FUNDAMENTALS OF
STRUCTURAL ANALYSIS

With Computer Analysis & Applications

FUNDAMENTALS OF
STRUCTURAL ANALYSIS

With Computer Analysis & Applications

[For B.E./B.Tech. in Civil Engineering and also Useful for M.E./M.Tech. Students]

Dr. SUJIT KUMAR ROY
B.E., M.E., Ph.D. (IIT, Kharagpur), MIE (India)
Assistant Professor of Civil Engineering

and

Dr. SUBRATA CHAKRABARTY
B.E., M.Tech, Ph.D. (IIT, Kharagpur)
Bengal Engineering College
Howrah - 711 103

S Chand And Company Limited
(ISO 9001 Certified Company)

S Chand And Company Limited

(ISO 9001 Certified Company)

Head Office: Block B-1, House No. D-1, Ground Floor, Mohan Co-operative Industrial Estate, New Delhi – 110 044 | Phone: 011-66672000

Registered Office: A-27, 2nd Floor, Mohan Co-operative Industrial Estate, New Delhi – 110 044 Phone: 011-49731800

www.**schandpublishing.com**; e-mail: **info@schandpublishing.com**

Branches

Chennai	:	Ph: 23632120; chennai@schandpublishing.com
Guwahati	:	Ph: 2738811, 2735640; guwahati@schandpublishing.com
Hyderabad	:	Ph: 40186018; hyderabad@schandpublishing.com
Jalandhar	:	Ph: 4645630; jalandhar@schandpublishing.com
Kolkata	:	Ph: 23357458, 23353914; kolkata@schandpublishing.com
Lucknow	:	Ph: 4003633; lucknow@schandpublishing.com
Mumbai	:	Ph: 25000297; mumbai@schandpublishing.com
Patna	:	Ph: 2260011; patna@schandpublishing.com

© S Chand And Company Limited, 2003

All rights reserved. No part of this publication may be reproduced or copied in any material form (including photocopying or storing it in any medium in form of graphics, electronic or mechanical means and whether or not transient or incidental to some other use of this publication) without written permission of the copyright owner. Any breach of this will entail legal action and prosecution without further notice.

Jurisdiction: *All disputes with respect to this publication shall be subject to the jurisdiction of the Courts, Tribunals and Forums of New Delhi, India only.*

First Edition 2003

Revised Edition 2009; Reprints 2010 (Twice), 2011; Reprint with Corrections 2011; Reprints 2012, 2014

Reprint 2016

ISBN : 978-81-219-2195-4 **Product Code:** H3STR46STRE10ENAB09O

PRINTED IN INDIA

By Vikas Publishing House Private Limited, Plot 20/4, Site-IV, Industrial Area Sahibabad, Ghaziabad – 201 010 and Published by S Chand And Company Limited, A-27, 2nd Floor, Mohan Co-operative Industrial Estate, New Delhi – 110 044.

PREFACE TO THE SECOND EDITION

In this edition the book has been thoroughly revised and corrected. A chapter on Plastic Analysis has been added. Hope the students will find the book more useful.

Kolkata

Dr. S.K. ROY
Dr. S. CHAKRABARTY

PREFACE TO THE FIRST EDITION

The book "Fundamentals of Structural Analysis" is the outcome of the authors' long experience of teaching structural engineering to undergraduate and post-graduate students in Civil Engineering. The book takes an integral look at structural engineering starting with fundamentals and ending with computer analysis. Computer applications have been introduced early though the full fledged analysis has been postponed to later stages when physical conception on structural behaviour is clear.

The book is suitable for 5th, 6th and 7th semesters of undergraduate course. The sixteenth chapter on Matrix Analysis and Computers can be taught in the 7th semester which will prepare the students for extensive study on Computer Analysis in the 8th semester and will be useful transition from matrix method to more general finite element study at P.G. level. Similarly the seventeenth chapter on "Dynamic Analysis" is introductory in character.

In writing this book we have consulted large number of Indian and foreign books on structural analysis. A select bibliography is given at the end of the book. We are grateful to all authors whose works form the background of this book. The senior author is grateful to Dr. N.C. Sinha, former Professor and Head of C.E. Deptt. of Jalpaiguri Govt. Engineering College and B.E. College (D.U.) for initiating him into the field of textbook writing.

A large number of examples have been worked out in the book so that students can master the subject by practising the examples and problems. It is quite possible to have some computational and typographical errors. Readers will earn our gratitude by pointing them to us.

We are grateful to Mr. Ravindra Kumar Gupta, Managing Director of M/s S. Chand and Co. Ltd. for publishing the book with great care and also in a very short time.

Our grateful thanks are due to Mr. R.S. Saxena, Advisor, Delhi office and to Mr. R.M. Nath, Manager, Calcutta Branch for coordinating the publication of the book.

Kolkata

Dr. S.K. ROY
Dr. S. CHAKRABARTY

Disclaimer : While the authors of this book have made every effort to avoid any mistake or omission and have used their skill, expertise and knowledge to the best of their capacity to provide accurate and updated information. The author and S. Chand does not give any representation or warranty with respect to the accuracy or completeness of the contents of this publication and are selling this publication on the condition and understanding that they shall not be made liable in any manner whatsoever. S.Chand and the author expressly disclaim all and any liability/responsibility to any person, whether a purchaser or reader of this publication or not, in respect of anything and everything forming part of the contents of this publication. S. Chand shall not be responsible for any errors, omissions or damages arising out of the use of the information contained in this publication.
Further, the appearance of the personal name, location, place and incidence, if any; in the illustrations used herein is purely coincidental and work of imagination. Thus the same should in no manner be termed as defamatory to any individual.

CONTENTS

1

BASIS OF STRUCTURAL ANALYSIS

1.1. INTRODUCTION

Engineering structures must carry load with small deformation. Examples of common engineering structures are : buildings, bridges, dams, offshore oil platforms, water and gas tanks, piping systems, nuclear and chemical reactor vessels, ships, air planes, buses and other transportation systems.

From engineering point of view

(*i*) a *Framed structure* is an assembly of members inter connected by joints. Other types of structural systems are shells, domes, dams, walls etc.

(*ii*) *Structural behaviour* is the response of a structure to applied loads and environmental effects (wind, earthquake and temperature changes).

(*iii*) *Structural analysis* is the determination of reactions, member forces, deformation of the structure and deflection of the joints due to applied loads and environmental effects.

1.2. RELATION BETWEEN ANALYSIS AND DESIGN

The primary work of an engineer is to design. But analysis of a structure is required to check whether the design is safe either of a new or old structure. Usually design consists of preliminary design and checking by analysis, redesign with suitable modifications which is repeated a number of times till final design is obtained. At this stage, all the design constraints are fulfilled, that is, stresses and deflections are within permissible limits and cost is usually minimum.

1.3. STRUCTURAL IDEALISATION

Structural analysis is conducted on an analytical (or mathematical) model which is an idealisation of the actual structure. Engineering judgment must be used in defining the idealised structure such that it represents the actual structural behaviour as accurately as is practically possible.

Depending on the problem in hand, structures are idealised using line, plate or brick elements. Line elements are used to model the structural components that have one dimension much greater than other two. Typical examples are truss members, beams, columns etc. Plate elements are used to model structural components such as slabs and shells of which two dimensions are much greater than the third. Brick elements are used to model structures such as thick slabs and machine components where the structural action is truly three dimensional. Assembling the elements in accordance with the geometry of the structure produces a model of the complete structure that can be used to investigate the effect of the applied loading.

Fig. 1.1. (*a*) shows an actual structure.

Fig. 1.1 (*b*) shows the idealisation.

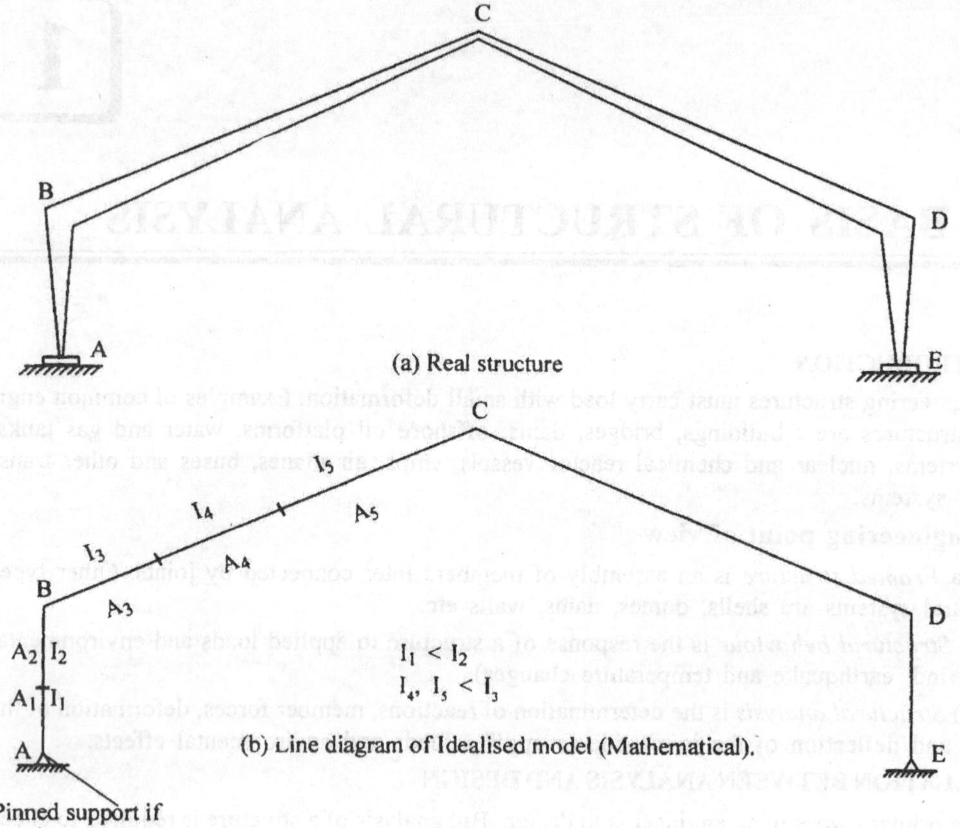

(a) Real structure

(b) Line diagram of Idealised model (Mathematical).

Fig. 1.1. Structure and its idealisation.

The following points can be observed :

(*i*) The actual structure is a welded frame of I section. The section is variable *i.e.* it is thicker at *B* and *D* and thinner at *A*, *C* and *E*.

(*ii*) It is supported at the ground level by means of concrete foundation. Hence in the idealised model represented by the line diagram we should have variable I and Area, large I and Area near *B* and *D* while small I and Area near the supports *A* and *E*. and mid region *C*. If the supports allow rotations it should be pinned, if no rotation occurs it should be considered as fixed support.

1.4. CLASSIFICATION OF STRUCTURES

Structures can be classified in many different ways. One way of classification is based on dimensions, such as

(*i*) One dimensional, for example, trusses, beams, frames, grids, arch ribs and cable structures. Here one dimension, say length, is very large compared to other two dimensions for the components of the structure.

Some of these structures are shown in Fig. 1.2. (*a*) to (*g*).

(*ii*) Two dimensional or surface structures *e.g.* slab, deep beams, shells or walls. Here length and breadth are large in comparison to thickness. Fig. 1.2 (*i*) and (*j*).

(*iii*) Three dimensional or solid structures where all the three dimensions of length, breadth or height are equally prominent *e.g.* foundations, retaining walls, gravity dams. Fig. 1.2. (*k*) (*l*) (*m*) and (*n*).

They can also be classified from the way they carry load. For example,

(*i*) Beams carry their loads by developing bending moments and shear forces at different sections.

(*ii*) Trusses carry their loads by developing axial forces in the bars.

(*iii*) Frames carry loads by axial force, bending moment and shear.

(*iv*) Arches carry load by compression and bending.

(*v*) Cables carry loads by developing axial tension.

(*vi*) Two way grids are subjected to both bending and twisting

(*vii*) Thin plates are subjected to both bending and twisting.

(*viii*) Thin shells transmit forces mainly by in plane or membrane stresses.

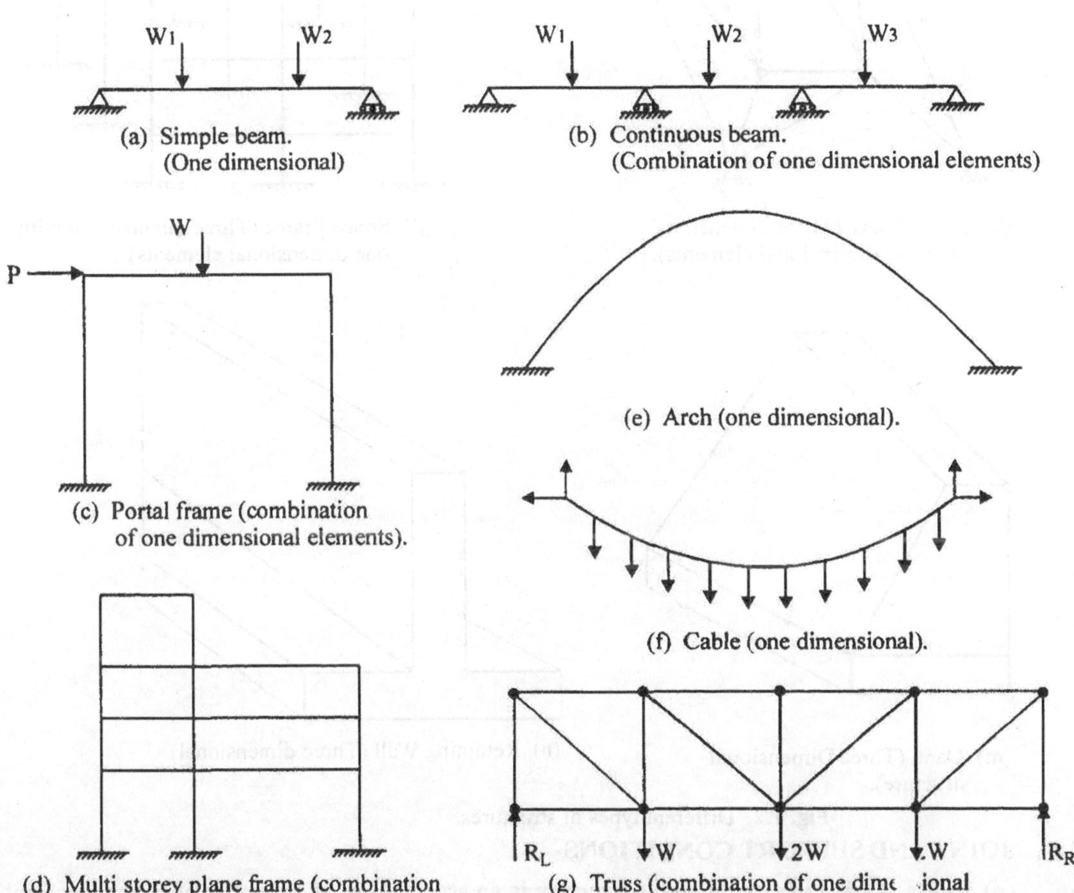

(a) Simple beam. (One dimensional)

(b) Continuous beam. (Combination of one dimensional elements)

(c) Portal frame (combination of one dimensional elements).

(e) Arch (one dimensional).

(f) Cable (one dimensional).

(d) Multi storey plane frame (combination of one dimensional elements)

(g) Truss (combination of one dimensional elements)

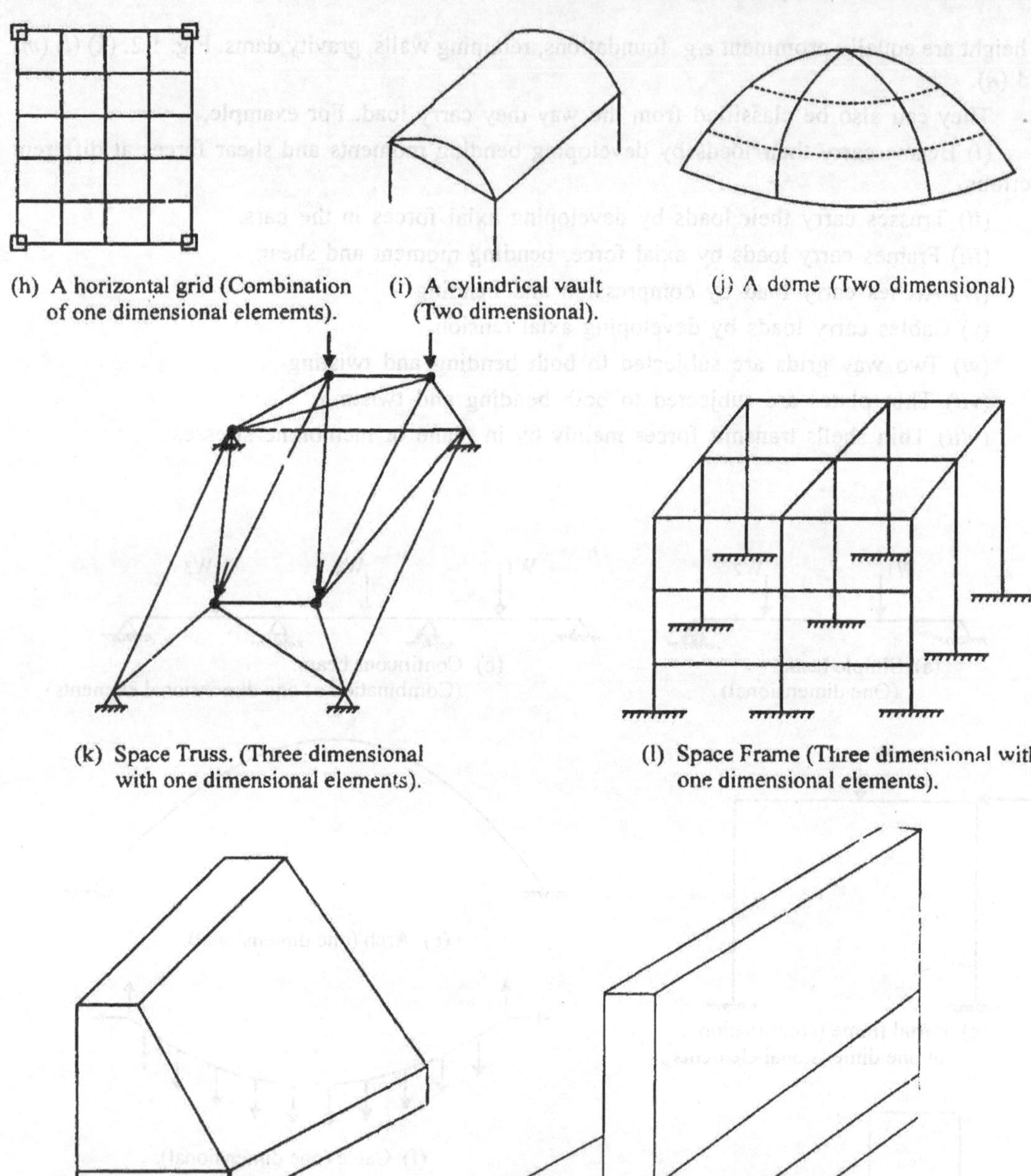

(h) A horizontal grid (Combination of one dimensional elememts).

(i) A cylindrical vault (Two dimensional).

(j) A dome (Two dimensional)

(k) Space Truss. (Three dimensional with one dimensional elements).

(l) Space Frame (Three dimensional with one dimensional elements).

(m) Dam. (Three Dimensional structure).

(n) Retaining Wall (Three dimensional).

Fig. 1.2. Different types of structures.

1.5. JOINT AND SUPPORT CONDITIONS

(a) Joints. As already explained, a structure is an assemblage of a number of elements. Joint is connection between these elements. Joints may be (*i*) Rigid or (*ii*) Pinned. Fig. 1.3. (*a*) shows rigid joint. A rigid joint maintains the angles between the centre lines of members framing into the joint. A rigid joint will carry any moment assigned to it and the joint will rotate as a whole and there will be no change in the angle between the members. Fig. 1.3. (*b*) shows a pinned joint.

This joint allows the members joining it independent rotation. Hence angle between the member changes. Since the joint is pinned it cannot resist any moment.

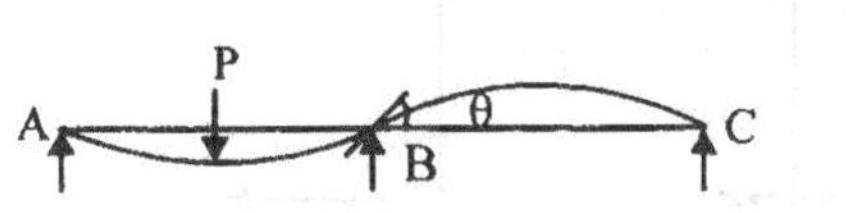

(i) 180° rigid joint between AB and BC. Joint rotates as a whole.

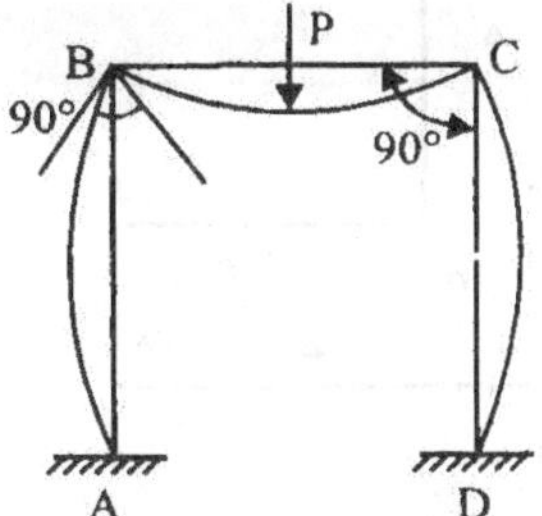

(ii) 90° rigid joint between AB and BC under load. Joint rotates as a whole.

(*a*) Behaviour of Rigid Joints.

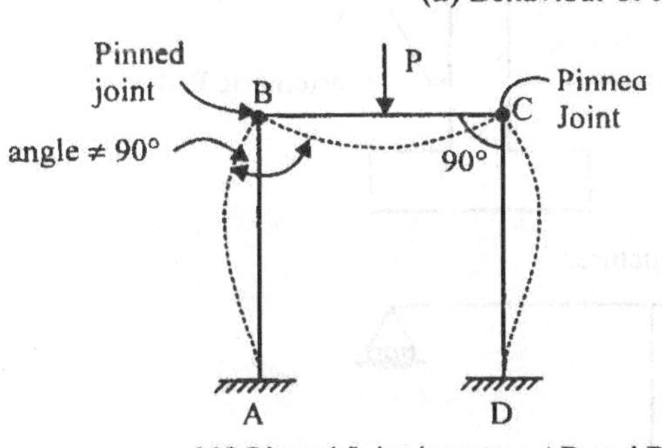

90° Pinned Joint between AB and BC.

(*i*) Angle between AB and BC does not remain 90° under load.

B'
A
B
C
D
Pinned

(ii) 180° pinned joint between AB and BC Angle between AB' and B'C does not remain 180°.

(*b*) Behaviour of pinned Joints.

Fig. 1.3. Behaviour of Joints.

(*b*) **Supports.** The loads applied to a structure must be transferred to the ground by the supports, that is, the support will generate reactions to counteract the actions of the loads. The type and number of reactions depend on the support conditions. Support may be roller, hinge or pin and fixed. Fig. 1.4. shows actual supports, how it is symbolically represented and forces and moments which act on the support. The bridge deck constructed as monolithic with the columns is shown in Fig. 1.4. (*d*) (*i*) and the corresponding support conditions that might be taken for analysis is shown in Fig. 1.4. (*d*) (*ii*).

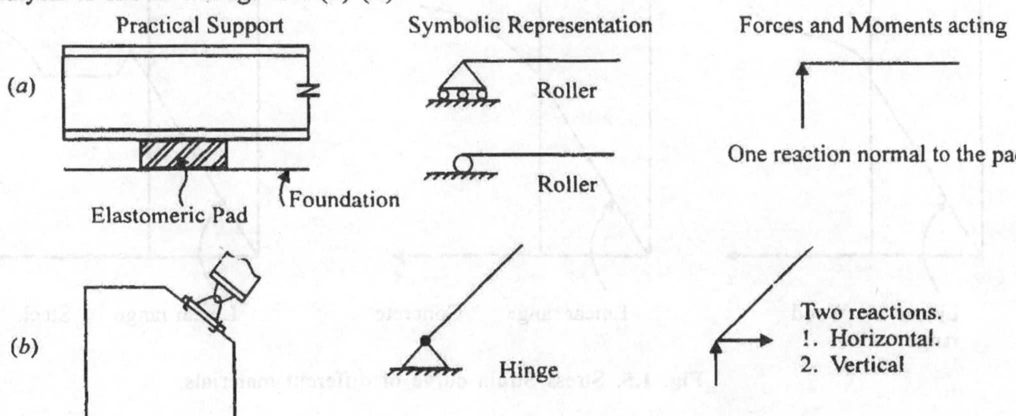

(c)

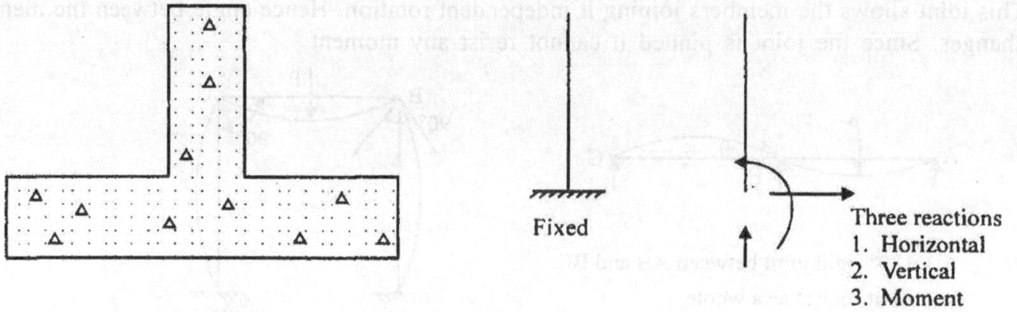

Fixed
Three reactions
1. Horizontal
2. Vertical
3. Moment

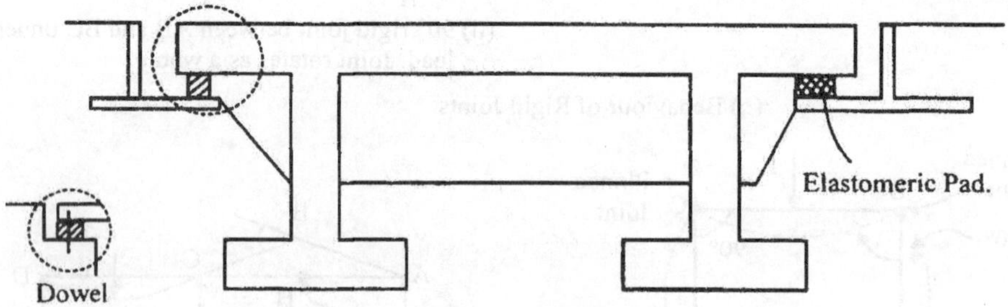

Elastomeric Pad.
Dowel

(i) Actual structure.

(ii) Mathematical Model.

(d) Actual structure and mathematical model of a bridge deck.

Fig. 1.4. Different types of supports *(a)* Roller support *(b)* Hinged support *(c)* Fixed support *(d)* Actual structure and mathematical model of a bridge deck.

1.6. MATERIAL CHARACTERISTICS

Structures are usually made of different materials like wood, concrete or steel. Their stress-strain curves are all different as shown in Fig. 1.5.

σ
τ
Linear range
Wood
σ
τ
Linear range
Concrete
σ
τ
Linear range
Steel.

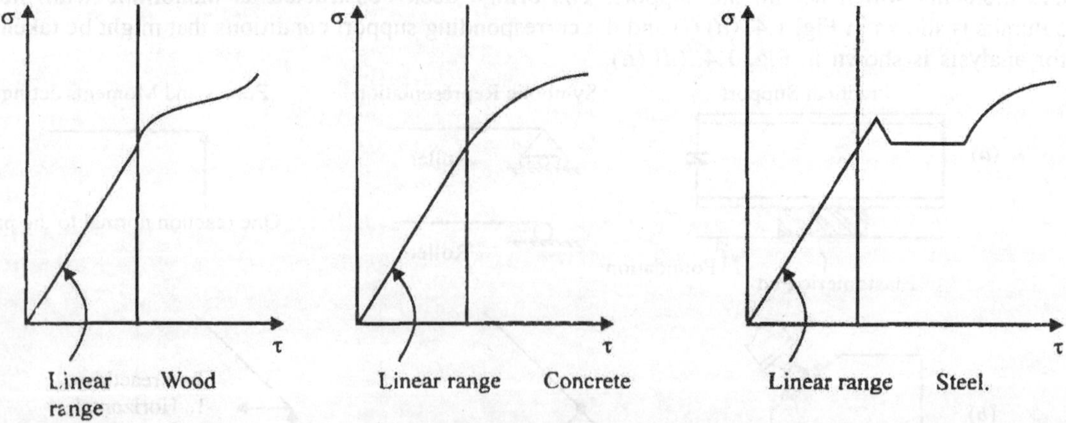

Fig. 1.5. Stress Strain curve of different materials.

Upto some level of load, stress is proportional to strain and *E*, the slope of the stress strain curve is constant and hence stress strain curve is linear. Another characteristic of member is its elasticity. Upon unloading, a material may come back to its original position, then it will be called elastic. When it does not come back to its original position it is termed inelastic, hence considering (*i*) linearity (*ii*) elasticity four combinations of curves are possible.

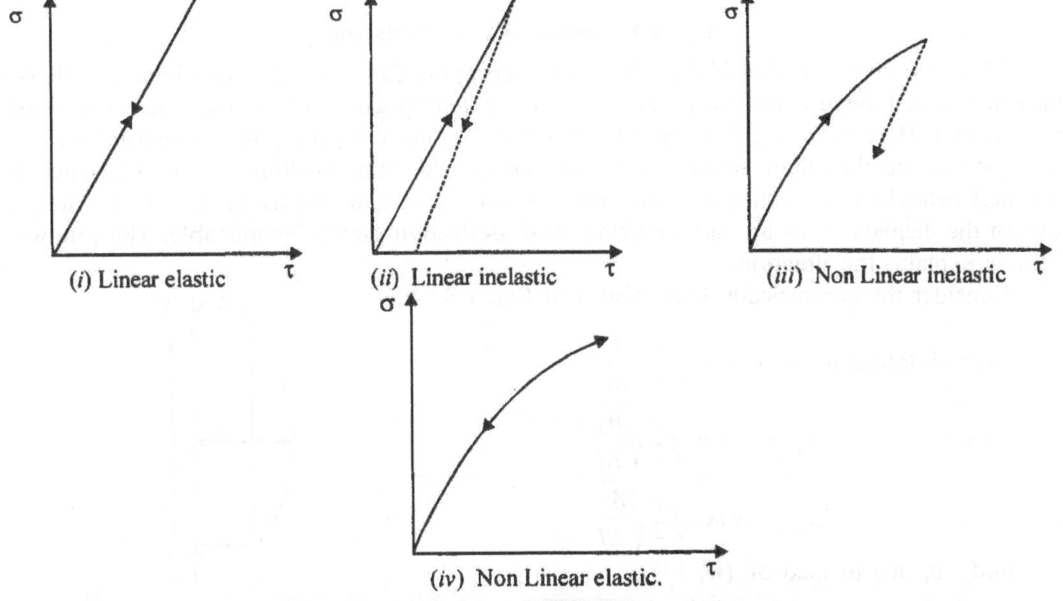

Fig. 1.6 Different types of stress strain curve.

Steel upto proportional limit is linear and elastic; steel beyond its yield point is nonlinear and inelastic. Rubber is nonlinear and elastic.

1.7. SUPER POSITION

The deformations of structures are generally small. In most cases, the geometry of the structure is sensibly unaffected by deformation. In such a case, the conditions of equilibrium and conditions of compatibility can be written down for the undisturbed structure without loss of accuracy. The principle of supper position states that

if	cause A	produces Effect A
and	cause B	produces Effect B
then	cause A + cause B produce Effect A + Effect B.	

Three categories of super position will be considered—

(*i*) Super position of force systems.

(*ii*) Super position of displacements.

(*iii*) Super position of linear elastic structures.

(*i*) **Super position of force systems** : For statically determinate structures the principle of super position is valid with only restriction that distortion of structure is small. This means internal forces and moments can be added together for two different loading systems. Under such conditions super position of forces is valid for statically indeterminate structures also with the additional restriction that material of the structure must obey Hooke's law because the internal forces depend on the deformation of the members.

(*ii*) Super position of displacements : Superposition of displacements is valid both for statically determinate and indeterminate structure provided deflection of the structure is not gross.

The propped cantilever shown in Fig. 1.7. illustrates the principle of superposition.

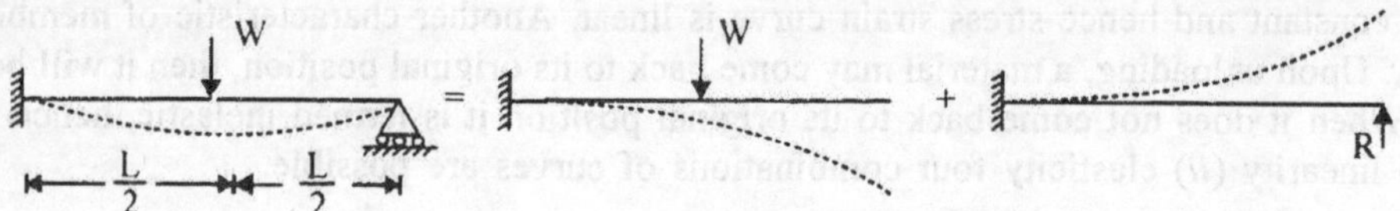

Fig. 1.7 Superposition of displacements.

The diagram shows that if the effect of the propping force (*i.e.* the right hand reaction) is superimposed on the effect of the applied load on the unpropped cantilever then the actual results are obtained. Here propping force and the applied load have bean treated as two independent load systems and then their effects have been superposed. Superposition is not valid when the structural behaviour is not linear. There are two sources of nonlinearity in structures. The *first* is when the displacements are large making small deflection theory inapplicable. The following example explains the situation.

Consider the eccentrically loaded strut of Fig. 1.8

Central deflection $d = e \sec \frac{l}{2}\sqrt{\frac{W}{EI}}$

Here $d_1 = e \sec \frac{l}{2}\sqrt{\frac{W_1}{EI}}$

$d_2 = e \sec \frac{l}{2}\sqrt{\frac{W_2}{EI}}$

and d, due to load of $(W_1 + W_2)$

$= e \sec \frac{l}{2}\sqrt{\frac{(W_1 + W_2)}{EI}}$

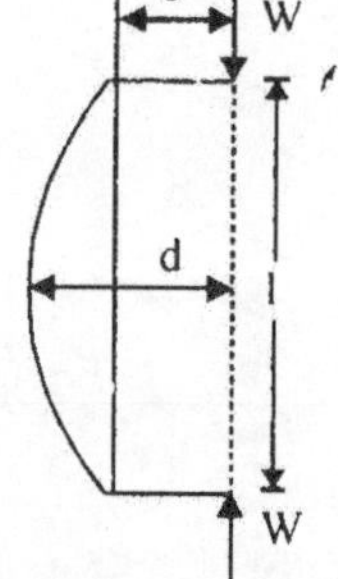

Fig. 1.8 Gross deformation of structure.

Obviously $d \neq d_1 + d_2$. This is because deformation is gross and conditions of equilibrium are appreciably affected by the distortion of the structure. The *Second* case when method of superposition is not valid occurs when the structural behaviour is nonlinear due to material nonlinearity.

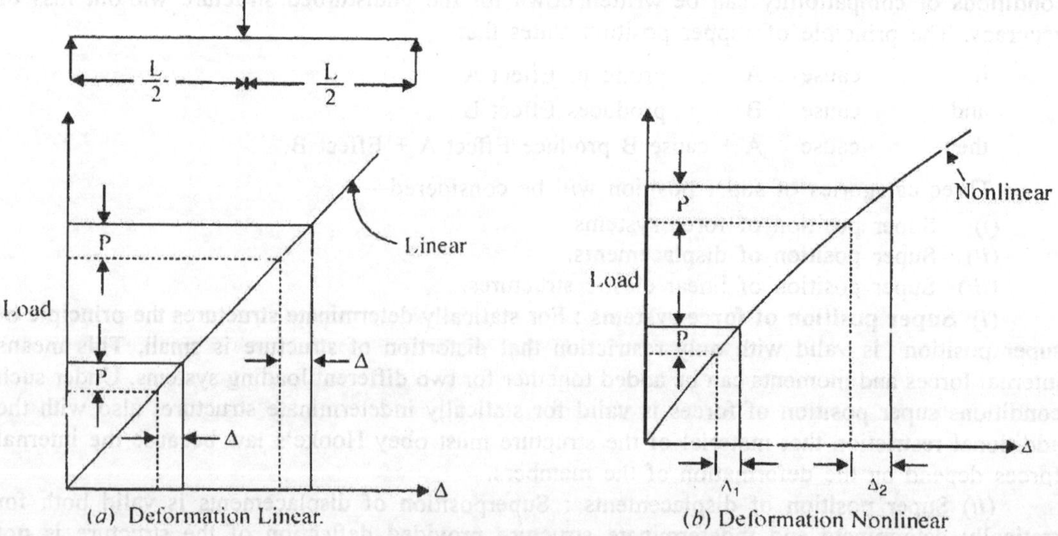

(*a*) Deformation Linear. (*b*) Deformation Nonlinear

Fig. 1.9 Load deformation graphs.

Fig. 1.9 shows load deformation graphs for a linear and a nonlinear structure respectively. We can see that for the linear structure an applied load P will give the same deformation Δ regardless of position of the application of the loading process. However, for the nonlinear structure, the same load P will give different deformations Δ_1 and Δ_2 depending upon the loading previously applied to the structure.

(iii) **Linear Elastic Structure** : Here the principle of superposition states that the changes in the deflections and the internal forces of a linear elastic structure due to applied loading, temperature change, movement of supports can all be superposed. It is based on two already explained conditions —

(i) The deformation must not be gross so that conditions of equilibrium and compatibility can be written down without loss of accuracy using the original data.

(ii) Member characteristics must be linear.

1.8. LOADS

It is difficult to accurately estimate the loads coming on a structure. They can be caused by gravity, wind pressure, hydrostatic pressure, accelerations, temperature changes, shrinkage, friction, earthquake, impacts and vibrations. Loads are usually modelled as concentrated point loads, line loads, surface or volume loads. A concentrated load or couple is applied at a point while a distributed load is applied over a length or area. Fig. 1.10 gives idealised representation of the forces.

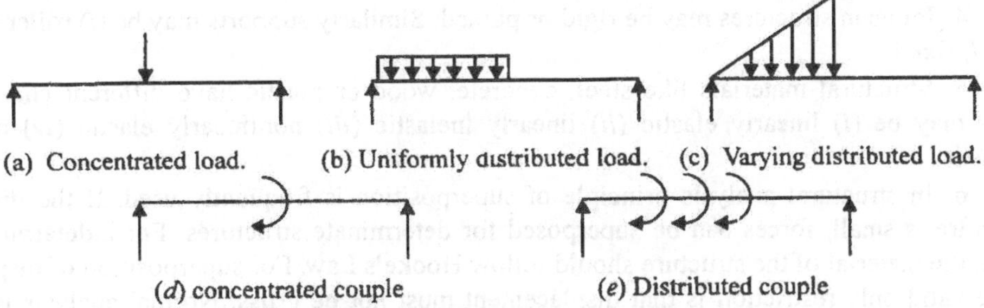

Fig.1.10. Different types of load.

Loads are generally classified in two groups

1. Dead load
2. Live load.

Dead loads are weight of the structural system and are computed on the basis of weight density and dimensions of the structures. In a building this includes the self weight of all beams, columns, trusses, walls etc and other service equipments supported by the structure.

Live loads, on the other hand, are —

(i) forces of nature *e.g.* wind, earthquake.

(ii) not permanent *e.g.* snow loads.

(iii) associated with intended use of the area *e.g.* occupancy loads as specified by various codes.

It is difficult to specify live load for a structural system accurately as

(i) The magnitude of the load itself is uncertain.

(ii) The location over which the loading acts at any instant is also uncertain.

Previously, uncertainty about loads and material properties was accounted for by using factors of safety in the design of structure. This allowed a deterministic analysis. At present, however, emphasis is on probabilistic design where uncertainty about the loading and the structure's properties are incorporated in design methods. These are given in codes. IS 875 – 1987. "Code of practice for structural safety of building" and IS 1893 – 2002 "Criteria for Earthquake Resistant Design of Structures" give guidelines on computation of Dead loads, Imposed loads, Snow loads, Windloads and Earthquake loads. There are also special publications by Bureau of Indian Standards, New Delhi giving explanations and examples on the use of these codes.

SUMMARY

1. A structure must carry loads with small deformation.
2. The real structure is idealised by its mathematical models.
3. Structure can be classified as
   (*i*) One dimensional
   (*ii*) Two dimensional
   (*iii*) Three dimensional

Most of the structures whether plane or space consist of one dimensional elements joined together. Typical two dimensional structures are plates and shells. Three dimensional structures are retaining walls and dams.

4. Joints in structures may be rigid or pinned. Similarly supports may be (*i*) roller (*ii*) pinned or (*iii*) fixed.

5. Structural materials like steel, concrete, wood or plastic have different characteristics. They may be (*i*) linearly elastic (*ii*) linearly inelastic (*iii*) nonlinearly elastic (*iv*) nonlinearly inelastic.

6. In structural analysis principle of superposition is frequently used. If the distortion of structure is small, forces can be superposed for determinate structures. For indeterminate structures, the material of the structure should follow Hooke's Law. For superposition of displacements to be valid only restriction is that displacement must not be gross. Normal analysis is based on linear elastic structure when all superposition is valid.

7. Loads acting on structure may be broadly divided into two groups :
   1. Dead loads
   2. Live loads.

Dead loads on a structure consist of the weight of the structure itself and all non movable items. They are usually known with greater certainty and are deterministic in nature. Liveload, on the other hand, is not permanent. They are associated with the intended use of the area. They may be caused by natural causes such as wind, earthquake, snowfall etc. Because of the uncertainty in calculating loads and material characteristics probabilistic design based on statistical concepts is gradually replacing deterministic designs.

PROBLEMS

A. Theoretical.

1.1. (*i*) Explain the difference in behaviour of (*a*) rigid joint (*b*) pinned joint.
(*ii*) Explain different types of structural supports and associated reaction components.
(*iii*) Explain with examples materials which are (*a*) linear (*b*) nonlinear (*c*) elastic (*d*) inelastic.

(*iv*) Explain with examples.

(*a*) Superposition of force systems.

(*b*) Superposition of displacements.

(*c*) Superposition of linear elastic structures.

Explain when superposition is not valid.

3. **Objective. Tick the correct answer.**

1.2. Principle of superposition is applicable when

(*a*) Deflections are linear functions of applied forces.

(*b*) Materials obey Hooke's law.

(*c*) None of the above [AMIE Winter'97].

2

EXTERNAL EQUILIBRIUM OF STRUCTURES

2.1. INTRODUCTION

A structure may be at rest or in motion. The usual civil engineering structures say, a building, a dam or a bridge are all at rest and hence may be said to be in statical equilibrium. An in-flight aeroplane is in motion and hence is in dynamic equilibrium. This book deals with structures which are all in statical equilibrium. The equations of statical equilibrium are based on Sir Isaac Newton's laws governing the motion of bodies (1687) which are —

(*a*) The summation of all forces in any axis is zero.

(*b*) The summation of all moments about any axis is zero.

If summation is taken about *x*, *y*, and *z* axes, symbolically this becomes —

$$\Sigma F_x = 0 \qquad \Sigma F_y = 0 \qquad \Sigma F_z = 0 \qquad \text{... (2.1)}$$

$$\Sigma M_x = 0 \qquad \Sigma M_y = 0 \qquad \Sigma M_z = 0 \qquad \text{... (2.2)}$$

For planar structures and forces acting in the same plane the equations reduce to

$$\Sigma F_x = 0 \quad \Sigma F_y = 0 \quad \Sigma M_z = 0 \qquad \text{... (2.3)}$$

Obviously the analysis becomes simpler when the structure as well as load is coplanar and Eq. (2.3) can be applied. Since analysis of a structure starts with computation of reactions, in this chapter, calculation of reactions of different types of structures has been explained.

2.2. FREE BODY DIAGRAM

This is one of the most useful tools at the disposal of the structural analyst. A free body diagram is simply a sketch of a structural component with all the appropriate forces acting on it. It may be of an entire structure or it may be a subsection of a larger structure. A free body diagram of an entire structure leads to relationships between the external forces on the structure. This is drawn by isolating the structure from its supports and showing it acted upon by all external loads and all independent component of the reactions. Such a diagram is shown in Fig. 2.1.

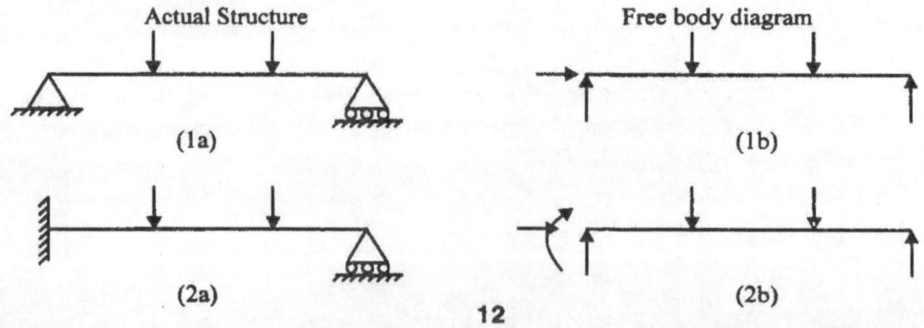

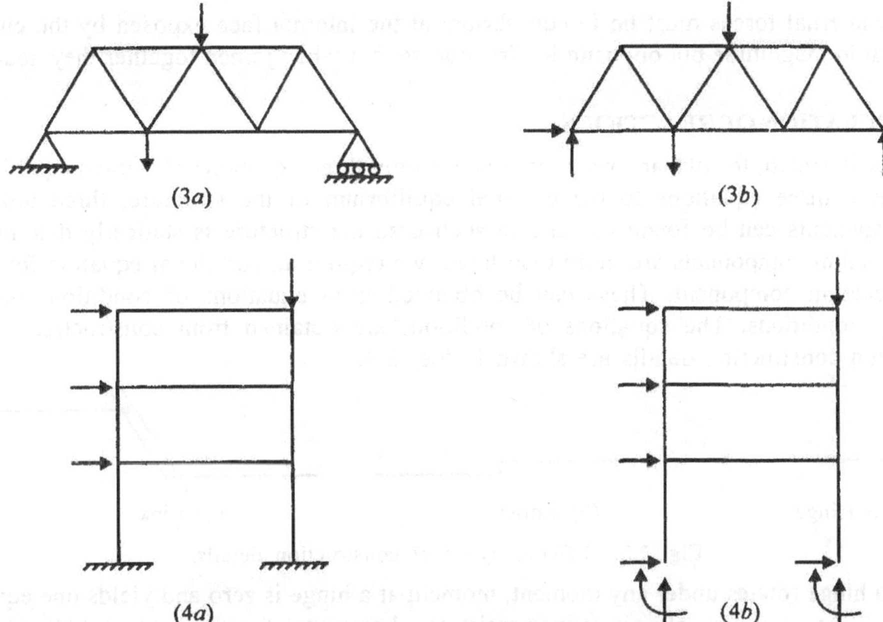

Fig. 2.1. Free body diagrams.

The reactions are generally shown acting in the positive coordinate directions. If the final calculations give positive sign, the assumed direction is correct. If the answer is negative the direction of force should be reversed.

Sometimes the free body diagram of part of a structure is required. In such a case part of the structure should be drawn separately and all the external forces acting on the structure should be shown in proper places and directions. The internal forces exerted by one part on the other will now become external forces on the corresponding part.

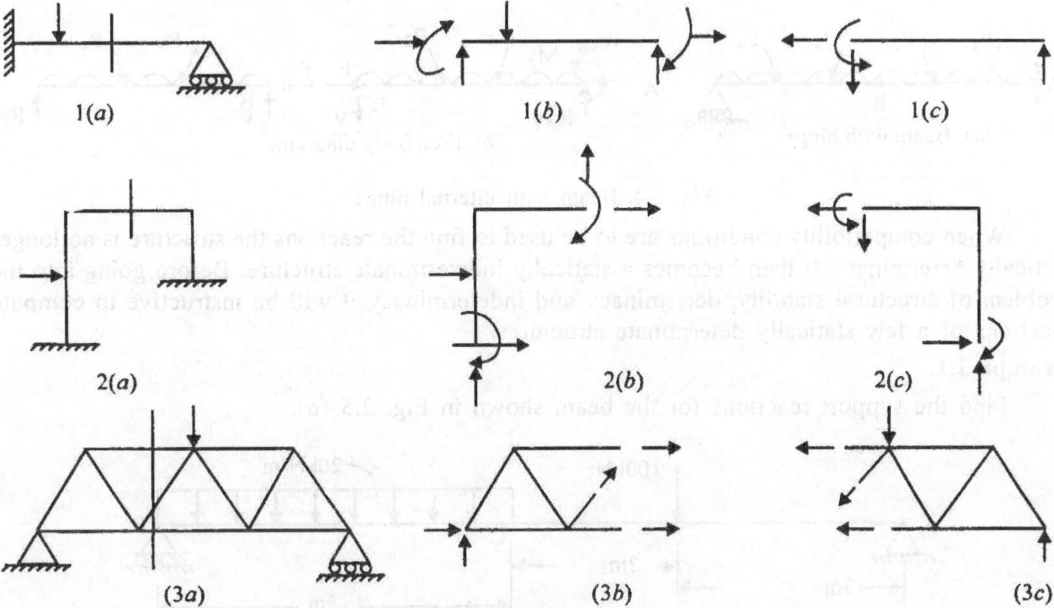

Fig. 2.2. Free Body Diagrams of Part of Structure.

As the internal forces must be in equilibrium at the internal face exposed by the cut they must be equal in magnitude but opposite in direction so that when joined together they neutralise each other.

2.3. CALCULATION OF REACTIONS

As already stated, for planar structures we have only three equations of statical equilibrium. Applying these three equations to the external equilibrium of the structure, three unknown reaction components can be found out and in such case the structure is statically determinate. When the reaction components are more than three, we require an additional equation for every additional reaction component. These can be obtained from equations of conditions or from compatibility conditions. The equations of conditions are obtained from construction details. Three common construction details are shown in Fig. 2.3.

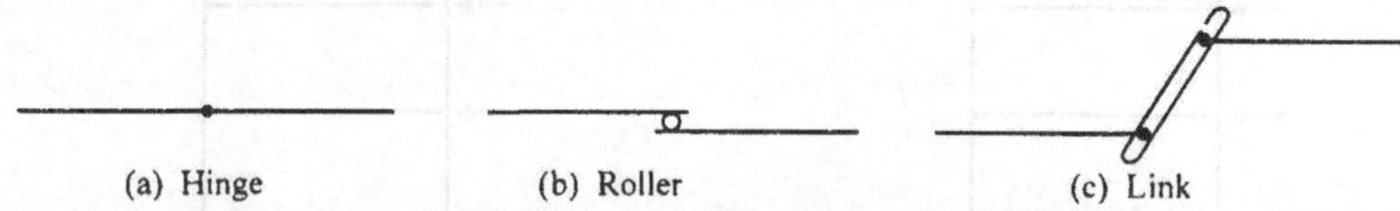

Fig. 2.3. Different types of construction details.

Since a hinge rotates under any moment, moment at a hinge is zero and yields one equation of condition. Since a roller rolls, it cannot resist any horizontal force or moment. It can only transmit vertical force and as such yields two equations of conditions.

The link in Fig. 2.3 (*c*) can transmit force only along its own axis and introduces two equations of conditions.

Fig. 2.4. a shows a beam with an internal hinge. Fig. 2.4 b shows the free body diagrams of the two beam segments. It can be seen that there are two internal components of force at point B, one parallel to the axis of the member (*F*) and another perpendicular to the axis (*V*). Since no moment is transmitted through the hinge, the equation $\Sigma M_B = 0$ can be imposed which gives the required equation of condition.

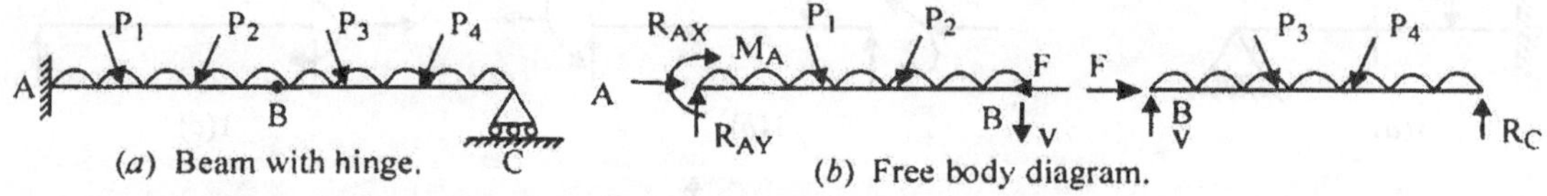

Fig. 2.4. Beam with internal hinge.

When compatibility conditions are to be used to find the reactions the structure is no longer statically determinate. It then becomes a statically indeterminate structure. Before going into the problem of structural stability, determinacy and indeterminacy it will be instructive to compute reactions of a few statically determinate structures.

Example 2.1.

Find the support reactions for the beam shown in Fig. 2.5 (*a*).

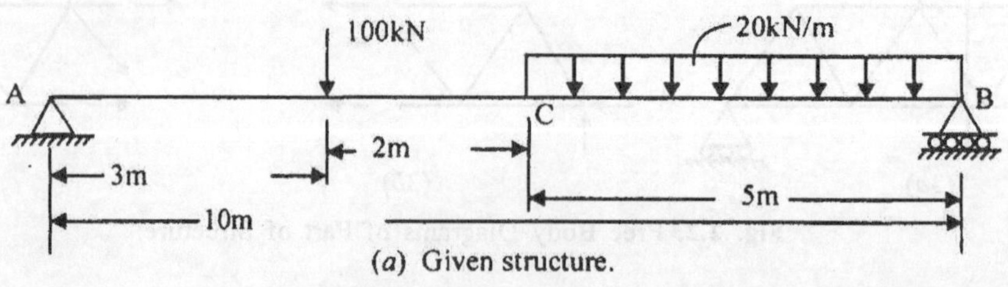

(*a*) Given structure.

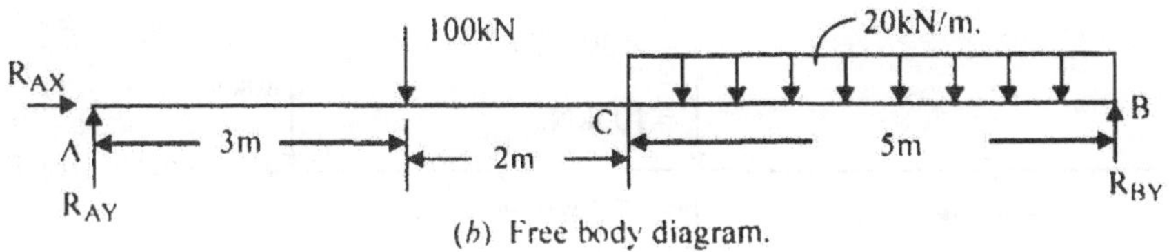

(b) Free body diagram.

Fig. 2.5. Example 2.1.

Solution.

1. **Free body diagram** — At the hinge A we have two reaction components R_{AX} and R_{AY}. At the roller B we have only the vertical reaction R_{BY}.
2. **Calculation of Reactions** — For equilibrium

$\Sigma F_x = 0$ Hence $R_{AX} = 0$, R_{AX} being the only force in the x direction.

Similarly as $\Sigma F_y = 0$, taking upward direction as positive

$$R_{AY} \uparrow - 100 \downarrow - 20 \times 5 \downarrow + R_{BY} \uparrow = 0$$

or $$R_{AY} + R_{BY} = 200$$

Again as $\Sigma M_z = 0$, taking moments about B of all forces acting on the beam and taking clockwise moment as positive

$$R_{AY} \times 10 - 100 \times 7 - \frac{20 \times 5 \times 5}{2} = 0$$

$$R_{AY} \times 10 = 700 + 250 = 950$$

$$R_{AY} = 95 kN \uparrow.$$

$$R_{BY} = 200 - R_{AY}$$

$$= 200 - 95 = 105 kN \uparrow.$$

Check. As moment of all forces about any point is zero taking moment of all forces about C.

$$R_{AY} \times 5 - 100 \times 2 + \frac{20 \times 5 \times 5}{2} - R_{BY} \times 5$$

$$= 95 \times 5 - 100 \times 2 + 250 - 105 \times 5 = 0.$$

Example 2.2.

Find the support reactions for the beam shown in Fig. 2.6*a*.

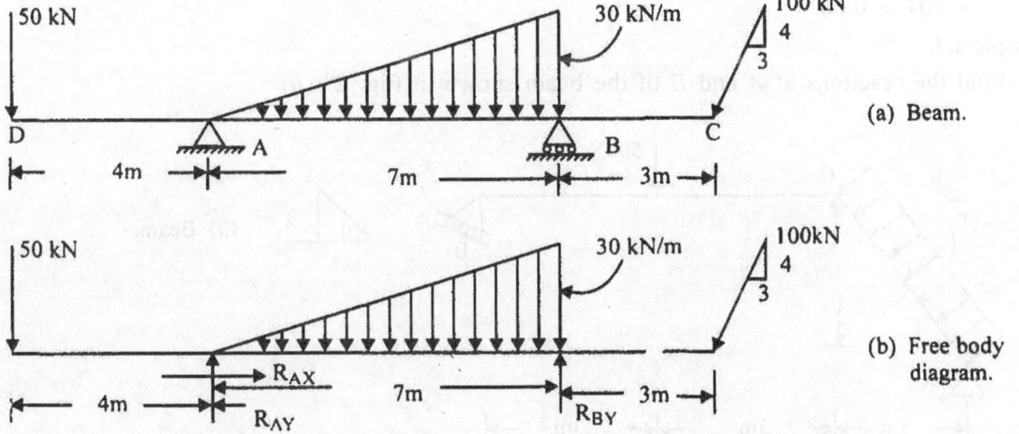

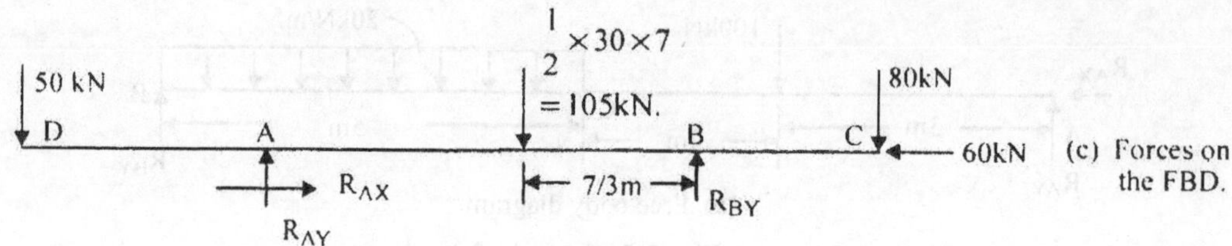

Fig. 2.6. Example 2.2.

Solution.

1. **Free body diagram.** Fig. 2.6 (*b*) gives the free body diagram. As usual hinge at *A* has two reaction components R_{AX} and R_{AY} while roller at *B* has only one reaction component, R_{BY}. Fig. 2.6(*c*) show the free body diagram with the resultant of the triangular loading acting at its centroid and the inclined 100kN load resolved into its horizontal and vertical components of 60kN and 80kN respectively.

Reactions.

Applying $\Sigma F_x = 0$

$$\overrightarrow{R_{AX}} - \overleftarrow{60} = 0 \quad \text{or} \quad R_{AX} = 60kN.$$

Applying $\Sigma F_y = 0$

$$-50\downarrow + R_{AY}\uparrow - 105\downarrow + R_{BY}\uparrow - 80\downarrow = 0.$$

or $R_{AY} + R_{BY} = 235$

Applying $\Sigma M_z = 0$ about *B*.

$$-50 \times 11 + R_{AY} \times 7 - \frac{105 \times 7}{3} + 80 \times 3 = 0$$

$$\text{or } R_{AY} = 79 \cdot 28kN.$$

$$R_{BY} = 235 - 79 \cdot 28 = 155 \cdot 72kN.$$

Check. Moment of all forces about *C*.

$$50 \times 14 - 79 \cdot 28 \times 10 + \frac{105 \times 16}{3} - 155 \cdot 72 \times 3$$

$$= 0 \cdot 04 \approx 0 \cdot 00.$$

Example 2.3.

Find the reactions at *A* and *B* of the beam shown in Fig. 2.7(*a*).

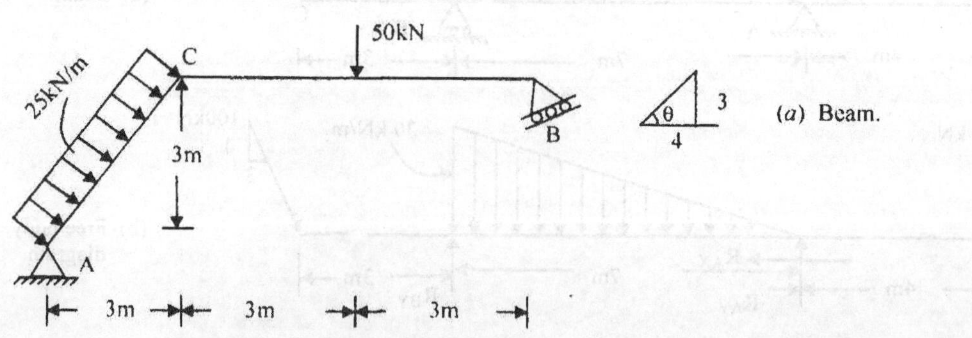

(*a*) Beam.

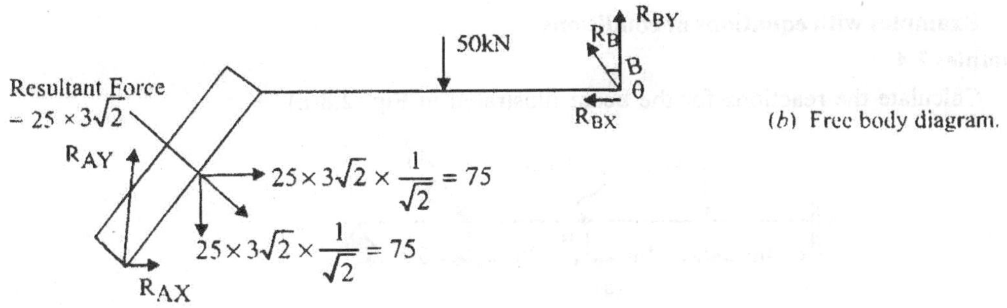

(b) Free body diagram.

Fig. 2.7. Example 2.3.

Solution.

1. *Free body diagram* — Fig. 2.7(*b*) shows the free body diagram. At the hinge A, there are two reactions R_{AX} and R_{AY}. At *B*, the reaction R_B will be at right angles to the roller. It has two components R_{BX} and R_{BY}.

   $R_{BX} / R_{BY} = \tan\theta = 3/4$.

   Hence $R_{BX} = 3/4\, R_{BY}$.

2. *Reactions* —

   With $\Sigma F_X = 0$, $\qquad \vec{R}_{AX} + \vec{75} - \overleftarrow{R}_{BX} = 0$

   $\Sigma F_Y = 0$, $\qquad R_{AY}\uparrow -75\downarrow -50\downarrow + R_{BY}\uparrow = 0$

   $\Sigma M_A = 0$

   $$25 \times 3\sqrt{2} \times 1{\cdot}5\sqrt{2} + 50 \times 6$$
   $$- R_{BX} \times 3 - R_{BY} \times 9 = 0$$

   Then $\qquad R_{AX} - R_{BX} = -75$

   $$R_{AY} + R_{BY} = 125$$

   or $\qquad 3\,R_{BX} + 9\,R_{BY} = 225 + 300 = 525.$

   or $\qquad 3 \times \frac{3}{4} R_{BY} + 9 R_{BY} = 525$

   or $\qquad R_{BY} = 46{\cdot}67$

   $$R_{BX} = \frac{3}{4} \times 46{\cdot}67 = 35.00$$

   $$R_{AX} = -75 + R_{BX} = -40kN.$$

   $$R_{AY} = 125 - 46{\cdot}67 = 78{\cdot}33kN.$$

Check. Moment of all forces about *B*.

$$R_{AY} \times 9 - R_{AX} \times 3 - 75 \times 1{\cdot}5 - 75 \times 7{\cdot}5 - 50 \times 3$$
$$= 78{\cdot}33 \times 9 - 40 \times 3 - 75 \times 1{\cdot}5 - 75 \times 7{\cdot}5 - 50 \times 3$$
$$= -\,{\cdot}03 \approx 0.$$

Examples with equations of conditions.

Examples 2.4

Calculate the reactions for the beam illustrated in Fig. 2.8(*a*).

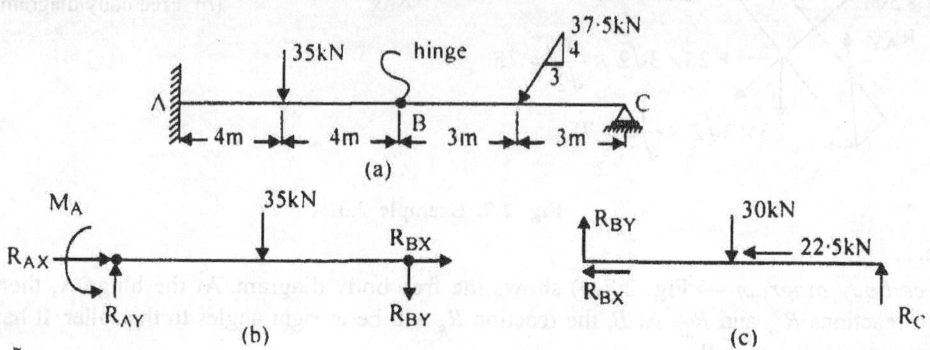

Fig. 2.8. Example 2.4.

Solution.

Considering Fig. 2.7(*c*)

$$\Sigma M_B = 0 \qquad R_C \times 6 - 30 \times 3 = 0.$$

$$R_C = 15kN\ .$$

From $\Sigma\ F_Y = 0$

$$R_C \uparrow + R_{BY} \uparrow - 30\downarrow = 0$$

$$15\uparrow + R_{BY} \uparrow - 30\downarrow = 0$$

$$R_{BY} = 15\uparrow$$

From $\Sigma F_X = 0$

$$\overleftarrow{22\cdot 5} + \overleftarrow{R_{BX}} = 0.$$

$$R_{BX} = -22\cdot 5kN.$$

Considering Fig. 2.7(*b*)

$$R_{AY} \uparrow - 35\downarrow - 15\downarrow = 0$$

$$R_{AY} = 50\uparrow.$$

As $\Sigma M_A = 0.$

$$R_{BY} \times 8 + 35 \times 4 - M_A = 0$$

or $M_A = 15 \times 8 + 35 \times 4$

$$= 120 + 140 = 260kNm.$$

Check. Considering equilibrium of the beam as a whole

$$\Sigma M_A = R_C \times 14 - 30 \times 11 - 35 \times 4 + 260$$

$$= 15 \times 14 - 30 \times 11 - 35 \times 4 + 260$$

$$= 0.$$

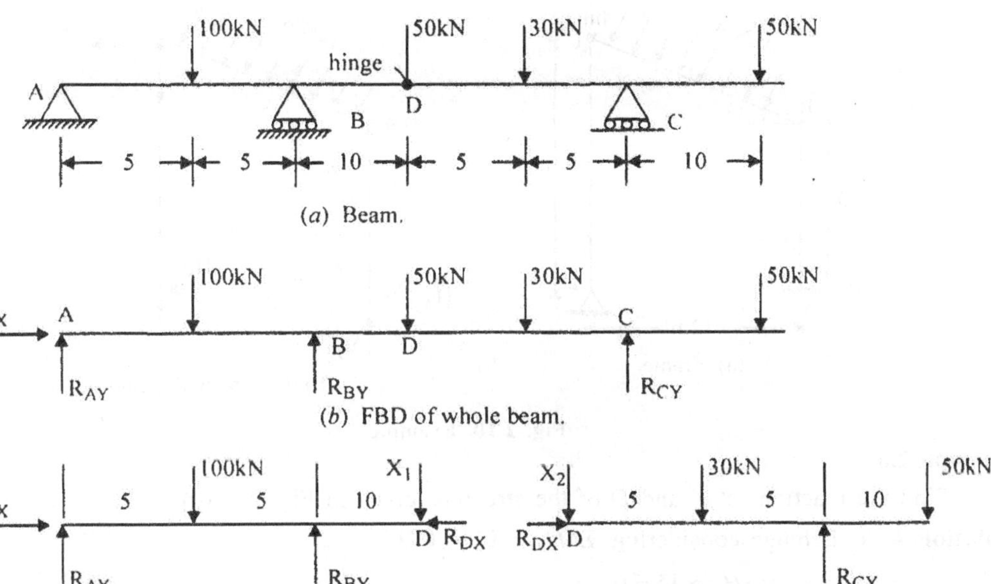

(a) Beam.

(b) FBD of whole beam.

(c) FBD of left part.

(d) FBD of right part.

Fig. 2.9 Example 2.5.

Example 2.5

Calculate the reactions at A, B and C for the beam given in Fig. 2.9(a).

Solution.

Taking moment of all forces about point D in Fig. 2.9(d)

$$R_{CY}\times 10-30\times 5-50\times 20=0$$

$$\text{or}\quad R_{CY}=(150+1000)/10$$

$$=115\text{kN}.$$

Putting $\Sigma F_X=0$ $\quad X_2\downarrow+30\downarrow+50\downarrow-115=0$

$$\text{or}\quad X_2=35\downarrow$$

Then $\quad X_1=15\downarrow$

Now Considering Fig. 2.9(c).

$$X_1\times 20+100\times 5-R_{BY}\times 10=0.$$

$$\text{or}\quad R_{BY}=80\uparrow$$

$$\therefore\quad R_{AY}=100+15-80=35\text{kN}.$$

$$R_{AX}=0.$$

Check. Considering equilibrium of the beam as a whole

$$\Sigma M_A=100\times 5-80\times 10+50\times 20+30\times 25-115\times 30+50\times 40=0.$$

$$\Sigma F_Y=R_{AY}\uparrow+R_{BY}\uparrow+R_{CY}\uparrow-100-50-30-50$$

$$=35+80+115-100-50-30-50=0.$$

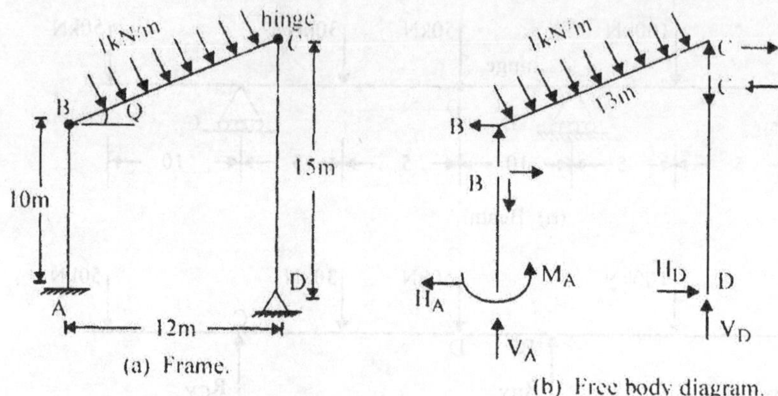

(a) Frame. (b) Free body diagram.

Fig. 2.10. Example 2.6.

Example 2.6

Find the reactions at A and D of the structure given in Fig. 2.10(a)

Solution As C is hinge considering $\Sigma M_{(D)} = 0$ in CD.

$$H_D \times 15 = 0$$

$$\text{or} \quad H_D = 0.$$

and hence $H_C = 0$ as there is no load in CD.

Hence reaction in C must be vertical (no horizontal component) to balance H_D.

Taking moments about B as B is a hinge

$$\Sigma M_B = 0, \quad \overset{\curvearrowleft}{R_C \times 12 - 13 \times 1 \times 6.5 = 0}$$

$$\text{or} \quad R_C = 7.04 \text{ kN} \uparrow = V_D.$$

From $\quad \Sigma V = 0 \quad$ in BC

$$V_B \uparrow + 7\cdot04 \uparrow = 13 \times 1 \cos\theta$$

$$V_B = 13 \cos\theta - 7\cdot04$$

$$= \frac{13 \times 12}{13} - 7\cdot04$$

$$= 4\cdot96 \uparrow.$$

$$H_B = 13 \times 1 \sin\theta = \frac{13 \times 5}{13} = 5 \leftarrow$$

$$V_A = V_B = 4\cdot96 \uparrow \quad H_A = H_B = 5 \leftarrow$$

$$M_A = 5 \times 10 = 50 \text{ kNm} \circlearrowright.$$

Check. $\quad V_A + V_D = 4\cdot96 + 7\cdot04 = 12\text{kN} \uparrow.$

$$\text{Downward vertical load} = 13 \times 1 \cos\theta = \frac{13 \times 1 \times 12}{13} = 12\text{kN} \downarrow.$$

$$\Sigma = 0.$$

$$\Sigma H = \overleftarrow{H_A} + H_D - 13 \times \overrightarrow{1} \sin\theta$$

$$13 \times 5$$

2.4. MATRIX FORMULATION OF REACTION PROBLEMS

Matrix formulation of structural problems has become a necessity these days. Matrix formulation facilitates use of electronic computers which enable very fast and efficient solution of structural problems. With matrix formulation and use of computers structural analysis becomes very general, compact and consistent. Moreover, the analysis becomes systematic. In addition, rules of matrix algebra can be applied to check structural stability, determinacy or indeterminacy. Hence matrix formulation of reaction problems is presented here in an elementary form. However, an exhaustive study on matrix and its application in modern structural analysis is presented in chapter 16.

Fig. 2.11(*a*) shows a simply supported beam. The equations of equilibrium of the beam are first formulated.

As $\Sigma M_A = 0$, taking moments about A.

$$-R_{BY} \times 9 + P_{1Y} \times 3 + P_{2y} \times 6 = 0$$

$$\text{or}\quad 3P_{1Y} + 6P_{2Y} - 9R_{BY} = 0$$

$$\text{or}\quad \frac{1}{3}P_{1Y} + \frac{2}{3}P_{2Y} - R_{BY} = 0.$$

Similarly putting $\Sigma F_Y = 0$

$$P_{1Y} + P_{2Y} - R_{AY} - R_{BY} = 0$$

Putting $\Sigma F_X = 0$

$$P_{1X} + P_{2X} - R_{AX} = 0.$$

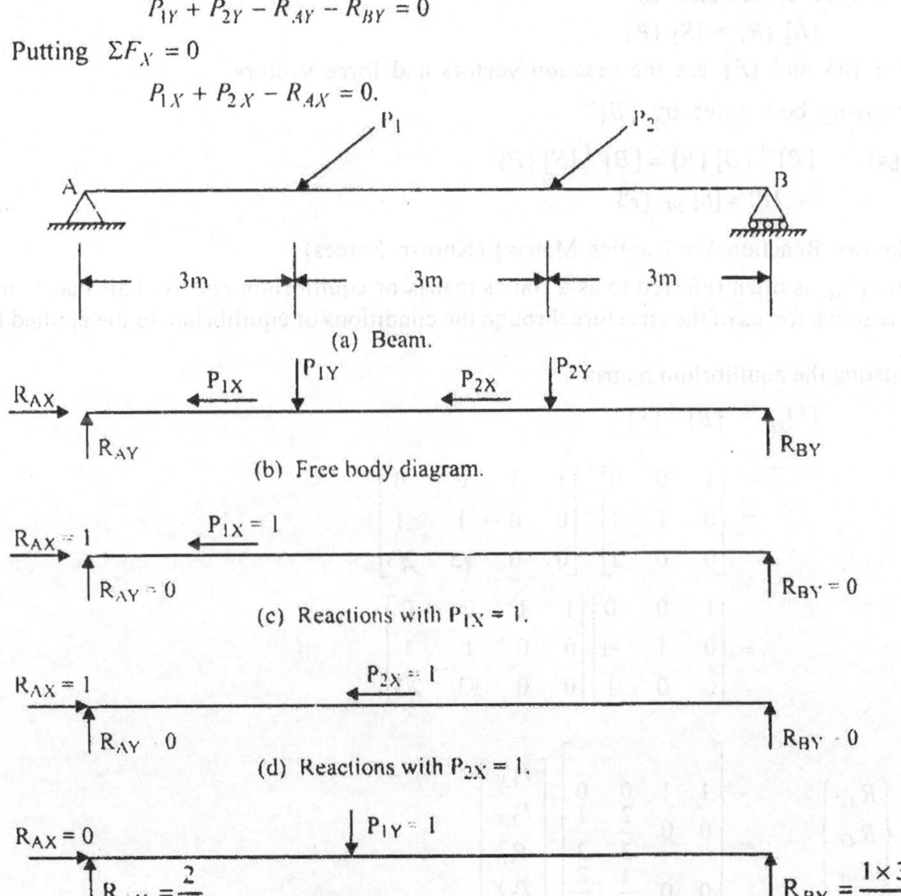

$R_{AX} = 0$ $P_{2Y} = 1$

$R_{AY} = \frac{1}{3}$ $R_{BY} = \frac{1 \times 6}{9} = \frac{2}{3}$

(f) Reactions with $P_{2Y} = 1$.

Fig. 2.11. Matrix Formulation of Reaction Problem.

Putting all reactions on the left side.

$$R_{BY} = \frac{1}{3} P_{1Y} + \frac{2}{3} P_{2Y}$$

$$R_{BY} + R_{AY} = P_{1Y} + P_{2Y}$$

$$R_{AX} = P_{1X} + P_{2X}$$

These can be easily expressed in the matrix form as —

$$\begin{bmatrix} 1 & 0 & 0 \\ 0 & 1 & 1 \\ 0 & 0 & 1 \end{bmatrix} \begin{Bmatrix} R_{AX} \\ R_{AY} \\ R_{BY} \end{Bmatrix} = \begin{bmatrix} 1 & 1 & 0 & 0 \\ 0 & 0 & 1 & 1 \\ 0 & 0 & \frac{1}{3} & \frac{2}{3} \end{bmatrix} \begin{Bmatrix} P_{1X} \\ P_{2X} \\ P_{1Y} \\ P_{2Y} \end{Bmatrix}$$

This can be abbreviated as

$$[B]\{R\} = [S]\{P\} \qquad \text{... (2.4)}$$

Where $\{R\}$ and $\{P\}$ are the reaction vectors and force vectors.

Multiplying both sides by $[B]^{-1}$

We get, $[B]^{-1}[B]\{R\} = [B]^{-1}[S]\{P\}$

or $\{R\} = [b]_{RP}\{P\}$ ... (2.5)

(Unknown Reactions) = (Statics Matrix) (Known Forces)

Where $[b]_{RP}$ is often referred to as a statics matrix or equilibrium matrix. This matrix merely relates the reactive forces of the structure through the conditions of equilibrium to the applied loads.

Evaluating the equilibrium matrix

$$[b]_{RP} = [B]^{-1}[S]$$

$$= \begin{bmatrix} 1 & 0 & 0 \\ 0 & 1 & 1 \\ 0 & 0 & 1 \end{bmatrix}^{-1} \begin{bmatrix} 1 & 1 & 0 & 0 \\ 0 & 0 & 1 & 1 \\ 0 & 0 & 1/3 & 2/3 \end{bmatrix}$$

$$= \begin{bmatrix} 1 & 0 & 0 \\ 0 & 1 & -1 \\ 0 & 0 & 1 \end{bmatrix} \begin{bmatrix} 1 & 1 & 0 & 0 \\ 0 & 0 & 1 & 1 \\ 0 & 0 & 1/3 & 2/3 \end{bmatrix}$$

and

$$\begin{Bmatrix} R_{AX} \\ R_{AY} \\ R_{BY} \end{Bmatrix} = \begin{bmatrix} 1 & 1 & 0 & 0 \\ 0 & 0 & \frac{2}{3} & \frac{1}{3} \\ 0 & 0 & \frac{1}{3} & \frac{2}{3} \end{bmatrix} \begin{bmatrix} P_{1X} \\ P_{2X} \\ P_{1Y} \\ P_{2Y} \end{bmatrix}$$

$$= \begin{Bmatrix}1\\0\\0\end{Bmatrix} P_{1X} + \begin{Bmatrix}1\\0\\0\end{Bmatrix} P_{2X} + \begin{Bmatrix}0\\2/3\\1/3\end{Bmatrix} P_{1Y} + \begin{Bmatrix}0\\1/3\\2/3\end{Bmatrix} P_{2Y} \qquad \ldots (2.6)$$

It can be observed from the right hand side of Eqn (2.6) that each column vector and associated load term gives the reaction induced by that load. Therefore the vector itself corresponds to the reactions induced by a unit value of the associated force. This can be seen from Figs. 2.11(*c*) to (*f*). In general any element of the statics or equilibrium matrix bij is the *i*th reaction quantity due to unit value of the *j*th force. This is expressed in matrix form as —

$$\begin{Bmatrix}\cdot\\R_i\\\cdot\end{Bmatrix} = \begin{bmatrix} & & \\ & bij & \\ & & \end{bmatrix} \begin{Bmatrix}\cdot\\P_J\\\cdot\end{Bmatrix}$$

or $\quad R_i = \cdots bij\, P_J + \cdots$

The problem given in Example 2.3 is now solved by Matrix Method.

The equilibrium equations are :

$$R_{AX} - R_{BX} = -75$$

$$R_{AY} + R_{BY} = 125$$

$$3R_{BX} + 9R_{BY} = 525$$

$$R_{BX} - 3/4\, R_{BY} = 0$$

In matrix form this is equal to

$$\begin{bmatrix} 1 & -1 & 0 & 0 \\ 0 & 0 & 1 & 1 \\ 0 & 3 & 0 & 9 \\ 0 & 1 & 0 & -3/4 \end{bmatrix} \begin{Bmatrix} R_{AX} \\ R_{BX} \\ R_{AY} \\ R_{BY} \end{Bmatrix} = \begin{Bmatrix} -75 \\ 125 \\ 525 \\ 0 \end{Bmatrix}.$$

To avoid zero at a diagonal the 2nd and 3rd equations are inter changed giving the new matrix equations as follows.

$$\begin{bmatrix} 1 & -1 & 0 & 0 \\ 0 & 3 & 0 & 9 \\ 0 & 0 & 1 & 1 \\ 0 & 1 & 0 & -3/4 \end{bmatrix} \begin{Bmatrix} R_{AX} \\ R_{BX} \\ R_{AY} \\ R_{BY} \end{Bmatrix} = \begin{Bmatrix} -75 \\ 525 \\ 125 \\ 0 \end{Bmatrix}$$

or

$$\begin{Bmatrix} R_{AX} \\ R_{BX} \\ R_{AY} \\ R_{BY} \end{Bmatrix} = \begin{bmatrix} 1 & -1 & 0 & 0 \\ 0 & 3 & 0 & 9 \\ 0 & 0 & 1 & 1 \\ 0 & 1 & 0 & -3/4 \end{bmatrix}^{-1} \begin{Bmatrix} -75 \\ 525 \\ 125 \\ 0 \end{Bmatrix}$$

$$= \begin{bmatrix} 1 & \frac{1}{15} & 0 & 12/15 \\ 0 & 1/15 & 0 & 12/15 \\ 0 & -4/45 & 1 & 4/15 \\ 0 & +4/45 & 0 & -4/15 \end{bmatrix} \begin{Bmatrix} -75 \\ 525 \\ 125 \\ 0 \end{Bmatrix}$$

Then $R_{AX} = -75 + \frac{525}{15} + 0 + 0$

$= -40$ kN.

$$R_{BX} = \frac{525}{15} = 35\,kN.$$

$$R_{AY} = -75 \times 0 - \frac{4}{45} \times 525 + 125 + 0$$

$$= 78 \cdot 33 \text{ kN}$$

$$R_{BY} = \frac{4}{45} \times 525 = 46 \cdot 67 \text{ kN as before.}$$

2.5. MATRIX SOLUTION AND STRUCTURAL BEHAVIOUR

(*i*) **Independent Equations and Equilibrium Matrices** — A valid set of equilibrium equations is known as an independent set. An invalid set is known as a set of dependent equations. It can be shown that a dependent set of equations cannot be solved for a unique set of unknowns. Dependence arises when any of the equations can be written as a linear combination of any or all of the remaining equations. This condition may also represent a condition of geometric instability. A set of equations is singular when the coefficient matrix of the unknown has a determinant equal to zero. In such a case the matrix cannot be inverted and no solution of the unknown is possible.

(*ii*) **Geometric Instability** — Another condition that leads to a singular set of equations arises when the body of the structure is improperly restrained against motion Although there may be an adequate number of support conditions, their arrangements may be such that they cannot resist motion due to an arbitrarily applied load. This condition does not merely represent a poor choice of equations but is indicative of the actual physical condition of the system. A rigid body will be unstable if the reactive forces acting on it are all concurrent at one point. As a result, the system cannot react against a moment. This is shown in Fig. 2.12.

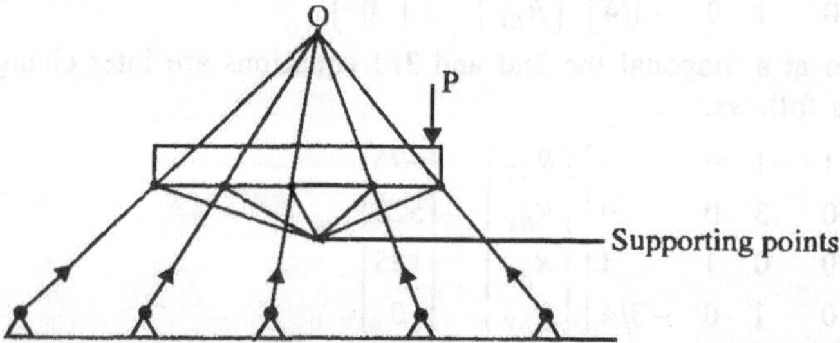

Fig. 2.12. Concurrent reactions of a rigid body. Cannot resist moment about O.

This is obvious if we take moments of external load about *O*. Reactions cannot offer any resistance against moment about *O*.

It is also important to note that the instability of this structure can be established independently of the loads. That is, the statics matrix does not involve applied loads, it only reflects the geometry of the structure and the location and orientation of the reactive forces. By examining the matrix alone we may find state of stability under any arbitrary set of loads. If any arbitrary loading cannot be resisted by a structure, the structure is unstable.

Geometric instability also arises when there is not a sufficient number of reactive forces to resist the applied loads. The situation is easily identified by comparing the number of equations of equilibrium available with the number of unknown reactive forces. Since there are three equations of equilibrium for a two dimensional body, there must be at least three independent reactive forces to prevent motion. These reactive forces should neither be concurrent nor parallel. Consider the equilibrium equations of Example 2.3.

$$R_{AX} - R_{BX} = -75$$

$$R_{AY} + R_{BY} = 125$$

$$3R_{BX} + 9R_{BY} = 525$$

$$R_{BX} - 3/4\, R_{BY} = 0.$$

Here we have four equations with four unknowns and hence the unknowns can be found out out. Whether a unique set of solution can be obtained or not, can be easily checked when the equations are presented in matrix form and we try to calculate its inverse. Inverse of a matrix exists only when the determinant of the matrix is nonzero and then unique solution exists. When the determinant of a matrix has a zero value, the matrix is said to be "Singular" and it has no inverse.

Consider again the Example 2.3. The supports are roller and one hinge and there are three unknown reactions, one due to roller and two due to hinge. As there are three equations of statics for planar structures number of equation matches with the number of unknowns and hence unique solution is possible. The fourth equation occurs due to the inclination of the roller to the horizontal and gives a relation between R_{BX} and R_{BY}. Now consider two alternate support conditions.

1. Both the supports are rollers : Obviously the structure is unstable as rollers will roll due to lateral load on AC. The new equations with R_{AX} equal to zero are —

$$-R_{BX} = -75$$

$$R_{AY} + R_{BY} = 125$$

$$3R_{BX} + 9R_{BY} = 525$$

$$R_{BX} - 3/4\, R_{BY} = 0.$$

Thus we have four equations with three unknowns and no unique solution of the unknowns is possible.

2. When both the supports are hinges — In this case the equation $R_{BX} - 3/4\, R_{BY} = 0$ is not valid as at a hinge, the reaction can take any direction.

The new equations are —

With $\quad \Sigma F_X = 0 \qquad R_{AX} - R_{BX} = -75.$

$\qquad\quad \Sigma F_Y = 0 \qquad R_{AY} + R_{BY} = 125.$

$\qquad\quad \Sigma M_A = 0 \qquad 3R_{BX} + 9R_{BY} = 525.$

The matrix expression becomes

$$\begin{bmatrix} 1 & -1 & 0 & 0 \\ 0 & 3 & 0 & 9 \\ 0 & 0 & 1 & 1 \end{bmatrix} \begin{Bmatrix} R_{AX} \\ R_{BX} \\ R_{AY} \\ R_{BY} \end{Bmatrix} = \begin{Bmatrix} -75 \\ 525 \\ 125 \end{Bmatrix}$$

For inversion of matrix to exist, it should be square. But this is 3 × 4 matrix, hence inversion is not possible and unique solution does not exist. With four unknown forces and three equations of statical equilibrium, the structure is statically indeterminate. Fig. 2.13 gives some examples of stable, unstable and indeterminate structure.

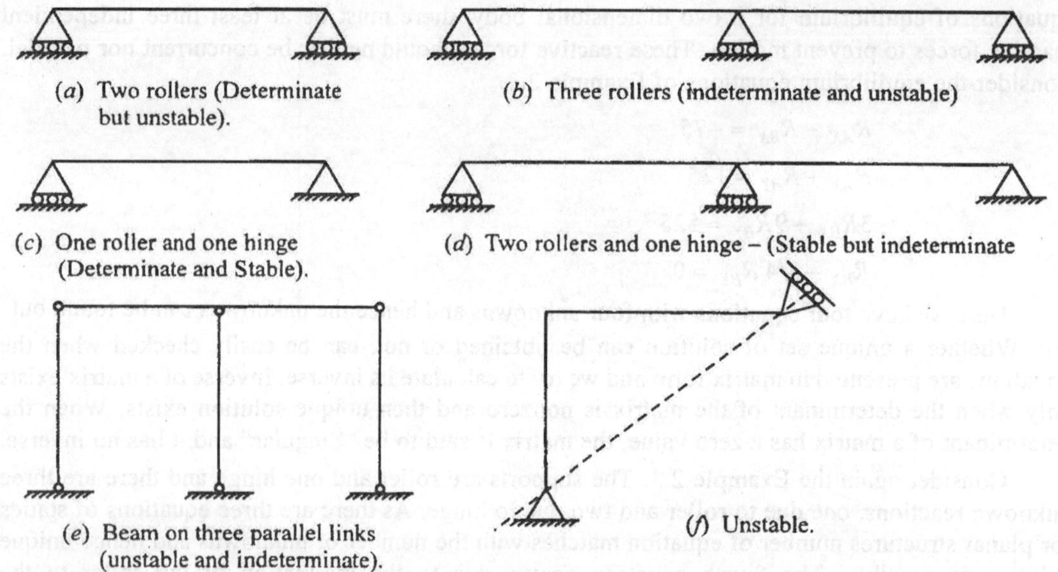

(a) Two rollers (Determinate but unstable).

(b) Three rollers (indeterminate and unstable)

(c) One roller and one hinge (Determinate and Stable).

(d) Two rollers and one hinge – (Stable but indeterminate

(e) Beam on three parallel links (unstable and indeterminate).

(f) Unstable.

Fig. 2.13. Stable, Unstable, Determinate and Indeterminate Structure.

2.6. GENERAL CRITERIA FOR DETERMINING STABILITY, DETERMINACY AND INDETERMINACY OF A STRUCTURAL MEMBER

Very broadly, a planar structure may be a beam, a frame or a truss. A structure may be externally indeterminate or internally indeterminate. A structure is said to be externally indeterminate if the available equations of equilibrium are not sufficient to determine the external reactions. Internal indeterminacy occurs when the internal forces or stress resultants cannot be determined from available equations of equilibrium. A frame consists of rigid joints, angles between the members do not change under load and members are subjected to bending moment, shear force and axial forces due to external loading. As a result if a member is cut there will be three unknowns. On the other hand a truss is pin jointed and members of a truss are subjected to axial compression or tension only. Hence if a truss member is cut, there will be one internal force only.

(*i*) **Plane Truss** — The simplest plane truss is a triangle with three joints and three bars. (Fig. 2.14). Each additional joint can be obtained by adding two bars to the joints already formed. It can be observed that for the first three joints we require three members and for each subsequent joint we require two members. If m be the number of members and j number of joints, then

$$m = 3 + 2\,(j - 3) = 2j - 3.$$

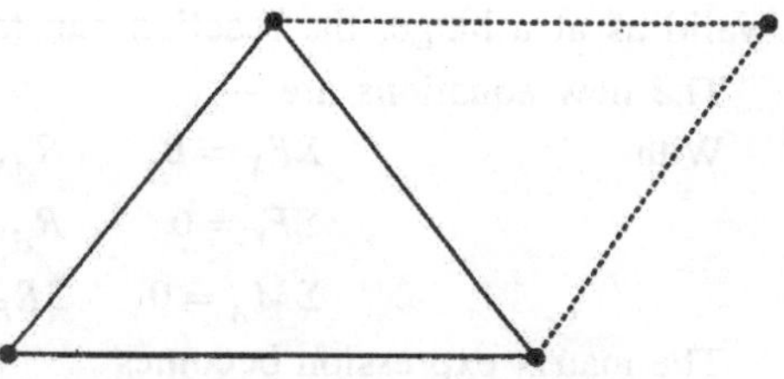

Fig. 2.14. Plane Truss formation.

If a simple plane truss has $m < (2j - 3)$, the truss is unstable. If $m > (2j - 3)$, the truss should be examined for stability and internal determinacy. For external determinacy, the supports of the truss should be examined. As there are three equations of equilibrium there should be three reactive forces which are neither parallel nor concurrent. This will ensure a stable and determinate structure. Otherwise the structure will be unstable or externally indeterminate as in the case of beams. The condition $m = 2j - 3$ is a necessary but not sufficient condition for stability or determinacy of a truss. Fig. 2.15 shows two similar trusses with different arrangement of bars.

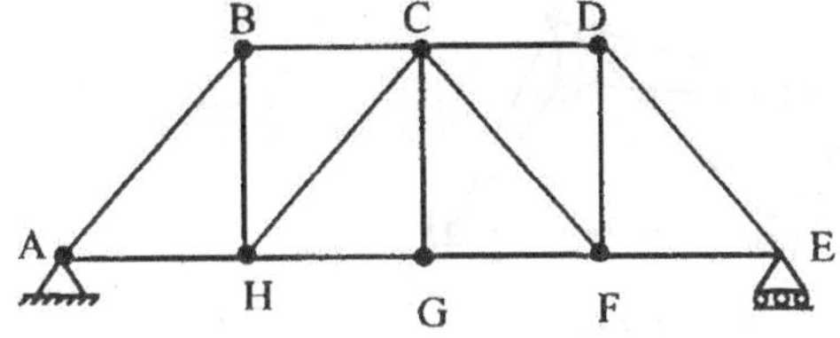

No. of joints j = 8
No. of members = 13
No of reactions = 3
Externally Determinate.
i = (2j – 3) – m
= (8 × 2 – 3) – 13
= 0.

(*a*) Stable and Determinate Structure.

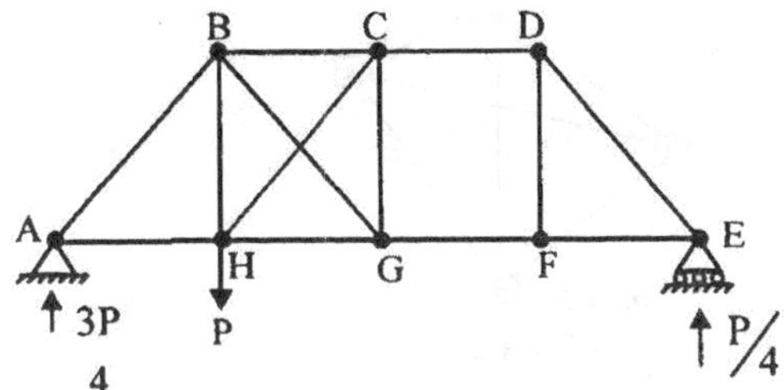

No. of joints j = 8
No. of members = 13
No of reactions = 3
Externally Determinate.
i = (8 × 2 – 3) – 13
= 0.
But panel BCGH is indeterminate and panel CDFG is unstable.

(*b*) Internally indeterminate and unstable.

Fig. 2.15. Plane Truss.

The panel *CDFG* cannot take the shear $P/4$. With removal of member *CF* the panel has become a mechanism and therefore unstable. A structure is considered geometrically stable if for any incipient movement due to any conceivable system of applied load, an elastic resistance to this movement is immediately developed. In the panel *CDFG* no such immediate resistance will be developed when there will be lateral joint movement.

Furthermore because of the additional bar *BG*, the panel *BCGH* has become over stiff that is indeterminate as all member forces cannot now be found out from statics alone. Provided the structure is stable some interchange between the number of members and number of reaction components is possible so that degree of indeterminacy.

$$i = (m + r) - 2j$$

Even if $(m + r)$ is equal to $2j$ structure may not be stable and determinate as *m* may be less than that required for stability whereas number of reactions may be more than that required for stability and determinacy. This is shown in Example 2.7(4), (7) (9) and (10).

***(ii)* Space Truss** — In the case of pin jointed space trusses there are three equations of statics at each joint namely,

$$\Sigma F_X = 0 \qquad \Sigma F_Y = 0 \qquad \text{and} \qquad \Sigma F_Z = 0$$

If there are *j* joints (including supports) there are 3*j* equations of equilibrium. If *m* is the number of members, *r* the number of reactions, for determinacy and stability necessary but not sufficient condition is

$$3j = m + r.$$

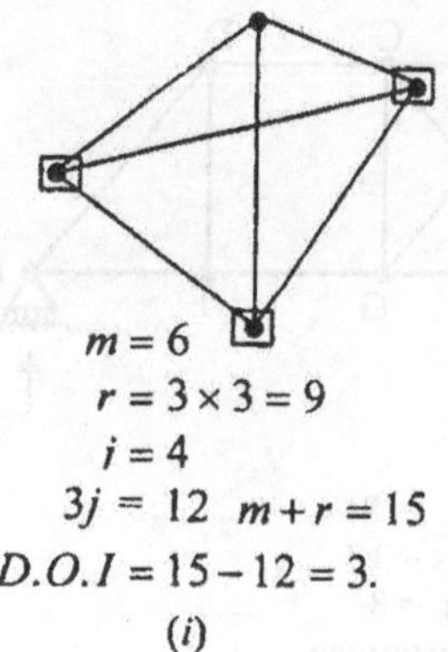

$m = 6$

$r = 3 \times 3 = 9$

$j = 4$

$3j = 12 \quad m + r = 15$

$D.O.I = 15 - 12 = 3.$

(*i*)

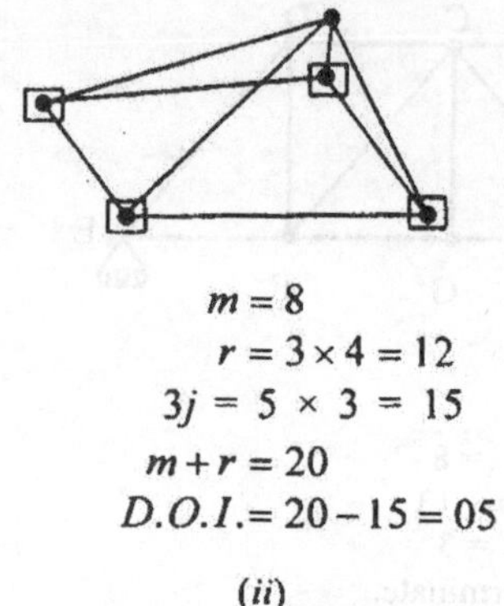

$m = 8$

$r = 3 \times 4 = 12$

$3j = 5 \times 3 = 15$

$m + r = 20$

$D.O.I. = 20 - 15 = 05$

(*ii*)

Fig. 2.16. Space Truss.

(*iii*) **Plane Frame** — A plane frame consists of members rigidly connected at the joints. The supports may be fixed, hinged or rollers. Depending on the type of supports the plane frame will be unstable (if there are two roller supports), determinate, (one roller and one hinge) or indeterminate (two hinges or two ends fixed), as shown in Fig. 2.17. If there is an internal hinge the moment at the hinge will be zero giving an equation of construction or condition as already explained in section 2.3. Fig. 2.17 shows some unstable, externally determinate and externally indeterminate frame structures.

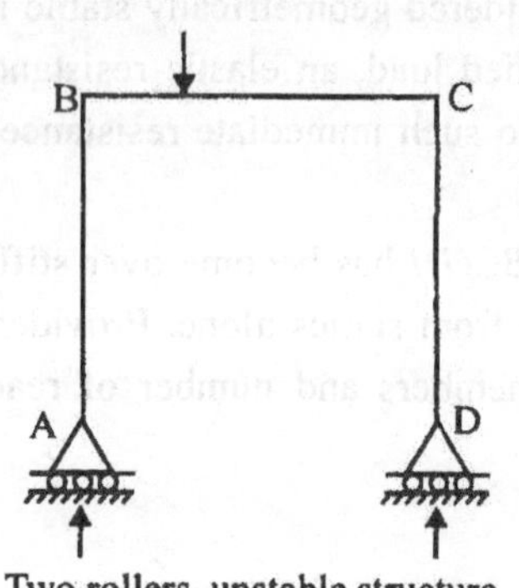

Two rollers, unstable structure.

(*i*)

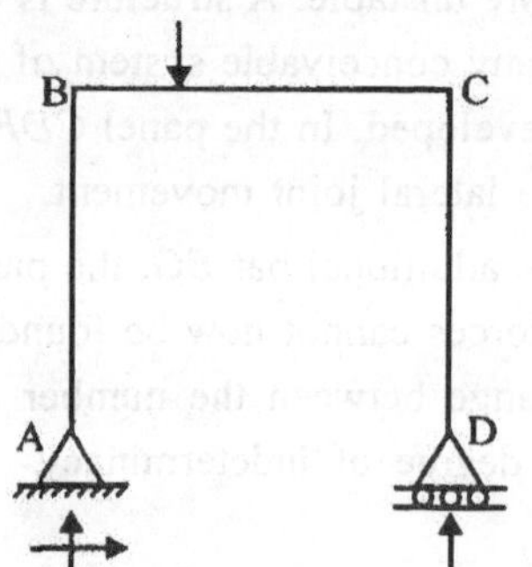

One roller 1 hinge, stable determinate structure.

(*ii*)

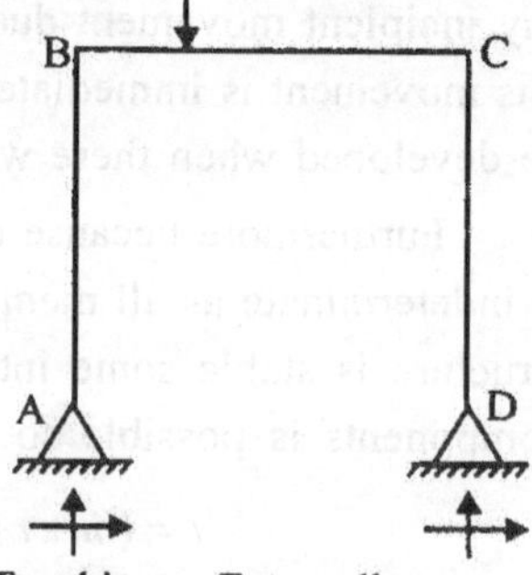

Two hinges, Externally Indeterminate.

(*iii*)

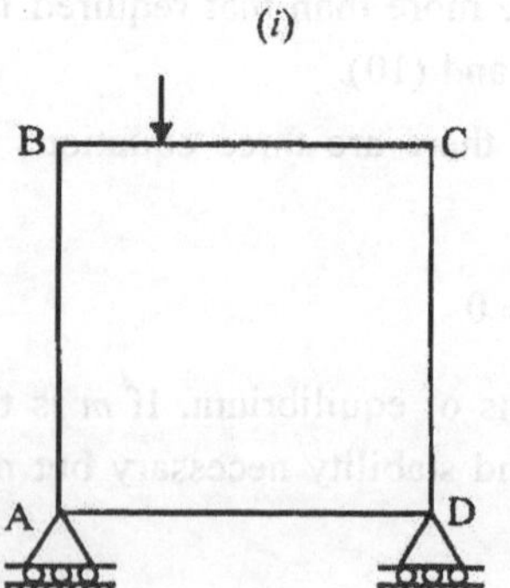

Externally unstable (−1) Internally Indeterminate (3) net $i = 3 + (-1) = 2$.

(*iv*)

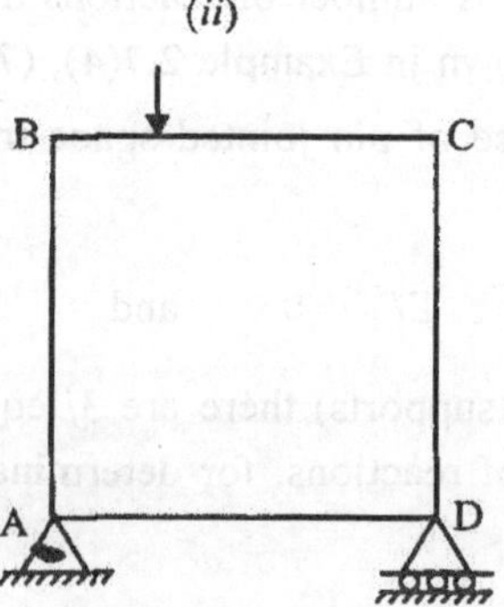

Externally stable – 0 Internally Indeterminate = 3 $i = 3 + 0 = 3$.

(*v*)

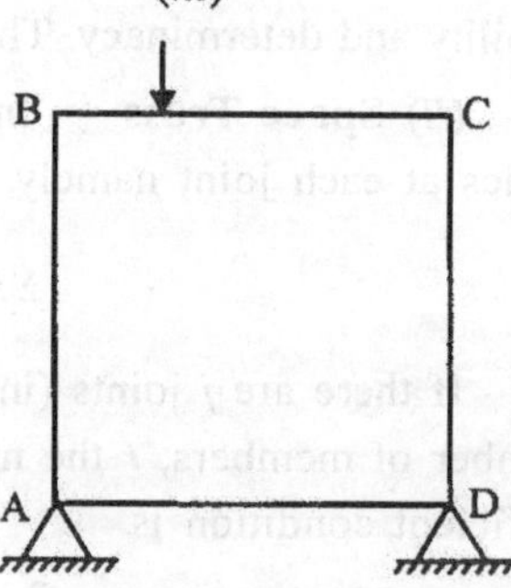

Externally Indeterminate = 1 Internally Indeterminate = 3 $i = 1 + 3 = 4$.

(*vi*)

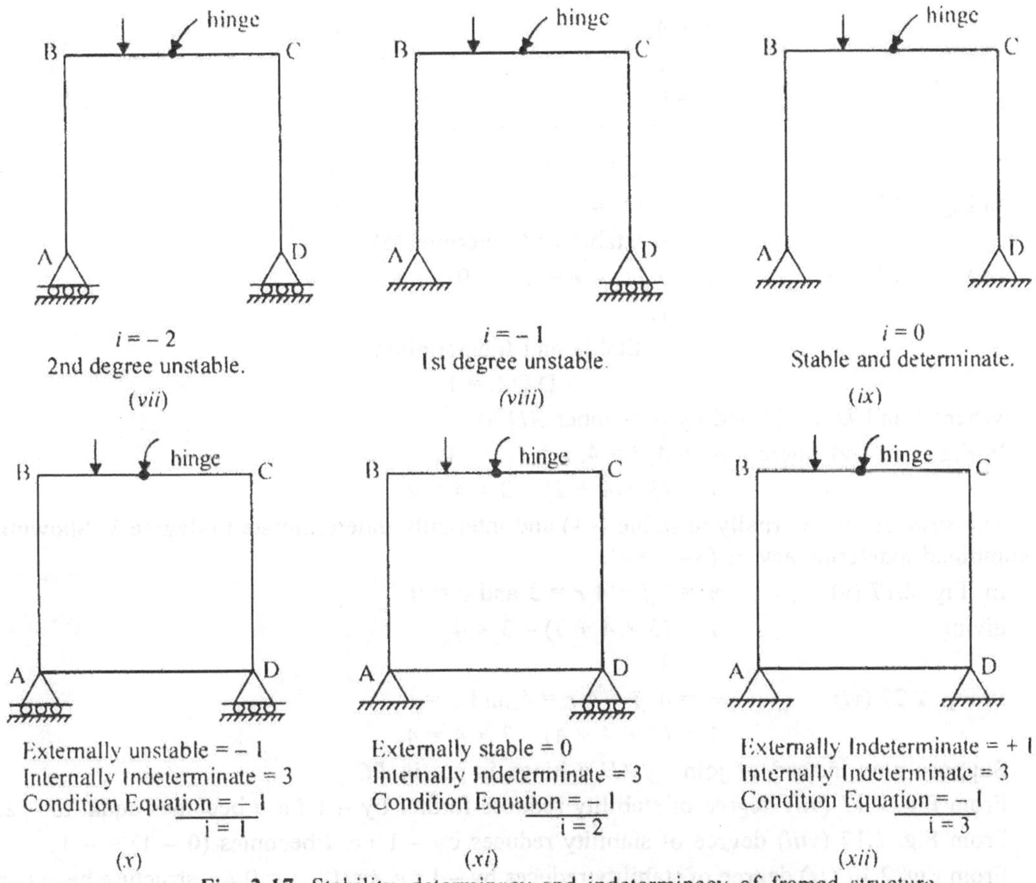

Fig. 2.17. Stability, determinacy and indeterminacy of framed structure.

Internal indeterminacy of a structure occurs when internal forces in a member (three in number) cannot be found out though reactions can be computed by applying the equations of equilibrium. If in Fig. 2.17 (*i*) *AD* is joined by a member, the frame still remains unstable though it becomes internally indeterminate. If in Fig. 2.17 (*ii*) *AD* is joined by a member, the beam though externally determinate and stable becomes again internally indeterminate. Similarly joining *AD* by a member in Fig. 2.17 (*iii*) makes the frame both internally and externally indeterminate.

To find the degree of indeterminacy of a frame we compute the number of joints *j*, no. of members *m*, no of reaction components *r* and number of equation of condition *c*.

In the case of a truss each member had one unknown stress resultant, in the case of a frame member there are three unknown stress resultants *i.e. B.M*, Axial Force and Shear Force. Hence total number of unknowns in members will be $3m$. Similarly each joint (including the supports) will give three equations. Hence the rigid plane frame is statically determinate if

$$3j = 3m + r.$$

If there are c condition equations then number of available equations increases to $3j + c$ which for determinacy should be equal to $3m + r$

Hence degree of indeterminacy

$$i = (3m + r) - (3j + c)$$

If we apply the above formula to

Fig. 2.17 (*i*) $m = 3$

$j = 4$
$r = 2$
$c = 0.$
$i = (3 \times 3 + 2) - 3 \times 4$
$= -1$ (unstable)

In Fig. 2.17 (*ii*) $m = 3\ j = 4\ r = 3\ c = 0$
$i = 0$ (stable and determinate).

In Fig. 2.17 (*iii*) $m = 3\ j = 4\ r = 4\ c = 0$
$i = 1.$

Stable and Indeterminate
D.O.I. = 1

When A and D are joined by a member AD as
In Fig. 2.17 (*iv*) where $m = 4, j = 4, r = 2, c = 0.$
$i = (3 \times 4 + 2) - 3 \times 4 = 2$

The structure is externally unstable (–1) and internally indeterminate to degree 3. Showing the combined indeterminacy of (3–1) = 2.

In Fig. 2.17 (*v*) $m = 4, j = 4\ r = 3$ and $c = 0$
giving $i = (3 \times 4 + 3) - 3 \times 4$
$= 3.$

In Fig. 2.27 (*vi*) $m = 4, j = 4\ r = 4$ and $c = 0.$
$i = (3 \times 4 + 4) - 3 \times 4 = 4.$

Suppose now instead of joining AD a hinge is put in BC.

From Fig. 2.17 (*vii*) degree of stability reduces further by – 1 *i.e.* i becomes equal to – 2.

From Fig. 2.17 (*viii*) degree of stability reduces by – 1 *i.e.* i becomes $(0 - 1) = -1$

From Fig. 2.17 (*ix*) degree of stability reduces by – 1 *i.e.* $i = 1 - 1 = 0$ *i.e.* structure becomes stable and determinate.

With member AD and with hinge at BC as shown in Fig. 2.17 (*x*) degree of indeterminacy reduces by 1 and i becomes equal $(2 - 1) = 1$. For external stability and determinacy we should have 1 hinge and 1 roller. *i.e.* 1 roller should be made hinge and at the hinge we have two unknowns. So degree of indeterminacy = 2 – 1 = 1.

For Fig. 2.17 (*xi*) D.O.I = 2

For Fig. 2.17 (*xii*) D.O.I = 3

(*iv*) Space Frame — At each joint of a space frame six equations can be written

$\Sigma F_X = 0, \quad \Sigma F_Y = 0 \quad \Sigma F_Z = 0$

$\Sigma M_X = 0, \quad \Sigma M_Y = 0 \quad \Sigma M_Z = 0$

Each member has six unknown $(F_x, F_y, F_z, M_x, M_y, M_z)$ stress resultants. Hence for statical determinacy we should have

$$6j = 6m + r$$

and $i = (6m + r) - 6j.$

Fig. 2.18. Indeterminate space frame.

Fig. 2.18 shows a three dimensional frame. Here

No of members $= 13$ No of joints $= 12$

Each support has six unknown reaction components. Hence

$$r = 6 \times 6 = 36$$

$$i = (6m + r) - 6j$$
$$= 6 \times 13 + 36 - 6 \times 12$$
$$= 78 + 36 - 72 = 42.$$

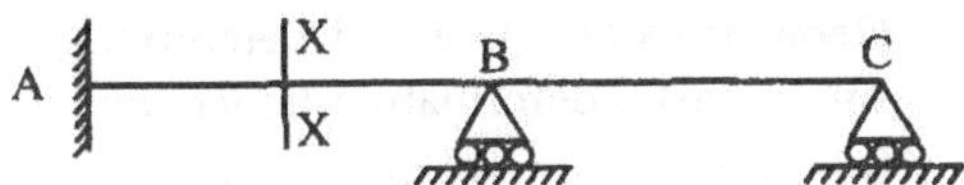

Fig. 2.19. Indeterminate Beam.

To determine the indeterminacy of a beam or frame, it is wise to remember two results. (Fig 2.19)

(1) A cantilever is a determinate structure

(2) When the member of a beam or frame is cut, there will be three stress resultants. If the frame is cut at X we have three unknown stress resultants. On the left the cantilever is a determinate structure. On the right we have an unstable structure with indeterminacy equal to (–1). Hence the D.O.I of the beam is $(3 - 1) = 2$.

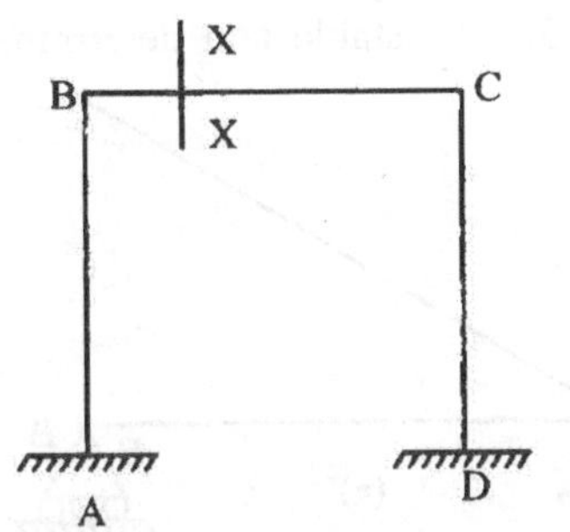

Fig. 2.20. Indeterminacy of Portal frame.

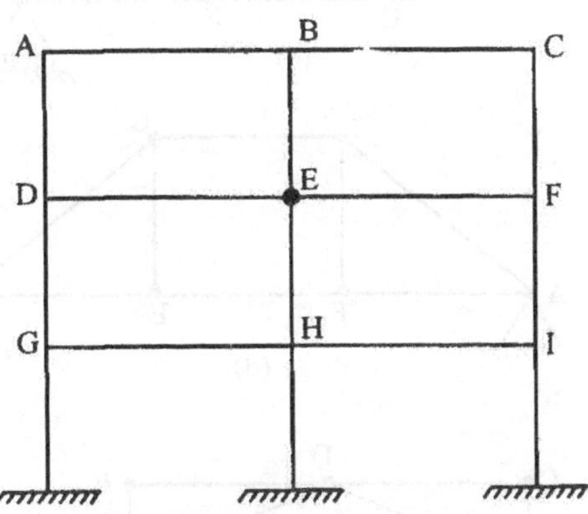

Fig. 2.21. Effect of hinge at a joint.

The portal frame if cut at X releases three stress resultants. Hence the degree of indeterminacy of the portal frame is 3. Any frame should be converted to a number of open cantilever trees which are all determinate structures. The addition subtraction or cut required to convert to on open ended cantilever will immediately give degree of indeterminacy.

(*v*) Hinges at a joint — It has already been explained that a hinge gives one condition equation when it is placed at any intermediate point of a beam or frame. Sometimes, however, it may be placed at a joint as in E as shown in Fig. 2.21. As a result moment at E of all the four members EB, ED, EH and EF will be zero. But as ΣM at E will be automatically zero, we get three additional condition equations. In general the net number of condition equations at a joint is one less than the number of members meeting at the joint. Hence the number of condition equations $= (4 - 1) = 3$.

Example 2.7 Check the external and internal conditions and determinacy of the plane trusses shown in Fig. 2.22(*a*) to (*k*)

(*a*) $m = 2$

$j = 3$ $m + r = 6$

$r = 4$ $2j = 6.$

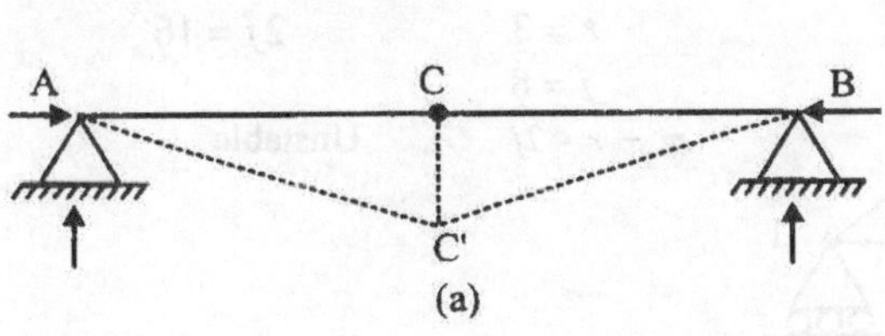

The truss is geometrically unstable as there will be substantial change of shape before equilibrium of joint C is obtained. This is known as critical form of the truss.

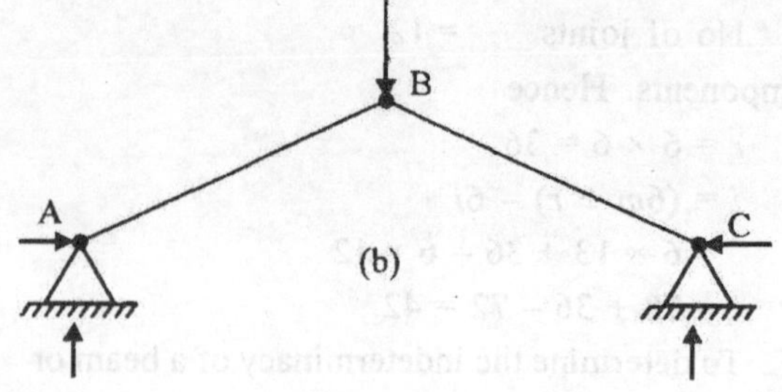

(*b*) $m = 2$

$j = 3$ $\quad m + r = 2j$

$r = 4$

Geometrically stable and determinate. Very small elastic deformation under load.

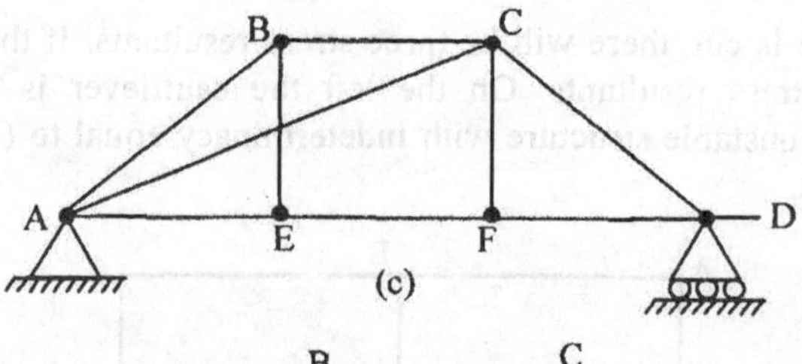

(*c*) $m = 9$ $\quad m + r = 12$

$j = 6$ $\quad 2j = 12$

$r = 3.$

Stable and determinate structure

(*d*) $m = 8$ $\quad m + r = 12$

$j = 6$ $\quad 2j = 12$

$r = 4$

Internally unstable. Externally Indeterminate

(*e*) $m = 3$ $\quad m + r = 6$

$j = 3$ $\quad 2j = 6$

$r = 3.$ $\quad$ **stable and determinate**

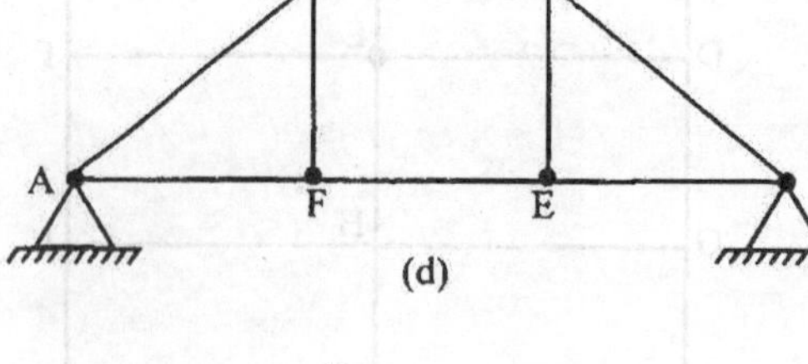

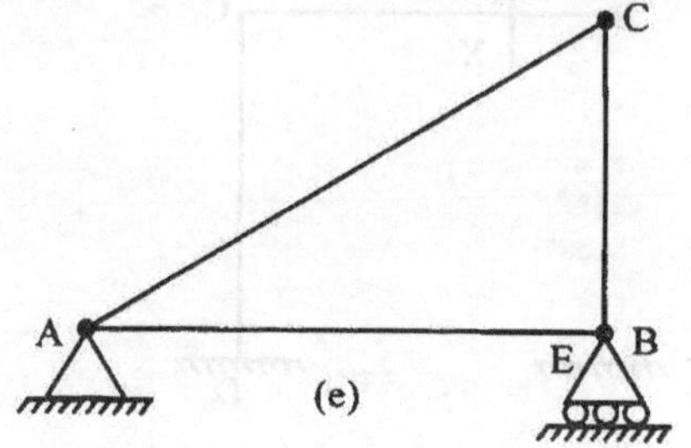

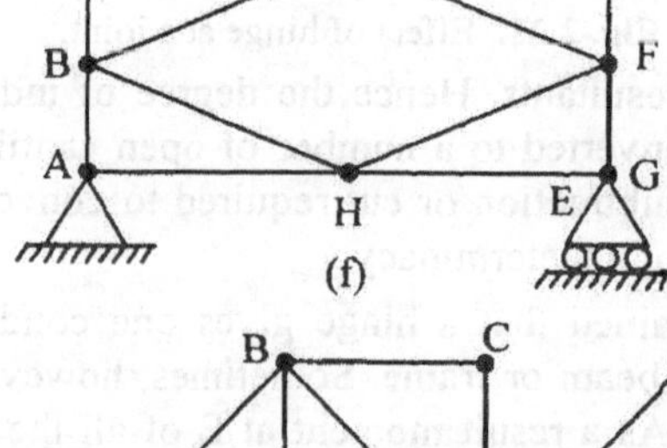

(*f*) $m = 12$ $\quad m + r = 15$

$j = 8$ $\quad 2j = 16$

$r = 3$ $\quad$ unstable

$m + r < 2j$

(g)

(*g*) $m = 12$ $\quad m + r = 16$

$j = 8$ $\quad 2j = 16$

$r = 4$

Externally Indeterminate
Internally unstable.

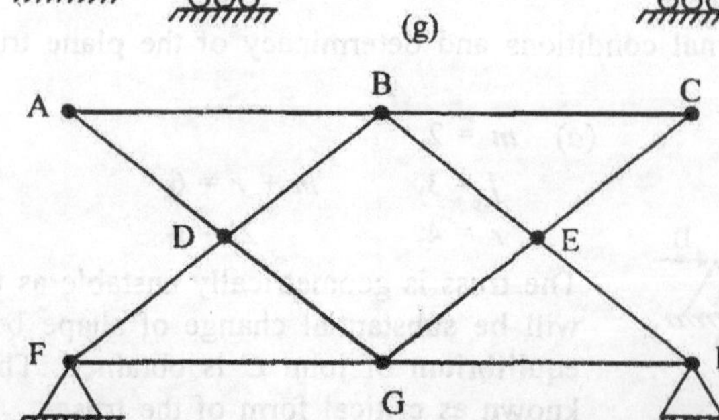

(*h*) $m = 12$ $\quad m + r = 15$

$r = 3$ $\quad 2j = 16$

$j = 8$

$m + r < 2j$ $\quad$ Unstable

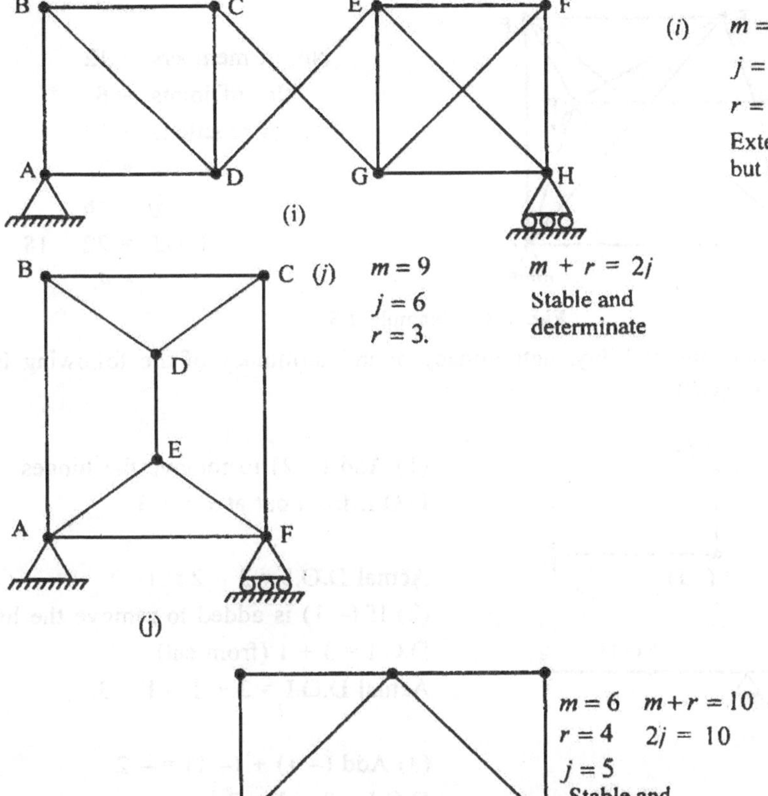

(*i*) $m = 13$, $j = 8$, $r = 3$; $m + r = 16$, $2j = 16$

Externally determinate but internally unstable.

(*j*) $m = 9$, $j = 6$, $r = 3$. $m + r = 2j$

Stable and determinate

$m = 6$, $r = 4$, $j = 5$; $m + r = 10$, $2j = 10$

Stable and Determinate

Fig. 2.22. Example 2.7.

Example 2.8 — Check the stability, determinacy or indeterminacy of the following space trusses given in Fig. 2.23.

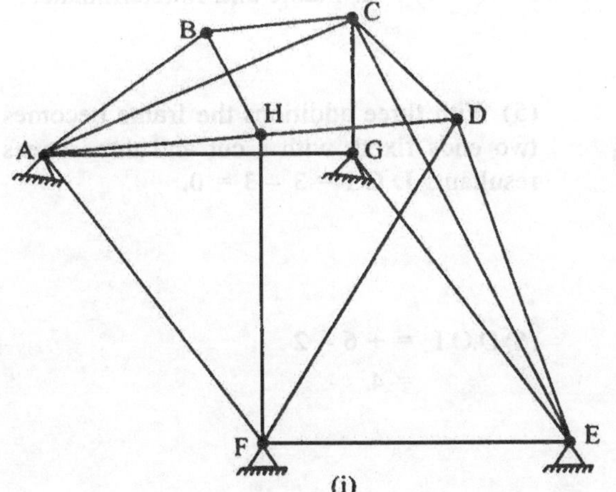

No. of members $= 16$

No. of joints $= 8$

No. of reactions $= 3 \times 4 = 12$

$m + r = 16 + 12 = 28$

$3j = 24$

No. of indeterminacy $= 4$.

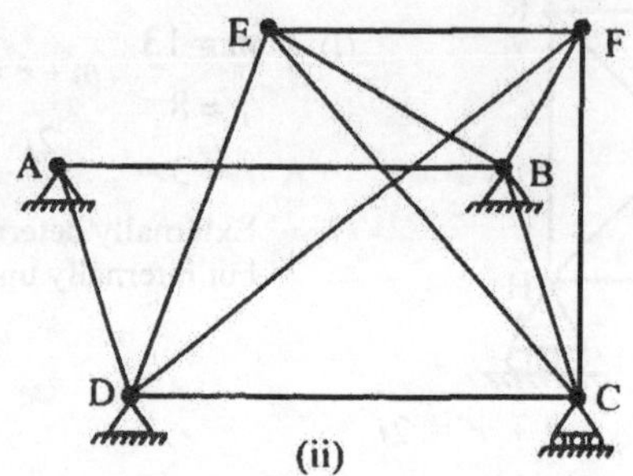

No. of members = 12
No. of joints = 6
No. of reactions = 10
$m + r = 22$
$3j = 18$
D.O.I = 22 – 18
= 4.

Fig. 2.23. Example 2.8.

Example 2.9 — Compute the stability, determinacy or indeterminacy of the following frames shown in Fig. 2.24 (*i*) to (*viii*).

(1) 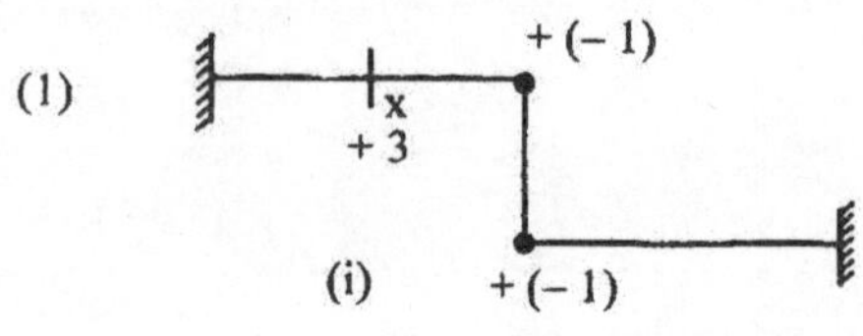

(1) Add (– 2) to remove the hinges.
D.O.I. from cut at x = + 3.

Actual D.O.I = 3 – 2 = 1

(2) 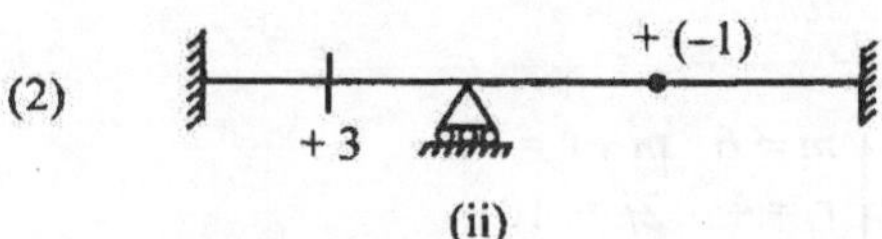

(2) If (– 1) is added to remove the hinge.
D.O.I = 3 + 1 (from cut)
Actual D.O.I = 3 + 1 – 1 = 3.

(3) 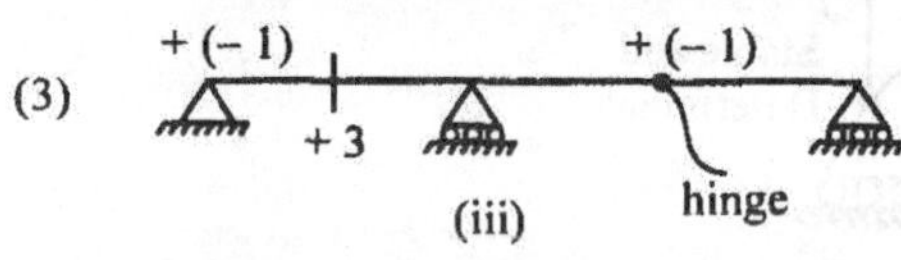

(3) Add (– 1) + (– 1) = – 2
D.O.I = 3 – 1 = 2 from cut.
Actual 2 – 2 = 0.
Externally and Internally Determinate.

(4) 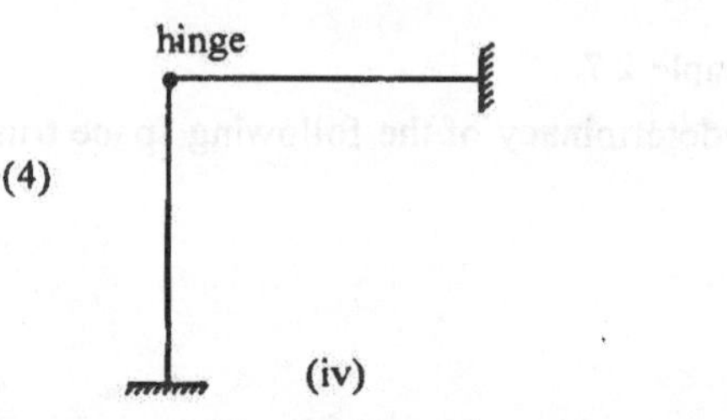

(4) Adding (– 1) to remove the hinge, the structure becomes rigid frame. With a cut we have three stress resultants. Actual D.O.I. = 3 + (– 1) = 2. Stable and Indeterminate.

(5) 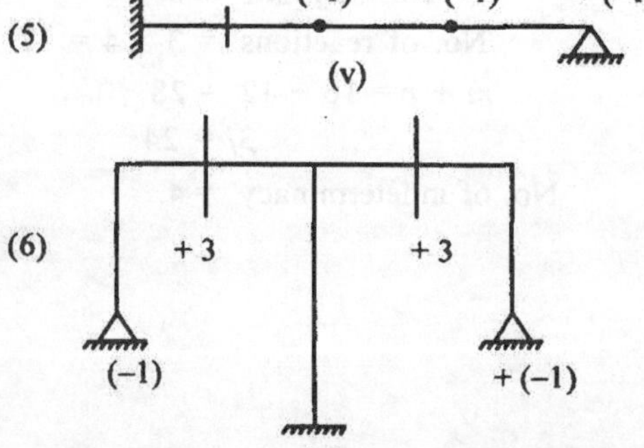

(5) With three additions the frame becomes two ends fixed. with a cut and three stress resultants D.O.I = 3 – 3 = 0.

(6) D.O.I = + 6 – 2
= 4.

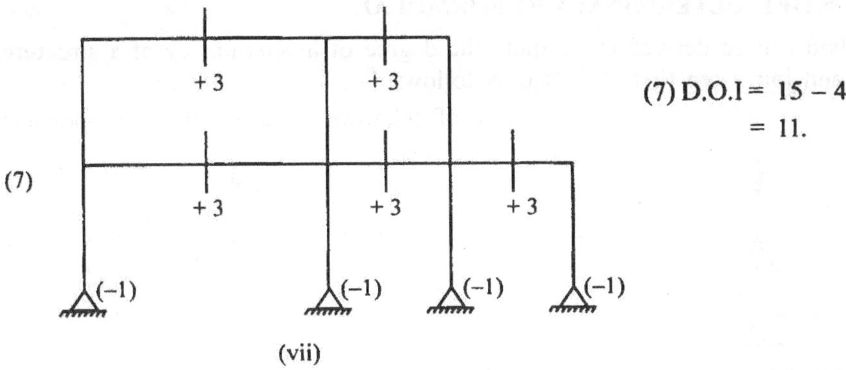

(7) D.O.I = 15 − 4
= 11.

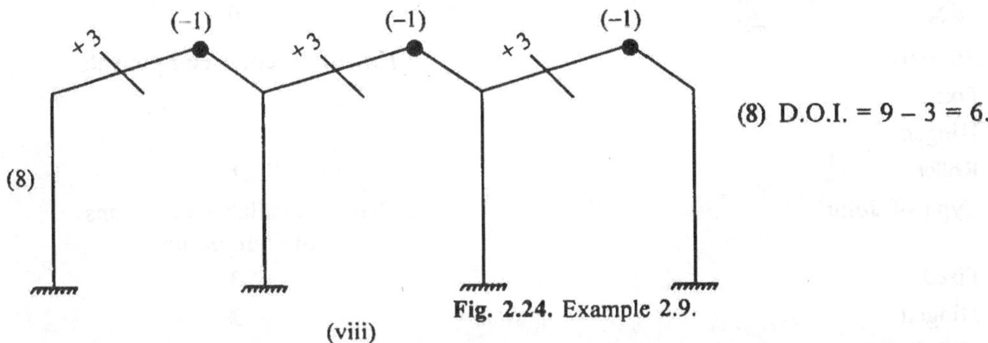

(8) D.O.I. = 9 − 3 = 6.

Fig. 2.24. Example 2.9.

Example 2.10 — Check the stability, determinacy or indeterminacy of the following space frames shown in Fig. 2.25.

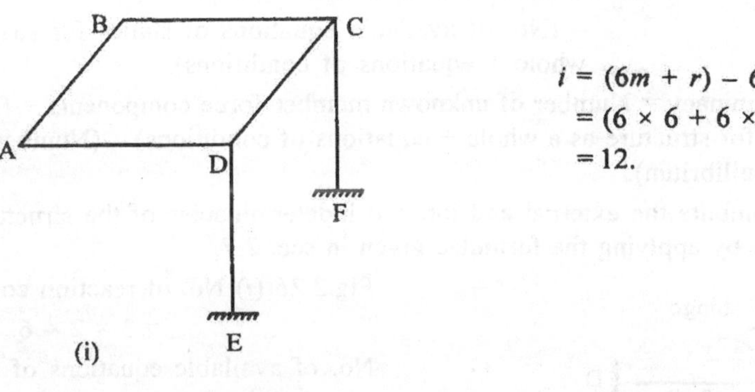

$$i = (6m + r) - 6j$$
$$= (6 \times 6 + 6 \times 2) - 6 \times 6$$
$$= 12.$$

(ii)

$$i = (6m + r) - 6j$$
$$= (6 \times 3 + 9) - 6 \times 4$$
$$= 3.$$

Fig. 2.25. Example 2.10.

2.7. CALCULATION OF INDETERMINACY BY FORMULAE

A general method can be derived to compute the degree of indeterminacy of a structure. Members, reactions and joints are first classified as follows —

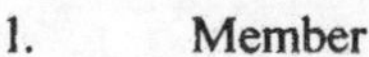

1. Member	No. of unknown Member Force Components.

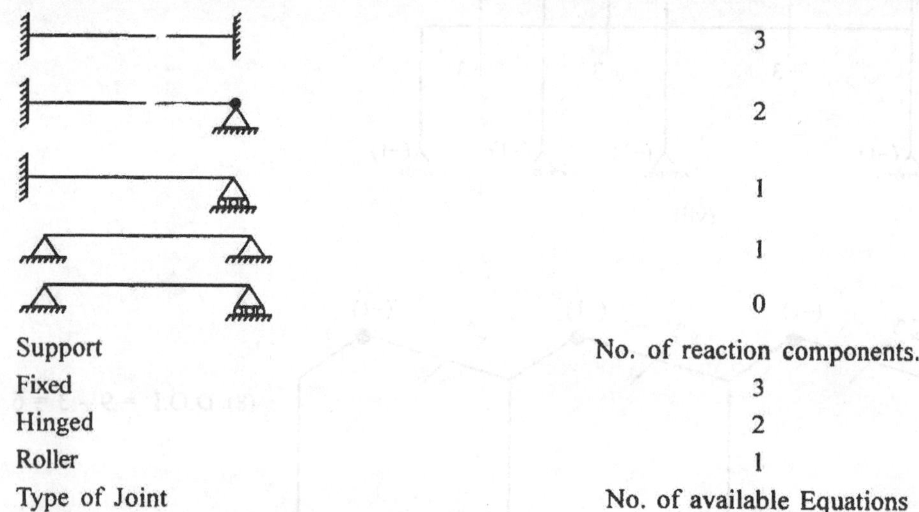

Member	No. of unknown Member Force Components
(fixed – fixed)	3
(fixed – hinged)	2
(fixed – roller)	1
(hinged – hinged)	1
(hinged – roller)	0

2. Support	No. of reaction components.
Fixed	3
Hinged	2
Roller	1

3. Type of Joint	No. of available Equations of Equilibrium.
Fixed	3
Hinged	2
sliding	1

Then External indeterminacy

= No. of reaction components

– (No. of available equations of statics for structure as a whole + equations of conditions).

Internal Indeterminacy = Number of unknown member force components + (No. of Equations of statics for structure as a whole + equations of conditions) – (Number of available equations of equilibrium).

Example 2.11 — Compute the external and internal indeterminacies of the structures (*i* to *vii*) given in Example 2.9 by applying the formulae given in sec. 2.7.

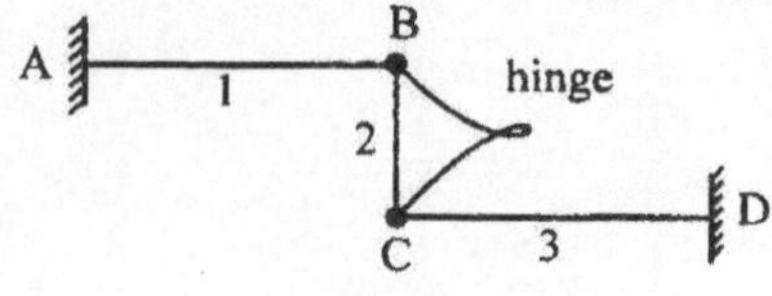

Fig. 2.26 (*i*)

Fig.2.26 (*i*) No. of reaction components

$= 3 \times 2 = 6.$

No. of available equations of statics = 3.

Equations of conditions = 2.

External indeterminacy $= 6 - (3 + 2)$

$= 1.$

Internal indeterminacy = No of unknown member force components

1 = 2

2 = 1

3 = 2

5

No. of available equations of equilibrium

[A + B + C + D] = (3 + 2 + 2 + 3)

= 10.

I.I = 5 + (3 + 2) − 10 = 0.

T.I = 1.

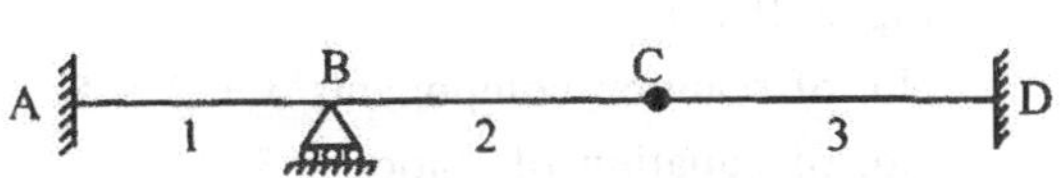

Fig. 2.26 (ii)

Fig. 2.26 (*ii*)

E.I. = (3 + 1 + 3) − (3 + 1)

= 3

A B C D

↓ ↓ ↓ ↓

I.I = 7 + (3 + 1) − (3 + 3 + 2 + 3)

= 0

No. of unknown member force components

Member 1 − 3

Member 2 − 2

Member 3 − 2

7

Fig. 2.26 (*iii*)

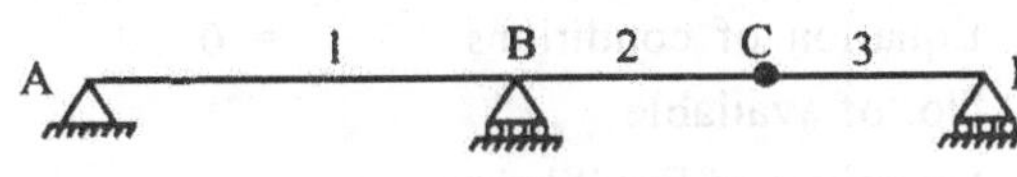

Fig. 2.26 (iii)

No. of reaction components

= 2 × 1 + 1 × 2 = 4

No. of Eqns. of statics = 3

Equations of condition = 1

External Indeterminacy = 4 − (3 + 1)

= 0.

No. of unknown member force components

Member 1 = 2

2 = 2

3 = 0

4.

No. of available Equations of equilibrium =

Joint A = 2

Joint B = 3

Joint C = 2

Joint D = 1

8.

Internal indeterminacy = 4 + (3 + 1) − 8 = 0.

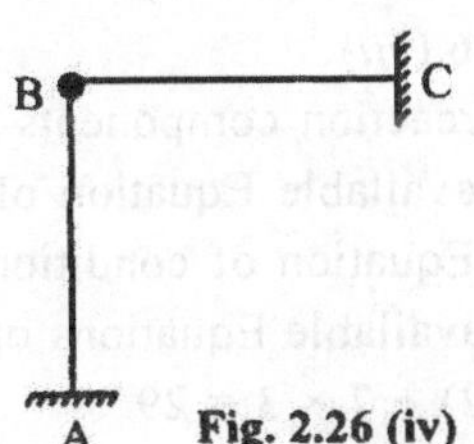

Fig. 2.26 (iv)

Fig. 2.26 (*iv*)

No. of reaction components = 3 × 2 = 6

No. of Equation of statics = 3

Equation of condition = 1

No of unknown Member force components

AB = 2

BC = 2

4.

No. of available Equations of Equilibrium

Joint

A = 3

B = 2

C = 3

8.

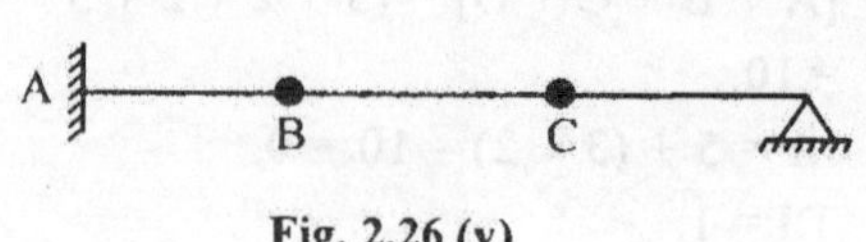

Fig. 2.26 (v)

External indeterminacy = (3 + 3) – (3 + 1)
= 2.

Internal indeterminacy = 4 + (3 + 1) – 8
= 0.

Fig. 2.26 (*v*)

No. of reaction components 3 + 2 = 5

No. of equation of statics = 3

Equations of conditions = 2.

No. of unknown member force components = AB = 2, BC = 1, CD = 1 Total = 4

No. of available Equations of equilibrium = 3 + 3 × 2 = 9

External Indeterminacy = 5 – (3 + 2) = 5 – 5 = 0.

Internal Indeterminacy = (2 + 1 + 1) + (3 + 2) – 9 = 0.

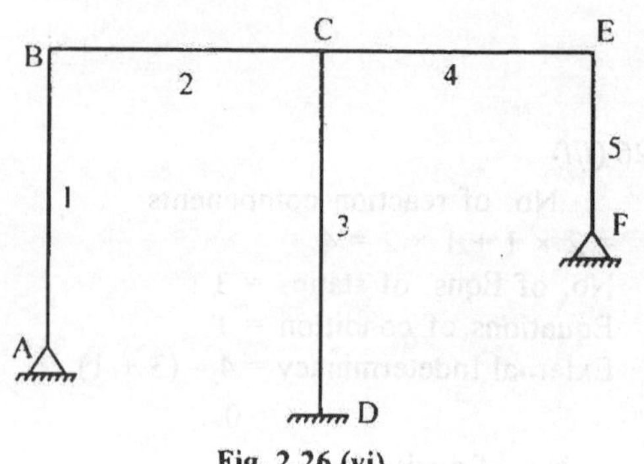

Fig. 2.26 (vi)

Fig. 2.26 (*vi*)

No. of reaction components = 3 + 2 + 2
= 7.

No. of available
Equations of statics = 3

Equation of conditions = 0

No. of available
Equations of Equilibrium

A = 2, B = 3, C = 3
D = 3, E = 3, F = 2
Total = 16.

No. of force components
1 = 2, 2 = 3, 3 = 3, 4 = 3, 5 = 2
Total = 13

Internal Indeterminacy
= 13 + (3 + 0) – (16) = 0.

External Indeterminacy
= 7 – (3 + 0) = 4

Fig. 2.26 (vii).

Fig. 2.26 Example 2.11.

Fig. 2.26 (*vii*)

No. of reaction components = 2 × 4 = 8

No. of available Equation of statics = 3

No. of Equation of conditions = 0

No. of available Equations of Equilibrium.
= (4 × 2) + 7 × 3 = 29.

No. of force components.
1 – 2, 2 – 3, 3 – 3, 4 – 3, 5 – 3,
6 – 2, 7 – 3 8 – 3, 9 – 3, 10 – 2,
11 – 3, 12 – 2 = 32.

External Indeterminacy
= 8 – (3 + 0) = 5.

Internal Indeterminacy
= 8 × 3 + 4 × 2 + (3 + 0) – 29 = 6.

Total indeterminacy = 5 + 6 = 11.

SUMMARY

1. A space structure in statical equilibrium must satisfy six equations of statics.

$$\Sigma F_x = 0 \qquad \Sigma F_y = 0 \qquad \Sigma F_z = 0$$

and

$$\Sigma M_x = 0 \qquad \Sigma M_y = 0 \qquad \Sigma M_z = 0.$$

For planar structure and when the forces act in the plane the equations reduce to

$$\Sigma F_x = 0 \qquad \Sigma F_y = 0 \qquad \Sigma M_z = 0$$

2. Free body diagram is very useful in structural analysis. It is simply a sketch of a structural component with all the appropriate forces acting on it.

3. If all the reactions of the structure can be found out from the equations of statics alone the structure is statically determinate. If not, it is statically indeterminate. We require an additional equation for every additional reaction component. These can be obtained from equations of conditions or from compatibility conditions.

4. Reactions can be computed by drawing free body diagram of a structure and then applying equations of statics and condition equations for statically determinate structures and compatibility conditions for statically indeterminate structures.

5. Matrix method is very useful in structural analysis. It can be used to find out stability, determinacy or indeterminacy of a structure.

6. A structure may be indeterminate

(*i*) externally (*ii*) internally.

As there are three equations of equilibrium, a planar structure will be externally determinate if there are only three reactive forces which are neither parallel nor concurrent. For internal determinacy of a truss a necessary but not sufficient condition is —

$m = 2j - 3.$ where m = no. of members

j = no. of joints

For space truss for determinacy and stability necessary but not sufficient condition is —

$3j = m + r$ where r is the number of reaction components.

For a plane frame degree of indeterminacy

$$i = (3m + r) - (3j + c)$$

where c = number of condition equations.

For a spaceframe $i = (6m + r) - 6j$.

PROBLEMS

A Theoretical

2.1 (*a*) How do you determine the degree of indeterminacy for a statically indeterminate beam?

(*b*) Explain how you can determine the degree of statical indeterminacy of a redundant plane truss. Distinguish between internal statical indeterminacy and external statical indeterminacy.

B. Objective

2.2 (*i*) Find the degree of statical indeterminacy in the structure below. (Fig. P. 2.1).

[AMIE S, 97]

(*ii*) A rectangular portal with one base fixed and the other base pinned

(*a*) is statically determinate.

(*b*) has a single degree of statical indeterminacy.

(*c*) has two degrees of statical indeterminacy.

(*d*) has three degrees of statical indeterminacy.

Tick the correct answer. **[AMIE Winter '98]**

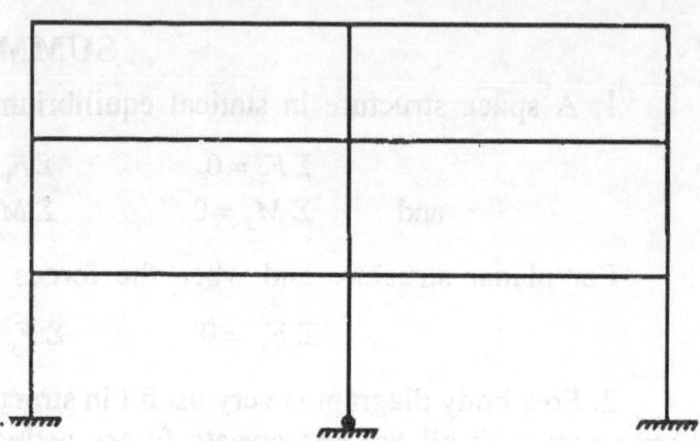

Fig. P 2.1. Problem 2.2 (i).

(*iii*) The number of independent equations to be satisfied for static equilibrium in a plane frame is

(*a*) 1 (*b*) 3 (*c*) 4 (*d*) 6

Tick the correct answer. **[AMIE Winter 97]**

(*iv*) The degree of statical indeterminacy of a propped cantilever is

(*a*) 0 (*b*) 1 (*c*) 2 (*d*) 3

[AMIE Summer 99]

Tick the correct answer.

2.3. Find the stability and determinacy or indeterminacy of the following structures.

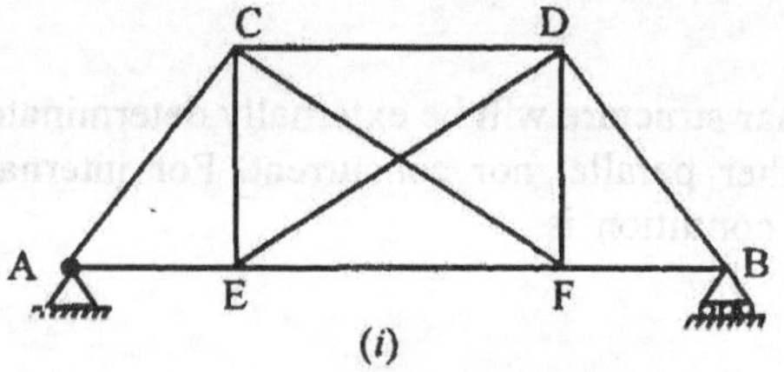

C D A B E F

(*i*)

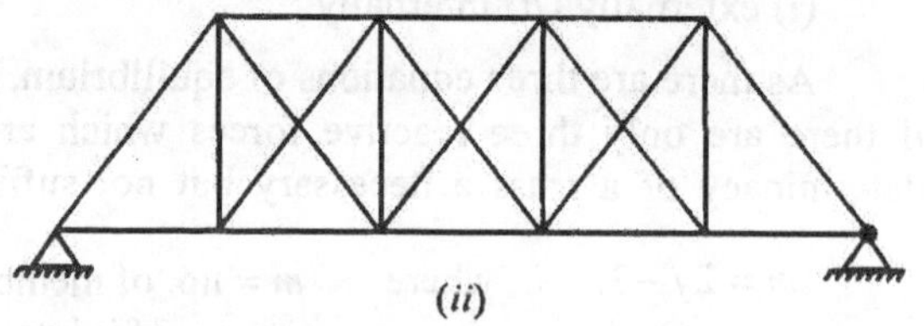

(*ii*)

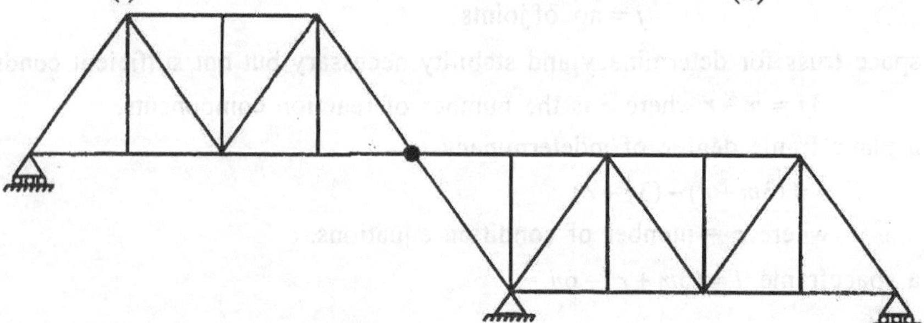

(*iii*)

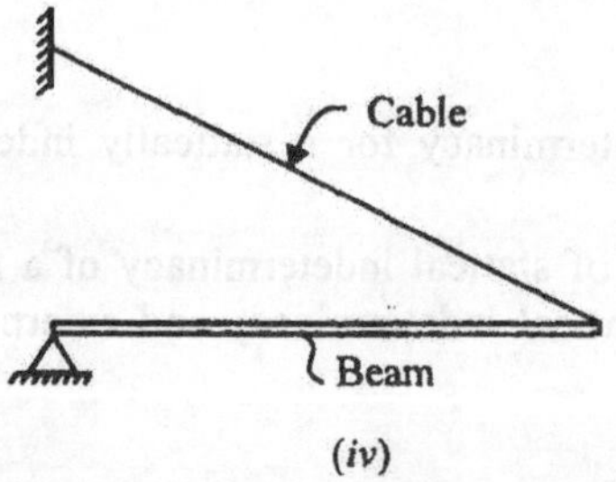

Cable Beam

(*iv*)

Cable Beam

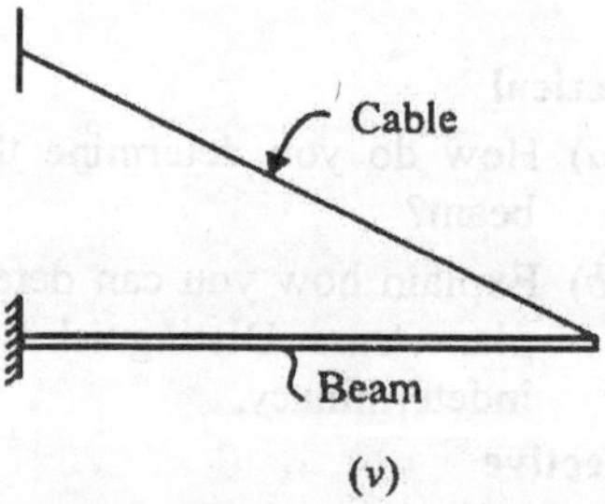

(*v*)

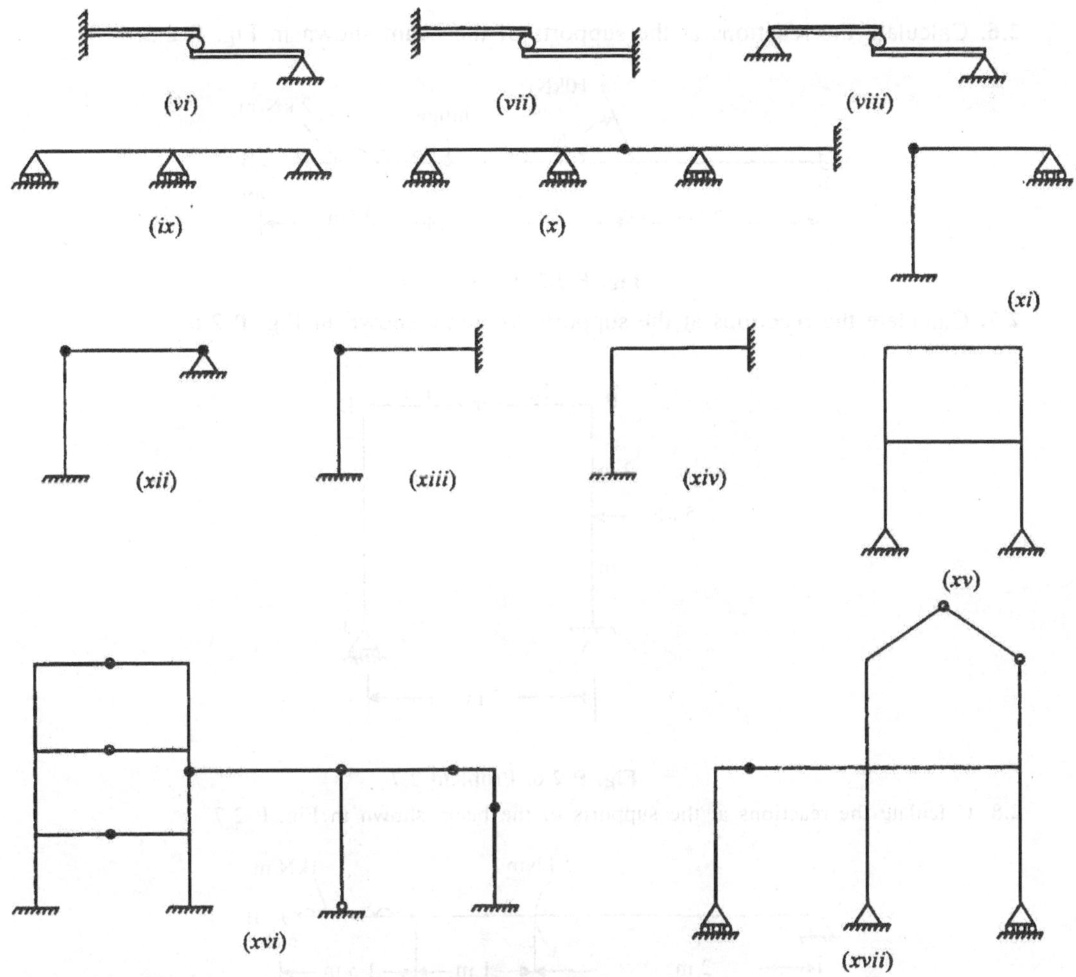

(vi)
(vii)
(viii)
(ix)
(x)
(xi)
(xii)
(xiii)
(xiv)
(xv)
(xvi)
(xvii)

Fig. P.2.2. Problem 2.3.

C. Mathematical

2.4. Calculate the Reactions at *A* and *D* for the beam shown in Fig. P.2.3.

4 kN
4kN.
1kN/m.
0·5m.
30°
A
B
C
D
E
2 m
2 m
2 m
1 m

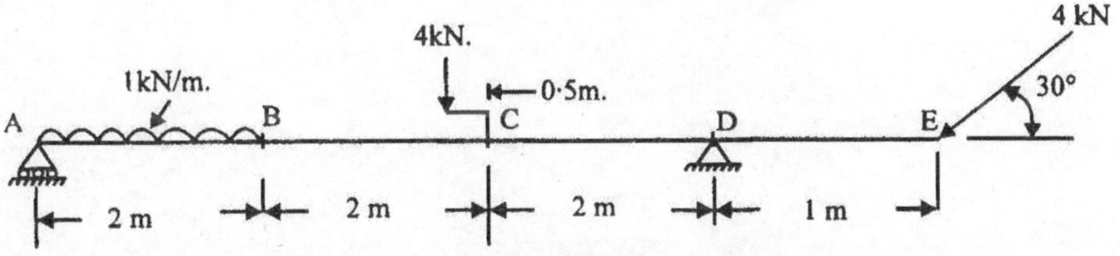

Fig. P 2.3. Problem 2.4.

2.5. Calculate the reactions at *A,C* and *D* for the beam shown in Fig. P 2.4.

Hinge
2P
P
A
B
C
D
E
L
L
L
L/2

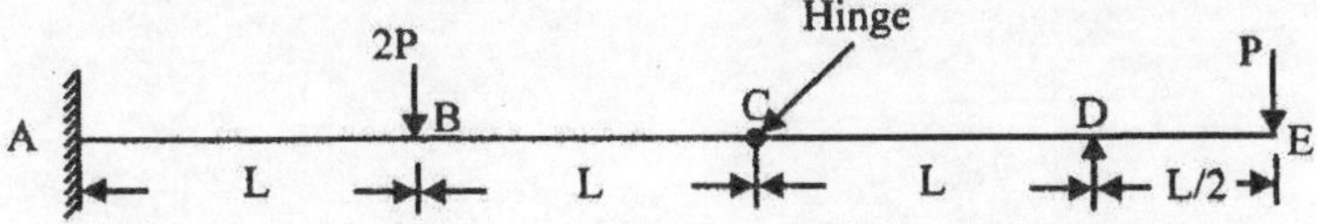

Fig. P. 2.4. Problem 2.5.

2.6. Calculate the reactions at the supports of the beam shown in Fig. P 2.5.

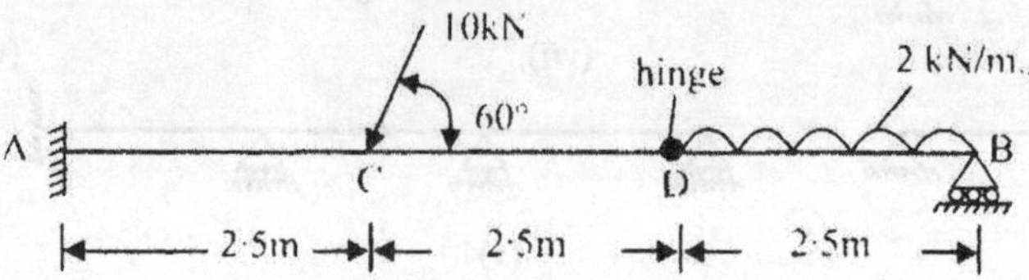

Fig. P 2.5. Problem 2.6.

2.7. Calculate the reactions at the supports of frame shown in Fig. P 2.6.

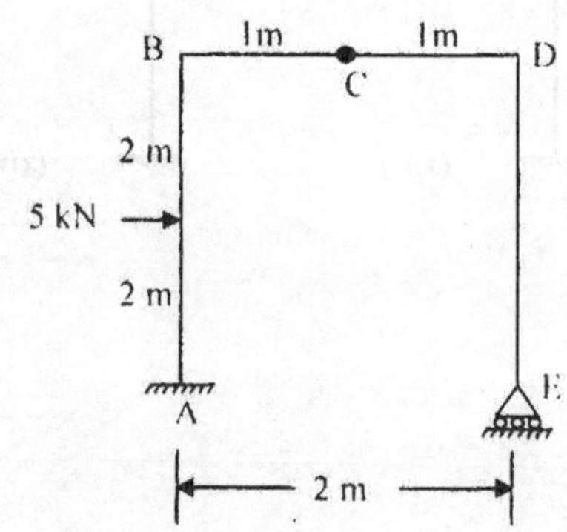

Fig. P 2.6. Problem 2.7.

2.8. Calculate the reactions at the supports of the beam shown in Fig. P 2.7.

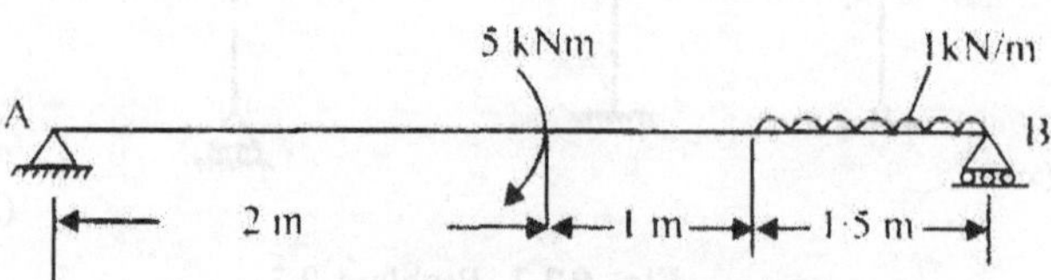

Fig. P 2.7. Problem 2.8.

3

TRUSS ANALYSIS

3.1. INTRODUCTION

A truss is a very common structure used in constructing bridges, building roofs, towers and lateral bracings of high rise buildings. With innovative use of reinforced concrete, prestressed concrete and cable stayed bridges, use of trusses in bridge building has declined. However, its other uses are still very important from structural point of view. Fig. 3.1 (*a*) shows different types of bridge trusses with approximate span range Fig. 3.1(*b*) shows different types of roof trusses. Roof trusses are usually inclined but bridge trusses may be plane. Fig. 3.2 shows details of an actual truss.

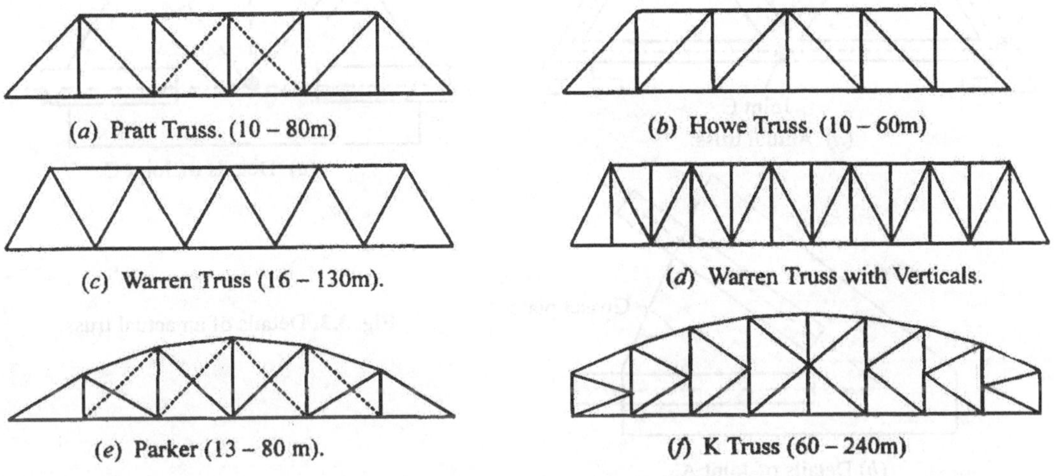

(*a*) Pratt Truss. (10 – 80m)

(*b*) Howe Truss. (10 – 60m)

(*c*) Warren Truss (16 – 130m).

(*d*) Warren Truss with Verticals.

(*e*) Parker (13 – 80 m).

(*f*) K Truss (60 – 240m)

Fig. 3.1. (*a*) Bridge Trusses.

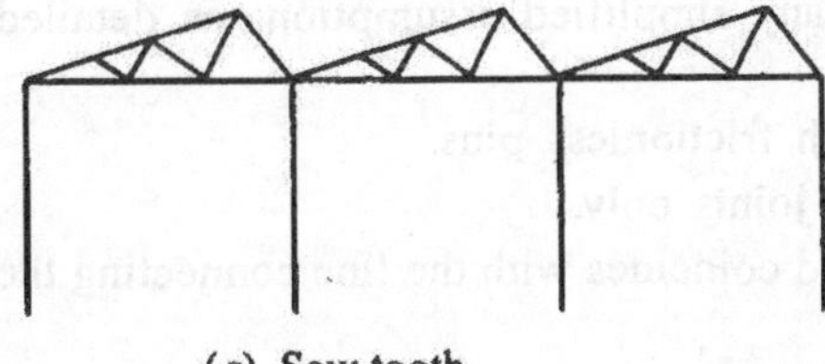

(*a*) Saw tooth.

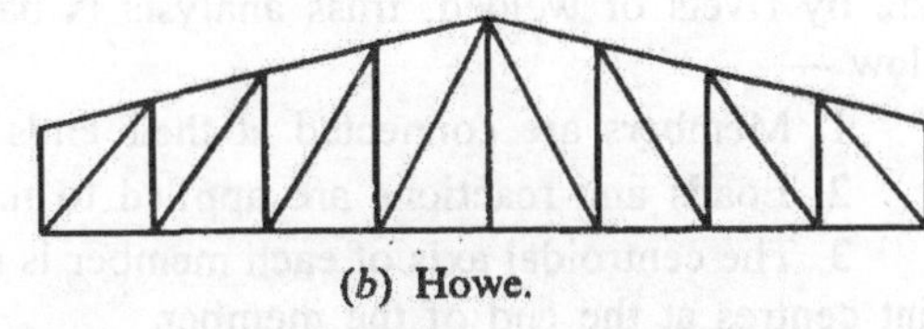

(*b*) Howe.

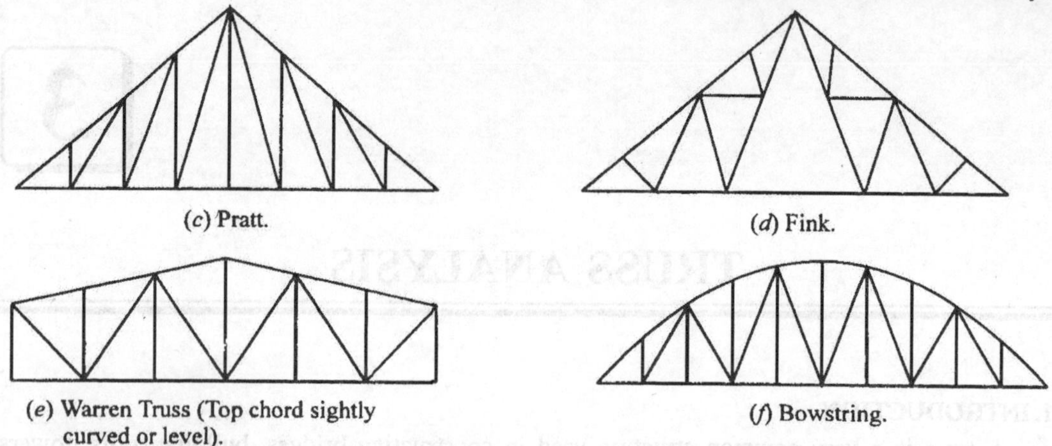

(c) Pratt.

(d) Fink.

(e) Warren Truss (Top chord sightly curved or level).

(f) Bowstring.

Fig. 3.1 (b) Roof Trusses.

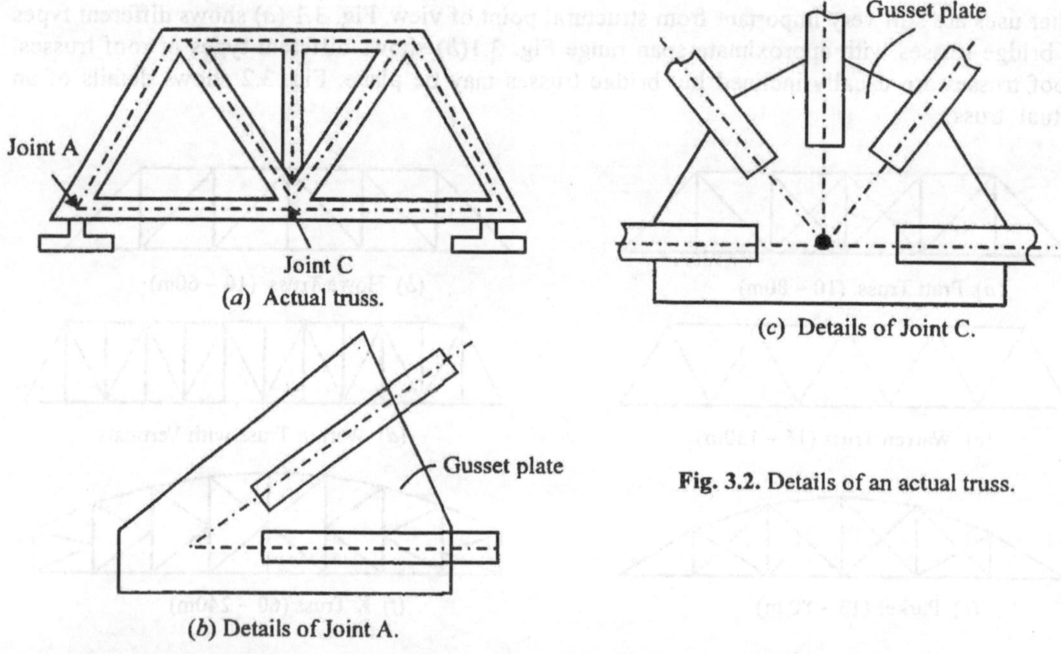

(a) Actual truss.

(b) Details of Joint A.

(c) Details of Joint C.

Fig. 3.2. Details of an actual truss.

3.2. BASIC ASSUMPTIONS OF TRUSS ANALYSIS :

While the actual truss is as shown in Fig. 3.2 and the joints are connected through gusset plate by rivets or welded, truss analysis is based on many simplified assumptions as detailed below —

1. Members are connected at their ends by smooth frictionless pins.
2. Loads and reactions are applied to the truss at joints only.
3. The centroidal axis of each member is straight and coincides with the line connecting the joint centres at the end of the member.
4. All members function as two-force members *i.e.* they are subjected to either tension or compression.

Fig. 3.3.(*a*) shows a truss model based on the above assumptions. As shown in Fig. 3.3.(*b*) the truss members are subjected to axial forces (compression or tension only).

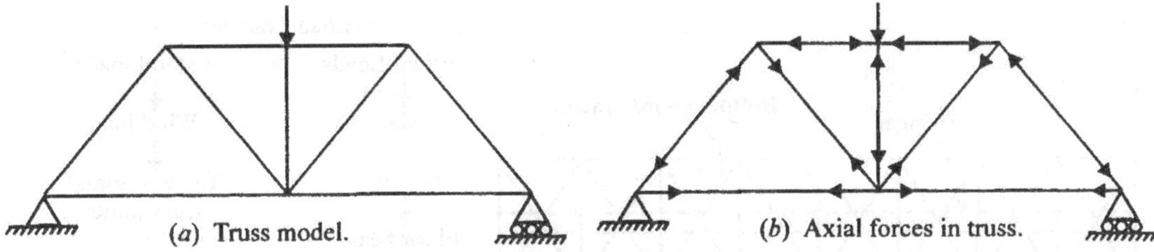

Fig. 3.3. Truss model.

The stresses which are developed in truss members due to idealised conditions is known as primary stresses. The stresses developed due to deviations from these idealised conditions in actual truss is known as "secondary stresses". The following secondary stresses may develop —

1. The joints being welded, riveted or bolted moments and shears will be developed at the ends of the members which will lead to secondary stresses.

2. A major contribution to secondary stresses occurs if the centroidal axis of the members meeting at a joint are not concurrent. Careful detailing can minimise these stresses.

3. If load is on the members and not at the joints, there will obviously be bending and shear in the members. Hence it is normally ensured that loads are transmitted only at the joints of the truss. In the case of roof truss this is done through purlins and cleats. However an elaborate arrangement is required in the case of bridge truss. Fig. 3.4(*a*) and (*b*) show how loads are transferred in the case of roof truss. Fig. 3.5 (*a*) to (*e*) explain the transmission in the case of bridges.

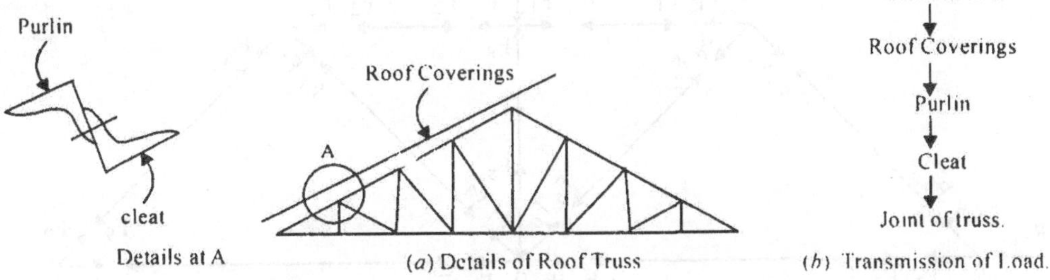

(*a*) Details of Roof Truss (*b*) Transmission of Load.

Fig. 3.4. Transfer of load of roof coverings to purlin, cleat and truss joints.

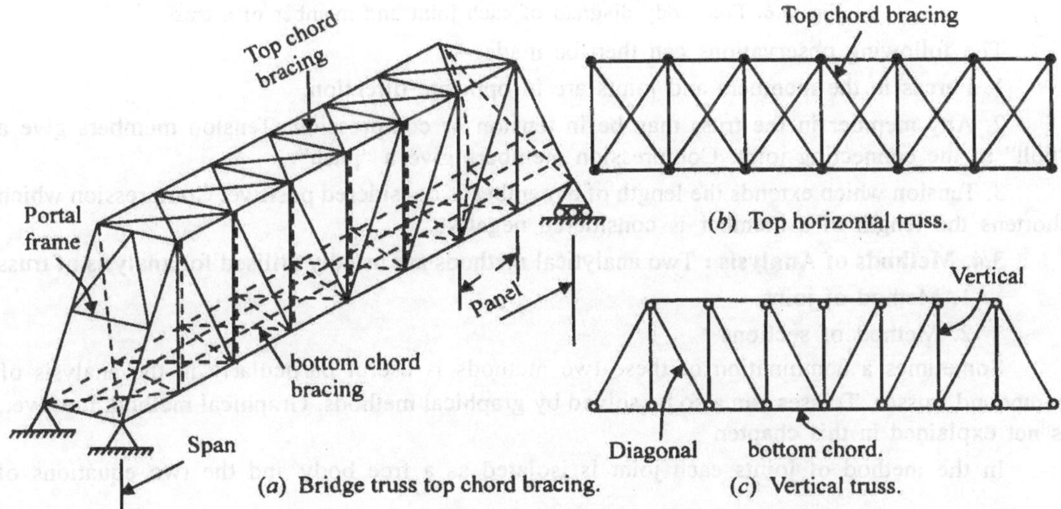

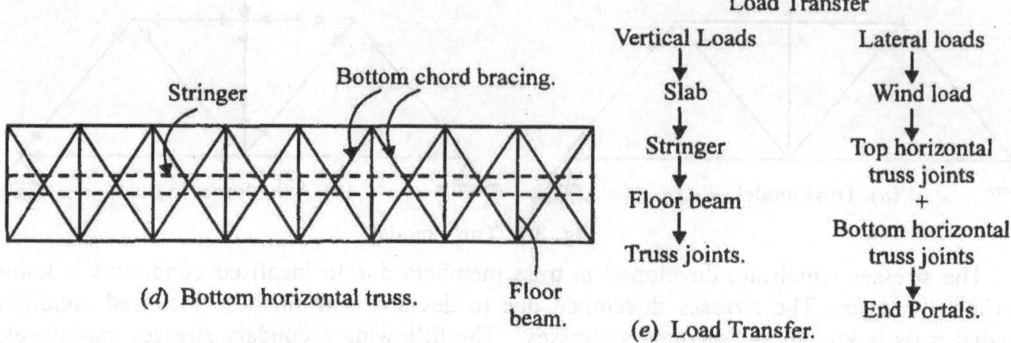

Fig. 3.5. Transmission of Load in bridges.

3.3. ANALYSIS OF PLANE TRUSS :

A plane truss is in equilibrium under the actions of applied loads and external reactions. Hence the three equations of statics, $\Sigma X = 0$ $\Sigma Y = 0$ $\Sigma M = 0$ can be applied to the truss as a whole to find the reactions at the supports when the truss is statically determinate. When the truss is externally indeterminate equations of compatibility have to be applied. Only axial forces act in the members of the truss. Since loads do not occur along the member, the forces on each end must be equal. Further more, since the member is straight, the end forces do not introduce bending. Fig. 3.3. (*b*) shows the forces in an idealised truss model. Fig. 3.6 shows the free body diagram of each member and each joint.

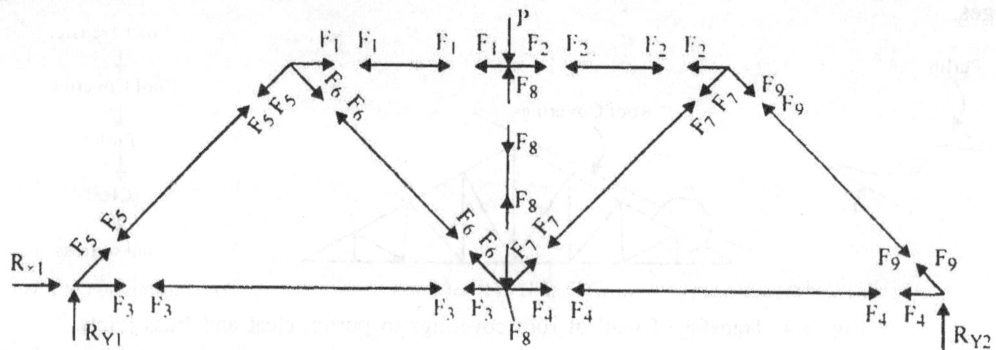

Fig. 3.6. Free body diagram of each joint and member of a truss.

The following observations can then be made —

1. Forces in the members and joints are in opposite direction.

2. Any member in the truss may be in tension or compression. Tension members give a "pull" to the connecting joint. Compression members give a "push".

3. Tension which extends the length of a member is considered positive. Compression which shortens the length of a member is considered negative.

3.4. Methods of Analysis : Two analytical methods are usually utilised for analysis of truss

1. Method of joint
2. Method of sections

Sometimes a combination of these two methods is useful particularly in the analysis of compound trusses. Trusses can also be solved by graphical methods. Graphical method, however, is not explained in this chapter.

In the method of joints each joint is isolated as a free body and the two equations of equilibrium $\Sigma X = 0$, $\Sigma Y = 0$ are applied at each joint sequentially. Since no moment is applied at the joint, the moment equation $\Sigma M = 0$ is automatically satisfied. At each joint there should not be more than two unknown bar forces as only two equations are available at each joint.

In the method of sections, the truss is usually divided into two parts by a cut through the entire truss. The entire truss is in equilibrium. Each part of the truss will also be in equilibrium if the bar forces cut by section are replaced. We can then use the three equations of equilibrium *i.e.* $\Sigma X = 0$ $\Sigma Y = 0$ $\Sigma M = 0$. It is to be noted that here the forces are nonconcurrent. In the method of joints unknown bar forces are to be found out sequentially. But in the method of section no such sequence is necessary. We will cut the section through the members whose force we want to find out. This is explained graphically in Fig. 3.7 (*a*), (*b*) and (*c*). In the isolated joint 2 we have four unknown forces F_1, F_3, F_4, F_5. With only two equations of statics $\Sigma F_x = 0$ $\Sigma F_y = 0$, we can find out only two unknowns. Hence before we go to joint 2, we have to find out F_1 and F_3 by considering equilibrium of joint 1 and 12. With known forces F_1 and F_3, equilibrium of joint 2 will give two equations to find out two unkowns F_4 and F_5.

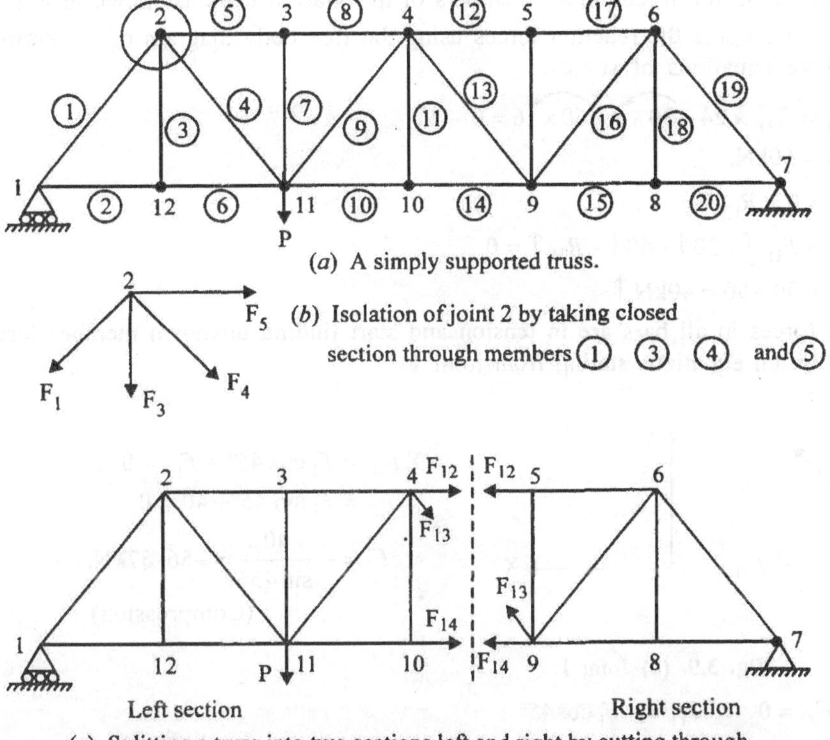

(*a*) A simply supported truss.

(*b*) Isolation of joint 2 by taking closed section through members (1) (3) (4) and (5)

(*c*) Splitting a truss into two sections left and right by cutting through a panel and replacing the forces in the cut members.

Fig. 3.7. Method of joints and method of sections.

In the method of section, however, we can straight way go to panel 4, 5, 9, 10 and cut a section through members 12, 13, 14 to find forces F_{12}, F_{13} and F_{14} in corresponding members. We can consider either the left or right part of the section. Usually the part which will involve less calculation is considered. To get the force F_{12}, it is obvious that the right section should be considered and moment should be taken about point 9. Similarly to find force F_{14} moment should be taken about point 4. To get F_{13} vertical equilibrium of the right portion should be considered. The methods are explained through a number of examples.

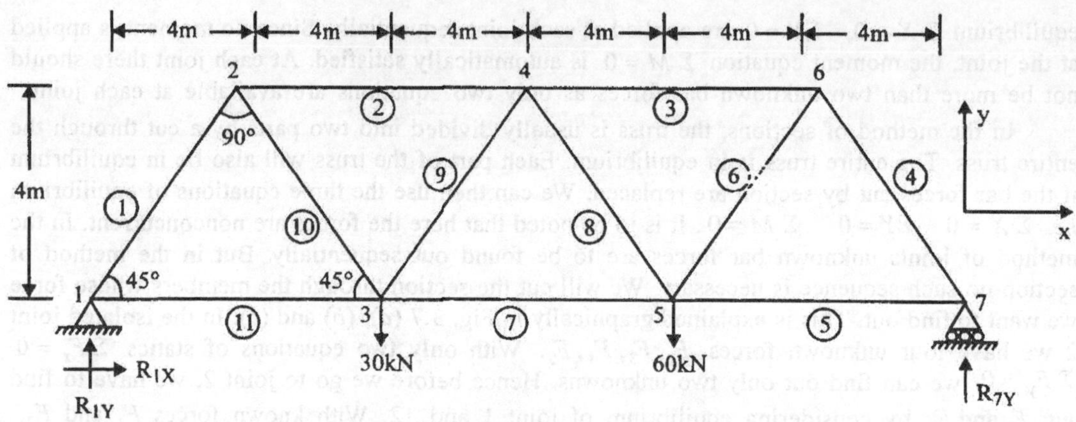

Fig. 3.8 (Example 3.1).

Example 3.1. Find the forces in all members of the Warren truss as shown in Fig. 3.8.

1. **Step 1** — Compute the reaction forces using the free body diagram of the entire truss and using the three equations of statics.

$$\Sigma M_1 = R_{7Y} \times 24 - 30 \times 8 - 60 \times 16 = 0$$

or $\quad R_{7Y} = 50\text{kN}.$

$$\Sigma F_X = 0 = R_{1X}$$

$$\Sigma F_Y = R_{1Y}\uparrow - 30\downarrow - 60\downarrow + R_{7Y}\uparrow = 0$$

or $\quad R_{1Y} = 90 - 50 = 40\text{kN}\uparrow$

2. Assume forces in all bars are in tension and start finding unknown member forces by using joint equilibrium equations stating from joint 1.

Joint 1.

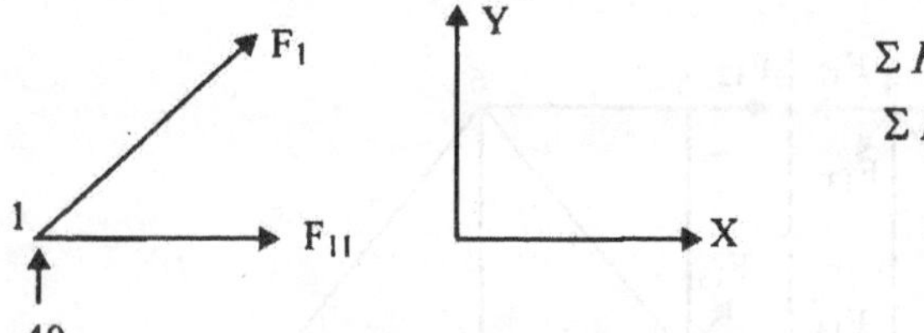

$$\Sigma F_X = F_1 \cos 45° + F_{11} = 0$$

$$\Sigma F_Y = F_1 \sin 45° + 40 = 0$$

$$F_1 = -\frac{40}{\sin 45°} = -56 \cdot 57\text{kN}.$$

(Compression)

Fig. 3.9. **(*a*)** Joint 1.

Solving $\Sigma F_X = 0 \quad F_{11} = -F_1 \cos 45°$

$= -(-56{\cdot}57) \cos 45°$

$= 40$ **KN. (tension).**

Note When the equation $\Sigma F_X = 0$ is solved a value of $(-56{\cdot}57)$ (with the minus sign) is substituted for F_1.

Joint 2 $\quad \Sigma F_X = F_2 + F_{10} \cos 45° - F_1 \cos 45° = 0$

$$\Sigma F_Y = F_1 \cos 45° + F_{10} \cos 45° = 0$$

Solving for $\Sigma F_Y = 0$

$$F_{10} = -\frac{F_{10} \cos 45°}{\cos 45°} = -F_1$$

$= +56 \cdot 57$ (tension).

Fig. 3.9(*b*) Joint 2.

Solving $\Sigma F_x = 0$

$$F_2 = -F_{10}\cos 45° + F_1 \cos 45°$$
$$= -56\cdot 57 \times 0\cdot 907) + (-56\cdot 57)(0\cdot 707)$$
$$= -80 kN. \text{ (compression)}$$

Note The initially assumed direction of the force F_1 was not changed in this free body diagram. The negative value of F_1 was substituted.

Joint 3

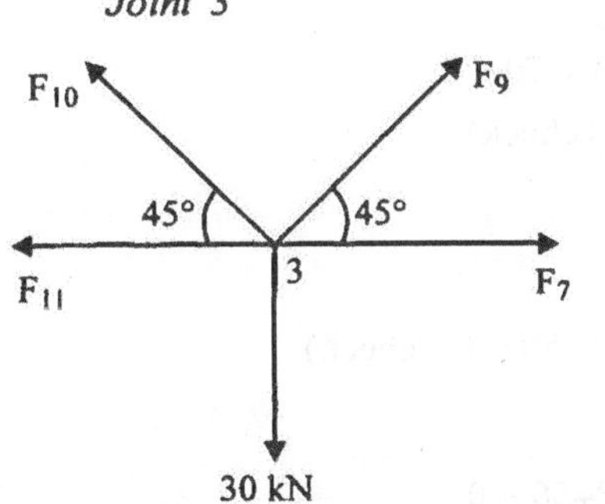

Fig. 3.9 (*c*) Joint 3.

$$\Sigma F_x = -F_{10}\cos 45° + F_9 \cos 45° - F_{11} + F_7 = 0.$$
$$\Sigma F_y = F_{10}\cos 45° + F_9 \cos 45° - 30 = 0$$

From $\Sigma F_y = 0.$

$$F_9 (0\cdot 707) = 30 - \cdot 707\,(56\cdot 57)$$

or $F_9 = -\ 14\cdot 14$ (compression).

From $\Sigma F_x = 0.$

$$F_7 = F_{10}\cos 45° - F_9 \cos 45° + F_{11}$$
$$= 56\cdot 57(0\cdot 707) + 14\cdot 14(0\cdot 707) + 40$$
$$= 90 kN \text{ (tension).}$$

Joint 4

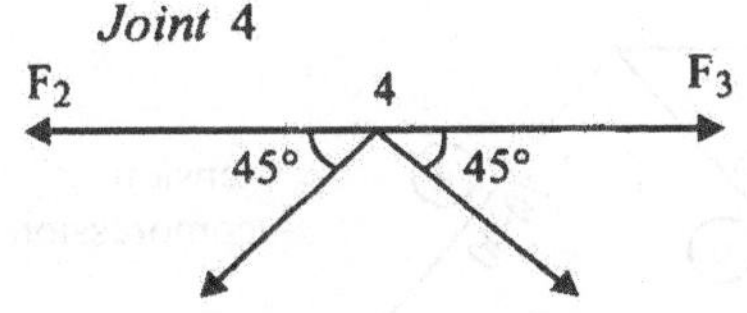

Fig. 3.9 (*d*) Joint 4.

$\Sigma F_x = 0$ gives

$$F_3 + F_8 \cos 45° - F_9 \cos 45° - F_2 = 0$$

$\Sigma F_y = 0$ gives

$$F_9 \cos 45° + F_8 \cos 45° = 0$$

or $F_8 = -F_9$

$$= -\ (-\ 14\cdot 14) = 14\cdot 14 \text{ (tension)}$$

From $\Sigma F_x = 0$

$$F_3 = +F_9 \cos 45° + F_2 - F_8 \cos 45°$$
$$= -14\cdot 14\,(0\cdot 707) + (-80) - 14\cdot 14\,(0\cdot 707)$$
$$= -100 \text{ kN (compression).}$$

Joint 5

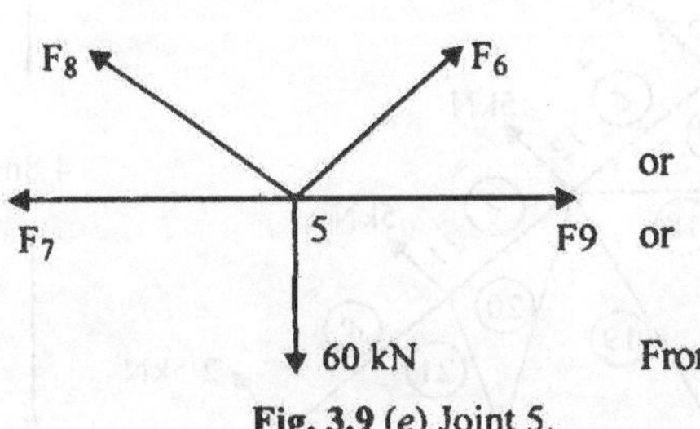

Fig. 3.9 (*e*) Joint 5.

$\Sigma F_x = 0$ gives

$$F_6 \cos 45° + F_5 - F_8 \cos 45° - F_7 = 0$$

$\Sigma F_y = 0$ gives

$$F_8 \cos 45° + F_6 \cos 45° - 60 = 0$$

or $F_6 \cos 45° = 60 - F_8 \cos 45°$

or $F_6 = 60\sqrt{2} - F_8.$

$$= 70\cdot 70 \text{ kN. (tension)}$$

From $\Sigma F_x = 0$

$$F_5 = F_7 + F_8 \cos 45° - F_6 \cos 45°$$
$$= 90 + 14\cdot 14\,(0\cdot 707)$$
$$-70\cdot 70\,(0\cdot 707)$$
$$= +\ 50 kN \quad \text{(tension)}$$

Joint 6

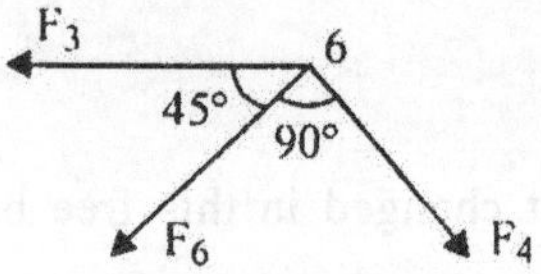

$$\Sigma F_x = F_4 \cos 45° - F_6 \cos 45° - F_3 = 0$$

$$\Sigma F_y = F_6 \cos 45° + F_4 \cos 45° = 0$$

From $\Sigma F_y = 0$

$$F_4 = -F_6 = -70 \cdot 70\text{kN}$$

Substituting in $\Sigma F_x = 0$

$$F_4 = 1 \cdot 414 F_3 + F_6$$

$$= 1 \cdot 414\,(-100) + 70 \cdot 70$$

$$= -70 \cdot 70 \text{ KN (check)}$$

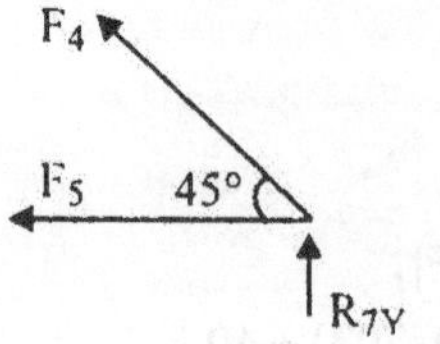

Fig. 3.9 (*f*) Joint 6.

Joint 7

$$\Sigma F_x = 0$$

or $F_4 \cos 45° + F_5 = 0$

or $(-70 \cdot 70)\,(0 \cdot 707) + 50 = 0$ (check)

$$\Sigma F_y = 0$$

$$F_4 \sin 45° + 50 = 0$$

$$(-70 \cdot 70)\,(0 \cdot 707) + 50 = 0 \text{ (check)}$$

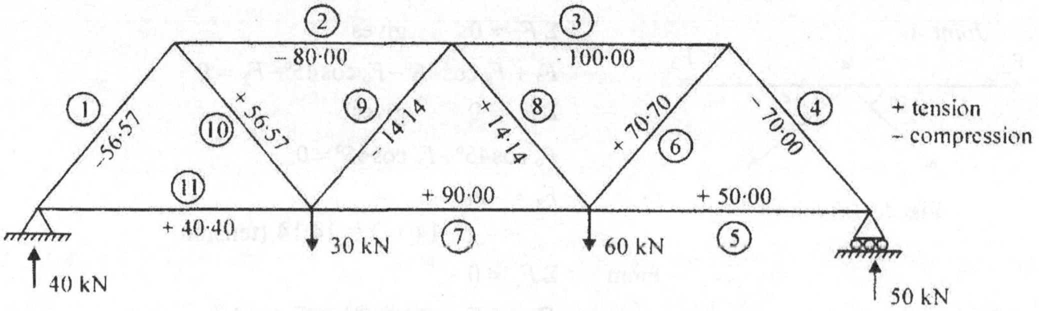

Fig. 3.9 (*g*) Joint 7.

Fig. 3.10. Forces in the members of the truss.

Example 3.2. Find forces in all members of the Fink roof truss subjected to wind loading. On the windward side there is pressure, on the leeward side there is suction as shown in Fig. 3.11.

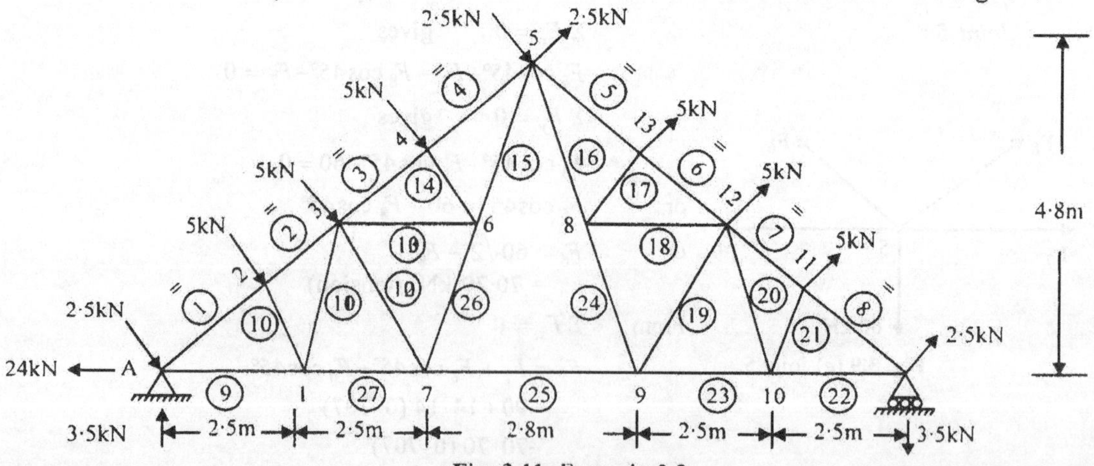

Fig. 3.11. Example 3.2.

Solution. Since all the dimensions of the truss are given, it is best to compute the coordinates of points taking coordinates of A, (0,0) as given below.

Point	x coordinate	y coordiante	Remark
A	0	0	
1	2.5	0	
2	1.6	1.2	
3	3.2	2.4	
4	4.8	3.6	
5	6.4	4.8	
6	5.7	2.4	$\frac{1}{2}$ of $pt5$ and $pt7$
7	5.0	0.0	
8	7.1	2.4	$\frac{1}{2}$ of $pt5$ and $pt9$
9	7.8	0.0	
10	10.3	0.0	
B	12.8	0.0	
11	11.2	1.2	
12	9.6	2.4	
13	8.0	3.6	

Slope of line $A - 5$, $\sin\theta = \dfrac{4\cdot 8}{8} = 0\cdot 6$

$$\cos\theta = \frac{6\cdot 4}{8} = 0\cdot 8$$

Resolving the forces in the horizontal and vertical directions

Horizontal component

of $2\cdot 5 = 2\cdot 5 \times 0\cdot 6 = 1\cdot 5$ kN.

Vertical component $= 2\cdot 5 \times 0\cdot 8 = 2\cdot 0$ kN.

Similarly horizontal component of 5KN $= 5 \times 0\cdot 6$

$= \mathbf{3\cdot 0}$ **kN.**

Vertical component of 5 KN $= 5 \times 0\cdot 8$

$= 4\cdot 0$ kN.

Taking moments of all forces about A.

$$R_{BY} \times 12\cdot 8 - (5\times 2 + 5\times 4 + 5\times 6 + 2\cdot 5\times 8)$$
$$+2\times 6\cdot 4 + 4\times 8\cdot 0 + 4\times 9\cdot 6$$
$$+4\times 11\cdot 2 + 2\times 12\cdot 8$$
$$-(1\cdot 5\times 4\cdot 8 + 3\cdot 0\times 3\cdot 6 + 3\cdot 0\times 2\cdot 4$$
$$+3\cdot 0\times 1\cdot 2 + 1\cdot 5\times 0) = 0$$

or $R_{BY} = \dfrac{1}{12\cdot 8}(80 + 28\cdot 8 - 153\cdot 6)$

$= -\ 3\cdot 5$ *i.e.* acting in opposite direction *i.e.* $\downarrow$.

Since $\Sigma V = 0$

$R_{AY} = -R_{BY} = 3\cdot 5\uparrow$.

As $\Sigma H = 0$

$$H_{AX} + (1{\cdot}5 + 3{\cdot}0 + 3{\cdot}0 + 3{\cdot}0 + 1{\cdot}5 + 1{\cdot}5 + 3{\cdot}0 + 3{\cdot}0 + 3{\cdot}0 + 3{\cdot}0 + 1{\cdot}5) = 0$$

or $H_{AX} = -24$ *i.e.* towards left.

Method of joint cannot be applied at joint 3 and 7 as the number of unknowns becomes more than 2 as we proceed through joints *A*, 1 and 2. Hence force in member 25 should initially be found out by considering only the right or left half of the body. Considering left half of the body and passing a section through joint 5 and member 25, the free body diagram of the left half of the truss with the section is shown in Fig. 3.12.

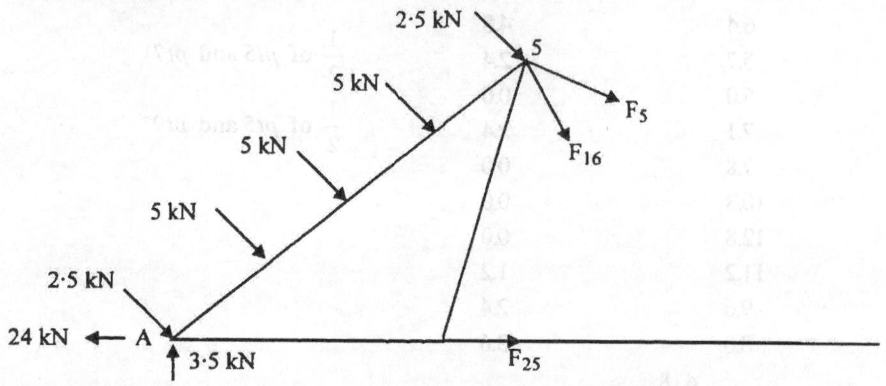

Fig. 3.12. Example 3.2.

$$\overset{\curvearrowleft}{+}\quad \Sigma M_5 = 2{\cdot}5\times{\cdot}8 + 5\times 6 + 5\times 4 + 5\times 2$$

$$+ F_{25}\times 4{\cdot}8 - 24\times 4{\cdot}8 - 3{\cdot}5\times 6{\cdot}4 = 0$$

or $F_{25} = 12.$

Joint A

$$\Sigma F_x = \vec{F_1}\frac{1{\cdot}6}{2{\cdot}0} + \vec{F_9} - 24 + 1{\cdot}5 = 0$$

$$\Sigma F_y = F_1\cdot\frac{1{\cdot}2}{2{\cdot}2{\cdot}0}\uparrow + 3{\cdot}5\uparrow - 2\downarrow = 0$$

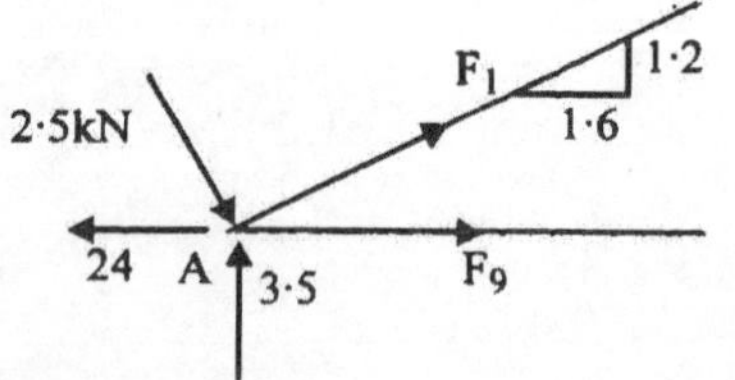

Fig. 3.13 (*a*) Joint A.

Solving $\Sigma F_y = 0$

$$F_1 = -\frac{1{\cdot}5\times 2{\cdot}0}{1{\cdot}2} = -2{\cdot}5.$$

From $\Sigma F_x = 0$

$$-2{\cdot}5\times\frac{1{\cdot}6}{2{\cdot}0} + F_9 - 24 + 1{\cdot}5 = 0$$

or $F_9 =$ **24·50 kN.**

Joint 2

Resolving along and perpendicular to force F_1 and F_2

$$5 + F_{10} = 0$$

$$F_1 - F_2 = 0$$

$F_2 = F_1 = -2{\cdot}5$ (compression)

$F_{10} = -5$ (compression)

Fig. 3.13 (*b*) Joint 2.

Joint 1

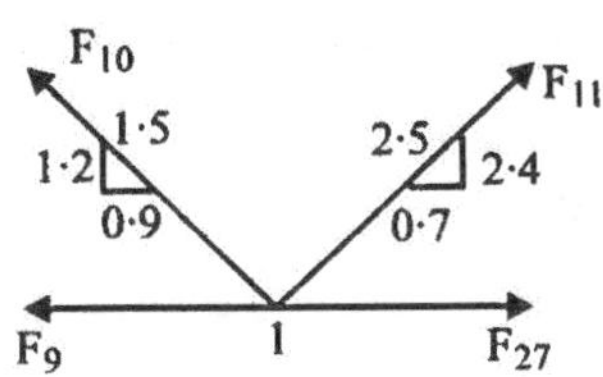

Fig. 3.13 (*c*) Joint 1.

$$\Sigma F_x = F_{11} \times \frac{0 \cdot 7}{2 \cdot 5} + F_{27}$$

$$-F_{10} \times \frac{9}{1 \cdot 5} - F_9 = 0$$

$$\Sigma F_y = F_{10} \times \frac{1 \cdot 2}{1 \cdot 5} + F_{11} \times \frac{2 \cdot 4}{2 \cdot 5} = 0$$

From $\Sigma F_y = 0$

$$F_{11} = -F_{10} \times \frac{1 \cdot 2}{1 \cdot 5} \times \frac{2 \cdot 5}{2 \cdot 4}$$

$$= +5 \times \frac{2 \cdot 5}{3} = +4 \cdot 166 \text{ kN.}$$

From $\Sigma F_x = 0$

$$\frac{4 \cdot 166 \times 0 \cdot 7}{2 \cdot 5} + F_{27} + \frac{5 \times 0 \cdot 9}{1 \cdot 5} - 24 \cdot 50 = 0$$

or $F_{27} = 20 \cdot 33$ kN.

Joint 7

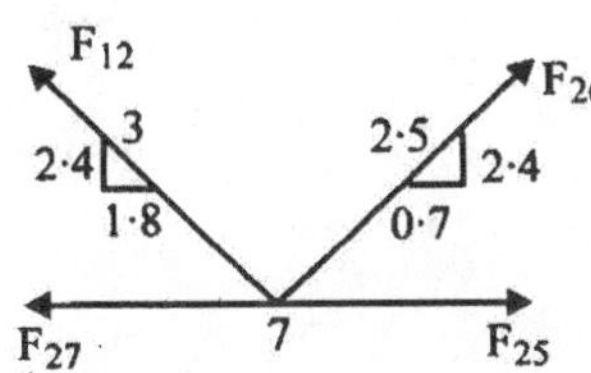

Fig. 3.13 (*d*) Joint 7.

$$\Sigma F_x = F_{25} + F_{26} \cdot \frac{0 \cdot 7}{2 \cdot 5} - F_{12} \times \frac{1 \cdot 8}{3}$$

$$- F_{27} = 0$$

$$\Sigma F_y = F_{26} \cdot \frac{2 \cdot 4}{2 \cdot 5} + F_{12} \cdot \frac{2 \cdot 4}{3} = 0$$

or $F_{26} = -F_{12} \cdot \frac{2 \cdot 5}{3}$

From $\Sigma F_x = 0$

$$12 + \left(-F_{12} \cdot \frac{2 \cdot 5 \times 0 \cdot 7}{3 \times 2 \cdot 5} \right)$$

$$-F_{12} \times \frac{1 \cdot 8}{3} - 20 \cdot 33 = 0$$

$$F_{12} = -9 \cdot 996$$

$$F_{26} = + \frac{9 \cdot 996 \times 2 \cdot 5}{3} = +8 \cdot 33$$

Joint 3

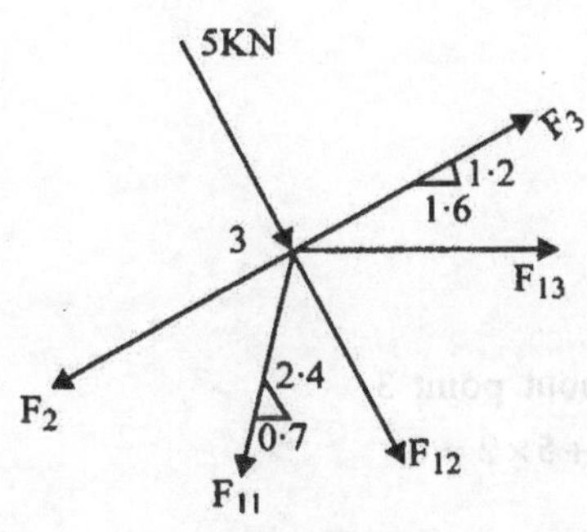

Fig. 3.13 (*e*) Joint 3.

Resolving perpendicular to $F_2 - F_3$

$$5 + F_{12} + F_{13} \cdot \frac{1 \cdot 5}{2 \cdot 5} + F_{11} \cdot \frac{1 \cdot 5}{2 \cdot 5} = 0$$

Resolving along $F_2 - F_3$.

$$F_3 + F_{13} \cdot \frac{2}{2 \cdot 5} - F_2 - F_{11} \frac{2}{2 \cdot 5} = 0$$

From 1st equation.

$$5 + (-9 \cdot 996) + F_{13} \cdot 0 \cdot 6 + \frac{4 \cdot 166 \times 1 \cdot 5}{2 \cdot 5} = 0$$

or $F_{13} = \mathbf{4 \cdot 16kN.}$

$$F_3 = F_2 + F_{11} \frac{2}{2 \cdot 5} - F_{13} \cdot \frac{2}{2 \cdot 5}$$

$$= -2\cdot 5 + \frac{4\cdot 166 \times 2}{2\cdot 5} - 4\cdot 16 \times \cdot 8$$

$$= -2\cdot 5.$$

Joint 4

$$F_3 - F_4 = 0$$

$$F_4 = F_3 = -2\cdot 50 \text{ (Comp)}$$

$$5 + F_{14} = 0$$

$$F_{14} = -5 \text{ (Comp)}$$

Fig 3.13 (*f*) Joint 4.

Joint 6

$$F_{15} - F_{26} + F_{14} \cdot \frac{1\cdot 5}{2\cdot 5} - \frac{F_{13} \times 1\cdot 4}{5} = 0$$

$$F_{14} \cdot \frac{2}{2\cdot 5} + F_{13} \times \frac{4\cdot 8}{5} = 0$$

$$\text{or } F_{13} = -F_{14} \cdot \frac{2}{2\cdot 5} \times \frac{5}{4\cdot 8}$$

$$= \mathbf{4\cdot 166 kN.}$$

Fig 3.13 (*g*) Joint 4.

$$F_{15} = F_{26} - F_{14} \times \frac{1\cdot 5}{2\cdot 5} + \frac{F_{13} \times 1\cdot 4}{5}$$

$$= \mathbf{12\cdot 50 \ kN.}$$

Checking of results. Fig. 3.14.

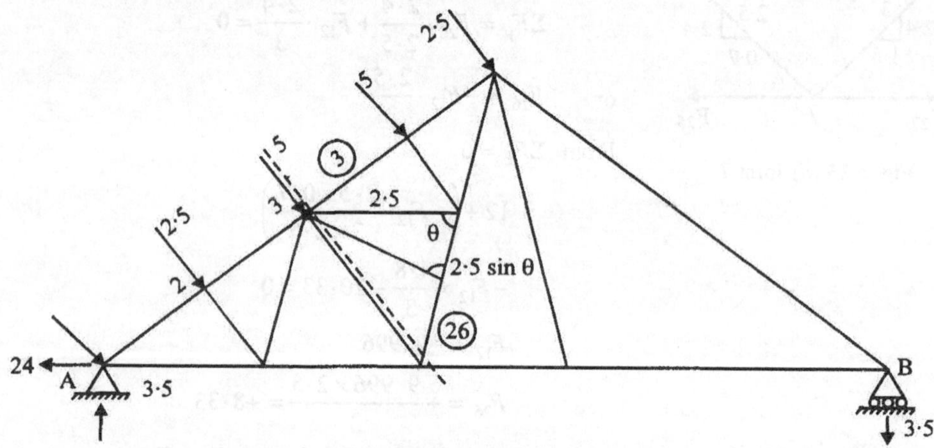

Fig. 3.14. Example 3·2.

Taking moments about 2

$$F_9 \times \frac{1}{4} \times 4\cdot 8 + 2\cdot 5 \times 2 - 2\cdot 4 \times \frac{1}{4} \times 4\cdot 8 - 3\cdot 5 \times 1\cdot 6 = 0$$

$$\text{or} \quad F_9 = \frac{1}{12}\left(3\cdot 5 \times 1\cdot 6 + 24 \times \frac{1}{4} \times 4\cdot 8 - 2\cdot 5 \times 2\right)$$

$$= 24\cdot 50.$$

Cutting section through 3 and 26 and taking moments about point 3.

$$F_{26} \times 2\cdot 5 \sin\theta + 12 \times 2\cdot 4 - 24 \times 2\cdot 4 - 3\cdot 5 \times 3\cdot 2 + 2\cdot 5 \times 4 + 5 \times 2 = 0$$

$$\text{or } F_{26} = \frac{1}{2\cdot 4}[20] = \mathbf{8\cdot 33 \ kN.}$$

We now start with right half of the frame

Joint B

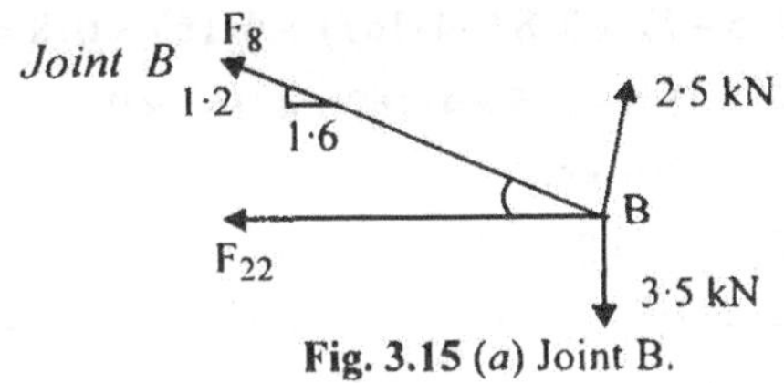

Fig. 3.15 (*a*) Joint B.

$$\Sigma F_x = \frac{F_8 \times 1 \cdot 6}{2} + F_{22} - \frac{2 \cdot 5 \times 1 \cdot 2}{2} = 0$$

$$\Sigma F_y = \frac{F_8 \times 1 \cdot 2}{3} - 3 \cdot 5 + \frac{2 \cdot 5 \times 1 \cdot 6}{2} = 0$$

$$F_8 = \frac{3 \cdot 5 \times 2}{1 \cdot 2} - \frac{2 \cdot 5 \times 1 \cdot 6}{1 \cdot 2}$$

$$= 2 \cdot 5 \text{ kN.}$$

$$F_{22} = -\frac{F_8 \times 1 \cdot 6}{2} + \frac{2 \cdot 5 \times 1 \cdot 2}{2}$$

$$= -0 \cdot 5 \text{ kN.}$$

Joint 11

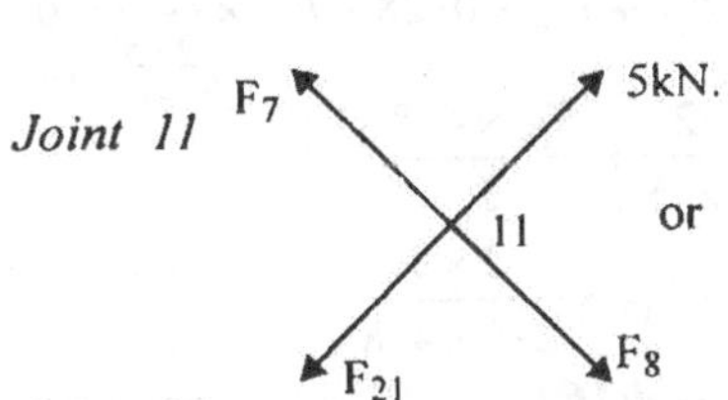

Fig. 3.15 (*b*) Joint 11.

or

$$F_{21} - 5\text{kN} = 0$$

$$F_{21} = 5\text{kN.}$$

$$F_8 = F_7 = 2 \cdot 5 \text{ kN.}$$

Joint 10

Fig. 3.15 (*c*) Joint 10.

$$\Sigma F_x = F_{21} \times \frac{0 \cdot 9}{1 \cdot 5} + F_{22} - F_{20} \times \frac{0 \cdot 7}{2 \cdot 5} - F_{23} = 0$$

$$\Sigma F_y = F_{20} \times \frac{2 \cdot 4}{2 \cdot 5} + F_{21} \times \frac{1 \cdot 2}{1 \cdot 5} = 0$$

or

$$F_{20} = -F_{21} \times \frac{1 \cdot 2}{1 \cdot 5} \times \frac{2 \cdot 5}{2 \cdot 4} = -4 \cdot 167$$

$$F_{23} = F_{21} \times \frac{0 \cdot 9}{1 \cdot 5} + F_{22} - \frac{F_{20} \times 0 \cdot 7}{2 \cdot 5}$$

$$= \frac{5 \times 0 \cdot 9}{1 \cdot 5} + (-0 \cdot 5) - \frac{4 \cdot 167 \times 0 \cdot 7}{2 \cdot 5}$$

$$= \mathbf{3 \cdot 667 \text{ kN.}}$$

Joint 9

$$\Sigma F_x = \frac{F_{19} \times 0 \cdot 9}{1 \cdot 5} + F_{23} - F_{24} \times \frac{0 \cdot 7}{2 \cdot 5} - F_{25} = 0$$

$$\Sigma F_y = F_{19} \times \frac{1 \cdot 2}{1 \cdot 5} + F_{24} \times \frac{2 \cdot 4}{2 \cdot 5} = 0$$

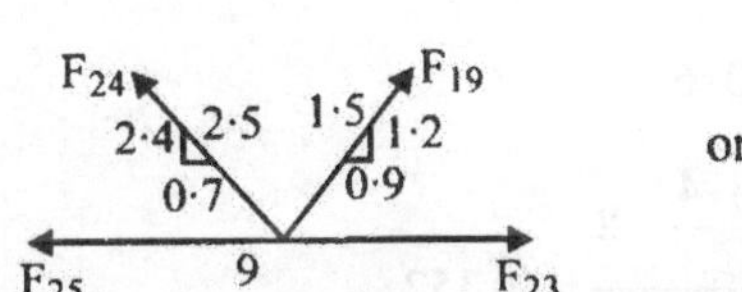

Fig. 3.15 (*d*) Joint 9.

or

$$F_{19} \times \frac{0 \cdot 9}{1 \cdot 5} + 3 \cdot 667 - F_{24} \times \frac{0 \cdot 7}{2 \cdot 5} - 12 = 0$$

$$F_{19} \times \frac{1 \cdot 2}{1 \cdot 5} + F_{24} \times \frac{2 \cdot 4}{2 \cdot 5} = 0$$

Solving $F_{24} = -8 \cdot 333$ kN.

$F_{19} = 10 \cdot 000$ kN.

Joint 12

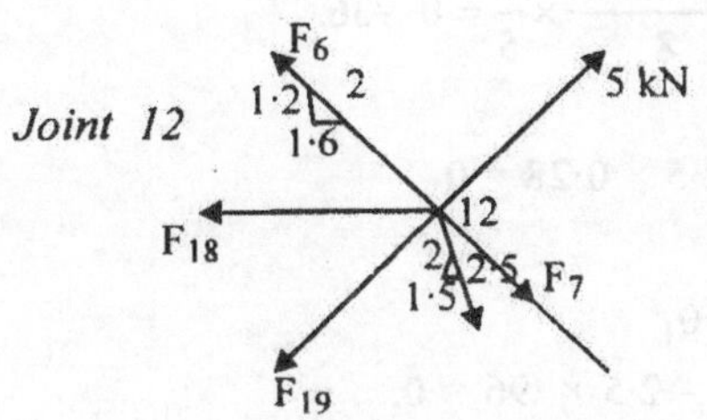

Fig. 3.15 (*e*) Joint 12.

Resolving perpenduclar to $F_6 - F_7$.

$$5 - F_{19} - F_{18} \times \frac{1 \cdot 2}{2} - F_{20} \times \frac{2 \cdot 5}{2 \cdot 5} = 0$$

Along $F_6 - F_7$,

$$F_7 - F_6 + F_{20} \times \frac{1 \cdot 6}{2} - F_{18} \times \frac{1 \cdot 6}{2} = 0$$

or

$$5 - 10 \cdot 00 - 0 \cdot 6 \ F_{18} + 4 \cdot 167 \times 0 \cdot 6 = 0$$

giving $F_{18} = -4 \cdot 167$

Similarly $2 \cdot 5 - F_6 + 0 \cdot 8\,(-4 \cdot 167) + 4 \cdot 167 \times 0 \cdot 8 = 0$

or $F_6 = 2 \cdot 5 - 0 \cdot 8 \times 4 \cdot 167 + 4 \cdot 167 \times 0 \cdot 8$

$= \mathbf{2 \cdot 5\ kN.}$

Joint 13

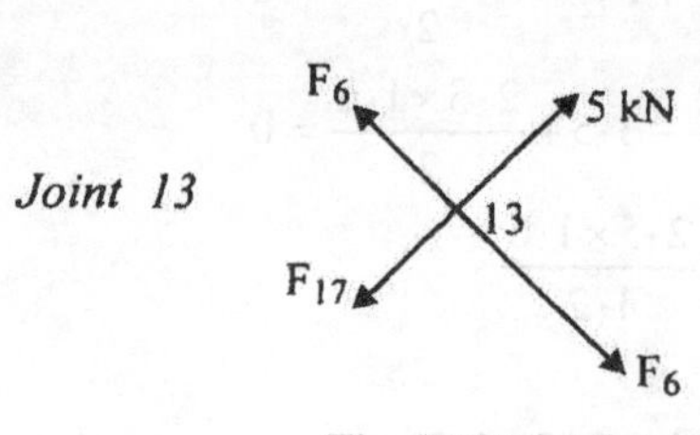

Fig. 3.15 (*f*) Joint 13.

$F_5 - F_6 = 0$

or $F_5 = F_6$

$5 - F_{17} = 0$

or $F_{17} = 5.$

Joint 8

$$F_{24} - F_{16} + F_{18} \times 0 \cdot 28 - F_{17} \times \frac{1 \cdot 5}{2 \cdot 5} = 0$$

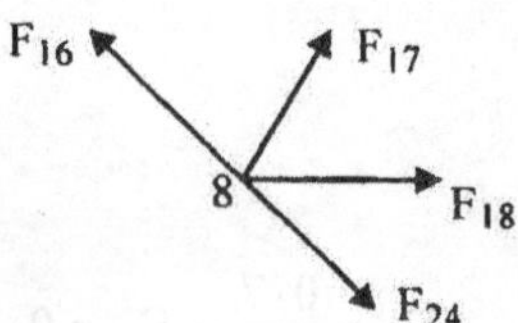

Fig. 3.15 (*g*) Joint 8.

$$F_{17} \times \frac{2}{2 \cdot 5} + \frac{F_{18} \times 4 \cdot 8}{5} = 0$$

or $$F_{18} = -F_{17} \times \frac{2 \times 5}{2 \cdot 5 \times 4 \cdot 8}$$

$= -\ \mathbf{4 \cdot 167}.$

$$-8 \cdot 333 - F_{16} + (-4 \cdot 167)\,(0 \cdot 28) - \frac{5 \times 1 \cdot 5}{2 \cdot 5} = 0.$$

$$F_{16} = -8 \cdot 33 - (4 \cdot 167)\,(0 \cdot 28) - 3 \cdot 0$$

$= -12 \cdot 50$

Joint 5

$$\cos\theta_1 = \frac{\frac{12 \cdot 8 \times 6 \cdot 4}{8} - 8}{8} = 0 \cdot 28$$

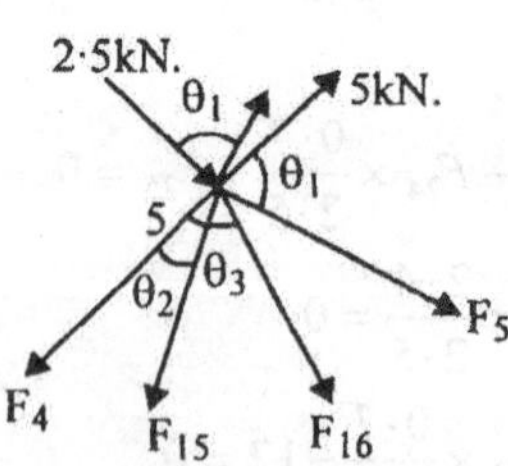

Fig. 3.15 (*h*) Joint 5.

$$\sin\theta_1 = \frac{\frac{12 \cdot 8 \times 4 \cdot 8}{8}}{8} = 0 \cdot 96$$

$$\cos\theta_2 = \frac{2}{2 \cdot 2} = 0 \cdot 8$$

$$\sin\theta_2 = \frac{1 \cdot 5}{2 \cdot 5} = 0 \cdot 6$$

$$\cos\theta_3 = \frac{\frac{7 \cdot 8 \times 6 \cdot 4}{8} - 8}{5} = 0 \cdot 352$$

$$\sin\theta_3 = \frac{7 \cdot 8 \times 4 \cdot 8}{8} \times \frac{1}{5} = 0 \cdot 936.$$

Check Resolving perpendicular to F_4

$2 \cdot 5 + 12 \cdot 5 \times 0 \cdot 6 - 12 \cdot 5 \times 0 \cdot 936 + 2 \cdot 5 \times \cdot 96 - 2 \cdot 5 \times 0 \cdot 28 = 0$

Resolving along F_4

$F_4 + F_{15}\cos\theta_2 + F_{16}\cos\theta_3 + F_5\cos\theta_1 - 2 \cdot 5 \sin\theta_1$

$= -\ 2 \cdot 5 + 12 \cdot 5 \times 0 \cdot 8 - 12 \cdot 5 \times \cdot 352 - 2 \cdot 5 \times \cdot 28 - 2 \cdot 5 \times \cdot 96 = 0.$

Example 3.3. Find the forces in the members of the truss in Example 3.2 through method of sections.

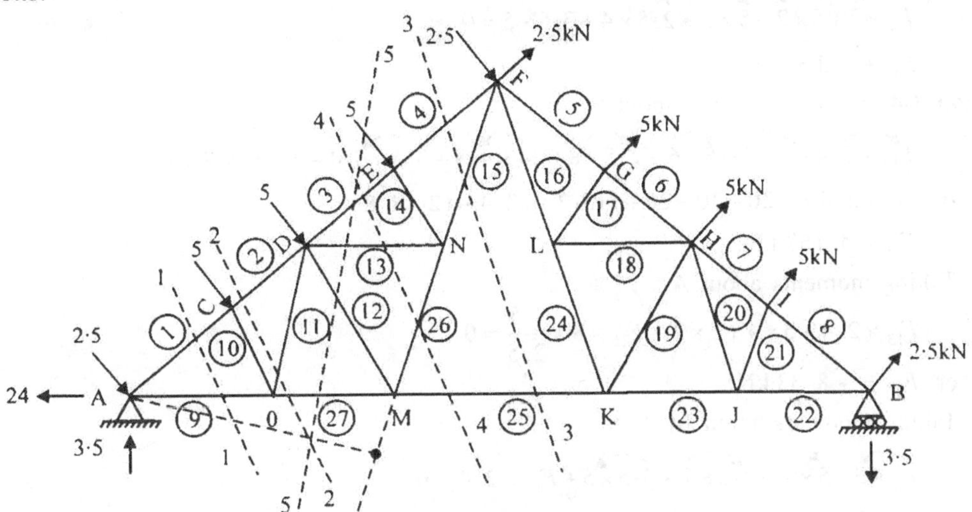

Fig. 3.16. Example 3.3.

Solution. Reactions and coordinates of all points can be computed as before.

Section 11. It cuts two members (1) and (9). Taking moments about 0.

$$F_1 \times 1 \cdot 5 + 3 \cdot 5 \times 2 \cdot 5 - 2 \cdot 5 \times 2 = 0 \quad \text{or} \quad F_1 = \frac{1}{15}(2 \cdot 5 \times 2 - 3 \cdot 5 \times 2 \cdot 5) = -2 \cdot 5.$$

Taking moments about C.

$$-F_9 \times 1 \cdot 2 - 2 \cdot 5 \times 2 + 24 \times 1 \cdot 2 + 3 \cdot 5 \times 16 = 0 \quad \text{or } F_9 = \mathbf{24 \cdot 50 \ kN.}$$

Section 22. Taking moments about 0.

$$F_2 \times 1 \cdot 5 + 3 \cdot 5 \times 2 \cdot 5 - 2 \cdot 5 \times 2 = 0$$

$$F_2 = -2 \cdot 5 \text{ kN.}$$

Taking moments about D.

$$-F_{27} \times 2 \cdot 4 - 2 \cdot 5 \times 4 - 5 \times 2.0 + 3 \cdot 5 \times 3 \cdot 2 + 24 \times 2 \cdot 4 = 0$$

or $\quad F_{27} = \mathbf{20 \cdot 33 kN.}$

Taking moment about A.

$$\Sigma M_A = 5 \times 2 - F_{11} \times \frac{4 \times 1 \cdot 5}{2 \cdot 5} = 0$$

$$F_{11} = \frac{5 \times 2 \times 2 \cdot 5}{4 \times 1 \cdot 5} = 4 \cdot 167 \text{ kN.}$$

Section 33. Taking moment about F.

$$-F_{25} \times 4 \cdot 8 + 24 \times 4 \cdot 8 + 3 \cdot 5 \times 6 \cdot 4 - 2 \cdot 5 \times 8 - 5 \times 6 - 5 \times 4 - 5 \times 2 = 0$$

or $F_{25} = \mathbf{12 \ kN.}$

Taking moments about M.

$$F_4 \times 3 + 5 \times 2 - 5 \times 2 - 2{\cdot}5 \times 4 + 3{\cdot}5 \times 5 = 0$$

or $F_4 = -$ **2·5 kN.**

Section 4-4. Taking moments about F.

$$F_{13} \times 2{\cdot}4 - 5 \times 4 - 5 \times 6 - 2{\cdot}5 \times 8 + 24 \times 4{\cdot}8 + 3{\cdot}5 \times 6{\cdot}4 - F_{25} \times 4{\cdot}8 = 0.$$

or $F_{13} \times 2{\cdot}4 = -20 - 30 - 20 + 115{\cdot}2 + 22{\cdot}4 - 12 \times 4{\cdot}8$

$F_{13} = 4{\cdot}167$ kN.

Taking moments about A.

$$F_{13} \times 2{\cdot}4 + 5 \times 4 + 5 \times 2 - F_{26} \times 8 \times \frac{1{\cdot}5}{2{\cdot}5} = 0$$

or $F_{26} = +8{\cdot}33$ kN.

Taking moments about M.

$$F_3 \times 3 - 5 \times 2 - 2{\cdot}5 \times 2 + 3{\cdot}5 \times 5 + F_{13} \times 2{\cdot}4 = 0$$

or $F_3 \times 3 = 10 + 10 - 17{\cdot}5 - 4{\cdot}167 \times 2{\cdot}4$

$= -7{\cdot}50$

$F_3 = -2{\cdot}50$ kN

Section 5-5. Taking moments about A.

$$F_{12} \times 4 + 5 \times 4 + 5 \times 2 + F_{13} \times 2{\cdot}4 = 0$$

or $F_{12} \times 4 = -(20 + 10 + 4{\cdot}167 \times 2{\cdot}4)$

or $F_{12} = -10$ kN.

Taking moments about A.

$$F_{10} \times 2 + 5 \times 2 = 0$$

or $F_{10} = -5$ kN.

$F_{14} = -5$ kN.

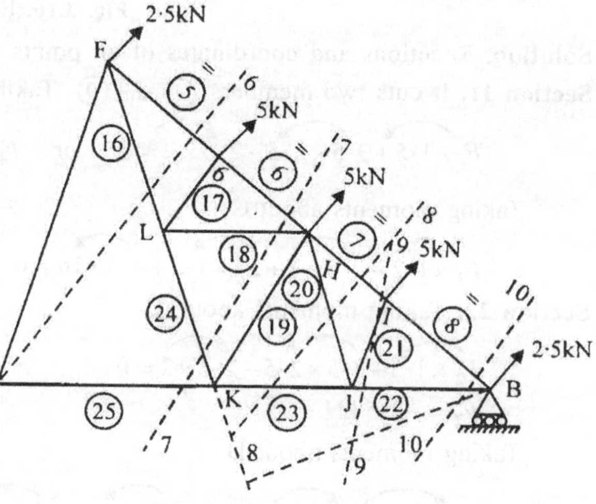

Fig. 3.17 Example 3.3.

Section 6-6. Taking moments about K.

Taking moments about K.

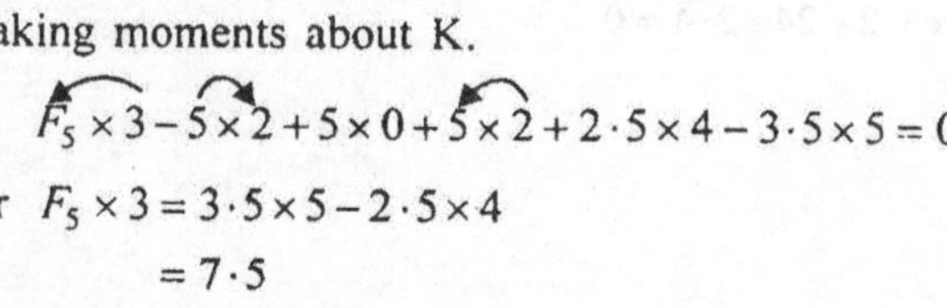

$$F_5 \times 3 - 5 \times 2 + 5 \times 0 + 5 \times 2 + 2{\cdot}5 \times 4 - 3{\cdot}5 \times 5 = 0$$

or $F_5 \times 3 = 3{\cdot}5 \times 5 - 2{\cdot}5 \times 4$

$= 7{\cdot}5$

or $F_5 = 2{\cdot}5$ kN.

Taking moments about B.

$$F_{16} \times \frac{8 \times 1{\cdot}5}{2{\cdot}5} + 5 \times 6 + 5 \times 4 + 5 \times 2 = 0$$

or $F_{16} = -\frac{60 \times 2{\cdot}5}{8 \times 1{\cdot}5} =$ **−12·5 kN.**

Section 7-7. Taking moments about F of all forces on the right of section 7-7.

$$F_{18} \times 2{\cdot}4 - 5 \times 4 - 5 \times 6 - 2{\cdot}5 \times 8 + 3{\cdot}5 \times 6{\cdot}4 + F_{25} \times 4{\cdot}8 = 0$$

or $F_{18} \times 2 \cdot 4 = 20 + 30 + 20 - 3 \cdot 5 \times 6 \cdot 4 - 12 \times 4 \cdot 8$

or $F_{18} = -4 \cdot 167$

Taking moments of all forces on the right section about K.

$$F_6 \times 3 + F_{18} \times 2 \cdot 4 + 5 \times 2 + 2 \cdot 5 \times 4 - 3 \cdot 5 \times 5 = 0$$

$$F_6 \times 3 = 7 \cdot 5$$

$$F_6 = \mathbf{2 \cdot 5\ kN.}$$

Taking moments of all forces about H.

$$F_{24} \times \frac{3 \times 4}{5} - 5 \times 2 - 2 \cdot 5 \times 4 + F_{25} \times 2 \cdot 4 + 3 \cdot 5 \times 3 \cdot 2 = 0$$

$$F_{24} \times \frac{12}{5} = 10 + 10 - 12 \times 2 \cdot 4 - 3 \cdot 5 \times 3 \cdot 2$$

or $F_{24} = \mathbf{-\,8 \cdot 33\ kN.}$

Section 8-8. Taking moments about H.

$$F_{23} \times 2 \cdot 4 - 5 \times 2 - 2 \cdot 5 \times 4 + 3 \cdot 5 \times 3 \cdot 2 = 0$$

or $F_{23} = \mathbf{3 \cdot 667\ kN.}$

Taking moments about J.

$$F_7 \times 1 \cdot 5 + 2 \cdot 5 \times 2 - 3 \cdot 5 \times 2 \cdot 5 = 0$$

or $F_7 = \mathbf{2 \cdot 5\ kN.}$

Taking moments about B.

$$F_{20} \times \frac{4 \times 1 \cdot 5}{2 \cdot 5} + 5 \times 2 = 0$$

$$F_{20} = \mathbf{-4 \cdot 167\ kN.}$$

Section 9-9. Taking moments about B.

$$F_{21} \times 2 - 5 \times 2 = 0$$

$$F_{21} = 5.$$

Section 10-10.

Taking moments about I.

$$F_{22} \times 1 \cdot 2 - 2 \cdot 5 \times 2 + 3 \cdot 5 \times 1 \cdot 6 = 0$$

$$F_{22} = \mathbf{-\,0 \cdot 5\ kN.}$$

Taking moments about J.

$$F_8 \times 1 \cdot 5 + 2 \cdot 5 \times 2 - 3 \cdot 5 \times 2 \cdot 5 = 0$$

or $F_8 = 2 \cdot 5.$

It can be observed that of the two hand methods for calculation of forces in a truss the method of section is very elegant and straight forward. Section should be passed through not more than three members and moments should be taken about points where two of the of the three members meet. In this way forces in almost all members can be obtained directly.

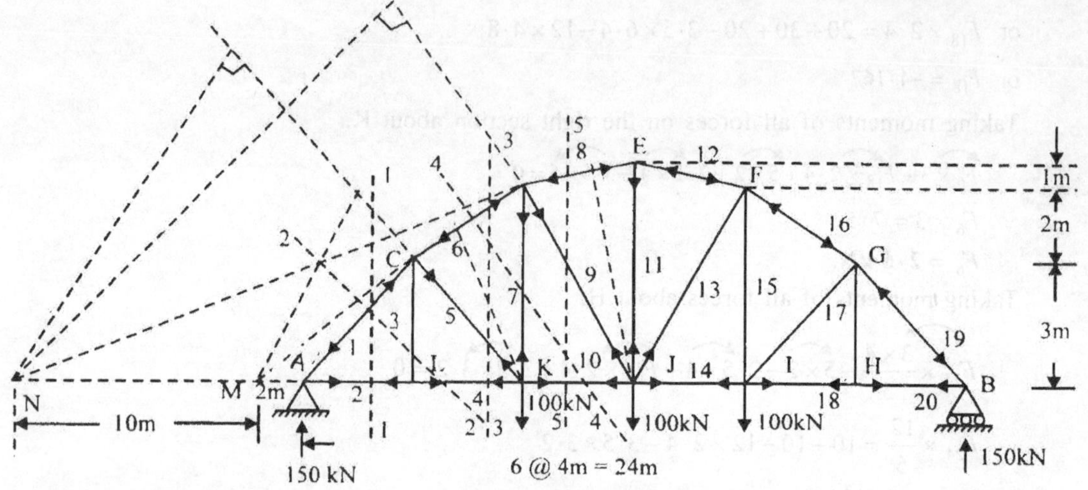

Fig. 3.18. Example 3.4.

Example 3.4. Find forces in all members of the Pratt truss by method of sections.

Solution. Since the truss is symmetrical and symmetrically loaded only left half of the truss is considered.

Coordinates of Points

Points	x	y
A	0	0
C	4	3
D	8	5
E	12	6
J	12	0
K	8	0
L	4	0

Computation of Reactions.

From symmetry $R_B = R_A = 150\text{kN}$.

Section 11. Taking moments about C.

$$F_2 \times 3 - 150 \times 4 = 0$$

or F_2 = 200 KN (tension).

Taking moments about L.

4,3

F_1

0,0 4,0

Fig. 3.19 (*a*) Example 3.4

$$F_1 \times 4 \times \frac{3}{5} + 150 \times 4 = 0$$

or F_1 = – 250 (compression)

Section 2-2. Taking moments about A.

$$F_3 = 0$$

Section 3-3.

$$\frac{CL}{LM} = \frac{DK}{RM}.$$

$$\frac{3}{4+AM} = \frac{5}{8+AM}.$$

or $\quad$ AM = 2.

Taking moments about M.

$$F_5 \times \frac{10\times 3}{5} - 150\times 2 = 0$$

$$F_5 = 50 \text{ kN (Tension)}$$

$$DM = \sqrt{5^2+10^2} = 11.19.$$

Taking moments about 'K'

$$F_6 \times 10 \times \frac{5}{11\cdot 19} + 150\times 8 = 0$$

or $\quad F_6 = -268\cdot 56$ (compression)

Taking moments about C.

$$F_4 \times 3 - 150\times 4 = 0$$

F_4 = **200 kN**. (tension)

Section 4-4. Taking moments about D.

$$\overset{\frown}{150\times 8} - F_{10}\times 5 = 0$$

F_{10} = **240 kN.**

Taking moments about M.

$$F_7 \times 10 + 150\times 2 - 100\times 10 = 0$$

or $\quad F_7$ = **70 kN**. (tension)

Section 5-5.

$$\frac{6}{JN} = \frac{1}{4}$$

$$JN = 24$$

$$NE = \sqrt{24^2+6^2}$$

= **24·74 kN.**

Taking moments about J.

$$F_8 \times 24 \times \frac{6}{24\cdot 74} + 150\times 12 - 100\times 4 = 0.$$

or $\quad F_8 = -240\cdot 33$ (compression)

$$F_9 \times 24 \times \frac{5}{6\cdot 40} + \overset{\frown}{100\times 20} - 150\times 12 = 0$$

or $\quad F_9$ = **10·67** (tension)

From symmetry $F_8 = F_{12} = -240\cdot 53$ (compression)

Considering Joint E.

$$F_8 \times \frac{1}{\sqrt{17}} + F_{12}\frac{1}{\sqrt{17}} + F_{11} = 0$$

or $F_{11} = \mathbf{116{\cdot}67\ kN.}$

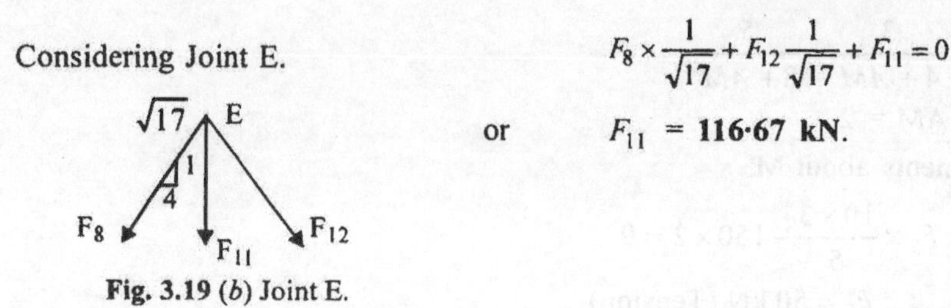

Fig. 3.19 (*b*) Joint E.

Example 3.5. Find forces in all members of the compound Warren truss as shown in Fig. 3.20.

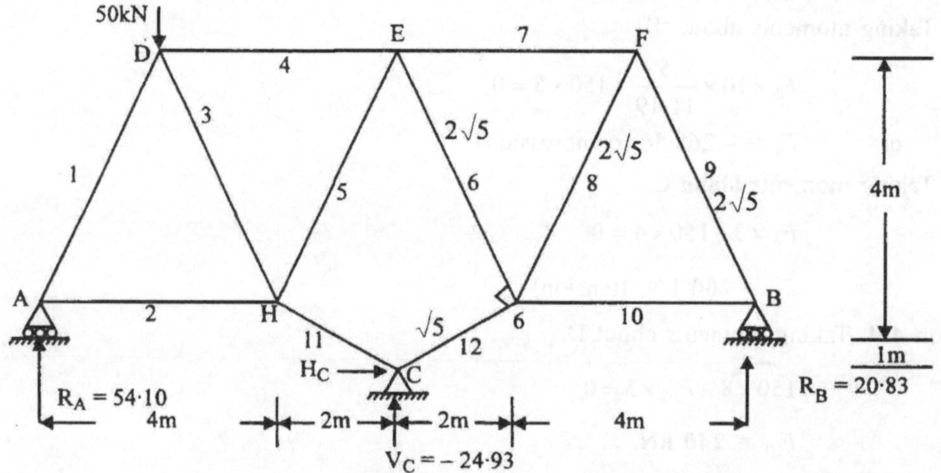

Fig. 3.20. Example 3.5.

Solution. Consider equilibrium of the whole truss. Three independent Equations of the truss are

$$\Sigma H = 0 \qquad H_C = 0 \qquad \text{... (3.1)}$$

$$\Sigma V = 0 \qquad R_A \uparrow + V_C \uparrow + R_B \uparrow - 50 = 0 \qquad \text{... (3.2)}$$

Taking moment of all forces about A

$$50 \times 2 - V_C \times 6 - R_B \times 12 = 0 \qquad \text{... (3.3)}$$

As there are only three equations with 4 unknowns complete solution is not possible. Isolating the joint *C*, we can write the joint equilibrium equations.

With $H_C = 0$

$F_{11} = F_{12} = F.$

Summing vertically

$$F_{11} \times \frac{1}{\sqrt{5}} + F_{12} \times \frac{1}{\sqrt{5}} + V_C = 0$$

$$\frac{2F}{\sqrt{5}} = -V_C$$

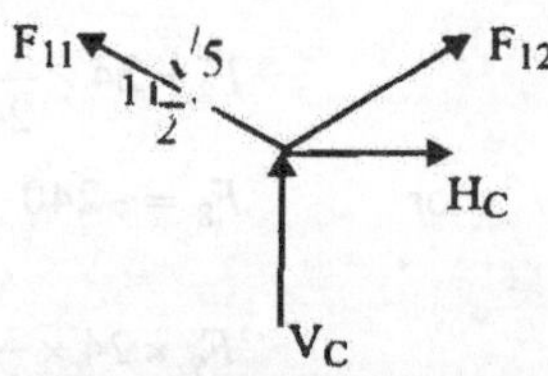

Fig. 3.21 (*a*) Example 3.5.

or $$V_C = -\frac{2}{\sqrt{5}}.F.$$

Substituting in Eqn. (3.3)

$$100 + \frac{2}{\sqrt{5}}.F.6 - R_B \times 12 = 0$$

Now pass a section through E and cut the member 12. Taking moments of all forces on the right of the section about E.

$$R_B \times 6 - F \times 2\sqrt{5} = 0$$

or $$F = \frac{R_B \times 6}{2\sqrt{5}} = \frac{3R_B}{\sqrt{5}}$$

$$100 + \frac{2}{\sqrt{5}} \cdot \frac{3R_B}{\sqrt{5}} \times 6 - R_B \times 12 = 0$$

or $$R_B = 20 \cdot 83.$$

$$F_{12} = \frac{3R_B}{\sqrt{5}} = \frac{3 \times 20 \cdot 83}{\sqrt{5}} = 27 \cdot 946 \text{ kN}.$$

Taking moments of all forces about A.

$$R_B \times 12 + V_C \times 6 - 50 \times 2 = 0$$

or $$V_C \times 6 = 100 - R_B \times 12$$
$$= 100 - 249 \cdot 6$$
$$= -149 \cdot 6$$
$$V_C = -24 \cdot 93 \text{ KN.}\downarrow$$
$$R_A \uparrow + V_C \uparrow + R_B \uparrow = 50$$
$$R_A = 50 - V_C - R_B$$
$$= 50 - (-24 \cdot 93) - 20 \cdot 83.$$
$$= 74 \cdot 93 - 20 \cdot 83 = \mathbf{54 \cdot 10 \text{ kN}}.$$

Solution by Method of Joint

Joint A (Fig. 3.21 (*b*))

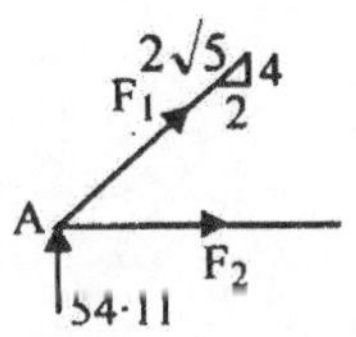

$2\sqrt{5}$ 4 F_1 2 A F_2 54·11

Fig. 3.21 (*b*) Joint A.

$$F_1 \times \frac{4}{2\sqrt{5}} + 54 \cdot 10 = 0$$

or $$F_1 = -\frac{54 \cdot 10 \times 2\sqrt{5}}{4}$$
$$= -60 \cdot 50 \text{ kN}.$$

$$F_1 \times \frac{2}{2\sqrt{5}} + F_2 = 0$$

$$F_2 = -\frac{F_1 \times 2}{2\sqrt{5}} = +\frac{60 \cdot 5}{\sqrt{5}} = 27 \cdot 05 \text{ kN}.$$

Joint D (Fig. 3.21 (*c*))

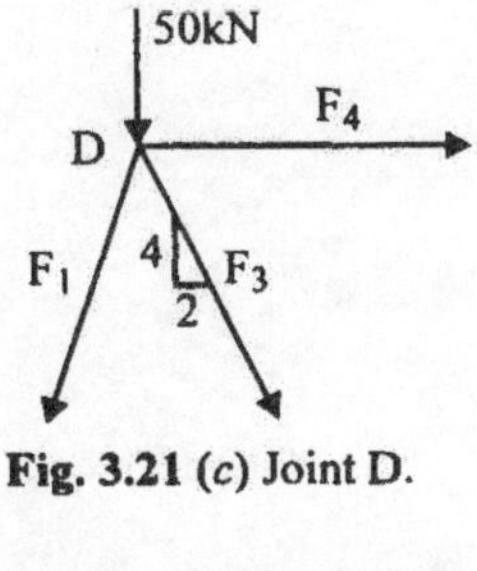

50kN F_4 D F_1 4 F_3 2

Fig. 3.21 (*c*) Joint D.

$$50\downarrow + F_3 \times \frac{4}{2\sqrt{5}} + F_1 \times \frac{4}{2\sqrt{5}} = 0$$

or $$F_3 = -F_1 - \frac{50 \times 2\sqrt{5}}{4}$$
$$= +60 \cdot 50 - 55 \cdot 90$$
$$= 4 \cdot 60 \text{ kN}.$$

$$F_4 + F_3 \times \frac{2}{2\sqrt{5}} - F_1 \times \frac{2}{2\sqrt{5}} = 0$$

$$\text{or} \qquad F_4 = +F_1 + \frac{2}{2\sqrt{5}} - F_3 \times \frac{2}{2\sqrt{5}}$$

$$= -\frac{60 \cdot 50}{\sqrt{5}} - \frac{4 \cdot 60}{\sqrt{5}}$$

$$= -29 \cdot 10 \text{ KN.}$$

Table 3.1.

Coordinates of Points (origin *A* (0,0))

Joint	*X*	*Y*
A	0	0
D	2	4
E	6	4
F	10	4
B	12	0
G	8	0
C	6	–2
H	4	0

Joint H. (Fig. 3.21 (*d*))

$$F_3 \times \frac{2}{2\sqrt{5}} + F_2 - F_5 \times \frac{2}{2\sqrt{5}} - F_{11} \times \frac{2}{\sqrt{5}} = 0$$

$$F_3 \times \frac{4}{2\sqrt{5}} + F_5 \times \frac{4}{2\sqrt{5}} - F_{11} \times \frac{1}{\sqrt{5}} = 0$$

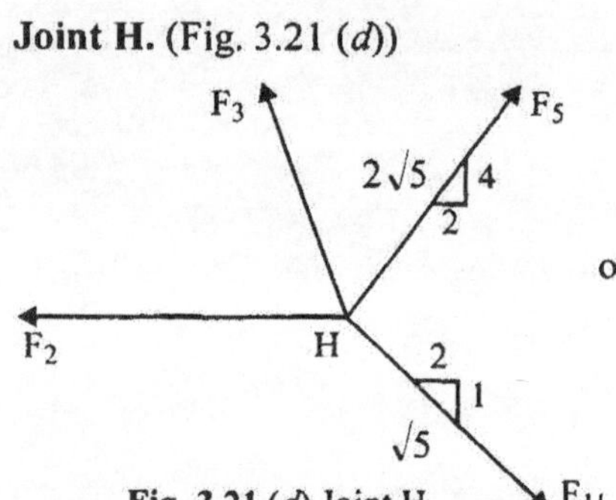

Fig. 3.21 (*d*) Joint H.

$$\text{or} \qquad F_5 \times \frac{1}{\sqrt{5}} + F_{11} \times \frac{2}{\sqrt{5}} = \frac{4 \cdot 60}{\sqrt{5}} + 27 \cdot 05$$

$$F_5 \times \frac{2}{\sqrt{5}} - F_{11} \times \frac{1}{\sqrt{5}} = -4 \cdot 60 \times \frac{2}{\sqrt{5}}$$

Solving $\qquad F_5 = +9 \cdot 34 \text{ KN}$

$\qquad F_{11} = 27 \cdot 88 \text{ KN.}$

Joint B (Fig. 3.21 (*e*))

$$F_9 \times \frac{2}{2\sqrt{5}} + F_{10} = 0$$

$$F_9 \times \frac{4}{2\sqrt{5}} + 20 \cdot 83 = 0$$

$$\text{or} \qquad F_9 = -\frac{20 \cdot 83 \times 2\sqrt{5}}{4}$$

$$= -23 \cdot 28.$$

$$F_{10} = -F_9 \times \frac{1}{\sqrt{5}}$$

$$= +23 \cdot 28 \times \frac{1}{\sqrt{5}} = +10 \cdot 41$$

Fig. 3.21 (*e*) Joint B.

Joint F (Fig. 3.21 (*f*))

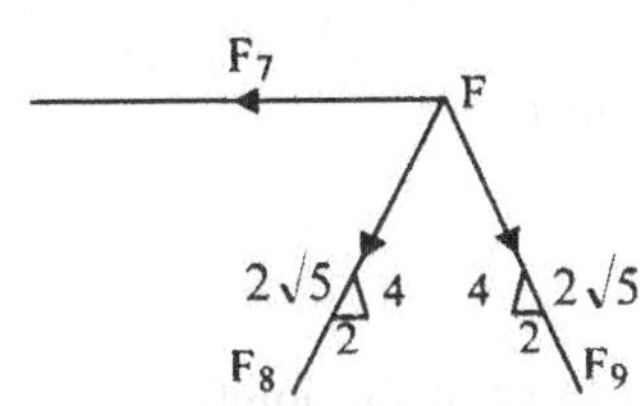

Fig. 3.21 (*f*) Joint F.

$$F_7 + F_8 \times \frac{2}{2\sqrt{5}} - F_9 \times \frac{2}{2\sqrt{5}} = 0$$

$$F_8 \times \frac{4}{2\sqrt{5}} + F_9 \times \frac{4}{2\sqrt{5}} = 0$$

or $\quad F_8 = -F_9 = 23 \cdot 28$

$$F_7 + 23 \cdot 38 \times \frac{1}{\sqrt{5}} + \frac{23 \cdot 28}{\sqrt{5}} = 0$$

or $\quad F_7 = -\dfrac{2 \times 23 \cdot 28}{\sqrt{5}} = -20 \cdot 82$

Joint G (Fig. 3.21 (*g*))

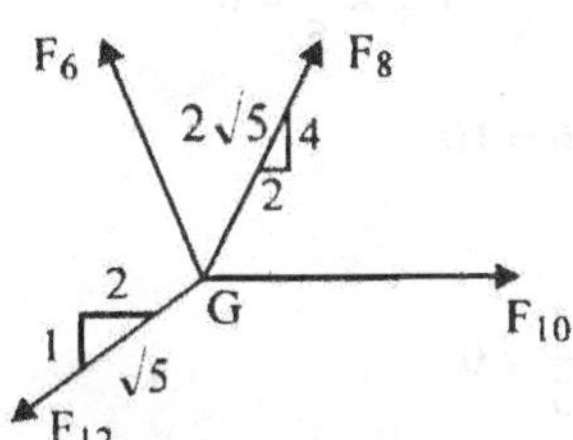

Fig. 3.21 (*g*) Joint G.

Resolving horizontally

$$F_{10} + F_8 \times \frac{2}{2\sqrt{5}} - F_6 \times \frac{2}{2\sqrt{5}} - F_{12} \times \frac{2}{\sqrt{5}} = 0$$

or $\quad 10 \cdot 14 + \dfrac{F_8}{\sqrt{5}} - \dfrac{F_6}{\sqrt{5}} - 27 \cdot 946 \times \dfrac{2}{\sqrt{5}} = 0$

or $\quad F_8 - F_6 = 32 \cdot 64$

Similarly Resolving vertically

$$F_8 \times \frac{4}{2\sqrt{5}} + F_6 \times \frac{4}{2\sqrt{5}} - F_{12} \times \frac{1}{\sqrt{5}} = 0$$

$$F_8 + F_6 = F_{12} \times \frac{1}{\sqrt{5}} \times \frac{\sqrt{5}}{2} = \frac{27 \cdot 946}{2} = 13 \cdot 973$$

Solving, $F_8 = 23 \cdot 30$, $F_6 = -9 \cdot 34$

Joint C (Fig. 3.21 (*h*))

Fig. 3.21 (*h*) Joint C.

$$F_{11} \frac{1}{\sqrt{5}} + F_{12} \times \frac{1}{\sqrt{5}} = 24 \cdot 99$$

or $\quad 2F_{11} = 24 \cdot 99 \times \sqrt{5}$

or $\quad F_{11} = 27 \cdot 94$ KN

Joint E (Fig. 3.21 (*i*))

Fig. 3.21 (*i*) Joint E.

$$F_4 + F_5 \times \frac{2}{2\sqrt{5}} - F_7 - F_6 \times \frac{2}{2\sqrt{5}} = 0$$

or $\quad -29 \cdot 10 + \dfrac{9 \cdot 34}{\sqrt{5}} + 20 \cdot 82$

$$+ \frac{9 \cdot 34}{\sqrt{5}} = 0 \cdot 07 \approx 0 \text{ (check).}$$

Example 3.6 Compute forces in all members of the truss given in Fig. 3.22.

Solution. At Joint *D*. (Fig. 3.22).

Summing horizontally

$$6 + F_{18} = 0$$

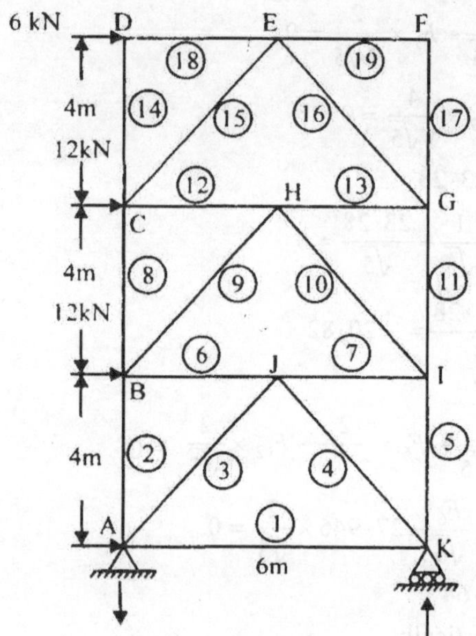

Fig. 3.22 Example 3.6.

Resolving vertically $F_{14} = 0$.

or $F_{18} = -6$.

Considering joint F.

$F_{19} = 0$

$F_{17} = 0$.

As there is no force at the joint.

Considering Joint E (Fig. 3.23 (*a*))

From $\Sigma H = 0$

$$F_{18} - F_{19} + F_{15} \times \frac{3}{5} - F_{16} \times \frac{3}{5} = 0$$

or $F_{15} - F_{16} = \frac{5}{3} \times 6 = 10$.

From $\Sigma V = 0$

$$F_{15} \times \frac{4}{5} + F_{16} \times \frac{4}{5} = 0$$

or $F_{15} = -F_{16}$

or $2F_{15} = 10$

$F_{15} = 5$ kN.

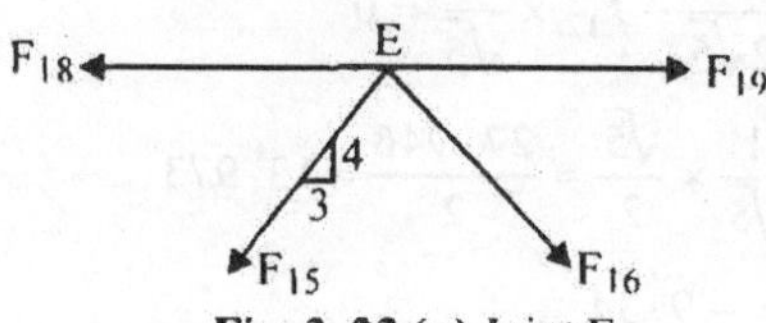

Fig. 3. 23 (*a*) Joint E.

Joint C (Fig. 3.23 (*b*))

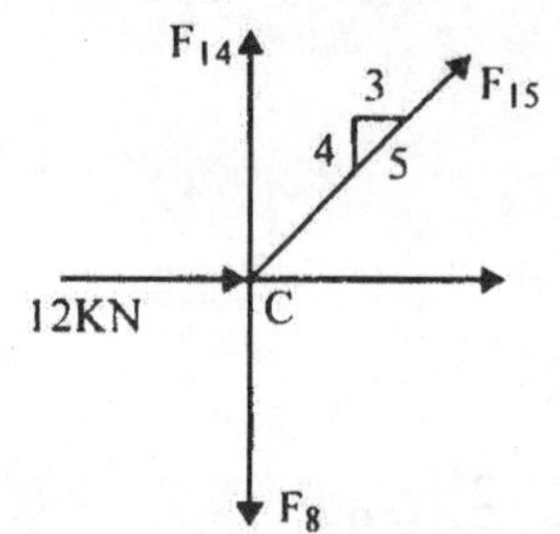

Fig. 3.23 (*b*) Joint C.

$$F_{14} + F_{15} \times \frac{4}{5} - F_8 = 0$$

$$F_{15} \times \frac{3}{5} + 12 + F_{12} = 0$$

or $5 \times \frac{3}{5} + 12 + F_{12} = 0$

or $F_{12} = -15$ kN.

$$F_{14} + F_{15} \times \frac{4}{5} - F_8 = 0$$

$$0 + 5 \times \frac{4}{5} - F_8 = 0$$

or $F_8 = 4$ kN.

Joint H (Fig. 3.23 (*c*))

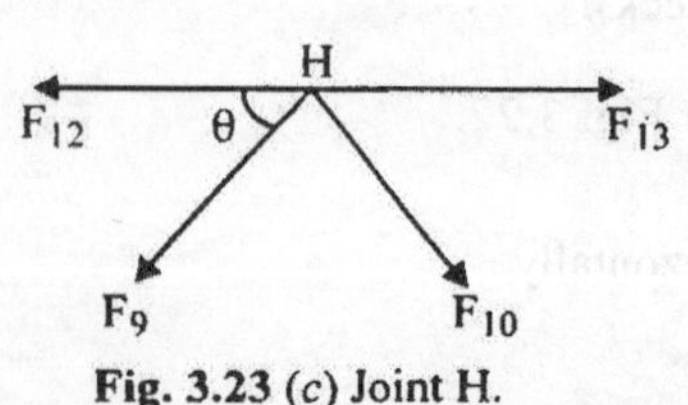

Fig. 3.23 (*c*) Joint H.

From symmetry and as there is no vertical force at H.

$$F_9 = -F_{10}$$

Resolving horizontally

$$F_{12} - F_{13} + F_9 \cos\theta - F_{10} \cos\theta = 0$$

or $-15 - F_{13} - 2F_{10} \cos\theta = 0$

or $2F_{10} \cos\theta = -18$

or $F_{10} = -15$kN.

Joint G (Fig. 3.23 (*d*))

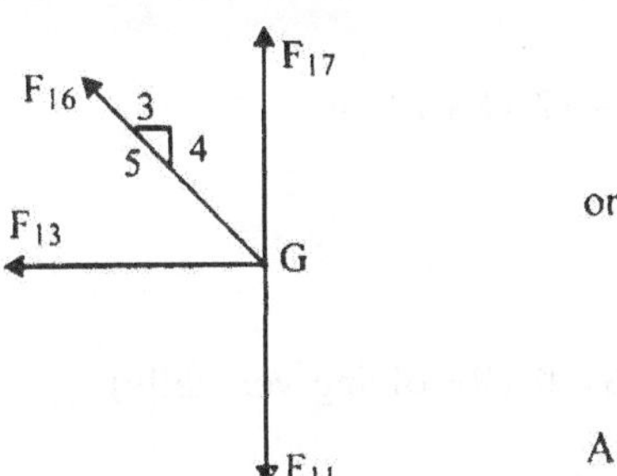

Fig. 3.23 (*d*) Joint G.

$$F_{16} \times \frac{4}{5} + F_{17} - F_{11} = 0 \text{ (Resolving vertically)}$$

$$F_{16} \times \frac{3}{5} + F_{13} = 0 \text{ (Resolving Horizontally)}$$

or
$$F_{13} = -\frac{3}{5} \times F_{16}$$
$$= -\tfrac{3}{5} \times (-5)$$
$$= +3 \text{ kN.}$$

As
$$F_{17} = 0$$
$$F_{11} = F_{16} \times \frac{4}{5} = -4 \text{ kN.}$$

Joint I (Fig. 3.23 (*e*))

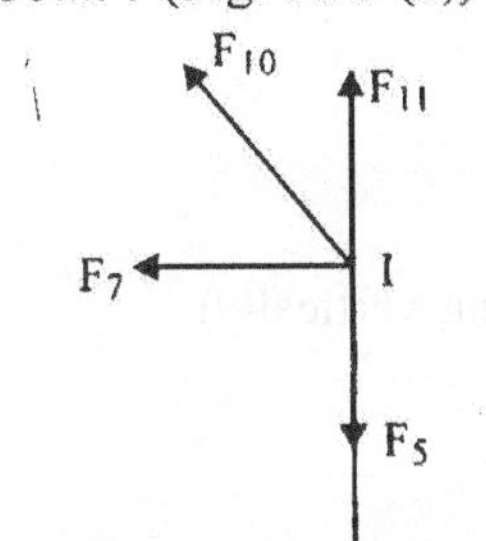

Fig. 3.23 (*e*) Joint I.

$$F_{10} \times \frac{4}{5} + F_{11} - F_5 = 0 \text{ (Resolving vertically)}$$

$$F_{10} \times \frac{3}{5} + F_7 = 0 \text{ (Resolving horizontally)}$$

or
$$-15 \times \frac{4}{5} + (-4) - F_5 = 0$$

or
$$F_5 = -16$$
$$F_7 = -F_{10} \times \frac{3}{5} = +\frac{15 \times 3}{5}$$
$$= 9 \text{ kN.}$$

Joint B (Fig. 3.23 (*f*))

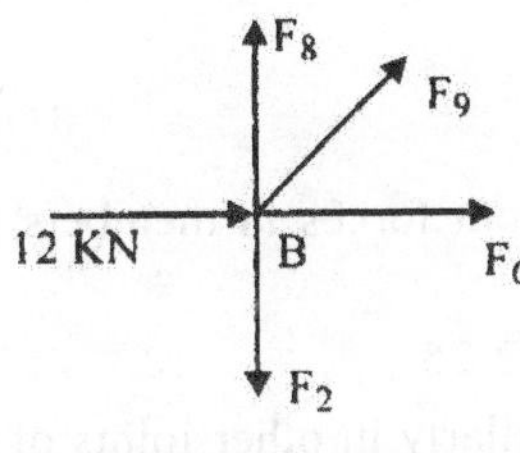

Fig. 3.23 (*f*) Joint B.

$$F_8 + F_9 \times \tfrac{4}{5} - F_2 = 0 \text{ (Resolving vertically)}$$

$$12 + F_9 \times \frac{3}{5} + F_6 = 0 \text{ (Resolving horizontally)}$$

Substituting values,

$$4 + F_9 \times \frac{4}{5} - F_2 = 0$$

or
$$F_2 = 4 + 15 \times \frac{4}{5}$$
$$= +16 \text{ kN.}$$
$$F_6 = -12 - 15 \times \frac{3}{5}$$
$$= -21 \text{ kN.}$$

Joint J (Fig. 3.23 (*g*))

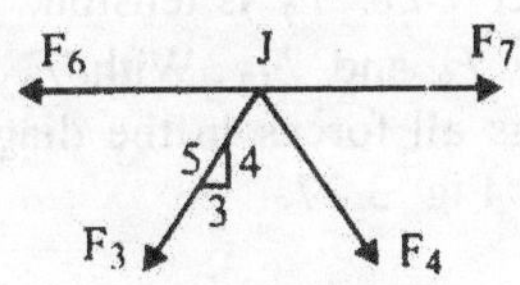

Fig. 3.23 (*g*) Joint J.

$$F_3 = -F_4 \text{ (Resolving vertically)}$$

$$F_6 - F_7 + F_3 \times \frac{3}{5} - F_4 \times \frac{3}{5} = 0 \text{ (Resolving horizontally)}$$

Substituting values,

$$-21 - 9 + 2F_3 \times \frac{3}{5} = 0$$

$$\text{or} \qquad F_3 = +\frac{30\times 5}{6}$$

$$= +\,25\text{kN}.$$

Joint A (Fig. 3.23 (*h*))

$$R_K \times 6 = 6\times 12 + 12\times 8 + 12\times 4$$

$$= 216$$

$$R_K = 36.$$

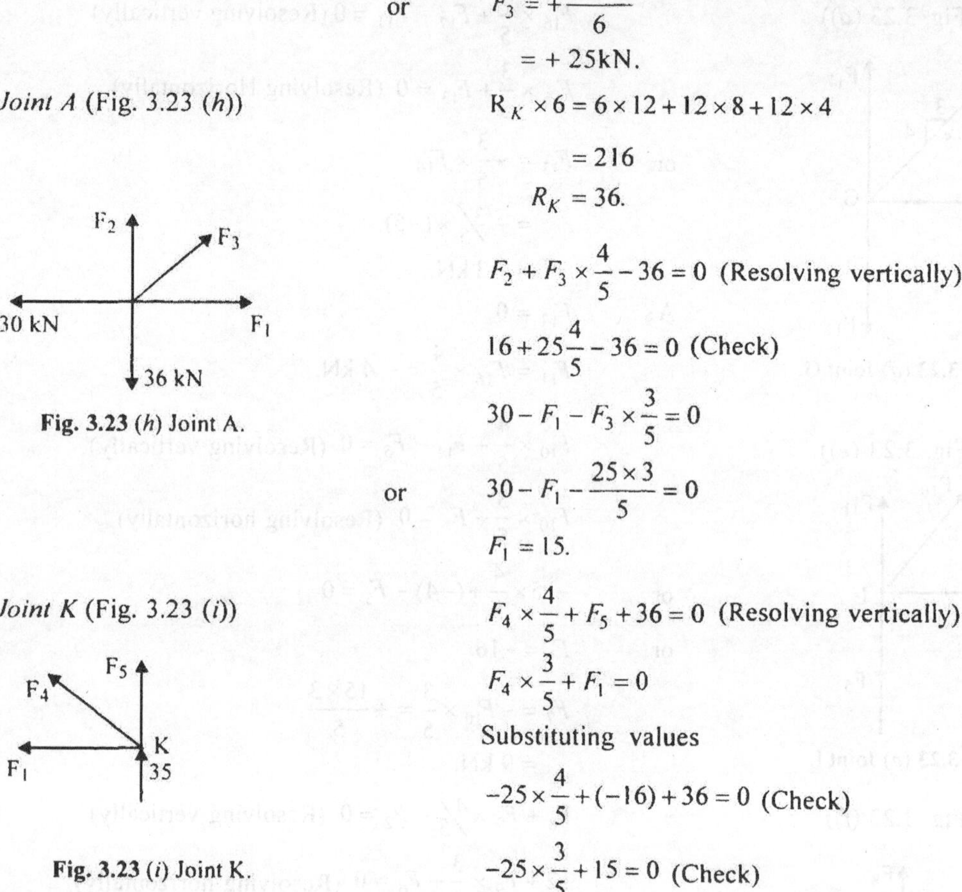

$$F_2 + F_3 \times \frac{4}{5} - 36 = 0 \text{ (Resolving vertically)}$$

$$16 + 25\frac{4}{5} - 36 = 0 \text{ (Check)}$$

Fig. 3.23 (*h*) Joint A.

$$30 - F_1 - F_3 \times \frac{3}{5} = 0$$

$$\text{or} \qquad 30 - F_1 - \frac{25\times 3}{5} = 0$$

$$F_1 = 15.$$

Joint K (Fig. 3.23 (*i*))

$$F_4 \times \frac{4}{5} + F_5 + 36 = 0 \text{ (Resolving vertically)}$$

$$F_4 \times \frac{3}{5} + F_1 = 0$$

Substituting values

$$-25 \times \frac{4}{5} + (-16) + 36 = 0 \text{ (Check)}$$

$$-25 \times \frac{3}{5} + 15 = 0 \text{ (Check)}$$

Fig. 3.23 (*i*) Joint K.

Example 3.7. Find Forces in all members of truss given in Fig. 3.24.

Solution. From the configuration of the truss certain observations about forces in members can be made.

1. Joints *E*, *G*, *H*, *I*, *K*, *B* and *C* are of the type shown in Fig. 3.25.

From consideration of equilibrium of joints $F_4 = -F_{10} = F$. Similarly in other joints of the same type *i.e.* if one member is in tension, other member will be in compression.

2. Joints of the type *L*, *M*, *N*, *O*, *P* are shown in Fig. 3.26. Since two forces are in one line, they must be equal and opposition direction.

Considering equilibrium of the truss as a whole (Fig. 3.24) $R_j \times 12 - 100\times 24 = 0$
$R_j = 200$ kN.

From the consideration of cantilever behaviour force in member 7 *i.e.* F_7 is tension. Then force in member 6 *i.e.* F_6 will be in compression. Say $-F$. Similarly F_9 and F_{12}. With $F_{12} = -F$ F_{18} is equal to F. Similarly F_{21} and F_{24}. Using these reasonings all forces in the diagonal members can be computed in terms of F and $-F$. This is shown in Fig. 3.27.

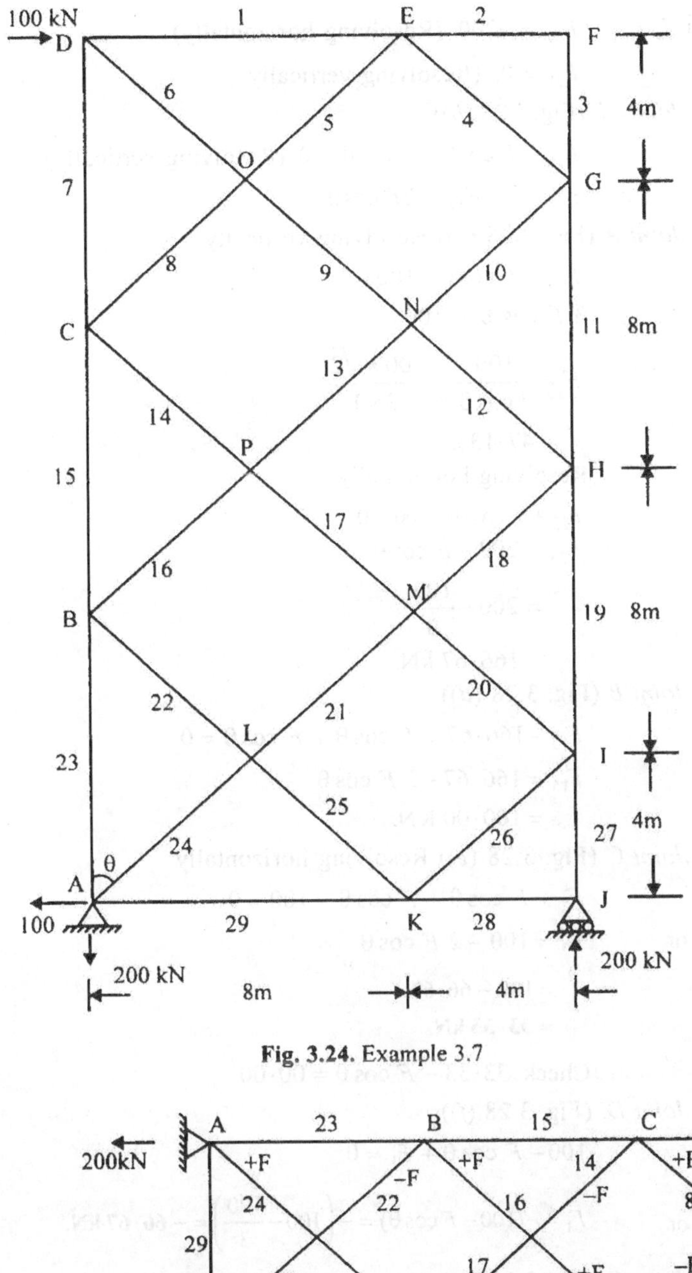

Fig. 3.24. Example 3.7

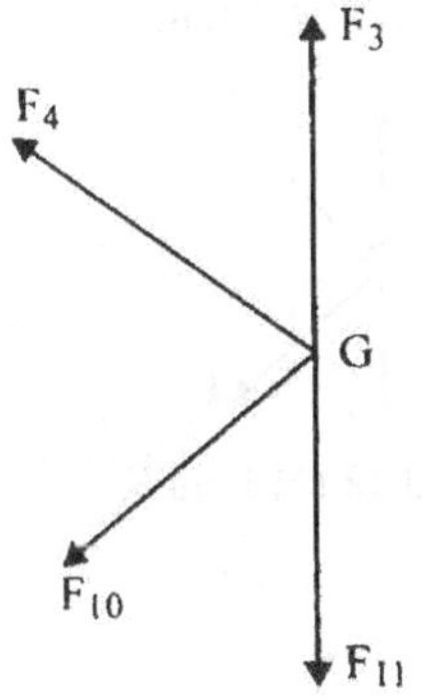

Fig. 3.25. Joint G.

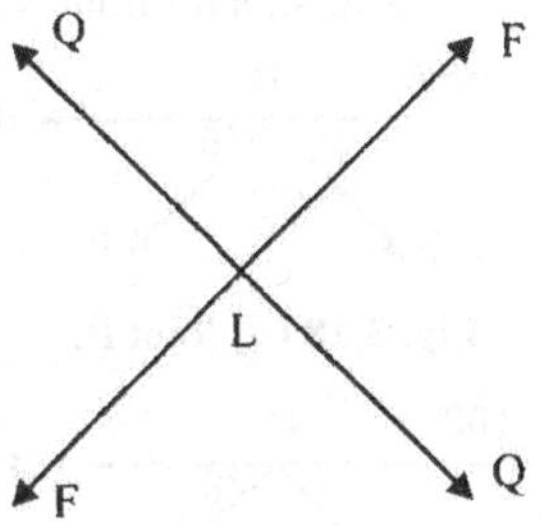

Fig. 3.26. Joint L.

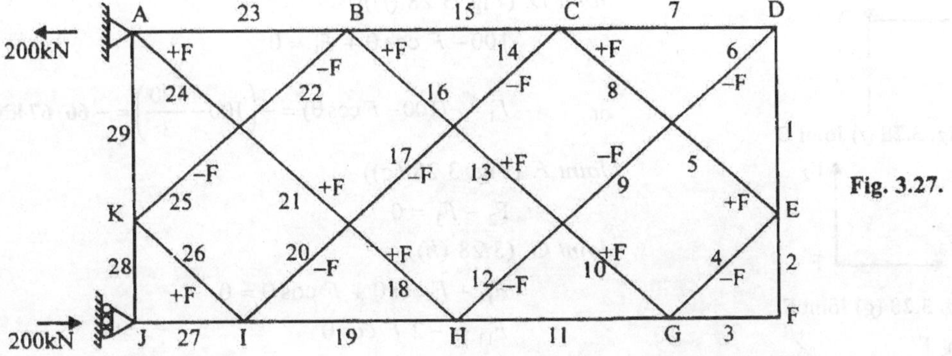

Fig. 3.27.

Consider equilibrium of joint *J*.

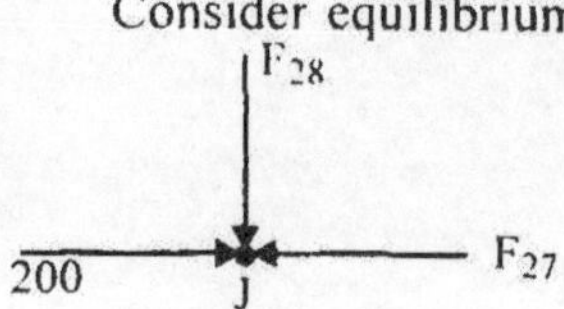

Fig. 3.28 (*a*) Joint J.

$F_{27} = -200$ (Resolving horizontally)

$F_{28} = 0$. (Resolving vertically

Joint K (Fig. 3.28 (*b*))

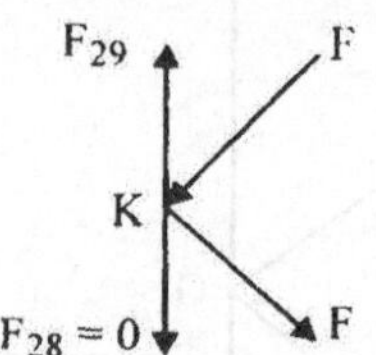

Fig. 3.28 (*b*) Joint K.

$F_{29} - F\cos\theta - F\cos\theta = 0$ (Resolving vertically)

or $F_{29} = 2F\cos\theta$.

Joint A (Fig. 3.28 (*c*)) Resolving vertically

$$F_{29} + F\cos\theta - 100 = 0$$

$$3\,F\cos\theta = 100$$

$$F = \frac{100}{3\cos\theta} = \frac{100 \times \sqrt{2}}{3 \times 1}$$

$$= 47 \cdot 133.$$

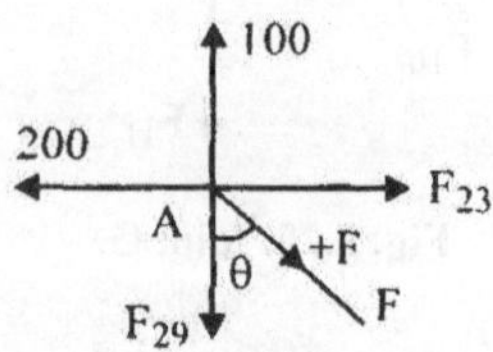

Fig. 3.28 (*c*) Joint A.

Resolving horizontally

$$F_{23} + F\cos\theta - 200 = 0$$

$$F_{23} = 200 - F\cos\theta$$

$$= 200 - \frac{100}{3}$$

$$= 166 \cdot 67 \text{ kN}.$$

Joint B (Fig. 3.28 (*d*))

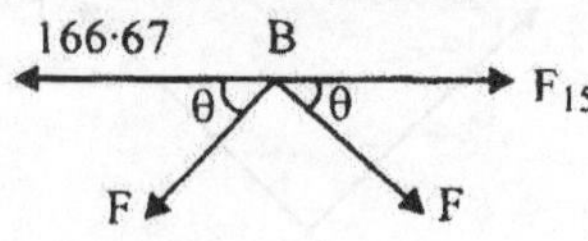

Fig. 3.28 (*d*) Joint B.

$$F_{15} - 166 \cdot 67 + F\cos\theta + F\cos\theta = 0$$

$$F_{15} = 166 \cdot 67 - 2\,F\cos\theta$$

$$= 100 \cdot 00 \text{ KN}.$$

Joint C (Fig. 3.28 (*e*)) Resolving horizontally

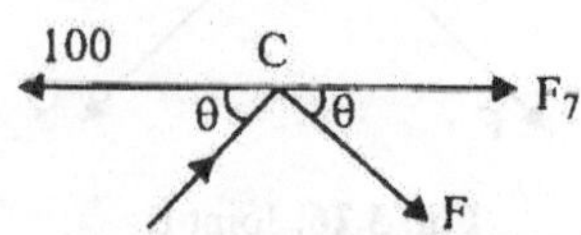

Fig. 3.28 (*e*) Joint C.

$$F_7 + F\cos\theta + F\cos\theta - 100 = 0$$

or $F_7 = 100 - 2\,F\cos\theta$

$$= 100 - 66 \cdot 67$$

$$= \mathbf{33 \cdot 33 \text{ kN}}.$$

Check $33 \cdot 33 - F\cos\theta = 00 \cdot 00$

Joint D. (Fig. 3.28 (*f*))

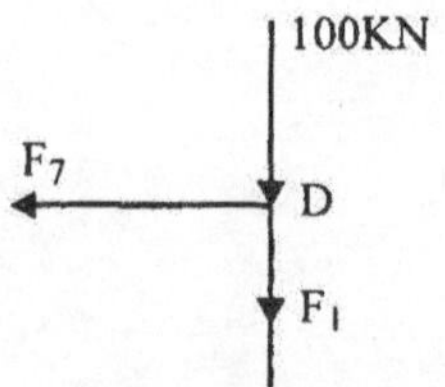

Fig. 3.28 (*f*) Joint D.

$$100 - F\cos\theta + F_1 = 0$$

or $F_1 = -(100 - F\cos\theta) = -\left(100 - \frac{100}{3}\right) = \mathbf{-\,66 \cdot 67 \text{ kN}}.$

Joint F. (Fig. 3.28 (*g*))

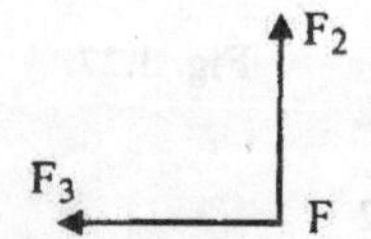

Fig. 3.28 (*g*) Joint F.

$$F_2 - F_3 = 0$$

Joint G. (3.28 (*h*))

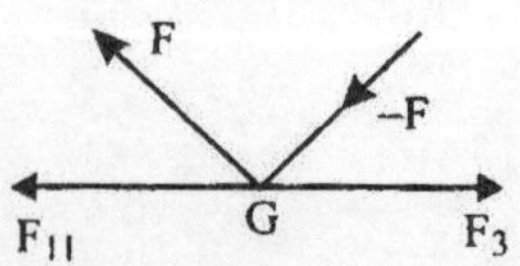

Fig. 3.28 (*h*) Joint G.

$$F_{11} + F\cos\theta + F\cos\theta = 0$$

$$F_{11} = -2\,F\cos\theta$$

$$= -2\,\frac{100}{3}$$

$$= \mathbf{-\,66 \cdot 67 \text{ kN}}.$$

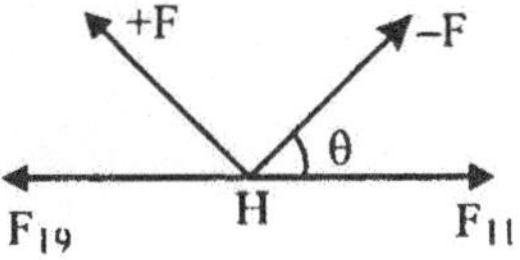

Fig. 3.28 (*i*) Joint H.

Joint H. (Fig. 3.28 (*i*))

$$F_{19} + F\cos\theta + F\cos\theta - (-66 \cdot 67) = 0$$

$$F_{19} = -2\,F\cos\theta - 66 \cdot 67$$

$$= -\frac{2 \times 100}{3} - 66 \cdot 67$$

$$= -\mathbf{133 \cdot 34\ kN.}$$

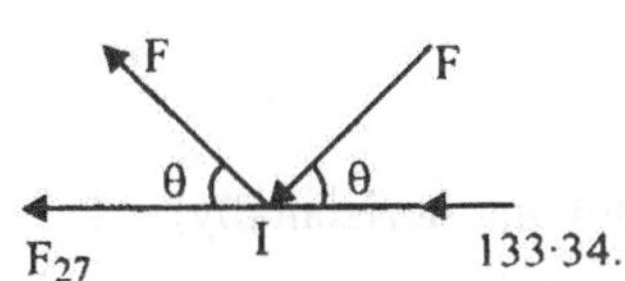

Fig. 3.28 (*j*) Joint I.

Joint I. (Fig. 3.28 (*j*))

$$F_{27} + 2\,F\cos\theta + 133 \cdot 34 = 0$$

$$F_{27} = -2\,F\cos\theta - 133 \cdot 34$$

$$= -66 \cdot 67 - 133 \cdot 34$$

$$= -\mathbf{200\ kN.}$$

Example 3.8. Find forces in all members of the *K*-truss shown in the Fig. 3.29.

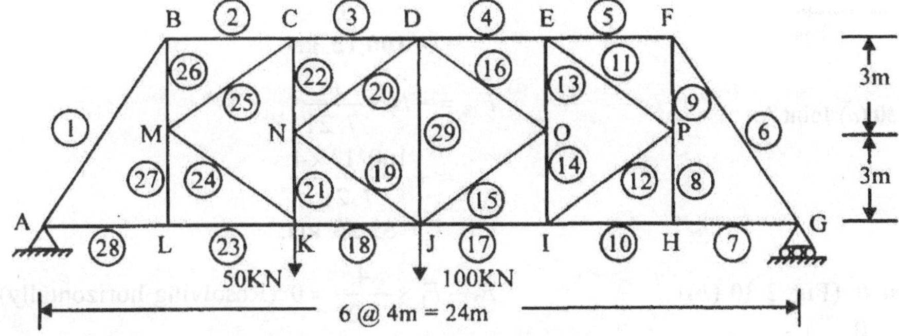

Fig. 3.29 Example 3.8.

Solution. Taking moments about A.

$$50 \times 8 + 100 \times 12 - R_G \times 24 = 0$$

or $$R_G = {(400 + 1200)}/{24}$$

$$= \mathbf{66 \cdot 67\ KN.}$$

Taking moments about G.

$$50 \times 16 + 100 \times 12 - R_A \times 24 = 0$$

or $$R_A = 83 \cdot 33\ \text{kN}$$

Check $$R_G + R_A = 66 \cdot 67 + 83 \cdot 33$$

$$= 150\ \text{kN}.$$

Coordinates of Points with *A* as origin.

Point	*X*	*Y*	*Point*	*X*	*Y*
A	0	0	L	4	0
B	4	6	M	4	3
C	8	6	N	8	3
D	12	6	O	16	3
E	16	6	P	20	3

Point	X	Y
F	20	6
G	24	0
H	20	0
I	16	0
J	12	0
K	8	0

Joint A. (Fig. 3.30 (*a*))

$$F_1 \times \frac{4}{7 \cdot 21} + F_{28} = 0 \text{ (Resolving horizontally)}$$

$$F_1 \times \frac{6}{7 \cdot 21} + 83 \cdot 33 = 0 \text{ (Resolving vertically)}$$

or $$F_1 = -\frac{83 \cdot 33 \times 7 \cdot 21}{6}$$

$$= -\mathbf{100 \cdot 13\ kN.}$$

$$F_{28} = -F_1 \times \frac{4}{7 \cdot 21}$$

$$= +\frac{100 \cdot 13 \times 4}{7 \cdot 21}$$

$$= +\mathbf{55 \cdot 55\ kN.}$$

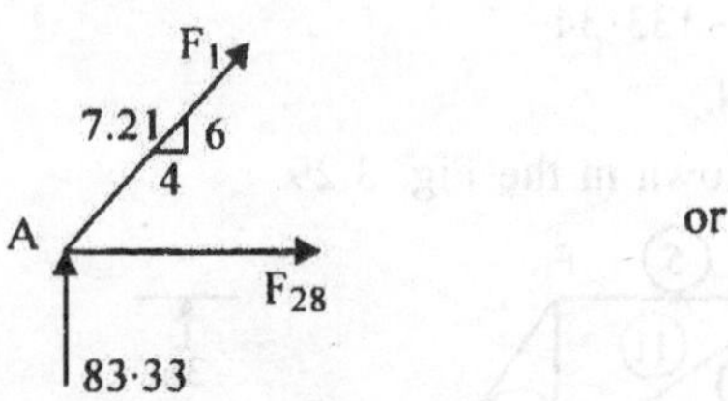

Fig. 3.30 (*a*) Joint A.

Joint B. (Fig. 3.30 (*b*))

$$F_2 - F_1 \times \frac{4}{7 \cdot 21} = 0 \text{ (Resolving horizontally)}$$

$$F_1 \times \frac{6}{7 \cdot 21} + F_{26} = 0 \text{ (Resolving vertically)}$$

or $$F_2 = F_1 \times \frac{4}{7 \cdot 21} = -\frac{100 \cdot 13 \times 4}{7 \cdot 21}$$

$$= -\mathbf{55 \cdot 55\ kN.}$$

Fig. 3.30 (*b*) Joint B.

$$F_{26} = -\frac{F_1 \times 6}{7 \cdot 21} = +\frac{100 \cdot 13 \times 6}{7 \cdot 21} = \mathbf{83 \cdot 33\ kN.}$$

Joint L. (Fig. 3.30 (*c*))

$$F_{27} = 0$$

Resolving horizontally

$$F_{23} - F_{28} = 0$$

or $$F_{23} = F_{28} = +\mathbf{55 \cdot 55\ kN.}$$

Joint M. (Fig. 3.30 (*d*))

$$\frac{4}{5}(F_{25} + F_{24}) = 0 \text{ (Resolving horizontally)}$$

$$F_{26} + \tfrac{3}{5} F_{25} - \tfrac{3}{5} F_{24} - F_{27} = 0 \text{ (Resolving vertically)}$$

or $$F_{26} + \frac{3}{5}(F_{25} + F_{25}) - F_{27} = 0$$

or $$\frac{3}{5} \times 2\, F_{25} = -F_{26} = -83 \cdot 33$$

or $$F_{25} = -\mathbf{69 \cdot 44\ kN.}$$

$$F_{24} = +69 \cdot 44 \text{ kN.}$$

Fig. 3.30 (*d*) Joint M.

Joint K. (Fig. 3.30 (*e*))

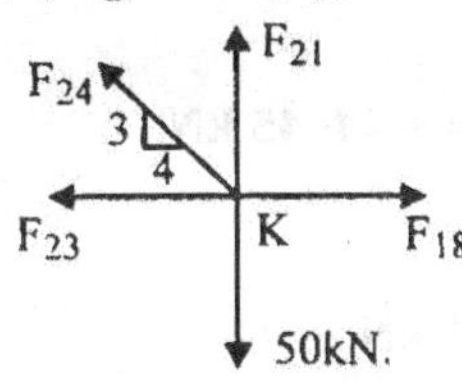

Fig. 3.30 (*e*) Joint K.

$$F_{18} - F_{23} - F_{24} \times \frac{4}{5} = 0 \text{ (Resolving horizontally)}$$

$$F_{21} + \frac{3}{5} \times F_{24} - 50 = 0$$

or $$F_{21} = 50 - \frac{3}{5} \times (69 \cdot 44)$$

$$= \mathbf{8 \cdot 34 kN.}$$

Joint C.

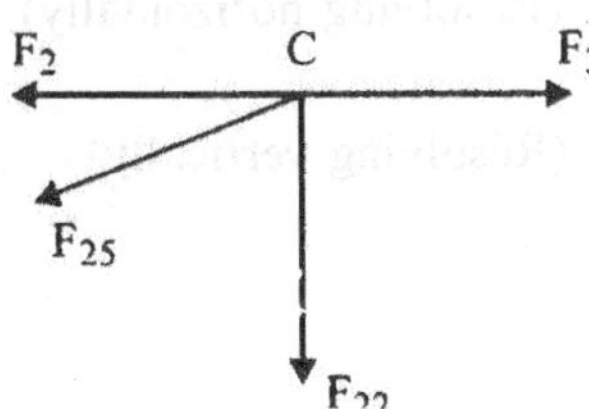

Fig. 3.30 (*f*) Joint C.

$$F_{18} = F_{23} + F_{24} \times \frac{4}{5} \text{ (Resolving horizontally)}$$

$$= 55 \cdot 55 + (69 \cdot 44) \times \frac{4}{5}$$

$$= \mathbf{111 \cdot 102 \ kN.}$$

$$F_2 + F_{25} \times \frac{4}{5} - F_3 = 0 \text{ (Resolving horizontally)}$$

$$F_{22} + F_{25} \times 3/5 = 0 \text{ (Resolving vertically)}$$

or $$F_{22} = -F_{25} \times 3/5$$

$$= \frac{69 \cdot 44 \times 3}{5} = +41 \cdot 66 \text{ kN.}$$

$$-55 \cdot 55 - 69 \cdot 44 \times \frac{4}{5} - F_3 = 0$$

or $$F_3 = -\mathbf{111 \cdot 102 \ kN.}$$

Joint N. Fig. 3.30 (*g*)

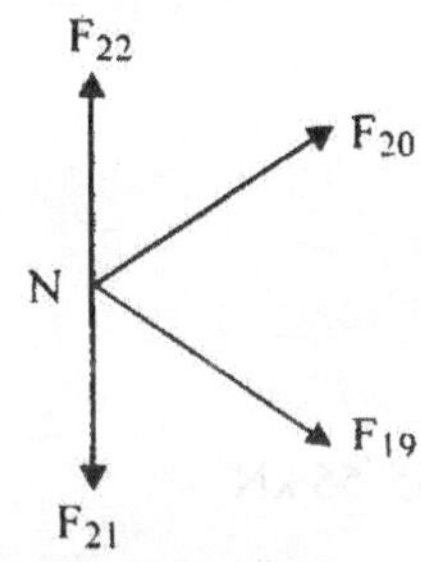

Fig. 3.30 (*g*) Joint N.

$$F_{22} - F_{21} + F_{20} \times \frac{3}{5} - F_{19} \times \frac{3}{5} = 0$$

(Resolving vertically)

$$F_{20} \times \frac{4}{5} + F_{19} \times \frac{4}{5} = 0 \text{ (Resolving horizontally)}$$

or $$F_{20} = -F_{19}.$$

Substituting

$$2 F_{20} \times \frac{3}{5} = -F_{22} + F_{21}$$

$$= -41 \cdot 66 + 8 \cdot 34$$

$$= -\mathbf{33 \cdot 32 \ kN.}$$

$$F_{20} = 5/6 \times (-33 \cdot 32) = -27 \cdot 76 \text{ kN.}$$

$$F_{19} = +27 \cdot 76 \text{ kN.}$$

Joint G Fig. 3.30 (*h*)

Fig. 3.30 (*h*) Joint G.

$$F_6 \cdot \frac{4}{7 \cdot 21} + F_7 = 0 \text{ (Resolving horizontally)}$$

$$F_6 \times \frac{6}{7 \cdot 21} + 66 \cdot 67 = 0 \text{ (Resolving vertically)}$$

or $$F_6 = -\frac{66 \cdot 67 \times 7 \cdot 21}{6} = -\mathbf{80 \cdot 12 \ kN.}$$

$$F_7 = -F_6 \times \frac{4}{7 \cdot 21}$$

$$= +\frac{80 \cdot 12 \times 4}{7 \cdot 21} = \mathbf{+\,44 \cdot 45\ kN.}$$

Joint H. (Fig. 3.30(*i*))

$$F_{10} - F_7 = 0$$

$$\text{or} \quad F_{10} = F_7 = +44 \cdot 45 \text{ kN}$$

Fig. 3.30 (*i*) Joint H.

Joint F. (Fig. 3.30(*j*))

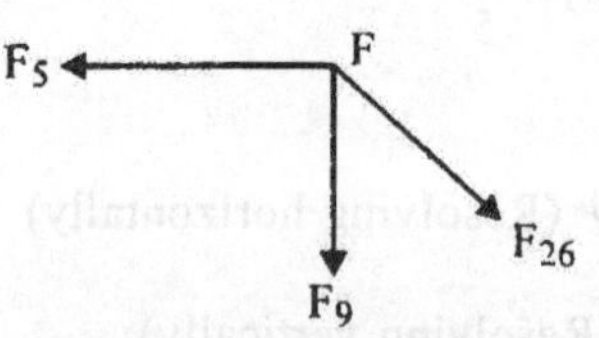

Fig. 3.30 (*j*) Joint F.

$$F_6 \times \frac{4}{7 \cdot 21} - F_5 = 0 \quad \text{(Resolving horizontally)}$$

$$F_6 \times \frac{4}{7 \cdot 21} + F_9 = 0 \quad \text{(Resolving vertically)}$$

$$\text{or} \quad F_5 = F_6 \times \frac{4}{7 \cdot 21}$$

$$= -\frac{80 \cdot 12 \times 4}{7 \cdot 21} = -\,44 \cdot 45 \text{ kN.}$$

$$F_9 = -F_6 \times \frac{4}{7 \cdot 21}$$

$$= \frac{80 \cdot 12 \times 6}{7 \cdot 21} = \mathbf{-\,66 \cdot 67\ kN.}$$

Joint P. (Fig. 3.30(*k*))

Resolving vertically,

$$F_{11} \times \frac{3}{5} + F_9 - F_8 - F_{12} \times \frac{3}{5} = 0$$

Resolving horizontally

$$F_{11} \times \frac{4}{5} + F_{12} \times \frac{4}{5} = 0$$

$$\text{or} \quad F_{11} = -F_{12}$$

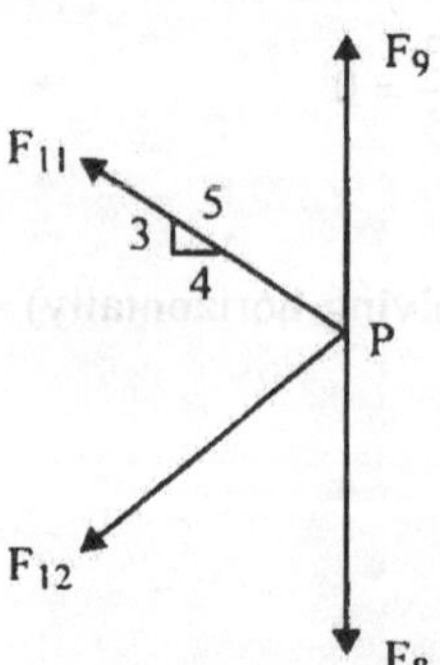

Fig. 3.30 (*k*) Joint P.

Substituting.

$$2\,F_{11} \times \frac{3}{5} = -F_9 + F_8$$

$$= -\,66 \cdot 67$$

$$\text{or} \quad F_{11} = -66 \cdot 67 \times \frac{5}{6} = -55 \cdot 55 \text{ kN}$$

$$F_{12} = +\,55 \cdot 55 \text{ kN}$$

Joint E (Fig. 3.30 (*l*))

$$F_4 - F_5 - F_{11} \times \frac{4}{5} = 0 \quad \text{(Resolving horizontally)}$$

$$F_{13} + F_{11} \times \frac{3}{5} = 0 \quad \text{(Resolving vertically)}$$

$$\text{or} \quad F_4 = F_5 + F_{11} \times \frac{4}{5}$$

$$= -44 \cdot 45 + (-55 \cdot 55) \times \frac{4}{5}$$

$$= -88 \cdot 90 \text{kN.}$$

Fig. 3.30 (*l*) Joint E.

$$F_{13} = -F_{11} \times \frac{3}{5} = +\frac{55 \cdot 55 \times 3}{5}$$

$$= \mathbf{33 \cdot 33 kN.}$$

Joint D. (Fig. 3.30 (*m*))

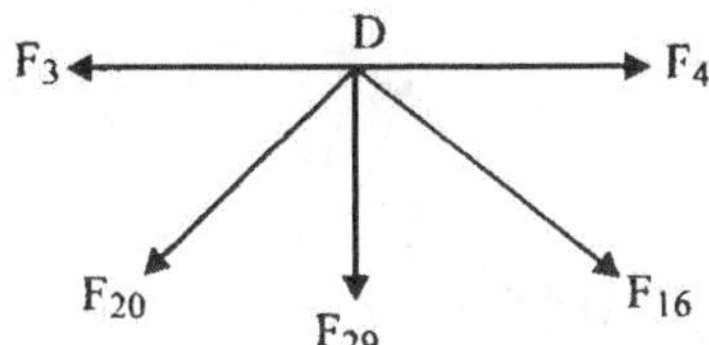

Fig. 3.30 (*m*) Joint D.

Resolving horizontally

$$F_3 - F_4 - F_{16} \times \frac{4}{5} + F_{20} \times \frac{4}{5} = 0$$

Resolving vertically

$$F_{29} + F_{16} \times \frac{3}{5} + F_{20} \times \frac{3}{5} = 0.$$

Substituting values.

$$F_{16} \times \frac{4}{5} = -111 \cdot 102 + 88 \cdot 90 + (-27 \cdot 76) \times \frac{4}{5}$$

or $F_{16} = -55 \cdot 51$

$$F_{29} = -F_{16} \times \frac{3}{5} - F_{20} \times \frac{3}{5}$$

$$= 49 \cdot 96.$$

Joint J (Fig. 3.30 (*n*))

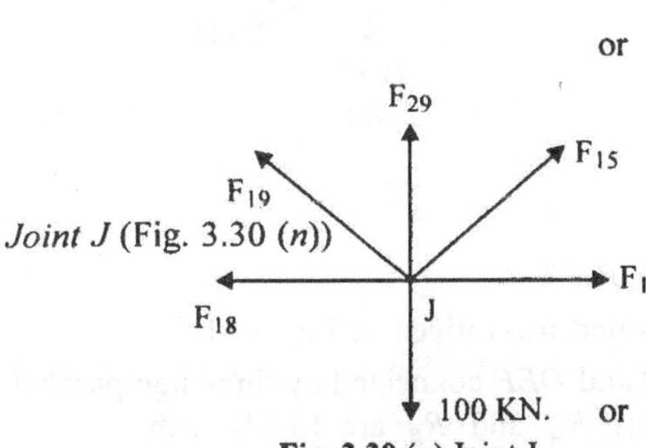

Fig. 3.30 (*n*) Joint J.

Resolving vertically

$$F_{29} + F_{15} \times \frac{3}{5} + F_{19} \times \frac{3}{5} - 100 = 0$$

or $49 \cdot 96 + F_{15} \times \frac{3}{5} + 27 \cdot 76 \times \frac{3}{5} 100 = 0$

or $F_{15} = 55 \cdot 64$

Joint I. (Fig. 3.30(*o*))

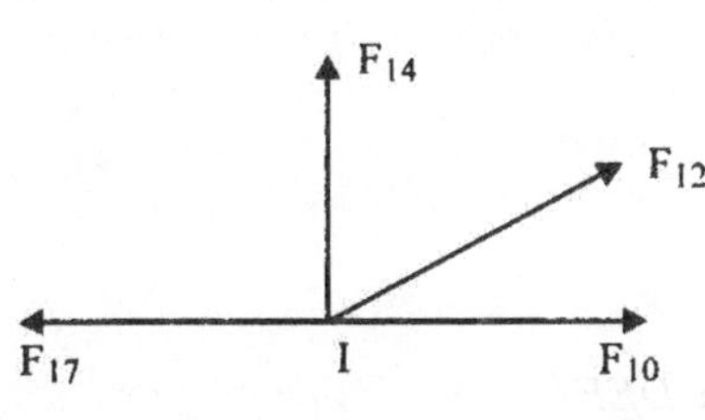

Fig. 3.30 (*o*) Joint I.

Resolving vertically

$$F_{14} + F_{12} \times \frac{3}{5} = 0$$

(Resolving horizontally)

$$F_{10} + F_{12} \times 4/5 - F_{17} = 0$$

$$F_{14} = -F_{12} \times \frac{3}{5}$$

$$= -\frac{55 \cdot 55 \times 3}{5}$$

$$= \mathbf{33 \cdot 33 kN.}$$

$$F_{17} = F_{10} + F_{12} \times \frac{4}{5}$$

$$= 44 \cdot 45 + \frac{55 \cdot 55 \times 4}{5}$$

$$= \mathbf{88 \cdot 89 kN.}$$

Joint J. $F_{29} + F_{15} \times \frac{3}{5} + F_{19} \times \frac{3}{5} - 100 = 0$

Substituting $F_{15} = 55 \cdot 64$kN.

Resolving horizontally at joint *J*.

$$F_{17} + F_{15} \times \frac{4}{5} - F_{19} \times \frac{4}{5} - F_{18} = 0$$

$$88 \cdot 89 + 55 \cdot 64 \times \frac{4}{5} - 27 \cdot 76 \times \frac{4}{5} - 111 \cdot 102 \approx 0. \text{ (check)}$$

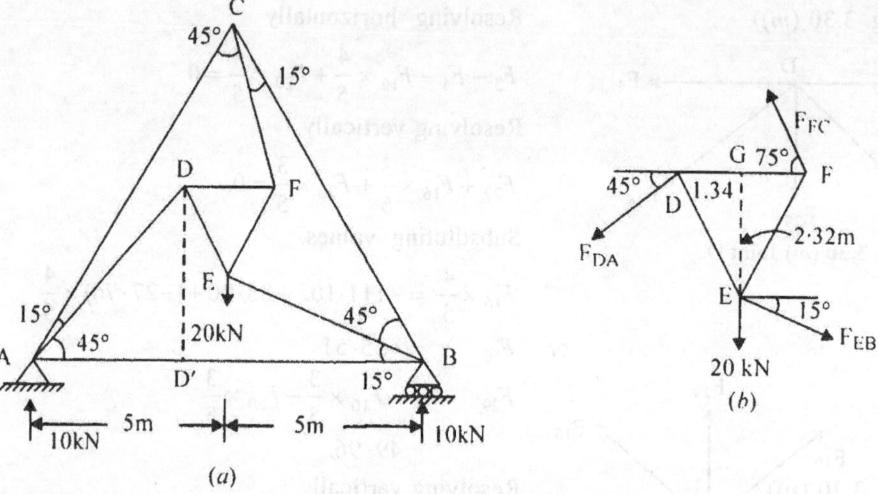

Fig. 3.31. Example 3.9.

Example 3.9. Find forces in all members of the compound truss given in Fig. 3.31.

Solution. The truss consists of two simple trusses *ABC* and *DEF* connected by three non parallel and non concurrent bars *AD*, *BE* and *CF*. The reactions R_A and R_B are 10 kN each. From geometry of figure *ABC* and *DEF* the following computations can be made.

$$AD = BE = CF = \frac{5}{\cos 15°} = 5 \cdot 176m.$$

$$AD' = 5 \cdot 176 \cos 45° = 3 \cdot 660m.$$

$$DF = AB - 2\,AD' = 10 - 2 \times 3 \cdot 660$$

$$= 2 \cdot 68m = DE = EF.$$

$$EG = DE \sin 60° = \frac{2 \cdot 68 \times \sqrt{3}}{2} = 2 \cdot 320.$$

Applying Equations of the Equilibrium to the triangle *DEF*

(*i*) $\Sigma X = 0$ $\quad F_{FC} \cos 75° + F_{DA} \cos 45° - F_{EB} \cos 15° = 0$

(*ii*) $\Sigma Y = 0$ $\quad F_{DA} \sin 45° - F_{FC} \sin 75°$

$\quad + F_{EB} \sin 15° + 20 = 0$

(*iii*) $\Sigma M = 0$ $\quad$ about point *E*.

$$F_{FA} \cos 45° \times 2 \cdot 32 + F_{DA} \sin 45° \times 1 \cdot 34 + F_{FC} \cos 75° \times 2 \cdot 32 + F_{FC} \sin 45° \times 1 \cdot 34 = 0$$

With $\cos 75° = 0{\cdot}259$ $\cos 45° = 0{\cdot}707$

and $\cos 15° = 0{\cdot}966$, $\sin 75° = 0{\cdot}966$

and $\sin 15° = 0{\cdot}259$

The equations reduce to

$$0 \cdot 259\, F_{FC} + 0 \cdot 707\, F_{FA} - 0 \cdot 966\, F_{EB} = 0$$

$$0 \cdot 707\, F_{FA} - 0 \cdot 966\, F_{FC} + 0 \cdot 259\, F_{FB} = -\,20$$

$$2 \cdot 587\, F_{FA} + 1 \cdot 548\, F_{FC} = 0$$

Solving $F_{DA} = -\,9 \cdot 426$ (compression)

$F_{EB} = -\,3 \cdot 450$ (compression)

$F_{FC} = +\,12 \cdot 880$ (tension)

Forces in other members are computed by method of joints.

Joint D. (Fig. 3.32 (*a*)).

Resolving horizontally

$$F_{DF} + F_{DE}\cos 60° - (-9\cdot426)\cos 45° = 0$$

Resolving vertically

$$F_{DE}\sin 60° + F_{DA}\sin 45° = 0$$

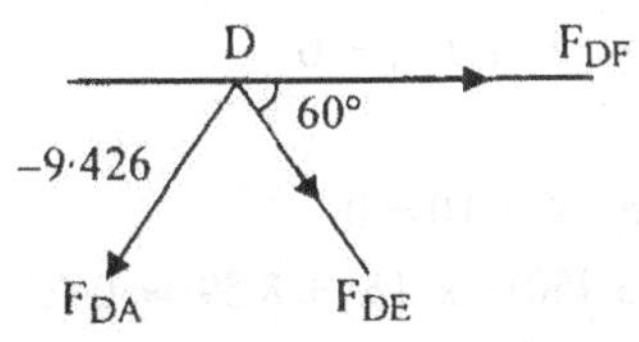

Fig. 3.32 (*a*) Joint D.

$$\text{or} \quad F_{DE} = -\frac{F_{DA}\sin 45°}{\sin 60°}$$

$$= +\frac{9\cdot426\times 0\cdot707}{0\cdot866}$$

$$= \mathbf{7\cdot695kN.}$$

$$F_{DF} = -F_{DE}\cos 60° - 9\cdot426\cos 45°$$

$$= -7\cdot695\cos 60° - 9\cdot426\cos 45°$$

$$= -\mathbf{10\cdot5126kN.}$$

Joint E (Fig. 3.32(*b*))

$$F_{DE}\cos 60° - F_{EF}\cos 60° - F_{EB}\cos 15° = 0.$$

(Resolving horizontally)

$$F_{DE}\sin 60° + F_{EF}\sin 60° - F_{EB}\sin 15° - 20 = 0$$

(Resolving vertically)

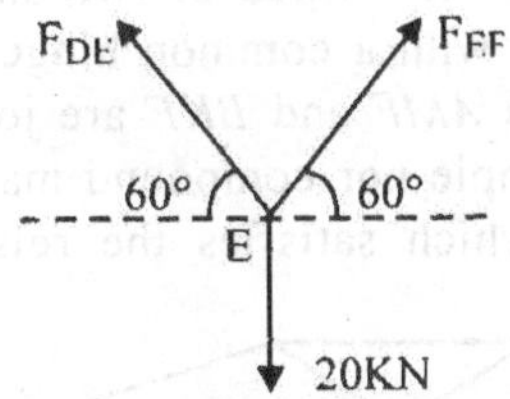

$$F_{EF}\cos 60° = F_{DE}\cos 60° - F_{EB}\cos 15°$$

$$= 7\cdot695\cos 60° + 3\cdot45\cos 15°$$

$$F_{EF} = 7\cdot695 + 3\cdot45\frac{\cos 15°}{\cos 60°}$$

Fig. 3.32 (*b*) Joint E.

$$= \mathbf{14\cdot359kN.}$$

Check $7\cdot695\sin 60° + 14\cdot359\sin 60° + 3\cdot450\sin 15° - 20 \approx 0.$

Joint A (Fig. 3.32 (*c*))

Resolving horizontally

$$F_{AB}\cos 60° + F_{AB} + F_{AD}\cos 45° = 0$$

$$F_{AC}\sin 60° + F_{AD}\sin 45° + 10 = 0$$

(Resolving vertically)

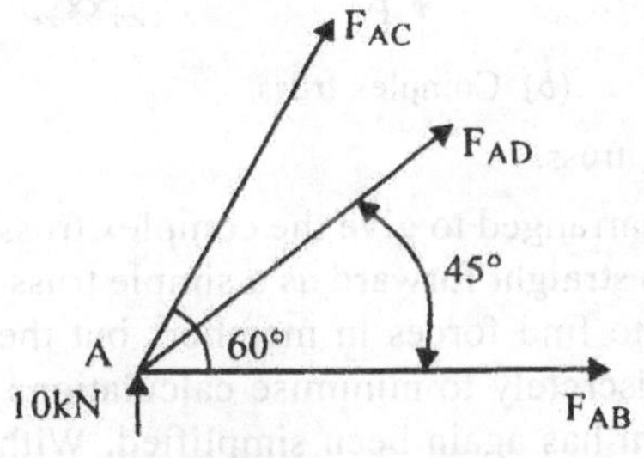

$$\text{or} \quad F_{AC}\sin 60° = -10 - F_{AD}\sin 45°$$

$$\text{or} \quad F_{AC} = \frac{-10}{\sin 60°} - F_{AD}\frac{\sin 45°}{\sin 60°}$$

$$= \mathbf{-3\cdot85kN.}$$

$$F_{AB} = -F_{AC}\cos 60° - F_{AD}\cos 45°$$

$$= +3\cdot85\cos 60° + 9\cdot426\cos 45°$$

Fig. 3.32 (*c*) Joint A.

$$= \mathbf{8\cdot59kN.}$$

Joint C (Fig. 3.32 (*d*))

$$F_{CB}\cos 60° + F_{FC}\cos 75° - F_{AC}\cos 60° = 0$$

(Resolving horizontally)

$$F_{AC}\sin 60° + F_{FC}\sin 75° + F_{CB}\sin 60° = 0$$

(Resolving vertically)

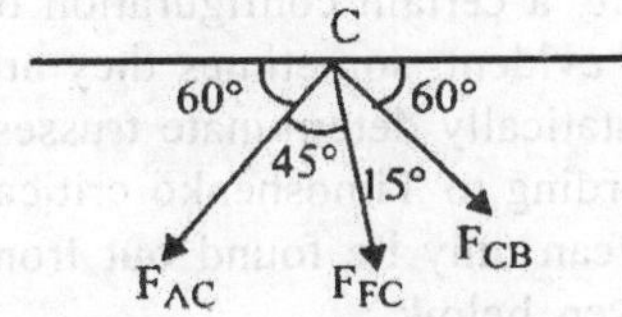

$$F_{CB}\cos 60° = F_{AC}\cos 60° - F_{FC}\cos 75°$$

$$= -3\cdot85\cos 60° - 12\cdot28\cos 75°$$

Fig. 3.32 (*d*) Joint C.

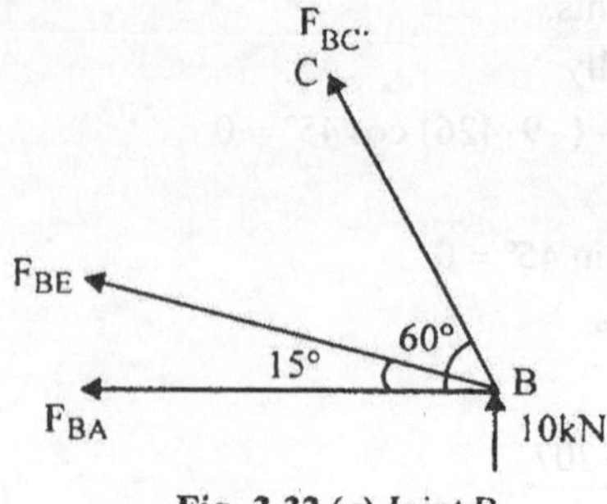

Fig. 3.32 (*e*) Joint B.

$$F_{CB} = -3 \cdot 85 - \frac{12 \cdot 88 \cos 75^\circ}{\cos 60^\circ}$$

$$= -\ \mathbf{10{\cdot}517kN}.$$

Resolving horizontally,

$$F_{BC} \cos 45^\circ + F_{BE} \cos 15^\circ + F_{BA} = 0$$

Resolving vertically,

$$F_{BC} \sin 60^\circ + F_{BE} \sin 15^\circ + 10 = 0$$

$$-10{\cdot}517 \cos 60^\circ + (-3{\cdot}450) \cos 15^\circ + 8{\cdot}59 \approx 0 \text{ (check)}$$

$$-10{\cdot}517 \sin 60^\circ - 3{\cdot}45 \sin 15^\circ + 10 \approx 0 \text{ (check).}$$

3.5. FURTHER COMMENTS ON TRUSS

It is possible to classify trusses as

(*i*) Simple (*ii*) Compound

(*iii*) Complex

A simple truss is composed of a number of triangles developed by adding two bars to the basic triangle. Example 3.1 shows a simple truss. A compound truss is formed of two simple trusses joined together by three non parallel non concurrent bars or with a common hinge and bar. Example 3.3 shows a compound truss where two simple trusses *AMF* and *BKF* are joined together at the hinge *F* and by bar *MK*. A truss which is neither simple nor compound may be classified as complex truss. Fig. 3.33(*a*) shows a simple truss which satisfies the relation $b = 2n - 3$.

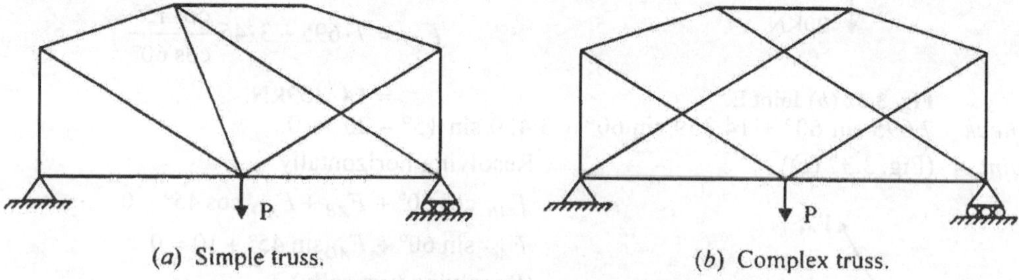

(*a*) Simple truss. (*b*) Complex truss.

Fig. 3.33. Simple and complex truss.

With the same number of members, the members can be rearranged to give the complex truss as shown in Fig. 3.33(*b*). The solution of complex truss is not as straight forward as a simple truss. The method of joints and the method of sections can be used to find forces in members but the sections and the sequence of joints analysed must be chosen discretely to minimise calculations. But with the use of computers in structural analysis the problem has again been simplified. With number of unknowns and number of equations equal, solution of simultaneous equations involve inversion of a matrix and if the structure is not unstable, inversion exists and hence the solution.

As a triangle is a stable structure, a simple truss obtained by adding triangle is usually stable. But with arbitrary change of members a complex truss may not always be stable. There is a possibility of accidentally obtaining a so-called critical form. *i.e.* a certain configuration of the truss which is unstable. Sometimes these critical forms are self evident, sometimes they are not. Two tests are available for investigating the critical forms of statically determinate trusses, the zero load test and the coefficient matrix determinant test. According to Timoshenko critical form of a truss is always indeterminate as forces in the members can only be found out from deformation considerations. Detailed steps of the two tests are given below :

***a.* The Zero Load Test:** This test was first suggested by *A.F.* Mobins (1790 – 1868). The steps are :

1. Assume zero external load and hence zero external reactions.

2. Assume a certain non zero force in any one member.

3. Determine the forces in all other members of the truss considering the equilibrium of all the joints in the truss.

4. If equilibrium can be established at all the joints, it indicates that a non-zero solution exists with zero external load, which leads to the conclusion that the structure is not geometrically stable.

5. If equilibrium cannot be established at all the joints and a non zero solution does not exist, the structure is stable.

***b.* Coefficient Matrix Determinant Test :** Here the stability or instability of a truss is determined by computing the determinant of the coefficient matrix of the joint equilibrium equations.

The steps are :

1. Write the joint equilibrium equations for all the joints.

2. Determine the coefficient matrix of the equilibrium equations and find its determinant.

3. If the determinant is positive and non zero, the truss is stable; otherwise the truss is unstable.

3.6. MATRIX METHOD OF SOLUTION OF TRUSSES

With the help of computer it is possible to analyse a truss in a straight forward manner as analysis of truss actually involves solution of simultaneous equations. In the hand method these equations are solved joint by joint whereas in the computer method the joint equations are written all at a time in matrix form. The number of unknowns is equal to the number of equations and they are then solved by inversion of the matrix. If the inversion exists, the system has a unique solution which indicates that the system is stable. On the other hand if the determinant is zero and the inversion does not exist, the system is unstable, since many solutions are possible. Fig. 3.35 explains the positive directions of the external forces, reactions and member forces. While external forces are positive in the positive coordinate directions, reactions and member forces are positive if they exert pull on the joints.

The procedure is illustrated through the simple truss shown in Fig. 3.34.

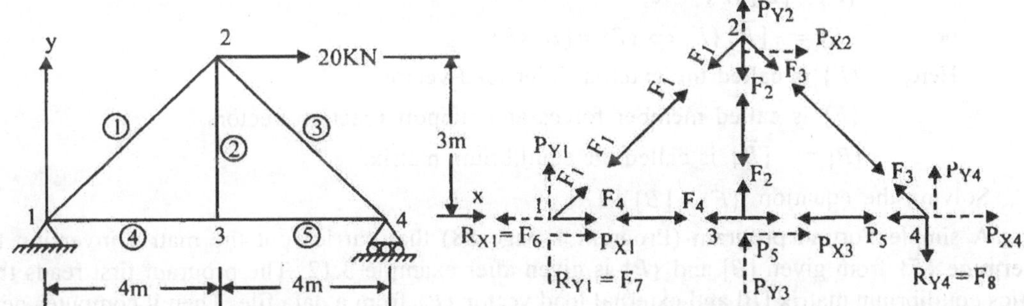

Fig. 3.34. Simple truss. **Fig. 3.35.** Free body diagram.

Here $m = 5$, $r = 3$ and $j = 4$ Hence $m + r = 5 + 3 = 2j = 2 \times 4 = 8$. Hence the truss is determinate with eight unknowns. The sign conventions just defined are followed *i.e.* all members are assumed in tension and taken positive, reactions are positive if they exert pull on the joints and external forces are assumed positive in the positive directions of x and y axes.

Two equations of equilibrium can then be written in the positive x and y direction at each joint giving $\Sigma F_x = 0$ and $\Sigma F_y = 0$ respectively. We adopt the convention of writing $\Sigma F_x = 0$ as the first equation and $\Sigma F_y = 0$ as the 2nd equation. Thus for

Joint 1 $\quad \Sigma F_x = 0$ gives $\quad P_{x_1} + \frac{4}{5} F_1 + F_4 - F_6 = 0$

$\Sigma F_y = 0$ gives $\quad P_{y_1} + \frac{3}{5} F_1 - F_7 = 0$

Joint 2 $\quad \Sigma F_x = 0$ gives $\quad P_{x_2} - \frac{4}{5} F_1 + \frac{4}{5} F_3 = 0$

$\Sigma F_y = 0$ gives $\quad P_{y_2} - \frac{3}{5} F_1 - F_2 - \frac{3}{5} F_3 = 0$

Joint 3 $\quad \Sigma F_x = 0$ gives $\quad P_{x_3} - F_4 + F_5 = 0$

$\Sigma F_y = 0$ gives $\quad P_{y_3} + F_2 = 0$

Joint 4 $\quad \Sigma F_x = 0$ gives $\quad P_{x_4} - \frac{4}{5} F_3 - F_5 = 0$

$\Sigma F_y = 0$ gives $\quad P_{y_4} + \frac{3}{5} F_3 - F_8 = 0$

The above eight equations can be written in matrix form as —

$$\begin{Bmatrix} P_{x_1} \\ P_{y_1} \\ P_{x_2} \\ P_{y_2} \\ P_{x_3} \\ P_{y_3} \\ P_{x_4} \\ P_{y_4} \end{Bmatrix} + \begin{bmatrix} 0{\cdot}8 & 0{\cdot}0 & 0{\cdot}0 & +1{\cdot}0 & 0{\cdot}0 & -1{\cdot}0 & 0{\cdot}0 & 0{\cdot}0 \\ 0{\cdot}6 & 0{\cdot}0 & 0{\cdot}0 & 0{\cdot}0 & 0{\cdot}0 & 0{\cdot}0 & -1{\cdot}0 & 0{\cdot}0 \\ -0{\cdot}8 & 0{\cdot}0 & +0{\cdot}8 & 0{\cdot}0 & 0{\cdot}0 & 0{\cdot}0 & 0{\cdot}0 & 0{\cdot}0 \\ -0{\cdot}6 & -1{\cdot}0 & -0{\cdot}6 & 0{\cdot}0 & 0{\cdot}0 & 0{\cdot}0 & 0{\cdot}0 & 0{\cdot}0 \\ 0{\cdot}0 & 0{\cdot}0 & 0{\cdot}0 & -1{\cdot}0 & +1{\cdot}0 & 0{\cdot}0 & 0{\cdot}0 & 0{\cdot}0 \\ 0{\cdot}0 & 1{\cdot}0 & 0{\cdot}0 & 0{\cdot}0 & 0{\cdot}0 & 0{\cdot}0 & 0{\cdot}0 & 0{\cdot}0 \\ 0{\cdot}0 & 0{\cdot}0 & -0{\cdot}8 & 0{\cdot}0 & 0{\cdot}0 & 0{\cdot}0 & 0{\cdot}0 & 0{\cdot}0 \\ 0{\cdot}0 & 0{\cdot}0 & +0{\cdot}6 & 0{\cdot}0 & 0{\cdot}0 & 0{\cdot}0 & 0{\cdot}0 & -1{\cdot}0 \end{bmatrix} \begin{Bmatrix} F_1 \\ F_2 \\ F_3 \\ F_4 \\ F_5 \\ F_6 \\ F_7 \\ F_8 \end{Bmatrix} = \begin{Bmatrix} 0 \\ 0 \\ 0 \\ 0 \\ 0 \\ 0 \\ 0 \\ 0 \end{Bmatrix}$$

Symbolically the above matrix can be written as

$$\{P\} + [\bar{B}]\{F\} = \{0\}$$

or $\quad \{P\} = -[\bar{B}]\{F\} \Rightarrow \{P\} = [B]\{F\}$

Here, $\{P\}$ is called the external joint load vector.

$\{F\}$ is called member forces and support reaction vectors

$[B] = -[\bar{B}]$ is called the equilibrium matrix.

Solving the equation $\{F\} = [B]^{-1}\{P\}$

A simple Fortran program (Program 1 page 88) that carries out the matrix inversion to determine $\{F\}$ from given $[B]$ and $\{P\}$ is given after example 3.12. The program first reads the statics equilibrium matrix $[B]$ and external load vector $\{P\}$ from a data file. Then it computes new $[B]$ as $-[B]$. Next it calls the inverse subroutine to get inverse of $[B]$ *i.e.* $[BI]$. Finally multiplication subroutine is called to get $\{F\} = [B]^{-1}\ \{P\}$.

The equilibrium matrix, the joint load vector, the inversion matrix, member forces and reactions of this problem are given in Example problem 1 (page 92).

The computer results are

$$
\begin{Bmatrix} F_1 \\ F_2 \\ F_3 \\ F_4 \\ F_5 \\ F_6 \\ F_7 \\ F_8 \end{Bmatrix} = \begin{bmatrix} 0 & 0 & 0\cdot625 & 0\cdot833 & 0 & 0\cdot833 & 0 & 0 \\ 0 & 0 & 0 & 0 & 0 & -1 & 0 & 0 \\ 0 & 0 & -0\cdot625 & 0\cdot833 & 0 & 0\cdot833 & 0 & 0 \\ 0 & 0 & 0\cdot5 & -0\cdot667 & 1 & -\cdot667 & 1 & 0 \\ 0 & 0 & 0\cdot5 & 0\cdot667 & 0 & -\cdot667 & 1 & 0 \\ 1 & 0 & 1\cdot0 & 0 & 1 & 0 & 1 & 0 \\ 0 & 1 & 0\cdot375 & 0\cdot5 & 0 & \cdot5 & 0 & 0 \\ 0 & 0 & -0\cdot375 & 0\cdot5 & 0 & \cdot5 & 0 & 1 \end{bmatrix} \begin{Bmatrix} 0 \\ 0 \\ 20 \\ 0 \\ 0 \\ 0 \\ 0 \\ 0 \end{Bmatrix} = \begin{Bmatrix} 12\cdot5 \\ 0 \\ -12\cdot5 \\ 10\cdot0 \\ 10\cdot0 \\ 20\cdot0 \\ 7\cdot5 \\ -7\cdot5 \end{Bmatrix}
$$

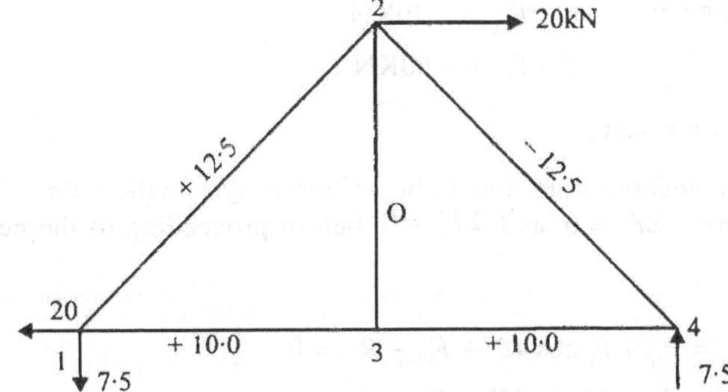

Fig. 3.36 Forces in the members.

Forces in various members and support reactions are shown in Fig. 3.36.

Example 3.10. Example 3.1. by matrix method

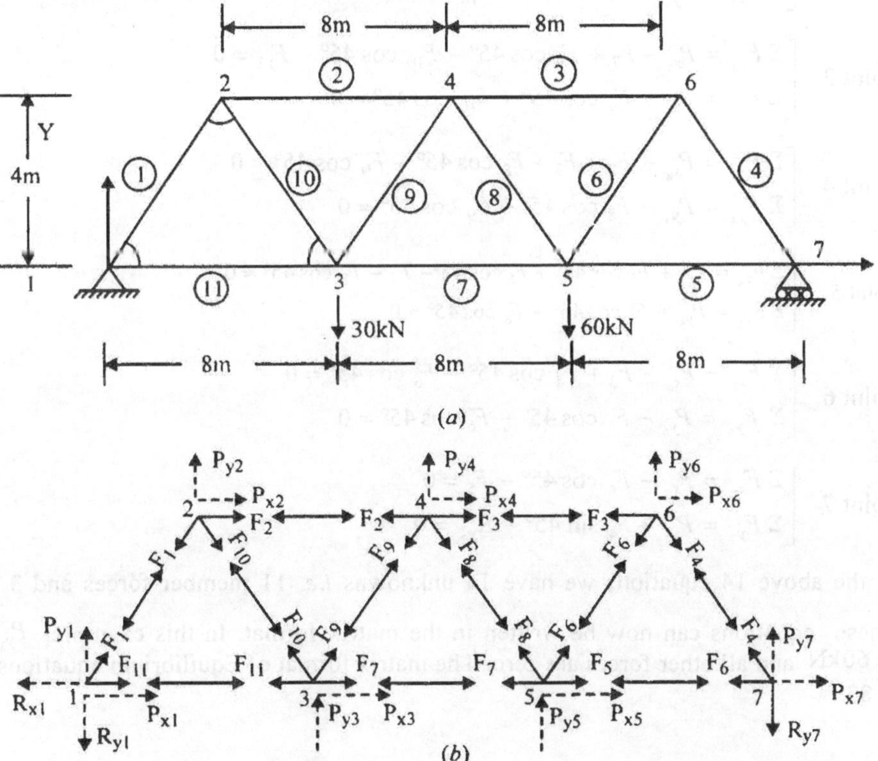

P_{x_1}, x_2 ... = Loads in the joints in the positive x direction.

P_{y_1}, y_2 ... = Loads in the joints in the positive y direction

R_{y_1} → Reactions in joint 1 in the y direction

R_{x_1} → Reaction in joint 1 in the x direction (positive if the joint is pulled)

Fig. 3.37. Example 3.10.

In the above example $P_{y_3} = -30\text{KN}$

$P_{y_5} = -60\text{KN}.$

All other forces are zero.

The equilibrium equations are now to be written in systematic order of joint 1, 2, 3, etc. In each joint we will write $\Sigma F_x = 0$ and $\Sigma F_y = 0$ before proceeding to the next joint.

Thus,

Joint 1 $\begin{bmatrix} \Sigma F_{x_1} = P_{x_1} + F_1 \cos 45° + F_{11} - R_{x_1} = 0 \\ \Sigma F_{y_1} = P_{y_1} + F_1 \sin 45° - R_{y_1} = 0 \end{bmatrix}$

Joint 2 $\begin{bmatrix} \Sigma F_{x_2} = P_{x_2} - F_1 \cos 45° + F_2 + F_{10} \cos 45° = 0 \\ \Sigma F_{y_2} = P_{y_2} - F_1 \cos 45° - F_{10} \cos 45° = 0 \end{bmatrix}$

Joint 3 $\begin{bmatrix} \Sigma F_{x_3} = P_{x_3} + F_7 + F_9 \cos 45° - F_{10} \cos 45° - F_{11} = 0 \\ \Sigma F_{y_3} = P_{y_3} + F_9 \cos 45° + F_{10} \cos 45° = 0 \end{bmatrix}$

Joint 4 $\begin{bmatrix} \Sigma F_{x_4} = P_{x_4} - F_2 + F_3 + F_8 \cos 45° - F_9 \cos 45° = 0 \\ \Sigma F_{y_4} = P_{y_4} - F_8 \cos 45° - F_9 \cos 45° = 0 \end{bmatrix}$

Joint 5 $\begin{bmatrix} \Sigma F_{x_5} = P_{x_5} + F_5 \cos 45° + F_6 \cos 45° - F_7 - F_8 \cos 45° = 0 \\ \Sigma F_{y_5} = P_{y_5} + F_6 \cos 45° + F_8 \cos 45° = 0 \end{bmatrix}$

Joint 6 $\begin{bmatrix} \Sigma F_{x_6} = P_{x_6} - F_3 + F_4 \cos 45° - F_6 \cos 45° = 0 \\ \Sigma F_{y_6} = P_{y_6} - F_4 \cos 45° - F_6 \cos 45° = 0 \end{bmatrix}$

Joint 7 $\begin{bmatrix} \Sigma F_{x_7} = P_{x_7} - F_4 \cos 45° - F_5 = 0 \\ \Sigma F_{y_7} = P_{y_7} + F_4 \sin 45° - R_{y_7} = 0 \end{bmatrix}$

In the above 14 equations we have 14 unknowns *i.e.* 11 member forces and 3 reactions.

These equations can now be written in the matrix format. In this example, $P_{y_3} = -30\text{kN}$ $P_{y_5} = -60\text{kN}$ and all other forces are zero. The matrix format of Equilibrium equations are given in Page 83.

Matrix Format of Equilibrium Equations. Example 3.10.

$$
\underbrace{\begin{Bmatrix} 0 \\ 0 \\ 0 \\ 0 \\ 0 \\ -30 \\ 0 \\ 0 \\ 0 \\ -60 \\ 0 \\ 0 \\ 0 \\ 0 \end{Bmatrix}}_{\{P\}} + \underbrace{\begin{bmatrix}
\cdot707 & 0 & 0 & 0 & 0 & 0 & 0 & 0 & 0 & 0 & 1 & -1 & 0 & 0 \\
\cdot707 & 0 & 0 & 0 & 0 & 0 & 0 & 0 & 0 & 0 & 0 & 0 & -1 & 0 \\
-\cdot707 & 1 & 0 & 0 & 0 & 0 & 0 & 0 & 0 & \cdot707 & 0 & 0 & 0 & 0 \\
-\cdot707 & 0 & 0 & 0 & 0 & 0 & 0 & 0 & 0 & \cdot707 & 0 & 0 & 0 & 0 \\
0 & 0 & 0 & 0 & 0 & 0 & 1 & 0 & \cdot707 & -\cdot707 & -1 & 0 & 0 & 0 \\
0 & 0 & 0 & 0 & 0 & 0 & 0 & 0 & \cdot707 & \cdot707 & 0 & 0 & 0 & 0 \\
0 & -1 & 1 & 0 & 0 & 0 & 0 & \cdot707 & -\cdot707 & 0 & 0 & 0 & 0 & 0 \\
0 & 0 & 0 & 0 & 0 & 0 & 0 & -\cdot707 & -\cdot707 & 0 & 0 & 0 & 0 & 0 \\
0 & 0 & 0 & 0 & 1 & \cdot707 & -1 & -\cdot707 & 0 & 0 & 0 & 0 & 0 & 0 \\
0 & 0 & 0 & 0 & 0 & \cdot707 & 0 & \cdot707 & 0 & 0 & 0 & 0 & 0 & 0 \\
0 & 0 & -1 & \cdot707 & 0 & -\cdot707 & 0 & 0 & 0 & 0 & 0 & 0 & 0 & 0 \\
0 & 0 & 0 & -\cdot707 & 0 & -\cdot707 & 0 & 0 & 0 & 0 & 0 & 0 & 0 & 0 \\
0 & 0 & 0 & -\cdot707 & -1 & 0 & 0 & 0 & 0 & 0 & 0 & 0 & 0 & 0 \\
0 & 0 & 0 & \cdot707 & 0 & 0 & 0 & 0 & 0 & 0 & 0 & 0 & 0 & -1
\end{bmatrix}}_{[B]} \underbrace{\begin{Bmatrix} F_1 \\ F_2 \\ F_3 \\ F_4 \\ F_5 \\ F_6 \\ F_7 \\ F_8 \\ F_9 \\ F_{10} \\ F_{11} \\ F_{x1} \\ F_{y1} \\ F_{y7} \end{Bmatrix}}_{\{F\}} = \begin{Bmatrix} 0 \\ 0 \\ 0 \\ 0 \\ 0 \\ 0 \\ 0 \\ 0 \\ 0 \\ 0 \\ 0 \\ 0 \\ 0 \\ 0 \end{Bmatrix}
$$

Here $\{P\}$ = Joint or structure load vector

$\{F\}$ = Member and support force vector

$[B]$ = Statics Matrix

The relation $\{F\} = [B]^{-1}\{P\}$ is already derived. Hence to get member forces and reactions at supports we only require $[B]$ and $\{P\}$.

The computer output (Example problem 2 page 93) gives the equilibrium matrix, the joint load vector, the member forces and reactions.

Example 3.11. Solve Example 3.5 by matrix method.

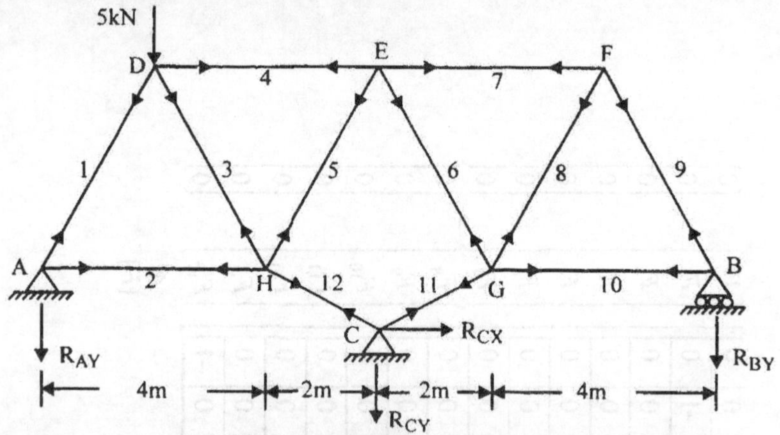

Fig. 3.38. Example 3.11.

Fig. 3.38 gives the details about the truss, direction of forces in members, support reactions. First the equilibrium equations at each joint are written sequentially.

Joint A

$$P_{AX} + F_1 \times \frac{2}{2\sqrt{5}} + F_2 = 0$$

$$P_{AY} + F_1 \times \frac{4}{2\sqrt{5}} - R_{AY} = 0$$

Joint D

$$P_{DX} + F_4 + F_3 \times \frac{2}{2\sqrt{5}} - \frac{F_1 \times 2}{2\sqrt{5}} = 0$$

$$P_{DY} - F_3 \times \frac{4}{2\sqrt{5}} - F_1 \times \frac{4}{2\sqrt{5}} = 0$$

Joint H

$$P_{HX} - \frac{F_3}{\sqrt{5}} - F_2 + \frac{F_5}{\sqrt{5}} + F_{11} \times \frac{2}{\sqrt{5}} = 0$$

$$P_{HY} + F_3 \times \frac{2}{\sqrt{5}} + F_5 \times \frac{2}{\sqrt{5}} - F_{11} \times \frac{1}{\sqrt{5}} = 0$$

Joint C

$$P_{CX} + F_{11} \times \frac{2}{\sqrt{5}} - F_{12} \times \frac{2}{\sqrt{5}} + R_{CX} = 0$$

$$P_{CY} + F_{11} \times \frac{1}{\sqrt{5}} + F_{12} \times \frac{1}{\sqrt{5}} - R_{CY} = 0$$

Joint E

$$P_{EX} - F_4 - F_5 \times \frac{1}{\sqrt{5}} + F_7 + F_6 \times \frac{1}{\sqrt{5}} = 0$$

$$P_{EY} - F_5 \times \frac{4}{2\sqrt{5}} - F_6 \times \frac{4}{2\sqrt{5}} = 0$$

Joint G

$$P_{GX} + F_{10} + F_8 \times \frac{1}{\sqrt{5}} - F_6 \times \frac{1}{\sqrt{5}} - F_{12} \times \frac{2}{\sqrt{5}} = 0$$

$$P_{GY} + F_8 \frac{2}{\sqrt{5}} + F_6 \times \frac{2}{\sqrt{5}} - F_{12} \times \frac{1}{\sqrt{5}} = 0$$

$\{P\}$	+	1	2	3	4	5	6	7	8	9	10	11	12	13	14	15	16	$[F]$
0	1	·450	1·00	·00	·00	·00	·00	·00	·00	·00	·00	·00	·00	·00	·00	·00	·00	F_1
0	2	0·89	·00	·00	·00	·00	·00	·00	·00	·00	·00	·00	·00	−1·00	·00	·00	·00	F_2
0	3	−·45	·00	·45	1·00	·00	·00	·00	·00	·00	·00	·00	·00	·00	·00	·00	·00	F_3
−50	4	−·89	·00	−·89	·00	·00	·00	·00	·00	·00	·00	·00	·00	·00	·00	·00	·00	F_4
0	5	·00	−1·00	−·45	·00	·45	·00	·00	·00	·00	·00	·89	·00	·00	·00	·00	·00	F_5
0	6	·00	·00	·89	·00	·89	·00	·00	·00	·00	·00	−·45	·00	·00	·00	·00	·00	F_6
0	7	·00	·00	·00	·00	·00	·00	·00	·00	·00	·00	·89	−·89	·00	·00	1·00	·00	F_7
0	8	·00	·00	·00	·00	·00	·00	·00	·00	·00	·00	·45	·45	·00	·00	·00	−1·00	F_8
0	9	·00	·00	·00	−1·0	−·45	·45	1·00	·00	·00	·00	·00	·00	·00	·00	·00	·00	F_9
0	10	·00	·00	·00	·00	−·89	−·89	·00	·00	·00	·00	·00	·00	·00	·00	·00	·00	F_{10}
0	11	·00	·00	·00	·00	·00	−·45	·00	·45	·00	1·00	·00	−·89	·00	·00	·00	·00	F_{11}
0	12	·00	·00	·00	·00	·00	·89	·00	·89	·00	·00	·00	−·45	·00	·00	·00	·00	F_{12}
0	13	·00	·00	·00	·00	·00	·00	−1·0	−·45	·45	·00	·00	·00	·00	·00	·00	·00	R_{AY}
0	14	·00	·00	·00	·00	·00	·00	·00	−·89	−·89	·00	·00	·00	·00	·00	·00	·00	R_{BY}
0	15	·00	·00	·00	·00	·00	·00	·00	·00	−·45	−1·00	·00	·00	·00	·00	·00	·00	R_{CX}
0	16	·00	·00	·00	·00	·00	·00	·00	·00	·89	·00	·00	·00	·00	−1·00	·00	·00	R_{CY}

$\{P\}$ $[B]$ $[F]$

Fig. Matrix Format Example 3.11.

Joint F

$$\begin{cases} P_{FX} + F_9 \times \dfrac{1}{\sqrt{5}} - F_7 - F_8 \times \dfrac{1}{\sqrt{5}} = 0 \\ P_{EY} - F_8 \times \dfrac{2}{\sqrt{5}} - F_9 \times \dfrac{2}{\sqrt{5}} = 0 \end{cases}$$

Joint B

$$\begin{cases} P_{BX} - F_9 \times \dfrac{1}{\sqrt{5}} - F_{10} = 0 \\ F_{BY} + F_9 \times \dfrac{2}{\sqrt{5}} - R_{BY} = 0 \end{cases}$$

In the above equations substituting numerical values for the joint loads it is observed that $P_{DY} = -50$. All the other joint loads are zero. The equations can now be written in matrix format as shown in page 85. The computer results are shown in Example problem 3 where the equilibrium matrix, the joint load vector, member forces and reactions are given. (pages 94 and 95)

Example 3.12. In the structure shown in Fig. 3.39 the members are vertical, horizontal and inclined at an angle of 30° to the horizontal and they cross without being connected. Show that the frame is statically determinate and find forces in all members.

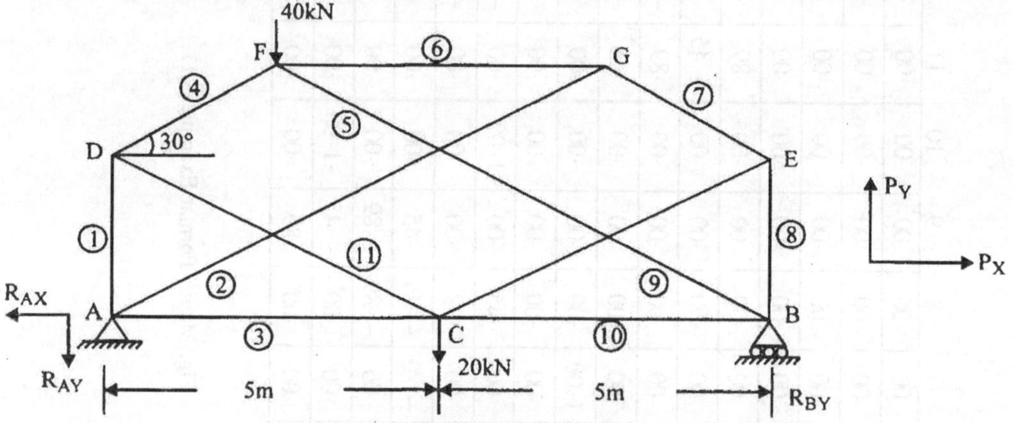

Fig. 3.39 (*a*) Example 3.12.

Solution. The truss is externally determinate with three reaction components. No. of joints = 7, No. of members = 11, $m = 2 \times 7 - 3 = 11$. Hence it is also internally determinate.

Reactions

$$R_B \times 10 = 40 \times 2{\cdot}5 + 20 \times 5$$

$$= 200\text{kN}$$

$$R_B = 20\text{kN}\uparrow \; R_A = 40\uparrow$$

Method of Joints *Joint A* (Fig. 3.39 (*b*)) Assuming $F_2 = T$

Fig. 3.39 (*b*) Joint A.

$$T \cos 30° + F_3 = 0$$

$$T \sin 30° + F_1 + 40 = 0$$

or $\quad F_1 = -40 - T/2$

$$F_3 = -T\frac{\sqrt{3}}{2}.$$

Joint D (Fig. 3.39 (*c*))

$$F_1 + F_{11} \sin 30° - F_4 \sin 30° = 0$$

$$F_4 \cos 30° + F_{11} \cos 30° = 0$$

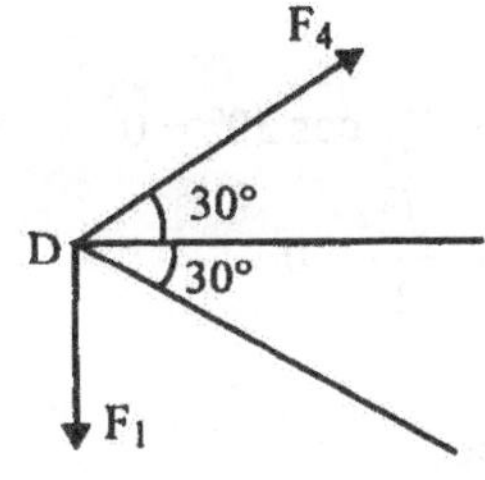

Fig. 3.39 (*c*) Joint D.

or $\quad -40 - T/2 + F_{11}/2 - F_4/2 = 0$

$$F_{11}\frac{\sqrt{3}}{2} + F_4\frac{\sqrt{3}}{2} = 0$$

or $\quad F_{11} = -F_4$

Substituting,

$$F_4 = -40 - T/2.$$

Joint F. (Fig. 3.39 (*d*))

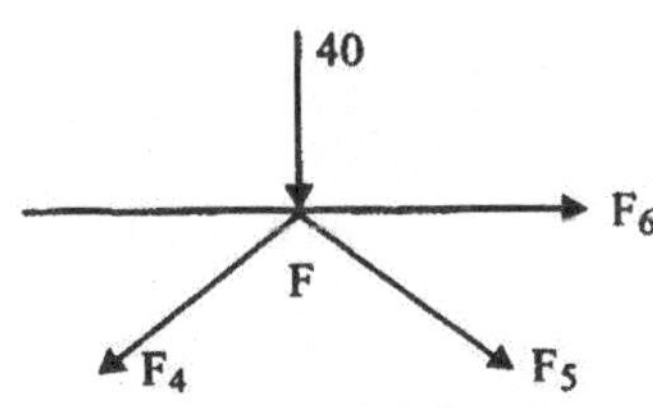

Fig. 3.39 (*d*) Joint F.

$$F_6 + F_5 \cos 30° - F_4 \cos 30° = 0$$

$$40 + F_5 \sin 30° + F_4 \sin 30° = 0$$

or $\quad 40 + F_5/2 + F_4/2 = 0$

or $\quad 40 + F_5/2 - 20 - T/4 = 0$

$$F_5 = T/2 - 40$$

$$F_6 = F_4 \cos 30° - F_5 \cos 30°$$

$$= \left(-40 - T/2\right)\cdot\frac{\sqrt{3}}{2} - \left(T/2 - 40\right)\frac{\sqrt{3}}{2}$$

$$= -20\sqrt{3} - \frac{T\sqrt{3}}{4} - \frac{T\sqrt{3}}{4} + 20\sqrt{3}$$

$$= -\frac{T\sqrt{3}}{2}$$

Joint G (Fig. 3.39 (*e*))

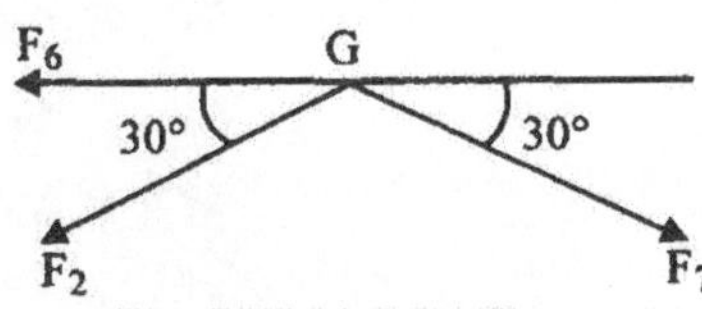

Fig. 3.39 (*e*) Joint G.

$$F_6 + F_2 \cos 30° - F_7 \cos 30° = 0$$

$$F_2 \sin 30° + F_7 \sin 30° = 0$$

$$F_2 = -F_7$$

Substituting numerical values

$$-T\frac{\sqrt{3}}{2} + T\frac{\sqrt{3}}{2} - F_7\frac{\sqrt{3}}{2} = 0$$

or $\quad F_7 = 0 \quad$ Hence, $F_2 = 0 \quad$ or $T = 0$.

Joint E (Fig. 3.39 (*f*))

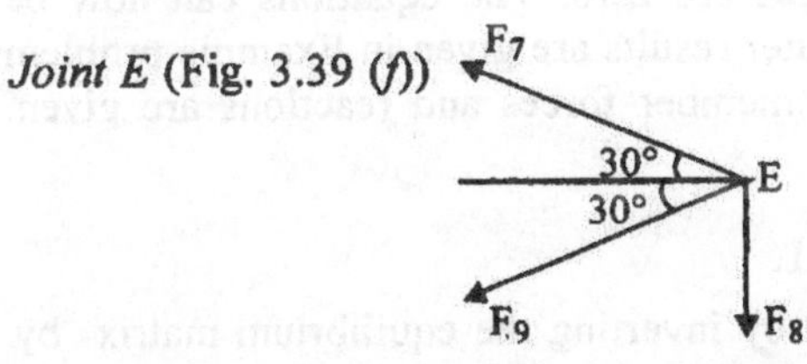

Fig. 3.39 (*f*) Joint E.

$$F_7 \cos 30° + F_9 \cos 30° = 0$$

$$F_7 \sin 30° - F_8 - F_9 \sin 30° = 0$$

or $\quad F_7 + F_9 = 0$

As $F_7 = 0$ $F_9 = 0$

and Hence $F_8 = 0$.

Joint B (Fig. 3.39 (*g*))

Fig. 3.39 (*g*) Joint B.

$$F_5 \sin 30° + F_8 + 20 = 0$$

$$F_5 \times \frac{1}{2} = -20$$

or $\quad F_5 = -40$

$$F_5 \cos 30° + F_{10} = 0$$

$$-40 \times \frac{\sqrt{3}}{2} + F_{10} = 0$$

Joint C (Fig. 3.39 (*h*))

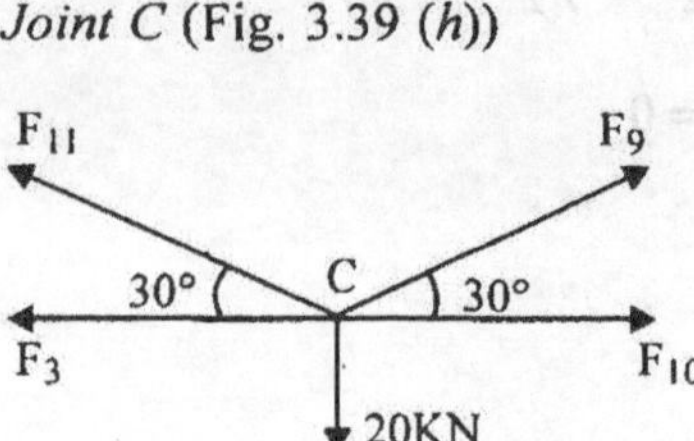

Fig. 3.39 (*h*) Joint C.

or $F_{10} = 20\sqrt{3}.$

$$F_9 \cos 30° + F_{10} - F_3 - F_{11} \cos 30° = 0$$

$$0 + 20\sqrt{3} + T\frac{\sqrt{3}}{2} - F_{11} \cdot \frac{\sqrt{3}}{2} = 0$$

or $F_{11} = 40$ as $T = 0$

as obtained before (check).

$$F_{11} \sin 30° + F_9 \sin 30° - 20$$

$$= 40 \times \frac{1}{2} + 0 - 20 = 0 \text{ (check)}$$

Solution by matrix method

The joint equations are :

Joint *A* $\begin{bmatrix} P_{XA} + F_2 \cos 30° + F_3 - R_{AX} = 0 \\ P_{YA} + F_1 + F_2 \sin 30° - R_{AY} = 0 \end{bmatrix}$

Joint *D* $\begin{bmatrix} P_{XD} + F_4 \cos 30° + F_{11} \cos 30° = 0 \\ P_{YD} - F_1 + F_4 \sin 30° - F_{11} \sin 30° = 0 \end{bmatrix}$

Joint *F* $\begin{bmatrix} P_{XF} - F_4 \cos 30° + F_5 \cos 30° + F_6 = 0 \\ P_{YF} - F_4 \sin 30° - F_5 \sin 30° = 0 \end{bmatrix}$

Joint *G* $\begin{bmatrix} P_{XG} - F_2 \cos 30° - F_6 + F_7 \cos 30° = 0 \\ F_{YC} - F_2 \sin 30° - F_7 \sin 30° = 0 \end{bmatrix} x$

Joint *E* $\begin{bmatrix} P_{XE} - F_7 \cos 30° - F_9 \cos 30° - 0 \\ P_{YE} + F_7 \sin 30° - F_8 - F_9 \sin 30° - 0 \end{bmatrix}$

Joint *B* $\begin{bmatrix} P_{XB} - F_5 \cos 30° - F_{10} = 0 \\ P_{YB} + F_9 \sin 30° + F_8 - R_{BY} = 0 \end{bmatrix}$

Joint *C* $\begin{bmatrix} P_{XC} - F_3 + F_9 \cos 30° + F_{10} - F_{11} \cos 30° = 0 \\ P_{YC} + F_9 \sin 30° + F_{11} \cos 30° = 0 \end{bmatrix}$

In the above equations substituting numerical values for joint loads it is observed that $P_{FY} = -40\text{kN}$ and $P_{CY} = -20\text{kN}$. All other joint loads are zero. The equations can now be written in matrix format as shown in Page 89 the computer results are given in Example problem 4 where the equilibrium matrix, the joint load vector, member forces and reactions are given. (page 95)

PROGRAM 1.

```
c-----> Program to obtain the member forces in a truss by inverting the equilibrium matrix by.
c       Gauss elimination
        parameter(nsize=12)
        dimension b(nsize, nsize),bi(nsize,nsize),p(nsize),f(nsize)
        character file1*10,file2*10
        write(*,'(a\) ' Input file name ? '
        read(*,'(a)') file1
        write(*,'(a\)') ' Output file name ? '
```

conted to Page 90

{P}	+	[B]														{F}	=	
0		·00	·86	1·00	·00	·00	·00	·00	·00	·00	·00	·00	−1·00	·00	·00	F_1		0
0		1·00	·50	·00	·00	·00	·00	·00	·00	·00	·00	·00	·00	−1·00	·00	F_2		0
0		·00	·00	·00	·86	·00	·00	·00	·00	·00	·00	·86	·00	·00	·00	F_3		0
0		−1·00	·00	·00	·50	·00	·00	·00	·00	·00	·00	−·50	·00	·00	·00	F_4		0
0		·00	·00	·00	−·86	·86	1·0	·00	·00	·00	·00	·00	·00	·00	·00	F_5		0
−40·00		·00	·00	·00	−·50	−·50	·00	·00	·00	·00	·00	·00	·00	·00	·00	F_6		0
0		·00	−·86	·00	·00	·00	−1·00	·86	·00	·00	·00	·00	·00	·00	·00	F_7		0
0		·00	−·50	·00	·00	·00	·00	−·50	·00	·00	·00	·00	·00	·00	·00	F_8		0
0		·00	·00	·00	·00	·00	·00	−·86	·00	−·86	·00	·00	·00	·00	·00	F_9		0
0		·00	·00	·00	·00	·00	·00	·50	1·00	−·50	·00	·00	·00	·00	·00	F_{10}		0
0		·00	·00	·00	·00	−·86	·00	·00	·00	·00	−1·00	·00	·00	·00	·00	F_{11}		0
0		·00	·00	·00	·00	·50	·00	·00	1·00	·00	·00	·00	·00	·00	−1·00	R_{AX}		0
0		·00	·00	−1·00	·00	·00	·00	·00	·00	·86	1·00	−·86	·00	·00	·00	R_{AY}		0
−20·00		−·00	·00	·00	·00	·00	·00	·00	·00	·50	·00	·50	·00	·00	·00	R_{BY}		0

Matrix Format Example 3.12.

```
        read(*,'(a)') file2
c
        open(10,file=file1)
        open(11,file=file2)
        write(*,*) 'give size of the matrix'
        read(*,*) n
c       Read the equilibrium matrix
        read(10,*)((b(i,j),j=1,n),i=1,n)
        write(11,*) 'The equilibrium matrix'
        write(11,1)((b(i,j),j=1,n),i=1,n)
c       Read the joint load vector
         read(10,*)(p(i), i=1,n)
        write(11,*) 'The joint load vector'
        write(11,1)(p(i),i=1,n)
          do 100 i=1,n
          do 100 j=1,n
        100 b(i,j)=-b(i,j)

        call inversion (b,n,n,bi)
          call mult(bi,p,f,n,n,1)

          write(11,*)' Member forces and reactions'
          write(11,*)
          write(11,*)'Member number','        ', 'Forces'
          do 300 i=1,n
          write(11,*) i,'     ',f(i)
        300  continue
        1 format(12f6.2/)
             stop
             end
c*****************************************************************************
c-----> Subroutine for inversion of a general matrix [B]
        subroutine inversion (b,n,nrhs,bi)
        dimension b(n,n),bi(n,n)
c*****************************************************************************

        ep=1.0e-6
        nl=n-1
             do 99 i=1,n
                  bi(i,i)=1.0
        99   continue
```

```
c------> Check zero diagonal
        do 100 k=1,nl
          kl=k+1
          c=b(k,k)
          cc=abs(c)-ep
          if (cc.gt.0.0) go to 1870
c------> Find row to swap if zero found
        do 110 j = kl,n
          cc = abs(b(j,k))-ep
          if(cc.le.0.0) go to 110
c------> row found, now swap
        do 112 1=k,n
          c=b(k,l)
          b(k,1)=b(j,l)
          b(j,l)=c
112     continue
c------> Swaping rows of B
          do 114 ir=1,nrhs
              c=bi(k,ir)
              bi(k,ir)=bi(j,ir)
              bi(j,ir)=c
114       continue
              c=b(k,k)
              go to 1870
110     continue
        write(*,*) 'singularity at row', k
        d=0.0
1860    go to 2140
1870      c=b(k,k)
        do 200 j=kl,n
              b(k,j)=b(k,j)/c
200       continue
        do 210 ir=1,nrhs
          bi(k,ir)=bi(k,ir)/c
210     continue
        do 212 i=k1,n
          c=b(i,k)
        do 214J=k1,n
          b(i,j)=b(i,j)c*b(kj)
214       continue
        do 216 ir=1,nrhs
```

```
            bi(i,ir)=bi(i,ir)–c*bi(k,ir)
216             continue
212             continue
100             continue
            cc=abs(b(n,n))-ep
            if(cc.le.0.0)go to 1840
            do 218 ir=1,nrhs
                bi(n,ir)=bi(n,ir)/b(n,n)
218             continue
            do 2201=1,n1
            k=n-1
            k1=k+1
            do 222 j=k1,n
            do 224 ir=1,nrhs
                bi(k,ir)=bi(k,ir)-b(k,j)*bi(j,ir)
224             continue
222         continue
220       continue
            d=1.0
            do 230 i=1,n
            d=d*b(i,i)
230         continue
2140        return
            end
c-------------------------------------------------------------------------------
            subroutine mult(a,b,c,m,k,n)
            dimension a(m,k),b(k,n),c(m,n)
c-------------------------------------------------------------------------------
          do 101 i=1,m
            do 101 j=1,n
            c(i,j)=0.0
            do 101 index=1,k
            c(i,j)=c(i,j)+a(i,index)*b(index,j)
101         continue
            return
            end
c-------------------------------------------------------------------------------
```

Example Problem 1 (Truss of Fig. 3.34)

The equilibrium matrix

```
.800E+00  .000E+00  .000E+00  .100E+01  .000E+00  -.100E+01  .000E+00   .000E+00
.600E+00  .000E+00  .000E+00  .000E+00  .000E+00  .000E+00   -.100E+01  .000E+00
```

-.800E+00 .000E+00 .800E+00 .000E+00 .000E+00 .000E+00 .000E+00 .000E+00

-.600E+00 -.100E+01 -.600E+00 .000E+00 .000E+00 .000E+00 .000E+00 .000E+00

.000E+00 .000E+00 .000E+00 -.100E+01 .100E+01 .000E+00 .000E+00 .000E+00

.000E+00 .000E+01 .000E+00 .000E+00 .000E+00 .000E+00 .000E+00 .000E+00

.000E+00 .000E+00 -.800E+00 .000E+00 -.100E+01 .000E+00 .000E+00 .000E+00

.000E+00 .000E+00 .600E+00 .000E+00 .000E+00 .000E+00 .000E+00 -.000E+01

The joint load vector

.000E+00 .000E+00 .200E+02 .000E+00 .000E+00 .000E+00 .000E+00 .000E+00

The inversion of [B] matrix

.000E+00 .000E+00 .625E+00 .833E+00 .000E+00 .833E+00 .000E+00 .000E+00

.000E+00 .000E+00 .000E+00 -.447E+07 .000E+00 -.100E+01 .000E+00 .000E+00

.000E+00 .000E+00 -.625E+00 .833E+00 .000E+00 .833E+00 .000E+00 .000E+00

.000E+00 .000E+00 .500E+00 -.667E+00 .100E+01 -.667E+00 .100E+01 .000E+00

.000E+00 .000E+00 .500E+00 -.667E+00 .000E+00 -.667E+00 .100E+01 .000E+00

.100E+00 .000E+00 .100E+01 .000E+00 .100E+01 .000E+00 .100E+01 .000E+00

.000E+00 .100E+01 .375E+00 .500E+00 .000E+00 .500E+00 .000E+00 .000E+00

.000E+00 .000E+00 -.375E+00 .500E+00 .000E+00 .500E+00 .000E+00 .100E+01

Member forces and reactions

Member number

Member	Forces
1	12.500000
2	0.000000E+00
3	-12.500000
4	10.000000
5	10.000000

(Horizontal reaction at 1) 20.000000

(Vertically reaction at 1) 7.500000

(Vertical reaction at 4) -7.500000

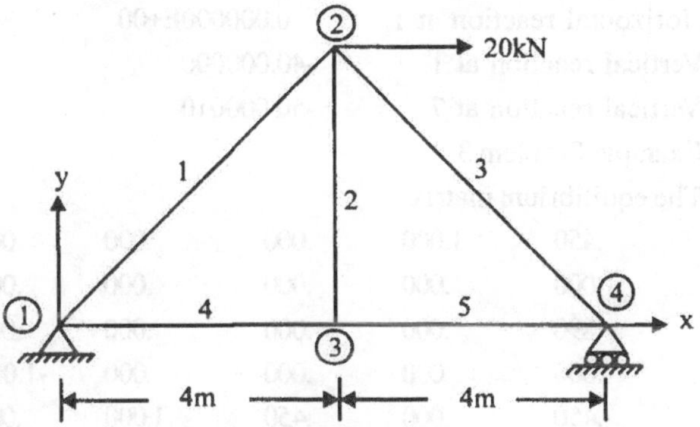

Fig. 3.34 (repeated).

Example Problem 2 (Truss of Example 3.10) Fig.3.37(*a*).

The equilibrium matrix.

.71 .00 .00 .00 .00 .00 .00 .00 .00 .00 1.00 -1.00 .00 .00

.71 .00 .00 .00 .00 .00 .00 .00 .00 .00 .00 .00 -1.00 .00

-.71 1.00 .00 .00 .00 .00 .00 .00 .00 .71 .00 .00 .00 .00

-.71 .00 .00 .00 .00 .00 .00 .00 .00 -.71 .00 .00 .00 .00

.00 .00 .00 .00 .00 .00 1.00 .00 .71 -.71 -1.00 .00 .00 .00

.00 .00 .00 .00 .00 .00 .00 .00 .71 .71 .00 .00 .00 .00

.00 -1.00 1.00 .00 .00 .00 .00 .71 .71 .00 .00 .00 .00 .00

.00 .00 .00 .00 .00 .00 .00 -.71 -.71 .00 .00 .00 .00 .00

.00 .00 .00 .00 1.00 .71 -1.00 -.71 .00 .00 .00 .00 .00 .00

.00 .00 .00 .00 .00 .71 .00 .71 .00 .00 .00 .00 .00 .00
.00 .00 -1.00 .71 .00 -.71 .00 .00 .00 .00 .00 .00 .00 .00
.00 .00 .00 -.71 .00 -.71 .00 .00 .00 .00 .00 .00 .00 .00
.00 .00 .00 -.71 -1.00 .00 .00 .00 .00 .00 .00 .00 .00 .00
.00 .00 .00 .71 .00 .00 .00 .00 .00 .00 .00 .00 .00 -1.00

The joint load vector

.00 .00 .00 .00 .00 -30.00 .00 .00 .00 -60.00 .00 .00 .00 .00

Member number and reactions

Member number	Forces
1	-56.577090
2	-80.000000
3	-100.000000
4	-70.721360
5	50.000000
6	70.721360
7	90.000000
8	14.144270
9	-14.144270
10	56.577090
11	40.000000

Fig. 3.37 *(a)* (repeated).

Horizontal reaction at 1 0.000000E+00
Vertical reaction at 1 -40.000000
Vertical reaction at 7 -50.000010

Example Problem 3

The equilibrium matrix

.450 1.000 .000 .000 .000 .000 .000 .000
.000 .000 .000 .000 .000 .000 .000 .000
.890 .000 .000 .000 .000 .000 .000 .000
.000 .000 .000 .000 -1.000 .000 .000 .000
-.450 .000 .450 1.000 .000 .000 .000 .000
.000 .000 .000 .000 .000 .000 .000 .000
-.890 .000 -.890 .000 .000 .000 .000 .000
.000 .000 .000 .000 .000 .000 .000 .000
.000 -1.000 -.450 .000 .450 .000 .000 .000
.000 .000 .890 .000 .000 .000 .000 .000
.000 .000 .890 .000 .890 .000 .000 .000
.000 .000 -.450 .000 .000 .000 .000 .000
.000 .000 .000 .000 .000 .000 .000 .000
.000 .000 .890 -.890 .000 .000 1.000 .000
.000 .000 .000 .000 .000 .000 .000 .000
.000 .000 .450 .450 .000 .000 .000 -1.000

.000	.000	.000	-1.000	-.450	.450	1..000	.000
.000	.000	.000	.000	.000	.000	.000	.000
.000	.000	.000	.000	-.890	-.890	.000	.000
.000	.000	.000	.000	.000	.000	.000	.000
.000	.000	.000	.000	.000	-.450	.000	.450
.000	1.000	.000	-.890	.000	.000	.000	.000
.000	.000	.000	.000	.000	.890	.000	.890
.000	.000	.000	-.450	.000	.000	.000	.000
.000	.000	.000	.000	.000	.000	-1.000	-.450
.450	.000	.000	.000	.000	.000	.000	.000
.000	.000	.000	.000	.000	.000	.000	-.890
-.890	.000	.000	.000	.000	.000	.000	.000
.000	.000	.000	.000	.000	.000	.000	.000
-.450	-1.000	.000	.000	.000	.000	.000	.000
.000	.000	.000	.000	.000	.000	.000	.000
.890	.000	.000	.000	.000	-1.000	.000	.000

The joint load vector

.000	.000	.000	-50.000	.000	.000	.000	.000
.000	.000	.000	.000	.000	.000	.000	.000

Member forces and reactions

Member number	Forces	
1	-61.510870	
2	27.679890	
3	5.331100	
4	-30.078880	
5	9.363305	
6	-9.363298	
7	-21.651910	
8	24.057690	
9	-24.057670	
10	10.825960	
11	29.062240	
12	29.062240	
13	-54.744670	(R at 1)
14	-21.411330	(R at 8)
15	6.501552E-07	(R_H at 4)
16	26.156020	(R_V at 4)

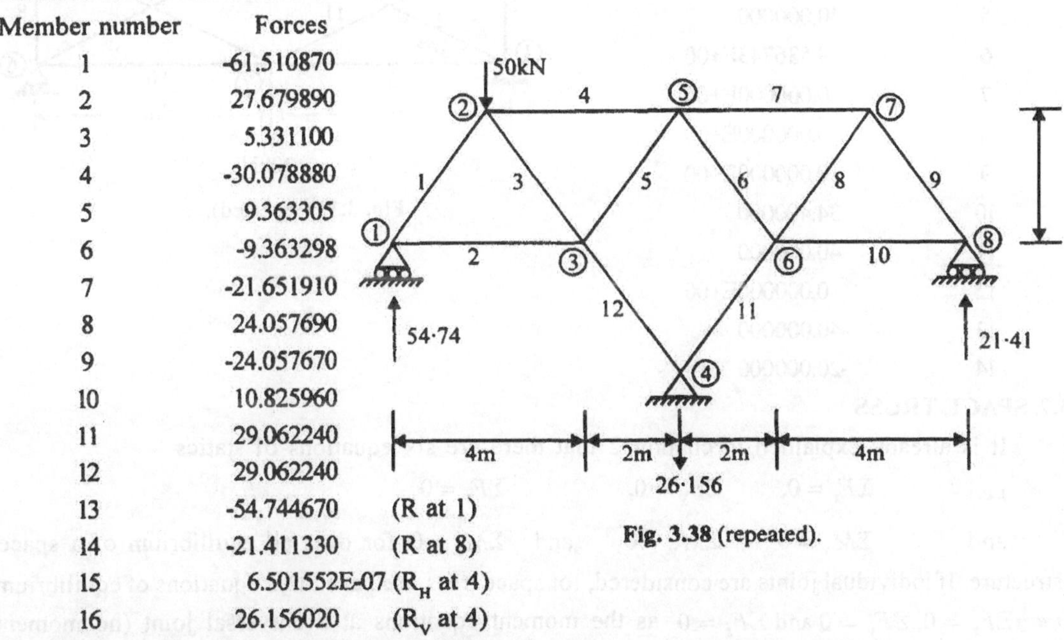

Fig. 3.38 (repeated).

Example problem 4 Example 3.12 (Fig. 3.39)

The equilibrium matrix

.00	.86	1.00	.00	.00	.00	.00	.00	.00	.00	.00	-1.00	.00	.00
1.00	.50	.00	.00	.00	.00	.00	.00	.00	.00	.00	.00	-1.00	.00
.00	.00	.00	.86	.00	.00	.00	.00	.00	.00	.86	.00	.00	.00

-1.00	.00	.00	.50	.00	.00	.00	.00	.00	.00	-.50	.00	.00	.00
.00	.00	.00	-.86	.86	1.00	.00	.00	.00	.00	.00	.00	.00	.00
.00	.00	.00	-.50	-.50	.00	.00	.00	.00	.00	.00	.00	.00	.00
.00	-.86	.00	.00	.00	-1.00	.86	.00	.00	.00	.00	.00	.00	.00
.00	-.50	.00	.00	.00	.00	-.50	.00	.00	.00	.00	.00	.00	.00
.00	.00	.00	.00	.00	.00	-.86	.00	-.86	.00	.00	.00	.00	.00
.00	.00	.00	.00	.00	.00	.50	1.00	-.50	.00	.00	.00	.00	.00
.00	.00	.00	.00	-.86	.00	.00	.00	.00	-1.00	.00	.00	.00	.00
.00	.00	.00	.00	.50	.00	.00	1.00	.00	.00	.00	.00	.00	-1.00
.00	.00	-1.00	.00	.00	.00	.00	.00	.86	1.00	-.86	.00	.00	.00
.00	.00	.00	.00	.00	.00	.00	.00	.50	.00	.50	.00	.00	.00

The joint load vector

.00	.00	.00	.00	.00	-40.00	.00	.00	.00	.00	.00	.00	.00	-20.00

Member forces and reactions

Member number	Forces
1	-40.000000
2	0.000000E+00
3	4.768372E-07
4	-40.000000
5	-40.000000
6	9.536743E+00
7	0.000000E+00
8	0.000000E+00
9	0.000000E+00
10	34.400000
11	40.000000
12	0.000000E+00
13	-40.000000
14	-20.000000

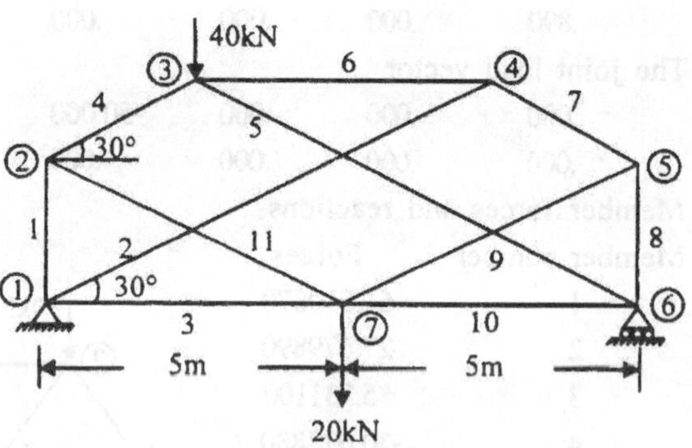

Fig. 3.39 (repeated).

3.7. SPACE TRUSS

It is already explained in chapter 2 that there are six equations of statics

i.e. $\Sigma F_x = 0, \quad \Sigma F_y = 0, \quad \Sigma F_z = 0$

and $\Sigma M_x = 0 \quad \Sigma M_y = 0$ and $\Sigma M_z = 0$ for over all equilibrium of a space structure. If individual joints are considered, for space truss, we have three equations of equilibrium *i.e.* $\Sigma F_x = 0, \Sigma F_y = 0$ and $\Sigma F_z = 0$ as the moment equations at a universal joint (no moment resistance capacity) are automatically satisfied. The members of a space truss can, therefore, carry only tension or compression.

For space frames also the overall equilibrium can be checked by six equations of statics as above. However, as the joints are rigid each bar has six unknown stress resultants 3 direct forces and three moments. It has already been shown in section 2.6(*ii*) that necessary but not sufficient condition for determinacy and stability of a space truss is

$3_j = m + r$ where j = no. of joints m = no. of members and r = no. of reactions.

3.7.1. Analysis of Space Truss : Three dimensional space trusses are generally analysed assuming that ——

1. The system is composed of two forces members *i.e.* members can carry only tension or compression.
2. The members are connected with ideal spherical hinges.
3. The axes of all members meeting at a joint intersect at a point.
4. All loads are applied at the points of connection.

For space trusses, as stated before, at each joint the three equations to be satisfied are

$$\Sigma F_x = 0 \qquad \Sigma F_y = 0 \qquad \Sigma F_z = 0$$

The following *conclusions* can be drawn when the above three equations are applied to forces meeting at a joint.

1. Any three concurrent forces that do not lie in a single plane will not be in equilibrium unless all three forces are zero.
2. If two forces of a system of four concurrent forces that are in equilibrium in space are collinear, the other two forces will be zero unless these forces are themselves collinear. Collinear forces will be equal in magnitude and opposite in direction.
3. If all but one force of a system of concurrent forces that are in equilibrium in space are coplanar, the out of plane force must be zero.
4. If all but two forces of a system of concurrent forces in space are in equilibrium and are coplanar and the magniiude of one of these out-of-plane forces is known, the magnitude of the other can be calculated by projecting all forces on to an axis perpendicular to the coplanar forces.
5. If all but two bars at a joint have no bar forces and these two are not collinear and if no external force acts at that joint, the bar force in each of these two bars is zero.

Explanations

(*i*) If P is the resultant force and it is in equilibrium, projections of P in three coordinate axes P_{xi}, P_{yj} and P_{zk} is equal to zero. This means P_x, P_y, P_z must be zero which is possible only if all the forces are zero.

(*ii*) By summing forces normal to a plane containing the two collinear forces, and since the other two forces are not collinear, they must be equal to zero.

(*iii*) By summing forces normal to the plane containing all the coplanar forces, it can be easily shown that the non-planar force is zero.

(*iv*) Two non collinear forces cannot balance. Hence for equilibrium, both of them must be zero.

Analysis of space truss problems will obviously be extension of plane truss problems. There are hand methods like method of joints, method of sections or computer methods based on matrix solutions of joint equilibrium equations. Initially some small problems are solved by hand methods before using computer for solution of simultaneous equations obtained from joint equations. All members are assumed to be in tension and joint equations are written taking direction of forces into account.

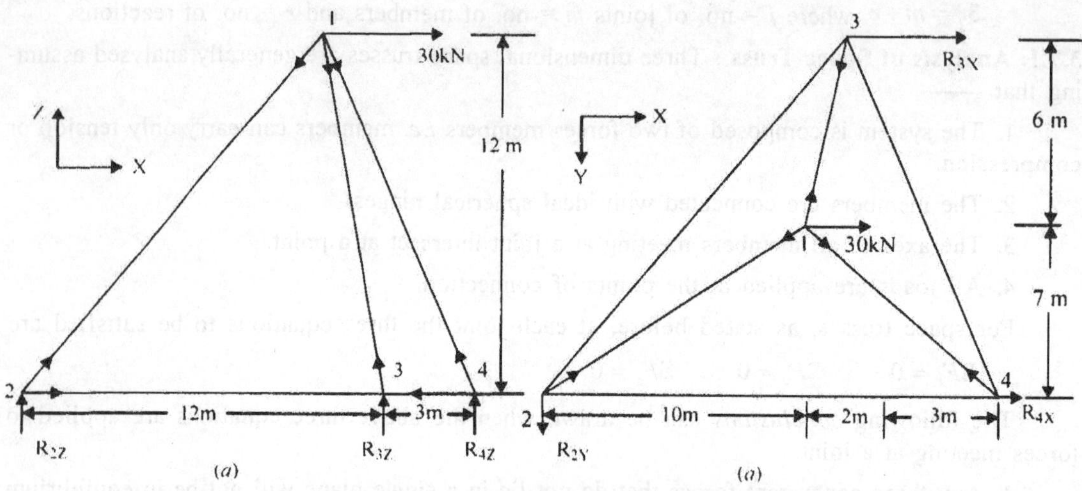

Fig. 3.40 (Example 3.13)

Example 3.13. Using the over all equations of equilibrium for a three dimensional structure, solve for the reactions of the truss. Compute the forces in the members by joint equilibrium.

Solution. The frame is statically determinate and stable externally with six external reactions R_{2z}, R_{3z}, R_{4z}, R_{2y}, R_{3x}, R_{4x} (1st suffix shows the point, the 2nd suffix shows the direction). The truss is statically determinate. $j = 4$, $m = 6$ and $r = 6$. Hence the equation $3j = m + r$ is satisfied

Overall Equilibrium Equations

Applying

$\Sigma M_x = 0$, and taking moments of all forces about 2–4, $R_{3z} = 0$.

From $\Sigma M_y = 0$ about line of action of R_{2y}, $-\overset{\frown}{R_{4z}} \times 15 - \overset{\frown}{R_{3z}} \times 12 + \overset{\frown}{30 \times 12} = 0$

or $$R_{4z} = +\frac{30 \times 12}{15} = +24\text{kN}\uparrow$$

From $\Sigma F_z = 0$

$$R_{2z} + R_{3z} + R_{4z} = 0$$

or $24 + 0 + R_{2z} = 0$ or $R_{2z} = -24\downarrow$

From $\Sigma F_x = 0$

$$R_{3x} + R_{4x} + 30 = 0$$

or $$R_{3x} + R_{4x} = -30$$

Taking moments about 3

and from $\Sigma M_z = 0$

$$\overset{\frown}{30 \times 6} + \overset{\frown}{R_{4x}} \times 13 + \overset{\frown}{R_{2y}} \times 12 = 0$$

From $\Sigma F_y = 0$, $R_{2y} = 0$, ...

Then $$R_{4x} = -\frac{30 \times 6}{13} = -13{\cdot}85.$$

Substituting in the equation.

$$R_{3x} + R_{4x} = -30$$

Forces in the members by method of joints. At each joint the three joint equilibrium equations are, $\Sigma F_x = 0, \quad \Sigma F_y = 0, \quad \Sigma F_z = 0$

If $\quad F$ = Force in any member which is axial only

F_x = Force in the x direction.

$$= \frac{F \cdot L_x}{L}$$

Similarly $\quad F_y = \dfrac{F \cdot L_y}{L} \qquad F_z = \dfrac{F \cdot L_z}{L}$

where L_x, L_y and L_z are the projections of the member length in the x, y, and z directions respectively. L_x, L_y, L_z and L can be easily obtained once the coordinates of two ends of the member are known. Taking point 2 as origin (0,0,0) the coordinates, projections and length of different members are given below. Projections are always taken as positive.

Table 3.2. Example 3.13.

Member	Coordinates of		Projections			Length
	End 1	End 2	X	Y	Z	
21	0, 0, 0	10, –7, 12	10	7	12	17·11
23	0, 0, 0	12, –13, 0	12	13	0	17·69
24	0, 0, 0	15, 0, 0	15	0	0	15·00
13	10, –7, 12	12, –13, 0	2	6	12	13·56
14	10, –7, 12	15, 0, 0	5	7	12	14·76
34	12, –13, 0	15, 0, 0	3	13	0	13·34

Joint 1 Assuming all bars are in tension

$\Sigma F_x = 0$ gives ($\rightarrow$ +ve)

$$30 + F_{14} \times \frac{5}{14 \cdot 76} + F_{13} \times \frac{2}{13 \cdot 56} - F_{12} \times \frac{10}{17 \cdot 11} = 0$$

$\Sigma F_y = 0$ gives ($\downarrow$ + ve)

$$F_{14} \times \frac{7}{14 \cdot 76} + F_{12} \times \frac{7}{17 \cdot 11} - F_{13} \times \frac{6}{13 \cdot 56} = 0$$

$\Sigma F_z = 0$ gives ($\uparrow$ + ve)

$$F_{14} \times \frac{12}{14 \cdot 76} - F_{12} \times \frac{12}{17 \cdot 11} - F_{13} \times \frac{12}{13 \cdot 6} = 0$$

or $\quad F_{14} \times 0 \cdot 3387 + F_{13} \times 0 \cdot 1474 - F_{12} \times \cdot 5845 = -30$

$$F_{14} \times 0 \cdot 4742 - F_{13} \times 0 \cdot 4424 + F_{12} \times 0 \cdot 4091 = 0$$

$$F_{14} \times (-0 \cdot 8130) - F_{13} \times (0 \cdot 8823) - F_{12} \times (0 \cdot 7013) = 0$$

Solving $F_{14} = -29 \cdot 52$ **(compression)**

$F_{13} = \cdot 00175 \approx 0$

$F_{12} = 34 \cdot 23$ **(tension)**

F_{13} will be zero can also be visualized from observation 3 as all forces except F_{13} meeting at joint 1 are coplanar.

Joint 2 $\quad \Sigma F_x = 0, \quad F_{24} \times \frac{15}{15} + F_{21} \times \frac{10}{17 \cdot 11} + F_{23} \times \frac{12}{17 \cdot 69} = 0$

$\Sigma F_y = 0 \quad F_{24} \times \frac{0}{15} - F_{21} \times \frac{7}{17 \cdot 11} - F_{23} \times \frac{13}{17 \cdot 69} = 0$

or $\quad -F_{21} \times \frac{7}{17 \cdot 11} - F_{23} \times \frac{13}{17 \cdot 69} = 0$

or $\quad F_{23} = -34 \cdot 23 \times \frac{7}{17 \cdot 11} \times \frac{17 \cdot 69}{13}$

$= -19{\cdot}05$ (compression)

Substituting $\quad F_{24} = -\left(\frac{34 \cdot 23 \times 10}{17 \cdot 11} + (-19 \cdot 05) \times \frac{12}{17 \cdot 69}\right)$

$= -7{\cdot}08$ (compression)

Joint 4 $\quad$ From $\Sigma F_y = 0$

$-F_{41} \times \left(+\frac{7}{14 \cdot 76}\right) - F_{43}\left(\frac{13}{13 \cdot 34}\right) = 0$

$F_{43} = -\frac{F_{41} \times 7}{14 \cdot 76} \times \frac{13 \cdot 34}{13}$

$= \frac{29 \cdot 52 \times 7 \times 13 \cdot 34}{14 \cdot 76 \times 13}$

$= 14 \cdot 37$ (tension)

Example 3.14. Using the overall equations of equilibrium for a three dimensional structure, solve for the reactions of the truss shown in Fig. 3.41. Compute the bar forces by joint equilibrium.

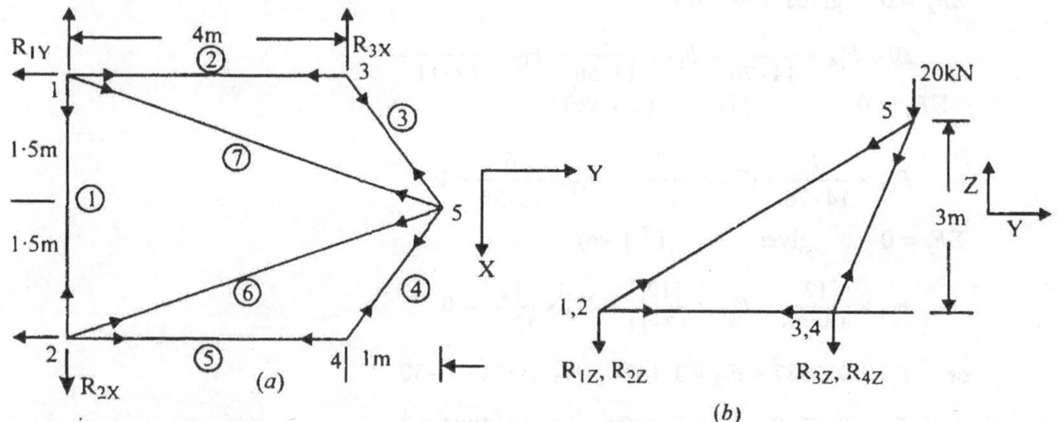

Fig. 3.41. Example 3.14.

The frame is statically indeterminate externally as there are seven reaction components and only six equations of statics. However, it is statically determinate internally as it satisfies the relation $3_j = m + r$. Here $j = 5$, $m = 8$ and $r = 7$. The frame can therefore be analysed statically. Bars are assumed to be in tension and reactions are assumed to be additional bars under tension as in the case of plane truss.

Taking $\quad \Sigma M_z = 0$ about line 1–2

$$\overset{\frown}{20 \times 5} + \overset{\frown}{R_{3z} \times 4} + \overset{\frown}{R_{4z} \times 4} = 0$$

or $R_{3z} + R_{4z} = -25$

From $\Sigma F_z = 0$ $R_{1z} + R_{2z} + R_{3z} + R_{4z} + 20 = 0$

or $R_{1z} + R_{2z} = -20 - (R_{3z} + R_{4z})$

$= -20 + 25 = +5.$

From $\Sigma M_y = 0$ about 1 3.

$$\overset{\frown}{R_{2z} \times 3} + \overset{\frown}{R_{4z} + 3} + \overset{\frown}{20 \times 1 \cdot 5} = 0$$

or $R_{2z} + R_{4z} = -10$

Since we have only three equations for four unknowns.

$R_{1z}, R_{2z}, R_{3z}, and\ R_{4z}.$

We assume $R_{1z} = R$ then $R_{2z} = +5 - R.$

$R_{4z} = -10 - R_{2z}$

$= -10 - 5 + R$

$= -$ **15 + R.**

and $R_{3z} = -25 - R_{4z}$

$= -25 + 15 - R$

$= -$ **10 − R.**

The member forces may now be determined in terms of R. These computations are continued until the forces at both ends of one bar are determined in terms of R. These two values must be equal and will give the value of R.

At joint 4 putting $\Sigma F_z = 0$

$$F_{45} \times \frac{3}{3 \cdot 5} - R_{4z} = 0$$

or $F_{45} = \frac{3 \cdot 5}{3} \times R_{4z}.$

But $R_{4z} = -15 + R$

Therefore $F_{45} = (-15 + R) \times \frac{3 \cdot 5}{3}$

Putting $\Sigma F_x = 0$ $F_{43} + F_{45}\left(+\frac{1 \cdot 5}{3 \cdot 5}\right) = 0$

or $F_{43} = -F_{45}\left(\frac{1 \cdot 5}{3 \cdot 5}\right) = -(-15 + R)\frac{1 \cdot 5}{3}$

$= -0 \cdot 5\ (-15 + R)$

Table 3.3 (Example 3.14)

Member		Coordinates of		Projections			Length
Points	No.	End 1	End 2	X	Y	Z	
1 – 2	1	0,0,0	3,0,0	3	0	0	3
1 – 3	2	0,0,0	0,4,0	0	4	0	4
1 – 5	3	0,0,0	1·5,5,3	1·5	5	3	6·02
2 – 1	1	3,0,0	0,0,0	3	0	0	3

2 – 4	5	3,0,0	3,4,0	0	4	0	4
2 – 5	6	3,0,0	1·5,5,3	1·5	5	3	6·02
3 – 1	2	0,4,0	0,0,0	0	4	0	4
3 – 4	8	0,4,0	3,4,0	3	0	0	3
3 – 5	3	0,4,0	1·5,5,3	1·5	1	3	3·5
4 – 2	5	3,4,0	3,0,0	0	4	0	4
4 – 3	8	3,4,0	0,4,0	3	0	0	3·0
4 – 5	4	3,4,0	1·5,5,3	1·5	1	3	3·5

At Joint 3 $\quad -R_{3z} + F_{35} \times \dfrac{3}{3 \cdot 5} = 0$

$$F_{35} = R_{3z} \times \frac{3 \cdot 5}{3}$$

$$= +(-10 - R) \times \frac{3 \cdot 5}{3}$$

From $\quad \Sigma M_z = 0 \quad R_{3x}(4) = 0$

or $\quad R_{3x} = 0$

$$F_{34} + F_{35} \times \frac{1 \cdot 5}{3 \cdot 5} = 0$$

$$F_{34} = -F_{35} \times \frac{1 \cdot 5}{3 \cdot 5}$$

$$= -(-10 - R) \cdot \frac{3 \cdot 5}{3} \times \frac{1 \cdot 5}{3 \cdot 5}$$

$$= -(-10 - R)/2.$$

As $\quad F_{43} = F_{34}$

$$-\frac{1}{2}(-15 + R) = -\frac{1}{2}(-10 - R)$$

$$-15 + R = -10 - R$$

or $\quad 2R = +5$

$$R = 2 \cdot 5$$

Substituting $\quad R_{1z} = +2 \cdot 5, \quad R_{2z} = +2 \cdot 5$

$$R_{3z} = -12 \cdot 5, \quad R_{4z} = -12 \cdot 5.$$

$$F_{45} = +(-15 + R)\frac{3 \cdot 5}{3}$$

$$= +(15 + 2 \cdot 5) \times \frac{3 \cdot 5}{3}$$

$$= -14 \cdot 58 \text{ (compression)}$$

$$F_{43} = (-10 - R) \cdot \frac{3 \cdot 5}{3}$$

$$= (-10 - 2 \cdot 5)\left(\frac{3 \cdot 5}{3}\right)$$

$$= -14 \cdot 58 \text{ (compression)}$$

$$F_{34} = -(-10 - R)\cdot\frac{1}{2}$$

$$= +12\cdot5 \times \frac{1}{2} = 6\cdot25 \text{ (tension)}$$

From $\quad \Sigma F_z = 0$ at joint 1

$$F_{15} \times \frac{3}{6\cdot02} - R_{1z} = 0$$

or $\quad F_{15} \times \frac{3}{6\cdot02} = R_{1z} = +2\cdot5$

or $\quad F_{15} = +5\cdot01$

From $\quad \Sigma F_x = 0$ at joint 1

$$F_{15} \times \frac{1\cdot5}{6} + F_{12} = 0$$

or $\quad F_{12} = -F_{15} \times \frac{1\cdot5}{6\cdot02}$

$$= -5\cdot01 \times \frac{1\cdot5}{6\cdot02} = -1\cdot25.$$

From $\quad \Sigma F_y = 0 \; R_{1y} = 0.$

$$F_{13} + F_{15} \times \frac{5\cdot0}{6\cdot02} = 0$$

or $\quad F_{13} = -\frac{5\cdot01 \times 50}{6\cdot02} = -4\cdot18$

From $\Sigma F_z = 0$ at joint 2

$$F_{25} \times \frac{3}{6\cdot02} - R_{2z} = 0$$

or $\quad F_{25} \times \frac{3}{6\cdot02} - (2\cdot5) = 0$

or $\quad F_{25} = 5\cdot02$ (tension)

$$F_{45} \times \frac{1}{3\cdot5} - F_{42} = 0$$

or $\quad F_{42} = \frac{F_{45}}{3\cdot5} = \frac{-14\cdot58}{3\cdot5} = -4\cdot17$ (compression)

Results

$R_{1y} = 0\cdot00$	$F_{13} = -4\cdot18$	(comp)
$R_{2x} = 0\cdot00$	$F_{15} = +5\cdot01$	(tension)
$R_{3x} = 0\cdot00$	$F_{12} = -1\cdot25$	(comp)
$R_{1z} = +2\cdot50$	$F_{25} = +5\cdot02$	(tension)
$R_{2z} = +2\cdot50$	$F_{24} = -4\cdot17$	(comp)
$R_{3z} = -12\cdot50$	$F_{45} = -14\cdot58$	(comp)
$R_{4z} = -12\cdot50$	$F_{35} = -14\cdot58$	(comp)
	$F_{34} = +6\cdot25$	(tension)

The problem is solved again by writing the joint equilibrium equations. All the conventions for joint equilibrium equations for plane truss are followed *i.e.* all members are assumed tensile, reactions are assumed to be additional members under tension. Forces are assumed positive in positive coordinate axes.

Joint 1 $\quad \Sigma F_x = 0 \quad F_1 + F_7 \times \dfrac{1 \cdot 5}{6 \cdot 02} = 0$ ...(1)

$\Sigma F_y = 0 \quad 0 + F_7 \cdot \dfrac{5}{6 \cdot 02} + F_2 - F_{1 \cdot y} = 0$ ...(2)

$\Sigma F_z = 0 \quad 0 + 0 + F_7 \cdot \dfrac{3}{6 \cdot 02} - F_{1 \cdot z} = 0$ ...(3)

Joint 2 $\quad \Sigma F_x = 0 \quad -F_1 - F_6 \times \dfrac{1 \cdot 5}{6 \cdot 02} + R_{2x} = 0$ ...(4)

$\Sigma F_y = 0 \quad F_5 + F_6 \times \dfrac{5}{6 \cdot 02} = 0$ ...(5)

$\Sigma F_z = 0 \quad 0 + F_6 \times \dfrac{3}{6 \cdot 02} - R_{2z} = 0$ ...(6)

Joint 3 $\quad \Sigma F_x = 0 \quad F_8 + F_3 \dfrac{1 \cdot 5}{3 \cdot 5} - R_{3x} = 0$ ...(7)

$\Sigma F_y = 0 \quad F_3 \times \dfrac{1}{3 \cdot 5} - F_2 = 0$ ...(8)

$\Sigma F_z = 0 \quad F_3 \times \dfrac{3}{3 \cdot 5} - R_{3z} = 0$ ...(9)

Joint 4 $\quad \Sigma F_x = 0 \quad -F_8 - F_4 \dfrac{1 \cdot 5}{3 \cdot 5} = 0$ ...(10)

$\Sigma F_y = 0 \quad F_4 \times \dfrac{1 \cdot 0}{3 \cdot 5} - F_5 = 0$ ...(11)

$\Sigma F_z = 0 \quad F_4 \times \dfrac{3}{3 \cdot 5} - R_{4z} = 0$ ...(12)

Joint 5 $\quad \Sigma F_x = 0 \quad -F_3 \times \dfrac{1 \cdot 5}{3 \cdot 5} - F_7 \times \dfrac{1 \cdot 5}{6 \cdot 02} + F_6 \times \dfrac{1 \cdot 5}{6 \cdot 02} + F_4 \times \dfrac{1 \cdot 5}{3 \cdot 5} = 0$ ...(13)

$\Sigma F_y = 0 \quad -F_3 \times \dfrac{1}{3 \cdot 5} - F_7 \times \dfrac{5}{6 \cdot 02} - F_6 \times \dfrac{5}{6 \cdot 02} - F_4 \times \dfrac{1}{3 \cdot 5} = 0$ ...(14)

$\Sigma F_z = 0 \quad -F_3 \times \dfrac{3}{3 \cdot 5} - F_7 \times \dfrac{3}{6 \cdot 02} - F_6 \times \dfrac{3}{6 \cdot 02} - F_4 \times \dfrac{3}{3 \cdot 5} - 20 = 0$ ...(15)

These 15 equations with 15 unknowns can be expressed in matrix form as shown in page 105. The equations are solved through computer with program given in Page 88 the results are given in Page 106 and they tally with the hand solutions.

Computer output of Example 3.14.

1·0	0·0	0·0	0·0	0·0	0·0	0·25	0·0	0·0	0·0	0·0	0·0	0·0	0·0	0·0
0·0	1·0	0·0	0·0	0·0	0·0	0·83	0·0	−1·0	0·0	0·0	0·0	0·0	0·0	0·0
0·0	0·0	0·0	0·0	0·0	0·0	0·5	0·0	0·0	0·0	0·0	−1·0	0·0	0·0	0·0
−1·0	0·0	0·0	0·0	0·0	−0·25	0·0	0·0	0·0	1·0	0·0	0·0	0·0	0·0	0·0
0·0	0·0	0·0	0·0	1·0	0·83	0·0	0·0	0·0	0·0	0·0	0·0	0·0	0·0	0·0
0·0	0·0	0·0	0·0	0·0	0·5	0·0	0·0	0·0	0·0	0·0	0·0	−1·0	0·0	0·0

Conted pg. 106

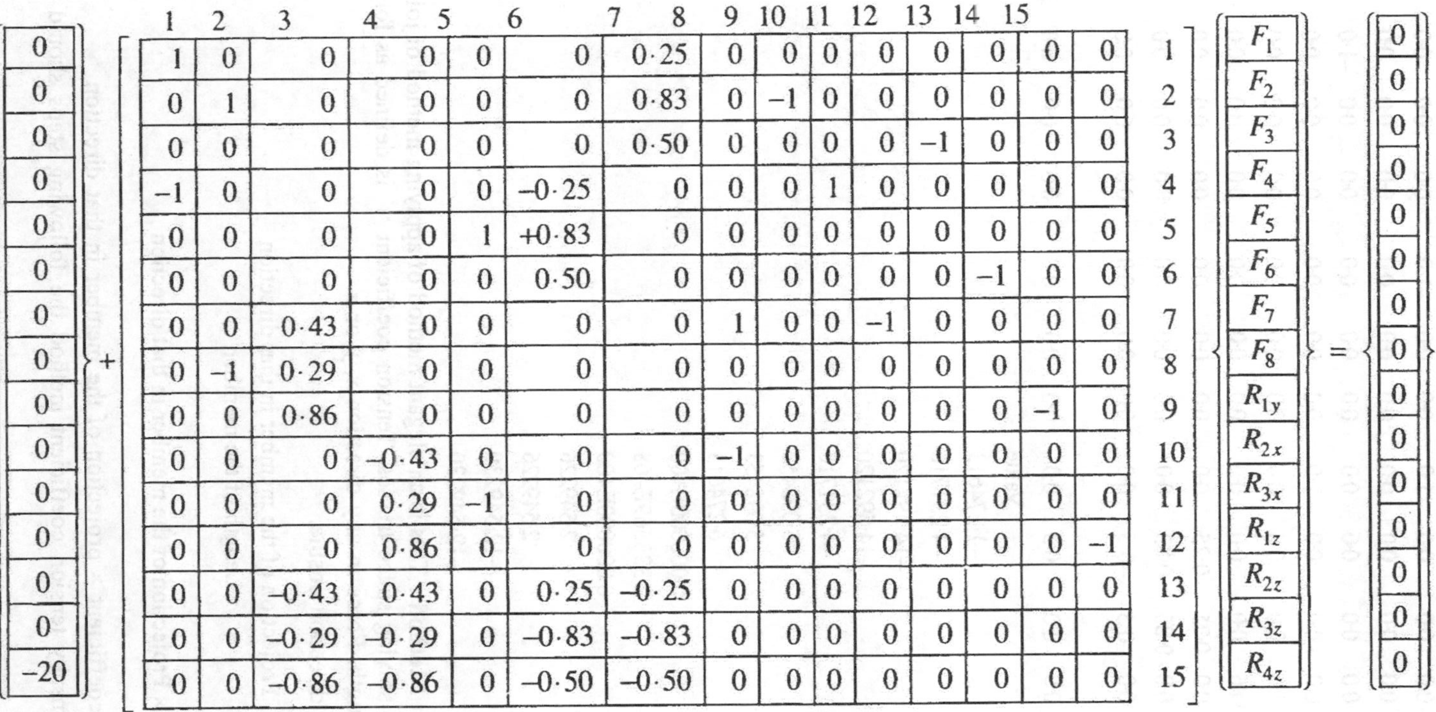

Load		1	2	3	4	5	6	7	8	9	10	11	12	13	14	15	Row	Force		RHS
0		1	0	0	0	0	0	0·25	0	0	0	0	0	0	0	0	1	F_1		0
0		0	1	0	0	0	0	0·83	0	–1	0	0	0	0	0	0	2	F_2		0
0		0	0	0	0	0	0	0·50	0	0	0	0	–1	0	0	0	3	F_3		0
0		–1	0	0	0	0	–0·25	0	0	0	1	0	0	0	0	0	4	F_4		0
0		0	0	0	0	1	+0·83	0	0	0	0	0	0	0	0	0	5	F_5		0
0		0	0	0	0	0	0·50	0	0	0	0	0	0	–1	0	0	6	F_6		0
0		0	0	0·43	0	0	0	0	1	0	0	–1	0	0	0	0	7	F_7		0
0	+	0	–1	0·29	0	0	0	0	0	0	0	0	0	0	0	0	8	F_8	=	0
0		0	0	0·86	0	0	0	0	0	0	0	0	0	0	–1	0	9	R_{1y}		0
0		0	0	0	–0·43	0	0	0	–1	0	0	0	0	0	0	0	10	R_{2x}		0
0		0	0	0	0·29	–1	0	0	0	0	0	0	0	0	0	0	11	R_{3x}		0
0		0	0	0	0·86	0	0	0	0	0	0	0	0	0	0	–1	12	R_{1z}		0
0		0	0	–0·43	0·43	0	0·25	–0·25	0	0	0	0	0	0	0	0	13	R_{2z}		0
0		0	0	–0·29	–0·29	0	–0·83	–0·83	0	0	0	0	0	0	0	0	14	R_{3z}		0
–20		0	0	–0·86	–0·86	0	–0·50	–0·50	0	0	0	0	0	0	0	0	15	R_{4z}		0

Matrix Format of Example 3.14.

0·0	0·0	0·43	0·0	0·0	0·0	0·0	1·0	0·0	0·0	–1·0	0·0	0·0	0·0	0·0
0·0	–1·0	0·29	0·0	0·0	0·0	0·0	0·0	0·0	0·0	0·0	0·0	0·0	0·0	0·0
0·0	0·0	0·86	0·0	0·0	0·0	0·0	0·0	0·0	0·0	0·0	0·0	0·0	–1·0	0·0
0·0	0·0	0·0	–0·43	0·0	0·0	0·0	–1·0	0·0	0·0	0·0	0·0	0·0	0·0	0·0
0·0	0·0	0·0	0·29	–1·0	0·0	0·0	0·0	0·0	0·0	0·0	0·0	0·0	0·0	0·0
0·0	0·0	0·0	0·0	0·86	0·0	0·0	0·0	0·0	0·0	0·0	0·0	0·0	0·0	–1·0
0·0	0·0	–0·43	0·43	0·0	0·25	–0·25	0·0	0·0	0·0	0·0	0·0	0·0	0·0	0·0
0·0	0·0	–0·29	–0·29	0·0	–0·83	–0·83	0·0	0·0	0·0	0·0	0·0	0·0	0·0	0·0
0·0	0·0	–0·86	–0·86	0·0	–0·5	–0·5	0·0	0·0	0·0	0·0	0·0	0·0	0·0	0·0

Force vector

0·0	0·0	0·0	0·0	0·0	0·0	0·0	0·0	0·0	0·0	0·0	0·0	0·0	0·0	20·0

Force Number	Value
1	–1·274613
2	–4·231716
3	–14·592120
4	–14·592120
5	–4·231716
6	5·098453
7	5·098453
8	6·274613
9	0·000000E+00
10	2·333569E–08
11	0·000000E+00
12	2·549226
13	2·549226
14	–12·549230
15	–12·549230

3.7.2. Tension Coefficient Method : This is an elegant method of applying method of joints and equilibrium equations for solving space trusses. Tension coefficient '*t*' is defined as force in a member divided by its length. Force in any direction *x*, *y* or *z*

$$= \text{Force} \times \text{direction cosine.}$$

$$= \text{Force} \times \frac{\text{Projection of the member in that direction}}{\text{Length of the member}}.$$

$$= \frac{\text{Force}}{\text{Length}} \times \text{Projection of the member in that direction}.$$

$$= \text{tension coefficient} \times \text{projection of the member in that direction}$$

For solving problems by tension coefficient method the following steps should be followed :—

1. Any joint can be taken as origin. Coordinates of all other joints should then be computed.

2. Projections of all the members should then be written. Projection of any member is equal to the difference of coordinates of further end and near end.

3. Force in each member is assumed to be tensile and positive. If it comes to be negative, it means it is compressive. Unlike method of joints, initially no direction should be assigned to the member forces.

4. External forces are positive in the positive direction of coordinate axes. Equilibrium equations are then written for all joints in terms of the tension coefficients (which are the unknowns) and the coordinates of members with proper signs. Solution of the equilibrium equations gives the values of tension coefficients. When multiplied by length, the forces in the members are obtained.

Example 3.15. Find forces in all members of the truss by tension coefficient method.

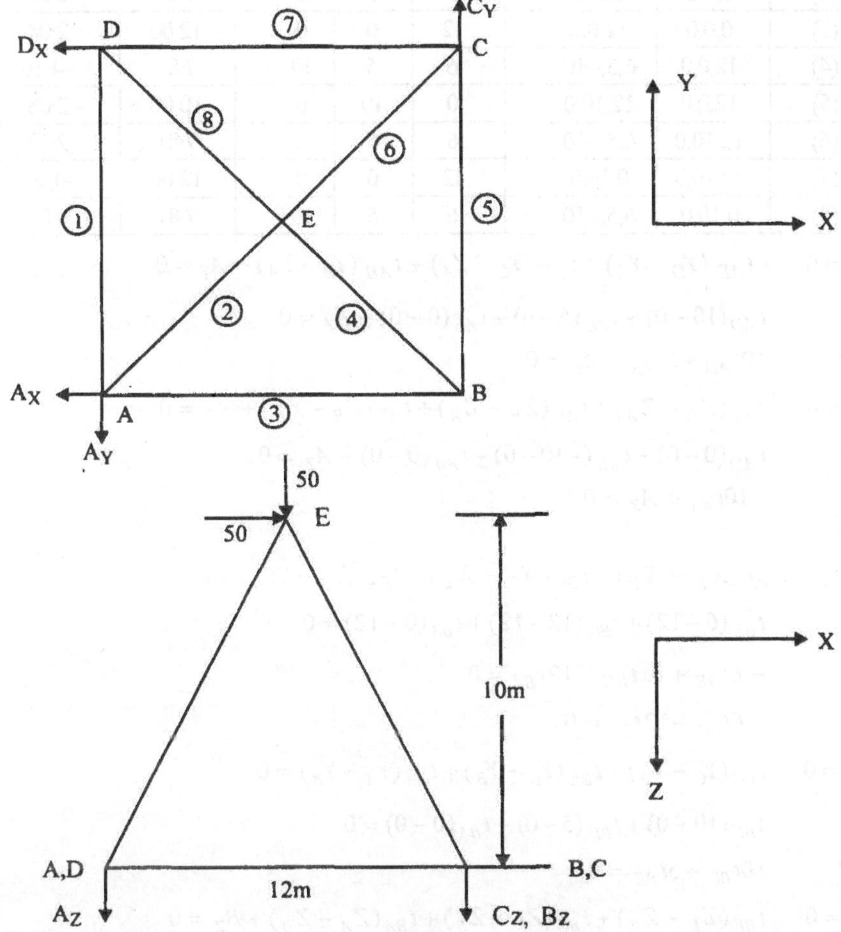

Fig. 3.42. Example 3.15.

Solution. The truss is determinate as it satisfies the relation $m + r = 3_j$.

Here $m = 8$, $r = 7$, $3j = 15$.

With A as origin 0,0,0 the coordinates of member ends, projections and member lengths are given in tabular form.

Equations of Equilibrium

(1) *Joint A.* $\Sigma F_x = 0 \qquad t_{AD}(x_D - x_A) + t_{AE}(x_E - x_A) + t_{AB}(x_B - x_A) - A_x = 0$

or $\quad t_{AD}(0-0)+t_{AE}(6-0)+t_{AB}(12-0)-A_x=0$

or $\quad 6t_{AE}+12t_{AB}-A_x=0.$

Table 3.4. Example 3.15

Members		Coordinates at ends		Projections neglecting signs			Length L	Tension Coefficient t	Force
Name	Number	end 1	end 2	X	Y	Z			
AD	(1)	0,0,0	0,10,0	0	10	0	10·00	0·00	0·00
AE	(2)	0,0,0	6,5,–10	6	5	10	7·81	6·66	52·01
AB	(3)	0,0,0	12,0,0	12	0	0	12·00	2·08	24·96
BE	(4)	12,0,0	6,5,–10	6	5	10	7·81	–4·16	–32·49
BC	(5)	12,0,0	12,10,0	0	10	0	10·00	2·08	20·80
CE	(6)	12,10,0	6,5,–10	6	5	10	7·81	2·50	19·53
CD	(7)	12,10,0	0,10,0	12	0	0	12·00	–1·25	–15·00
DE	(8)	0,10,0	6,5,–10	6	5	10	7·81	0·00	0·00

$\Sigma F_y=0 \quad t_{AD}(Y_D-Y_A)+t_{AE}(Y_E-Y_A)+t_{AB}(Y_B-Y_A)-A_Y=0$

or $\quad t_{AD}(10-0)+t_{AE}(5-0)+t_{AB}(0-0)-A_Y=0$

or $\quad 10t_{AD}+5t_{AE}-A_Y=0$

$\Sigma F_z=0 \quad t_{AD}(Z_D-Z_A)+t_{AE}(Z_E-Z_A)+t_{AB}(Z_B-Z_A)+A_Z=0$

or $\quad t_{AD}(0-0)+t_{AE}(-10-0)+t_{AB}(0-0)+A_Z=0$

or $\quad -10t_{AE}+A_Z=0$

(2) *Joint B.*

$\Sigma F_x=0 \quad t_{BE}(X_E-X_B)+t_{BC}(X_C-X_B)+t_{BA}(X_A-X_B)=0$

or $\quad t_{BE}(6-12)+t_{BC}(12-12)+t_{BA}(0-12)=0$

$\quad -6t_{BE}+oxt_{BC}-12t_{BA}=0$

or $\quad -6t_{BE}-12t_{BA}=0.$

$\Sigma F_y=0 \quad t_{BC}(Y_C-Y_B)+t_{BE}(Y_E-Y_B)+t_{BA}(Y_A-Y_B)=0$

$\quad t_{BC}(10-0)+t_{BE}(5-0)+t_{BA}(0-0)=0$

or $\quad 10t_{BC}+5t_{BE}=0.$

$\Sigma F_z=0 \quad t_{BE}(Z_E-Z_B)+t_{BC}(Z_C-Z_B)+t_{BA}(Z_A-Z_B)+B_Z=0$

or $\quad t_{BE}(-10-0)+t_{BC}(0-0)+t_{BA}(0-0)+B_Z=0$

or $\quad -10t_{BE}+B_Z=0.$

(3) *Joint C*

$\Sigma F_x=0 \quad t_{CD}(X_D-X_C)+t_{CE}(X_E-X_C)+t_{CB}(X_B-X_C)=0$

or $\quad t_{CD}(0-12)+t_{CE}(6-12)+t_{CB}(12-12)=0$

or $\quad 12t_{CD}+6t_{CE}=0.$

$\Sigma F_y = 0 \quad t_{CD}(Y_D - Y_C) + t_{CE}(Y_E - Y_C) + t_{CB}(Y_B - Y_C) + C_Y = 0$

or $\quad t_{CD}(10-10) + t_{CE}(5-10) + t_{CB}(0-10) + C_Y = 0$

or $\quad -5t_{CE} - 10t_{CB} + C_Y = 0.$

$\Sigma F_z = 0 \quad t_{CD}(Z_D - Z_C) + t_{CE}(Z_E - Z_C) + t_{CB}(Z_B - Z_C) + C_Z = 0$

or $\quad t_{CD}(0-0) + t_{CE}(-10-0) + t_{CB}(0-0) + C_Z = 0$

or $\quad -10t_{CE} + C_Z = 0.$

(4) *Joint D.*

$\Sigma F_x = 0 \quad t_{DC}(X_C - X_D) + t_{DE}(X_E - X_D) + t_{DA}(X_A - X_D) - D_X = 0$

or $\quad t_{DC}(12-0) + t_{DE}(6-0) + t_{DA}(0-0) - D_X = 0$

or $\quad 12t_{DC} + 6t_{DE} - D_X = 0.$

$\Sigma F_y = 0 \quad t_{DC}(Y_C - Y_D) + t_{DE}(Y_E - Y_D) + t_{DA}(Y_A - Y_D) = 0$

or $\quad t_{DC}(10-10) + t_{DE}(5-10) + t_{DA}(0-10) = 0$

or $\quad 5t_{DE} - 10t_{DA} = 0.$

$\Sigma F_z = 0 \quad t_{DC}(Z_C - Z_D) + t_{DE}(Z_E - Z_D) + t_{DA}(Z_A - Z_D) = 0$

or $\quad t_{DC}(0-0) + t_{DE}(-10-0) + t_{DA}(0-0) = 0$

$-10t_{DE} = 0 \quad$ or $\quad t_{DE} = 0.$

(5) *Joint E.*

$\Sigma F_x = 0 \quad t_{EC}(X_C - X_E) + t_{EB}(X_B - X_E) + t_{EA}(X_A - X_E) + t_{ED}(X_D - X_E) + 50 = 0$

or $\quad t_{EC}(12-6) + t_{EB}(12-6) + t_{EA}(0-6) + t_{ED}(0-6) + 50 = 0$

or $\quad 6t_{EC} + 6t_{EB} - 6t_{EA} - 6t_{ED} + 50 = 0.$

$\Sigma F_y = 0 \quad t_{EC}(Y_C - Y_E) + t_{EB}(Y_B - Y_E) + t_{EA}(Y_A - Y_E) + t_{ED}(Y_D - Y_E) = 0$

or $\quad t_{EC}(10-5) + t_{EB}(0-5) + t_{EA}(0-5) + t_{ED}(10-5) = 0$

or $\quad 5t_{EC} - 5t_{EB} - 5t_{EA} + 5t_{ED} = 0.$

$\Sigma F_z = 0 \quad t_{EC}(Z_C - Z_E) + t_{EB}(Z_B - Z_E) + t_{EA}(Z_A - Z_E) + t_{ED}(Z_D - Z_E) + 50 = 0$

or $\quad t_{EC}(0-10) + t_{EB}(0-10) + t_{EA}(0-10) + t_{ED}(0-10) + 50 = 0$

or $\quad -10t_{EC} - 10t_{EB} - 10t_{EA} - 10t_{ED} + 50 = 0.$

The equations then are :

$$6t_{AE} + 12t_{AB} - A_x = 0 \quad \text{...(1)}$$

$$10t_{AD} + 5t_{AE} - A_y = 0 \quad \text{...(2)}$$

$$-10t_{AE} + A_z = 0 \quad \text{...(3)}$$

$$-6t_{BE} - 12t_{BA} = 0 \quad \text{...(4)}$$

$$+5t_{BE} + 10t_{BC} = 0 \quad \text{...(5)}$$

$$-10t_{BE} + B_Z = 0 \quad \text{...(6)}$$

$$12t_{CD} + 6t_{CE} = 0 \quad \text{...(7)}$$

$$-5t_{CE} - 10t_{CB} + C_y = 0 \quad \text{...(8)}$$

$$-10t_{CE} + C_Z = 0 \quad \text{...(9)}$$

$$12t_{DC} + 6t_{DE} - D_x = 0 \quad \text{...(10)}$$

$$-5t_{DE} - 10t_{DA} = 0 \quad \text{...(11)}$$

$$t_{DE} = 0 \quad \text{...(12)}$$

$$6t_{EC} + 6t_{EB} - 6t_{EA} - 6t_{ED} + 50 = 0 \quad \text{...(13)}$$

$$5t_{EC} - 5t_{EB} - 5t_{EA} + 5t_{ED} = 0 \quad \text{...(14)}$$

$$-10t_{EC} - 10t_{EB} - 10t_{EA} - 10t_{ED} + 50 = 0 \quad \text{...(15)}$$

With $t_{DE} = 0$ from eqn. (12)

$t_{DA} = 0$ from eqn. (11)

The equations (13), (14) and (15) reduce to

$$6t_{EC} + 6t_{EB} - 6t_{EA} + 50 = 0 \quad \text{...(13a)}$$

$$5t_{EC} - 5t_{EB} - 5t_{EA} = 0 \quad \text{...(14a)}$$

$$-10t_{EC} - 10t_{EB} - 10t_{EA} + 50 = 0 \quad \text{...(15a)}$$

Solving, $t_{EA} = 6{\cdot}55$ KN

$t_{EB} = -4{\cdot}15$ KN

$t_{EC} = 2{\cdot}50$ KN

From eqn. (3) $A_Z = 10t_{AE} = 10 \times 6{\cdot}66 = 66{\cdot}66$

$$t_{CD} = -\frac{1}{2}t_{CE} = -\frac{2{\cdot}50}{2{\cdot}00} = \mathbf{-1{\cdot}25\ kN.}$$

$C_Z = 10 \times 2{\cdot}5 =$ **25kN.**

$B_Z = +10t_{BE} = 10 \times (-4{\cdot}16) = -$ **41·6 kN.**

Check : $A_Z + B_Z + C_Z = 66{\cdot}66 + 25{\cdot}00 - 41{\cdot}60 = 50.$

From Eqn. (4) $12t_{BA} = -6t_{BE}$

$= -6 \times (-4{\cdot}16)$

$= 24{\cdot}96$

or $t_{BA} = 2{\cdot}08$

From eqn. (5) $10t_{BC} = -5t_{BE}$

$= -5 \times (-4{\cdot}16) = 20{\cdot}80$

$t_{BC} = 2{\cdot}08$ kN.

From Eqn. (7) $12t_{CD} = -6t_{CE}$

$= 6 \times (2{\cdot}50) = -15.$

$t_{CD} = -1{\cdot}25$ kN.

From Eqn. (8) $C_y = 5t_{CE} + 10t_{CB}$

$= 5 \times 2{\cdot}50 + 10 \times 2{\cdot}08$

Equilibrium Equations in Matrix form (t_{DA} and t_{DE} are zero)

(Example 3.15)

	1	2	3	4	5	6	7	8	9	10	11	12	13			
1	6	12					−1							t_{AE}		0
2	5							−1						t_{AB}		0
3	−10								+1					t_{BE}		0
4		−12	−6											t_{BC}		0
5			5	10										t_{CE}		0
6			−10							1				t_{CD}		0
7					6	12								A_x	=	0
8				−10	−5						1			A_y		0
9					−10							1		A_z		0
10						12							−1	B_z		0
11	−6		6		6									C_y		−50
12	−5		−5		5									C_z		0
13	−10		−10		−10									D_x		−50

Values of the tension coefficients and the reactions are given in computer output on page 112.

$= 33{\cdot}30$ kN.

From Eqn. (1) $\quad A_x = 6t_{AE} + 12t_{AB}$

$= 6 \times 6{\cdot}66 + 12 \times 2{\cdot}08 = 64{\cdot}92$

$A_y = 5t_{AE} = 5 \times 6{\cdot}66 = 33{\cdot}30$ KN.

Check : $\quad A_y - C_y = 33{\cdot}30 - 33{\cdot}30 = 0$

$D_x = 12t_{DC} = 12 \times (-1{\cdot}25) = -15{\cdot}00$

Check : $\quad A_x + D_x - 50 = 64{\cdot}92 - 15{\cdot}00 - 50{\cdot}00 \approx 0$

Values of forces are given in Table.

Computer output of Example 3.15.

6·0	12·0	0·0	0·0	0·0	0·0	–1·0	0·0	0·0	0·0	0·0	0·0	0·0
5·0	0·0	0·0	0·0	0·0	0·0	0·0	–1·0	0·0	0·0	0·0	0·0	0·0
–10·0	0·0	0·0	0·0	0·0	0·0	0·0	0·0	1·0	0·0	0·0	0·0	0·0
0·0	–12·0	–6·0	0·0	0·0	0·0	0·0	0·0	0·0	0·0	0·0	0·0	0·0
0·0	0·0	5·0	10·0	0·0	0·0	0·0	0·0	0·0	0·0	0·0	0·0	0·0
0·0	0·0	–10·0	0·0	0·0	0·0	0·0	0·0	0·0	1·0	0·0	0·0	0·0
0·0	0·0	0·0	0·0	6·0	12·0	0·0	0·0	0·0	0·0	0·0	0·0	0·0
0·0	0·0	0·0	–10·0	–5·0	0·0	0·0	0·0	0·0	0·0	1·0	0·0	0·0
0·0	0·0	0·0	0·0	–10·0	0·0	0·0	0·0	0·0	0·0	0·0	1·0	0·0
0·0	0·0	0·0	0·0	0·0	12·0	0·0	0·0	0·0	0·0	0·0	0·0	–1·0
–6·0	0·0	6·0	0·0	6·0	0·0	0·0	0·0	0·0	0·0	0·0	0·0	0·0
–5·0	0·0	–5·0	0·0	5·0	0·0	0·0	0·0	0·0	0·0	0·0	0·0	0·0
–10·0	0·0	–10·0	0·0	–10·0	0·0	0·0	0·0	0·0	0·0	0·0	0·0	0·0

Force vector

0·0	0·0	0·0	0·0	0·0	0·0	0·0	0·0	0·0	0·0	–50·0	0·0	–50·0

Force No	Value
1 (t_{AE})	6·666667
2 (t_{AB})	2·083333
3 (t_{BE})	–4·166667
4 (t_{BC})	2·083333
5 (t_{CE})	2·500000
6 (t_{CD})	–1·250000
7 (Ax)	65·000000
8 (Ay)	33·333330
9 (Az)	66·666660
10 (Bz)	–41·666660
11 (Cy)	33·333330
12 (Cz)	25·000000
13 (Dx)	–15·000000

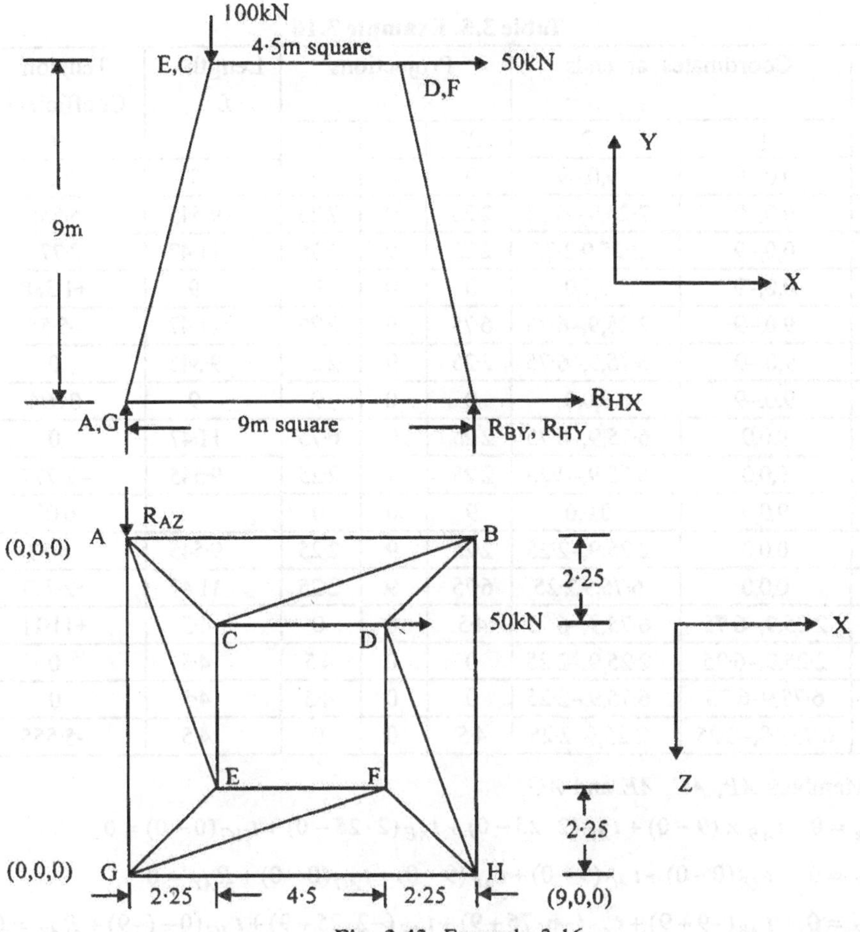

Fig. 3.43. Example 3.16.

Example 3.16. Find forces in all members of the space truss. Compute the reactions. Use the theorems of space truss given in section 3.7 as much as possible. (Fig. 3.43)

Solution. External Reactions are

$$R_{AY}, R_{AZ} = 2$$

$$R_{BX}, R_{BY}, R_{BZ} = 3$$

$$R_{GY}, R_{HX}, R_{HY} = \frac{3}{8}$$

No. of members = 16

No. of joint = 8

$$3j = 24 = 16 + 8$$

Hence the truss is statically determinate. From Fig. 3·43 it can be observed that at joint C F_{CA}, F_{CB}, and F_{CD} are in one plane while F_{CE} is out of plane similarly, at joint F, F_{FE} F_{FG} and F_{FH} are in one plane while F_{FD} is out of plane. Hence from conclusion 3 of section 3.7.1 F_{CE} and F_{FD} are zero.

Now consider joint D. at joint D. F_{DB}, F_{DH} and F_{DF} are coplanar. F_{CD} and 50 kN. load are out of plane but collinear. Hence from conclusion 2 of section 3.7.1. $F_{CD} = 50$ kN. Now F_{DF} is zero as already derived. Hence from conclusion 5, F_{DB} and F_{DH} must be zero. Hence, F_{CE}, F_{FD}, F_{DB} and F_{DH} are all zero and $F_{CD} = 50$ kN.

Table 3.5. Example 3.16

Members	Coordinates at ends		Projections			Length L	Tension Coefficient t	Force = $t \times L$
	1	2	X	Y	Z			
AB	0,0,–9	9,0,–9	9	0	0	9	–0·694	–6·25
AC	0,0,–9	2·25,9,–6·75	2·25	9	2·25	9·545	5·555	43·02
AE	0,0,–9	2·25,9,2·25	2·25	9	6·75	11·47	2·777	–31·86
AG	0,0,–9	0,0,0	0	0	9	9	+1·388	+12·50
BC	9,0,–9	2·25,9,–6·75	6·75	9	2·25	11·47	–5·55	–63·72
BD	9,0,–9	6·75,9,–6·75	2·25	9	2·25	9·545	0	0
BH	9,0,–9	9,0,0	0	0	9	9	0·694	6·246
HD	9,0,0	6·75,9,–6·75	2·25	9	6·75	11·47	0	0
HF	9,0,0	6·75,9,–2·25	2·25	9	2·25	9·545	–2·777	26·51
HG	9,0,0	0,0,0	9	0	0	9	0·00	0·00
GE	0,0,0	2·25,9,–2·25	2·25	9	2·25	9·545	–8·33	–79·54
GF	0,0,0	6·75,9,2·25	6·75	9	2·25	11·47	+2·777	31·85
CD	2·25,9,–6·75	6·75,9,–6·75	4·5	0	0	4·5	+11·11	50·00
CE	2·25,9,–6·75	2·25,9,–2·25	0	0	4·5	4·5	0	0
DF	6·75,9–6·75	6·75,9,–2·25	0	0	4·5	4·5	0	0
FE	6·75,+9,–2·25	2·25,9–2·25	4·5	0	0	4·5	–5·555	–25·00

Joint A Members *AB*, *AC*, *AE* and *AG*

$$\Sigma F_x = 0 \quad t_{AB} \times (9-0) + t_{AC}(2\cdot 25 - 0) + t_{AE}(2\cdot 25 - 0) + t_{AG}(0-0) = 0 \tag{1}$$

$$\Sigma F_Y = 0 \quad t_{AB}(0-0) + t_{AC}(9-0) + t_{AE}(9-0) + t_{AG}(0-0) + R_{AY} = 0 \tag{2}$$

$$\Sigma F_Z = 0 \quad t_{AB}(-9+9) + t_{AC}(-6\cdot 75+9) + t_{AE}(-2\cdot 25+9) + t_{AG}(0-(-9)) + R_{AZ} = 0 \tag{3}$$

Joint B Members *AB*, *BC*, *BD*, *BH*

$$\Sigma F_x = 0 \quad t_{BA}(0-9) + t_{BC}(2\cdot 25-9) + t_{BD}(6\cdot 75-9) + t_{BH}(9-9) + R_{BX} = 0 \tag{4}$$

$$\Sigma F_y = 0 \quad t_{BA}(0-0) + t_{BC}(9-0) + t_{BD}(9-0) + t_{BH}(0-0) + R_{BY} = 0 \tag{5}$$

$$\Sigma F_z = 0 \quad t_{BA}(-9+9) + t_{BC}(-6\cdot 75+9) + t_{BD}(-6\cdot 75+9) + t_{BH}(0+9) + R_{BZ} = 0 \tag{6}$$

Joint C Members *CA*, *CB*, *CD*, *CE*.

$$\Sigma F_x = 0 \quad t_{CA}(0-2\cdot 25) + t_{CB}(9-2\cdot 25) + t_{CD}(6\cdot 75-2\cdot 25) + t_{CE}(2\cdot 25-2\cdot 25) = 0 \tag{7}$$

$$\Sigma F_y = 0 \quad t_{CA}(0-9) + t_{CB}(0-9) + t_{CD}(9-9) + t_{CE}(9-9) = 0 \tag{8}$$

$$\Sigma F_z = 0 \quad t_{CA}(-9+6\cdot 75) + t_{CB}(-9+6\cdot 75) + t_{CD}(-6\cdot 75+6\cdot 75) + t_{CE}(-2\cdot 25-2\cdot 25) = 0 \tag{9}$$

Joint D Members *DC*, *DB*, *DF*, *DH*.

$$\Sigma F_x = 0 \quad t_{DC}(2\cdot 25-6\cdot 75) + t_{DB}(9-6\cdot 75) + t_{DF}(6\cdot 75-6\cdot 75) + t_{DH}(9-6\cdot 75) + 50 = 0 \tag{10}$$

$$\Sigma F_y = 0 \quad t_{DC}(9-9) + t_{DB}(0-9) + t_{DF}(9-9) + t_{DH}(0-9) = 0 \tag{11}$$

$$\Sigma F_z = 0 \quad t_{DC}(-6\cdot 75+6\cdot 75) + t_{DB}(-9+6\cdot 75) + t_{DF}(-2\cdot 25+6\cdot 75) + t_{DH}(0+6\cdot 75) = 0 \tag{12}$$

Joint E Members *EC*, *EA*, *EG* and *EF*.

$$\Sigma F_x = 0 \quad t_{EC}(2\cdot 25-2\cdot 25) + t_{EA}(0-2\cdot 25) + t_{EG}(0-2\cdot 25) + t_{EF}(6\cdot 75-2\cdot 25) = 0 \tag{13}$$

$$\Sigma F_y = 0 \quad t_{EC}(9-9)+t_{EA}(0-9)+t_{EG}(0-9)+t_{EF}(9-9)-100=0 \tag{14}$$

$$\Sigma F_z = 0 \quad t_{EC}(-6\cdot 75+2\cdot 25)+t_{EA}(-9+2\cdot 25)+t_{EG}(0+2\cdot 25)+t_{EF}(-2\cdot 25+2\cdot 25)=0 \tag{15}$$

Joint F Members *FE, FG, FH* and *FD*.

$$\Sigma F_y = 0 \quad t_{EF}(2\cdot 25-6\cdot 75)+t_{FG}(0-6\cdot 75)+t_{FH}(9-6\cdot 75)+t_{FD}(6\cdot 75-6\cdot 75)=0 \tag{16}$$

$$\Sigma F_y = 0 \quad t_{FE}(9-9)+t_{FG}(0-9\cdot 00)+t_{FH}(0-9)+t_{FD}(9-9)=0 \tag{17}$$

$$\Sigma F_z = 0 \quad t_{FE}(-2\cdot 25+2\cdot 25)+t_{FG}(0+2\cdot 25)+t_{FH}(0+2\cdot 25)+t_{FD}(-6\cdot 75+2\cdot 25)=0 \tag{18}$$

Joint G Members *GA, GE, GF, GH.*

$$\Sigma F_x = 0 \quad t_{GA}(0-0)+t_{GE}(2\cdot 25-0)+t_{GF}(6\cdot 75-0)+t_{GH}(9-0)=0 \tag{19}$$

$$\Sigma F_y = 0 \quad t_{GA}(0-0)+t_{GE}(9\cdot 00-0)t_{GF}(9\cdot 00-0)+t_{GH}(0-0)+R_{GY}=0 \tag{20}$$

$$\Sigma F_z = 0 \quad t_{GA}(-9-0)+t_{GE}(-2\cdot 25-0)+t_{GF}(-2\cdot 25-0)+t_{GH}(0-0)=0 \tag{21}$$

Joint H Members *HG, HF, HD* and *HB*.

$$\Sigma F_x = 0 \quad t_{HG}(0-9)+t_{HF}(6\cdot 75-9)+t_{HD}(6\cdot 75-9)+t_{HB}(9-0)+R_{HX}=0 \tag{22}$$

$$\Sigma F_y = 0 \quad t_{HG}(0-0)+t_{HF}(9\cdot 00-0)+t_{HD}(9-0)+t_{HB}(0-0)+R_{HY}=0 \tag{23}$$

$$\Sigma F_z = 0 \quad t_{HG}(0-0)+t_{HF}(-2\cdot 25-0)+t_{HD}(-6\cdot 75-0)+t_{HB}(-9-0)=0 \tag{24}$$

From eqn. (7) $\quad -2\cdot 25\, t_{CA}+6\cdot 75\, t_{CB} = -\dfrac{50}{4\cdot 5}\times 4\cdot 5$

eqn. (8) $\quad -9\, t_{CA}-9t_{CB}=0$

$t_{CA}=-t_{CB}$

Substituting, $\quad +2\cdot 25t_{CB}+6\cdot 75t_{CB}=-50$

$9t_{CB}=-50$

$t_{CB}=50/9$

$F_{CB}=-\dfrac{50}{9}\times 11\cdot 47=-63\cdot 72\text{kN}.$

$t_{CA}=\dfrac{50}{9}\; F_{CA}=\dfrac{50}{9}\times 9\cdot 545=$ **53·02kN.**

From eqn. (14) $\quad -9t_{EA}-9t_{EG}=100$

From eqn. (15) $\quad -6\cdot 75t_{EA}+2\cdot 25t_{EG}=0$

Solving $\quad -36t_{EA}=100$

or $\quad t_{EA}=-\dfrac{100}{36}$

$F_{EA}=-\dfrac{100}{36}\times 11\cdot 47=$ **– 31·86kN.**

Substituting $\quad t_{EG}=-\dfrac{75}{9}$

$F_{EG}=-\dfrac{75}{9}\times 9\cdot 545$

= **79·54kN.**

From eqn. (13) $\quad 4\cdot 50t_{EF}=2\cdot 25(t_{EA}+t_{EG})$

$$=2\cdot 25\left(-\frac{100}{36}-\frac{75}{9}\right)$$

or $\quad t_{EF} = -5 \cdot 555$

$F_{EF} = -25\text{kN}.$

From eqn. (17) $\quad -9(t_{FG} + t_{FH}) = 0$

$t_{FG} = -t_{FH}.$

From eqn. (16) $\quad -t_{FE}(4 \cdot 45) - 6 \cdot 75 t_{FG} + 2 \cdot 25 t_{FH} = 0$

or $\quad -6 \cdot 75 t_{FG} - 2 \cdot 25 t_{FG} = t_{FE}(4 \cdot 50)$

or $\quad t_{FG} = \dfrac{5 \cdot 555}{2}$

or $\quad F_{FG} = \dfrac{5 \cdot 555}{2} \times 11 \cdot 47 = \mathbf{31 \cdot 85kN.}$

From eqn. (19) $\quad 9t_{GH} = -(2 \cdot 25 t_{GE} + 6 \cdot 75 t_{GF})$

$$= -\left(2 \cdot 25 \times \left(-\frac{75}{9}\right) + 6 \cdot 75 \times \frac{5 \cdot 555}{2}\right)$$

$$= 0$$

or $\quad t_{GH} = 0 = F_{GH}$

From eqn. (24) $\quad -2 \cdot 25 t_{HF} - 9 t_{HB} = 0$

or $\quad 9t_{HB} = -2 \cdot 25\, t_{HF}$

or $\quad t_{HB} = -\dfrac{2 \cdot 25 \times 5 \cdot 555}{2 \times 9}$

or $\quad t_{HB} = \mathbf{0 \cdot 694kN.}$

From eqn. (21) $\quad -9t_{GA} = 2 \cdot 25(t_{GE} + t_{GF})$

or $\quad t_{GA} = -\dfrac{2 \cdot 25}{9}\left(-\dfrac{75}{9} + \dfrac{5 \cdot 555}{2}\right)$

$F_{GA} = 12 \cdot 50$

$9t_{AB} = -2 \cdot 25(t_{AC} + t_{AE})$

or $\quad t_{AB} = -\dfrac{2 \cdot 25}{9}\left(\dfrac{50}{9} - \dfrac{100}{36}\right)$

or $\quad t_{AB} = -0 \cdot 694$

$F_{AB} = -6 \cdot 25\text{kN}$

Reactions

$$R_{AY} = -9(t_{AC} + t_{AE})$$

$$= -9\left(\frac{50}{9} - \frac{100}{36}\right)$$

$$= -25\text{ kN}.$$

$$-R_{AZ} = t_{AC}(-6 \cdot 75 + 9) + t_{AE}(-2 \cdot 25 + 9) + 9 \times t_{AG}$$

$$= \frac{50}{9} \times 2 \cdot 25 + \left(-\frac{100}{36}\right) \times 6 \cdot 75 + 9 \times 1 \cdot 388$$

$$= 6 \cdot 242$$

or $R_{AZ} = -\mathbf{6 \cdot 242.}$

$$R_{BX} = 9t_{BA} + 6 \cdot 75 t_{BC}$$

$$= 9 \times (-0 \cdot 694) + 6 \cdot 75\left(-\frac{50}{9}\right)$$

$$= \mathbf{-43 \cdot 746kN.}$$

$$R_{BY} = -9t_{BC} = -9 \times \left(-\frac{50}{9}\right) = +50\text{kN}$$

$$-R_{BZ} = 2 \cdot 25 t_{BC} + 9t_{BH}$$

$$= 2 \cdot 25 \times \left(-\frac{50}{9}\right) + 9 \times 0 \cdot 694$$

$$= -6 \cdot 254$$

or $R_{BZ} = \mathbf{6 \cdot 254kN.}$

$$-R_{GY} = 9(t_{GE} + t_{FG})$$

$$= 9\left(-\frac{75}{9} + \frac{5 \cdot 555}{2}\right)$$

$$= -50 \cdot 00$$

$$R_{GY} = \mathbf{50 \cdot 00kN.}$$

$$-R_{HX} = -9t_{HG} - 2 \cdot 25 t_{HF}$$

$$= 0 - 2 \cdot 25 \times \left(-\frac{5 \cdot 555}{2}\right)$$

$$= +6 \cdot 25\text{kN.}$$

$$R_{HY} = -9t_{HF} = -9 \times \left(-\frac{5 \cdot 555}{.2}\right)$$

$$= \mathbf{25kN.}$$

Check the Equilibrium equations

$$\Sigma Y = 0 \quad R_{AY} + R_{GY} + R_{BY} + R_{HY} - 100$$

$$= -25 + 50 + 50 + 25 - 100$$

$$= 0.$$

$$R_{HX} + R_{BX} = -6 \cdot 25 + (-43 \cdot 746) + 50$$

$$= 0.$$

$$R_{AZ} + R_{BZ} = -6 \cdot 242 + 6 \cdot 25 \approx 0.$$

SUMMARY

1. A truss is a very common structure for constructing bridges, roofs, towers and lateral bracings of high rise buildings.

2. A truss is analysed by making three basic assumptions :—

(*a*) Members are connected at their ends by smooth friction less pins.

(*b*) Loads and reactions are applied to the truss at joints only.

(*c*) The centroidal axis of each member is straight and coincides with the line connecting the joint centres at the end of the member.

3. Two methods are usually utilised for analysis of truss —

1. Method of Joints
2. Method of sections

Both are useful hand methods for solution of truss.

4. A truss may be classified as

1. Simple
2. Compound
3. Complex

A simple truss is composed of a number of triangles developed by adding two bars to a basic triangle. A compound truss is formed by two simple trusses joined together by three non-parallel non-concurrent bars or with a common hinge and bar. Complex truss does not fall in any of the above two categories. A complex truss may not always be rigid. Critical form of a complex truss is a certain configuration which is non-rigid.

Two tests

1. Zero load test
2. Coefficient matrix determinant test

are available for checking the stability of a critical form of a truss.

5. Matrix methods can be utilised in finding the forces in different members of a truss. They can also be utilised to check the determinacy, indeterminacy and stability of a truss. Computer solution of truss problems is based on matrix methods. It can be semi-automatic when computer is used only for inversion of matrix. Fully automated procedure is explained in later chapters.

6. Space structures are three dimensional. Hence methods of plane truss solution can be extended to solve space trusses. For space trusses there are three equations of equilibrium, $\Sigma F_x = 0, \Sigma F_y = 0$ and $\Sigma F_z = 0$. For space frames, there are six equations of equilibrium $\Sigma F_x = 0$, $\Sigma F_y = 0$ and $\Sigma F_z = 0$ and $\Sigma M_x = 0, \Sigma M_y = 0$ and $\Sigma M_z = 0$. Each member has three unknown stress resultants, three direct forces and three moments.

Space truss can be solved by method of joints. Tension coefficient method is a very elegant tool for solving space trusses.

PROBLEMS

A. Theoretical

3.1. (*a*) What is a plane truss ? Obtain relationship between the number of joints and number of members of a stable plane truss.

(*b*) What are simple, compound and complex trusses? Illustrate by suitable examples.

(*c*) What is a space truss? Obtain a relationship between the number of members and number of joints of a space truss.

(*d*) What is meant by tension coefficient? Obtain equations for solution of a space truss by tension coefficient method.

B. Objective

3.2 (*i*) In a pin jointed truss the members may be subjected to bending if.

(*a*) The material of the truss does not obey Hooke's law.

(*b*) The truss is statically indeterminate.

(*c*) The loads are not applied at the nodes.

(*d*) There is support settlement. **[AMIE Winter'98]**

(*ii*) A pinjointed truss cannot deflect unless

(*a*) There is load on it.

(*b*) The members are stressed.

(*c*) There is change in lengths of members.

(*d*) The material is elastic **[AMIE Winter'98]**

(*iii*) Find the force in truss member 1 (Fig. P3.1)

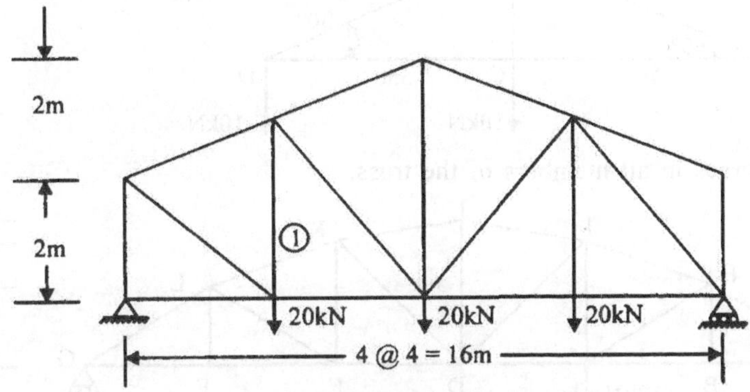

Fig. P 3.1 Problem 3.2 (*iii*).

C. Mathematical

3.3. Find forces in all members of the truss given in Fig. P. 3.2.

[5th Semester Final 1996 B.E. College (D.U.)]

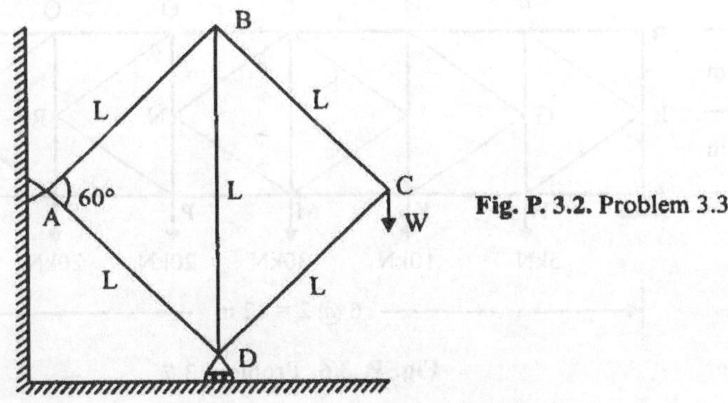

Fig. P. 3.2. Problem 3.3.

3.4. A Warren girder of 24m span has four equal panels of 6m each. All the triangles are equilateral. Find forces in all members of the truss. **[5th Semester Final 1996 B.E. College (D.U.)]**

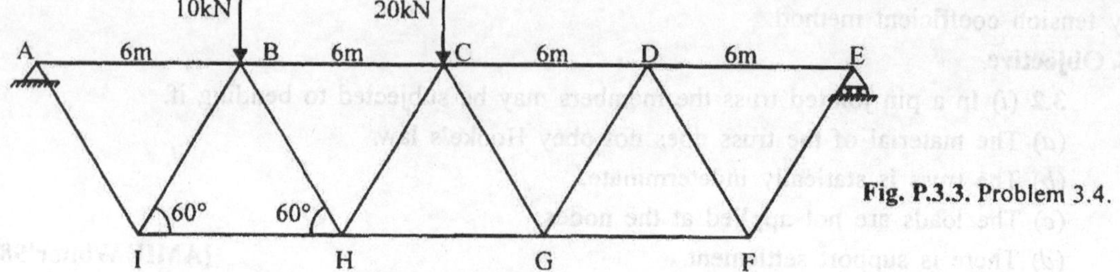

Fig. P.3.3. Problem 3.4.

3.5. Find forces in all members of the truss.

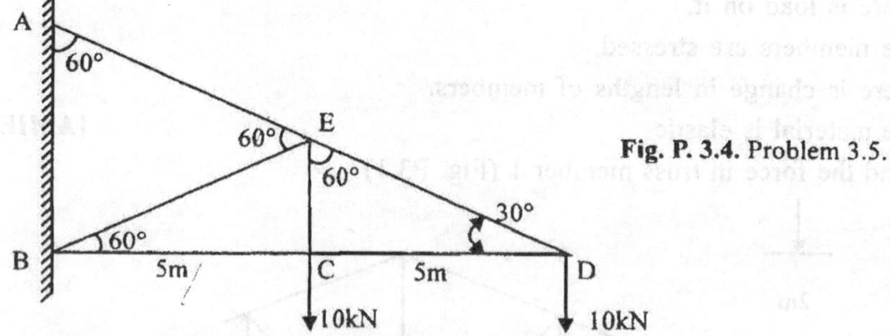

Fig. P. 3.4. Problem 3.5.

3.6. Find forces in all members of the truss.

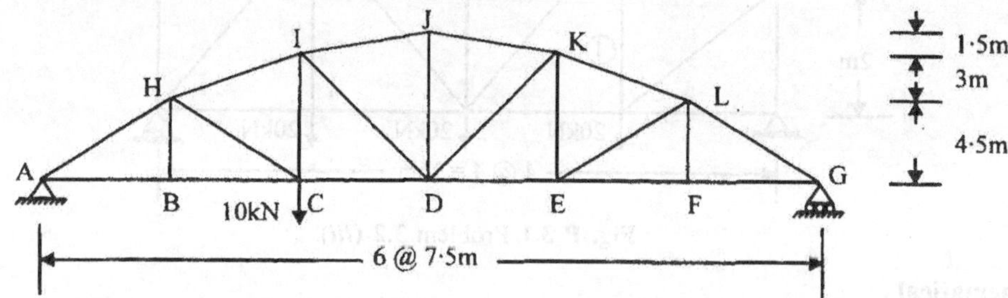

Fig. P. 3.5. Problem 3.6.

3.7. Find forces in the members *a*, *b*, *c*, and *d*. Indicate whether they are tensile or compressive

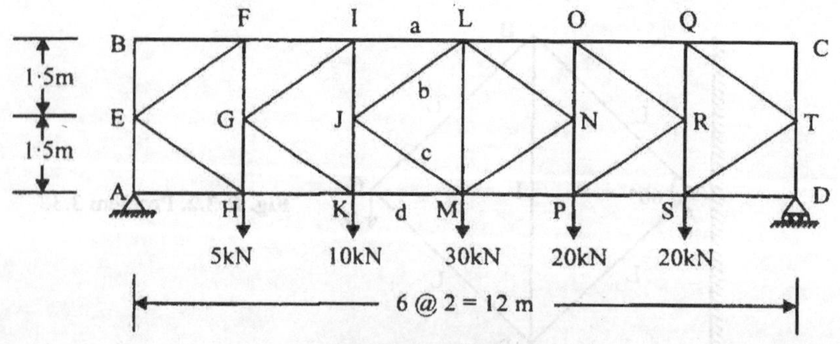

Fig. P. 3.6. Problem 3.7.

3.8. Find forces in all members of the truss shown in Fig. P. 3.7 and indicate whether they are tensile or compressive.

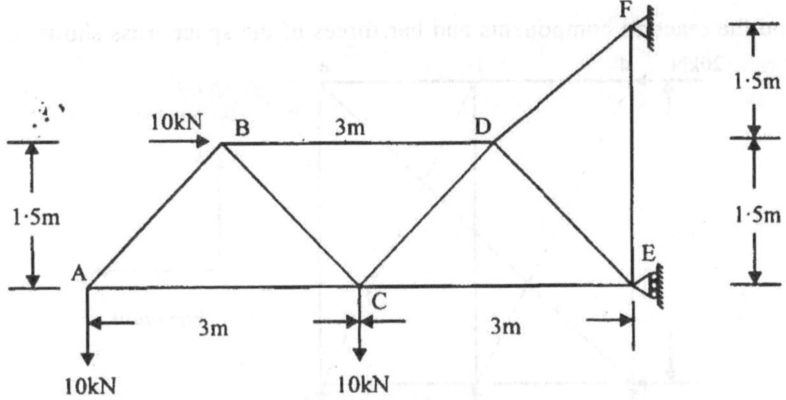

Fig. P. 3.7. Problem 3.8.

3.9. Fig. P. 3.8 shows two views of a tripod bracket formed of three bars *OA*, *OB* and *OC* connected by spherical hinges. Find the forces in the three bars due to a load of 20kN acting in the direction at *O* (downward).

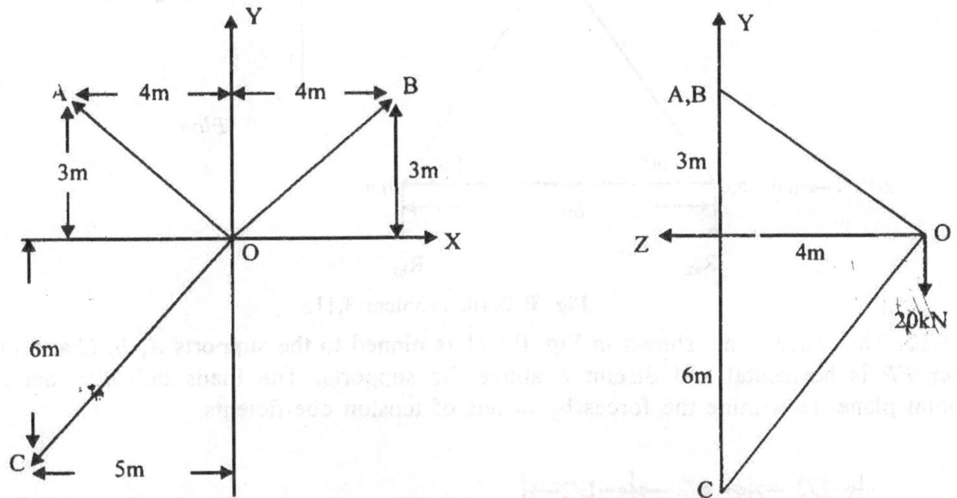

Fig. P 3.8. Problem 3.9.

3.10. Find P. 3.9 shows plan of a tripod the foot *A*, *B* and *C* being in the same horizontal plane and apex *D* being 7·5m above the plane. Horizontal loads 100kN and 150kN are applied at *D*. Find forces in the members.

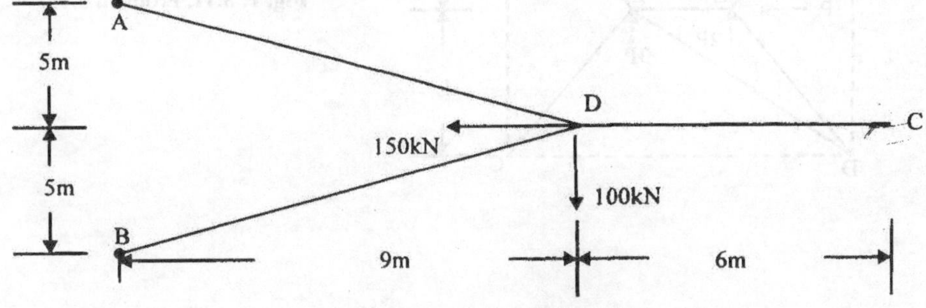

Fig. P. 3.9. Problem 3.10.

3.11. Find the reaction components and bar forces of the space truss shown in Fig. P. 3.10.

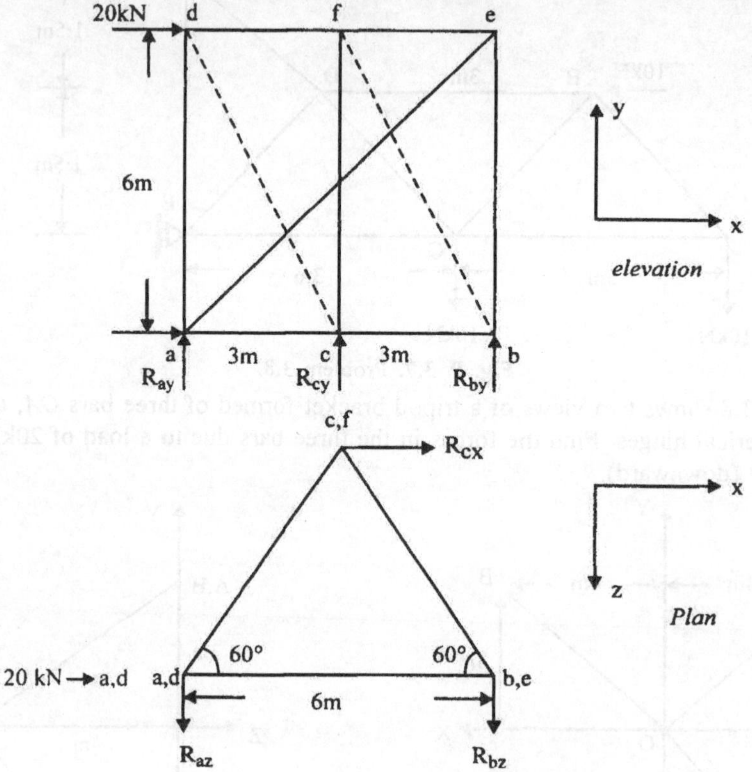

Fig. P. 3.10. Problem 3.11.

3.12. The space frame shown in Fig. P.3.11 is pinned to the supports *A*, *B*, *C* and *D*. The member *EF* is horizontal and distant *L* above the supports. The loads indicated act in the horizontal plane. Determine the forces by means of tension coefficients.

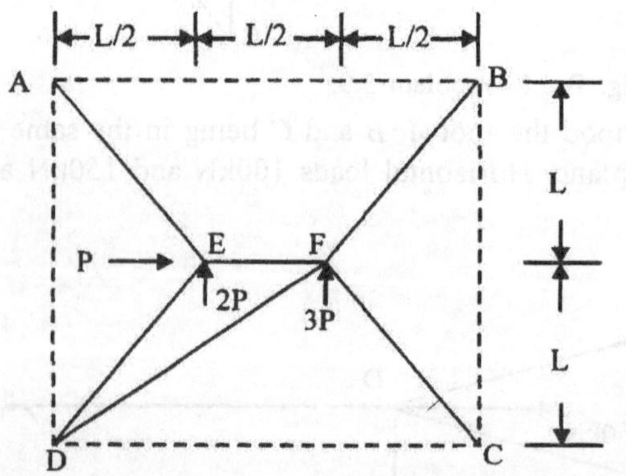

Fig. P. 3.11. Problem 3.12.

3.13. Find all reactions and forces in all members of the space truss shown in Fig. P. 3.12.

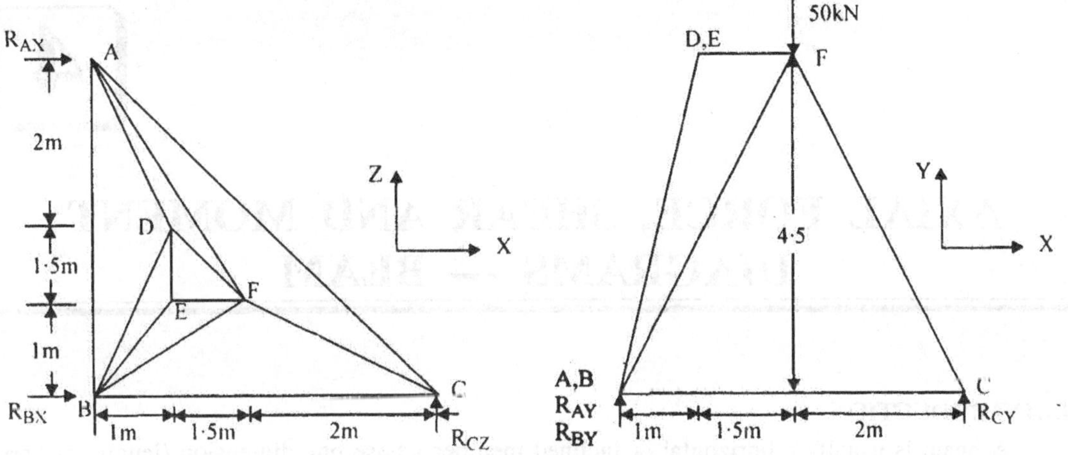

Fig. P. 3.12. Problem 3.13.

4

AXIAL FORCE, SHEAR AND MOMENT DIAGRAMS — BEAM

4.1. INTRODUCTION

A beam is usually a horizontal or inclined member whose one dimension (length) is very large in comparison to other two dimensions (cross section). Unlike truss whose members are assumed to be subjected to only axial forces, a beam is subjected to axial force, bending moment and shear. However, bending moment is predominant in beam analysis. In simplified analysis, it is assumed that a beam is

1. prismatic
2. of uniform cross section.
3. of such shape that shear centre and centroid are coincident. It is also assumed that the beam is acted on by only transverse loading and moment loading and that all the loads and reactions lie in a single plane which contains the centroidal axis of the member and one of the principal axis for every cross section. Under these conditions the beam will deflect in the plane of loading without twisting.

4.2 INTERNAL EQUILIBRIUM OF A FLEXURAL MEMBER

The external reactive forces acting on a flexural member have already been analysed considering external equilibrium. But the external reactions do not give a full picture of the internal forces acting on different points of a structure. It is therefore, necessary to cut the structure at different points and consider the equilibrium of the resulting free body diagram to obtain internal forces at different points.

Fig. 4.1 (*a*) Shows a simply supported beam which has been cut at point C.

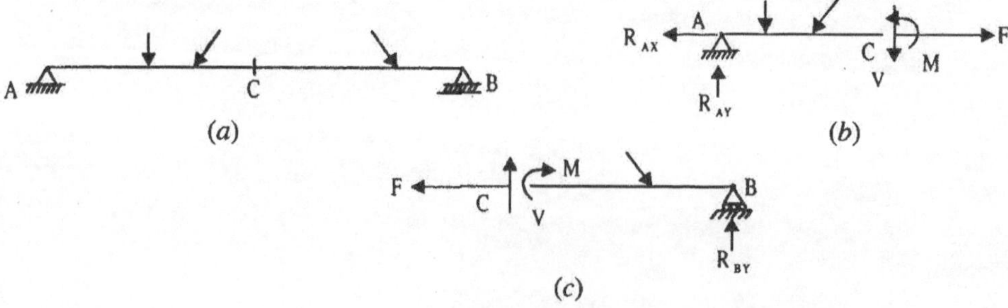

Fig. 4.1. Internal Equilibrium of Beam.

Fig. 4.1 (*b*) shows the left half of the cut while Fig. 4.1 (*c*) shows the right half.

As it is a planar structure, the three unknowns, shear force *V*, bending moment M and axial force F can be obtained from three equations of equilibrium applied to the portion AC or CB. i.e.

$\Sigma F_y = 0$ $\Sigma F_x = 0$ and $\Sigma M = 0$

As the two portions AC and CB are merged the internal forces cancel and we are left with original structure and loading.

From equilibrium considerations it is clear that

(*i*) F will be equal and opposite to the sum of horizontal forces to the left of point C.

(*ii*) V will be equal and opposite to the sum of vertical forces to be left of point C.

(*iii*) M will be equal and opposite to the sum of moments of all forces to the left of point C about C.

Similar reasoning can be applied to the right section. Hence,

(*i*) Axial force can be defined as the algebraic sum of all forces acting parallel to the longitudinal axis of the member on either side of the section. Tensile force is taken as positive.

(*ii*) Shear force at any section is the algebraic sum of all forces either to the left or to the right of the section. A shear force is considered positive when the left hand of the section goes up or the right hand goes down.

(*iii*) Bending moment at any cross section is the algebraic sum of moments of all the forces either to the left or to the right of the cross section about the section. A bending moment is taken to be positive when it produces compression on the top of the cross section and tension at the bottom. The cross section then appears to sag and it is known as sagging moment. Similarly bending moment which produces tension on the top and compression at the bottom and produces a hogging shape is negative moment and is also called hogging moment Fig. 4.2 shows all the sign conventions.

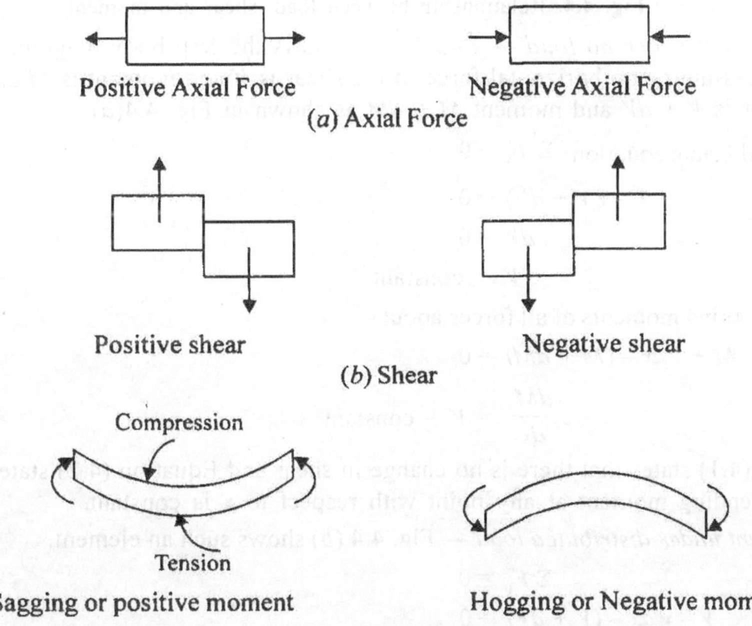

Fig. 4.2. Sign convention for A.F., B.M. and Shear.

4.3 RELATIONSHIP BETWEEN LOAD, SHEAR AND MOMENT

A beam may be loaded by different types of loading at different points. Fig 4.3 shows a simply supported beam subjected to different types of loading at different parts.

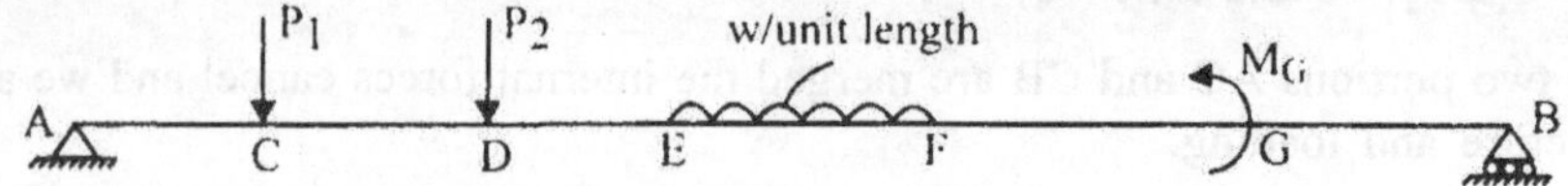

Fig. 4.3. Beam with different types of loads.

If we consider an elementary length '*dx*' between C and D, it is under no transverse load. An element *dx* between E and F is subjected to distributed load. An element at C or D is subjected to a concentrated load. Similarly an element at G is subjected to moment M. Free body diagrams of these elements are shown in Fig. 4.4 (*a*), (*b*), (*c*) and (*d*) respectively.

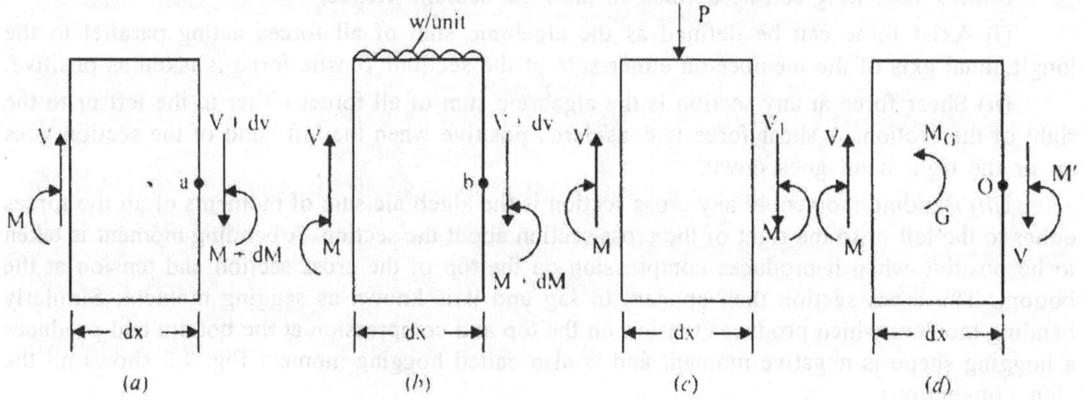

Fig. 4.4. Relationship between load, shear and moment.

(*i*) *An element under no load* — Fig. 4.4 (*a*) shows the free body diagram of an element under no load. Assuming no horizontal force, if the shear is V and moment is M on the left side, at $x + dx$, shear is $V + dV$ and moment $M + dM$ as shown in Fig. 4.4(a)

From equilibrium equation $\Sigma F_y = 0$

$$V - (V + dV) = 0$$

or $$dV = 0 \quad \text{...(4.1)}$$

That is, $$V = \text{constant}$$

Similarly taking moments of all forces about

'*a*', $$M + Vdx - (M + dM) = 0$$

or $$\frac{dM}{dx} = V = \text{constant} \quad \text{...(4.2)}$$

Equation (4.1) states that there is no change in shear and Equation (4.2) states that the rate of change of bending moment at any point with respect to x is constant.

(*ii*) *Element under distributed load* — Fig. 4.4 (*b*) shows such an element.

From $$\Sigma F_y = 0$$

$$V - wdx - (V + dV) = 0$$

or $$dV = -wdx$$

or $$\frac{dV}{dx} = -w \qquad ...(4.3)$$

From $\Sigma M = 0$ about point 'b'

$$M + V.dx - w.dx.\frac{dx}{2} - (M + dM) = 0$$

Neglecting the small term $\frac{wdx^2}{2}$

we get $$\frac{dM}{dx} = V \qquad .... (4.4)$$

Equation (4.3) states that the rate of change of shear with respect to x is equal to the intensity of load but with opposite sign. Similarly Equation (4.4) states that the rate of change of B.M. at any point is equal to the shear force at that point.

(*iii*) *Element under concentrated load* — Fig. 4.4 (*c*) As the load is concentrated and assumed to act on an infinitesimal area there will be no moment difference between the left and right of the concentrated load. This means same moment M acts on the left and right of load P. As regards shear there will be an abrupt change because of the load.

Putting $$\Sigma F_y = 0$$

$$V - P - V_1 = 0$$

or $$V_1 = V - P$$

As $\frac{dM}{dx} = V$, there will be an abrupt change in the derivative $\frac{dM}{dx}$ at the point of application of concentrated load. Fig. 4.4. (*c*)

(*iv*) *Element under moment load* — Fig. 4.4 (*d*). Moment M_G is acting at the point G. As the distance between the two adjacent sections becomes smaller and smaller there will be no shear difference between the left and right of M_G. However, there will be an abrupt change of moment equal to M_G between the left and right of the section.

Putting $$\Sigma M = 0 \text{ about point 'O'}$$

$$M - M_G - M' = 0$$

or $$M' = M - M_G$$

Significance of the equations derived –

(*i*) $\frac{dV}{dx} = -w$. The relation implies that the slope of the shear curve at any point is equal to the negative value of the ordinate of the load diagram applied to the beam at that point. From this it follows that

(*a*) For a segment under no load, the slope of the shear curve is zero. The shear diagram is, therefore, a line parallel to the beam axis.

(*b*) When the load intensity is uniform the slope of the curve is constant. The shear curve is therefore, a sloping line.

(*c*) Under a concentrated load, the load intensity is infinite which means slope of the shear curve is infinite i.e. the line is vertical to the beam axis. There will be discontinuity in shear curve and a sudden change in shear between right and left side of the load point occurs by an amount equal to the load.

(*d*) With distributed load w, the change in shear between two points a and b, $dV = -wdx$

The difference between V_a and V_b

$$V_b - V_a = -\int_a^b w.dx$$

= (area of the load diagram between a and b)

(*iii*) $\frac{dM}{dx} = V$. The equation means that slope of the bending moment curve at any point is equal to the ordinate of the S.F. diagram at that point. From this it follows that

(*a*) If the shear force is constant, the B.M. will be a straight line.

(*b*) If the shear force varies in any way the B.M. will be a curved line.

(*c*) At a concentrated load there will be an abrupt change in S.F. and hence an abrupt change in the slope of the bending moment curve at that point. The moment curve will have two different slopes on two sides of the point.

(*d*) When S.F. is zero $\frac{dM}{dx} = 0$ which is a test of maximum or minimum value of the bending moment. The maximum occurs when shear changes from positive to negative and the minimum in the reverse manner.

(*e*) For concentrated load, the maximum bending moment occurs under a particular load as only under the particular load will S.F. change sign.

(*f*) From the relation $\frac{dM}{dx} = V$

$$dM = V dx$$

or $$M_b - M_a = \int_a^b V\,dx$$

= area of the shear force diagram between b and a.

4.4 DRAWING AXIAL FORCE, S.F. AND B.M. DIAGRAMS

The easiest way of computing the axial force, B.M. and S.F. diagrams is to make cuts at appropriate sections, draw the free body diagram (FBD) and use equilibrium equations to compute the values. The draw the complete curve sometimes it is necessary to express the function in the form of an equation and derive necessary values at required poits. Axial force, S.F. and B.M. diagrams are now drawn for two fundamental cases of cantilever and simply supported beams under three basic types of loadings namely, moment, concentrated load and distributed load.

1. Cantilever (Fig. 4.5)

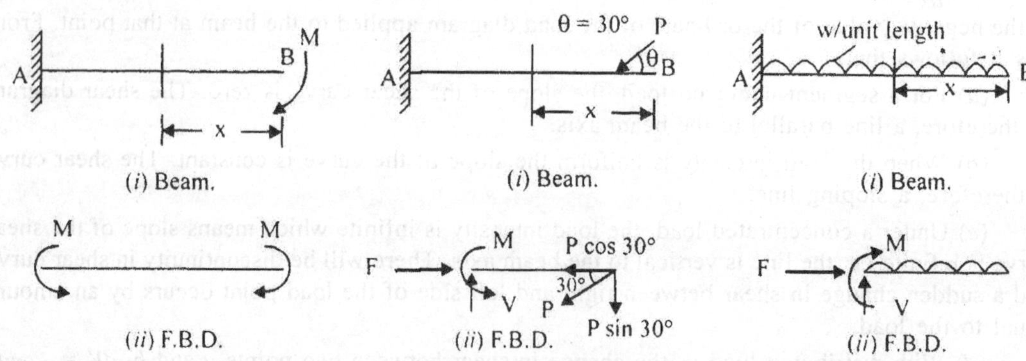

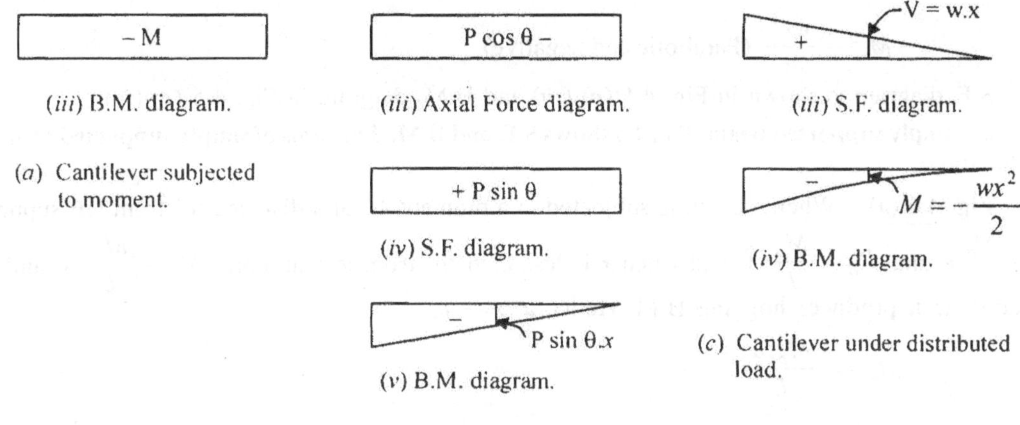

(b) Cantilever subjected to a concentrated load.

Fig. 4.5. Cantilever under different types of loading.

Explanations

Fig. 4.5 (*a*) (*i*) shows a cantilever subjected to moment M at the free end. Fig. 4.5 (*a*) (*ii*) shows the FBD of a section at a distance *x* from the right end. It is clear that for equilibrium of the section, an anticlockwise moment *M* equal to *M* is required and it is the same for any value of *x*. Hence there is no shear force or axial force in the section. It is subjected to only same hogging B.M. M_H in all sections.

Hence the diagram is rectangular and negative.

Fig. 4.5 (*b*) (*i*) shows a cantilever subjected to an inclined concentrated load at the free end. The free body diagram at a distance *x* is shown in Fig. 4.5 (*b*) (*ii*). From equilibrium considerations it is obvious, when θ = 30°.

$$F = P\cos\theta = \frac{P\sqrt{3}}{2}$$

$$V = P\sin\theta = \frac{P}{2}$$

$$M = P\sin\theta\cdot x = \frac{P}{2}\cdot x$$

The axial force diagram is shown in Fig. 4.5 (*b*) (*iii*)

It is negative as the section is under compression. Fig. 4.5 (*b*) (*iv*) gives the shear force diagram with a constant value of $\frac{P}{2}$. S.F. is positive as left hand goes up and right hand goes down. Fig. 4.5 (*b*) (*v*) gives the B.M. diagram.

$$M = \frac{P}{2}\cdot x, \text{ when } \theta = 30°$$

As the expression varies with *x*, the diagram is triangular. The B.M. tends to give tension on the top, hence negative.

Fig. 4.5 (*c*) (*i*) shows a cantilever under uniformly distributed load. Fig. 4.5 (*c*) (*ii*) shows the FBD from the right end at a distance '*x*'. From equilibrium considerations it is clear that

$F = 0$ (No. axial force)

$V = w.x$ (*S. F* is linear)

$$M = -\frac{wx^2}{2} \text{ (Parabolic and negative)}$$

S.F. diagram is shown in Fig. 4.5 (*c*) (*iii*) and B.M. diagram in Fig. 4.5 (*c*) (*iv*)

2. Simply supported beam. Fig. 4.6 shows S.F. and B.M. diagrams of simply supported beams.

Explanations

Fig. 4.6 (a) – When a beam is subjected to a moment M_a at a distance '*a*' from left support $R_A = \frac{M_a}{L} \downarrow$ and $R_B = \frac{M_a}{L} \uparrow$ and when *x* is less than '*a*' from left support. $M_x = \frac{M_a}{L} \cdot x$ and is negative as it produces hogging B.M. Hence at $x = a$,

$$M_x = -\frac{M_a . a}{L}$$

When $x > a$ we consider *FBD* of right half. Taking *B* as origin $M_x = R_B . x = +\frac{M_a}{L} \cdot x$ as upward reaction produces sagging moment. This is shown in Fig. 4.6 (*a*) (*v*).

At $x = b$ $\quad M_x = \frac{M_a \cdot b}{L}$. Observe that the sum of two moments

$$= M_a \cdot \frac{a}{L} + \frac{M_a \cdot b}{L}$$

$$= M_a$$

Similarly as shown in Fig. 4.6 (*a*) (*iv*)

for portion *a* S.F. is $-\frac{M_a}{L}$

for portion *b* S.F. is $-\frac{M_a}{L}$

Fig. 4.6 (*b*) When a load P acts at a distance '*a*' from *A* and *b* from B

$$R_A = \frac{Pb}{L} \quad \text{and} \quad R_B = \frac{P_a}{L}$$

F.B.D. when $x < a$ is shown in Fig. 4.6 (*b*) (*ii*), when $x < b$ with *B* as origin. F.B.D. is shown in Fig. 4.6 (*b*) (*iii*). From the F.B.D. it is clear that

Axial force $F = 0$

S.F. $V = \frac{Pb}{L}$ (positive) for $x < a$

$\frac{Pa}{L}$ (Negative) for $x < b$

This is shown in Fig. 4.6 (*b*) (*iv*)

B.M. $M = \frac{P.b.x}{L}$ at a distance *x* when $x < a$

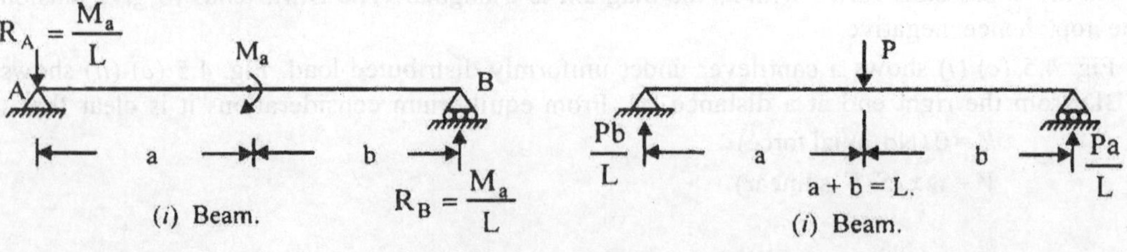

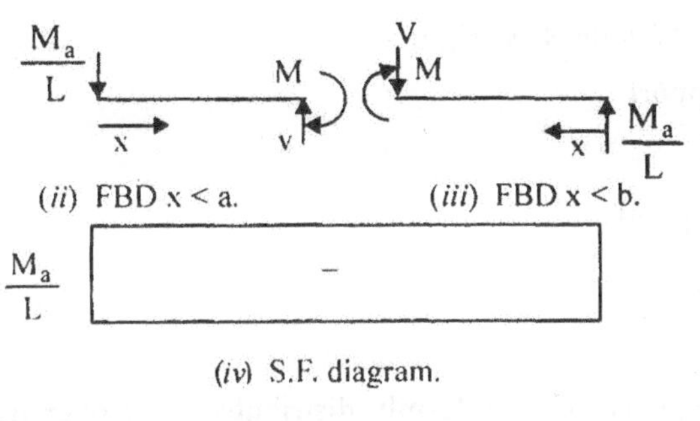

(*ii*) FBD $x < a$. (*iii*) FBD $x < b$.

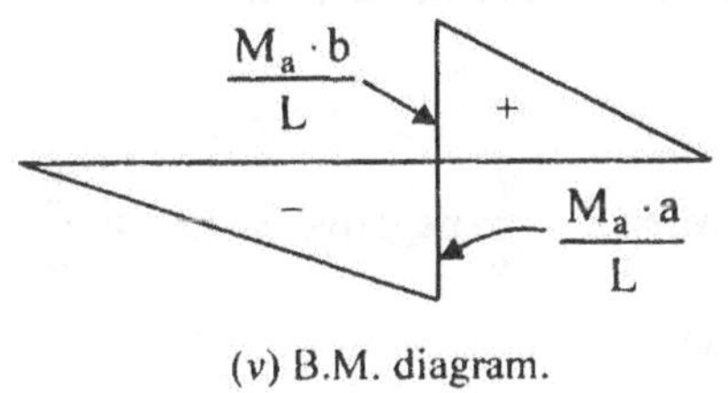

(*iv*) S.F. diagram.

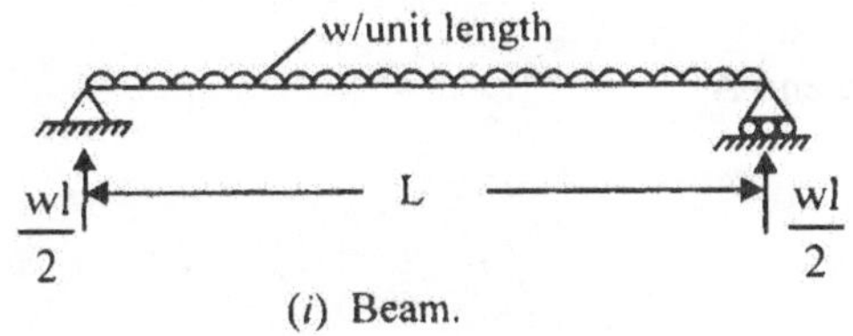

(*v*) B.M. diagram.

(*a*) Simply supported beam subjected to moment.

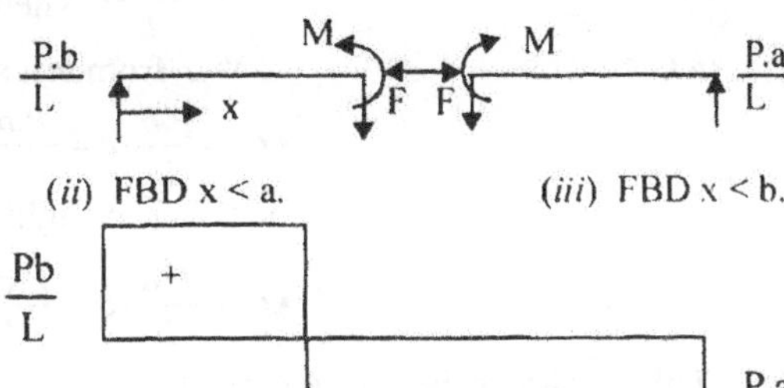

(*ii*) FBD $x < a$. (*iii*) FBD $x < b$.

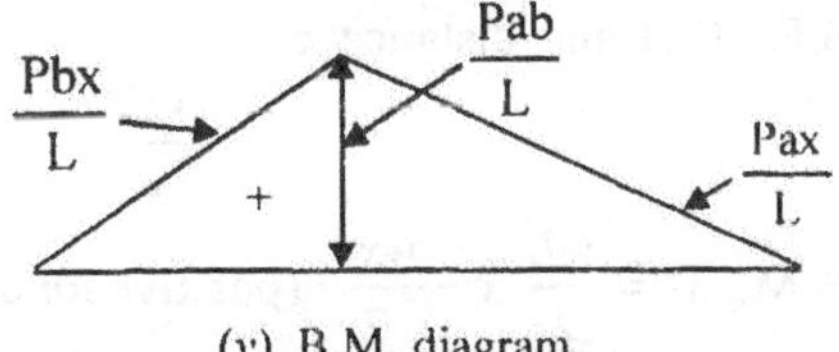

(*iv*) S.F. diagram.

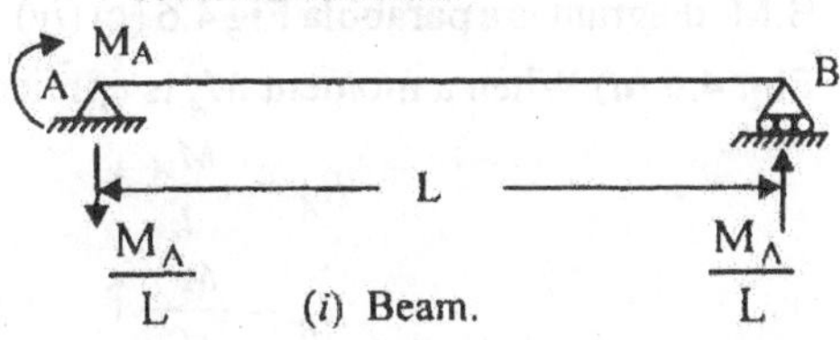

(*v*) B.M. diagram.

(*b*) Simply supported beam subjected to a concentrated load.

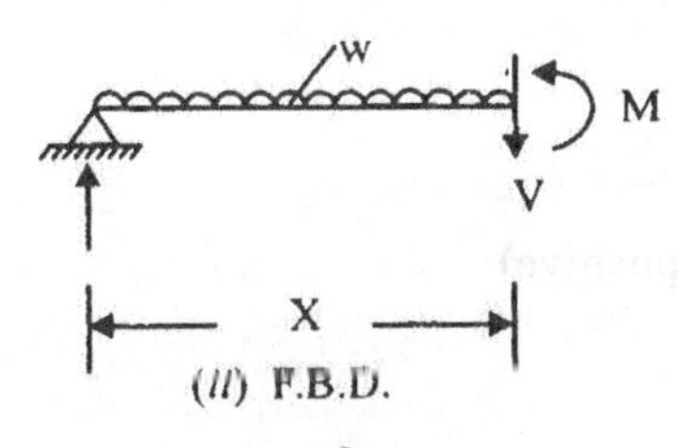

(*i*) Beam.

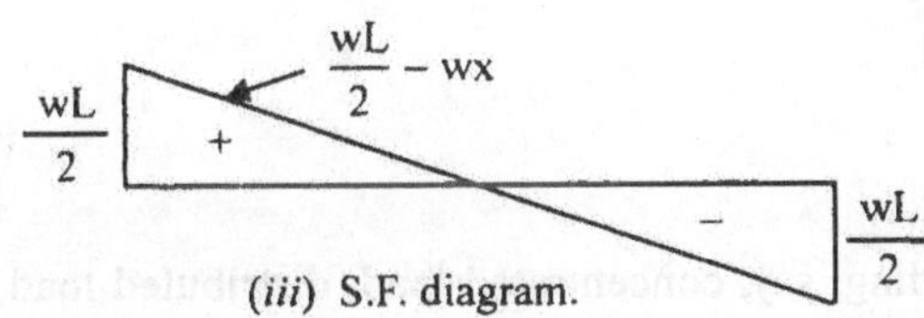

(*ii*) F.B.D.

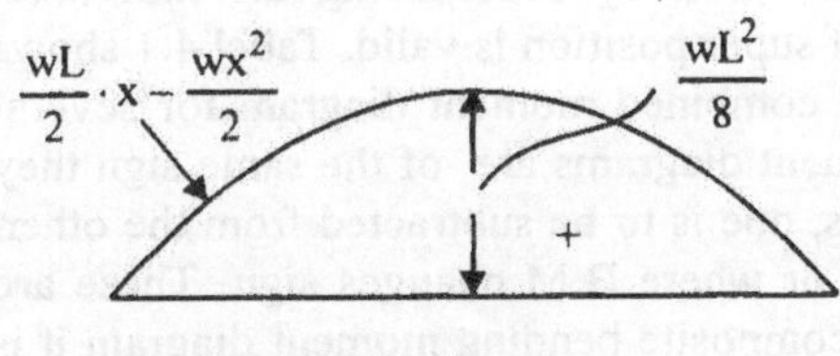

(*iii*) S.F. diagram.

$\frac{wL}{2} \cdot x - \frac{wx^2}{2}$ $\quad \frac{wL^2}{8}$ $\quad +$

(*iv*) B.M. diagram.

(*c*) Simply supported beam with uniformly distributed load.

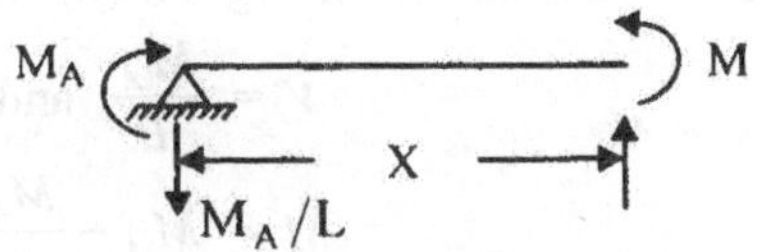

(*i*) Beam.

M_A $\quad M$ $\quad X$ $\quad M_A/L$

(*ii*) F.B.D.

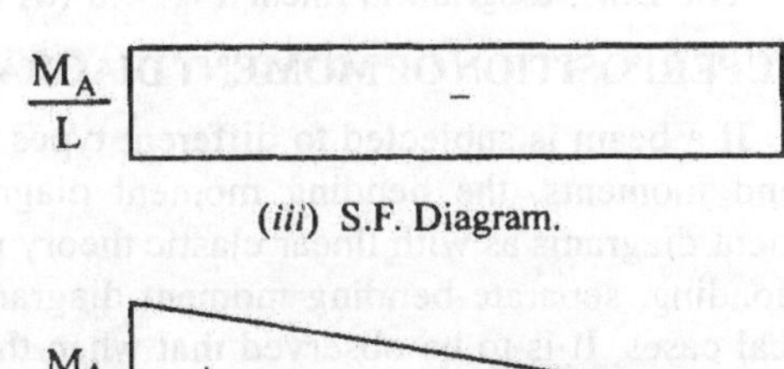

(*iii*) S.F. Diagram.

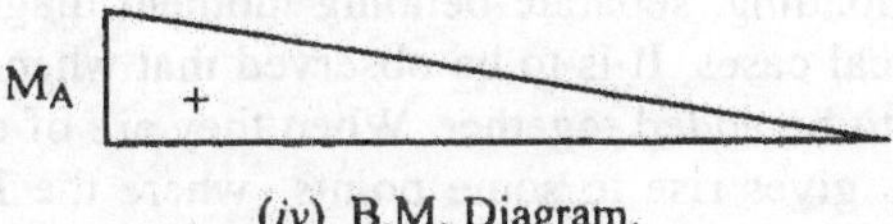

(*iv*) B.M. Diagram.

(*d*) Simply supported beam under end moment M_A.

Fig. 4.6. S.F. and B.M. diagrams of simply supported beams.

$$= \frac{P.a.x}{L} \text{ when } x < b \text{ with } B \text{ as origin.}$$

At $x = a$ from left support

$$M = \frac{P.b}{L} \cdot a = \frac{P.a.b}{L}$$

$x = b$ from right support

$$M = \frac{P.a}{L} \cdot b = \frac{P.a.b}{L}$$

This is shown in Fig. 4.6 (*b*) (*v*)

Fig. 4.6 (*c*). A simply supported beam subjected to a uniformly distributed load over its full span. FBD at a distance x is shown in Fig. 4.6 (*c*) (*ii*). It can be seen that

Axial force $F = 0$

S.F., V at any distance x

$$= \frac{wL}{2} - wx \quad \text{(Positive when } x < \frac{L}{2} \text{ negative when } x > \frac{L}{2})$$

B.M., $M = \frac{wL}{2}.x - \frac{wx^2}{2}$ (positive for all values of x)

From the equations it can be seen that

S.F. diagram is a straight line Fig. 4.6 (*c*) (*iii*)

B.M. diagram is a parabola Fig 4.6 (*c*) (*iv*)

Fig. 4.6 (*d*) When a moment M_A is applied at the end A

$$R_A = \frac{M_A}{L} \downarrow$$

$$R_B = \frac{M_A}{L} \uparrow$$

From the FBD shown in Fig 4.6 (*d*) (*ii*)

$$V = \frac{M_A}{L} \text{ and S.F. is negative}$$

$$M = M_A - \frac{M_A}{L} \cdot x \text{ (concave and positive)}$$

at $x = 0$ $M = M_A$

at $x = L$ $M = 0$

The S.F. diagram is shown in Fig. 4.6 (*d*) (*iii*)

The B.M. diagram is linear Fig. 4.6 (*d*) (*iv*)

4.5 SUPERPOSITION OF MOMENT DIAGRAMS

If a beam is subjected to different types of loading, say, concentrated load, distributed load or end moments, the bending moment diagram can be drawn by superposing the individual moment diagrams as with linear elastic theory principle of super position is valid. Tabel 4.1 shows the loading, separate bending moment diagrams and the combined moment diagram for several typical cases. It is to be observed that when the two moment diagrams are of the same sign they are to be added together. When they are of opposite signs, one is to be subtracted from the other. This gives rise to some points where the B.M is zero or where B.M changes sign. These are known as points of inflexion or contraflexure. From the composite bending moment diagram it is possible to find out the regions where B.M is positive, zero or negative.As the clastic deformations of beam are primarily caused by bending, it is possible to easily sketch the deflected elastic curve.

Table 4.1 Composite B.M. diagram

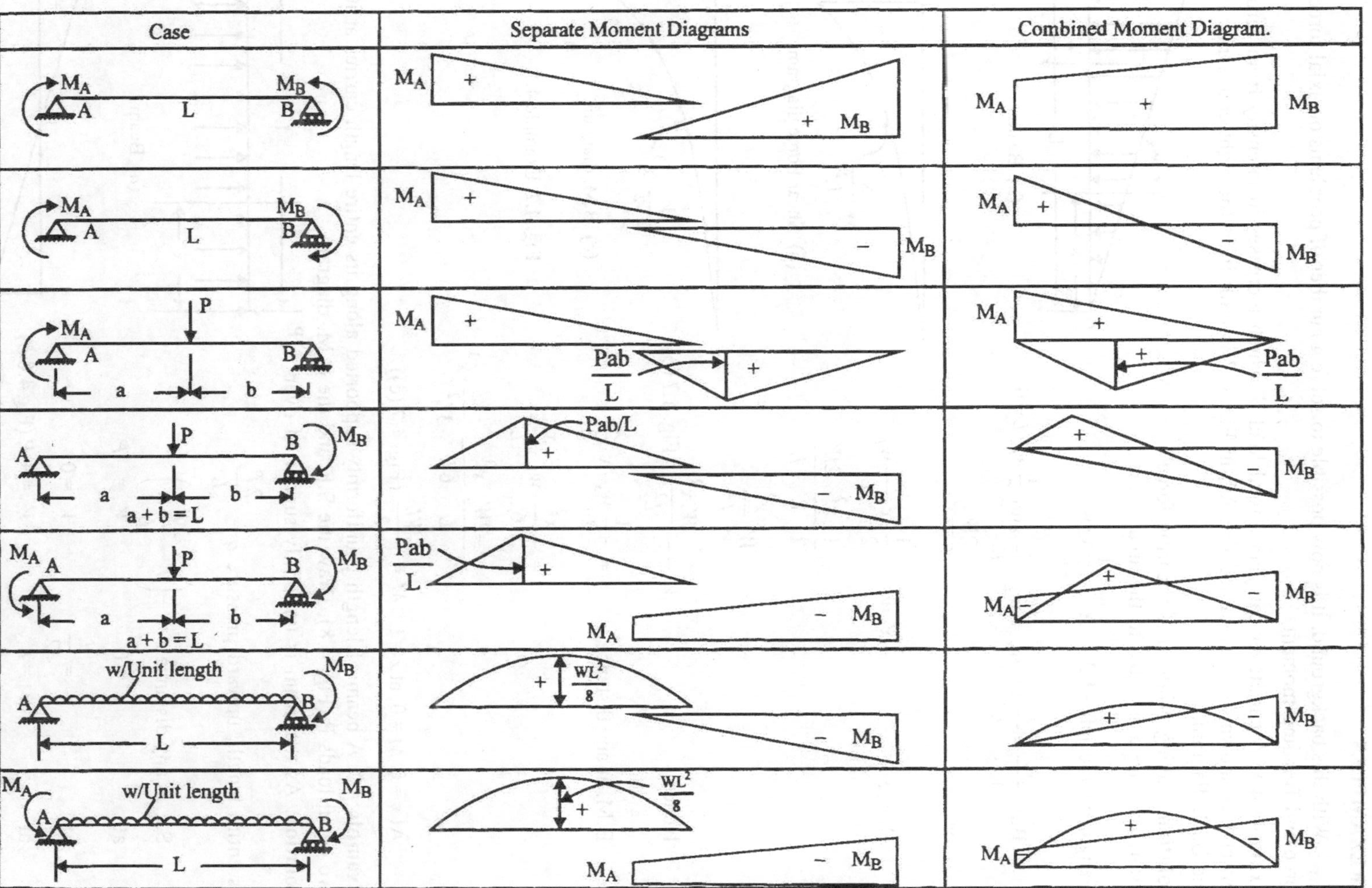

4.6 EXAMPLES

With this background, it is now possible to solve a number of problems on axial force, shear force and Bending moment.

Example 4.1. A cantilever carrying a total load W which increases in intensity from zero at the left end as shown by the inclined line AC in Fig. 4.7 is built in at the right end. Draw the S.F. and B.M. diagram.

Solution. Taking origin at A shearing force at any distance x is equal to the area of the triangle on the left of $X - X = \frac{1}{2} x \cdot w_x$

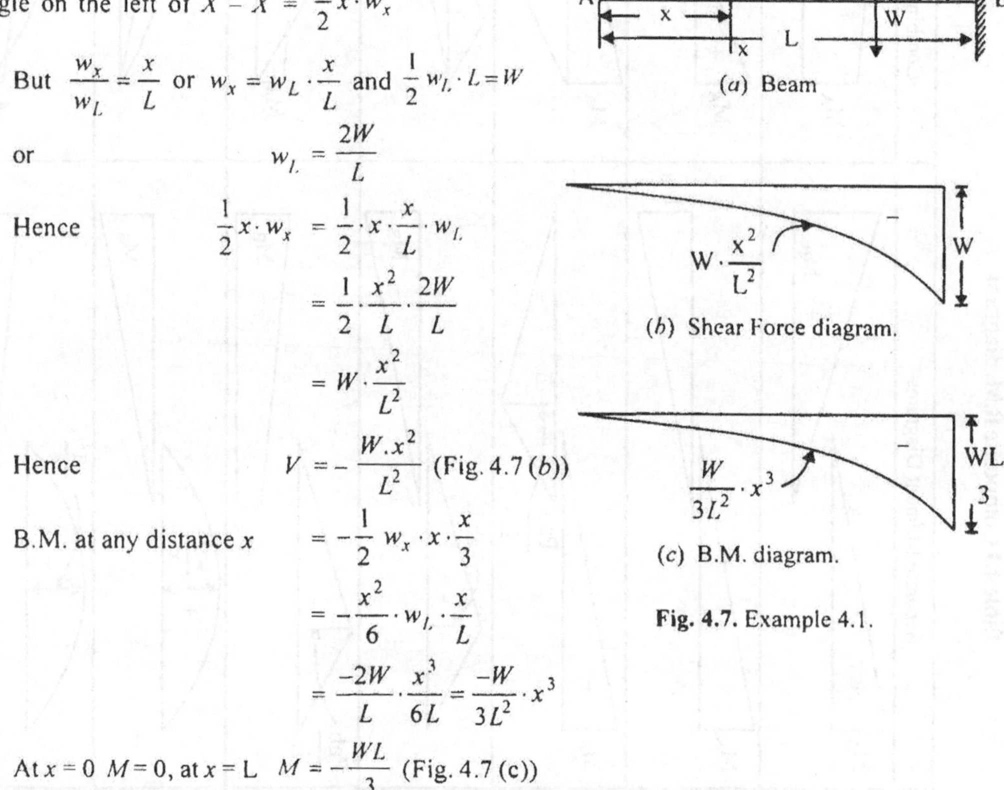

But $\frac{w_x}{w_L} = \frac{x}{L}$ or $w_x = w_L \cdot \frac{x}{L}$ and $\frac{1}{2} w_L \cdot L = W$

or $$w_L = \frac{2W}{L}$$

Hence $$\frac{1}{2} x \cdot w_x = \frac{1}{2} \cdot x \cdot \frac{x}{L} \cdot w_L = \frac{1}{2} \cdot \frac{x^2}{L} \cdot \frac{2W}{L} = W \cdot \frac{x^2}{L^2}$$

Hence $$V = -\frac{W.x^2}{L^2} \text{ (Fig. 4.7 (b))}$$

B.M. at any distance x $$= -\frac{1}{2} w_x \cdot x \cdot \frac{x}{3} = -\frac{x^2}{6} \cdot w_L \cdot \frac{x}{L} = \frac{-2W}{L} \cdot \frac{x^3}{6L} = \frac{-W}{3L^2} \cdot x^3$$

At $x = 0$ $M = 0$, at $x = L$ $M = -\frac{WL}{3}$ (Fig. 4.7 (c))

(a) Beam

(b) Shear Force diagram.

(c) B.M. diagram.

Fig. 4.7. Example 4.1.

Example 4.2. A beam of length L uniformly supported along its entire length carries at its ends two equal loads P (Fig. 4.8). Draw the S.F. and the B.M. diagram.

Solution. As the beam is uniformly supported over its entire length, upward pressure $q = \frac{2P}{L}$.

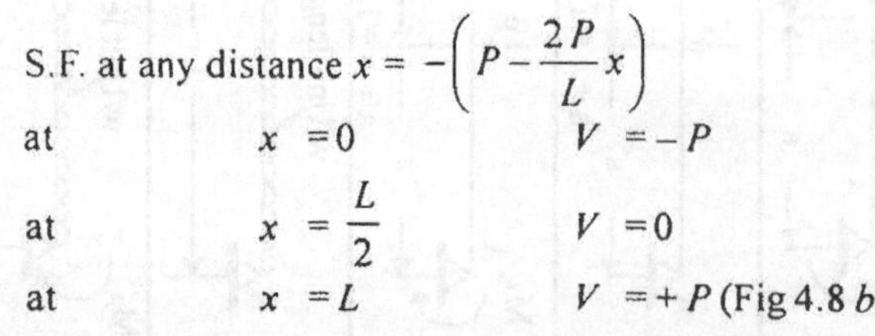

S.F. at any distance $x = -\left(P - \frac{2P}{L} x\right)$

at $x = 0$ $V = -P$

at $x = \frac{L}{2}$ $V = 0$

at $x = L$ $V = +P$ (Fig 4.8 b)

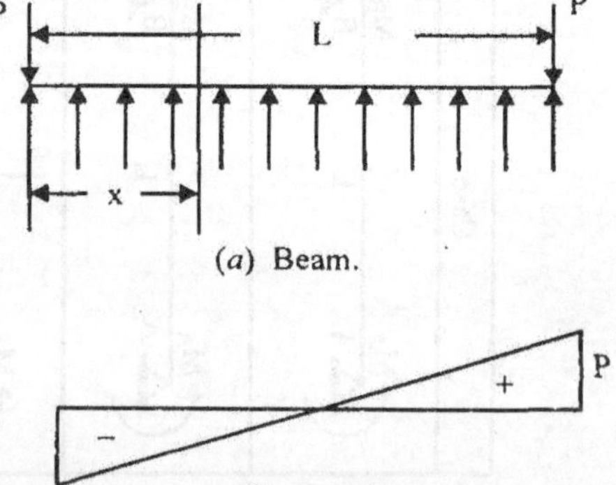

(a) Beam.

(b) S.F. diagram.

B.M. at any distance x,

$$M_x = -\left(P.x - \frac{2P}{L}\cdot\frac{x^2}{2}\right)$$

at $\quad x = 0 \quad M = 0$

at $\quad x = L \quad M = 0$

at $\quad x = \frac{L}{2} \quad M = \frac{PL}{4}$

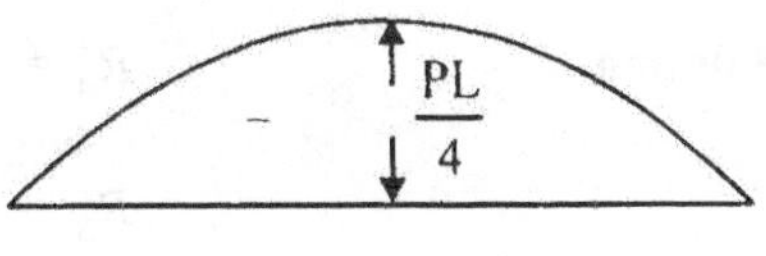

(c) B.M. diagram.

Fig. 4.8. Example 4.2.

Example 4.3. A simply supported beam of length L carries a total distributed load W which increase in intensity uniformly from zero at the left end. Draw the B.M. and S.F. diagrams.

Solution. If w is the rate of intensity of loading at B.

$$\frac{1}{2}w.L = W$$

or $\quad w = \frac{2W}{L}$

At any distance x $\quad w_x = \frac{w}{L}\cdot x$

$$= \frac{2W}{L^2}\cdot x$$

$$R_A = \frac{W}{3}$$

S.F. at $X-X$ $\quad = \frac{W}{3} - \frac{1}{2}w_x.x$

$$= \frac{W}{3} - \frac{1}{2}\cdot\frac{2W}{L^2}\cdot x \cdot x$$

$$= \frac{W}{3} - \frac{Wx^2}{L^2} \text{ (Fig. 4.9 (b))}$$

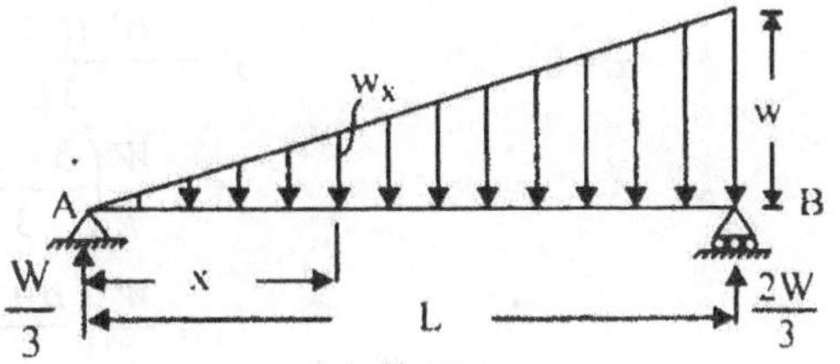

At $\quad x = 0 \;$ S.F. $= \frac{W}{3}$

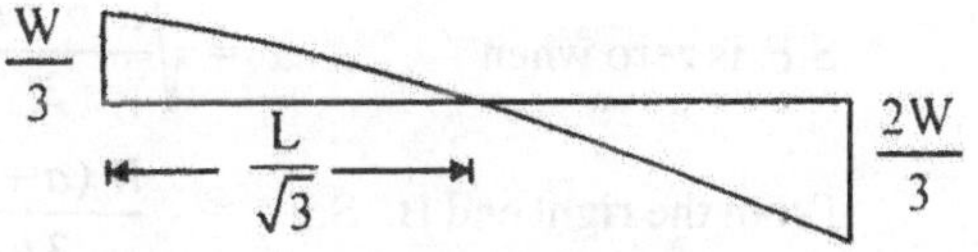

at $x = L \quad$ S.F. $= -\frac{2w}{3}$

at $x = \frac{L}{\sqrt{3}} \quad$ S.F. $= 0$

B.M. at X – X

$$\frac{W}{3}\cdot x - \frac{W}{L^2}\cdot\frac{x^3}{3}$$

at $\quad x = 0 \quad$ B.M. $= 0$

at $\quad x = L \quad$ B.M. $= 0$

at $\quad x = \frac{L}{2} \quad$ B.M. $= \frac{WL}{8}$ (Fig. 4.9 (c))

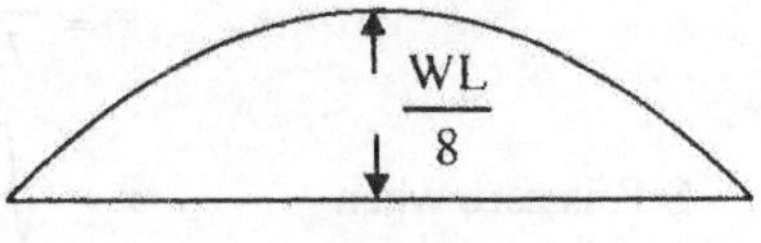

(a) Beam

(b) S.F. diagram.

(c) B.M. diagram.

Fig. 4.9. Example 4.3.

Example 4.4. A simply supported beam AB carries a triangular load as shown in Fig. 4.10 (a). Draw the SF and B.M. diagrams.

Solution.

$$R_A = \frac{W.(b+L)}{3L}$$

$$R_B = \frac{W\cdot(a+L)}{3L}$$

$$W = \frac{1}{2}CD\times L$$

$$CD = \frac{2W}{L}$$

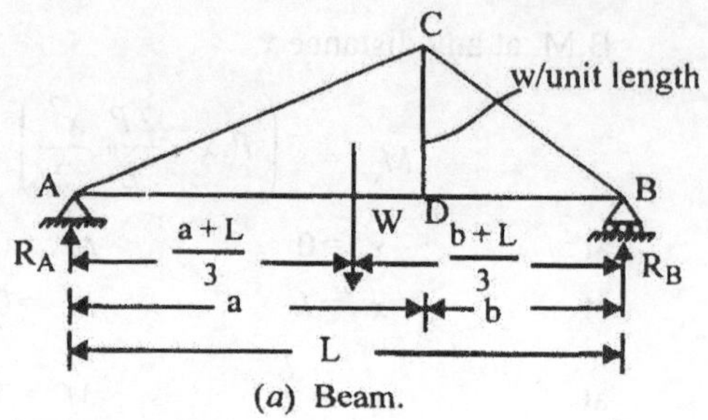

(*a*) Beam.

At any distance x the load intensity is $\frac{CD}{a}\times x = \frac{2W.x}{aL}$

(*i*) S.F. diagram

$$\text{S.F. at } x = R_A - \frac{1}{2}\times\frac{2W.x}{aL}\cdot x$$

$$= \frac{W\cdot(b+L)}{3L} - \frac{W.x^2}{aL}$$

$$= \frac{W}{L}\left(\frac{b+L}{3} - \frac{x^2}{a}\right)$$

which is equation of 2nd degree

At $x = 0$ $\qquad V = \frac{W\cdot(b+L)}{3L}$

At $x = a$ $\qquad V = \frac{W}{L}\left(\frac{b+L}{3} - a\right)$

$$= \frac{W}{L}\left(\frac{a+2b-3a}{3}\right)$$

$$= \frac{2W}{3}\cdot\frac{b-a}{L}$$

S.F. is zero when $\qquad x = \sqrt{\frac{a(b+L)}{3}}$

From the right end B, S.F. $= -\frac{W(a+L)}{3L} + \frac{W\cdot x^2}{bL}$

$$= \frac{W}{L}\left(\frac{x^2}{b} - \frac{a+L}{3}\right)$$

S.F. is zero when $\qquad x = \sqrt{\frac{(a+L)b}{3}}$

(*b*) S.F. diagram.

(*c*) B.M. diagram.

Fig. 4.10. Example 4.4.

(*ii*) B.M. diagram

When x lies between A and D

$$\text{B.M.} = R_A\cdot x - \frac{1}{2}\times\frac{2Wx}{aL}\cdot x\cdot\frac{x}{3}$$

$$= \frac{W(b+L)}{3L}\cdot x - \frac{Wx^3}{3aL}$$

At $\quad x = 0 \quad$ B.M. $= 0$

at $x = a$,

$$\text{B.M.} = \frac{W(b+L)\cdot a}{3L} - \frac{Wa^3}{3aL}$$

$$= \frac{a^2(b+L) - a^3}{3aL}\cdot W$$

When $a = b = \dfrac{L}{2}$

$$B.M. = W\left[\frac{L^2}{4}\left(\frac{3L}{2}\right) - \left(\frac{L}{2}\right)^3\right]\times\frac{2}{3L^2}$$

$$= W\cdot\left(\frac{3L^3}{8} - \frac{L^3}{8}\right)\times\frac{2}{3L^2}$$

$$= \frac{WL}{6} = \frac{wL^2}{12}$$

Example 4.5. Draw the shear force and bending moment diagram for the beam shown in Fig. 4.11. Give equation for each curve.

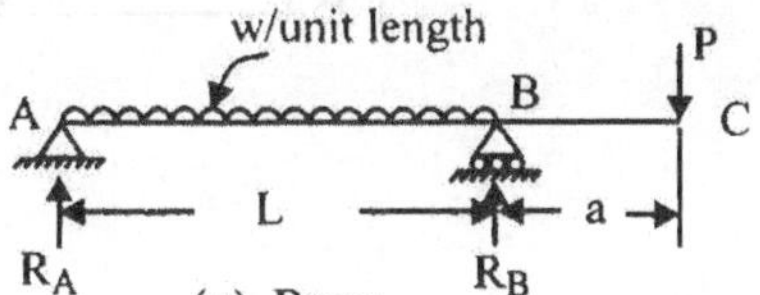

Solution. Taking moments about A, $P(L+a) + \dfrac{wL^2}{2} = R_B . L$

or $R_B = \dfrac{P(L+a)}{L} + \dfrac{wL}{2}$

Taking moments about 'B'

$$R_A \times L - \frac{wL^2}{2} + P.a. = 0$$

or $R_A = \dfrac{wL}{2} - \dfrac{Pa}{L}$

Taking origin at A

S.F. at any distance x from A,

$$V = R_A - w.x$$

$$= \frac{wL}{2} - \frac{Pa}{L} - w.x$$

at $x = L$, $\quad V = -\left(\dfrac{Pa}{L} + \dfrac{wL}{2}\right)$

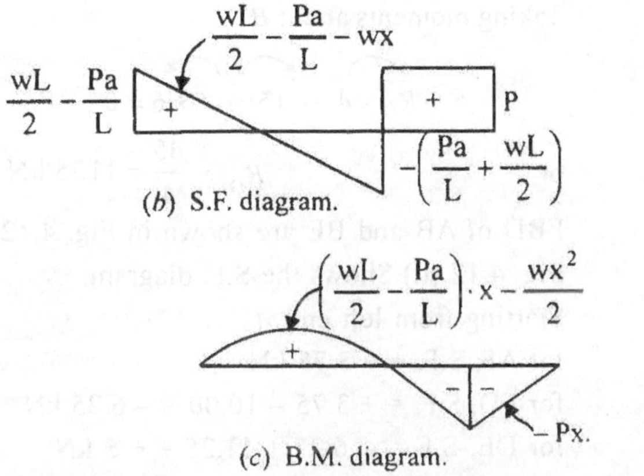

B.M. at a distance x

$$M = R_A.x - \frac{wx^2}{2} = \left(\frac{wL}{2} - \frac{Pa}{L}\right).x - \frac{wx^2}{2}$$

at $x = L$ $\quad M = \dfrac{wL^2}{2} - Pa - \dfrac{wL^2}{2} = -P.a$

For the portion BC taking C as origin

S.F. $= \text{P}$

B.M. $= -\text{P}x$

At $x = a$ $\quad$ B.M. $= -\text{P.a.}$

(*a*) Beam.

(*b*) S.F. diagram.

(*c*) B.M. diagram.

Fig. 4.11. Example 4.5.

Example 4.6. Draw S.F. and B.M. diagram for the beam shown in Fig. 4.12.

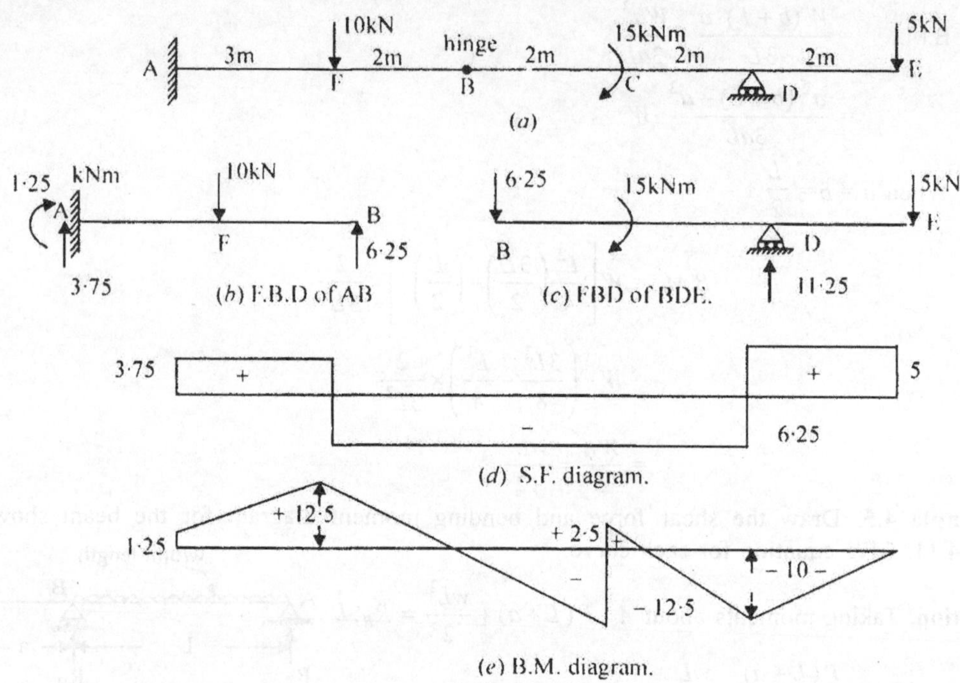

Fig. 4.12. Example 4.6.

Taking moments about B

$$R_D \times 4 - 15 - 5 \times 6 = 0$$

or $$R_D = \frac{45}{4} = 11.25 \text{ kN} \uparrow$$

FBD of AB and BE are shown in Fig. 4.12 (b) and (c) respectively.

Fig. 4.12 (d) Shows the S.F. diagram.

Starting from left end A

for AF, S.F. = + 3.75 kN.

for FD, S.F. = + 3.75 – 10.00 = – 6.25 kN

for DE, S.F. = – 6.25 + 11.25 = **+ 5 kN**

B.M. diagram

For portion AB, taking origin at 'A'

for AF, $M = 1.25 + 3.75x$

at $x = 0$ $M = 1.25$ kNm.

at $x = 3$ $M = 1.25 + 11.25$

$= \mathbf{12.5\ kNm}$

for FB $M = 1.25 + 3.75x - 10(x - 3)$

$= 31.25 - 6.25x$

at $x = 3$ $M = 31.25 - 6.25 \times 3$

$= \mathbf{12.50\ kNm}$

	$x = 5$	$M = 31.25 - 6.25 \times 5$
		$= 0$ as required.

For portion BE

For	BC (origin at B)	
		$M = -6.25x$
at	$x = 0$	$M = 0$
at	$x = 2$	$M = -12.50$ kNm
For	CD	$M = -6.25x + 15$
at	$x = 2$ m	$M = 2.50$ kNm
at	$x = 4$ m	$M = -10$ kNm
For	ED (origin at E)	
		$M = -5x$
at	$x = 2$	$M = -5 \times 2 = $ **– 10 kNm**

Example 4.7. Draw the axial force, shear force and bending moment diagram of the beam loaded as shown in Fig. 4.13.

Solution. The inclined force of 25 kN is resolved at D into 20 kN horizontal and 15 kN vertical downward. As a result of 20 kN horizontal force at *D*, there is an anticlockwise moment of 40 kNm at C along with a horizontal force of 20 kN.

Computation of Reactions

Taking moments of all forces about *A*,

$$R_F \times 25 + 10 \times 20 + 40 - 15 \times 15 - \frac{1}{2} \times 3 \times 12 \times 4 = 0$$

or $R_F = \frac{1}{25}[57] = 2.28 \text{ kN} \uparrow$

Similarly taking moments about *F*

$$R_A \times 25 - \frac{1}{2} \times 3 \times 12 \times 21 - 40 - 15 \times 10 + 10 \times 5 = 0$$

or $R_A = \frac{1}{25}[51.8] = 20.72 \uparrow$

$H_A = 20 \text{ kN} \rightarrow$

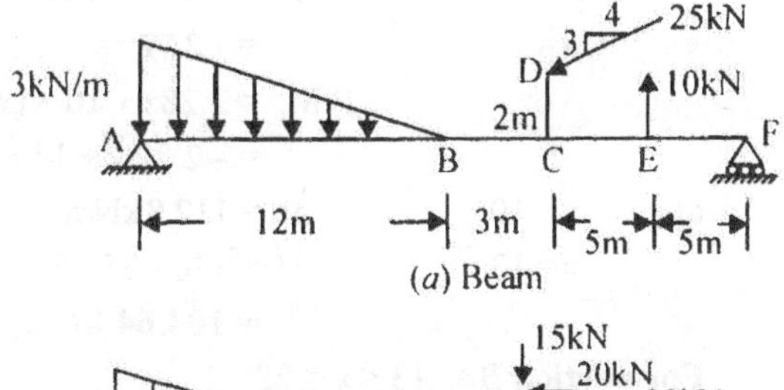

(*a*) Beam

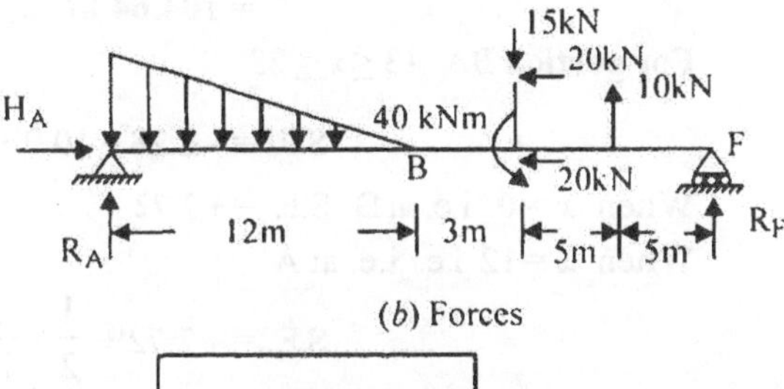

(*b*) Forces

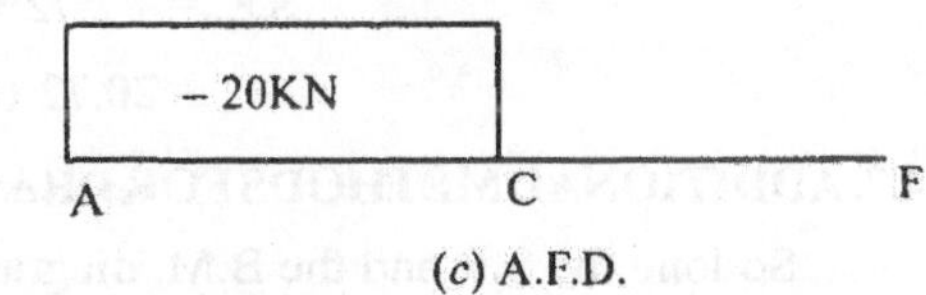

(*c*) A.F.D.

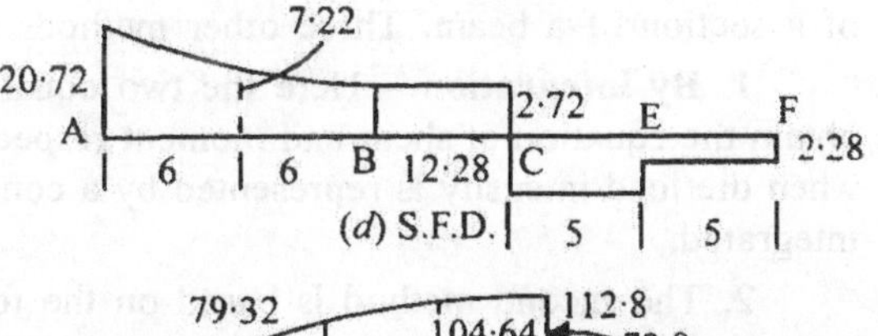

(*d*) S.F.D.

79·32
104·64
112·8
72·8
11·40
6 6 6 3 5 5

(*e*) B.M. diagram.

Fig. 4.13. Example 4.7

Axial force diagram

Axial force is constant from *A* to *C* and equal to 20 kN (compressive). From *C* to *F* axial force is zero.

S.F. and B.M. diagrams

For portion FE $0<x\leq 5$

$$\text{S.F.} = -2.28 \text{ kN}$$

$$\text{B.M.} = +2.28\,x \text{ (}F\text{ is origin)}$$

at $x=0$ B.M. $=0$

at $x=5$ B.M. $=2.28\times 5$

$$= \mathbf{+11.40 \text{ kNm}}$$

For portion EC $5\leq x\leq 10$

$$\text{S.F.} = -2.28-10 = -12.28 \text{ kN}$$

$$\text{B.M.} = 2.28\times x+10\times(x-5)$$

at $x=5$ $M=+11.40$ kNm

$x=10$

$$M = 22.8+50 = 72.8 \text{ kNm.}$$

For portion CB $10\leq x\leq 13$

$$\text{S.F.} = -2.28-10.0+15$$

$$= +2.72$$

$$\text{B.M.} = 2.28x+10\times(x-5)-15\times(x-10.0)+40$$

$$= -2.72x+140$$

at $x=10$ $M=112.8$ kNm

$x=13$ $M=140-35.36$

$$= \mathbf{104.64 \text{ kNm}}$$

For portion BA $13\leq x\leq 25$

$$\text{S.F.} = -2.28-10.0+15.0+\frac{1}{2}\times\frac{3}{12}.x.x \text{ (origin at B)}$$

When $x=0$ i.e. at B. S.F. $=+2.72$

When $x=12$ i.e. i.e. at A

$$\text{S.F.} = +2.72+\frac{1}{2}\times\frac{3}{12}\times 12^2$$

$$= 20.72 \text{ (check)}$$

4.7. ADDITIONAL METHODS FOR DRAWING THE S.F. AND B.M. DIAGRAMS

So long the S.F. and the B.M. diagrams have been drawn by considering free body diagrams of a section of a beam. Three other methods can also be used.

1. By integration – Here the two equations $dV = wdx$ and $dM = Vdx$ are integrated to obtain the equation of shear and moment respectively. Obviously this method is most appropriate when the load intensity is represented by a continuous mathematical function that can be readily integrated.

2. The second method is based on the following fundamental results which were derived in section 3.4.

(*a*) The slope of the shear diagram at any point is given by the load intensity at that point.

(*b*) The slope of the moment diagram at any point is given by the value of shear at that point.

(*c*) The change in shear between two points is equal to the area under the load intensity diagram between these two points.

(*d*) The change in moment between two points is equal to the area under the shear diagram between these two points. In this method changes in shear and moment between key points are determined integrating the load and shear diagrams. By adding or subtracting changes, values of shear or bending moment at selected points are then obtained.

3. Another method of drawing the B.M. diagram for a simply supported beam is to imagine one end of the beam as fixed and draw the cantilever B.M. diagram for different loads including the reactions with proper signs. The combined B.M. diagram will be the summation of individual B.M. diagrams which should satisfy the boundary conditions. Example 4.10 shows application of this method.

Example 4.8. Construct the shear and moment diagrams by the formal integration method for the beam loaded as shown in Fig. 4.14.

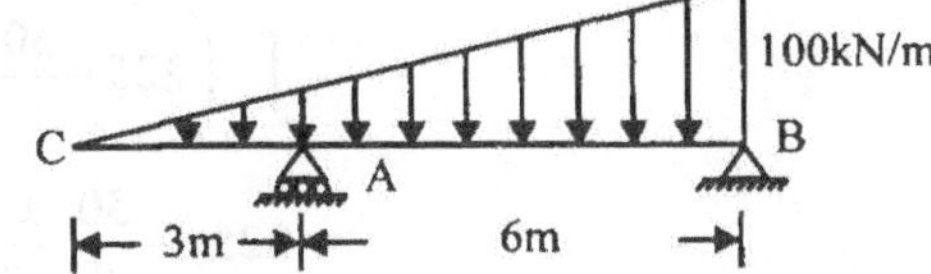

Fig. 4.14. Example 4.8.

Solution. Total load $= \frac{1}{2} \times 9 \times 100 = 450\,\text{kN}$ acting at 3 m from *A*.

Taking moments about *B*

$$R_A \times 6 - 450 \times 3 = 0$$

or $\qquad R_A = 225\,\text{kN}$

$\qquad R_B = 225\,\text{kN}$

S.F. diagram origin at C

For CA $\qquad V = -\int w\,dx = -\int \frac{100}{9} \cdot x \cdot dx$

$0 \le x \le 3$ $\qquad = -\frac{50}{9} \cdot x^2 + C_1$

at $\quad x = 0 \qquad V = 0 \quad \therefore\ C_1 = 0$

$\therefore \qquad V = -\frac{50}{9} \cdot x^2$

At $\quad x = 3 \qquad V = -50\,\text{kN}$

Fig. 4.15. Example 4.8.

For AB $3 < x \le 9$

$$V = R_A - \int w\,dx$$

$$= 225 - \int \frac{100}{9} \cdot x \cdot dx$$

$$= 225 - \frac{100}{9} \cdot \frac{x^2}{2} + C_2$$

At $\quad x = 3 \qquad V = +175$ Hence $C_2 = 0$

Equation then becomes

$$V = 225 - \frac{50}{9} \cdot x^2$$

B.M. diagram

For CA $\qquad M = \int V\,dx$

$$= \int \frac{-50}{9} \cdot x^2 \cdot dx$$

$$= \frac{-50}{9} \cdot \frac{x^3}{3} + C_3$$

at $x = 0$ $M = 0$ $\therefore C_3 = 0$

$$M = \frac{-50}{27} \cdot x^3$$

For AB $M = \int V.dx$

$$= \int \left(225 - \frac{50}{9} \cdot x^2\right) dx$$

$$= 225x - \frac{50}{9} \cdot \frac{x^3}{3} + C_4$$

At $x = 3$ $M = -\frac{50}{9} \times \frac{27}{3} = -50$

or $-50 = 225 \times 3 - 50 + C_4$

or $C_4 = -675$

Hence $M = 225x - \frac{50}{27} . x^3 - 675$

Fig. 4.16 shows the S.F. and B.M. diagrams

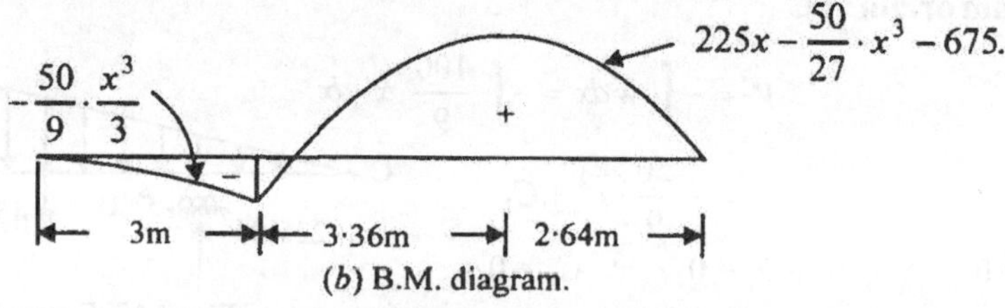

(a) S.F. diagram.

(b) B.M. diagram.

Fig. 4.16. Example 4.8

Example 4.9. Use fundamental principles of drawing S.F. and B.M. diagrams to construct the same for the structure shown in Fig. 4.17.

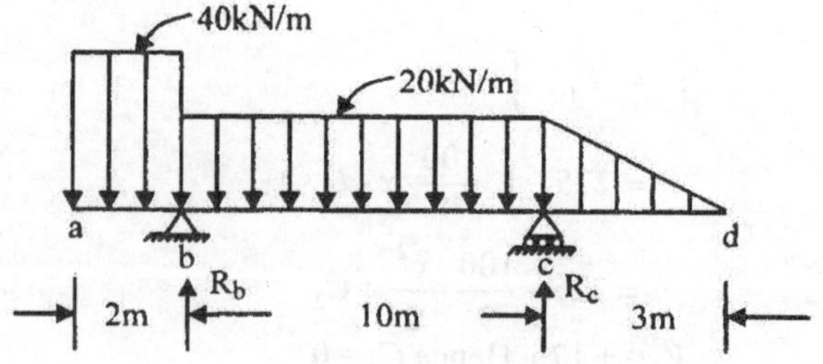

Fig. 4.17. Example 4.9.

Solution. The reactions are first computed. Taking moment about c of all forces

$$40 \times 2 \times 11 - R_b \times 10 + 20 \times 10 \times 5$$

$$-\frac{1}{2} \times 20 \times 3 \times \frac{1}{3} \times 3 = 0$$

or $R_b = 185$ kN

Taking moments about b

Taking moments about b.

$$40 \times 2 \times 1 - 20 \times 10 \times 5 + R_C \times 10$$

$$-\frac{1}{2} \times 20 \times 3 \times 11 = 0$$

or $\quad R_C = 125 \text{ kN}$

$$R_b + R_c = 185 + 125 = 310$$

$$= 40 \times 2 + 10 \times 20 + \frac{1}{2} \times 20 \times 3$$

$$= 310 \text{ (check)}$$

Fig. 4.18. Example 4.9.

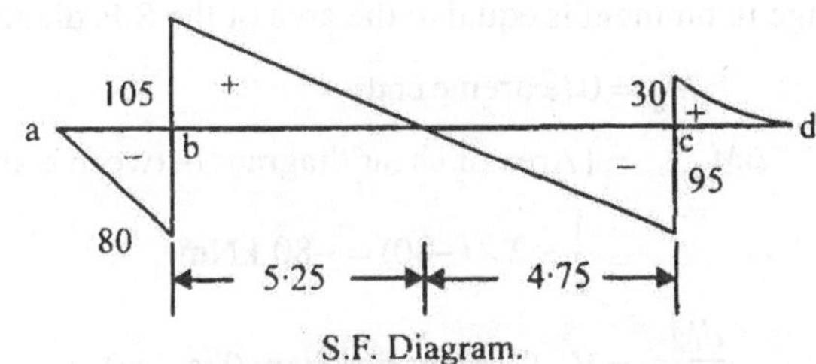

S.F. Diagram.

Fig. 4.19. Example 4.9.

Shear diagram

Shear diagram is determined by considering changes in shear from point to point and using the relation $\frac{dV}{dx} = -w$

Point a Shear force is zero.

Δv_{a-b} = {Area under load diagram between a and b} with a negative sign.

$= -40 \times 2 = \mathbf{-80 \text{ kN}}$

Shear at left half of $b = 0 + (-80) = -80$ kN

[Negative from sign convention for shear]

Slope of the shear diagram in $ab = -40$

Shear just to the right of $b = -80 + 185 = +105$ kN

b to c $\qquad \frac{dv}{dx} = -20$

Change in shear from b to c

$= -20 \times 10 = -200$ kN

Shear at left of c $\quad = +105 - 200$

$$= -95 \text{ kN}$$

Shear just to right of c $\quad = -95 + 125$

$$= +30 \text{ kN}$$

c to d $\quad \dfrac{dv}{dx} = -w(x) = -\dfrac{20}{3} \times (3-x)$

$$\Delta V = -\int_0^3 -\frac{20}{3} \times (3-x) \cdot dx$$

from c to d $\quad = -\dfrac{20}{3}\left[3x - \dfrac{x^2}{2}\right]_0^3$

$$= -\frac{20}{3}[9-4.5] = -30$$

Shear at $\quad d = +30 - 30 = 0$

Slope of shear diagram $\dfrac{dv}{dx}$ is negative and decreasing (w_x is decreasing) towards the right and hence it is concave upward.

Bending Moment Diagram

B.M. diagram is drawn by considering the change in moment between consecutive points and by using the relation.

$\dfrac{dM}{dx} = V$ or change in moment is equal to the area of the S.F. diagram.

Point a $\quad M_a = 0$ (extreme end)

a to b $\quad \Delta M_{a-b} = \{$Area of shear diagram between a and $b\}$

$$= \frac{1}{2} \times 2 \times (-80) = -80 \text{ kNm}$$

$$\frac{dM}{dx} = V \quad \text{Slope varies from 0 to } -80$$

$$M_b = 0 + (-80) = -80 \text{ kNm}$$

b to c $\quad \Delta M_{b-c} = \{$Area of the shear diagram between b and $c\}$

$= \{$Area of the shear diagram between b and $e\}$

$+ \{$Area of the shear diagram between e and $c\}$

$$= \frac{1}{2}[105] \times 5.25 + \frac{1}{2}[-95] \times 4.75$$

$$= [275.625] + [-225.625]$$

At e, change in moment $\quad = 275.62$

and value is $-80 + 275.62 = +195.625$

at c change in moment $\quad = -225.625$

and value of moment is $\quad = 195.625 - 225.625$

$= $ **− 30.00 kNm**

Slope of the moment diagram

at $\quad a, \quad \dfrac{dM}{dx} = V = 0 \quad$ Slope $= 0$

at b left $\quad \dfrac{dM}{dx} = -80 \quad$ Slope $= -80$ (clockwise)

at b right $\quad \dfrac{dM}{dx} = +105 \quad$ Slope $= +105$ (anticlockwise)

There is a sudden change in shear and hence sudden change in slope.

at $e \quad \dfrac{dM}{dx} = V = 0$, Slope is zero and tangent is paralel to x axis.

From e to left of c $\quad V =$ negative, hence slope is negative (clockwise) and with increase of negative shear negative slope is increasing.

From right of c to d shear is positive and hence slope is positive (anticlockwise).

Thereis a sudden change in shear from left of c to right of c. hence there is sudden change in slope from clockwise to anticlockwise.

At d $\quad$ S.F. $= 0$ Slope $= 0$.

B.M. diagram.

Fig. 4.20. Example 4.9.

Example 4.10. Imagining the point D as fixed draw individual B.M. diagram for different loads and then finally draw the combined B.M. diagram. (Fig. 4.21).

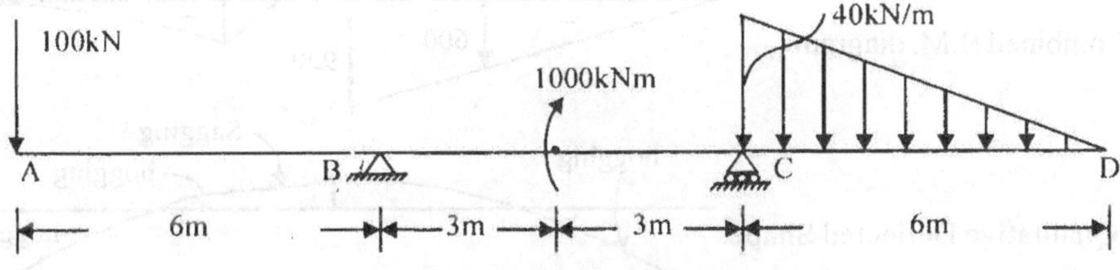

Fig. 4.21. Example 4.10.

Solution. First step is to compute reactions.

Taking moments about B

$$-R_C \times 6 - 100 \times 6 + 1000.00 + \frac{1}{2} \times 40 \times 6 \times 8 = 0$$

or $R_C = 226.67$ kN $\uparrow$

Taking moments about C

$$-100 \times 12 + R_B \times 6 + 1000 + \frac{1}{2} \times 40 \times 6 \times 2 = 0$$

or $R_B = -6.67$ kN

Check $\quad R_C + R_B = 226.67 + (-6.67) = 220.00 \uparrow$

$$= 100 + \frac{1}{2} \times 40 \times 6 = 220 \downarrow$$

B.M. diagrams

1. With AD **imagined** as cantlever and fixed at D load 100 kN downward and $R_B = 6.67 \downarrow$ will produce hogging B.M. as shown in Figs 4.22 (*a*) and (*b*) respectively.

2. Similarly, 1000 kNm (anticlockwise) and $R_C = 226.67$ (upward) will produce sagging (positive) B.M. Fig. 4.22 (*c*) and (*d*).

3. Finally, triangular downward load will producd hogging B.M. Fig. 4.22 (*e*)

4. Al gebraic sum of all B.Ms at *D* should be zero.

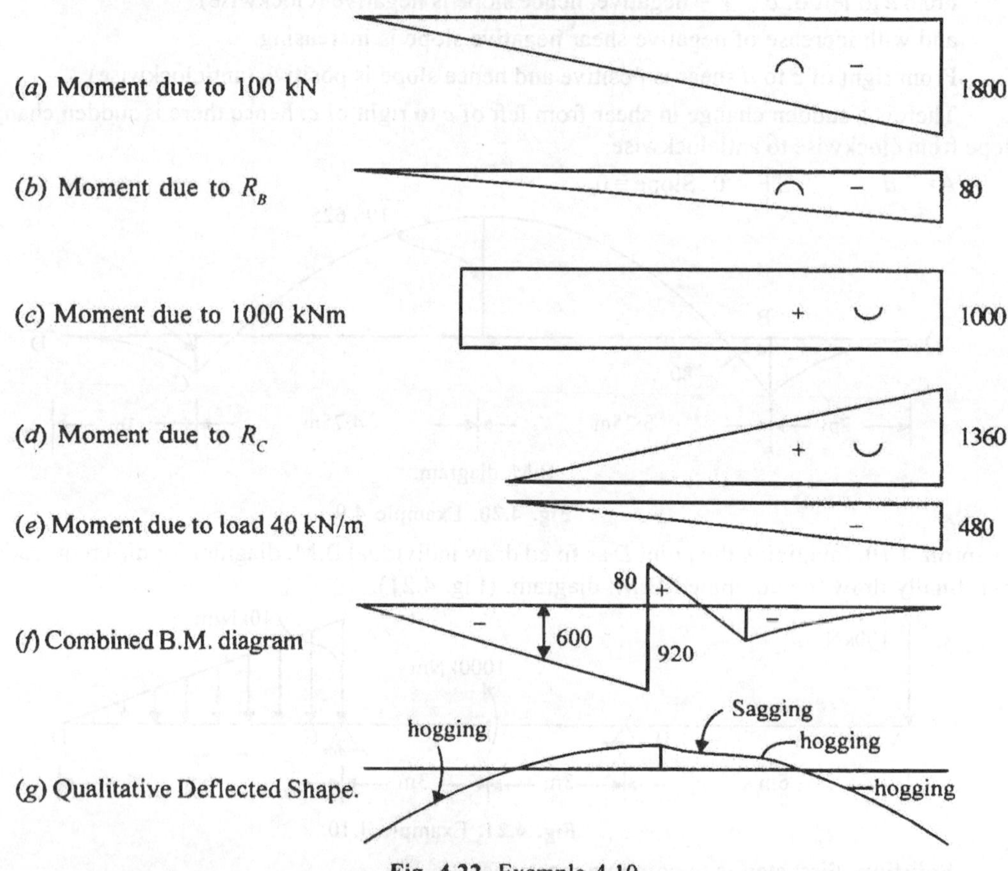

Fig. 4.22. Example 4.10.

4.8 LOADING DIAGRAM FROM S.F. AND/OR BENDING MOMENT DIAGRAM

It is often necessary to draw the load diagram from a S.F. and/or a bending moment diagram. This is necessary in a continuous prestressed concrete beam for deriving the support moments. The method becomes systematic if the following points are remembered.

1. Loading diagram from S.F. diagram. The S.F.D. will consist of
   (*a*) a number of rectangles if the loads are concentrated loads.
   (*b*) inclined lines if the loading is uniformly distributed.
   (*c*) parabolic lines if loading is triangular or trapezoidal.
   (*d*) cubical parabolas when the loading is parabolic.
2. Loading diagram from B.M. diagram. The B.M. diagram is
   (*a*) Linear if the loads are concentrated loads.
   (*b*) parabolic for distributed load.

(*c*) cubical parabola if the load is triangular
(*d*) Fourth degree parabola if the load is parabolic.

Example 4.11. Draw the S.F. diagram and the loading diagram corresponding to the given bending moment diagram shown in Fig. 4.23 (*a*)

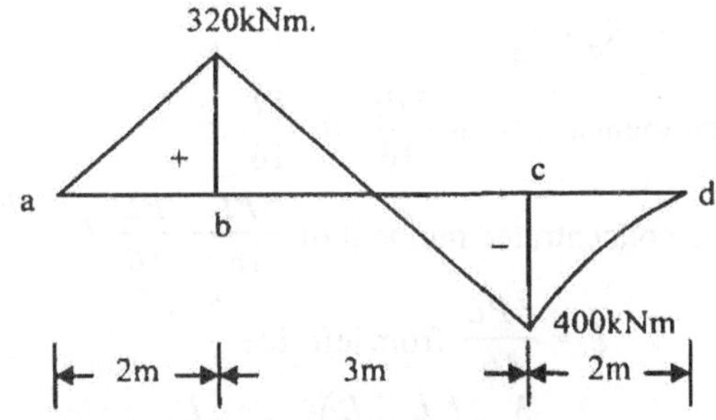

Fig. 4.23. (*a*) Bending Moment Diagram (Example 4.11).

Solution. From the figure it is apparent that support is at *a* and *c*. From *a* to *c*, B.M. diagram is linear, hence loads must be concentrated.

Reaction at *a* $= \dfrac{320}{2} = 160$ kN

Moment at *c* $= -400$ kNm

Therefore $R_a \times 5 - P_b \times 3 = -400$

$160 \times 5 - P_b \times 3 = -400$

or $P_b = \mathbf{400\ kN}$

Moment at $c = 400 = \dfrac{w \times 2^2}{2}$

or $w = 200$ kN/m

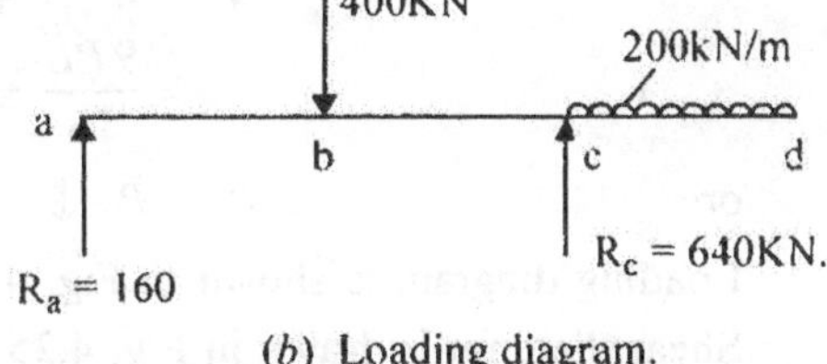

(*b*) Loading diagram.

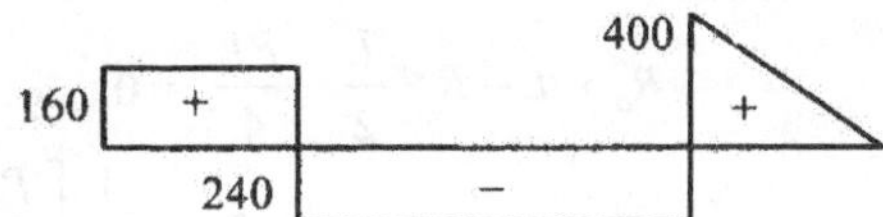

(*c*) Shear Force diagram.

Fig. 4.23. (*b*) and (*c*) (Example 4.11).

Example 4.12. Draw the S.F. and the loading diagram from the B.M. diagram shown in Fig. 4.24.

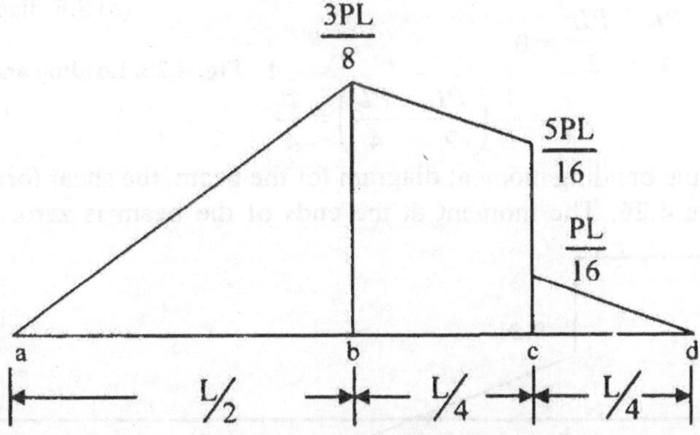

Fig. 4.24. Example 4.12.

Solution. Moments at *a* and *d* are zero. Hence supports must be at *a* and *d* as all moments are positive

From Fig 4.24

$$R_a \times \frac{L}{2} = \frac{3PL}{8}$$

or $$R_a = \frac{3}{4} \cdot P$$

$$R_d \times \frac{L}{4} = \frac{PL}{16}$$

or $$R_d = \frac{P}{4}$$

There is a drop of moment at c from $\frac{5PL}{16}$ to $\frac{PL}{16}$.

Hence there must be concentrated moment of $\frac{5PL}{16} - \frac{PL}{16}$ *i.e.* $\frac{PL}{4}$ at c (anticlockwise)

Moment at $$c = \frac{5PL}{16} \text{ from left side}$$

$$= \frac{3}{4} P\left(\frac{L}{2} + \frac{L}{4}\right) - P_b \times \frac{L}{4}$$

or $$P_b \times \frac{L}{4} = \frac{3}{4} \cdot P \cdot \frac{3}{4} \cdot L - \frac{5PL}{16}$$

$$= \frac{9PL}{16} - \frac{5PL}{16} = \frac{PL}{4}$$

or $$P_b = P. \downarrow$$

Loading diagram is shown in Fig. 4.25 (*a*)

Shear diagram is drawn in Fig. 4.25 (*b*)

check

$$R_a \times L - R \times \frac{L}{2} - \frac{PL}{4} = 0$$

or $$R_a = \frac{1}{L}\left[\frac{PL}{2} + \frac{PL}{4}\right] = \frac{3P}{4}$$

$$R_d \times L + \frac{PL}{4} - \frac{PL}{2} = 0$$

or $$R_d = \frac{1}{L}\left(\frac{PL}{2} - \frac{PL}{4}\right) = \frac{P}{4}$$

(*a*) Loading diagram.

(*b*) S.F. diagram.

Fig. 4.25. Loading and S.F. diagram.

Exampel 4.13. Draw the bending moment diagram for the beam, the shear force diagram for which is shown in the figure 4.26. The moment at the ends of the beam is zero.

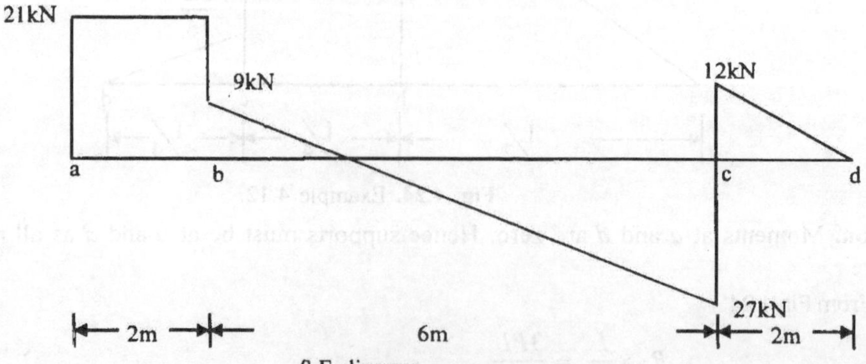

S.F. diagram.

Fig. 4.26. Example 4.13.

Solution. From the figure it is clear that supports are at *a* and *c*. The S.F. of 21 kN is due to support reaction at a (Positive) and there is a change in shear at *c* due to support reaction at *c*. As the S.F. is sloping from *b* to *c* there is a uniformly distributed load of $\frac{(27+9)}{6}$ i.e. 6 kN/m from *b* to *c*. From *c* to *d* the loading is $\frac{12}{2}$ = 6 kN/m. There is a change in S.F. at *c* of 12 – (–27) = 39 kN which is the reaction at *c*. Fig. 4.27 (*a*) gives the loading diagram and Fig. 4.27 (*b*) the corresponding B.M. diagram.

Check

$$R_a \times 8 - 6 \times 6 \times 3 + 6 \times 2 \times 1 - 12 \times 6 = 0$$

or $R_a = 21$ kN ↑

$$R_c \times 8 - 12 \times 2 - 6 \times 8 \times 6 = 0$$

or $R_c = 39$ kN ↑

$\Sigma V = 21 + 39 = 60 = 6 \times 8 + 12$ (check)

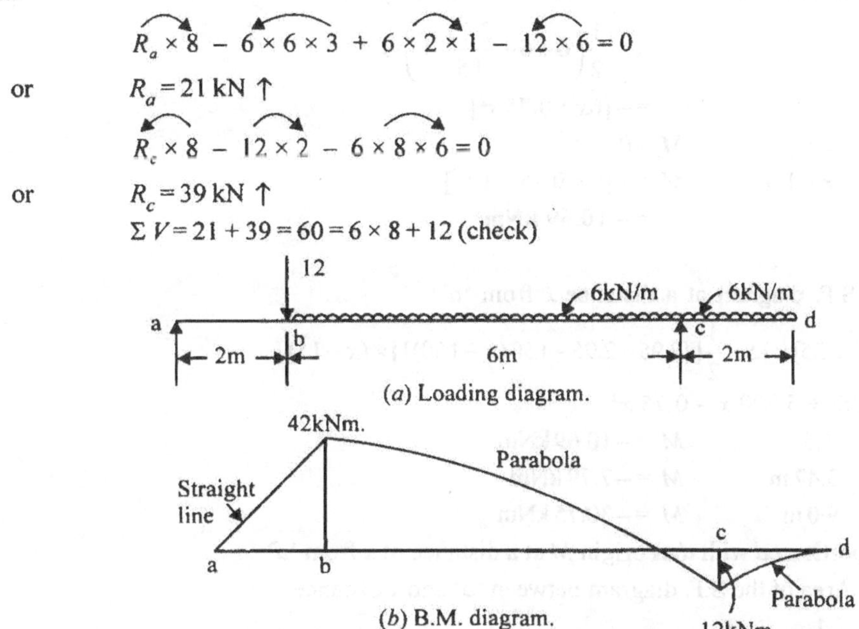

Fig. 4.27. Loading and B.M. diagram (Example 4.13).

Example 4.14. Draw directly from the S.F. diagram (Fig. 4.28 *a*) the B.M. diagram and the loading diagram for the beam.

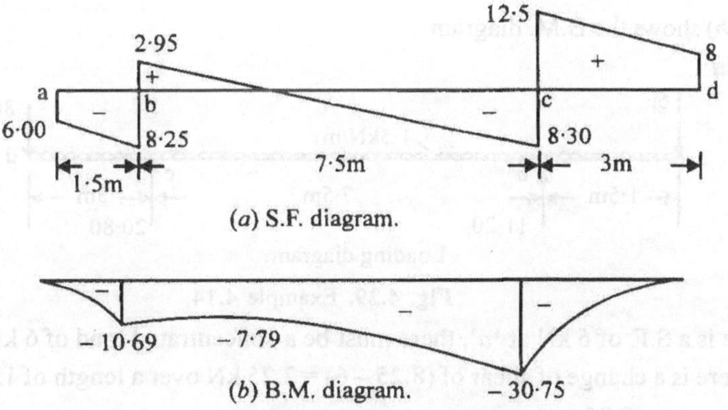

Fig. 4.28. Example 4.14.

Solution. 1. From the S.F. diagram it is obvious that the suppors are at b and c with overhangs ba and cd. Negative value of shear in ba indicates left portion going down with respect to the right because of downward load, similarly positive sign in cd indicates right portion going down with respect to the left due to downward vertical load.

The S.F. of 6 kN in 'a' indicates a concentrated load of 6 kN at 'a'. The linear increase of S.F. from 6 to 8.25 indicates presence of distributed load in ab.

2. B.M. diagram

For ab

At any distance x from 'a' BM = area of the S.F. diagram

$$= -\frac{1}{2}\left(6+6+\frac{2.25}{1.5}\cdot x\right)\cdot x$$

$$= -[6x + 0.75x^2]$$

at $x = 0$ $M = 0$

at $x = 1.5$ $M = -[9 + 0.75 \times 1.5^2]$

$$= \mathbf{-10.69\ kNm}$$

For bc

Area of S.F. diagram at a distance x from 'a'

$$-\frac{1}{2}\times[6+8.25]\times 1.5+\frac{1}{2}[\{2.95+2.95-1.50(x-1.50)\}]\times(x-1.5)$$

$$= -16.802 + 5.200x - 0.75x^2$$

at $x = 1.5$ $M = -10.69$ kNm

at $x = 3.47$ m $M = -7.79$ kNm

at $x = 9{\cdot}0$ m $M = -30.75$ kNm

For the portion cd with d as origin M at a distance of x from 'd'

= Area of the S.F. diagram between 'd' and x distance

$$= -\frac{1}{2}(8+8+1.5x)x$$

$$= -(8x + 0.75x^2)$$

At $x = 0$ $M = 0$

$x = 3$m $M = 30{\cdot}75$ kNm

Fig. 4.28 (b) shows the B.M. diagram

Loading diagram

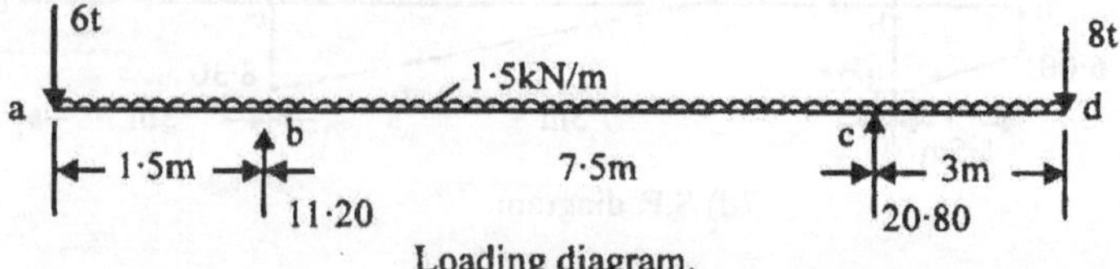

Loading diagram.

Fig. 4.29. Example 4.14.

1. As there is a S.F. of 6 kN at 'a', there must be a concentrated load of 6 kN at 'a'
2. In ab there is a change of shear of $(8.25 - 6) = 2.25$ kN over a length of 1.5 m

Hence $\frac{w}{\text{metre}} = \frac{2.25}{1.5} = 1.5$ kN / m

3. At b, because of reaction from support there is a change of shear from -8.25 to $+2.95$. Hence the reaction at $b = 2.95 - (-8.25) = 11.20$ kN
4. From b to c, change of shear is $8.30 - (-2.95) = 11.25$ kN

$$\text{Hence } \frac{w}{\text{metre}} = \frac{11.25}{7.5} = 1.5 \text{ kN / m}$$

5. Change of shear at $c = 8.30 - (12.5) = 20.80$ kN
6. Change in shear in cd

$$= 12.5 - 8 = 4.5 \text{ kN}$$

$$w = \frac{4.5}{3} = 1.5 \text{ kN/m}$$

7. 8 kN S.F. at d indicates a concentrated load of 8 kN at d.

Check Total down word load

$$= 1.5 \times 12 + 6 + 8 = 32 \text{ kN}$$

$$\text{Upward reaction } = 11.20 + 20.80 = 32 \text{ kN}$$

Fig. 4.29 shows the loading diagram.

Example 4.15. Draw S.F. and B.M. diagrams for the beam shown in Fig. 4.30. Determine location and magnitude of maximum shear and maximum bending moment.

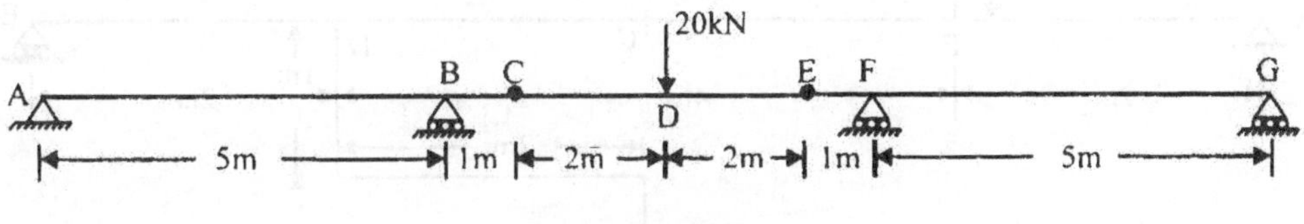

(*a*) Compound beam.

10kN 20kN 10kN

A B C D E F G

10kN 10kN

(*b*) F.B.D. diagrams.

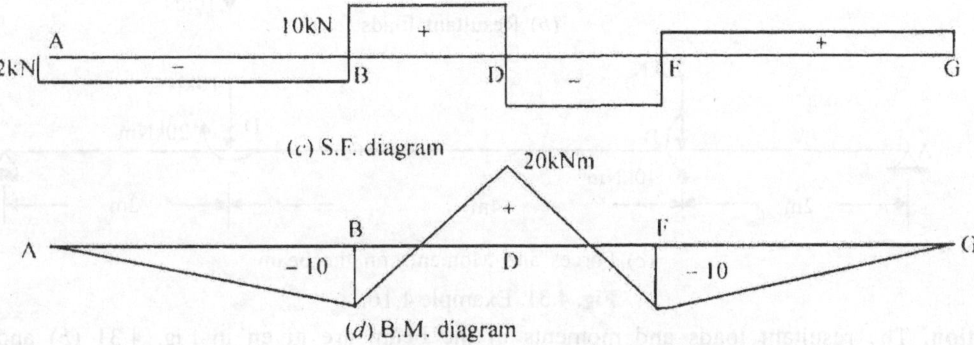

(*c*) S.F. diagram

(*d*) B.M. diagram

Fig. 4.30. Example 4.15.

Solution. The free body diagrams for the portion ABC, CE and EFG are shown in Fig. 4.30 (*b*). On CE 20 kN being central load the reactions at C and E are 10 kN each. The corresponding loads in span ABC at C and in span EFG at E are 10 kN downward.

$R_A = 2$ kN ↓ and $R_B = 12$ kN ↑

Similarly $R_F = 12$ kN ↑ and $R_G = 2$ kN ↓

As there is no horizontal load $H_A = 0$

The S.F. diagram is shown in Fig. 4.30 (*c*) which is self explanatory

B.M. diagram is drawn by noting

B.M. at $A = 0$, at $G = 0$

B.M. at $B = -10$ kNm, at $F = -10$ kNm

B.M. at $C = 0$ at $E = 0$

B.M. at $D = \dfrac{20 \times 4}{4} = +20$ kNm

B.M. diagram is shown in Fig. 4.30 (*d*)

Example 4.16. Draw the S.F. and B.M. diagrams for the beam shown in Fig. 4.31 (*a*).

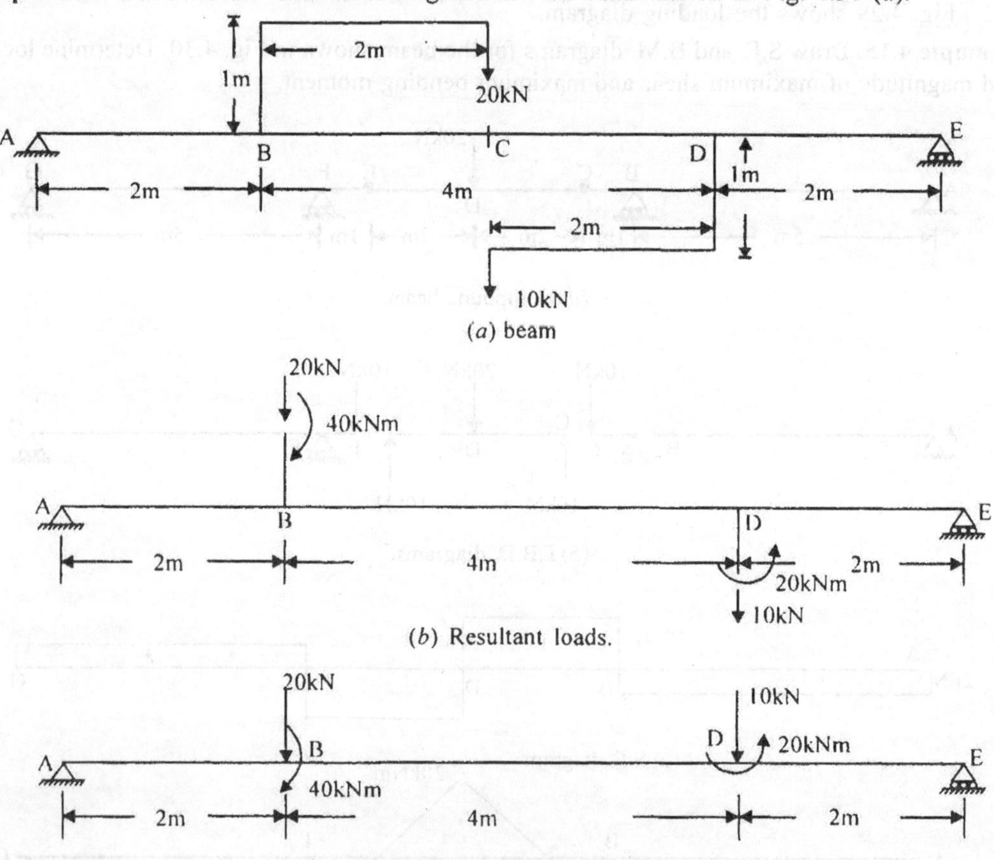

Fig. 4.31. Example 4.16.

Solution. The resultant loads and moments in the beam are given in Fig. 4.31 (*b*) and (*c*) respectively.

Calculation of Reactions

Taking moments about E

$$R_A \times 8 + 40 - 20 \times 6 - 20 - 10 \times 2 = 0$$

$$R_A = \frac{120}{8} = 15 \text{ kN} \uparrow$$

$$-R_E \times 8 - 20 + 10 \times 6 + 40 + 20 \times 2 = 0$$

or $\quad R_E \times 8 = 120 \quad R_E = 15 \text{ kN} \uparrow$

$$R_A + R_E = 15 + 15 = 30 = 20 + 10 \text{ (Check)}$$

S.F. diagram is shown in Fig. 4.32 (a). It is + 15 kN between A and B, – 5 kN between B and D and – 15 kN between D and E.

B.M. diagram is shown in Fig. 4.32 (b)

Moment at B (left side) = + 30 kNm

Moment at B (right side) = + 30 + 40 = + 70 kNm.

at $C = 15 \times 4 + 40 - 20 \times 2 = 60$ kNm

at D (right side) = $15 \times 2 = 30$ kNm

at D (left side) = $30 + 20 = 50$ kNm

at $E = 0$

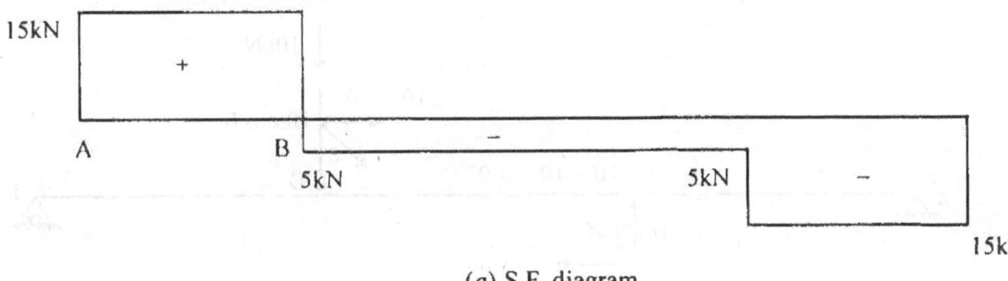

(a) S.F. diagram

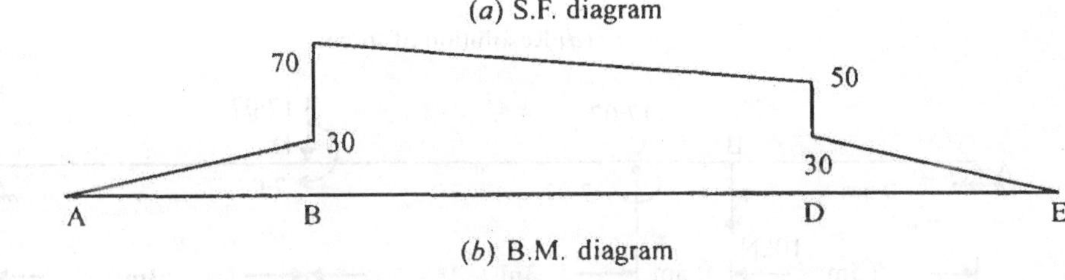

(b) B.M. diagram

Fig. 4.32. Example 4.16.

Example 4.17. Draw the S.F. and B.M. digrams of the beam loaded as shown in Fig. 4.33 (a)

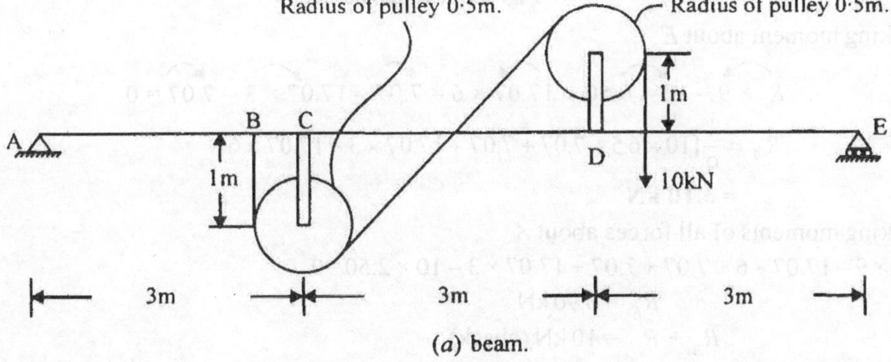

(a) beam.

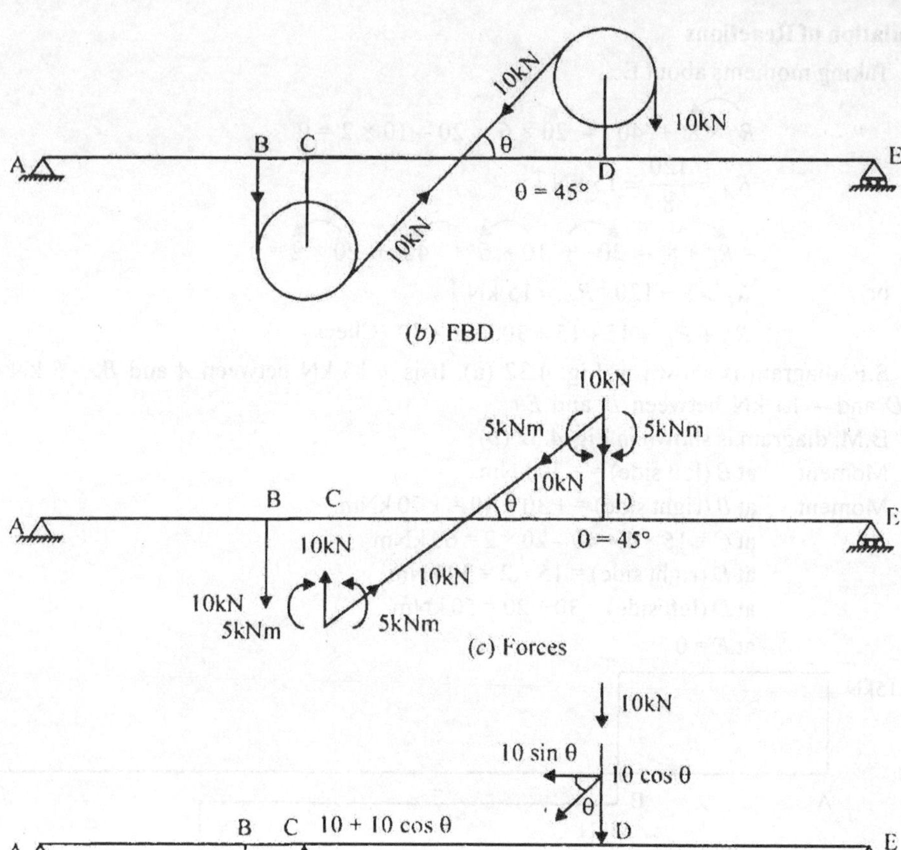

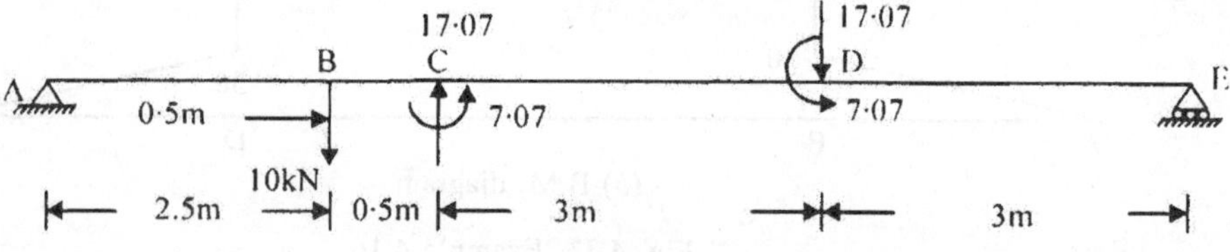

(e) Forces and moments acting.

Fig. 4.33. Example 4.17.

Taking moment about E

$$R_A \times 9 - 10 \times 6.50 + 17.07 \times 6 - 7.07 - 17.07 \times 3 - 7.07 = 0$$

or $$R_A = \frac{1}{9}[10 \times 6.5 + 7.07 + 7.07 + 17.07 \times 3 - 17.07 \times 6]$$

$$= \mathbf{3.10\ kN}$$

Taking moments of all forces about A

$$R_E \times 9 - 17.07 \times 6 + 7.07 + 7.07 + 17.07 \times 3 - 10 \times 2.50 = 0$$

$$R_E = 6.90\text{ kN}$$

$$R_A + R_E = 10\text{ kN (check)}$$

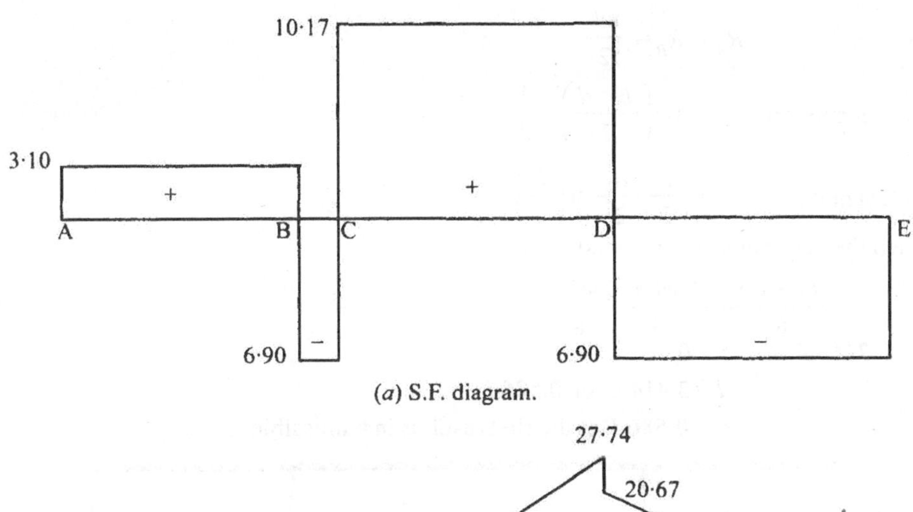

(*a*) S.F. diagram.

(*b*) B.M. diagram.

Fig. 4.34. Example 4.17.

S.F. at	$A = R_A = +3.10$
Right of	$B = 3.10 - 10.00 = -6.90$
Right of	$C = -6.90 + 17.07 = +10.17$
	$D = +10.17 - 17.07 = -6.90$
	$E = -6.90$
B.M. at	$A = 0$
at	$B = 3.1 \times 2.5 = 7.75$ kNm
Left of	$C = 4.30$
Right of	$C = 4.30 - 7.07 = -2.77$
Left of	$D = 3.1 \times 6.0 - 10 \times 3.5 + 17.07 \times 3 - 7.07$ $= 27.74$
Right of	$D = 27.74 - 7.07 = 20.67$
	$E = 0$

4.9 MISCELLANEOUS EXAMPLES

Example 4.18. A beam with two equal overhangs loaded by a uniformly distributed load has a length '*l*'. Find the distance '*d*' between the supports such that the bending moment at the middle is numerically equal to the moments at the supports. Draw the shearing force and bending moment diagrams for this case.

Solution.

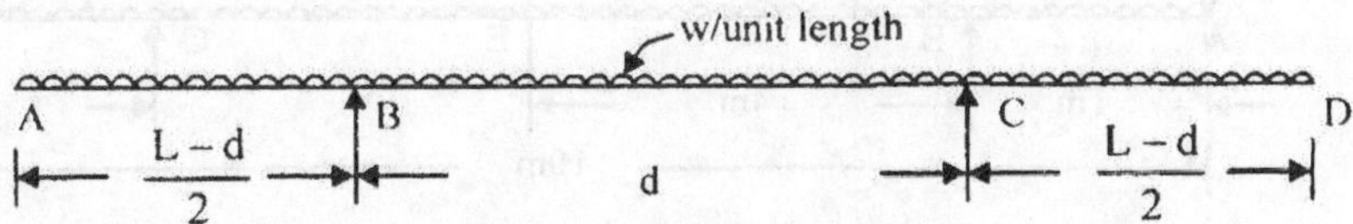

Fig. 4.35. (*a*) Example 4.18.

$$R_A = R_B = \frac{wL}{2}$$

B.M. at support $= w\cdot\left(\frac{L-d}{2}\right)^2\cdot\frac{1}{2}$

B.M. at middle $= \frac{wL}{2}\cdot\frac{d}{2} - w\left(\frac{L}{2}\right)^2\cdot\frac{1}{2}$

When they are numerically equal

$$\frac{w(L-d)^2}{8} = \frac{wLd}{4} - \frac{wL^2}{8}$$

or $\quad 2L^2 - 4Ld + d^2 = 0$

or $\quad d = 3.414\,L$ or $0.586\,L$

$d = 0.586\,L$ as the first result is in admissible

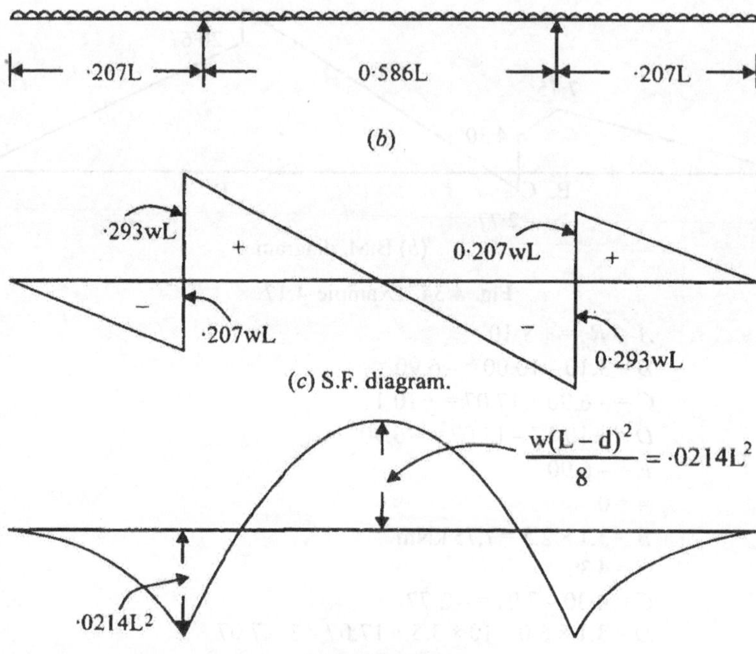

(d) B.M. diagram.

Fig. 4.35. (*b*), (*c*) and (*d*) Example 4.18.

Example 4.19. A horizontal beam *AD*, 10 metres long carries a uniformly distributed load of 180 *N* per metre run together with a concentrated load of 450 *N* at the left end *A*. The beam is supported at a point *B* which is 1 m from *A* and at *C* which is in the right half of the beam and *x* metres from the end *D*. Determine the value of *x* if the mid point of the beam is a point of inflexion and for this arrangement plot the bending moment diagram indicating the principal numerical values. Locate any other point of inflexion. [L.U]

450N 180N/m

A B E C D

Solution. 1m 4m x

10m

Fig. 4.36. Example 4.19.

Taking moment about C

$$R_B \times (10-1-x) - 450(10-x) - 180 \times 10 \times (5-x) = 0$$

or $$R_B = \frac{13500-2250x}{9-x}$$

Bending moment at E (mid-point)

$$= 450 \times 5 + 180 \times 5 \times 2.5 - \frac{13500-2250x}{9-x} \times 4 = 0$$

or $$x = 3 \text{ m}$$

$$R_B = \frac{13500-2250 \times 3}{9-3} = 1125\text{N}$$

B.M. at $$B = -(450 \times 1 + 180 \times 0.5) = -540 \text{ Nm}$$

B.M. at x from A

$$= -450 \times x - \frac{180 \times x^2}{2} + 1125 \times (x-1)$$

Points of inflexion B.M. = 0

or $$90x^2 - 675x + 1125 = 0$$

or $$x = 5 \text{ or } 2.5$$

Hence the 2nd inflexion point is at 2.5 m from A.

B.M. is maximum when $\frac{dM}{dx} = 0$

or $$2x - 7.5 = 0$$

or $$x = 3.75 \text{ m}$$

$$M_{x=3.75} = -90 \times 3.75^2 + 6.75 \times 3.75 - 1125$$
$$= +140.625 \text{ kNm}$$

$$M \text{ at } B = -450 - \frac{180}{2} = -540 \textbf{ Nm}$$

$$M \text{ at } C = \frac{-180 \times 3^2}{2} = -810 \text{ Nm}$$

The B.M. diagram is shown in Fig. 4.37

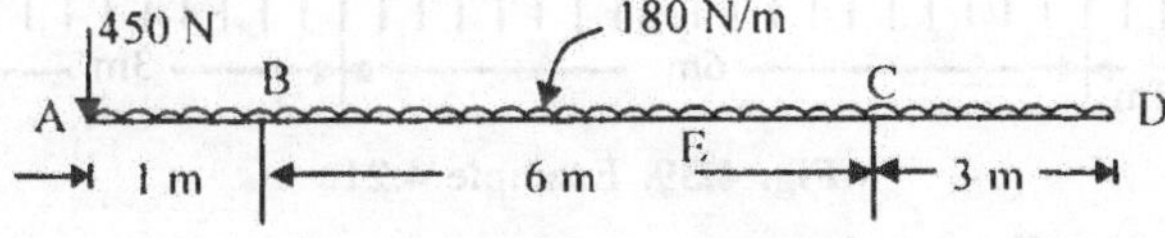

(a) Beam and Loading

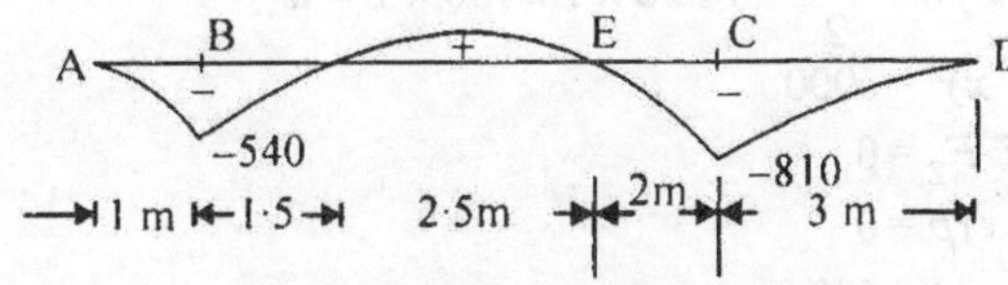

(b) B.M. diagram

Fig. 4.37. Example 4.19.

Example 4.20. A beam carrying a uniformly distributed load rests on two suppors 'b' meters apart with equal overhang 'a' meters at each end. Determine the ratio $\frac{b}{a}$ for the maximum bending moment as small as possible. Use this result to determine the most economical length for a sleeper if the rail centres are 1.6 m apart. [I.A.S. 1969]

Solution.

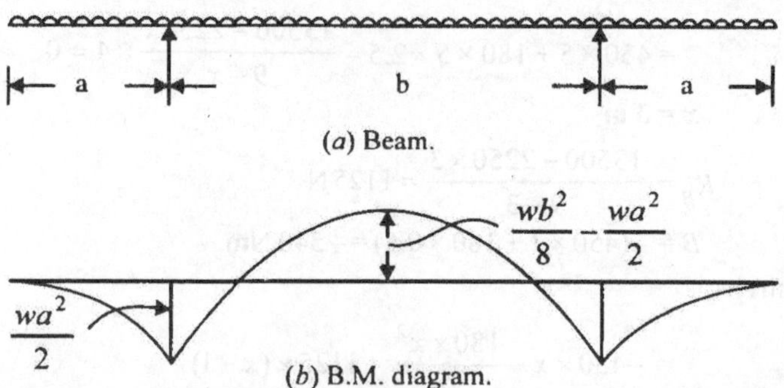

Fig. 4.38. Example 4.20.

Maximum B.M. will be minimum when negative moment at support is equal to positive moment at midspan. Hence

$$\frac{wa^2}{2} = \frac{1}{2}\left[\frac{wb^2}{8}\right]$$

or $$\frac{b^2}{a^2} = 8 \text{ or } \frac{b}{a} = 2\sqrt{2}$$

When $$b = 1.6\ m \quad a = \frac{1.6}{2\sqrt{2}}$$

$$\text{Total length} = 1.6 + \frac{1.6}{\sqrt{2}} = 2.73 \text{ m}$$

Example 4.21. For the foundation grade beam shown in Fig. 4.39. assuming uniform soil pressure, determine the value of 'x' and draw the S.F. and B.M. diagram.

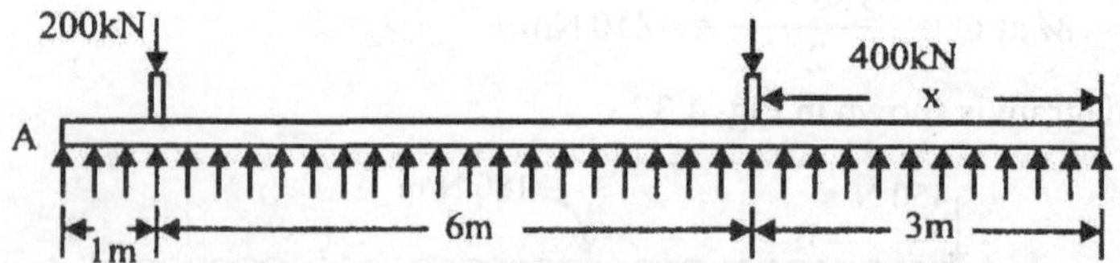

Fig. 4.39. Example 4.21.

Solution. Taking moments about 'A'

$$-p(7+x) \times \frac{(7+x)}{2} + 200 \times 1 + 400 \times 7 = 0$$

or $$p.(7+x)^2 = 6000$$

From $$\Sigma F_z = 0$$

$$200 + 400 - (7+x)p = 0$$

or $$p(7+x) = 600$$

or $$(7+x) = 10$$

$$x = 3 \text{ m}$$

$$p = \frac{600}{7+x} = 60 \text{ kN/m}$$

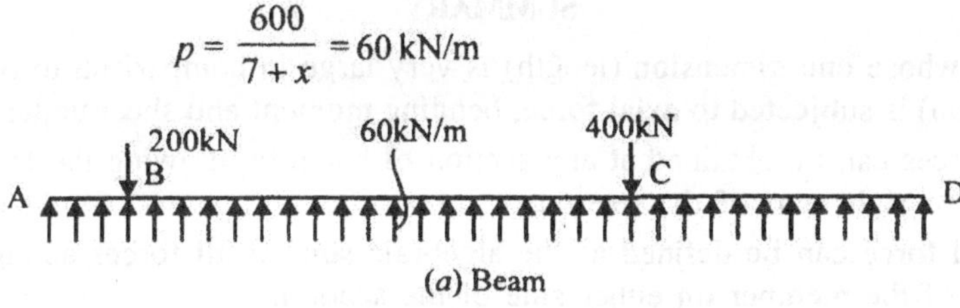

(a) Beam

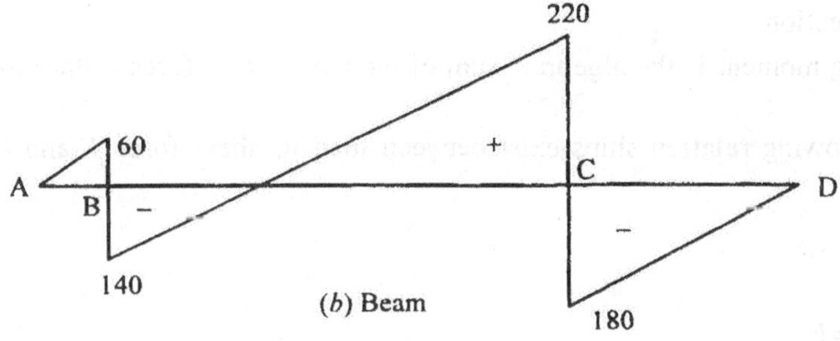

(b) Beam

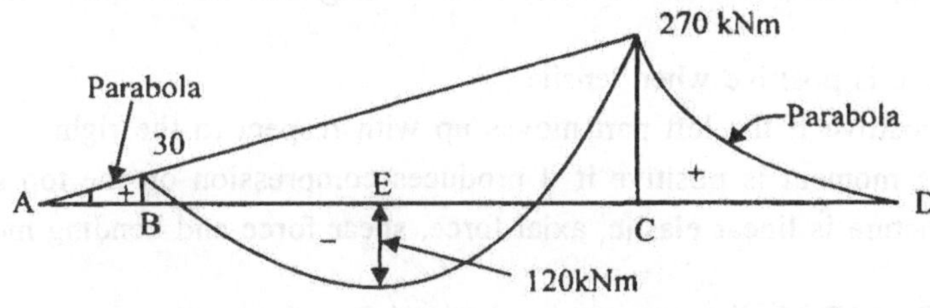

Fig. 4.40. Example 4.21.

S.F. diagram

S.F. at $A = 0$

At left of $B = 1 \times 60 = +60$

right of $B = +60 - 200 = -140$

at $C = -140 + 6 \times 60 = +220$

at right of $C = 220 - 400 = -180$

at $D = -180 + 60 \times 3 = 0$

B.M. diagram

At $A = 0$

at $B = \frac{60 \times 1^2}{2} = +30$ kNm

at E (midpoint of *BC*)

$$= \frac{60 \times 4^2}{2} - 200 \times 3$$

$$= 480 - 600 = -120 \text{ kNm}$$

at $C = \frac{60 \times 3^2}{2} = 270$ kNm

SUMMARY

1. A beam whose one dimension (length) is very large in comparison to other two dimensions (cross section) is subjected to axial force, bending moment and shear under transverse load.

2. Three forces can be obtained at any section of beam by drawing the free body diagram either to the left or right part of the section.

3. (*a*) Axial force can be defined as the algebraic sum of all forces acting parallel to the longitudinal axis of the member on either side of the section.

(*b*) Shear force can be defined as the algebraic sum of all normal forces either to the left or right of the section.

(*c*) Bending moment is the algebraic sum of moments of all forces either to the left or right of the section.

4. The following relation ships exist between load *w*, shear force *V* and bending moment *M*.

$$\frac{dV}{dx} = -w$$

$$\frac{dM}{dx} = V$$

5. Axial force, shear force and bending moment diagrams can be drawn with the following sign conventions:

(*i*) Axial force is positive when tensile

(*ii*) S.F. is positive if the left part moves up with respect to the right.

(*iii*) Bending moment is positive if it produces compression on the top surface.

6. If the structure is linear elastic, axial force, shear force and bending moment can all be superposed.

7. Just as S.F. and B.M. diagrams can be obtained from the loading diagram, loading diagram can be obtained when S.F. and/or B.M. diagram are given.

PROBLEMS

A. Objective

4.1 Tick the correct answer.

(*i*) A simply supported beam has an overhang of length less than the supported span. If the beam is uniformly loaded throughout the whole length, the point of contraflexure

(*a*) does not exist

(*b*) is more than one in number

(*c*) occurs within the over hang

(*d*) occurs within the supported span **[AMIE Winter 1998]**

(*ii*) A small settlement of the supports of a simply supported beam with an over hang will

(*a*) increase the value of the maximum sagging B.M.

(*b*) increase the value of maximum hogging B.M.

(*c*) increase the value of the maximum S.F.

(*d*) None of the above **[AMIE Winter 1998]**

(*iii*) In a vertically loaded propped cantilever any settlement of the prop would.

(*a*) reduce the hogging B.M. at the fixed end.

(*b*) increase the hogging B.M. at the fixed end.

(*c*) affect only S.F. and not B.M. values.

(*d*) affect neither B.M. nor S.F. values. [AMIE Winter 1998]

(*iv*) Point of contraflexure occurs in a structure when

(*a*) bending moment is zero

(*b*) bending moment changes sign

(*c*) shear force is zero

(*d*) All of the above [AMIE Winter '97]

(*v*) Due to settlement of support in a fixed beam midspan bending moment.

(*a*) increases

(*b*) decreases

(*c*) remains unchanged [AMIE Winter '97]

(*vi*) A two span continuous beam on simple supports carries a u.d.l over its entire length. If the central support sinks by a small amount, the maximum sagging moment in the beam will.

(*a*) increase (*b*) decrease (*c*) remain the same

[AMIE Summer '99]

B. Mathematical

4.2. Calculate the reactions at A and D for the beam shown in Fig. P 4.1. Draw the bending moment and shear force diagrams showing all the important values.

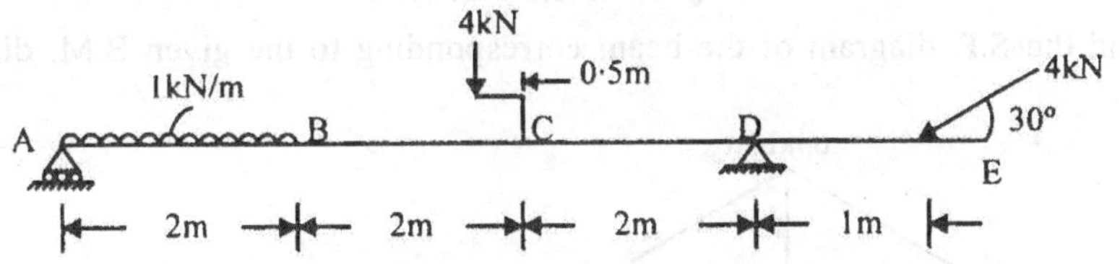

Fig. P4.1 for Prob. 4.2.

4.3. Draw the S.F. and B.M. diagrams for the beam shown below Fig. P. 4.2.

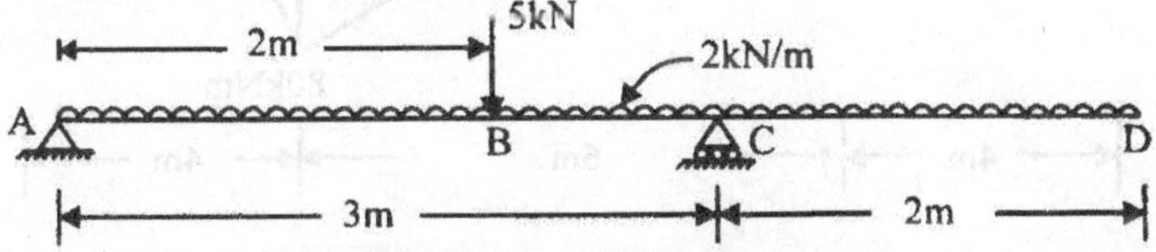

Fig. P 4.2. for Prob. 4.3.

4.4. Draw the S.F. and B.M. diagrams for the beam shown in Fig. P 4.3.

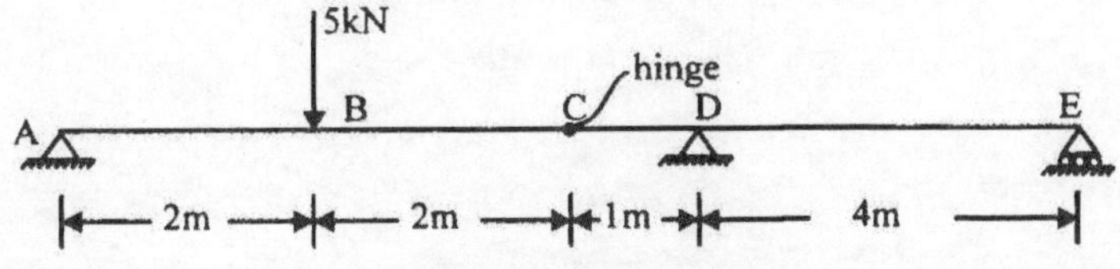

Fig. P 4.3. for Prob 4.4.

4.5. A beam having a span of 6 m is simply supported at its ends. If carries a distributed load which increases linearly from zero at one end to 10 t/m at the other, draw the shearing force and bending moment diagrams. Determine the maximum bending moment and its location.

[AM and S.O.M. Winter 1987]

4.6. A beam AB hinged at B and freely supported at A carries loads as shown in Fig. P 4.4. Estimate (*i*) The magnitude and direction of reaction at B (*ii*) the maximum B.M. occurring in the beam and (ii) draw the B.M, S.F. and thrust diagrams.

[A.M. and S.O.M. Summer 1987]

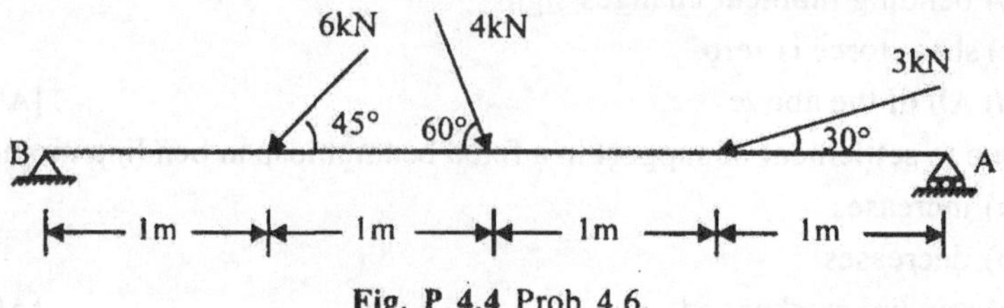

Fig. P 4.4 Prob 4.6.

4.7. Draw B.M. and S.F. for the beam given in Fig. P 4.5.

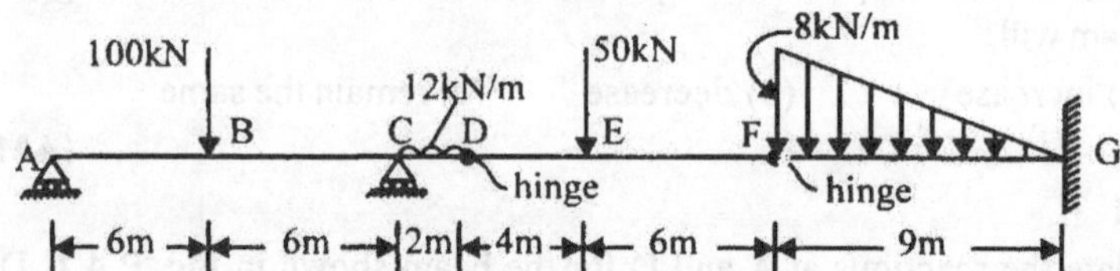

Fig. P 4.5 for Prob 4.7.

4.8. Find the S.F. diagram of the beam corresponding to the given B.M. diagram

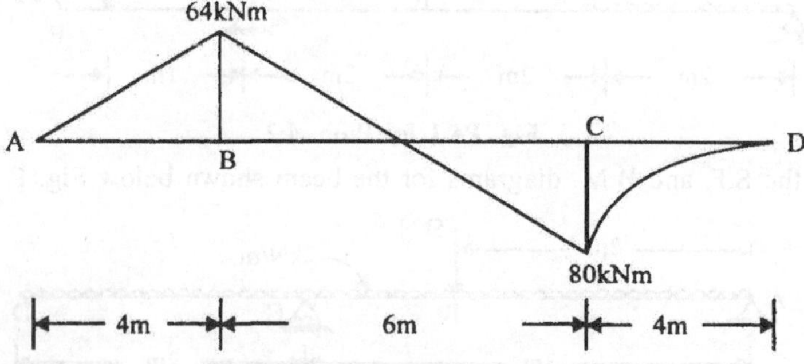

Fig. P 4.6 For Problem 4.8.

5

THRUST, SHEAR AND MOMENT DIAGRAMS – FRAMES

5.1 INTRODUCTION

A rigid frame consists of members (horizontal, vertical or inclined) rigidly connected at the joints. As a result when a joint translates, all the connected members at the joint translates the same amount. Similarly when a joint rotates, all the connected members rotate by the same amount. In steel structures rigidity of joints can be obtained by rivetting, bolting or welded connections. In reinforced concrete structure the rigidity of joint is obtained by casting the two members meeting at the joint together and extending reinforcements of one member to another member. Typical frames are shown in Fig. 5.1 (a), (b) and (c).

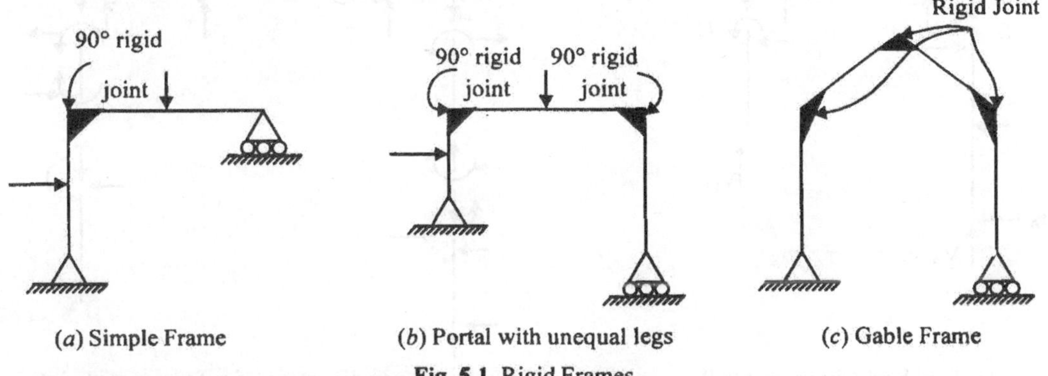

Fig. 5.1. Rigid Frames.

5.2 ANALYSIS OF RIGID FRAMES

Analysis of rigid frames usually involves three calculations

1. Reactions at supports.
2. Member end actions.
3. Internal member actions.

Reactions are computed by applying equations of statics for statically determinate problems. For plane frames when all the loads and members lie in one plane, there are three equations of statics, $\Sigma X = 0$, $\Sigma Y = 0$ and $\Sigma M_z = 0$. For space frame the equations are $\Sigma X = 0$, $\Sigma Y = 0$, $\Sigma Z = 0$ and $\Sigma M_x = 0$ $\Sigma M_y = 0$ and $\Sigma M_z = 0$.

Member end actions are obtained by drawing the free body diagrams (FBDs) of the members and applying the equations of statical equilibrium.

Internal member actions are obtained by

(*a*) Considering part of the member by making suitable cut.

(*b*) Drawing FBD of the part cut

(*c*) Applying equilibrium equations.

Fig. 5.2 shows all these steps.

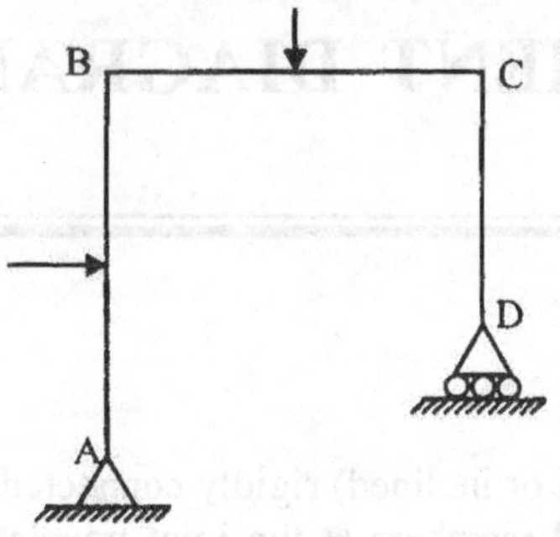

(*a*) The Portal frame

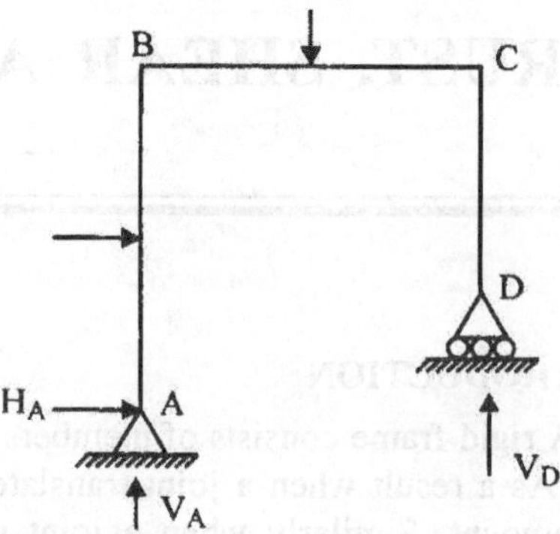

(*b*) Free body diagram for computing reactions H_A, V_A and V_D.

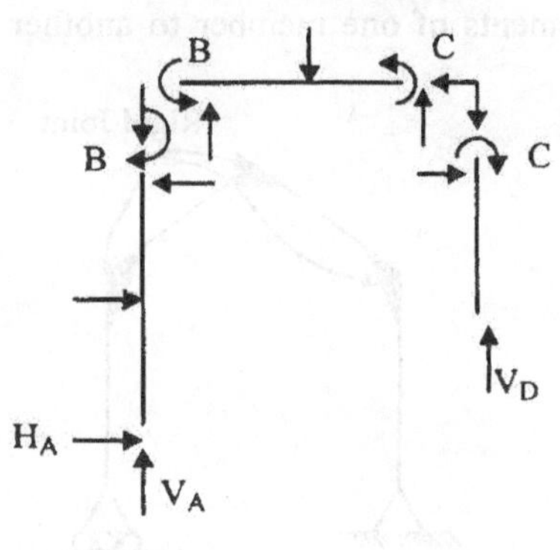

(*c*) Free body diagram of all members to compute end actions.

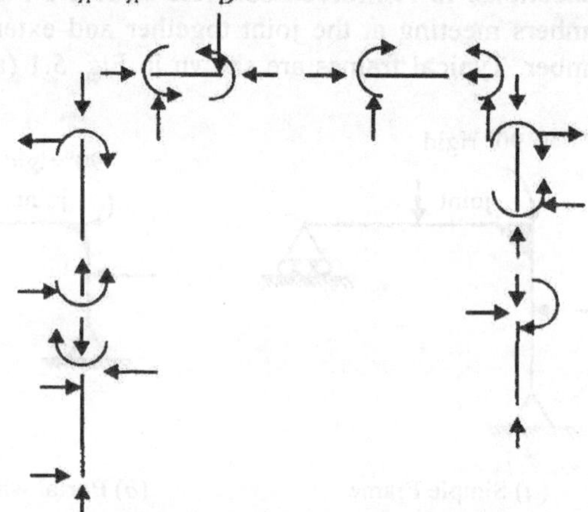

(*d*) Free body diagrams of cut to compute internal actions for computation of axial force, S.F. and B.M.

Fig. 5.2. Free body diagrams.

The substitution of proper number of unkown forces in a cut is important. If a cut is made in

(*i*) a member of a truss – unknown force is 1 i.e. tension or compression force along the member.

(*ii*) a hinge – unknown forces are 2 (horizontal and vertical) as hinge does not carry any moment.

(*iii*) a plane frame – unknown forces are three, horizontal and vertical forces and a moment.

(*iv*) a space frame – unknown force are six – three forces and three moments.

5.3 SIGN CONVENTION – A frame is subjected to

(*i*) Axial force

(*ii*) S.F.

(*iii*) B.M.

In axial force diagram tension is considered positive and compression negative.

In a beam, the idea of moving up and down as regards shear and sagging and hogging in B.M. can be visualised. Such a concept, however, is not valid for frames which may have vertical member. Hence we should look at a frame from inside. Looking from inside if the forces to the right are inward, shear is positive. Similarly for B.M. if the tension is inside the frame, it is positive. These are shown in Fig. 5.3.

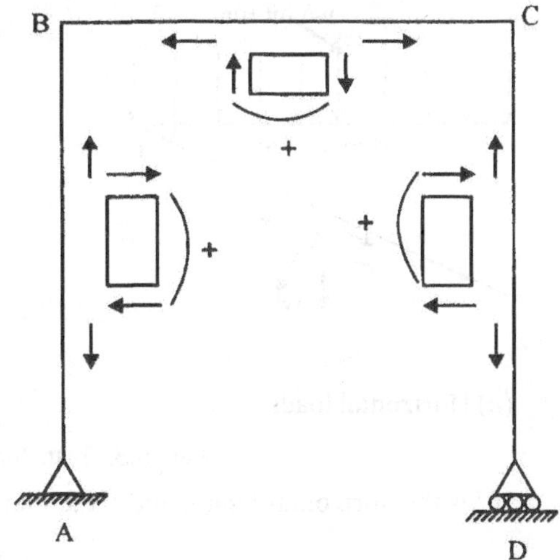

Fig. 5.3. Sign conventions of A.F, S.F. and B.M. diagrams of frames.

5.4 TRANSFORMATION OF CONCENTRATED AND DISTRIBUTED LOADS

Reactions are usually foundout along horizontal and vertical axes (global). All member axes may not coincide with the global axes. Axial force, S.F. and B.M. are to be computed with respect to member axes as

(*i*) Axial force is force acting along the axis of the member.

(*ii*) S.F. is force acting at right angle to the member axis.

(*iii*) B.M. is moment of forces at right angle to the axis of the member.

Hence to compute A.F., S.F. and B.M. it is necessary to transform forces from global axis to member axes.

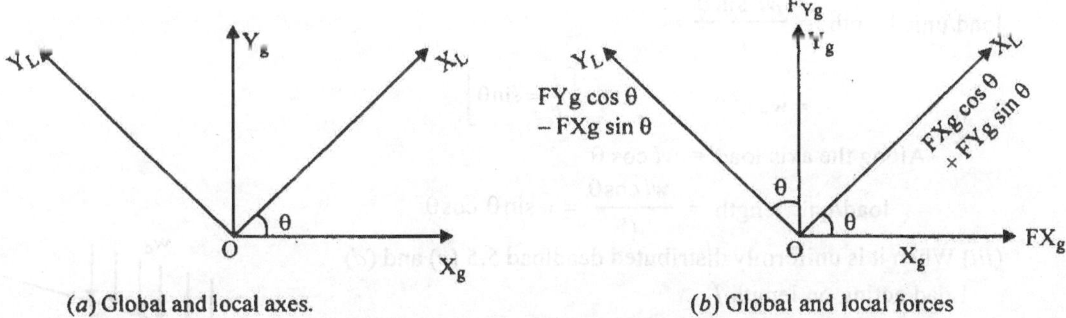

(*a*) Global and local axes. (*b*) Global and local forces

Fig. 5.4.

(*a*) Transformation of concentrated forces:

Fig. 5.4 (*a*) shows the global axes OX_g and OY_g and member axes (local) OX_L and OY_L

Fig. 5.4 (*b*) shows the global axes forces FX_g and FY_g and corresponding member axes forces $FX_g \cos\theta + FY_g \sin\theta$ along OX_L and $FY_g \cos\theta - FX_g \sin\theta$ along OY_L.

(*b*) Transformation of distributed forces:

(*i*) Fig. 5.5 (*a*) shows uniformly distributed horizontal loads acting on an inclined member.

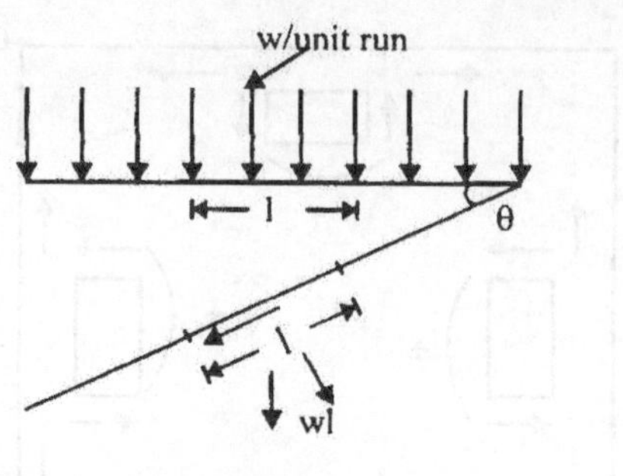

(*a*) Horizontal loads.

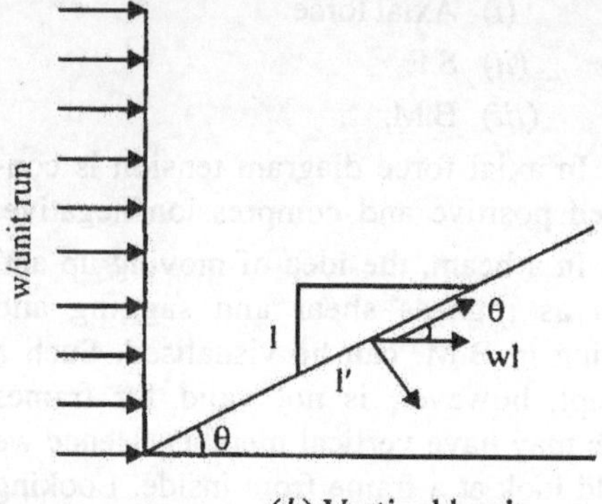

(*b*) Vertical loads

Fig. 5.5. Transformation of distributed forces.

If l is the horizontal length and l' the inclined length,

$$l' = \frac{l}{\cos\theta}$$

Load acting over length $l = wl$. which is acting now over the inclined length l'. Resolving along and perpendicular to the member axes.

Force perpendicular to the axis $= w.l.\ \cos\theta = w\,l'\cos^2\theta$

load/unit length, $w_n = \dfrac{wl'\cos^2\theta}{l'} = w\cos^2\theta$

Similarly force along the axis

$$= wl\ .\ \sin\theta$$

$$= wl'\ .\ \sin\theta\cos\theta$$

force/unit length $\quad w_a = w\sin\theta\cos\theta$

(*ii*) When the distributed load is vertical Fig. 5.5 (*b*), as before load acting over lenth $l = wl$. Resolving, load perpendicular to $l' = wl\sin\theta$

load/unit length $= \dfrac{wl.\sin\theta}{l'}$

$= w\sin^2\theta \qquad \left[\dfrac{l}{l'} = \sin\theta\right]$

Along the axis load $= wl\cos\theta$

load/unit length $= \dfrac{wl\cos\theta}{l'} = w\sin\theta\cos\theta$

(*iii*) When it is uniformly distributed deadload 5.5 (*c*) and (*d*)

Load acting on length l

$$= w_d\ .\ l$$

Load normal to length l

$$= w_d\ .\ l\ .\ \cos\theta$$

Normal load/unit length

$$= w_d\ .\ \cos\theta$$

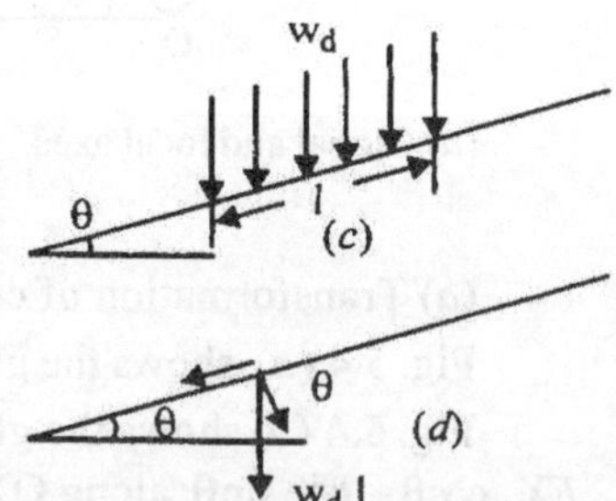

Fig. 5.5. (*c*) and (*d*)

Load along the axis of the member

$$= w_d \cdot \sin\theta$$

Example 5.1. Draw the free body diagrams of the following frame and its individual members. Compute the end actions.

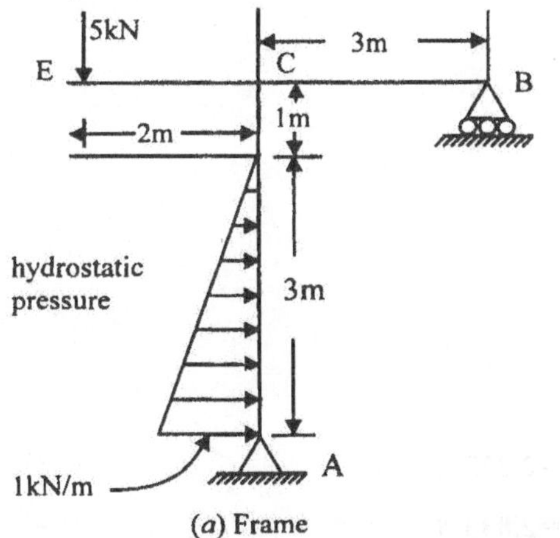

(*a*) Frame

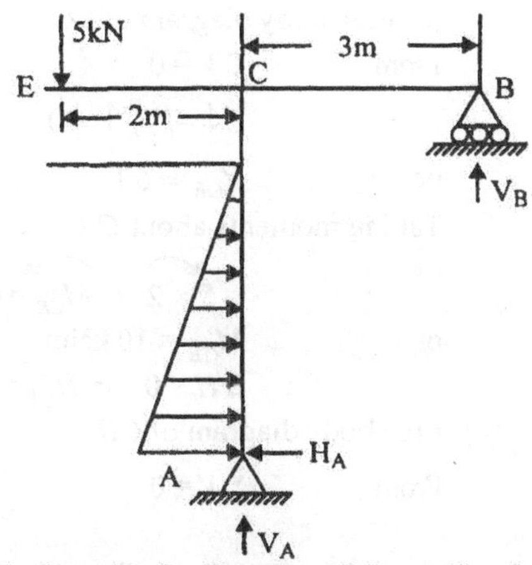

(*b*) Free body diagram of the entire frame for computing external reactions at the supports.

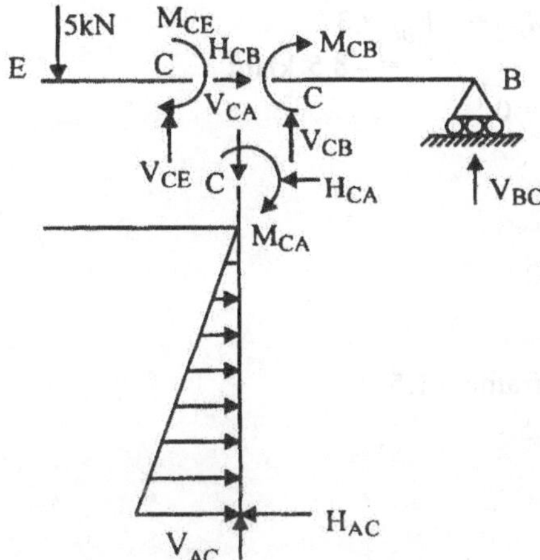

(*c*) Free body diagrams for computing member end forces

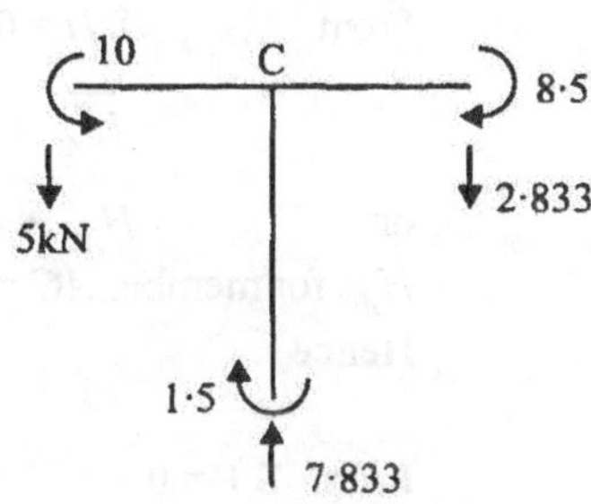

(*d*) Free body diagram of Joint *c*.

Fig. 5.6. Example 5.1.

Computations

(*i*) Reactions. Taking moments of all external forces about '*A*'

$$V_B \times 3 - \frac{1}{2} \times 1 \times 3 \times 1 + 5 \times 2 = 0$$

or $\quad V_B = \frac{1}{3}[-8.5] = -2.833 \text{ kN} \downarrow$

From $\quad \Sigma V = 0$

$$5\downarrow + 2.833\downarrow - V_A = 0$$

or $\quad V_A = 7.833\uparrow$

From $\Sigma H = 0 \quad \frac{1}{2} \times 1 \times 3 - H_A = 0$

or $H_A = 1.5 \leftarrow$

(*ii*) End Actions

(*a*) Free body diagram of *EC*

From $\Sigma V = 0$

$$5\downarrow - V_{CE}\uparrow = 0$$

or $V_{CE} = 5\uparrow$

Taking moments about *C*

$$-5 \times 2 + M_{CE} = 0$$

or $M_{CE} = 10 \text{ kNm}$

$\Sigma H = 0$ or $H_{CE} = 0$

(*b*) Free body diagram of *CB*

From $\Sigma V = 0 \qquad V_{CB}\uparrow - 2.833\downarrow = 0$

or $V_{CB} = 2.833\uparrow$

From $\Sigma M_C = 0 \qquad V_{BC} \times 3 + M_{CB} = 0$

or $M_{CB} = -V_{BC} \times 3$

$= -8.5 \text{ kNm}$

From $\Sigma H = 0, \qquad H_{CB} = 0$

(*c*) Freebody diagram of *AC*

From $\Sigma H = 0$

$$\overleftarrow{H}_{AC} + \overleftarrow{H}_{CA} - \frac{1}{2} \times 1 \times 3 = 0$$

or $H_{AC} + H_{CA} = 1.5$

H_{AC} for member $AC = H_A$ for the entire frame $= 1.5$

Hence $1.5 + H_{CA} = 1.5$

$H_{CA} = 0$

From $\Sigma V = 0$

$$V_{CA}\downarrow - V_{AC}\uparrow = 0$$

But $V_{AC} = V_A = 7.833$

$\therefore \quad V_{CA} - 7.833 = 0$

or $V_{CA} = 7.833\downarrow$

From $\Sigma M_C = 0$

$$H_{AC} \times 4 + M_{CA} - \frac{1}{2} \times 1 \times 3 \times 3 = 0$$

or $M_{CA} = 4.5 - 4H_{AC}$

$= 4.5 - 4 \times 1.5$

$= -1.5$

Free body diagram of joint C is given in Fig. 5.6 (*d*)

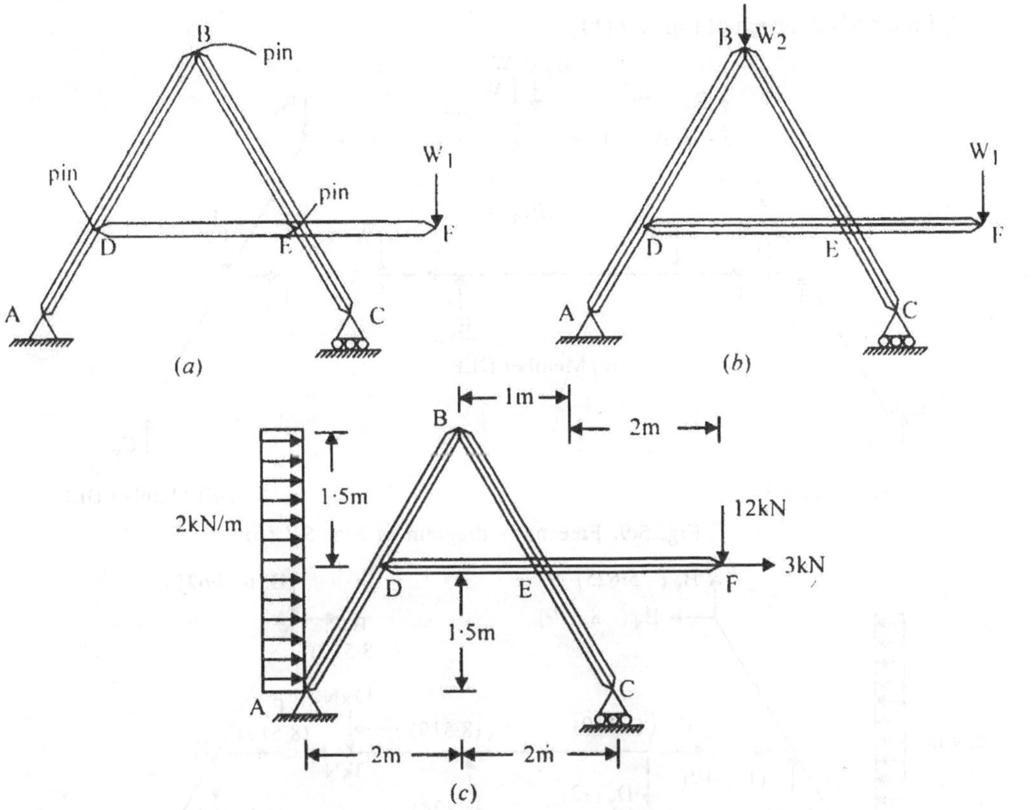

Fig. 5.7 Example 5.2.

Example 5.2. Draw the free body diagram of the following frames shown in Fig. 5.7 (*a*) and (*b*) and then solve the actions in the frame shown in Fig. 5.7 (*c*).

Solution. The free body diagrams of all members should be drawn first. At the pin connections two components of force must be shown acting in opposite directions on the connected members. As the pins are in equilibrium, the force exerted by one member must be equal and opposite to the force exerted by the other member provided no load acts on the pin. When a load acts on the pin forces exerted by members on the pin (which is equal and opposite to the forces exerted by the pin on the members) and the external load on the pin should keep the pin in equilibrium.

(*i*) Free body diagram of Fig. 5.7 (*a*)

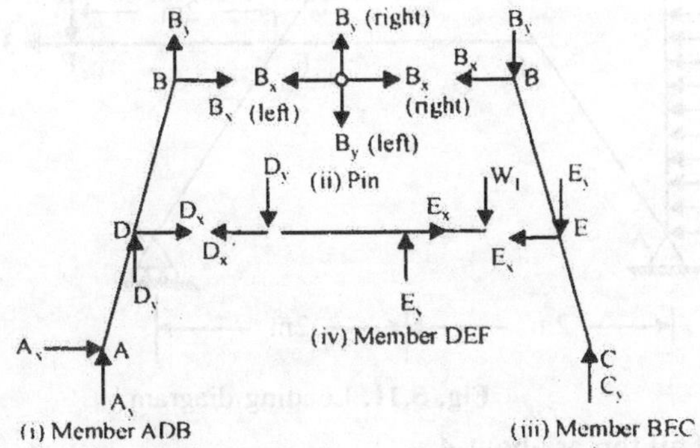

Fig. 5.8. Free body diagram of Fig. 5.7 (*a*).

(*ii*) Free body diagram of Fig. 5.7 (*b*)

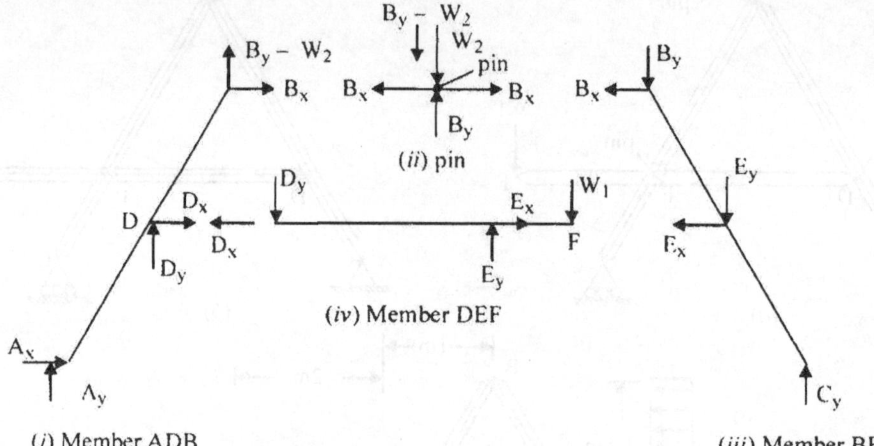

Fig. 5.9. Free body diagram of Fig. 5.7 (*b*).

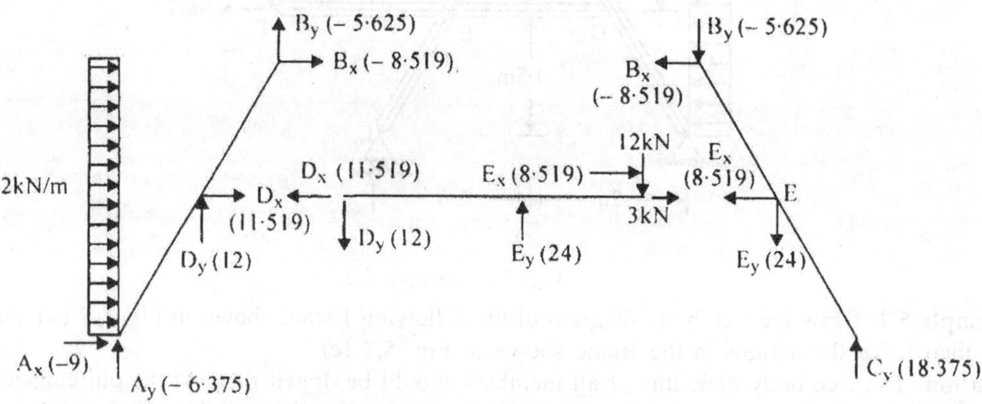

Fig. 5.10. FBD of Fig. 5.7 (*c*)

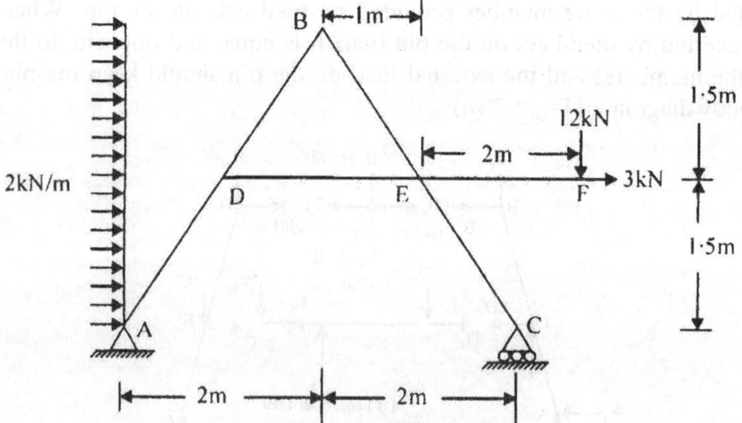

Fig. 5.11. Loading diagram.

Taking moments of all forces about *A*

$$2 \times 3 \times 1.5 + 12 \times 5 + 3 \times 1.5 - C_y \times 4 = 0$$

or $C_y = \frac{1}{4}(9+60+4.5)$

$= 18.375 \text{ kN} \uparrow$

From $\Sigma H = 0$

$\overrightarrow{A_x} + 2\times 3 + 3 = 0$

or $A_x = -9$ *i.e.* $\leftarrow$

From $\Sigma V = 0$

$A_y \uparrow + 18\cdot 375 \uparrow - 12 \downarrow = 0$

or $A_y = -6\cdot 6375 \downarrow$

From *FBD* of *ADB*

$\Sigma H = 0$ gives $\overrightarrow{2\times 3} + \overrightarrow{D_x} - \overleftarrow{9} + \overrightarrow{B_x} = 0$

or $D_x + B_x = 3.0$

$\Sigma V = 0$ gives. $-6.375 + D_v + B_v = 0$ or $D_y + B_y = 6.375$

$\Sigma M = 0 \quad 2\times 3\times 1.5 - B_y\times 2 + B_x\times 3 - D_y\times 1 + D_x\times 1.5 = 0$

or $3B_x + 1.5D_x - 2B_y - D_y = -9$

Substituting values,

$3B_x + 1.5(3 - B_x) - 2B_y - (6.375 - B_y) = -9$

or $1.5B_x - B_y = -9 - 4.5 + 6.375$

$= -7.125$

From *FBD* of *DE*

$D_x - E_x - 3 = 0$

$D_y - E_y + 12 = 0$

Adding $E_x + E_y = D_x + D_y + 9$

From *FBD* of *BEC*

$E_y = C_y - B_y = 18.375 + 5.625 = 24.00$

$B_x + E_x = 0$ *or* $E_x = -B_x = +8.519$

$D_x = 3 + E_x = 3 + 8.519 = 11.519$

$D_y = -12 + E_y = -12 + 24 = 12.00$

Adding four equations and substituting values

$$\begin{aligned} D_x - E_x - 3 &= 0 \\ D_y - E_y + 12 &= 0 \\ B_x + E_x &= 0 \\ B_y + E_y - C_y &= 0 \\ \hline B_x + B_y + (D_x + D_y) &= 9.375\,. \end{aligned}$$

In the free body diagram of *BEC* taking moments about *C*

$\overset{\curvearrowleft}{B_x} \times 3 + \overset{\curvearrowleft}{B_y} \times 2 + \overset{\curvearrowleft}{E_x} \times 1.5 + \overset{\curvearrowleft}{E_y} \times 1 = 0$

From *FBD* of *DE*, taking moments about *D*

$$-E_y \times 2 + 12 \times 4 = 0$$

or $E_y = 24$.

Substituting $3B_x + 2B_y + 1.5E_x = -24.00$

Adding $B_x + E_x = 0$ to $B_y + E_y - C_y = 0$

$$B_x + B_y + E_x + E_y = C_y = 18.375$$

$$B_x + B_y + E_x = 18.375 - 24 = -5.625$$

Solving $B_x = -8.519$ and $B_y = -5.625$

Example 5.3. Draw the *FBD* of each portion of the frame *ABCD*. Indicate the type of actions involved.

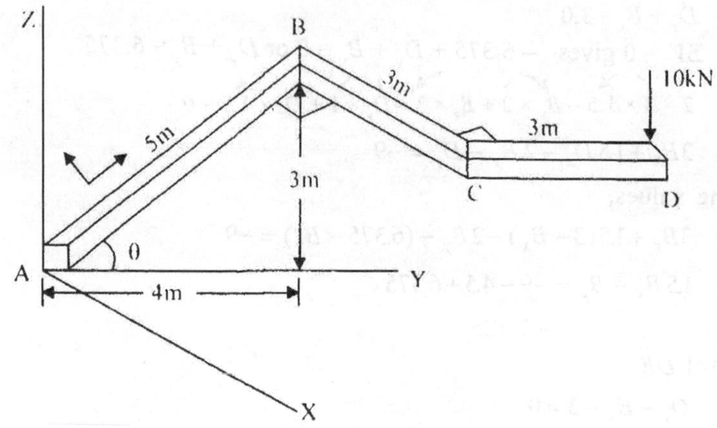

Fig. 5.12. Example 5.3.

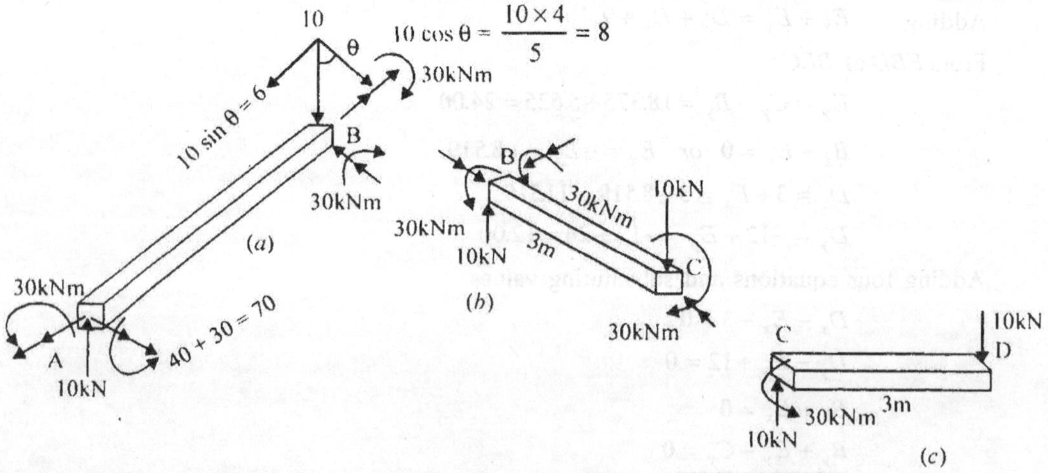

(*a*) Torsion, Shear B.M and Axial force (*b*) Torsion, Shear and and B.M. (*c*) Shear and B.M.

Fig. 5.13. Example 5.3.

Example 5.4. Analyse the composite structure shown in Fig. 5.14. Draw the shear, moment and axial force diagram of the members.

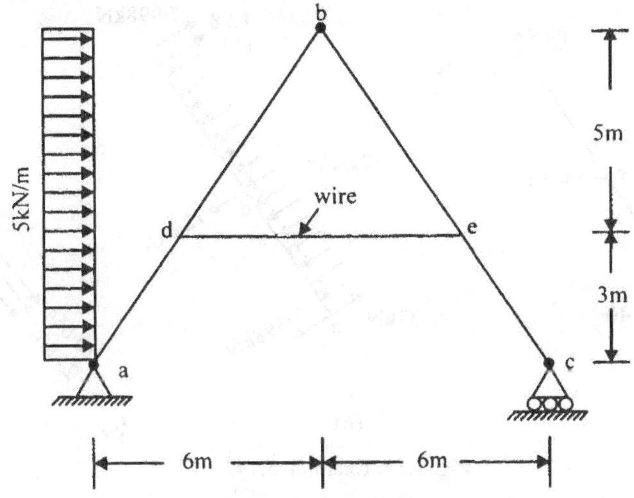

Fig. 5.14. Example 5.4.

Solution. Taking moments of all forces about '*a*'

$$V_c \times 12 - 5 \times 8 \times 4 = 0$$

or
$$V_c = 13.33 \text{ kN} \uparrow$$
$$V_a = 13.33 \text{ kN} \downarrow$$
$$H_a = 5 \times 8 = 40 \text{ kN} \leftarrow$$

Considering member adb (The wire de can exert tension only)

Taking moments about '*b*'

$$13.33 \times 6 - 40 \times 8 + 5 \times 8 \times 4 + H_d \times 5 = 0$$

or $H_d = 16 \text{ kN} \rightarrow$

From $\Sigma H = 0$

$$H_b + 5 \times 8 + 16 - 40 = 0$$

or $H_b = -16 \text{ kN} \leftarrow$

$V_b = 13.33 \text{ kN} \uparrow$

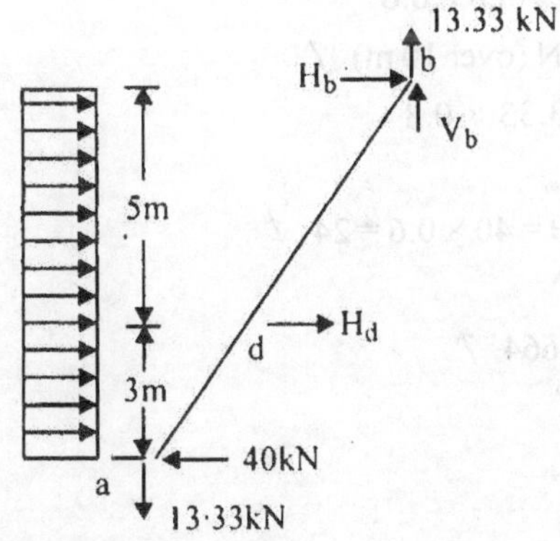

Fig. 5.15. FBD of ab (Example5.4).

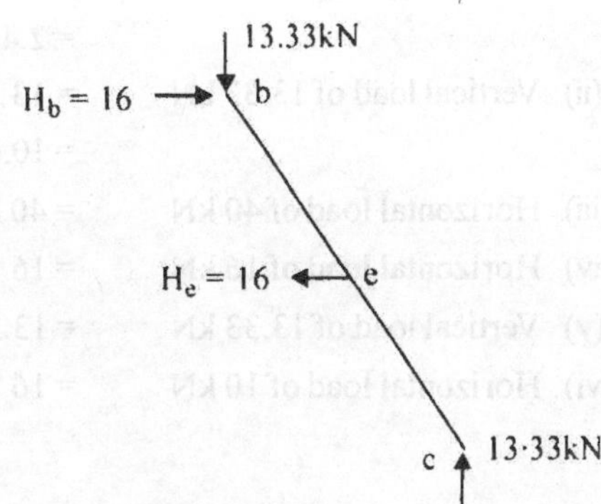

Fig. 5.16. FBD of cb (Example 5.4).

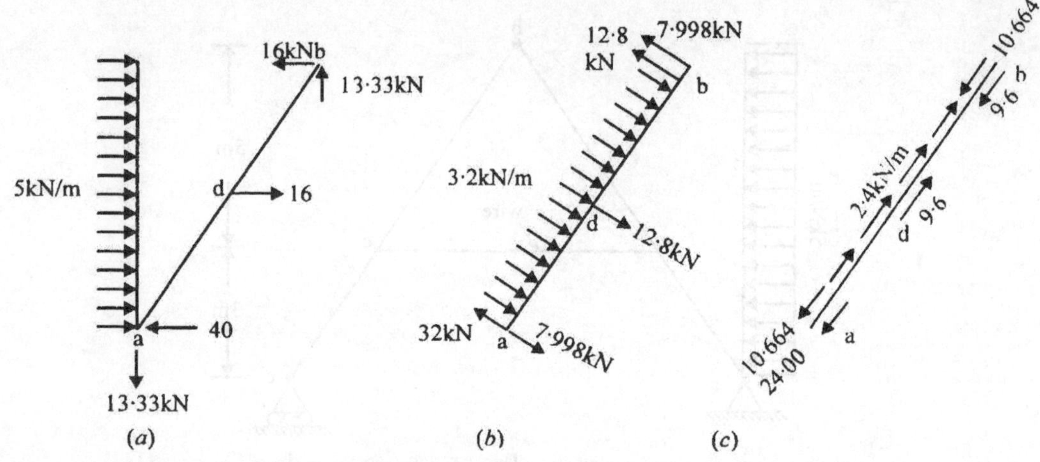

Fig. 5.17. Example 5.14.

$$\tan\theta = \frac{8}{6} = 1.333$$

$$\theta = 53.129°$$

$$\sin\theta = 0.8000$$

$$\cos\theta = 0.6000$$

Forces normal to the axis (member adb)

(i) Distributed load 5 kN/m = 5 $\sin^2\theta$ = 3.2 kN/m ↓ = 32kN (over 10m)

(ii) Vertical load of 13.33 kN = 13.33 $\cos\theta$ = 13.33 × 0.6 = 7.998 kN ↑

(iii) Horizontal load of 40 kN = 40 $\sin\theta$ = 40 × 0.8 = 32 kN ↑

(iv) Horizontal load of 16 kN = 16 $\sin\theta$ = 16 × 0.8 = 12.8 ↓

(v) Vertical load of 13.33 kN = 13.33 $\cos\theta$ = 7.9978 ↑

(vi) Horizontal load of 16 kN = 16 $\sin\theta$ = 12.8 ↑

Forces along the axis (member adb)

(i) Distributed load of 5 kN/m = 5.$\sin\theta\cos\theta$ = 5 × 0.8 × 0.6
= 2.4 kN/m = 24 kN (over 10 m) ↗

(ii) Vertical load of 13.33 kN = 13.33 $\sin\theta$ = 13.33 × 0.8
= 10.664 ↙

(iii) Horizontal load of 40 kN = 40 kN = 40 $\cos\theta$ = 40 × 0.6 = 24 ↙

(iv) Horizontal load of 16 kN = 16 × 0.6 = 9.6 ↗

(v) Vertical load of 13.33 kN = 13.33 × 0.8 = 10.664 ↗

(vi) Horizontal load of 10 kN = 16 × 0.6 = 9.6 ↙

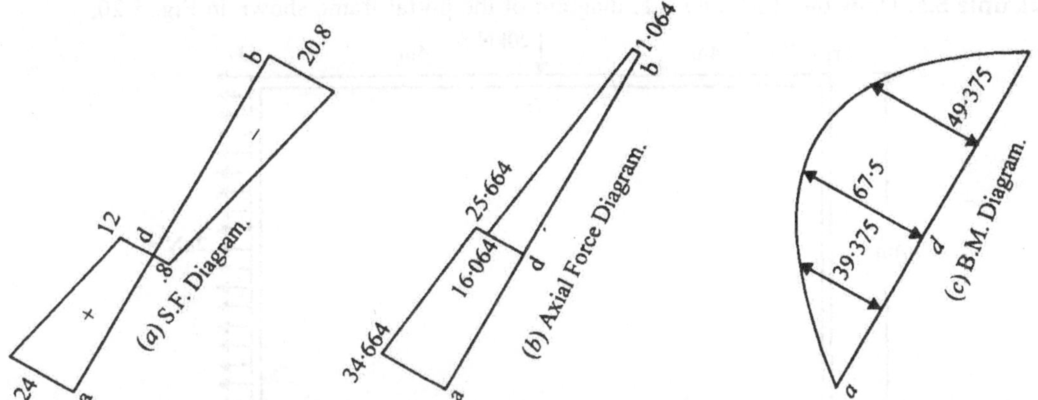

Fig. 5.18. S.F., A.F. and B.M. diagram for adb.

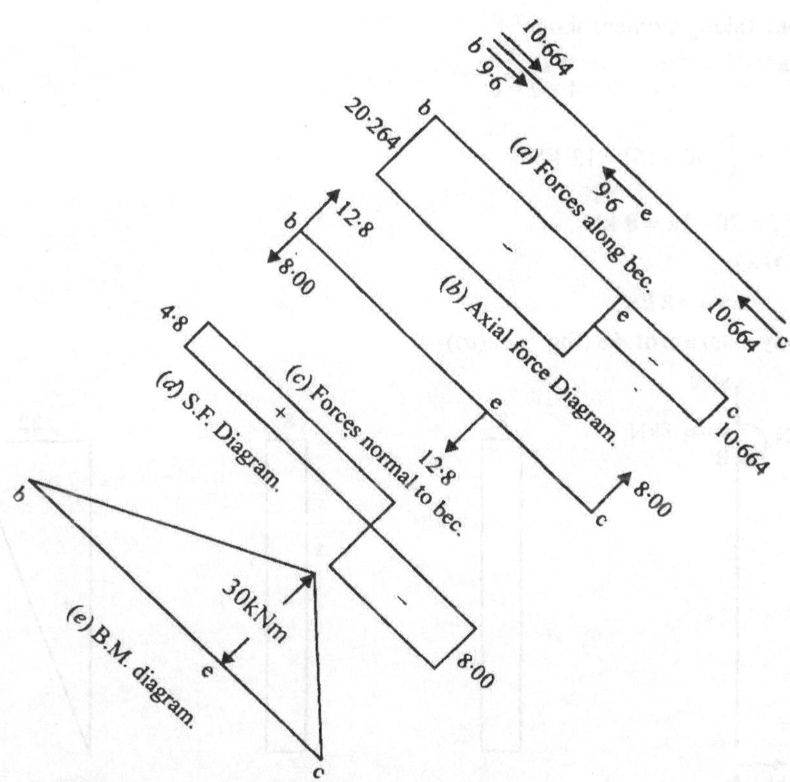

Fig. 5.19. S.F., A.F. and B.M. diagram for bec.

Example 5.5. Draw the B.M. and S.F. diagram of the portal frame shown in Fig. 5.20.

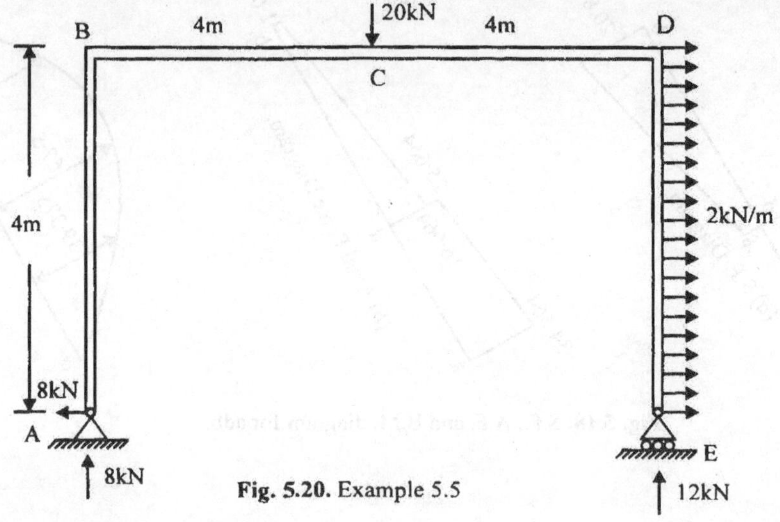

Fig. 5.20. Example 5.5

Solution: Taking moment about '*A*'

$$V_E \times 8 - 20 \times 4 - 2 \times 4 \times 2 = 0$$

or $$V_E = \frac{1}{8}[80 + 16] = 12 \text{ kN}$$

$$V_A = 20 - 12 = 8 \text{ kN}$$

From $\Sigma H = 0$

$$H_A = 2 \times 4 = 8 \text{ kN}.$$

Free body diagram of *AB* (Fig. 5.21 (*a*))

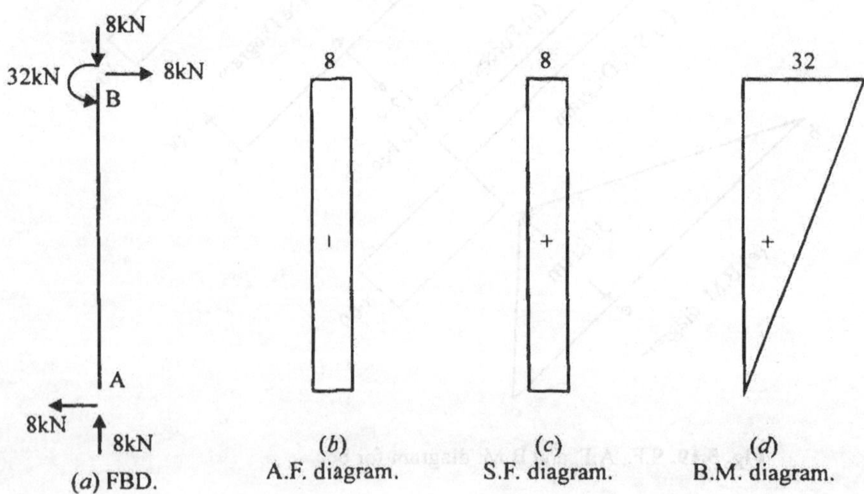

Fig. 5.21. *FBD, A.F., S.F., B.M.* diagram of *AB*

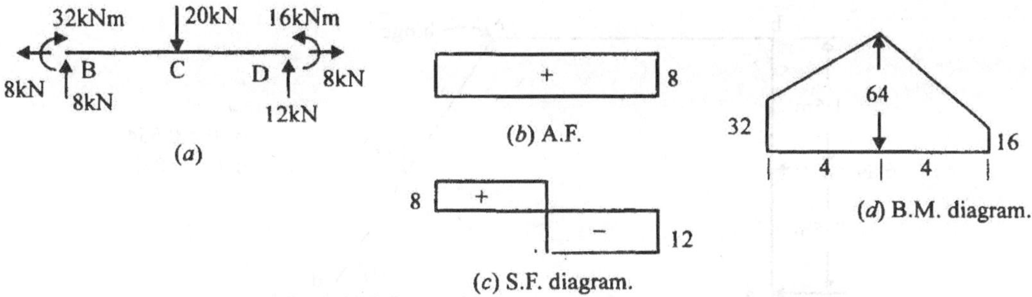

Fig. 5.22. *FBD, A.F., S.F.,* and *BM* diagrams of *BD*.

From Equilibrium considerations

$\Sigma H = 0 \qquad H_D = 8\text{ kN} \rightarrow$

$\Sigma V = 0 \qquad H_D = 12\text{ kN} \uparrow$

$\Sigma M_{DB} = 0 \qquad 32 + 8 \times 8 - 20 \times 4 - M_{DB} = 0$

or $\qquad M_{DB} = 16\text{ kNm}$

12, 16, 8kN, D, 2kN/m, 4m, E, 12

(a) FBD

12, 12

(b) A.F. diagram.

8, −

(c) S.F. diagram.

16kNm, −

(d) B.M. diagram.

Fig. 5.23. Example 5.5. (*FBD, AF, SF, B.M.* diagram of EB).

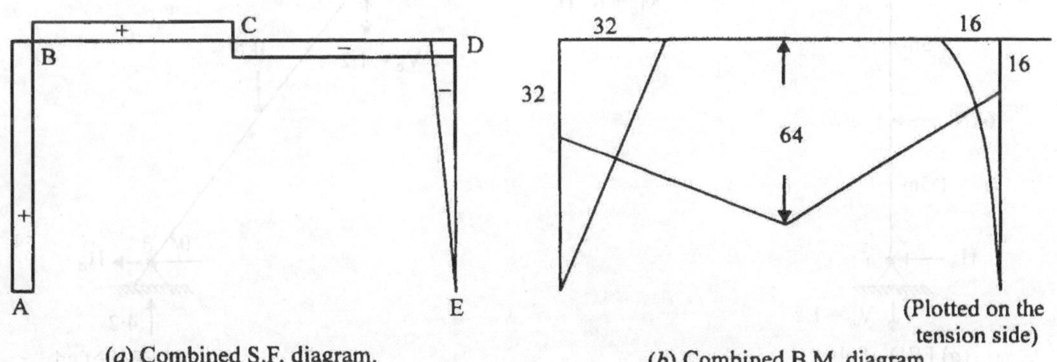

Fig. 5.24. Example 5.5

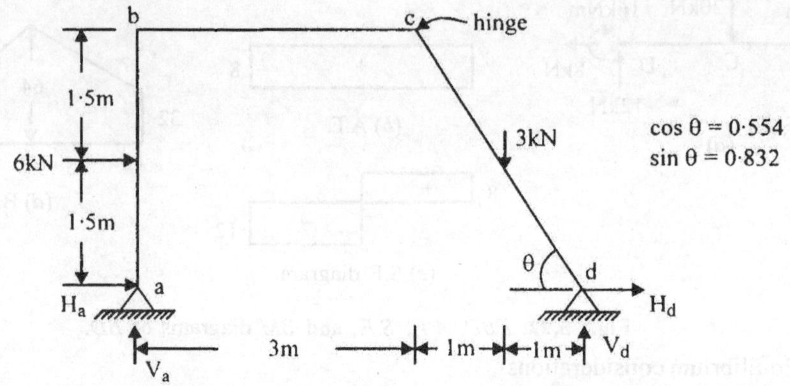

Fig. 5.25. Example 5.6

Example 5.6. Draw the A.F., S.F. and B.M. diagrams of the frame shown in Fig. 5.25.

From the equilibrium of the frame as whole

$$\Sigma H = 0 \text{ gives}$$

$$6 + H_a + H_d = 0$$

From $\Sigma V = 0$ $\quad V_a + V_d - 3 = 0$

From $\Sigma M_a = 0$ $\quad -V_d \times 5 + 3 \times 4 + 6 \times 1.5 = 0.$

$$\text{or } V_d = \frac{21}{5} = 4.2\uparrow$$

$\Sigma M_d = 0$ $\quad -V_a \times 5 + 6 \times 1.5 - 3 \times 1 = 0.$

$$\text{or } V_a = -\frac{6}{5} = -1.2\downarrow$$

Check $\quad 4.2\uparrow - 1.2\downarrow - 3\downarrow = 0$

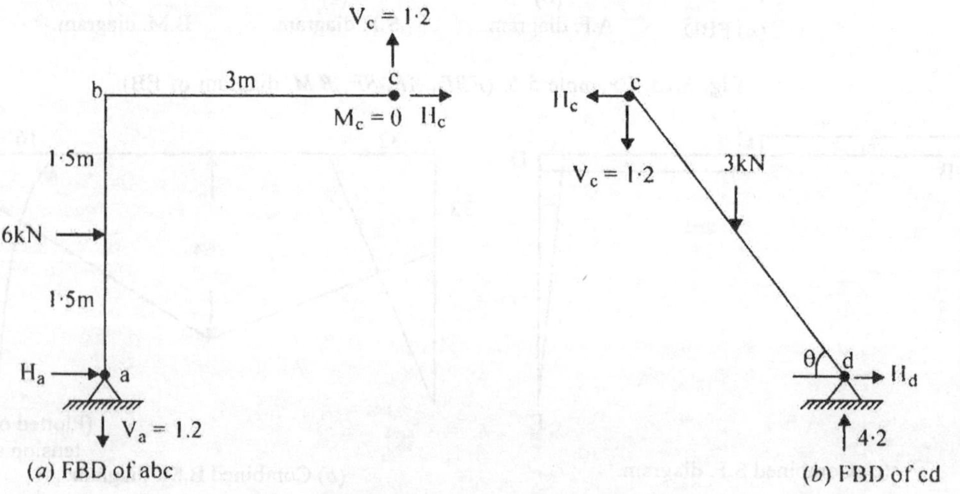

(*a*) FBD of abc (*b*) FBD of cd

Fig. 5.26. FB Diagrams Example 5.6

From $M_c = 0$

$$-1.2 \times 3 - 6 \times 1.5 - H_a \times 3 = 0$$

$$\text{or } H_a = -\frac{12.6}{3} = -4.2$$

From $\quad \Sigma H = 0 \qquad 6 - 4.2 + H_c = 0$

$$\text{or } H_c = -1.8$$

From $\quad$ FBD of cd

$$-H_d \times 3 + 3 \times 1 - 4.2 \times 2 = 0$$

$$\text{or } H_d = -\frac{5.4}{3} = -1.8$$

(*a*) FBD of abc (*b*) Fbd of aeb. (*c*) A.F. diagram. (*d*) S.F. diagram. (*e*) B.M. diagram.

Fig. 5.27. FBD, A.F., S.F., B.M. of aeb (Example 5.6)

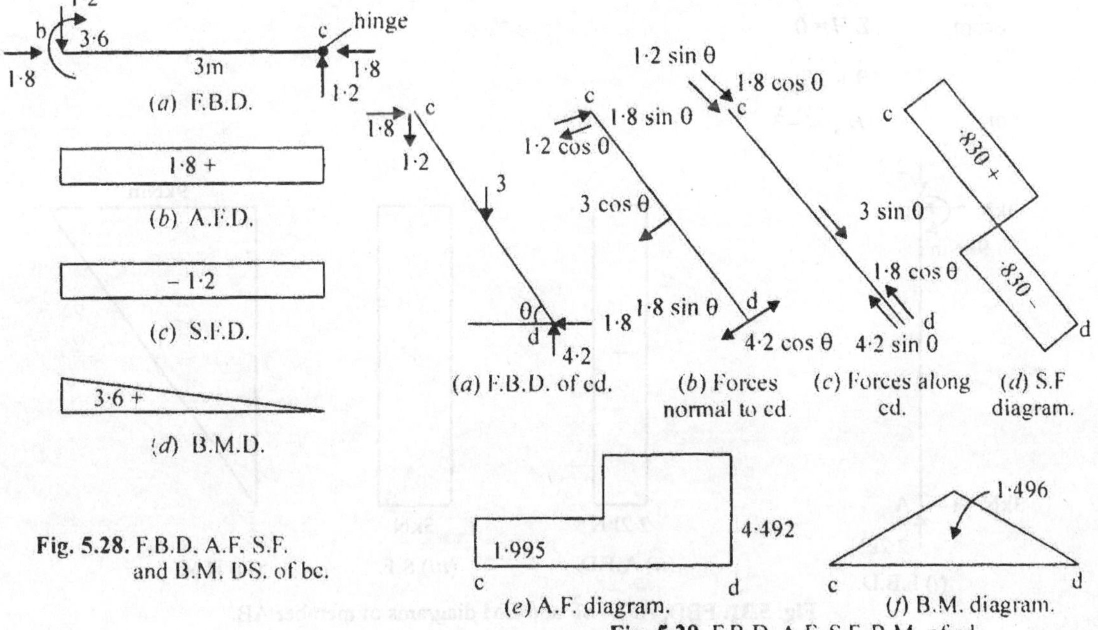

Fig. 5.28. F.B.D. A.F. S.F. and B.M. DS. of bc.

Fig. 5.29. F.B.D. A.F. S.F. B.M. of cd.

Example 5.7. Draw the axial force, S.F. and B.M. diagrams of the gable frame shown in Fig. 5.30.

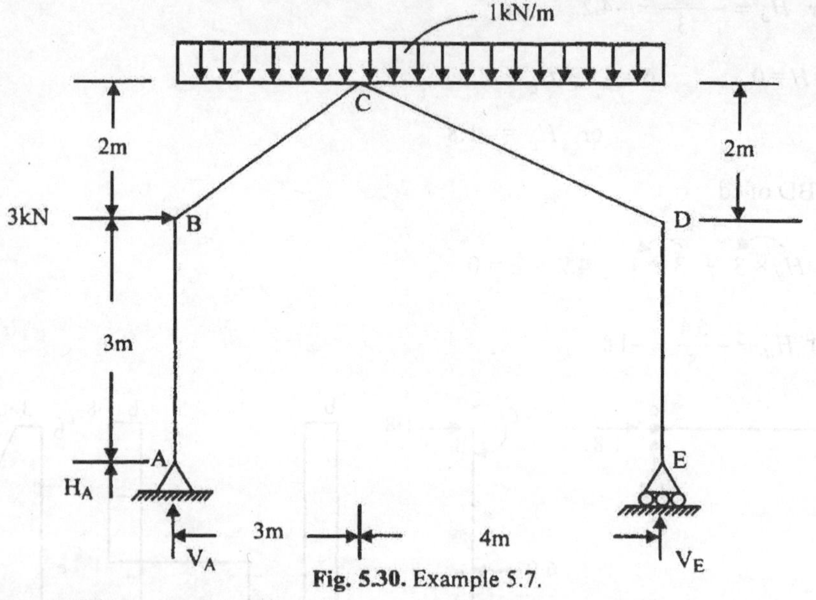

Fig. 5.30. Example 5.7.

Solution. Taking moments about A

$$-V_E \times 7 + 1 \times 7 \times \frac{7}{2} + 3 \times 3 = 0$$

or $\quad V_E = \frac{1}{7}(24.5 + 9) = 4.8 \text{ kN}$

Taking moments about E,

$$V_A \times 7 + 3 \times 3 - \frac{1 \times 7 \times 7}{2} = 0$$

or $\quad V_A = \frac{1}{7}(24.5 - 9) = 2.2 \text{ kN}$.

From $\quad \Sigma H = 0$

$$\vec{3} + \overrightarrow{H_A} = 0$$

or $\quad H_A = -\overleftarrow{3}$

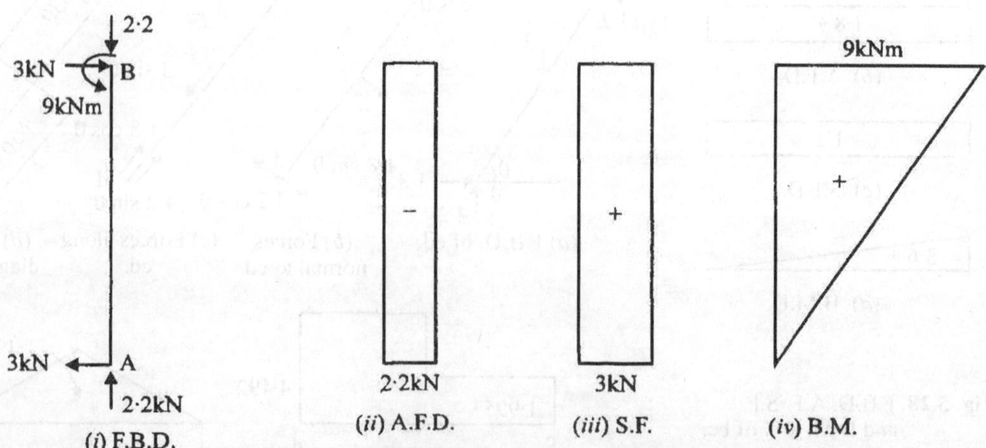

Fig. 5.31. FBD, AFD, SF and BM diagrams of member AB.

Taking moment about c

$$2.2 \times 3 + 9 - 1 \times 3 \times 1.5 - M_C = 0$$

or $$M_C = 11.1 \text{ kNm}$$

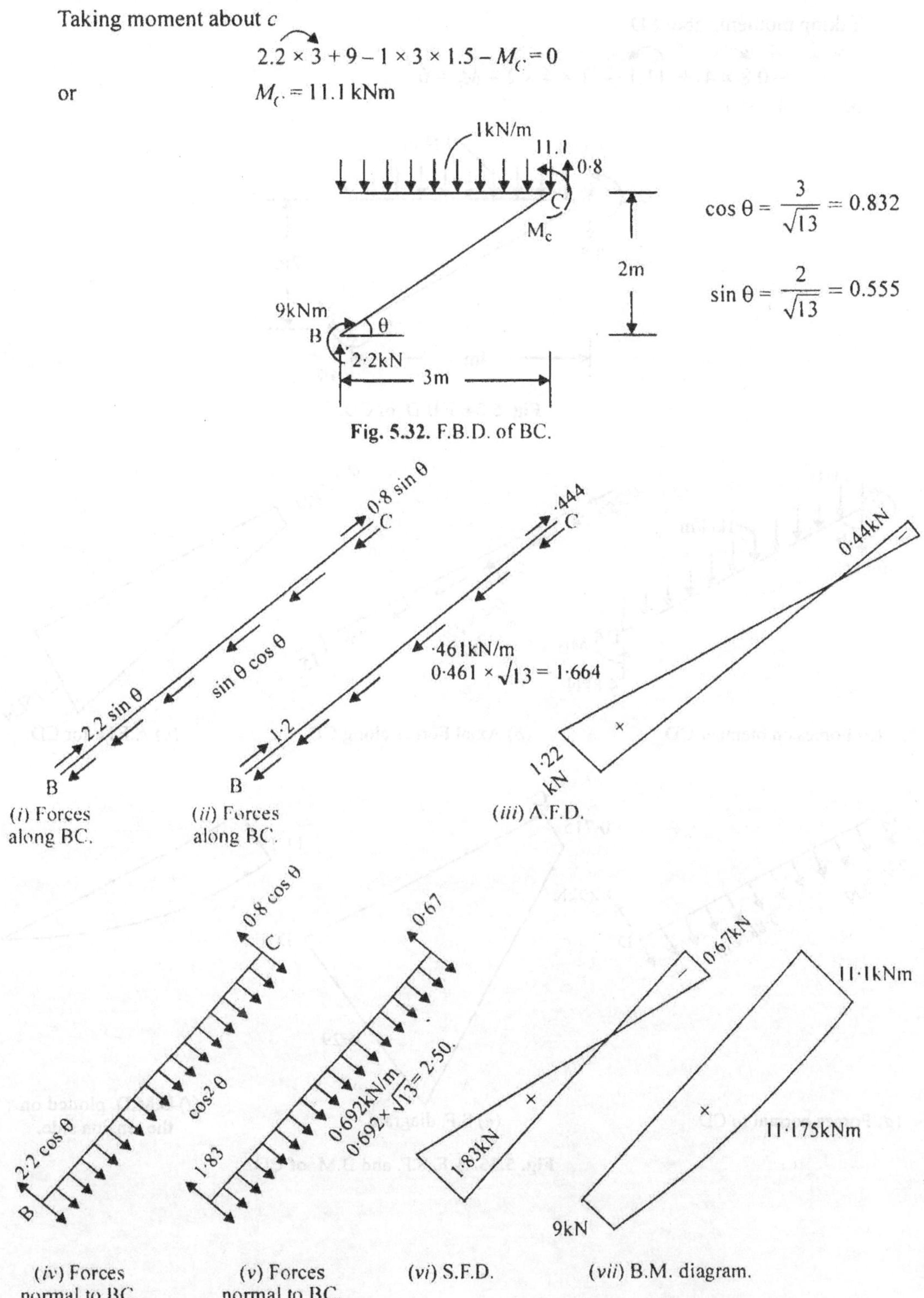

Fig. 5.32. F.B.D. of BC.

(*i*) Forces along BC. (*ii*) Forces along BC. (*iii*) A.F.D.

(*iv*) Forces normal to BC. (*v*) Forces normal to BC. (*vi*) S.F.D. (*vii*) B.M. diagram.

Fig. 5.33. F.B.D, A.F.D, S.F.D. and B.M.D of BC.

Taking moments about D

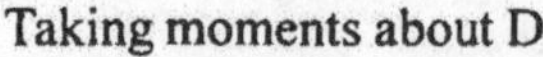

$$-0.8 \times 4 + 11.1 - 1 \times 4 \times 2 - M_D = 0$$

or $\quad M_D = 0$

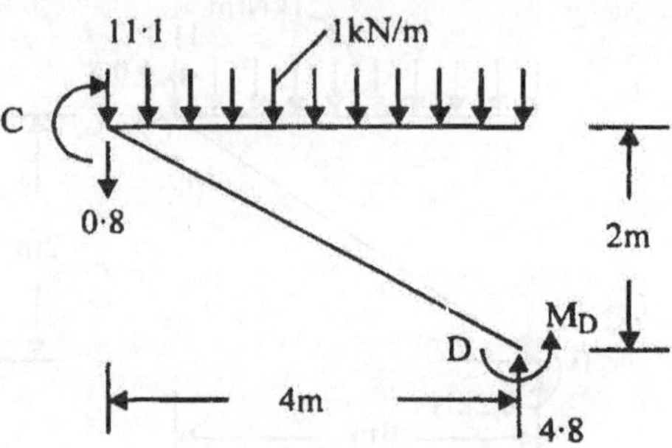

Fig. 5.34. F.B.D. of CD.

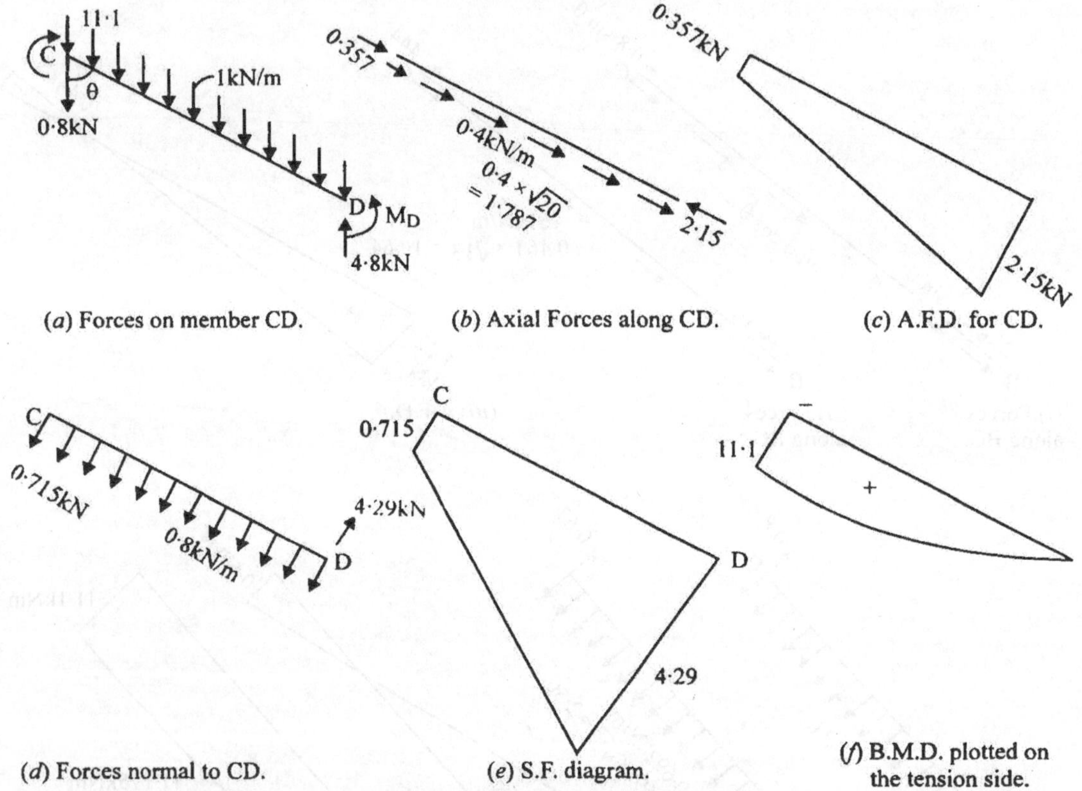

(*a*) Forces on member CD.

(*b*) Axial Forces along CD.

(*c*) A.F.D. for CD.

(*d*) Forces normal to CD.

(*e*) S.F. diagram.

(*f*) B.M.D. plotted on the tension side.

Fig. 5.35. A.F, S.F. and B.M. of CD.

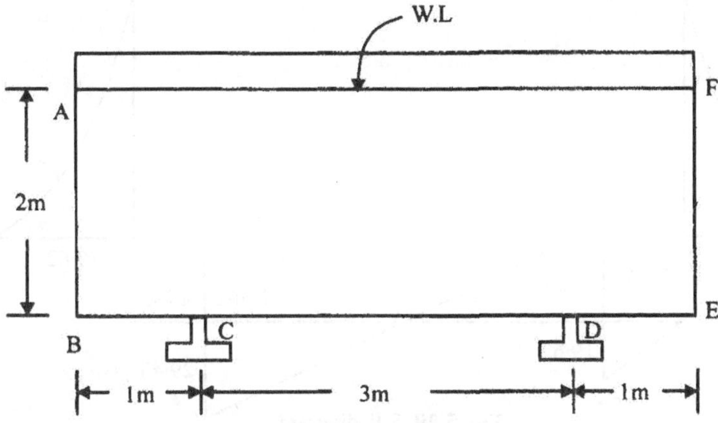

Fig. 5.36. (Example 5.8)

Example 5.8. Construct the shear and B.M. diagrams for the water retaining structure shown in Fig. 5.36.

Solution.

1. Water pressure exerts lateral forces on the vertical members which may be treated as cantilevers fixed at the lower end. The pressure distribution is linear as given by $p = wh$ where w is density and h, the depth. Consider $1m$ width of the water retaining structure.

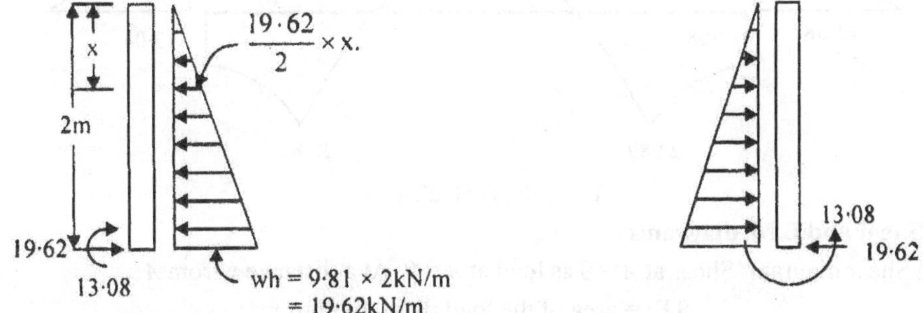

Fig. 5.37. Example 5.8.

Lateral force at the base of 1 m width $= \frac{1}{2} \times 19.62 \times 2 \times 1 = 19.62$ kN

Moment at the base $= 19.62 \times \frac{1}{3} \times 2 = 13.08$ kNm

2. F.B.D. of the base

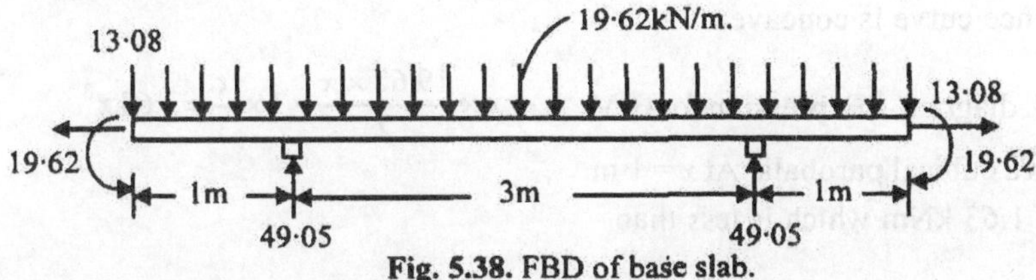

Fig. 5.38. FBD of base slab.

The free body diagram of the base slab for 1 m width is shown in Fig. 5.38. Forces at the ends are equal and opposite to those acting at the vertical walls. The downward vertical load is the water pressure at the base. Reactions at the supports are

$$\frac{1}{2} \times 19.62 \times 5 = 49.05 \text{ kN}$$

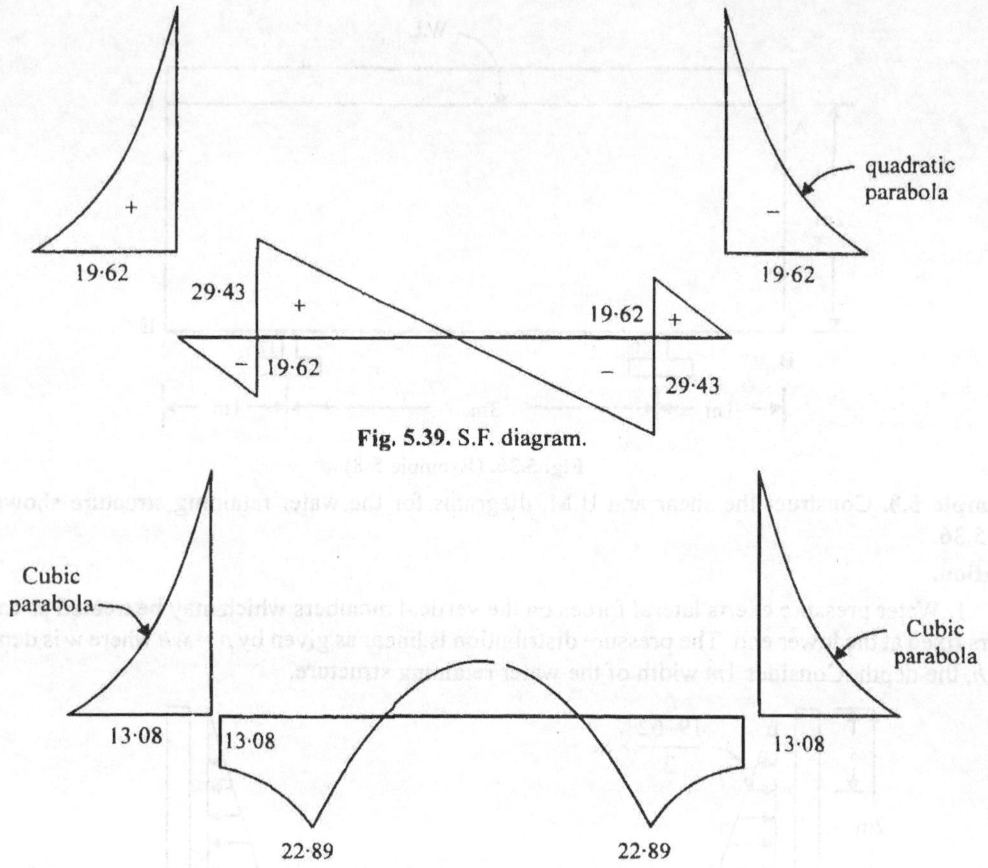

Fig. 5.39. S.F. diagram.

Fig. 5.40. B.M. diagram.

3. Shear and B.M. diagrams

(*a*) Shear diagram: Shear at $A = 0$ as load at $A = 0$. At a distance x from A

$$\begin{aligned} \text{S.F.} &= \text{area of the load diagram upto } x. \\ &= \frac{1}{2} \times x \times \frac{19.62 \times x}{2} \times 1 \\ &= 4.9\, x^2 \end{aligned}$$

which is a quadratic expression. Whether it will be concave or convex can be easily obtained by computing the value at mid depth i.e. 1 m. S.F. at $x = 1$ m from top = 4.9 kN/m which is less than $\frac{1}{2}(19.62)$. Hence curve is concave.

(*b*) B.M. diagram – Expression for B.M. $= \frac{1}{2} x \times \frac{19.62 \times x}{2} \times 1 \times \frac{x}{3} = 1.63\, x^3$

which is a cubical parobala. At $x = 1$ m.

B.M. = 1.63 kNm which is less than

$\frac{1}{2}(13.08)$. Hence the curve concaves inward.

(*c*) The B.M. diagram for the base is drawn like that of a simply supported beam with over hang. Critical values are given in Fig. 5.40.

Example 5.9. Draw the B.M. and Torsional moment diagram of the frame ABCD.

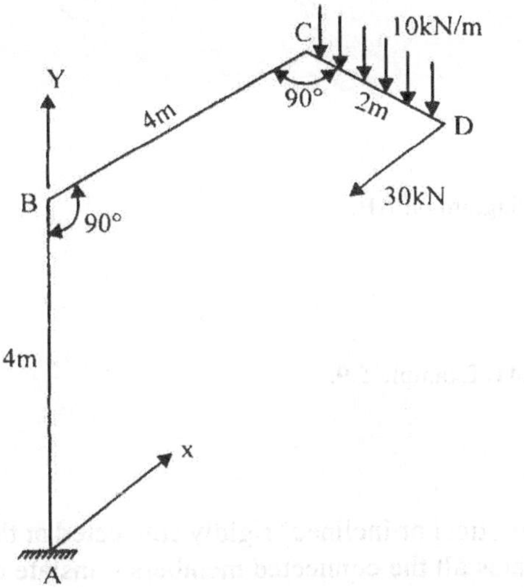

(*i*) The frame ABCD with loading

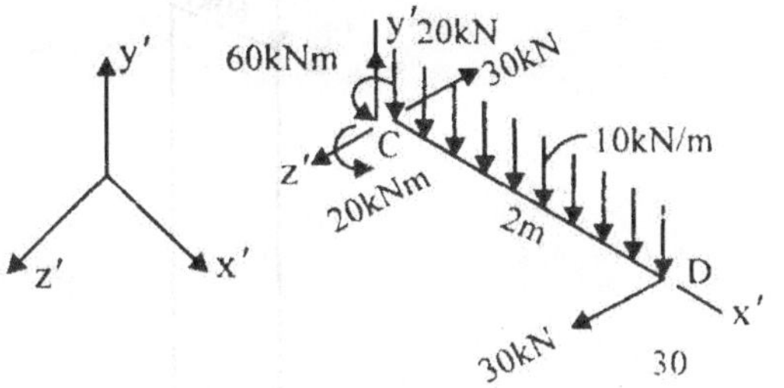

(*ii*) FBD of CD.

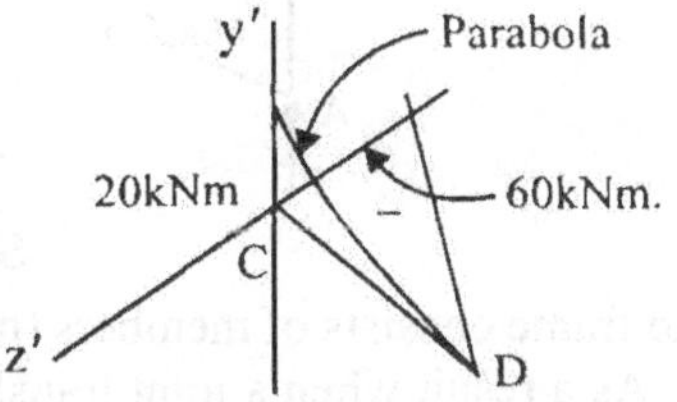

(*iii*) B.M. diagram on CD.
(*a*) about y' axis
(*b*) about z' axis (parabola)

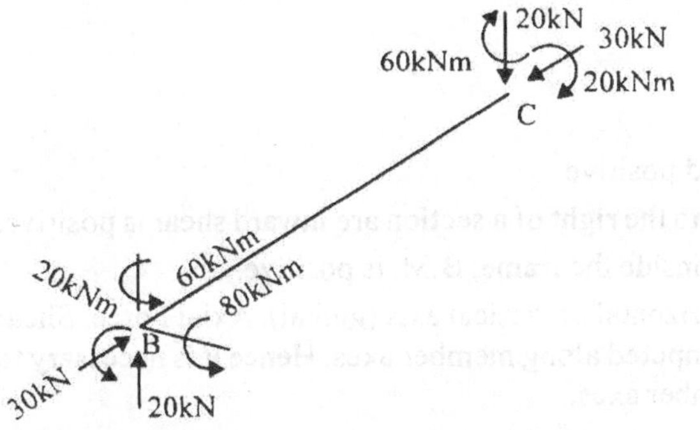

(*iv*) Forces on BC.

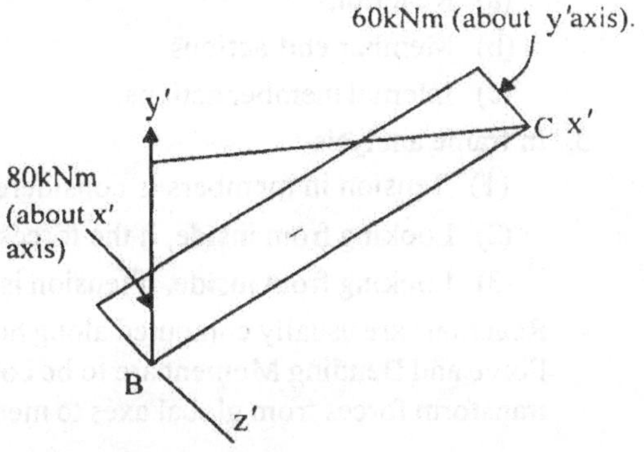

(*v*) B.M. diagram on BC (about y' and z' axes)

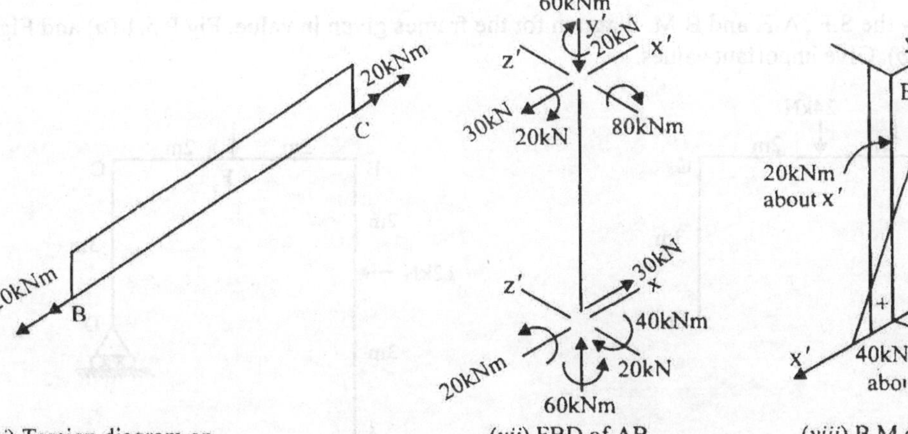

(*vi*) Torsion diagram on BC (about x' axis)

(*vii*) FBD of AB

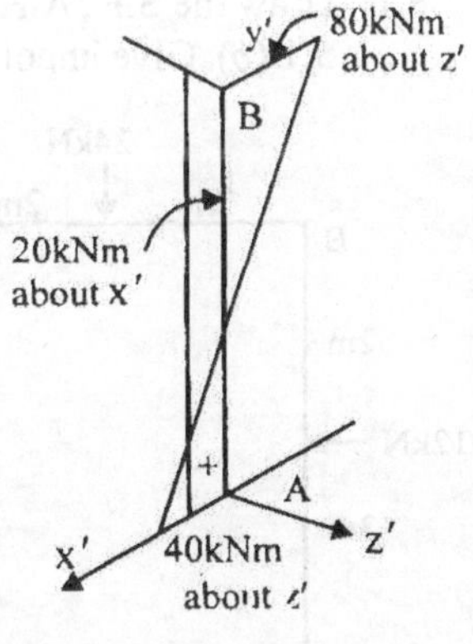

(*viii*) B.M diagram on AB

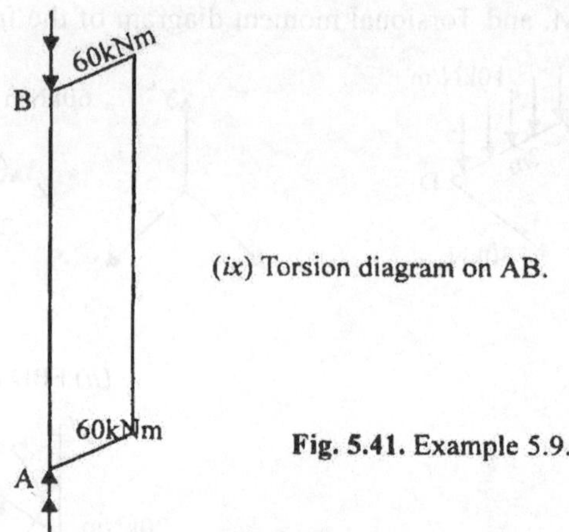

(*ix*) Torsion diagram on AB.

Fig. 5.41. Example 5.9.

SUMMARY

1. A rigid frame consists of members (horizontal, vertical or inclined) rigidly connected at the joints. As a result when a joint translates or rotates all the connected members translate or rotate by the same amount.
2. Analysis of frames requires calculation of
   (a) Reactions
   (b) Member end actions
   (c) Internal member actions.
3. In frame analysis
   (1) Tension in members is considered positive.
   (2) Looking from inside, if the forces to the right of a section are inward shear is positive.
   (3) Looking from inside, if tension is inside the frame, B.M. is positive.
4. Reactions are usually computed along horizontal or vertical axis (global). Axial Force, Shear Force and Bending Moment are to be computed along member axes. Hence it is necessary to transform forces from global axes to member axes.

PROBLEMS

5.1. Draw the S.F., A.F. and B.M. diagram for the frames given in value. Fig P 5.1 (*a*) and Fig. P 5.1 (*b*). Give important values.

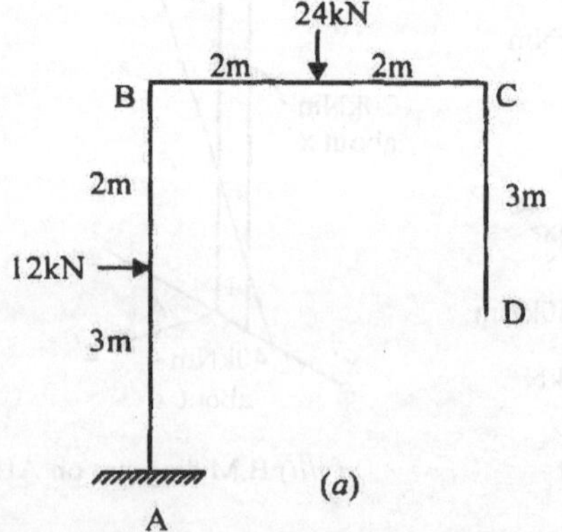

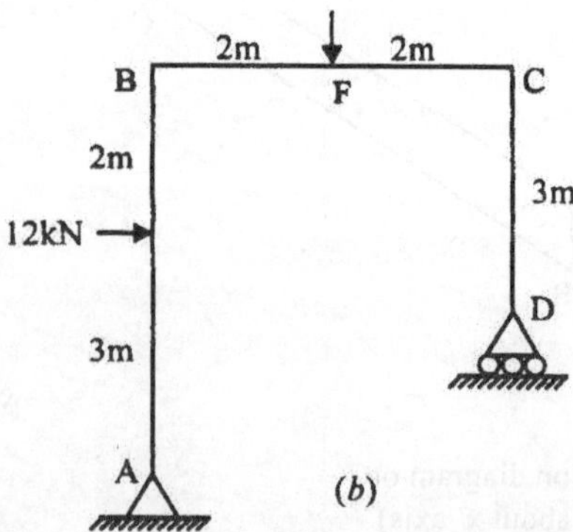

Fig. P 5.1.

5.2. Draw the S.F. and B.M. diagrams for the frames given in Fig P 5.2 (*a*) and Fig. P 5.2 (*b*).

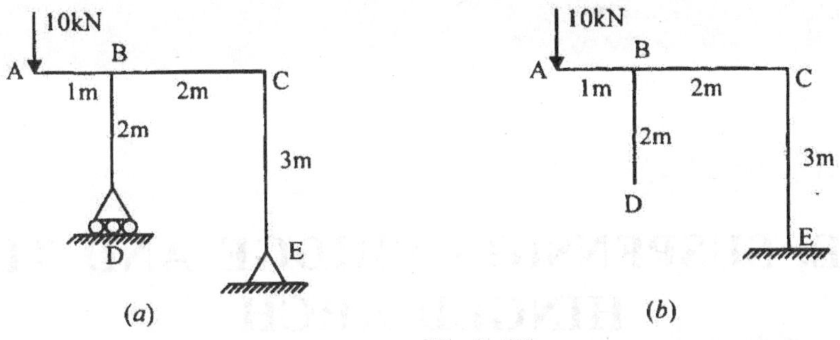

Fig. P. 5.2

5.3. Draw S.F., A.F. and B.M. diagrams for the frame given below. in Fig. P 5.3 (*a*) and Fig. P 5.3 (*b*)

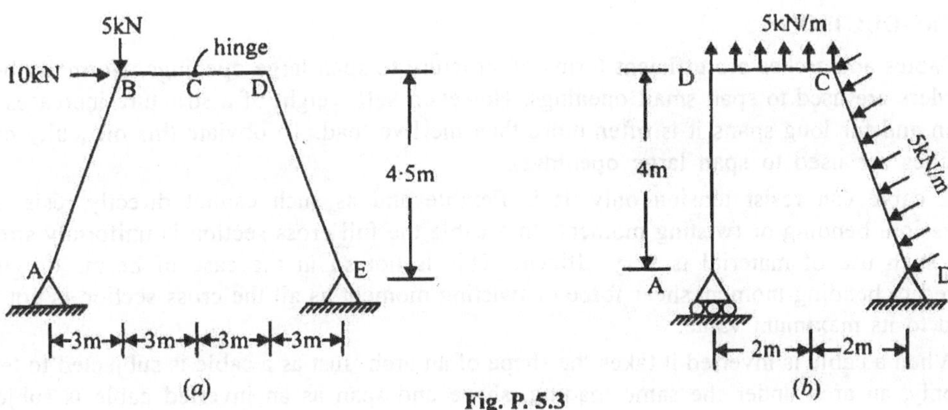

Fig. P. 5.3

6

CABLE, SUSPENSION BRIDGE AND THREE HINGED ARCH

6.1 INTRODUCTION

Cables and arches are efficient forms of structure to span large openings. Normally beams and girders are used to span small openings. However, self weight of a structure increases with the span and for long spans it is often more than the live load. To obviate this difficulty cables and arches are used to span large openings.

A cable can resist tension only. It is flexible and as such cannot directly resist axial compression, bending or twisting moment. In a cable the full cross-section is uniformly stressed and as such use of material is very efficent. This is not so in the case of beams or girders subjected to bending moment,shear force or twisting moment as all the cross-section is not fully stressed to its maximum value.

When a cable is inverted it takes the shape of an arch. Just as a cable is subjected to tensile force only, an arch under the same loading, shape and span as an inverted cable is subjected to only compressive force. However, unlike a cable an arch is a rigid structure and cannot change its shape with change of loading. Hence except in particular cases when shape of the arch and loading corespond such that it is subject to compressive stress only, normally an arch is subjected to bending moment (reduced), shear force and axial force (predominant).

6.2. ANALYTICAL SOLUTION OF CABLE STRUCTURE

A cable is used to carry concentrated loads, uniformly distributed load along its length and uniformly distributed load along the horizontal length. These are shown in Fig. 6.1 (a), (b) and (c).

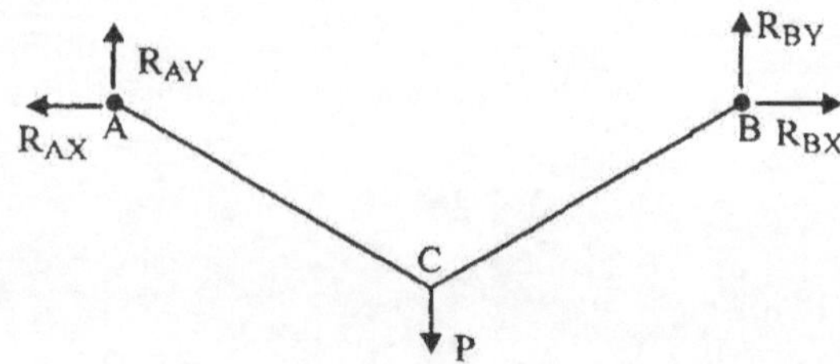

(*a*) A cable with single concentrated vertical load

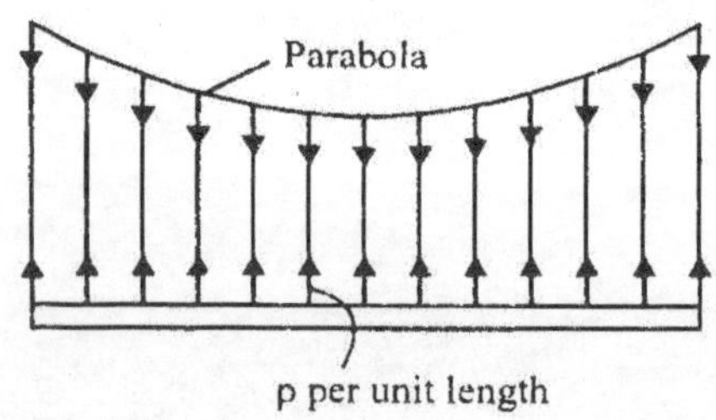

(*b*) Cable with uniform load/horizontal length

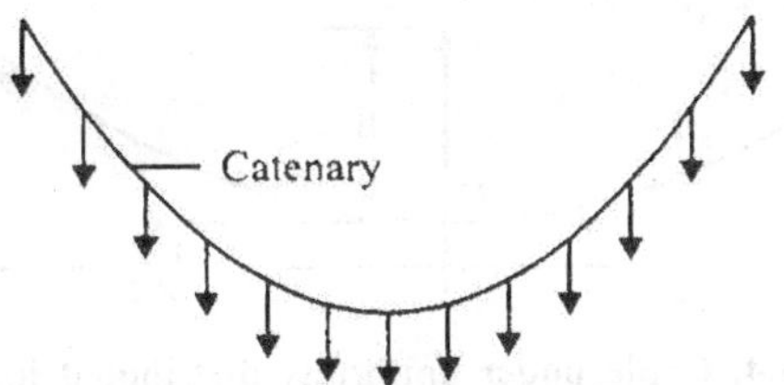

(*c*) Cable with uniform load along its length.

Fig. 6.1. Different shapes of cables.

There is not much difference between a catenary and a parabola (Fig. 6.2) and to simplify analysis often a catenary is approximated by a parabola.

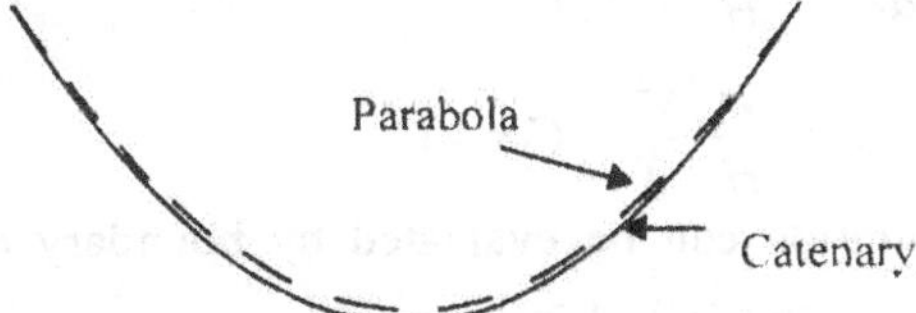

Fig. 6.2. A caterary and a parabola.

6.2.1 General Equation of a cable

Consider an elementary length of a cable subjected to a uniformly distributed vertical load w/unit horizontal length. The free body diagram of the elementary length *ds* of the cable is shown in Fig. 6.3.

From $\Sigma F_x = 0$

$$T\cos\theta = (T + dT)\cos(\theta + d\theta)$$

$= H$. as there is no horizontal load on the cable.

From $\Sigma F_y = 0$

$$V\downarrow + w.dx \downarrow - (V + dV) = 0$$

or $$w = \frac{dV}{dx}$$

At A $$V = H\tan\theta$$

$\therefore$ $$w = \frac{d}{dx}(H\tan\theta)$$

But $$\tan\theta = \frac{dy}{dx}$$

or $$w = \frac{d}{dx}\left(H.\frac{dy}{dx}\right)$$

As H does not change with x.

$$w = H.\frac{d^2y}{dx^2}$$

Fig. 6.3. FBD of Elementary length of a cable.

or $$\frac{d^2y}{dx^2} = \frac{w}{H} \qquad ...(6.1)$$

The equation has been derived with x positive towards right and y positive upwards.

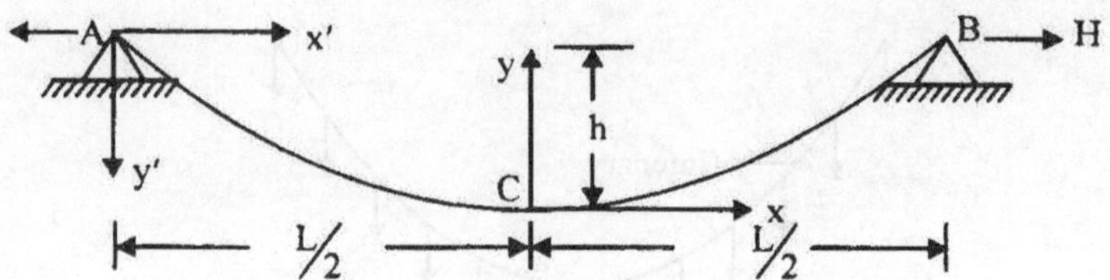

Fig. 6.4. Cable under uniformly distributed load.

Case 1. Load uniform along the horizontal span and supports at the same level.

With
$$\frac{d^2y}{dx^2} = \frac{w}{H}$$
$$\frac{dy}{dx} = \frac{w}{H}\cdot x + C_1$$
$$y = \frac{w}{H}\cdot\frac{x^2}{2} + C_1x + C_2$$

The constants of integration can be evaluated by boundary conditions.

With origin at C, at $x = 0$, $y = 0$

at $x = 0$ $\frac{dy}{dx} = 0$ (due to symmetry)

$\therefore C_1 = 0$ and $C_2 = 0$.

The equation of the cable then becomes.
$$y = \frac{w}{2H}\cdot x^2 \qquad \text{.... (6.2)}$$

This is equation of a parabola.

At $x = \frac{L}{2}$ $y = h$ Substituting

in the eqn 6.2 above
$$h = \frac{w}{2H}\cdot\left(\frac{L}{2}\right)^2$$

or
$$H = \frac{wL^2}{8h} \qquad \text{.... (6.3)}$$

Substituting this value in Equation 6.2, the equation $y = \frac{w}{2H}\cdot x^2$ can be rewritten as
$$y = \frac{8h.H}{L^2}\cdot\frac{1}{2H}\cdot x^2$$
$$= \frac{4hx^2}{L^2}$$

with C as origin and positive axes as shown in Fig. 6.4

When the origin is shifted to A
$$x' = x + \frac{L}{2}$$
$$y' = h - y$$

or
$$x = x' - \frac{L}{2}$$
$$y = h - y'$$

Substituting in the equation

$$y = \frac{4hx^2}{L^2}$$

$$h - y' = \frac{4h}{L^2}\left(x' - \frac{L}{2}\right)^2$$

$$= \frac{4h}{L^2}\left(x'^2 - 2x'.\frac{L}{2} + \left(\frac{L}{2}\right)^2\right)$$

or $$y' = \frac{4h}{L^2}(-x'^2 + x'L) - \frac{4h}{L^2}\cdot\frac{L^2}{4} + h$$

$$= \frac{4h}{L^2}\cdot x'(L - x')$$

or $$y = \frac{4h}{L^2}\cdot x(L - x) \qquad \text{...(6.4)}$$

The tension in the cable at any point can be obtained as,

$$T\cos\theta = H$$

or $$T = H\sec\theta$$

$$= H\sqrt{1 + \tan^2\theta}$$

$$= H\cdot\sqrt{1 + \left(\frac{dy}{dx}\right)^2}$$

$$= \frac{wL^2}{8h}\sqrt{1 + \frac{64h^2.x^2}{L^4}} \qquad \text{...(6.5)}$$

Case II. **Load uniform along cable length. Let the intensity of weight be w_l/unit length of cable.**

Weight in length $ds = w_l.ds$

Weight per unit horizontal length = $w_l.\frac{ds}{dx}$

The equation $\frac{d^2y}{dx^2} = \frac{w}{H}$

now becomes $$\frac{d^2y}{dx^2} = \frac{w_l\cdot ds}{dx\cdot H}$$

$$= \frac{w_l}{H}\cdot\sec\theta$$

$$= \frac{w_l}{H}\cdot\sqrt{1 + \tan^2\theta}$$

$$= \frac{w_l}{H}\sqrt{1 + \left(\frac{dy}{dx}\right)^2}$$

with $\frac{dy}{dx} = p$ and $\frac{d^2y}{dx^2} = \frac{dp}{dx}$

The equation becomes

$$\frac{dp}{dx} = \frac{w_l}{H} \cdot \sqrt{1+p^2}$$

or
$$\frac{dp}{\sqrt{1+p^2}} = \frac{w_l}{H} \cdot dx$$

Integrating both sides,

$$\int \frac{dp}{\sqrt{1+p^2}} = \frac{w_l}{H} \int dx$$

or
$$\sin h^{-1} p = \frac{w_l}{H} \cdot x + C_1$$

with C as origin,

at
$$x = 0 \quad \frac{dy}{dx} = 0 \text{ due to symmetry or } C_1 = 0$$

or
$$p = \sin h . \frac{w_l \cdot x}{H} = \frac{dy}{dx}$$

Integrating again

$$y = \frac{H}{w_l} \cos h \frac{w_l . x}{H} + C_2$$

at $x = 0 \quad y = 0$ which gives $C_2 = -\frac{H}{w_l}$.

The equation then becomes

$$y = \frac{H}{w_l} \left[\cos h \frac{w_l \cdot x}{H} - 1 \right]$$

H cannot be determined directly by putting the values of coordinates of A or B but is obtained by iterative method. It can be seen that the sag of the cable greatly influences the tension in the cable segments. The behaviour of cable is significantly different from other structures where the deflection of a structure has negligible effect on the structural analysis. For example in rigid structures change in geometry is ignored. However, for a cable structure, change in geometry has a pronounced effect on the tension in the cable.

Case III. Load uniform along the horizontal span and cable supported at different levels.

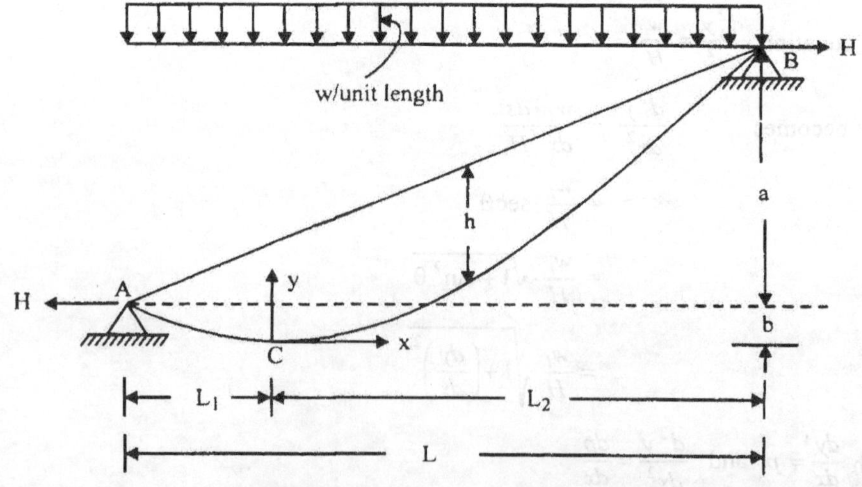

Fig. 6.5. Cable supported at different levels.

The equation of a cable with C as origin as already derived,

$$y = \frac{wx^2}{2H}$$

at A $x = -L_1$ $y = b$

or
$$b = \frac{wL_1^2}{2H}$$

at B $x = L_2$ $y = a + b$

or
$$a + b = \frac{wL_2^2}{ZH}$$

Dividing the two equations

$$\frac{b}{a+b} = \frac{L_1^2}{L_2^2} \text{ or } \frac{L_1}{L_2} = \sqrt{\frac{b}{a+b}}$$

or
$$L_2 = \frac{L}{\left[\sqrt{\frac{b}{a+b}} + 1\right]}$$

Horizontal thrust is given by

$$H = \frac{wL_1^2}{2b} = \frac{w}{2(a+b)} \cdot L_2^2$$

$$T = H \sec \theta = H\sqrt{1 + \left(\frac{dy}{dx}\right)^2}$$

Differentiating

$$\frac{dy}{dx} = \frac{w.x}{H} \text{ or } T = H\sqrt{1 + \left(\frac{wx}{H}\right)^2}$$

H can be obtained directly by applying equations of statics. Taking moment of all forces in the span CB about B,

$$H(a+b) - w \cdot L_2 \cdot \frac{L_2}{2} = 0$$

or
$$H = \frac{wL_2^2}{2(a+b)}$$

When the sag is Y_p at any distance x from A,

$$H.Y_P = \left(\frac{wL}{2}\right) \cdot x - wx \cdot \left(\frac{x}{2}\right) = \frac{wx}{2}(L - x)$$

If the sag is h at midspan i.e. when $x = \frac{L}{2}$

$$H.h = \frac{W}{2} \cdot \frac{L}{2}\left(L - \frac{L}{2}\right) \text{ or } H = \frac{wL^2}{8h}$$

Then
$$\frac{wL^2}{8h} \cdot Y_p = \frac{wx}{2}(L - x)$$

or
$$Y_P = \frac{4hx}{L^2} \cdot (L - x)$$

With origin at A, y upwards positive and x positive towards right

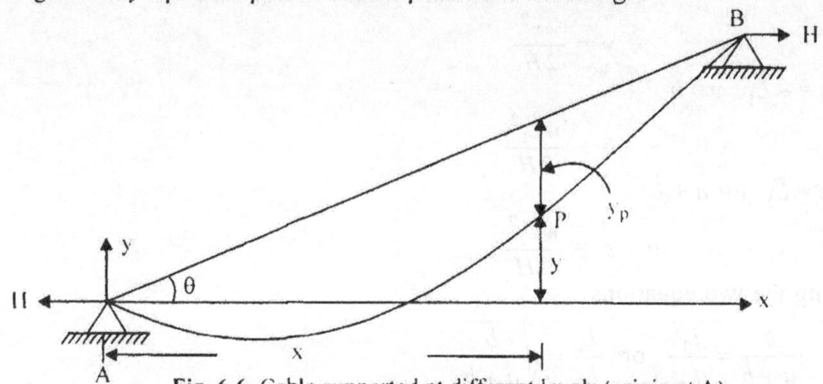

Fig. 6.6. Cable supported at different levels (origin at A).

$$Y_P = x\tan\theta - y$$

or $$x\tan\theta - y = \frac{4hx}{L^2}(L-x)$$

or $$y = x\tan\theta - \frac{4hx}{L^2}(L-x) \qquad \text{....(6.7)}$$

6.3. GRAPHICAL SOLUTION OF CABLE PROBLEMS

Before taking up graphical solution of cable problems fundamentals of graphics statics are reviewed. Basic points of graphical analysis are:

1. A force is completely specified when its magnitude, direction and point of application are known.

2. Resultant of a number of forces acting on a body can be obtained by drawing the force polygon. If the resultant is zero, that is, the forces are in equilibrium, the force polygon will close.

3. While the force polygon gives the resultant in magnitude and direction, it does not give the point of application. This can be obtained by drawing the funicular polygon.

4. A diagram which shows the point of application of forces with respect to a structure is called a space diagram.

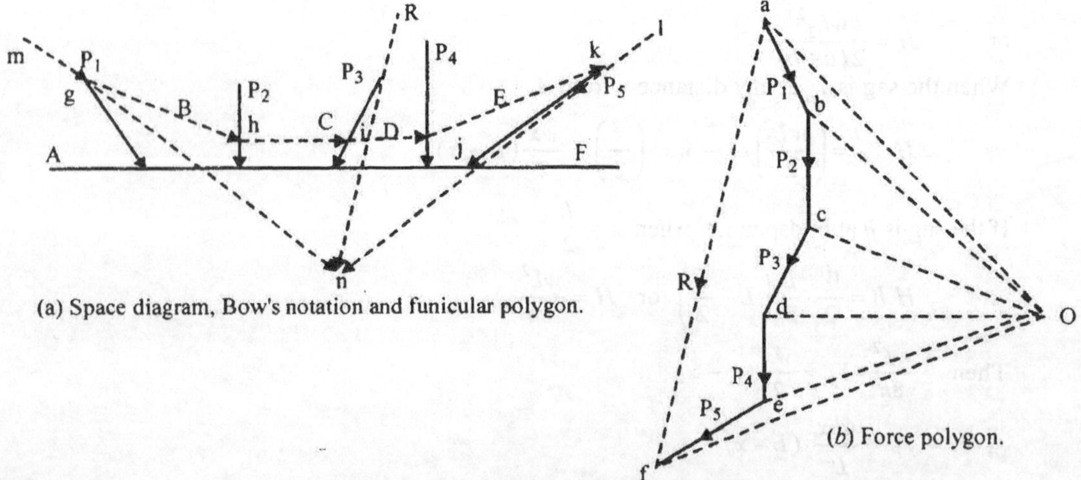

Fig. 6.7. Space diagram and Bow's notation.

The space on either side of a force is given a notation by means of which the force can be identified. The notation is known as Bow's notation.

These are explained in Fig. 6.7. A body is subjected to a number of forces P_1, P_2, P_3, P_4 and P_5. To find the resultant graphically,

(*i*) Space diagram is drawn.

(*ii*) Space on either side of a force is given designation *A, B, C, D, E, F.*

(*iii*) Bow's notation is applied. i.e. P_1 is designated by *ab*, P_2 by *bc* and so on.

(*iv*) Then the force polygon *abcdef* is drawn where *ab* gives the value of P_1 to some scale in magnitude and direction. Similarly *bc* gives P_2 and so on.

(*v*) Joining *af* gives the resultant in magnitude and direction.

(*vi*) To get the point of application, it is necessary to draw the funicular polygon.

This is drawn by choosing a point O as pole and joining *ao, bo, co* and *do* and so on. From triangle of forces *ab* can be resolved into two components *ao* and *ob*.

bc can be resolved into *bo* and *oc*

cd can be resolved into *co* and *od* and so on.

Since *ob* is component of both P_1 and P_2 *gh* is drawn parallel to *ob* cutting P_1 and P_2.

As *oc* is component of both P_2 and P_3 *hi* is drawn parallel to *oc* cutting P_2 and P_3.

Similarly as *od* is component of P_3 and P_4 *ij* is drawn parallel to *od* cutting both P_3 and P_4. *As oe* is component of both P_4 and P_5 , *jk* is drawn parallel to *oe* cutting P_4 and P_5 . As *ao* is component of both P_1 and resultant *af, mg* drawn parallel to *ao* will cut P_1 and the resultant. Similarly *of* is a component of both P_5 and resultant *af*. Hence *kl* drawn parallel to *of* will cut P_5 and the resultant. In other words, intersection of *mg* and *kl* when produced will give the point through which the resultant force will pass. Hence the resultant will be

(*i*) equal to *af* converted to force (magnitude)

(*ii*) in the direction from *a* to *f* (direction)

(*iii*) passing through point *n* (point of application)

As regards force and funicular polygons, there are three possibilities.

(*i*) Force and funicular polygon both close – Forces are in equilibrium. No resultant force and no resultant couple.

(*ii*) Force polygon closes but funicular polygon does not close. Forces are equivalent to a couple.

(*iii*) Force polygon does not close. Forces are equivalent to a resultant.

In a cable subjected to concentrated loads it is necessary to find

(*i*) the cable shape

(*ii*) the end reactions

(*iii*) tensions at different parts of the cable.

Consider a cable supported at *A* and *B* and subjected to loads P_1 , P_2, P_3 and P_4 . Assuming for the present that *AB* is a rigid body with *A* as hinge and *B* as roller,

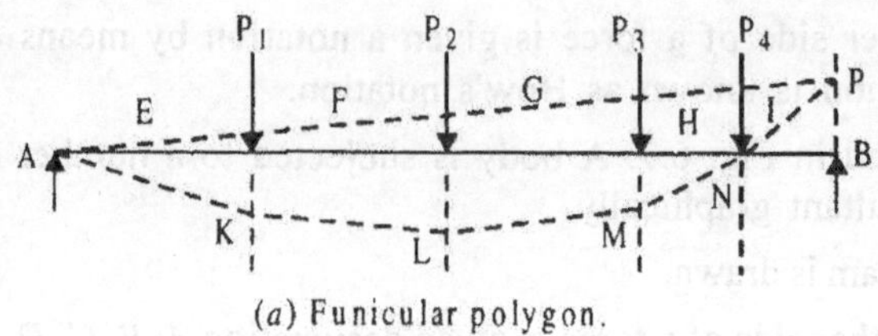

(a) Funicular polygon.

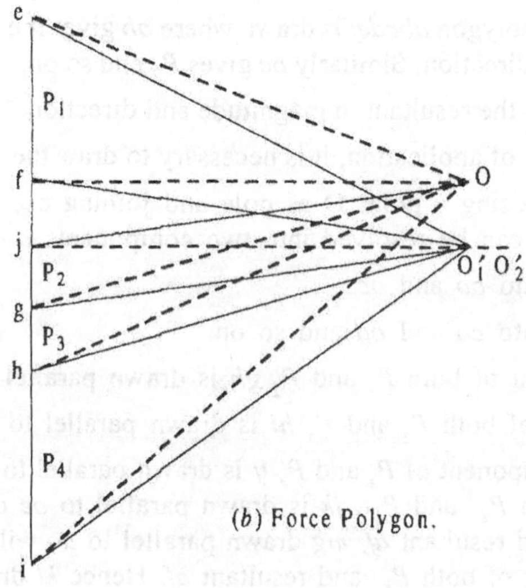

(b) Force Polygon.

Fig. 6.8. Funicular polygon and force polygon.

the funicular polygon can be drawn starting from point *A* through which the reaction at A must pass and ending at *P* cutting the vertical reaction at *P*. The line *oj* drawn parallel to *AP* cuts the vertical line *ei* at *j* giving the upward reactions *ij* at *B* and *je* at *A* respectively. But as the cable is supported at *A* and *B*, the funicular polygon must pass through *A* and *B*. This is because the points of application of support reactions only are known and their magnitudes and directions are unknown.

As the vertical component of reactions at *A* and *B* are fixed by point *j* which remains unchanged, a new pole $o_1{}'$ is chosen any where on the line passing through point *j* and parallel to the chord joining support points *A* and *B*. The funicular polygon thus drawn will pass through the points, *A* and *B*. It can be seen that position of $o_1{}'$ can be chosen any where ($o_1{}'$, $o_2{}'$ ...).

The distance of $o_1{}'$ from the vertical line *ei* gives the value of *H* and the shape of the funicular polygon so obtained is the shape of the cable. It can be observed that with different values of *H* (i.e. position of $o_1{}'$), the sag of the cable changes or more clearly with different cable profiles value of *H* will change. But as the product of *H* and sag is constant for any given loading and span length, once *H* × sag is known from anyone solution, different values of *H* from different cable profiles can be easily obtained. If the supports are not at the same level final vertical reactions will be sum of vertical reactions obtained from above construction and the vertical component of the inclined line $o_1{}'j$.

6.4. ANALYTICAL SOLUTION OF CABLES UNDER CONCENTRATED LOADS

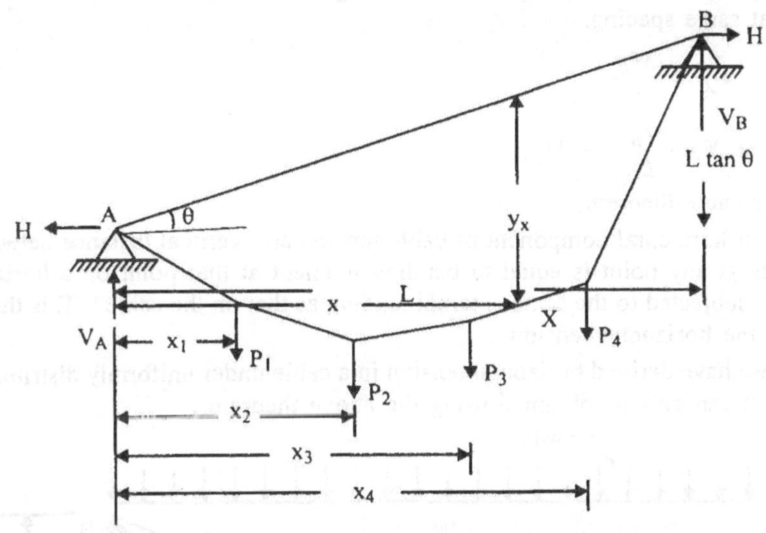

Fig. 6.9. Analytical solution of cable under concentrated loads.

Fig. 6.9 shows a cable under concentrated loads with supports at different levels.

Since all the loads are vertical the horizontal reactions at A and B must be equal and may be designated by H.

Taking moment of all forces about B.

$$+\curvearrowright \; V_A \times L - P_1(L-x_1) - P_2(L-x_2) - P_3(L-x_3) - P_4(L-x_4) + H \times L\tan\theta = 0$$

Now the summation of terms

$P_1(L-x_1)+P_2(L-x_2)+P_3(L-x_3)+P_4(L-x_4)$ can be designated as M_B and is equal to moment of all the external vertical forces about B.

Therefore $V_A \times L - M_B + H \times L\tan\theta = 0$

or $$V_A = \frac{M_B}{L} - H\tan\theta \qquad \text{...(6.8)}$$

Since the cable is assumed to be perfectly flexible, the bending moment is zero at all points. Applying that condition to any arbitrary point X.

$$V_A(x) - H(y_x - x\tan\theta) - M_x = 0.$$

Substituting V_A from above,

$$\left(\frac{M_B}{L} - H\tan\theta\right)x - H(y_x - x\tan\theta) - M_x = 0$$

or $$\frac{M_B}{L}\cdot x - M_x - H\cdot y_x = 0$$

or $$H.y_x = M_B\cdot\frac{x}{L} - M_x$$

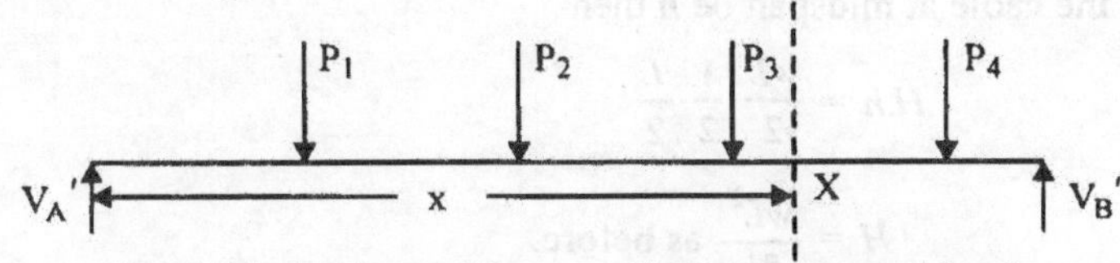

Fig. 6.10. Horizontal beam under concentrated loads.

If instead of a cable, a horizontal beam of length L is subjected to the same vertical loads P_1, P_2, P_3, P_4 at same spacing.

$$V_A' = \frac{M_B}{L}$$

and B.M. at $X = \frac{M_B}{L} \cdot x - M_x$

Hence the cable theorem,

"Product of horizontal component of cable tension and vertical distance between the cable chord and cable at any point is equal to bending moment at that point on a horizontal simply supported beam subjected to the same external loading as that on the cable". This theorem enable us to compute the horizontal tension.

Though we have derived horizontal tension in a cable under uniformly distributed load from first principle, it can also be obtained using the above theorem.

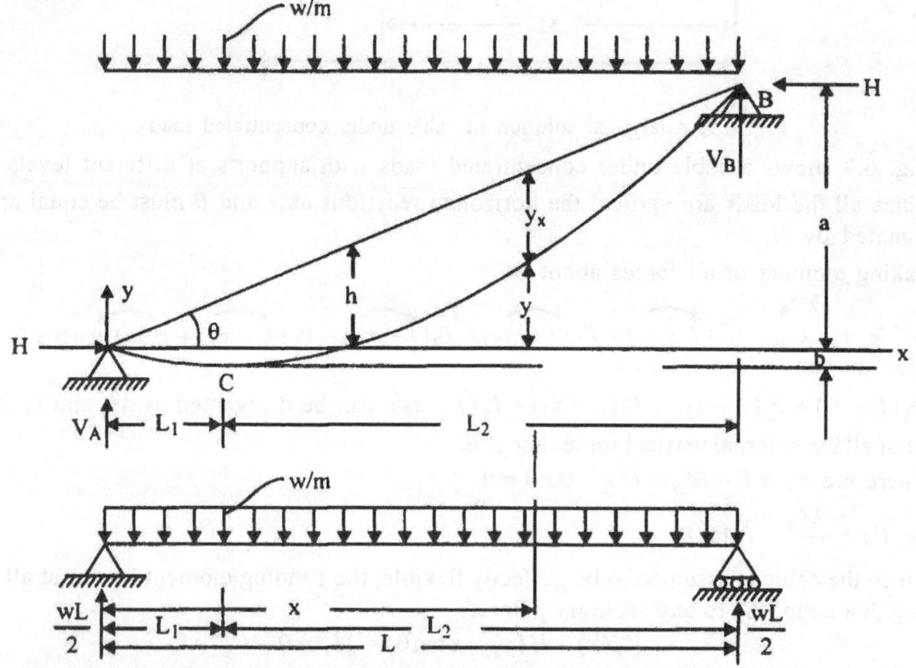

Fig. 6.11. Cable under uniformly distributed load.

Applying the above theorem to the case of distributed load as shown in Fig. 6.11.

$$H(y_x) = \frac{wL}{2} \cdot x - (wx) \cdot \frac{x}{2}$$

$$= \frac{wx}{2}(L - x)$$

If the sag of the cable at midspan be h then

$$H.h = \frac{wL}{2} \cdot \frac{1}{2} \cdot \frac{L}{2}$$

or $$H = \frac{wL^2}{8h} \text{ as before.}$$

The sag of the cable at the lowest point C

$$= a \cdot \frac{L_1}{L} + b$$

Hence $$H\left(\frac{aL_1 + bL}{L}\right) = w \cdot \frac{L}{2} \cdot L_1 - \frac{wL_1^2}{2}$$

or $$H = w \cdot \frac{LL_1 - L_1^2}{2} \times \frac{L}{aL_1 + bL}$$

$$= w \cdot \frac{L_1 \cdot L_2}{2} \cdot \frac{L}{aL_1 + bL} \qquad \text{...(6.9)}$$

6.5. THE LENGTH OF A CABLE

The length of the cable can be obtained using the basic relation

$$dS^2 = dx^2 + dy^2$$

or $$dS = \sqrt{dx^2 + dy^2}$$

$$= \sqrt{1 + \left(\frac{dy}{dx}\right)^2} \cdot dx$$

Integrating both sides.

$$S = \int_0^L \sqrt{1 + \left(\frac{dy}{dx}\right)^2} \cdot dx$$

Particular Cases

(*i*) Parabolic cable with level supports with mid-point C as origin, equation of a parabolic cable

$$y = \frac{4hx^2}{L^2}$$

$$\frac{dy}{dx} = \frac{8hx}{L^2}$$

$$S = 2\int_0^{L/2} \sqrt{1 + \frac{64h^2x^2}{L^4}} \cdot dx.$$

Expanding by Binomial theorem the expression under root

$$S = 2\int_0^{L/2} \left[1 + \frac{1}{2} \cdot \frac{64h^2x^2}{L^4} - \frac{1}{8}\left(\frac{64h^2x^2}{L^4}\right)^2\right] dx$$

Neglecting higher order terms

$$= 2\int_0^{L/2} \left[1 + \frac{32h^2x^2}{L^4}\right] dx$$

$$= L + \frac{8}{3}\frac{h^2}{L} = L\left[1 + \frac{8}{3}\frac{h^2}{L^2}\right] \qquad \text{... (6.10)}$$

(*ii*) Catenary at level support. The equation of a catenary

$$y = \frac{H}{w_l}\left[\cosh\cdot\frac{w_l . x}{H} - 1\right]$$

$$\frac{dy}{dx} = \sinh . \frac{w_l . x}{H}$$

$$\text{Length of cable } = 2\int_0^{L/2} \sqrt{1 + \sin h^2 \frac{w_l \cdot x}{H}}\, dx$$

$$= \frac{2H}{w_l}\cdot \sin h \frac{w_l . x}{2H} \qquad \text{... (6.11)}$$

(*iii*) Parabolic cable with supports at different levels.

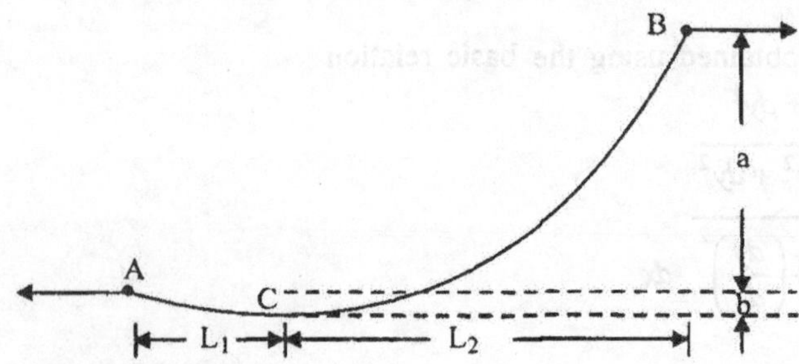

Fig. 6.12. Parabolic cable with supports at different levels.

The total length $S = AC + CB$

Considering AC as half parabola with central dip equal to b

$$\text{length of } AC = L_1\left[1 + \frac{8}{3}\cdot\frac{b^2}{4L_1^2}\right]$$

$$\text{Similarly length of } CB = L_2\left[1 + \frac{8}{3}\cdot\frac{(a+b)^2}{4L_2^2}\right]$$

$$\text{Total Length } = L_1 + \frac{2}{3}\frac{b^2}{L_1} + L_2 + \frac{2}{3}\cdot\frac{(a+b)^2}{L_2}$$

$$= L_1 + L_2 + \frac{2}{3}\left[\frac{b^2}{L_1} + \frac{(a+b)^2}{L_2}\right]$$

$$= L + \frac{2}{3}\left[\frac{b^2}{L_1} + \frac{(a+b)^2}{L_2}\right] \qquad \text{... (6.12)}$$

6.6. ELASTIC STRETCH OF A CABLE

A cable subjected to concentrated load consists of a number of straight segments each subjected to constant tension T_1, T_2, T_3 etc.

Applying Hooke's law, elastic stretch due to $T_1 = \frac{T_1 l_1}{AE}$. Similarly for other segments. The total elongation is sum of all the segmental elongations.

For a distributed loading the tension varies from point to point. Hence instead of segmental addition, there will be integration.

Elastic stretch $= S = \int_0^S \frac{T_x\, ds}{AE}$

where T_x = tension at any distance x.

$$T_x = \sqrt{H^2 + V^2} = H\sqrt{1+\left(\frac{V}{H}\right)^2} = H\sqrt{1+\left(\frac{dy}{dx}\right)^2}$$

$$ds = dx\sqrt{1+\left(\frac{dy}{dx}\right)^2}$$

Hence $$\delta = \frac{H}{AE}\int_0^L \sqrt{1+\left(\frac{dy}{dx}\right)^2}\cdot\sqrt{1+\left(\frac{dy}{dx}\right)^2}\cdot dx$$

$$= \frac{H}{AE}\int_0^L \left[1+\left(\frac{dy}{dx}\right)^2\right]\cdot dx.$$

For a parabolic cable $y = 4h\cdot\frac{x^2}{L^2}$

$$\therefore \quad \frac{dy}{dx} = \frac{8h.x}{L^2}$$

$$\delta = \frac{2H}{AE}\int_0^{L/2}\left[1+\frac{64h^2x^2}{L^4}\right]dx$$

$$= \frac{H.L}{AE}\left[1+\frac{16}{3}\cdot\frac{h^2}{L^2}\right]. \quad \text{...(6.13)}$$

Example 6.1. A light cable with negligible weight is suspended between two points spaced 300 m apart horizontally with the right support 12 m higher than the left. Four vertical gravity loads of 400, 200, 400 and 1200 kN (from left to right) centred between the supports and spaced 60 m apart horizontally are applied to the cable. If the largest sag is 24 m, calculate (*a*) the sag at each applied load; (*b*) the *y* coordinate of the cable at each load with the coordinate origin at the left support (*c*) the maximum cable tension. (*d*) the unstressed length of the cable.

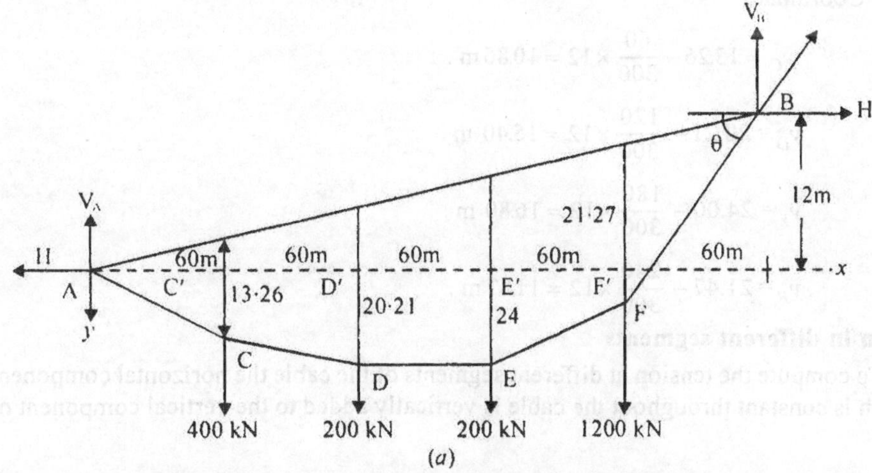

(*a*)

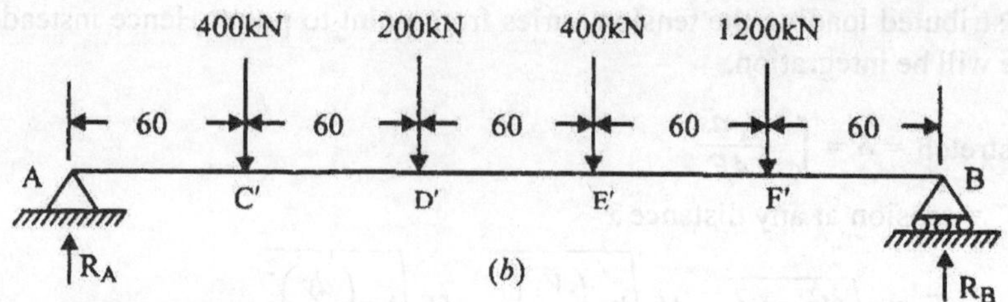

(*b*)

Taking moments about *B*,

$$R_A \times 300 = 400 \times 240 + 200 \times 180 + 400 \times 120 + 1200 \times 60$$

or $$R_A = 840 \text{ kN}$$

Taking moments about *A*

$$R_B \times 300 = 1200 \times 240 + 400 \times 180 + 200 \times 120 + 400 \times 60$$

or $$R_B = 1360, R_A + R_B = 2200 \text{ kN (check)}$$

B.M. at

$$C' = 840 \times 60 = 50400 \text{ kNm}$$

$$D' = 840 \times 120 - 400 \times 60 = 76800 \text{ kNm}$$

$$E' = 840 \times 180 - 400 \times 120 - 200 \times 60 = 91200 \text{ kNm.}$$

$$F' = 1360 \times 60 = 81{,}600 \text{ kNm}$$

Max^m B.M. at $E' = 91{,}200$ kNm.

$\therefore$ Max^m sag of 24 m will occur at E

$$\text{Horizontal force} = \frac{91{,}200}{24} = 3800 \text{ kN}$$

$$\text{Sag at } C = \frac{50400}{3800} = 13.26 \text{ m}$$

$$\text{at } D = \frac{76800}{3800} = 20.21 \text{ m}$$

$$\text{at } E = \frac{91200}{3800} = 24 \text{ m}$$

$$\text{at } F = \frac{81600}{3800} = 21.47 \text{ m}$$

Y Coordinate

$$y_C = 13.26 - \frac{60}{300} \times 12 = 10.86 \text{ m}.$$

$$y_D = 20.21 - \frac{120}{300} \times 12 = 15.40 \text{ m}.$$

$$y_E = 24.00 - \frac{180}{300} \times 12 = 16.80 \text{ m}.$$

$$y_F = 21.47 - \frac{240}{300} \times 12 = 11.87 \text{ m}.$$

Tension in different segments

To compute the tension at different segments of the cable the horizontal component of tension *H* which is constant throughout the cable is vertically added to the vertical component of tension in that cable segment. The vertical component of tension is equal to the algebraic sum of vertical forces acting either to the left or to the right of cable segment. This is done in Tabular form.

Taking moments of all forces about B,

$$H \times 12 + V_A \times 300 - 400 \times 240 - 200 \times 180 - 400 \times 120 - 1200 \times 60$$

or $V_A = 688\text{ kN}.\ V_B = 2200 - 688 = 1512\text{ kN}$

Segment	Horizontal Component kN	Vertical Component kN	Resultant Tension, KN
AC	3800	688	$\sqrt{3800^2 + 688^2} = 3861.78$
CD	3800	$688 - 400$	$\sqrt{3800^2 + 288^2} = 3810.89$
DE	3800	$688 - 400 - 200$	$\sqrt{3800^2 + 88^2} = 3801.01$
EF	3800	$688 - 400 - 200 - 400$	$\sqrt{3800^2 + (-312)^2} = 3812.78$
FB	3800	$688 - 400 - 200 - 400 - 1200$	$\sqrt{3800^2 + (-1512)^2} = 4089.76$

Unstressed length of cable

$$\sqrt{60^2 + 10.86^2} + \sqrt{60^2 + (15.40 - 10.86)^2} + \sqrt{60^2 + (16.80 - 15.40)^2}$$
$$+ \sqrt{60^2 + (16.80 - 11.87)^2} + \sqrt{60^2 + 23.87^2}$$
$$= 60.97 + 60.17 + 60.02 + 60.20 + 64.57$$
$$= 305.933\text{ m}$$

Check

$$F_{AC} \cos\theta = H = 3800\text{ kN}$$

$$F_{AC} = \frac{H}{\cos\theta} = \frac{3800}{\cos\theta} = \frac{3800}{\dfrac{60}{\sqrt{60^2 + 10.86^2}}}$$

$$= \frac{3800}{60} \times \sqrt{60^2 + 10.86^2} = \mathbf{3861.78\ kN.}$$

Example 6.2. Show that the equation for the shape of a cable with uniformly distributed load along its horizontal projection can be written.

$H\bar{y} = \frac{1}{2} q\bar{x}^2 + H(\bar{y}_b - \bar{y}_a)\frac{\bar{x}}{L} - \frac{1}{2} q L\bar{x} + H_{\bar{y}_a}$ where the cartesian coordinates $\bar{x}$ and $\bar{y}$ are as shown in fig. 6.14 and $\bar{y}_a$ and $\bar{y}_b$ are the $\bar{y}$ coordinates of points a and b respectively [Sachs]

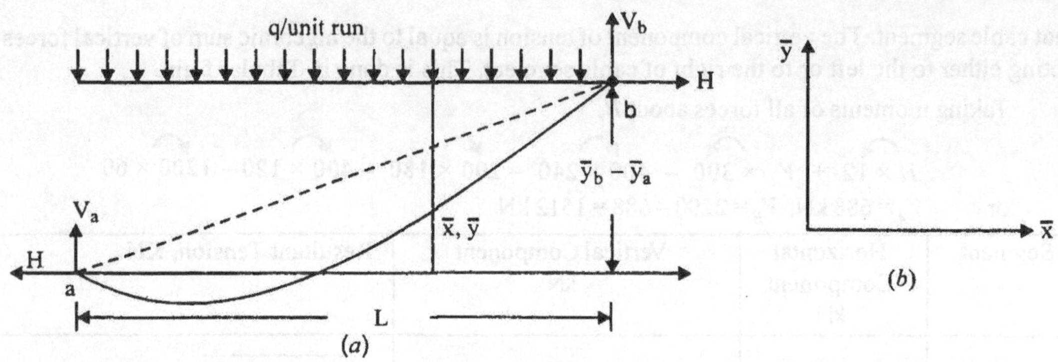

Fig. 6.14. Example 6.2

Applying the principle

Horizontal force × sag = B.M. on horizontal span due to vertical load.

$$H \times \left[\frac{\bar{y}_b - \bar{y}_a}{L} \cdot \bar{x} - (\bar{y} - \bar{y}_a)\right] = \frac{1}{2} qL\bar{x} - \frac{1}{2} q\,\bar{x}^2$$

or $$H\bar{y} = \frac{1}{2} q\,\bar{x}^2 - \frac{1}{2} q\,L.\bar{x} + H(\bar{y}_b - \bar{y}_a).\frac{\bar{x}}{L} + H\bar{y}_a.$$

Example 6.3. Determine (*a*) the reactions and (*b*) the location and magnitude of maximum tension in the cable shown in Fig. 6.15.

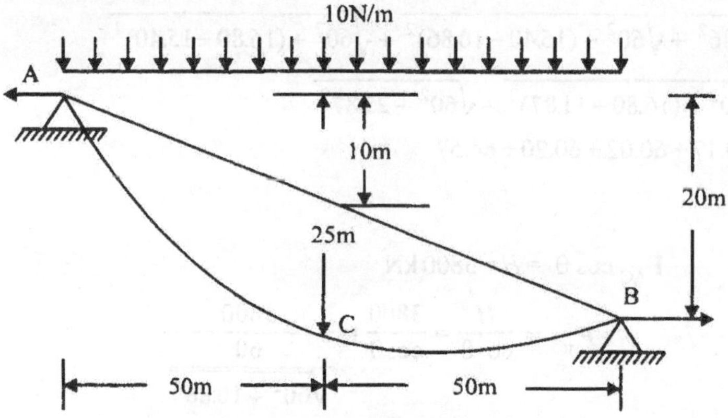

Fig. 6.15. Example 6.3.

Solution: $H \times y$ = B.M. at the section

or $$H \times 15 = \frac{10 \times 100^2}{8}$$

or $$H = \frac{10^5}{120} = 833.33\,\text{N}.$$

Taking moments about *A*

$$R_B \times 100 + 833.33 \times 20 - \frac{10 \times 100^2}{2} = 0$$

or $$R_B = \frac{1}{100}\left[\frac{10 \times 100^2}{2} - 833.33 \times 20\right]$$

$$= \mathbf{333.33\ N.}$$

$$R_A \times 100 - 833.33 \times 20 - \frac{10 \times 100 \times 100}{2} = 0$$

or $$R_A = 666.67$$

$$R_B + R_A = 333.33 + 666.67 = 1000 \text{ (check)}$$

Maximum cable tension occurs at the higher end

$$= \sqrt{833.33^2 + 666.67^2}$$

$$= \mathbf{1067.186\ N}$$

Example 6.4. For the structure shown in Fig. 616 fino

(*a*) max^m tension in the main cable

(*b*) max^m tension in the anchor cable

(*c*) the compressive force in the mast.

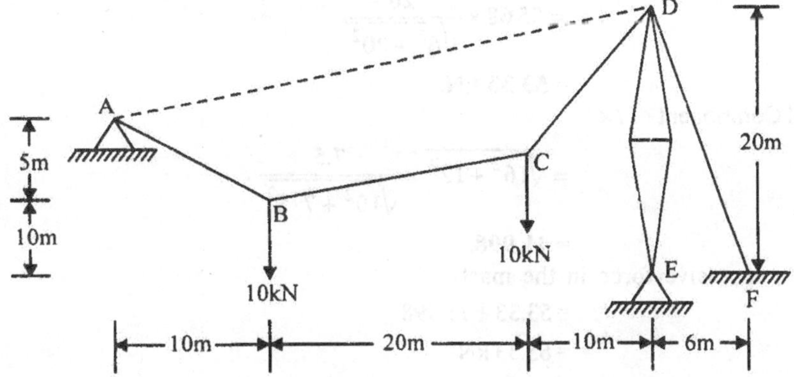

Fig. 6.16. Example 6.4

Solution:

$$H \cdot y_B = 100 \text{ kNm (B.M for Simply supported span)}.$$

$$H = \frac{100}{y_B} = \frac{100}{5 + \frac{10 \times 5}{40}} = 16 \text{ kN}$$

$$y_C = \frac{100}{16} = 6.25 \text{ m}$$

Vertical distance of C from $D = 6.25 + 1.25 = 7.50$

Taking moment about D as the tower is pinned at both D and E.

$$V_A \times 40 + H_A \times 5 - 10 \times 30 - 10 \times 10 = 0$$

or $$V_A = \frac{1}{40}[300 + 100 - 80] = 8 \text{ kN}$$

At A $H = 16$ $V_A = 8$ $T_A = \sqrt{8^2 + 16^2} = 17.885$ kN

At B $H = 16$ $V_B = 8 - 10 = -2$ $T_B = \sqrt{16^2 + 2^2} = 16.125$ kN

At C $H = 16$ $V_C = 12$ $T_C = \sqrt{16^2 + 12^2} = 20$ kN

As the tower is pinned at both ends the horizontal component of forces in DC and DF must be equal. Hence

Horizontal Component of DF = Horizontal Component of DC = H

$$F_{DF} \times \frac{6}{\sqrt{6^2+20^2}} = 16$$

or
$$F_{DF} = \frac{16 \times \sqrt{6^2+20^2}}{6} = \mathbf{55.68\ kN}$$

$$\text{Vertical Component of } DF = F_{DF} \times \frac{20}{\sqrt{6^2+20^2}}$$

$$= 55.68 \times \frac{20}{\sqrt{6^2+20^2}}$$

$$= \mathbf{53.33\ kN.}$$

Vertical Component of DC

$$= \sqrt{16^2+12^2} \times \frac{7.5}{\sqrt{10^2+7.5^2}}$$

$$= \mathbf{11.998.}$$

Total Compressive force in the mast

$$= 53.33 + 11.998$$

$$= 65.33 \text{ kN}$$

Hence (*i*) Maximum Tension in the main cable = 20 kN.

(*ii*) Maximum tension in the anchor cable = 55.68 kN.

(*iii*) Compressive force in the mast = 65.33 kN.

Example 6.5. For the cable shown in Fig. 6.17 determine the

(*a*) Maximum tension

(*b*) Location of the lowest point

(*c*) Magnitude of sag at mid span.

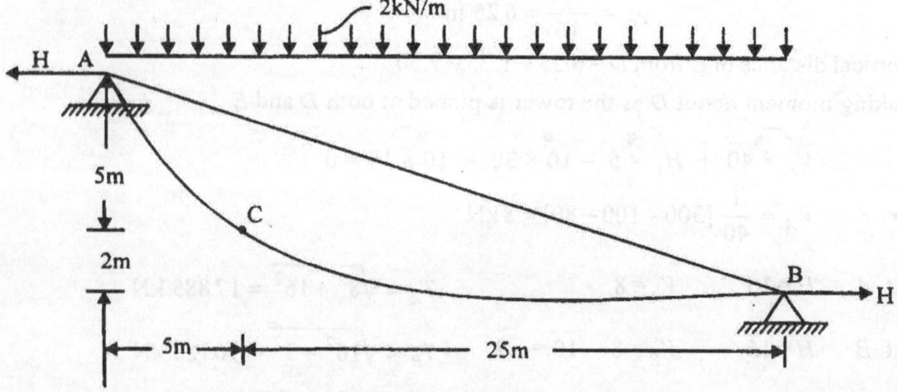

Fig. 6.17. Example 6.5

Solution:

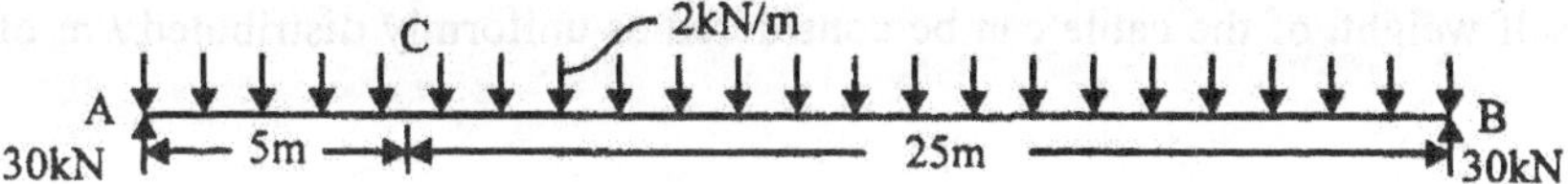

Fig. 6.18. Beam diagram (Example 6.5)

$$\text{BM. at } C = 30 \times 5 - \frac{2 \times 5^2}{2} = 150 - 25 = \mathbf{125\ kNm}$$

$$H = \frac{\text{B.M at } C}{\text{sag at } C} = \frac{125}{5 - \frac{5 \times 7}{30}} = 32.60\,\text{N.}$$

Taking moments of all forces about B

$$V_A \times 30 - 32.60 \times 7 - 2 \times \frac{30^2}{2} = 0.$$

or $$V_A = \frac{1}{30}[32.60 \times 7 + 30^2] = \mathbf{37.60\ kN.}$$

Maximum Tension occurs at the highest point and is equal to

$$\sqrt{32.60^2 + 37.60^2} = 49.76\ \text{kN}\cdot$$

General Equation of a parabola

$y = Ax^2 + Bx + C$ (origin at A, y positive upward)

at $x = 0$ $y = 0$ $\therefore c = 0$

at $x = 30$ $y = -7$

or $-7 = A \times 30^2 + B \times 30$

At $x = 5$ $y = -5$

$-5 = A \times 5^2 + B \times 5$

giving $5A + B = -1$

$30A + B = -0.23.$

Solving $A = 0.0308$

$B = -1.154.$

At the lowest point tangent is horizontal or $\frac{dy}{dx} = 0$

with $y = Ax^2 + Bx$

$$\frac{dy}{dx} = 2Ax + B = 0$$

or $$x = -\frac{B}{2A}$$

$$= +\frac{1.154}{2 \times \cdot 0308}$$

$= 18.73$ m from A

Sag at mid span.

$$H \times y_{\frac{L}{2}} = \frac{2 \times 30^2}{8} \quad \text{or} \quad y_{\frac{L}{2}} = 6.90\ \text{m.}$$

Example 6.6. A steel suspension bridge cable having a central sag equal to one tenth of its span carries a superimposed distributed load of 20 kN/m. The density of the steel cable is 76 kN/m³. The self weight. of the cable can be considered as uniformly distributed / m of horizontal span.

Calculate

(*i*) Maximum possible span of the cable if the maximum permissible stress in the cable is 200 MPa.

(*ii*) If the anchorage difficulties limit the maximum tension in the cable to 50 MN determine the maximum possible span.

Solution. Length of parabolic cable $= L+\frac{8}{3}\frac{h^2}{L}$

Total weight of cable $=(A)\left(L+\frac{8}{3}\cdot\frac{h^2}{L}\right).76$

$$\frac{\text{weight}}{\text{m}}=\frac{\text{Total weight}}{L}=76\,A\left(1+\frac{8}{3}\frac{h^2}{L^2}\right)$$

$$=76\,A\,(1+0.0267)$$

$$= \mathbf{78.03\ A\ kN.}$$

Total uniformly distributed load

$$w=(20+78.03\,A)\,\text{kN/m}$$

$$H=\frac{wL^2}{8h}=\frac{wL^2}{8(0.1L)}=1.25\,wl$$

$$\text{Vertical reaction }=\frac{wL}{2}=0.50\ wL$$

$$\text{Maximum tension }=\sqrt{(1.25\,wL)^2+(0.50\,wL)^2}$$

$$=1.3463\ wL$$

Substituting $w=(20+78.03\,A)\,\text{kN}$

$$T_{max}=(1.346)(20+78.03\text{ A}).\text{L kN}$$

$$\text{Max}^{m}\text{ stress}=\frac{T_{max}}{A}=\frac{(1.346)(20+78.03\,A)}{A}\text{ L kN/m}^2$$

This should not exceed 200 N/mm² $=\frac{200}{1000}\times10^6\ \text{kN/m}^2$

$$=2\times10^5\ \text{kN/m}^2$$

or $$\frac{1.346(20+78.03\,A)\,L}{A}\not>2\times10^5$$

or $$\left(\frac{20}{A}+78.03\right)L\not>1.485\times10^5$$

For maximum possible span A should be maximum. When $A\to\infty$

$$78.03\text{ L}=1.485\times10^5$$

$$L = \frac{1.485 \times 10^5}{78.03} = \mathbf{1903 \cdot 114 \ m}$$

(*ii*) When the maximum tension is restricted to 50 MN

$$1.346\,(20 + 78.03\,A)\,L \not> 50 \times 10^3$$

Again, $$A = \frac{50 \times 10^3}{2 \times 10^5} = 0.25\ m^2$$

$$1.346\,(20 + 78.03 \times 0.25)\,L \not> 50 \times 10^3$$

or $$L \not> \frac{50 \times 10^3}{53.177}$$

or $$L \not> \mathbf{940.25\ m.}$$

6.7. SUSPENSION BRIDGE

Suspension bridges are used in highways, when span is more than 200 m. Fig. 6.19 shows a suspension bridge and the main elements which are

(*i*) Main cable
(*ii*) Hangers or suspenders
(*iii*) Anchor cable
(*iv*) Bridge deck and
(*v*) Pier

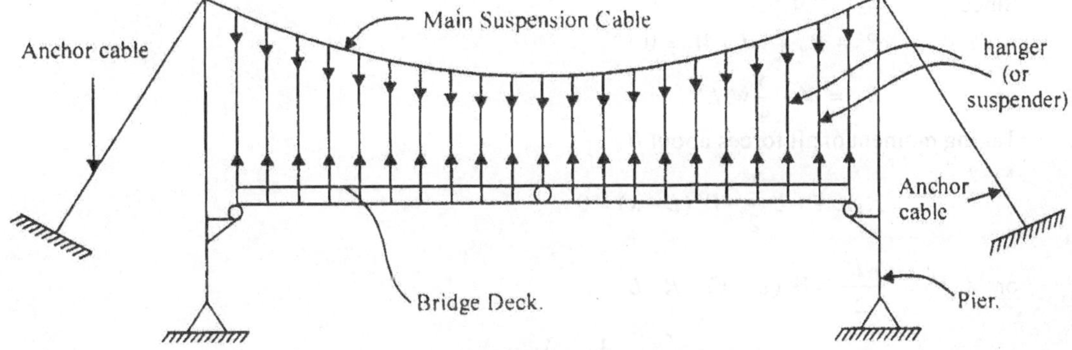

Fig. 6.19. Suspension bridge.

The load from the bridge deck is transferred to the main cable by means of hangers which are in direct tension. As the load in the bridge can be any where the shape of the cable and the deck level changes with load position which is undesirable. To overcome this the deck is stiffened by two hinged or three hinged stiffened girders or truss on either side of the roadway. The purpose of the stiffening girder or truss is to provide sufficient rigidity to the bridge deck such that all the hangers carry equal load for all position of the live load. The main cable thus retains its parabolic shape for all load positions.

The girders are suspended from cables through hanger cables. Hence the uniformly distributed dead load of the roadway and the stiffening girders is transmitted to the cables through hanger cables and is taken up entirely by the tension in the cables. The stiffening girders do not suffer any S.F. or B.M. under dead load as girders are supported by closely spaced hanger cables throughout.

Any live load on the bridge will be transmitted to the girders as point loads. The stiffening girders transmit the live load to the cable as uniformly distributed load. While doing so the stiffening girders will be subjected to S.F. and B.M. throughout their length. The analysis of stiffened grider can be carried out as follows. (Fig. 6.20)

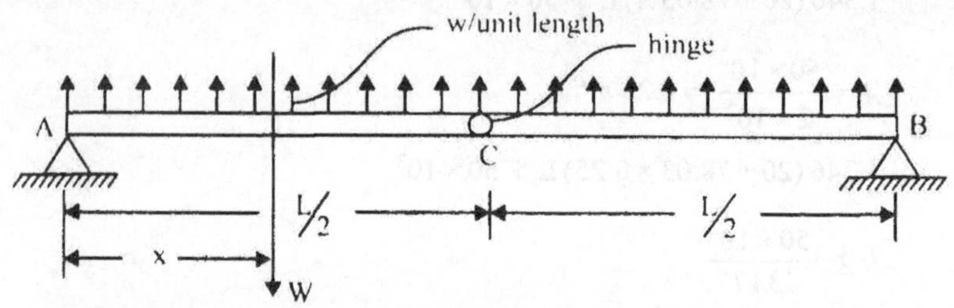

Fig. 6.20. Analysis of stiffened girder (Three hinged)

Fig. 6.20. shows the forces acting on the girder when the live load is transferred to the girder as a concentrated load W at a distance x from left support A. The stiffening girder transfers the load to the main cable as a uniformly distributed load w per unit length thus enabling the main cable to retain its parabolic shape. (The magnitude of 'w' varies with the position of load W). Since the bending moment at hinge C is zero, considering moment of all forces to the right of C.

$$R_B \cdot \frac{L}{2} + w\left(\frac{L}{2}\right).\left(\frac{L}{4}\right) = 0$$

or $$R_B = -\frac{wL}{4}$$

Since $$\Sigma F_y = 0$$

$$R_A + R_B + wL - W = 0$$

or $$R_A = W - \frac{3}{4}w.L.$$

Taking moment of all forces about B,

$$R_A.L + wL.\frac{L}{2} - W\,(L - x) = 0$$

or $$\frac{wL^2}{2} = W\,(L - x) - R_A.L$$

$$= W\,(L - x) - \left(W - \frac{3}{4}w.L\right).L$$

$$= \boldsymbol{WL - Wx - WL + \frac{3}{4}\,w^{\rho.c}.L^2}$$

or $$wL^2\left(\frac{1}{2} - \frac{3}{4}\right) = -W.x$$

or $$w = \frac{4W.x}{L^2} \qquad \text{...(6.14)}$$

and $$H = \frac{wL^2}{8h} = \frac{4W.x}{L^2}\cdot\frac{L^2}{8h} = \frac{Wx}{2h} \qquad \text{...(6.14a)}$$

Knowing W, w, R_A and R_B, the bending moment and shear force at any section can be found out. In between the suspension cable and the anchor cable there is a guide pulley or a saddle mounted on roller over the pier as shown in Fig. 6.21. The different arrangements will introduce different forces in the anchor cable or in the pier. The anchor cable is fixed inside a huge mass of concrete.

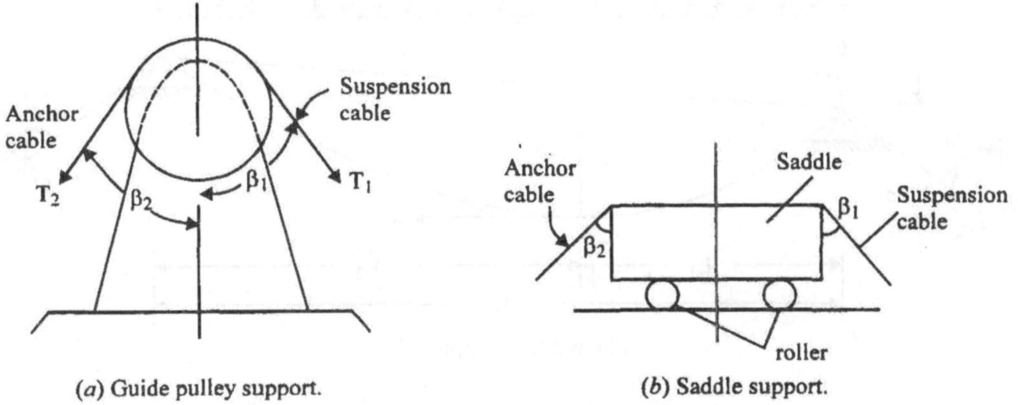

(*a*) Guide pulley support.

(*b*) Saddle support.

Fig. 6.21. Different types of cable supports.

Case 1. Guide pulley support (Fig. 6.21 *a*). Because of friction less pulley, force T_2 in anchor cable is equal to force T_1 in suspension cable

Downward Thrust on Pier

$$= T_1 \cos\beta_1 + T_2 \cos\beta_2$$

$$= T_1 (\cos\beta_1 + \cos\beta_2)$$

Net horizontal force on the top of pier

$$= T_1 \sin\beta_1 - T_2 \sin\beta_2$$

$$= T_1 (\sin\beta_1 - \sin\beta_2).$$

This horizontal force will cause bending in the tower.

Case II. Roller support (Fig. 6.21*b*).

As the rollers cannot resist any horizontal force, the horizontal component of tension on both sides must be equal. Hence

$$T_1 \sin\beta_1 = T_2 \sin\beta_2$$

or

$$T_2 = T_1 \cdot \frac{\sin\beta_1}{\sin\beta_2}$$

The bending moment in the tower is then zero.

Vertical thrust on tower

$$= T_1 \cos\beta_1 + T_2 \cos\beta_2.$$

Example 6.7. A suspension cable is suspended from two piers 44 m apart, one support being 1.1 m above the other. (Fig. 6.22.). The cable carries a uniformly distributed horizontal load of 1/2 t/m and has its lowest point 2.5 m below the lower support. The ends of the cables are attached to saddles on rollers at the top of the piers and the back stays (may be assumed straight) are inclined at 60° to the vertical. Determine the maximum tension in the cable, the tension in the back stays and the thrust on each pier.

Solution:

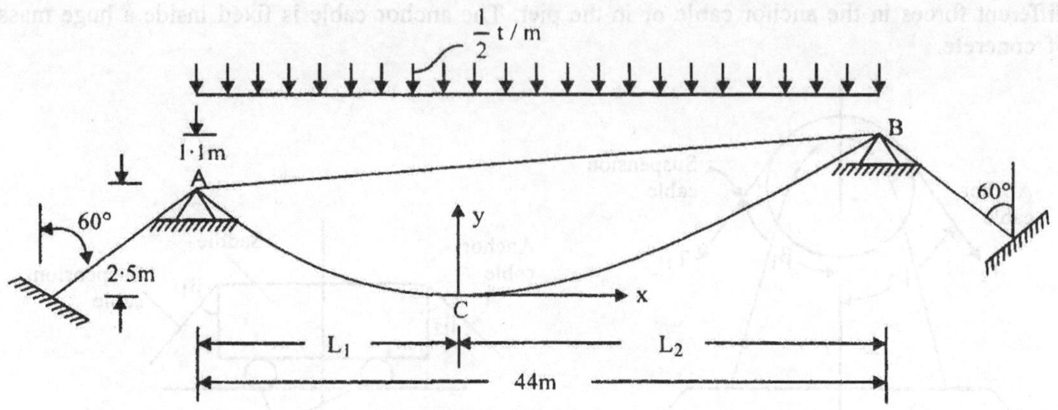

Fig. 6.22. Example 6.7

$$L_2 = \frac{L}{\left[\sqrt{\frac{b}{a+b}}+1\right]}$$

Case III. Ref case III (Page 192).

Here $a = 1.1$ m $b = 2.5$ m and $L = 44$ m

$$\therefore \quad L_2 = \frac{44}{\left[\sqrt{\frac{2.5}{1.1+2.5}}+1\right]} = 24 \text{ m}$$

$$L_1 = 44 - 24 = 20$$

$$H = \frac{wL_1^2}{2b} = \frac{1}{2} \times \frac{20^2}{2 \times 2.5} = 40.0$$

$$= \frac{w}{2(a+b)} \cdot L_2^2 = \frac{1}{2} \times \frac{24^2}{2 \times 3.6} = 40.00 \text{ (check)}$$

$$T_B = H\sqrt{1+\left(\frac{wx}{H}\right)^2} = 40\sqrt{1+\left(\frac{1}{2}\cdot\frac{24}{40}\right)^2} = 41.76t$$

$$T_A = H\sqrt{1+\left(\frac{wx}{H}\right)^2} = 40\sqrt{1+\left(\frac{1}{2}\frac{(-20)}{40}\right)^2}$$

$$= 41.23$$

Maximum Cable Tension $= 41.76\ t$

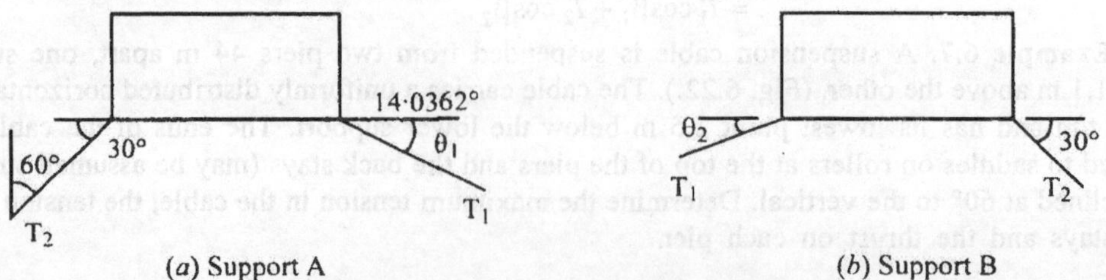

Fig. 6.23. Supports (Example 6.7)

$$\frac{dy}{dx} = \frac{w.x}{H}$$

at $x = 24$ $\quad \frac{dy}{dx} = \tan\theta_2 = \frac{1}{2} \times \frac{24}{40} = 0.3$

$$\theta_2 = 16.6992°$$
$$\cos\theta_2 = 0.9578$$
$$\sin\theta_2 = 0.2873$$

at $x = -20$ $\quad \frac{dy}{dx} = \tan\theta_1 = \frac{1}{2}\left(\frac{-20}{40}\right) = -0.25$

$$\theta_1 = -14{\cdot}0362$$
$$\sin\theta_1 = -0{\cdot}2425$$
$$\cos\theta_1 = 0.97014$$

Support A. At $\quad x = -20$

$$T_1 \cos\theta_1 = T_2 \cos 30°$$

or $\quad 41.23 \times ({\cdot}9701) = T_2 \cdot \frac{\sqrt{3}}{2}$

or $\quad T_2 = 46{\cdot}1848t$

(tension at back stay at A)

Pressure on Pier A

$$T_1 \sin\theta_1 + T_2 \sin 30°$$

$$= 4123 \times 0.2425 + 46.1848 \times \frac{1}{2}$$

$$= \mathbf{33.0906t}$$

Support B

$$x = 24$$
$$T_1 \cos\theta_2 = T_2 \cos 30°$$

or $\quad T_2 = \frac{T_1 \cos\theta_2}{\cos 30°}$

(tension at back stay at B)

$$= \frac{41.76 \times 0.9578 \times 2}{\sqrt{3}}$$

$$= \mathbf{46.1854t}$$

Pressure on Pier B

$$T_1 \sin\theta_2 + T_2 \sin 30°$$

$$= 41.76 \times 0.2873 + 46.1854 \times \frac{1}{2}$$

$$= 35.0903t$$

2nd Solution (From 1st Principle)

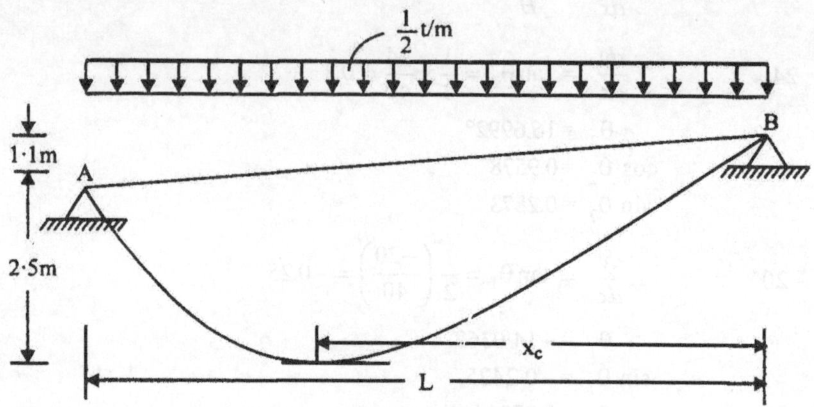

Fig. 6.24. Example 6.7

General Equation of a Cable

$$\frac{d^2y}{dx^2} = \frac{w}{H}$$

or $$\frac{dy}{dx} = -\frac{w}{H}\cdot x + C_1$$

or $$y = -\frac{w}{H}\cdot\frac{x^2}{2} + C_1x + C_2$$

Boundary Conditions (origin at B)

at $x=0$ $y=0$ ∴ $C_2 = 0$ At $x=44$ $y=1.1$

or $$1.1 = -\frac{1}{2}\cdot\frac{44^2}{2H} + 44\,C_1$$

or $$C_1 = \frac{1.1}{44} + \frac{44}{4H} = \frac{1}{40} + \frac{11}{H}$$

The equation then becomes,

$$y = -\frac{x^2}{4H} + \left(\frac{1}{40} + \frac{11}{H}\right)x$$

Check $x=44$ $$y = -\frac{44\times 44}{4H} + \left(\frac{1}{40} + \frac{11}{H}\right)44 = \mathbf{1.1\ m.}$$

Position of Lowest Point

At the lowest point $\frac{dy}{dx} = 0$.

or $$-\frac{x}{2H} + \frac{1}{40} + \frac{11}{H} = 0$$

or $$x_c = \left(\frac{1}{40} + \frac{11}{H}\right)\times 2H = \frac{H}{20} + 22$$

But $H = \frac{wL^2}{8h} = \frac{44^2}{16.h}$

or $h = 3.6 - \frac{1.1}{44} \times \left(\frac{H}{20} + 22\right)$

or $H = \frac{44^2}{16\left(3.6 - \frac{1.1H}{880} - \frac{1.1}{2}\right)}$

or $3.6H - \frac{1.1H^2}{880} - 0.55H = \frac{44^2}{16}$

Solving $H = 40.34t$ Rest as before.

Example 6.8. The two cables of a suspension bridge each has a span of 50 m and a central dip of 5 m. Each cable is siffened by a girder hinged at the ends and at mid span to constrain the cable to retain its parabolic shape. There is uniform dead load of 1t/horizontal meter span over the whole girder and a live load of 3t/horizontal meter and 12 m long. These D.L. and L.L. are for each girder. Determine the maximum cable tension when the live load is situated on the left hand half of the stiffening girder with right end near the central hinge. Sketch the S.F.D. and B.M.D. for the girder showing important values. **[AMIE Summer 1979]**

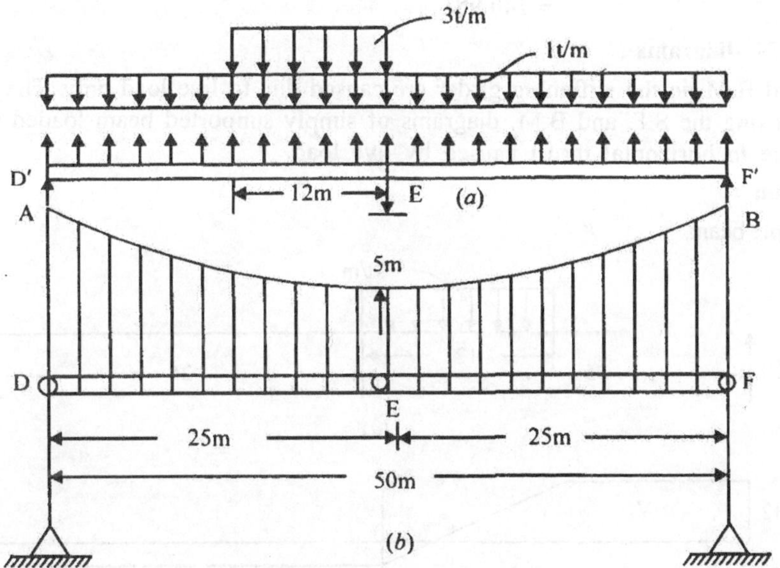

Fig. 6.25. Example 6.8.

1. Calculation of *H*

In the three hinged stiffened girder,

$$M_E = 0$$

But $M_E = \text{Beam B.M.} - Hy_c$

From Fig. 6.25 (*a*)

$$M_{E_1} = \frac{3 \times 12 \times 19}{50} \times 25 = 342 \text{ tm (due to live load)}$$

$$\therefore \quad H_l = \frac{342}{5} = 68.4\ t$$

$$M_{E_2} = \frac{wL^2}{8} \text{ due to D.L}$$

$$= \frac{1 \times 50^2}{8} = 312.50$$

$$H_d = \frac{312.50}{5} = 62.5t$$

2. Calculation of Maximum Tension

Maximum tension in the cable for this condition of loading shall occur at A

$$T_{max} = \sqrt{V^2 + H^2}$$

$$H = 68.4 + 62.5 = 130.9t$$

Equivalent *u.d.l* due to L.L

$$w_e = \frac{8 \cdot y_C \cdot H_l}{L^2} = \frac{8 \times 5 \times 68 \cdot 4}{50^2}$$

$$= 1.0944 \text{ t/m}$$

$$V = \frac{w_d \times L}{2} + \frac{w_e \times L}{2} = \frac{1 \times 50}{2} + \frac{1.0944 \times 50}{2}$$

$$= 25 + 27 \cdot 36$$

$$= \mathbf{52.36t}$$

$$T_{max} = \sqrt{52.36^2 + 130.9^2}$$

$$= \mathbf{140.98t.}$$

3. S.F. and B.M. diagrams

S.F. and B.M. in the stiffening girder are caused due to live load only. They are obtained by superimposing the S.F. and B.M. diagrams of simply supported beam loaded with live load and those due to horizontal thrust caused by live load.

4. S.F. diagram

(*i*) Simple beam.

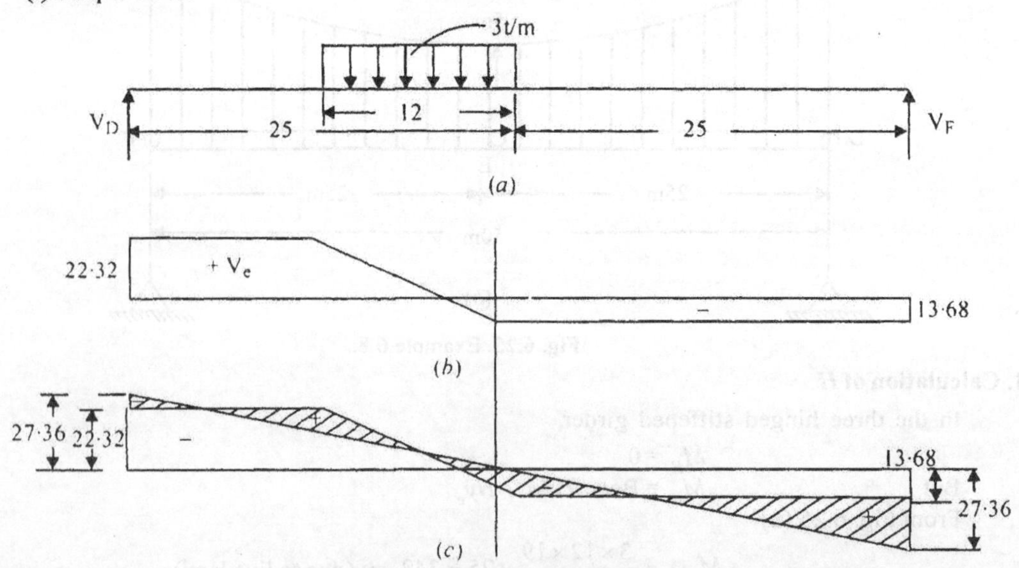

Fig. 6.26. S.F. diagram (Example 6.8).

(*ii*) S.F. due to Horizontal Thrust.

The shear force due to horizontal thrust H_l at any point $= H_l \cdot \tan\phi$ where $\tan\phi$ represents the slope equation of the cable at any point. Equation of the cable with origin at A.

$$y = \frac{4y_C}{L^2}(Lx - x^2)$$

$$\tan\phi = \frac{dy}{dx} = \frac{4y_C}{L^2}(L - 2x)$$

This is equation of a straight line.

At $\quad x = 0, \tan\phi = \frac{4y_C}{L} = \frac{4 \times 5}{50} = 0.4$

$H_l.\tan\phi = 68.4 \times 0.4 = 27.36t\ (-ve)$

At $\quad x = 13 \quad \tan\phi = \frac{4 \times 5}{50^2}(50 - 2 \times 13) = 0.192$

$H_l.\tan\phi = 68.40 \times 0.192 = 13.133\ (-ve)$

At $\quad x = 25 \quad \tan\phi = \frac{4 \times 5}{50^2} \times (50 - 50) = 0$

$H_l \cdot \tan\phi = 0$

At $\quad x = 50 \quad \tan\phi = \frac{4 \times 5}{50^2}(50 - 100)$

$= -0{\cdot}4.$

$H_l \tan\phi = 68{\cdot}4 \times (-0{\cdot}4) = 27{\cdot}36(+ve)$

B.M. diagram

(*i*) Simple beam bending moment.

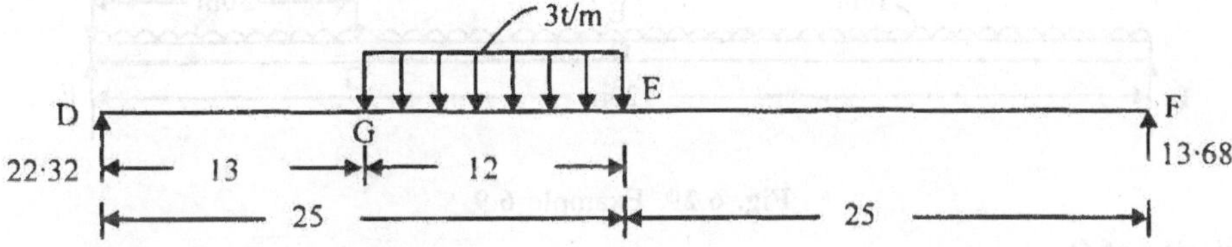

Fig. 6.27. Beam bending moment (Example 6.8).

$M_G = 22{\cdot}32 \times 13 = 290{\cdot}16$

$M_E = 13{\cdot}68 \times 25 = 342{\cdot}00$

(*ii*) Horizontal thrust due to live load.

B.M. at any pont due to H

$$M_H = H_l.y = H_l\left[\frac{4y_C}{L^2}(Lx - x^2)\right]$$

at $\quad x = 0 \qquad M_H = 0$

$x = 13 \qquad M_H = 68.4\left[\frac{4 \times 5}{50^2}(50 \times 13 - 13^2)\right]$

$= 263.20 \text{ tm}$

$$x = 25 \qquad M_H = 68.4\left[\frac{4\times 5}{50^2}(50\times 25 - 25^2)\right]$$

$$= \mathbf{342\ tm.}$$

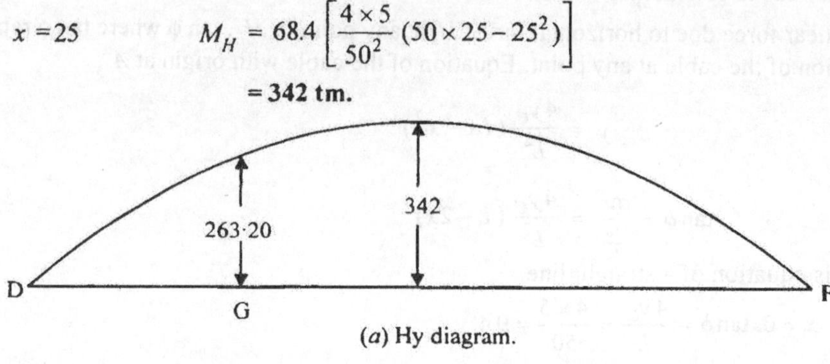

(a) Hy diagram.

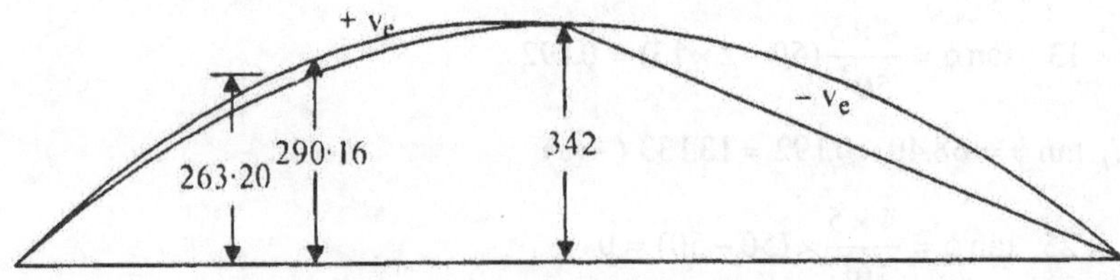

(b) Combined B.M. diagram.

Fig. 6.28. B.M. diagram (Example 6.8).

Example 6.9. A three hinged suspension girder bridge has a span of 120 m and a central dip of 15 m. it has to carry a uniformly distributed load of 1ᵗ/m. A concentrated load of 20ᵗ is placed at distance of 30 m from the right support. Determine the maximum bending moment and shear force at a section 30 m from the left hand support. **[AMIE Winter 1987]**

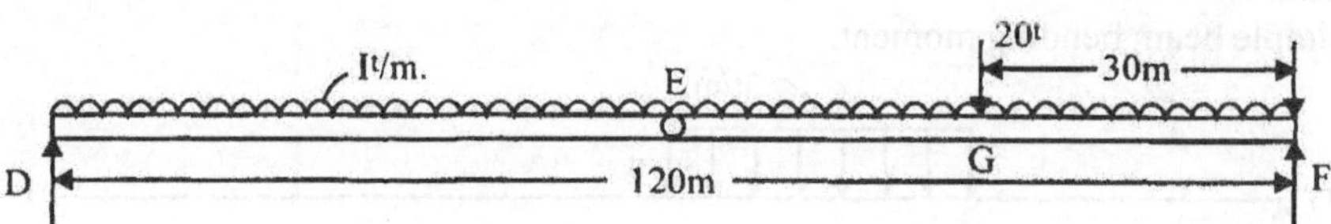

Fig. 6.29. Example 6.9.

1. Calculation of H.

In the three hinged stiffened girder $M_E = 0$

But $M_E = \text{Beam B.M.} - Hy_C = 0$

or $$H = \frac{\text{Beam B.M.}}{y_C}$$

Assuming 1ᵗ/m is the uniformly distributed dead load and 20ᵗ is the live load.

$$V_D \times 120 - 1\times 120\times 60 - 20\times 30 = 0$$

or $V_D = 65^t$

$V_F = 75^t$

Moment at $$E = 65\times 60 - \frac{1\times 60^2}{2}$$

$$= 2100.$$

$$H = \frac{2100}{15} = \mathbf{140^t.}$$

2. The maximum tension in the cable for this condition of loading shall occur at support.

Reaction at F due to live load

$$= \frac{20 \times 90}{120} = \mathbf{15t}$$

Reaction at $$D = \frac{20 \times 30}{120} = 5t$$

Horizontal thrust due to live load

$$H_l = \frac{5 \times 60}{15} = 20^t$$

w_e for live load

$$= \frac{8 \times H_l \times y_C}{L^2} = \frac{8 \times 20 \times 15}{120^2}$$

$$= \mathbf{0.1666t/m}$$

$$V = (w_d + w_e) \cdot \frac{L}{2}$$

$$V = (1 + 0.1666) \times 60$$

$$= 69.996$$

$$T_{max} = \sqrt{V^2 + H^2} = \sqrt{69{,}996^2 + 140^2}$$

$$= 156.52^t$$

3. Shear Force diagram

Shear force due to horizontal thrust H_l at any point $= H_l \, . \, \tan\theta$

$$\tan\theta = \frac{4y_C}{L^2}(L - 2x)$$

At $x = 0$ $$\tan\theta = \frac{4\,y_C}{L} = \frac{4 \times 15}{120} = 0.5$$

$$H_l \tan\theta = 20 \times 0.5 = 10^t(-ve)$$

at $x = 30$ $$\tan\theta = \frac{4 \times 15}{120^2} \times 60 = 0.25$$

$$H_l \, . \tan\theta = 20 \times 0.25 = 5.$$

at $x = 60$ $$\tan\theta = 0$$

at $x = 90$ $$\tan\theta = \frac{4 \times 15}{120^2} \cdot (120 - 180)$$

$$= -6.25$$

$$H_l \, . \tan\theta = 20 \times (-0.25) = -5^t$$

$x = 120$ $$\tan\theta = \frac{4 \times 15}{120^2} \cdot (120 - 240)$$

$$= -0.5^t \; (+ve)$$

$$H_l \, . \tan\theta = 10^t(+ve)$$

S.F. diagram is obtained by superposing S.F.D. of a simply supported beam on S.F.D. due to H_l.

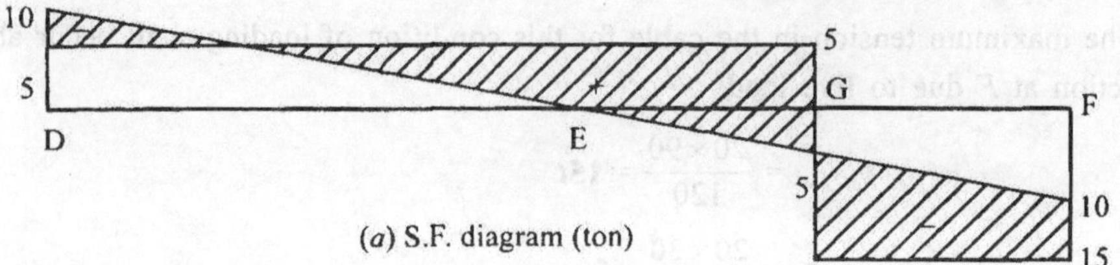

(a) S.F. diagram (ton)

B.M.D. is obtained by superposing B.M.D. due to H_l on simply supported B.M. diagram.

B.M. due to $H_l = H_l \cdot y$

$$= H_l \left[\frac{4 y_C}{L^2} (Lx - x^2) \right]$$

at $x = 0$ $\quad M_H = 0$

$x = 30$ $\quad M_H = 20 \left[\frac{4 \times 15}{120^2} (120 \times 30 - 30^2) \right]$

$= \mathbf{225\ tm.}$

$x = 60$ $\quad M_H = 20 \left[\frac{4 \times 15}{120^2} (120 \times 60 - 60^2) \right]$

$= 300$tm.

$x = 90$ $\quad M_H = 20 \left[\frac{4 \times 15}{120^2} (120 \times 90 - 90^2) \right]$

$= +225$tm.

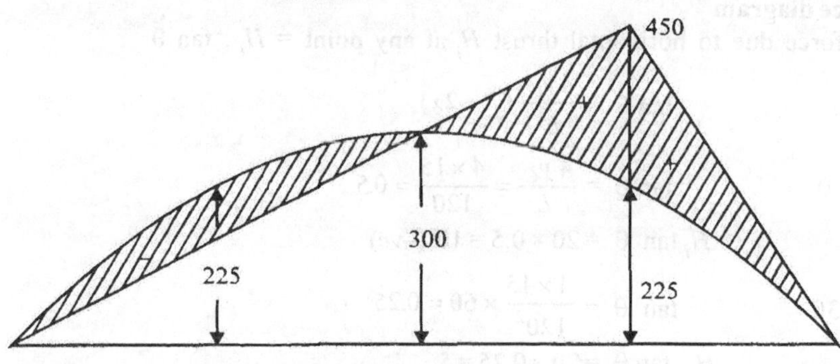

(b) B.M. diagram. in tm.

Fig. 6.30. S.F. and B.M. diagrams Example 6.9

Example 6.10. A suspension cable of 36 m span and 6 m maximum dip at centre carries a dead load of 1t/m through out and a uniformly distributed live load of 18 m length of an intensity of 1t/m which covers left hand half span. The cable is stiffened by a three hinged girder with two hinges at the abutments and the third at the centre.

(a) Draw the B.M.D. and S.F.D. for the girder giving values at salient points

(b) Calculate the maximum tension in the cable **[AMIE Summer 1988]**

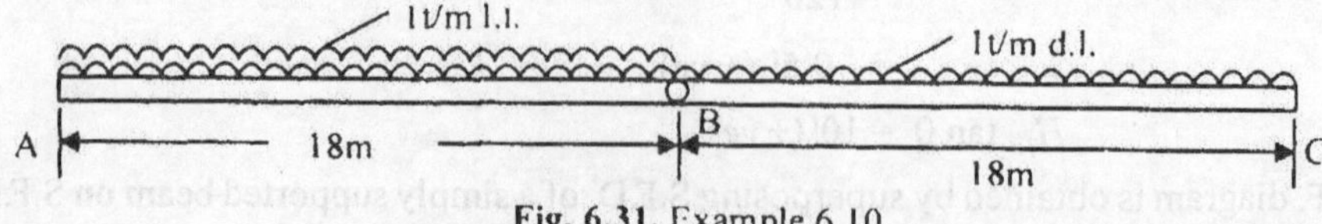

Fig. 6.31. Example 6.10.

1. Calculation of H

$$H = \frac{\text{B.M. at } B \text{ of the simply supported girder}}{y_C}$$

$$R_C \times 36 - 1 \times 36 \times 18 - 1 \times 18 \times 9 = 9$$

or $$R_C = \frac{1}{36}(18 \times 36 + 18 \times 9)$$

$$= 22.5^t$$

B.M. at $$B = 22.5 \times 18 - 1 \times 18 \times 9$$

$$= 243 \text{ tm}$$

$$H = \frac{243}{6} = 40.5^t$$

2. S.F. and B.M. diagrams.

Only live load produces Shear and Bending moment in the girder.

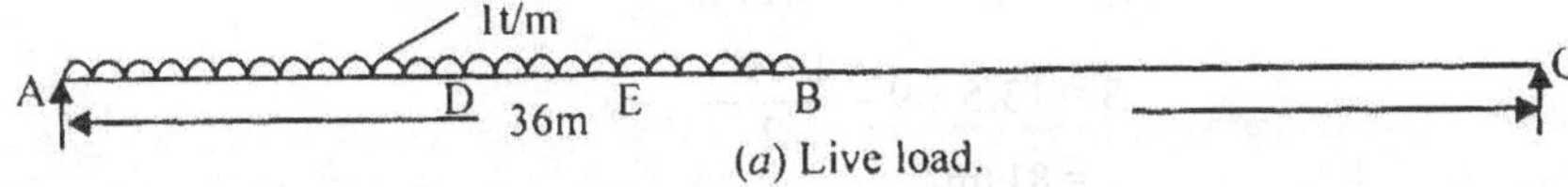

(a) Live load.

Fig. 6.32. Example 6.10.

$$R_C \times 36 - 1 \times 18 \times 9 = 0$$

or $$R_C = \frac{18 \times 9}{36} = 4.5^t$$

$$R_A = 13.5^t$$

Horizontal thrust due to liveload

$$H_l = \frac{W.x}{2h} = \frac{18 \times 9}{2 \times 6} = 13.5^t$$

$$w_e \text{ for live load} = \frac{8\, y_C . H_l}{L^2} = \frac{8 \times 6 \times 13.5}{36^2}$$

$$= 0.5 \ t/m$$

S.F. due to Horizontal thrust at any point

$$H_l \tan\theta = 13.5 \tan\theta$$

$$\tan\theta = \frac{4\, y_C}{L^2}(L - 2x)$$

at $x = 0$ $$\tan\theta = \frac{4\, y_C}{L} = \frac{4 \times 6}{36} = 0.667$$

$$H_l . \tan\theta = 13.5 \times 0.667$$

$$= 9^t (-ve)$$

at $x = 9$ $$\tan\theta = \frac{4 \times 6}{36 \times 36}(36 - 18)$$

$$= 0.333$$

$$H_l . \tan\theta = 13.5 \times 0.333 = 4.50$$

at $x = 18$ $$\tan\theta = 0 \quad H_l \tan\theta = 0$$

at $x = 36$ $\tan\theta = +\frac{4y_C}{L^2} - L = -0.667$

$H_l \tan\theta = 9t\ (+ve)$

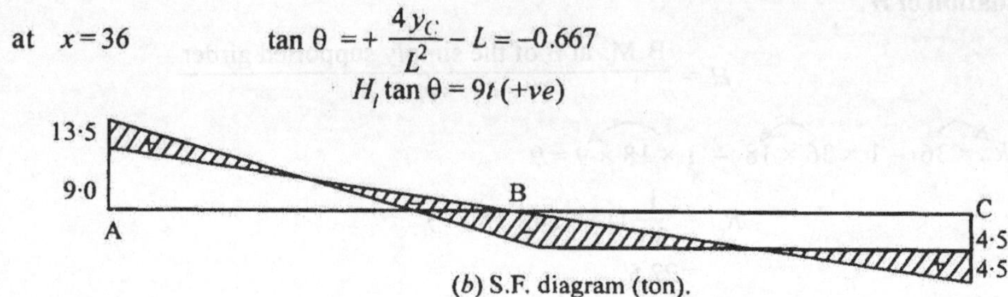

(b) S.F. diagram (ton).

B.M.D.

(i) Due to *u.d.l.l.*

$$\text{B.M.} = 13.5\,x - \frac{x^2}{2}$$

At $A = 0$

$C = 0$

$D = 4.5 \times 18 = +81$ tm

$$B = 13.5 \times 9 - \frac{1 \times 9^2}{2}$$

$= 81$ tm.

$$E = 13.5 \times 13.5 - \frac{1. \times 13.5^2}{2} = 91.125 \text{ tm}.$$

(ii) Due to H_l

$$\text{B.M.} = H_l \times \frac{4y_C}{L^2}(Lx - x^2)$$

at A $x = 0$ B.M. $= 0$

D $x = 9$ $\text{B.M.} = \frac{13.5 \times 4 \times 6}{36^2}(36 \times 9 - 9^2)$

$= 60.75$ tm

B $x = 18$ $\text{B.M.} = \frac{13.5 \times 4 \times 6}{36^2}(36 \times 18 - 18^2)$

$= 81$ tm

C $x = 36$ B.M. $= 0$

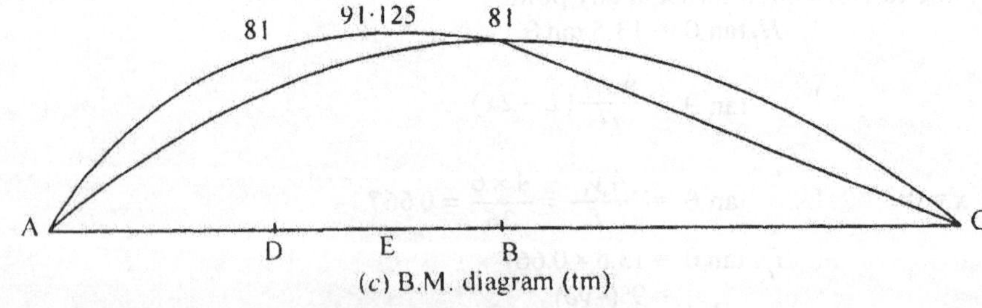

(c) B.M. diagram (tm)

Fig. 6.32. (a), (b) and (c) Example 6.10.

3. Maximum tension in cable

$$V = (w_d + w_l)\frac{L}{2} = (1 + 0.5)\,18 = 27'$$

$$T_{max} = \sqrt{27^2 + 40.5^2}$$

$= 48.67'$

Example 6.11. The cables of a suspension bridge have span of 60 m and central dip of 7.5 m. Each cable is stiffened by a girder hinged at the ends and at the middle to retain a parabolic shape. There is a uniformly distributed load of 1 t/m of the horizontal span over the whole girder and a live load of 2 t/m and 15 m long. Determine the maximum tension in the cable when the live load is placed on the left hand half of the sliffening girder in such a way that right hand end of the live load is just over the central hinge. **[AMIE Winter 1989]**

Solution:

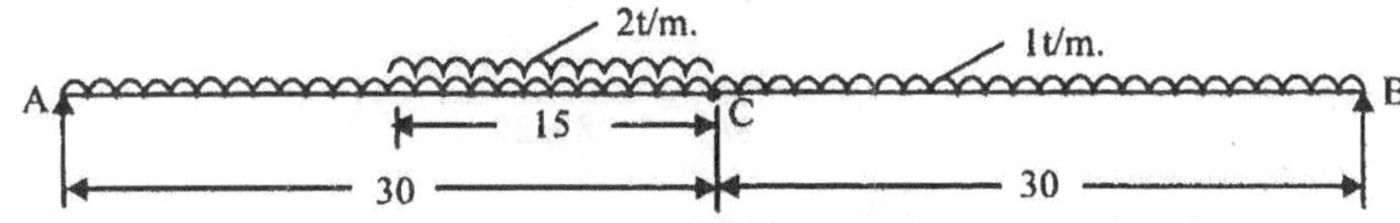

Fig. 6.33. Example 6.11.

$$V_B \times 60 - 1 \times 60 \times 30 - 2 \times 15 \times 22.5 = 0$$

or
$$R_B = \frac{1}{60}[60 \times 30 + 30 \times 22.5]$$
$$= 41.25^t$$

$$R_A = \frac{1}{60}[60 \times 30 + 30 \times 37.5]$$
$$= 48.75$$

$$H = \frac{M_C}{y_C} = \frac{41.25 \times 30 - \frac{1 \times 30^2}{2}}{7.5}$$
$$= \mathbf{105^t.}$$

H due to uniformly distributed dead

$$\text{load} = \frac{w_d . l^2}{8y_C} = \frac{1 \times 60^2}{8 \times 7.5} = 60^t$$

$$H_l \text{ due to live load} = 105 - 60 = 45^t$$

$$w_e \text{ for live load} = \frac{8\, y_C . H_l}{L^2} = \frac{8 \times 7.5 \times 45}{60^2}$$
$$= 0.75 \text{ t/m}$$

$$V_A = (w_l + w_d)\frac{l}{2} = (1 + 0.75) \times 30$$
$$= 52.5^t$$

Max^m tension in cable

$$= \sqrt{H^2 + V_A{}^2} = \sqrt{105^2 + 52.5^2}$$
$$= \mathbf{117.39^t.}$$

Example 6.12. A suspension bridge of 200 m span has a dip of 20 m. If carries a uniformly distributed load (spread horizontally) of 500 tonnes. If the chains are of mild steel the allowable stress of which is 1 ton per square cm, calculate the required sectional area of the chain. The anchor makes an angle of 45° with the piers. Calculate the over turning force on the pier.

(*i*) When the chains run over a pulley device

(*ii*) When the chains are attached to a saddle resting on rollers at the top of the pier. **[AMIE. Winter '97]**

Solution: Total load $= 500^t$

$$\text{load/cable} = \frac{500}{2} = 250^t$$

$$\text{load/horizontal m run of cable} = \frac{250}{200} = 1.25\,t/m$$

$$H = \frac{wL^2}{8h} = \frac{1.25 \times 200^2}{8 \times 20} = 312.5^t$$

$$V_A = V_B = \frac{250}{2} = 125.0\,t$$

$$T_{max} = \sqrt{312.5^2 + 125.0^2} = 336.572^t$$

$$\text{Area of cable} = \frac{336.572}{1} = 336{\cdot}57\ \text{cm}^2$$

Over turning force

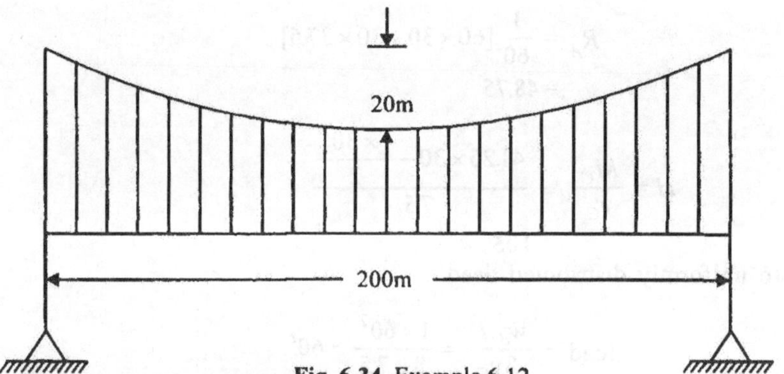

Fig. 6.34. Example 6.12.

Equation of the cable with A as origin (Fig. 6.34)

$$y = \frac{4y_C}{L^2}(Lx - x^2)$$

$$\frac{dy}{dx} = \tan\theta = \frac{4y_C}{L^2}(L - 2x)$$

At $x = 0$ $\tan\theta = \frac{4y_c}{L} = \frac{4 \times 20}{200} = 0{\cdot}4$ $\theta = 21.8°$

$$\sin\theta = 0.37139 \quad \cos\theta = 0.92847$$

Over turning force

(*i*) Pulley Support

Over turning force

$$= T\cos 21.80° - T\cos 45°$$
$$= T(0.92847 - 0.70710)$$
$$= 336.572 \times 0.2213$$
$$= 74.50^t$$

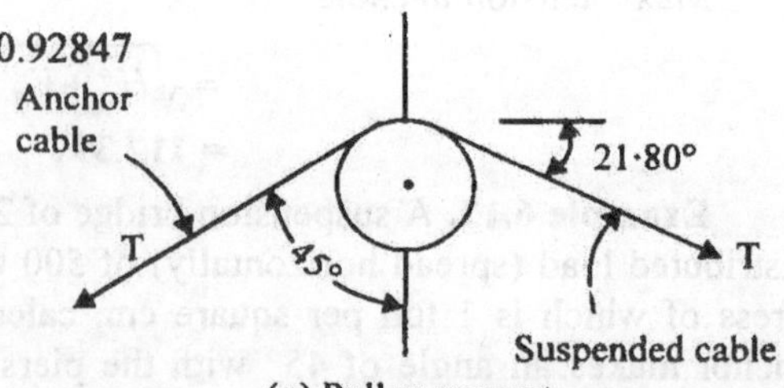

(*a*) Pulley support

***(ii)* Saddle Support**

Overturning force

$$= T\cos 21.8° - T_1 \cos 45° = 0$$

or

$$T_1 = \frac{T\cos 21.8}{\cos 45°}$$

$$= \frac{336.572 \times .92847}{0.70710}$$

$$= 441.9417$$

Overturning force is zero.

(*b*) Saddle Support.

Fig. 6.35. Over turning force Example 6.12.

6.8. THREE HINGED ARCH

While cable is a tension member, arch behaves in a reverse fashion and it is basically a compression member. There is, however, some difference. *A* cable is flexible and it changes its shape with different types and positions of loads. If cannot resist any bending moment hence moment everywhere is zero. On the other hand on arch is a rigid structure. It cannot change its shape corresponding to different types of loading. Hence through primarily it is a compression member it is often subjected to B.M. and shear though of small magnitude An arch is of three types as per structure configuration.

1. Three hinged arch
2. Two hinged arch
3. Hingeless or fixed arch.

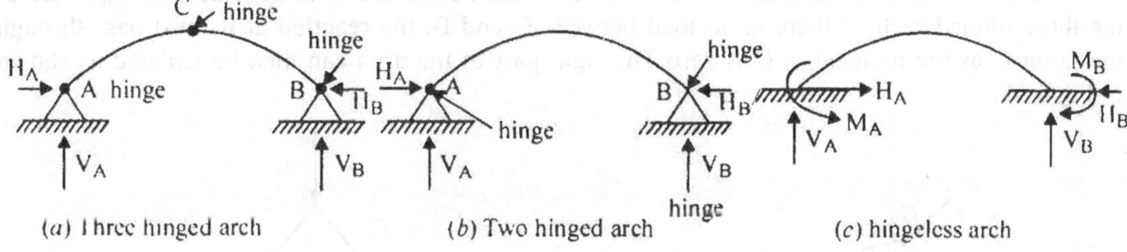

Fig. 6.36. Different types of arches

Structurally

(*a*) a three hinged arch is statically determinate. There are 4 unknowns and there are three equations of equilibrium and one condition equation.
(Moment at the third hinge is Zero)

(*b*) The two hinged arch is statically indeterminate to the first degree : there are four unnowns and three equations of static equilibrium. Two hinged arch is dealt separately in chapter 12.

(*c*) The hingeless arch is statically indeterminate to the third degree as there are six unknows (Three at each support) and only three equations of equilibrium.

6.8.1. Comparison of the Behaviour of an Arch and beam.

Fig 6.3 shows the difference in analysis of a beam and a three hinged arch.

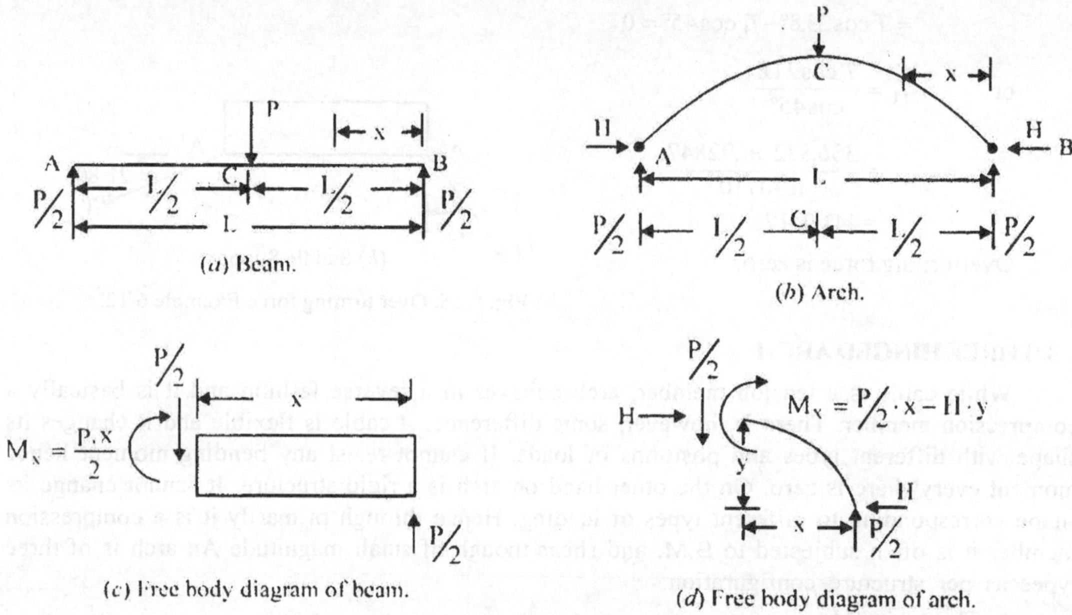

Fig. 6.37. Comparison between an arch and a beam.

From the study of the figures above it is clear that the moment at the arch section gets reduced by the factor H.y.

Another way to understand the behaviour of arch structure is to consider the rightside of the three pinned arch. If there is no load between C and B, the reaction at B must pass through the point C as the moment at C is zero.The right part of the arch can then be isolated as shown in Fig 6.38.

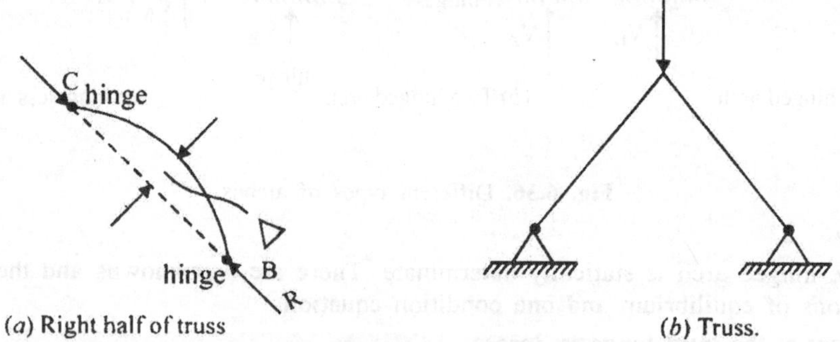

Fig. 6.38. Behaviour of an arch and a truss.

The moment at any section is then R.Δ. If Δ is made zero, the moment is then zero and we have a truss type structure. The arch, therefore, lies between a moment resistant strcture and a truss type structure. It is, however, possible , to design the shape of an arch in such a way that it is subjected to compression only. Such a shape of the arch is called the "funicular shape". For a given span , location of hinges and loading there is only one funicular shape of the arch. When loading changes, the shape is no longer funicular and there will be bending moments at different sections of the arch rib.

6..9. ANALYSIS OF THREE HINGED ARCH

A three hinged arch is statically determinate as the four unknowns can be solved by three

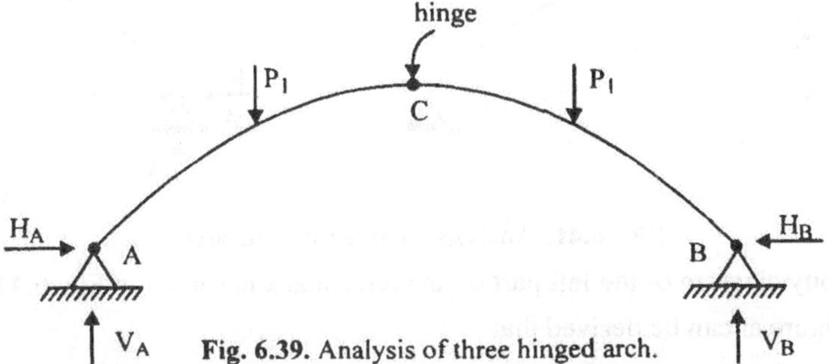

Fig. 6.39. Analysis of three hinged arch.

equations of equilibrium and the fourth equation of condition, namely moment at C is zero. For a symmetrical parabolic arch, the equation can be written as

$$y = \frac{4h}{L^2} \cdot x(L-x)$$

where h is the rise and origin is at one end.

$$\frac{dy}{dx} = \frac{4h}{L^2}(L-2x) = \tan\theta_x .$$

From the above two equations it is possible to calculate y when x is known or to calculate the slope θ_x at any point x. (Fig. 6.40 *a*). If the arch is semi circular or part of a circle its radius R can be obtained from the relation. (Fig. 6.40*b*)

$$\frac{L}{2} \times \frac{L}{2} = h(2R-h)$$

or

$$R = \frac{1}{2h}\left(\frac{L^2}{4} + h^2\right).$$

(*a*) Parabolic. (*b*) Circular

Fig. 6.40. Different types of arches.

With D as origin and x and y coordinates as shown the equation of the circular arch can be derived as

$$R^2 = x^2 + [R-(h-y)]^2$$

where $x = R\sin\alpha$ and $y = R\cos\alpha - OD = R\cos\alpha - R\cos\beta$.
and α is the angle of inclination of radius OQ with vertical (Fig. 6.40 *b*).

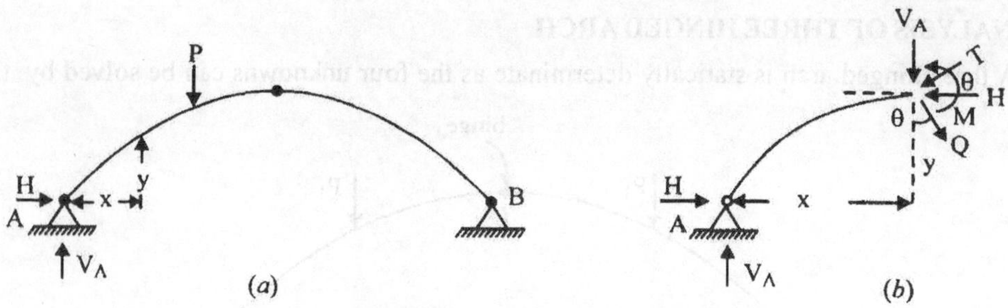

Fig. 6.41. Analysis of three hinged arch.

The free body diagram of the left part of an arch cut at x is shown in Fig. 6.41 *b*.

From the figure at can be derived that

$$T = H\cos\theta + V_A \sin\theta \qquad \text{...(6.15)}$$

$$Q = V_A \cos\theta - H\sin\theta \qquad \text{...(6.16)}$$

$$M = V_A . x - H . y \qquad \text{...(6.17)}$$

Shear force is taken as positive if it is acting upward on the left. A study of the above equations reveal that compared to a beam, in an arch

(*a*) Axial force increases

(*b*) Shear force decreases

(*c*) B.M. decreases

Example 6.13. Obtain the equation of the symmetrical three hinged arch using the coordinate system illustrated if each structure is to perform as a compressive arch and support the loadings shown.

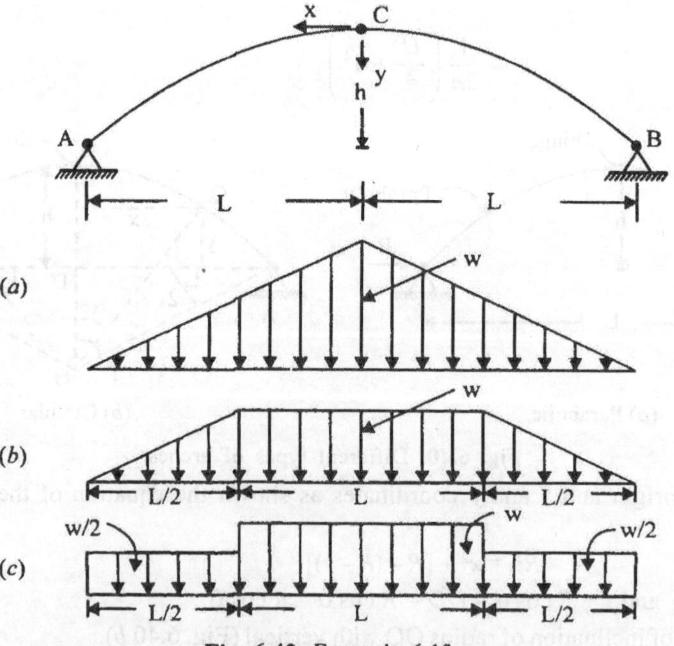

Fig. 6.42. Example 6.13.

Solution:

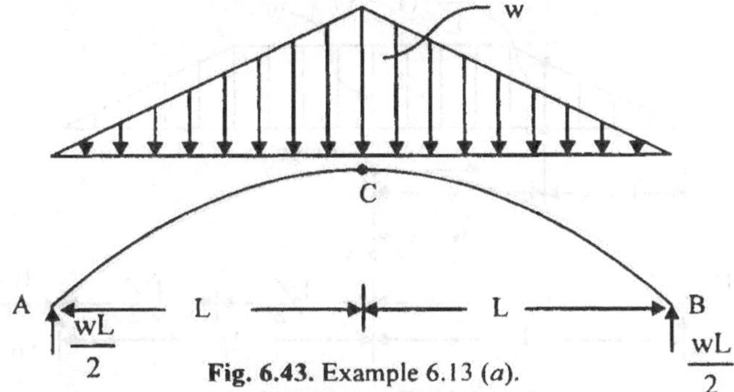

Fig. 6.43. Example 6.13 (*a*).

Reactions $V_A = V_B = \dfrac{wL}{2}$

Taking moments about C

$$\frac{wL}{2} \times L - \frac{1}{2} wL \times \frac{L}{3} - H.h = 0$$

or $$H = \frac{1}{h}\left[\frac{wL^2}{2} - \frac{wL^2}{6}\right]$$

$$= \frac{1}{h} \cdot \frac{wL^2}{3}.$$

At any distance x moment is zero for a compressive arch.

Hence $$\frac{wL^2}{3h}(h-y) - \frac{wL}{2}(L-x) + \frac{1}{2}\frac{w}{L} \cdot (L-x)^2 \times \frac{1}{3}(L-x) = 0$$

or $$y = \frac{3h}{wL^2}\left[\frac{w(L-x)^3}{6L} - \frac{wL^2}{6} + \frac{wL.x}{2}\right]$$

$$= \frac{3h}{wL^2}\left[\frac{wx^2}{2} - \frac{wx^3}{6L}\right]$$

$$= \frac{hx^2}{2L^2}\left[3 - \frac{x}{L}\right] \text{ when } x > 0$$

$$y = \frac{hx^2}{2L^2}\left[3 + \frac{x}{L}\right] \text{ where } x < 0.$$

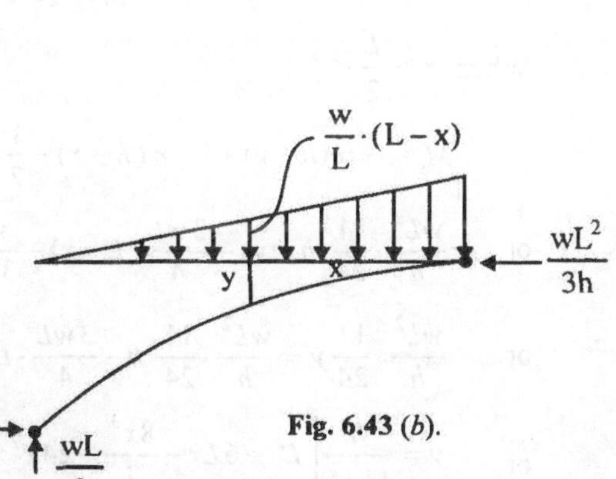

Fig. 6.43 (*b*).

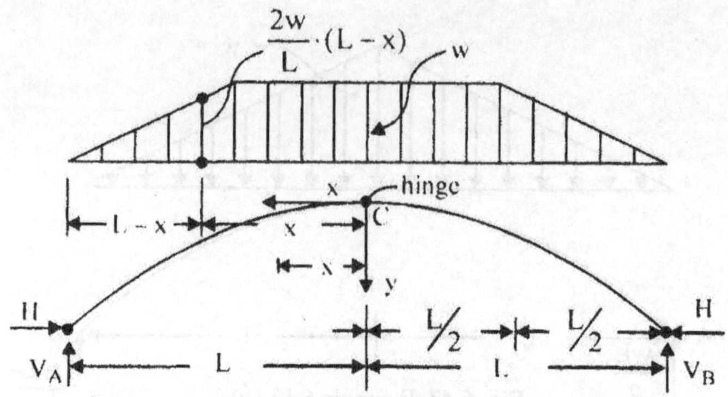

Fig. 6.44. Example 6.13 (*b*).

(*b*) Reactions $V_A = V_B = \frac{1}{2} w \cdot \frac{L}{2} + \frac{wL}{2}$

$$= wL \cdot \frac{3}{4}$$

Taking moments about C

$$V_A \times L - \frac{1}{2} w \cdot \frac{L}{2} \times \left(\frac{L}{2} + \frac{1}{3} \cdot \frac{L}{2}\right)$$

$$- w \cdot \frac{L}{2} \cdot \frac{L}{4} - H.h = 0$$

or
$$H = \frac{1}{h}\left[-\frac{wL}{4} \cdot \frac{2}{3} \cdot L - \frac{wL^2}{8} + \frac{3}{4} wL^2\right]$$

$$= \frac{wL^2}{h} \cdot \frac{11}{24}$$

When $x > \frac{L}{2} < L$

$$M_x = -H(h-y) + V_A \times (L-x) - \frac{1}{2} \cdot \frac{2w}{L} \cdot \frac{(L-x)(L-x)(L-x)}{3} = 0$$

or
$$\frac{wL^2}{h} \cdot \frac{11}{24}(h-y) - \frac{3wL}{4}(L-x) + \frac{w}{3L}(L-x)^3 = 0$$

or
$$\frac{wL^2}{h} \cdot \frac{11}{24} y = \frac{wL^2}{h} \cdot \frac{11}{24} \cdot h - \frac{3wL}{4} \cdot L + \frac{3wL}{4} \cdot x + \frac{w}{3L}(L^3 - 3L^2 x + 3x^2 L - x^3)$$

or
$$y = \frac{h}{11L^2}\left[L^2 - 6Lx - \frac{8x^3}{L} + 24x^2\right]$$

when $x > 0 < \frac{L}{2}$

$$M_x = -H(h-y) + V_A \times (L-x) - \frac{L}{2} w \cdot \frac{1}{2}\left(\frac{L}{2} + \frac{L}{6} - x\right) - w.\left(\frac{L}{2} - x\right)\left(\frac{L}{2} - x\right) \cdot \frac{1}{2} = 0$$

or $$y = \frac{24h}{11wL^2}\left[\frac{wx^2}{2}\right]$$

$$= \frac{h}{11L^2}\cdot[12x^2]$$

(c)

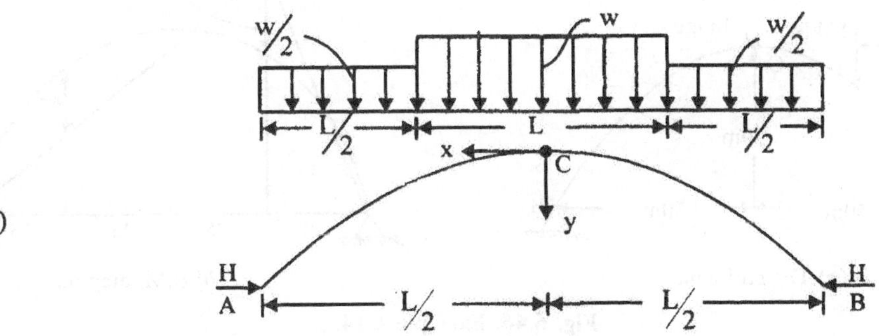

Fig. 6.45. Example 6.13 (c).

$$V_A = V_B = \frac{w}{2}\cdot\frac{L}{2}+\frac{wL}{2} = \frac{3}{4}wL$$

$$-H\times h + V_A \times L - \frac{w}{2}\cdot\frac{L}{2}\left(\frac{L}{2}+\frac{L}{4}\right) - \frac{wL}{2}\times\frac{L}{4} = 0$$

or $$H = \frac{1}{h}\left[\frac{3}{4}wL^2 - \frac{wL^2}{4}\cdot\frac{3}{4} - \frac{wL^2}{8}\right]$$

$$= \frac{7wL^2}{16h}$$

when $x > \frac{L}{2} < L$

$$M_x = \frac{3}{4}wL(L-x) - \frac{7wL^2}{16h}(h-y) - \frac{w}{2}(L-x)^2\cdot\frac{1}{2} = 0$$

or $$y = \frac{16h}{7wL^2}\left[\frac{7wL^2}{16} + \frac{wL^2}{4} - \frac{3}{4}wL^2 + \frac{3}{4}wLx - \frac{wLx}{2} + \frac{wx^2}{4}\right]$$

$$= \frac{h}{7L^2}[4x^2 + 4Lx - L^2]$$

when $x > 0 < \frac{L}{2}$

$$M_x = \frac{3}{4}wL(L-x) - \frac{7wL^2}{16h}(h-y) - \frac{w}{2}\cdot\frac{L}{2}\cdot\left(\frac{L}{4}+\frac{L}{2}-x\right) - w\left(\frac{L}{2}-x\right)^2\cdot\frac{1}{2} = 0$$

or $$y = \frac{16h}{7wL^2}\left[\frac{7wL^2}{16} - \frac{3}{4}wL^2 + \frac{wL^2}{8} + \frac{3}{4}wLx + \frac{3wL^2}{16} - \frac{1}{4}wLx - \frac{wLx}{2} + \frac{wx^2}{2}\right]$$

or $$y = \frac{8h}{7} \cdot \frac{x^2}{L^2}.$$

Example 6.14. A three hinged semicircular arch of 100 m span has a rise of 50 m and is loaded as shown in Fig. 6.46. Determine the reactions at the support and draw the bending moment diagram.

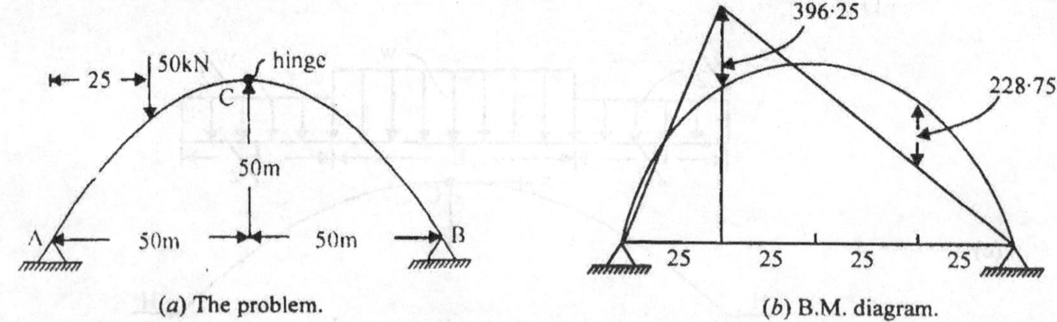

(*a*) The problem. (*b*) B.M. diagram.

Fig. 6.46. Example 6.14.

Solution: $$V_A \times 100 - 50 \times 75 = 0$$

or $$V_A = 37.5 \uparrow$$

$$V_B \times 100 - 50 \times 25 = 0$$

$$V_B = \frac{50 \times 25}{100} = 12.5 \uparrow$$

$$V_A + V_B = 50 \text{ (check)}$$

Taking moments about C

$$V_A \times 50 - 50 \times 25 - H \times 50 = 0$$

or $$H = 12.5 \rightarrow$$

For a circle $x^2 + y^2 = R^2$

At 5 m from A $x = 45$, $y = \sqrt{50^2 - 45^2} = 21.79$ m

At 10 m from A $y = \sqrt{50^2 - 40^2} = 30.00$ m

15 m from A $y = \sqrt{50^2 - 35^2} = 35.70$ m

20 m from A $y = \sqrt{50^2 - 30^2} = 40.00$ m

25 m from A $y = \sqrt{50^2 - 25^2} = 43.30$ m

At 5 m from B $y = 21.79$

10 m from B $y = 30.00$

15 m from B $y = 35.70$

20 m from B $y = 40.00$

25 m from B $y = 43.30$

30 m from A or B $y = \sqrt{50^2 - 20^2} = 45.82$ m

35 m from A or B $y = \sqrt{50^2 - 15^2} = 47.69$ m

40 m from A or B $y = \sqrt{50^2 - 10^2} = 48.98$ m

45 m from A or B $y = \sqrt{50^2 - 5^2} = 49.75$ m

B.M. under the load $= V_A \times 25 - H \times y$

$= 37.5 \times 25 - 12.5 \times 43.30 = 396.25$ kNm.

B.M. diagram

Distance from A	Distance from B	x'	y'	$V_A x' - Hy'$
0	0	0	0	0
5		5	21.79	−84.875
10		10	30.00	−75.00
15		15	35.70	116.25
20		20	40.00	250.00
25		25	43.30	396.25
				$V_B x' - Hy'$
30	70	30	45.82	302.250
35	65	35	47.69	216.375
40	60	40	48.98	137.750
45	55	45	49.75	65.625
50	50	50	50.00	0.000
	0	0	0	0
	5	5	21.79	−209.875
	10	10	30.00	−250.000
	15	15	35.70	−258.750
	20	20	40.00	−250.000
	25	25	43.30	−228.75
	30	30	45.82	−197.75
	35	35	47.69	−158.625
	40	40	48.98	−112.250
	45	45	49.75	−59.375
	50	50	50.00	0.00

Example 6.15. For the three hinged arch shown in Fig. 6.47 determine (*a*) the support reactions (*b*) axial force, shear force and bending moment at *D*.

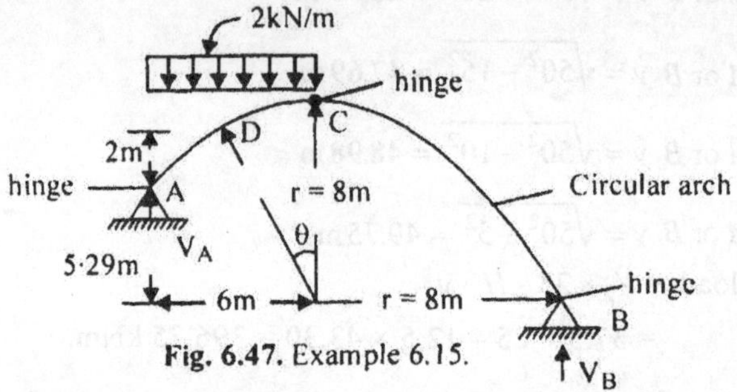

Fig. 6.47. Example 6.15.

Solution: From the geometry of a circle

$$8^2 = y^2 + 6^2$$

or $$y^2 = 8^2 - 6^2 = 28$$

or $$y = 5.29.$$

Taking moments about C

$$V_A \times 6 - H \times 2.71 - 2 \times 6 \times 3 = 0$$

or $$6V_A - 2.71H = 36.$$

Taking moments about B

$$V_A \times 14 + H \times 5.29 - 2 \times 6 \times 11 = 0$$

or $$14V_A + 5.29H = 132$$

Solving the two equations

$$V_A - 0.451H = 6$$

$$V_A + 0.377H = 9.43$$

$$H = 4.14 \text{ kN}$$

$$V_A = 7.867 \text{ kN}$$

$$V_B = 4.133 \text{ kN}.$$

Taking moments about C

$$V_B \times 8 - H \times 8 = 0$$

or $$H = V_B = 4.133 \text{ kN (check)}$$

At D $$x = \sqrt{8^2 - 7.29^2} = 3.29$$

$$\sin\theta = \frac{3.29}{8} = 0.412$$

$$\cos\theta = \frac{7.29}{8} = 0.911$$

$$\text{Axial force} = V_A \sin\theta - 2 \times (6 - 3.29)\sin\theta + 4.133\cos\theta$$

$$= (7.867)(0.412) - 2(2.71)(0.412) + 4.133(0.911)$$

$$= \mathbf{4.773\ kN.}$$

$$\begin{aligned} \text{S.F.} &= V_A \cos\theta - 2\times 2.71 \cos\theta - 4.133 \sin\theta \\ &= 7.867\,(0.911) - 2\,(2.71)(0.911) - 4.133\,(0.412) \\ &= \mathbf{0.526\ kN.} \end{aligned}$$

B.M at D

$$V_A \times 2.71 - H\times 2 - \frac{2\times 2.71^2}{2}$$

$$= 7.867\,(2.71) - 4.133\times 2 - 2{\cdot}71^2$$

$$= \mathbf{5.71\ kNm.}$$

Example 6.16. A three hinged parabolic arch of span 100 m and rise 20 m carries a uniformly distributed load of 2 kN/m length on the right half as shown in the figure 6.48. Determine the maximum bending moment in the arch.

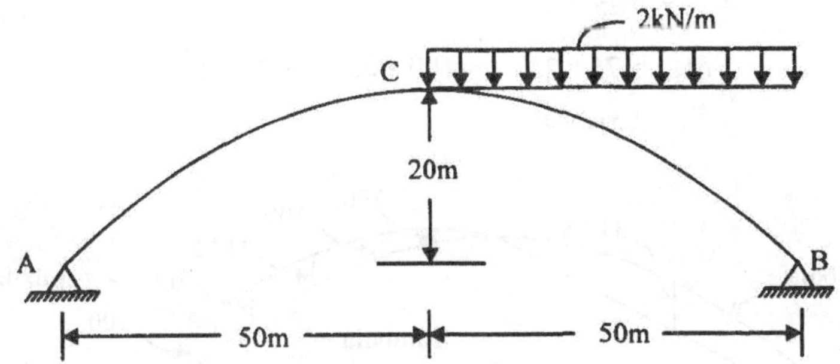

Fig. 6.48. Fxample 6.16.

Solution:

$$V_A \times 100 - 2\times 50\times 25 = 0$$

or $$V_A = \frac{2\times 50\times 25}{100} = 25\,\text{kN.}$$

$$V_B = 75\,\text{kN.}$$

Taking moments about C

$$V_A \times 50 - H\times 20 = 0$$

or $$H = \frac{25\times 50}{20} = 62.5\ \text{kN} \rightarrow$$

Equation of a parabola with origin at A

$$y = \frac{4h}{L^2}(L-x)(x)$$

At any distance x from A

$$M_x = V_A \times x - H.y$$

$$= 25x + 62.5\times \frac{4\times 20}{100^2}(x^2 - 100.x)$$

$$\frac{d\,M_x}{dx} = 25 + \frac{1}{2}\times 2x - 50 = 0$$

or $$x = 25$$

$$M_{25} = 25 \times 25 - 25 \times 75$$
$$= 625 - 937.5 = -312.5.$$
$$M_{50} = 25 \times 50 - 62.5 \times 20 = 0$$

At any distance x from B,

$$M_x = V_B \times x - H.y - \frac{2.x^2}{2}$$

$$= 75x - \frac{1}{2}(100x - x^2) - x^2$$

$$= 25x - \frac{x^2}{2}$$

$$\frac{dM_x}{dx} = 25 - x = 0$$

or $x = 25$

$$M_{25} = 75 \times 25 - \frac{1}{2}(100 \times 25 - 25^2) - 25^2$$
$$= 312.50$$

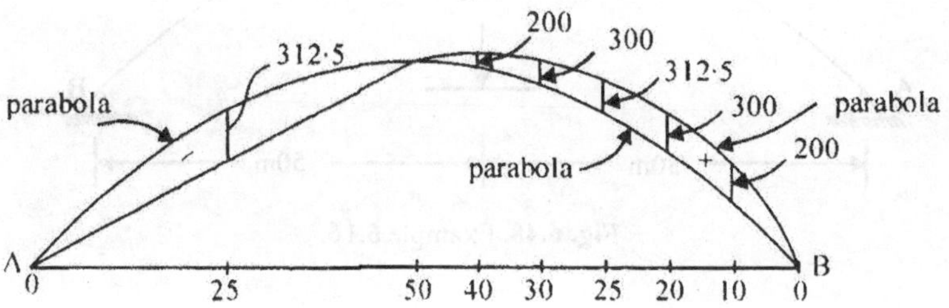

Fig. 6.49. B.M. diagram (kNm).

Example 6.17. Compute the vertical shear forces and bending moments for the three hinged parabolic arch shown in Fig. 6.50 and indicate their values at locations D and E.

[AMIE Summer '94]

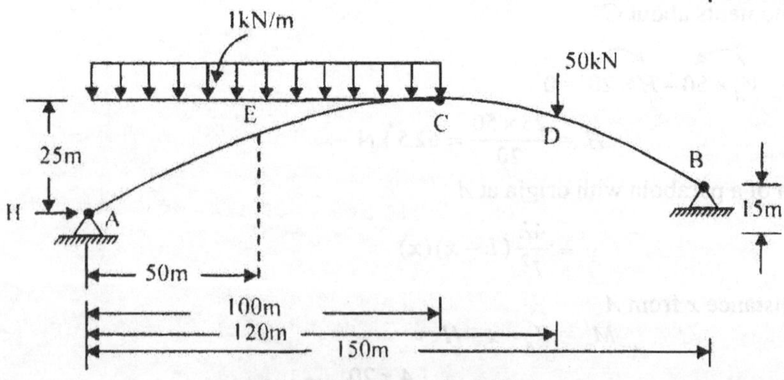

Fig. 6.50. Example 6.17.

General Equation of Parabola.

$$y = Ax^2 + Bx + C$$

at $x = 0$ $y = 0$ $\therefore C = 0$

at $x = 100$ $y = 25$

$$\therefore 25 = A \cdot 100^2 + B \cdot 100$$

at $x = 150$ $y = 15$

or $15 = A \cdot 150^2 + B \cdot 150$

Solving $A = -.003$

$B = 0.55$

∴ Equation of the parabola

$$y = -.003\,x^2 + 0.55\,x.$$

Taking moments of all forces to the left of C about C

$$V_A \times 100 - H \times 25 - 1 \times 100 \times 50 = 0 \quad ...(1)$$

Taking moments of all forces to the right of C about C

$V_B \times 50 - 50 \times 20 - H \times 10 = 0$

or $H = 5\,V_B - 100$ ...(2)

From $\Sigma V = 0$ $V_A - 1 \times 100 - 50 + V_B = 0$

or $V_A + V_B = 150$

or $V_A = 150 - V_B$ ...(3)

Substituting (2) and (3) in (1)

$$(150 - V_B) \times 100 - (5\,V_B - 100) \times 25 - 5000 = 0$$

Solving $V_B = 55.55$

$H = 5V_B - 100 = 177.77$

$$V_A = \frac{1}{100}[25H + 50 \times 100]$$

$$= 94.44 \text{ kN}$$

$$V_A + V_B = 55.55 + 94.44 = 149.99 = 150 \text{ (check)}$$

S.F. and B.M. at E

$$y = -\cdot003\,x^2 + 0.55\,x$$

at $x = 50$ $y = -\cdot003 \times 50^2 + 0.55 \times 50$

$= 20$

Vertical shear force at $E = V_A - 1 \times 50 = 44.44$

B.M. at $E = V_A \times 50 - H \times 20 - 1 \times 50 \times 25$

$= 94.44 \times 50 - 177.77 \times 20 - 50 \times 25$

$= -83.4$ kNm

S.F. and B.M. at D

$$y = -.003x^2 + 0.55\,x$$

at $x = 120$ $y = -.003\,(120)^2 + 0.55 \times 120$

$= 22.8$ m

Hence vertical distance from $B = 22.8 - 15 = 7.8$ m

Vertical S.F. at $D = V_B = 55.55$

B.M. at $D = V_B \times 30 - H \times 7.8$

$= 55.55 \times 30 - 177.77 \times 7.8$

$$= 279.894 \text{ kNm}.$$

Example 6.18. (*a*) **In a three hinged symmetrical parabolic arch loaded with uniformly distributed load calculate the bending moment and radial shear at any section.**

(*b*) An arch in the form of a parabola with axis vertical has hinges at the abutments and at the vertex. The abutments are at different levels, the horizontal span being l and the heights of the vertex above abutments being h_1 and h_2. Show that the horizontal thrust due to load w per unit length distributed across the span is

$$\frac{wl^2}{2\left(\sqrt{h_1}+\sqrt{h_2}\right)^2}$$

[AMIE Winter '97]

Solution:

(*a*)

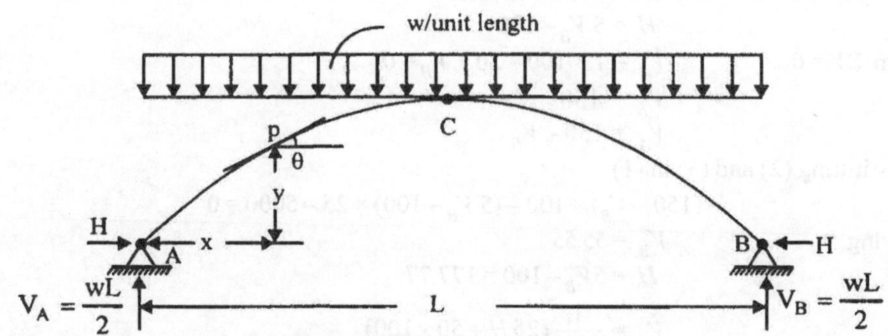

Fig. 6.51. Example 6.18 (*a*).

Taking moments about C

$$\frac{wL}{2} \times \frac{L}{2} - H \times y_C - \frac{w \times L}{2} \times \frac{L}{4} = 0$$

or $$H = \frac{wL^2}{8y_C}$$

At any point $P\ (x, y)$

$$y = \frac{4y_C}{L^2} x(L-x)$$

$$M_x = \frac{wL}{2} \cdot x - H \cdot \frac{4y_C}{L^2} \cdot x(L-x) - \frac{wx^2}{2}$$

$$= \frac{wLx}{2} - \frac{wL^2}{8y_C} \times \frac{4y_C}{L^2} \cdot x(L-x) - \frac{wx^2}{2}$$

$$= \frac{wLx}{2} - \frac{wLx}{2} + \frac{wx^2}{2} - \frac{wx^2}{2} = 0$$

Shear force at any point p

$$\text{S.F.} = H\sin\theta - V_A\cos\theta + wx\cos\theta$$

$$y = \frac{4y_C}{L^2}(Lx - x^2)$$

$$\frac{dy}{dx} = \frac{4y_C}{L^2}(L-2x) = \tan\theta$$

$$\therefore \quad \cos\theta = \frac{L^2}{[L^4 + 16y_C^{\,2}(L-2x)^2]^{1/2}}$$

$$\sin\theta = \frac{4y_C(L-2x)}{[L^4 + 16y_C^{\,2}(L-2x)^2]^{1/2}}$$

$$\text{S.F.} = \frac{wL^2}{8y_C} \cdot \frac{4y_C(L-2x)}{[L^4 + 16y_C^{\,2}(L-2x)^2]^{1/2}}$$

$$-\frac{wL}{2} \cdot \frac{L^2}{[L^4 + 16y_C^{\,2}(L-2x)^2]^{1/2}}$$

$$+\, w.x \frac{L^2}{[L^4 + 16y_C^{\,2}(L-2x)^2]^{1/2}}$$

$$= 0$$

Hence radial shear is zero at any point

(b)

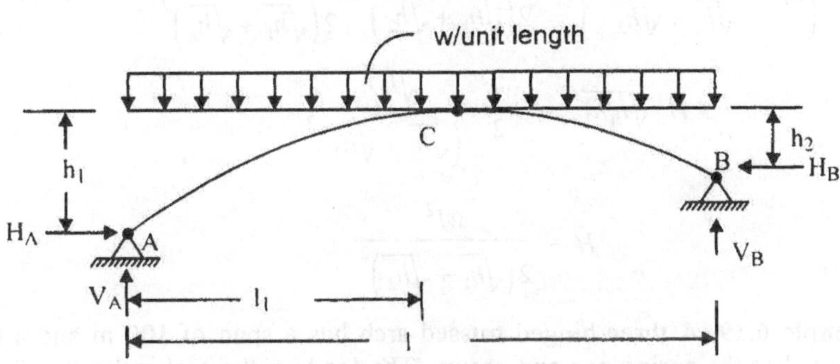

Fig. 6.52. Example 6.18 (*b*).

With C as origin, the equation of the parabola $y = Kx^2$. Substituting for the portion AC.

$$h_1 = Kl_1^{\,2} \quad \text{or} \quad K = \frac{h_1}{l_1^{\,2}}$$

For the portion CB

$$h_2 = K(l-l_1)^2 \quad \text{or} \quad K = \frac{h_2}{(l-l_1)^2}$$

Equating the value of K

$$\frac{h_1}{l_1^{\,2}} = \frac{h_2}{(l-l_1)^2}$$

or $$l_1^{\,2}h_2 = (l-l_1)^2 \,.\, h_1$$

or
$$l_1\sqrt{h_2} = (l-l_1)\cdot\sqrt{h_2}$$

or
$$l_1 = \frac{l\sqrt{h_1}}{\sqrt{h_1}+\sqrt{h_2}}$$

Takinig moments of all forces to the left of C about C. $[H_A = H_B = H]$

$$H\times h_1 + \frac{wl_1^2}{2} - V_A\times l_1 = 0 \qquad ...(1)$$

Similarly moments of all forces to the right of C about C

$$H\times(h_1-h_2)+\frac{wl^2}{2}-V_A.l = 0$$

or
$$V_A = \frac{H\times(h_1-h_2)}{l}+\frac{wL}{2} \qquad ...(2)$$

Substituting the value of V_A from Eqn. (2) in Eqn. (1)

$$Hh_1 = \left[\frac{H(h_1-h_2)}{l}+\frac{wl}{2}\right]l_1 - \frac{wl_1^2}{2}$$

$$= \left[\frac{H(h_1-h_2)}{l}+\frac{wl}{2}\right]\frac{l\sqrt{h_1}}{\sqrt{h_1}+\sqrt{h_2}} - \frac{w}{2}\frac{h_1l^2}{\left(\sqrt{h_1}+\sqrt{h_2}\right)^2}$$

or
$$H\left[h_1 - \frac{(h_1-h_2)\sqrt{h_1}}{\sqrt{h_1}+\sqrt{h_2}}\right] = \frac{wl^2\sqrt{h_1}}{2\left(\sqrt{h_1}+\sqrt{h_2}\right)} - \frac{wh_1l^2}{2\left(\sqrt{h_1}+\sqrt{h_2}\right)^2}$$

or
$$H\cdot\sqrt{h_1h_2} = \frac{wl^2}{2}\frac{\sqrt{h_1h_2}}{\left(\sqrt{h_1}+\sqrt{h_2}\right)^2}$$

or
$$H = \frac{wl^2}{2\left(\sqrt{h_1}+\sqrt{h_2}\right)^2}$$

Example 6.19. A three hinged trussed arch has a span of 100 m and a central rise of 20 m. It is hinged at the springings and crown. The load on the arch consists of a dead load of 20 kN/m. Determine the forces in members 1, 2, and 3.

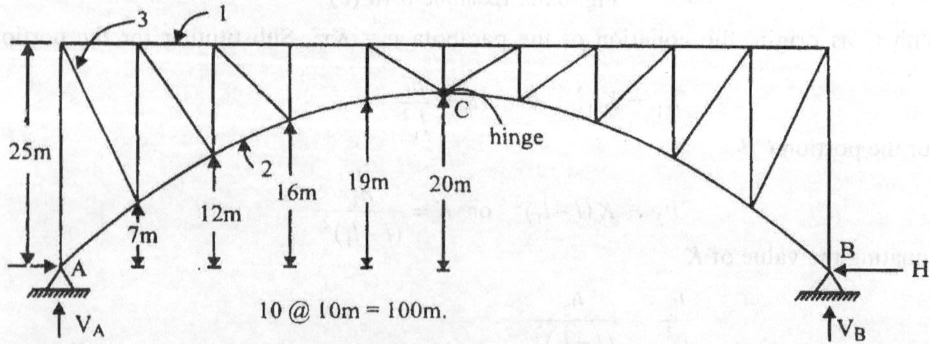

Fig. 6.53. Example 6.19.

Solution. $V_A = V_B = \dfrac{20 \times 100}{2} = 1000\,\text{kN}$

Taking moments about C

$$V_A \times 50 - 20 \times 50 \times 25 - H_A \times 20 = 0$$

or $$H_A = \frac{1}{20}[1000 \times 50 - 20 \times 50 \times 25]$$

$$= \mathbf{1250\ kN.}$$

Passing a section through member 1 and taking moments of all forces at the opposite joint D (Fig. 6.54)

$$1000 \times 20 - 1250 \times 12 - 100 \times 20 - 200 \times 10 + F_1 \times 13 = 0$$

or $$F_1 \times 13 = -1000$$

$$F_1 = -76.92\ \text{kN (comp)}$$

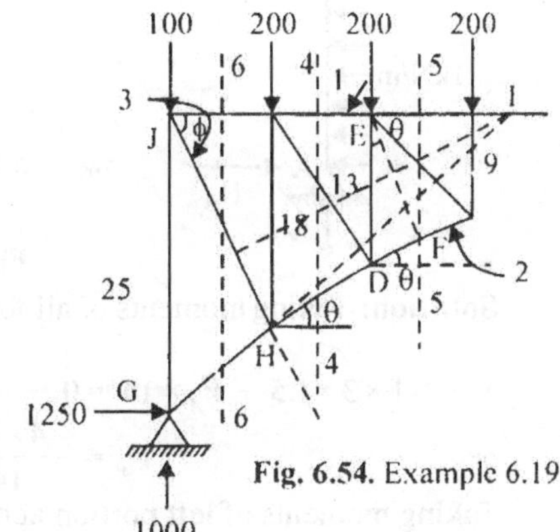

Fig. 6.54. Example 6.19.

Force in member 2

$$\tan\theta = \frac{4}{10} = \frac{2}{5}$$

$$\theta = \tan^{-1}\frac{2}{5} = 21.8°$$

$$\cos\theta = 0.9284$$

Perpendicular distance $EF = 13\cos\theta$

$$= 13 \times 0.9284$$

$$= \mathbf{12.07\ m.}$$

Taking mpments of all forces on the left of the section 5 – 5 about E.

$$1000 \times 20 - 1250 \times 25 - 100 \times 20 - 200 \times 10 - F_2 \times 12{\cdot}07 = 0$$

or $$F_2 = \frac{1}{12.07}[-15250]$$

$$= -1263.46\ \text{kN (compressive)}$$

Force in member 3

GH rises 7 m in 10 m.

Hence rise is 25 m in $\dfrac{10}{7} \times 25 = 35.71\,\text{m}$

$$\tan\phi = \frac{18}{10} = 1.8$$

$$\sin\phi = 0.8741$$

Perpendicular distance from I on JH

$$= 35.71\sin\phi = 35.71 \times 0.8741$$

$$= \mathbf{31.21\ m}$$

Taking moments of all forces on the left of section 6-6 about I

$$1000 \times 35.71 - 1250 \times 25 - 100 \times 35.71 - F_3 \times 31.21 = 0$$

or $$F_3 = \frac{1}{31.21}[35710 - 31250 - 3571]$$

$$= 28.48 \text{ kN. (tension)}$$

Example 6.20. For the three hinged gabled frame shown determine (a) Support reactions. (b) axial force, shear force and bending moment at *B*.

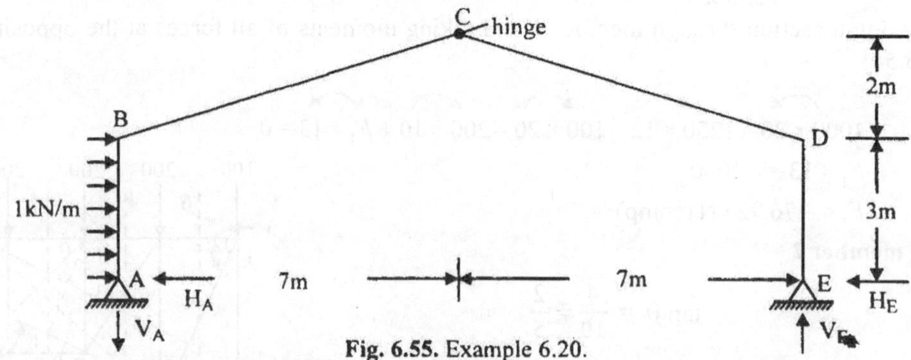

Fig. 6.55. Example 6.20.

Solution: Taking moments of all forces about *E*

$$1 \times 3 \times 1.5 + V_A \times 14 = 0$$

or $$V_A = -\frac{4 \times 1.5}{14} = -0.3214 \text{ kN.}$$

Taking moments of left portion about hinge *C*

$$H_A \times 5 - 1 \times 3 \times 3.5 - V_A \times 7 = 0$$

or $5H_A - 7V_A = 10.5$

Substituting the value of V_A H_A **= 2.55 kN.**

From $$\Sigma V = 0$$

$$-0.3214 + V_E = 0$$

or $$V_E = 0.3214$$

From $$\Sigma H = 0$$

$$H_A - 1 \times 3 + H_E = 0$$

$$H_E = 3 - H_A$$

$$= 3 - 2.55 = \mathbf{0.45 \text{ kN.}}$$

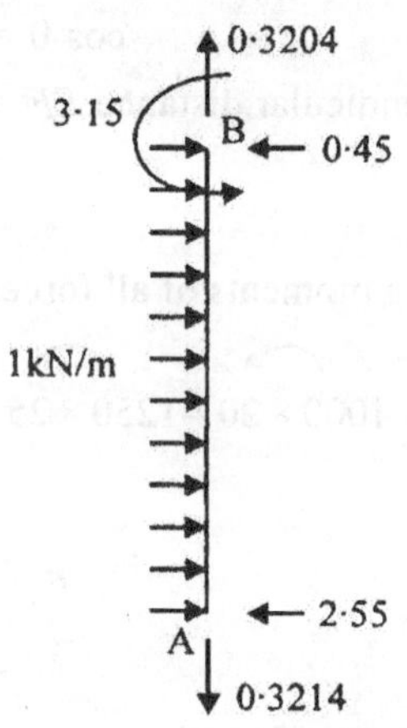

Fig. 6.56. Example 6.20.

In AB

Axial force at *B* = 0.3214 (tension)

S.F. at *B* = 2.55 − 3 = − 0.45 kN

$$\text{B.M. at } B = 2.55 \times 3 - 1 \times \frac{3^2}{2}$$

= 3.15 kNm (tension inside)

In BC (Fig. 6.57)

$$\sin\theta = \frac{2}{\sqrt{2^2 + 7^2}} = 0.27$$

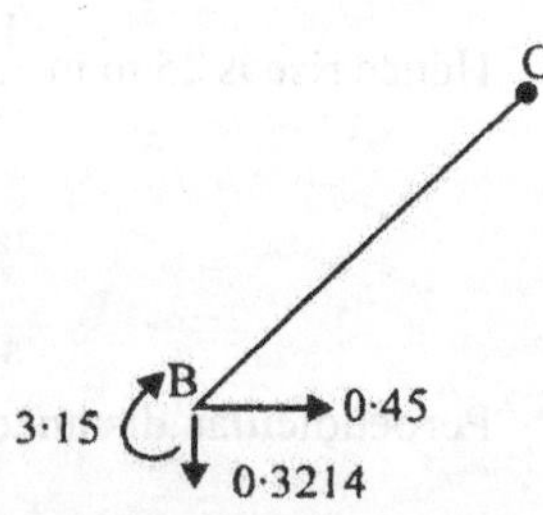

Fig. 6.57. Example 6.20.

$$\cos\theta = \frac{7}{\sqrt{2^2+7^2}} = 0.96$$

$$\text{Axial force in } BC = 0.45 \times 0.96 - 0.3214 \times 0.27$$

$$= 0.3452 \text{ kN (Compressive)}$$

$$\text{S.F. in BC at } B = 0.45 \times 0.27 + 0.3214 \times 0.96$$

$$= 0.43 \text{ kN}$$

SUMMARY

1. **Cables and arches are used to span large openings. A cable can resist tension only. In a cable the full cross section is uniformly stressed and this results in efficient use of materials.**
2. **An arch is a rigid structure and resists loads predominantly through compression. It may also be subjected to bending and shear.**
3. **A cable with uniform load along its length takes the shape of a catenary. However, in most cases, it is approximated as a parabola.**
4. **A cable subjected to a number of concentrated loads can be analysed**

   (*i*) Graphically (*ii*) Analytically

   Analytical solution of cables leads to the following cable theorem:

   "Product of horizontal component of cable tension and vertical distance between the cable chord and cable at any point is equal to bending moment at that point on a horizontal simply supported beam subjected to the same external loading as that on the cable". This theorem enables computation of horizontal tension.
5. **Suspension bridges are used in highways when the span is more than 200 m. Here the bridge deck is suspended from cables by means of hangers.It is also stiffened by stiffened girders or truss on either side of the road way. The stiffened girders are not subjected to any S.F. and B.M. due to deadload as the girders are supported by closely spaced hangers.**

   Any live load on the bridge will be transmitted to the girders as point loads and will induce S.F. and B.M. throughout their length.
6. **In between the suspension cable and the anchor cable there is a guide pulley or saddle mounted on roller over the pulley. When the support is pulley, the pier will be subjected to both direct thrust and bending. When it is a saddle, pier is subjected to direct thrust only.**
7. **Three hinged arch is a statically determinate structure. Though primarily a compression member it is often subjected to B.M. and Shear which can be computed by the formulae,**

$$T = H\cos\theta + V_A \sin\theta$$

$$Q = V_A \cos\theta - H\sin\theta$$

$$M = V_A \cdot x - H \cdot y$$

where T**= Axial thrust**

H**= Horizontal Reaction**

V_A**= Vertical reaction at** A

Q**= Shear Force**

M**= Bending Moment**

x, y**= coordinates of the point**

θ**= angle between the tangent and horizontal at the point.**

PROBLEMS

A. Theoretical

6.1 Distinguish between parabolic, segmental and semi circular arches. **[AMIE Winter 1986]**

B. Objective. Tick the correct answers.

6.2. (*i*) A three hinged arch is (*a*) determinate (*b*) indeterminate (*c*) unstable. **[AMIE Winter '97]**

(*ii*) If the supports of a three hinged arch yield horizontally by a small amount, the horizontal support reaction (*a*) decreases (*b*) increases (*c*) remains the same. **[AMIE Summer '99]**

C. Mathematical

6.3 **A suspension cable is suspended from two piers 200 m apart, the left support being 5 m above the right support. The cable carries a uniformly distributed load of 1.5 *t* f/m and has its lowest point 10 m below the lower support. The ends of the cable are attached to saddles on rollers at the top of the piers and the back stays which may be assumed straight and inclined at 60° to the vertical. Determine the maximum tension in the cable, the tension in the back stays and the thrust on each pier. [AMIE Summer 1983]**

6.4. **A symmetrical suspension bridge has two cables each of 200 m span and central dip of 20 m. It carries a total load of 500 tf which is uniformly distributed over the entire span. Calculate the horizontal component of the tension and the maximum tension in each cable. If the allowable stress for the materials of the cable is 1.2 t/cm^2, calculate the required sectional area of each. Also calculate the overturning force at the top of the pier where the cable runs over pulley device given that the anchor cables make an angle of 45° with the vertical. [AMIE Winter 1985]**

6.5. **The stiffening girder of a suspension bridge of span 100 m is pinned at its supports and at centre. Calculate the reactions at the supports and draw the S.F. and B.M. diagrams for the girder due to a load of 5 tf acting at 25 m from left support. The dead load of the girder is 1 t/m and dip of cable 7.5 m.**

6.6. **A suspension bridge of 100 m span has two stiffening girders with hinged supports at the ends. They are further supported by two cables each having a central dip of 10 m. The deadload of the bridge per girder is 0.5 tf per meter run and the live load is 1 tf per meter run which covers left half of the span. Determine for each girder the shear force and bending moment at 25 meters from the left hand. Find also the maximum tension in each cable. [AMIE Summer 1982]**

6.7 **The three hinged stiffening girder of a suspension bridge of span 120 m is subjected to two point load of 24 tf and 30 tf at distances of 25 and 80 m from the left end. Find the shear force and bending moment for the girder at a distance of 40 m from the left end. The supporting cable has a central dip of 12 m. Find also the maximum tension in the cable and draw the bending moment diagram for the girder. [AMiE Winter 1982]**

6.8. **A suspension bridge stiffened by a three hinged girder has a span of 100 m. Its cables are supported at the same level and they have a central dip of 10 m. Find the maximum bending moment in the girder if a live load of 3 tf/m and 20 m long lies with its centre at 20 m from the left support. Determine the required section modulus of the girder if the permissible stress in the girder is 1200 kgf./cm^2. If the dead load per girder is 0.5 tf/m, find the maximum tension in the cable. [AMIE Winter 1983]**

6.9. **A three hinged circular arch has a span of 20 m between hinged supports at the same level and rise of 4 m to its centre where there is a hinge. It carries a uniformly distributed load**

of 2.5 t/m on the left half of the span. Calculate the B.M. at the left quarter span and the position and magnitude of maximum B.M. on right half of the span.

[AMIE Summer 1979]

6.10. A semicircular arch of span 2R has hinged supports at the springings and an internal hinge at the crown. Determine the vertical and horizontal components of the reactions at the supports when it is subjected to a uniformly distributed load = *W* covering left hand half of span. Compute the values of maximum B.M. in each half spans and their locations and sketch the bending moment diagram. Also determine the radial SF and normal thrust at left quarter point. **[AMIE Winter 1986]**

6.11. A three hinged arch consists of two segments of circular arcs as shown in Fig. P 6.1 and is loaded by a concentrated load P at the crown. Determine the horizontal thrust at the supports and draw the bending moment diagram showing the maximum bending moment.

[AMIE Summer '97]

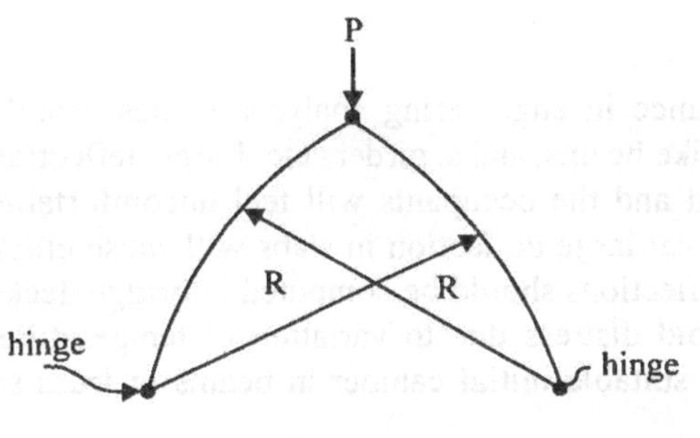

Fig. P 6.1. Problem 6.11.

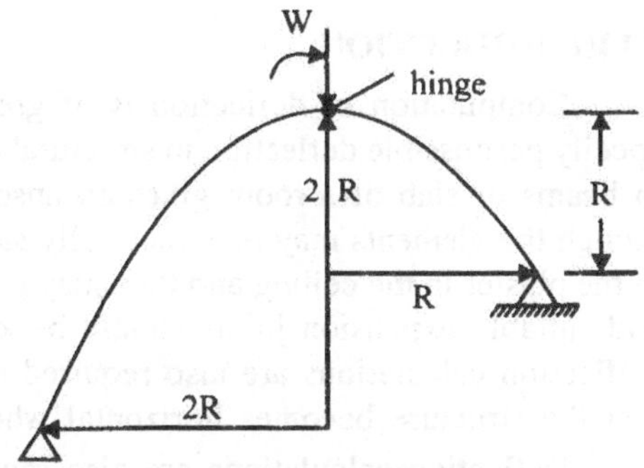

Fig. P 6.2. Problem 6.12.

6.12. A three hinged arch shown in Fig. P 6.2 consists of two segments, first one is a quadrant of a circle of radius 2R and the second one is a quadrant of a circle of radius R joined together by a pin connection at the crown where a point load *W* is placed. Draw the bending moment diagram. Also determine the radial shear force and normal thrust at a section situated at a horizontal distance R from left hand side. **[AMIE Summer 1987]**

6.13. An unsymmetrical parabolic arch of 20 m span has two hinges at the springings and one at the highest point of the arch which is 9 m and 4 m above the left hand and right hand supports respectively. The arch is loaded with a uniformly distributed load of 30000 N/m run covering a horizontal distance from the left hand support to the highest point of the arch. Find the reactions at the supports and normal thrust and radial shear at a section 4 m from the right hand support. **[AMIE Summer 1990]**

6.14. A parabolic tied arch rib ABC has its left support A hinged to a rigid abutment and the right support C connected to a roller resting on a rigid horizontal base. It was a span equal to 16 m between these supports and a central rise of 4 m to the crown B. where an internal hinge is provided. The arch is tied by means of horizontal tie EF placed 2m below the crown. The left half span AB of the arch is loaded with a uniformly distributed load of intensity 3 t/m measured horizontally. Assuming the tie to be inextensible, find the force in it. Also calculate the B, M, S.F. and Normal Thrust on the rib section at a distance of 4 m from the left hand support A. **[AMIE Winter 1980]**

7

FORCE DISPLACEMENT RELATIONS : GEOMETRIC METHODS

7.1 INTRODUCTION

Computation of deflection is of great importance in engineering analysis. Codes usually specify permissible deflection in structural elements like beams, slabs, girders etc. Large deflection in beams or slab of a room gives an unseemly sight and the occupants will feel uncomfortable though the elements may be structurally safe. Moreover large deflection in slabs will cause crack in the plaster in the ceiling and they may peel off. Deflections should be computed in bridge deeks and suitable expansion joints should be kept to avoid distress due to variation of temperature. Deflection calculations are also required to provide suitable initial camber in beams or truss so that the structure becomes horizontal when loaded.

Deflection calculations are also required for statically indeterminate problems where the available equations of statics are not enough to solve the unknowns. Additional equations are required which are provided by the compatibility conditions of the structure based on slopes and deflections. Hence a large number of methods have been developed to calculate deflections of elastic structures. These can be broadly classified as.

1. Geometrical Methods
2. Energy Methods

In the geometrical method the displacements are computed through the direct application of equilibrium, compatibility and material constitutive laws. In the energy method, however, one of the first two concepts is replaced by an energy principle.

7.2 GEOMETRICAL METHOD

The advantage of this method is that it enables us to determine deflections at several points of the structure. In elastic structure loads produce stresses, stresses produce strain and strain produce deflection. The different geometrical methods are :

1. Double Integral Method
2. Area Moment Method
3. Conjugate Beam Method
4. Method of concentrated Angle changes.

Of the above four methods, method of concentrated angle changes is very suitable for computing deflection of frames.

7.3 DIFFERENTIAL EQUATION OF THE DEFLECTION CURVE :

Let us consider a beam subjected to pure bending. Unless it is a very deep beam with short span $\left(\text{normally } \frac{d}{l} < \frac{1}{10}\right)$ deflection due to shear is negligible, hence effect of shear is not considered.

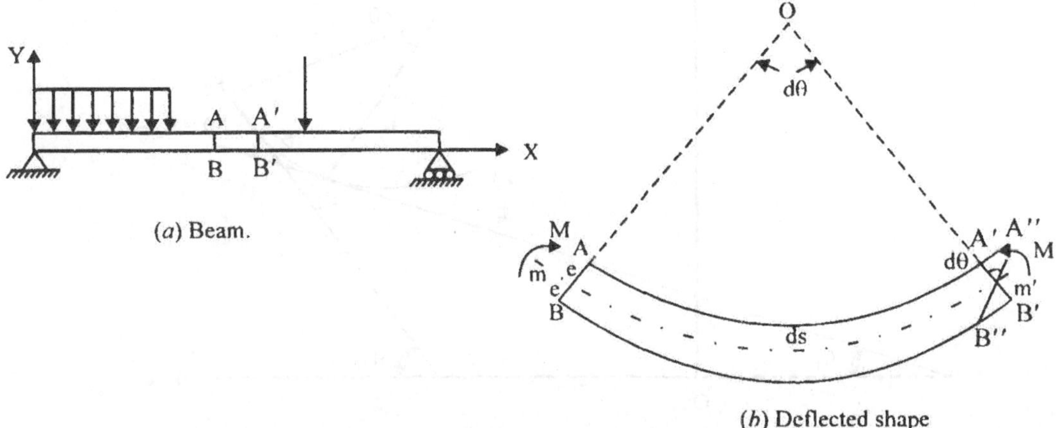

Fig. 7.1. Differential Equation of Deflection Curve.

Fig. 7.1 (*b*) shows the bent shape of a small deflected element of the beam.

Two originally plane parallel sections *ds* distance apart AB and A'B' will now make a small angle $d\theta$. A'' B'' is drawn parallel to AB. From Fig. 7.1 (b) it is seen that

A'A'' = B'B'' = $e d\theta$.

If f is the maximum stress in the extreme top and bottom fibres, E is the young's modulus of elasticity strain = $\frac{f}{E}$ and extension or contraction over length $ds = \frac{f}{E} \cdot ds$.

Hence
$$e d\theta = \frac{f}{E} \cdot ds$$

or
$$\frac{d\theta}{ds} = \frac{f}{eE}.$$

Using the familiar strength of materials formula.

$$\frac{M}{I} = \frac{f}{y} \text{ or } f = \frac{My}{I}$$

Here $y = e$

$$\therefore \quad \frac{d\theta}{ds} = \frac{Me}{I} \cdot \frac{1}{eE} = \frac{M}{EI} \qquad \text{.....(7.1)}$$

Since the formula $\frac{M}{I} = \frac{f}{y}$ has been used, the derivation is valid when the following assumption are fulfilled.

1. Small deflection of beam.
2. Elastic material, E same for tension and compression.
3. Only bending moment is considered significant.
4. Plane sections remain plane and normal to the longitudinal fibres after bending.

5. Cross sections has a longitudinal plane of symmetry and is bent by end couples acting in this plane.

$\frac{d\theta}{ds}$ is called curvature and is defined as the rate at which curve is changing direction.

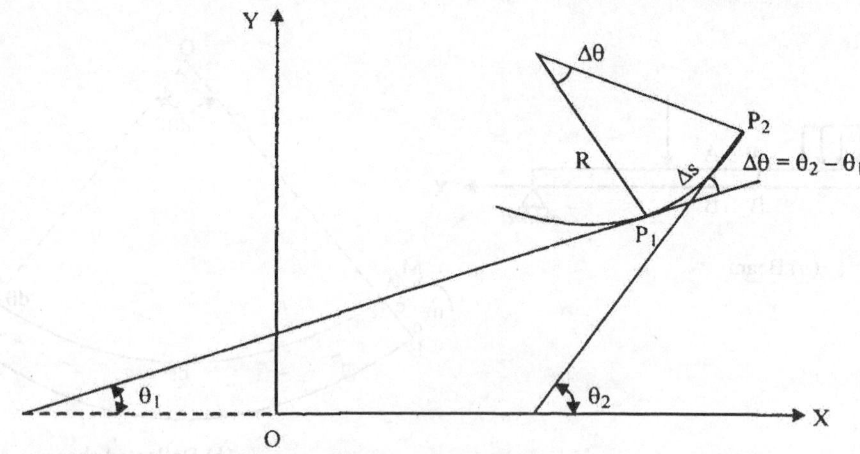

Fig. 7.2. Curvature Equation.

Let P_1 and P_2 be two points on a curve a small distance Δs apart. The change of direction from P_1 to $P_2 = \theta_2 - \theta_1 = \Delta\theta$. The average rate of change of direction between P_1 and $P_2 = \frac{\Delta\theta}{\Delta s}$. The limiting value of this ratio as Δs approaches zero is called curvature and the radius of curvature is reciprocal to this curvature.

Hence curvature $\rho = \frac{1}{R}$

$$\text{Lim} \frac{\Delta\theta}{\Delta s} = \frac{d\theta}{ds}$$

To convert $\frac{d\theta}{ds}$ to x, y coordinates as shown in Fig. 7.2.

$$\tan\theta = \frac{dy}{dx}$$

$$\frac{d}{dx}(\tan\theta) = \frac{d^2y}{dx^2}$$

or $$\frac{d}{d\theta}(\tan\theta)\cdot\frac{d\theta}{dx} = \frac{d^2y}{dx^2}$$

or $$(1+\tan^2\theta)\cdot\frac{d\theta}{dx} = \frac{d^2y}{dx^2}$$

or $$\frac{d\theta}{dx} = \frac{\frac{d^2y}{dx^2}}{1+\tan^2\theta} = \frac{\frac{d^2y}{dx^2}}{1+\left(\frac{dy}{dx}\right)^2}$$

Now,
$$\frac{d\theta}{ds} = \frac{d\theta}{dx}\cdot\frac{dx}{ds}$$

$$= \frac{\frac{d^2y}{dx^2}}{1+\left(\frac{dy}{dx}\right)^2}\cdot\frac{dx}{\sqrt{dx^2+dy^2}}$$

$$= \frac{\frac{d^2y}{dx^2}}{\left\{1+\left(\frac{dy}{dx}\right)^2\right\}^{\frac{3}{2}}} \quad ....(7.2)$$

Equating Eqn. 7.1 with eqn. 7.2

$$\frac{M}{EI} = \frac{\frac{d^2y}{dx^2}}{\left\{1+\left(\frac{dy}{dx}\right)^2\right\}^{\frac{3}{2}}}$$

or
$$M = EI\cdot\frac{\frac{d^2y}{dx^2}}{\left\{1+\left(\frac{dy}{dx}\right)^2\right\}^{\frac{3}{2}}}$$

Usually the elastic deflection of a beam is small. $\frac{dy}{dx}$ is then small and $\left(\frac{dy}{dx}\right)^2$ is still smaller and can be neglected in comparison to 1.

Hence,
$$\frac{M}{EI} \approx \frac{d^2y}{dx^2}$$

The derivation is based on coordinate axes as shown. However, usually, y axis is taken as positive down ward instead of upward. In that case the equation will be

$$\frac{d^2y}{dx^2} = -\frac{M}{EI} \quad ....(7.3)$$

Solution of the above differential equation involves integrating twice and hence the method is known as double integration method.

Example 7.1. For the beam shown in Fig. 7.3 find the expression for the elastic curve. What is the maximum deflection of the beam and where does it occur?

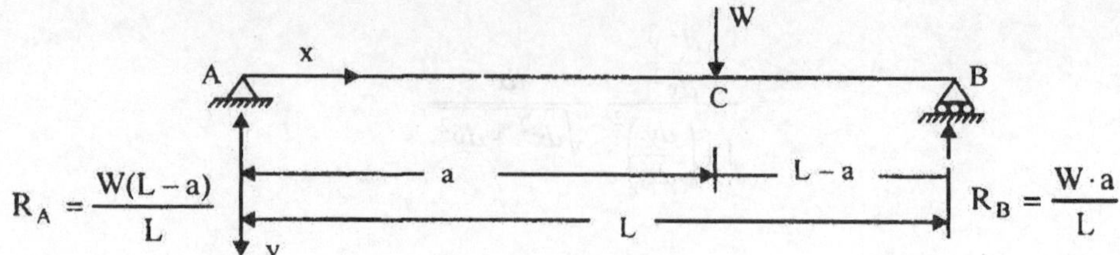

Fig. 7.3. Example 7.1

Solution:

For the portion AC expression for BM. is $\frac{W(L-a)}{L} \cdot x$

For the portion CB B.M. is

$$\frac{W(L-a)}{L} \cdot x - W(x-a)$$

Hence,
$$EI_z \cdot \frac{d^2y}{dx^2} = -\frac{W(L-a)}{L} \cdot x \qquad \text{for } x \le a \qquad \text{....(7.4 a)}$$

and
$$EI_z \cdot \frac{d^2y}{dx^2} = -\frac{W(L-a)}{L} \cdot x + W(x-a) \qquad \text{for } x \ge a \qquad \text{....(7.4 b)}$$

Integrating once

$$EI_z \cdot \frac{dy}{dx} = -W\frac{(L-a)}{L} \cdot \frac{x^2}{2} + C_1$$

$$EI_z \cdot \frac{dy}{dx} = -W\frac{(L-a)}{L} \cdot \frac{x^2}{2} + W\frac{(x-a)^2}{2} + C_2$$

At point C, $x = a$ the tangent must have a single slope *i.e.* $\frac{dy}{dx}$ must be the same for both the branches. Hence, $C_1 = C_2 = C$ (say).

Similarly integrating once more

$$EI_z y = \frac{-W(L-a)}{L} \cdot \frac{x^3}{6} + C \cdot x + C_3 \qquad \text{for } x \le a$$

$$EI_z y = \frac{-W(L-a)}{L} \cdot \frac{x^3}{6} + \frac{W(x-a)^3}{6} + C \cdot x + C_4 \qquad \text{for } x \ge a$$

At $x = a$ deflection for both branches must be the same. Hence $C_3 = C_4$. From the boundary conditions at $x = 0, y = 0$. Hence $C_3 = 0$ and C_4 also is zero.

Putting $y = 0$ at $x = L$

$$C = \frac{W(L-a) \cdot L}{6} - \frac{W(L-a)^3}{6L}$$

$$= \frac{W(L-a)\{L^2 - (L-a)^2\}}{6L}$$

The deflection equations then become

$$EI_z y = \frac{W(L-a)}{6L} \cdot x\left\{L^2 - (L-a)^2 - x^2\right\} \qquad \text{for } x \le a \text{ ....(7.5 a)}$$

and
$$EI_z y = \frac{W(L-a)x}{6L} \cdot \left\{L^2 - (L-a)^2 - x^2\right\} + \frac{W(x-a)^3}{6} \qquad \text{for } x \ge a \text{ ....(7.5 b)}$$

Similarly
$$EI_z \frac{dy}{dx} = \frac{W(L-a)}{6L}\left\{L^2 - (L-a)^2 - 3x^2\right\} \qquad \text{for } x \le a \text{ ....(7.6 a)}$$

and
$$EI_z \frac{dy}{dx} = \frac{W(L-a)}{6L}\left\{L^2 - (L-a)^2 - 3x^2\right\} + \frac{W(x-a)^2}{2} \qquad \text{for } x \ge a \text{ ....(7.6 b)}$$

Putting $x = 0$ in the first equation

$$\theta_A = \left(\frac{dy}{dx}\right)_{x=0} = \frac{W(L-a)}{6L\, EI_z} \cdot \left\{L^2 - (L-a)^2\right\}$$

Putting $x = L$ in the 2nd equation

$$\theta_B = \left(\frac{dy}{dx}\right)_{x=L} = -\frac{Wa\,(L-a)}{6L\, EI_z} \cdot (L+a)$$

Here, θ_A is clockwise, hence positive.
θ_B is anticlockwise, hence negative.

The maximum deflection occurs where $\frac{dy}{dx} = 0$.

or tangent is horizontal.

Assuming $a > \frac{L}{2}$, maximum deflection occurs within the left span. Hence putting 1st expression of $\frac{dy}{dx}$ (Eq. 7.6a) equal to zero.

$$L^2 - (L-a)^2 - 3x^2 = 0$$

or
$$x = \frac{\sqrt{L^2 - (L-a)^2}}{\sqrt{3}}$$

Substituting the value of x in Eq. 7.5 a.

$$y_{max} = \frac{W(L-a)\left\{L^2 - (L-a)^2\right\}^{\frac{3}{2}}}{9\sqrt{3} \cdot L \cdot EI_z}$$

when $a = \frac{L}{2}$, $x = \frac{L}{2}$ *i.e.*, for central load maximum deflection occurs under the load and the value is

$$\frac{W \cdot \frac{L}{2}\left(L^2 - \frac{L^2}{4}\right)^{\frac{3}{2}}}{9\sqrt{3} \cdot L \cdot EI} = W \cdot \frac{L}{2} \cdot \frac{3^{\frac{3}{2}} \cdot L^3}{8.9 \cdot \sqrt{3}} \cdot \frac{1}{L \cdot EI}$$

$$= \frac{W \cdot L^3 \cdot 3\sqrt{3}}{16.9 \cdot \sqrt{3} \cdot EI} = \frac{WL^3}{48\ EI}$$

When $\quad a \to L \quad$ (Load very close to the right support)

$$x \to \frac{L}{\sqrt{3}} = 0.577\ L$$

Hence the position of maximum deflection changes by a very small distance as the load changes its position from mid point to the end of the beam. Hence for all practical purposes deflection at the middle is taken as the maximum deflection.

When there are a large number of concentrated loads the expression for B.M. will vary from segment to segment and the solution by equating slopes and deflections at common points will be very tedious. To overcome this, singularity function and Macaulay's notation have been developed which are explained in the next section.

7.4. SINGULARITY FUNCTION AND MACAULAY'S NOTATION

Singularity function was first introduced by A. Clebsch in 1862. In 1919 W. H. Macaulay specifically suggested the use of special brackets for beam problems. Hence the name Macaulay's method. To obtain the slopes and deflections by double - integration method the expression,

$\frac{d^2y}{dx^2} = -\frac{M}{EI}$ has to be integrated twice.

If the loading is a continuous function between the supports solution of the differential equation is convenient. But with a large number of concentrated loads integration has to be done for each part separately and constants of integration evaluated by considering same slope and deflection at common points.

To over come this difficulty Macaulay's notation will be used. Macaulay suggested the use of special brackets for beam problems. Fig. 7.4 shows a simply supported beam with concentrated loads and moments.

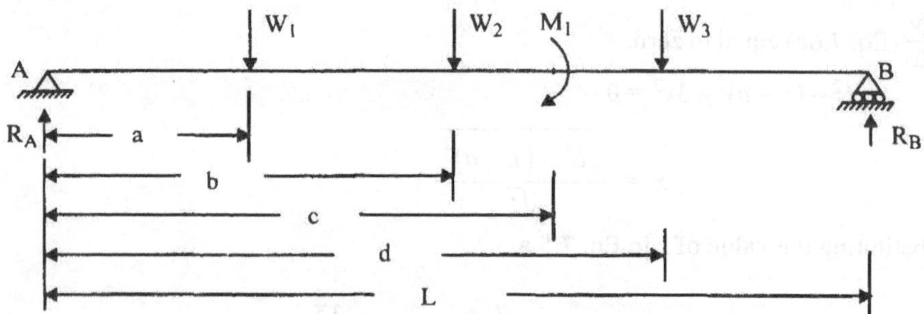

Fig. 7.4. Beam with concentrated loads and moments.

Since the applied loads are point (concentrated) loads five distinet regions exist to which different bending moment expressions apply.

$$M = R_A \cdot x \qquad \text{when } 0 \le x \le a.$$

$$M = R_A \cdot x - W_1(x-a) \qquad \text{when } a \le x \le b.$$

$$M = R_A \cdot x - W_1(x-a) - W_2(x-b) \qquad \text{when } b \le x \le c.$$

$$M = R_A \cdot x - W_1(x-a) - W_2(x-b) + M_1 \qquad \text{when } c \le x \le d.$$

$$M = R_A \cdot x - W_1(x-a) - W_2(x-b) + M_1 - W_3(x-d) \qquad \text{when } d \le x \le L.$$

All five equations can be written as one providing we define the following symbolic functions.

$$\langle x-p\rangle^n = \begin{cases} 0 & \text{for } 0 < x < p \\ (x-p)^n & \text{for } p < x < \infty \end{cases}$$

where $\qquad n \geq 0 \qquad (n = 0, 1, 2, ...)$

The expression within pointed brackets is non existent until x reaches p. For x beyond p, the expression becomes an ordinary binomial. For $n = 0$ and for $x > p$ the function is unity. Using this artefact we can write only one expression for all the five expressions above :

$$M = R_A\langle x-0\rangle^1 - W_1\langle x-a\rangle^1 - W_2\langle x-b\rangle^1 + M_1\langle x-c\rangle^0 - W_3\langle x-d\rangle^1$$

Here values of p are 0, a, b, c and d respectively.

The method is now explained with the help of an example.

Example 7.2. Find the slopes at A, C, and D and deflections at B and D of the beam given below. EI constant.

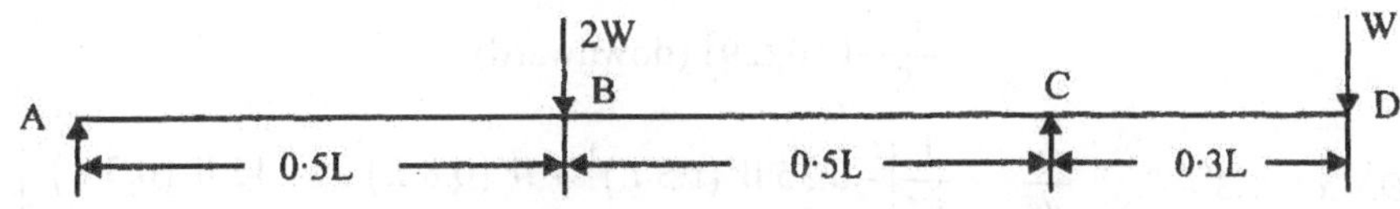

Fig. 7.5. Example 7.2

Solution:

$$R_A \times L - 2\,W \times 0.5\,L + W \times 0.3\,L = 0$$

or $\qquad R_A = 0.7\,W$

$$-R_C \times L + 2\,W \times 0.5\,L + W \times 1.3\,L = 0$$

or $\qquad R_C = 2.3\,W$

$$R_A + R_C = 3\,W \text{ (check)}$$

$$\frac{d^2y}{dx^2} = -\frac{M}{EI}$$

$$M = 0.7\,Wx - 2\,W\langle x-0.5\,L\rangle + 2.3\,W\langle x-L\rangle$$

or $$\frac{d^2y}{dx^2} = \frac{1}{EI}\left[-0.7\,Wx + 2\,W\langle x-0.5L\rangle - 2.3\,W\langle x-L\rangle\right]$$

Integrating $$\frac{dy}{dx} = \frac{1}{EI}\left[-0.7\,W\cdot\frac{x^2}{2} + \frac{2\,W}{2}\langle x-0.5L\rangle^2 - \frac{2.3\,W}{2}\langle x-L\rangle^2\right] + C_1$$

$$y = \frac{1}{EI}\left[-0.35\,W\cdot\frac{x^3}{3} + \frac{W}{3}\langle x-0.5\,L\rangle^3 - \frac{1.15\,W}{3}\langle x-L\rangle^3\right] + C_1x + C_2$$

At $\qquad x = 0 \quad y = 0 \;\therefore\; C_2 = 0$

At $\qquad x = L \quad y = 0$ or $0 = \dfrac{1}{EI}\left[-0.35\dfrac{W\cdot L^3}{3} + \dfrac{W}{3}\cdot(0.5\,L)^3\right] + C_1L$

or $$C_1L = \frac{1}{EI}\left[0.35\frac{W\cdot L^3}{3} - \frac{W}{3}\cdot(0.5\,L)^3\right]$$

or $$C_1 = 0.075\frac{WL^2}{EI}$$

At A, $x = 0$ $$\frac{dy}{dx} = 0.075\frac{WL^2}{EI} \text{ (Clockwise)}$$

At C, $x = L$ $$\frac{dy}{dx} = \frac{1}{EI}\left[-0.7\,W\frac{L^2}{2} + W(0.5\,L)^2\right] + \left[0.075\frac{WL^2}{EI}\right]$$

$$= -0.025\frac{WL^2}{EI} \text{ (anticlockwise)}$$

At B, $x = 0.5\,L$ $$y = \frac{1}{EI}\left[\frac{-0.35\,W(0.5\,L)^3}{3}\right] + \frac{0.075\,WL^2}{EI}\cdot\frac{L}{2}$$

$$= \frac{WL^3}{EI}[0.0229] \text{ (downward)}$$

At D, $x = 1.3\,L$ $$\frac{dy}{dx} = \frac{1}{EI}\left[-0.35\,W(1.3\,L)^2 + W(0.8\,L)^2 - 1.15\,W(0.3\,L)^2\right]$$

$$+ 0.075\frac{WL^2}{EI}$$

$$= .02\frac{WL^2}{EI} \text{ (clockwise).}$$

Deflection at D.

$$\frac{d^2y}{dx^2} = \frac{1}{EI}\left[-\frac{0.35\,W}{3}(1.3\,L)^3 + \frac{W}{3}(0.8\,L)^3 - \frac{1.15\,W}{3}(0.3\,L)^3\right] + \frac{0.075\,WL^2}{EI}\cdot 1.3\,L$$

$$= 0.0015\frac{WL^3}{EI} \text{ (downward)}$$

Macaulay's method has been used in solving this problem. It is important to remember that the expression under the pointed brackets is valid only when it is positive, otherwise it is to be ignored. When evaluating constants of integration from the conditions $x = 0$, $y = 0$ both the expressions under printed brackets become negative and hence ignored. However, when using the condition $y = 0$ at $x = L$, none of the two expressions under bracket becomes negative and hence both are valid expressions.

Example 7.3. Compute the rotations of points A and C and the deflection at B due to the applied moment M at B. EI constant.

Solution:

$$R_A \times L - M = 0$$

$$R_A = \frac{M}{L}$$

$$M = \frac{M}{L}\cdot x - M < x - a >^0$$

$$EI\frac{d^2y}{dx^2} = -M$$

Fig. 7.6. Example 7.3.

or
$$\frac{d^2y}{dx^2} = \frac{1}{EI}\left[-\frac{M}{L}\cdot x + M < x-a >^0\right]$$

$$\frac{dy}{dx} = \frac{1}{EI}\left[-\frac{M}{L}\cdot\frac{x^2}{2} + M < x-a >\right] + C_1$$

$$y = \frac{1}{EI}\left[-\frac{M}{2L}\cdot\frac{x^3}{3} + \frac{M}{2} < x-a >^2\right] + C_1x + C_2$$

At $x = 0$ $y = 0$ $\therefore C_2 = 0$

At $x = L$ $y = 0$

$$0 = \frac{1}{EI}\left[-\frac{M}{2L}\cdot\frac{L^3}{3} + \frac{M}{2}\cdot b^2\right] + C_1 \cdot L$$

or
$$C_1 = \frac{M}{EI}\cdot\frac{1}{L}\left[\frac{L^2}{6} - \frac{b^2}{2}\right]$$

$$= \frac{M}{6\,EI}\cdot\frac{1}{L}\left[L^2 - 3b^2\right]$$

At $x = 0$
$$\frac{dy}{dx} = C_1 = \frac{M}{6\,EI\,L}\left[L^2 - 3b^2\right] \text{ (clockwise)}$$

At $x = L$
$$\frac{dy}{dx} = \frac{1}{EI}\left[-\frac{M.L^2}{2L} + M.b\right] + \frac{M}{6EIL}[L^2 - 3b^2]$$

$$= +\frac{M}{6EIL}[L^2 - 3a^2] \text{ clockwise}$$

when $a = L$ *i.e.*, when moment is applied at the end

$$\text{slope} = +\frac{M}{6\,EI\,L}\cdot\left(-2L^2\right) = -\frac{ML}{3\,EI} \text{ (anticlockwise)}$$

deflection at $x = a$

$$y = \frac{1}{EI}\left[-\frac{M}{6L}\cdot x^3 + \frac{M}{2} < x-a >^2\right] + \frac{M}{6\,EI\,L}\left(L^2 - 3b^2\right)\cdot x$$

$$= \frac{1}{EI}\left[-\frac{M}{6L}\cdot a^3\right] + \frac{M}{6\,EI\,L}\left(L^2 - 3b^2\right)\cdot a$$

$$= -\frac{M}{3\,EI\,L}\cdot ab(a-b).$$

Example 7.4. A cantilever of length L and constant flexural rigidity EI carries a uniformly distributed load of intensity ω per unit length in the middle half of its length. Determine the ratio of the slopes and the ratio of deflections at the centre and free end of the cantilever. (London University).

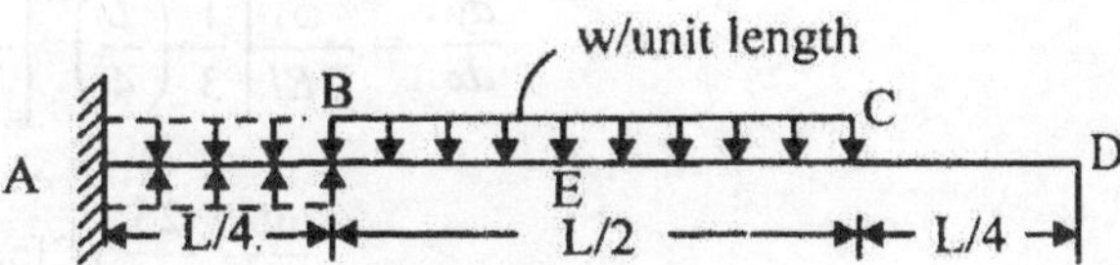

Fig. 7.7. Example 7.4.

Solution:

To apply Macaulay's method the load should countinue upto the end. Hence downward load/ unit length w is continued upto the end and upward load of same magnitude is introduced from $\frac{3}{4}L$ to L.

Taking origin at D

$$M = -\frac{\omega}{2}\left\langle x-\frac{L}{4}\right\rangle^2 + \frac{\omega}{2}\left\langle x-\frac{3L}{4}\right\rangle^2$$

$$\frac{d^2y}{dx^2} = -\frac{M}{EI} = \frac{1}{EI}\left[\frac{\omega}{2}\left\langle x-\frac{L}{4}\right\rangle^2 - \frac{\omega}{2}\left\langle x-\frac{3L}{4}\right\rangle^2\right]$$

$$\frac{dy}{dx} = \frac{\omega}{2EI}\left[\frac{1}{3}\left\langle x-\frac{L}{4}\right\rangle^3 - \frac{1}{3}\left\langle x-\frac{3L}{4}\right\rangle^3\right] + C_1$$

$$y = \frac{\omega}{2EI}\left[\frac{1}{12}\left\langle x-\frac{L}{4}\right\rangle^4 - \frac{1}{12}\left\langle x-\frac{3L}{4}\right\rangle^4\right] + C_1x + C_2$$

at $\quad x = L, \ \frac{dy}{dx} = 0$

$$0 = \frac{w}{2EI}\left[\frac{1}{3}\cdot\left(\frac{3L}{4}\right)^3 - \frac{1}{3}\cdot\left(\frac{L}{4}\right)^3\right] + C_1$$

or

$$C_1 = \frac{\omega L^3}{2EI}\left[\frac{1}{192} - \frac{27}{192}\right] = -\frac{\omega L^3}{192}\cdot\frac{13}{EI}$$

Hence

$$\frac{dy}{dx} = \frac{\omega}{2EI}\left[\frac{1}{3}\left\langle x-\frac{L}{4}\right\rangle^3 - \frac{1}{3}\left\langle x-\frac{3L}{4}\right\rangle^3\right] - \frac{13}{192}\cdot\frac{\omega L^3}{EI}$$

at $x = 0$ *i.e.*, at the free end,

$$\frac{dy}{dx} = -\frac{13}{192}\cdot\frac{\omega L^3}{EI} \text{ (Clockwise)}$$

At $\quad x = \frac{L}{2}$

$$\frac{dy}{dx} = \frac{\omega}{2EI}\left[\frac{1}{3}\cdot\left(\frac{L}{4}\right)^3\right] - \frac{13}{192}\cdot\frac{\omega L^3}{EI}$$

$$= -\frac{\omega L^3}{EI}\cdot\frac{25}{384} \text{ (Clockwise)}$$

Ratio of

$$\frac{\text{Slope at centre}}{\text{Slope at free end}} = \frac{25}{26}$$

Deflection
$$y = \frac{\omega}{2EI}\left[\frac{1}{12}\left\langle x-\frac{L}{4}\right\rangle^4 - \frac{1}{12}\left\langle x-\frac{3L}{4}\right\rangle^4\right] - \frac{13}{192}\cdot\frac{\omega L^3}{EI}\cdot x + C_2$$

$y = 0$ at $x = L$

or
$$0 = \frac{\omega}{2EI}\left[\frac{1}{12}\cdot\left(\frac{3L}{4}\right)^4 - \frac{1}{12}\left(\frac{L}{4}\right)^4\right] - \frac{13}{192}\cdot\frac{\omega L^3}{EI}\cdot L + C_2$$

or
$$C_2 = \frac{7}{128}\cdot\frac{\omega L^4}{EI}$$

Hence
$$y = \frac{\omega}{2EI}\left[\frac{1}{12}\left\langle x-\frac{L}{4}\right\rangle^4 - \frac{1}{12}\left\langle x-\frac{3L}{4}\right\rangle^4\right] - \frac{13}{192}\cdot\frac{\omega L^3}{EI}\cdot x + \frac{\omega L^4}{EI}\cdot\frac{7}{128}$$

at $x = 0$ $y = +\frac{\omega L^4}{EI}\cdot\frac{7}{128}$

At $x = \frac{L}{2}$
$$y = \frac{\omega}{2EI}\left[\frac{1}{12}\cdot\frac{L^4}{256}\right] - \frac{13}{192}\cdot\frac{\omega L^3}{EI}\cdot\frac{L}{2} + \frac{\omega L^4}{EI}\cdot\frac{7}{128}$$

$$= \frac{wL^4}{EI}\cdot\frac{129}{24\times 256}.$$

Ratio of deflections

$$\frac{\text{At midspan}}{\text{At free end}} = \frac{129}{24\times 256}\times\frac{128}{7} = \frac{43}{112}.$$

Example 7.5. Find slopes at A, B, C and D and deflections at B and D of the beam loaded as shown below. EI is constant.

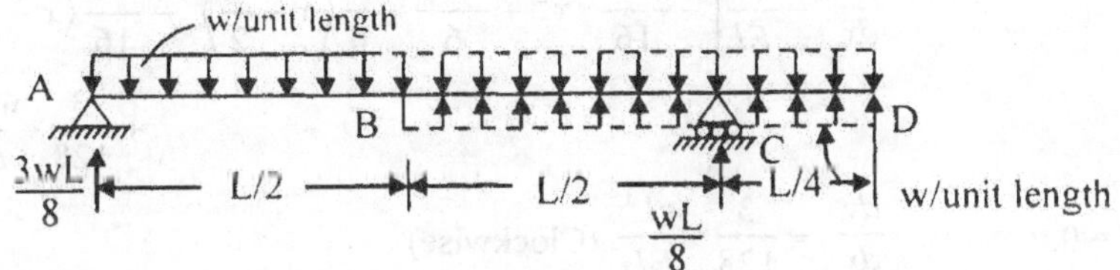

Fig. 7.8. Example 7.5

Solution:

To apply Macaulay's method, the distributed load should continue upto the end and a negative upward loading should be introduced from B to D.

$$R_A \times L - w\times\frac{5L}{4}\left(\frac{3L}{8}\right) + \frac{w\times 3L}{4}\times\frac{L}{8} = 0$$

or
$$R_A = wL\cdot\frac{3}{8}$$

$$R_C\times L - w\times\frac{5L}{4}\cdot\frac{5L}{8} + w\times\frac{3L}{4}\times\frac{7L}{8} = 0$$

or $$R_C = \frac{wL}{8}$$

$$M = \frac{3wL}{8} \times x - \frac{wx^2}{2} + \frac{w}{2}\left\langle x - \frac{L}{2}\right\rangle^2 + \frac{wL}{8}\langle x - L\rangle$$

$$EI\frac{d^2y}{dx^2} = -M = -\frac{3wL}{8}\cdot x + \frac{wx^2}{2} - \frac{w}{2}\left\langle x - \frac{L}{2}\right\rangle^2 - \frac{wL}{8}\langle x - L\rangle$$

$$\frac{d^2y}{dx^2} = \frac{1}{EI}\left[-\frac{3wL}{8}\cdot x + \frac{wx^2}{2} - \frac{w}{2}\left\langle x - \frac{L}{2}\right\rangle^2 - \frac{wL}{8}\langle x - L\rangle\right]$$

$$\frac{dy}{dx} = \frac{1}{EI}\left[-\frac{3wL}{16}\cdot x^2 + \frac{wx^3}{6} - \frac{w}{6}\left\langle x - \frac{L}{2}\right\rangle^3 - \frac{wL}{16}\langle x - L\rangle^2\right] + C_1$$

$$y = \frac{1}{EI}\left[-\frac{3wL}{16}\cdot\frac{x^3}{3} + \frac{wx^4}{24} - \frac{w}{24}\left\langle x - L/2\right\rangle^4 - \frac{wL}{48}\langle x - L\rangle^3\right] + C_1 x + C_2$$

At $x = 0$ $y = 0$ $\therefore C_2 = 0$

At $x = L$ $y = 0$

or $$0 = \frac{1}{EI}\left[-\frac{wL^4}{16} + \frac{wL^4}{24} - \frac{w}{24}\left\langle x - \frac{L}{2}\right\rangle^4 - \frac{wL}{48}\langle x - L\rangle^3\right] + C_1 x + C_2$$

or $$C_1 = \frac{3}{128}\cdot\frac{wL^3}{EI}$$

Slopes at *A, B, C* and *D*.

$$\frac{dy}{dx} = \frac{1}{EI}\left[-\frac{3wL}{16}\cdot x^2 + \frac{wx^3}{6} - \frac{w}{6}\left\langle x - \frac{L}{2}\right\rangle^3 - \frac{wL}{16}\langle x - L\rangle^2\right] + \frac{3}{128}\cdot\frac{wL^3}{EI}$$

At A $x = 0$ $$\frac{dy}{dx} = \frac{3}{128}\cdot\frac{wL^3}{EI} \text{ (Clockwise)}$$

At B $x = \frac{L}{2}$ $$\frac{dy}{dx} = \frac{1}{EI}\left[-\frac{3wL}{16}\cdot\frac{L^2}{4} + \frac{wL^3}{48}\right] + \frac{3}{128}\cdot\frac{wL^3}{EI}$$

$$= -\frac{wL^3}{EI}\cdot\frac{1}{384} \text{ (anticlockwise)}$$

At $x = L$ $$\frac{dy}{dx} = \frac{1}{EI}\left[-\frac{3wL}{16}\cdot L^2 + \frac{wL^3}{6} - \frac{wL^3}{48}\right] + \frac{3}{128}\cdot\frac{wL^3}{EI}$$

$$= -\frac{wL^3}{EI}\cdot\frac{7}{128 \times 3} \text{ (anticlockwise)}$$

At $x = \frac{5L}{4}$ $$\frac{dy}{dx} = \frac{1}{EI}\left[-\frac{3wL}{16} \times \frac{25L^2}{16} + \frac{wL^3 \times 125}{64 \times 6} + \frac{3}{128}\cdot wL^3\right.$$

$$-\frac{w}{6}\cdot\frac{27L^3}{64}-\frac{wL^3}{16\times16}\Bigg]$$

$$=-.018229\frac{wL^3}{EI}\text{ (anticlockwise)}$$

Deflection at *B* and *D*.

At $B,\ x=\dfrac{L}{2}$

$$y=\frac{1}{EI}\left[-\frac{3wL}{48}\cdot x^3+\frac{wx^4}{24}-\frac{w}{24}\left(x-\frac{L}{2}\right)^4-\frac{wL}{48}(x-L)^3+\frac{3}{128}wL^3\cdot x\right]$$

Putting $x=\dfrac{L}{2}$

$$y=\frac{1}{EI}\left[-\frac{3wL}{48}\cdot\frac{L^3}{8}+\frac{wL^4}{16\times24}+\frac{3}{128}wL^3\cdot\frac{L}{2}\right]$$

$$=0.0065104\frac{wL^4}{EI}\text{ (downward)}$$

At $x=\dfrac{5L}{4}$

$$y=\frac{1}{EI}\left[-\frac{3wL}{48}\left(\frac{5L}{4}\right)^3+w\left(\frac{5L}{4}\right)^4\cdot\frac{1}{24}-\frac{w}{24}\left(\frac{3L}{4}\right)^4-\frac{wL}{48}\left(\frac{L}{4}\right)^3+\frac{3}{128}\cdot wL^3\cdot\frac{5L}{4}\right]$$

$$=\frac{wL^4}{EI}[-0.004557]\text{ upward.}$$

7.5. MOMENT AREA METHOD

Normally it is required to compute slopes and deflections at selected points of a beam. It is then not necessary to integrate the differential equation to get the elastic equation of the complete curve. Instead, semi graphical method like Moment Area Method can be used. The method is due to Charles E. Green who taught it in his classes in 1873. However from records it appears that Otto Mohr of Dresden, Germany developed this method in 1868.

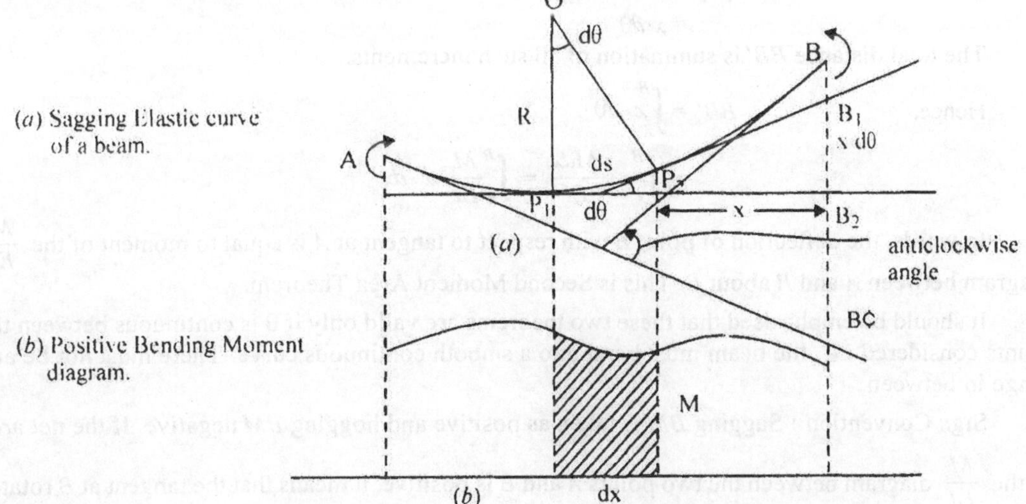

Fig. 7.9. Derivation of Moment Area Method.

It is already derived that for a bent beam as shown in Fig. 7.9.

$$\frac{M}{EI} = \frac{d^2 y}{dx^2} = \frac{d}{dx}\left(\frac{dy}{dx}\right)$$

Now $\frac{dy}{dx}$ = tan θ and for flat curves, when θ is very small, tan θ ≈ θ.

Hence,
$$\frac{dy}{dx} \approx \theta$$

Substituting`
$$\frac{M}{EI} = \frac{d}{dx}\left(\frac{dy}{dx}\right) = \frac{d\theta}{dx}$$

or
$$d\theta \approx \frac{M}{EI}\cdot dx$$

Graphically this means that the change in angle between the radii or the tangent at two points P_1 and P_2, *ds* distance apart in the deflection curve is equal to the shaded elemental area M.*dx* of the bending moment diagram divided by the flexural rigidity *EI*.

To get the total change of angle between tangents at *A* and *B*, it is necessary to integrate the expression between *A* and *B*.

$$\int_A^B d\theta = \int_A^B \frac{M dx}{EI}$$

or
$$\theta_B - \theta_A = \int_A^B \frac{M dx}{EI}$$

$$= \text{Area of the } \frac{M}{EI} \text{ diagram between } A \text{ and } B.$$

In words, the change of angle between the tangents at two points *A* and *B* of the deflection curve of a beam initially straight equals the area of the $\frac{M}{EI}$ diagram between the corresponding verticals. This is first Area Moment Theorem.

Let the tangent at *A* cut the vertical through *B* at *B'*. Then *BB'* is the vertical displacement of *B* above the tangent *AB'* at *A*. If the distance of P_2 from *BB'* is 'x' then

B_1B_2 = incremental distance along *BB'* due to small rotation $d\theta$

$= x\cdot d\theta$

The total distance *BB'* is summation of all such increments.

Hence,
$$BB' = \int_A^B x\cdot d\theta$$

$$= \int_A^B x\cdot\frac{M dx}{EI} = \int_A^B \frac{M}{EI}\cdot x\cdot dx$$

In words, the deflection of point *B* with respect to tangent at *A* is equal to moment of the $\frac{M}{EI}$ diagram between *A* and *B* about *B*. This is Second Moment Area Theorem.

It should be emphasised that these two theorems are valid only if θ is continuous between the points considered *i.e.*, the beam must bend into a smooth continuous curve. There must not be any hinge in between.

Sign Convention : Sagging *BM* is taken as positive and hogging *BM* negative. If the net area of the $\frac{M}{EI}$ diagram between the two points *A* and *B* is positive, it means that the tangent at *B* rotates anticlockwise with respect to tangent at A. If negative, rotation will be in opposite direction *i.e.*, clockwise.

As regards deflection, moment of the $\frac{M}{EI}$ diagram is considered positive if the BM is positive and negative if BM is negative. If the net moment is positive, B lies above the tangent at A if negative, B lies below the tangent at A. The above sign conventions can be checked from Fig. 7.9. A few simple examples are now solved by moment area method.

Example 7.6. Find the slope and deflection at B for the cantilever beam AB. EI is constant.

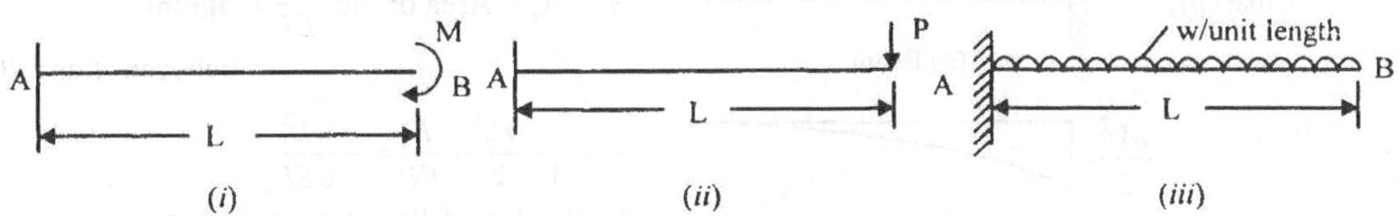

Fig. 7.10. Example 7.6

Solution:

Case (*i*)

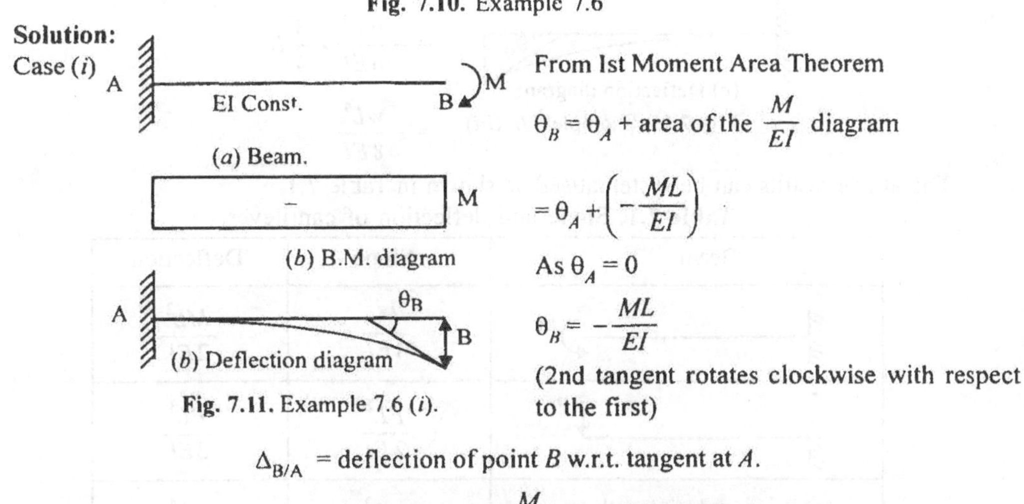

(*a*) Beam.

(*b*) B.M. diagram

(*b*) Deflection diagram

Fig. 7.11. Example 7.6 (*i*).

From Ist Moment Area Theorem

$$\theta_B = \theta_A + \text{area of the } \frac{M}{EI} \text{ diagram}$$

$$= \theta_A + \left(-\frac{ML}{EI}\right)$$

As $\theta_A = 0$

$$\theta_B = -\frac{ML}{EI}$$

(2nd tangent rotates clockwise with respect to the first)

$\Delta_{B/A}$ = deflection of point B w.r.t. tangent at A.

= Moment of the $\frac{M}{EI}$ diagram between A and B about B.

$$= -\frac{ML}{EI} \cdot \frac{L}{2} = \frac{-ML^2}{2EI}$$ (Point B moves downward w.r.t. 1st tangent)

Case (*ii*)

From 1st Moment Area Theorem.

$$\theta_B = \theta_A + \text{area of the } \frac{M}{EI} \text{ diagram}$$

$$= -\frac{1}{2} \cdot \frac{PL}{EI} \cdot L = -\frac{PL^2}{2EI}$$

with $\theta_A = 0$

$$\theta_B = -\frac{PL^2}{2EI}$$

(2nd tangent rotates clockwise w.r.t. the first).

A P B

(*a*) Beam.

– M

(*b*) B.M. diagram.

A θB B

(*c*) Deflection diagram.

Fig. 7.12. Example 7.6. (*ii*).

Deflection of point B w.r.t. tangent at A

$$= \text{Moment of the } \frac{M}{EI} \text{ diagram between } A \text{ and } B \text{ about } B.$$

$$= -\frac{1}{2} \times \frac{PL^2}{EI} \times \frac{2}{3} \times L = -\frac{PL^3}{3EI}$$ (Point B moves downward w.r.t. tangent at A)

Case (*iii*)

w

(*a*) Beam.

$\frac{wL^2}{2}$ –

(*b*) B.M. diagram.

θ_B

(*c*) Deflection diagram.

Fig. 7.13. Example 7.6. (*iii*)

$$\theta_B - \theta_A = \text{Area of the } \frac{M}{EI} \text{ diagram between } A \text{ and } B$$

$$= -\frac{1}{3} \cdot \frac{wL^2}{2} \cdot \frac{L}{EI} = -\frac{wL^3}{6EI}$$

Deflection of B w.r.t. tangent at A

$$= -\frac{wL^3}{6EI} \times \frac{3}{4} L$$

$$= -\frac{wL^4}{8EI}$$

The above results can be systematised as shown in Table 7.1.

Table 7.1. Slope and deflection of cantilever

Beam	Slope	Deflection
M	$\frac{ML}{1EI}$	$\frac{ML^2}{2EI}$
P	$\frac{PL^2}{2EI}$	$\frac{PL^3}{3EI}$
w/unit length	$\frac{wL^3}{6EI}$	$\frac{wL^4}{8EI}$

The numbers in the denominator 12 23 68 will help in memorising the results.

Example 7.7. Find by moment area method slopes at ends and deflections at the middle of the following beams. EI constant.

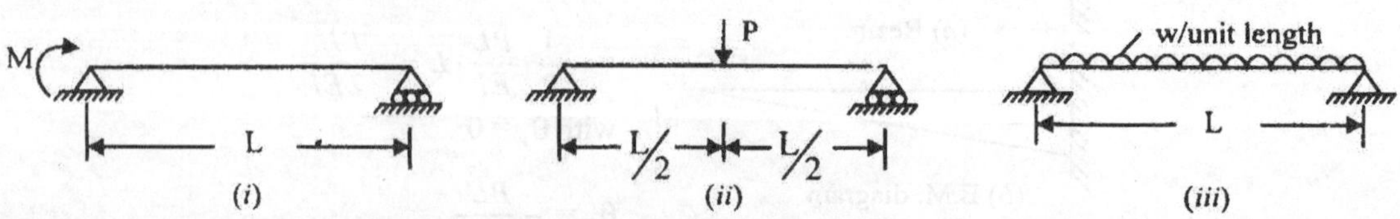

Fig. 7.14. Example 7.7

Solution: case (*i*)

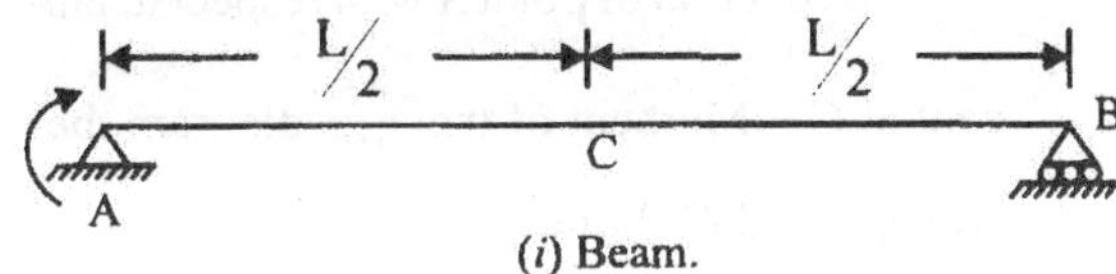

(*i*) Beam.

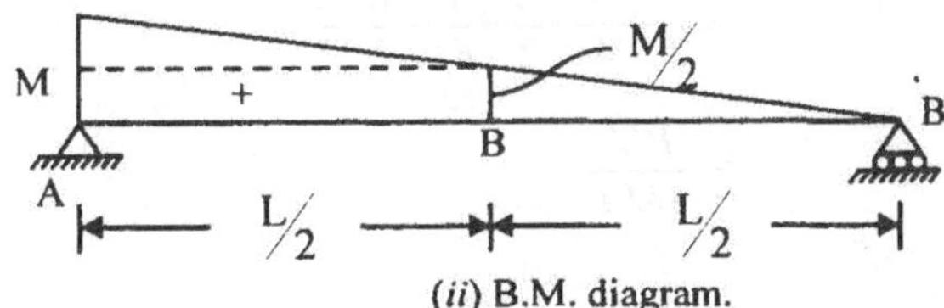

(*ii*) B.M. diagram.

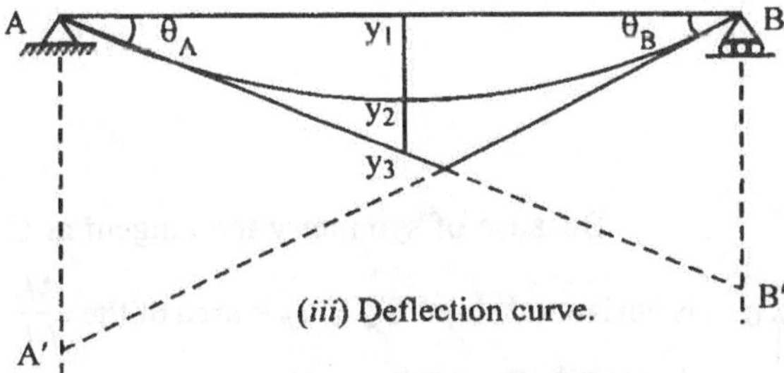

(*iii*) Deflection curve.

Fig. 7.15. Example 7.7 (*i*)

As there is no horizontal tangent, it is not possible to directly compute θ_A or θ_B. However, they can be computed in an indirect way. (Fig. 7.15).

BB' = deflection of point B w.r.t. tangent at A

= Moment of the $\dfrac{M}{EI}$ diagram between A and B about B.

$$= \frac{1}{2} \cdot \frac{ML}{EI} \cdot \frac{2}{3} \cdot L = \frac{ML^2}{3EI}$$

As θ_A is small, taking $AB = AB' = L$

$$\theta_A = \frac{ML}{3EI}$$

Similarly, taking moment about A,

$$AA' = \frac{1}{2} \cdot \frac{ML}{EI} \cdot \frac{1}{3} \cdot L = \frac{ML^2}{6EI}$$

Deflection at $C \left(x = \dfrac{L}{2} \right)$

$$y_1y_2 = y_1y_3 - y_2y_3$$

y_2y_3 = deflection of point C with respect to tangent at A.

= Moment of the $\dfrac{M}{EI}$ diagram between A and C about C

$$= \frac{1}{EI}\left[\frac{M}{2} \cdot \frac{L}{2} \cdot \frac{L}{4} + \frac{1}{2} \cdot \frac{M}{2} \cdot \frac{L}{2} \cdot \frac{2}{3} \cdot \frac{L}{2}\right]$$

$$= \frac{ML^2}{EI} \cdot \frac{5}{48}$$

Actual deflection $\quad y_1y_2 = y_1y_3 - y_2y_3$

$$= \theta_A \cdot \frac{L}{2} - \frac{ML^2}{EI} \cdot \frac{5}{48}$$

$$= \frac{ML}{3EI} \cdot \frac{L}{2} - \frac{ML^2}{EI} \cdot \frac{5}{48}$$

$$= \frac{ML^2}{16EI}.$$

Case (*ii*) (Fig. 7.16)

As the load is applied at the middle point, the tangent under the load is horizontal. Change of slope between

$$A \text{ and } C = \frac{1}{2} \cdot \frac{L}{2} \cdot \frac{PL}{4} \cdot \frac{1}{EI} = \frac{PL^2}{16EI}$$

As tangent at C is horizontal, $\theta_A = \dfrac{PL^2}{16EI}$ and 2nd tangent at C makes an anticlockwise rotation with respect to the 1st tangent at A.

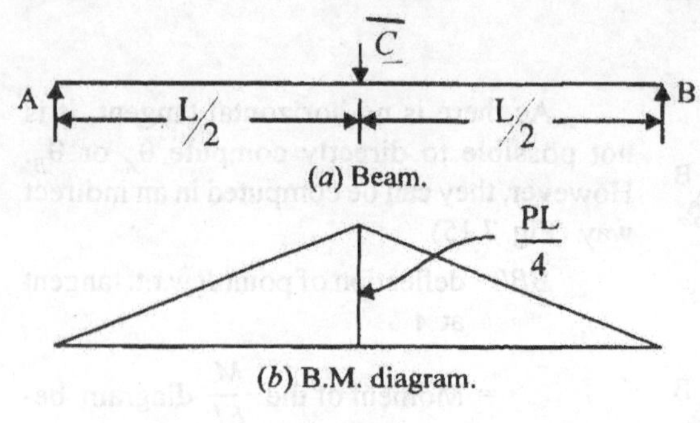

(a) Beam.

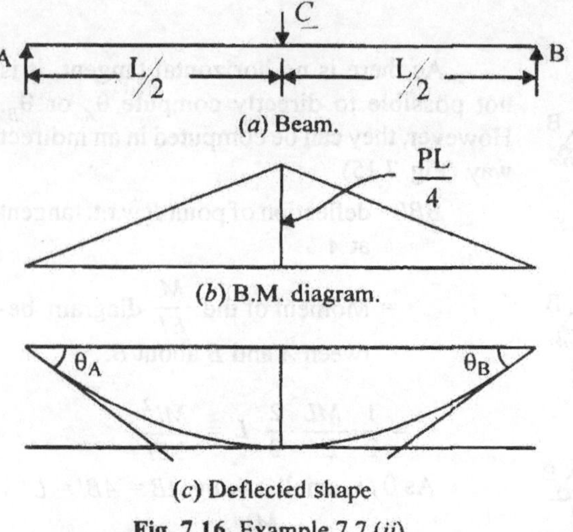

(b) B.M. diagram.

(c) Deflected shape

Fig. 7.16. Example 7.7 (*ii*)

Deflection of point C:

Delfection of point A with respect to tangent at C = Moment of the $\frac{M}{EI}$ diagram between A and C about A.

$$= \frac{1}{2}\cdot\frac{PL}{4}\cdot\frac{L}{2}\cdot\frac{1}{EI}\cdot\frac{L}{2}\cdot\frac{2}{3}$$

$$= \frac{PL^3}{48EI}.$$

Case (*iii*) (Fig. 7.17)

(*i*) Beam.

(*ii*) B.M. diagram.

(*iii*) Deflection curve.

Fig. 7.17. Example 7.7 (*iii*)

Because of symmetry the tangent at C is horizontal slope $\theta_A = \theta_B$ = area of the $\frac{M}{EI}$ diagram between.

$$A \text{ and } C = \frac{2}{3}\times\frac{L}{2}\times\frac{wL^2}{8EI}$$

$$= \frac{wL^3}{24EI} = \theta_A = \theta_B$$

Deflection of point A w.r.t. tangent at C

= Moment of the $\frac{M}{EI}$ diagram between A and C about A.

$$= \frac{wL^3}{24EI}\times\frac{5}{8}\times\frac{L}{2} = \frac{5wL^4}{384EI}.$$

7.6. AREAS AND CENTRE OF GRAVITIES OF STANDARD FIGURES

Moment Area Methods require areas of bending moment diagrams and their centre of gravities to compute moment of areas. While this can always be obtained by integration formulae areas and C.G's of standard figures are also available. These are given in Table 7.2 for ready reference.

Table 7.2 Areas and Centre of gravities of Standard Figures.

Figure	Area	Centroid
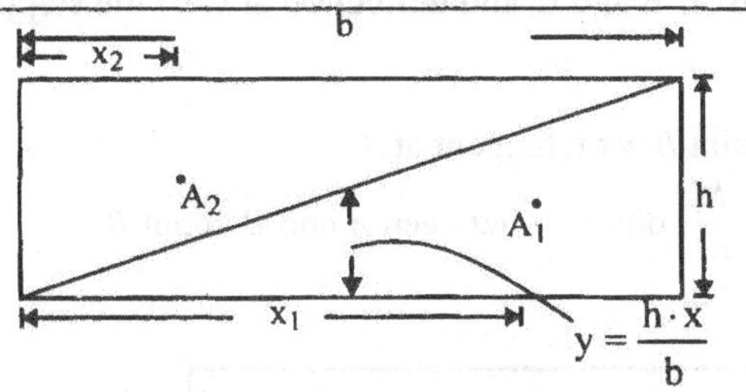 Rectangle of width b and height h divided by a diagonal into areas A_1 and A_2; $y = \frac{h \cdot x}{b}$; x_1, x_2	$A_1 = \frac{bh}{2}$ $A_2 = \frac{bh}{2}$	$x_1 = \frac{2}{3} \cdot b$ $x_2 = \frac{1}{3} \cdot b$
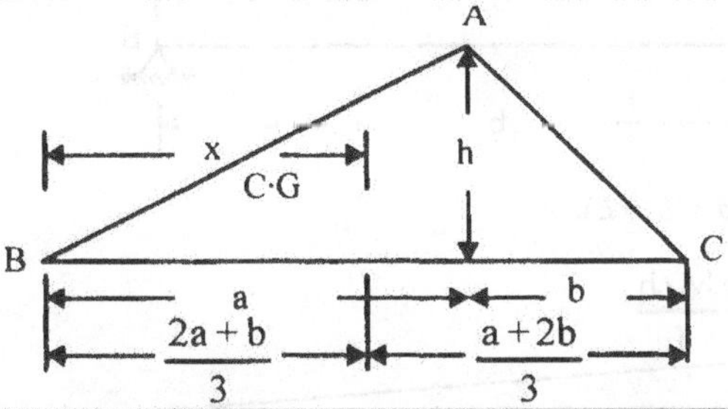 Triangle ABC, height h, base divided into a and b; C·G at x; $\frac{2a+b}{3}$, $\frac{a+2b}{3}$	$A = \frac{1}{2}(a+b) \cdot h$	$x = \frac{2a+b}{3}$
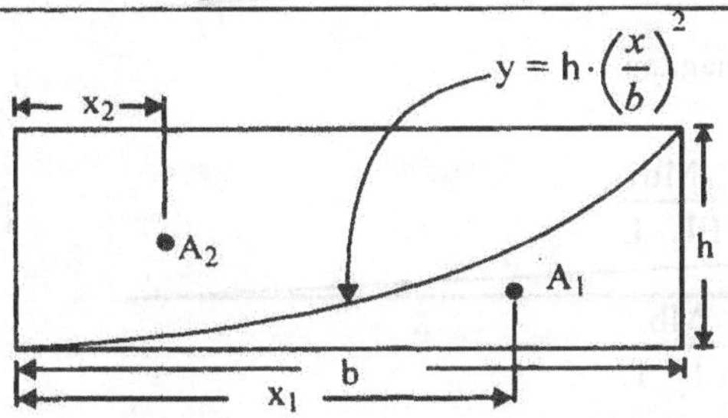 Rectangle b × h divided by the curve $y = h \cdot \left(\frac{x}{b}\right)^2$ into A_1 and A_2; x_1, x_2	$A_1 = \frac{bh}{3}$ $x_1 = \frac{3}{4} b$	$A_2 = \frac{2bh}{3}$ $x_2 = \frac{3}{8} b$
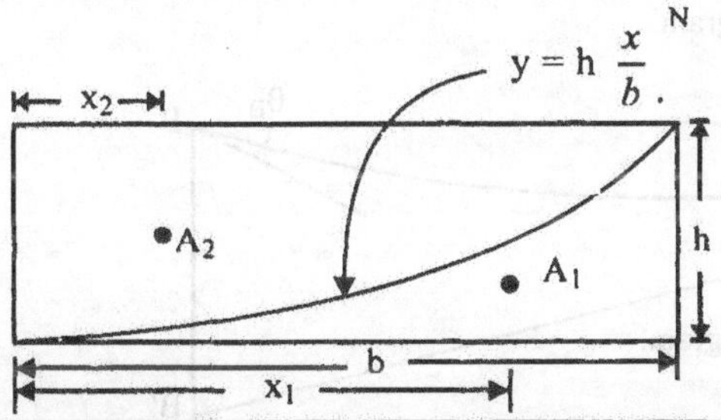 Rectangle b × h divided by the curve $y = h \frac{x}{b}$ (power N) into A_1 and A_2; x_1, x_2	$A_1 = \frac{bh}{N+1}$ $x_1 = \frac{N+1}{N+2} . b$	$A_2 = \frac{N\, bh}{N+1}$ $x_2 = \frac{N+1}{N+2} \cdot \frac{b}{2}$

7.7 FURTHER EXAMPLES ON MOMENT AREA METHODS

In the previous examples, simple cantilevers and simply supported beams with constant *EI* and subjected to moment, concentrated load and uniformly distributed load were studied. Moment Area Methods can also be applied to beams with overhangs, beams with internal hinges, beams with variable *EI*. In such cases moment area methods should be applied to different parts of a beam. (It cannot be applied straight way to a beam with internal hinge) separately and the part results superposed to get the final results. When there are a large number of loads, the B.M. diagram should be drawn in parts. One way of doing this will be to imagine one end of the beam to be a cantilever and to draw cantilever B.M. diagrams for all the loads. Total bending moment will be summation of all these part bending moments which should satisfy the boundary conditions of bending moments.

Example 7.8. Compute the rotations of points *A*, *B* and *C* and deflection at *C* of the stepped beam shown due to the applied end moment *M*.

Solution:

$$BB' = \text{Deflection of point } B \text{ w.r.t. tangent at } A.$$

$$= \text{Moment of the } \frac{M}{EI} \text{ diagram between } A \text{ and } B \text{ about } B.$$

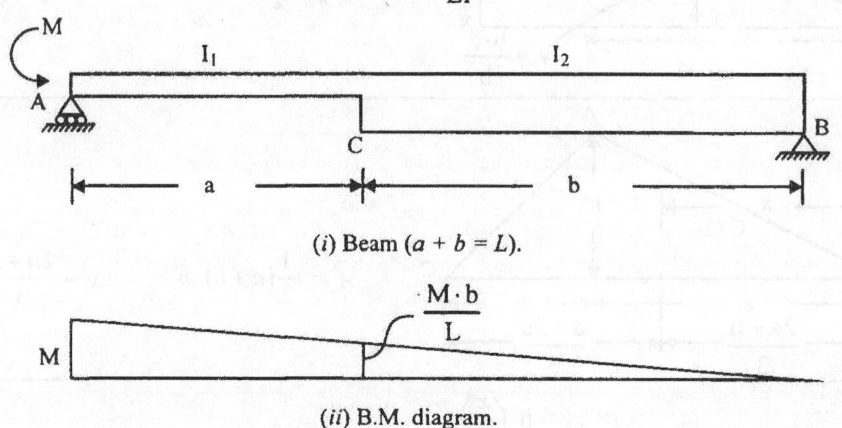

(*i*) Beam ($a + b = L$).

(*ii*) B.M. diagram.

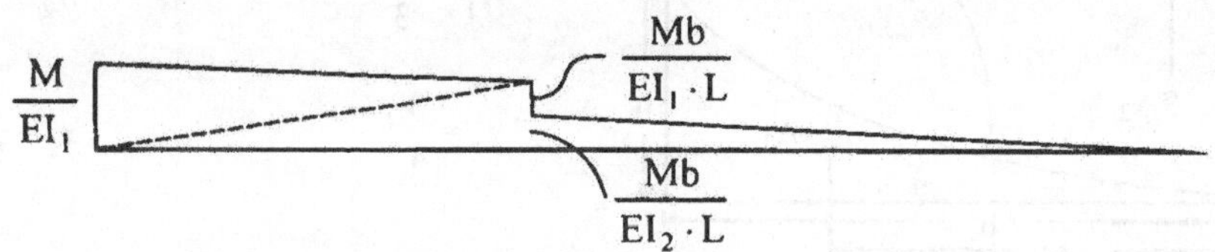

(*iii*) M/EI diagram.

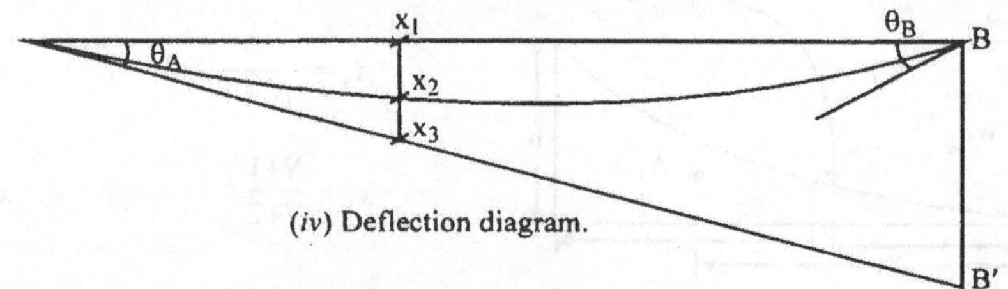

(*iv*) Deflection diagram.

Fig. 7.18. Example 7.8

$$= \frac{1}{2} \cdot \frac{Mb}{EI_2 \cdot L} \cdot b \times \frac{2}{3} \cdot b + \frac{1}{2} \cdot \frac{Mb}{EI_1 \cdot L} \cdot a \cdot \left(b + \frac{a}{3}\right)$$

$$+ \frac{1}{2} \cdot \frac{M}{EI_1} \cdot a \cdot \left(b + \frac{2a}{3}\right)$$

$$\theta_A = \frac{1}{L}\left[\frac{Mb^3}{3EI_2 \cdot L} + \frac{Mba}{6EI_1 \cdot L}(3b + a) + \frac{Ma}{6EI_1}(3b + 2a)\right]$$

Deflection of point A w.r.t. tangent at B.

$$= \text{Moment of the } \frac{M}{EI} \text{ diagram between } A \text{ and } B \text{ about } A.$$

$$= \frac{1}{2}\cdot\frac{M}{EI_1}\cdot a\cdot\frac{a}{3}+\frac{1}{2}\cdot\frac{Mb}{EI_1\cdot L}\cdot a\cdot\frac{2a}{3}+\frac{1}{2}\cdot\frac{Mb}{EI_2\cdot L}\cdot b\left(a+\frac{b}{3}\right)$$

$$\theta_B = \frac{1}{L}\left[\frac{Ma^2}{6EI_1}+\frac{Ma^2b}{3EI_1\cdot L}+\frac{Mb^2(3a+b)}{6EI_2\cdot L}\right]$$

Rotation at C

$$\theta_C = \theta_A - \text{area of the } \frac{M}{EI} \text{ diagram between } A \text{ and } C$$

$$= \frac{1}{L}\left[\frac{Mb^3}{3EI_2\cdot L}+\frac{Mab(3b+a)}{6EI_1\cdot L}+\frac{Ma(3b+2a)}{6EI_1}\right] - \frac{1}{2}\left[\frac{M}{EI_1}+\frac{Mb}{EI_1 L}\right]\cdot a$$

Deflection at C [Fig. 7.8 (*iv*)]

$$\Delta_c = x_1x_3 - x_2x_3$$

$$= \theta_A\cdot a - \left[\frac{1}{2}\cdot\frac{M}{EI_1}\cdot a\cdot\frac{2a}{3}+\frac{1}{2}\cdot\frac{Mb}{EI_1\cdot L}\cdot a\cdot\frac{a}{3}\right]$$

$$= \frac{a}{L}\left[\frac{Mb^3}{3EI_2\cdot L}+\frac{Mab}{6EI_1\cdot L}(3b+a)+\frac{Ma}{6EI_1}(3b+2a)\right] - \left[\frac{Ma^2}{3EI_1}+\frac{Ma^2b}{6EI_1\cdot L}\right].$$

Example 7.9. Compute the rotations of points A, B, C and D and displacements of points B and D of the beam given in Fig. 7.19. $E = 200 \times 10^6$ kN/m² $\quad I = 12 \times 10^{-6}$ m⁴.

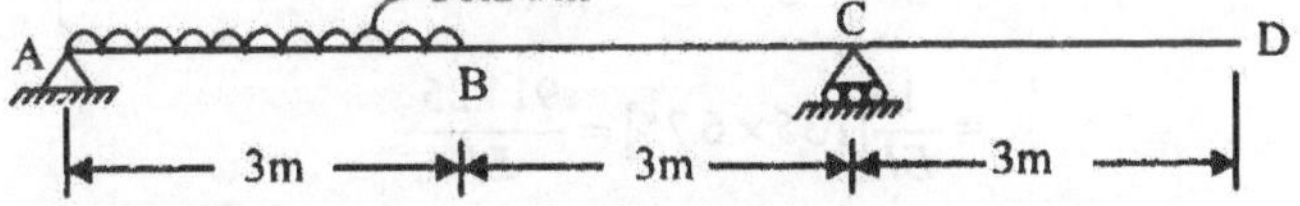

30kN/m, A, B, C, D, 3m, 3m, 3m

Fig. 7.19. Example 7.9

Taking moments about C

$$R_A \times 6 - 30\times 3\times 4.5 = 0$$

or $\quad R_A = \dfrac{30\times 3\times 4.5}{6} = 67.5$ kN.

$$R_C \times 6 - 30\times 3\times 1.5 = 0$$

or $\quad R_C = \dfrac{30\times 3\times 1.5}{6} = 22.5$ kN.

Check $\quad 67.5 + 22.5 = 90.00$ kN.

Imagining point A as the fixed end of a cantilever $ABCD$, the BM diagram is drawn by parts.

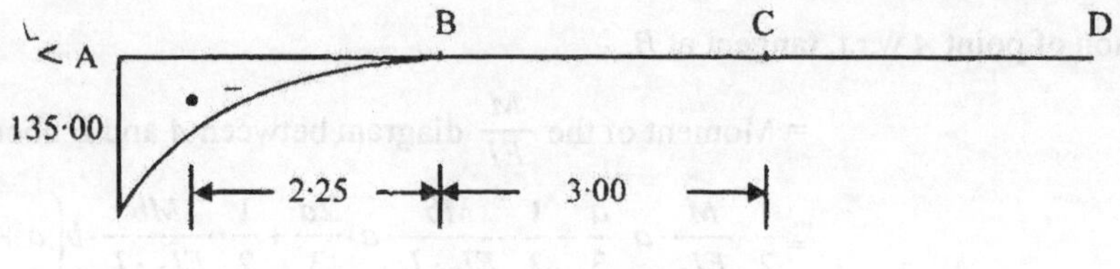

(*i*) B.M. diagram due to 30kN/m.

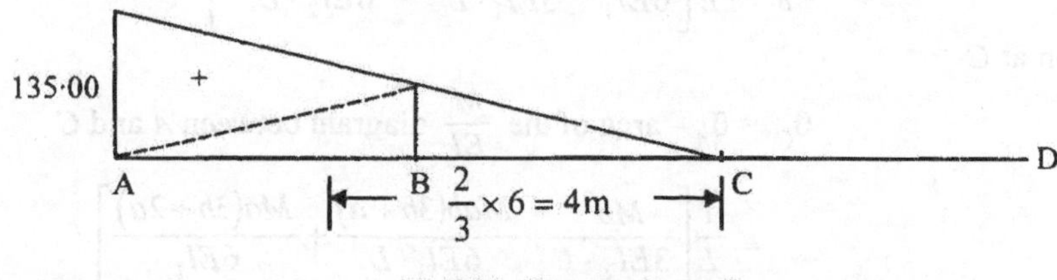

(*ii*) B.M. diagram due to R_C.

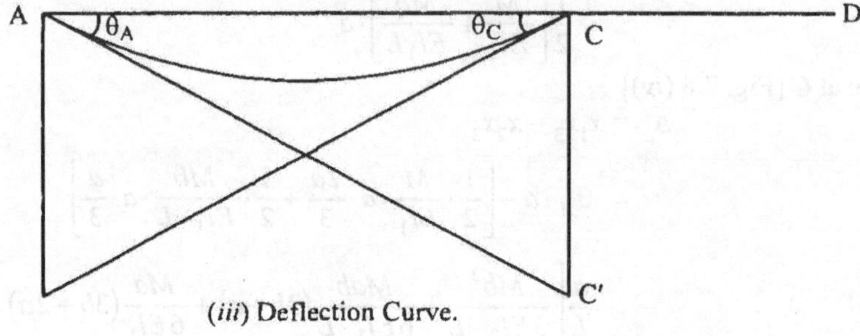

(*iii*) Deflection Curve.

Fig. 7.20. Part *BM* diagrams and deflection curve. (Example 7.9)

Deflection of point *C* w.r.t. tangent at *A*

$$= \text{Moment of the } \frac{M}{EI} \text{ diagram between } A \text{ and } C \text{ about } C$$

$$= \frac{1}{EI}\left[-\frac{1}{3}\times 135\times 3\times 5.25+\frac{1}{2}\times 135\times 6\times 4\right]$$

$$= \frac{1}{EI}[135\times 6.75] = \frac{911.25}{EI}$$

$$\theta_A = \frac{911.25}{6EI} = \frac{911.25}{6\times 200\times 10^6\times 12\times 10^{-6}}$$

$$= 0.06328 \text{ radian (clockwise)}$$

Deflection of *A* w.r.t. tangent at *C*

$$= \text{Moment of the } \frac{M}{EI} \text{ diagram between } A \text{ and } C \text{ about } A$$

$$= \frac{1}{EI}\left(-\frac{1}{3}\times 135\times 3\times 0.75+\frac{1}{2}\times 135\times 6\times 2\right)$$

$$= \frac{1}{EI}(708.75).$$

$$\theta_C = \frac{708.75}{6EI} = \frac{708.75}{6 \times 200 \times 10^6 \times 12 \times 10^{-6}}$$
$$= 0.049218 \text{ radian (anticlokwise)}$$

$$\theta_B = \theta_A - \text{Area of the } \frac{M}{EI} \text{ diagram between } A \text{ and } B$$

$$= 0.06328 - \left(-\frac{1}{3} \times \frac{135 \times 3}{200 \times 10^6 \times 12 \times 10^{-6}} + \frac{101.25 \times 3}{200 \times 12}\right)$$
$$= 0.06328 - (-0.05625 + 0.12656)$$
$$= 0.007030 \text{ radian (anticlockwise)}$$

$$\theta_C = 0.06328 - \left(-0.05625 + \frac{1}{2} \times \frac{135 \times 6}{200 \times 10^6 \times 12 \times 10^{-6}}\right)$$
$$= 0.06328 + 0.05625 - 0.16875$$
$$= -0.049218 \text{ (anticlockwise)}$$

$$\text{Area of the } \frac{M}{EI} \text{ diagrams} = \left(-\frac{1}{3} \times 135 \times 3 + \frac{1}{2} \times 135 \times 6\right)\frac{1}{EI}$$
$$= (-135 + 135 \times 3)\frac{1}{EI}$$
$$= \frac{2 \times 135}{200 \times 10^6 \times 12 \times 10^{-6}} = 0.1125$$

$$\text{Sum of } \theta_A \text{ and } \theta_C = 0.06328 + 0.049218$$
$$= 0.112498 \text{ (check)}$$

Deflection of B and D.

Deflection of B w.r.t. tangent at A

$$= \text{Moment of the } \frac{M}{EI} \text{ diagram between } A \text{ and } B \text{ about } B.$$

$$= \frac{1}{EI}\left(-\frac{1}{3} \times 135 \times 3 \times \times 2.25 + \frac{1}{2} \times 135 \times 3 \times 2 + \frac{1}{2} \times 67.5 \times 3 \times 1\right)$$

$$= \frac{1}{EI}(202.5)$$

$$\text{Actual Deflection} \quad \Delta_B = \theta_A \times 3 - \frac{1}{EI}(202.5)$$
$$= \frac{911.25 \times 3}{6EI} - \frac{202.5}{EI}$$
$$= \frac{1}{EI}(455.625 - 202.5)$$
$$= 0.10546\, m.$$

$$\text{Deflection at} \quad D = \theta_C \times 3\, m$$
$$= 0.049218 \times 3$$
$$= 0.14765\, m \text{ (upward)}$$

Example 7.10. Determine the slope and vertical deflections at C and B of the cantilever beam.

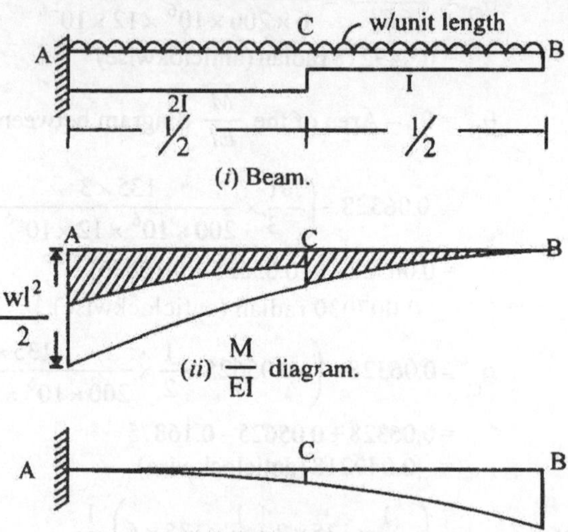

Fig. 7.21. Example 7.10

Solution:

Change of slope between A and C (origin at B)

$$\theta_C - \theta_A = \text{Area of the } \frac{M}{EI} \text{ diagram between } A \text{ and } C$$

$$= -\int_{l/2}^{l} \frac{wx^2}{2 \times 2EI} \cdot dx$$

$$= -\frac{w}{4EI}\left(\frac{x^3}{3}\right)_{l/2}^{l}$$

$$= -\frac{wl^3}{EI} \cdot \frac{7}{96} \qquad \text{As } \theta_A = 0 \qquad \theta_C = -\frac{wl^3}{EI} \cdot \frac{7}{96} \text{ (Clockwise)}$$

Change of slope between A and B

$$\theta_B - \theta_A = \text{Area of the } \frac{M}{EI} \text{ diagram between } A \text{ and } B$$

$$= -\frac{wl^3}{EI} \cdot \frac{7}{96} + \int_0^{l/2} -\frac{wx^2}{2EI} \cdot dx$$

$$= -\frac{wl^3}{EI} \cdot \frac{7}{96} + (-) \cdot \frac{w}{2EI}\left(\frac{x^3}{3}\right)_0^{l/2}$$

$$= -\frac{wl^3}{EI} \cdot \frac{3}{32}.$$

As $\theta_A = 0$ $\qquad \theta_B = -\frac{wl^3}{EI} \cdot \frac{3}{32}$ (clockwise).

Deflection of *C*.

Deflection of *C* w.r.t. tangent at *A*

$$= \text{Moment of the } \frac{M}{EI} \text{ diagram between } A \text{ and } C \text{ about } C.$$

$$= -\int_{l/2}^{l} \frac{wx^2}{2\times 2EI}\left(x-\frac{l}{2}\right)\cdot dx$$

$$= -\frac{w}{4EI}\left(\int_{l/2}^{l} x^3 \cdot dx\right) - \left[(-)\frac{wl}{8EI}\int_{l/2}^{l} x^2 \cdot dx\right]$$

$$= -\frac{w}{4EI}\left(\frac{x^4}{4}\right)_{l/2}^{l} + \frac{wl}{8EI}\left(\frac{x^3}{3}\right)_{l/2}^{l}$$

$$= -\frac{wl^4}{EI}\cdot\frac{15}{256} + \frac{wl^4}{EI}\cdot\frac{7}{192}$$

$$= \frac{wl^4}{EI}\left(\frac{-45+28}{12\times 64}\right) = -\frac{wl^4}{EI}\cdot\frac{17}{12\times 64} \text{ (downward)}$$

Deflection of *B*.

Deflection of *B* w.r.t. tangent at *A*

$$= \text{Moment of the } \frac{M}{EI} \text{ diagram between } A \text{ and } B \text{ about } B.$$

$$= -\int_{l/2}^{l} \frac{wx^2}{4EI}\cdot x \cdot dx + \int_{0}^{l/2} -\frac{wx^2}{2EI}\cdot x \cdot dx$$

$$= -\left[\frac{wl^4}{EI}\cdot\frac{15}{256} + \frac{w}{2EI}\left(\frac{x^4}{4}\right)_0^{l/2}\right]$$

$$= -\frac{wl^4}{EI}\left(\frac{15}{256} + \frac{1}{128}\right)$$

$$= -\frac{wl^4}{EI}\cdot\frac{17}{256} \text{ (downward).}$$

Example 7.11. Calculate the critical displacements and rotations of the frame shown in Fig. 7.22. Draw the slope and deflection diagrams. *EI* constant.

Solution:

Taking moments about *B*.

$$R_D \times 6 - 30 \times 4 = 0$$

or $$R_D = \frac{30\times 4}{6} = 20 \text{ kN}.$$

Putting $\Sigma V = 0$ $\quad R_B + R_D = 30$

$$R_B = 30 - 20 = 10 \text{ kN}.$$

The B.M. diagram is shown in Fig. 7.22 (iii)

$$\Delta_{BA} = \text{displacement of point } B \text{ with respect to tangent at } A$$

$$= BB_1$$

$$= \text{Moment of the } \frac{M}{EI} \text{ diagram between } A \text{ and } B \text{ about } B.$$

$$= -\frac{1}{2}\times 20\times 2\times \frac{2}{3}\times \frac{2}{EI} + (-)\frac{1}{2}\times 10\times 1\times \frac{5}{3}$$

$$= \frac{1}{EI}\left(-\frac{80}{3} - \frac{25}{3}\right) = -\frac{105}{3EI} = -\frac{35}{EI}$$

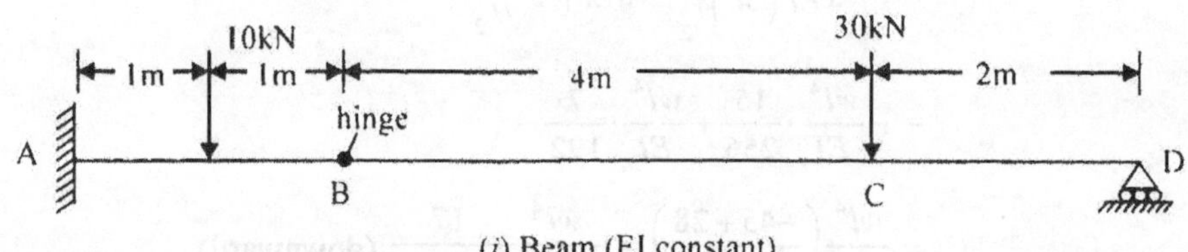

(*i*) Beam (EI constant).

(*ii*) F.B.D.

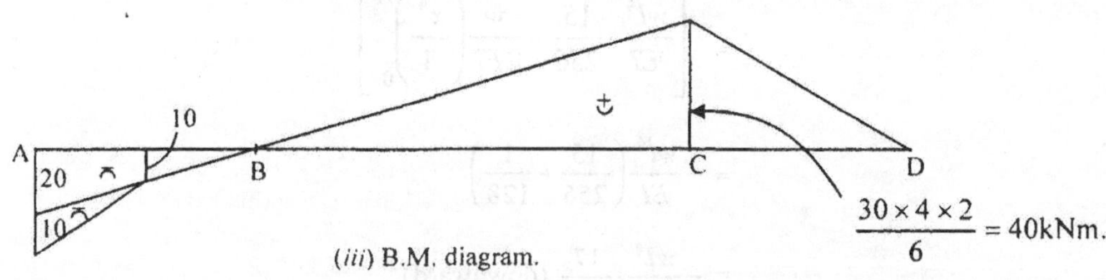

(*iii*) B.M. diagram.

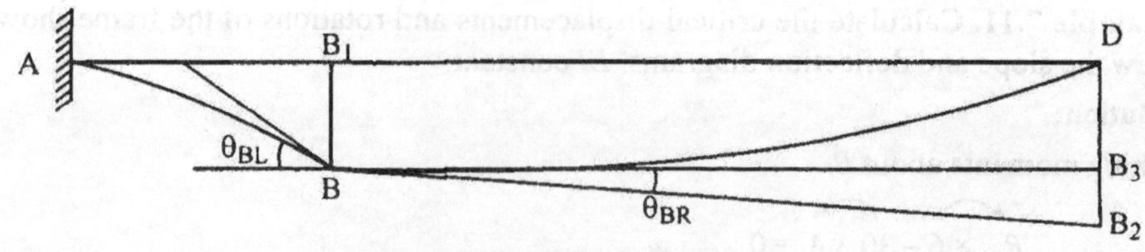

(*iv*) Deflection diagram.

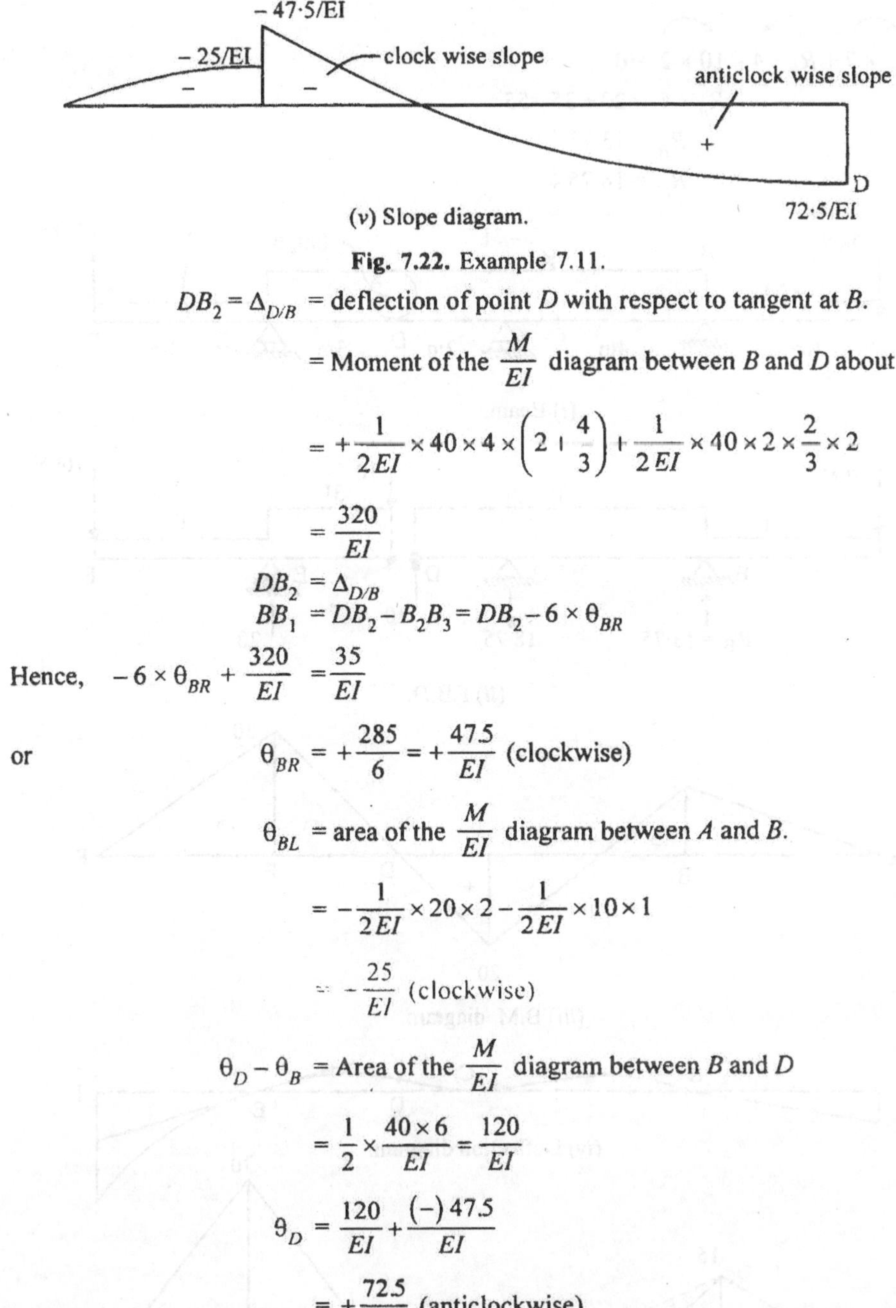

(v) Slope diagram.

Fig. 7.22. Example 7.11.

$$DB_2 = \Delta_{D/B} = \text{deflection of point } D \text{ with respect to tangent at } B.$$

$$= \text{Moment of the } \frac{M}{EI} \text{ diagram between } B \text{ and } D \text{ about } D.$$

$$= +\frac{1}{2EI} \times 40 \times 4 \times \left(2 + \frac{4}{3}\right) + \frac{1}{2EI} \times 40 \times 2 \times \frac{2}{3} \times 2$$

$$= \frac{320}{EI}$$

$$DB_2 = \Delta_{D/B}$$

$$BB_1 = DB_2 - B_2B_3 = DB_2 - 6 \times \theta_{BR}$$

Hence, $$-6 \times \theta_{BR} + \frac{320}{EI} = \frac{35}{EI}$$

or $$\theta_{BR} = +\frac{285}{6} = +\frac{47.5}{EI} \text{ (clockwise)}$$

$$\theta_{BL} = \text{area of the } \frac{M}{EI} \text{ diagram between } A \text{ and } B.$$

$$= -\frac{1}{2EI} \times 20 \times 2 - \frac{1}{2EI} \times 10 \times 1$$

$$= -\frac{25}{EI} \text{ (clockwise)}$$

$$\theta_D - \theta_B = \text{Area of the } \frac{M}{EI} \text{ diagram between } B \text{ and } D$$

$$= \frac{1}{2} \times \frac{40 \times 6}{EI} = \frac{120}{EI}$$

$$\theta_D = \frac{120}{EI} + \frac{(-)\,47.5}{EI}$$

$$= +\frac{72.5}{EI} \text{ (anticlockwise).}$$

Example 7.12. Calculate the dispalcements and slopes at the critical points of the beam shown in Fig. 7.23.

Solution:

$$-R_E \times 3 + 10 \times 6 = 0$$

or $$R_E = 20 \text{ kN} \uparrow$$

$$-5 \times 7 + R_B \times 4 - 10 \times 2 = 0$$

or $$R_B \times 4 = 20 + 35 = 55$$

$$R_B = 13.75 \uparrow$$

$$R_C = 18.75 \downarrow.$$

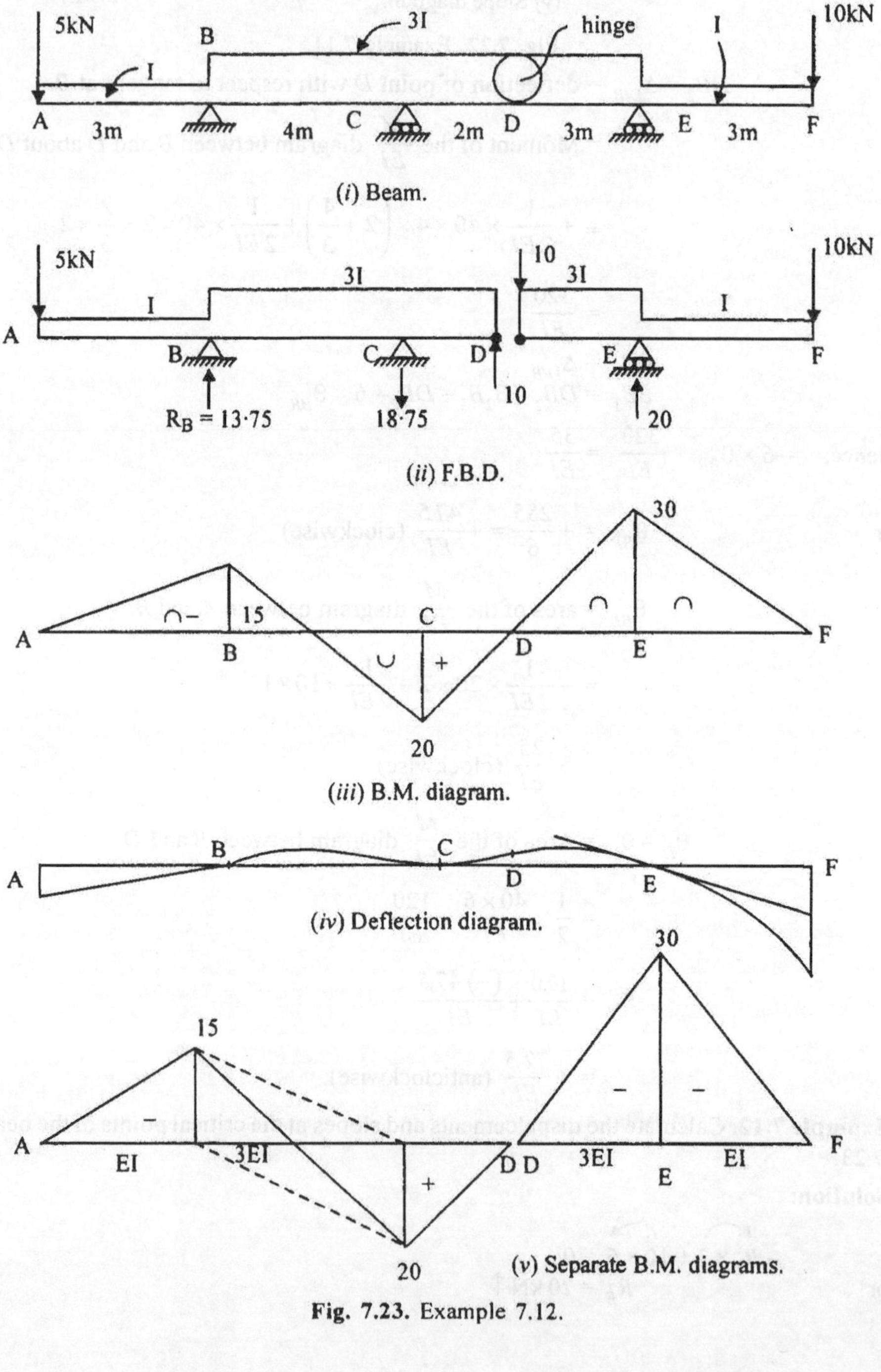

(*i*) Beam.

(*ii*) F.B.D.

(*iii*) B.M. diagram.

(*iv*) Deflection diagram.

(*v*) Separate B.M. diagrams.

Fig. 7.23. Example 7.12.

$$\text{Slope at } B = \frac{1}{L}\text{ [deflection of } C \text{ w.r.t. tangent at } B]$$

$$= \frac{1}{4}\left[-\frac{1}{2}\times 15\times 4\times\frac{8}{3}\times\frac{1}{3EI}+\frac{1}{2}\times 20\times 4\times\frac{4}{3}\times\frac{1}{3EI}\right]$$

$$= \frac{1}{3\times 4EI}\left[-80+\frac{160}{3}\right] = -\frac{20}{9EI}\text{ (anticlockwise)}$$

$$\text{Deflection at } A = \frac{1}{12EI}\left[-\frac{80}{3}\right]\times 3-\frac{1}{2}\times\frac{3\times 15}{EI}\times\frac{2}{3}\times 3$$

$$= -\frac{51.67}{EI} = \frac{51.67}{EI}\text{ downward}$$

$$\theta_A = -\frac{20}{9EI}-\frac{1}{2}\times\frac{15}{EI}\times 3$$

$$= -\frac{24.72}{EI}\text{ (anticlockwise)}$$

$$\theta_C = \frac{1}{L}\text{ [deflection of point B w.r.t. tangent at } C]$$

$$= \frac{1}{3\times 4\,EI}\left[-\frac{1}{2}\times 15\times 4\times\frac{4}{3}+\frac{1}{2}\times 20\times 4\times\frac{8}{3}\right]$$

$$= \frac{5.56}{EI}\text{ (anticlockwise)}$$

$$\theta_{DL} = \theta_C+\frac{1}{2}\times 20\times\frac{2}{3\,EI}$$

$$= \frac{5.56}{EI}+\frac{6.67}{EI} = \frac{12.23}{EI}\text{ (anticlockwise)}$$

Deflection at $D = \theta_C\times 2$ + deflection of D with respect to tangent at C

$$= \frac{5.56\times 2}{EI}+\frac{1}{2}\times\frac{20\times 2}{3\,EI}\times\frac{2}{3}\times 2$$

$$= \frac{20.00}{EI}\text{ upward}$$

Deflection at D left = Deflection at D right

$= \theta_E\times 3$ + deflection of point D w.r.t. tangent at E

$$= \theta_E\times 3+\left(-\frac{1}{2}\times\frac{30\times 3}{3\,EI}\times\frac{2}{3}\times 3\right)$$

$$= \theta_E\times 3-\frac{30}{EI}$$

Hence
$$\frac{20.00}{EI} = \theta_E\times 3-\frac{30}{EI}$$

or $$\theta_E = \frac{16 \cdot 67}{EI} \text{ (clockwise)}$$

$$\text{Slope at } F = \frac{16.67}{EI} + \frac{1}{2} \times \frac{30 \times 3}{EI}$$

$$= \frac{1}{EI}[16.67 + 45] = \frac{61.67}{EI} \text{ (clockwise)}$$

$$\text{Deflection at } F = \theta_E \times 3 + \text{deflection of } F \text{ w.r.t. tangent at } E$$

$$= \frac{16.67 \times 3}{EI} + \frac{1}{2} \times 30 \times \frac{3}{EI} \times 2$$

$$= \frac{1}{EI}[50.00 + 90]$$

$$= \frac{1}{EI}[140.00] \text{. (downward)}$$

Example 7.13. For the statically determinate beam of Fig. 7.24 find the deflections at B and D and slopes at B and C. EI constant .

Solution. Taking moment of all forces about B for the part on the right of B

$$R_C \times 6b - 3P \times 4b = 0$$

or $$R_C = \frac{12\,Pb}{6b} = 2P$$

Hence $$R_B = P$$

Deflection at $$B = \text{Moment of the } \frac{M}{EI} \text{ diagram between } A \text{ and } B \text{ about } B.$$

$$= \frac{1}{EI}\left[\frac{1}{2} \cdot 2Pb \cdot 2b \times \frac{4b}{3} + \frac{1}{2} \cdot Pb \cdot b \cdot \left(b + \frac{2b}{3}\right)\right]$$

$$= \frac{1}{EI} \frac{7Pb^2}{2} = \frac{3.5\,Pb^3}{EI}$$

(*i*) Beam.

(*ii*) Beam.

$\frac{3P \times 4b \times 2b}{6\,b} = 4Pb$

(*iii*) B.M. diagram.

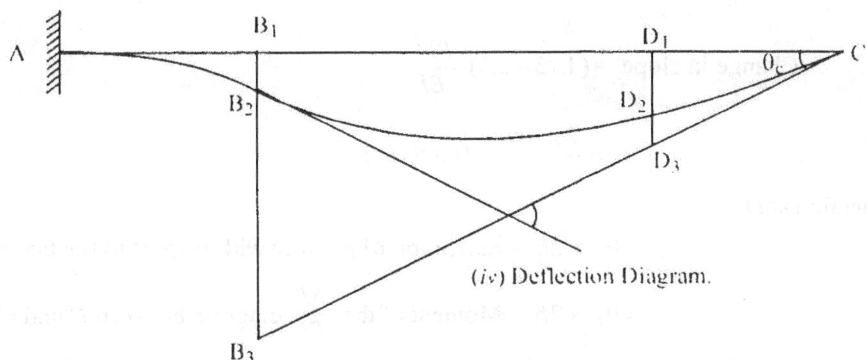

Fig. 7.24. Example 7.13.

$$\text{Slope at left of } B = \text{area of the } \frac{M}{EI} \text{ diagram between } A \text{ and } B$$

$$= \frac{1}{EI}\left[\frac{1}{2}\cdot 2Pb\cdot 2b + \frac{1}{2}Pb\cdot b\right]$$

$$= \frac{2.5\,Pb^2}{EI} \text{ (clockwise)}$$

$$6b(\theta_C) = \text{Deflection of } B \text{ with respect to tangent at } C + B_1B_2$$

$$= \text{Moment of the } \frac{M}{EI} \text{ diagram between } B \text{ and } C \text{ about } B + B_1B_2$$

$$= \frac{1}{EI}\left[\frac{1}{2}\cdot 4\,Pb \times 4b \times \frac{8b}{3} + \frac{1}{2} \times 4\,Pb \times 2b \times \left(4b + \frac{2b}{3}\right) + 3.5\,Pb^3\right]$$

$$= \frac{130.5}{3}\cdot Pb^3$$

$$\theta_C = \frac{Pb^3}{EI}\cdot\frac{130.5}{3} \times \frac{1}{6b}$$

$$= \frac{Pb^2}{EI}\cdot 7.25 \text{ (anticlockwise)}$$

Slope at right of B, $\theta_{BR} + \theta_C$

$$= \text{Area of the } \frac{M}{EI} \text{ diagram between } B \text{ and } C.$$

$$= \frac{1}{2}\cdot\frac{4\,Pb \times 6b}{EI}$$

$$= 12\cdot\frac{Pb^2}{EI}$$

$$\theta_{BR} = \frac{12\,Pb^2}{EI} - \frac{7.25\,Pb^2}{EI}$$

$$= \frac{4.75\,Pb^2}{EI} \text{ (clockwise)}$$

$$\text{Change in slope} = (4.75 - 2.5)\,\frac{Pb^2}{EI}$$

$$= \frac{2.25\,Pb^2}{EI} \text{ (increase)}$$

Deflection at D

$= \theta_C \times 2b$ – deflection of point D with respect to the tangent at C.

$= \theta_C \times 2b$ – Moment of the $\frac{M}{EI}$ diagram between D and C about D.

$$= \frac{7.25\,Pb^2}{EI} \times 2b - \frac{1}{2}\,\frac{4\,Pb \cdot 2b}{EI} \cdot \frac{1}{3} \cdot 2b$$

$$= \frac{Pb^3}{EI} \cdot \frac{35.5}{3}.$$

Example 7.14. Apply Moment Area method to compute the vertical, horizontal and rotational displacement at the free end of the cantilever frame shown in Fig. 7.25.

$EI = 2000 \text{ kN m}^2$

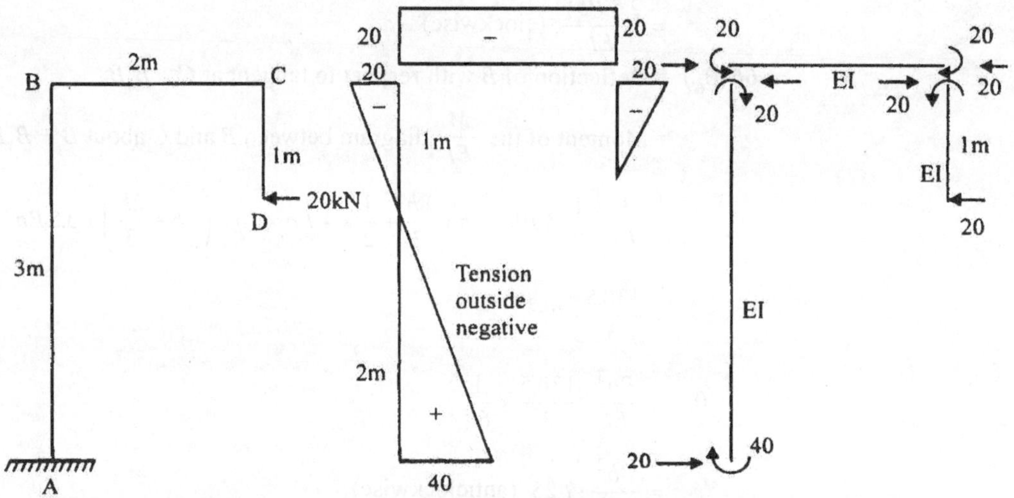

(i) The frame. *(ii)* B.M. diagram. *(iii)* F.B.D.

Fig. 7.25. Example 7.14.

Solution: The free body diagram and the Bending Moment diagrams are shown in Fig. 7.25 (i), (ii) and (iii).

Starting with fixed end as deflection and rotation at fixed end is zero.

$$\theta_B = 0 + \text{Area of the } \frac{M}{EI} \text{ diagram between } A \text{ and } B.$$

$$= \frac{1}{EI}\left[\frac{1}{2} \times 2 \times 40 - \frac{1}{2} \times 1 \times 20\right]$$

$$= \frac{30}{EI} = +\frac{30}{2000} = +0.015 \text{ radian (anticlockwise).}$$

y_B = deflection of B with respect to tangent at A

= Moment of the $\frac{M}{EI}$ diagram between A and B about B.

$$= \frac{1}{EI}\left[-\frac{1}{2}\times 20\times 1\times\frac{1}{3}+\frac{1}{2}\times 2\times 40\left(1+\frac{4}{3}\right)\right]$$

$$= \frac{90}{EI} = \frac{90}{2000} = \cdot 045\,\text{m (towards left)}$$

$$\theta_C = \underset{\text{(anticlockwise rotation)}}{+.015} - \underset{\text{(clockwise rotation)}}{\frac{20\times 2}{EI}}$$

$$= +.015 - \frac{40}{2000} = -.005 \text{ radian (clockwise)}$$

$$y_C = \theta_B \cdot \overline{BC} - \text{Moment of the } \frac{M}{EI} \text{ diagram between } B \text{ and } C \text{ about } C$$

$$= +.015\times 2 - \frac{20\times 2}{EI}\times 1$$

$= .01$ m upward.

Deformation at D

(a) Horizontal movement

(i) Due to horizontal movement of C

= Horizontal movement of B

$= -.045$ m (left)

(ii) Horizontal movement due to rotation θ_C

$= -.005 \times 1 = -.005$ (left)

(iii) Due to B.M. on the arm CD.

$$= \frac{PL^3}{3\,EI} = \frac{20\times 1^3}{3\times 2000} = -\ .00333 \text{ (left)}$$

Total displacement = .05333m (left)

(b) Vertical movement of D = Vertical movement of C

= .01 m upward

(c) Rotation of D = rotation at C + rotation due to moment at the portion CD

$$= -.005 + \left(-\frac{20\times 1^2}{2\times 2000}\right)$$

$= -.01$ radian.

7.8 METHOD OF CONCENTRATED ANGLE CHANGES

Example 7.14 shows application of Moment Area method to frame problems. However, for easy application to any type of frame problem, the method needs systematizations as regards rotation (clockwise or anticlockwise) and horizontal movement (right or left) and vertical movement (up or down). This is done with the help of the method of concentrated angle changes. The method is systematic, efficient and can be organised in a tabular form. To recapitulate the Moment Area method, Fig. 7.26 shows a simply supported beam AB and C any intermediate point.

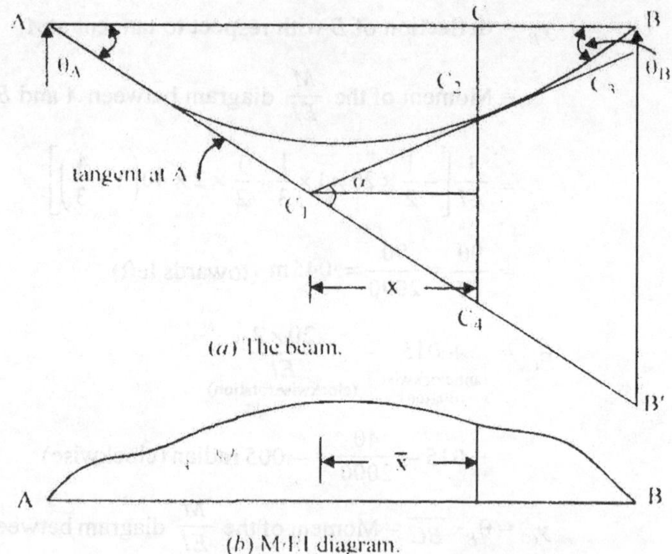

(a) The beam.

(b) M/EI diagram.

Fig. 7.26. Method of Concentrated Angle Changes.

From the 1st Area Moment Theorem.

$$\theta_C = \theta_A + \text{Area of the } \frac{M}{EI} \text{ diagram between } A \text{ and } C.$$

The previous convention of sagging moment and anticlockwise angle positive still holds.

Applying the 2nd Moment Area Theorem, deflection of point C with respect to tangent at A =

C_2C_4 = Moment of the $\frac{M}{EI}$ diagram between A and C about C.

If symbolically A_r = Area of the $\frac{M}{EI}$ diagram, $\bar{x}$ is the distance of the centroid of area A_r from C, we can write the two area moment theorems as

$$\theta_C = \theta_A + A_r \quad + \circlearrowleft \qquad ...(7.7)$$

$$\Delta_{C/A} = \bar{x}\ A_r$$

$$\Delta = CC_2 = \overline{AC}\ \theta_A + \bar{x}\ A_r \qquad ...(7.8)$$

In this equation A_r is positive and hence $\bar{x}\ A_r$ is positive. θ_A is clockwise, hence negative and $\overline{AC}\ \theta_A$ is negative Δ, therefore, is negative with the sign convention Δ is negative downward and positive upward.

Considering the geometry of the figure (Fig. 7.26).

We can write

$$\theta_C = \theta_A + \alpha \qquad ...(7.9)$$

where $\alpha = \angle C_2C_1C_4$

Similarly

$$\Delta = \overline{AC}\ \theta_A + x.\alpha \qquad ...(7.10)$$

where x is the distance of the point C_1 from C_2C.

Comparing Equation 7.7 and 7.9 we get $A_r = \alpha$

and from Equations 7.8 and 7.10 $\bar{x} = x$.

indicating that the centroid of the $\frac{M}{EI}$ diagram between A and C is at the intersection of the tangents of the elastic curve at A and C. The analogy leads to the method of concentrated angle changes where the actual elastic curve is replaced by a series of straight lines and concentrated angle changes. The concentrated angle change A_r is thus located on the elastic curve at the point that coincides with the centroid of the $\frac{M}{EI}$ diagram between A and C. Let us consider a cantilever beam subjected to a uniformly distributed load w/unit length. (Fig. 7.27). The elastic curve is replaced by two straight line segments ao and ob'. The concentrated angle change is located at the centroid of the $\frac{M}{EI}$ diagram. The displacement of point B is then found by multiplying the angle change by the distance from o to b.

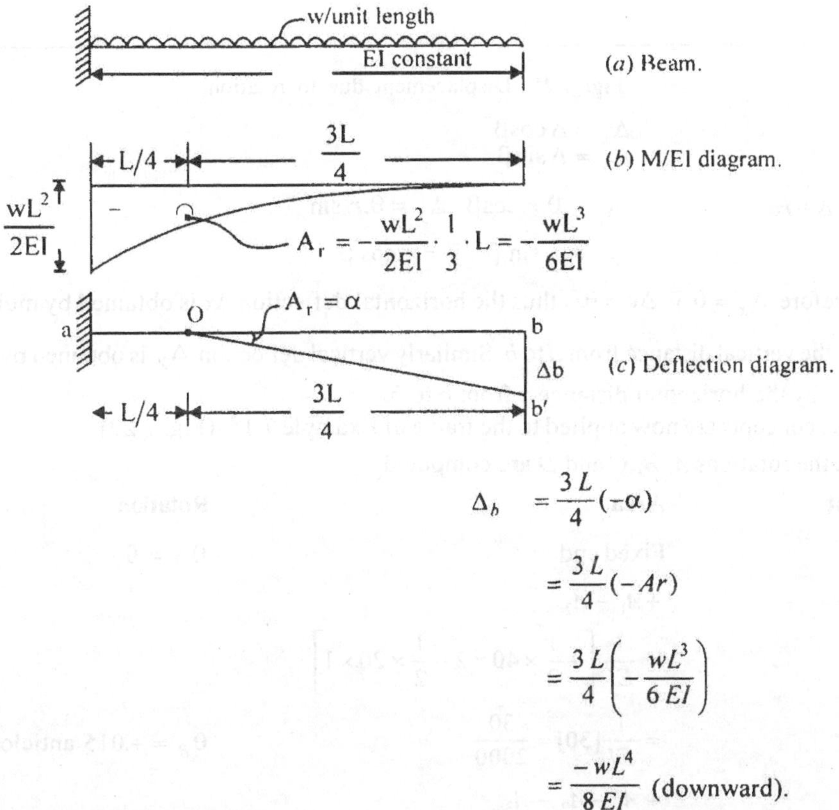

$$\Delta_b = \frac{3L}{4}(-\alpha)$$

$$= \frac{3L}{4}(-Ar)$$

$$= \frac{3L}{4}\left(-\frac{wL^3}{6EI}\right)$$

$$= \frac{-wL^4}{8EI} \text{ (downward).}$$

Fig. 7.27. Example 7.15 on method of concentrated angle change.

To apply the method of concentrated angle change to a frame (for which the method is most appropriate) it is necessary to study displacement of a rigid body subjected to a rotation about some point O. Fig. 7.28, shows a rigid body subjected to a rotation θ about point O. As a result a point b at a distance r from O moves to b' and $bb' = r.\theta = \Delta$ (say)

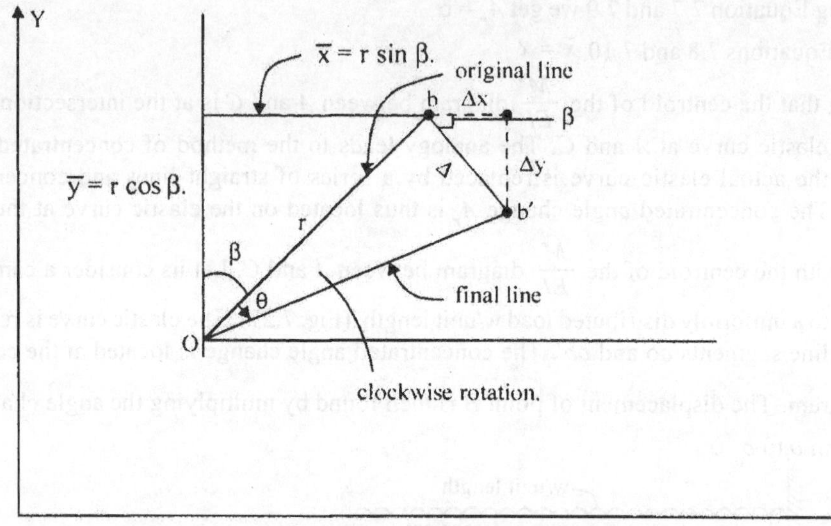

Fig. 7.28. Displacement due to rotation.

$$\Delta_x = \Delta \cos\beta$$
$$\Delta_y = \Delta \sin\beta$$

with $\Delta = r\theta$ $\quad \Delta_x = \theta.r.\cos\beta \quad \Delta_y = \theta.r \sin\beta$

$$\overline{x} = r \sin\beta \quad \overline{y} = r\cos\beta$$

Therefore $\Delta_x = \theta.\overline{y}$ $\Delta y = \theta\overline{x}$ thus the horizontal deflection Δx is obtained by multiplying the rotation by the vertical distance from o to b. Similarly vertical deflection Δ_y is obtained by multiplying the rotation by the horizontal distance x from o to b.

These concepts are now applied to the frame of Example 7.14. (Fig. 7.29)

First the rotations A, B, C and D are computed.

Point	**Area**	**Rotation**
A	Fixed end	$\theta_A = 0$
B	$+A_1 - A_2$	
	$= \dfrac{1}{EI}\left[+\dfrac{1}{2}\times 40\times 2 - \dfrac{1}{2}\times 20\times 1\right]$	
	$= \dfrac{1}{EI}[30] = \dfrac{30}{2000}$	$\theta_B = +.015$ anticlockwise
C	$+A_1 - A_2 - A_3$	
	$= +.015 - \dfrac{20\times 2}{2000}$	$\theta_C = -.005$
	$= +.015 - .020$	(clockwise)
	$= -.005$	
D	$+A_1 - A_2 - A_3 - A_4$	
	$= -.005 - \dfrac{1}{2\,EI}\times 20\times 1$	

$$= -.005 - \frac{1}{2} \times \frac{20 \times 1}{2000} \qquad -.01 \text{ (clockwise)}$$

$$= -.01 \text{ (clockwise)}$$

Calculation of Deflections

Points	Area	$\bar{x}$	$\bar{y}$	$\Delta_y = A\bar{x}$	$\Delta_x = A\bar{y}$	Deflection
B	$A_1 = +40$ ↺	0	$\frac{7}{3}$	0	$+40 \times \frac{7}{3}$	$\Delta x_B = +\frac{90}{2000}$
	$A_2 = -10$ ↻ +30	0	$\frac{1}{3}$	0	$-10 \times \frac{1}{3}$ +90	$= \cdot 045 \leftarrow$
	$A_1 = +40$ ↺	2	$\frac{7}{3}$	$+40 \times 2 \uparrow$	$+40 \times \frac{7}{3} \leftarrow$	$\Delta y_C = \frac{20}{EI}$ $= \frac{20}{2000}$
	$A_2 = -10$ ↻	2	$\frac{1}{3}$	$-10 \times 2 \downarrow$	$-10 \times \frac{1}{3} \rightarrow$	$= \cdot 01 \text{ m} \uparrow$
C	$A_3 = -40$ -10	1	0	$-40 \times 1 \downarrow$ $+20 \uparrow$	$-40 \times 0 \rightarrow$ $+90 \leftarrow$	$\Delta x_C = \frac{90}{2000}$ $= .045 \text{ m} \leftarrow$
	$A_1 = +40$ ↺	2	$\frac{4}{3}$	$40 \times 2 = +80 \uparrow$	$+40 \times \frac{4}{3} \leftarrow$	$\Delta y_D = \frac{20}{2000}$
	$A_2 = -10$ ↻	2	$-\frac{2}{3}$	$-10 \times 2 = -20 \downarrow$	$-10 \times \left(-\frac{2}{3}\right) \leftarrow$	$= \cdot 01 \text{ m} \uparrow$
D	$A_3 = -40$ ↻	1	-1	$-40 \times 1 = -40 \downarrow$	$-40 \times (-1) \leftarrow$	$\Delta x_D = \frac{320}{3 \times 2000}$
	$A_4 = -10$ ↻	0	$-\frac{2}{3}$	$-10 \times 0 = 0 \downarrow$ $+20 \uparrow$	$-10 \times \left(-\frac{2}{3}\right) \leftarrow$ $\frac{320}{3} \leftarrow$	$= \cdot 0533 \text{ m} \leftarrow$

In this analysis the following points are of great importance.

1. Type of B.M. diagram, sagging or hogging and the corresponding direction of angular rotation.
2. Points of concentrated angle change.
3. Direction of the point whose deflection is being measured from the point of concentrated angle change.
4. Direction of lateral movement of the point due to rotation right or left.

Hence the steps of the method are:

1. Draw the moment diagram on each member taking special care to find the direction of curvature. Divide by the appropriate *EI* values to obtain the curvature diagram.
2. Locate the centroid of the B.M. diagram of each member. Here the B.M. diagrams may be taken by parts and the C.G. located separately. These are the points of concentrated

angle changes.

3. Determine the horizontal and vertical distances x and y from each concentrated angle change to the point in question.
4. Determine the areas of the $\frac{M}{EI}$ diagrams.
5. Arrange the areas and x and y distances in Tabular form.
6. Compute the results by super posing the part results with proper signs foı rıght or left, up or down.

7.9. CORRECTION FOR PINNED STRUCTURES

When one end of a structure is pinned and the other is roller supported. There will be rotations at both supports and it is not possible to get a fixed line of reference. Hence the *hinged* support is assumed to be fixed and deflections computed at the roller support. Because of change of support conditions there will be both horizontal and vertical movements in the roller support. Since we have violated the support condition at the hinge by assuming it to be fixed we now satisfy the support conditions by allowing the frame to rotate by θ_{cr} as a rigid body so that the vertical moment at the roller is brought back to zero.

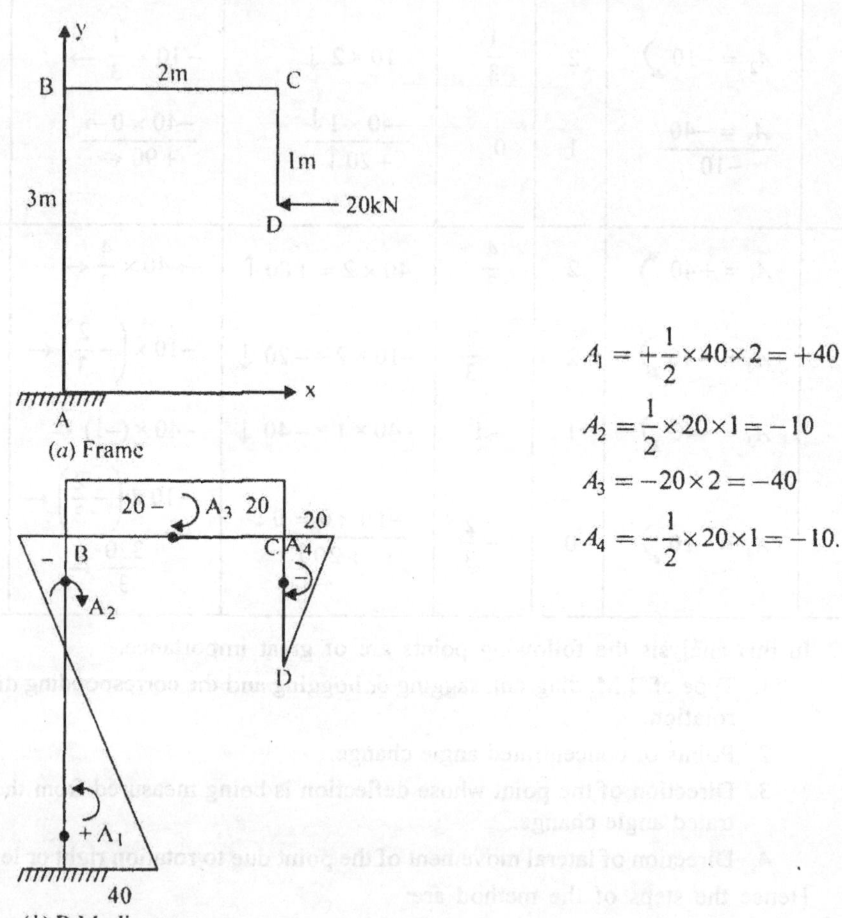

$$A_1 = +\frac{1}{2} \times 40 \times 2 = +40$$

$$A_2 = \frac{1}{2} \times 20 \times 1 = -10$$

$$A_3 = -20 \times 2 = -40$$

$$A_4 = -\frac{1}{2} \times 20 \times 1 = -10.$$

Fig. 7.29. Solution of Example 7.14 by Concentrated Angle Changes

7.10. CONJUGATE BEAM METHOD

The conjugate beam method is based on an analogy that can be made between the operations performed to compute shear and bending moment diagrams from load diagrams and the operations required to compute slopes and deflections from curvature diagrams. The equation

$$\frac{d^2y}{dx^2} = -\frac{M}{EI} \text{ has already been derived.}$$

or $$\frac{d}{dx}\left(\frac{dy}{dx}\right) = -\frac{M}{EI}$$

But $$\frac{dy}{dx} = \text{slope} = \tan\theta \approx \theta \text{ for flat curves.}$$

Hence $$\frac{d\theta}{dx} = -\frac{M}{EI}$$

or $$d\theta = -\frac{M}{EI}\cdot dx.$$

Integrating $$\theta = -\int \frac{M}{EI}\cdot dx \qquad ....(7.11)$$

or $$\frac{dy}{dx} = -\int \frac{M}{EI}\cdot dx$$

Integrating $$y = -\iint \frac{M}{EI}\cdot dx\cdot dx \qquad ....(7.12)$$

For a beam under distributed load of *w*/unit run it is also already derived that (section 4.3).

$$\frac{dV}{dx} = -w \qquad ....(4.3)$$

and $$\frac{dM}{dx} = V \qquad ....(4.4)$$

Thus, $$dV = -w\,dx$$

or $$V = -\int w\cdot dx \qquad .....(7.13)$$

Similarly $$dM = V\,dx = -w\int dx\cdot dx$$

or $$M = -\iint w\cdot dx\cdot dx \qquad .....(7.14)$$

If Eqn. 7.11 is compared with 7.13, it can be observed that if in Eqn. 7.11. *w* is replaced by $\frac{M}{EI}$, the shear force *V* will correspond with the slope θ of the beam. Similarly when 7.12 is compared with Eqn. 7.14, it can be seen that if $\frac{M}{EI}$ replaces *w* then the B.M. at any point will give the deflection *y*. The beam, however, has to be replaced by a conjugate beam whose length will be equal to the actual beam but support conditions will change such that shear and bending moment in the conjugate beam will conform to the slopes and deflections of the actual beam. The two conjugate beam theorems, therefore, are :

1. The slope at any point of an actual beam subjected to any type of loading is given by the shear force at the corresponding section of the conjugate beam subjected to the elastic load of $\frac{M}{EI}$ where M, E and I correspond to actual beam conditions.

2. Similarly deflection at any section is given by the corresponding B.M. at that section of the conjugate beam subjected to elastic load of $\frac{M}{EI}$.

7.11. CHANGE OF SUPPORT CONDITIONS

As already explained in the conjugate beam support conditions are modified compared to the actual beam. The modification can be summarised as

Fixed end ↔ Free end.

Simple end ↔ Simple end.

Interior Connection ↔ Interier Support.

The reasons for the above conjugation are obvious. In the fixed end of an actual beam there is no rotation or displacement. As such there can be no shear or moment at that end of the conjugate beam, which is possible only if the end is free. Similarly in an intermediate support of an actual beam deflection is zero. In the conjugate beam, therefore, bending moment should be zero which is possible if there is a hinge at that point. These are summarised in Table 7.3.

Table 7.3. Actual beam vs Conjugate beam.

Actual beam		Conjugate beam	
Support Condition	Slope and Deflection at support	Required shear and moment	Required support condition
Fixed end	$\theta = 0$, $y = 0$	$V = 0$, $M = 0$	Free end
Free end	θ and y both exist	V and M should exist	Fixed end
Simple end (hinge or roller)	θ exists $y = 0$	V should exist $M = 0$	Simple end (hinge or roller)
Interior Support (hinge or roller)	θ exists $y = 0$	V should exist $M = 0$	Interior connection (hinge or roller)
Interior connection (hinge or roller)	θ exists y exists	V should exist M should exist	Interior support (hinge or roller)
Overhang (free end)	θ exists y exists	V should exist M should exist	Fixed end
Overhang (support)	θ exists $y = 0$	V should exist $M = 0$	Interior support (hinge)

Fig. 7.30 Acutal beam and corresponding conjugate beam.

Type of beam	Actual	⟷ Conjugate
	Statically Determinate	
Cantilever		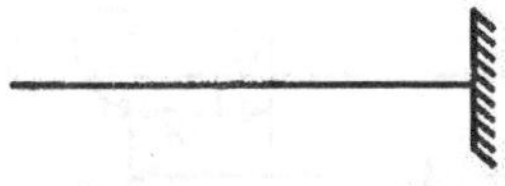
Simply Supported		
Beam with overhang.		 hinge
Beam with hinge		 hinge
Beam with fixed end and hinges		
Beam with fixed end and hinge.		
	Statically Indeterminate	
Continuous beam		
Propped Cantilever	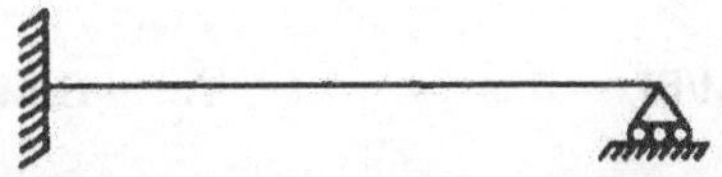	
Fixed end	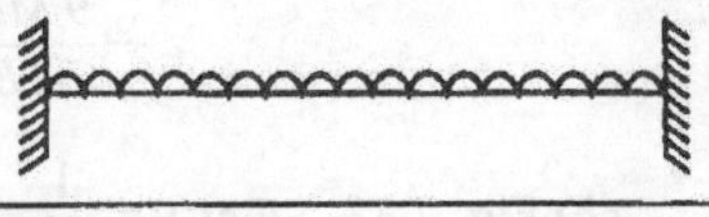	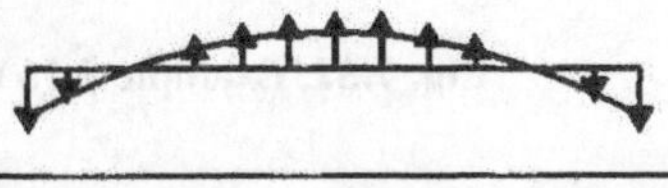

From the study of Fig. 7.30 it can be observed that the conjugate beam of a statically determinate beam is also statically determinate. Conjugate beam of a statically indeterminate beam appears to be unstable. However, it is stabilised by elastic loading. This is because beam is in equilibrium and thus yields a moment diagram that reflects this condition.

7.12. SIGN CONVENTION

The sign convention depends on the direction of coordinate axes. With coordinate axes y downward and x towards right positive, downward deflection and clockwise slope is positive. With the bending moment sign convention, downward load, sagging moment and left up and right down shear positive, sagging moment should be taken as downward load and hogging moment upward load. In such a case positive bending moment will produce downward deflection and positive shear clockwise rotation.

Example 7.16. Find θ_B, θ_C and Δ_B, Δ_C for cantilever beams and θ_A, θ_B, θ_C and Δ_C for simply supported beams. EI constant.

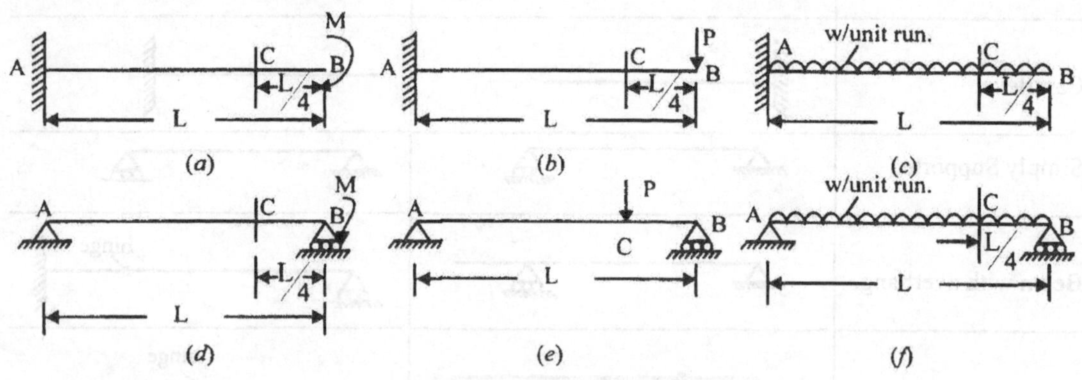

Fig. 7.31. Example 7.16.

Solution:

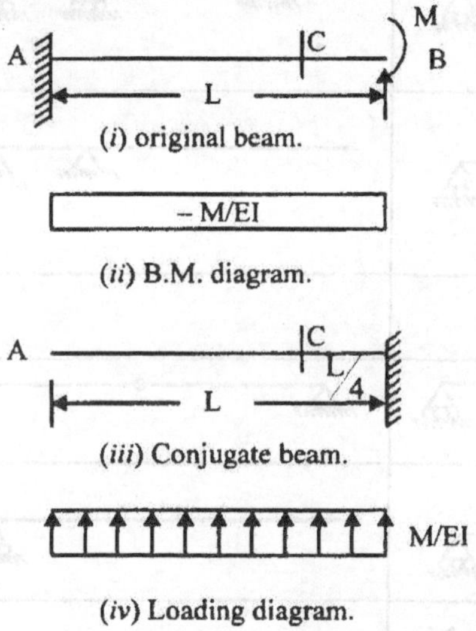

(*a*) Fig. 7.32

Shear at $B = +\dfrac{M}{EI}\cdot L$

Slope at $B = \dfrac{M}{EI}$ (clockwise)

Shear at $C = \theta_C = \dfrac{M}{EI}\cdot\dfrac{3L}{4} = \dfrac{3ML}{4EI}$ (clockwise)

B.M. at $B = \dfrac{M}{EI}\cdot L\cdot\dfrac{L}{2} = +\dfrac{ML^2}{2EI}$ (sagging)

Deflection at $B = \dfrac{ML^2}{2EI}$ ↓ downward

Δ_C = B.M. at $C = \dfrac{M}{EI}\cdot\dfrac{3L}{4}\cdot\dfrac{3}{8}\cdot L$

$= \dfrac{9ML^2}{32EI}$ ↓.

(*i*) original beam.

(*ii*) B.M. diagram.

(*iii*) Conjugate beam.

(*iv*) Loading diagram.

Fig. 7.32. Example 7.16 (*a*)

(*b*) Fig. 7.33 SF at $B = \dfrac{1}{2}\times\dfrac{PL.L}{EI} = +\dfrac{PL^2}{2EI}$

Hence $\theta_B = \dfrac{PL^2}{2EI}$ (clockwise)

B.M. at $B = \dfrac{1}{2}\cdot\dfrac{PL}{EI}\cdot L\cdot\dfrac{2}{3}\cdot L = +\dfrac{PL^3}{3EI}$

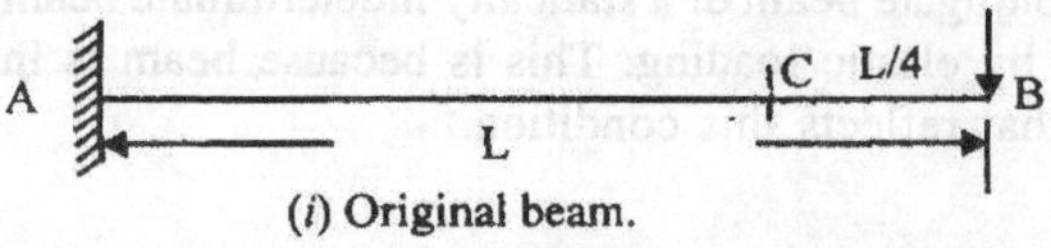

(*i*) Original beam.

(*b*) Fig. 7.33.

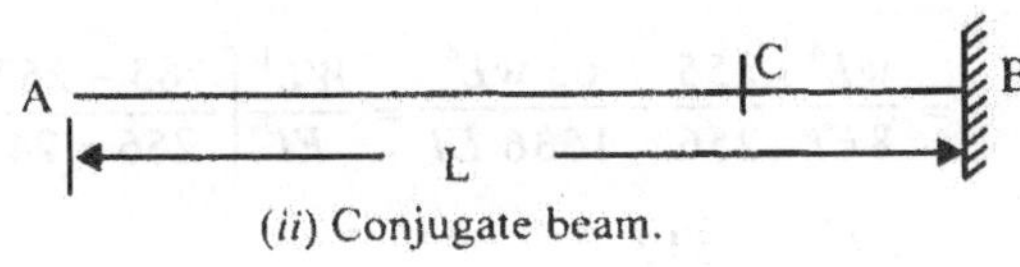

(*ii*) Conjugate beam.

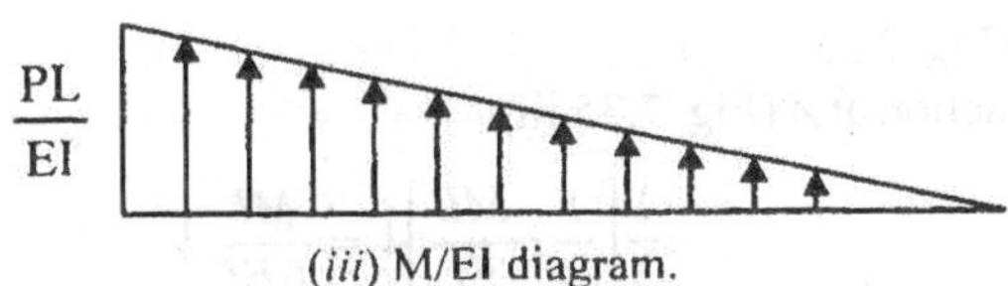

(*iii*) M/EI diagram.

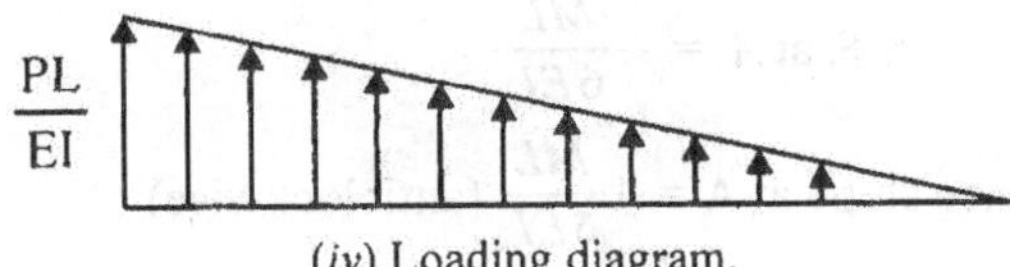

(*iv*) Loading diagram.

Fig. 7.33. Example 7.16 (*b*)

Deflection at $B = +\frac{PL^3}{3EI}$ (downward)

$$\theta_C = \text{S.F. at } C$$

$$= \frac{1}{2}\left(\frac{PL}{EI} + \frac{PL}{4EI}\right)\frac{3}{4}L$$

$$= \frac{15}{32} \cdot \frac{PL^2}{EI}$$

$$\Delta_C = \frac{1}{2} \cdot \frac{PL}{EI} \cdot \frac{3L}{4} \cdot \frac{2}{3} \cdot \frac{3L}{4} + \frac{1}{2} \cdot \frac{PL}{4EI} \cdot \frac{3L}{4} \cdot \frac{1}{3} \cdot \frac{3L}{4}$$

$$= \frac{PL^3}{EI} \cdot \frac{27}{128}$$

(*c*) Fig. 7.34

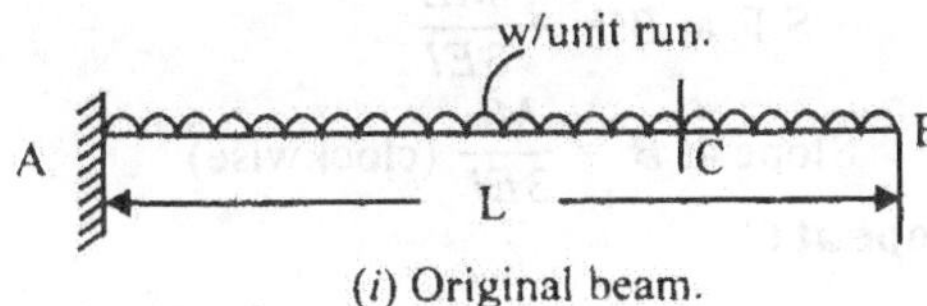

(*i*) Original beam.

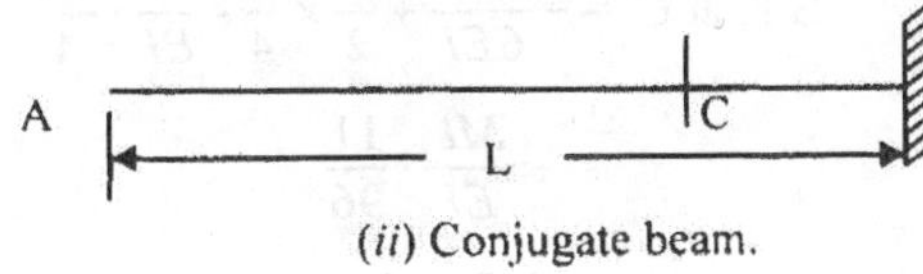

(*ii*) Conjugate beam.

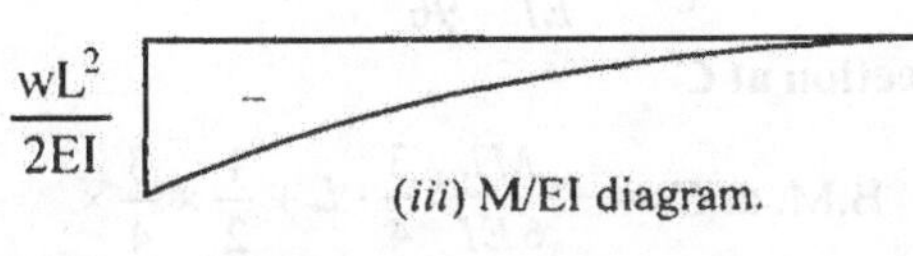

(*iii*) M/EI diagram.

wL²/2EI

(*iv*) Loading diagram.

Fig. 7.34. Example 7.16 (*c*)

$$\text{S.F. at } B = +\frac{1}{3} \cdot \frac{wL^2}{2EI} \cdot L = \frac{wL^3}{6EI}$$

Hence $\theta_B = \frac{wL^3}{6EI}$ (clockwise)

$$\text{B.M. at } B = +\frac{1}{3} \cdot \frac{wL^2}{2EI} \cdot L \cdot \frac{3}{4} \cdot L = \frac{wL^4}{8EI}$$

Hence deflection $= \frac{wL^4}{8EI} \downarrow$

θ_C = area of the $\frac{M}{EI}$ diagram between A and C

$$= \int_{L/4}^{L} \frac{wx^2}{2EI} \cdot dx = \frac{w}{6EI}\left[x^3\right]_{L/4}^{L}$$

$$= \frac{w}{6EI}\left[L^3 - \frac{L^3}{64}\right] = \frac{wL^3}{EI} \cdot \frac{63}{384}$$

Δ_C = B.M at C.

$$= \int_{L/4}^{L} \frac{wx^2}{2EI} \cdot (x - L/4)dx$$

$$= \int_{L/4}^{1} \frac{wx^3}{2EI} \cdot dx - \frac{L}{4} \cdot \frac{wL^3}{EI} \cdot \frac{63}{384}$$

$$= \frac{w}{8EI}\left[L^4 - \frac{L^4}{256}\right] - \frac{L}{4} \cdot \frac{wL^3}{EI} \cdot \frac{63}{384}$$

$$= \frac{wL^4}{8EI} \cdot \frac{255}{256} - \frac{63\,wL^4}{1536\,EI} = \frac{WL^4}{EI}\left[\frac{765-252}{256 \times 24}\right]$$

$$= \frac{wL^4}{EI} \cdot \frac{513}{256 \times 24}$$

(*d*) Fig. 7.35

Reaction at A (Fig. 7.35 iii)

$$= \frac{1}{3}\left[\frac{1}{2} \cdot \frac{ML}{EI}\right] \downarrow = \frac{ML}{6EI} \downarrow$$

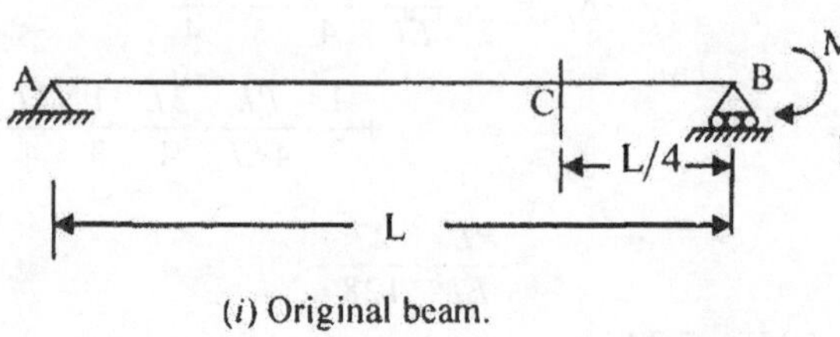

(*i*) Original beam.

(*ii*) Conjugate beam.

$$\text{S.F. at } A = -\frac{ML}{6EI}$$

$$\text{Hence slope at A} = -\frac{ML}{6EI} \text{ (anticlockwise)}$$

$$\text{Reaction at } B = \frac{2}{3}\left[\frac{1}{2} \cdot \frac{M}{EI} \cdot L\right] = \frac{ML}{3EI} \uparrow$$

$$\text{S.F. at } B = +\frac{ML}{3EI}$$

$$\text{Slope at } B = \frac{ML}{3EI} \text{ (clockwise)}$$

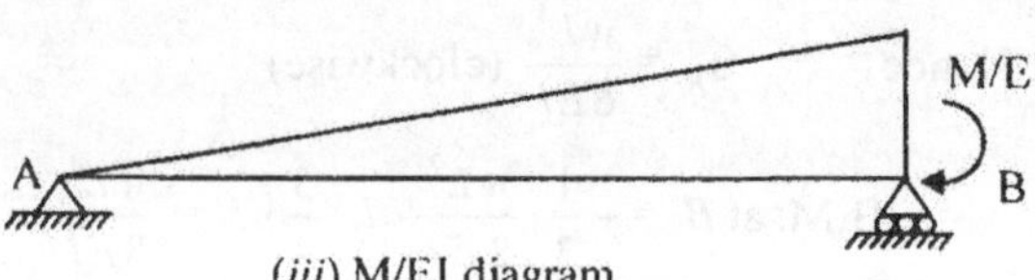

(*iii*) M/EI diagram.

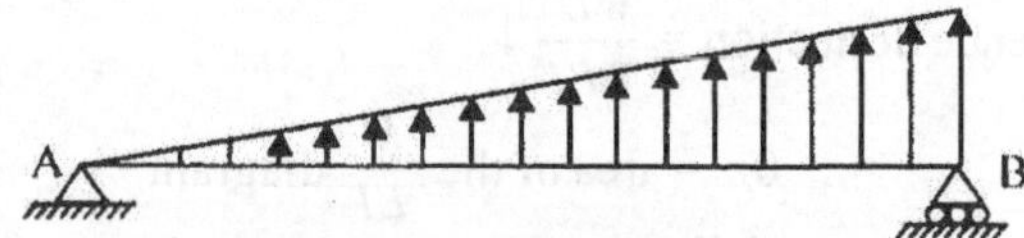

Fig. 7.35. Example 7.16 (*d*).

Slope at C

$$\text{S.F. at } C = -\frac{ML}{6EI} + \frac{1}{2} \times \frac{3}{4} \cdot \frac{M}{EI} \cdot \frac{3L}{4}$$

$$= +\frac{ML}{EI} \cdot \frac{11}{96}$$

$$\theta_C = \frac{ML}{EI} \cdot \frac{11}{96} \text{ (clockwise)}$$

Deflection at C

$$\text{B.M. at } C = -\frac{ML}{6EI} \cdot \frac{3}{4} \cdot L + \frac{1}{2} \times \frac{3}{4} \times \frac{M}{EI} \cdot \frac{3L}{4} \times \frac{1}{3} \times \frac{3}{4} \cdot L$$

$$= -\frac{ML^2}{EI} \cdot \frac{7}{128}$$

$$\text{Hence deflection} = \frac{ML^2}{EI} \cdot \frac{7}{128} \text{ (upward)}$$

(*e*) Fig. 7.36

$$\theta_A = \text{S.F. at } A$$

$$= R_A = +\frac{5PL^2}{128EI} \text{ (clockwise)}$$

$$\theta_B = \text{S.F at } B$$

$$= R_B = -\frac{7PL^2}{128EI} \text{ (anticlockwise)}$$

(*i*) Original beam.

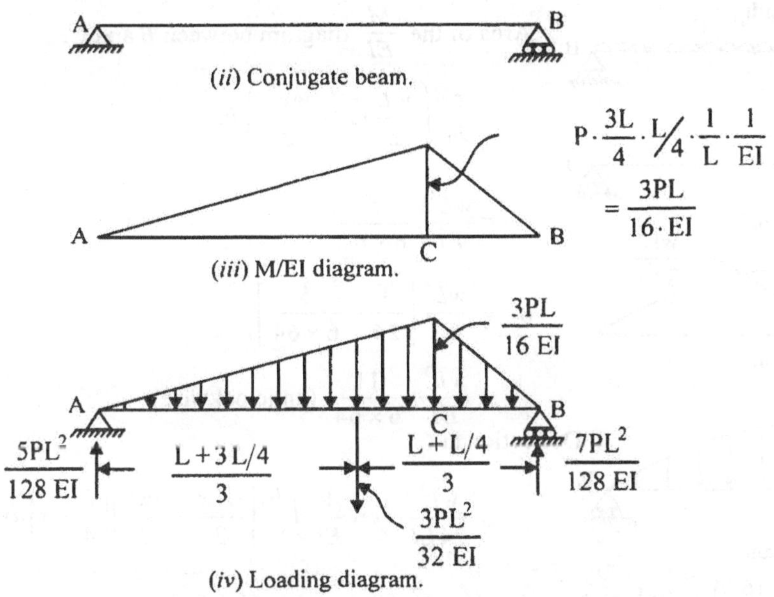

(ii) Conjugate beam.

(iii) M/EI diagram.

(iv) Loading diagram.

Fig. 7.36. Example 7.16 (e).

$$\theta_C = -\frac{7PL^2}{128EI} + \frac{1}{2} \cdot \frac{3PL}{16EI} \cdot \frac{L}{4}$$

$$= -\frac{PL^2}{32EI} \text{ (anticlockwise)}$$

Δ_C = deflection at C
= B.M. at C

$$= \frac{7PL^2}{128} \times \frac{L}{4EI} - \frac{1}{2} \cdot \frac{3PL}{16EI} \cdot \frac{L}{4} \cdot \frac{1}{3} \cdot \frac{L}{4}$$

$$= \frac{PL^3}{EI} \cdot \frac{3}{256}$$

(f) Fig. 7.37

θ_A = Reaction at A

$$= \frac{1}{2} \times \frac{2}{3} \times \frac{wL^2}{8EI} \times L$$

$$= \frac{wL^3}{24EI} \text{ (clockwise)}$$

$$\theta_B = \frac{wL^3}{24EI} \text{ (anticlockwise)}$$

θ_C = Shear force at C

$= \frac{wL^3}{24EI}$ − area of the $\frac{M}{EI}$ diagram between B and C

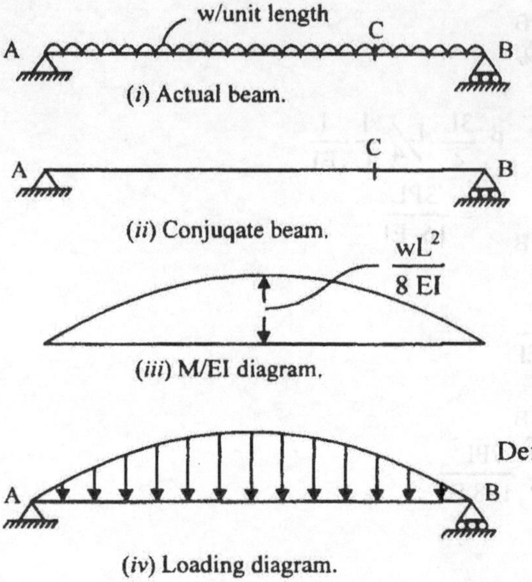

w/unit length
A
C
B
(i) Actual beam.
A
C
B
(ii) Conjuqate beam.
wL²
8 EI
(iii) M/EI diagram.
A
B
(iv) Loading diagram.

Fig. 7.37. Example 7.16 (f).

Area of the $\frac{M}{EI}$ diagram between B and C.

$$= \int_0^{L/4}\left(\frac{wL}{2}\cdot x - \frac{wx^2}{2}\right)dx$$

$$= \frac{wL^3}{EI}\cdot\frac{5}{6\times 64}$$

$$\theta_C = \frac{wL^3}{EI}\left[\frac{1}{24} - \frac{5}{6\times 64}\right]$$

$$= \frac{wL^3}{EI}\cdot\frac{11}{6\times 64} \text{ (anticlockwise)}$$

Deflection at C

$$= \frac{wL^3}{24EI}\cdot\frac{L}{4} - \frac{1}{EI}\int_0^{L/4}\left(\frac{wLx}{2} - \frac{wx^2}{2}\right)\left(\frac{L}{4} - x\right)dx$$

$$= \frac{wL^4}{EI}\times\frac{1}{96} - \frac{1}{EI}\int_0^{L/4}\frac{wL^2}{8}\cdot x\,dx + \frac{1}{EI}\int_0^{L/4}\frac{wL}{8}\cdot x^2\,dx$$

$$+\frac{1}{EI}\int_0^{L/4}\frac{wL}{2}\cdot x^2\,dx - \frac{1}{EI}\int_0^{L/4}\frac{wx^3}{2}\cdot dx$$

$$= \frac{wL^4}{EI}\cdot\frac{1}{96} - \frac{wL^2}{8}\cdot\frac{L^2}{32EI} + \frac{wL}{8}\cdot\frac{L^3}{192EI}$$

$$+\frac{wL}{2}\cdot\frac{L^3}{192EI} - \frac{w}{2}\cdot\frac{L^4}{256\times 4EI}$$

$$= \frac{wL^4}{EI}\left[\frac{1}{96} - \frac{1}{256} + \frac{1}{1536} + \frac{1}{384} - \frac{1}{2048}\right]$$

$$= \frac{wL^4}{EI}\cdot\frac{57}{1536\times 4}.$$

Example 7.17. Find the slopes and deflections at B and C of the beam given below. $E = 20{,}000$ kN/cm^2 and $I = 5{,}000$ cm^4.

20kN
hinge
A
B
C
D
E
2m
2m
2m
2m

(a) Beam.

20kNm.
+
20kNm.
–

(b) B.M. diagram.

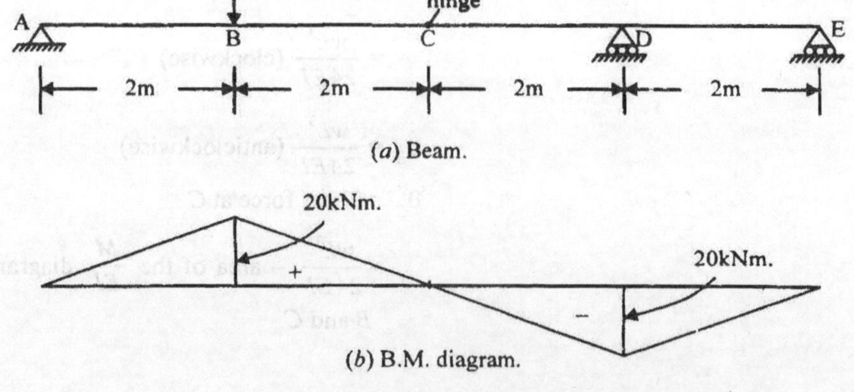

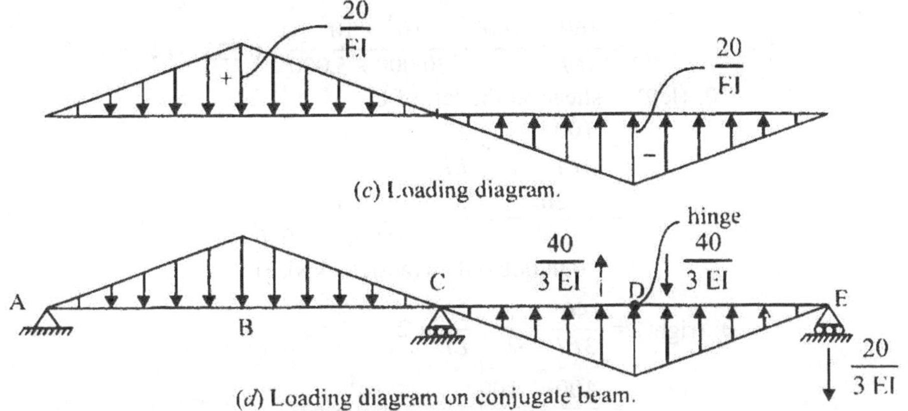

(d) Loading diagram on conjugate beam.

Fig. 7.38. Example 7.17 (by Conjugate beam method)

Solution:

(1) By conjugate beam method

As moment at $C = 0,$

$$R_A \times 4 - 20 \times 2 = 0$$

$$R_A = 10 \text{ kN} \uparrow$$

$$R_C = 10 \text{ kN} \uparrow$$

The B.M. diagram is then drawn in Fig. 7.38 (b).

The conjugate beam and loading diagram is shown in Fig. 7.38 (d).

As $M_D = 0$ $\quad R_E \times 2 + \frac{1}{2} \times \frac{20}{EI} \times 2 \times \frac{2}{3} = 0$

or $\quad R_E = -\frac{20}{3EI} \downarrow$ (+ve shear clockwise rotation)

$$R_D = \frac{1}{2} \times \frac{20 \times 2}{EI} - \frac{20}{3EI} = \frac{40}{3EI} \uparrow \text{ (-- ve shear anticlockwise rotation)}$$

Taking moments about C.

$$R_A \times 4 - \frac{1}{2} \times \frac{20}{EI} \times 4 \times 2 - \frac{1}{2} \times \frac{20}{EI} \times 2 \times \frac{2}{3} \times 2 - \frac{40}{3EI} \times 2 = 0$$

or $\quad R_A = \frac{100}{3EI} \uparrow$ (positive shear clockwise rotation)

Slope at *B*

$$\frac{100}{3EI} - \frac{1}{2} \times \frac{20}{EI} \times 2 = \frac{40}{3EI} \text{ (clockwise)}$$

$$= \frac{40 \times 10^4}{3 \times 20{,}000 \times 5{,}000} = 0.00133 \text{ radian}$$

Deflection at *B*

$$\text{B.M. at } B = \frac{100}{3EI} \times 2 - \frac{1}{2} \times \frac{20}{EI} \times 2 \times \frac{2}{3}$$

$$= \frac{160}{3EI} = \frac{160}{3} \times \frac{10^4 \times 10^2}{20{,}000 \times 5{,}000} = 0.533\ cm \downarrow$$

θ_C (left) = shear on the left of C

$$= \frac{100}{3EI} - \frac{1}{2} \times \frac{20}{EI} \times 4$$

$$= -\frac{20}{3EI} = -\frac{20}{3} \times \frac{10^4}{20{,}000 \times 5{,}000}$$

= – 0.00066 radian (anticlockwise)

$$\theta_C \text{ (right)} = \frac{40}{3EI} + \frac{1}{2} \times \frac{20}{EI} \times 2$$

$$= \frac{100}{3EI} = \frac{100}{3} \times \frac{10^4}{20{,}000 \times 5{,}000}$$

= 0.00333 radian (anticlockwise)

Check

$$R_C + R_A - \frac{1}{2} \times \frac{20}{EI} \times 4 + \frac{1}{2} \times \frac{20}{EI} \times 2 + \frac{40}{3EI} = 0$$

or
$$R_C = +\frac{40}{EI} - \frac{20}{EI} - \frac{40}{3EI} - \frac{100}{3EI} = -\frac{80}{3EI}$$

Shear on right of C
$$= -\frac{20}{3EI} - \frac{80}{3EI} = -\frac{100}{3EI}$$

= – 0.00333 (anticlockwise)

Deflection at C

$$\text{B.M. at } C = \frac{100 \times 4}{3EI} - \frac{1}{2} \times \frac{20}{EI} \times 4 \times 2$$

$$= \frac{1}{EI} \cdot \frac{160}{3} = 0.533\ cm \downarrow$$

(2) By Area Moment method.

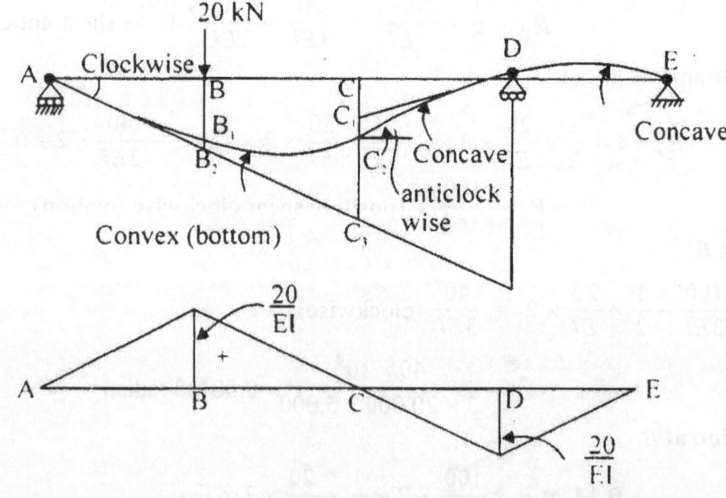

Fig. 7.39. Example 7.17 (Solution by Moment Area Method)

Deflection of point E w.r.t. tangent at D..

$$= \text{Moment of the } \frac{M}{EI} \text{ diagram between } D \text{ and } E \text{ about } E.$$

$$= \frac{1}{2} \times \frac{20}{EI} \times 2 \times \frac{4}{3} = \frac{80}{3EI}$$

$$\theta_D = \frac{80}{3EI} \times \frac{1}{2} = \frac{40}{3EI} \text{ (anticlockwise)}$$

$$CC_1 = \frac{40}{3EI} \cdot 2 = \frac{80}{3EI}$$

$$C_1C_2 = \text{deflection of point } C \text{ w.r.t. tangent at } D$$

$$= \text{Moment of the } \frac{M}{EI} \text{ diagram between } C \text{ and } D \text{ about } C$$

$$= \frac{1}{2} \times \frac{20}{EI} \times 4 \times 2 = \frac{80}{EI}$$

$$\therefore \quad \text{Deflection at } C = \frac{160}{3EI} = 0.533 \; cm \downarrow$$

$$\text{Slope on right side of } C = \frac{40}{3EI} + \text{area of the } \frac{M}{EI} \text{ diagram between } C \text{ and } D$$

$$= \frac{40}{3EI} + \frac{1}{2} \times \frac{20}{EI} \times 2 = \frac{100}{3EI} \text{ (anticlockwise)}$$

$$C_2C_3 = \text{deflection of point } C \text{ w.r.t. tangent at } A$$

$$= \text{Moment of the } \frac{M}{EI} \text{ diagram between } A \text{ and } C \text{ about } C$$

$$= \frac{1}{2} \times \frac{20}{EI} \times 4 \times 2 = \frac{80}{EI}$$

$$CC_3 = \frac{80}{EI} + \frac{160}{3EI} = \frac{400}{3EI}$$

$$\theta_A = \frac{400}{3EI} \times \frac{1}{4} = \frac{100}{3EI} \text{ (clockwise)}$$

$$\text{Slope at } B = \frac{100}{3EI} - \frac{1}{2} \times \frac{20}{EI} \times 2 = \frac{1}{EI} \cdot \frac{40}{3} \text{ (clockwise)}$$

Deflection at B

$$= BB_2 - B_1B_2$$

$$= \theta_A \times 2 - \text{Moment of the } \frac{M}{EI} \text{ diagram between } A \text{ and } B \text{ about } B.$$

$$= \frac{100 \times 2}{3EI} - \frac{1}{2} \times \frac{20}{EI} \times 2 \times \frac{2}{3} = \frac{160}{3EI}.$$

Example 7.18. For the beam and loading shown in Fig. 7.40 determine the slope at A, B, C and D and deflections at A and D.

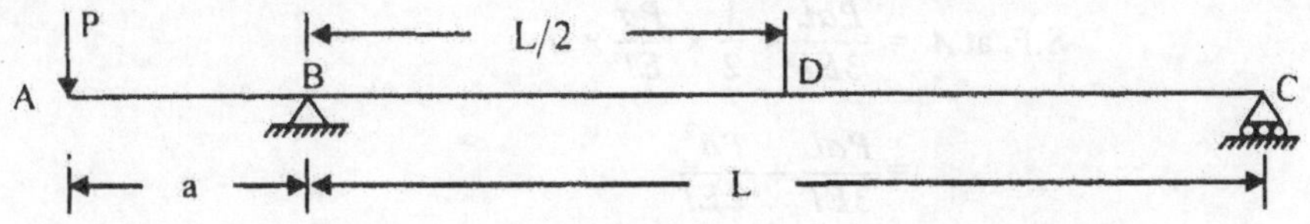

Fig. 7.40. Example 7.18

Solution:

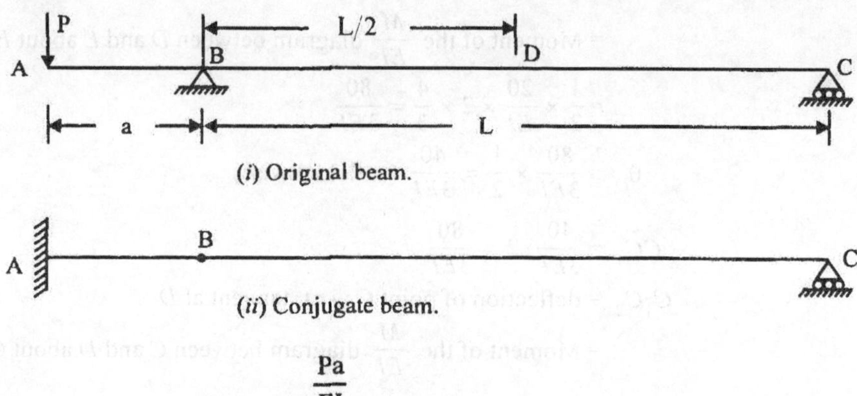

(*i*) Original beam.

(*ii*) Conjugate beam.

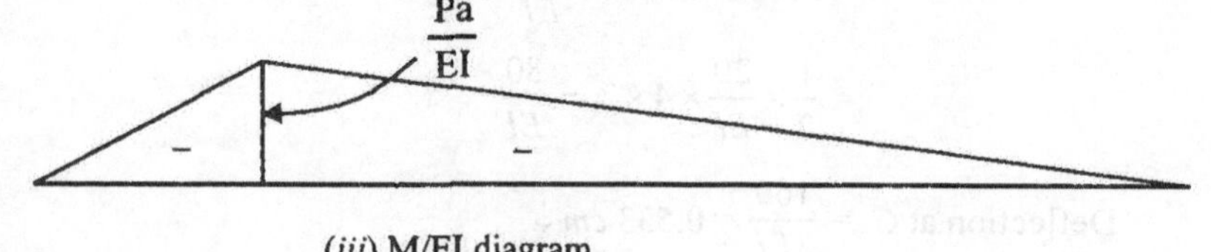

(*iii*) M/EI diagram.

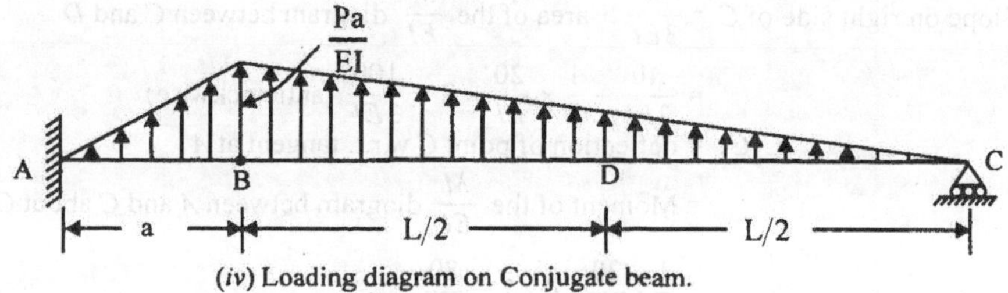

(*iv*) Loading diagram on Conjugate beam.

Fig. 7.41. Example 7.18

As B is a hinge in the conjugate beam $M_B = 0$

Therefore $$R_C \times L + \frac{1}{2} \times \frac{Pa}{EI} \times L \times \frac{L}{3} = 0$$

or $$R_C = -\frac{PaL}{6EI} \downarrow$$

From equilibrium of forces on BC

$$R_{BR} \uparrow + \frac{1}{2} \cdot \frac{PaL}{EI} \uparrow - \frac{PaL}{6EI} \downarrow = 0.$$

or $$R_{BR} = \frac{PaL}{EI}\left(\frac{1}{6} - \frac{1}{2}\right) = -\frac{PaL}{EI} \cdot \frac{1}{3} \downarrow$$

Hence $$R_B \text{ (left side)} = \frac{PaL}{3EI} \uparrow$$

$$\text{S.F. at } A = \frac{PaL}{3EI} + \frac{1}{2} \times \frac{Pa}{EI} \times a$$

$$= \frac{PaL}{3EI} + \frac{Pa^2}{2EI}$$

$$\text{B.M. at } A = \frac{PaL}{3EI}\cdot a + \frac{1}{2}\cdot\frac{Pa}{EI}\cdot a\cdot\frac{2a}{3}$$

$$= \frac{Pa^2}{3EI}(a+L)$$

$$\text{Slope at } C = \text{shear at } C$$

$$= +\frac{PaL}{6EI} \text{ (clockwise)}$$

$$\text{Slope at } D = \text{Shear at } D$$

$$= \frac{PaL}{6EI} - \frac{1}{2}\cdot\frac{Pa}{2EI}\cdot\frac{L}{2}$$

$$= \frac{PaL}{EI}\left(\frac{1}{6}-\frac{1}{8}\right) = \frac{PaL}{EI}\cdot\frac{1}{24} \text{ (clockwise)}$$

Deflection at D

$$= \text{B.M. at } D$$

$$= -\frac{PaL}{6EI}\cdot\frac{L}{2} + \frac{1}{2}\cdot\frac{Pa}{2EI}\cdot\frac{L}{2}\cdot\frac{1}{3}\cdot\frac{L}{2} = -\frac{PaL^2}{EI}\cdot\frac{1}{16} \text{ (upward)}$$

$$\text{Slope at } B \text{ (right)} = \text{shear on right of } B$$

$$= \frac{PaL}{6EI} - \frac{1}{2}\cdot\frac{Pa}{EI}\cdot L = -\frac{PaL}{3EI} \text{ (anticlockwise)}$$

$$\text{Slope at } A = \text{S.F. at } A$$

$$= -\left[\frac{PaL}{3EI} + \frac{Pa^2}{2EI}\right] = -\frac{Pa}{EI}\left[\frac{L}{3}+\frac{a}{2}\right] \text{ (anticlockwise)}$$

$$\text{Deflection at } A = \text{B.M. at } A$$

$$= +\frac{Pa^2}{3EI}(a+L) \text{ (downward).}$$

Example 7.19. Find slopes and deflections at critical points of the beam given below. Draw the deflected shape. Fig. 7.42.

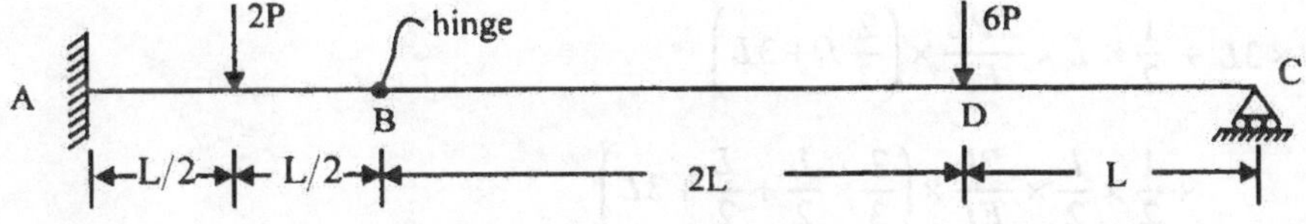

Fig. 7.42. Example 7.19

Solution:

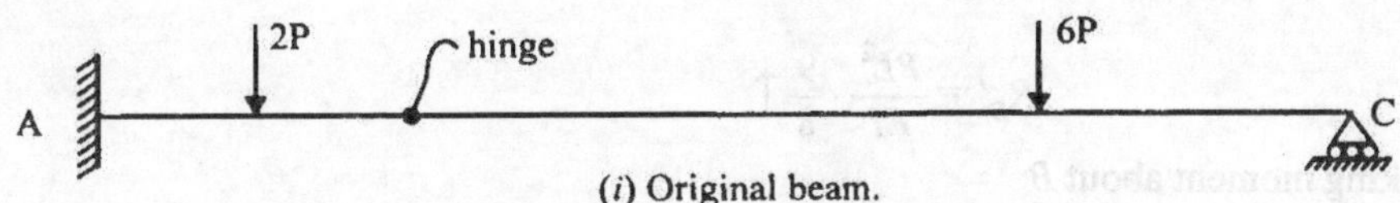

(*i*) Original beam.

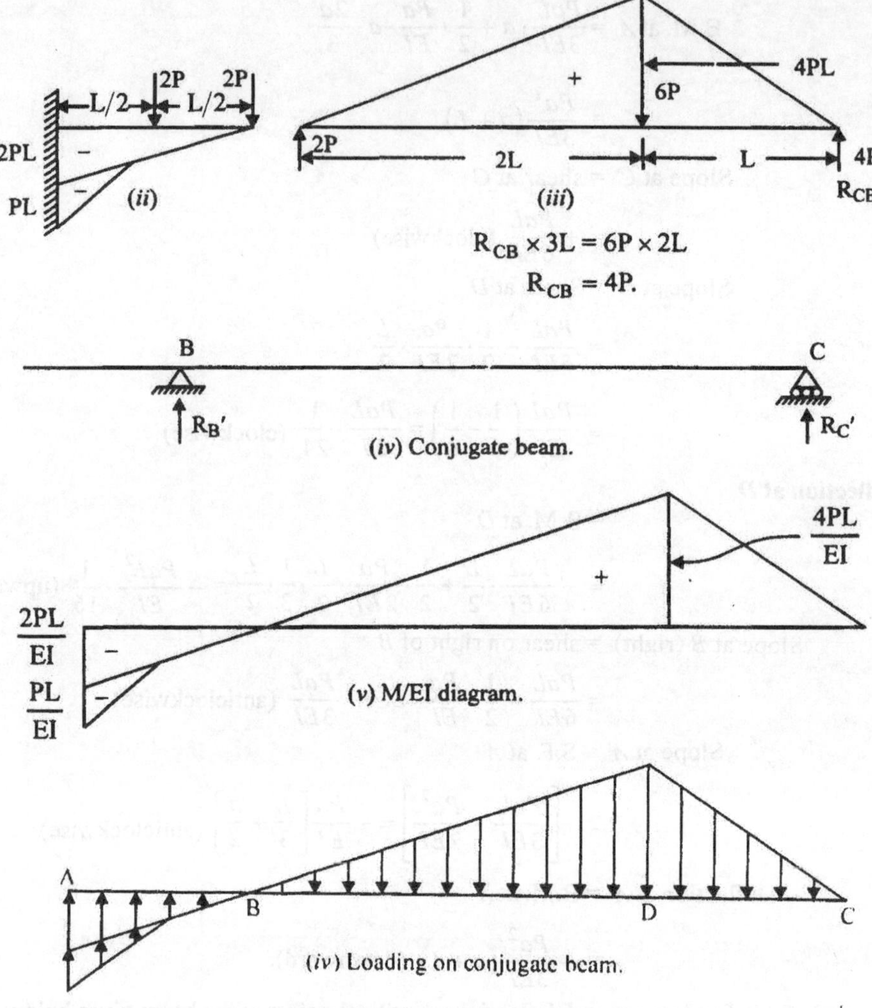

$$R_{CB} \times 3L = 6P \times 2L$$

$$R_{CB} = 4P.$$

Fig. 7.43. Example 7.19

In the conjugate beam taking moments of all forces about C

$$R_B' \times 3L + \frac{1}{2} \times L \times \frac{2PL}{EI} \times \left(\frac{2}{3}L + 3L\right)$$

$$+\frac{1}{2} \times \frac{L}{2} \times \frac{PL}{EI} \times \left(\frac{2}{3} \times \frac{L}{2} + \frac{L}{2} + 3L\right)$$

$$-\frac{1}{2} \times \frac{4PL}{EI} \times \frac{3L + L}{3} \times 3L = 0$$

or
$$R_B' = \frac{PL^2}{EI} \cdot \frac{9}{8} \uparrow$$

Taking moment about B

$$R_C' \times 3L = \frac{1}{2} \times \frac{4PL}{EI} \times 3L \times \frac{5L}{3} + \frac{1}{2} \times \frac{2PL}{EI} \times L \times \frac{2}{3} \cdot L$$

$$+\frac{1}{2}\times\frac{PL}{EI}\times\frac{L}{2}\times\left(\frac{L}{2}+\frac{2}{3}\times\frac{L}{2}\right)$$

or $$R_C' = \frac{PL^2}{EI}\cdot\frac{29}{8}\uparrow$$

$$R_B' + R_C' = \text{Total vertical load on the conjugate beam}$$

$$= \frac{1}{2}\times\frac{4\,PL}{EI}\times 3L - \frac{1}{2}\times\frac{2\,PL}{EI}\times L - \frac{1}{2}\times\frac{PL}{EI}\times\frac{L}{2}$$

$$= \frac{PL^2}{EI}\cdot\frac{19}{4}$$

$$R_B' + R_C' = \frac{PL^2}{EI}\left[\frac{9}{8}+\frac{29}{8}\right] = \frac{PL^2}{EI}\cdot\frac{19}{4} \text{ (check)}$$

Slope at C = shear force at C

$$= R_C' = -\frac{PL^2}{EI}\cdot\frac{29}{8} \text{ (anticlockwise)}$$

Slope to the right at B = S.F. to the right of B

$$= -\frac{PL^2}{EI}\cdot\frac{29}{8}+\frac{1}{2}\times\frac{4\,PL}{EI}\times 3L$$

$$= +\frac{PL^2}{EI}\cdot\frac{19}{8} \text{ (clockwise)}$$

Slope to the left of B $$= +\frac{PL^2}{EI}\cdot\frac{19}{8} - \frac{PL^2}{EI}\cdot\frac{9}{8}$$

$$= +\frac{PL^2}{EI}\cdot\frac{5}{4} \text{ (clockwise)}$$

Deflection at D = B.M. at D

$$= \frac{PL^2}{EI}\cdot\frac{29}{8}\times L - \frac{1}{2}\times\frac{4\,PL}{EI}\times L\times\frac{1}{3}L$$

$$= +\frac{PL^3}{EI}\cdot\frac{71}{24} \text{ (downward)}$$

Deflection at B = B.M. at B

$$= \frac{PL^2}{EI}\cdot\frac{29}{8}\times 3L - \frac{1}{2}\times\frac{4\,PL}{EI}\times 3L\times\frac{5L}{3}.$$

$$= \frac{PL^3}{EI}\cdot\frac{7}{8} \text{ (downward).}$$

Example 7.20. Find the deflection at C in the statically indeterminate beam with constant EI shown in Fig. 7.44.

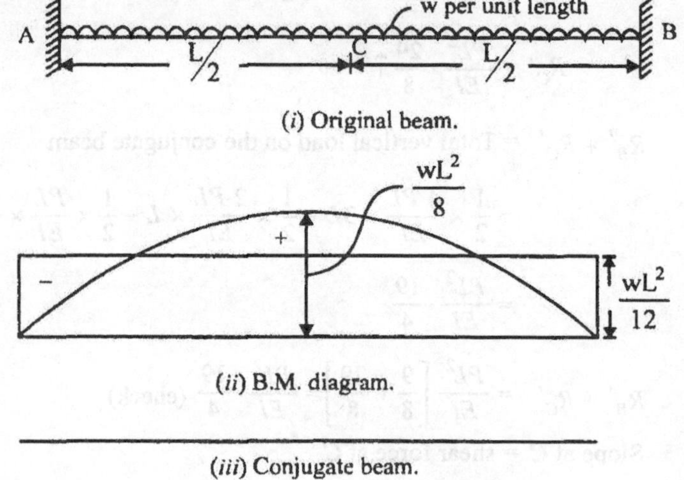

(*i*) Original beam.

(*ii*) B.M. diagram.

(*iii*) Conjugate beam.

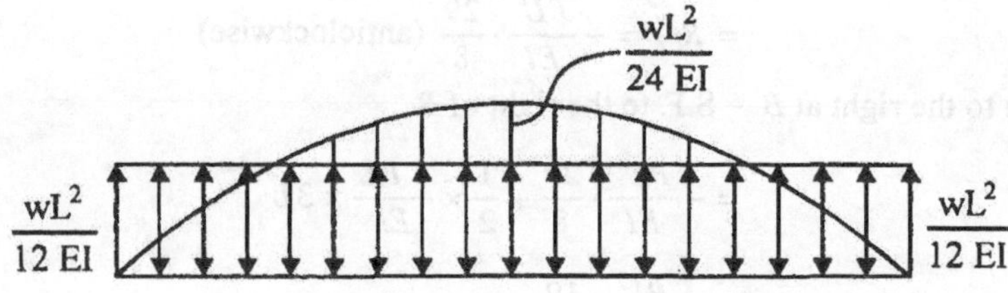

(*iv*) Conjugate beam with elastic loading.

Fig. 7.44. Example 7.20.

Deflection at C = B.M. at C

$$= \frac{wL^2}{12\,EI} \times \frac{L}{2} \times \frac{L}{4} - \frac{2}{3} \times \frac{wL^2}{8\,EI} \times \frac{L}{2} \times \frac{3}{8} \times \frac{L}{2}$$

$$= \frac{wL^4}{EI} \cdot \frac{1}{384} \text{ (downward)}$$

7.13. METHOD OF SUPERPOSITION

It is already explained in chapter 1 that method of superposition is valid for linear elastic analysis which forms the major part of this book. It is already shown how bending moment diagrams can be super posed i.e. how to draw the composite bending moment diagrams by superposing individual part bending moment diagrams. Similarly, it is possible to calculate total deflection by superposing deflection due to various causes. This is particularly useful in the case of a beam with overhang. Fig. 7.45 shows a beam with over hang.

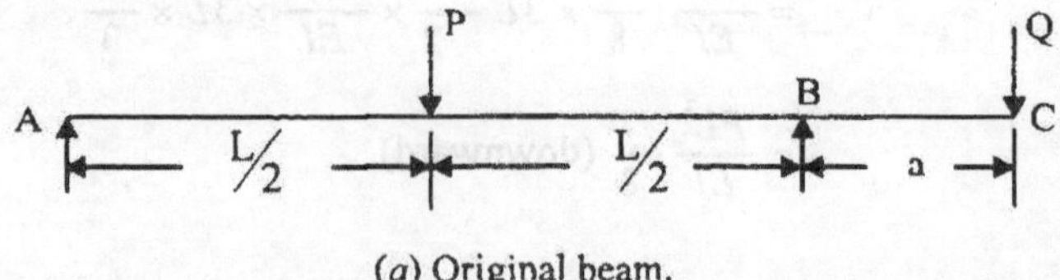

(*a*) Original beam.

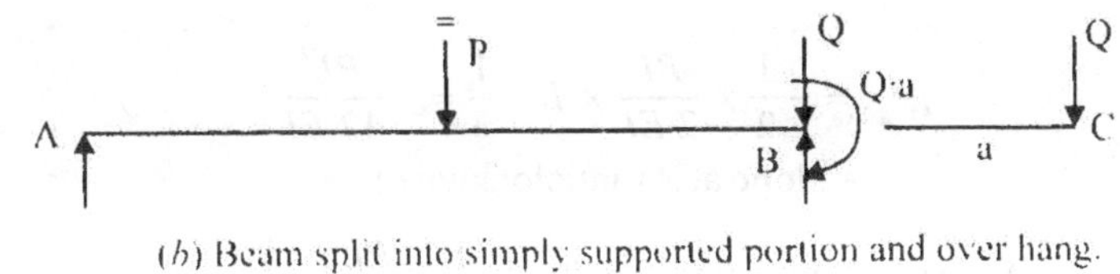

(*b*) Beam split into simply supported portion and over hang.

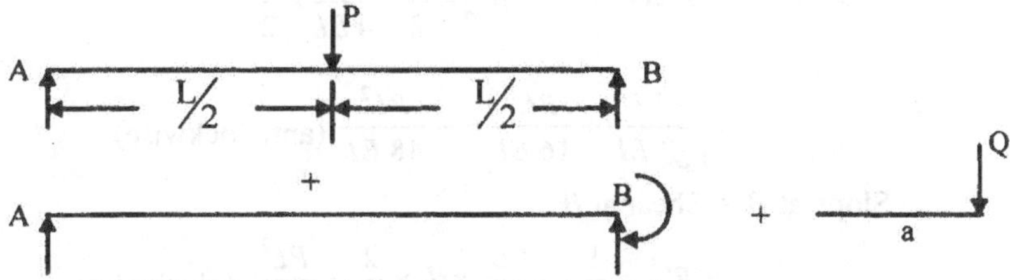

(*c*) Forces acting on different parts shown separately.

Fig. 7.45. Super position of forces.

The effect of the force Q on the support B is a force Q directly on the support and a moment Q.a as shown in Fig. 7.45 (b). For computing slopes and deflections for the simply supported portion the effects of load P and moment Q.a. are to be considered. To compute slope and deflection in the over hang portion slope at *B* from the simply supported portion should be first computed. On this slope the over hang is considered as a cantilever. These are explained by means of examples.

Example 7.21. Find the slopes at *A, C, B* and *D* and deflections at *C* and *D* of the beams with over hang. (Fig. 7.46.

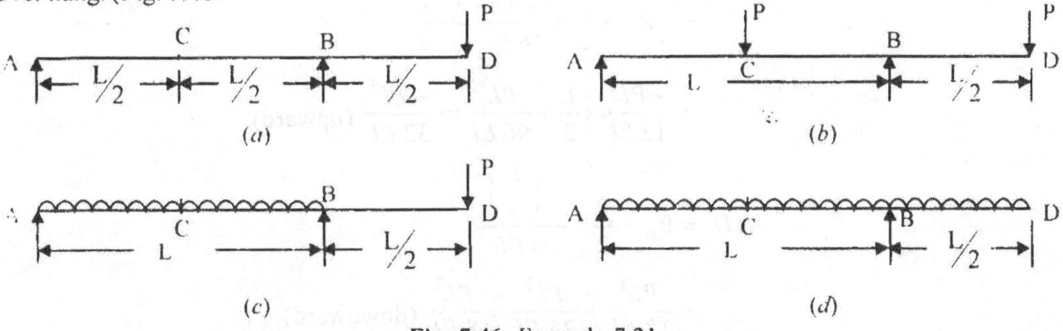

Fig. 7.46. Example 7.21

Solution : (*a*)

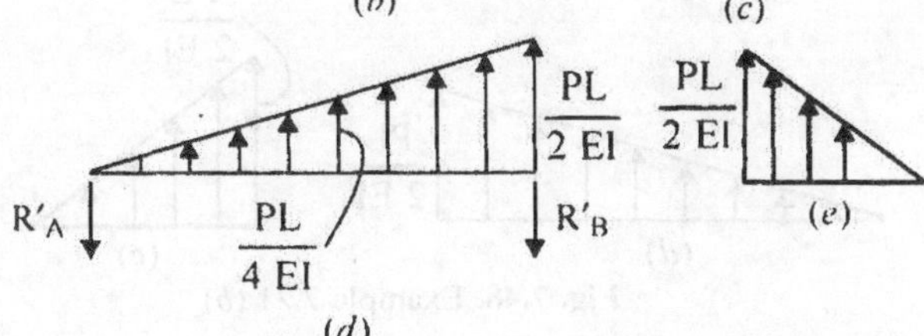

Fig. 7.47. Example 7.21. (*a*)

$$R'_A = -\frac{1}{2} \times \frac{PL}{2\,EI} \times L \times \frac{1}{3} = -\frac{PL^2}{12\,EI}$$

= slope at A (anticlockwise)

Slope C = Shear at $C = R'_A - \frac{1}{2} \times \frac{PL}{4\,EI} \times \frac{L}{2}$

$$= \frac{-PL^2}{12\,EI} + \frac{PL^2}{16\,EI} = -\frac{PL^2}{48\,EI} \text{ (anticlockwise)}$$

Slope at B = Shear at B

$$= R'_B = \frac{1}{2} \times \frac{PL}{2\,EI} \times L \times \frac{2}{3} = \frac{PL^2}{6\,EI} \text{ (clockwise)}$$

Slope at D = Slope at B + Slope at D due to loading between B and D.

$$= \frac{PL^2}{6\,EI} + \frac{P\left(\frac{L}{2}\right)^2}{2\,EI} = \frac{7\,PL^2}{24\,EI} \text{ (clockwise)}$$

Deflection

At C = Moment at C of the conjugate beam

$$= -R'_A \times \frac{L}{2} + \frac{PL^2}{16\,EI} \cdot \frac{1}{3} \cdot \frac{L}{2}$$

$$= \frac{-PL^2}{12\,EI} \times \frac{L}{2} + \frac{PL^3}{96\,EI} = \frac{-PL^3}{32\,EI} \text{ (upward)}$$

At D $= \theta_B \times \frac{L}{2} + \frac{P\left(\frac{L}{2}\right)^3}{3\,EI}$

$$= \frac{PL^3}{12\,EI} + \frac{PL^3}{24\,EI} = \frac{PL^3}{8\,EI} \text{ (downward)}$$

(*b*)

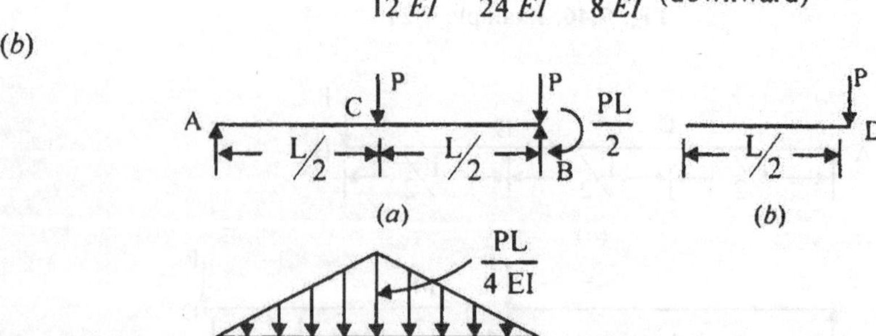

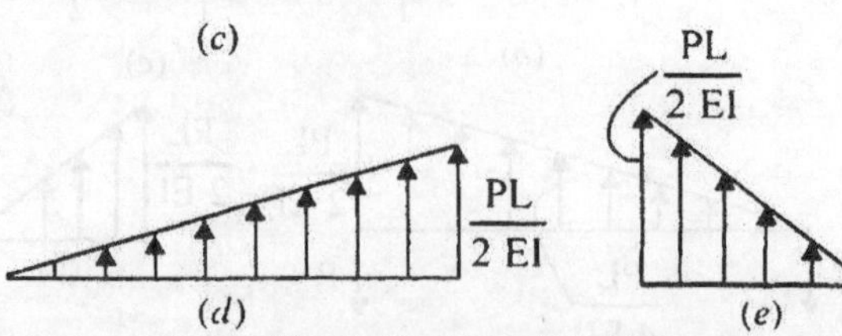

Fig. 7.48. Example 7.21 (*b*)

Slope at A θ_A due to concentrated load $= \frac{1}{2} \times \frac{1}{2} \times \frac{PL}{4\,EI} \cdot L = +\frac{PL^2}{16\,EI}$

$$\theta_A \text{ due to moment } = -\frac{PL^2}{12\,EI}$$

Net $$\theta_A = -\frac{PL^2}{12\,EI} + \frac{PL^2}{16\,EI} = -\frac{PL^2}{48\,EI} \text{ (anticlockwise)}$$

Slope at C

(*i*) Due to concentrated load = 0

(*ii*) Due to moment $= -\frac{PL^2}{48\,EI}$

Net Slope $= -\frac{PL^2}{48\,EI}$ (anticlockwise)

Slope at B

(*i*) Due to concentrated load $= -\frac{PL^2}{16\,EI}$

(*ii*) Due to moment $= +\frac{PL^2}{6\,EI}$

Net $\theta_B = +\frac{PL^2}{6\,EI} - \frac{PL^2}{16\,EI}$

$= \frac{5}{48} \cdot \frac{PL^2}{EI}$ (clockwise)

Slope at D Slope at B + slope due to elastic loading between B and D.

$$= \frac{5}{48} \cdot \frac{PL^2}{EI} + \frac{P\left(\frac{L}{2}\right)^2}{2\,EI} = +\frac{11}{48}\,\frac{PL^2}{EI} \text{ (clockwise)}$$

Deflection at C = Deflection due to moment + Deflection due to concentrated load at centre

$$= -\frac{PL^3}{32\,EI} + \frac{PL^3}{48\,EI} = -\frac{PL^3}{EI} \cdot \frac{1}{96} \text{ (upward)}$$

Deflection at D

$$= \theta_B \times \frac{L}{2} + \frac{P\left(\frac{L}{2}\right)^3}{3\,EI}$$

$$= \frac{5}{48} \cdot \frac{PL^2}{EI} \cdot \frac{L}{2} \downarrow + \frac{PL^3}{24\,EI} \downarrow$$

$$= \frac{PL^3}{EI} \cdot \frac{9}{96} \downarrow$$

(c) (Fig. 7.49)

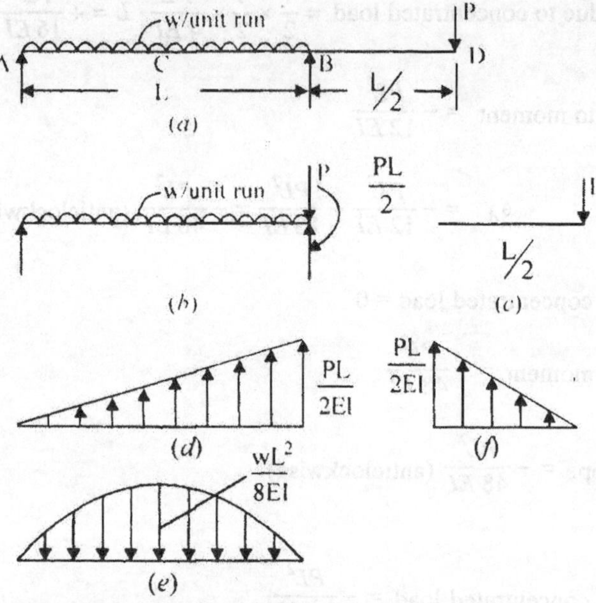

Fig. 7.49. Example 7.21 (c).

Slope at A,
$$\theta_A = -\frac{PL}{2}\cdot\frac{L}{6\,EI} + \frac{1}{2}\cdot\frac{2}{3}\cdot\frac{wL^2}{8\,EI}\cdot L$$

$$= -\frac{PL^2}{12\,EI} + \frac{wL^3}{24\,EI}$$

$$\theta_C = -\frac{PL^2}{48\,EI} + 0 = -\frac{PL^2}{48\,EI}$$

$$\theta_B = -\frac{PL^2}{6\,EI} + \frac{wL^3}{24\,EI}$$

θ_D = slope at B + slope in cantilever due to load P

$$= \frac{-PL^2}{6\,EI} + \frac{wL^3}{24\,EI} - \frac{P\left(\frac{L}{2}\right)^2}{2\,EI}$$

$$= \frac{L^2}{EI}\left[\frac{wL}{24} - \frac{7P}{24}\right]$$

Deflection at C

= Deflection due to uniformly distributed load

+ Deflection due to moment.

$$= \frac{5}{384}\cdot\frac{wL^4}{EI} - \frac{PL^3}{32\,EI}$$

$$= \frac{L^3}{32\,EI}\left[\frac{5\,wL}{12} - P\right]$$

$$\text{Deflection at } D = \text{Slope at } B \times \frac{L}{2} + \text{Deflection due to concentrated load } P$$

$$= \left(-\frac{PL^2}{6\,EI} + \frac{wL^3}{24\,EI}\right) \times \frac{L}{2} - \frac{P\left(\frac{L}{2}\right)^3}{3\,EI}$$

$$= \frac{wL^4}{48\,EI} - \frac{PL^3}{8\,EI}$$

(*d*)

(*a*)

(*b*) (*c*)

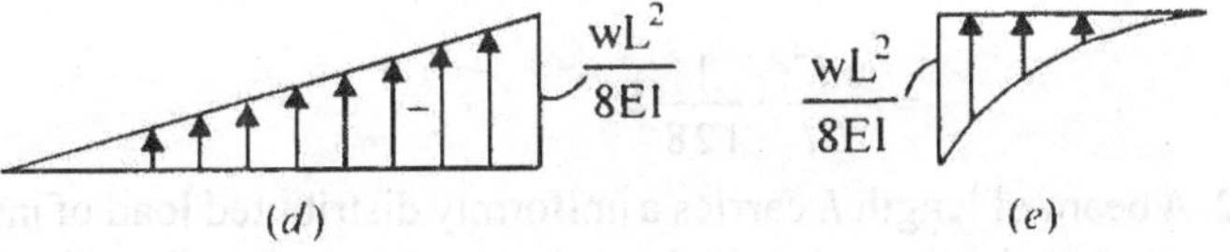

(*d*) (*e*)

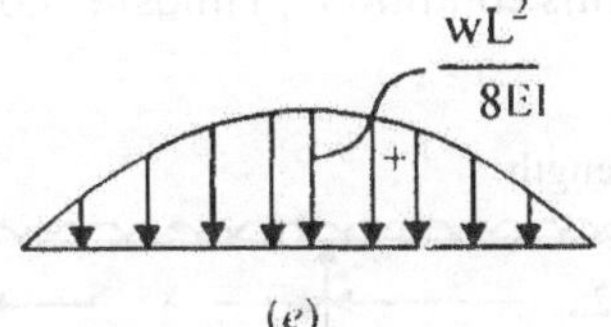

(*e*)

Fig. 7.50. Example 7.21 (*d*).

$$\text{Slope at } A,\ \theta_A = \frac{wL^3}{24\,EI} - \frac{wL^2 \times L}{8 \times 6\,EI}$$

$$= \frac{wL^3}{48\,EI} \text{ (clockwise)}$$

$$\theta_C = 0 - \left[\frac{1}{2} \cdot \frac{wL^2}{8\,EI} \cdot L \times \frac{1}{3} - \frac{1}{2} \cdot \frac{wL^2}{16\,EI} \cdot \frac{L}{2}\right]$$

$$= -\frac{wL^3}{EI} \cdot \frac{1}{192} \text{ (anticlockwise)}$$

$$\theta_B = \frac{wL^2}{8} \cdot \frac{L}{3\,EI} - \frac{wL^3}{24\,EI} = 0$$

$$\theta_D = \text{Slope at } B + \text{Slope at } D \text{ due to cantilever portion}$$

$$= 0 + \frac{w\left(\frac{L}{2}\right)^3}{6\,EI} = \frac{wL^3}{48\,EI} \quad \text{(clockwise)}$$

Deflection C

$$\Delta_C = \text{Deflection due to uniformly distributed load} + \text{deflection due to moment}$$

$$= \frac{5\,wL^4}{384\,EI} - \frac{wL^2}{8} \times \frac{L^2}{16\,EI}$$

$$= \frac{wL^4}{EI} \cdot \frac{1}{192}$$

Deflection at D

$$\Delta_C = \text{Slope at } B \times \frac{L}{2} + \text{deflection of cantilever}$$

$$= 0 \times \frac{L}{2} + \frac{w\left(\frac{L}{2}\right)^4}{8\,EI}$$

$$= \frac{wL^4}{EI} \cdot \frac{1}{128} \downarrow$$

Example 7.22. A beam of length L carries a uniformly distributed load of intensity w. Determine the length of overhangs to make the numerical maximum value of the bending moment as small as possible. Determine the deflection at the middle for this condition. [Timoshenko]

Solution;

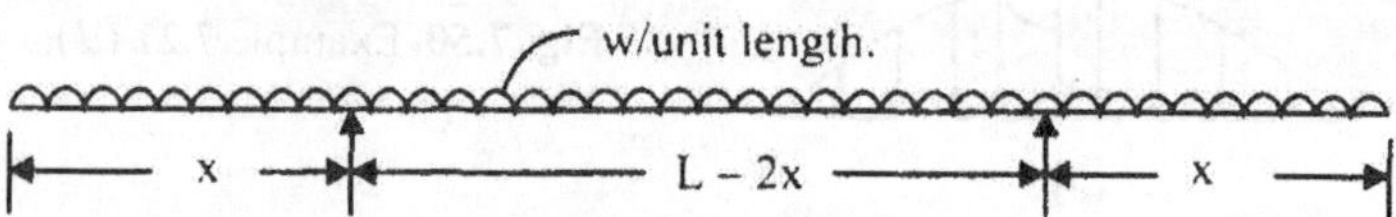

Fig. 7.51. Example 7.22

Numerical maximum value will be as small as possible when negative moment at the support will be equal to positive moment at the mid span.

$$\text{Negative moment at support} = \frac{wx^2}{2}$$

$$\text{Positive moment at mid span} = \frac{wL}{2} \cdot \frac{(L-2x)}{2} - w\left(\frac{L}{2}\right)^2 - \frac{1}{2}$$

$$= \frac{wL^2}{8} - \frac{wLx}{2}$$

Equating numerically negative and positive moments

$$\frac{wx^2}{2} = \frac{wL^2}{8} - \frac{wLx}{2}$$

or $\qquad x = 0.207\,L$

Deflection at middle

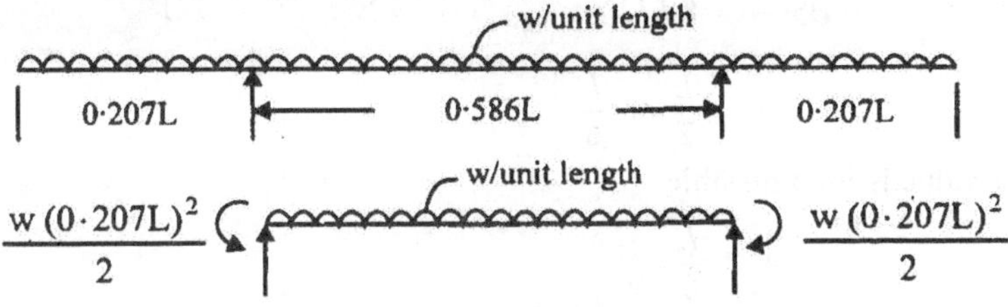

Fig. 7.52. Example 7.22

$$\delta\,\text{middle} = \frac{5w}{384}\frac{(0.586\,L)^4}{EI_Z} - \frac{w(0.207\,L)^2}{2}\cdot\frac{(0.586\,L)^2}{8\,EI}$$

$$= 0.00355\frac{wL^4}{EI_Z}$$

Example 7.23. A beam of length L with two equal overhangs is loaded by two equal forces at the ends. Determine the ratio $\frac{x}{L}$ at which

(i) the deflection at the middle is equal to the deflection at either end.

(ii) the deflection at the middle has its maximum value. [Timoshenko]

Solution:

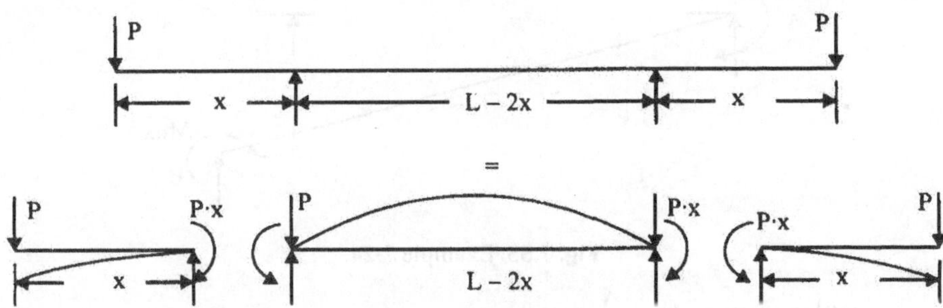

Fig. 7.53. Example 7.23.

$$\text{Deflection at middle} = \frac{P\cdot x\cdot(L-2x)^2}{8\,EI}$$

$$\text{Deflection at end} = \frac{P\cdot x(L-2x)\cdot x}{2\,EI} + \frac{Px^3}{3\,EI}$$

Equating the two,

$$\frac{Px(L-2x)^2}{8\,EI} = \frac{Px^2(L-2x)}{2\,EI} + \frac{Px^3}{3\,EI}$$

Solving $x = 0.705\,L$ or $0.152\,L$

As the first value is not admissible $x = 0.152\,L$

$$\text{Deflection at middle, } \Delta_C = \frac{Px(L-2x)^2}{8\,EI}$$

$$\frac{\partial \Delta_C}{\partial x} = \frac{P}{8\,EI}\left[(L-2x)^2 + 2x(L-2x)(-2)\right] = 0$$

or $$x = \frac{L}{2} \text{ or } \frac{L}{6}$$

As the first value is inadmissible

$$x = \frac{L}{6}.$$

Example 7.24. Find the fixed end moments generated when there is a relative displacement Δ between the two supports A and B of a fixed end beam as shown in Fig. 7.54.

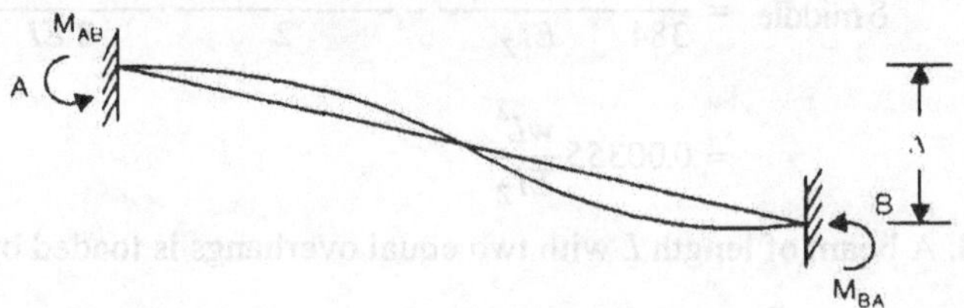

Fig. 7.54.

Solution: We can visualise the fixed end beam as an inclined simply supported beam with moments at A and B so that slopes at ends A and B are reduced to zero as shown in Fig. 7.55

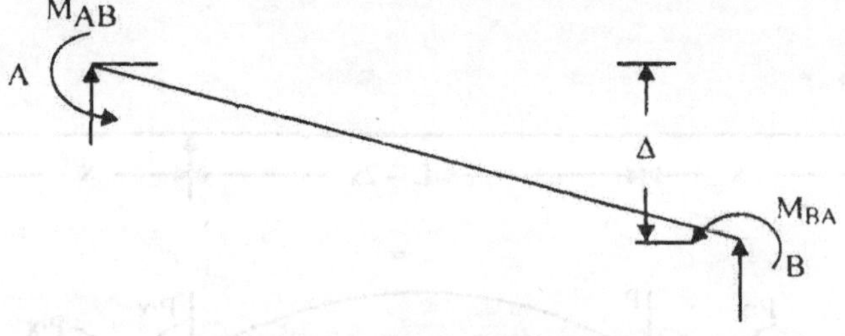

Fig. 7.55. Example 7.24.

$$\text{Slope at } A \text{ or } B = \frac{\Delta}{L}$$

$$\text{Slope at } A = \frac{M_{AB}\cdot L}{3\,EI} - \frac{M_{BA}\cdot L}{6\,EI} = \frac{\Delta}{L}$$

Similarly,

$$\text{Slope at } B = \frac{M_{BA}\cdot L}{3\,EI} - \frac{M_{AB}\cdot L}{6\,EI} = \frac{\Delta}{L}$$

$$\text{Solving } M_{AB} = M_{AB} = \frac{6\,EI\,\Delta}{L^2}$$

From B.M sign convention

M_{AB} is hogging i.e. negative

M_{BA} is sagging i.e. positive

SUMMARY

1. Calculation of deflection is of great importance in structural analysis. Deflection computations are required in solving statically indeterminate structures. Methods for computing deflections can be broadly classified as

   1. Geometrical Methods.

   2. Energy Methods.

   This chapter deals with geometrical method.

2. In this chapter the following geometrical methods are explained.

   1. Double Integral Method

   2. Area Moment Method

   3. Method of Concentrated Angle Changes.

   4. Conjugate Beam Method.

3. When there are a large number of concentrated loads and moments, double integration method is very tedious. To overcome this difficulty singularity function and Macaulay's method can be utilised.

4. Moment Area method can be utilised to compute slopes and deflections of beams and frames. Though it does not straight way gives absolute slopes or deflections of structures and there should not be any discontinuity between the points considered it can be used to compute absolute slopes and deflections at any point.

5. Method of concentrated angle changes is derived from moment area methods and is suitable for solution of frame problems. The method is systematic, efficient and can be organised in a tabular form.

6. Conjugate beam method, on the other hand gives straight way the slope and deflection at any point. However, support conditions should be suitably modified to get the correct result.

PROBLEMS

A. Theoretical

7.1 (*i*) State the conjugate beam theorems **[AMIE Summer '94]**

(*ii*) State Moment Area Theorems for obtaining slopes and deflections in cases of beams. How can you obtain the deflection at any point of a simply supported beam by moment area theorems? **[AMIE Winter 94]**

(*iii*) State the two moment area theorems with illustrative sketches. **[AMIE Summer 1981]**

(*iv*) Explain the "double integration" method of finding deflection and slope at a point of a simply supported beam. **[AMIE Winter 1985]**

(*v*) Explain the advantages and disadvantages of "Moment Area Method" **[AMIE Summer 1986]**

(*vi*) Explain with example the Macaulay's method of double integration for determining slopes and deflections of beams. **[AMIE Summer 1990]**

B. Objective. Tick the correct answers.

7.2 (*i*) A simply beam of 8 m span is loaded by a uniformly distributed load and has maximum deflection of 16 mm. If the span is halved (i.e. 4 m) and the loading is doubled. The maximum deflection will be

(*a*) 2 mm (*b*) 4 mm (*c*) 8 mm (*d*) 16 mm **[AMIE Summer, '97]** (*ii*) If $\frac{M}{EI}$ diagram is loaded on a conjugate beam, the shear force at any section gives.

(*a*) the end reaction in the real beam.

(*b*) the slope at that section in the real beam.

(*c*) the deflection at that section in the real beam.

(*d*) none of the above. **[AMIE Summer '97]**

(*iii*) The maximum deflection of a cantilever beam loaded uniformly over its entire length is 1.2 cm. If the whole load is concentrated at the free end, the maximum deflection would be

(*a*) 2.4 cm (*b*) 3.2 cm (*c*) 0.45 cm (*d*) 3 cm **[AMIE Winter 1998]**

(iv) Bending moment at any section in a conjugate beam gives in the actual beam.,

(*a*) slope (*b*) deflection (*c*) curvature **[AMIE Winter '97]**

(v) The fixed support in a real beam becomes in the conjugate beam

(*a*) roller support (*b*) hinged support (*c*) free support **[AMIE Winter '97]**

(vi) If fixity of support increases in a simply supported beam central deflection

(*a*) increases (*b*) decreases (*c*) no change. **[AMIE Winter '97]**

(*vii*) For a cantilever beam subjected to a point load at its mid length the maximum slope (a) is at its free end (*b*) is at the point of application of the load (*c*) has a constant value over the length between the load point and the free end **[AMIE Summer '99]**

(viii) A point load of 20 kN acting at the quarter span point of a simply supported beam produces a central deflection of 2 mm. For a central load of 40 kN. The deflection at the quarter span point will be.

(*a*) 4 mm (*b*) 8 mm (*c*) 2 m (*d*) 1 m **(AMIE Summer '99]**

(*ix*) At the location of an intermediate simple support of a continuous beam, the corresponding conjugate beam will be

(*a*) a simple support (*b*) a fixed support (*c*) a free end (*d*) internal hinge

[AMIE Summer 99]

(*x*) A point load of 50 kN acting centrally on a simply supported beam produces the same maximum deflection in the beam as caused by a *u.d.l* of 20 kN/m over the whole span. The span of the beam is

(*a*) 10 m (*b*) 8 m (*c*) 6 m (*d*) 4 m

C. Mathematical

7.3. (*i*) A simply supported beam of span L is loaded with two equal loads W at points distant $\frac{L}{4}$ from each end. Show that the central deflection is $\frac{11}{384}\cdot\frac{WL^3}{EI}$

(*ii*) A beam so loaded is to be propped at the centre so that the maximum positive and negative moments are numerically equal. Determine the central reaction and central deflection interms of w, L and EI **[AMIE Summer '97]**

7.4. A propped cantilever AB of length = 10 m is fixed at A and propped at B to the same level of A. The beam is loaded with a load of 2 t/m run over the entire length. Take EI = constant.

(*a*) Find the reaction at A and B (by Area Moment Theorem)

(*b*) Draw the bending moment diagram, shear force diagram showing the point of contra flexure and also determine the slope at B. Draw the deflected shape of the beam.

[AMIE Summer '78]

7.5. A simply supported beam with overhang is subjected to a moment M at the support A and a concentrated load P at the centre of the span. Compute (*i*) the slope and deflection at the point C (*ii*) deflection under load P. (Fig. P.7.1.) **[BE college (DU) 1996]**

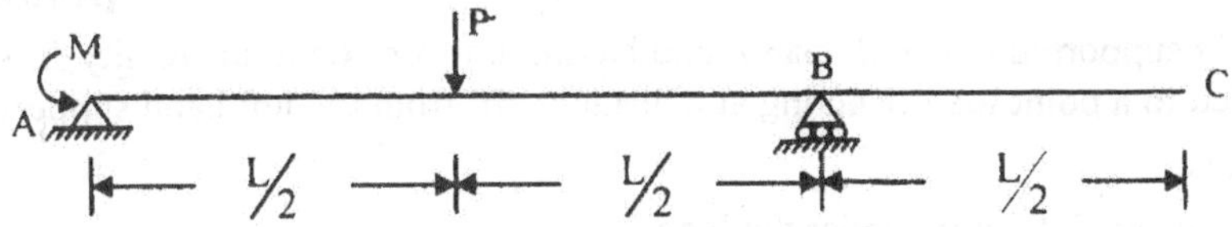

Fig. P 7.1. Prob. 7.5.

7.6. A propped cantilever with overhang hinged at C is loaded with 5 kN at A as shown in Fig. P 7.2. Find the slope and deflection at A. [B.E. College (DU) 1996]

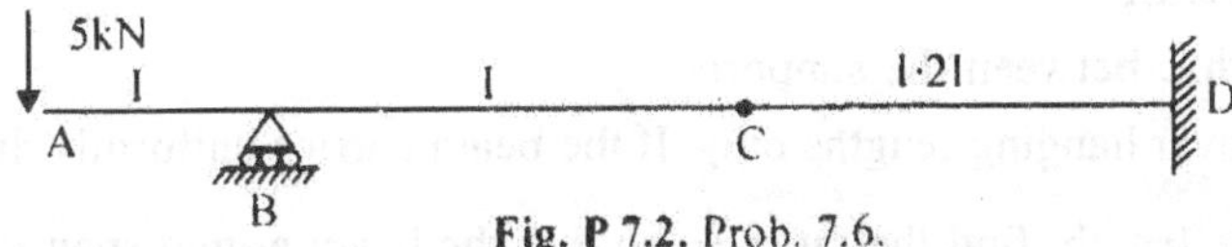

Fig. P 7.2. Prob. 7.6.

7.7. A beam built in at A and B of span L of constant EI carries a load W at a distance $\frac{L}{3}$ from A. Find (*i*) the bending moments and reactions at supports. (*ii*) deflection under the load, the position and magnitude of the maximum deflection (*iii*) the positions of the points of contraflexure. Apply area moment theorems for analysis. **[AMIE Winter 1978]**

7.8. A simply supported beam of uniform flexural rigidity ($EI = 2000$ tm^2) and span 8 m is subjected to a couple (clockwise) of 10 tm at a distance of 6 m from the left support. Calculate with the help of moment area theorems (*i*) the slope at the left support (*ii*) the deflection at the point where the couple is applied (*iii*) the maximum deflection in the beam.

[AMIE Summer 1981]

7.9. A simply supported beam AB of span 6 m carries a point load of 10t at a distance of 4 m from left hand support A. The flexural rigidity of the beam (EI) is uniform throughout the span and is equal to 3000 tm^2. Calculate by moment area theorems the deflection of the beam under the load and the maximum deflection of the beam and the location where it would occur.

[AMIE Winter 1981]

7.10. AB is a cantilever of span 4 m having uniform flexural rigidity 4000 tm^2 and it carries a triangular load having maximum intensity of 8 t/m at the fixed at A and zero intensity at the end B. Using the method of successive integration of the differential equation from bending of beams (*a*) Find the slope at B (*b*) Find the deflection at B. (*c*) If the cantilever is provided with an unyielding prop at B, determine the magnitude of prop reaction (*d*) In the latter case locate the point of inflection. **[AMIE Winter 1984]**

7.11. A vertical member ABC of total height h has relative moment of inertia 2I for the bottom $\frac{2}{3}$ rd height AB and I for top $\frac{1}{3}$ rd height of BC. It is fixed at its base A and free at the top C and is subjected to three loads as follows. A uniformly distributed horizontal load of intensity $\frac{P}{2h}$

over the full height AC, a clockwise couple of magnitude $\frac{Ph}{4}$ and a horizontal load of 1.2P from left to right both acting at B. Find with the help of area moment theorems the slope at B and horizontal deflection at C interms of P, h, E and I, E being young's modulus of the material.

[AMIE Summer 85]

7.12. **A simply supported beam of span L and having uniform flexural rigidity EI through out is subjected to a point load W acting at a distance 'a' from the left hand support.**

Determine

(*i*) Slope and deflection under the load.

(*ii*) Deflection at mid span.

(*iii*) The support reaction if an unyielding support is provided at mid span.

7.13. **A beam of uniform section and total length $(2l + L)$ is simply supported on span L with two over hanging length l. Calculate the deflection at mid span section due to a uniformly distributed load covering.**

(*i*) the length L between the supports

(*ii*) the two over hanging lengths only. If the beam carries uniformly distributed load over the whole length, find the ratio $\frac{l}{L}$ so that the beam at mid span is just level with the supports.

[AMIE Summer 1990]

7.14. (*a*) **A beam 6 m long, simply supported at its ends, carries a point load at its centre. If the slope at its ends is not to exceed 1°, find the maximum deflection of the beam due to the load.**

(*b*) **A beam AB of span L is fixed at A, the other end B resting freely at the centre of a simply supported cross beam CD of span L and same cross section. If the beam is loaded with a uniformly distributed load w per unitlength, find the reaction at the end B from the cross beam without using any standard formula.**

[AMIE Summer 1991]

7.15. **A simply supported beam AB of span L with an over hang BC of length $\frac{l}{4}$ is subjected to a clockwise moment M at a point D at a distance $\frac{L}{3}$ from A and a concentrated load P at the free end C of the over hang. If the slope at the point D is zero, what is the value of M and the deflection at D interms of P, L and EI, EI being uniform flexural rigidity of the beam.**

[AMIE Summer 99]

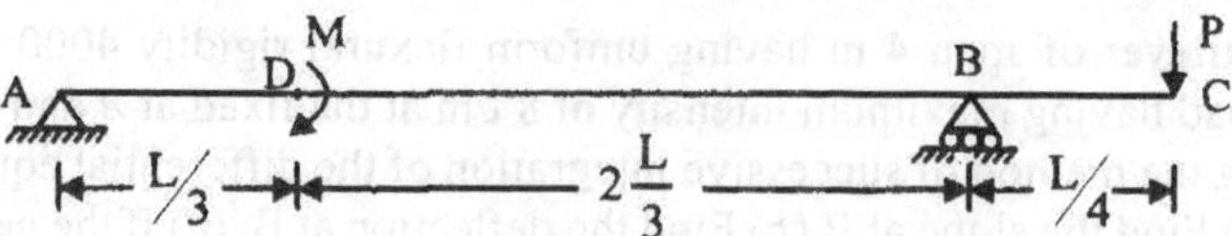

Fig. P 7.3 Prob. 7.15

8

FORCE DISPLACEMENT RELATIONS — ENERGY METHOD

8.1 INTRODUCTION

In the last chapter the geometric method of computing deflections was explained. The method is direct. The basic equations of equilibrium, compatibility, boundary conditions and the material stress-strain relations (Constitutive laws) were used to generate the governing differential equations. These equations were then solved by analytic (double integration), graphic or combined method (Moment Area). Deflection can also be calculated by energy method where internal energy is equated with external work. Euler called it as the "means of final causes". Significant pioneering work in this field was done by Euler, Bernoulli, Lagrange, Castigliano, Raleigh, Engesser, Westergaard and Ritz. In the broad generalised application of Energy methods two general principles (*i*) Virtual work (virtual displacements) and (*ii*) complementary virtual work (virtual forces) play a significant part.

8.2 CONSERVATION OF ENERGY, WORK AND STRAIN

Conservation of Energy is one of the basic law of physics and a "closed system" consisting of a structure and the applied forces must obey this law. Mathematically the work energy balance can be written as

$$W = E_S + E_L \qquad \text{... (8.1)}$$

where W = Work performed.

E_S = Energy stored in the body.

E_L = Energy Loss.

Now in a structure work is performed by the external load moving through a distance and energy is stored due to elastic deformation of the members. If the structure is static there is no Kinetic energy in the system. With no energy loss due to heat, permanent set etc. the equation reduces to

$$W = E_S$$

The usual terminology for E_S in a structure is elastic strain energy with the symbol U. Hence for a conservative structural system (no energy loss) $W = U$ ... (8.1*a*)

8.3 WORK AND COMPLEMENTARY WORK

In a structure the total external load cannot be applied all at once. It is applied gradually and hence the load displacement curve is not parallel to the X-axis but generally follows the non linear path as shown in Fig. 8.1.

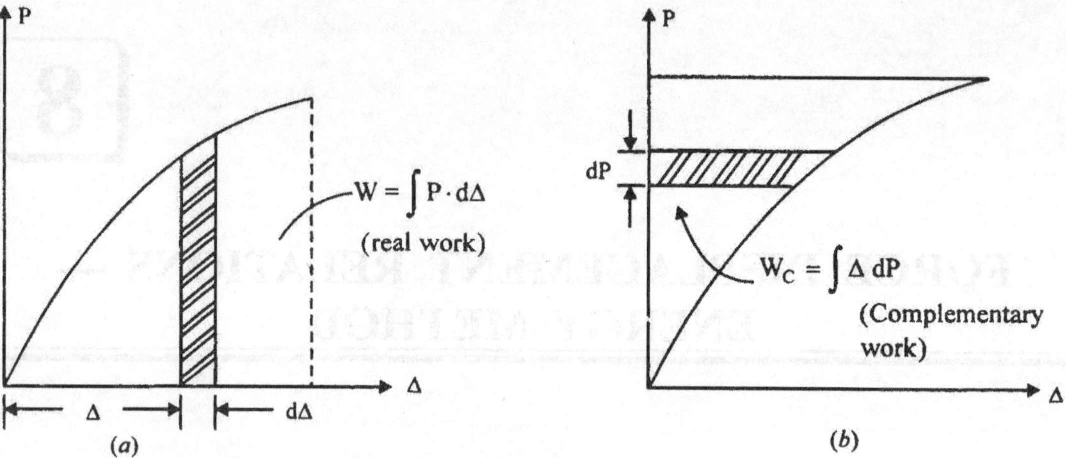

Fig. 8.1. Load deflection curve.

As the definition of work is force × displacement,

the work done as the force P moves through a distance $d\Delta$, $\Delta W = P \cdot d\Delta$,

and the total work done, (Fig. 8.1 a)

$$W = \int P\, d\,\Delta \qquad \text{...(8.2)}$$

If force P is three dimensional with components P_x, P_y, P_z the total work done will be

$$W = \int P_x \cdot d\Delta_x + \int P_y\, d\Delta_y + \int P_Z d\Delta_Z \qquad \text{...(8.2 a)}$$

This work is known as real work and graphically it is below the load-deflection curve. It can be physically understood.

Complementary work W_c is defined as $\int \Delta \cdot dP$ and as shown in Fig. 8.1 b it is the area above the load deflection curve. It has no physical meaning. However it is very useful in structural analysis.

Since in this book we are dealing with linear elastic analysis, load deflection curve should be linear. This is correct if the material constitutive law is linear and small displacement theory is applicable. The linear load-displacement curve in shown in Fig. 8.2.

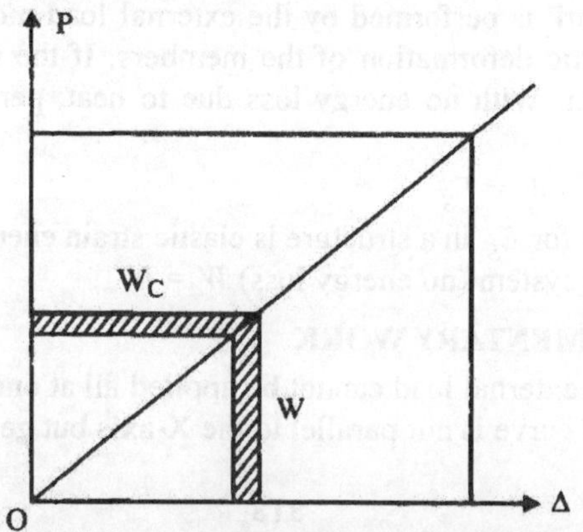

Fig. 8.2. Linear Load Deflection Curve.

As can be easily seen from Fig. 8.2 for linear load deflection curve

$$W = W_C = \frac{1}{2} P \cdot \Delta \text{ i.e. area below the curve}$$

$$= \text{area above the curve.}$$

8.4. EXPRESSIONS FOR STRAIN ENERGY AND COMPLEMENTARY ENERGY

General expression for strain energy in a deformable body subjected to generalised stress is now derived. Fig 8.3. shows the coordinate axes and the generalised stresses acting on an element of sides *dx, dy* and *dz.*

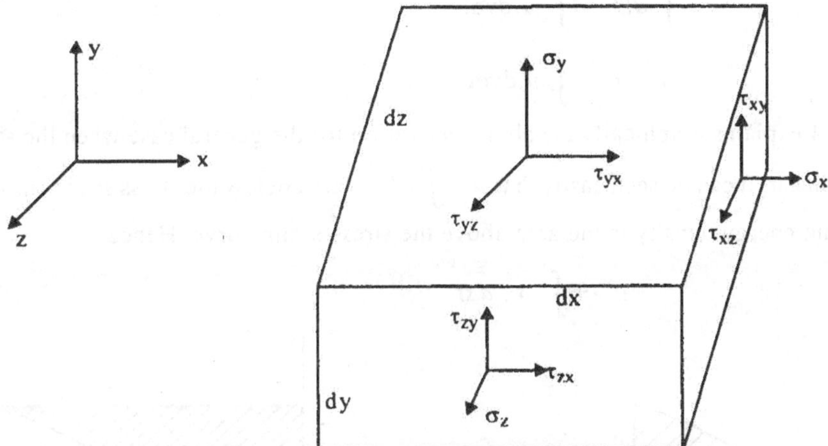

Fig. 8.3. Generalised stresses.

Because of these stresses there will be corresponding strains. Stresses multiplied by areas will give forces. Strains multiplied by corresponding length will give displacement. Force multiplied by displacement will give internal work done or strain energy stored.

Symbolically force acting in the *x* direction.

$$= \sigma_x \ (dy\, dz)$$

Incremental strain in the *x* direction $= d \in_x$

Displacement increment $= dx\ d \in_x$

Hence incremental work done

$$= \sigma_x . (dx\ dy\ dz)\ (d \in_x)$$

$$= \sigma_x . d \in_x \cdot \text{dvol}$$

As the material strains from $\in_x = 0$ to any arbitrary value $\in_x = \in_{\bar{x}}$ work done on the elementary volume.

$$= \int_0^{\in_{\bar{x}}} \sigma_x \cdot d \in_x \cdot \text{dvol}$$

This is strain energy stored in the elementary volume and is designated dU.

Hence $$dU = \int_0^{\in_{\bar{x}}} \sigma_x \cdot d \in_x \cdot \text{dvol} \qquad \text{... (8.3)}$$

and $$\frac{dU}{dV} = \int_0^{\in_{\bar{x}}} \sigma_x \cdot d \in_x \qquad \text{... (8.3 a)}$$

$$= \text{strain energy density}$$

$$= u \text{ (small)}$$

Strain energy density measures the energy stored per *unit* volume.

If all the stresses are considered total energy will be the sum of energy stored due to all causes. Hence energy per unit volume.

$$u = \int \sigma \cdot d\epsilon_x + \int \sigma_y . d\epsilon_y + \int \sigma_z . d\epsilon_z$$

$$+ \int \tau_{xy} \, dr_{xy} + \int \tau_{xz} d\, r_{xz} + \tau_{yz} . dr_{yz} \quad ...(8.3b)$$

Total strain energy in a given volume

$$\int dU = \int u.\text{dvol}$$

or

$$U = \int u.\text{dvol}$$

Fig. 8.4 explains graphically the above derivation for the general case when the stress strain curve is nonlinear. It can be seen easily that $u = \int \sigma d\epsilon$ = area below the stress strain curve Complementary strain energy density is the area above the stress-strain curve. Hence

$$u_c = \int \varepsilon . d\sigma \quad ...(8.4)$$

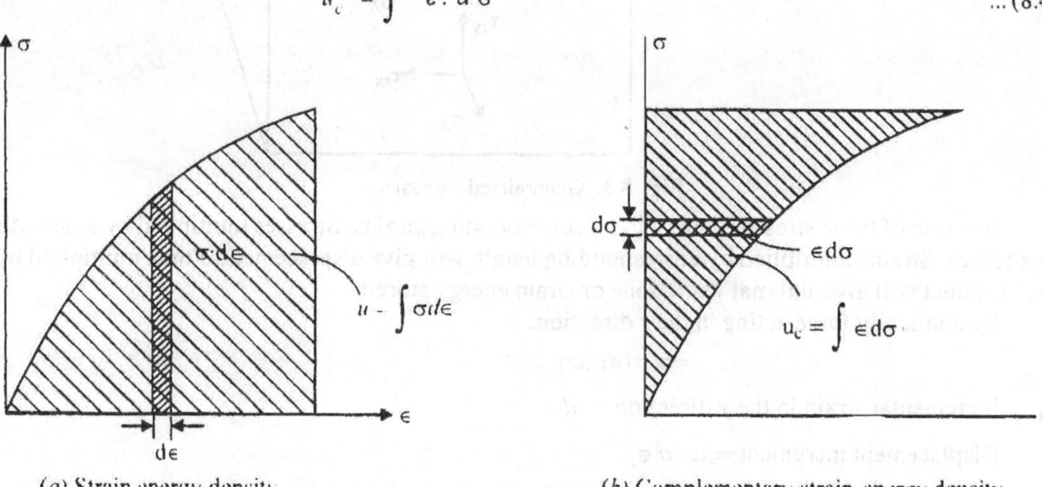

(*a*) Strain energy density (*b*) Complementary strain energy density

Fig. 8.4. Strain energy density and complementary strain energy density.

and total complementary strain energy

$$U_C = \int u_c . dvol$$

$$= \int \varepsilon . d\sigma . dvol \quad ...(8.6)$$

When the stress strain curve is linear

$$u = u_c = \frac{1}{2} \sigma . \varepsilon . \quad ...(8.6)$$

When energy stored due to all stresses are taken into account the expression becomes.

$$u = u_c = \frac{1}{2}(\sigma_x \varepsilon_x + \sigma_x \varepsilon_y + \sigma_z \varepsilon_z + \tau_{xy} \gamma_{xy} + \tau_{xz} . \gamma_{xz} + \tau_{yz} . \gamma_{yz}) \quad ...(8.6\,a)$$

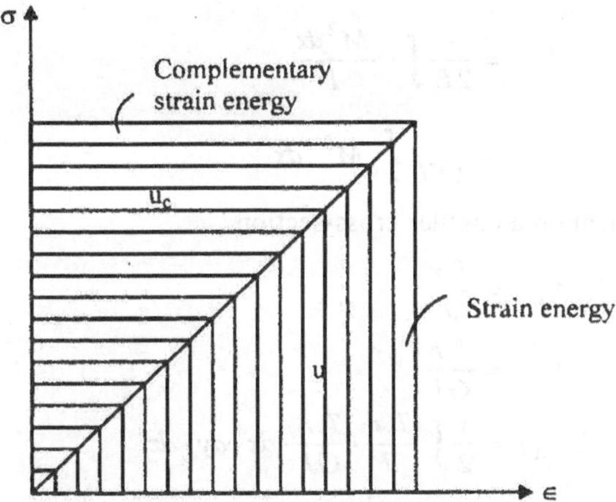

Fig. 8.5. Strain energy and complementary strain energy for linear elastic structure.

8.5. EXPRESSIONS FOR STRAIN ENERGY FOR LINEAR ELASTIC MEMBER

It is now possible to compute expressions for strain energy for different effects from the fundamental expressions already derived for linear elastic member of prismatic section.

(*i*) Axial Force

$$u = \frac{1}{2} \cdot \sigma \cdot \varepsilon$$

$$U = \frac{1}{2} \int \sigma . \varepsilon . dV$$

$$= \frac{1}{2} \int \sigma . \frac{\sigma}{E} \, dx . dy . dz$$

For member of length L and constant cross section and constant force

$$\sigma = \frac{F}{A}, \quad \int dy . dz = A$$

and

$$U = \frac{1}{2} \int \frac{F}{A} \cdot \frac{F}{AE} \cdot dx \cdot A$$

$$= \frac{1}{2} \int \frac{F^2 \, dx}{AE} \qquad \text{...(8.7)}$$

$$= \frac{1}{2} \frac{F^2 L}{AE} \qquad \text{as } \int dx = L.$$

(*ii*) Bending moment M.

$$\sigma = \frac{M}{I} \cdot y$$

$$U = \frac{1}{2} \int \frac{M}{I} \cdot y \frac{M.y}{I \cdot E} \cdot dx \, dy \cdot dz$$

$$= \frac{1}{2E} \int \frac{M^2 . dx}{I^2} \cdot \int y^2 \, dy . dz$$

$$= \frac{1}{2E} \cdot \int \frac{M^2}{I^2} \cdot I \cdot dx$$

$$= \frac{1}{2E}\int \frac{M^2 dx}{I}$$

$$= \frac{1}{2\,EI}\int M^2 .dx \qquad \text{... (8.8)}$$

(*iii*) Torsional moment on a circular cross-section.

$$\tau_{xy} = \frac{T.r}{J}$$

$$\gamma_{xy} = \frac{T.r}{GJ}$$

$$U = \frac{1}{2}\int \frac{T.r}{J}\cdot\frac{T.r}{GJ}\cdot dx\cdot dy\cdot dz$$

$$\frac{1}{2G}\int \frac{T^2}{J^2}dx\cdot\int r^2 dy.dz$$

$$= \frac{1}{2\,GJ^2}\cdot J\int T^2\cdot dx \qquad \text{as } \int r^2 dy\cdot dz = J$$

$$= \frac{1}{2\,GJ}\cdot\int T^2\cdot dx \qquad \text{...(8.9)}$$

Expression for torsional stresses are derived based on elementary theory of torsion, neglecting warping effect. It is true only for circular cross sections. But for other cross-sections J is to be derived. For example for a square cross-section $J = 0.1406\ a^4$. where a is the side of a square.

(*iv*) Shear force on a rectangular section.

$$\tau_{xy} = \frac{VQ}{Ib}\quad \gamma_{xy} = \frac{VQ}{Ib.G}$$

$$U = \frac{1}{2}\int \frac{VQ}{Ib}\cdot\frac{VQ}{Ib.G}\cdot dx\cdot dy\cdot dz$$

$$= \frac{1}{2G}\int V^2\ dx\cdot\int\frac{Q^2.dy.dz}{I^2b^2}$$

Now $\int \frac{Q^2.dy.dz}{I^2b^2}$ **can be written as** $\frac{1}{I^2}\int\left(\frac{Q}{b}\right)^2.dA = \frac{1}{A_S}$

A_s is called the effective shear area

Then $$U = \frac{1}{2G}\cdot\int\frac{V^2\,dx}{A_S} \qquad \text{...(8.10)}$$

8.5.1. Computation of $\frac{1}{I^2}\int\left(\frac{Q}{b}\right)^2.dA$

(*i*) For a rectangular section $b \times d$.

$$\frac{1}{I^2}\int\left(\frac{Q}{b}\right)^2\cdot dA = \frac{1}{\left(\frac{bd^3}{12}\right)^2}\cdot\int\left[\frac{b}{2}\left(\frac{d^2}{4}-y^2\right)\cdot\frac{1}{b}\right]^2 b.dy$$

$$= \frac{144}{b^2.d^6}\cdot\frac{b}{4}\int_{-d/2}^{d/2}\left(\frac{d^2}{4}-y^2\right)^2\cdot dy$$

$$= \frac{36d^5}{bd^6} \cdot \frac{16}{480} = \frac{1.2}{bd} = \frac{1}{A_s}$$

Hence for a rectangular section $A_s = \frac{bd}{1.2}$. This factor is termed as shape factor whose value is dependent on actual shear stress distribution over the cross-section.

(*ii*) Thin circular section of thickness '*t*' and mean radius r_m.

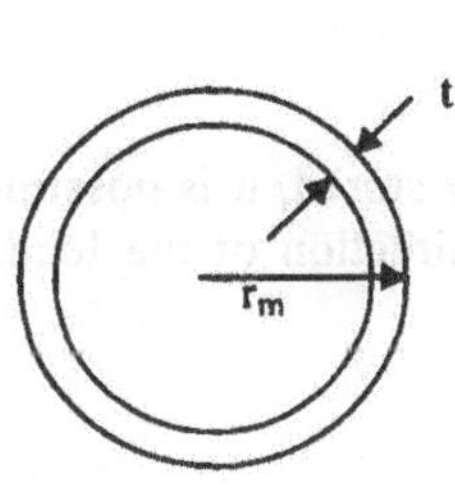

(*a*) Thin circular tube

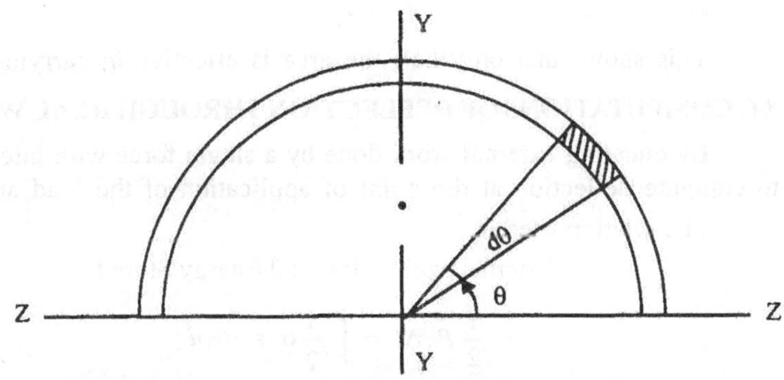

(*b*) Part of circular tube

Fig. 8.6. Circular tube.

Area of element of tube $= r_m . d\theta . t$

distance from Z axis $= r_m . \sin\theta$

Q = Moment of area above θ about Z axis

$$= 2\int_{\theta}^{\pi/2} r_m . d\theta . t . r_m . \sin\theta = 2\, r_m^{\ 2} . t \cos\theta$$

I = Moment of inertia of a thin tube about the centroid axis.

$$= \int_0^{\pi/2} 4\, r_m . d\theta . t . (r_m . \sin\theta)^2$$

$$= 4\, r_m^{\ 3} . t \int_0^{\pi/2} \sin^2\theta . d\theta$$

$$= 4 r_m^{\ 3} . t . \frac{\pi}{4} = \pi t . r_m^{\ 3}$$

$$\frac{1}{A_S} = \frac{1}{I^2} \int \left(\frac{Q}{b}\right)^2 . dA$$

$$= \frac{1}{\pi^2 t^2 r_m^6} \cdot \int_0^{2\pi} \left(\frac{2r_m^{\ 2} . t . \cos\theta}{2t}\right)^2 \cdot r_m . d\theta . t$$

$$= \frac{r_m^{\ 5} . t . \int_0^{2\pi} \cos^2\theta . d\theta}{\pi^2 . t^2 . r_m^{\ 6}}$$

$$= \frac{\pi . t . r_m^{\ 5}}{\pi^2 . t^2 . r_m^{\ 6}}$$

$$= \frac{1}{\pi . t . r_m}$$

or $\quad A_S = \pi . t . r_m$

or $\quad \frac{A_S}{A} = \frac{\pi . t . r_m}{2\pi r_m . t} = \frac{1}{2}$

or $\quad A_S = \frac{A}{2}$

This shows that only half the area is effective in carrying shear.

8.6. COMPUTATIONS OF DEFLECTION THROUGH REAL WORK

By equating external work done by a single force with internal energy stored, it is possible to compute deflection at the point of application of the load and in the direction of the load.

The relation used is

$$\text{External work} = \text{Internal Energy Stored}$$

$$\frac{1}{2} P . \Delta = \int \frac{1}{2} \sigma . \varepsilon . dvol .$$

However the method has two major limitations.

(*i*) If the applied loads are more than one energy balance is still valid. However, the equation will not lead to finding out deflection. For example, with twoloads P_1 and P_2 the energy balance is

$$\frac{1}{2} P_1 \Delta_1 + \frac{1}{2} P_2 \Delta_2 = \int \frac{1}{2} \sigma . \varepsilon . dvol .$$

With this one equation neither Δ_1 nor Δ_2 can be computed.

(*ii*) The second limitation is that P and Δ must correspond i.e. we can compute deflection only at the point of application of P and in the direction of P. We cannot compute deflection at any other point or at any other direction at P.

Because of these two limitations, method of real work is of very little use. However, in the next example, its limited use is demonstrated.

Example 8.1. Find

(*i*) The rotation of point B due to applied moment M.

(*ii*) Deflection of point B due to load P. Take shear deflection into account. Examine the magnitude of shear deflection compared to that due to bending.

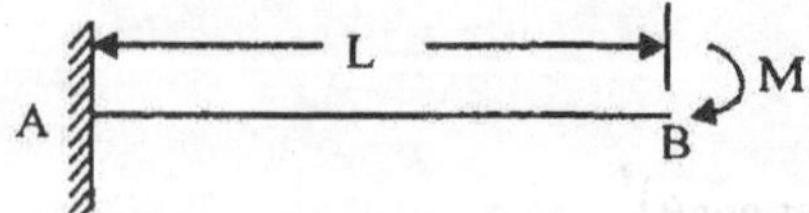

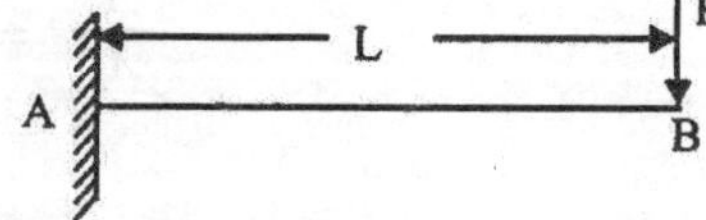

Fig. 8.7. Example 8.1.

Solution:

(*i*) Strain Energy stored due to bending

$$= \int_0^L \frac{M^2 dx}{2 EI} = \frac{M^2 L}{2 EI}$$

External work done by $\quad M = \frac{M . \theta}{2}$

Equating $$\frac{M^2 L}{2EI} = \frac{M\theta}{2}$$

or $$\theta = \frac{ML}{EI}$$

We cannot find out the vertical deflection at B as it does not correspond with the direction of moment M.

(*ii*) Strain energy stored due to bending and shear.

$$= \int_0^L \frac{M^2 dx}{2\,EI} + \int \frac{V^2\,dx}{2\,A_S.G}$$

$$= \int_0^L \frac{(-Px)^2\,.dx}{2\,EI} + \frac{P^2.L \times 1.2}{2 \times A.G.}$$

$$= \frac{P^2 L^3}{6\,EI} + \frac{1.2\,P^2 L}{2\,AG}$$

External work done $= \dfrac{P.\Delta}{2}$

Equating work done with energy stored.

$$\frac{P.\Delta}{2} = \frac{P^2.L^3}{6\,EI} + \frac{1.2\,P^2 L}{2\,AG}$$

or $$\Delta = \underbrace{\frac{PL^3}{3EI}}_{\text{Deflection due to B.M.}} + \underbrace{\frac{1\cdot 2\,PL}{AG}}_{\text{Deflection due to S.F.}}$$

Usually deflection due to B.M. is very large compared to that due to S.F.

$$\text{Ratio of } \frac{\text{deflection due to shear}}{\text{deflection due to B.M.}} = \frac{1.2\,PL}{AG} \times \frac{3\,EI}{PL^3}$$

$$= \frac{1.2 \times 3 \times E \times bd^3 \times L}{12 \times b \times d \times G \times L^3}$$

$$= 0.3\,\frac{E}{G} \times \left(\frac{d}{L}\right)^2$$

If $\dfrac{E}{G} = 2.5$ Ratio $= 0.75 \times \left(\dfrac{d}{L}\right)^2$

If $\dfrac{d}{L} = 1$, it is a case of deep beam and deflection due to shear is 0.75 times deflection due toB.M. and hence significant.

If $\dfrac{d}{L} = \dfrac{1}{10}$, shear deflection = 0.0075 of bending deflection and hence can be neglected.

8.7. VIRTUAL WORK AND COMPLEMENTARY VIRTUAL WORK

The limitations of real work in computing deflections are obvious. This is over come by using the virtual work principle. The word "virtual" means imaginery. This type of work may not physically exist but is a very useful concept in structural analysis. The load displacement curve of a structure is shown in Fig. 8.8.

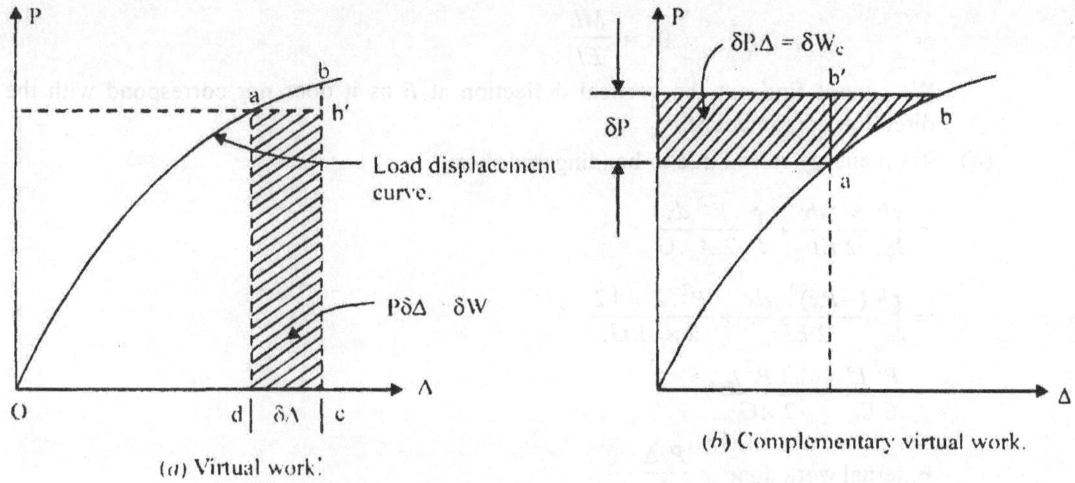

Fig. 8.8. Virtual work and Complementary Virtual Work.

If we now give the structure a small imaginery (virtual) displacement $\delta\Delta$ (The prefix δ indicates virtual quantity). The virtual work $\delta W = P\delta\Delta$. Here P is real whereas $\delta\Delta$ is virtual. It is important to observe that the virtual displacement $\delta\Delta$ is along *ab'* and *not* along *ab* i.e. the load displacement curve. $\delta\Delta$ is independent of P and not caused by P.

Similarly virtual force is defined as a small imaginery force and is designated as δP. (δ indicates virtual quantity). The hatcned area in Fig. 8.8 b indicates complementary virtual work $\delta W_c = \delta P.\Delta$. Here force δP is imaginary (vitual) but displacement Δ is real. However the displacement Δ is not caused by δP. Corresponding to the load deflection curve we can draw the associated internal stress-strain curve (Fig. 8.9).

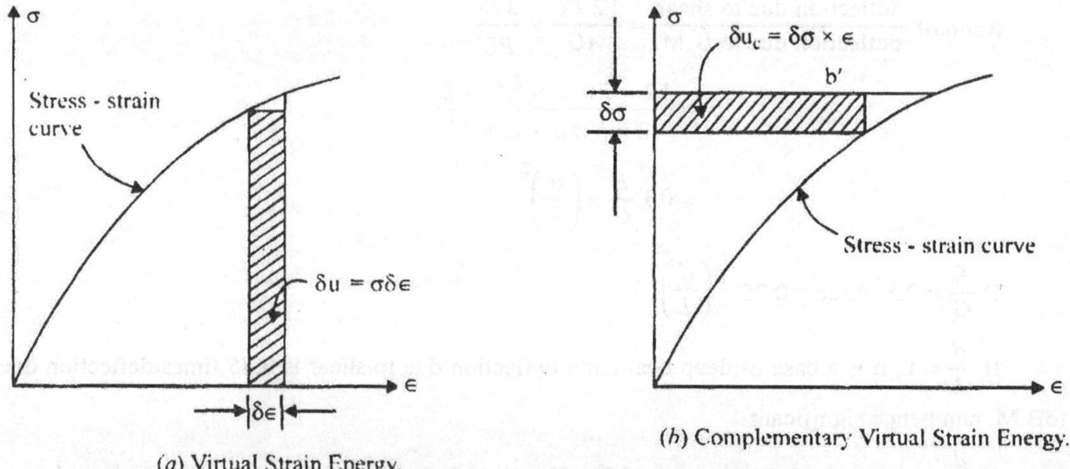

Fig. 8.9. Virtual Strain Energy and Complementary Virtual Strain Energy.

Instead of virtual displacement, there will be virtual strain and instead of virtual work there will be virtual strain energy density defined as $\delta u = \sigma\delta\varepsilon$. The prefixed symbol δ again indicates virtual strains and virtual strain energy density. Similarly if virtual stress $\delta\sigma$ is applied to real strain ε the product $\delta\sigma \times \varepsilon = \delta u_c$ i.e. complementry virtual work density.

Total virtual strain energy

$$\delta U = \int \sigma \delta \varepsilon . \, dvol \qquad ...(8.11)$$

and total complementary virtual.

$$\text{Strain energy } \delta U_C = \int \epsilon . \delta \sigma . dvol \qquad ...(8.12)$$

For ready reference the basic definitions can be summarised as shown in Table 8.1.

Table 8.1. Basic Work and Energy Definitions.

Symbol	Definition	Symbol	Definition
$W = \int P d\Delta$	Real Work (area below the P. Δ curve)	$W_C = \int \Delta dP$	Complementary work. (Area above P.Δ curve)
u (small) $= \int \sigma d\varepsilon$	Strain Energy density (Area below σ – ε curve)	u_c (small) $= \int \varepsilon d\sigma$	Complementary strain energy density (Area above σ – ε curve)
$U = \int u dvol$	Total Strain energy	$U_C = \int u_c dvol$	Total Complementary Energy
$\delta\Delta$	Virtual displacement	δP	Virtual Force
$\delta w = P\delta\Delta$	Virtual work	$\delta W_C = \Delta\delta P$	Complementary virtual work.
$\delta\varepsilon$	Virtual strain	$\delta\sigma$	Virtual stress
$\delta u = \sigma\delta\varepsilon$	Virtual strain energy density	$\delta u_c = \varepsilon\delta\sigma$	Virtual Complementary strain energy density
$\delta U = \int \sigma\delta\epsilon \, dvol$	Total virtual strain energy	$\delta U_C = \int \varepsilon\delta\sigma \, dvol$	Total Complementary Strain Energy.

8.8 PRINCIPLE OF VIRTUAL WORK AND COMPLEMENTARY VIRTUAL WORK (RIGID BODIES)

There are two principles. Where force is real and displacement virtual, the product of real force and virtual displacement leads to virtual work. Where force is virtual and displacement real the product leads to complementary virtual work. The first principle of virtual work or more properly the principle of virtual displacement states.

"If a body in equilibrium is given a small virtual displacement consistent with boundary conditions, the virtual work done by the real forces going through this virtual displacement is zero." Symbolically,

$\delta W = \Sigma P_i \, \delta \Delta_i = 0$ where P_i's are the real forces and $\delta\Delta_i$'s are the corresponding virtual displacements. To fix ideas let us consider a simply supported beam loaded as shown in Fig. 8.10.

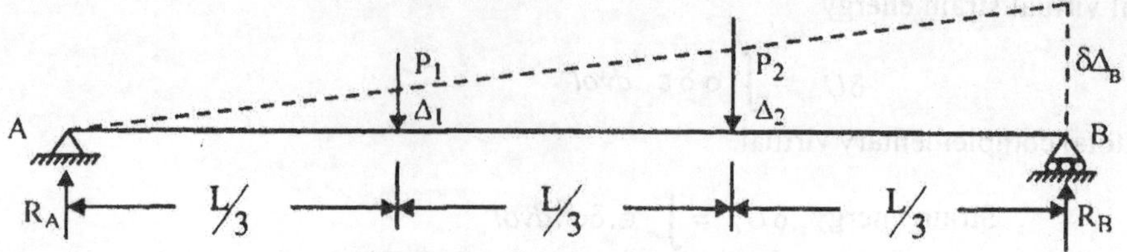

Fig. 8.10. Virtual displacement of rigid body.

If the rigid body is given a small virtual displacement $\delta\Delta_B$ at B, it will rotate about the hing point A. To be consistent with boundary conditions it cannot leave the point A. As the body is rigid, it cannot change its shape. Hence the displacement at point $1 = \Delta_1 = \dfrac{\delta\Delta_B}{3}$

at point 2 = $\Delta_2 = \dfrac{2\,\delta\Delta_B}{3}$

$$\delta W = \Sigma P_i\,\delta\Delta_i = (-P_1)\frac{\delta\Delta_B}{3} + (-P_2)\frac{2\delta\Delta_B}{3} + R_B.\delta\Delta_B = 0$$

(upward force and upward displacement are positive)

or $R_B - \dfrac{P_1}{3} - \dfrac{2P_2}{3} = 0$

or $R_B = \dfrac{1}{3}[P_1 + 2P_2]$

This is nothing but the equilibrium equation of all forces acting on the body about point A or the equation $\Sigma M_A = 0$.

Another equilibrium equation $\Sigma M_B = 0$ can be obtained by making small virtual displacement Δ_A at A.

The second principle is principle of complementary virtual work. Here force is virtual and displacement real. The principle is:

"If a set of virtual forces in equilibrium moves along the real displacements consistent with boundary conditions of a rigid body caused by a set of real forces, the total complementary virtual work given by the product of virtual force and corresponding real displacement is zero"

Symbolically $\delta W_C = \Sigma\,\delta P_i\,\Delta_i = 0$

Let us consider again the simply supported *rigid beam* hinged at A and on the roller at B. If by any force the end B is moved upward through Δ_B consistent with boundary conditions, Δ_1 will be displacement at 1 and Δ_2 at 2. If unit virtual force is applied at 1, the reactions at A and B are $\dfrac{2}{3}$ and $\dfrac{1}{3}$ respectively. The principle of complementary virtual work gives.

$$\delta W_C = \Sigma\,\delta P_i.d_i = (-1)\times\Delta_1 + \left(\frac{1}{3}\right)(\Delta_B) = 0 \text{ (Fig. 8.11)}$$

or $\Delta_1 = \dfrac{\Delta_B}{3}$

The condition $\Delta_1 = \dfrac{\Delta_B}{3}$ is the required compatibility for a rigid body, that is, the body cannot elongate or shorten but it simply rotates about point A. It can, therefore, be seen that principle of complementary virtual work produces equations of compability and can be used to compute displacement.

8.9. PRINCIPLE OF VIRTUAL WORK AND COMPLEMENTARY VIRTUAL WORK (DEFORMABLE BODIES)

In the case of deformable bodies, it is subjected to stresses and strains. Real external force will produce real internal stresses and real internal deformations or strains. Virtual force, on the other hand will produce internal virtual stresses and strains. Principle of virtual work relates real external force and coresponding external virtual deformation with real internal streses and corresponding internal virtual strains.

Let us take a simple problem.

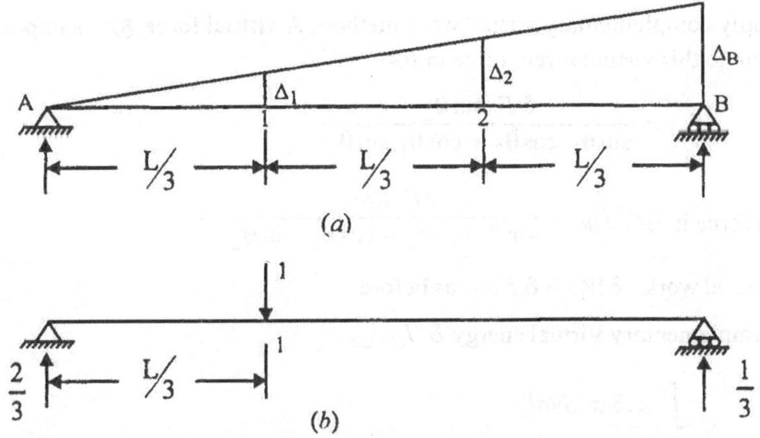

Let two bars *AB* and *BC* intersect at *B* and be subjected to a load *P* at *B*. (Fig. 8.12)

Resolving vertically,

$$F_{BA}\cos\theta_1 + F_{BC}\cos\theta_2 = P$$

This is equation of equilibrium in the vertical direction If the principle of vertual work is to be applied, an imaginery virtual displacement is imposed at *B*. The external virtual work $= P\delta\Delta$.

The internal total virtual strain energy $\delta U = \int \sigma\,\delta\,\varepsilon\; dvol$

Here because of discreet members $\delta U = \Sigma\sigma.\delta\,\varepsilon.dvol$ for two members *AB* and *BC*.

$$= \frac{F_{BA}}{a_1}\cdot\frac{\delta\Delta.\cos\theta_1}{l_1}\cdot a_1 l_1 + \frac{F_{BC}}{a_2}\cdot\frac{\delta\,\Delta.\cos\theta_2}{l_2}\cdot a_2 l_2$$

$$= (F_{BA}.\cos\theta_1 + F_{BC}.\cos\theta_2)\,\delta\Delta$$

Equating external virtual work δW with internal virtual strain energy δU equation of equilibrium $F_{BA}\cos\theta_1 + F_{BC}\cos\theta_2 = P$ is obtained which can be used for computing forces.

Suppose we are now intersected in computing vertical deflection Δ at the point *B*.

Resolving vertically

$$F_{BA}\cos\theta_1 + F_{BC}\cos\theta_2 = P$$

Resolving horizontally

$$F_{BA}\sin\theta_1 = F_{BC}\sin\theta_2$$

Substituting

$$F_{BA}\cos\theta_1 + F_{BA}\frac{\sin\theta_1}{\sin\theta_2}\cos\theta_2 = P$$

or $$F_{BA} = \frac{P\sin\theta_2}{\sin\theta_1\cdot\cos\theta_2 + \cos\theta_1\cdot\sin\theta_2}$$

Similarly $$F_{BC} = \frac{P\sin\theta_1}{\sin\theta_1\cdot\cos\theta_2 + \cos\theta_1\cdot\sin\theta_2}$$

We then apply complementary virtual work method. A virtual force δF is imposed at B in the direction of P. Due to this virtual force, force in BA,

$$f_{BA} = \frac{\delta F.\sin\theta_2}{\sin\theta_1\ \cos\theta_2 + \cos\theta_1.\sin\theta_2}$$

and virtual force in BC $$f_{BC} = \frac{\delta F\ \sin\theta_1}{\sin\theta_1\cos\theta_2 + \cos\theta_1\cdot\sin\theta_2}$$

External virtual work $\delta W_C = \delta F.\Delta.$ as before.

Internal Complementary Virtual energy δU_C

$$= \int \varepsilon.\delta\sigma.dvol$$

For discreet members this is equal to

$$= \Sigma\ \varepsilon.\delta\sigma.dvol$$

For member BA.

$$\in = \frac{P\sin\theta_2}{\sin\theta_1\cos\theta_2 + \cos\theta_1\sin\theta_2}\cdot\frac{1}{a_1}\cdot\frac{1}{E}$$

$$\delta\sigma = \frac{\delta F\sin\theta_2}{\sin\theta_1\cos\theta_2 + \cos\theta_1\sin\theta_2}\cdot\frac{1}{a_1}$$

$$dvol = a_1 l_1$$

$$\Sigma\text{ Product} = \frac{P\delta F.\sin^2\theta_2}{(\sin\theta_1\cos\theta_2 + \cos\theta_1\cos\theta_2)^2}\cdot\frac{a_1 l_1}{a_1^2.E}$$

$$+\frac{P\delta F.\sin^2\theta_1}{(\sin\theta_1\cos\theta_2 + \cos\theta_1.\sin\theta_2)^2}\cdot\frac{a_2\ l_2}{a_2^2.E}$$

Equating this expression of U_C with external virtual work $\delta W_C = \delta F.\Delta$.

$$\Delta = \frac{P\sin^2\theta_2}{(\sin\theta_1\cos\theta_2 + \cos\theta_1\cdot\sin\theta_2)^2}\cdot\frac{l_1}{a_1 E}$$

$$+\frac{P\sin^2\theta_1}{(\sin\theta_1\cos\theta_2 + \cos\theta_2\sin\theta_2)^2}\cdot\frac{l_2}{a_2 E}$$

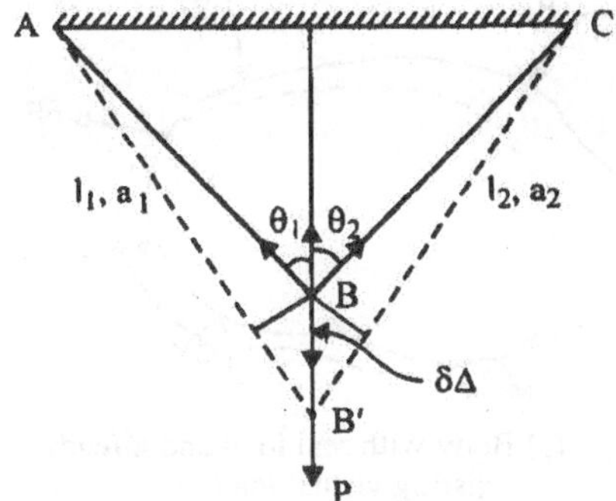

Fig. 8.12. Application of Virtual Work Principle

As only one deflection is required and that too at the load point, method of real work can also be applied giving.

$$\frac{P.\Delta}{2} = \frac{1}{2}\frac{F_{AB^2}.l_1}{a_1 E} + \frac{1}{2}\frac{F_{BC^2}l_2}{a_2 E}$$

$$= \frac{1}{2E}\left[\left(\frac{P\sin\theta_2}{\sin\theta_1\cos\theta_2 + \cos\theta_1\sin\theta_2}\right)^2 \cdot \frac{l_1}{a_1} + \left(\frac{P\sin\theta_1}{\sin\theta_1\cos\theta_2 + \cos\theta_1\sin\theta_2}\right)^2 \cdot \frac{l_2}{a_2}\right]$$

or $$\Delta = \frac{P}{E}\left[\frac{\sin^2\theta_2}{(\sin\theta_1\cos\theta_2 + \cos\theta_1\sin\theta_2)^2}\cdot\frac{l_1}{a_1} + \frac{\sin^2\theta_1}{(\sin\theta_1\cos\theta_2 + \cos\theta_1\sin\theta_2)^2}\cdot\frac{l_2}{a_2}\right]$$

as before.

8.10. DERIVATION OF PRINCIPLE OF COMPLEMENTARY VIRTUAL WORK FOR A GENERAL ELASTIC BODY

As shown in the last section this principle can be used to compute deflections of elastic body and is very frequently used. As such the principle merits a closer study.

We consider an elastic body subjected to a general real loading of $P_1, P_2, \dots P_r$. Suppose we are interested in computing the deflection at any point A due to these real loads. To compute the deflection at A we apply a virtual (or imaginary) point load δF at A in the direction in which deflection is desired. This is done before the real loads $P_1, P_2 \dots P_r$ are applied to the body.

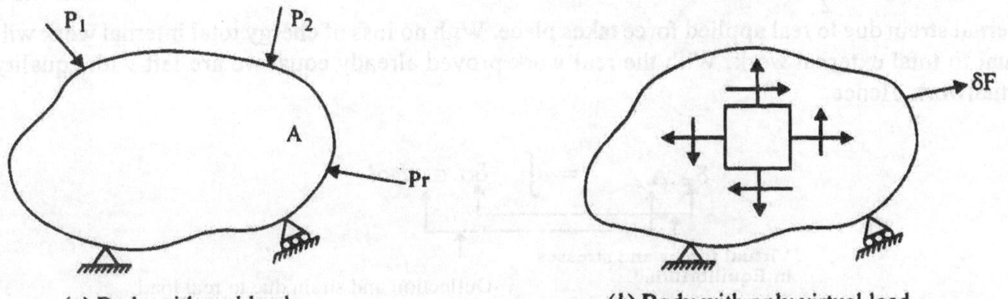

(a) Body with real load. (b) Body with only virtual load.

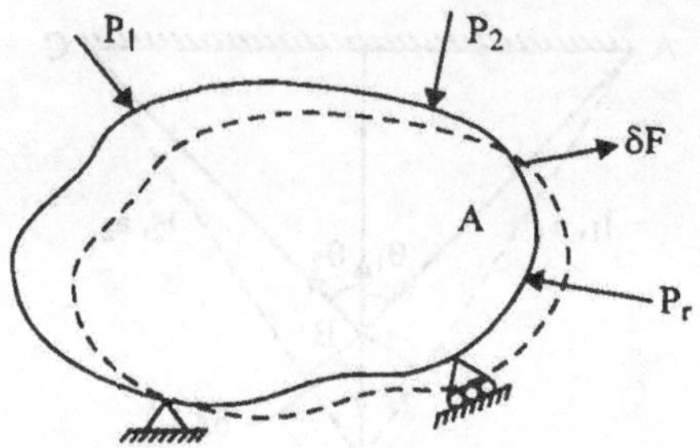

(c) Body with real load and already existing virtual load.

Fig. 8.13. Derivation of principle of complementary energy.

As a result of the virtual load δF, there will be virtual stresses and strains in each element of the body as shown in Fig. 8.13. (*b*). The body is in equilibrium and the external load and internal stresses balance.

Now, the real loads $P_1, P_2, \ldots P_r$ are applied and displacements in the body occur due to this real loading. Two types of work result:

1. Real work due to real loading and real displacement.

This is given by the expression

$$\underbrace{\frac{1}{2}\Sigma\,\Delta_i\,P_i}_{\text{External Real Work}} = \underbrace{\frac{1}{2}\int \epsilon\cdot\sigma\cdot \text{dvol}}_{\text{Internal Real Work.}} \qquad \ldots(8.13)$$

Here the body is assumed to be linearly elastic. Hence the factor $\frac{1}{2}$. However, linearity is not a restriction and the final result will be applicable to non-linear structure also.

2. Because of real external load, there will be real displacement at the point and direction where virtual force is applied and this will result in virtual work given by

$$\delta F\,.\,\Delta_A$$

The $\frac{1}{2}$ factor will not come as the virtual force δF is already there when the displacement occurs. Similarly internal virtual work due to already existing virtual stress is given by.

$$\int \delta\,\sigma\,.\,\varepsilon\, dvol$$

Here again $\frac{1}{2}$ factor is not there as the virtual stresses are already existing when the real internal strain due to real applied force takes place. With no loss of energy total internal work will be equal to total external work. With the real work proved already equal we are left with equality of virtualwork. Hence

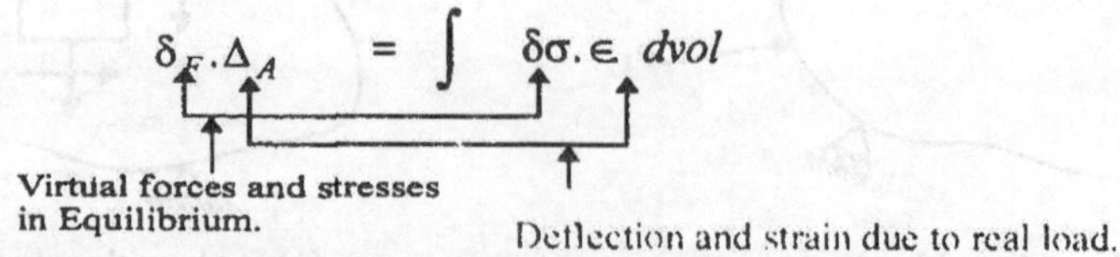

The principle may be stated in words as. "The external virtual work done by a virtual force moving through a real displacement due to applied real loads is equal to the internal virtual work

done by internal virtual stresses moving through compatible virtual strains."

The important points in the principle are

1. Virtual force and virtual stresses are in equilibrium i.e. virtual stresses are caused by virtual force.
2. Real applied force causes real external displacement and compatible internal strains.

Therefore steps in the method for computing displacements by applying virtual work principle are:

1. Apply a virtual force at the point and in the direction where deflection is required.
2. Compute internal virtual stresses.
3. Apply the real loads and compute real stresses and strains.
4. Apply the virtual work equation.

We now apply the above steps to the truss problem given in Fig. 8.14. We are interested in finding the horizontal and vertical displacement of point c. Instead of applying an arbitrary virtual force δF it is usual to apply a unit load (1 kN, 1 ton or 1 kg) as a virtual load which gives the name to the method "Dummy Load" or "unit load method".

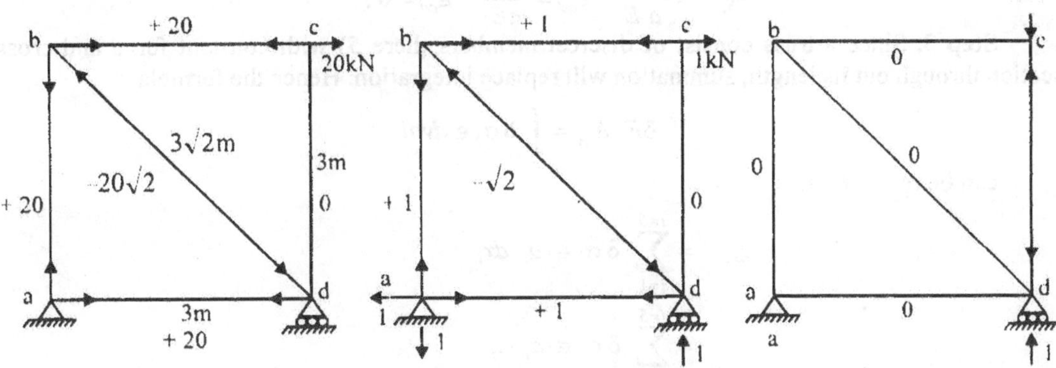

(*a*) Real force and loading.

(*b*) Unit horizontal force of 1 kN to compute horizontal displacement.

(*c*) Unit vertical force of 1 kN to compute vertical displacement at C.

Fig. 8.14. Truss Problem

Computation of Horizontal Displacement at *c*

Step 1. Apply 1 kN as virtual force in the horizontal direction. Due to this load forces in the members are (tension +, compression)

$$\delta F_{bc} = +1$$
$$\delta F_{bd} = -\sqrt{2}$$
$$\delta F_{ba} = +1$$
$$\delta F_{ad} = +1$$
$$\delta F_{cd} = 0$$

Step 2. If a is the area of all the numbers virtual stresses.

$$\delta \sigma_{bc} = +\frac{1}{a}$$
$$\delta \sigma_{bd} = -\frac{\sqrt{2}}{a}$$
$$\delta \sigma_{ba} = +\frac{1}{a}$$

$$\delta\,\sigma_{ad} = +\frac{1}{a}$$

$$\delta\,\sigma_{cd} = 0.$$

Step 3. The 20 kN horizontal force is now applied to the structure. Forces in the members due to this load are

$$F_{bc} = +20 \text{ kN}$$

$$F_{bd} = -20\sqrt{2} \text{ kN}$$
$$F_{ba} = +20 \text{ kN}$$
$$F_{ad} = +20 \text{ kN}$$
$$F_{cd} = +00 \text{ kN}$$

The strain in the members are

$$\epsilon_{bc} = +\frac{20}{aE} \quad \epsilon_{bd} = -\frac{20\sqrt{2}}{a\,E}$$

$$\epsilon_{ba} = \frac{+20}{a\,E} \quad \epsilon_{ad} = \frac{+20}{aE} \quad \epsilon_{cd} = 0\,.$$

Step 3. Since a truss consist of discreet members (here 5) with constant force and cross-section through out its length, summation will replace integration. Hence the formula

$$\delta F \cdot \Delta_A = \int \delta\sigma.\epsilon\, dvol$$

can be rewritten as

$$1 \cdot \Delta_C = \sum_{i=1}^{i=5} \delta\sigma \cdot \epsilon \cdot a_i\ dx_i$$

$$= \sum_{i=1}^{i=5} \delta\sigma \cdot \epsilon \cdot a_i \cdot l_i$$

$$= +\frac{1}{a} \cdot \frac{20}{a\,E} \cdot a \cdot 3 - \frac{\sqrt{2}}{a} \cdot \left(\frac{-20\sqrt{2}}{a\,E}\right) \cdot a \cdot 3\sqrt{2}$$

$$+2 \times \frac{1}{a} \times \frac{20}{a\,E} \cdot a \times 3 + 0$$

$$= \frac{180}{a\,E} + \frac{120\sqrt{2}}{a\,E} = \frac{1}{aE}(180 + 169.68)$$

$$= 349.68 \times \frac{1}{a\,E}$$

When $a = 1000\ \text{mm}^2$ and $E = 200$ GPa

$$\Delta_C = \frac{349.68 \times 1000}{1000 \times 200} = \mathbf{1.7484\ mm}$$

+ result indicates deflection is in the direction of unit horizontal load i.e. towards right.

Computation of vertical deflection at *c*

Step 1. Apply 1 kN as virtual force in the vertical direction at *c*. Due to this load forces in the members are :

$$\delta\,F_{bc} = 0$$

$$\delta F_{bd} = 0$$
$$\delta F_{ba} = 0$$
$$\delta F_{ad} = 0$$
$$\delta F_{cd} = -1.$$

Step 2. Virtual stresses in members.

$$\delta \sigma_{bc} = 0$$
$$\delta \sigma_{bd} = 0$$
$$\delta \sigma_{ba} = 0$$
$$\delta \sigma_{ad} = 0$$
$$\delta \sigma_{cd} = -\frac{1}{a}$$

Step 3. Forces in members due to applied load are as before

$$F_{bc} = +20$$
$$F_{bd} = -20\sqrt{2}$$
$$F_{ba} = +20$$
$$F_{ad} = +20$$
$$F_{cd} = 0.00$$

Step 4. Apply the formula for truss

$$1 . \Delta = \sum_{i=1}^{5} \delta \sigma_i \in_i \cdot a_i \cdot l_i$$

or $$\Delta = \left(-\frac{1}{a}\right) \times 0 \times a \times 3 = 0 .$$

Hence the vertical displacement at $c = 0$.

8.11. EXPRESSIONS FOR INTERNAL COMPLEMENTARY VIRTUAL WORK

The basic complementary virtual work equation is in terms of virtual stress and real strains and volume of members. Another way of looking at the equation is to equate internal virtual work i.e. internal virtual force × real internal displacement with external virtual force × real external displacement. Expressions are now derived from this point of view.

1. Axial Force—Let δP be the virtual force and P real axial force. If

L = Length of member
A = Area,
E = Young's modulus

then real internal displacement

$$= \frac{PL}{AE}$$

Virtual work equation then becomes

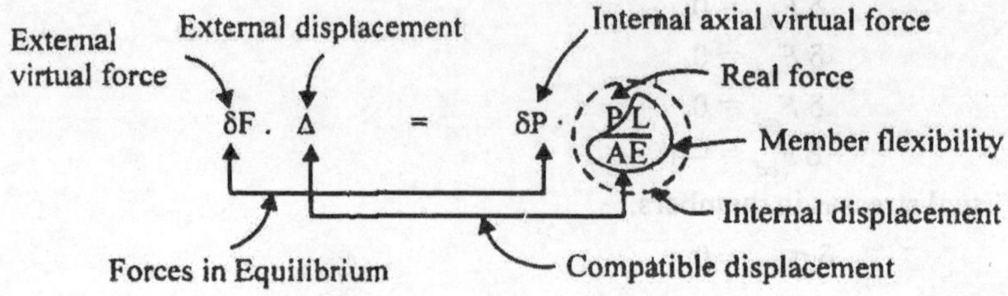

If there are a number of axial forces in members

$$\delta F.\Delta \;=\sum \delta\, P_i \cdot \frac{P_i L_i}{A_i\, E_i} \qquad \text{....(8.15)}$$

2. Bending Moment — Let $\delta\, M$ be the virtual moment, M real bending moment due to applied force, then external virtual work.

$$\delta\, F\cdot\Delta \;= \int \delta\sigma\cdot \in dvol$$

Internal stress due to B.M., $\delta\, M$

$$\delta\,\sigma \;= \frac{\delta\, M.Z.}{EI}$$

$$\text{Real strain } C \;= \frac{M.Z.}{EI}$$

$$\int \delta\,\sigma.\in dvol \;= \int \frac{\delta\, M.Z.}{I}\cdot\frac{M.Z.}{EI}\cdot dvol$$

$$= \int_x \int_A \frac{\delta\, M.Z.}{I}\cdot\frac{M.Z.}{EI}\, dA\,.dx$$

$$= \int_x \frac{\delta\, M.M.}{EI^2}\cdot dx\cdot \int Z^2.dA$$

But $\int z^2 dA \;= I$

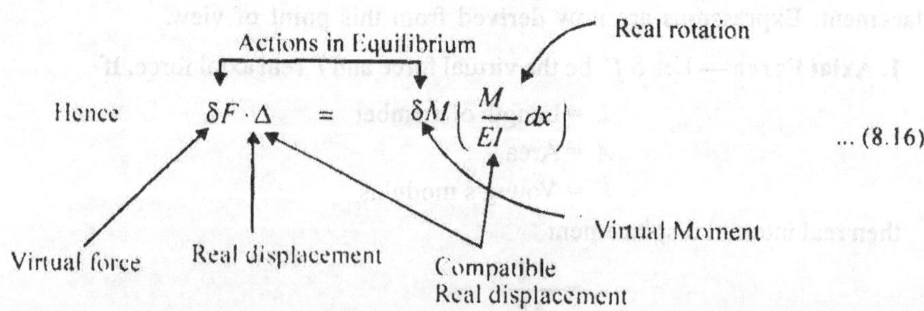

$$\text{Hence} \quad \delta F\cdot\Delta \;= \delta M\left(\frac{M}{EI}dx\right) \qquad \text{... (8.16)}$$

3. Shear — The basic equation

$$\Delta F.\Delta = \int \delta\sigma.\epsilon.dvol$$

remains valid but $\delta\sigma$ changes to shear stress $\delta\tau$ and ϵ changes to shearing strain γ.

If V is the applied shear, δV is the virtual shear force.

$$\delta\tau = \text{virtual shear stress due to shear force } \delta V$$

$$= \frac{\delta V.Q.}{Ib}$$

where Q = 1st moment of area above a distance Z about the centroidal axis

I = Moment of inertia about Z-axis

Similarly, $$\tau = \frac{VQ}{Ib}$$

$$\epsilon = \frac{VQ}{GIb}$$

$$\delta F.\Delta = \int\int \frac{\delta V.Q}{I.b.} \cdot \frac{VQ}{GIb} \cdot dA \cdot dx.$$

$$= \int \frac{\delta V.V.}{G} \cdot \frac{1}{I^2} \cdot dx \int \left(\frac{Q}{b}\right)^2 \cdot dA$$

or (External Virtual Force, External displacement) $$\delta F.\Delta = \int \delta V \frac{V}{A_S G} \cdot dx \quad \text{... (8.17)}$$

(Internal Virtual Shear δV; Internal Real displacement $\frac{V}{A_S G} \cdot dx$)

where $$A_S = \frac{I^2}{\int_A \left(\frac{Q}{b}\right)^2 \cdot dA}$$

Torsion

Torsional Shear stress $\tau = \frac{Tr}{J}$ for a circular member.

Torsional Strain $\gamma = \frac{\tau}{G} = \frac{Tr}{GJ}$

$$\delta\tau = \frac{\delta T.r.}{J}$$

Hence

$$\Delta.\delta F = \int \delta\tau.\gamma.dvol$$

$$= \int\int \frac{\delta Tr}{J} \cdot \frac{Tr}{GJ} \cdot dA \cdot dx$$

(External displacement Δ; External Virtual Force δF)

$$= \int_x \frac{\delta T.T.}{GJ^2} \cdot dx \int_A r^2 \cdot dA$$

$$= \int_x \frac{\delta T.T.}{GJ^2} \cdot J \cdot dx$$

$$= \int \delta T \cdot \left(\frac{T\,dx}{GJ}\right) \quad \text{...(8.18)}$$

Internal virtual torisonal force ↗ ↑ Real internal rotation

8.12. APPLICATION OF COMPLEMENTARY VIRTUAL WORK

Virtual work method is a versatile tool for structural analysis. Though deflection due to applied load is a general problem to be solved, deflection may be due to many other causes like.

(i) temperature variation

(ii) support settlement

(iii) fabrication error.

Generalised deflection includes rotation also. A few examples are given now to indicate different types of problems which can be solved by virtual work method.

Example 8.1. Find the angular displacement of bar *AB* of the structure shown in Fig 8.15. *A* and *E* for each member are the same.

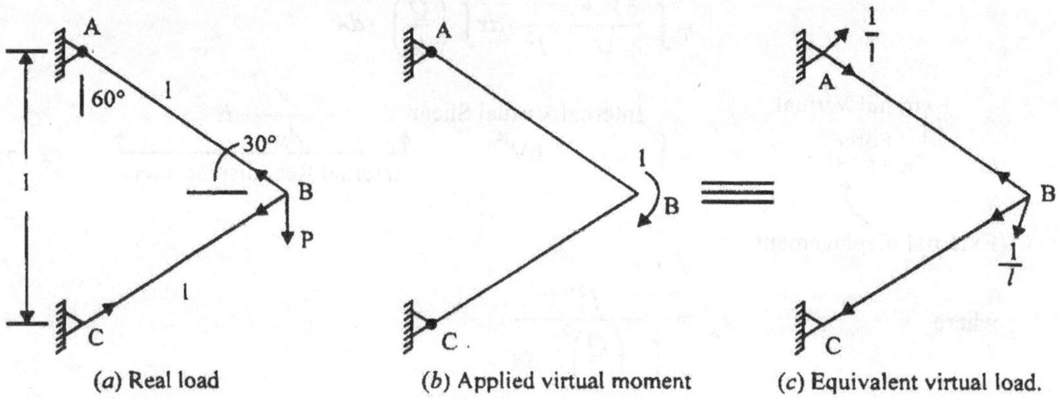

Fig. 8.15. Example 8.1

Soluton:

(*i*) Compute real load in the members

From equilibrium of Joint B (Fig. 8.15 (*i*))

$$F_{BA} \cdot \cos 30° + F_{BC} \cos 30° = 0$$

or $$F_{BA} = -F_{BC}$$

$$F_{BA} \cdot \sin 30° - F_{BC} \sin 30° - P = 0$$

or $$F_{BA} = P$$

$$F_{BC} = -P$$

(*ii*) To get the rotation of member *AB* a unit moment is applied at *B*. This is equivalent to applying a force $\frac{1}{l}$ at *A* and *B* perpendicular to *AB* in opposite directions

Resolving horizontally and vertically

$$f_{BA}\cos 30° + f_{BC}\cos 30° + \frac{1}{l}\sin 30° = 0$$

and $$f_{BA}\sin 30° - f_{BC}\sin 30° - \frac{1}{l}\cos 30° = 0$$

or $$f_{BA} + f_{BC} = -\frac{1}{l}\cdot\frac{1}{\sqrt{3}}$$

$$f_{BA} - f_{BC} = \frac{1}{l}\cdot\sqrt{3}$$

solving $$f_{BA} = \frac{1}{\sqrt{3}\,l},\ f_{BC} = -\frac{2}{\sqrt{3}\,l}.$$

(*iii*) Applying virtual work principle

$$1.\theta_B = \frac{Pl}{AE}\cdot\frac{1}{\sqrt{3}\,l} + \left(-\frac{Pl}{AE}\right)\left(-\frac{2}{\sqrt{3}\,l}\right)$$

$$= \frac{3P}{\sqrt{3}\,AE} = \frac{\sqrt{3}\,P}{AE}$$

when P = 10 kN

A = 1000 mm^2

E = 200 × 10^3 N/mm^2

$$\theta_B = \frac{1.712 \times 10 \times 1000}{1000 \times 200 \times 10^3} = 0.856 \times 10^{-4} \text{ radian.}$$

Example 8.2. For the truss shown, determine (*a*) the vertical deflection of joint C (*b*) the horizontal deflection at joint *B* (*c*) the relative deflection between joint *B* and *E* along the line joining them. (*d*) the rotation of member *EF*. *AE* for all members is same.

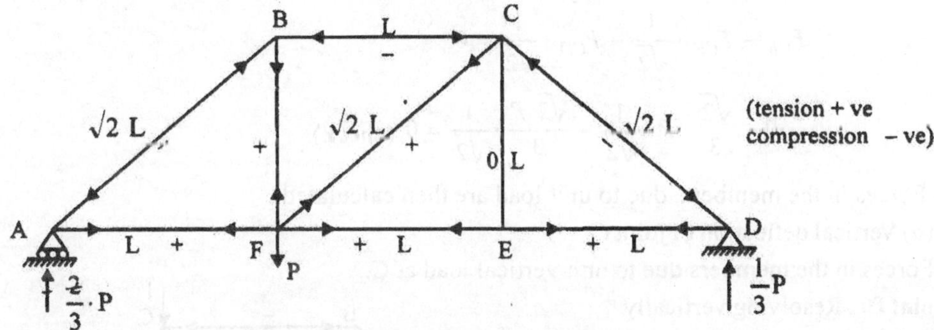

Fig. 8.16. (Example 8.2)

Solution: Forces in all members of the truss under real loading are calculated first

Joint A Resolving vertically,

$$F_{AB}\cdot\frac{1}{\sqrt{2}} - \frac{2}{3}P = 0$$

or $$F_{AB} = \frac{2\sqrt{2}}{3}P(c)$$

Resolving horizontally

$$\frac{2\sqrt{2}P}{3}\cdot\frac{1}{\sqrt{2}} = F_{AF}$$

or $$F_{AF} = +\frac{2}{3}P\,(t)$$

Joint B Resolving vertically

$$F_{BA}\cos 45° - F_{BF} = 0$$

$$\text{or } F_{BF} = F_{BA}\cdot\frac{1}{\sqrt{2}} = \frac{2\sqrt{2}}{3}P\cdot\frac{1}{\sqrt{2}} = +\frac{2}{3}P\ (t)$$

Resolving horizontally

$$F_{BA}\cdot\frac{1}{\sqrt{2}} - F_{BC} = 0$$

$$\text{or } F_{BC} = F_{BA}\cdot\frac{1}{\sqrt{2}} = \frac{2\sqrt{2}}{3}\cdot P\cdot\frac{1}{\sqrt{2}} = \frac{2}{3}P\ (c)$$

Joint D Resolving vertically,

$$F_{DC}\frac{1}{\sqrt{2}} = \frac{1}{3}P$$

$$\text{or } F_{DC} = \frac{\sqrt{2}}{3}\cdot P\ (c)$$

Resolving horizontally,

$$F_{DC}\cdot\frac{1}{\sqrt{2}} = F_{DE} = \frac{\sqrt{2}}{3}\cdot P\cdot\frac{1}{\sqrt{2}} = +\frac{P}{3}(t) = F_{EF}.$$

Joint C Resolving vertically

$$F_{CF}\cdot\frac{1}{\sqrt{2}} - F_{CD}\cdot\frac{1}{\sqrt{2}} = 0$$

$$\text{or } F_{CF} = F_{CD} = +\frac{\sqrt{2}}{3}\cdot P\ (t)$$

Resolving horizontally,

$$F_{CB} - F_{CF}\cdot\frac{1}{\sqrt{2}} - F_{CD}\cdot\frac{1}{\sqrt{2}} = 0$$

$$\frac{2}{3}P - \frac{\sqrt{2}}{3}\cdot P\cdot\frac{1}{\sqrt{2}} - \frac{\sqrt{2}\,P}{3}\cdot\frac{1}{\sqrt{2}} = 0 \text{ (check).}$$

(*ii*) **Forces in the members, due to unit load are then calculated.**

(*a*) **Vertical deflection of joint *C*.**

Forces in the members due to unit vertical load at *C*.

Joint D. Resolving vertically

$$f_{CD}\frac{1}{\sqrt{2}} - \frac{2}{3} = 0$$

$$f_{CD} = \frac{2\sqrt{2}}{3}\text{ (comp)}$$

Resolving horizontally,

$$f_{DE} = \frac{2}{3} = f_{FE}\text{ (tension)}$$

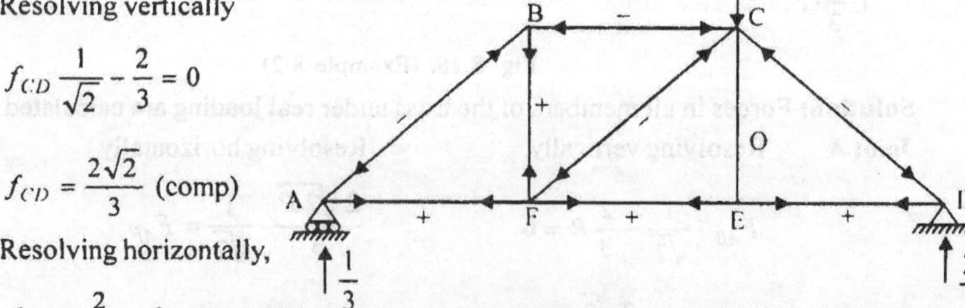

Fig. 8.17. Forces in members due to unit vertical load at c.

Joint A Resolving vertically

$$f_{AB}\cdot\frac{1}{\sqrt{2}}=\frac{1}{3}$$

$$f_{AB}=\frac{\sqrt{2}}{3}\ \text{(compression)}$$

Jonit B Resolving vertically

$$f_{AB}\cdot\frac{1}{\sqrt{2}}=f_{BF}$$

$$\text{or } f_{BF}=\frac{\sqrt{2}}{3}\cdot\frac{1}{\sqrt{2}}=\frac{1}{3}\ \text{(tension)}$$

Joint A **Resolving horizontally**

$$f_{AF}=\frac{\sqrt{2}}{3}\cdot\frac{1}{\sqrt{2}}=\frac{1}{3}\ \text{(tension)}$$

Resolving horizontally at joint B

$$f_{BC}=f_{BA}\cdot\frac{1}{\sqrt{2}}=\frac{\sqrt{2}}{3}\cdot\frac{1}{\sqrt{2}}=\frac{1}{3}\ \text{(compression)}$$

Resolving vertically at Joint F

$$f_{CF}\cdot\frac{1}{\sqrt{2}}=f_{FB}=\frac{1}{3}$$

$$f_{CF}=\frac{\sqrt{2}}{3}\ \text{(compression)}$$

Check equilibrium of joint C

$$f_{CB}+f_{CF}\cdot\frac{1}{\sqrt{2}}-f_{CD}\cdot\frac{1}{\sqrt{2}}=0$$

$$\frac{1}{3}+\frac{\sqrt{2}}{3}\cdot\frac{1}{\sqrt{2}}-\frac{2\sqrt{2}}{3}\cdot\frac{1}{\sqrt{2}}=0$$

$$f_{CF}\cdot\frac{1}{\sqrt{2}}+f_{CD}\cdot\frac{1}{\sqrt{2}}-1$$

$$=\frac{\sqrt{2}}{3}\cdot\frac{1}{\sqrt{2}}+\frac{2\sqrt{2}}{3}\cdot\frac{1}{\sqrt{2}}-1=0\ \text{(check)}$$

(*b*) Horizontal Deflection of joint B

Forces in the members due to unit horizontal load at B.

Jont A Resolving vertically

$$f_{AB}\cdot\frac{1}{\sqrt{2}}=\frac{1}{3}$$

$$f_{AB}=\frac{\sqrt{2}}{3}\ \text{(tension)}$$

Resolving horizontally

$$f_{AB}\cdot\frac{1}{\sqrt{2}}=f_{AF}$$

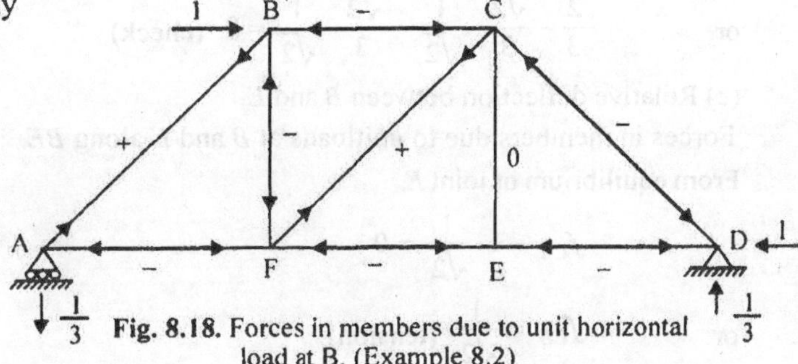

Fig. 8.18. Forces in members due to unit horizontal load at B. (Example 8.2)

$$f_{AF} = \frac{\sqrt{2}}{3} \cdot \frac{1}{\sqrt{2}} = \frac{1}{3} \text{ (comp)}$$

Joint B Resolving vertically.

$$f_{BA} \cdot \frac{1}{\sqrt{2}} = f_{BF} = \frac{\sqrt{2}}{3} \cdot \frac{1}{\sqrt{2}}$$

$$= \frac{1}{3} \text{ (comp)}$$

Resolving horizontally

$$1 - f_{BA} \cdot \frac{1}{\sqrt{2}} + f_{BC} = 0$$

$$\text{or } f_{BC} = f_{BA} \frac{1}{\sqrt{2}} - 1$$

$$= \frac{\sqrt{2}}{3} \cdot \frac{1}{\sqrt{2}} - 1 = -\frac{2}{3} \text{ (comp)}$$

Joint D Resolving vertically

$$f_{DC} \frac{1}{\sqrt{2}} = \frac{1}{3} \qquad f_{DC} = +\frac{\sqrt{2}}{3} \text{ (comp)}$$

Resolving horizontally

$$f_{DE} + F_{DC} \cdot \frac{1}{\sqrt{2}} - 1 = 0$$

$$\text{or } f_{DE} + \frac{\sqrt{2}}{3} \cdot \frac{1}{\sqrt{2}} - 1 = 0$$

$$\text{or } f_{DE} = \frac{2}{3} \text{ (comp.)}$$

Joint F Resolving vertically

$$f_{FC} \cdot \frac{1}{\sqrt{2}} = f_{FB} = \frac{1}{3}$$

$$\text{or } f_{FC} = \frac{\sqrt{2}}{3} \text{ (tension).}$$

Check joint C

$$f_{CB} - f_{CF} \cdot \frac{1}{\sqrt{2}} - f_{CD} \cdot \frac{1}{\sqrt{2}}$$

or
$$\frac{2}{3} - \frac{\sqrt{2}}{3} \cdot \frac{1}{\sqrt{2}} - \frac{\sqrt{2}}{3} \cdot \frac{1}{\sqrt{2}} = 0 \text{ (check)}$$

(*c*) Relative deflection between *B* and *E*

Forces in members due to unitloads at *B* and *E* along *BE*.

From equilibrium of joint *E*.

$$f_{CE} - 1 \cdot \frac{1}{\sqrt{2}} = 0$$

or
$$f_{CE} = \frac{1}{\sqrt{2}} \text{ (tension)}$$

$$f_{EF} - \frac{1}{\sqrt{2}} = 0$$

or $$f_{EF} = \frac{1}{\sqrt{2}} (t)$$

Similarly

$$f_{BF} = \frac{1}{\sqrt{2}} (t) \quad f_{BC} = \frac{1}{\sqrt{2}} (t)$$

$f_{CF} = 1$ (compression)

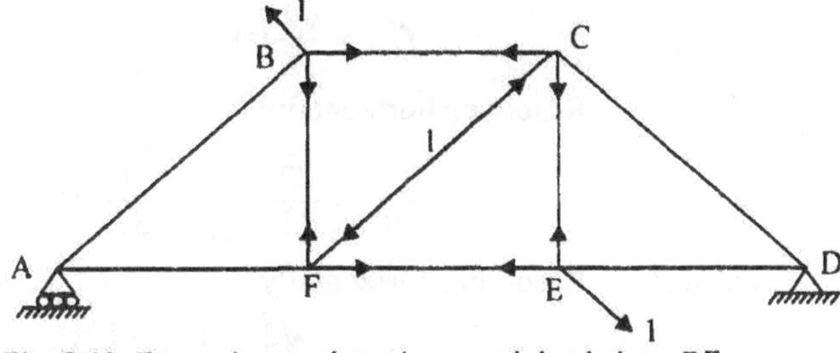

Fig. 8.19. Forces in members due to unit load along BE. (Example 8.2)

(*d*) Rotation of Member *EF*. Forces in members when unit moment is applied at *E*.

Reactions at A and D

$$R_A \times 3L + \frac{1}{L} \times 2L - \frac{1}{L} \times L = 0$$

or $$R_A = -\frac{1}{3L}$$

and $$R_D \times 3L - \frac{1}{L} \times 2L + \frac{1}{L} \times L = 0$$

or $$R_D = \frac{1}{3L}$$

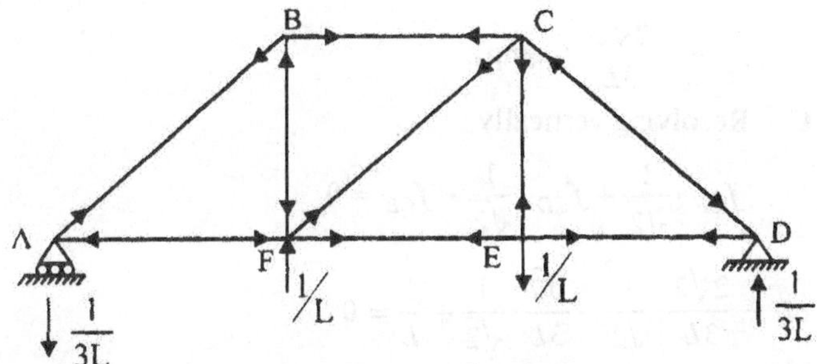

Fig. 8.20. Forces in members with $\frac{1}{L}$ load at E and F. (Example 8.2).

Joint D Resolving vetically

$$f_{DC} \frac{1}{\sqrt{2}} = \frac{1}{3L}$$

$$f_{DC} = \frac{\sqrt{2}}{3L} \text{ (comp)}$$

Resolving horizontally

$$f_{DE} = f_{CD} \cdot \frac{1}{\sqrt{2}}$$

$$= \frac{\sqrt{2}}{3L} \cdot \frac{1}{\sqrt{2}} = \frac{1}{3L} \text{ (tension)}$$

$$= f_{EF} \text{ (tension)}$$

Joint A Resolving vertically

$$f_{AB} \cdot \frac{1}{\sqrt{2}} = \frac{1}{3L}$$

or $\quad f_{AB} = \dfrac{\sqrt{2}}{3L}(t).$

Resolving horizontally

$$f_{AF} = \frac{\sqrt{2}}{3L}\cdot\frac{1}{\sqrt{2}} = \frac{1}{3L}(c)$$

Joint *B* Resolving horizontally

$$f_{BA}\cdot\frac{1}{\sqrt{2}} = f_{BC} = \frac{\sqrt{2}}{3L}\cdot\frac{1}{\sqrt{2}} = \frac{1}{3L}(t)$$

Resolving vertically

$$f_{BA}\cdot\frac{1}{\sqrt{2}} = f_{BF} = \frac{\sqrt{2}}{3L}\cdot\frac{1}{\sqrt{2}} = \frac{1}{3L}\text{ (comp)}$$

Joint *F* Resolving horizontally

$$f_{FA} + \frac{1}{3L} + f_{FC}\cdot\frac{1}{\sqrt{2}} = 0$$

or
$$f_{FC}\cdot\frac{1}{\sqrt{2}} = -f_{FA} - \frac{1}{3L}$$
$$= -\frac{1}{3L} - \frac{1}{3L} = -\frac{2}{3L}$$

or $\quad f_{FC} = -\dfrac{2\sqrt{2}}{3L}$ (comp).

Check Joint *C* Resolving vertically

$$f_{CF}\cdot\frac{1}{\sqrt{2}} + f_{CD}\cdot\frac{1}{\sqrt{2}} - f_{CE} = 0$$

or
$$-\frac{2\sqrt{2}}{3L}\cdot\frac{1}{\sqrt{2}} - \frac{\sqrt{2}}{3L}\cdot\frac{1}{\sqrt{2}} + \frac{1}{L} = 0$$

Resolving vertically at joint *F*

$$f_{FB} + f_{FC}\cdot\frac{1}{\sqrt{2}} - \frac{1}{L}$$
$$= \frac{1}{3L} + \frac{2\sqrt{2}}{3L}\cdot\frac{1}{\sqrt{2}} - \frac{1}{L} = 0\text{ (check)}$$

Calculations of deflections are shown in Table 8.2.

From summation of corresponding columns

(*a*) Vertical deflection of $C = \dfrac{FLU_c}{AE} = \dfrac{1}{AE}\left[\dfrac{6\sqrt{2}\,PL}{9} + \dfrac{10\,PL}{9}\right] = \dfrac{PL}{9AE}\left[6\sqrt{2} + 10\right]$

(*b*) Horizontal deflection of $B = \dfrac{F.LU_B}{AE} = \text{O}$

(*c*) Relative displacement between *B* and $E = \dfrac{F.LU_{BE}}{AE} = \dfrac{1}{AE}\left[\dfrac{PL}{3\sqrt{2}} - \dfrac{2PL}{3}\right]$

Table 8.2. Calculation of Deflections (Example 8.2)

Member	Length	F	U_C	U_B	U_{BE}	U_{EF}	$F \cdot LU_c$	FLU_B	FLU_{BE}	FLU_{EF}
AB	$\sqrt{2}L$	$-\frac{2\sqrt{2}}{3}P$	$-\frac{\sqrt{2}}{3}$	$+\frac{\sqrt{2}}{3}$	0	$+\frac{\sqrt{2}}{3L}$	$\frac{4\sqrt{2}PL}{9}$	$-\frac{4\sqrt{2}PL}{9}$	0	$-\frac{4\sqrt{2}P}{9}$
BC	L	$-\frac{2}{3}P$	$-\frac{1}{3}$	$-\frac{2}{3}$	$+\frac{1}{\sqrt{2}}$	$+\frac{1}{3L}$	$\frac{2PL}{9}$	$\frac{4PL}{9}$	$-\frac{2PL}{3\sqrt{2}}$	$-\frac{2}{9}P$
CD	$\sqrt{2}L$	$-\frac{\sqrt{2}}{3}P$	$-\frac{2\sqrt{2}}{3}$	$-\frac{\sqrt{2}}{3}$	0	$-\frac{\sqrt{2}}{3L}$	$+\frac{4\sqrt{2}}{9}PL$	$+\frac{2\sqrt{2}PL}{9}$	0	$+\frac{2\sqrt{2}P}{9}$
DE	L	$+\frac{P}{3}$	$+\frac{2}{3}$	$-\frac{2}{3}$	0	$+\frac{1}{3L}$	$+\frac{2PL}{9}$	$-\frac{2PL}{9}$	0	$+\frac{P}{9}$
EF	L	$+\frac{P}{3}$	$+\frac{2}{3}$	$-\frac{2}{3}$	$+\frac{1}{\sqrt{2}}$	$+\frac{1}{3L}$	$+\frac{2PL}{9}$	$-\frac{2PL}{9}$	$\frac{PL}{3\sqrt{2}}$	$+\frac{P}{9}$
FA	L	$+\frac{2}{3}P$	$+\frac{1}{3}$	$-\frac{1}{3}$	0	$-\frac{1}{3L}$	$+\frac{2PL}{9}$	$-\frac{2PL}{9}$	0	$-\frac{2P}{9}$
BF	L	$+\frac{2}{3}P$	$+\frac{1}{3}$	$\frac{1}{3}$	$+\frac{1}{\sqrt{2}}$	$-\frac{1}{3L}$	$+\frac{2PL}{9}$	$+\frac{2PL}{9}$	$+\frac{2PL}{3\sqrt{2}}$	$-\frac{2P}{9}$
FC	$\sqrt{2}L$	$+\frac{\sqrt{2}}{3}P$	$-\frac{\sqrt{2}}{3}$	$+\frac{\sqrt{2}}{3}$	−1	$-\frac{2\sqrt{2}}{3L}$	$-\frac{2\sqrt{2}PL}{9}$	$+\frac{2\sqrt{2}PL}{9}$	$-\frac{2PL}{3}$	$-\frac{4\sqrt{2}P}{9}$
CE	L	0	0	0	$+\frac{1}{\sqrt{2}}$	$+\frac{1}{L}$	0	0	0	0
							Σ	Σ	Σ	Σ
							(a)	(b)	(c)	(d)

(*d*) Rotation of $EF = \frac{F.L.U_{EF}}{AE} = \frac{1}{AE}\left|\frac{6\sqrt{2}P}{9} - \frac{4P}{9}\right| = \frac{P}{AE}\left|\frac{2\sqrt{2}}{3} - \frac{4}{9}\right|$

Example 8.3. Calculate the length change in members *AE*, *BC* and *ED* of the truss shown in Fig. 8.21 in order to induce an initial vertical upward deflection at *D* equal to 5 mm. This is accomplished by shortening *BC* and lengthening *AE* and *ED* by the same amount.

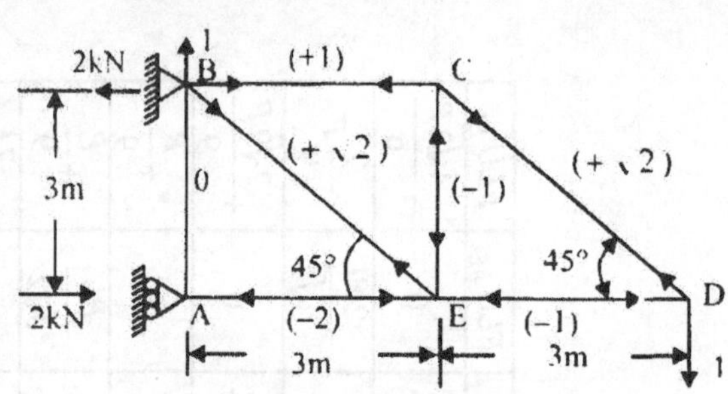

Fig. 8.21. Example 8.3.

Real displacement for $BC = -\delta$

$AE = +\delta$

$ED = +\delta$

Forces induced in members due to unit load acting downward at *D* are shown in Fig. 8.21. Applying virtual work principle.

$$1 \times v_D = (+1)(-\delta) + (-2)(+\delta) + (-1)(+\delta)$$
$$= -4\delta$$

Here $v_D = -5$

Hence $-5 = -4\delta$ or $\delta = \mathbf{1.25\ mm.}$

Example 8.4. Calculate the vertical and horizontal displacement of joint *D* in the truss shown (Fig. 8.22). Support A moves horizontally 2 mm to the left; support *B* moves 3 mm vertically downward and 3 mm to the right. Assume $A = 1000\ mm^2$. E = 200 GPa. Use virtual work method.

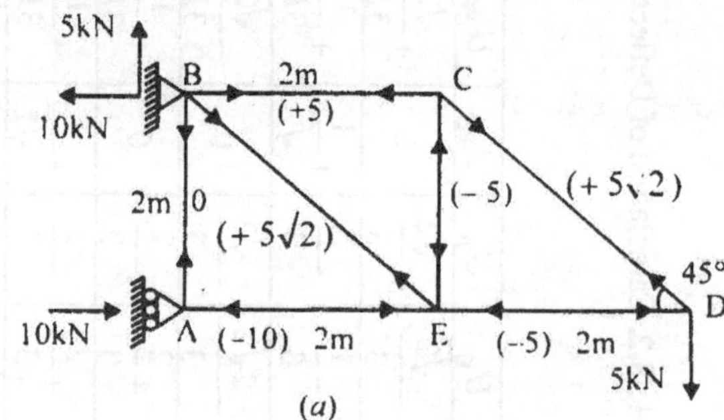

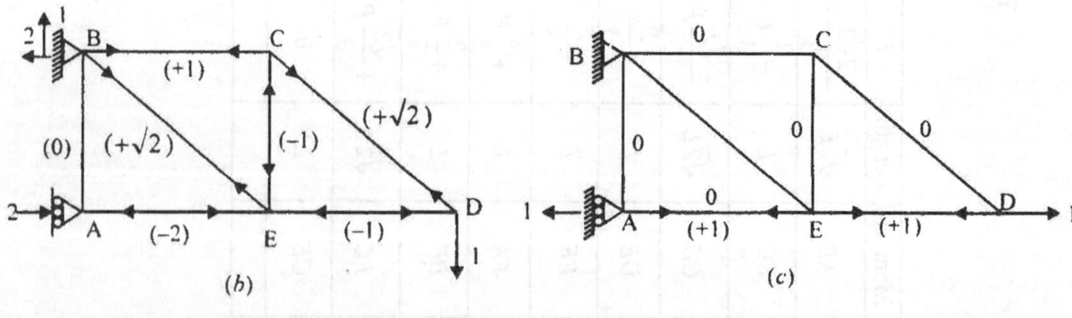

Fig. 8.22. Example 8.4.

Solution: Forces in the members due to real applied load, virtual load of 1 kN acting vertically at *D* and 1 kN acting horizontally at *D* are shown in Figs 8.22 (a), (b) and (c).

Table 8.3 (Example 8.4)

Member	Length (m)	F (kN)	u_v	u_h	$Fu_v.L$	$Fu_h.L$
AB	2	0	0	0	0	0
BC	2	+5	+1	0	+10	0
CD	$2\sqrt{2}$	$+5\sqrt{2}$	$+\sqrt{2}$	0	$+20\sqrt{2}$	0
DE	2	−5	−1	+1	+10	−10
EA	2	−10	−2	+1	+40	−20
EC	2	−5	−1	0	+10	0
EB	$2\sqrt{2}$	$+5\sqrt{2}$	$\sqrt{2}$	0	$+20\sqrt{2}$	0
					$\Sigma 70+40\sqrt{2}$	$\Sigma -30$

External virtual work $= 1 \times v_d + 2 \times (-2) + 1 \times (-3) + 2 \times (-3)$

From Table 8.3, Internal virtual work

$$= \frac{70 + 40\sqrt{2}}{AE}$$

$$= \frac{126.56 \times 10 \times 10 \times 10}{1000 \times 200}$$

$$= 0.6328 \text{ kN mm}$$

Equating external and internal virtual work

$$v_d - 4 - 3 - 6 = 0.6328$$

or $\quad v_d = \mathbf{13.6328\ mm}$

Horizontal deflection

External virtual work $= 1 \times h_d + 1 \times 2$

From table 8.3, Internal virtual work $= \dfrac{-30}{200} = -0.15$

Equating external and internal virtual work

$$h_d + 2 = -0.15$$

$$h_d = -2.15 \text{ mm (towards left)}$$

Example 8.5. Compute the horizontal deflection of joint *f* if (*a*) member cd is shortened by 10 mm (*b*) members *ab* and *ef* both experience a temperature rise of 25°C ($\alpha = 1.2 \times 10^{-5}$ mm/mm/°C)

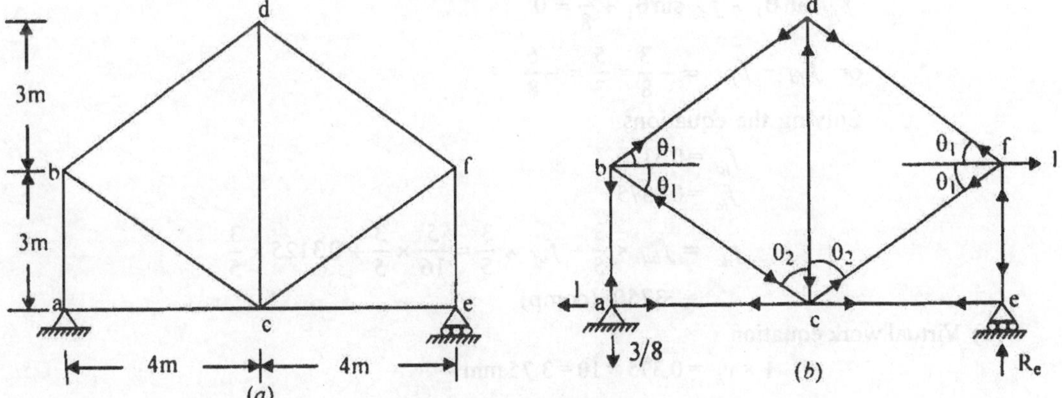

Fig. 8.23. Example 8.5.

Solution: Apply unit load at f

$$R_e \times 8 = 1 \times 3$$

$$R_e = \frac{3}{8}$$

Joint e Resolving vertically

$$f_{ef} - R_e = 0$$

or $$f_{ef} = \frac{3}{8} \text{ (comp)}$$

$$f_{ec} = 0$$

$$f_{ac} = 1 \text{ (tension)}$$

$$f_{ab} = \frac{3}{8} \text{ (tension)}$$

Joint b Resolving horizontally

$$f_{bd} \cos\theta_1 - f_{bc} \cos\theta_1 = 0$$

or $$f_{bd} = f_{bc}$$

Resolving vertically

$$f_{bd} \sin\theta_1 + f_{bc} \sin\theta_1 - \frac{3}{8} = 0$$

$$\text{or } 2 f_{bd} \sin\theta_1 = \frac{3}{8}$$

or $$f_{bd} = \frac{3}{8 \times 2} \times \frac{1}{\sin\theta_1}$$

$$= \frac{3}{8 \times 2} \times \frac{5}{3} = \frac{5}{16} \text{ (tension)}$$

Joint f Resolving horizontally

$$f_{fd} \cos\theta_1 + f_{fc} \cos\theta_1 - 1 = 0$$

$$\text{or } (f_{fd} + f_{fc}) \cos\theta_1 = 1$$

$$\text{or } f_{fd} + f_{fc} = \frac{1}{\cos\theta_1} = \frac{5}{4}$$

Resolving vertically,

$$f_{fd} \sin\theta_1 - f_{fc} \sin\theta_1 + \frac{3}{8} = 0$$

$$\text{or } f_{fd} - f_{fc} = -\frac{3}{8} \times \frac{5}{3} = -\frac{5}{8}$$

Solving the equations

$$f_{fd} = 0.3125$$

$$f_{fc} = 0.9375$$

$$f_{dc} = f_{bd} \times \frac{3}{5} + f_{df} \times \frac{3}{5} = \frac{5}{16} \times \frac{3}{5} + 0.3125 \times \frac{3}{5}$$

$$= \cdot 3750 \text{ (comp)}$$

(*a*) Virtual work equation

$$1 \times v_h = 0.375 \times 10 = 3.75 \text{ mm}$$

or $\quad v_h = 3.75 \text{ mm} \rightarrow$

(*b*) Virtual work equation

$$1 \times v_h = \frac{3}{8} \times 25 \times 1.2 \times 10^{-5} \times 3000$$

$$+\left(-\frac{3}{8}\right) \times 25 \times 1.2 \times 10^{-5} \times 3000$$

$$= 0$$

$$v_h = 0$$

Example 8.6. For the beam illustrated in Fig. 8.24. Calculate deflection at a due to (i) given loading (ii) if the upper and lower beam surfaces are 20°C and 0°C respectively with no applied load.

$$I = 100 \times 10^6 \text{ mm}^4 \qquad E = 200 \text{ GPa}$$

$$\alpha = 1.2 \times 10^{-5} \text{ mm/mm/°C} \qquad h = 250 \text{ mm}$$

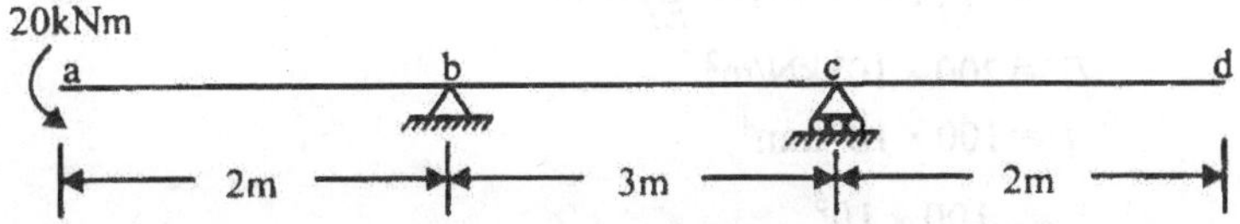

Fig. 8.24. Example 8.6

Solution: A 1 kN load is applied at 'a' downward. The B.M. diagram due to applied load and virtual load are drawn. Table 8.4 gives equation of M, δM, origin and limits within which the equations are valid.

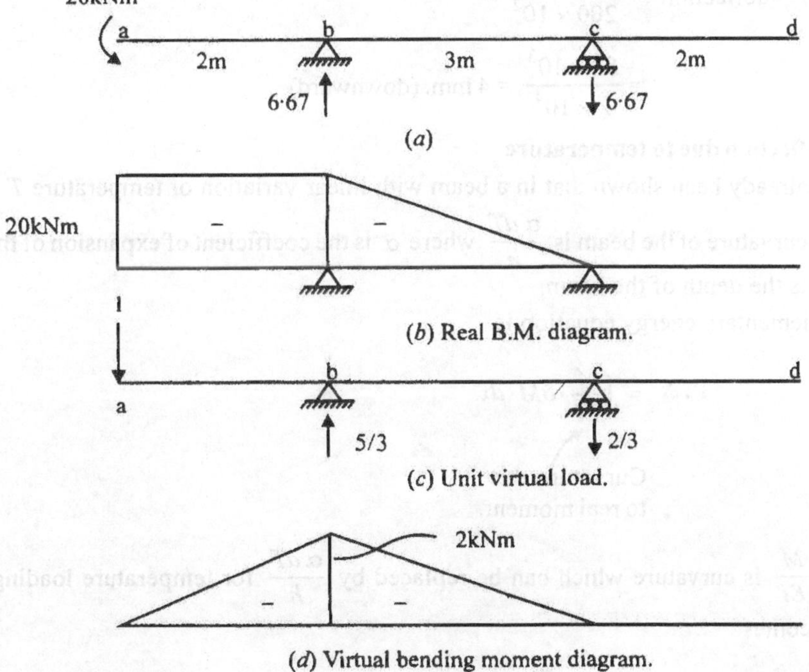

Fig. 8.25. Example 8.6

Table 8.3

Member	M	δM	Origin	Limit
ab	-20	$-x$	a	$0-2$
cb	$-6.67x$	$-\frac{2}{3}\cdot x$	c	$0-3$

(i) Deflection at a due to given loading

$$= \int_0^2 \frac{(-20)(-x)\,dx}{EI} + \int_0^3 \frac{-20}{3}\cdot x\left(-\frac{2}{3}x\right)\cdot dx$$

$$= \frac{1}{EI}\left[\frac{20}{2}\cdot 2^2 + \frac{40}{9}\cdot\frac{3^3}{3}\right]$$

$$= \frac{1}{EI}[40+40] = \frac{80}{EI}\text{ m.}\downarrow$$

$$E = 200\times 10^6\text{ kN/m}^2$$

$$I = 100\times 10^6\text{ mm}^4$$

$$= \frac{100\times 10^6}{10^{12}} = 10^{-4}\text{ m}^4$$

$$EI = 200\times 10^6\times 10^{-4}$$

$$= 200\times 10^2$$

$$\text{deflection} = \frac{80}{200\times 10^2}\text{ m}$$

$$= \frac{80\times 10^3}{2\times 10^4} = 4\text{ mm. (downward)}$$

(ii) Deflection due to temperature

It has already been shown that in a beam with linear variation of temperature T between top and bottom, curvature of the beam is $\frac{\alpha\, dT}{h}$ where α is the coefficient of expansion of the material of the beam, h is the depth of the beam.

Complementary energy equation is

$$1.\Delta = \int\frac{M}{EI}\cdot\delta M\cdot dx.$$

Curvature due to real moment.

Now $\frac{M}{EI}$ is curvature which can be replaced by $\frac{\alpha\, dT}{h}$ for temperature loading. Hence the equation becomes

$$1.\Delta = \int\frac{\alpha\ dT}{h}\cdot\delta M\cdot\delta x$$

With 20°C at top and 0°C at bottom the curvature is convex on top, hence negative. Hence

$$\Delta_1 = \int_0^{2000} -\frac{(1.2\times10^{-5})(20)}{250}(-x)\,dx$$

$$= \frac{1.2\times10^{-5}}{250}\times 20\times\frac{2^2\times10^6}{2}$$

$$= 1.92 \text{ mm.}$$

$$\Delta_2 = \int_0^{3000} -\frac{(1.2\times10^{-5})(20)}{250}\times\left(-\frac{2}{3}x\right)\cdot dx$$

$$= \frac{(1.2\times10^{-5})(20)}{250}\times\frac{2}{3}\times\frac{1}{2}\times 3^2\times10^6 = \mathbf{2.88\ mm.}$$

Total deflection $\Delta = \Delta_1 + \Delta_2 = 1.92 + 2.88 =$ **4.80 mm.** (downward)

Example 8.7. Calculate the vertical deflection and slope at point C and D of the beam shown in Fig. 8.26. EI contant.

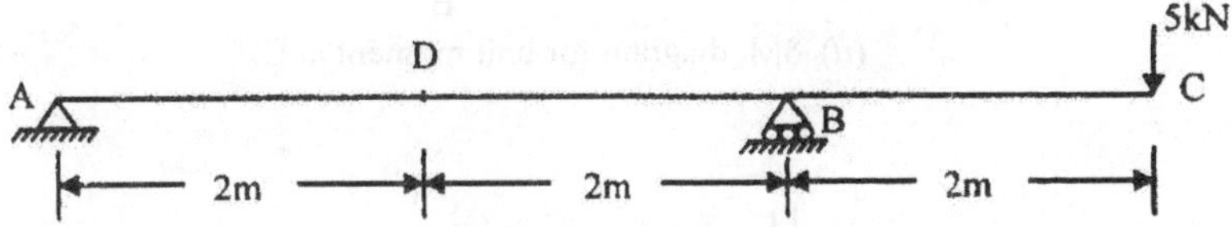

Fig. 8.26. Example 8.7

Solution: Fig. 8.27 gives the M diagram and δM diagrams for (*i*) unit vertical load (*ii*) unit moment at C and (*iii*) unit vertical load and (*iv*) unit moment at D. Table 8.4 gives the equations for (*i*) M diagrams (*ii*) δM diagrams and the validity limits for different parts of the beam.

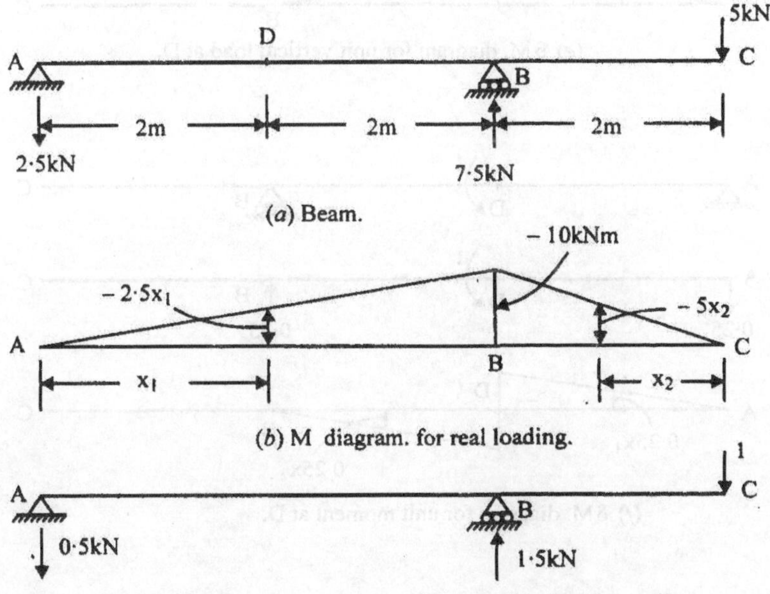

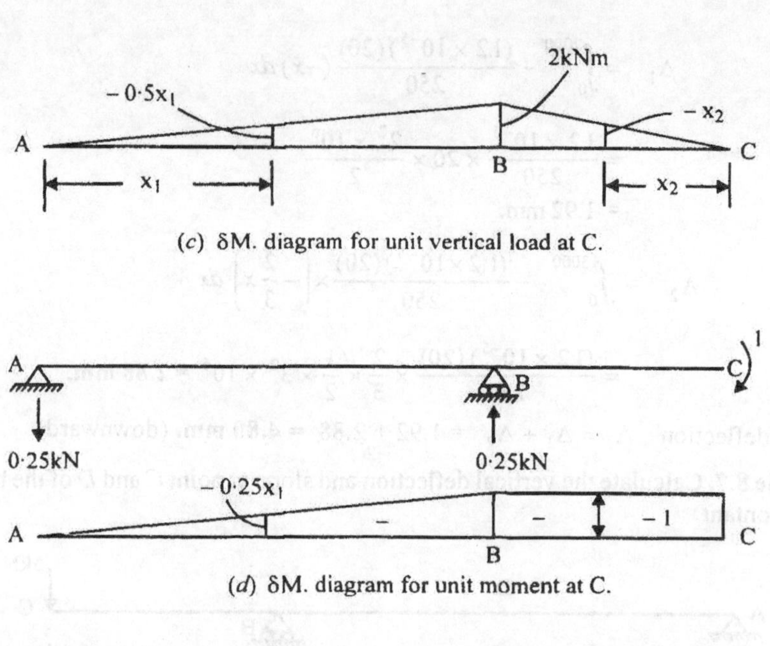

(c) δM. diagram for unit vertical load at C.

(d) δM. diagram for unit moment at C.

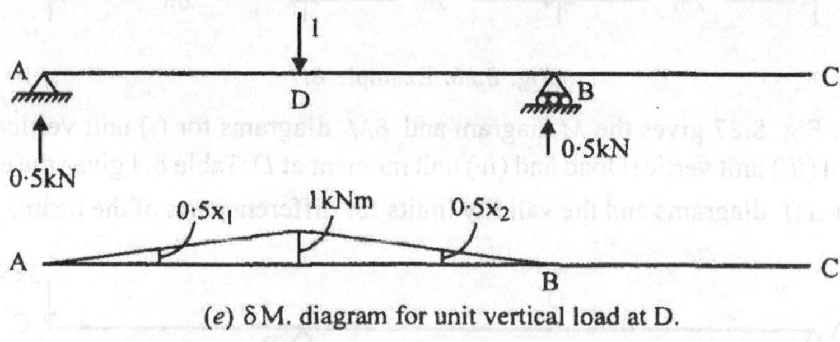

(e) δM. diagram for unit vertical load at D.

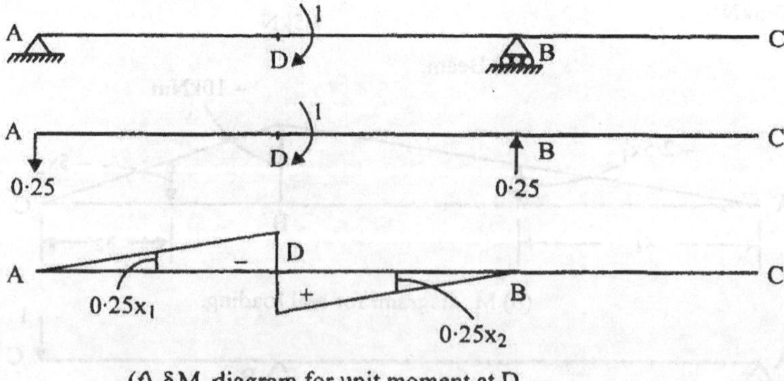

(f) δM. diagram for unit moment at D.

Fig. 8.27. *M* diagram and δM diagrams (Example 8.7)

Table 8.4

	Member	Origin	M	δM	Limit
Deflection of C	AB	A	$-2.5\,x_1$	$-0.5\,x_1$	$0-4$
	BC	C	$-5\,x_2$	$-x_2$	$0-2$
Rotation at C	AB	A	$-2.5\,x_1$	$-0.25\,x_1$	$0-4$
	BC	C	$-5\,x_2$	-1	$0-2$
Deflectiom of D	AD	A	$-2.5\,x_1$	$+0.5\,x_1$	$0-2$
	BD	B	$-10+\frac{10}{4}\cdot x_2$	$+0.5\,x_2$	$0-2$
Rotation at D	AD	A	$-2.5\,x_1$	$-0.25\,x_1$	$0-2$
	BD	B	$-10+\frac{10}{4}\cdot x_2$	$+0.25\,x_2$	$0-2$

Vertical deflection C

$$\Delta_C = \frac{1}{EI}\int_0^4 (-2.5\,x_1)(-0.5\,x_1)\,dx_1 + \frac{1}{EI}\int_0^2 (-5\,x_2)(-x_2)\,dx_2$$

$$= \frac{1}{EI}\times 1.25\left[\frac{x_1^{\,3}}{3}\right]_0^4 + \frac{5}{EI}\left[\frac{x_2^{\,3}}{3}\right]_0^2$$

$$= \frac{40}{EI}\downarrow$$

Rotation at C

$$\theta_C = \frac{1}{EI}\int_0^4 (-2.5\,x_1)(-0.25\,x_1).\,dx_1 + \frac{1}{EI}\int_0^2 -5x_2\,(-1).\,dx_2$$

$$= \frac{1}{EI}\left[\frac{0.625\times 4^3}{3} + \frac{5\times 2^2}{2}\right]$$

$$= \frac{23.33}{EI}$$

Deflection at D

$$\Delta_D = \frac{1}{EI}\int_0^2 (-2.5\,x_1)(+0.5\,x_1)\,dx_1 + \frac{1}{EI}\int_0^2 (-10+2.5\,x_2)(+0.5\,x_2)\,dx_2$$

$$= \frac{1}{EI}\left[\frac{-1.25\times 2^3}{3} - \frac{5\times 2^2}{2} + \frac{1.25\times 2^3}{3}\right]$$

$$= -\frac{10}{EI}$$

Rotation at D

$$\theta_D = \frac{1}{EI}\int_0^2 (-2.5\,x_1)(-0.25\,x_1)\,dx_1 + \frac{1}{EI}\int_0^2 (-10+2.5\,x_2)(0.25\,x_2)\,dx_2$$

$$= -\frac{1.67}{EI}.$$

Example 8.8. For the frame shown, compute the horizontal displacement of point D and the rotation of point A. Use $E = 200.00$ GPa. What value of I limits this displacement to less than 5 cm? Consider bending deformation only.

Solution: In a portal only bending deformation is usually considered as deflection due to axial load and shear is very small.

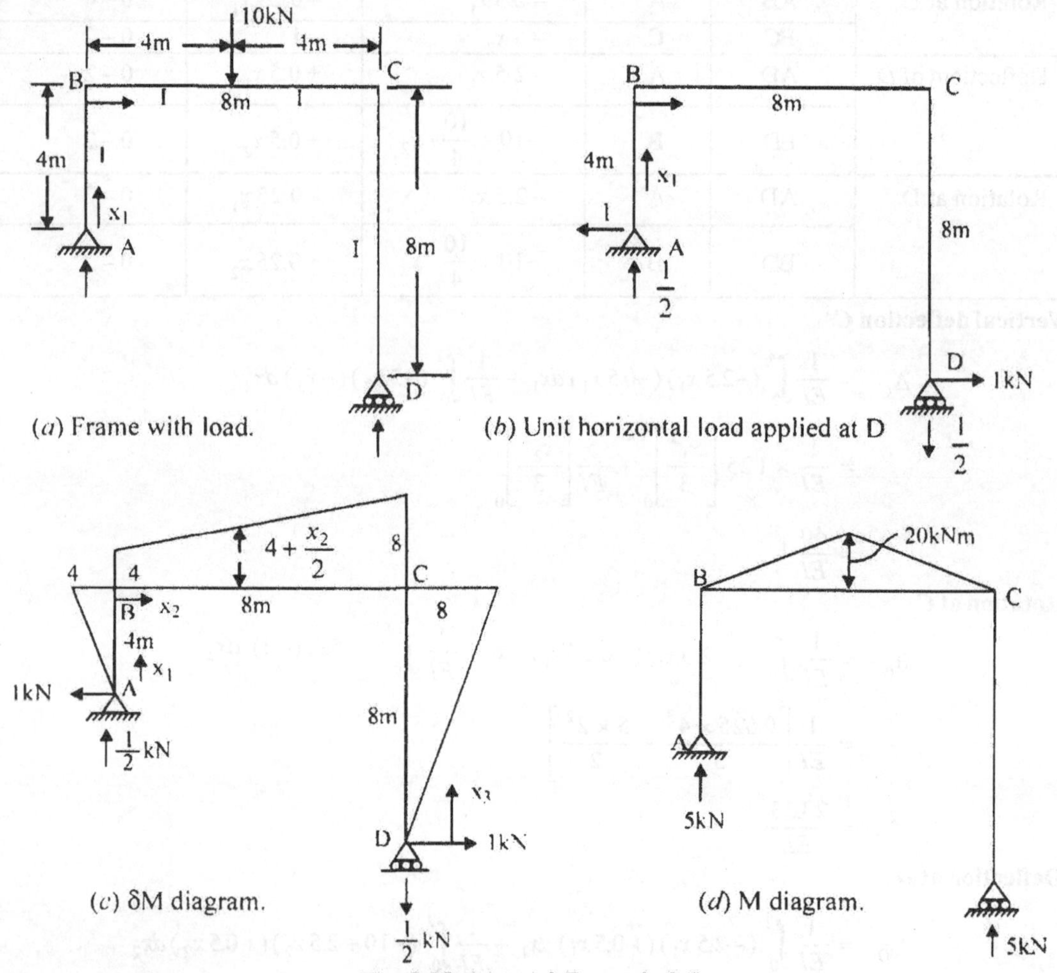

Fig. 8.28. (a) to (d) Example 8.8

Deflection of point D

Step 1. To compute deflection of point D a unit horizontal force is applied at D, as shown in Fig. 8.28 (b).

Step 2. The internal moment due to this virtual load is shown in Fig. 8.28 (c). For each member a local coordinate is defined x_1, x_2, x_3 as shown in Fig. 8.28 (b). The equation for virtual moment δM is then written on the diagram. The sign for B.M. (out side compression is positive) is followed.

Step 3 and Step 4. Real loads are now applied and the bending moment diagram with same coordinate axes are computed and drawn. The $\frac{M}{EI}$ diagram is then drawn.

Step 5. The basic equation

$$1.\Delta = \int \delta M.\left(\frac{M}{EI}\right).dx$$

is now applied. Since there are three members with three different δM and M diagram integration has to be done for all three members separately and the results added.

Horizontal deflection at D

$$1\,.\,\Delta_{xD} = \int \delta M.\frac{M\,dx}{EI}$$

$$= \frac{1}{EI}\int_0^4 \left(4+\frac{x_2}{2}\right)5x_2\,.dx_2 + \frac{1}{EI}\int_4^8 \left(4+\frac{x_2}{2}\right)[20-(x_2-4).5]\,dx_2$$

$$= \frac{1}{EI}\left[20\cdot\frac{x_2^2}{2}+\frac{2.5}{3}\cdot x_2^3\right]_0^4 + \frac{1}{EI}\int_4^8\left(4+\frac{x_2}{2}\right)(40-5x_2)\,dx_2$$

$$= \frac{1}{EI}\left[160+\frac{2.5\times 64}{3}\right]+\frac{1}{EI}\left[160\,x_2+10x_2^2-10\,x_2^2-\frac{2.5}{3}x_2^3\right]_4^8$$

$$= \frac{1}{EI}[213.33]+\frac{1}{EI}[640-373.33]$$

$$= \frac{480}{EI}\rightarrow$$

When $\Delta_{xD} = 5$ cm

$$\frac{480}{200\times 10^6}\cdot\frac{1}{I} = \frac{5}{100}\,\text{m}$$

or
$$I = \frac{480\times 100}{200\times 5\times 10^6}$$

$$= 48\times 10^{-6}\ m^4$$

$$= 48\times 10^{-6}\times 10^8\ \text{cm}^4$$

$$= 4800\ \text{cm}^4.$$

Rotation of Point A [Fig. 8.28 (e) and (f)]

Taking moments about A

$$R_D\times 8-1 = 0$$

or
$$R_D = +\frac{1}{8}$$

$$\theta_A = \int\frac{\delta M.M.dx}{EI}$$

$$= \frac{1}{EI}\int_0^4 -\frac{1}{8}x_4\,(5x_4)\,dx_4 + \frac{1}{EI}\int_4^8 -\frac{1}{8}\cdot x_4\,(20-(x_4-4)5]\times dx_4$$

$$= \frac{1}{EI}\left[-\frac{5}{8}\cdot\left(\frac{x_4^3}{3}\right)\right]_0^4+\left(-\frac{1}{8}\right)\left[\frac{40.x_4^{\,2}}{2}-5\cdot\frac{x_4^{\,3}}{3}\right]_4^8$$

$$= -\frac{40}{EI}\ \text{rotation clockwise.}$$

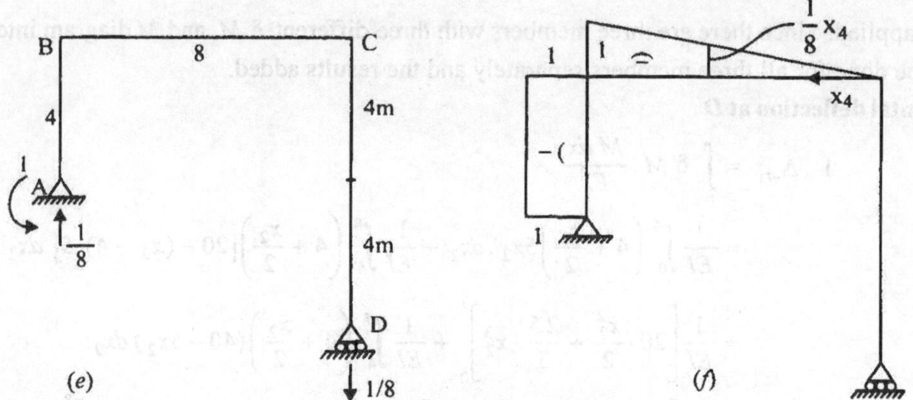

Fig. 8.28. (e) and (f) Example 8.8

Example 8.9. The figure 8.29. shows a frame subjected to a uniform load of 1 kN/m. *EI* is constant. Compute horizontal displacement and rotation at *C*.

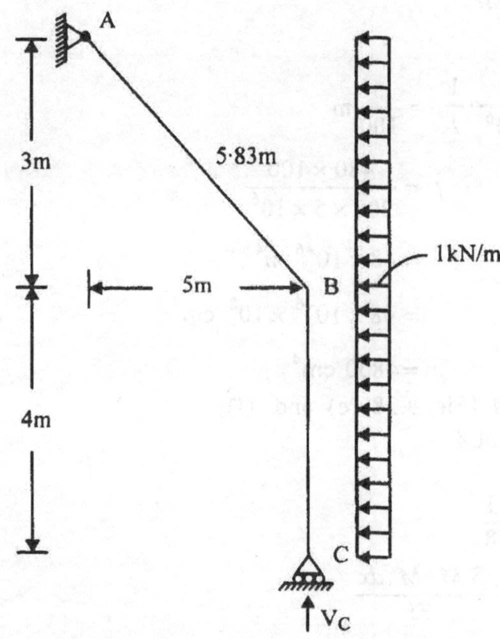

Fig. 8.29. Frame with load (Example 8.9)

Solution: Reaction at *C* is vertical. Taking moments about *A*

$$V_C \times 5 - 1 \times 7 \times \frac{7}{2} = 0$$

or $$V_C = 4.9\,\text{kN}.$$

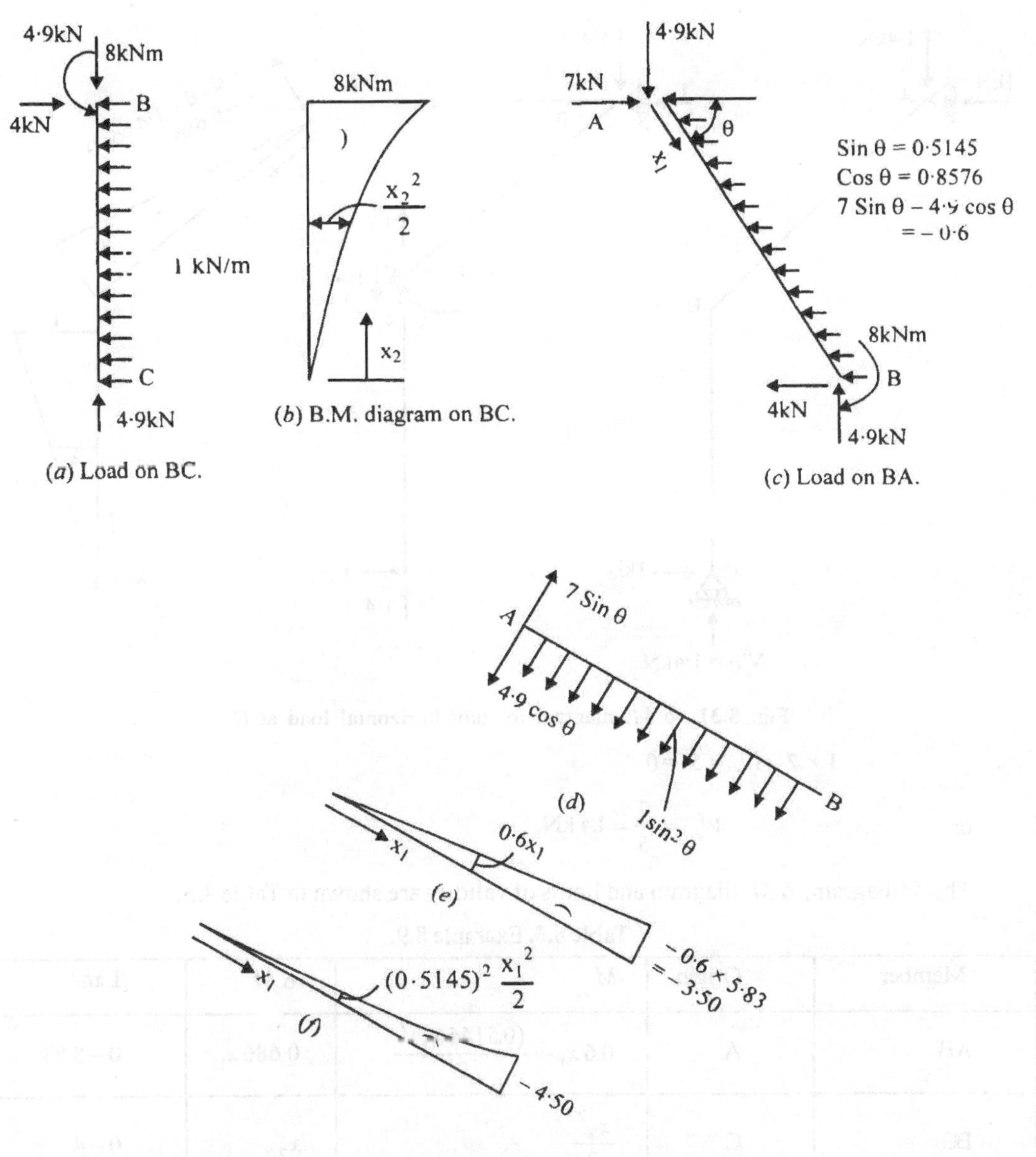

Fig. 8.30. *M* diagram of the frame.

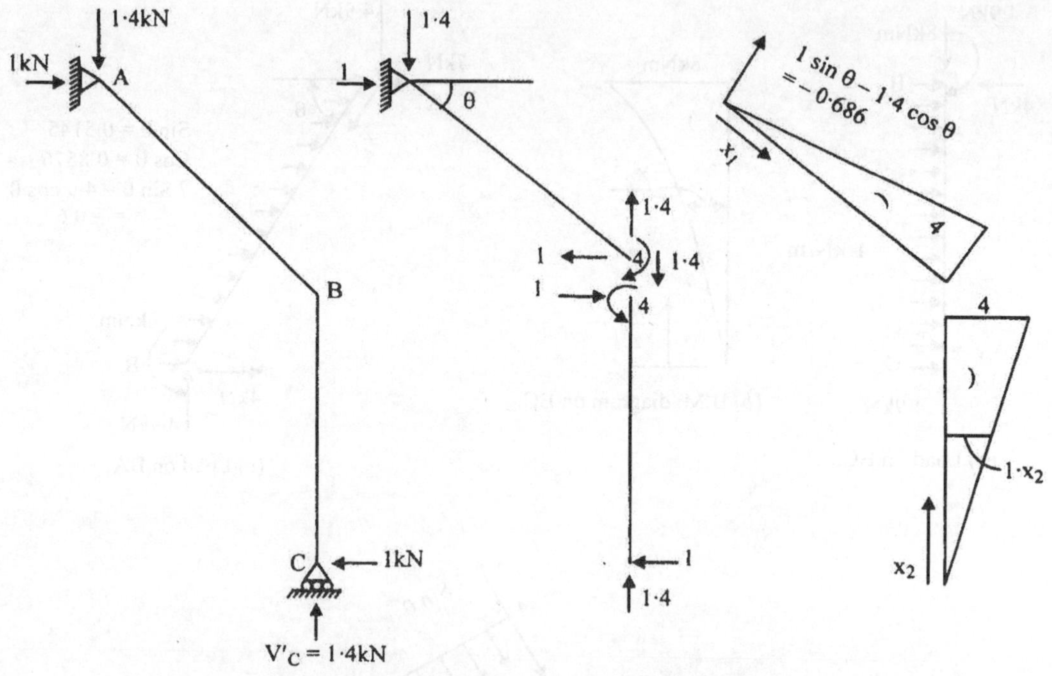

Fig. 8.31. δ *M* diagram for unit horizontal load at *C*.

$$1 \times 7 - V'_C \times 5 = 0$$

or
$$V'_C = \frac{7}{5} = 1.4 \text{ kN}.$$

The *M* diagram, δ *M* diagram and limits of validity are shown in Table 8.5.

Table 8.5. Example 8.9.

Member	Origin	M	δM	Limit
AB	A	$0.6\,x_1 + \dfrac{(0.5145\,x_1)^2}{2}$	$0.686\,x_1$	0 – 5.83
BC	C	$\dfrac{x_2^{\,2}}{2}$	x_2	0 – 4

Horizontal displacement of point C

1.
$$\Delta_h = \frac{1}{EI}\int_0^{5.83} (0.6\,x_1 + 0.1323\,x_1^2)\cdot 686\,x_1\,dx_1 + \frac{1}{EI}\int_0^4 \frac{x_2^{\,2}}{2}\cdot x_2 dx_2$$

$$= \frac{1}{EI}\left[0.411\cdot\frac{x_1^{\,3}}{3} + \cdot 09\,\frac{x_1^{\,4}}{4}\right]_0^{5.83} + \frac{1}{EI}\cdot\frac{1}{2}\left[\frac{x_2^{\,4}}{4}\right]_0^4$$

$$= \frac{1}{EI}[27.14 + 25.99 + 32] = \frac{85.13}{EI}$$

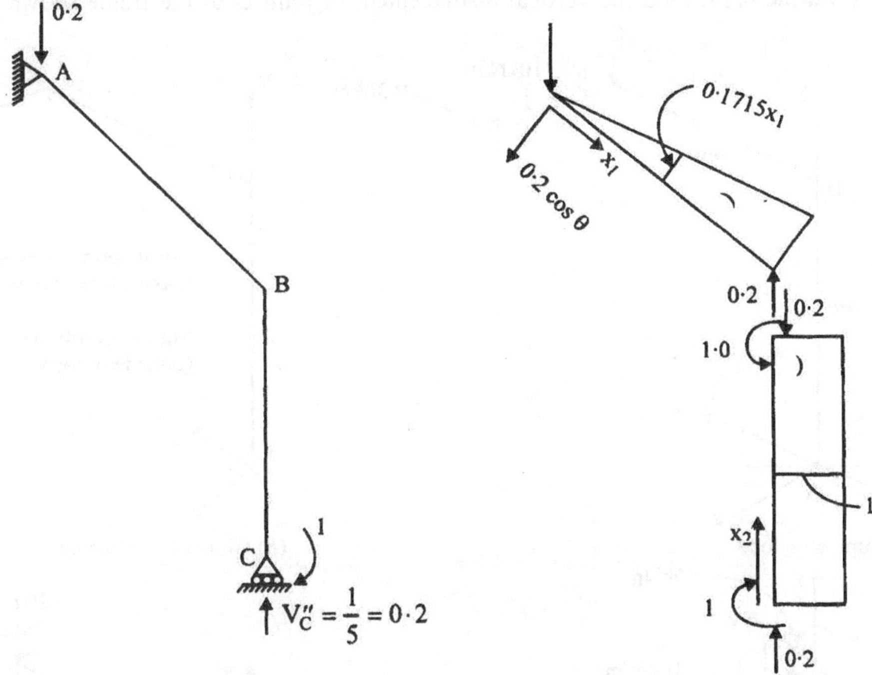

Fig. 8.32. δM diagram for unit moment at C.

Rotation at *C*

Table 8.6 Example 8.9

Member	Origin	M	δM	Limit
AB	A	$0.6\,x_1 + \dfrac{0.5145^2 \cdot x_1{}^2}{2}$	$0.1715\,x_1$	$0-5.83$
BC	C	$\dfrac{x_2{}^2}{2}$	1	$0-4$

$$1.\ \Delta\theta = \frac{1}{EI}\int_0^{5.83} (0.6\,x_1 + 0.1323\,x_1^2)(0.1715\,x_1)\,dx_1 + \frac{1}{EI}\int_0^4 \frac{x_2{}^2}{2} \cdot 1 \cdot dx_2$$

$$= \frac{1}{EI}[6.80 + 6.55 + 10.66]$$

$$= \frac{24}{EI}.$$

Example 8.10. Find the vertical displacement of joint D of the frame shown in Fig. 8.33.

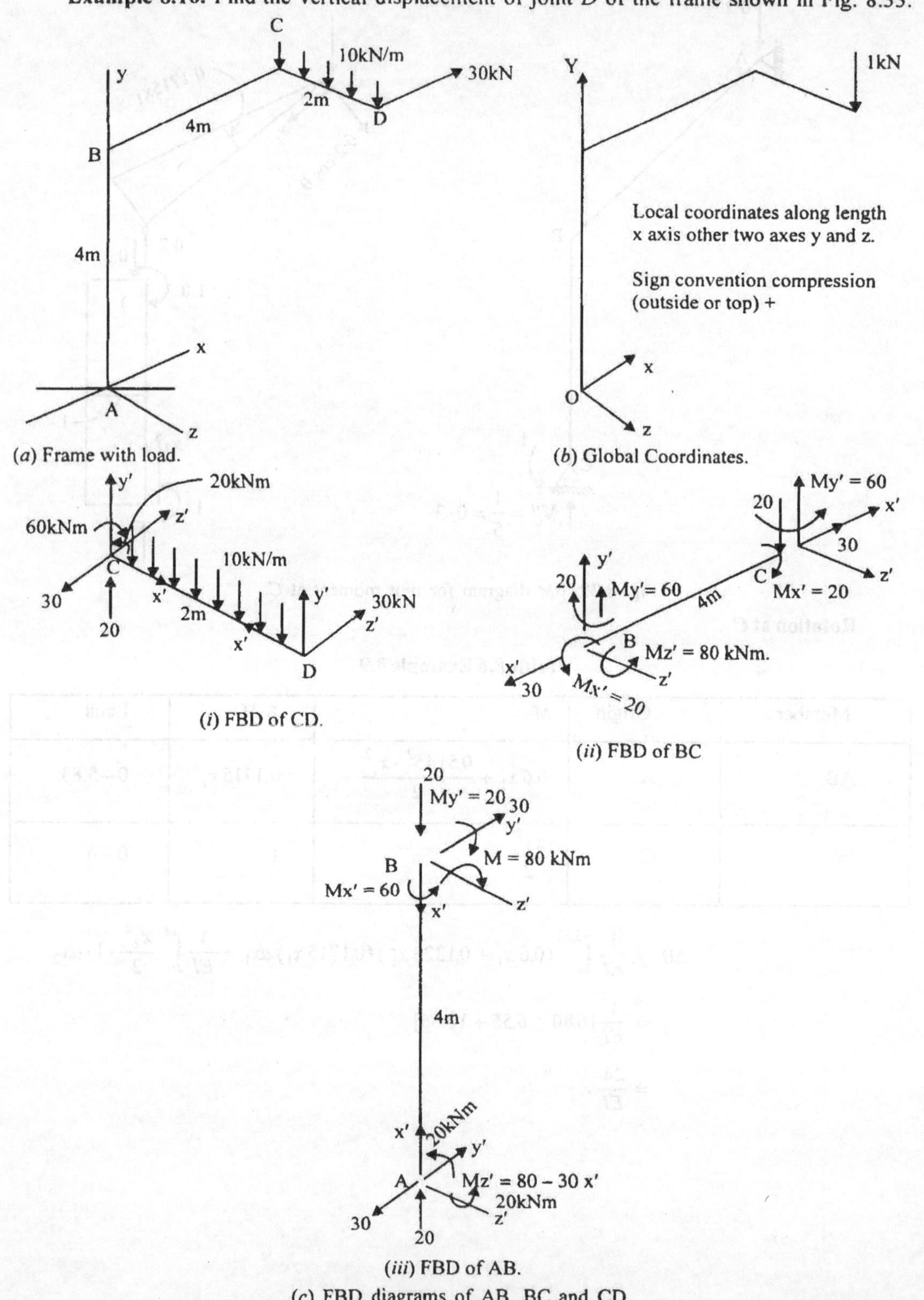

(*a*) Frame with load.

(*b*) Global Coordinates.

(*i*) FBD of CD.

(*ii*) FBD of BC

(*iii*) FBD of AB.

(*c*) FBD diagrams of AB, BC and CD.

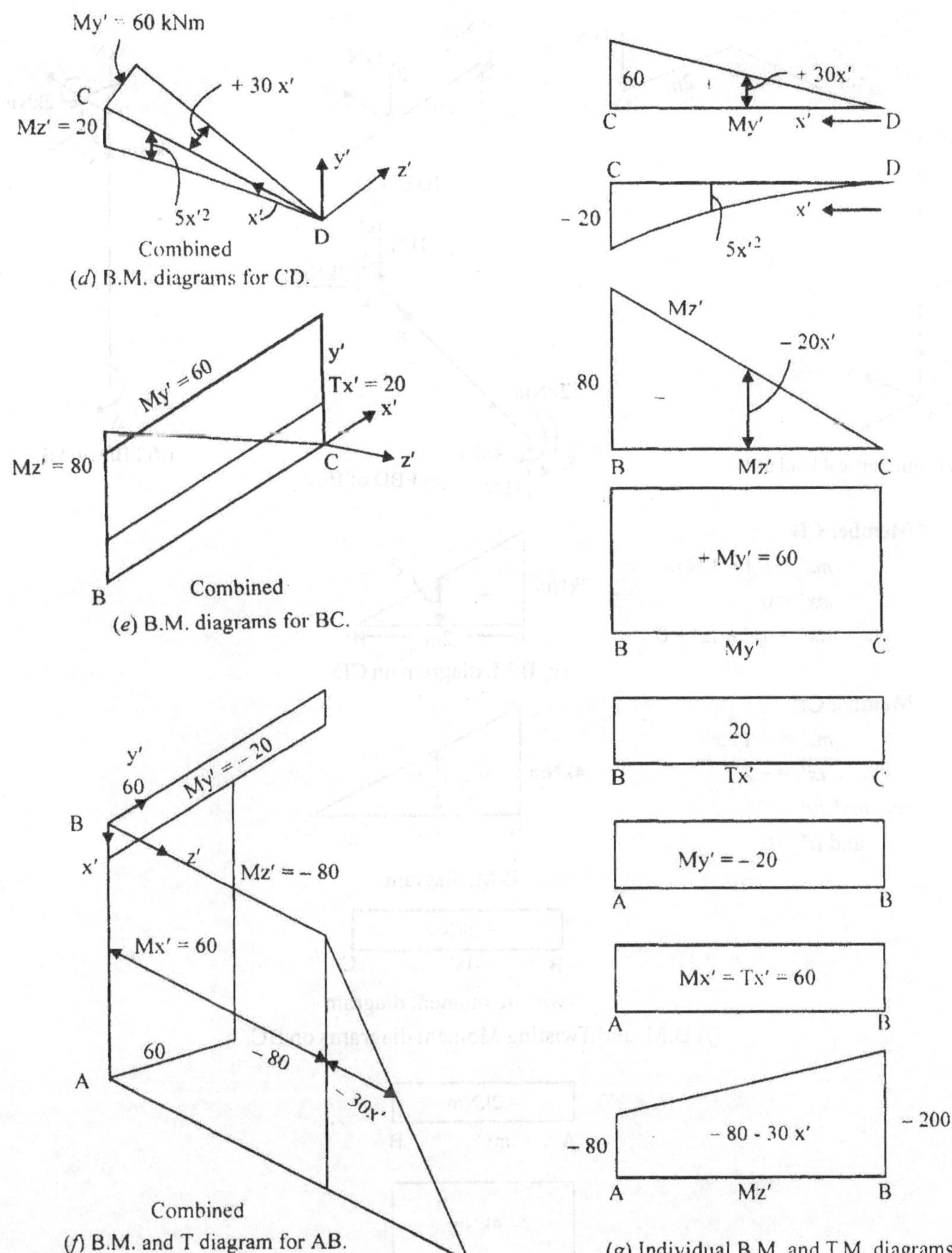

Fig. 8.33. FBD, B M & T diagrams (Example 8.10).

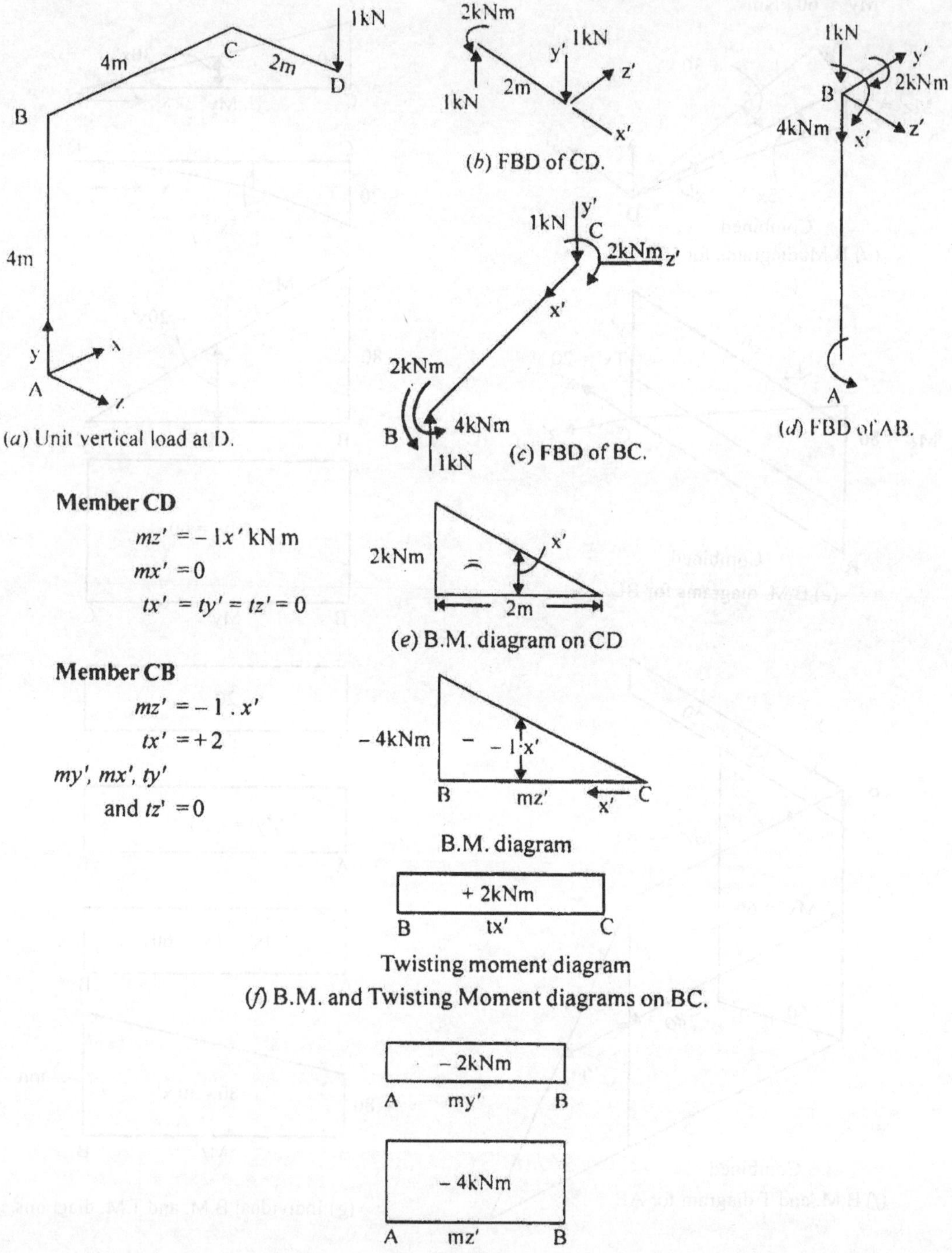

Member CD

$mz' = -1x'$ kN m

$mx' = 0$

$tx' = ty' = tz' = 0$

Member CB

$mz' = -1 \, . \, x'$

$tx' = +2$

my', mx', ty' and $tz' = 0$

(g) B.M. diagrams on AB.

Fig. 8.34. FBD, B.M. and Torsional diagram with unit vertical load at *D*.

Vertical deflection at D

$$1.\ \Delta_{yD} = \int_0^2 \frac{-5x'^2(-x').dx'}{EI_Z} + \int_0^4 \frac{-20x'(-x')dx'}{EI_Z} + \int_0^4 \frac{+20\times 2\times dx'}{GI_x} + \int_0^4 \frac{(-20)(-2)dx'}{EI_y}$$

$$+\int_0^4 \frac{(-80-30x')}{EI_Z}(-4)\cdot dx'$$

$$= \frac{5\times 2^4}{4EI_Z} + \frac{20\times 4^3}{3EI_Z} + \frac{40\times 4}{GI_x} + \frac{40\times 4}{EI_Y} + \frac{4}{EI_Z}\left[80\times 4 + \frac{30\times 4^2}{2}\right]$$

$$= \frac{160}{EI_y} + \frac{2686.66}{EI_Z} + \frac{160}{GI_x}.$$

8.13. IMPORTANT ENERGY THEOREMS

A number of energy theorems have been profitably used in structural analysis for a long time. It will be useful to have an integrated study of them. Virtual work principle with appropriate assumptions helps us in this study.

The main theorems are:

1. Principle of Stationary Potential Energy and Minimum Potential energy.
2. Castigliano's First Theorem.
3. Principle of Stationary and Minimum Complementary Potential Energy.
4. Castigliano's Second Theorem and Engesser's Theorem.

(*a*) Principle of Stationary Potential Energy and Minimum potential Energy — This principle states. "If an elastic structure is given a small virtual displacement and as a result the variation (change) in the total potential energy is zero, the structure will be in static equilibrium."

Total potential energy is the sum of strain energy U of a structure and the potential energy of the external forces so that total potential (energy)

$$\pi_P = U + V \qquad ...(8.19)$$

where V is the potential of the applied loads.

Taking potential energy of loads in initial condition as zero, $V = -PD = -W$ where P is the load and D is the displacement in the direction of P as force × displacement is equal to work done.

Now we apply virtual displacement principle which equates external virtual work δW with internal virtual work δU.

But from $\qquad V = -W$

$$\delta V = -\delta W = -\delta U$$

or $\qquad \delta V + \delta U = 0$

or $\qquad \delta(V + U) = 0$

$$\delta\pi_P = 0 \qquad ...(8.20)$$

In accordance with the virtual work principle loads and stresses remain constant during the virtual displacement.

The theorem of Minimum Potential Energy states that amongst all possible sets of deformations, that which ensures that all the equilibrium conditions are fulfilled will lead to minimisation of total potential π_P ie $\dfrac{\partial \pi_P}{\partial D_i}\cdot\delta D_i = 0$

in which we would sum our indices $i = 1, 2, ... n$. The δD_i are independent first variations of displacements. Since δD_i can have any value, $\delta \pi_P$ will be zero only if all $\frac{\partial \pi_P}{\partial D_i}$ is zero. The theorem can be used to form stiffness relations.

It should be noted that the method of minimum total potential energy is valid for linear and non linear structures as well as for structures experiencing gross distortion. The method is valuable when the number of degrees of freedom of the joints is less than the number of redundant bars i.e. when there are many members but few joints such as the spoke wheel problem. It should be noted that in order to satisfy the condition of continuity of displacements over the entire body, the strain must first be expressed in terms of the displacement and the variation of strain energy U must be carried out with respect to the displacements. This theorem is widely used in mechanics and is one of the major approaches used in finite element analysis for developing stiffness relations.

(b) Castigliano's 1st Theorem — In 1879, Alberto Castigliano published two theorems known as Casligliano's first and 2nd theorems. The 1st theorem states that if for any structure with independent displacements ($D_1, D_2, D_3 ... D_n$) corresponding to externally applied loads ($P_1, P_2, P_3 ... P_n$) along their lines of action strain energy U is expressed interms of displacements D then n equilibrium equations can be written in the form.

$$\frac{\partial U}{\partial D_i} = P_i$$

Proof: Potential of a series of concentrated loads or moments.

$$V = -\Sigma P_i D_i$$

As strain energy U is expressed in terms of displacements D, $U = U(D_i)$

Hence $\pi_P = U + V = U(D_i) - \Sigma P_i D_i$

$$\text{Now } \frac{\partial \pi_P}{\partial D_i} = \frac{\partial U(D_i)}{\partial D_i} - \frac{\partial}{\partial D_i} \Sigma P_i D_i = 0$$

$$\text{or } \frac{\partial U}{\partial D_i} = P_i \qquad ...(8.21)$$

The above theorem is applicable to both linear and non-linear elastic structures including gross distortions. Another way of expressing the theorem is

"The partial derivative of the strain energy with respect to displacement D_i is equal to the load P_i in the direction of D_i"

(c) Principle of stationary and minimum complementary potential energy: Complementary potential energy has no physical interpretation but it is an important mathematical concept. It can be visualised as the ability of *displacement* to do some work when the load varies from one level to other but the displacement remains constant.

$$V_C = -P_i D_i = -W_C$$

Total complementary energy then

$$\pi_C = U_C + V_C \qquad ...(8.22)$$

Principle of stationary complementary energy then states that. "If the first variation of the total complementary energy is equal to zero of an elastic structure, it will satisfy compatibility."

From principle of complementary virtual work.

$$\delta W_C = \delta U_C \qquad \text{But} \qquad V_C = -W_C$$

$$\text{or} \qquad \delta V_C = -\delta W_C$$

$$\text{or} \qquad \delta V_C = -\delta U_C$$

or $\delta(U_C + V_C) = \delta \pi_C = 0$ ...(8.23)

To utilise this principle we should have a system of forces in equilibrium and keeping displacement constant there should be first variation of load. The principle of minimum complementary potential energy states that "for a stable equilibrium condition of an elastic system the complementary potential energy should be minimum."

We have already derived the relation

$$\delta \pi_C = 0 \qquad ...(8.23)$$

But $\delta \pi_C$ for multiple load

$$= \frac{\partial \pi_C}{\partial P_i} \cdot \delta P_i$$. As δP_i is arbitrary and can taken any value

$$\frac{\partial \pi_C}{\partial P_1} = 0$$

$$\frac{\partial \pi_C}{\partial P_2} = 0$$

$$\frac{\partial \pi_C}{\partial P_n} = 0$$

As $\delta \pi_C = 0$, $\partial \pi_C = 0$ which leads to the minimisation principle. This principle is used to form flexibility relations and is also an energy approach for directly solving indeterminate structures.

(d) Engesser's Theorem and Castigliano's 2nd Theorem

Engesser's theorem states that if the complementary energy U_C of any elastic structure can be expressed in terms of the loads acting on the structure then the partial differentation of the complementary energy with respect to a load gives the displacement in the direction of the load or

$$\frac{\partial U_C}{\partial P_i} = D_i \qquad ...(8.24)$$

We already known

$$V_C = -\Sigma\, P_i\, D_i$$

$$U_C = U(P_i) \text{ (assumption)}$$

From principle of stationary potential energy

$$\frac{\partial \pi_C}{\partial P_i} = \frac{\partial}{\partial P_i}(U_C + V_C)$$

$$= \frac{\partial}{\partial P_i} \cdot (U_C(P_i) - \Sigma\, P_i - D_i)$$

$$= \frac{\partial U_C.(P_i)}{\partial P_i} - D_i = 0$$

or $$\frac{\partial U_C}{\partial P_i} = D_i$$

This is known as Engesser's Theorem. The theorem is applicable to any elastic structure linear or non-linear excluding gross distortion. This is true because we are using complementary energy.

Castigliano's theorem is valid only for linear elastic structure and when effect of environmental cause is zero i.e. the load deflection curve must start from origin.

In such a case $U_C = U$

and we can write $$\frac{\partial U}{\partial P_i} = D_i \qquad \text{... (8.25)}$$

Castigliano's 2nd Theorem can now be stated as. "For a linear elastic structure, if the strain energy is expressed in terms of the load, partial derivative of the strain energy with respect to any load will give the deflection in the direction of load."

This principle is utilised in solving indeterminate structure. Where support displacement is zero or a known prescribed value the theorem can be straight way applied to compute support reactions. In other cases, Castigliano's Theorem of least work which follows from Castigliano's 2nd theorem can be used. The theorem of least work states that "the work done in stressing an indeterminate structure under a given system of loads is the *least possible* consistent with maintenance of equilibrium" Like Castigliano's 2nd theorem it is applicable only to linearly elastic structures.

It will be shown in this and subsequent chapters that energy methods can be utilised to

(*i*) Calculate deflections by Engessers' theorem and Casligliano's 2nd theorem.

(*ii*) Compute compatible displacements, form flexibility relations, and Flexibility matrices to directly solve redundants of statically indeterminate structures (force method) through stationary Complementary Potential energy. Engessers' theorem and Casligliano's theorem.

The principles of virtual displacements, stationary potential energy and Castigliano's first theorem lead to a stiffness analysis. They may be used to compute forces in equilibrium, form stiffness relations and solve for the displacements of a structure.

As indeterminate structures will be dealt separately in later chapters, a number of examples are solved on deflections of statically determinate structures only by energy principles.

8.14 LINEAR AND NONLINEAR PROBLEMS

For linear problems Castigliano's 2nd theorem gives

$$\frac{\partial U}{\partial P_i} = D_i$$

where U is the strain energy.

Total strain energy of a beam or frame subjected to axial force F, bending moment M, Torsional Moment T and shear force V is given by,

$$U = \int \frac{F^2 dx}{2AE} + \int \frac{M^2 dx}{2EI} + \int \frac{T^2 dx}{2GJ} + \int \frac{V^2 dx}{2GA_S}$$

These expressions for individual cases have been derived in sec. 8.5. Total energy will be the sum of energies due to individual cases. Applying Castigliano's theorem.

$$D_i = \frac{\partial U}{\partial P_i} = \int \frac{F \cdot \frac{\partial F}{\partial P_i} \cdot dx}{AE} + \int \frac{M \cdot \frac{\partial M}{\partial P_i} \cdot dx}{EI} + \int \frac{T \cdot \frac{\partial T}{\partial P_i} \cdot dx}{GJ} + \int \frac{V \cdot \frac{\partial V}{\partial P_i} \cdot dx}{GA_s}$$

For nonlinear problems Engesser's theorem $\frac{\partial U_C}{\partial P_i} = D_i$ has to be applied where complementary energy replaces strain energy.

Example 8.11. Find the deflection at the centre of the simply supported beam due to flexure and shear. Find their ratio. *EI* constant.

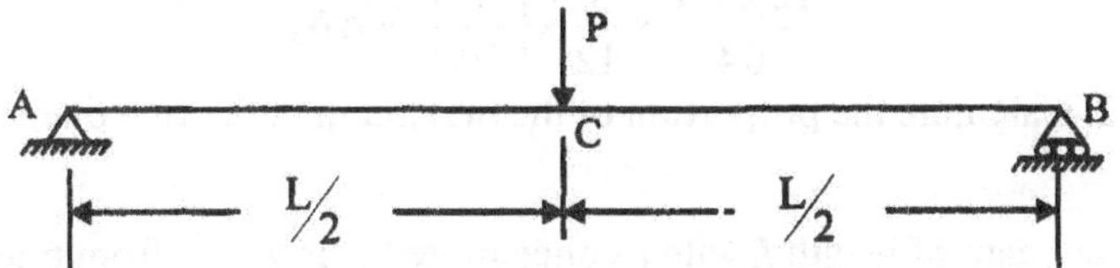

Fig. 8.35. Example 8.11

Solution:

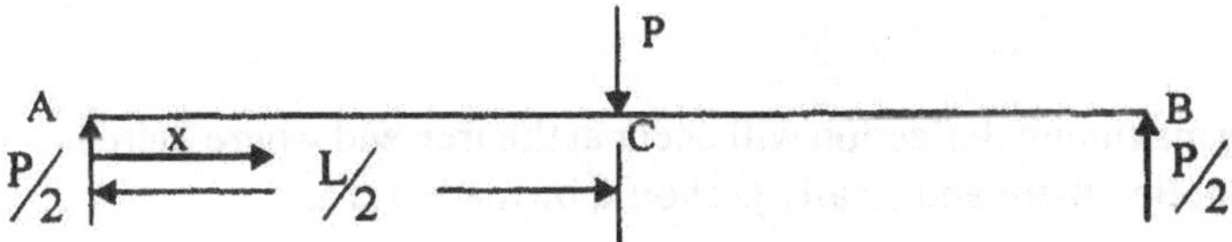

Fig. 8.36. Calculatations of B.M.

For $0 \le x \le \frac{L}{2}$ and with origin at *A*

$$M = \frac{P}{2} \cdot x$$

$$\frac{\partial M}{\partial P} = \frac{x}{2}$$

Deflection due to flexure (factor 2 because of symmetry)

$$= 2\int_o^{L/2} \frac{P}{2EI} \cdot x \cdot \frac{x}{2} \cdot dx = \frac{P}{2EI}\left[\frac{x^3}{3}\right]_0^{L/2}$$

$$= \frac{P}{2EI} \cdot \frac{L^3}{24} = \frac{PL^3}{48EI}$$

For $0 \le x \le \frac{L}{2}$ and with origin at *A*

$$V = \frac{P}{2}$$

$$\frac{\partial V}{\partial P} = \frac{1}{2}$$

Deflection due to shear $= \frac{2}{As.G} \cdot \int_0^{L/2} \frac{P}{2} \cdot \frac{1}{2} \cdot dx$

$$= \frac{P}{2As \cdot G} \cdot [x]_0^{L/2} = \frac{PL}{4As \cdot G}$$

With $\frac{d}{L} = \frac{1}{10}$ G = 0.4 *E* ratio of $\frac{\text{deflection due to shear}}{\text{deflection due to B.M.}}$

$$= \frac{PL}{4AsG} \times \frac{48EI}{PL^3}$$

$$= \frac{12 \cdot L}{0.4} \times \frac{1.2}{b \times d} \times \frac{bd^3}{12} \times \frac{1}{L^3}$$

$$= \frac{12 \times 1 \cdot 2}{0.4} \times \frac{1}{12} \times \left(\frac{1}{10}\right)^2 = 0.03.$$

Example 8.12. Calculate the proportion of the maximum deflection due to shear and bending moment in

(*a*) A cantilever beam of length L with a concentrated load P, $\frac{L}{2}$ from free end.

(*b*) A simply supported beam with uniformly distributed load w/unit length on the full span.

Assume the beams to be rectangular section with $d = 2b$ and $\frac{d}{L} = \frac{1}{10}$. Also assume $G = 0.4\,E$. EI is contant.

Solution:

(*a*) Since the maximum deflection will occur at the free end where there is no load, an inaginary load Q has to be applied there and finally Q should be made zero.

Fig. 8.37. Example 8.12(*a*).

With origin at C.

for $0 \le x \le \frac{L}{2}$ $\qquad M = -Qx, \frac{\partial M}{\partial Q} = -x$

$\frac{L}{2} \le x \le L$ $\qquad M = -Qx - P\left(x - \frac{L}{2}\right)$

$$\frac{\partial M}{\partial Q} = -x$$

$0 \le x \le \frac{L}{2}$ $\qquad V = Q$

$$\frac{\partial V}{\partial Q} = 1$$

$\frac{L}{2} \le x \le L$ $\qquad V = Q + P$

$$\frac{\partial V}{\partial Q} = 1.$$

Deflection due to Moment

$$\Delta = \int_0^{L/2} \frac{(-Qx)(-x)dx}{EI} + \int_{L/2}^{L} \frac{\{-Qx - P(x - L/2)\}(-x)dx}{EI}$$

$$= \left[\frac{Q \cdot x^3}{3EI}\right]_0^{L/2} + \left[\frac{Q \cdot x^3}{3EI}\right]_{L/2}^{L} + \left[\frac{Px^3}{3EI}\right]_{L/2}^{L} - \frac{PL}{2EI}\left[\frac{x^2}{2}\right]_{L/2}^{L}$$

As Q is zero first two terms are zero.

Hence $\qquad \Delta = \frac{P}{3EI}\left[L^3 - \left(\frac{L}{2}\right)^3\right] - \frac{PL}{4EI}\left[L^2 - \left(\frac{L}{2}\right)^2\right]$

$$= \frac{7PL^3}{24EI} - \frac{3PL^3}{16EI} = \frac{5}{48EI} \cdot PL^3$$

Check

$$\frac{P\left(\frac{L}{2}\right)^3}{3EI} + \frac{P\left(\frac{L}{2}\right)^2}{2EI} \cdot \frac{L}{2} = \frac{PL^3}{EI}\left(\frac{1}{24} + \frac{1}{16}\right) = \frac{PL^3}{EI}\frac{5}{48}.$$

Deflection due to shear.

$$= \int_0^{L/2} \frac{Q \cdot dx}{GA_s} + \int_{L/2}^{L} \frac{(Q+P)\,dx}{GA_s}$$

$$= \frac{Q \cdot L/2}{GA_s} + \frac{(Q+P) \cdot L}{2GA_s}$$

As Q is zero, deflection due to shear $= \dfrac{PL}{2 \cdot G \cdot A_s}$

Ratio of $\dfrac{\text{Deflection due to shear}}{\text{Deflection due to B.M.}} = \dfrac{PL}{2 \cdot G \cdot A_s} \cdot \dfrac{48EI}{5PL^3}$

$$= \frac{48}{10} \times \frac{1}{12} \cdot \frac{L \cdot Ebd^3 \times 1.2}{G \cdot b \cdot d \cdot L^3}$$

$$= \frac{48 \times 1.2 \times 1}{0.4 \times 10 \times 12 \times 10^2} = \cdot 012$$

(*b*) Distributed load on simply supported span.

Deflecton will be maximum at centre since there is no load at centre we will have to apply an imaginery load Q at centre which should finally be put to zero.

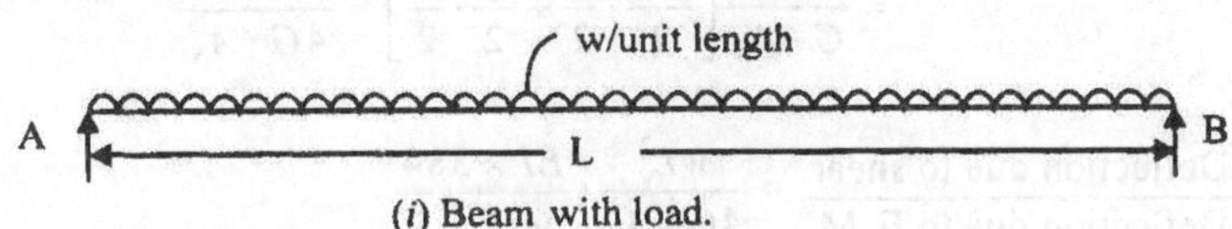

(*i*) Beam with load.

Q
C
w/unit length
A
B
L/2
L/2

$R_A = \dfrac{wL}{2} + \dfrac{Q}{2}$ (*ii*) Imaginary load at C. $R_B = \dfrac{wL}{2} + \dfrac{Q}{2}$

Fig. 8.38. Example 8.12(*b*)

Deflection due to flexure.

For $0 \le x \le \dfrac{L}{2}$ and origin at A

$$M = \left(\frac{wL}{2} + \frac{Q}{2}\right) \cdot x - \frac{wx^2}{2}$$

$$\frac{\partial M}{\partial Q} = \frac{x}{2}$$

$$\Delta = \frac{2}{EI}\int_0^{L/2}\left[\left(\frac{wL}{2}+\frac{Q}{2}\right)\cdot x-\frac{wx^2}{2}\right]\cdot\frac{x}{2}\cdot dx$$

[The factor 2 is because of symmetry]

$$= \frac{2}{2EI}\int_0^{L/2}\left(\frac{wL}{2}\cdot x^2+\frac{Q}{2}\cdot x^2-\frac{wx^3}{2}\right)\cdot dx$$

$$= \frac{1}{EI}\left[\frac{wL}{2}\cdot\frac{x^3}{3}+\frac{Q}{2}\cdot\frac{x^3}{3}-\frac{wx^4}{8}\right]_0^{L/2}$$

As Q is zero, the expression reduces to $\frac{1}{EI}\left[\frac{wL}{6}\cdot\frac{L^3}{8}-\frac{wL^4}{8\cdot 16}\right]$

$$= \frac{5}{384}\cdot\frac{wL^4}{EI}$$

Deflection due to shear

For $0 \le x \le \frac{L}{2}$ $\qquad V = \frac{wL}{2}+\frac{Q}{2}$

$$\frac{\partial V}{\partial Q} = \frac{1}{2}$$

Deflection due to shear $= 2\int_0^{L/2}\frac{\left(\frac{wL}{2}+\frac{Q}{2}\right)\left(\frac{1}{2}\right)\cdot dx}{G\cdot A_S}$

$$= 2\times\frac{1}{2}\times\frac{1}{G\cdot A_S}\int_0^{L/2}\left(\frac{wL}{2}+\frac{Q}{2}\right)dx$$

$$= \frac{1}{G\cdot A_S}\left[\frac{wL}{2}\cdot\frac{L}{2}+\frac{Q}{2}\cdot\frac{L}{2}\right] = \frac{wL^2}{4G\cdot A_S}$$

as $Q = 0$.

Ratio of $\frac{\text{Deflection due to shear}}{\text{Deflection due to B.M.}} = \frac{wL^2}{4G\cdot A_S}\cdot\frac{EI\times 384}{wL^4\times 5}$

$$= \frac{E}{4G}\cdot\frac{1.2}{bd}\times\frac{bd^3}{12}\times\frac{384}{5}\times\frac{1}{L^2}$$

$$= \frac{1}{4\times 0.4}\times\frac{1.2\times 384}{12\times 5}\times\frac{1}{100} = 0.048$$

Example 8.13. Find the deflection under the load for the beam shown in Fig. 8.39.

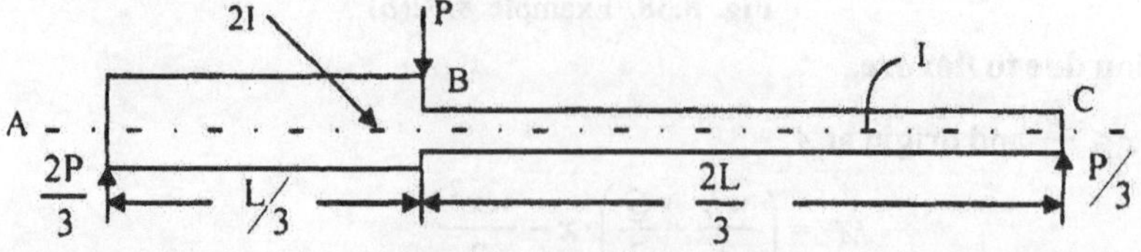

Fig. 8.39. Example 8.13

Solution:

Deflection under load

For portion AB, origin at A

$$0 \le x \le \frac{L}{3} \qquad M = \frac{2P}{3} \cdot x$$

$$\frac{\partial M}{\partial P} = \frac{2x}{3}$$

$$I_{AB} = 2I$$

For portion BC, origin at C

$$0 \le x \le \frac{2L}{3} \qquad M = \frac{P}{3} \cdot x$$

$$\frac{\partial M}{\partial P} = \frac{x}{3}$$

$$I_{CB} = I$$

$$\Delta = \frac{1}{2EI}\int_0^{L/3} \frac{2P}{3} \cdot x \cdot \frac{2x}{3} \cdot dx + \frac{1}{EI}\int_0^{2L/3} \frac{P}{3} \cdot x \cdot \frac{x}{3} \cdot dx$$

$$= \frac{2P}{9EI}\left[\frac{x^3}{3}\right]_0^{L/3} + \frac{P}{9EI}\left[\frac{x^3}{3}\right]_0^{2L/3}$$

$$= \frac{PL^3}{EI}\left[\frac{2}{9 \times 27 \times 3} + \frac{8}{9 \times 3 \times 27}\right]$$

$$= \frac{10PL^3}{9 \times 27 \times 3EI} = 0.0137\frac{PL^3}{EI}$$

Example 8.14. **A quadrantal ring AB with flexural rigidity EI is fixed at B and is loaded by a vertical load P acting in the plane of the ring at the free end A. The support B may be taken to be horizontal. The radius of the ring is R. Calculate the vertical and horizontal displacement of the end.**

Solution:

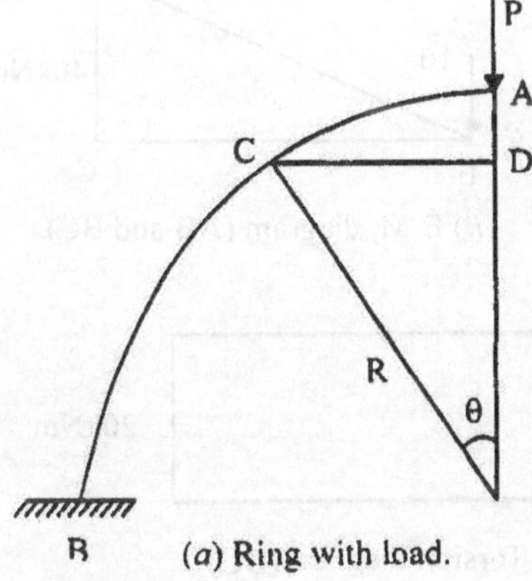

(*a*) Ring with load.

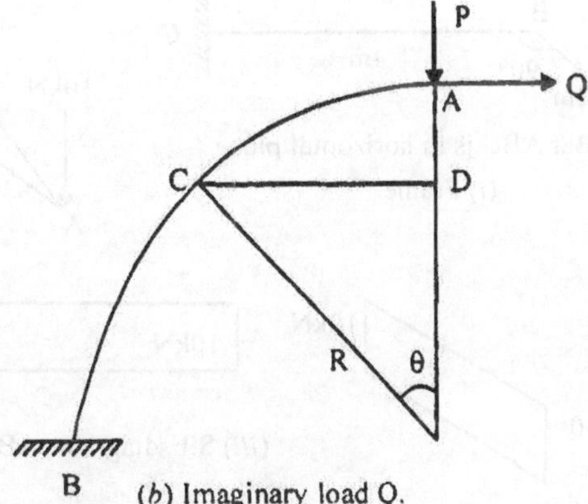

(*b*) Imaginary load Q.

Fig. 8.40. Example 8.14

For vertical deflection

for $0 \le \theta \le \frac{\pi}{2}$

$$M = -P \cdot CD$$

$$= -PR \sin\theta$$

$$\frac{\partial M}{\partial P} = -R \sin\theta$$

$$\Delta_V = \frac{1}{EI}\int_0^{\pi/2} -PR\sin\theta\,(-R)\,(\sin\theta)R.\,d\theta$$

$$= \frac{PR^3}{EI}\int_0^{\pi/2} \sin^2\theta \cdot d\theta = \frac{\pi}{4}\cdot\frac{PR^3}{EI}$$

For horizontal deflection

Apply imaginery horizontal load Q

$$M = -P \times CD - Q \times AD$$

$$= -P\,R\sin\theta - QR\,(1-\cos\theta)$$

$$\frac{\partial M}{\partial Q} = -R\,(1-\cos\theta).$$

$$\Delta_h = \frac{1}{EI}\int_0^{\pi/2}\left[-PR\sin\theta - QR(1-\cos\theta)\right]\times -R(1-\cos\theta)\cdot R d\theta$$

$$= \frac{R^3}{EI}\int_0^{\pi/2}\left[P\sin\theta + Q(1-\cos\theta)\right](1-\cos\theta)\cdot d\theta$$

$$= \frac{R^3}{EI}\left|-P\cos\theta + Q\cdot\theta - Q\sin\theta - \frac{P\sin^2\theta}{2} - Q\sin\theta\right|_0^{\pi/2} + Q\cdot\frac{\pi}{4}$$

Putting $Q = 0$ $$= \frac{R^3}{EI}\cdot\frac{P}{2} = \frac{PR^3}{2EI}$$

Example 8.15. The figure 8.41 shows a hollow thin circular pipe in a horizontal plane loaded by a vertical load of 10 kN at A. Determine the vertical deflection at A. Cross sectional area = 1200 mm^2, E = 200 GPa, G = 80 GPa, $I = 250 \times 10^6$ mm^4, $J = 500 \times 10^6$ mm^4.

Solution:

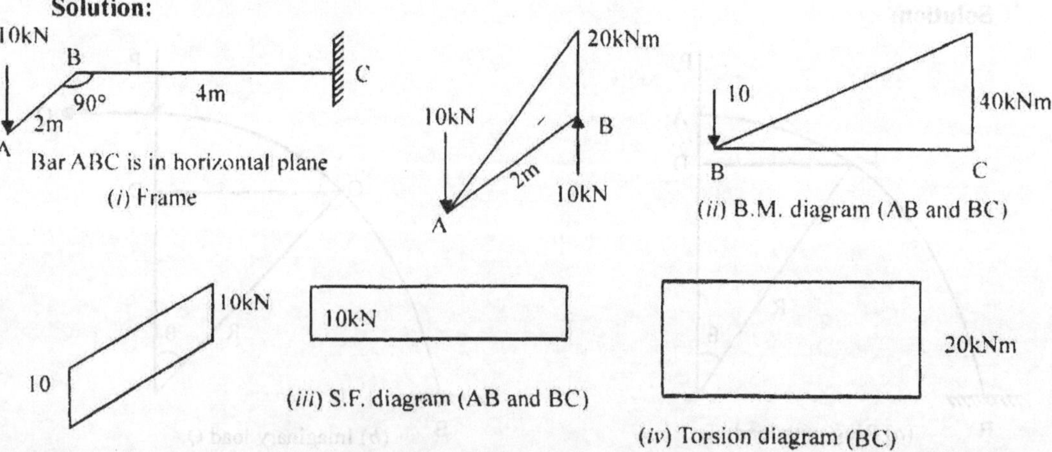

Fig. 8.41. Example 8.15

Imaginary Vertical load.

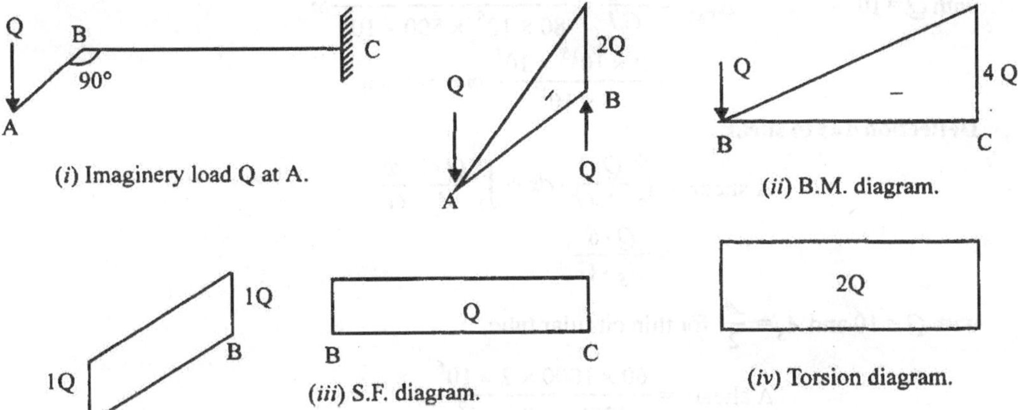

(i) Imaginery load Q at A. (ii) B.M. diagram. (iii) S.F. diagram. (iv) Torsion diagram.

Fig. 8.42. B.M., S.F. and Torsion diagram due to load Q.

$$\frac{\partial U}{\partial Q} = \text{deflection at } A$$

$$= \frac{1}{EI}\int M \cdot \frac{\partial M}{\partial Q} \cdot dx + \frac{1}{A_S \cdot G}\int V \cdot \frac{\partial V}{\partial Q} \cdot dx + \frac{1}{GJ}\int T \cdot \frac{\partial T}{\partial Q} \cdot dx$$

For *AB*

Origin at A $\quad M = -Q.x \quad \frac{\partial M}{\partial Q} = -x$ Limit 0 to 2 m

$V = +Q \quad \frac{\partial V}{\partial Q} = 1$ Limit 0 to 2 m

$T = 0$

For *BC*

Origin at B $\quad M = -Q.x \quad \frac{\partial M}{\partial Q} = -x$ Limit 0 to 4 m

$V = +Q \quad \frac{\partial V}{\partial Q} = 1$ Limit 0 to 4 m

$T = +2Q \quad \frac{\partial T}{\partial Q} = 2$ Limit 0 to 4 m

Deflection due to B.M.

$$\int_0^2 \frac{Qx^2 dx}{EI} + \int_0^4 \frac{Qx^2 dx}{EI} = \frac{Q}{EI}\left[\frac{2^3}{3} + \frac{4^3}{3}\right] = \frac{24Q}{EI}$$

with $Q = 10$ kN $\quad \Delta_{B.M} = \frac{10 \times 24 \times 1000 \times 1000}{200 \times 10^9 \times 250 \times 10^{-6}} = 4.8\ mm$

Deflection due to torsion.

$$\int_0^4 \frac{2(Q) \times 2}{GJ} \cdot dx = \frac{4Q}{GJ} \cdot [x]_0^4 = \frac{16Q}{GJ}$$

with $Q = 10$ $$\Delta_{T.M} = \frac{16Q}{GJ} = \frac{160 \times 1000 \times 10^{12}}{80 \times 10^9 \times 500 \times 10^6} m$$

$$= \frac{2 \times 10^{15} \times 10^3}{5 \times 10^{17}} mm = 4 mm.$$

Deflection due to shear.

$$\Delta \text{ shear} = \int_0^2 \frac{Q \cdot 1}{A_S G} \cdot dx + \int_0^4 \frac{Q \cdot 1 \cdot dx}{A_S \cdot G}$$

$$= \frac{Q \cdot 6}{A_S \cdot G}$$

with $Q = 10$ and $A_S = \frac{A}{2}$ for thin circular tube,

$$\Delta \text{ shear} = \frac{60 \times 1000 \times 2 \times 10^6}{1200 \times 80 \times 10^9} m$$

$$= \frac{12 \times 10^{10} \times 10^3}{12 \times 10^{12} \times 8} mm = 1.25 mm.$$

Total deflection $= 4.8 + 4.0 + 1.25 = 10.05$ *mm.*

It can be observed that for a thin circular cross section deflection due to shear is comparable to that due to B.M. and Torsion.

Example 8.16. Determine (*a*) the horizontal deflection and (*b*) the vertical deflection of joint *C* in the pin jointed truss shown. Each member has a cross sectional area of 1000 mm^2. $E = 200$ GPa.

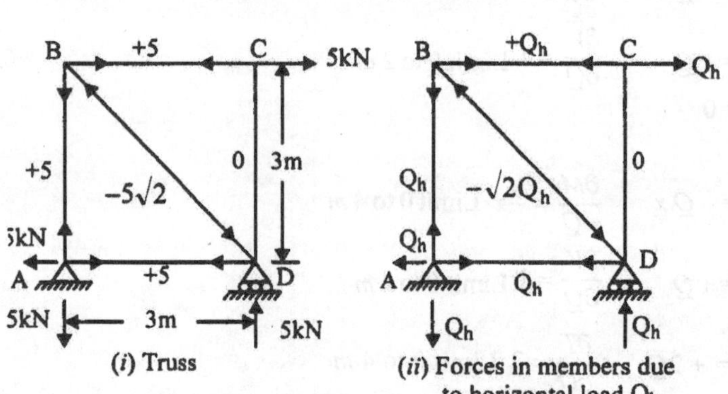

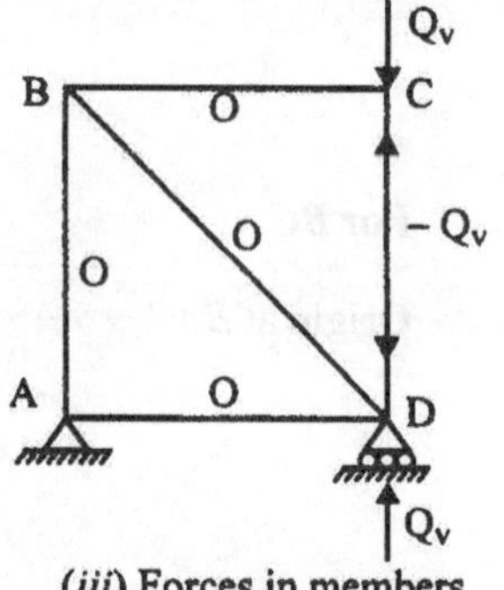

Fig. 8.43. Example 8.16.

Solution:

Forces in the members due to applied load

$$P_{AB} = +5 \, kN$$

$$P_{AD} = +5 \, kN$$

$$P_{BD} = -5\sqrt{2} \, kN$$

$$P_{BC} = +5 \, kN$$

$$P_{CD} = 0.00$$

For imaginary horizontal load Q_h at C

$$f_h AB = +Q_h$$
$$f_h AD = +Q_h$$
$$f_h BD = -\sqrt{2}\,Q_h$$
$$f_h BC = +Q_h$$
$$f_h CD = +0.00$$

For imaginery vertical load Q_V at C

$$f_V CD = -Q_V$$
$$f_V AB = 0$$
$$f_V AD = 0$$
$$f_V BD = 0$$
$$f_V BC = 0$$

Strain Energy due to axial forces

$$= \sum \frac{F^2 L}{2AE} \qquad \text{where } F = \text{total force} = P + f.$$

$$\text{deflection} = \sum F \cdot \frac{\partial F}{\partial Q} \cdot \frac{L}{AE}$$

The calculations are shown in Table 8.7.

Table 8.7 (Example 8.16)

Member	L in m	$A\ mm^2$	F	$\frac{\partial F}{\partial Q_h}$	$\frac{F \cdot \partial F}{\partial Q_h} \cdot L$	$\frac{\partial F}{\partial Q_V}$	$\frac{F \cdot \partial F}{\partial Q_V} \cdot L$
AB	3	1000	$5 + Q_h$	1	$3(5 + Q_h)$	0	0
BC	3	1000	$5 + Q_h$	1	$3(5 + Q_h)$	0	0
CD	3	1000	0.00	0	0	–1	$3Q_V$
DA	3	1000	$5 + Q_h$	1	$3(5 + Q_h)$	0	0
BD	$3\sqrt{2}$	1000	$-\sqrt{2}(5 + Q_h)$	$-\sqrt{2}$	$6\sqrt{2}(5 + Q_h)$	0	0
					$\Sigma 15(3+2\sqrt{2}) + (9+6\sqrt{2})Q_h$		$\Sigma\, 3Q_V$

With $Q_h = 0$ horizontal deflection

$$\Delta_h = \frac{15(3+2\sqrt{2}) \times 10^6 \times 10^3}{1000 \times 200 \times 10^9}$$

$$= 43.7 \times 10^{-5}\, m = 0.437\, mm$$

with $Q_V = 0$ $\qquad \Delta_V = 0$

Example 8.17. Use (i) Engesser's theorem and (ii) principle of complementary virtual work to compute the horizontal displacement of joint three of the truss shown in Fig. 8.44

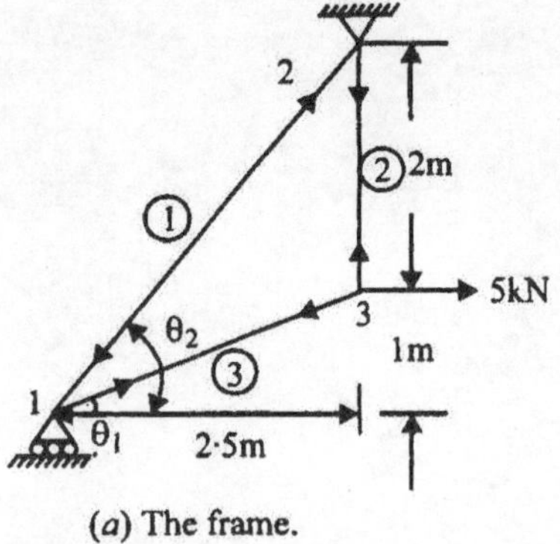

(a) The frame.

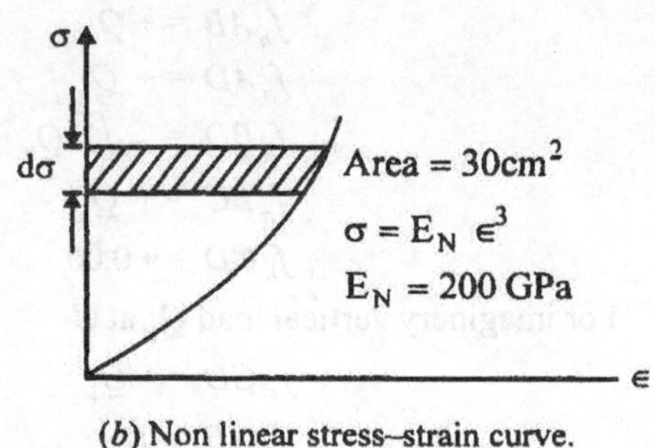

(b) Non linear stress–strain curve.

Fig. 8.44. Example 8.17

Solution:

Forces in the truss members due to applied load of *P kN*.

Taking moments about point 2

$$R_1 \times 2.5 - P \times 2 = 0$$

or $$R_1 = \frac{2P}{2.5} = 0.8\,P$$

Consider equilibrium of joint 3

$$F_3 \cos \theta_1 = P$$

or $$F_3 = \frac{P}{\cos\theta_1} = \frac{P}{0.928} \quad \left[\cos\theta_1 = \frac{2.5}{\sqrt{2.5^2 + 1^2}} = 0.928\right]$$

$$= 1.078\,P$$

$$F_2 = 1.078\,P \sin \theta_1$$

$$= 1.078P \cdot \frac{1}{\sqrt{2.5^2 + 1^2}} = 0.40\,P$$

Joint 1

$$F_3 \cos \theta_1 - F_1 \cos \theta_2 = 0$$

or $$F_1 \cos \theta_2 = F_3 \cos \theta_1 = 1.078\,P \times 0.928$$

or $$F_1 = \frac{1.078 \times 0.928P}{0.64} \quad \left[\cos\theta_2 = \frac{2.5}{\sqrt{2.5^2 + 3.0^2}}\right]$$

$$= 1.563\,P$$

(i) By Engesser's theorem $\dfrac{\partial U_C}{\partial P_i} = D_i$

$$U_C = \int \in \cdot d\sigma \cdot dvol$$

Here $$\in = \left(\frac{\sigma}{E_n}\right)^{1/3}$$

Hence $$U_C = \int \left(\frac{\sigma}{E_n}\right)^{1/3} \cdot d\sigma . dvol$$

$$= \frac{3}{E_n^{1/3}} \cdot \frac{\sigma^{4/3}}{4} \cdot dvol$$

For a truss of 3 members with constant cross section.

$$U_C = \frac{3}{4} \cdot \frac{1}{E_n^{1/3}} \left[A_1 L_1 \cdot \sigma_1^{4/3} + A_2 L_2 \cdot \sigma_1^{4/3} + A_3 L_3 \cdot \sigma_3^{4/3} \right]$$

$$= \frac{3}{4} \cdot \frac{1}{E_n^{1/3}} \left[A_1 L_1 \cdot \left(\frac{1.563P}{A_1} \right)^{4/3} + A_2 L_2 \cdot \left(\frac{0.40P}{A_2} \right)^{4/3} + A_3 L_3 \cdot \left(\frac{1.078P}{A_3} \right)^{4/3} \right]$$

$$= \frac{3}{4} \cdot \frac{1}{E_n^{1/3}} \left[\frac{A_1 L_1}{A_1^{4/3}} \cdot (1.563)^{4/3} \cdot P^{4/3} + \frac{A_2 L_2}{A_2^{4/3}} \cdot (0.40)^{4/3} \cdot P^{4/3} \right.$$

$$\left. \frac{A_3 L_3}{A_3^{4/3}} \cdot (1.078)^{4/3} \cdot P^{4/3} \right]$$

$$\frac{\partial U_C}{\partial P} = \frac{3}{4} \cdot \frac{1}{E_n^{1/3}} \left[\frac{A_1 L_1}{A_1^{4/3}} \cdot (1.563)^{4/3} \cdot \frac{4}{3} \cdot P^{1/3} + \frac{A_2 L_2}{A_2^{4/3}} \cdot (0.40)^{4/3} \cdot \frac{4}{3} \cdot P^{1/3} \right.$$

$$\left. \frac{A_3 L_3}{A_3^{4/3}} \cdot (1.078)^{4/3} \cdot \frac{4}{3} \cdot P^{1/3} \right]$$

with $P = 5\ kN$

$$\frac{\partial U_C}{\partial P} = \frac{3}{4} \cdot \frac{1}{E_n^{1/3}} \cdot \frac{4}{3} \cdot P^{1/3} \left[\frac{L_1}{A_1^{1/3}} \cdot 1.563^{4/3} + \frac{L_2}{A_2^{1/3}} \cdot 0.40^{4/3} + \frac{L_3}{A_3^{1/3}} \cdot 1.078^{4/3} \right]$$

$$= \frac{1}{(200 \times 10^6)^{1/3}} \cdot \frac{5^{1/3}}{(0.003)^{1/3}} \left[3.905 \times 1.563^{4/3} + 2.00 \times 0.40^{4/3} + 2.69 \times 1.078^{4/3} \right]$$

$$= \frac{1.710}{584.80} \cdot \frac{1}{0.144} [10.646] = 0.216\ m = 216\ mm.$$

(ii) By principle of complementary virtual work.

Due to unit horizontal load at joint 3 forces in members are

$$f_3 = 1.078$$
$$f_2 = 0.40$$
$$f_1 = 1.563$$

By the principle of complementary virtual work

$$\Delta_3 = \int \delta\sigma \cdot \epsilon \cdot dvol$$

For the truss

$$\int \delta\sigma \cdot \epsilon \cdot dvol = \sum_{i=1}^{3} \delta\sigma_i \cdot \epsilon_i \cdot dvol = \sum_{i=1}^{3} \delta\sigma_i \cdot \epsilon_i \cdot A_i L_i$$

$$= \sum_{i=1}^{3} f_i \cdot \epsilon_i \cdot L_i$$

$$\sigma = E_n \epsilon^3 \text{ or } \epsilon = \left(\frac{\sigma}{E_n}\right)^{1/3}$$

$$\epsilon_1 = \left(\frac{7.81}{0.003} \times \frac{1}{200 \times 10^6}\right)^{1/3} = 0.0235$$

$$\epsilon_2 = \left(\frac{2.00}{0.003} \times \frac{1}{200 \times 10^6}\right) = 0.0149$$

$$\epsilon_3 = \left(\frac{5.38}{0.003} \times \frac{1}{200 \times 10^6}\right)^{1/3} = 0.0208$$

$$\delta_3 = 1.563 \times 0.0235 \times 3.905 + 0.40 \times 0.0149 \times 2.00 + 1.076 \times 0.0208 \times 2.69$$
$$= 0.216\, m = 216\, mm$$

Example 8.18. Find the horizontal deflection at D, rotation at C and the vertical deflection of point E taking into account, axial, bending and shear effects. I for columns $200 \times 10^6\ mm^4$, I for beam $= 400 \times 10^6\ mm^4$. Cross sectional area of each column $= 9000\ mm^2$. Cross section area of beam $= 15{,}000\ mm^2$. $E = 200\ GPa$. Use Castigliano's theorem.

Solution:

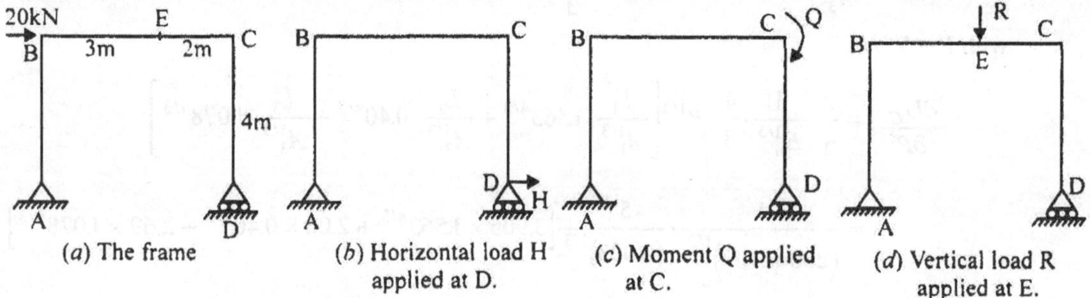

(a) The frame (b) Horizontal load H applied at D. (c) Moment Q applied at C. (d) Vertical load R applied at E.

Fig. 8.45. Example 8.18.

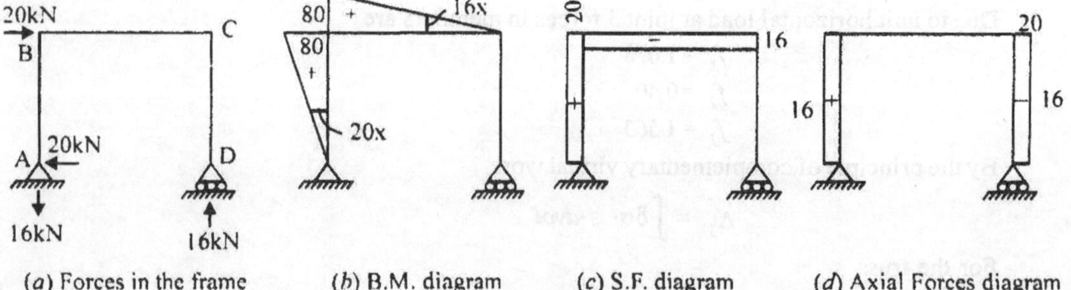

(a) Forces in the frame (b) B.M. diagram (c) S.F. diagram (d) Axial Forces diagram

Fig. 8.46. Example 8.18.

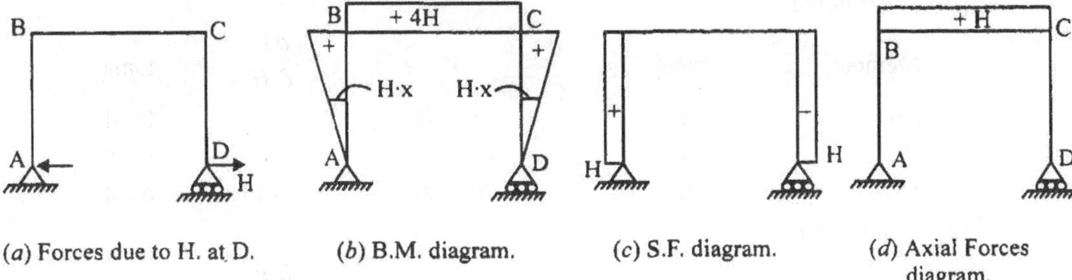

(*a*) Forces due to H. at D. (*b*) B.M. diagram. (*c*) S.F. diagram. (*d*) Axial Forces diagram.

Fig. 8.47. B.M., S.F. and Axial Force due to H.

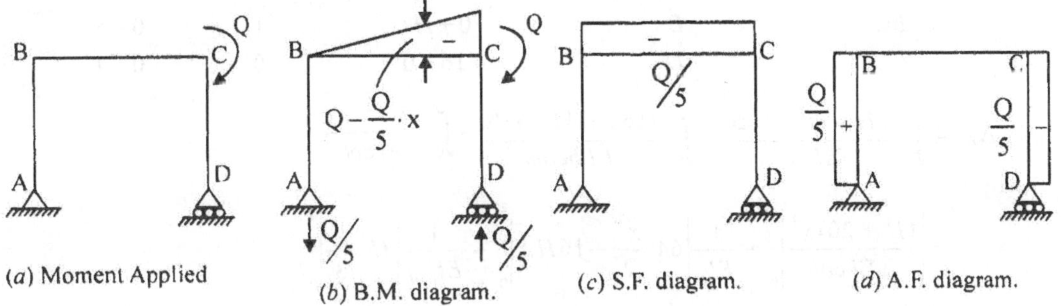

(*a*) Moment Applied (*b*) B.M. diagram. (*c*) S.F. diagram. (*d*) A.F. diagram.

Fig. 8.48. B.M, S.F. and A.F.D. due to Q.

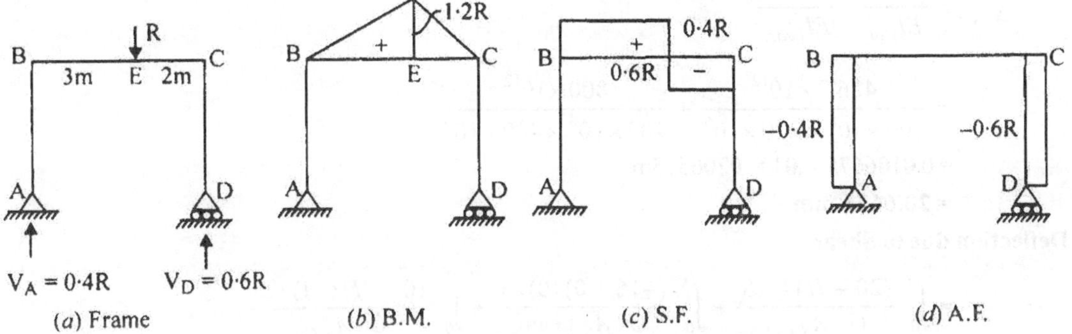

(*a*) Frame (*b*) B.M. (*c*) S.F. (*d*) A.F.

Fig. 8.49. B.M, S.F. and A.F. diagrams due to R.

Horizontal Deflection at *D*

$$\Delta_h = \frac{1}{EI}\int M \cdot \frac{\partial M}{\partial H} \cdot dx + \frac{1}{A_S G}\int V \cdot \frac{\partial V}{\partial H} \cdot dx + \frac{1}{AE}\int P \cdot \frac{\partial P}{\partial H} \cdot dx$$

Forces in members

(*i*) B.M.

Member	Origin	M	$\frac{\partial M}{\partial H}$	Limit
AB	A	$(H+20)x$	x	0 – 4
BC	C	$16x+4H$	4	0 – 5
CD	D	$0+H\cdot x$	x	0 – 4

(*ii*) Shear Force

Member	Origin	V	$\frac{\partial V}{\partial H}$	Limit
AB	A	$20 + H$	1	0 – 4
BC	B	$-16 + 0$	0	0 – 5
CD	D	$0 - H$	– 1	0 – 4

(*iii*) Axial Force

Member	Origin	P	$\frac{\partial P}{\partial H}$	Limit
AB	A	$16 + 0$	0	0 – 4
BC	B	$0 + H$	1	0 – 5
CD	D	$-16 + 0$	0	0 – 4

$$\Delta_h = \int_0^4 \frac{(H+20)\,x^2\,dx}{EI\,col} + \int_0^5 \frac{(16x + 4H)4\,dx}{EI\,beam} + \int_0^4 \frac{H \cdot x^2 \cdot dx}{EI\,col}$$

$$= \left[\frac{(H+20)\,x^3}{3\,EI\,col}\right]_0^4 + \frac{1}{EI_b}\left[64 \cdot \frac{x^2}{2} + 16H.x\right]_0^5 + \frac{1}{EI_{Col}}\left[H \cdot \frac{x^3}{3}\right]_0^4$$

With $H = 0$ deflection due to bending

$$\Delta_{hb} = \frac{426.7}{EI_{Col}} + \frac{800}{EI_{beam}}$$

$$= \frac{426.7 \times 10^{12}}{200 \times 10^6 \times 200 \times 10^6} + \frac{800 \times 10^{12}}{200 \times 10^6 \times 400 \times 10^6}$$

$= 0.0106675 + .01 = .0206675\text{m}$

= 20.6675 mm

Deflection due to Shear

$$= \int_0^4 \frac{(20 + H)1 \cdot dx}{G \cdot A_{S\,Col}} + \int_0^5 \frac{(-16 + 0)(0)\,dx}{G\,.\,A_S \text{ beam}} + \int_0^4 \frac{(0 - H)(-1) \cdot dx}{G\,.\,A_{S\,Col}}$$

$$= \frac{1}{G}\left[\left[\frac{(20 + H)\,x}{A_{S\,Col}}\right]_0^4 + \left[\frac{H\,.\,x}{A_{S\,Col}}\right]_0^4\right]$$

With $H = 0$

$$\Delta_{hS} = \frac{1}{G} \cdot \frac{20 \times 4}{A_{S\,Col}} = \frac{20 \times 4 \times 10^6 \times 1.2}{0.4 \times 200 \times 10^6 \times 9000} = 0.000133\text{m}$$

$= 0.133$ mm

Due to Axial Force

$$= \frac{1}{A_b E}\int_0^5 (0 + H) \cdot 1\,dx$$

$$= \frac{1}{A_b \cdot E}[H \cdot x]_0^5$$

With $H = 0$

$$\text{deflection due to axial force} = 0$$

$$\text{Total deflection} = 20.6675 + 0.133$$

$$= \mathbf{20.8005\ mm}$$

Rotation at C

$$\theta_C = \frac{1}{EI}\left[\int M \cdot \frac{\partial M}{\partial Q} \cdot dx\right] + \left[\frac{1}{A_S G}\int V \cdot \frac{\partial V}{\partial Q} \cdot dx\right] + \left[\frac{1}{AE}\int P \cdot \frac{\partial P}{\partial Q} \cdot dx\right]$$

Due to BM

$$\theta_{CM} = \frac{1}{EI}\int_0^5 \left(16x - Q + \frac{Q}{5} \cdot x\right)\left(\frac{x}{5} - 1\right) dx$$

$$= \frac{1}{EI}\left[\frac{16}{5} \cdot \frac{x^3}{3} - \frac{Qx^2}{2 \times 5} + \frac{Q.x^3}{3 \times 25} - \frac{16x^2}{2} + Qx - \frac{Q}{5} \cdot \frac{x^2}{2}\right]$$

With $Q = 0$

$$\theta_{CM} = \frac{1}{EI_b}\left[\frac{16}{15} \times 125 - 8 \times 25\right]$$

$$= \frac{1}{EI_b} \cdot [133.33 - 200] = -\frac{66.67}{EI_{beam}}$$

$$= -\frac{66.67 \times 10^{12}}{200 \times 10^6 \times 400 \times 10^6} = -8.33 \times 10^{-4} \text{ radian (anticlockwise)}$$

Due to Shear

$$= \frac{1}{A_S G} \cdot \int V \cdot \frac{\partial V}{\partial Q} \cdot dx$$

$$= \frac{1}{A_S G} \cdot \int_0^5 + \left(16 + \frac{Q}{5}\right) \cdot \frac{1}{5} \cdot dx$$

$$= \frac{1}{A_S G}\left[\frac{16}{5} x + \frac{Q}{25} x\right]_0^5$$

With $Q = 0$

$$\theta_C \text{ shear} = \frac{16}{5} \cdot 5 \cdot \frac{1}{A_S G}$$

$$= \frac{16 \times 1.2 \times 10^6}{15{,}000 \times 0.4 \times 200 \times 10^6}$$

$$= 1.6 \times 10^{-5} \text{ radian. (clockwise)}$$

Due to Axial Force

$$\theta_C \text{ axial} = \frac{1}{AE}\int P\cdot\frac{\partial P}{\partial Q}\cdot dx$$

$$= \frac{1}{A_C\cdot E}\int_0^4\left(16+\frac{Q}{5}\right)\cdot\frac{1}{5}\cdot dx + \frac{1}{A_C\cdot E}\int_0^4\left(-16-\frac{Q}{5}\right)\left(-\frac{1}{5}\right)dx$$

$$= \frac{1}{A_C\cdot E}\left[\frac{x}{5}\left(16+\frac{Q}{5}\right)+\frac{x}{5}\left(16+\frac{Q}{5}\right)\right]_0^4$$

With $Q = 0$

$$\theta_C \text{ axial} = \frac{1}{A_C\cdot E}\cdot\frac{2\times 16}{5}\times 4$$

$$= \frac{10^6\times 128}{9000\times 200\times 10^6}\times\frac{1}{5}$$

$$= 1.42\times 10^{-5} \text{ radian}$$

$$\text{Total rotation} = -83.3\times 10^{-5}+1.6\times 10^{-5}+1.42\times 10^{-5}$$

$$= -80.28\times 10^{-5} \text{ radian (anticlockwise).}$$

Vertical Deflection at *E*

Bending Moment

Member	Origin	M	$\frac{\partial M}{\partial R}$	Limit
AB	A	$20x+0$	0	0 – 4
BE	B	$80-16x+0.4R\cdot x$	$0.4x$	0 – 3
CE	C	$16x+0.6R\cdot x$	$0.6x$	0 – 2
CD	D	0	0	0 – 4

Shear Force

Member	Origin	V	$\frac{\partial V}{\partial R}$	Limit
AB	A	$+20+0$	0	0 – 4
BE	B	$-16+0.4R$	0.4	0 – 3
EC	C	$-16-0.6R$	−0.6	0 – 2
CD	D	$0+0$	0	0 – 4

Axial Force

Member	Origin	P	$\frac{\partial P}{\partial R}$	Limit
AB	A	$+16-0.4R$	−0.4	0 – 4
BC	B	0	0	
CD	D	$-16-0.6R$	−0.6	0 – 4

Deflection due to B.M.

$$\Delta_{B.M.} = \frac{1}{EI_b}\int_0^3 (80-16x+0.4R\cdot x)(0.4x)\,dx$$

$$+\frac{1}{EI_b}\int_0^2 (16x+0.6R\cdot x)\,(0.6x)\,dx$$

$$=\frac{1}{EI_b}\int_0^3 (32x-6.4x^2+0.16R\cdot x^2)\,dx+\frac{1}{EI_b}\int_0^2 (9.6x^2+0.36\,R\cdot x^2)\,dx$$

With $R=0$

$$\Delta_{B.M.}=\frac{1}{EI_b}\left[32\cdot\frac{x^2}{2}-6.4\frac{x^3}{3}\right]_0^3+\frac{1}{EI_b}\left[9.6\frac{x^3}{3}\right]_0^2$$

$$=\frac{1}{EI_b}\left[144-\frac{6.4}{3}\cdot 27+\frac{9.6\times 8}{3}\right]$$

$$=\frac{112\times 10^{12}\times 10^3}{200\times 10^6\times 400\times 10^6}\text{ mm}$$

$$=\mathbf{1.4\ mm}$$

Deflection due to shear

$$\Delta_{Shear}=\frac{1}{A_S.G}\left[\int_0^3 (-16+0.4\,R)(0.4)\,dx+\int_0^2 (-16-0.6\,R)(-0.6)\,dx\right]$$

$$=\frac{1}{A_SG}[-6.4\,x+0.16R.x]_0^3+\frac{1}{A_sG}[+9\cdot 6x+36R.x]_0^2$$

With $R=0$

$$\Delta_{Shear}=\frac{1}{A_SG}[-6\cdot 4\times 3+9\cdot 6\times 2]$$

$$=0$$

$$\Delta_{Axial\ Force}=\frac{1}{EA_C}\cdot\int_0^4 [(16-0.4R)(-0.4)+(16+0.6R)(0.6)]\,dx$$

$$=\frac{1}{EA_C}[-6.4\,x+0.16\,R\cdot x+9.6\,x+0.36\,R\cdot x]_0^4$$

With $R=0$ $\quad\Delta_{Axial\ Force}=\dfrac{3.2\times 4}{EA_C}$

$$=\frac{3.2\times 4\times 10^6\times 10^3}{200\times 10^6\times 9000}$$

$$=0.711\times 10^{-2}=0.00711\text{ mm}.$$

Fig. 8.50. Example 8.19.

Total deflection = 1.4 + 0.00711 = 1.40711 mm.

Example 8.19. Use Castigliano's 2nd Theorem to directly compute the displacement in the direction of load P on the propped beam shown in Fig. 8.50. Express your answers in terms of L, I, A, E and P.

Solution: Taking moment about A

$$T \times 2L = P \times 3L$$

or

$$T = \frac{3}{2}P$$

$$R_A = \frac{P}{2} \downarrow \text{ (downward)}$$

For beam

AB

$$M = -\frac{P}{2} \cdot x \text{ (origin at } A)$$

$$\frac{\partial M}{\partial P} = -\frac{x}{2}$$

BC

$$M = -P \,.\, x \text{ (origin at } C)$$

$$\frac{\partial M}{\partial P} = -x$$

For cable

$$T = \frac{3}{2} \cdot P$$

$$\frac{\partial T}{\partial P} = \frac{3}{2}$$

$$\Delta = \int_0^{2L} \frac{-\frac{P}{2} \cdot x\left(-\frac{x}{2}\right)dx}{EI} + \int_0^{L} \frac{-P.x \cdot (-x)\,dx}{EI} + \int_0^{L} \frac{\frac{3}{2}P \cdot \frac{3}{2}\,dx}{AE}$$

$$= \frac{P}{4EI} \cdot \left[\frac{x^3}{3}\right]_0^{2L} + \frac{P}{EI}\left[\frac{x^3}{3}\right]_0^{L} + \frac{9P}{4AE}[x]_0^L$$

$$= \frac{PL^3}{EI}\left[\frac{2}{3}\right] + \frac{P}{EI} \cdot \frac{L^3}{3} + \frac{9PL}{4AE}$$

$$= \frac{PL^3}{EI} + \frac{9PL}{4AE}$$

8.15. MISCELLANEOUS EXAMPLES ON APPLICATION OF ENERGY PRINCIPLES

Example 8.20. A horizontal beam AB 100 cm long is fixed at A and unsupported at B. The beam connected at centre point C by a vertical string of length 100 cm and area 2 cm^2. Find the tension in the string when a vertical load of 1000 N is placed at B. Assume elasticity of the string and that of the beam as 2×10^6 N/mm^2 and I for the beam as 1000 cm^4.

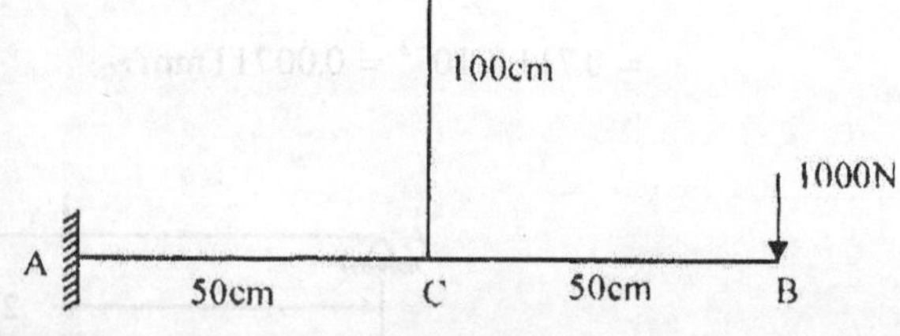

Fig. 8.51. Example 8.20.

Solution: This is a statically indeterminate problem. The theory of least work.

$\frac{\partial U}{\partial T} = 0$ where T is the tension in the string, is applied.

Let T be the tension in the string

$$U = U_{beam} + U_{rod}$$

$$= \int \frac{M^2\,dx}{2\,EI} + \frac{T^2 L}{2\,AE}$$

$$\frac{\partial U}{\partial T} = \frac{1}{EI}\int M \cdot \frac{\partial M}{\partial T} \cdot dx + \frac{TL}{AE}$$

For BC $0 < x < 50$

$$M = -1000\,x$$

$$\frac{\partial M}{\partial T} = 0$$

For CA, $50 \le x \le 100$

$$M = -1000\,x + T(x-50)$$

$$\frac{\partial M}{\partial T} = x - 50$$

$$\frac{\partial U}{\partial T} = \int_{50}^{100} \frac{\{-1000\,x + T(x-50)\}}{EI} \cdot (x-50) \cdot dx + \frac{T \times 100}{2 \times E}$$

$$= \frac{1}{EI}\left[-1000 \cdot \frac{x^3}{3} + \frac{Tx^3}{3} - 50T \cdot \frac{x^2}{2} + 50{,}000 \cdot \frac{x^2}{2} - 50T \cdot \frac{x^2}{2}\right.$$

$$\left. + 2500\,T.x \right]_{50}^{100} + \frac{50\,T}{E}$$

or
$$E\frac{\partial U}{\partial T} = \frac{1}{1000}\left[\frac{-1000}{3} \cdot (100^3 - 50^3) + \frac{T}{3}(100^3 - 50^3)\right.$$

$$\left. -25T(100^2 - 50^2) + 25{,}000(100^2 - 50^2) - 25T(100^2 - 50^2) + 2500 \times 50T\right] + 50T$$

Simplifying and putting $\frac{\partial U}{\partial T} = 0$

$$91.66\,T = 1.0416 \times 10^5$$

or
$$\mathbf{T = 1136.37\ N}$$

Example 8.21. A continuous beam ABC consist of two equal spans $AB = BC = L$. The beam is simply supported on two equal springs of stiffness k each at the two endsupports A and C and on a rigid knife edge at the mid-point B. The flexural rigidity EI of the beam is same all through. If the beam carries a uniformly distributed load w per unit length over its entire length, calculate the reaction at A

Solution:

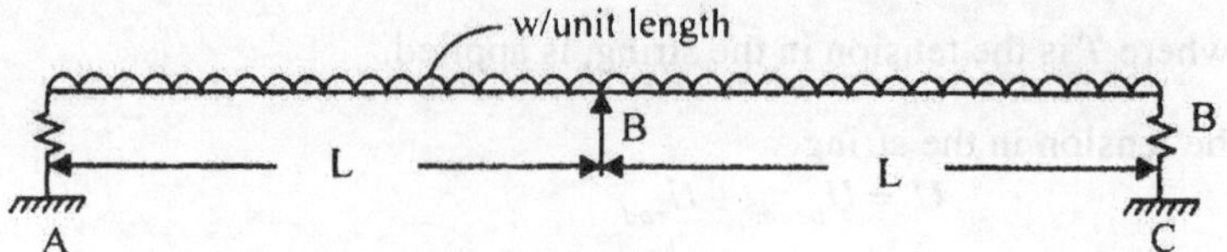

Fig. 8.52. Example 8.21.

This is again an indeterminate problem.

Let $R_A = R_C = R$ (by symmetry)

By theorem of least work $\frac{\partial U}{\partial R} = 0$

$$\frac{\partial U}{\partial R} = \frac{2}{EI}\int_0^L M \cdot \frac{\partial M}{\partial R} \cdot dx + 2 \cdot \frac{\partial}{\partial R} \text{ (Energy of spring)}$$

If R is the reaction of the spring K, the stiffness, deflection of the spring $\Delta = \frac{R}{K}$

$$\text{Work done or Energy Stored in Spring} = \frac{1}{2} \cdot R \cdot \frac{R}{K} = \frac{R^2}{2K}$$

$$M = R.x - \frac{wx^2}{2}$$

$$\frac{\partial M}{\partial R} = x$$

$$\frac{\partial U}{\partial R} = \frac{2}{EI}\int_0^L \left(R.x - \frac{wx^2}{2}\right) \cdot x \cdot dx + 2 \cdot \frac{R}{K}$$

$$= \frac{2}{EI}\left[\frac{R.x^3}{3} - \frac{w}{2}\frac{x^4}{4}\right]_0^L + \frac{2R}{K}$$

$$= 0$$

or $$\frac{1}{EI} \cdot \frac{RL^3}{3} + \frac{R}{K} = \frac{wL^4}{8\,EI}$$

or $$R \cdot \left[\frac{L^3}{3\,EI} + \frac{1}{K}\right] = \frac{wL^4}{8\,EI}$$

or $$R = \frac{\frac{wL^4}{8\,EI}}{\frac{L^3}{3EI} + \frac{1}{K}} = \frac{wL^4}{\frac{8\,EI}{K} + \frac{8\,L^3}{3}}.$$

Example 8.22. Find the reaction at B for the beam supported as shown in Fig. 8.53.

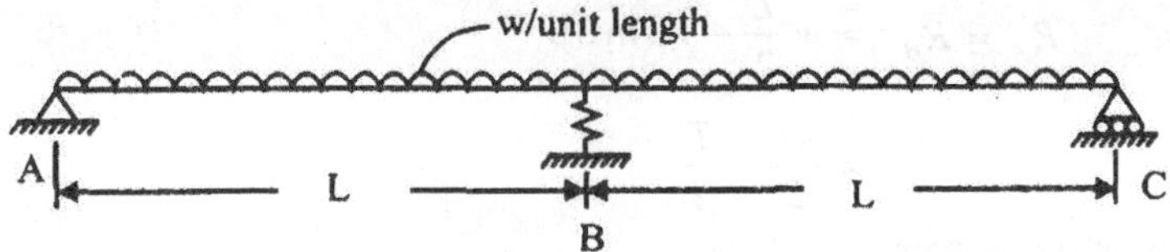

Fig. 8.53 Example 8.22

Solution: Let the reaction at B be R_B

From symmetry $R_A = R_C = \frac{1}{2}(2wL - R_B)$

For AB

$$M = \frac{1}{2}(2wL - R_B) \cdot x - \frac{wx^2}{2}$$

$$\frac{\partial M}{\partial R_B} = -\frac{x}{2}$$

$$\frac{\partial U}{\partial R_B} = \frac{2}{EI}\int_0^L \left[\frac{1}{2}(2\,wL - R_B)x - \frac{wx^2}{2}\right]\left(-\frac{x}{2}\right)dx + \frac{R_B}{K}$$

$$= 2 \cdot \frac{1}{2} \cdot \frac{1}{2} \cdot \frac{1}{EI}\left[-2\,wL \cdot \frac{x^3}{3} + R_B \cdot \frac{x^3}{3} + \frac{wx^4}{4}\right]_0^L + \frac{R_B}{K}$$

$$= \frac{1}{EI}\left[-\frac{5}{24}wL^4\right] + R_B\left(\frac{L^3}{6\,EI} + \frac{1}{K}\right) = 0$$

or
$$R_B = \frac{\frac{5}{24} \cdot \frac{wL^4}{EI}}{\left(\frac{L^3}{6\,EI} + \frac{1}{K}\right)}.$$

Example 8.23. A simply supported beam AB of span L and moment of Inertia I is strengthened by two tierods attached symmetrically to its mid span as shown in Fig. 8.54. Find tension in the tierod E same for tie rod and beam.

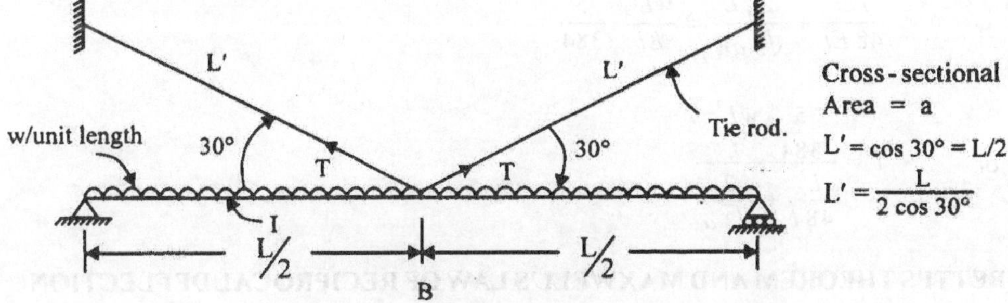

Fig. 8.54. Example 8.23.

Solution: We apply theory of least work i.e. $\frac{\partial U}{\partial T} = 0$ for the redundant tension T in the tie rod.

From equilibrium of the beam

$$R_A = R_B = w\frac{L}{2} - T\sin 30^\circ$$

$$= \frac{wL}{2} - \frac{T}{2}$$

Total Energy in the system

= Energy in the beam + Energy in the tie rod.

$$\text{Energy in the beam} = 2\int_0^{L/2} \frac{M^2 dx}{2EI}$$

$$M = \left(\frac{wL}{2} - \frac{T}{2}\right)\cdot x - \frac{wx^2}{2}$$

$$\frac{\partial M}{\partial T} = -\frac{x}{2}$$

$$\text{Energy of the Tie rods } E_T = 2\cdot\frac{1}{2}\cdot\frac{T^2 L}{2\cos 30^\circ}\cdot\frac{1}{aE}$$

$$= \frac{T^2 L}{\sqrt{3}\, aE}$$

$$\frac{\partial E_T}{\partial T} = \frac{2TL}{\sqrt{3}\, aE}$$

Hence

$$2\int_0^{L/2} -\frac{1}{EI}\left[\left(\frac{wL}{2} - \frac{T}{2}\right)x - \frac{wx^2}{2}\right]\cdot\frac{x}{2}dx + 2\frac{TL}{\sqrt{3}\, aE} = 0$$

or $$-\frac{1}{EI}\int_0^{L/2}\left[\left(\frac{wL}{2} - \frac{T}{2}\right)x^2 - \frac{wx^3}{2}\right]dx + \frac{2\,TL}{\sqrt{3}aE} = 0$$

or $$\frac{TL^3}{48\,EI} + \frac{2\,TL}{\sqrt{3}\, aE} = \frac{wL^4}{EI}\cdot\frac{5}{384}$$

or $$T = \frac{\frac{5}{384}\frac{wL^3}{I}}{\frac{L^2}{48\,I} + \frac{2}{\sqrt{3}\, a}}.$$

8.15. BETTI'S THEOREM AND MAXWELL'S LAW OF RECIPROCAL DEFLECTION

These two theorems are very useful for analysis of linear elastic structures and can be derived by the principle of virtual work. Betti's law states.

"For a linearly elastic structure with unyielding supports and no induced deformations if the forces are all increased from zero, the work done by a system of P_i' forces moving through the corresponding displacements caused by a second set of P_i'' forces is equal to the work done by the P_i'' forces acting through the displacements produced by the P_i' forces".

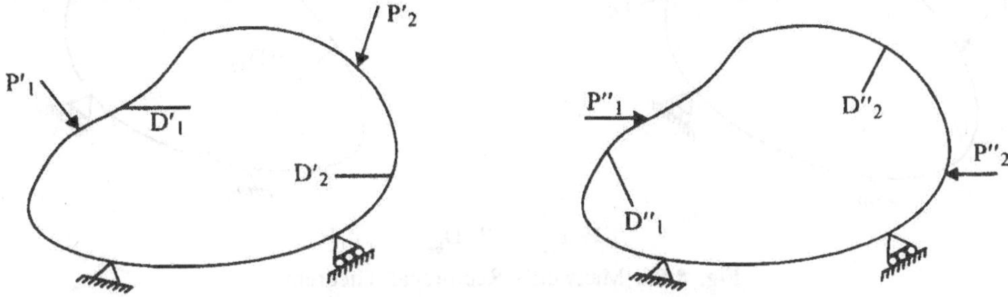

Fig. 8.55. Betti's law – $P_1' D_1'' + P_2' D_2'' = P_1'' D_1' + P_2'' D_2'$

Consider the first system of P_i' forces acting on an elastic body causing internal stresses and strains σ' and $\in'$ and displacements such as D_i'. Now suppose second set of P_i'' forces are now applied causing virtual external distortions D_i'' and internal virtual strains $\in''$ and stresses σ''. The principle of virtual work states that

$$\Sigma P_i D_i'' = \int \sigma' \in'' dvol$$

Now suppose we reverse the process. The loads P_i'' are applied first and the load P_i' are applied later. Again from the principle of virtual work.

$$\Sigma P_i'' D_i' = \int \sigma'' \in' dvol$$

If the structure is linearly elastic.

$$\sigma'' = E \in''$$

$$\sigma' = E \in'$$

Substituting these values in the two virtual work equations.

$$\Sigma P_i' D_i'' = \int E \in' \in'' dvol$$

$$\Sigma P_i'' D_i' = \int E \in'' \in' dvol$$

Since the right hand sides of these two expressions are the same

$$\Sigma P_i' D_i'' = \Sigma P_i'' D_i' \quad \text{...(8.25)}$$

In words the above equation states. "The work done by a set of loads P_i' riding through the virtual displacement D_i'' caused by a second set of forces P_i'' is equal to the work done by the second set of forces P_i'' moving through the virtual displacement D_i' caused by the 1st set of forces P_i' This is Betti's law.

Makwell's law of reciprocal deflection is a special case of Betti's law. It states. "For a linear elastic structure with unyielding supports and no induced deformations, if the forces are increased from zero, the deflection at point '*a*' in the direction of applied force *P* at point '*b*' is equal to the deflection at point '*b*' in the direction of the applied force at this point due to a force *P* at point '*a*'

Mathematically,

$$P \,.\, D_{ab} = P \,.\, D_{ba} \quad \text{...(8.25a)}$$

or

$$D_{ab} = D_{ba}$$

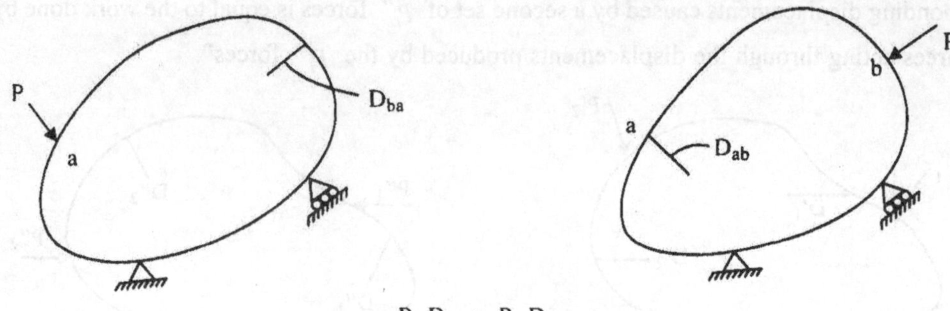

$P.\ D_{ab} = P.\ D_{ba}$

Fig. 8.56. Maxwell's Reciprocal Theorem

Reciprocal theorem is extensively used in the analysis of statically indeterminate structure. It

(*i*) finds broad applications in the analysis of structures using matrix methods.

(*ii*) can be used to construct influence lines for the deflection of structures.

(*iii*) can also be applied to yield the Müller Breslau principle which is used to construct influence lines for generalised forces in structures.

Example 8.24. *A beam ABCDE is supported at A and E. A forced vertical deflection δ_B at B causes vertical deflection δ_C at C. and δ_D at D. A forced vertical deflection δ_d at D causes vertical deflections δ_b and δ_c at B and C. Show that if supports are placed at B and D to the same level as A and E a vertical load W at C induced a reaction of*

$$\frac{\delta_B\,\delta_c - \delta_b\delta_C}{\delta_B\delta_d - \delta_b\delta_D}\cdot W \text{ at } D.$$

Solution: There are three system of forces shown in Fig 8.57 (*a*), (*b*) and (*c*) respectively with their corresponding deflections.

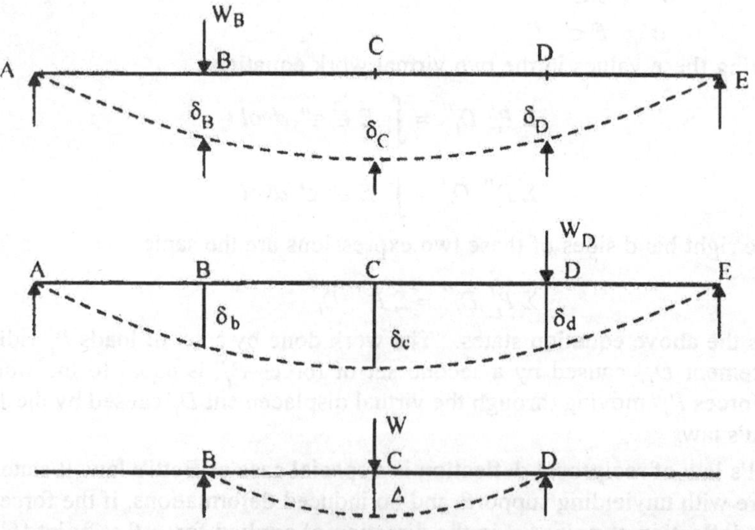

Fig. 8.57. Example 8.24.

System	Force in point			Deflection in point		
	(B)	(C)	(D)	(B)	(C)	(D)
1.	W_B	0	0	δ_B	δ_C	δ_D
2.	0	0	W_D	δ_b	δ_c	δ_d
3.	$-R_B$	W	$-R_D$	0	Δ	0

Applying Maxwell Bett's Theorem to systems 1 and 3.

$$W_B \times 0 + 0 \times \Delta + 0 \times 0 = -R_B \times \delta_B + W \times \delta_C - R_D.\delta_D$$

whence $R_B = \dfrac{1}{\delta_B}(W\delta_c - R_D\delta_D)$

Applying the theorem again to systems 2 and 3.

$$-R_B \times \delta_b + W\delta_c - R_D.\delta_d = 0 \times 0 + 0 \times \Delta + W_D \times 0 = 0$$

Substituting the value of R_B

$$-\frac{\delta_b}{\delta_B}(W\delta_C - R_D.\delta_D) + W\delta_c - R_D.\delta_d = 0$$

or $$R_D = \frac{\delta_B\delta_c - \delta_b.\delta_C}{\delta_B.\delta_d - \delta_b.\delta_D}.W$$

SUMMARY

1. Deflection of a structure can be calculated by energy method where internal energy is equated with external work. Two general principles of energy methods are :
   (1) Virtual work (virtual displacements)
   (2) Complementary virtual work (virtual forces)
2. Conservation of Energy is one of the basic laws of physics and a "closed system" consisting of a structure and the applied forces must obey this law. Mathematically the work energy balance can be written as
   $$W = E_S + E_L$$
   where W = work performed
   E_S = Energy stored
   E_L = Energy loss.
   when there is no energy loss due to heat, vibration etc.
   $$W = E_S$$
   The usual terminology for E_S in a structure is elastic strain energy with the symbol U. Hence for a conservative structural system. (no energy loss)
   $$W = U$$
3. Work is defined as Force × displacement and symbolically $W = \int P\,d\Delta$

   Complementary work, on the other hand is defined as $\int \Delta.dP$ and it is the area above the load deflection curve. Work on the other hand is the area below the load deflection curve. For linear load deflection curve
   $$W = W_C = \frac{1}{2}P.\Delta$$
4. Similarly strain energy density x is the area below the $\sigma - \epsilon$ curve whereas complementary strain energy density is the area above the $\sigma - \epsilon$ curve.
5. By equating external work done by a single force with internal energy stored it is possible to compute deflection at the point of application of the load and in the direction of the load. This is method of real work. But it is not very useful because,

(*i*) If the applied loads are more than one deflection cannot be found out.

(*ii*) Deflection cannot be computed at any arbitrary point or in any arbitrary direction.

6. The limitations of real work are over come by using virtual work principles. For a deformable body the principle of complementary virtual work as applied in structural analysis may be stated as "The external virtual work done by a virtual force moving through a real displacement due to applied loads is equal to the internal virtual work done by internal virtual forces moving through compatible virtual strains".
7. The important energy theorems used in structural analysis are:

   (*a*) Principle of stationary Potential Energy and Minimum Potential energy.

   (*b*) Castigliano's First Theorem.

   (*c*) Principle of Stationary and minimum complementary potential energy.

   (*d*) Castigliano's Second Theorem and Engesser's Theorem.

Of the above Theorems

(*i*) Castigliano's 2nd Theorem and Engesser's Theorem are used to calculate deflections.

(*ii*) Stationary Complementary Potential Energy, Engesser's Theorem and Castigliano's Theorems are used to compute compatible displacements, form flexibility relations and flexibility matrices to directly solve redundants of statically indeterminate structures (force method)

(*iii*) Principle of virtual displacements, stationary potential energy and Castigliano's First theorem lead to a stiffness analysis. They may be used to compute forces in equilibrium, form stiffness relations and solve for the displacements of a structure.

8. Betti's theorem and Maxwell's law of reciprocal deflection are very useful for analysis of linear elastic structures. Betti's law states. "For a linearly elastic structure with unyielding supports and no induced deformations, if the forces are all increased from zero, the work done by a system of P_i' forces moving through the corresponding displacements caused by a second set of P_i'' forces is equal to the work done by the P_i'' forces acting through the displacements produced by the P_i' forces."
9. Maxwell's law of reciprocal deflection is a special case of Betti's law. It states. "For a linear elastic structure with unyielding supports and no induced deformations, if the forces are increased from zero, the deflection at point '*a*' in the direction of applied force *P* at this point due to a force *P* at point *b* is equal to the deflection at point *b* in the direction of the applied force at this point due to a force *P* at point *a*"

PROBLEMS

A Theoretical

8.1 (*i*) (*a*) Define the principle of virtual work.

(*b*) Explain the unit load method. **[AMIE Summer '94]**

(*ii*) Briefly explain how the unit load method is based on the principle of virtual work. **[AMIE Winter '94]**

(*iii*) State and explain Castigliano's Theorems **[AMIE Summer '95]**

(*iv*) State and prove Clark-Maxwell Reciprocal Theorem **[AMIE Winter '97]**

(*v*) Enunciate Betti's Theorem. **[AMIE Summer '78]**

(*vi*) State Castigliano's First Theorem

(*vii*) Develop the expression for calculating the deflection of a joint of a pin connected truss using the dummy unit load method. **[AMIE Summer 1981]** (*ix*) State with illustrative sketches Maxwell's reciprocal theorem.

(*x*) Distinguish between "Strain energy" and Complementary Energy and state "Castigliano's First Theorem" **[AMIE Winter 1981]**

(*xi*) Explain Betti's law and Maxwell's law as applied to structures. **[AMIE Summer 1983]**

(*xii*) Explain the two theorems of Castigliano's as applied to structures. **[AMIE Winter 1983]**

(*xiii*) State and explain Castigliano's theorems pointing out the limitations of their use, if any **[AMIE Summer 1990]**

(*xiv*) What is understood by 'Complementary Energy'? State Engesser's Theorem of Compatibility. **[AMIE Advanced Structure 1991]**

(*xv*) Derive expressions for strain energy for

(*i*) Bending Moment (*ii*) Axial Force

B. Objective

Tick the correct answer.

8.2 (*i*) Castigliano's theorem for deflection i.e. $\frac{\partial U}{\partial P} = \delta$ (deflection) is true for

(*a*) linearly elastic material.

(*b*) rigid material

(*c*) non linearly elastic material

(*d*) any material - elastic (linear or non linear) or in elastic.

[AMIE Summer '97]

(*ii*) A two span beam simply supported at the two end supports and continuous over the intermediate support suffers a down ward deflection of 3 cm at the centreof the left span when subjected to a clockwise couple of 6 kNm at the right end. If a down ward load of 2 kN acts at the centre of the left span, the slope at the right end will be

(*a*) 0.01 rad clockwise

(*b*) 0.01 rad anti clockwise

(*c*) 0.09 rad clockwise

(*d*) 0.09 rad anticlockwise. **[AMIE Winter '98]**

(*iii*) The theorem which states that the partial derivative of strain energy with respect to an applied load is equal to the displacement of the load point in the direcion of load is valid for a

(*a*) statically determinate structure

(*b*) statically indeterminate structure

(*c*) structure made of material with linear stress - strain relation

(*d*) structure with no material and geometrical non linearity. **[AMIE Winter '98]**

(*iv*) Castigliano's theorem is applicable

(*a*) when the system behaves elastically

(*b*) only when principle of super position is valid

(*c*) None of the above. **[AMIE Winter '97]**

C. Mathematical

8.3 Determine the deflection of the point A of the truss shown in Fig. P 8.1. Take the stress in tension members as 140 N/mm^2 and in compression members as 90 N/mm^2.

Take $E = 2.0 \times 10^5$ N/mm^2 **[AMIE Winter '97]**

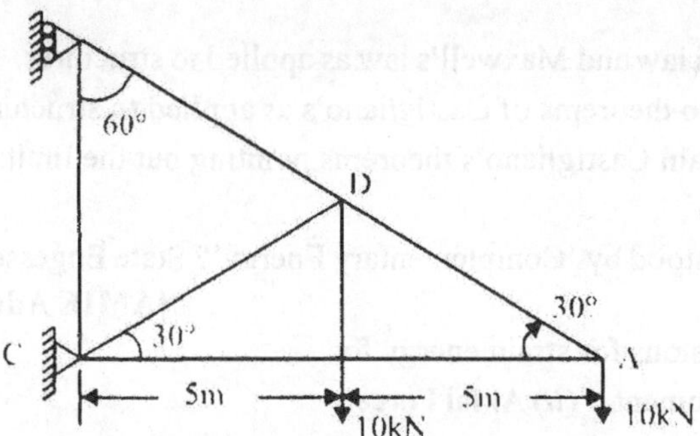

Fig. P. 8.1 (Prob. 8.3).

8.4. Analyse the propped cantilever shown in Fig. P. 8.2. Plot the S.F. and B.M. diagrams. **[AMIE Winter '97]**

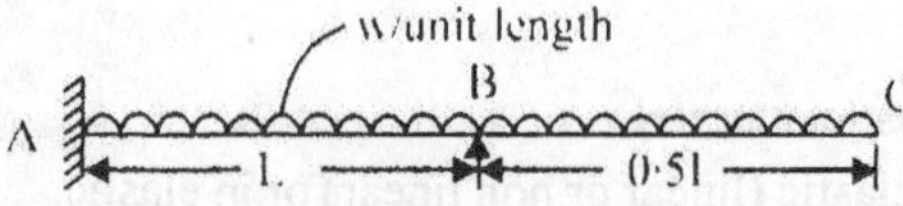

Fig. P. 8.2 (Prob. 8.4.).

8.5. A frame *ABCD* consisting of two equilateral triangles is hinged at *A* and supported on rollers at *D* as shown in Fig. P 8.3. Determine the vertical deflection of *C* due to a load applied vertically at *C*. All the members are of length *L*. All the tension members are of area '*a*' and compression members of area 2*a*. **[BE College (D.U.)]**

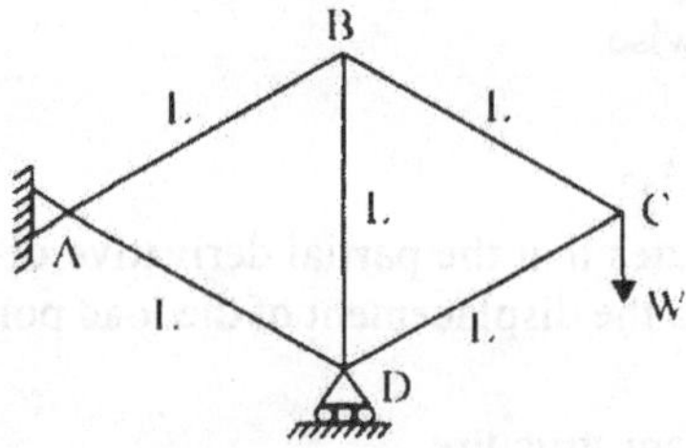

Fig. P. 8.3 (Prob. 8.5).

8.6. A right angled bent of constant cross-section fixed at the base *A* and free at *C*. The vertical member *AB* is 4 m long and the horizontal member BC is 1m in length, joint being rigid. Calculate the rotation, horizontal and vertical deflection of *C* when a vertical load of 3^t is suspended at *C*. The section has I = 10,000 cm^4 and E = 2000 t/cm^2. Use Castigliano's theorem. **[AMIE Winter, 1980]**

8.7. Using Castigliano's Theorem, find the deflection of the point of application of the force *P* on the beam of different cross-section as shown in Fig. P 8.4. Assume uniform value of Young's Modulus *E*. Moments of Inertia of the beam section are *I* and 2*I* as shown in the figure.

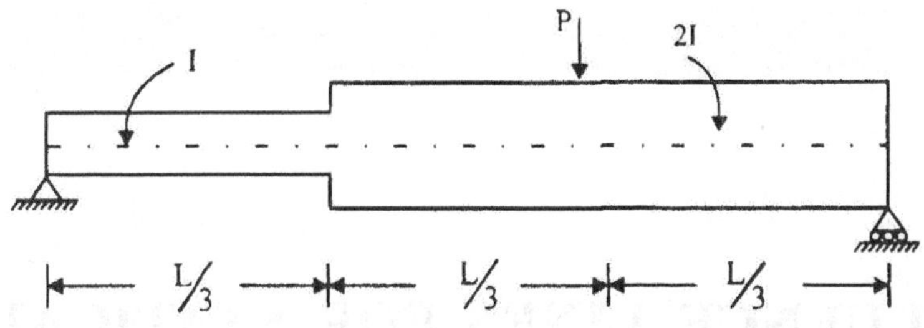

Fig. P. 8.4. (Prob. 8.7)

8.8 Analyse the frame with two pinned support shown in Fig. P 8.5. Calculate the reactions and draw the B.M. diagram. Take EI to be same for all members.

[AMIE Summer '99]

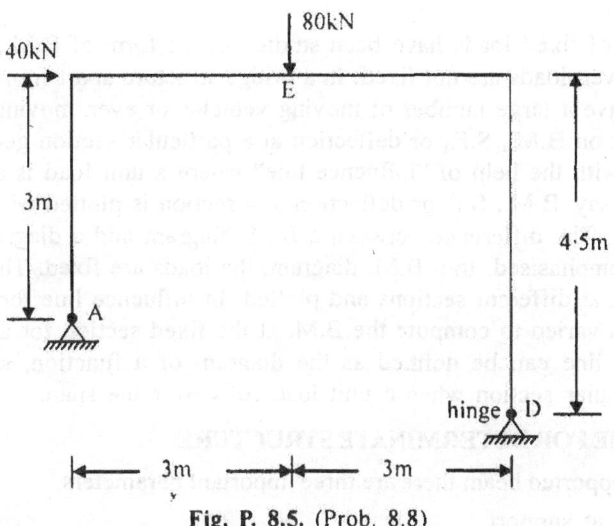

Fig. P. 8.5. (Prob. 8.8)

9

INFLUENCE LINES FOR STATICALLY DETERMINATE STRUCTURES

9.1 INTRODUCTION

So long effect of fixed loads have been studied in the form of B.M., S.F. and deflection. In many cases, however, loads are not fixed. In a bridge structure apart from its own dead weight which is fixed we have a large number of moving vehicles or even moving pedestrians. As the loads move its effect on B.M., S.F., or deflection at a particular section goes on changing. This variation is studied with the help of "influence line" where a unit load is allowed to move and a particular function say, B.M., S.F. or deflection at a section is plotted with respect to different positions of the load. The difference between a B.M. diagram and a diagram for influence line for B.M. should be emphasised. In a B.M. diagram, the loads are fixed. The sections are varied to compute the B.M. at different sections and plotted. In influence line for B.M., the section is fixed the unit load is varied to compute the B.M. at the fixed section for all the load positions. Therefore, influence line can be defined as the diagram of a function, say, of B.M., S.F. or deflection at a particular section when a unit load rolls over the span.

9.2 INFLUENCE LINE FOR DETERMINATE STRUCTURE

For a simply supported beam there are three important parameters.

1. Reaction at support.
2. S.F. at a section
3. B.M. at a section.

Fig. 9.1 (*a*) shows a simply supported beam AB with a specified section X-X at a distance a from *A*. We are interested in computing the influence line for R_A. Let the unit load which rolls over span *AB* be at a distance *x* from *A*.

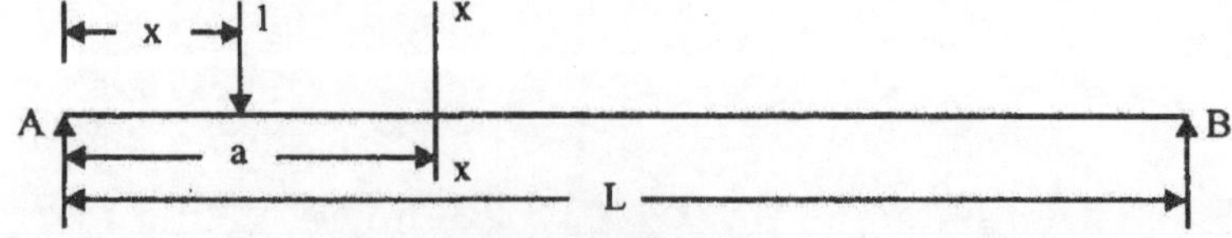

Fig. 9.1. (*a*) Simply supported beam with section *x-x*

$$R_A = \frac{L-x}{L} \qquad \text{...(9.1)}$$

at $\quad x=0 \qquad R_A = 1$

$\quad x=L \qquad R_A = 0$

The results are obvious as when unit load is over A R_A has its full value. When unitload is over B, R_A is zero as all the load goes to support B. As $L - x$ is linear, the variation of R_A is also linear. The plotting is done in Fig. 9.1 (*b*).

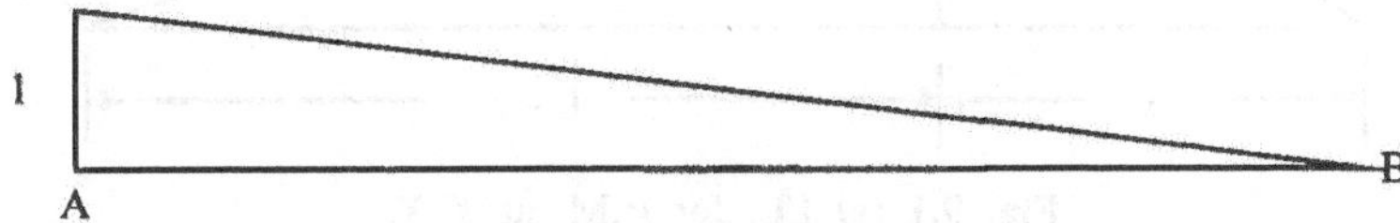

Fig. 9.1. (*b*) **I.L. for R_A**

Similarly Fig. 9.1 (c) shows the influence line for R_B.

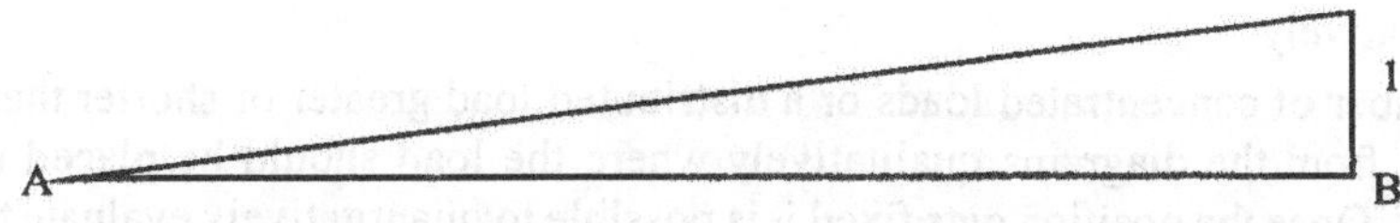

Fig. 9.1. (*c*) **I.L. for R_B**

Referring to fig 9.1 (*a*) S.F. at any section *X–X* is given by R_B when unit load is to the left of *X–X* i.e. when $x < a$. Similarly when load is on the right side of *X–X*, ie $x > a$ S.F at *X–X* is given by R_A. Hence influence line for shear can be obtained from influence line for R_A and R_B. With left hand upward taken as positive R_A gives +ve shear and R_B –ve shear Fig. 9.1 (*d*) shows the two influence lines for Reactions superposed.

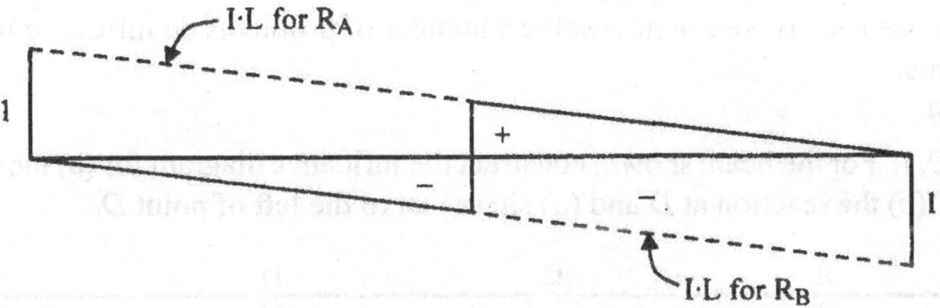

Fig. 9.1. (*d*) **I.L. for shear.**

With the same reasoning when the load is on the left of *X–X* B.M. at section is

$$R_B \times (L-a) \quad \text{or} \quad \frac{x}{L} \times (L-a) \qquad \text{...(9.2)}$$

When the load is on the right of section

$$\text{B.M.} = R_A \times x \quad \text{or} \quad \frac{L-x}{L} \cdot a \qquad \text{...(9.3)}$$

From Eq. (9.2)

At $\quad x=0 \qquad$ B.M. $=0$

at $\quad x=a \qquad$ B.M. $= \dfrac{a(L-a)}{L}$

Similarly from Eq. (9·3) at $x = a$.

$$\text{B.M.} = \frac{(L-a)a}{L}$$

at $x = L$ B.M. $= 0$

Fig. 9.1 (*e*) shows the plot of I.L. for B.M. with respect to *x*.

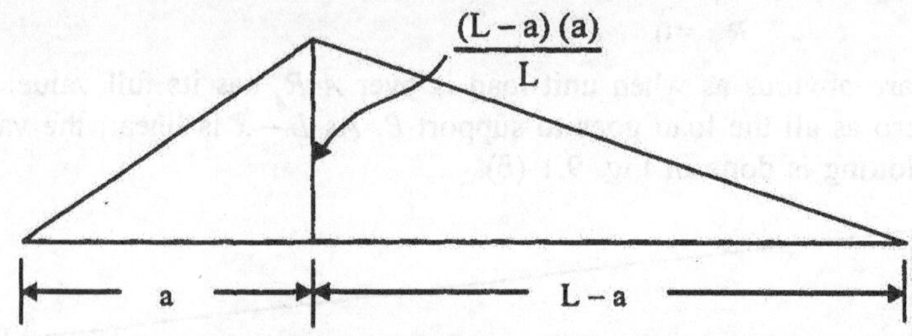

Fig. 9.1 (*e*) I.L. for B.M. at *X–X*.

It can be observed that the I.L. for Reaction, B.M. and S.F. for a determinate structure consists of a number of straight lines. Influence line is useful to structural analyst in two ways:

1. Qualitatively
2. Quantitatively

For a number of concentrated loads or a distributed load greater or shorter than the span it is possible to find from the diagrams qualitatively where the load should be placed to produce the maximum effect. Once the position gets fixed it is possible to quantitatively evaluate the value of the function e.g. B.M., S.F. or Reaction. From Fig. 9.1 above it can be seen that loads should be placed closer.

to *A* to get maximum R_A

to *B* to get maximum R_B

to section *X–X* - to get maximum positive or negative shear.

to section *X–X* – to get the maximum B.M.

With these basic ideas we can now solve a number of problems on influence line for statically determinate beams.

9.3. EXAMPLES

Example 9.1. For the beam shown, construct the influence diagram for (a) the reaction at *B* (b) the moment at *C* (*c*) the reaction at *D* and (*d*) shear just to the left of point *D*.

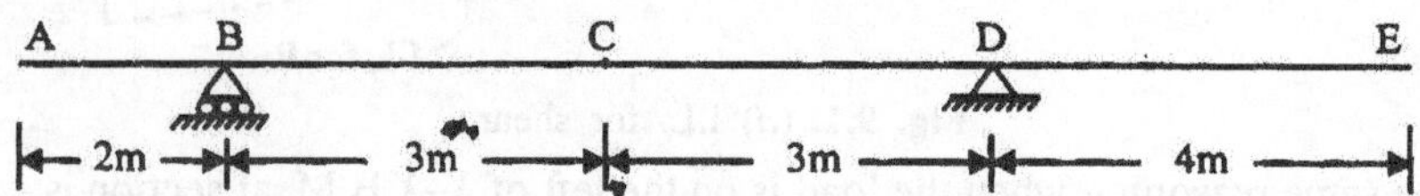

Fig. 9.2. (*a*) Example 9.1

Solution: (a) I.L. for Reaction at *B*

Load at a distance *x* from *A*

within span AB

$$-1 \times (8 - x) + R_B \times 6 = 0$$

or $\quad R_B = \dfrac{8 - x}{6}$

at $\quad x = 0 \qquad R_B = \dfrac{8}{6} = \dfrac{4}{3} \uparrow$

at $x = 2$ $R_B = 1 \uparrow$

In span BD

$$-1 \times (8 - x) + R_B \times 6 = 0$$

or $R_B = \frac{8 - x}{6}$

at $x = 2$ $R_B = 1$

$x = 8$ $R_B = 0$

In span DE

$$+1 \times (x - 8) + R_B \times 6 = 0$$

or $R_B = -\frac{(x - 8)}{6}$

when $x = 8$ $R_B = 0$

$x = 12$ $R_B = -\frac{2}{3}$

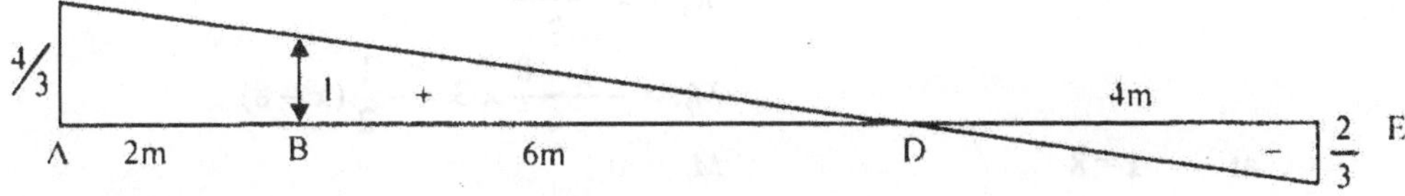

Fig. 9.2. (*b*) I.L. for R_B

(*b*) Moment at *C*

Load at a distance *x* from *A*.

(*i*) **Within span *AB***

Taking moments about *B*

$$1 \times (2 - x) + R_D \times 6 = 0$$

or $R_D = -\frac{2 - x}{6}$

Moment at $C = -\frac{2 - x}{6} \times 3$

$$= -\frac{2 - x}{2}$$

at $x = 0$ $M_C = -1$ (hogging)

$x = 2$ $M_C = 0$

Within span BC

$$-1 \times (x - 2) + R_D \times 6 = 0$$

or $R_D = \frac{x - 2}{6}$

$$M_C = \frac{x - 2}{6} \times 3 = \frac{x - 2}{2}$$

at $x = 2$ $M_C = 0$

$x = 5$ $M_C = \frac{3}{2}$

Within span CD

$$R_B = \frac{8 - x}{6}$$

$$M_C = \frac{8 - x}{6} \times 3 = \frac{8 - x}{2}$$

at	$x=5$	$M_C = \frac{3}{2}$
at	$x=8$	$M_C = 0$

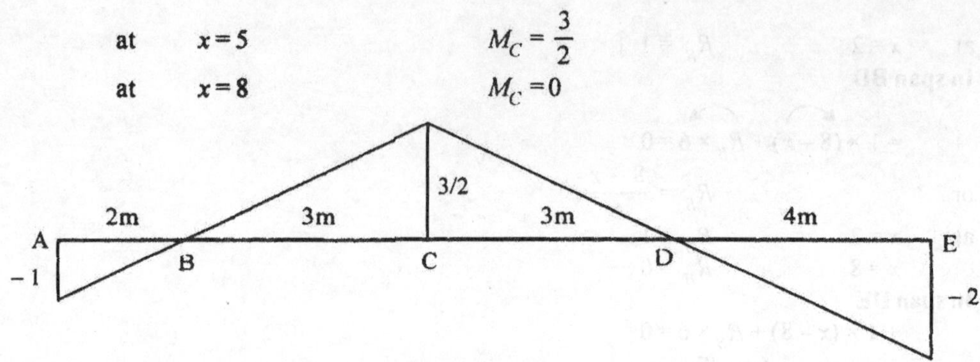

Fig. 9.2. (*c*) I.L. for M_c (example 9.1)

Within span DE

$$R_B \times 6 + 1 \times (x-8) = 0$$

$$R_B = -\frac{x-8}{6}$$

$$M_C = -\frac{x-8}{6} \times 3 = -\frac{1}{2}(x-8)$$

at	$x=8$	$M_C = 0$
	$x=12$	$M_C = -2$

I.L. diagram for M_C is shown in Fig. 9.2 (*c*).

Reaction at D.

Load in span *AB* (*x* is measured from *A*)

Taking moments about *B*

$$1 \times (2-x) + R_D \times 6 = 0$$

or $$R_D = -\frac{2-x}{6}$$

at	$x=0$	$R_D = -\frac{1}{3}\downarrow$
	$x=2$	$R_D = 0.$

In span BD

$$1 \times (x-2) - R_D \times 6 = 0$$

or $$R_D = \frac{x-2}{6}$$

at	$x=2$	$R_D = 0$
at	$x=8$	$R_D = 1$

Inspan DE

$$1 \times (x-2) - R_D \times 6 = 0$$

or $$R_D = \frac{x-2}{6}$$

at	$x=8$	$R_D = 1$
	$x=12$	$R_D = \frac{10}{6} = \frac{5}{3}.$

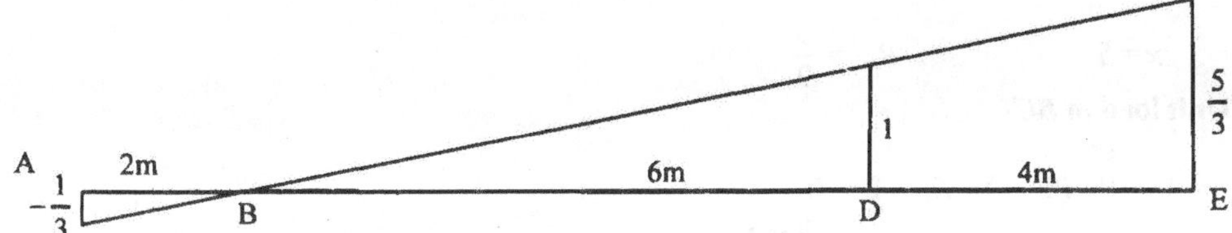

Fig. 9.2. (*d*) Influence Line for R_D

Shear just to the left of point D.

Load is in AD (Shear force sign convention left up and right down + ve) when load is in *DE* shear is R_B, when load in AD shear $= R_D$.

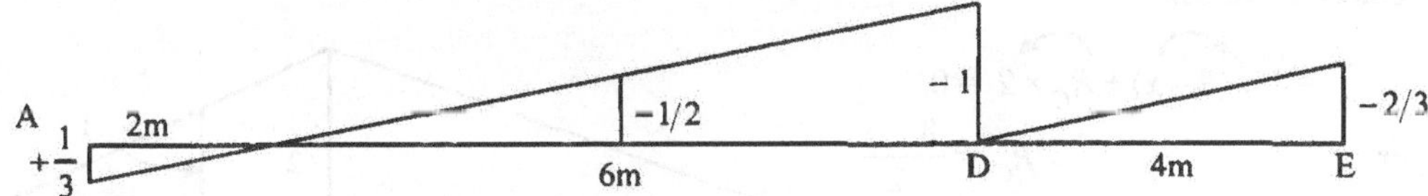

Fig. 9.2. (*e*) - I.L. for shear just to the left of point *D*.

Example 9.2. For the beam shown construct the influence diagram for (*a*) the reaction at *A* (*b*) the reaction at *C* (*c*) the reaction at *D* (*d*) the shear at *B* and (*e*) the moment at *C*.

Solution.

A B C D

5m 1m 2m

Fig. 9.3. (*a*) Example 9.2.

Reaction at *A*. Unit Load in *AB* (*x* measured from *A*)

$$R_A \times 5 - 1 \times (5 - x) = 0$$

or $$R_A = \frac{5-x}{5}$$

at $x = 0$ $\quad R_A = 1$

$x = 5$ $\quad R_A = 0.$

Unit load in *BC* – No load is carried to *AB*, hence $R_A = 0$

Unit load in *CD* – No load is carried to *AB*, hence $R_A = 0$

Reaction at *C*.

Unit load in AB

$$-R_B \times 5 + 1 \times x = 0$$

$$R_B = \frac{x}{5}$$

$$R_C \times 2 - R_B \times 3 = 0$$

$$R_C = \frac{R_B \times 3}{2} = \frac{x}{5} \times \frac{3}{2} = \frac{3x}{10}$$

At $x = 0$ $\quad R_C = 0$

$x = 5 \qquad R_C = \frac{3}{2}$

Unit load in BC

$$-1 \times (8 - x) + R_C \times 2 = 0$$

$$R_C = \frac{8 - x}{2}$$

at $x = 5 \qquad R_C = \frac{3}{2}$

$x = 6 \qquad R_C = 1.$

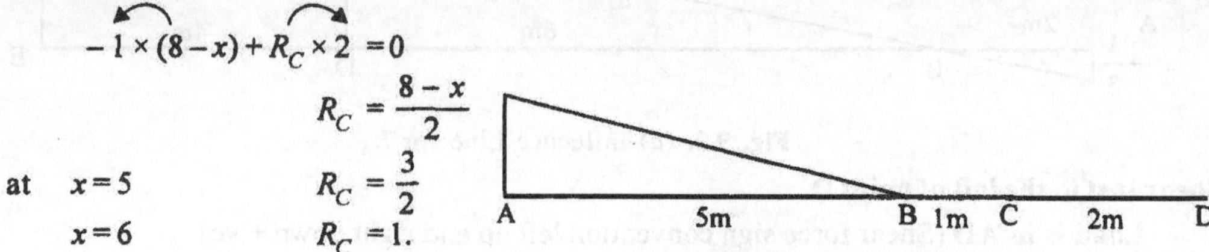

Fig. 9.3. (*b*) - I.L. for R_A

Unit load in CD

$$-1 \times (8 - x) + R_C \times 2 = 0$$

or $\qquad R_C = \frac{8 - x}{2}$

at $x = 6 \qquad R_C = 1$

at $x = 8 \qquad R_C = 0$

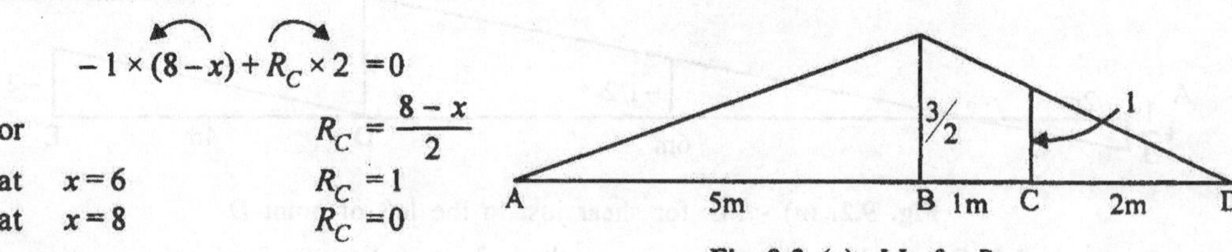

Fig. 9.3. (*c*) - I.L. for R_C

Reaction at D

Unit load in AB $\qquad R_B = \frac{x}{5}$

Taking moments about C

$$-R_B \times 1 - R_D \times 2 = 0$$

or $\qquad R_D = -\frac{R_B}{2} = -\frac{x}{10}$

at $x = 0 \qquad R_D = 0$

at $x = 5 \qquad R_D = -\frac{1}{2}$

Unit load in BC

$$1 \times (6 - x) + R_D \times 2 = 0$$

$$R_D = -\frac{6 - x}{2}$$

at $x = 5 \qquad R_D = -\frac{1}{2}$

$x = 6 \qquad R_D = 0.$

Unit load in CD

$$1 \times (x - 6) - R_D \times 2 = 0$$

or $\qquad R_D = \frac{x - 6}{2}$

at $x = 6 \qquad R_D = 0$

at $x = 8 \qquad R_D = 1$

Fig. 9.3. (*d*) – I.L. for R_D

Shear at B

When load is in *AB*

$$R_B = \frac{x}{5}$$

at $x = 0$ $\quad R_B = 0$

$x = 5$ $\quad R_B = 1$

Fig. 9.3. (*e*) I.L. for shear at *B*.

Moment at *C*.

Unit load in AB

Moment at $C = R_B \times 1$ (hogging)

$$= \frac{x}{5}$$

at $x = 0$ $\quad M_C = 0$

at $x = 5$ $\quad M_C = 1$

Unit load in *BC*

$$M_C = 1 \times (6 - x)$$

at $x = 5$ $\quad M_C = 1$

at $x = 6$ $\quad M_C = 0$

Unit load in *CD* $\quad M_C = 0$

Fig. 9.3. (*f*) Influence line for M_C

Example 9.3. For the frame shown construct the influence line diagram for (*a*) vertical reaction at *A* (*b*) The moment just to the right of *B* (*c*) the moment at *C* (*d*) the shear at *C* and (*e*) the reaction at *D*.

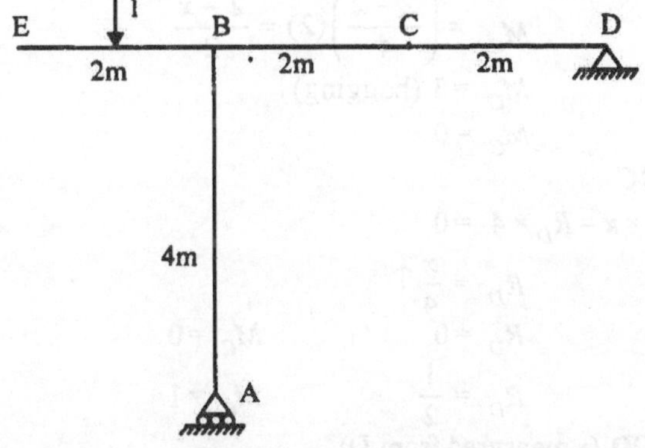

Fig. 9.4. (*a*) Example 9.3.

Solution:

(*a*) Vertical reaction at *A*

Unit load in EB (*x* measured from *E*).

$$-1 \times (6 - x) + R_A \times 4 = 0$$

$$R_A = \frac{1 \times (6-x)}{4}$$

at $x=0$ $\quad R_A = \frac{6}{4} = \frac{3}{2}$

$x=2$ $\quad R_A = 1.$

Unit load in *BD*

$$-1 \times (6-x) + R_A \times 4 = 0$$

$$R_A = \frac{6-x}{4}$$

at $x=2$ $\quad R_A = 1$

$x=6$ $\quad R_A = 0$

(*b*) Moment just to the right of *B*

Load in *EB*

$$-1 \times (2-x) + R_D \times 4 = 0$$

$$R_D = \frac{2-x}{4}$$

$$M_B = \frac{2-x}{4} \times 4 \text{ (hogging)}$$

at $x=0$ $\quad M_B = 2$ (hogging)

$x=2$ $\quad M_B = 0$ when load is in *BD*

Moment at *B* is zero.

(c) Moment at *C*

Unit load in *EB*

$$R_D = \frac{2-x}{4} \downarrow$$

$$M_C = \left(\frac{2-x}{4}\right)(2) = \frac{2-x}{2}$$

at $x=0$ $\quad M_C = 1$ (hogging)

at $x=2$ $\quad M_C = 0$

Unit load in *BC*

$$1 \times x - R_D \times 4 = 0$$

or $\quad R_D = \frac{x}{4} \uparrow$

at $x=0$ $\quad R_D = 0$ $\quad M_C = 0$

$x=2$ $\quad R_D = \frac{1}{2}$ $\quad M_C = 1$

Unit load in *CD* (*x* measured from *D*)

$$-1 \times x + R_A \times 4 = 0$$

or $\quad R_A = \frac{x}{4}$

at $x=0$ $\quad R_A = 0$ $\quad M_C = 0$

$x=2$ $\quad R_A = \frac{1}{2}$ $\quad M_C = 1$ (hogging)

(*d*) I.L. for shear at C

Unit load in *EB*

$$-1\times(2-x)-R_D\times 4=0$$

or $R_D=-\dfrac{(2-x)}{4}$

at $x=0$ $R_D=-\dfrac{1}{2}$

at $x=2$ $R_D=0$

as S.F. is given by R_D at $x=0$ $V=+\dfrac{1}{2}$

at $x=2$ $V=0$.

Unit in *BC*

$$1\times(x-2)-R_D\times 4=0$$

$R_D=\dfrac{x-2}{4}\uparrow$

at $x=2$ $R_D=0$

$x=4$ $R_D=\dfrac{1}{2}\uparrow$ S.F. $=-\dfrac{1}{2}$

Unit load in *CD*. S.F. is given by R_A

$$-1\times(6-x)+(R_A\times 4)=0$$

or $R_A=\dfrac{6-x}{4}$

at $x=4$ $R_A=\dfrac{1}{2}$, $V=\dfrac{1}{2}$ + shear

$x=6$ $R_A=0$

I.L. for R_D

Unit load in *EB*

$R_D=-\dfrac{2-x}{4}$

at $x=0$ $R_D=-\dfrac{1}{2}$

at $x=2$ $R_D=0$

Unit load in *BC*

$R_D=\dfrac{x-2}{4}$

at $x=2$ $R_D=0$

at $x=4$ $R_D=\dfrac{1}{2}$

Unit load in *CD*

$$1\times(x-2)-R_D\times 4=0$$

or $R_D=\dfrac{x-2}{4}$ at $x=4$

$R_D=\dfrac{1}{2}$

at $x=6$ $R_D=1$

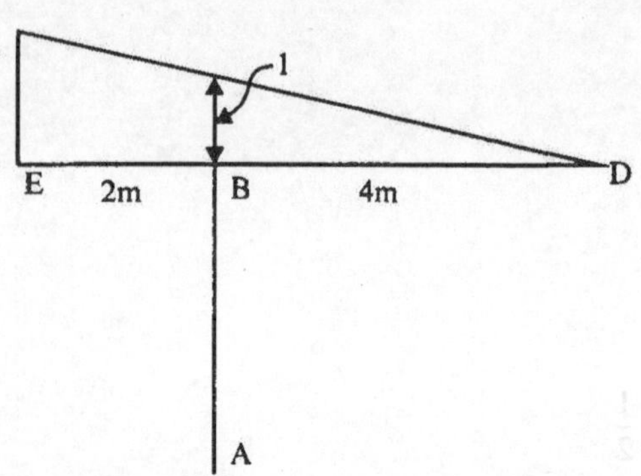

Fig. 9.4. (*b*) I.L. for R_A

Fig. 9.4. (*c*) I.L. for Moment just to the right of *B*.

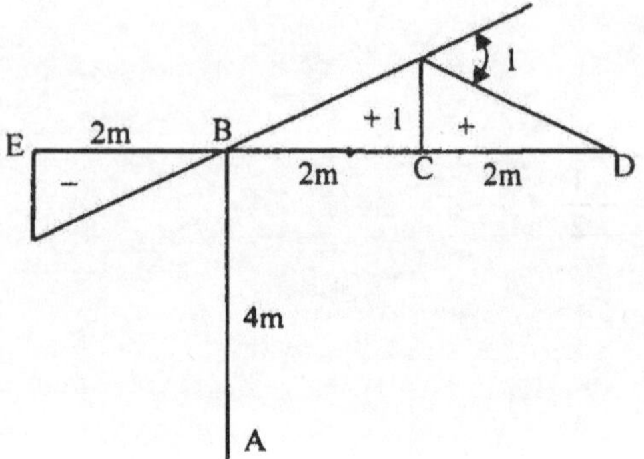

Fig. 9.4. (*d*) I.L. for moment at C

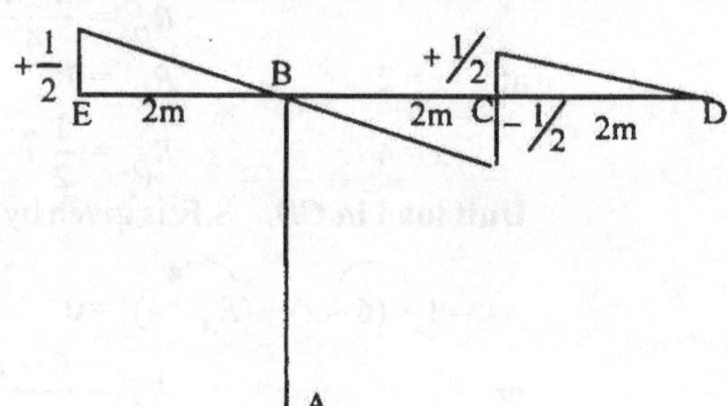

Fig. 9.4. (*e*) I.L. for shear at *C*

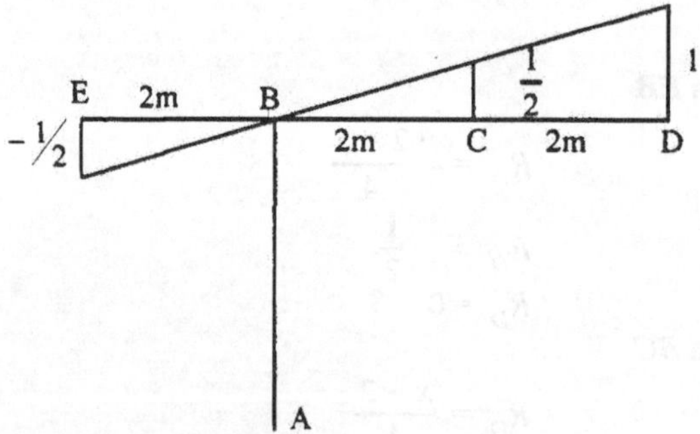

Fig. 9.4. (*f*) I.L. for R_D

9.4 INFLUENCE LINES BY VIRTUAL WORK:

There are two principles of virtual work.

(*i*) Principle of virtual force

(*ii*) Principle of virtual displacement.

Principle of virtual force enables us to determine real displacement. Principle of virtual displacement will enable us to determine real force. The virtual displacement principle states that

"If a system of forces acting on a rigid body is in equilibrium, the total virtual work done by the forces as the structure under goes a small virtual displacement is zero."

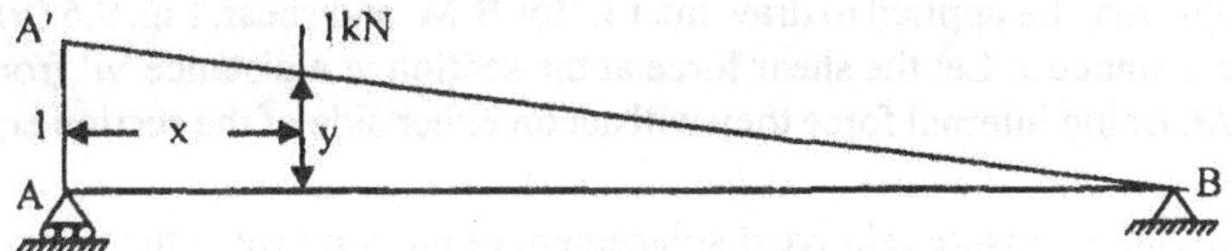

Fig. 9.5. Principle of virtual work.

The simply supported beam *AB* is in equilibrium under the action of three forces 1 kN, R_A and R_B. If we give a small virtual displacement to the beam the total work done will be zero. Since the virtual displacement is arbitrary consistent with the boundary condition, we move the end *A* through *unit* distance keeping *B* in its original position i.e. *AB* becomes *A'B*.

The total work done

$R_A \times 1 + 1 \times (-y) + R_B \times 0 = 0$

[*y* is taken as negative as it is in opposite direction to the unit load]

or $\quad R_A = y.$

As there is no displacement at *B*, work done by R_B is zero. Since *y* gives the value of R_A under the load *A'AB* is the influence line diagram for R_A. Hence if virtual displacement is made unity, the displaced shape of the beam is the influence line diagram for R_A. The steps in this method, therefore, are:

1. Select the action for which the Influence Line is required.
2. Allow this action to move through unit displacement with the structure subjected to all other constraints.
3. The resulting displacement shape will give to some scale the influence line for the action.

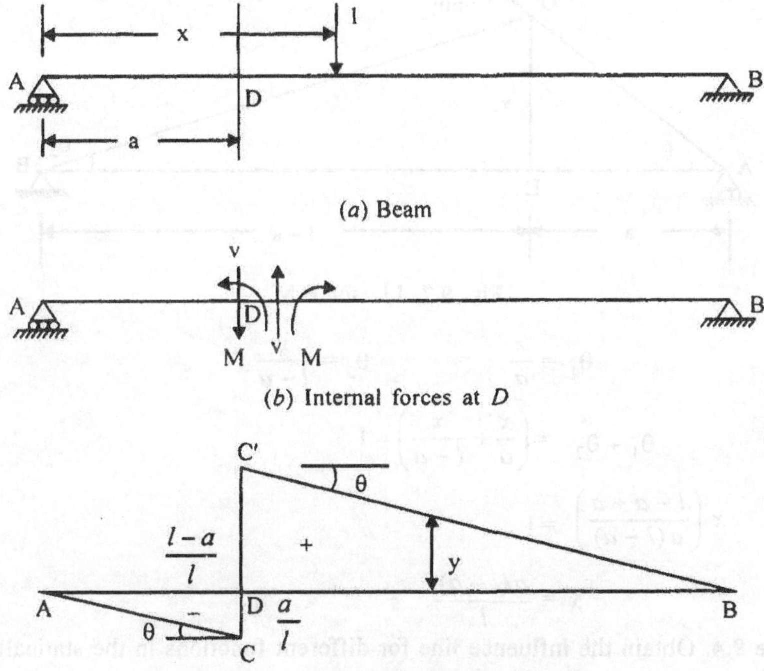

Fig. 9.6. I.L. for shear

Same principle may be applied to draw the I.L. for B.M. and shear. Fig. 9.6 (*a*) shows the beam with unit load at a distance x. Let the shear force at the section at a distance 'a' from A be V and the moment there be M. Being internal force they will act on either side of the section equal and opposite Fig. 9.6 (*b*).

For shear we should ensure relative displacement of unity between the two parts of the beam. At the same time, in order that the moment M gets eliminated, the parts should be parallel so that work done by moments on the two parts are equal. Fig. 9.6 (*c*).

Applying virtual work principle

$$V(CD) + V(C'D) + 1 \times (-y) + M(\theta) + M(-\theta) = 0$$

or $\quad V(CD + C'D) = y.$

If $\quad CD + C'D = \text{unity}, \; V = y$

As AC and BC' are parallel,

$$\frac{CD}{AD} = \frac{C'D}{BD} = \frac{CD + C'D}{AD + BD} = \frac{1}{l}$$

or
$$CD = \frac{AD}{l} = \frac{a}{l}$$

$$C'D = \frac{l-a}{l}$$

To get the influence line for B.M. we should give unit rotation at D'. At the same time vertical displacement at D of the two parts must be the same so that work done by shear equalises. This gives the diagram shown in Fig. 9.7.

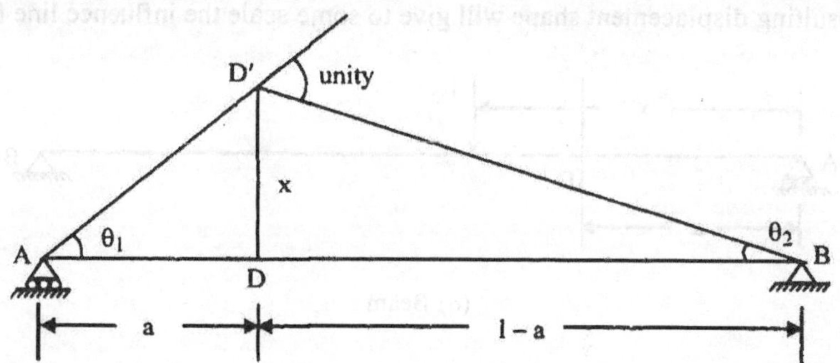

Fig. 9.7. I.L. for B.M.

$$\theta_1 = \frac{x}{a} \qquad \theta_2 = \frac{x}{l-a}$$

$$\theta_1 + \theta_2 = \left(\frac{x}{a} + \frac{x}{l-a}\right) = 1$$

or
$$x\left(\frac{l-a+a}{a(l-a)}\right) = 1$$

or
$$x = \frac{a(l-a)}{l}.$$

Example 9.4. Obtain the influence line for different functions in the statically determinate beam. (Fig. 9.8)

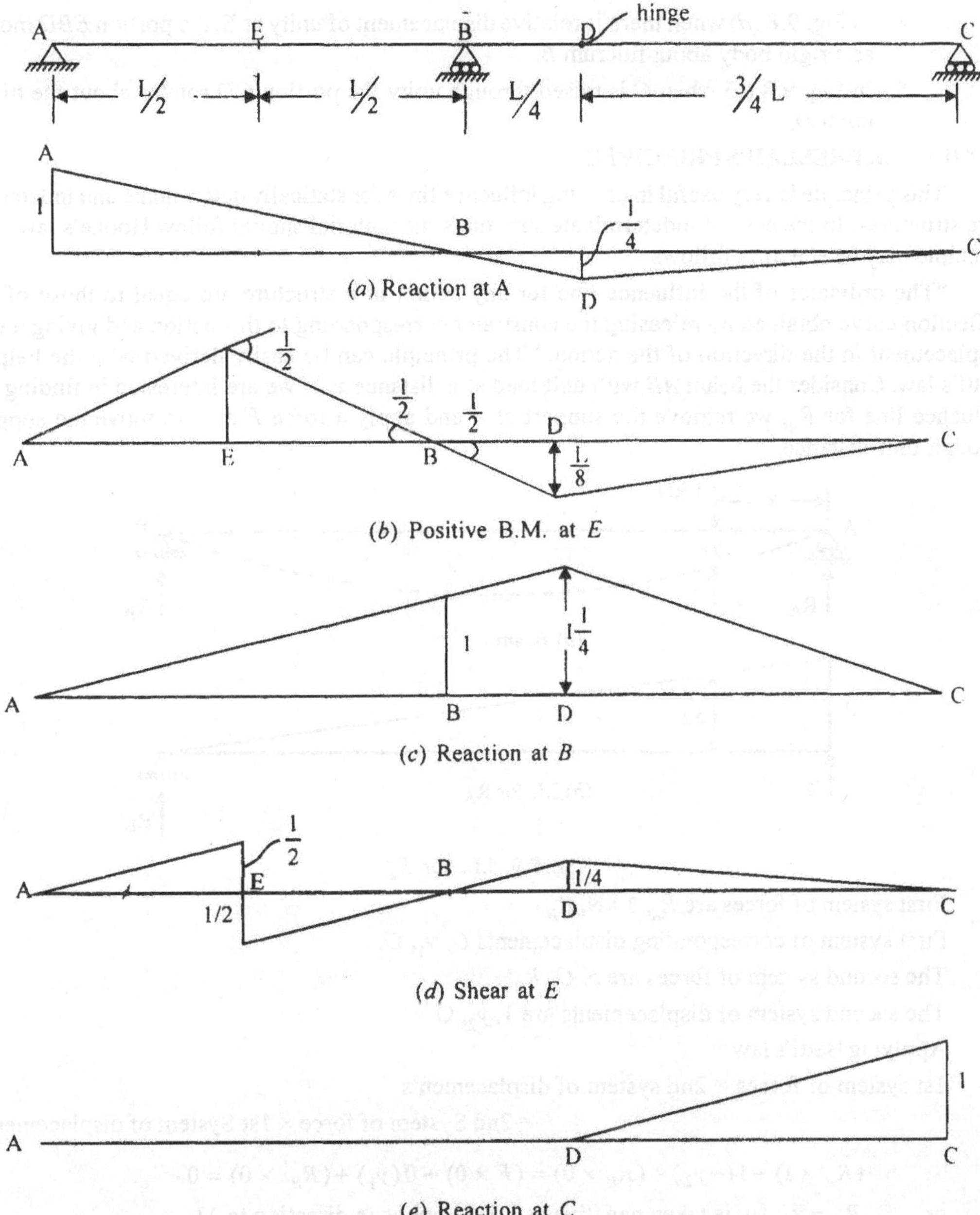

Fig. 9.8. Influence Lines for Example 9.4.

Explanation

1. In Fig. 9.8 (*a*) when *A* is raised the rigid beam rotates about *B* as fulcrum.
2. In Fig. 9.8 (*b*) when the resistance at *E* is removed and unit rotation is given at *E* the portion *EBD* rotates with *B* as fulcrum.
3. In Fig. 9.8 (*c*) when *B* is raised through unit distance the portion *ABD* moves as a rigid body and *D* is raised $1\frac{1}{4}$ units.

4. In Fig. 9.8 (*d*) when there is relative displacement of unity at *E*, the portion *EBD* moves as a rigid body about fulcrum *B*.
5. In Fig. 9.8 (*e*) when *C* is raised through unity the portion *CD* rotates about the hinge point *D*.

9.5 MÜLLER BRESLAU'S PRINCIPLE

This principle is very useful in drawing influence lines for statically determinate and indeterminate structures. In the case of indeterminate structures the material should follow Hooke's law. The principle may be stated as follows.

"The ordinates of the influence line for any action in a structure are equal to those of the deflection curve obtained by releasing the constraint corresponding to this action and giving a unit displacement in the direction of the action." The principle can be easily derived with the help of Betti's law. Consider the beam *AB* with unit load at a distance *x*. If we are interested in finding the influence line for R_A, we remove the support at *A* and apply a force *F* at *A* to move the support through unit distance.

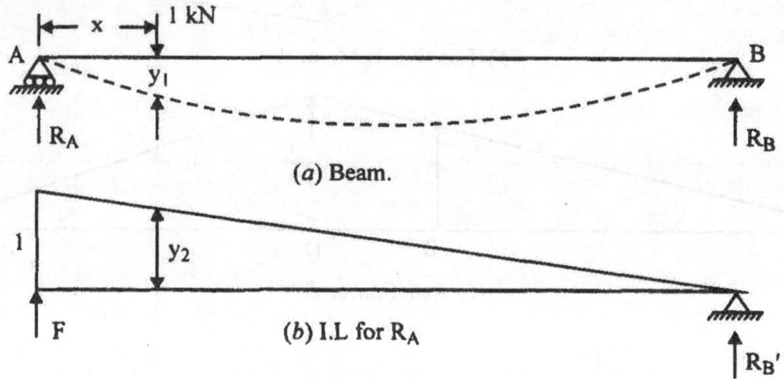

Fig. 9.9. I.L. for R_A

First system of forces are R_A, 1 kN, R_B

First system of corresponding displacements *O*, y_1, *O*.

The second system of forces are *F*, *O*, R_B'

The second system of displacements are 1, y_2, O

Applying Betti's law.

1st system of forces × 2nd system of displacements

= 2nd System of force × 1st System of displacements.

$$(R_A \times 1) + 1(-y_2) + (R_B \times 0) = (F \times 0) + 0(y_1) + (R_B' \times 0) = 0 \cdot$$

or $R_A = y_2$ [y_2 is taken negative as it is in opposite direction to 1]

Hence by displacing the end A by unity we get the I.L. for reaction R_A.

Similarly the I.L. for B.M. is obtained by removing the resistance at *C* and giving unit rotation at *C*.

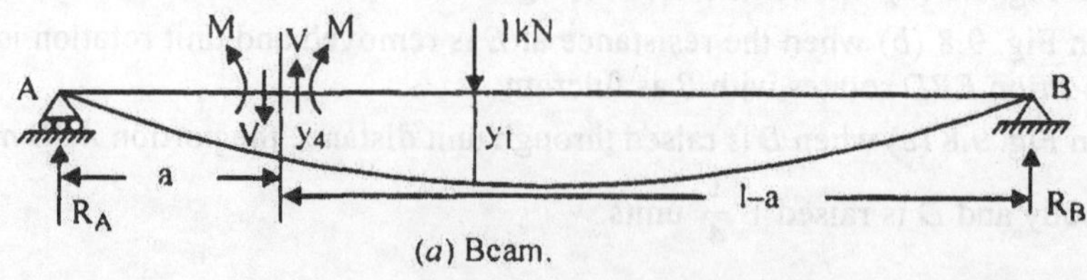

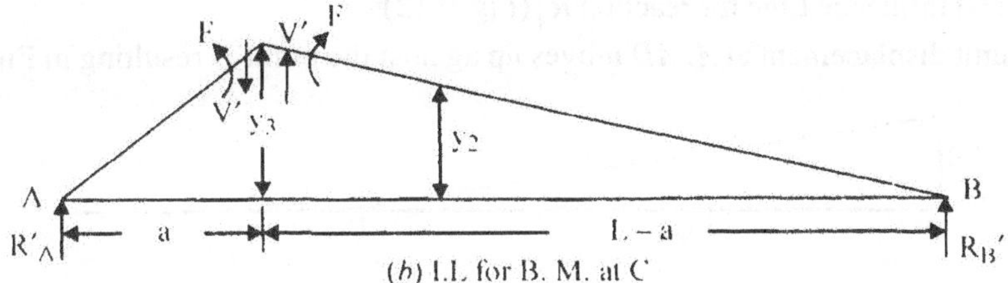

(*b*) I.L. for B. M. at C

Fig. 9.10. I.L. for B.M.

1st System of Forces

$$R_A \uparrow,\ M \circlearrowleft,\ V \downarrow\ \circlearrowright M,\ V \uparrow\ 1 \downarrow\ R_B \uparrow$$

2nd System of displacements

$$0,\ \frac{y_3}{a} \circlearrowleft,\ y_3 \uparrow\ \frac{y_3}{l-a} \circlearrowright,\ y_3 \uparrow\ y_2 \uparrow\ 0$$

2nd System of forces

$$R_A',\ F \circlearrowleft,\ V' \downarrow\ \circlearrowright F,\ V' \uparrow 0,\ R_B'$$

1st System of displacements

$$0, \theta_1 \circlearrowright, y_4 \downarrow,\ \circlearrowright \theta_1, y_4 \downarrow y_1 \downarrow 0$$

Applying Bett's law

$$R_A(0) + M \times \left(\frac{y_3}{a}\right) + V \times (-y_3) + M.\left(\frac{y_3}{l-a}\right) + V(+y_3) + 1 \times (-y_2) + R_B \times 0$$

$$= R_A' \times 0 + F \times (-\theta_1) + V' \times (y_4) + F \times (+\theta_1) + V' \times (-y_4) + 0 \times y_1 + R_B' \times 0$$

or $M\left(\frac{y_3}{a} + \frac{y_3}{l-a}\right) + V(y_3 - y_3) + 1 \times (-y_2) + 0 = F(\theta_1 - \theta_1) + V'(y_4 - y_4) = 0$

If unit rotation is given

or $M(1) - y_2 = 0$ or $M = y_2$

Hence by removing the restraint against B.M. at *C i.e.* putting a hinge at *C* gives the I.L. for B.M. Similarly, the I.L. for shear can be derived. Influence line for determinate structures consists of straight lines as reducing one restraint reduces a determinate structure to a mechanism and rigid body rotation is possible. I.L. for indeterminate structure will, however, consists of curved lines as removal of constraint and giving required displacement will involve elastic considerations. For determinate structure there is no added advantage of using Müller Breslau's principle over virtual work method. For indeterminate structures Müller Breslau's principle is especially useful.

Example 9.5. Apply Müller Breslau's principle to draw the influence lines for (*i*) Reaction at *A* (*ii*) Shears at *D* and *C* (*iii*) Moments at *B* and *C* for the beam shown in Fig. 9.11.

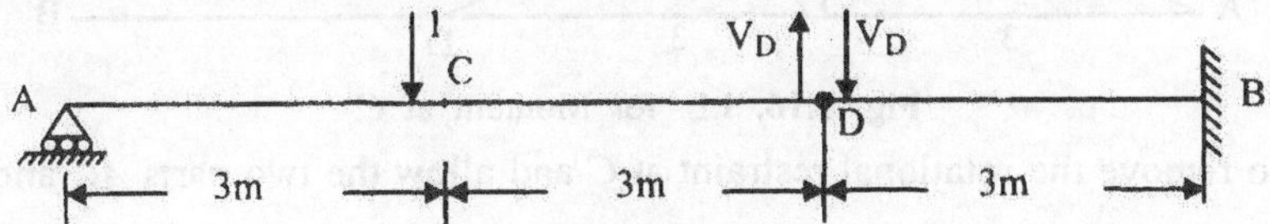

Fig. 9.11. Example 9.5

Solution: (*i*) Influence Line for reaction R_A (Fig. 9.12)

We give unit displacement at *A*. *AD* moves up against the point *D* resulting in Fig. 9.12.

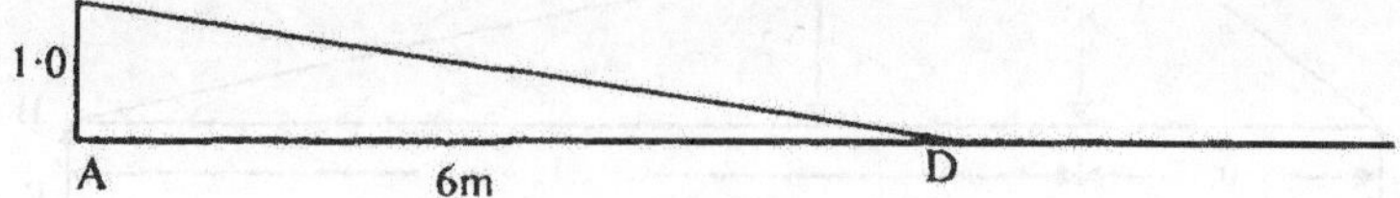

Fig. 9.12. I.L. for R_A. (Example 9.5)

(*ii*) Shear at *C* (Fig. 9.13)

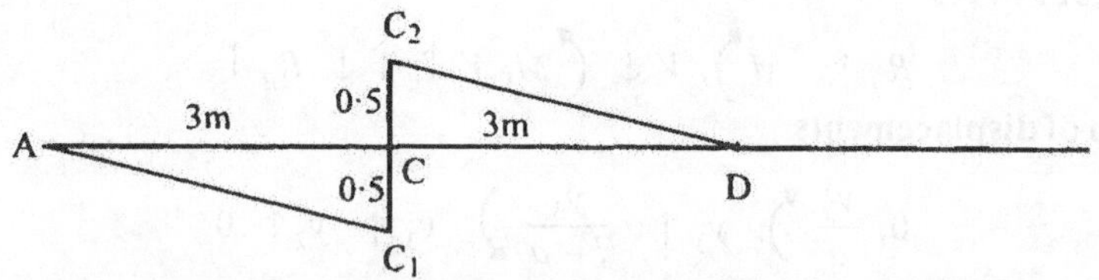

Fig. 9.13. I.L. for shear at *C*. (Example 9.5)

We give relative displacement of unity at *C*. making AC_1 parallel to C_2D. $C_2C = CC_1 = 0.5$.

(*iii*) Shear at *D* (Fig. 9.14)

V_D will occur when load is in *AD*. V_D will be zero when load crosses *D* as the load will be carried by reaction at *B*.

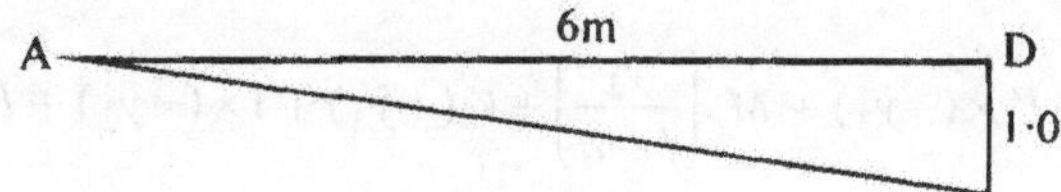

Fig. 9.14. I.L. for shear V_D. (Example 9.5)

(*iv*) I.L. for Moment at *B*. (Fig. 9.15)

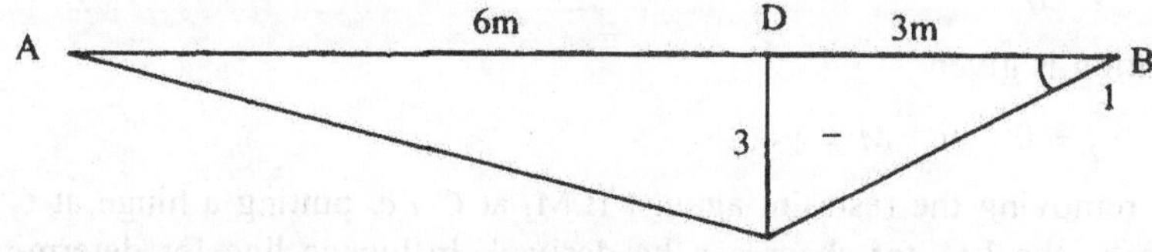

Fig. 9.15. I.L. for moment at *B*. (Example 9.5)

We have to remove the restraint at *B* and allow *DB* at rotate through unity.

(*v*) I.L. for Moment at *C* (Fig. 9.16)

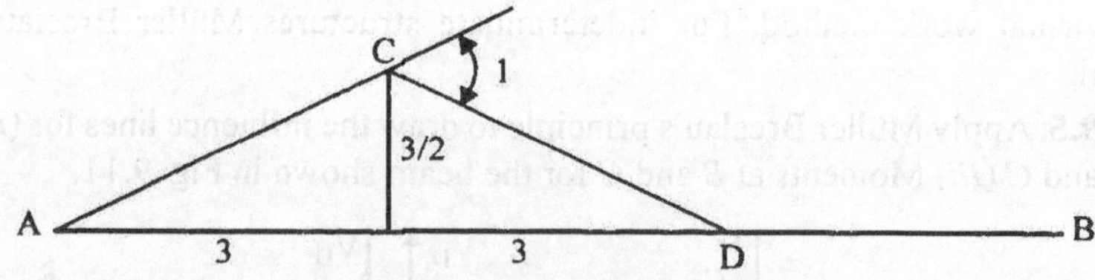

Fig. 9.16. I.L. for Moment at *C*

We are to remove the rotational restraint at *C* and allow the two parts *AC* and *CD* to rotate through unity.

Example 9.6. For the beam shown, construct the influence for (*a*) the moment at *A* (*b*) the moment at *B* (*c*) the shear at *B* (*d*) The reaction at *D* (*e*) the moment at *F* and (*f*) the shear just to the left of *D*.

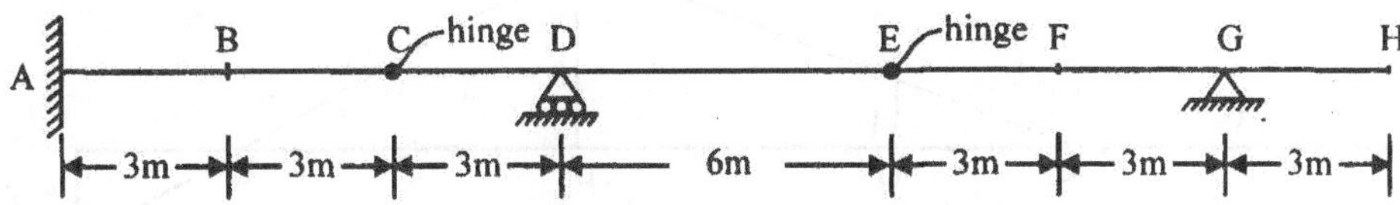

Fig. 9.17. Example 9.6

Solution:

(*a*) Moment at *A*

Remove the restraint at *A* and give unit rotation at *A*. The portion *CDE* rotates about *D*. *EFGH* rotates about *G*.

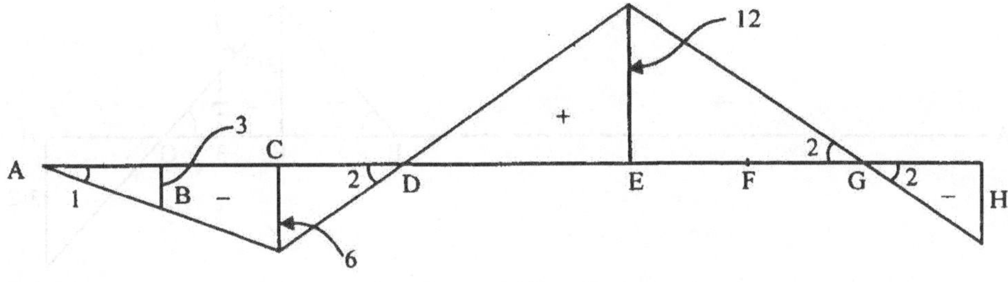

(*a*) I.L. for Moment at *A*

(*b*) Moment at B

Remove the rotational restraint at *B* and give unit rotation at *B*. *CDE* rotates about *D* and *EFGH* about *G*.

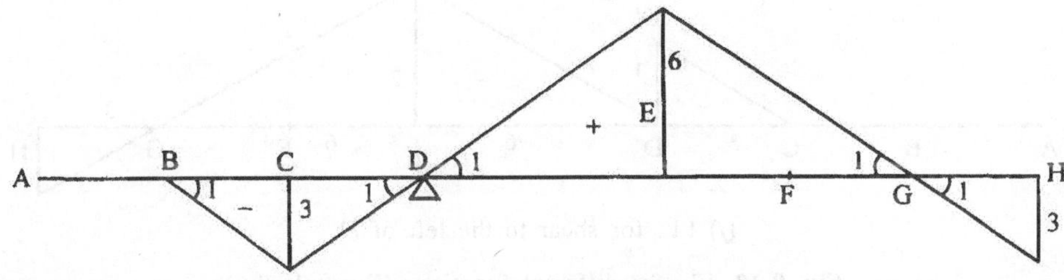

(*b*). I.L. for moment at *B*.

(c) Shear at *B*

Give unit relative displacement between *AB* and *BC*. As *AB* is fixed *BC* goes down by 1 with other consequential displacements.

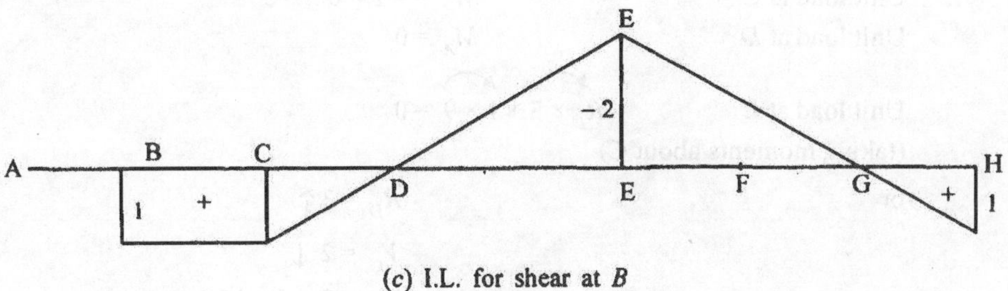

(*c*) I.L. for shear at *B*

(d) **Reaction at *D*.** Unit vertical displacement is given at *D*. *CDE* rotates about *C*. *EFGH* about *G*.

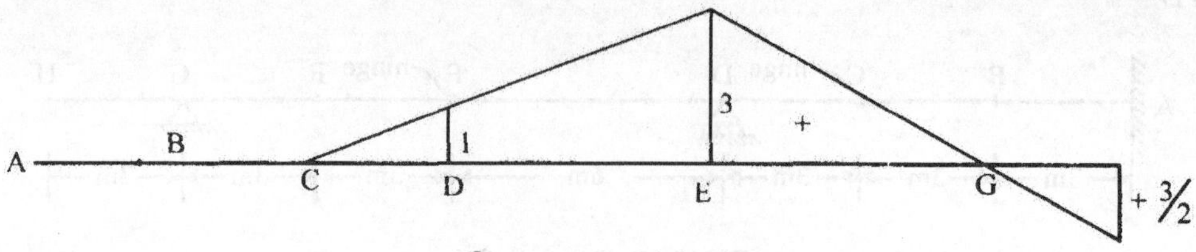

(*d*) I.L. for Reaction at *D*.

(*e*) Moment at *F*. Unit rotation is given at *F*.

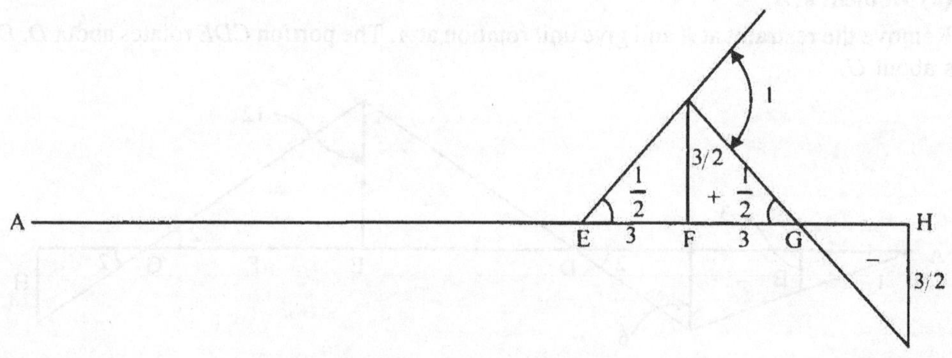

(*e*) I.L. for moment at *F*.

(*f*) Shear just to the left of D. Unit relative displacement is given just to the left of *D*. *D* moves up through unity. As the two parts should be parallel *DE'* is made parallel to *CD'*.

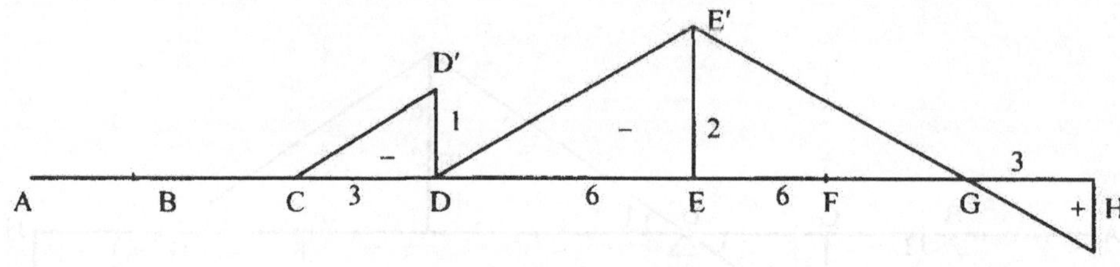

(*f*) I.L. for shear to the left of *D*.

Fig. 9.18. I.L. for different functions (Example 9.6)

Check

(*a*) Moment at A

Unit load at *A* $M_A = 0$

Unit load at *C* $M_A = -1 \times 6 = -6$

Unit load at *D* $M_A = 0$

Unit load at *E* $-R_D \times 3 + 1 \times 9 = 0$

(taking moments about *C*)

or $R_D = 3 \uparrow$

$V_C = 2 \downarrow$

$M_A = 2 \times 6 = 12$ (sagging)

Unit load at G $\qquad M_A = 0$

Unit load at H. (taking moments about E).

$$-R_G \times 6 + 1 \times 9 = 0$$

$$R_G = \frac{9}{6} = \frac{3}{2}$$

$$V_E = \frac{1}{2}\downarrow$$

Taking moments about C, $-V_E \times 9 + R_D \times 3 = 0$

or $\qquad R_D = \frac{3}{2}\downarrow$

$$V_C = 1$$

$$M_A = 1 \times 6 \text{ (Hogging)}$$

$$= 6$$

(*b*) **Unit load at** A $\qquad M_B = 0$

Unit load at B $\qquad M_B = 0$

Unit load at C $\qquad M_B = -3$

Unit load at D $\qquad M_B = 0$

Unit load at E $\qquad 1 \times 9 - R_D \times 3 = 0.$

or $R_D = 3$

$$V_C = 2\downarrow$$

$$M_B = 2 \times 3 = +6$$

Unit load at G $\qquad M_B = 0$

Unit load at H $\qquad V_C = 1 \qquad M_B = 1 \times 3 = 3$ (hogging)

(*c*) **Shear at** B

Unit load at A $\qquad$ Shear at $B = 0$

Unit load at B (left) $\qquad$ Shear at $B = 0$

Unit load at B (right) $\qquad$ Shear at $B = 1$

Unit load at D $\qquad$ Shear at $B = 0$

Unit load at E $\qquad V_C = 2$ Shear at $B = -2$

Unit load at G $\qquad$ Shear at $B = 0$

Unit load at H $\qquad V_C = 1$ Shear at $B = +1$

(*d*) **Reaction at** D

Unit load at A $\qquad R_D = 0$

Unit load [illegible] $\qquad R_D = 0$

Unit load at D $\qquad R_D = 1\uparrow$

Unit load at E $\qquad R_D = 3\uparrow$

Unit load at G $\qquad R_D = 0$

Unit load at H $\quad R_D = \frac{3}{2}\downarrow$

(*e*) **Moment at F**

Unit load at H

$$-R_G \times 6 + 1 \times 9 = 0$$

or $$R_G = \frac{9}{6} = \frac{3}{2}$$

$$M_F = \frac{3}{2} \times 3 - 1 \times 6 = -\frac{3}{2}$$

Unit load at F

$$-R_G \times 6 + 1 \times 3 = 0$$

$$R_G = \frac{1}{2} \quad M_F = \frac{1}{2} \times 3 = \frac{3}{2}$$

Unit load at E $\quad R_G = 0 \quad M_F = 0$

Similarly at D, C, B and A.

(*f*) **Shear just to the left of D**

(*i*) Unit load between A and C. Load is carried to CDE. Load is taken by R_A.
Hence shear at D is zero.

(*ii*) Unit load between C and D.

$R_D = \frac{x}{3}$ (x measured from C)

at $\quad x = 0 \quad R_D = 0$. Hence shear at $D = 0$

at $\quad x = 3 \quad R_D = 1$. Hence shear at $D = 1$

As right side of D tends to move up shear is negative.

(*iii*) When load is between D and E. R_D is again $\frac{x}{3}$

at $\quad x = 9 \quad R_D = 3$

Shear to the left of $D = 2$ (Shear is negative)

(*iv*) When load is at G. Shear at $D = a$

(*v*) When load is at H

$$R_G = \frac{3}{2}$$

$$V_E = \frac{1}{2}\downarrow$$

and $$R_D = \frac{3}{2}\downarrow$$

Hence shear to the left of $D = 1 \uparrow$

Hence shear is positive.

9.6 USES OF INFLUENCE LINES

With the help of influence lines, it is possible to compute the value of a function,

(*i*) When load position is known.

(*ii*) for a distributed load longer or shorter than the span.

(*iii*) for its maximum value due to

(*a*) moving concentrated load

(*b*) moving distributed load longer or shorter than the span.

(*c*) combination of distributed and concentrated load.

These are explained with the help of a simple example. Fig. 9.19. shows a simply supported beam *AB*.

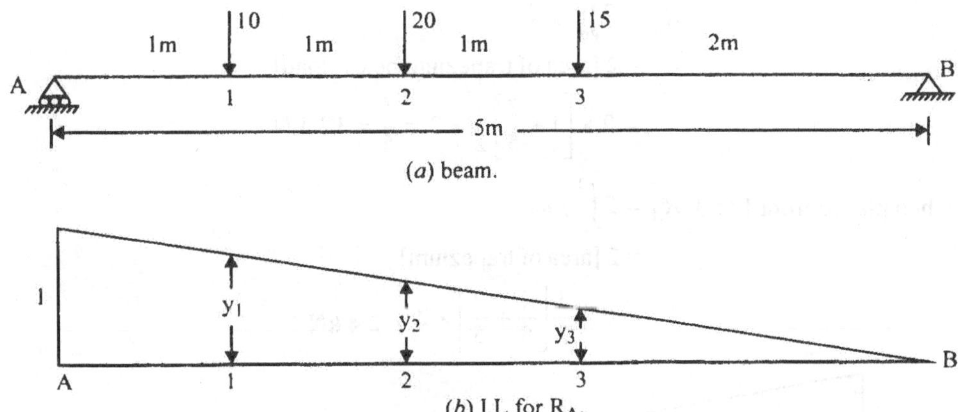

(*a*) beam.

(*b*) I.L for R_A.

Fig. 9.19. Calculation of ordinates of I.L.

The influence line for R_A is given in Fig. 9.19 (b). The beam is subjected to a series of concentrated loads of 10, 20 and 15 kN as shown in Fig. 9.19 (a). The value of reaction due to unit load when placed in section.

$$1 = y_1 = \frac{4}{5}$$

$$2 = y_2 = \frac{3}{5}$$

$$3 = y_3 = \frac{2}{5}$$

Therefore, for

10 kN load at section 1, $R_A = \frac{4}{5} \times 10 = 8 \text{ kN}$

20 kN load at section 2, $R_A = \frac{3}{5} \times 20 = 12 \text{ kN}$

15 kN load at section 3, $R_A = \frac{2}{5} \times 15 = 06 \text{ kN}$

Total $R_A = 26 \text{ kN}$

A distributed load of w kN/m may be thought of as a continuous series of concentrated load of *wdx* at a distance *x* from one end (Fig. 9.20). If the ordinate of the influence line at a distance *x* is *y*, the value of the reaction is *y* × concentrated load at a distance *x* which is equal to *w.dx*.

Hence the value of the function is *y* × *w dx*. As *w* is constant this is equal to w.y.dx. For the net effect the function is to be integrated. When the full span is loaded the integration is from 0 to *L*. For any other value it is from x_1 to x_2. Integration of *y dx* means area of the influence line diagram and hence value of the function is *w* times area of the influence line diagram below the load.

For example, when the beam is fully loaded with 2 kN/m of distributed load.

$$R_A = 2\int_0^5 y\,dx$$

$$= 2 \times \text{area of the triangle}$$

$$= 2 \times \frac{1}{2} \times 1 \times 5 = \mathbf{5\ kN}$$

When the load is of 2 m length and is placed from 0 to 2.

$$R_A = 2\int_0^2 y\,dx.$$

$$= 2\ [\text{area of trapezium below load}]$$

$$= 2 \times \left[1 + \frac{3}{5}\right]\frac{1}{2} \times 2 = \frac{8}{5} = 3.2\ \text{kN}$$

when placed from 1 to 3 $R_A = 2\int_1^3 y\,dx$

$$= 2\ [\text{area of trapezium}]$$

$$= 2 \times \frac{1}{2}\left[\frac{4}{5} + \frac{2}{5}\right] \times 2 = \mathbf{2.4\ kN}$$

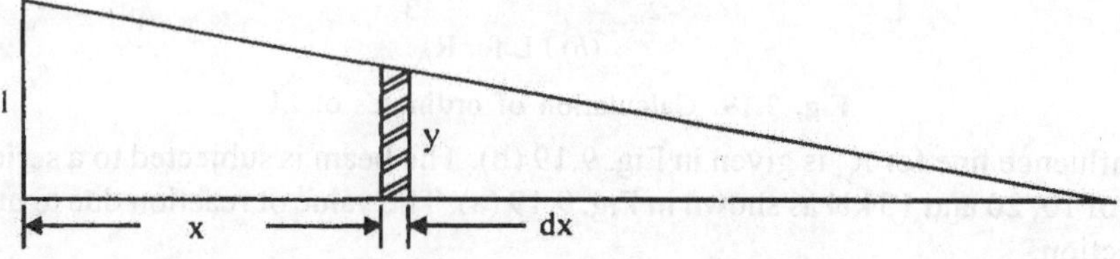

Fig. 9.20. Computation of effect of distributed force.

To get the maximum value due to

(*a*) Single concentrated load — The load should be placed at the maximum ordinate. In this case the load should be placed at the support where the ordinate is maximum i.e. unity.

(*b*) Distributed load shorter than the span — The distributed load should be placed as close to the maximum ordinates as possible. The distributed load of 2 kN/m of 2 m span should extend from the support into the span by 2m and the value is 3.2 kN as deduced above which is greater than the values for load in any other position.

(*c*) A number of concentrated load — In this case loads to be so placed so as to produce the maximum values. For this some guide lines are available as explained in the next section.

9.7. MOVING CONCENTRATED LOADS

Criteria for Maxima —

(*a*) Maximum value of Reaciton —

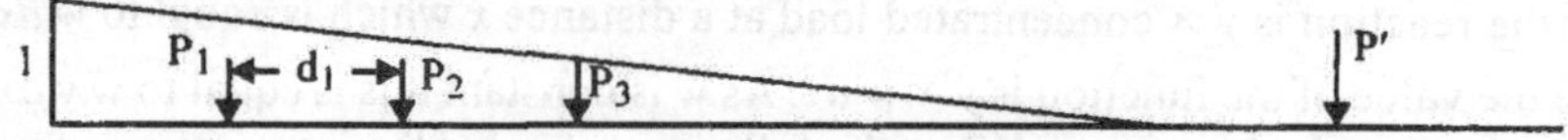

Fig. 9.21. Maximum Value of Reaction.

From a study of influence line diagram for reaction R_A it is clear that reaction from P_1 is maximum when load P_1 is placed over left support. For maximum reaction from P_2 load P_2 is to be placed over left support. As a result of this movement load P_1 will go off the span and some load P' may come over the span. We have to study the effect of this movement on R_A. If ΔR is the change,

$$\Delta R = \frac{\Sigma P d_1}{L} - P_1 + \frac{P'e}{L}$$

where ΣP = sum of all loads which are on the span and *stay* on during movement.

P_1 = the load which was over the reaction and has moved off the span.

d_1 = distance between P_1 and the following wheel.

L = length of span.

P' = a load which comes on during movement.

e = distance of P' from the right support.

If ΔR is positive, there is a gain and movement should continue.

(*b*) Maximum value of shear — Influence line for shear is shown in fig. 9.22(*a*) corresponding beam diagram is shown in fig. 9.22(*b*)

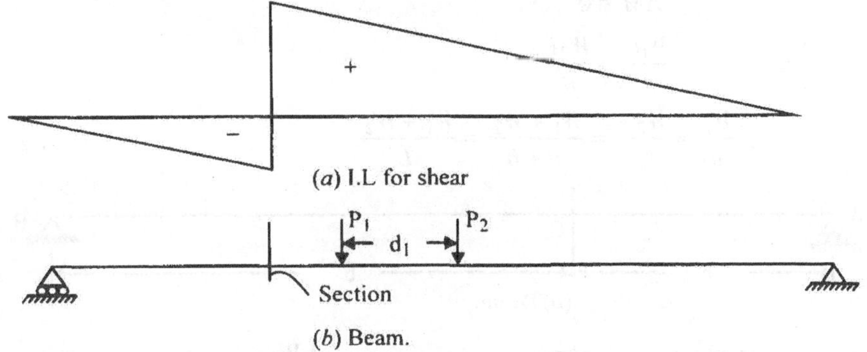

(*a*) I.L for shear

(*b*) Beam.

Fig. 9.22. Calculation of Maximum value of Shear Force.

As before maximum positive shear occurs when load P_1 is placed over the section. This is given by R_A. When load P_1 just crosses the section shear gets reduced by P_1. With further movement towards left shear again increases till P_2 is over the section and so on.

From P_1 on section to P_2 on section change in shear.

$$\Delta V = \frac{\Sigma P d_1}{L} - P_1 + \frac{P'e}{L}$$

where $\frac{\Sigma P d_1}{L}$ = Increase in reaction due to movement of d_1 of existing load.

P_1 = Load which crosses the section and reduces the shear.

$\frac{P'e}{L}$ = Value of additional reaction due to entry of new load P' on the span.

e = distance of P' from right support.

If it is positive the shear increases. When any load goes off the span its effect on shear should be considered.

(*c*) Maximum moment at a section—

The beam section and the corresponding influence line for moment is shown in Fig. 9.23. For maximum B.M. some load must be placed over the section. If now movement of all loads are made to the left moment due to loads on the right of the section will increase and that due to loads at the left of the section will decrease.

Hence change in B.M.

$$\Delta M = I - D$$

where I is the increase.

and D is the decrease.

For unit movement of the loads

$$I = W_2 \times \frac{i}{b} \times 1$$

$$D = W_1 \times \frac{i}{a} \times 1$$

and
$$\Delta M = \frac{W_2 \times i}{b} - \frac{W_1 \times i}{a}$$

Unless there is a change in W_2 or W_1 *i.e.*, W_2 becoming smaller or W_1 becoming larger ΔM cannot change from a positive quantity to a negative quantity and for this it must pass through a zero value *i.e.*, for a maximum.

$$\Delta M = 0$$

and
$$\frac{W_1 i}{a} = \frac{W_2 i}{b}$$

or
$$\frac{W_1}{a} = \frac{W_2}{b} = \frac{W_1 + W_2}{a+b} = \frac{W_1 + W_2}{L}$$

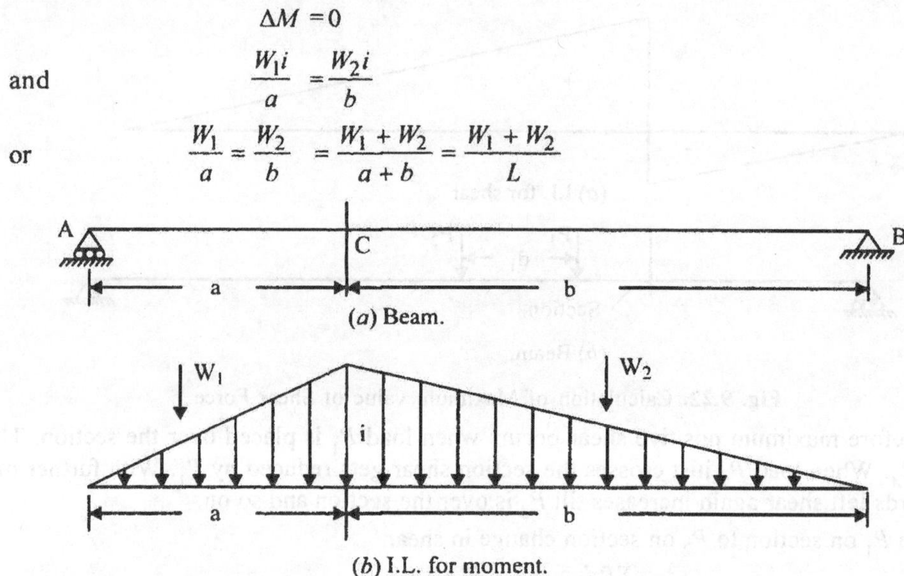

(*a*) Beam.

(*b*) I.L. for moment.

Fig. 9.23. Maximum Value of B.M.

This can be stated in words as.

"The moment is a maximum (at a section) when average load on the left of the section is equal to average load on the right of the section or average load on the span." While with distributed load shorter than the span this exact equalisation is possible with concentrated loads this is not always possible. With concentrated loads the criterion is :

With one load just to the left of the section if $\frac{W_1}{a}$ is greater than $\frac{W_2}{b}$ with that load just on the right of the section $\frac{W_1}{a}$ should be less than $\frac{W_2}{b}$

If the criterion is fulfilled, then the corresponding load should be placed over the section to get the maximum B.M.

(*d*) Maximum bending moment any where in the girder. (under a particular wheel load) Fig. 9.24 respresents a number of wheel loads over a simply supported beam of span L.

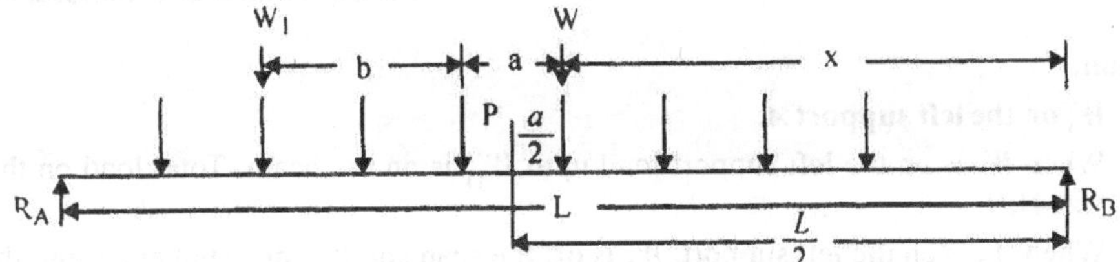

Fig. 9.24. Maximum B.M. any where in the girder.

Let W = sum of all wheel loads on the beam.

x = distance of centre of gravity of all loads (W) from the right support.

P = any one load.

a = distance of load P from W.

W_1 = sum of all loads to the left of P.

b = distance of W_1 from P.

It is required to find the positions of the loads which will make the moment at P a maximum.

Moment at $P = R_A(L-a-x) - W_1 b$

But $R_A = \dfrac{Wx}{L}$

$\therefore$ $M = \dfrac{W}{L}.\left(Lx - ax - x^2\right) - W_1 b.$

Differentiating w.r.t. x

$$\frac{dM}{dx} = \frac{W}{L}(L-a-2x)$$

For maximum value $\dfrac{dM}{dx} = 0$

or $L - a - 2x = 0$

or $x = \dfrac{L-a}{2} = \dfrac{L}{2} - \dfrac{a}{2}.$

In words, for maximum B.M. under a particular wheel load, that load and C.G. of the loads should be equidistant from the centre. Some guide lines regarding application of this rule are :

(*a*) Largest possible moment occurs under the load closest to the resultant provided that load is not less than the other adjacent load. When the closest load is less than the further load both positions should be checked.

(*b*) As a further aid, the load placed as per guide line above, must also satisfy the criterion for maximum moment at that point (*i.e.*, the average load to the left should equal the average on the span); if it does not it can be dropped from further consideration.

9.8 EXAMPLES

Example 9.7. A simply supported beam of 20 m span is subjected to a series of loads moving from right to left as given below.

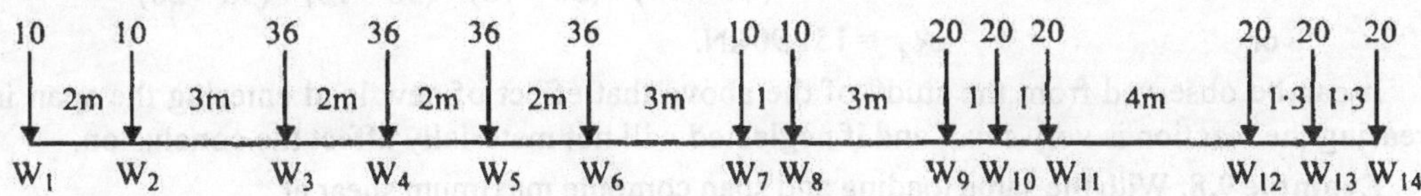

Fig. 9.25. Example 9.7.

Find the maximum reaction R_A.

Solution:

(*a*) ***W_1* on the left support *A*.**

When W_1 is on the left support load upto W_{11} is on the beam. Total load on the beam = 244 kN.

(*b*) When W_2 is on the left support, W_1 is off the span and W_{12} does not enter into the span. Hence load remaining in the structure = 234 kN.

$$\Delta R = \frac{234 \times 2}{20} - 10 = 23.4 - 10 = 13.4 \text{ (increase)}$$

(*c*) When W_3 is on the left support, W_2 is off the span and W_{12} enters from the right through a distance of 1 *m*.

$$\Delta R = \underset{\begin{bmatrix}\text{Existing load}\\ \text{movement}\end{bmatrix}}{\frac{224 \times 3}{20}} + \underset{\begin{bmatrix}\text{New load}\\ W_{12}\end{bmatrix}}{\frac{20 \times 1}{20}} - \underset{\begin{bmatrix}\text{Load off}\\ \text{the span}\end{bmatrix}}{10}$$

$$= 33.6 + 1 - 10 = 24.6$$

(*d*) ***W_4* on the left support**

W_3 is now off the span, W_{13} and W_{14} enters the span as W_4 moves through 2*m*.

$$\Delta R = \frac{208 \times 2}{20} + \underbrace{\frac{20 \times 1.7}{20}}_{\text{New load}} + \underbrace{\frac{20 \times 0.4}{20}}_{\text{New load}} - \underbrace{36}_{\text{Load off the span}}$$

$$= 20.8 + 1.7 + 0.4 - 36 = -13.1 \text{ (decrease).}$$

(*e*) ***W_5* on the support**

W_4 is now off the span.

$$\Delta R = \frac{212 \times 2}{20} - 36$$

$$= 21.2 - 36 = -14.8 \text{ (decrease)}$$

Hence Maximum value of reaction is obtained with W_3 on the left support.

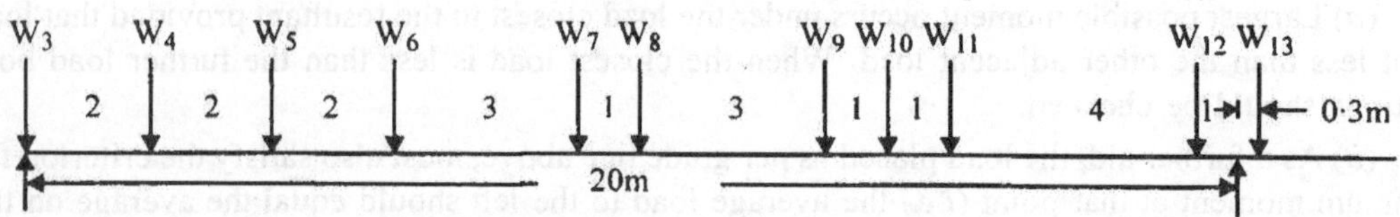

Fig. 9.26. Load W_3 on left support (Ex. 9.7)

With W_3 on the left support

$$R_A \times 20 = (20 \times 1) + (20 \times 5) + (20 \times 6) + (20 \times 7) + (10 \times 10) + (10 \times 11) + (36 \times 14) + (36 \times 16) + (36 \times 18) + (36 \times 20)$$

or $\quad R_A = 151.90 \text{ kN}.$

It can be observed from the study of the above that effect of new load entering the span in increasing the reaction is very small and if neglected will not materially affect the conclusion.

Example 9.8. With the same loading and span compute maximum shear at

(*a*) 6 *m* from left support.

(*b*) 3 *m* from left support.

Solution.

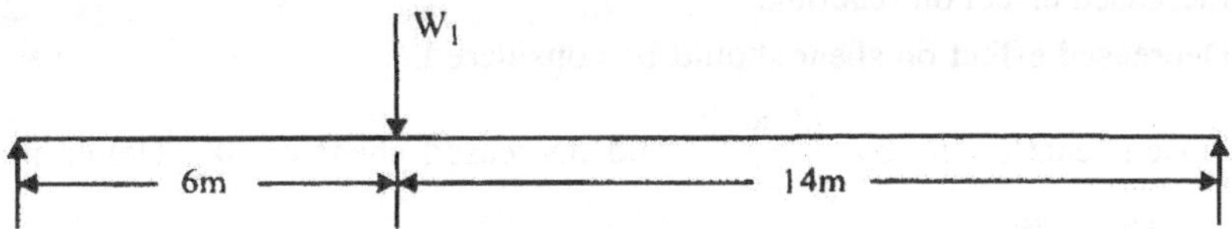

Fig. 9.27. Example 9.8

(*a*) W_1 on the section—Load on the right of section W_1 to W_7. When W_2 is on the section load on the right of section is W_2 to W_8. (distance between W_1 and W_2 is 2 *m*).

Change ΔV from W_1 to W_2 on section.

$$\underbrace{\frac{174\times 2}{20}}_{\text{Movement of original load}} + \underbrace{\frac{10\times 1}{20}}_{\text{Effect of New load } W_s} - \underbrace{10}_{W_1 \text{ left of section}}$$

$$= 17.4 + 0.5 - 10 = 7.9 \text{ (increase)}$$

Change due to W_2 to W_3 on section.

$$\Delta V = \underbrace{\frac{184\times 3}{20}}_{\text{Movement of Existing load}} + \underbrace{\frac{20\times 1}{20}}_{\text{Due to new load } W_9} + \underbrace{\frac{20\times 0}{20}}_{\text{Due to New load } W_{10}} - \underbrace{10}_{\text{Load } W_{10} \text{ left of section}}$$

$$= +18.6 \text{ (an increase)}$$

Change due to W_3 to W_4.

Load on right of section W_4 to W_{11}.

$$\text{Change } \Delta V = \underbrace{\frac{214\times 2}{20}}_{\text{Movement of Existing load}} + \underbrace{\frac{20\times 10}{20}}_{\text{New load } W_{11}} - \underbrace{36}_{W_3 \text{ leftof section}} + \underbrace{\frac{10\times 1}{20}}_{W_1 \text{ off the span}}$$

$$= 21.4 + 1 - 36 + 0.5 = -13.1 \text{ (decrease).}$$

Therefore, for maximum positive shear W_3 should be on section.

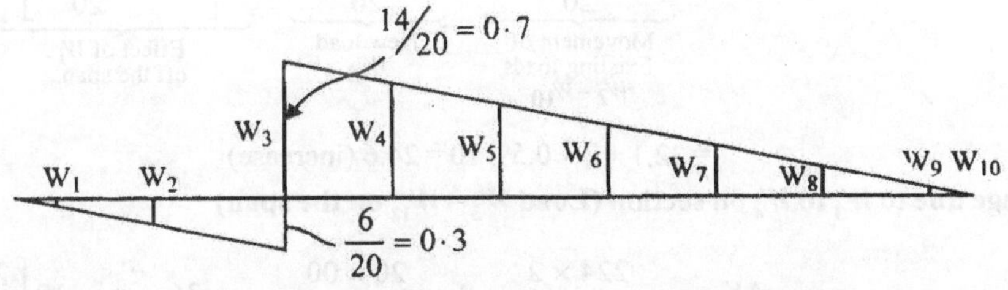

Fig. 9.28. Example 9.8 — Maximum positive shear.

$$\text{Maximum positive shear} = \frac{36\times 14}{20} + \frac{36\times 12}{20} + \frac{36\times 10}{20} + \frac{36\times 8}{20} + \frac{10\times 5}{20} + \frac{10\times 4}{20}$$

$$+ \frac{20\times 1}{20} + \frac{20\times 0}{20} - \frac{10\times 3}{20} - \frac{10\times 1}{20}$$

$$= \mathbf{84.7 - 2.00 = 82.7\ kN.}$$

When load goes off the section its

(i) Increased effect on reaction.

(ii) Decreased effect on shear should be considered.

W_1 increased reaction R_A by $\dfrac{W_1 \times 19}{20}$ and decreased shear by W_1. Hence net effect was decrease $W_1\left(1-\dfrac{19}{20}\right)=\dfrac{W_1}{20}$. This should be added when W_1 goes off the span.

Example 9.9. Same problem when the section is 3 *m* from left.

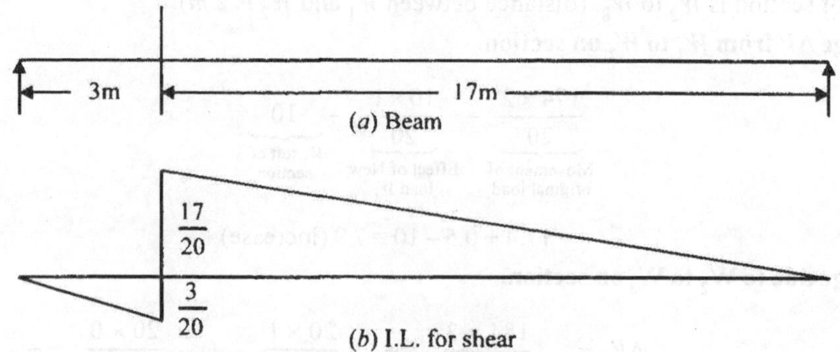

Fig. 9.29. Example 9.9

Solution:

Change due to wheel 1 to wheel 2 on section.

$$\Delta V = \underbrace{\frac{184 \times 2}{20}}_{\substack{\text{Movement of} \\ \text{Existing loads} \\ W_1 - W_8}} + \underbrace{\frac{20 \times 1}{20}}_{W_9} + \underbrace{\frac{20 \times 0}{20}}_{W_{10}} - \underbrace{10}_{W_1}$$

$$= 18.4 + 1 - 10 = 9.4 \text{ (increase).}$$

Change due to W_2 to W_3 on section

$$\Delta V = \underbrace{\frac{214 \times 3}{20}}_{\substack{\text{Movement of} \\ \text{Existing loads} \\ W_2 - W_{10}}} + \underbrace{\frac{20 \times 2}{20}}_{\substack{\text{New load} \\ W_{11}}} + \underbrace{\left[10 - \frac{19 \times 10}{20}\right]}_{\substack{\text{Effect of } W_1 \\ \text{off the span.}}} - \underbrace{10}_{W_2}$$

$$= 32.1 + 2 + 0.5 - 10 = 24.6 \text{ (increase)}$$

Change due to W_3 to W_4 on section (Load $W_3 - W_{12}$ on the span)

$$\Delta V = \underbrace{\frac{224 \times 2}{20}}_{\substack{\text{Movement of} \\ \text{Existing loads} \\ W_3 - W_{11}}} + \underbrace{\frac{20 \times 00}{20}}_{\substack{\text{New load} \\ W_{12}}} - \underbrace{36}_{W_3} + \underbrace{10\left[\frac{20}{20} - 1\right]}_{\substack{W_2 \\ \text{off the span.}}}$$

$$= 22.4 + 0 - 36 + 0 = -13.6 \text{ (decrease).}$$

Hence for maximum shear wheel load 3 should be placed at section.

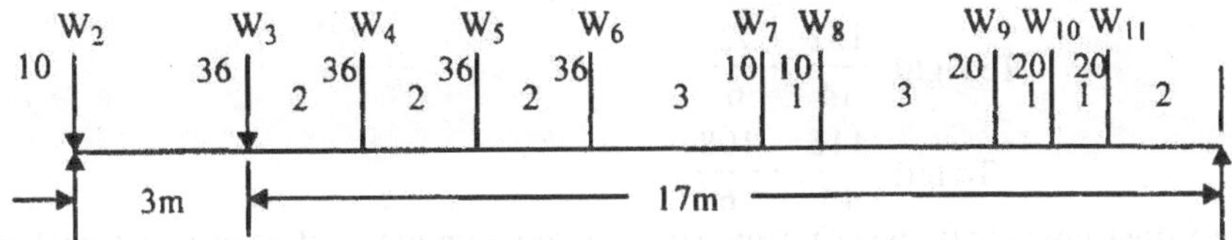

Fig. 9.30. Calculation for maximum shear.

$$R_A \times 20 = 10 \times 20 + 36 \times 17 + 36 \times 15 + 36 \times 13 + 36 \times 11 + 10 \times 8 + 10 \times 7 + 20 \times 4 + 20 \times 3 + 20 \times 2$$

or $$R_A = 127.3$$

$$\text{S.F.} = 127.3 - 10 = \mathbf{117.3\ kN.}$$

Example 9.10. With the same span and loading find maximum B.M. at 6 *m* from left end.

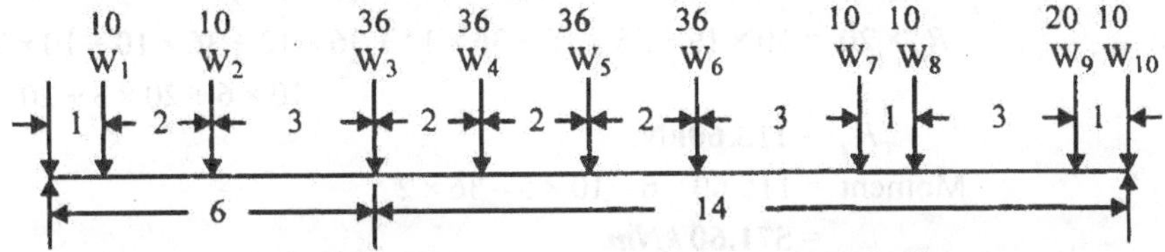

Fig. 9.31. Wheel 3 on section (Example 9.10)

Solution:

For maximum bending moment as many loads as possible should be in the span. With W_3 on the section wheel loads upto W_{10} will be in the span.

Wheel 3 To right $\frac{184}{14} > \frac{20}{6}$

To left $\frac{148}{14} > \frac{56}{6}$

Therefore wheel 3 does not satisfy the criterion.

Wheel 4.

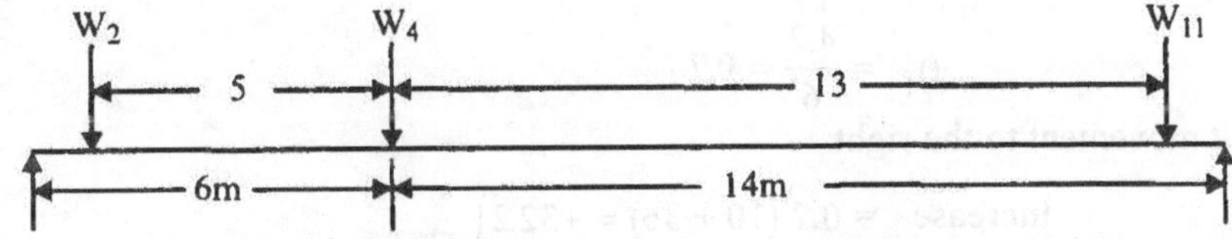

Fig. 9.32. Wheel 4 on section (Example 9.10)

To right $\frac{188}{14} > \frac{46}{6}$

To left $\frac{152}{14} < \frac{82}{6}$

Wheel 4 does satisfy the criterion.

Wheel 5.

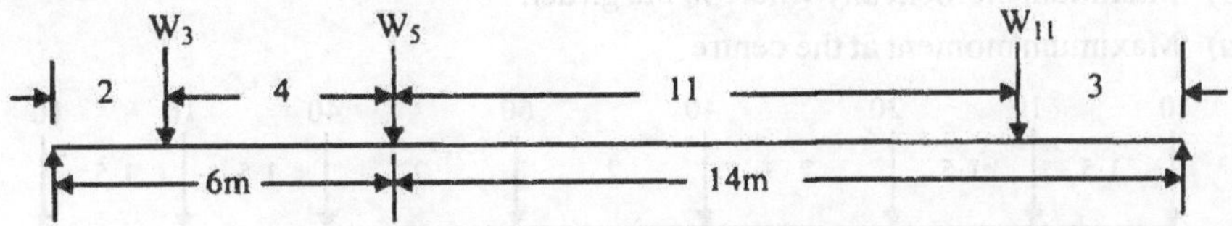

Fig. 9.33. Wheel 5 on section (Example 9.10)

$$\text{To right} \quad \frac{152}{14} > \frac{72}{6}$$

$$\text{To left} \quad \frac{116}{14} > \frac{108}{6}$$

Wheel 5 does not satisfy the criterion. Hence maximum moment occurs under wheel load 4.

Calculation of moment.

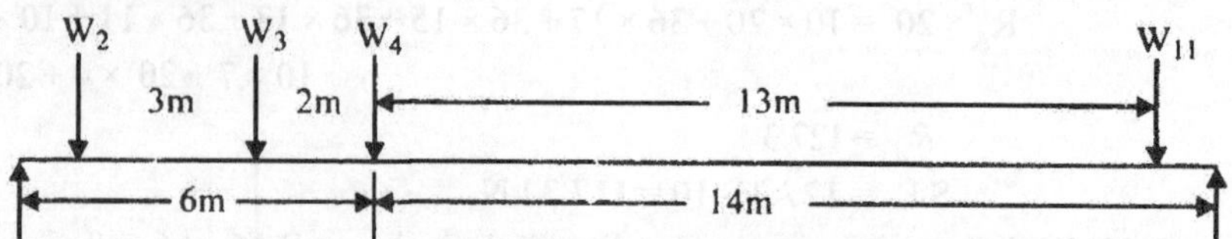

Fig. 9.34. Calculation of moment (Example 9.10)

$$R_A \times 20 = 10 \times 19 + 36 \times 16 + 36 \times 14 + 36 \times 12 + 36 \times 10 + 10 \times 7 + 10 \times 6 + 20 \times 3 + 20 \times 2 + 20 \times 1$$

or $$R_A = \mathbf{115.60 kN.}$$

$$\text{Moment} = 115.60 \times 6 - 10 \times 5 - 36 \times 2$$

$$= \mathbf{571.60\ \mathit{kNm}.}$$

Check

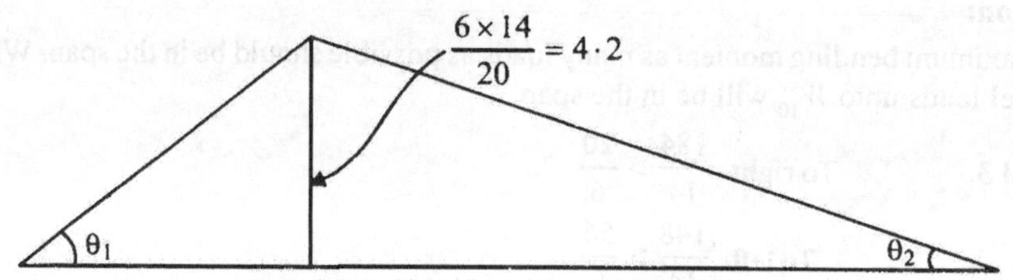

Fig. 9.35. Checking from I.L. for B.M.

$$\theta_2 = \frac{4.2}{14} = 0.3$$

$$\theta_1 = \frac{4.2}{6} = 0.7$$

For unit movement to the right

$$\left.\begin{array}{l} \text{Increase} = 0.7\,(10 + 36) = +32.2 \\ \text{Decrease} = 0.3\,(188) = -56.4 \end{array}\right\} \text{Decrease}$$

For unit movement to the left.

$$\left.\begin{array}{l} \text{Decrease} = 0.7\,(10 + 36 + 36) = -57.4 \\ \text{Increase} = 0.3\,(152) = +45.6 \end{array}\right\} \text{Decrease}$$

Example 9.11. For a span of 17 *m*, calculate the following due to the loading shown.

(*a*) Maximum moment at the centre.

(*b*) Maximum moment any where in the girder.

(*a*) Maximum moment at the centre

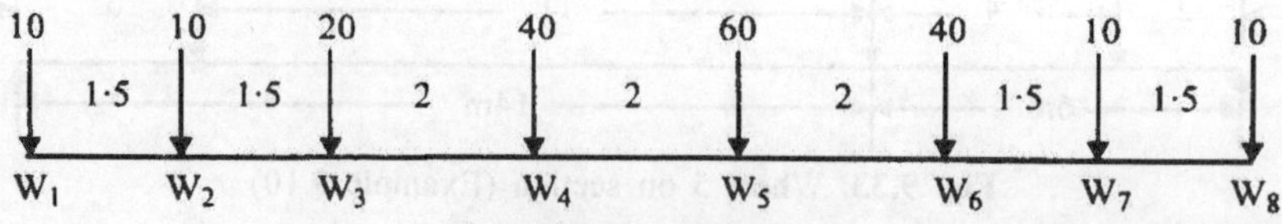

Fig. 9.36. Example 9.11

The maximum moment will occur with W_4 or W_5 over the section.

with W_4 just on the left of the section.

$$\frac{80}{8.5} < \frac{120}{8.5}$$

with W_4 on the right

$$\frac{40}{8.5} < \frac{160}{8.5}$$

The criterion for maximum moment is not satisfied.

with W_5 on the left of the section.

$$\frac{140}{8.5} > \frac{60}{8.5}$$

with W_5 on the right of the section.

$$\frac{80}{8.5} < \frac{120}{8.5}$$

Hence the criterion for maximum moment is satisfied. Hence for maximum moment W_5 should be on the section.

$$R_B \times 17 = 10 \times 1.5 + 10 \times 3.0 + 20 \times 4.5 + 40 \times 6.5 + 60 \times 8.5 + 40 \times 10.5 + 10 \times 12.0 + 10 \times 13.5$$

or $\quad R_B = \mathbf{92.94\ kN}$

$$\text{Moment at centre} = 92.94 \times 8.5 - 40 \times 2 - 10 \times 3.5 - 10 \times 5 = \mathbf{625\ kNm.}$$

(*b*) Maximum B.M. any where in the girder.

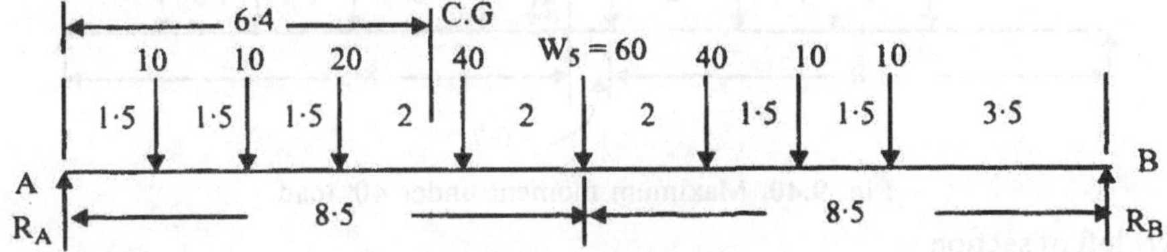

Fig. 9.37. Calculation of Maximum Moment with W_5 on the section.

Distance of C.G. from left end

$$\frac{(10 \times 1.5) + (20 \times 3) + (40 \times 5) + (60 \times 7) + (40 \times 9) + (10 \times 10.5) + (10 \times 12)}{10 + 10 + 20 + 40 + 60 + 40 + 10 + 10}$$

$$= \frac{1280}{200} = 6.4$$

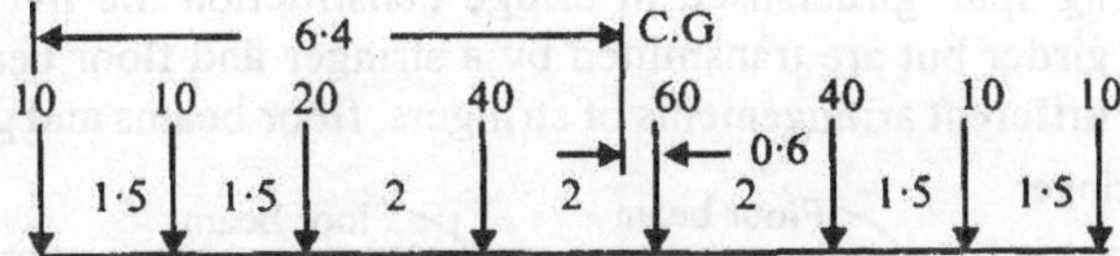

Fig. 9.38 C.G. from left end.

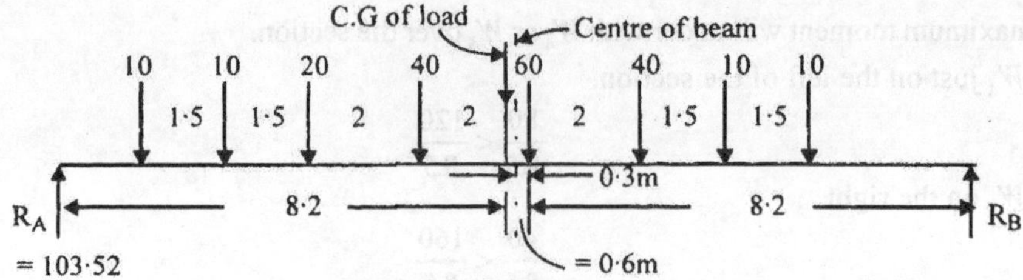

Fig. 9.39. Maximum moment any where in the girder.

$$R_A = \frac{200 \times 8.8}{17} = 103.52$$

Moment under 60ᵗ load $= 103.52 \times 8.8 - (10 \times 7) - (10 \times 5.5) - (20 \times 4) - (40 \times 2)$

$= 625.976 \text{ kNm.}$

Check

$$\frac{140}{8.8} > \frac{60}{8.7}$$

$$\frac{80}{88} < \frac{120}{8.2}$$

Hence maximum criterion of B.M. at a section is satisfied.

Check for maximum under 40ᵗ load.

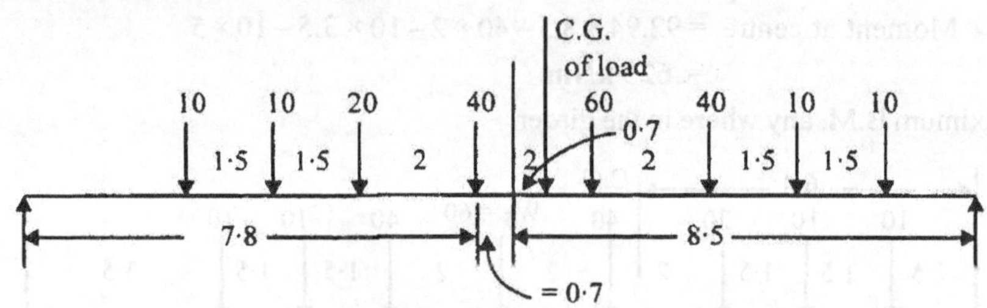

Fig. 9.40. Maximum moment under 40ᵗ load

40 kN left of section

$$\frac{80}{7.8} < \frac{120}{9.2}$$

40 kN right of section

$$\frac{40}{7.8} < \frac{160}{9.2}$$

Hence maximum criterion is not satisfied. The difference between maximum moment at mid span and that any where in the grider is almost negligible.

9.9. GIRDERS WITH FLOOR BEAMS

In the case of long span girder used in bridge construction the live loads are usually not applied directly to the girder but are transmitted by a stringer and floor beam system to the main girder. Fig. 9.39 shows different arrangements of stringers, floor beams and girders.

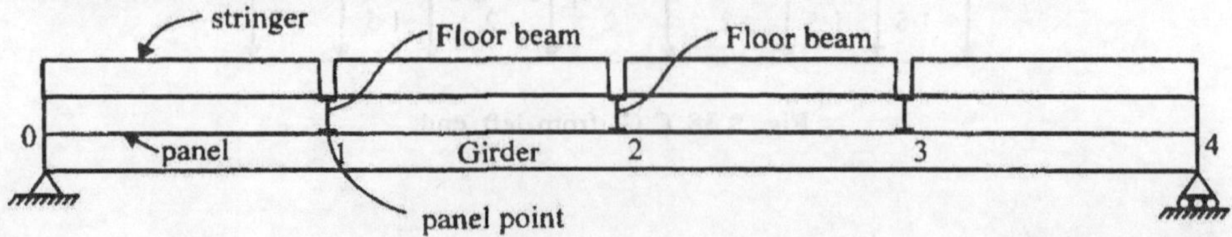

(a)

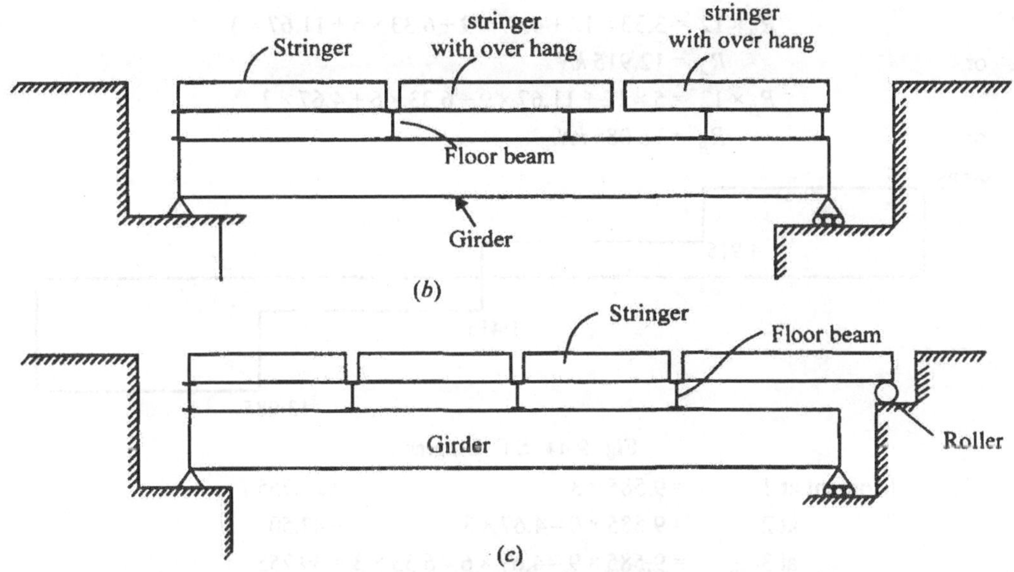

Fig. 9.41. Girders with floorbeams.

Transmission of load is (*i*) from the floor slab to stringer (*ii*) from the stringer to the floor beam (*iii*) from the floor beam to the girder at panel points.

9.10. STATICAL ANALYSIS OF A GRIDER WITH FLOOR BEAM

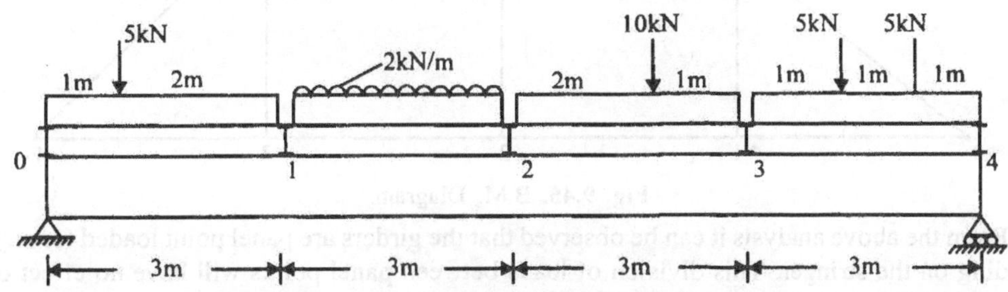

Fig. 9.42. Statical Analysis of girder with floor beams.

Assuming that stringers are simply supported over the floor beams loads transferred to the panel points are

Panel point

0. $\dfrac{5 \times 2}{3} = 3.33 \text{ kN}$

1. $\dfrac{5 \times 1}{3} + \dfrac{2 \times 3}{2} = 4.67 \text{ kN}$

2. $\dfrac{2 \times 3}{2} + \dfrac{10 \times 1}{3} = 6.33 \text{ kN}$

3. $\dfrac{10 \times 2}{3} + \dfrac{5 \times 2}{3} + \dfrac{5 \times 1}{3} = 11.67 \text{ kN}$

4. $\dfrac{5 \times 1}{3} + \dfrac{5 \times 2}{3} = 5 \text{ kN}$

3·33kN 4·67kN 6·33kN 11·67kN 5kN

4 @ 3m = 12m

R_A R_B

Fig. 9.43. Calculations of reactions

$$R_A \times 12 = 3.33 \times 12 + 4.67 \times 9 + 6.33 \times 6 + 11.67 \times 3$$

or $$R_A = 12.915\ kN$$

$$R_B \times 12 = 5 \times 12 + 11.67 \times 9 + 6.33 \times 6 + 4.67 \times 3$$

or $$R_B = 18.085\ kN.$$

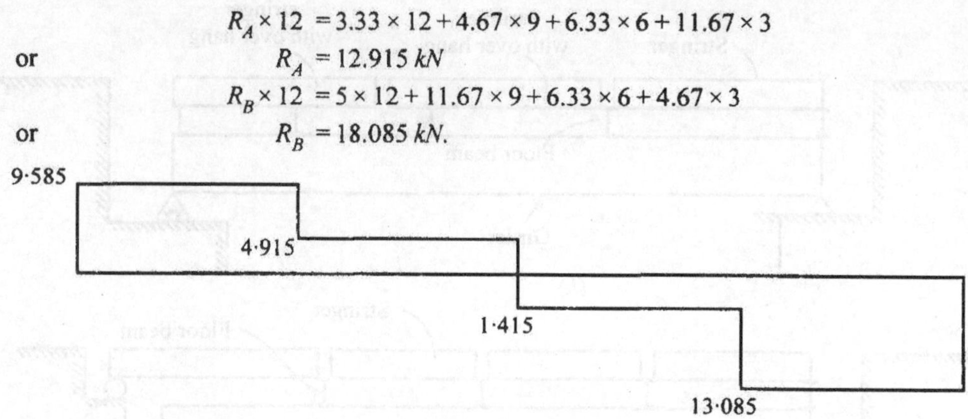

Fig. 9.44. S.F. Diagram.

Moment at 1 $= 9.585 \times 3 = 28.755$

at 2 $= 9.585 \times 6 - 4.67 \times 3 = 43.50$

at 3 $= 9.585 \times 9 - 4.67 \times 6 - 6.33 \times 3 = 39.255$

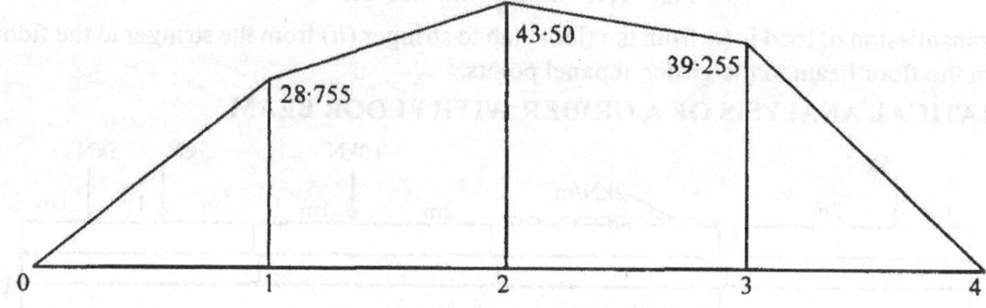

Fig. 9.45. B.M. Diagram.

From the above analysis it can be observed that the girders are panel point loaded for any type of loading on the stringer. This division of loads between panel points will have no effect on the reactions at A and B. However, as a result, the internal shear at a panel becomes constant. Hence it is appropriate to say panel shear instead of shear at a particular section.

9.11. INFLUENCE LINES FOR GIRDERS WITH FLOOR BEAMS

Fig. 9.46 gives the I.L. for different functions for girders with floorbeams. The figures can be explained as follows :

Fig. 9.46 (*b*) and (*c*). The end reactions R_A and R_B are not affected by the presence of the floo system. Hence they will be same as the reactions of a simply supported grider.

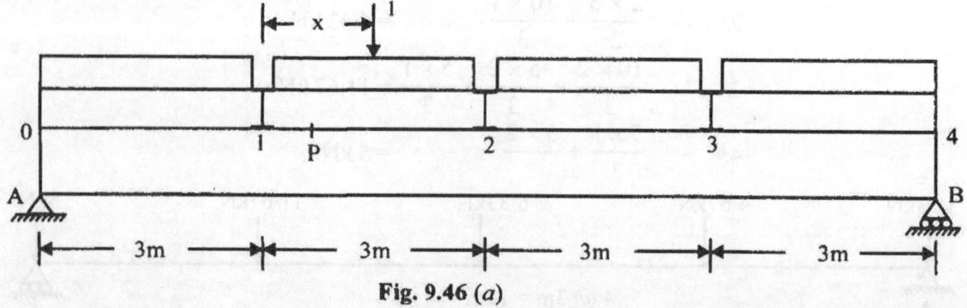

Fig. 9.46 (*a*)

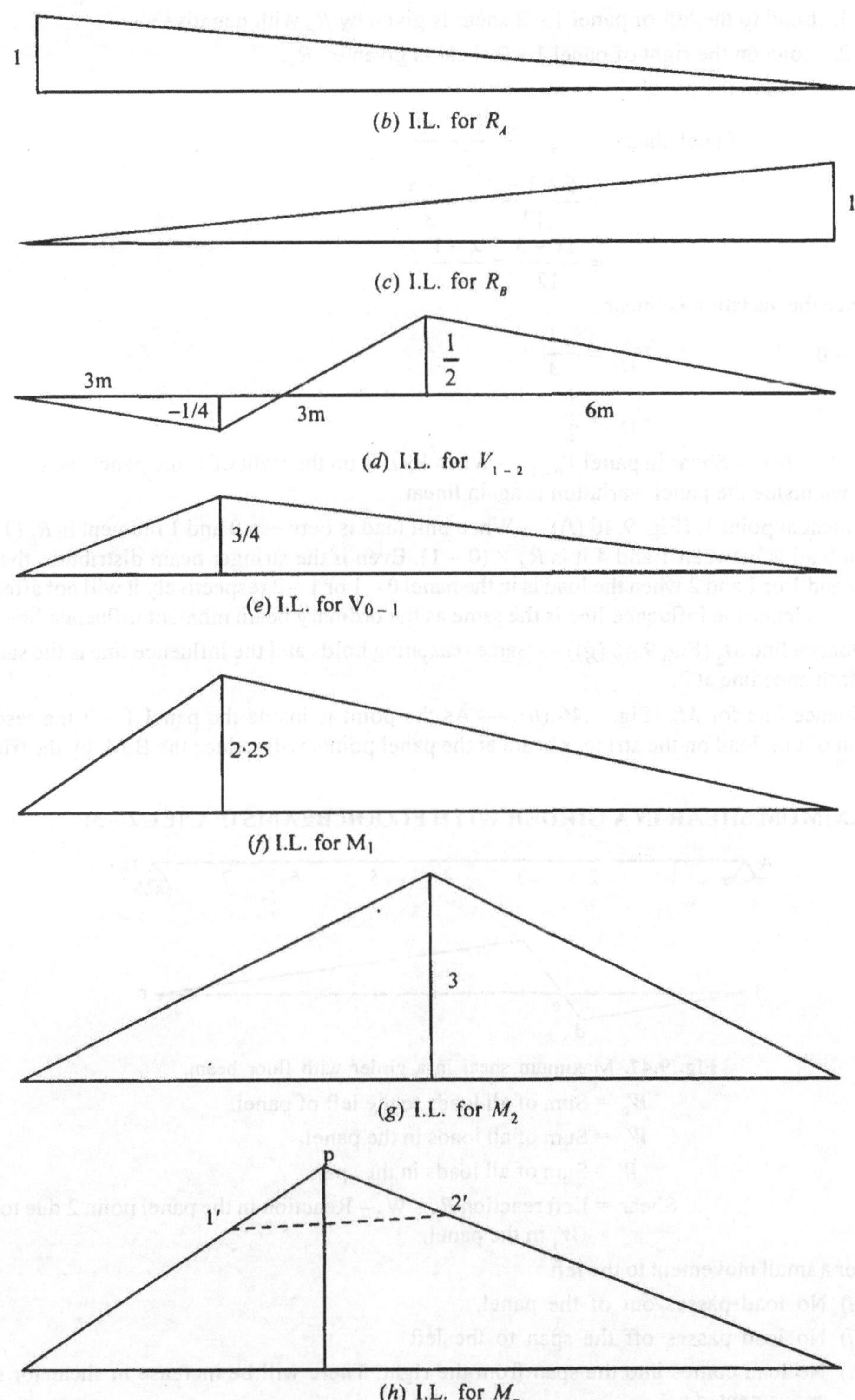

Fig. 9.46. I.L. for girder with floor beam.

Fig. 9.46 (*d*) — Influence line for shear in the panel should be drawn in three stages :

1. Load to the left of panel 1 – 2 shear is given by R_B with negative sign.
2. Load on the right of panel 1 – 2 shear is given by R_A.
3. Load in the panel.

$$\text{Panel shear } = R_A - \frac{1 \times (3 - x)}{3}$$

$$= \frac{6 + 3 - x}{12} - \frac{3 - x}{3}$$

$$= \frac{3x - 3}{12} = \frac{x - 1}{4}$$

Hence the variation is linear.

at $x = 0$ $\quad V_{12} = -\frac{1}{4}$

at $x = 3$ $\quad V_{12} = \frac{1}{2}$.

Fig. 9.46 (*e*) — Shear in panel V_{0-1} — When load is on the right of 1, the panel shear is given by R_A. When inside the panel, variation is again linear.

Moment at point 1. (Fig. 9.46 (*f*)) — When unit load is between 0 and 1 moment is $R_B(1 - 4)$. When unit load is between 1 and 4 it is $R_A \times (0 - 1)$. Even if the stringer beam distributes the load between 0 and 1 or 1 and 2 when the load is in the panel 0 – 1 or 1 – 2 respectively it will not affect the moment at 1. Hence the influence line is the same as the ordinary beam moment influence line.

Influence line M_2 (Fig. 9.46 (g)) — Same reasoning holds and the influence line is the same as Moment Influence line at 2.

Influence line for M_p (Fig. 9.46 (*h*)) — As the point is inside the panel 1 – 2 the resolved component of unit load on the stringer beam at the panel points will reduce the B.M. by the triangle *p*1′2′.

9.12. MAXIMUM SHEAR IN A GIRDER WITH FLOOR BEAMS (PANEL 2 – 3)

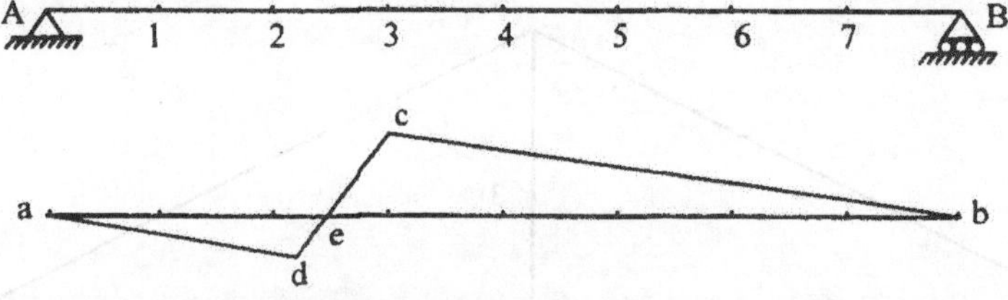

Fig. 9.47. Maximum shear in a girder with floor beam.

If $\quad W_0$ = Sum of all loads to the left of panel.

W_1 = Sum of all loads in the panel.

W = Sum of all loads in the span.

Shear = Left reaction R_A – W_0 – Reaction in the panel point 2 due to load W_1 in the panel.

If for a small movement to the left

(*i*) No load passes out of the panel.

(*ii*) No load passes off the span to the left

(*iii*) No load comes into the span from the right. There will be increase in shear for small movement *dx*,

$$\Delta V = \underbrace{\frac{W.dx}{L}}_{\substack{\text{Increase in} \\ \text{Reaction}}} - \underbrace{\frac{W_1.dx}{P}}_{\substack{\text{Decrease due to} \\ \text{increase of reaction} \\ \text{at panel point 2.}}}$$

The shear will have a maximum value when for small movement change of shear is zero.

or $$\frac{Wdx}{L} = \frac{W_1dx}{P}$$

or $$\frac{W}{L} = \frac{W_1}{P}$$

This expression states that shear in a panel is maximum when the average load in the panel is equal to average load in the span. This can only occur as a load passes from an infinetesimal distance to the right of the panel point to an infinetesimal distance to the left of the panel point i.e. only when load is at a panel point.

Example 9.12. Find the maximum shear in panel 2 – 3 in the girder given in Fig. 9.47 with the loading of Example 9.7. (Span 8 @ 5 = 40m)

Solution:

Wheel 1

To the right of panel point 3, $\frac{264}{40} > \frac{0}{5}$

To the left of panel point 3, $\frac{264}{40} > \frac{10}{5}$

Wheel 1 déos not satisfy the criterion.

Wheel 2

To the right of panel point 3, $\frac{284}{40} > \frac{10}{5}$

To the left of panel point 2, $\frac{284}{40} > \frac{20}{5}$

Wheel 3

To the right of panel point 3, $\frac{284}{40} > \frac{20}{5}$

To the left of panel point 3, $\frac{284}{40} < \frac{46}{5}$

Hence wheel 3 satisfies the criterion and it should be placed over point 3.

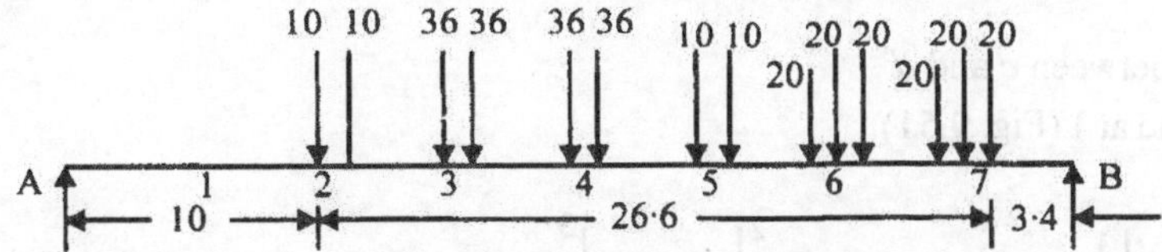

Fig. 9.48. Wheel 3 over point 3

$$R_A \times 40 = (10 \times 30) + (10 \times 28) + (36 \times 25) + (36 \times 23) + (36 \times 21) + (36 \times 19)$$
$$+ (10 \times 16) + (10 \times 15) + (20 \times 12) + (20 \times 11) + (20 \times 10)$$
$$+ (20 \times 60) + (20 \times 4.7) + (20 \times 3.4)$$

or $$R_A = \frac{5000}{40} = \mathbf{125\ kN.}$$

Reaction of stringer beam at 2

$$= \frac{10 \times 5}{5} + \frac{10 \times 3}{5} + \frac{36 \times 10}{5}$$

$$= 10 + 6 + 0 = 16$$

Maximum panel shear $= 125 - 16 = 109$

Check

With the next 36 kN on points 3

$$R_4 \times 40 = (10 \times 32) + (10 \times 30) + (36 \times 27) + (36 \times 25) + (36 \times 23) + (36 \times 21) + (10 \times 18) + (10 \times 17) + (20 \times 14) + (20 \times 13) + (20 \times 12) + (20 \times 8.0) + (20 \times 6.7) + (20 \times 5.4)$$

or $\quad R_4 = 140.2$ **kN.**

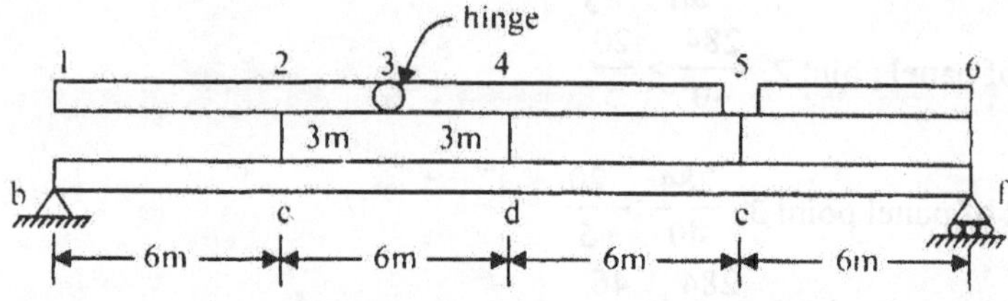

Fig. 9.49. Wheel 4 over point 3.

$$\text{Shear} = 140.2 - 10 - \left[\frac{36 \times 0}{5} + \frac{36 \times 2}{5} + \frac{10 \times 5}{5}\right]$$

$$= 140.2 - 10 - 24.4$$

$$= 105.8\ kN < 109\ kN.$$

Example 9.13. Draw the following influence lines for the girder (*a*) the shear between *c* and *d*. and (*b*) the bending moment at *d*.

Fig. 9.50. Example 9.13

Solution:

(*a*) Shear between *c* and *d*

(*i*) Unit load at 1 (Fig. 9.51)

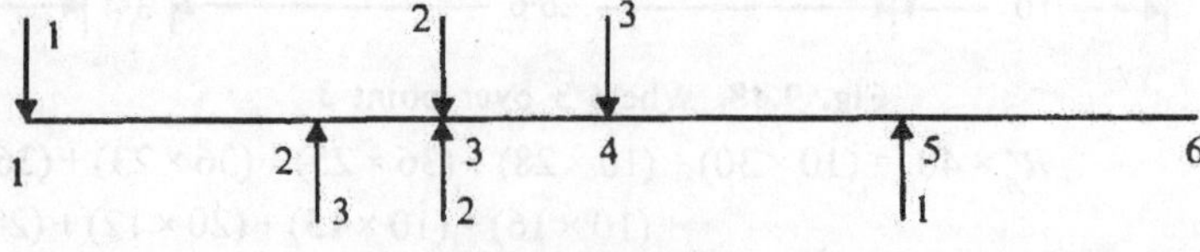

(*i*) Wheel load at 1.

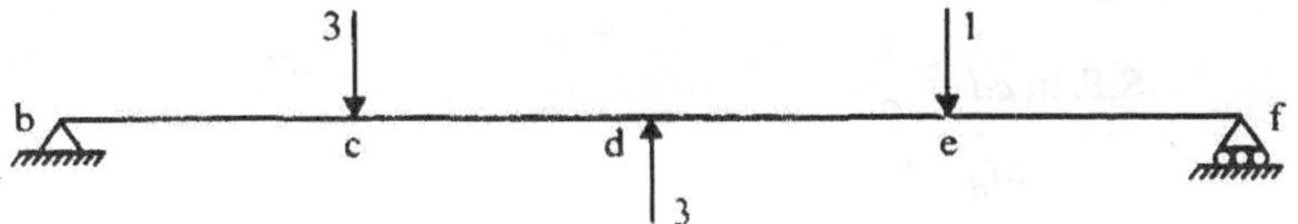

(*ii*) Forces in the girder due to wheel load at 1.

Fig. 9.51. Unit load at 1 (Example 9.13)

$$R_2 \times 3 - 1 \times 9 = 0$$

$$R_2 = 3 \uparrow$$

$$2 \times 9 - R_4 \times 6 = 0$$

$$R_4 = 3 \downarrow$$

Taking moments about *f*.

$$R_b \times 24 - 3 \times 18 + 3 \times 12 - 1 \times 6 = 0$$

or $$R_b = 1$$

S.F. in cd = –2 $$M_d = 1 \times 12 - 3 \times 6 = -6.$$

(*ii*) Unit load at 2 (Fig. 9.52)

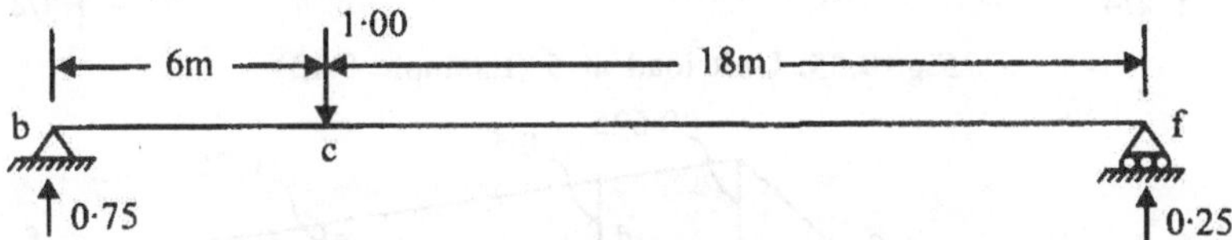

Fig. 9.52. Unit load at 2 (Example 9.13)

Taking moments about *f*.

$$R_b \times 24 - 1 \times 18 = 0$$

or $$R_b = \frac{18}{24} = 0.75$$

S.F. in *cd* = –0.25

$$M_d = 0.75 \times 12 - 1 \times 6 = 3.$$

Unit load to the right of 3 (Fig. 9.53).

Taking moments about 5

$$1 \times 9 - R_d \times 6 = 0$$

$$R_d = 1.5$$

$$R_b \times 24 - 1.5 \times 12 + 0.5 \times 6 = 0$$

or $$R_b = \frac{1}{24}(18 - 3) = \frac{5}{8}$$

S.F. in *cd* = 0.625

$$M_d = 7.5$$

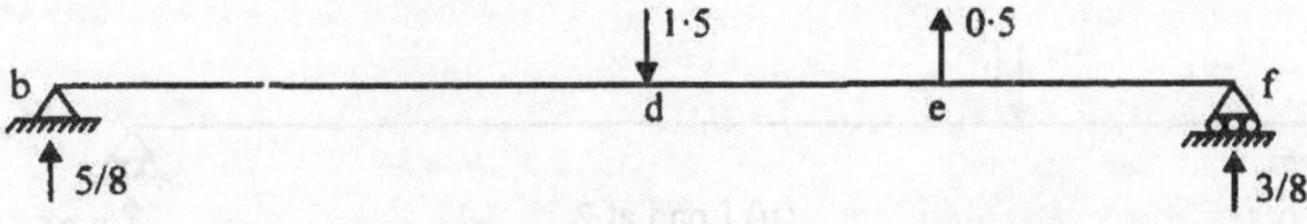

Fig. 9.53. Unit load to the right of point 3 (Example 9.13)

Unit load at 4. (Fig. 9.54)

$$\text{S.F. in } cd = \frac{1}{2}$$

$$M_d = 6.$$

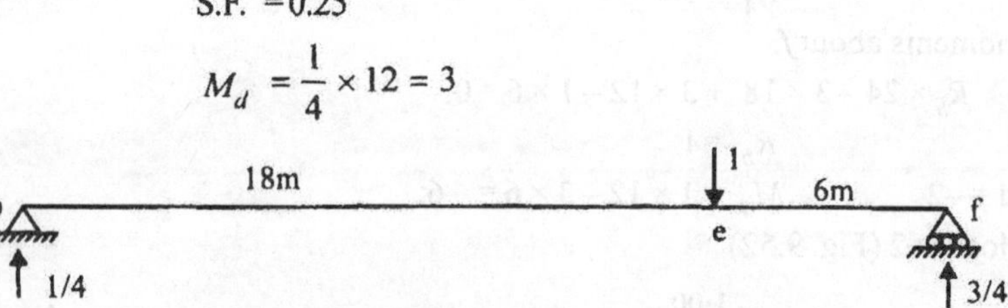

Fig. 9.54. Unit load at 4 (Example 9.13)

Unit load at 5.

$$\text{S.F.} = 0.25$$

$$M_d = \frac{1}{4} \times 12 = 3$$

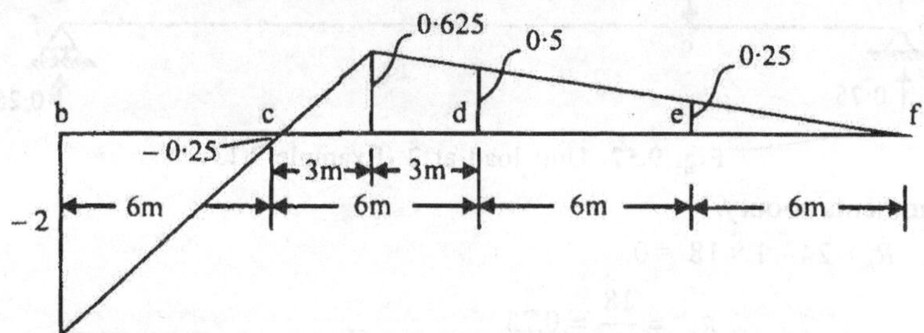

Fig. 9.55. Unit load at 5 (Example 9.13)

Fig. 9.56. I.L. for shear force in panel *cd* (Example 9.13)

Influence line for M_d

When load is at 1. Fig. 9.57 (*i*)

$$R_b = 1$$

$$M_d = 1 \times 12 - 3 \times 6$$

$$= -6.$$

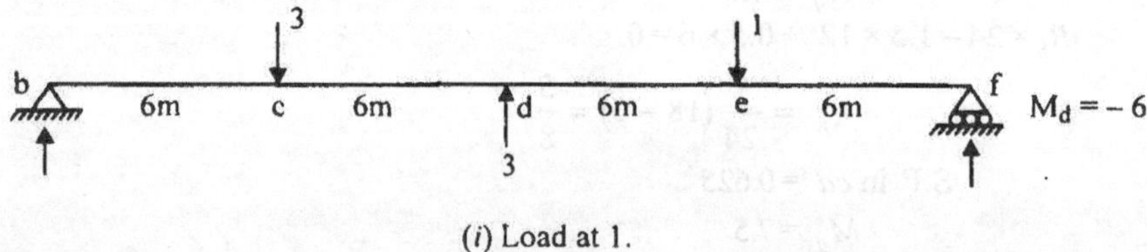

(*i*) Load at 1.

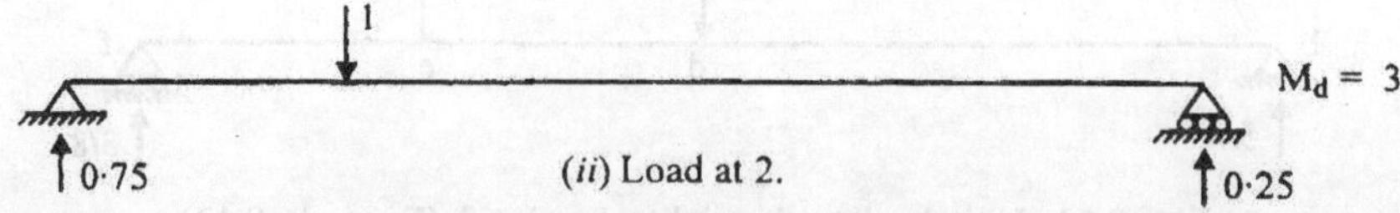

(*ii*) Load at 2.

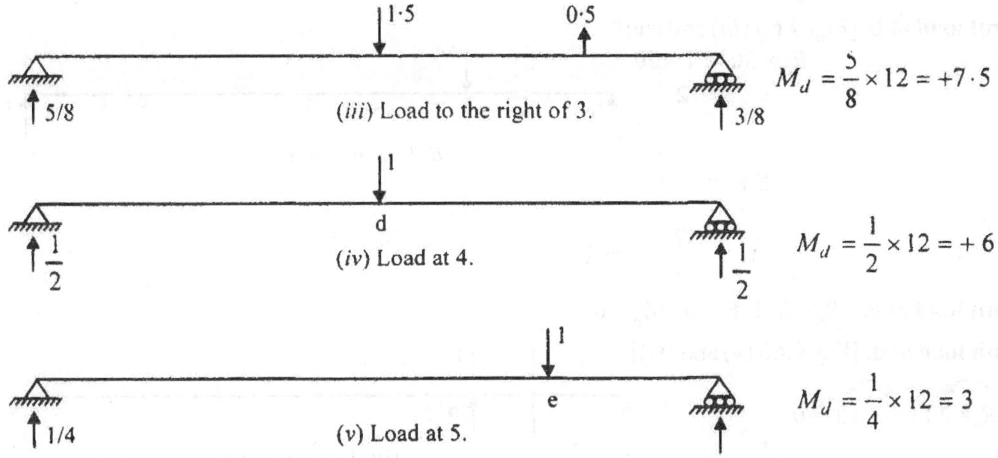

Fig. 9.57. Reactions of Girder with different load positions.

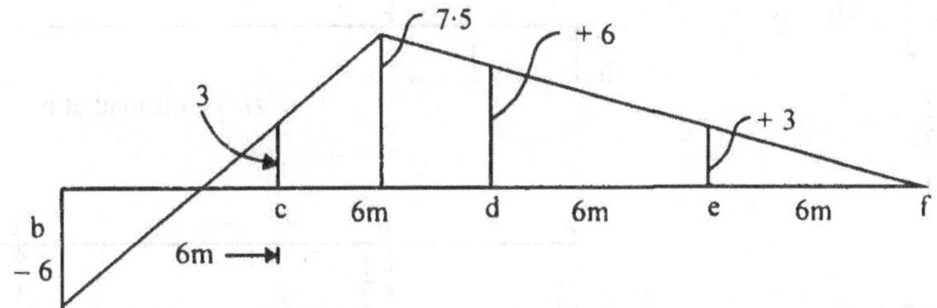

Fig. 9.58. I.L. for moment at *d*.

Example 9.14. Draw influence lines for the shear and moment at 'g' in the girder.

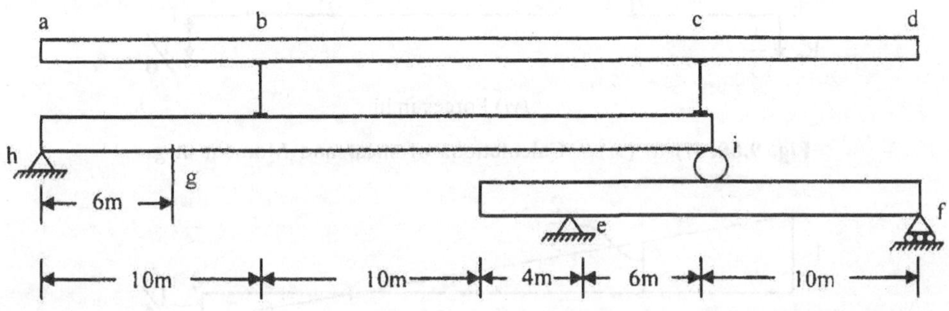

Fig. 9.59. Example 9.14

Solution:

Unit load at *a*. [Fig. 9.60 (*i*) and (*ii*)]

Taking moments about C

$$1 \times 30 - R_b \times 20 = 0$$

or

$$R_b = 1.5$$

$$R_h \times 30 = 1.5 \times 20$$

$$R_h = 1$$

$$\text{Shear at g} = 1$$

$$M_g = +6$$

1 a b c d 10m 20m 10m

(*i*) Unit load at a.

Unit load at b. [Fig. 9.60 (*iii*) and (*iv*)]

$$R_h \times 30 = 1 \times 20$$

$$R_h = \frac{2}{3}$$

$$\text{S.F.} = +\frac{2}{3}$$

$$M_g = \frac{2}{3} \times 6 = 4$$

(*ii*) Loading on hi

Unit load at c. $R_h = 0$ S.F. = 0 $M_g = 0$

Unit load at d. [Fig. 9.60 (*v*) and (*vi*)]

$$R_b \times 20 + 1 + 10 = 0$$

or $$R_b = -\frac{1}{2}$$

$$R_h \times 30 + \frac{1}{2} \times 20 = 0$$

$$R_h = -\frac{1}{3}$$

$$\text{S.F.} = -\frac{1}{3}$$

$$M_g = \frac{1}{3} \times 6 = -2.$$

(*iii*) Unit load at b

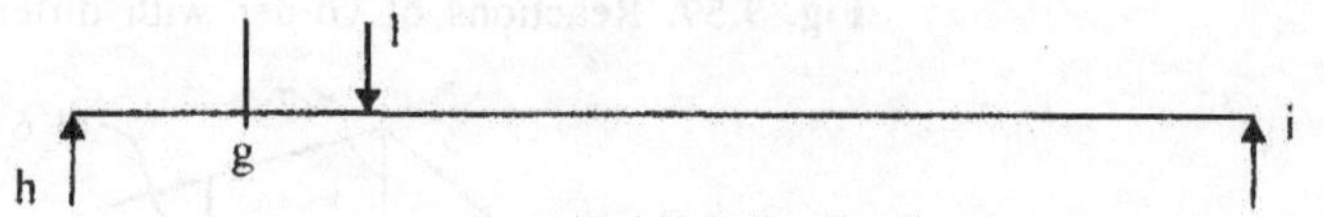

(*iv*) Unit load at b

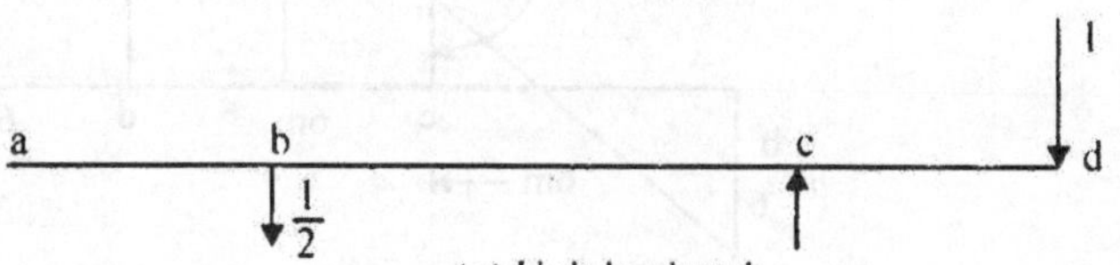

(*v*) Unit load at d.

(*vi*) Forces in hi.

Fig. 9.60. (*i*) to (*vi*) – Calculations of shear and Moment at *g*.

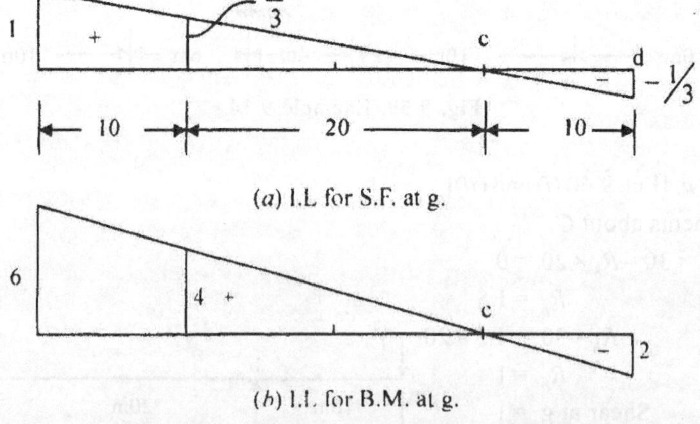

(*a*) I.L for S.F. at g.

(*b*) I.L. for B.M. at g.

Fig. 9.61. Influence lines for S.F. and B.M.

9.13. INFLUENCE LINE FOR TRUSS

The influence line diagram for a truss is particularly important since trusses are commonly used as bridge structure which must be analysed for moving loads. It has already been explained in chapter three how load is transmitted to the roadway slab and then to stringers, which in turn deliver their reactions to floor beams. The floor beams will transmit the loads to the joints or panel points of the trusses where they are attached. As the stringers are considered to be simply supported spanning between the panel points the influence lines for the truss will be linear functions between these points. This was also shown for the girder-floor-beam system. Method of section is usually very convenient for truss analysis. For this the whole truss is divided into three zones:

1. Portion of the truss to the left of the panel where the member is situated.
2. Portion of the truss to the right of the panel point.
3. Portion inside the panel point.

Hence the steps for drawing influence lines for trusses are :

1. Express the reactions as a function of the position of the unit load.
2. When the load is on the left of the panel containing the truss, consider the right portion of the section.
3. When the load is on the right of the panel consider the left part.
4. The influence line when the unit load is inside the panel is obtained by joining by means of a straight line the influence line ordinates at the two end panel points.

Example 9.15. Draw the influence line diagram for members 4, 5 and 6 of the truss shown. Assume that load moves along the bottom chord.

Solution:

Step 1. The reactions R_1 and R_5 are obtained interms of the x position of the load as

$$R_1 = \frac{30-x}{30} \qquad R_5 = \frac{x}{30}.$$

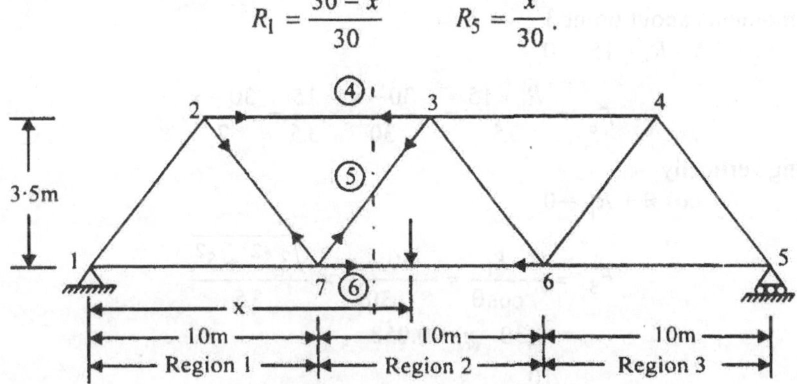

Fig. 9.62. Example 9.15.

Step 2. The structure is divided into three regions as shown.

Step 3. With unit load at region 1, forces are found considering the right hand portion. Taking moments of all forces about point 7.

$$F_4 \times 3.5 + R_5 \times 20 = 0$$

or

$$F_4 = -\frac{R_5 \times 20}{3.5} = -\frac{x}{30} \times \frac{20}{3.5}$$

At $x = 0$ $F_4 = 0$, at $x = 10\,m$ $F_4 = -\dfrac{10 \times 20}{30 \times 3.5} = -1.90$ (comp.)

Taking moments of all forces cut about point 3.

$$F_6 \times 3.5 - R_5 \times 15 = 0$$

or
$$F_6 = \frac{R_5 \times 15}{3.5} = \frac{x}{30} \times \frac{15}{3.5} = \frac{x}{7.0}$$

Resolving vertically

$F_5 \cos\theta = R_5$

or
$$F_5 = \frac{R_5}{\cos\theta} = \frac{x}{30}\,\frac{\sqrt{3.5^2 + 5.0^2}}{3.5}$$
$$= \mathbf{0.058\,x}$$

At $x = 0$ $F_6 = 0$, **at** $x = 10$ $F_6 = \dfrac{10}{7} = 1.429$ **(tension)**

At $x = 0$ $F_5 = 0$, **at** $x = 10$ $F_5 = 0.58$ **(tension)**

Step 4.

When unit load is a Region 3 considering left portion of the section and taking moments about point 7.

$$F_4 \times 3.5 + R_1 \times 10 = 0$$

$$F_4 = -\frac{R_1 \times 10}{3.5} = -\frac{30 - x}{30} \times \frac{10}{3.5} = -\frac{30 - x}{10.5}.$$

Taking moments about point 3

$$F_6 \times 3.5 - R_1 \times 15 = 0$$

or
$$F_6 = \frac{R_1 \times 15}{3.5} = -\frac{30 - x}{30} \times \frac{15}{3.5} = \frac{30 - x}{7}$$

Resolving vertically

$$F_5 \cdot \cos\theta + R_1 = 0$$

or
$$F_5 = -\frac{R_1}{\cos\theta} = -\frac{30 - x}{30} \times \frac{\sqrt{3.5^2 + 5^2}}{3.5}$$
$$= -(30 - x) \times 0.058$$

At $x = 20$ $F_6 = \dfrac{10}{7} = 1.43$ (tension)

$x = 30$ $F_6 = 0$

$x = 20$ $F_4 = -\dfrac{10}{10.5} = -0.95$ (comp.)

$x = 30$ $F_4 = 0$

$x = 20$ $F_5 = -10 \times 0.058 = -0.58$ (comp.)

$x = 30$ $F_5 = 0.$

Steps 5.

These values are plotted in Fig. 9.63 (*a*), (*b*), (*c*). The values of the ordinates of the influence line when the unit load is inside the panel can be obtained by joining the ordinates at 6 and 7.

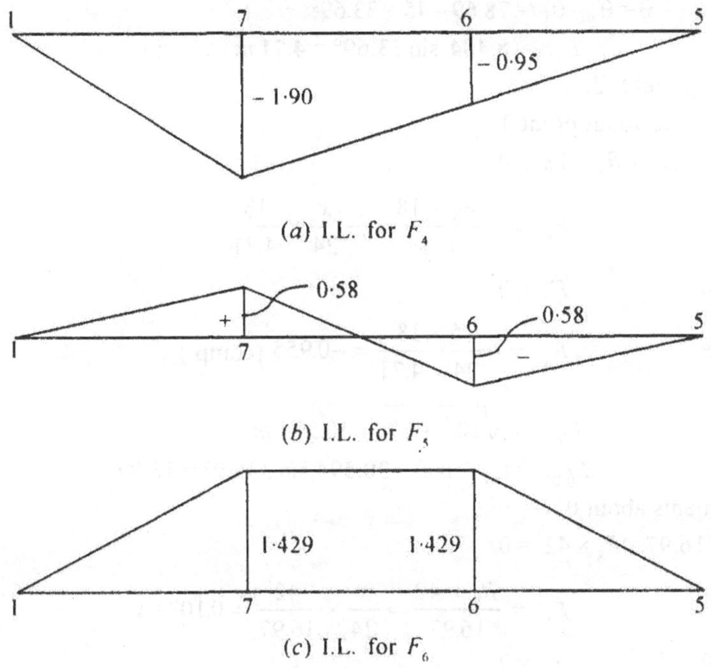

(a) I.L. for F_4

(b) I.L. for F_5

(c) I.L. for F_6

Fig. 9.63. I. Lines for members 4, 5, 6.

Example 9.16. Draw the influence line diagram for members 4, 5 and 6 of the truss shown in fig. 9.64. Assume the load moves accross the bottom chord.

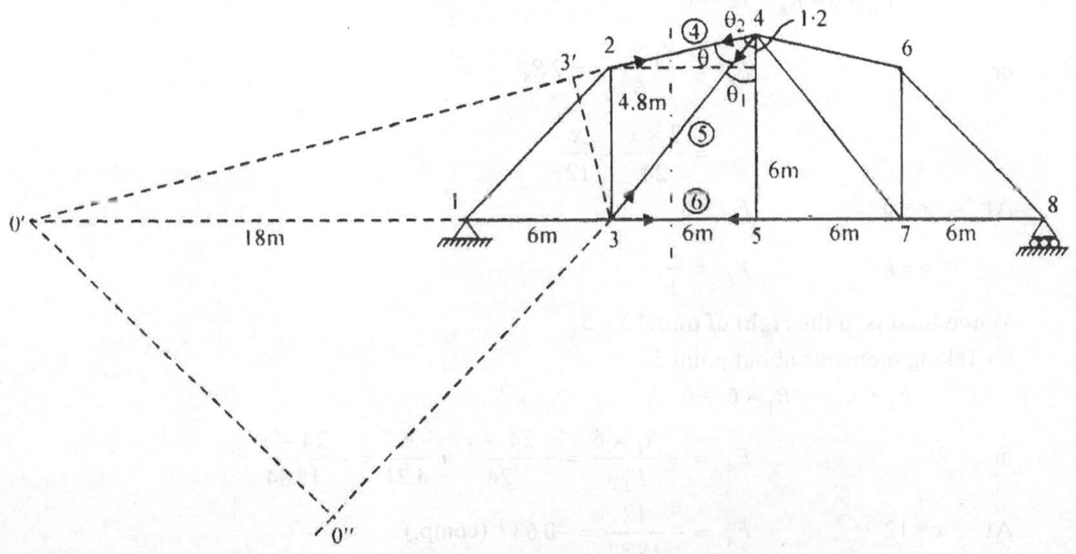

Fig. 9.64. Example 9.16

Solution: Distance 50′ can be obtained from the ratio $\frac{1.2}{6} = \frac{6}{50'}$ or 50′ = 30 *m*; $\tan\theta_2 = \frac{30}{6} = 5$

$\theta_2 = 78.69°$.

$$L_{34} = 6\sqrt{2} = 6 \times 1.414 = 8.484\ m$$

$$\theta = \theta_2 - \theta_1 = 78.69 - 45 = 33.69°$$

$$L_{33'} = 8.484\ \sin 33.69° = \mathbf{4.71\ m}$$

Unit load in panel 1–3.

Taking moments about point 3,

$$F_4 \times L_{33'} + R_8 \times 18 = 0$$

or
$$F_4 = -\frac{R_8 \times 18}{L_{33'}} = -\frac{x}{24} \times \frac{18}{4.71}$$

At $x = 0$ $\quad F_4 = 0$

$x = 6$ $\quad F_4 = -\dfrac{6 \times 18}{24 \times 4.71} = -0.955$ (comp.)

$$L_{0'4} = \sqrt{30^2 + 6^2} = 30.594\ m$$

$$L_{0'0''} = L_{o'4} \sin\theta = 30.594.\sin 33.69° = 16.97$$

Taking moments about 0′

$$F_5 \times 16.97 - R_8 \times 42 = 0$$

$$F_5 = \frac{R_8 \times 42}{16.97} = \frac{x}{24} \times \frac{42}{16.97} = 0.1031\,x$$

At $x = 0$ $\quad F_5 = 0$

$x = 6$ $\quad F_5 = 0.6187$ (tension)

Taking moments about 4.

$$F_6 \times 6 - R_8 \times 12 = 0$$

or
$$F_6 = \frac{R_8 \times 12}{6} = 2R_8$$

$$= \frac{2 \times x}{24} = \frac{x}{12}$$

At $x = 0$ $\quad F_6 = 0$

$x = 6$ $\quad F_6 = \dfrac{1}{2}$

When load is to the right of panel 3 – 5

(*i*) Taking moments about point 3

$$F_4 \times L_{33'} + R_1 \times 6 = 0$$

or
$$F_4 = -\frac{R_1 \times 6}{L_{33'}} = -\frac{24 - x}{24} \times \frac{6}{4.71} = -\frac{24 - x}{18.84}$$

At $x = 12$ $\quad F_4 = -\dfrac{12}{18.84} = -0.637$ (comp.)

$x = 24$ $\quad F_4 = 0$

(*ii*) Taking moments about 0′

$$F_5 \times L_{0'0''} + R_1 \times 18 = 0$$

or $\quad F_5 = -\dfrac{R_1 \times 18}{L_{0'0''}} = -\dfrac{24-x}{24} \times \dfrac{18}{16.97}$

$\quad = -(24-x) \times 0.04419$

at $x = 12 \quad F_5 = -12 \times 0.04419 = -0.530$

$x = 24 \quad F_5 = 0$

(*iii*) Taking moments about point 4

$$F_6 \times 6 - R_1 \times 12 = 0$$

or $\quad F_6 = 2R_1 = \dfrac{2(24-x)}{24} = \dfrac{24-x}{12}$

at $x = 12 \quad F_6 = 1$

$x = 24 \quad F_6 = 0$

Fig. 9.65 gives influence lines for members F_4, F_5 and F_6.

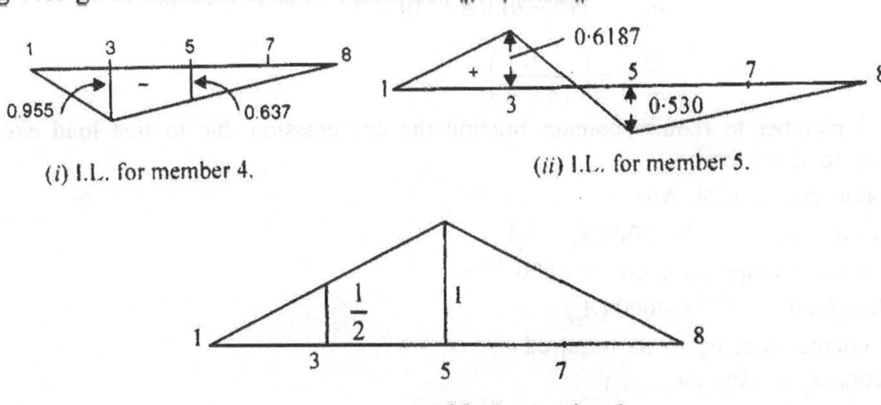

Fig. 9.65. Influence lines for forces in members 4, 5 and 6.

Example 9.17. An N girder consists of 15 square panels with a total span of 60 *m*. It weighs approximately 1500 *N/m*. The heaviest live loading is uniformly distributed load of 4000 *N/m*. If the diagonal bracing is unsuitable for compressive loads, how many bays must be counter braced and what is the maximum tensile stress in a diagonal?

Solution:

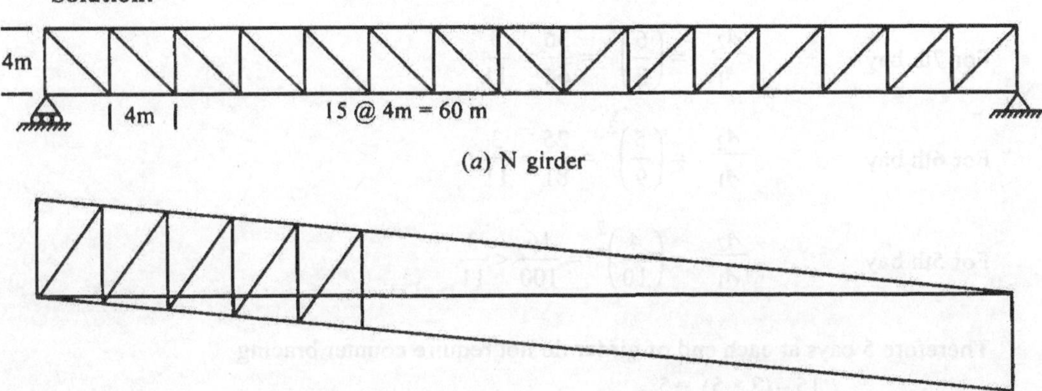

(*a*) N girder

(*b*) Family of Influence lines for diagonals.

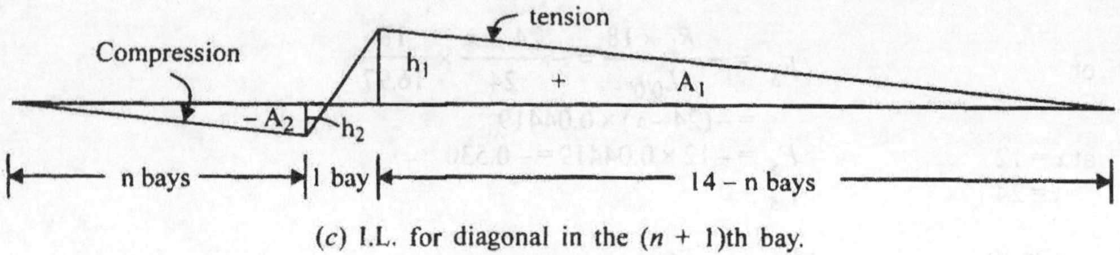

(c) I.L. for diagonal in the $(n + 1)$th bay.

Fig. 9.66. Example 9.17

From similar trangles

$$\frac{A_2}{A_1} = \frac{h_2^2}{h_1^2}$$

But
$$\frac{h_2}{h_1} = \frac{4n}{(14-n)\times 4} = \frac{n}{14-n}$$

$$\therefore \quad \frac{A_2}{A_1} = \left(\frac{n}{14-n}\right)^2$$

For a member to require counter bracing the compression due to live load exceeds the tension due to dead load.

Tension due to 1500 N/m.

dead load $= 1500\,(A_1 - A_2)$.

Maximum Compression due to 4000 N/m.

live load $= 4000\,(A_2)$

For counter bracing to be required

$4000\,A_2 > 1500\,(A_1 - A_2)$

or $5500\,A_2 > 1500\,A_1$

or $\dfrac{A_2}{A_1} > \dfrac{1500}{5500} = \dfrac{3}{11}$

For 8th bay $\quad \dfrac{A_2}{A_1} = \left(\dfrac{7}{7}\right)^2 = 1 > \dfrac{3}{11}$

For 7th bay $\quad \dfrac{A_2}{A_1} = \left(\dfrac{6}{8}\right)^2 = \dfrac{36}{64} > \dfrac{3}{11}$

For 6th bay $\quad \dfrac{A_2}{A_1} = \left(\dfrac{5}{9}\right)^2 = \dfrac{25}{81} > \dfrac{3}{11}$

For 5th bay $\quad \dfrac{A_2}{A_1} = \left(\dfrac{4}{10}\right)^2 = \dfrac{16}{100} < \dfrac{3}{11}$

Therefore 5 bays at each end of girder do not require counter bracing

$$15 - (2 \times 5) = 5$$

$\therefore$ 5 centre bays require counter bracing.

Maximum tension occurs in end diagonal with dead and live loads on whole span.

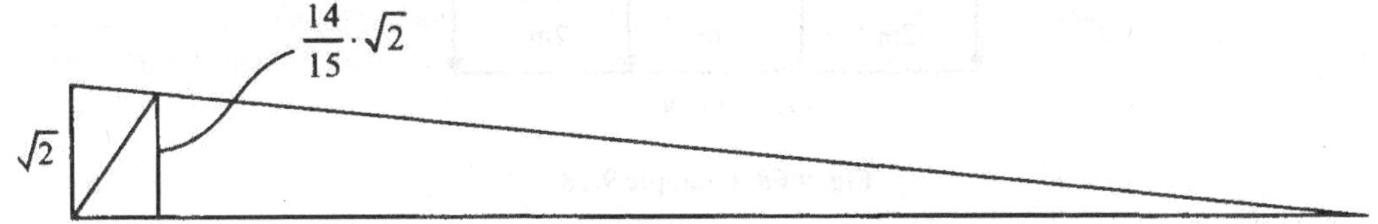

Fig. 9.67. Maximum tension in end diagonal (Example 9.17).

$$\text{Area of diagram} = \frac{1}{2} \times \frac{14\sqrt{2}}{15} \times 60$$

$$= 28\sqrt{2}$$

$$\therefore \quad \text{Maximum tensile force} = (1500 + 4000)28\sqrt{2}$$

$$= 28 \times 5500\sqrt{2} = \mathbf{2.1778 \times 10^5 N.}$$

Example 9.18. The warren girder of span 18 *m* as shown in Fig. 9.68 has six equal panels and all the triangles are equilateral. The dead load is 5 *t/m* run and the live load consists of four wheel loads 12′, 20′, 20′ and 18′ at 2 *m* spacings. Find the maximum and minimum forces in the members *CD* and *CE*, stating whether tension or compression.

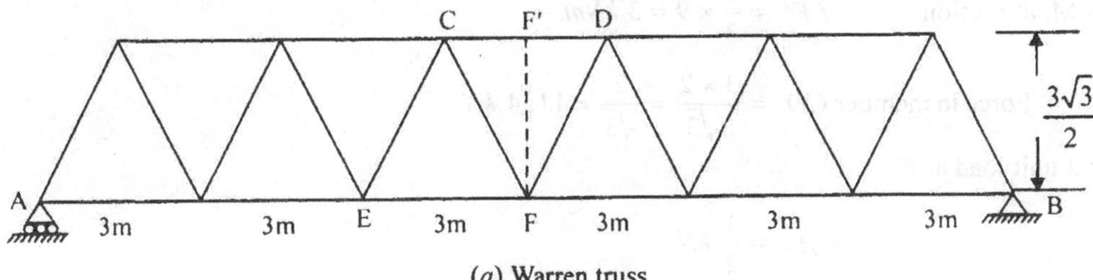

(*a*) Warren truss.

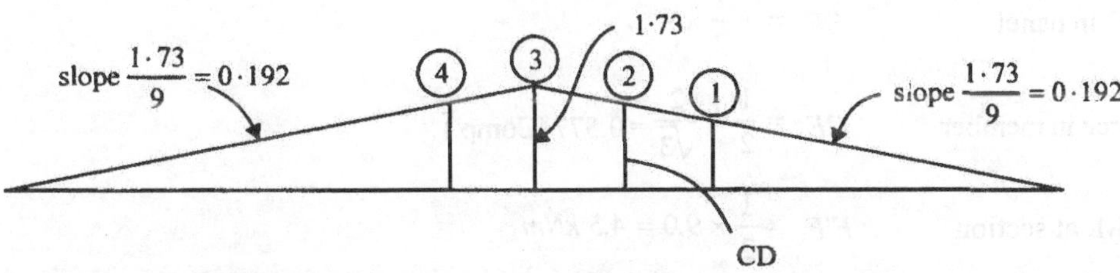

(*b*) I.L. for *CD*

Slope = ·064

0·385

0·57

y

x

6m

Slope ·064

$$\text{slope} = \frac{\cdot962}{3} \doteq 0\cdot321$$

(*c*) I.L. for *CE*

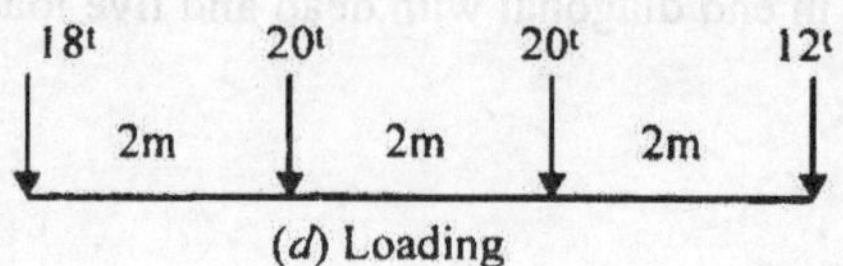

(*d*) Loading

Fig. 9.68. Example 9.18.

Solution:

Put unit load at *E*

$$R_A = \frac{4}{6} = \frac{2}{3}\ kN$$

$$R_B = \frac{1}{3}\ kN$$

S.F. at panel $EF = \frac{1}{3}\ kN$

$$R_{CE} \sin 60^\circ = \frac{1}{3}\ kN$$

$\therefore$ $F_{CE} = \frac{2}{\sqrt{3}} \cdot \frac{1}{3} = 0.385\ kN$ (tension)

B.M. at section $FF' = \frac{1}{3} \times 9 = 3\ kNm.$

$\therefore$ Force in member $CD = \frac{3 \times 2}{3\sqrt{3}} = \frac{2}{\sqrt{3}} = 1.154\ kN.$

Put unit load at *F*.

$$R_A = R_B = \frac{1}{2}\ kN$$

S.F. in panel $EF = +\frac{1}{2}\ kN$

Force in member $CE = \frac{1}{2} \times \frac{2}{\sqrt{3}} = 0.577$ (Comp.)

B.M. at section $F'F = \frac{1}{2} \times 9.0 = 4.5\ kNm$

Force in member $CD = \frac{4.5 \times 2}{3\sqrt{3}} = \sqrt{3}\ kN.$

Maximum force in *CD*

Try load 3 on F.

If load moves 1 *m* to left increase of force

$$= 32 \times 0.192 - 38 \times 0.192$$

$$= -6 \times 0.192 = 1.152 \text{ (decrease)}$$

If load moves 1 *m* to right increase of force

$$= 18 \times 0.192 - 52 \times 0.192, \text{ hence decrease.}$$

$\therefore$ Maximum value occurs with load 3 on *F*.

Maximum force in CD (compression) due to live load

$$18 \times \frac{1.73}{9} \times 7 = 24.22$$

$$20 \times 1.73 = 34.60$$

$$20 \times \frac{1.73}{9} \times 7 = 26.91$$

$$12 \times \frac{1.73}{9} \times 5 = \underline{11.53}$$

97.26 kN (compression)

$$\text{Max}^m \text{ force due to D.L.} = 5 \times \frac{1}{2} \times 1.73 \times 18 = \underline{77.85} \text{ (compression)}$$

$= 175.11\ kN$ (compression)

Max^m force with only D.L. acting = 77.85kN (compression)

For maximum force in CE (compression)

try load 4 at F

If load moves 1 m to the left, increase of force.

$52 \times 0.064 - 18 \times 0.321$ = negative.

Hence force decreases. When load moves towards right then also force decreases. Hence maximum value occurs with load 4 on F.

$$18 \times 0.577 = 10.386$$

$$20 \times \frac{7}{9} \times 0.577 = 8.976$$

$$20 \times \frac{5}{9} \times 0.577 = 6.411$$

$$12 \times \frac{3}{9} \times 0.577 = 2.308$$

28.08 kN (compression)

Hence maximum compression force in CE.

due to live load = 28.08 kN.

For maximum tensile force in CE put load 2 at E.

If load moves 1 m to the left.

$$\text{Increase} = 12 \times 0.231 - 58 \times 0.064$$
$$= 3.852 - 3.712$$

= + quantity, hence increase.

Also rate of increase increases when 18′ load leaves beam.

Hence, maximum tension due to live load is with load 1 at E.

Value is equal to $12 \times 0.385 = 4.620$

$$20 \times 0.385 \times \frac{4}{6} = 5.133$$

$$20 \times \frac{0.385 \times 2}{6} = \underline{2.566}$$

12.319t

Force due to D.L.

$$\frac{x}{y} = \frac{0.385}{0.577} = 0.667$$

$$\frac{x}{x+y} = \frac{0.667}{1.667}$$

or

$$x = \frac{3 \times 0.667}{1.667} = 1 \cdot 2\ m$$

$$y = 1.8\ m$$

$$\text{Net area} = \frac{1}{2} \times 7.2 \times 0.385 - \frac{1}{2} \times 10.8 \times 0.577$$

$$= -1.7298 \text{ (compression)}$$

Force in *CE* due to D.L. $= 5\,(1.7298) = 8.649$ **(compression)**

Maximum compressive force in *CE*.

$= 28.08 + 8.649 = $ **36.729 *kN*.**

Minimum force in *CE* $= 12.319 - 8.649 = 3.670$ **(tensile)**

Example 9.19. A Pratt truss (*N*) has a span of 40 *m*, a depth of 4 *m* and is divided into ten equal parts. Draw the influence lines for the loads in the top boom of the eighth panel from the left hand support and for the diagonal in the sixth panel from the same support when the bottom chord is loaded. Determine the loads in these members when a uniform load of 40 *kN/m* and 10 *m* long crosses the bridge.

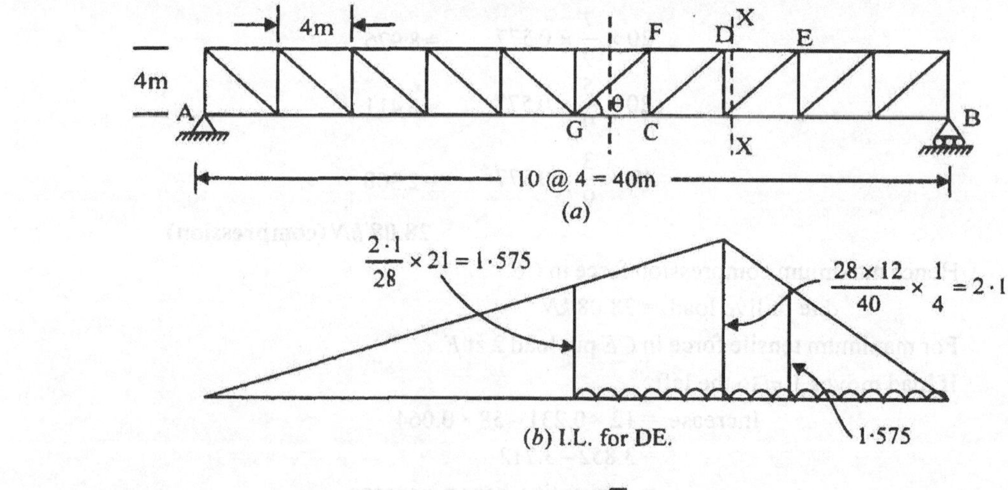

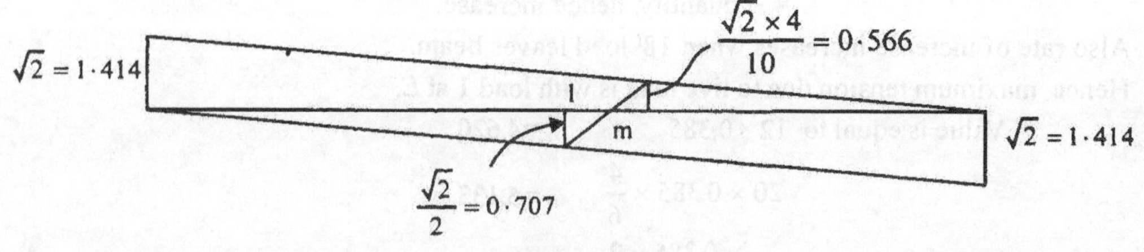

Fig. 9.69. Example 9.19.

Solution :

Force in DE can be obtained by taking moment about the opposite joint C. Hence I.L. for force in DE is influence line for

$$\frac{M_C}{h} = \frac{M_C}{4}$$

For maximum effect of live load of 10 m length the load should be so placed that the point C divides the load in the same ratio as the span. Then,

$$\frac{x}{y} = \frac{28}{12}$$

or $$\frac{x}{x+y} = \frac{28}{40}$$

or $$x = 7\,m.$$

Value of the force in DE $$= 40 \times \left[\frac{1}{2}(1.575 + 2.1) \times 10\right]$$

$$= \mathbf{735\ kN.}$$

Force in FG

The influence line diagram for force in FG is shown in Fig. 9.69 (*c*)

From Fig. $$\frac{l}{m} = \frac{0.707}{0.566}$$

or $$\frac{l}{l+m} = \frac{0.707}{0.707 + 0.566} = \frac{0.707}{1.273}$$

or $$l = \frac{4 \times 0.707}{1.273} = 2.22\ m$$

$$m = 1.78\,m$$

Value of Maximum tensile force in F_G (Fig. 9.70)

$$\frac{x}{y} = \frac{20}{2.22}$$

or $$\frac{x}{x+y} = \frac{20}{22.22}$$

or $$x = 9\,m.$$

Value of the force $$= \frac{1}{2}[0.707 + 0.388] \times 10 \times 40$$

$$= \mathbf{219\ kN.}$$

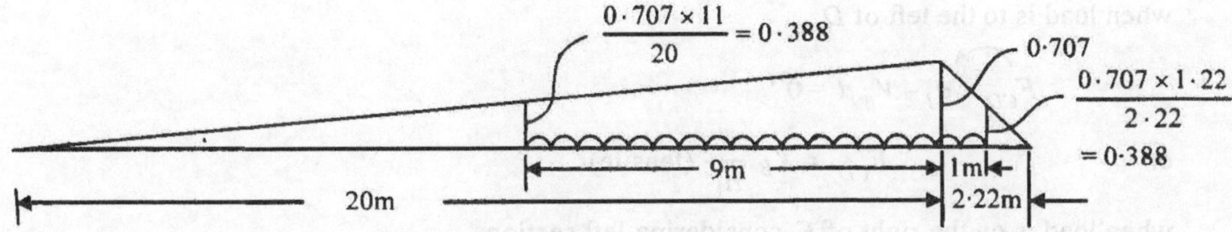

Fig. 9.70. Maximum tensile force in FG.

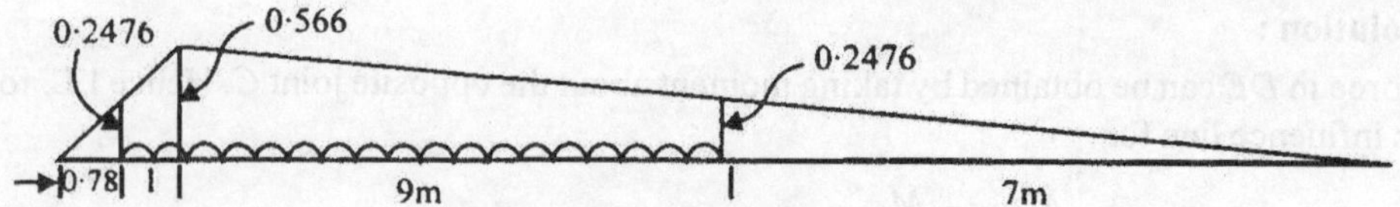

Fig. 9.71. Maximum compressive force in *FG*.

Value of Max^m compressive force in *FG* (Fig. 9.71)

$$\frac{x}{y} = \frac{1.78}{16.00}$$

or
$$\frac{x}{x+y} = \frac{16}{17.78}$$

or
$$x = \frac{16 \times 10}{17.78} = 9\ m$$

Value of compressive force in $FG = \frac{1}{2}[0.566 + 0.2476] \times 40 \times 10$

$$= \mathbf{162.72\ kN.}$$

Example 9.20. Assuming the load to be moving on the upper chord of the panelled arch truss, find the influence lines for the members *CD, DE, DI* and *CE*.

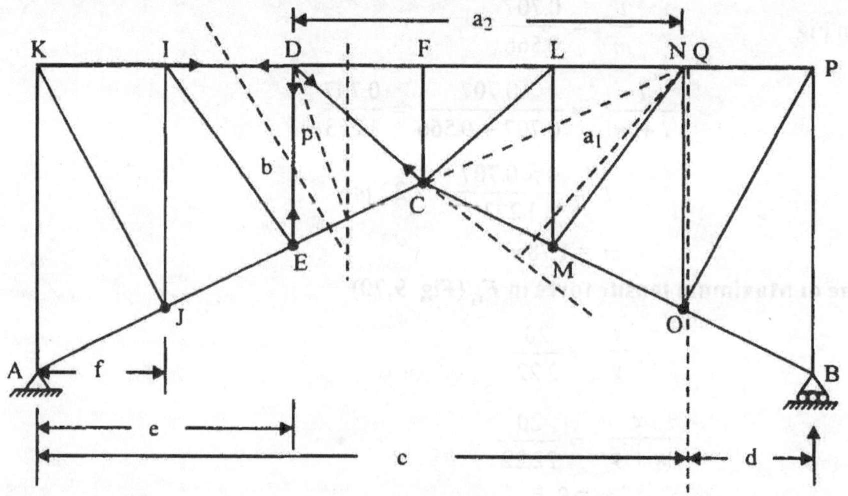

Fig. 9.72. Example 9.20.

Influence Line for member *CD*. (Fig. 9.73 (a))

We take a vertical section through three members *DF, DC* and *CE* in the panel *DFCE*. Taking moments about point *Q*, (where *EC* produced meets NP).

when load is to the left of *D*.

$$F_{CD} \times a_1 - V_B d = 0$$

or
$$F_{CD} = V_B \cdot \frac{d}{a_1} \text{ (tensile)}$$

when load is on the right of *F*, considering left section.

$$V_A \times C - F_{CD} \times a_1 = 0$$

or $$F_{CD} = V_A \cdot \frac{c}{a_1}$$

Fig. 9.73 (*a*). I.L. for *CD*

Influence line for *DE* (Fig. 9.73 (b))

A section is taken cutting members *ID*, *DE* and *EC*. When unit load is to the left of *I*.

$$F_{DE} \times a_2 + V_B \times d = 0$$

or $$F_{DE} = -V_B \times \frac{d}{a_2} \text{ (compression)}$$

when to the right of *D*

$$V_A \times C + F_{DE} \times a_2 = 0$$

or $$F_{DE} = -V_A \times \frac{c}{a_2} \text{ (compression)}$$

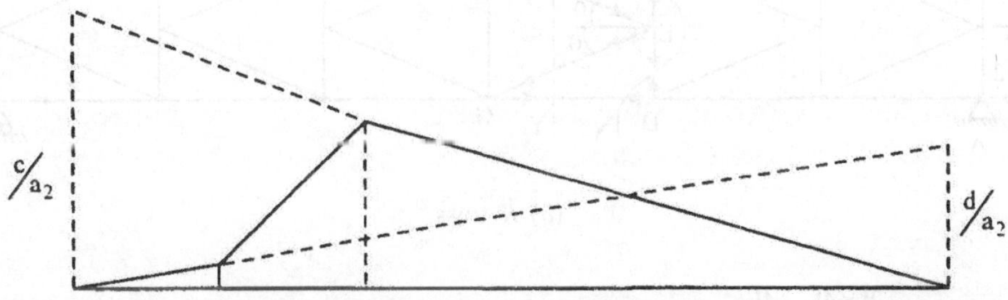

Fig. 9.73 (*b*). I.L. for *DE*

Influence line for *DI* (Fig. 9.73 (c))

We take moments about *E*. When unit load is to the right of *D* taking left section.

$$V_A \times e + F_{DI} \times b = 0$$

or $$F_{DI} = -V_A \times \frac{e}{b}$$

$$= \frac{\text{Moment at } E}{b} \text{ (compression)}$$

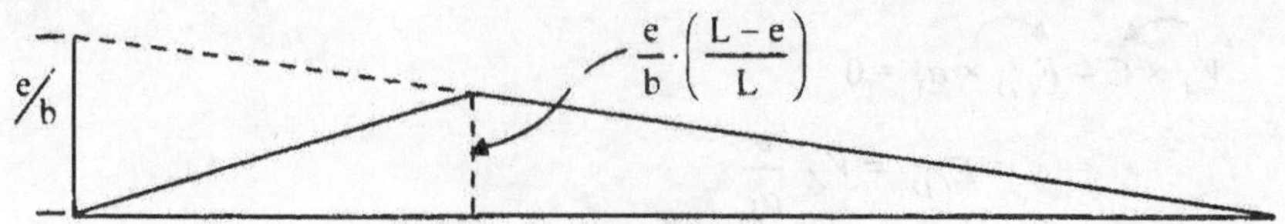

Fig. 9.73 (*c*). I.L. for force in *DI*

Influence line for *CE*. (Fig. 9.73 (d))

We take moments about *D*, the opposite joint

$$F_{CE} = \frac{M_D}{p} \text{ (tensile)}$$

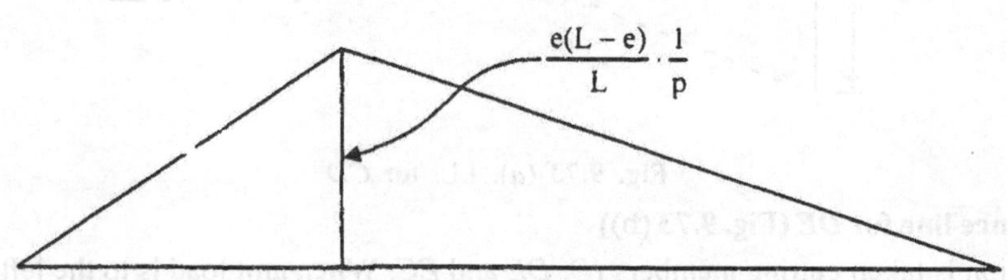

Fig. 9.73 (*d*). I.L. for *CE*.

Example 9.21. Draw the influence lines for different members of the panel *CDGF* in the *K* truss shown in Fig. 9.63 (*a*).

Solution:

The *K* truss and the influence lines for different members are shown in Fig. 9.63 (a) to (e).

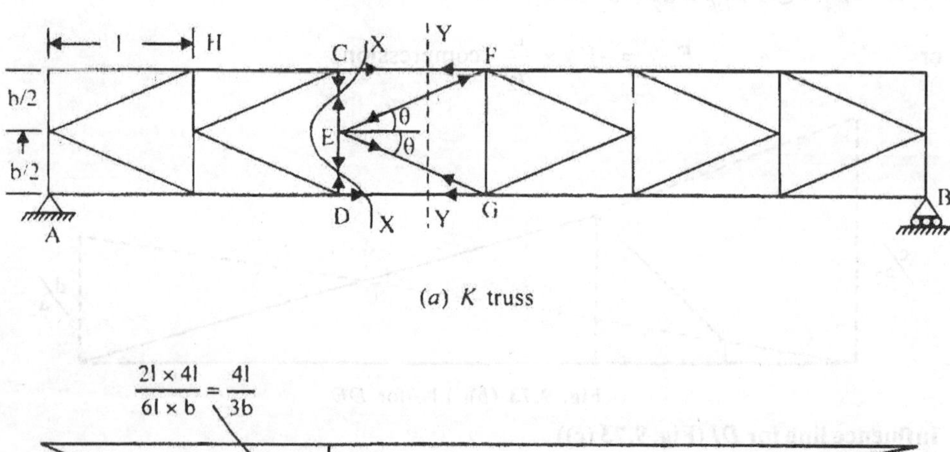

(*a*) *K* truss

(*b*) I.L. for F_{CF}

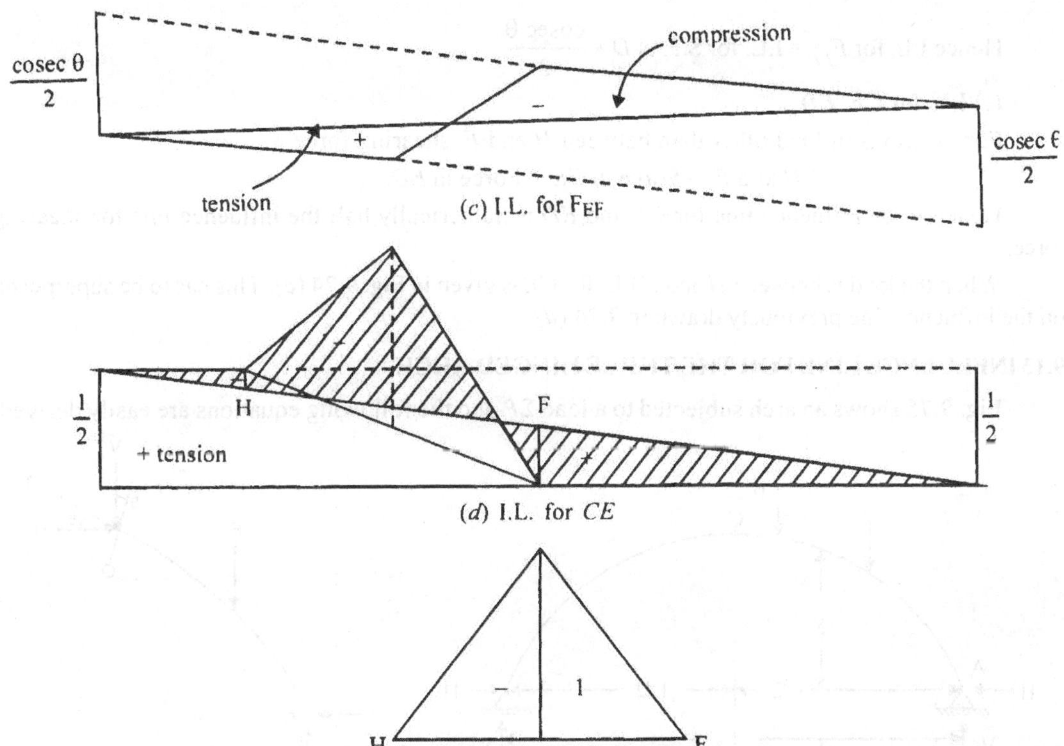

(c) I.L. for F_{EF}

(d) I.L. for CE

(e) I.L. for CE when load is between H and F.

Fig. 9.74. I.L. for K truss.

(a) I. Lines for forces in *CF* and *DG*.

We take X-X section. Moment centre for F_{CF} is D and for F_{DG} is C. Taking moments about D or C.

$$F_{CF} \times b + V_A \times AD = 0 \text{ (when load is on the right of } F\text{)}$$

$$F_{CF} \times b + V_B \times DB = 0 \text{ (when load is on the left of } C\text{)}$$

or $$F_{CF} = -\frac{M_D}{b} \text{ (compression) Fig. 9.74(}b\text{)}$$

$$F_{DG} = \frac{M_D}{b} \text{ (tension)}$$

(b) I.L. for forces in *EF* and *EG*. (Fig. 9.74 (*c*))

From horizontal equilibrium of joint E.

$$F_{EF} \cos\theta - F_{EG} \cos\theta = 0$$

or $$F_{EF} = F_{EG}.$$

Considering section YY and equating vertical forces.

$$R_A - 2F_{EF} \sin\theta = 0$$

i.e., $$\text{S.F. at } D - 2F_{EF} \sin\theta = 0$$

or $$F_{EF} = \frac{\text{S.F. at } D}{2\sin\theta} = \frac{\text{S.F.}}{2}.\text{cosec}\,\theta$$

$$= -F_{EG}$$

Hence I.L. for F_{EF} = I.L. for S.F. at $D \times \dfrac{\text{cosec }\theta}{2}$

(c) I.L. for *CE, ED*

For positions of load other than between *H* and *F*, shearing force

at *H* and *F* = Force in *CE* + Force in *ED*.

Hence force influence line for *CE* and *ED* is numerically half the influence line for shearing force.

When the load is between *H* and *F* I.L. for *CE* is given in Fig. 9.74 (*e*). This has to be superposed on the influence line previously drawn in 9.74 (*d*).

9.13 INFLUENCE LINE FOR THE THREE HINGED ARCH.

Fig. 9.75 shows an arch subjected to a load ΣP and the following equations are easily derived.

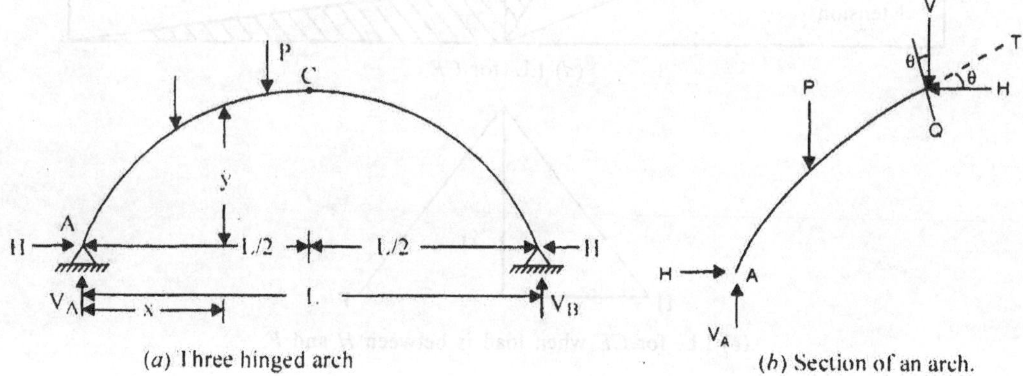

Fig. 9.75. I.L. for three hinged arch.

$$\text{Axial Thrust } T = H\cos\theta + V\sin\theta$$

$$\text{Shear force } Q = V\cos\theta - H\sin\theta$$

$$M = V_A \,.\, x - \Sigma P.x - H.y$$

$$= \mu - H.y \quad \text{where } \mu = \text{simple Beam B.M.}$$

The shear force is considered positive when the left end tends to move upward. Arch is a fovourite structure for bridges as it can be supported on rock formations at the ends of the span over a deep valley. An arch proves to be economical because if designed properly, the dead load moments are almost absent. It is therefore, essential to study influence lines for arches.

1. Influence line for Horizontal Reaction *H*.

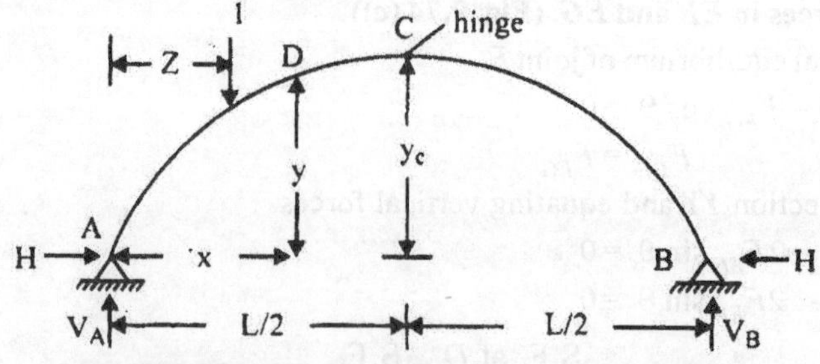

Fig. 9.76. I.L. for three hinged arch (Derivation).

When a unit load is at a distance *Z* from *a* (Fig. 9.76) $V_B = \dfrac{1.Z}{L}$

Taking moments about the point C.

$$V_B \times L/2 = H \times Y_C$$

or

$$H = V_B \times \frac{L}{2Y_C} = \frac{Z}{L} \times \frac{L}{2Y_C}$$

$$= \frac{Z}{2Y_C}$$

This is valid when $Z \leq L\backslash 2$

when $Z = L/2$ $\quad H = \frac{L}{4Y_C}$

when $Z > L/2$ $\quad V_A = \frac{L-Z}{L}$

and taking moments about C

$$H = V_A \times \frac{L}{2Y_C} = \frac{L-Z}{L} \times \frac{L}{2Y_C}$$

$$= \frac{L-Z}{2Y_C}.$$

when $Z = L/2$ $\quad H = \frac{L}{4Y_C}.$

The influence line is therefore, a triangle with mid ordinate $\frac{L}{4Y_C}$. (Fig. 9.77 b)

2. Bending Moment (Fig. 9.77 c) When the load is between A and D.

$$\text{B.M.} = V_B(L-x) - H.y.$$

The first term $V_B(L-x)$ is the simple beam B.M. μx

$$\text{Hence B.M.} = \mu x - H.y.$$

At a particular section x, y are constant

Hence I.L. for B.M. = I.L. for beam B.M. – (Influence line for H) $\times y$

3. Axial Thrust (Fig. 9.77 d) Influence line for axial thrust is obtained by adding cos θ times influence line for H to sin θ times influence line for vertical shear at D.

4. Shear (Fig. 9.77 e) Influence line for radial shear at a section is obtained by subtracting sin θ times influence line for H from cos θ times influence line for vertical shear at section with slope θ.

The influence lines are drawn in Fig. 9.77 (a) to (e).

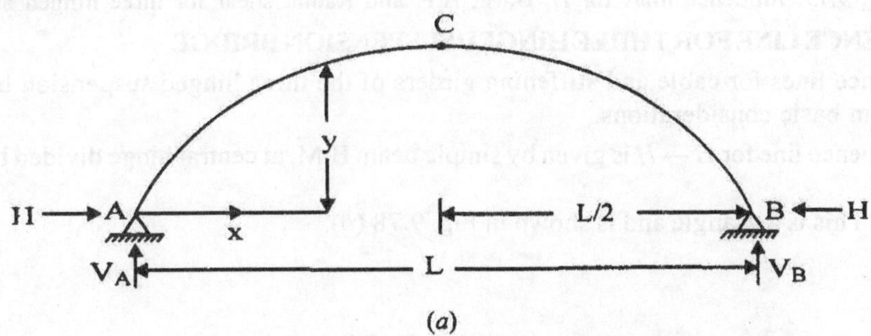

(a)

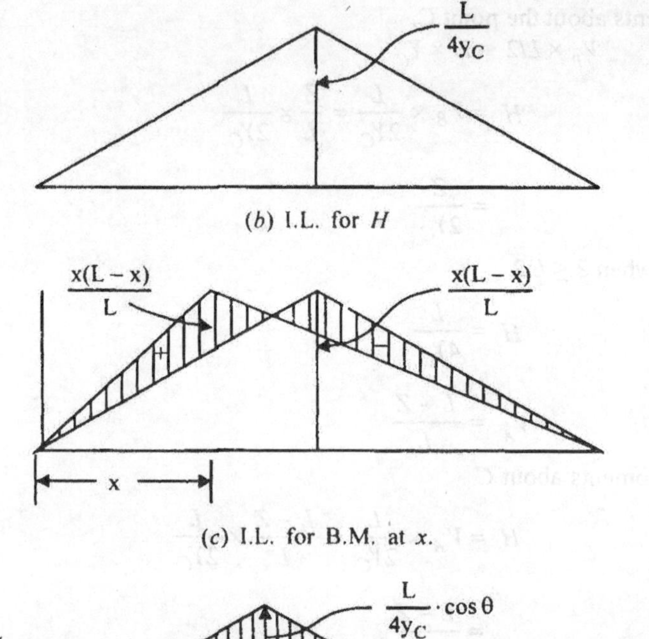

(b) I.L. for H

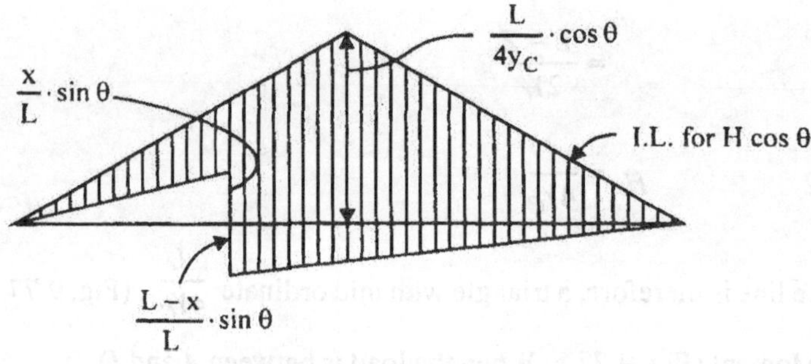

(c) I.L. for B.M. at x.

(d) I.L. for Axial Thrust at x.

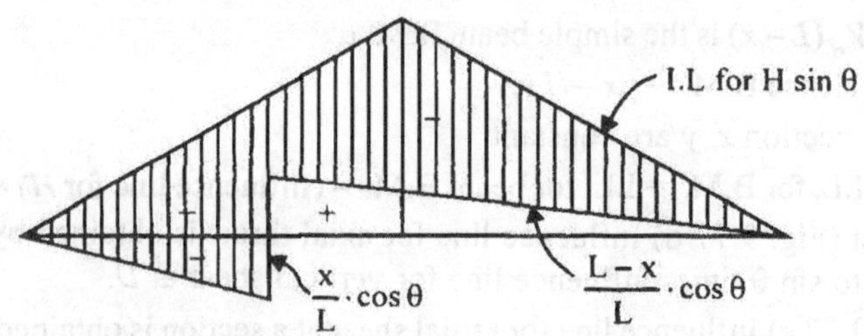

(e) I.L. for radial shear at x.

Fig. 9.77. Influence lines for H, B.M., A.F. and Radial shear for three hinged arch.

9.14 INFLUENCE LINE FOR THREE HINGED SUSPENSION BRIDGE

Influence lines for cable and stiffening girders of the three hinged suspension bridge can be obtained from basic considerations.

1. Influence line for H— H is given by simple beam B.M. at central hinge divided by Y_C. Hence

$H = \dfrac{1 . L}{4 . Y_C}$. This is a triangle and is shown in Fig. 9.78 (b).

2. Influence line for B.M. at any point X in the girder.

B.M. at any point X = Simple Beam B.M. – $H.y$

Hence influence line for B.M. = Influence line for beam B.M. – Influence line $H \times Y$. This is shown in Fig. 9.78 (*c*).

3. Influence line for shear at any point

$$X \text{ in the girder S.F.} = \text{Beam shear at any point } X - \text{Shear due to suspender forces.}$$
$$= \text{Beam shear at any point } X - H_l \,.\, \tan\phi.$$

Hence I.L. for shear in girder at X = I.L. for beam shear at any point x – Influence line for $H_l . \tan\phi$

Two cases will occur as the point X is < $L/4$ or > $L/4$. These are shown in Fig. 9.78 (*d*) and (*e*).

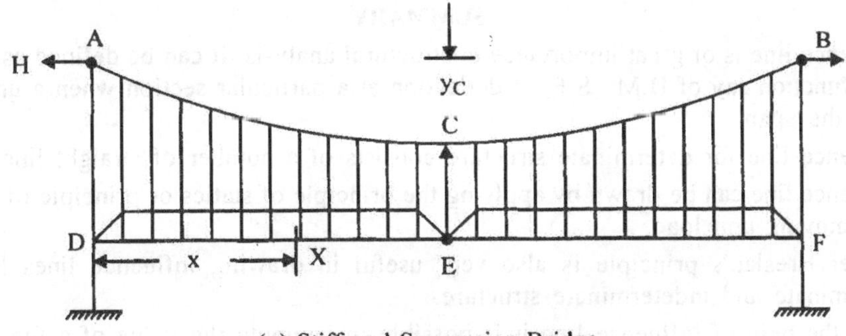

(*a*) Stiffened Suspension Bridge.

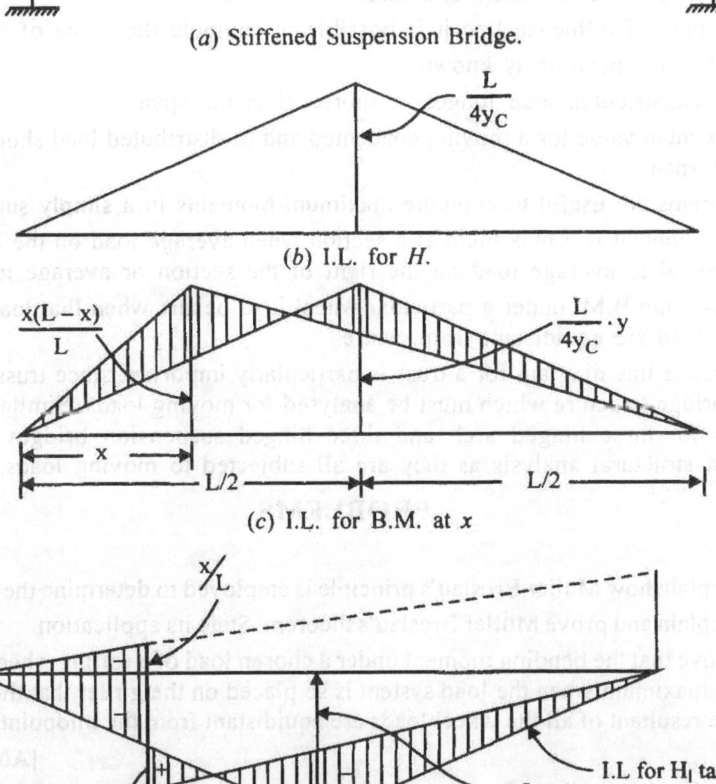

(*b*) I.L. for *H*.

(*c*) I.L. for B.M. at *x*

(*d*) I.L. for shear at *x* (*x* < *L*/4)

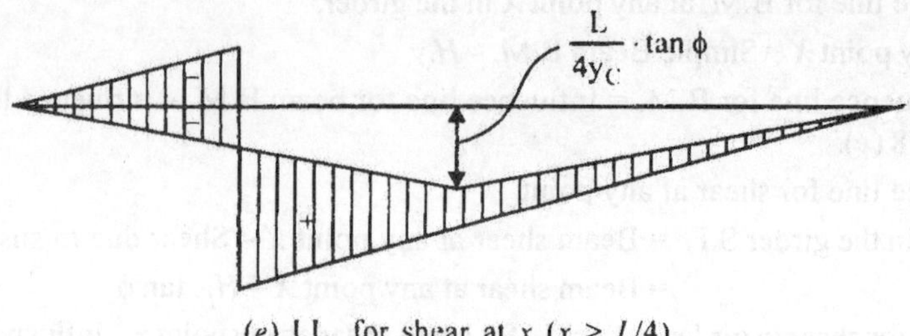

(e) I.L. for shear at x ($x > L/4$)

Fig. 9.78. Influence lines for three hinged stiffened girder.

SUMMARY

1. Influence line is of great importance in structural analysis. It can be defined as the diagram of a function say of B.M., S.F. or deflection at a particular section when a unit load rolls over the span.
2. Influence line for determinate structure consists of a number of straight lines.
3. Influence line can be drawn by applying the principle of statics or principle of virtual work to a moving unit load.
4. Muller Breslau's principle is also very useful in drawing influence lines for statically determinate and indeterminate structures.
5. With the help of influence line it is possible to compute the value of a function.
   (*i*) when load position is known.
   (*ii*) for a distributed load longer or shorter than the span.
   (*iii*) maximum value for a moving concented and or distributed load shorter or longer than the span.
6. Two theorems are useful to compute maximum moments in a simply supported beam.
   (*i*) The moment is a maximum at a section when average load on the left of the section is equal to average load on the right of the section or average load on the span.
   (*ii*) Maximum B.M. under a particular wheel load occurs when that load and the C.G. of the load are equidistant from centre.
7. The influence line diagram for a truss is particularly important since trusses are commonly used as bridge structure which must be analysed for moving loads. Similarly, influence line diagrams for three hinged arch and three hinged suspension bridges prove extremely helpful in structural analysis as they are all subjected to moving loads.

PROBLEMS

A. Theoretical

9.1 (*i*) Explain how Müller Breslau's principle is employed to determine the influence line.

(*ii*) Explain and prove Müller Breslau's theorem. State its application. **[AMIE 1978]**

(*iii*) Prove that the bending moment under a chosen load of a rolling wheel load system will be maximum when the load system is so placed on the girder that the chosen load and the resultant of all the wheel loads are equidistant from the midpoint of the girder. **[AMIE Winter 1981]**

(*iv*) Derive the condition for maximum bending moment under any given load in a system of loads rolling across a simply supported beam. Also indicate the method of calculating the absolute maximum bending moment any where in the span for the above system of rolling loads.

(*v*) Explain what you understand by "Influence Line diagrams."

B. Objective. Tick the correct answer.

9.2. (*i*) The force in a diagonal member of a truss under a moving load remains in

(*a*) tension only (*b*) compresssion only

(*c*) zeroforce (*d*) both tension and compression.

[AMIE Summer 97]

(*ii*) The value of the ordinate at quarter span of the influence line diagram for mid span B.M. of a simply supported beam is 0.6 *m*. If the span of the beam is doubled what will be the corresponding value?

(*a*) 0.6 m (*b*) 1.2 m

(*c*) 0.3 m (*d*) None of the above. **[AMIE Winter 98]**

(*iii*) A simply supported beam of span 3 *m* carries a point load of 30 *kN* which may be placed anywhere within the middle third of the span. The maximum B.M. in the beam with the load placed so as to minimise this valuc is

(*a*) 20 kNm (*b*) 30 kNm

(*c*) 22.5 kNm (*d*) None of the above. **[AMIE Winter 98]**

(*iv*) The ordinates of influence line diagram for bending moment always have the dimensions of

(*a*) force (*b*) length

(*c*) force × length (*d*) force/length. **[AMIE Winter 97]**

(*v*) The maximum bending moment due to train of wheel loads on a simply supported girder.

(*a*) always occur at the centre of span

(*b*) always occur under a wheel load

(*c*) all of above. **[AMIE Winter 97]**

(*vi*) The ordiniates of the influence line for the left support reaction of a simply supported beam with an over hang on the left are 1.0 and 1.2 at the left support point and at the free end respectively. If the span of the beam is 5.0 *m*, the length of the overhang is

(*a*) 1.2 m (*b*) 0.2 m

(*c*) 1.0 m (*d*) 6.00 m **[AMIE Summer 99]**

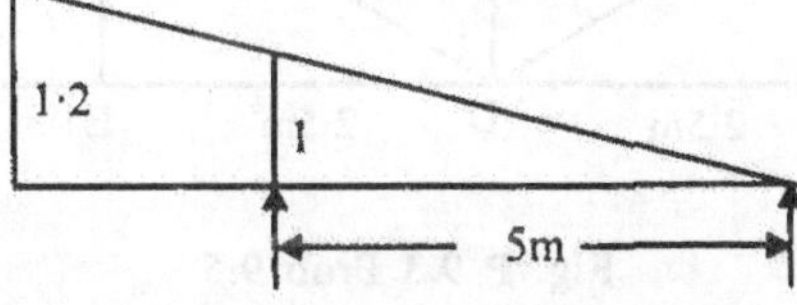

Fig. P 9.1 (Prob 9.1 (*vi*)

(*vii*) The ordinate of the influence line for the mid span B.M. of a simply supported beam is 1 *m*. If the beam carries a *u.d.l.* of 6 *kN/m* over the entire span, the maximum B.M. will be

(*a*) 8 kNm (*b*) 6 kNm

(*c*) 12 kNm (*d*) 24 kNm. **[AMIE Summer' 99]**

(*viii*) At a section 2 *m* from the left support of a simply supported beam of span 6 *m*, the B.M. due to a moving *u.d.l.* shorter than the span will be maximum when

(*a*) the left end of the load will be at the section.

(*b*) the section will divide the load in the ratio 1 : 2.

(*c*) the section will divide the load equally. **[AMIE Summer 99]**

C. Mathematical

9.3 A Pratt truss shown in Fig. P 9.2 is loaded by a dead load of 20 *kN/m* and a uniformly distributed live load of 100 *kN/m* of length 4 *m*. Find the maximum force in member *IJ*. Loading is in bottom chord. [AMIE Summer 97]

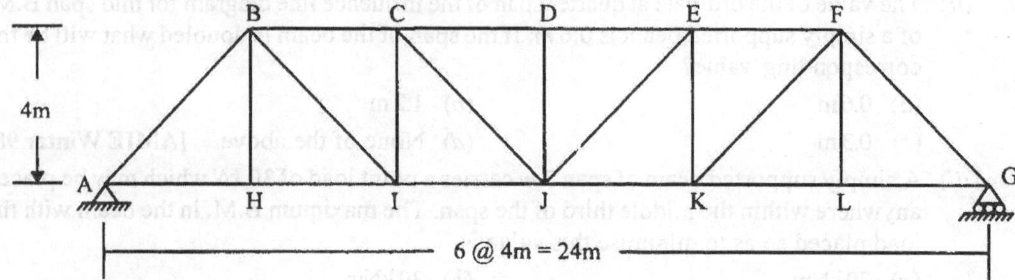

Fig. P 9.2. Prob 9.3.

9.4 Draw the Influence Line diagrams for B.M. and S.F. at a section 4 *m* from the left support of a simply supported beam of span 10 *m*. Explain how the diagrams were obteined. If a moving *u.d.l.* of intensity 18 *kN/m* and of length 8 *m* crosses the beam what will be the maximum values of S.F. and B.M. at the section 4 *m* from the left support? Also find the absolute maximum B.M. and S.F. any where in the span.

[AMIE Summer 99]

9.5 Construct the influence line for the force in the member *GD* of the pin jointed truss shown in Fig. P 9.3. The unit load moves along the bottom chord *AB*. [AMIE S 94]

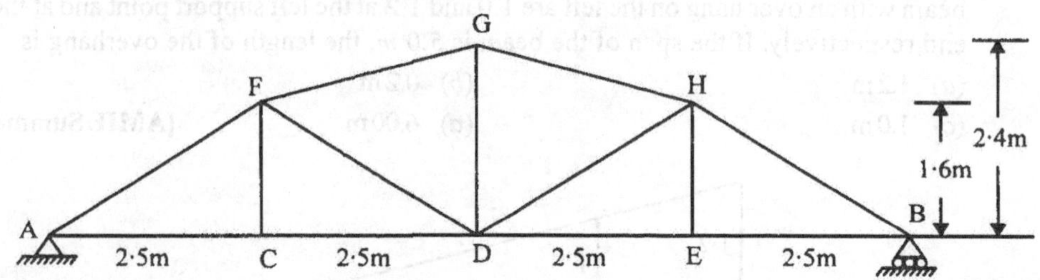

Fig. P 9.3 Prob 9.5

9.6 A three hinged structure *ABC* shown in Fig. P 9.4 comprises two beams *AB* and *BC*. The span *AC* is 20 *m* and the rise to the central hinge is 6 *m*. Vertical columns spaced 10/3 *m* centres transfer the force actions to the rib from simply supported beams. Draw influence line diagrams for (*i*) bending moment (*ii*) normal thrust at section 5 *m* from the left hand support. [AMIE Winter 78]

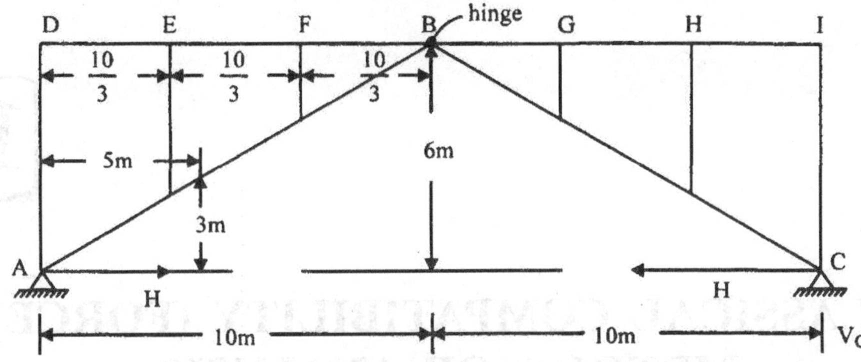

Fig. P 9.4 Prob 9.6.

9.7 A suspension bridge stiffened by three hinged symmetrical stiffening girders has two set of cables of span 200 *m* and central dip of 15 metres. The dead load of the bridge is 4*t* per metre which is carried by both the cables equally. The live load comprises two rolling loads each of magnitude 20*t* separated by a distance of 4 *m*. Assuming equal distribution of the live load to both the girders, find maximum values for (*i*) tension in each cable (*ii*) B.M. and S.F. at left quarter span of stiffening girder. (Fig. p 9.5)

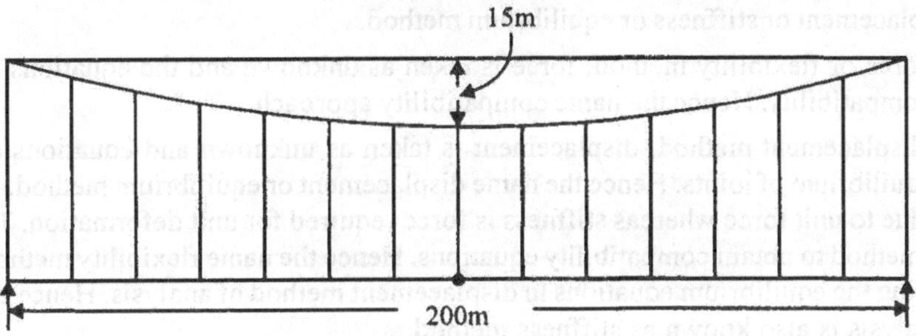

Fig. P 9.5. Prob 9.7

10

CLASSICAL COMPATIBILITY (FORCE) METHOD OF ANALYSIS

10.1 INTRODUCTION

Indeterminate structure can be solved by two methods.

(*a*) Force or flexibility or compatibility method

(*b*) Displacement or stiffness or equilibrium method.

In the force or flexibility method, force is taken as unknown and the equations are obtained considering compatibility. Hence the name compatibility approach.

In the displacement method, displacement is taken as unknown and equations are obtained considering equilibrium of joints. Hence the name displacement or equilibrium method. Flexibility is deformation due to unit force whereas stiffness is force required for unit deformation. The former is used in force method to obtain compatibility equations. Hence the name flexibility method. The latter is used in writing the equilibrium equations in displacement method of analysis. Hence displacement method of analysis is also known as stiffness method.

In this chapter classical force methods are considered. They are :

1. Consistent deformation method
2. Method of least work
3. Theorem of three moments

10.2 CONSISTENT DEFORMATION METHOD

This is the basic method for solving indeterminate structure taking force as unknown and enforcing compatibility. Fig 10.1 shows a cantilever under uniformly distributed load. The structure is statically determinate as we can calculate moment and shear at any section by equations of statics only. However, as soon as a support is given at B, the structure becomes statically indeterminate as reaction at B i e R_B cannot be obtained from equation of statics alone and as such B.M and S.F. at any section cannot be obtained. R_B, however, can be obtained by using the compatibility condition that displacement at B is zero. Taking downward displacement as positive the compatibility equation becomes (Fig. 10.2).

$$\frac{wL^4}{8EI} - \frac{R_B L^3}{3EI} = 0$$

or

$$R_B = \frac{3}{8}wL$$

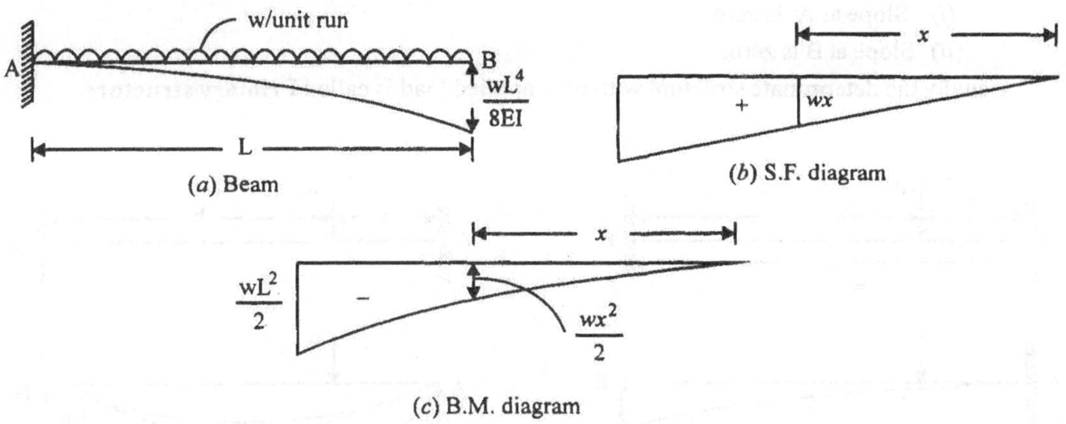

Fig. 10.1. Statically determinate cantilever.

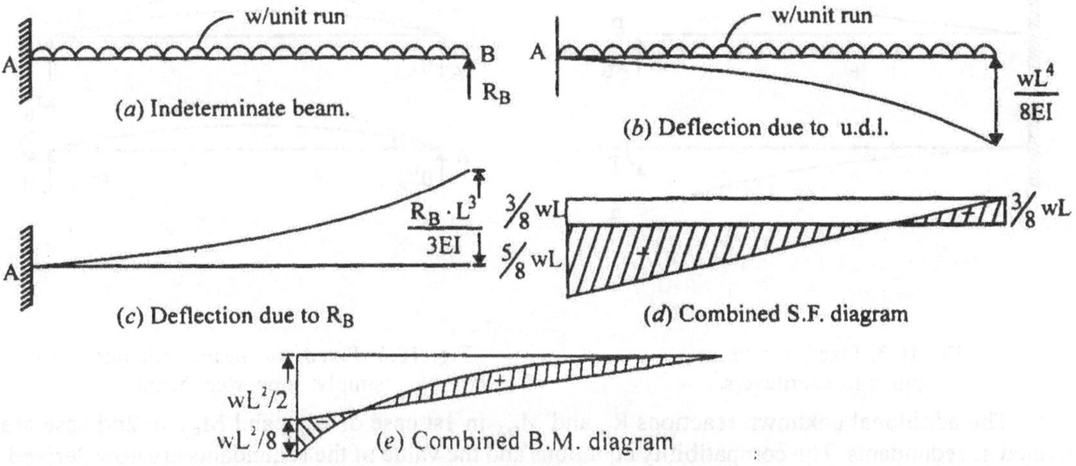

Fig. 10.2. Propped Cantilever.

Once R_B is known, the B.M and S.F. diagram can be drawn separately and then superposed to get the final B.M. and S.F. diagram. Here we found out only one unknown R_B by using compatibility condition. Hence degree of indeterminacy of the propped cantilever is one. The number of equations will depend on the number of unknowns to be found out from compatibility equations which again will depend on the degree of indeterminacy of the structure.

Next we consider a fixed end beam with say a concentrated load as shown in Fig 10.3. The structure is statically indeterminate to the 2nd degree. Neglecting horizontal reactions as only vertical load is acting, there will be four unknown reactions R_A and M_{AB} at A and R_B and M_{BA} at B. With two equations of statics $\Sigma Y = 0$ and $\Sigma M = 0$ there will be two unknown reactions to be determined from compatibility equations. If we split the beam into three cantilevers as in Fig 10.3 the compatibility conditions will be

(*i*) Slope at *B* is zero.

(*ii*) Deflection at *B* is zero.

If we split the fixed beam into three simply supported beams the compatible conditions will be.

(*i*) Slope at A is zero

(*ii*) Slope at B is zero.

Usually the determinate structure with only applied load is called **Primary structure.**

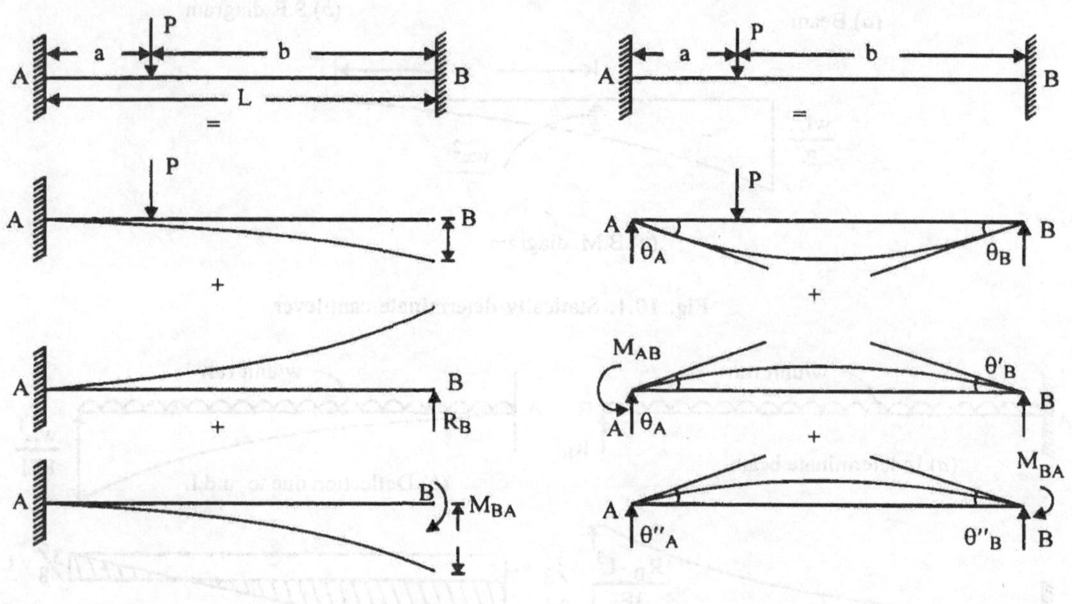

Fig 10.3. Fixed end beam split into cantilevers.

Fig. 10.4. Fixed end beam split into three simply supported beams.

The additional unknown reactions R_B and M_{BA} in 1st case or M_{AB} and M_{BA} in 2nd case are termed as redundants. The compatibility equations and the value of the redundants are now derived.

1st Case (Fig 10.3)

Slope at B is Zero or $\dfrac{Pa^2}{2EI} - \dfrac{R_B L^2}{2EI} + \dfrac{M_{BA}.L}{EI} = 0$

Deflection at B is Zero or $\dfrac{Pa^3}{3EI} + \dfrac{Pa^2}{2EI}.b - \dfrac{R_B L^3}{3EI} + \dfrac{M_{BA} L^2}{2EI} = 0$

Solving the above two equations

$$M_{BA} = \frac{Pa^2 b}{L^2},\ R_B = \frac{Pa^2(a+3b)}{L^2}$$

2nd case (Fig 10.4)

Slope at $A = 0$

or $\theta_A - \theta'_A - \theta''_A = 0$

$$\theta_A = \frac{Pab\,(L+b)}{6EI.L}, \theta'_A = \frac{M_{AB}.L}{3EI}, \theta''_A = \frac{M_{BA}.L}{6EI}$$

Substituting $\frac{Pab\,(L+b)}{6EI.L} - \frac{M_{AB}.L}{3EI} - \frac{M_{BA}.L}{6EI} = 0$

Similarly from 2nd condition slope at B = 0

or $\theta_B - \theta'_B - \theta''_B = 0$

Substituting $\frac{Pab\,(L+a)}{6EI.L} - \frac{M_{AB}.L}{6EI} - \frac{M_{BA}.L}{3EI} = 0$

Rearranging, $Pab\,(L+b) - 2M_{AB}L^2 - M_{BA}L^2 = 0$

$Pab\,(L+a) - M_{AB}L^2 - 2M_{BA}L^2 = 0$

Solving $M_{AB} = \frac{Pab^2}{L^2}$

$$M_{BA} = \frac{Pa^2b}{L^2}$$

10.3 General steps of the consistent deformation method - Based on the two illustrations just given, it is now possible to formulate the general steps to be followed in solving indeterminate structures by consistent deformation method.

Step 1 Find the degree of indeterminacy of the structure. Choose a determinate stable primary structure. Identify the redundants.

Step 2 Compute displacement of the primary structure at the point and in the direction of redundants due to applied loading only.

Step 3 Compute displacements of the primary structure in the direction of redundants due to a variable value of the redundants.

Step 4 The total displacements at the point and in the direction of redundants should be equal to the existing or prescribed value which gives the compatibility equations.

Step 5 Solve the equations and get the value of the redundants.

Step 6 Superpose the effects of the redundants on the primary structure to get final forces or B.M or S.F. as required.

The above steps are valid for beams, frames and trusses. The first step of

(*i*) finding the degree of indeterminacy.

(*ii*) determining the primary structure

and (*iii*) identifying the redundants is of great importance.

Fig. 10.5. shows the different types of structures with (*i*) degree of indeterminacy

(*ii*) appropriate stable primary structures.

(*iii*) corresponding redundants.

As compatibility equations involve displacements of determinate structures, Table 10.1 shows common beam displacements.

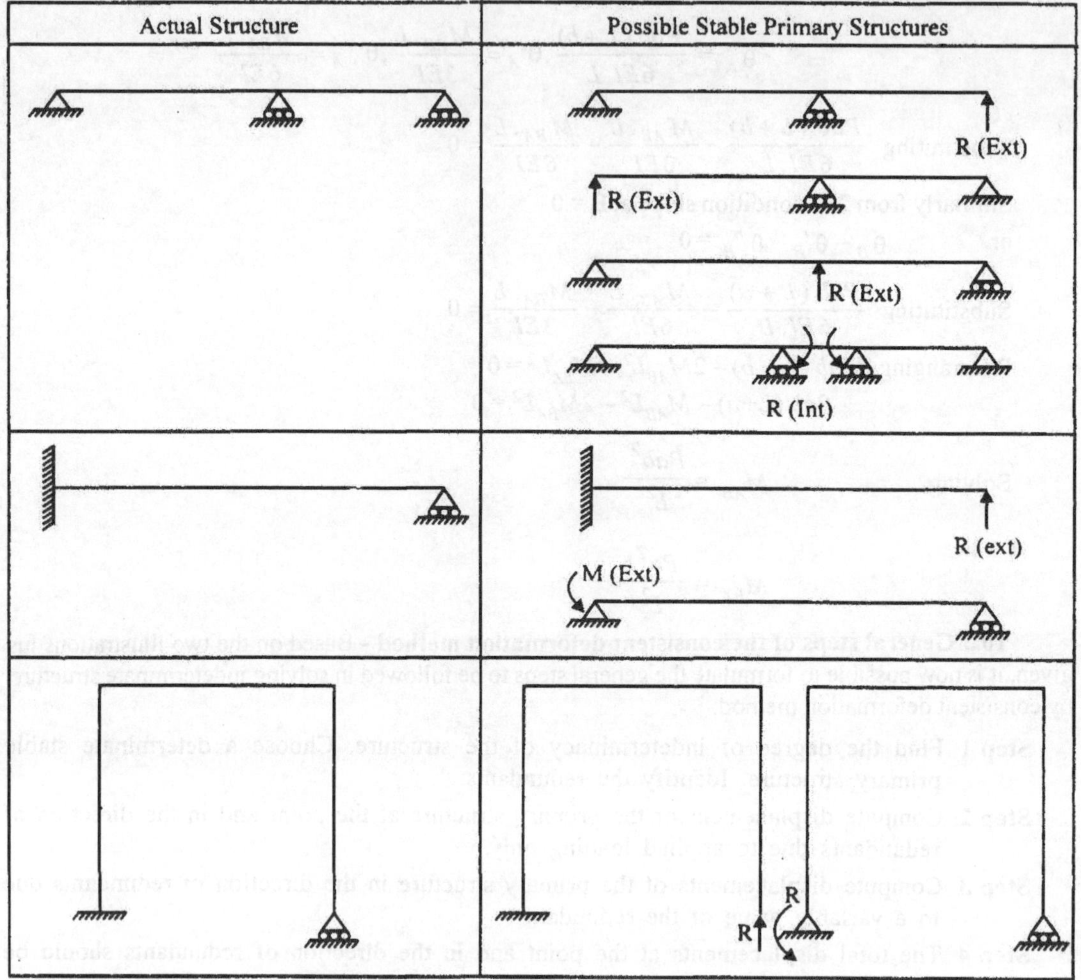

(*a*) Statically Indeterminate Structure of 1st Degree (contd)

Fig. 10.5 (Contd)

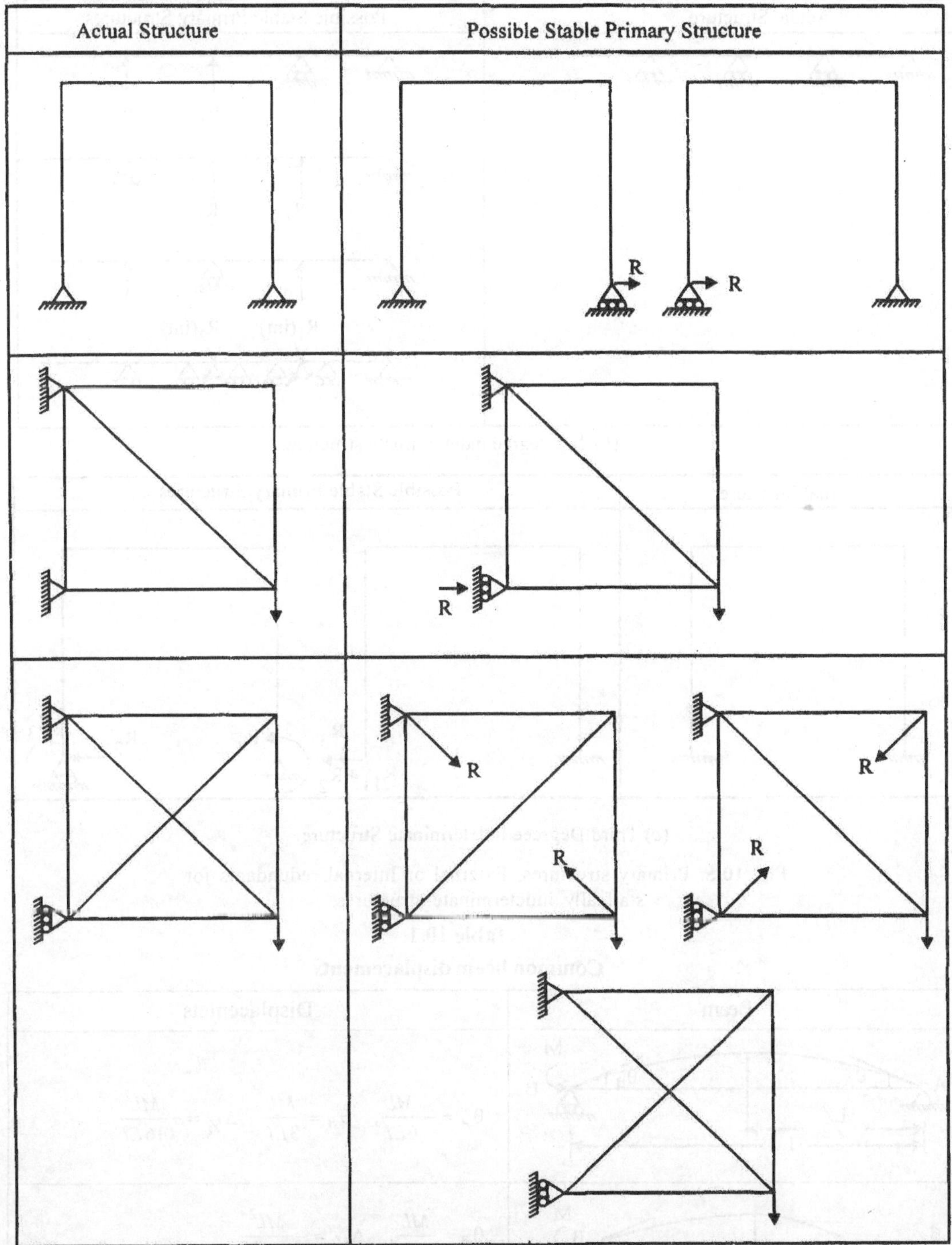

(a) Statically Indeterminate Structure of 1st degree.

Fig. 10.5 (Contd)

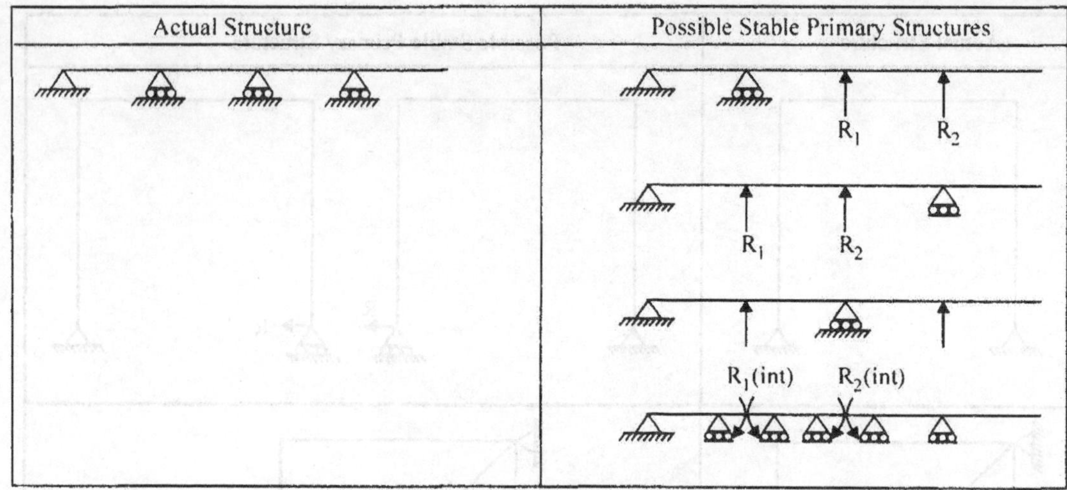

(*b*) 2nd degree indeterminate structure.

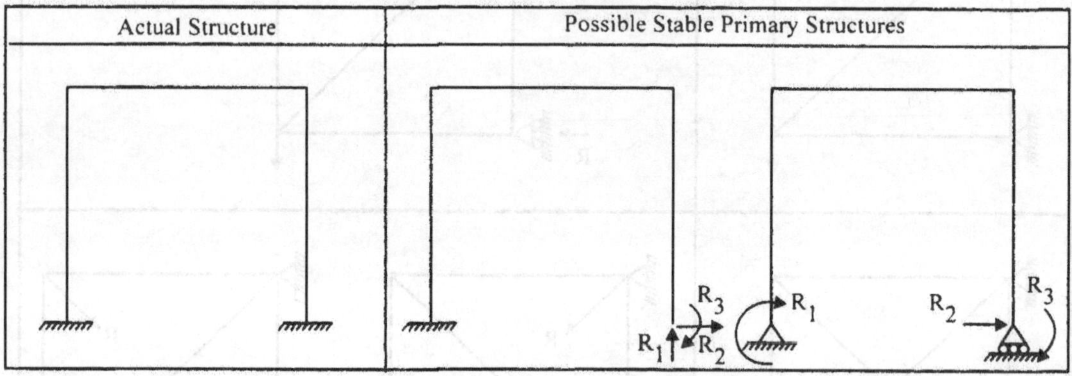

(*c*) Third Degreee Indeterminate Structure.

Fig. 10.5. Primary structures, External or Internal redundants for statically indeterminate structures.

Table 10.1

Common beam displacements

Beam	Displacemets
A, B, M, θ_A, θ_B, L/2, L	$\theta_A = -\frac{ML}{6EI}$, $\theta_B = \frac{ML}{3EI}$ $\Delta_{L/2} = -\frac{ML^2}{16EI}$
A, B, M, θ_B, L/2	$\theta_B = \frac{ML}{4EI}$, $\Delta_{L/2} = -\frac{ML^2}{32EI}$

(Contd)

Case	Results
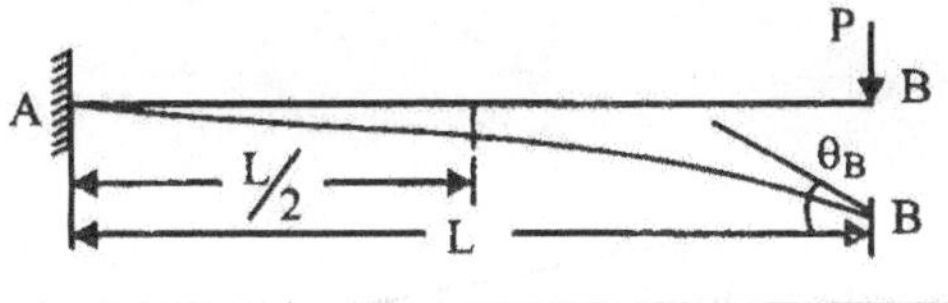 1	$\theta_B = \frac{PL^2}{2EI}, \quad \Delta_B = \frac{PL^3}{3EI} \quad \Delta_{L/2} = \frac{5PL^3}{48EI}$
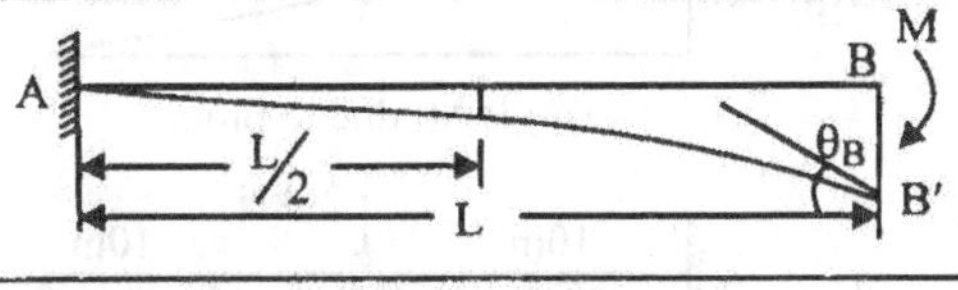 2	$\theta_B = \frac{ML}{EI}, \quad \Delta_B = \frac{ML^2}{2EI} \quad \Delta_{L/2} = \frac{ML^3}{8EI}$
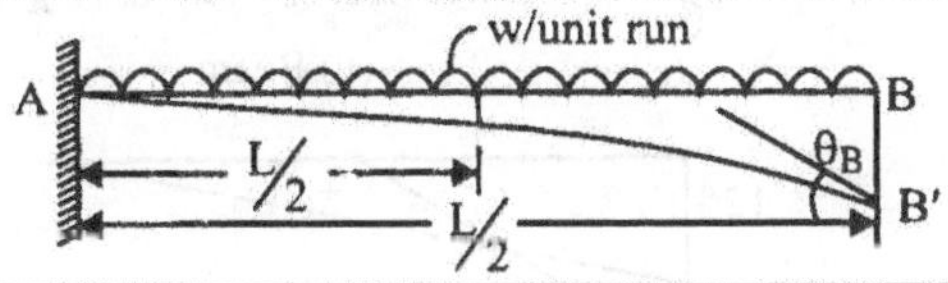 3	$\theta_B = \frac{wL^3}{6EI}, \quad \Delta_B = \frac{wL^4}{8EI} \quad \Delta_{L/2} = \frac{7wL^4}{192EI}$
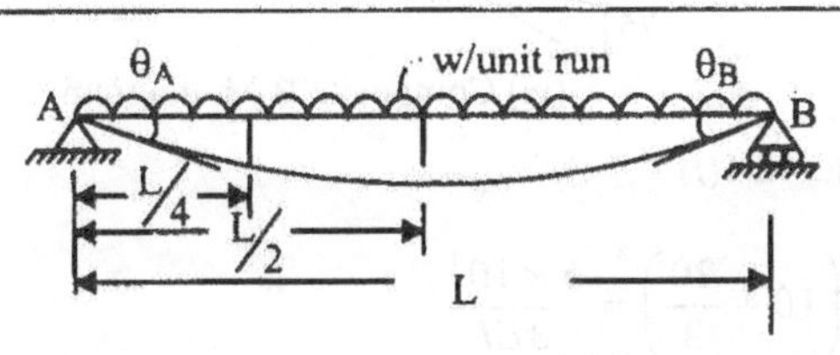 4	$\theta_A = -\theta_B = \frac{wL^3}{24EI} \quad \Delta_{L/4} = \frac{19wL^4}{2048EI} \quad \Delta_{L/2} = \frac{5wL^4}{384EI}$
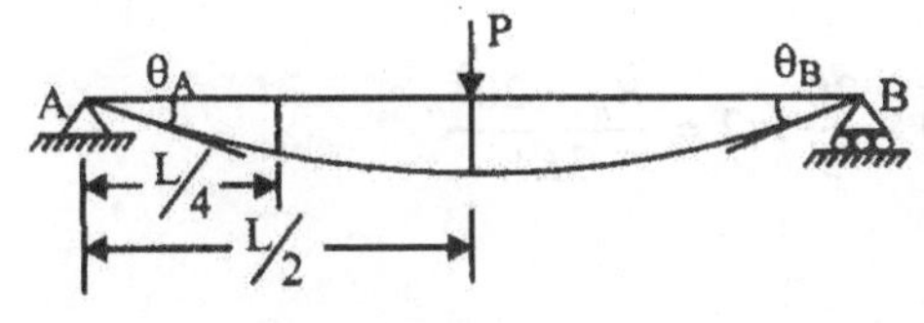 5	$\theta_A = -\theta_B = \frac{PL^2}{16EI} \quad \Delta_{L/4} = \frac{11PL^3}{768EI} \quad \Delta_{L/2} = \frac{PL^3}{48EI}$
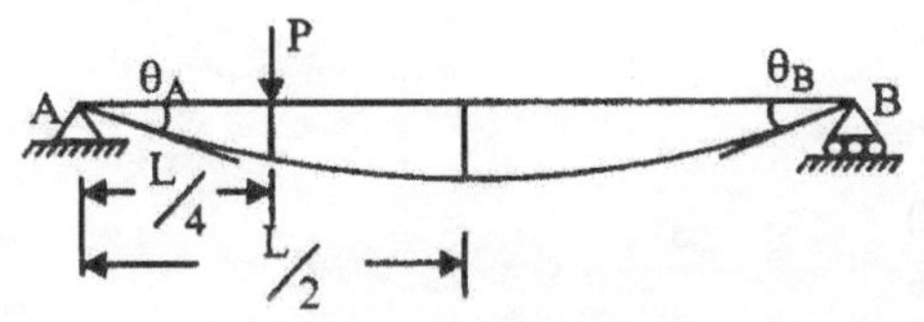 6	$\theta_A = \frac{21PL^2}{384EI} \quad \Delta_{L/4} = \frac{9PL^3}{768EI}$ $\theta_B = -\frac{15PL^2}{384EI} \quad \Delta_{L/2} = \frac{11PL^3}{768EI}$

Deflection downward + rotation clockwise +

Example 10.1. Analyse the propped cantilever beams shown. Use the reaction at the prop as the redundant. Use moment area method to compute the required displacements. Draw the final moment diagrams.

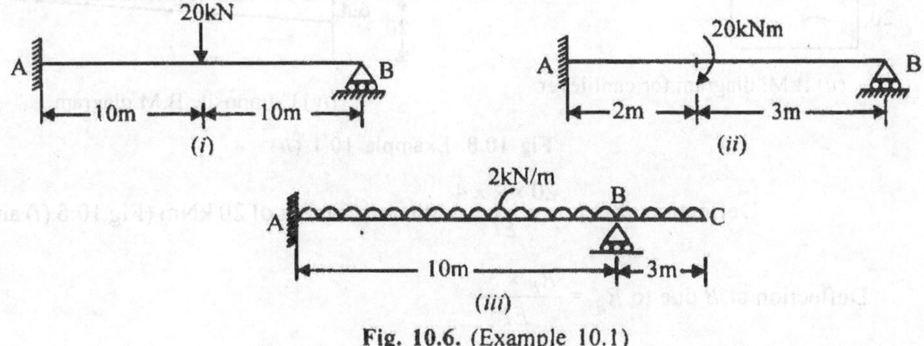

Fig. 10.6. (Example 10.1)

Solution:

Case 1

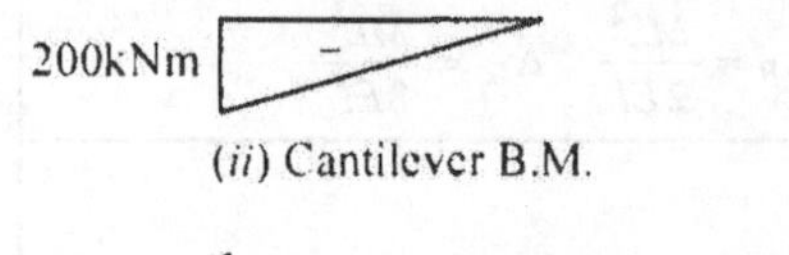

(*i*) Cantilever

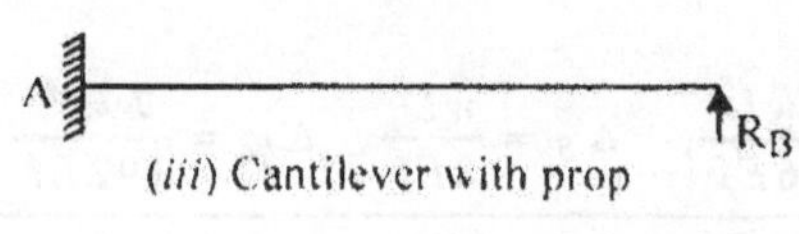

(*ii*) Cantilever B.M.

A R_B

(*iii*) Cantilever with prop

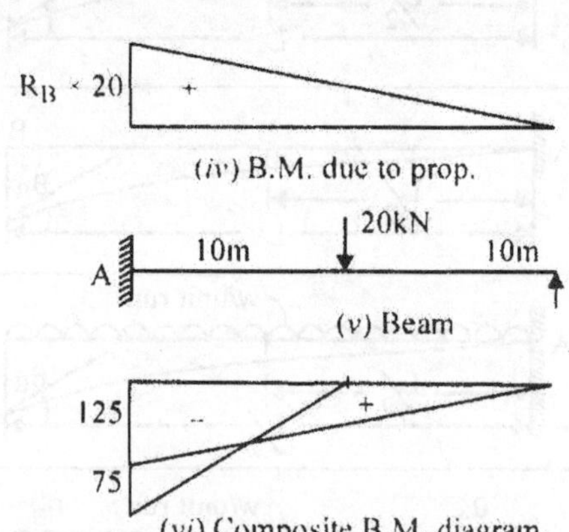

(*iv*) B.M. due to prop.

(*v*) Beam

(*vi*) Composite B.M. diagram.

Fig. 10.7 Example 10.1(*i*)

Deflection at B (Fig 10.7 (ii)) $= \frac{1}{2} \times 200 \times 10 \times \left(10 + \frac{20}{3}\right) = \frac{5 \times 10^4}{3EI}$

De. tion due to redundant R_B (Fig 10.7 (iv))

$$= \frac{1}{2} \times R_B \times 20 \times 20 \times \frac{20}{3} \times 2 = \frac{R_B \times 20^3}{3EI}$$

Equating $\quad R_B \times \frac{{}^{\neg}0^3}{3.} = \frac{5 \times 10^4}{3EI}$

or $\quad R_B = \frac{5 \times 10^4}{8 \times 10^3} = 6.25$ kN.

Composite B.M diagram is shown in Fig 10.7 (*vi*)

Case 2

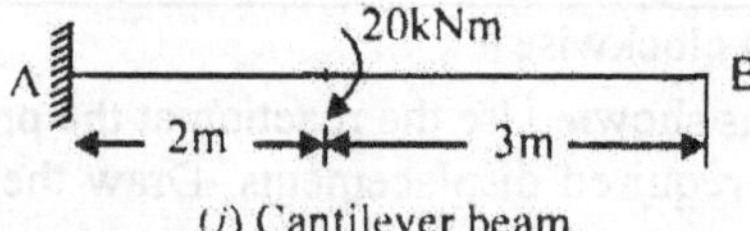

(*i*) Cantilever beam.

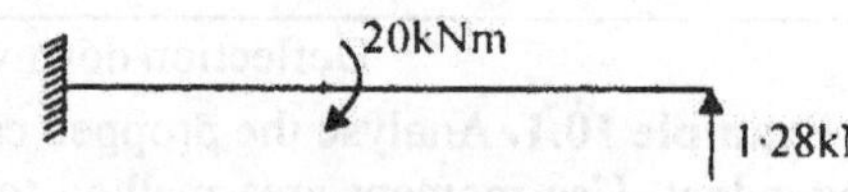

(*iii*) Beam with reaction at B.

(*ii*) B.M. diagram for cantilever

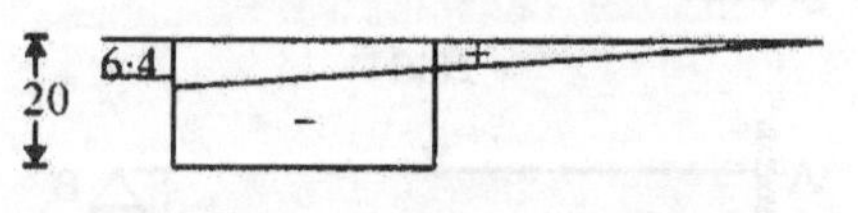

(*iv*) Composite B.M diagram.

Fig 10.8. Example 10.1 (*ii*)

Deflection at $B = \frac{20 \times 2 \times 4}{EI}$, due to moment of 20 kNm (Fig 10.8 (*i*) and (*ii*))

Deflection at B due to $R_B = \frac{R_B \times 5^3}{EI}$

Equating deflections,

$$R_B \times \frac{5^3}{EI} = \frac{160}{EI}$$

or $$R_B = \frac{160}{5^3} = 1.28 \text{ kN (Fig 10.8 } (iii))$$

Case 3

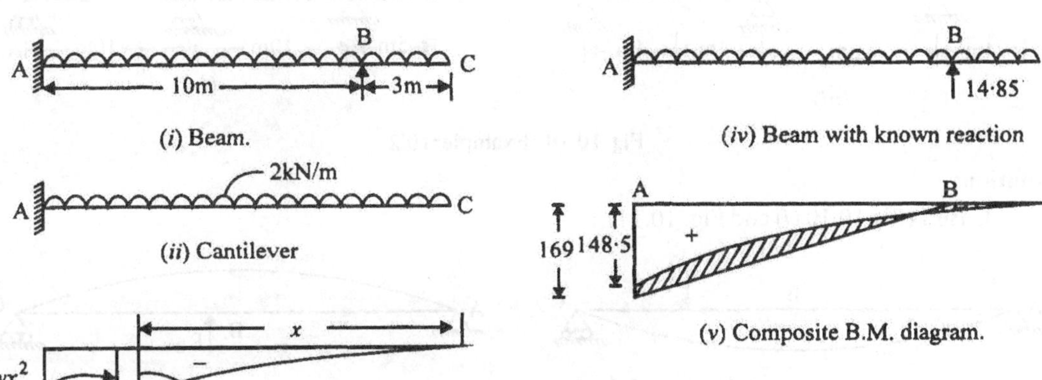

Fig 10.9 Example 10.1 (*iii*)

$$\text{Deflection at B due to load} = \int_3^{13} \frac{wx^2}{2EI}(x-3).dx$$

$$= \frac{w}{2EI}\int_3^{13}(x^3 - 3x^2)dx$$

$$= \frac{1}{EI}\left[\frac{x^4}{4} - \frac{3x^3}{3}\right]_3^{13}$$

$$= \frac{4950}{EI}$$

$$\text{Deflection due to } R_B = \frac{R_B \times 10^3}{3EI}$$

Equating deflections

$$\frac{R_B \times 1000}{3EI} = \frac{4950}{EI}$$

$$R_B = \mathbf{14.85\ kN}$$

Example 10.2. Calculate the reactions for the following continuous beams. Draw the complete B.M diagrams. Use data from Table 10.1 if available.

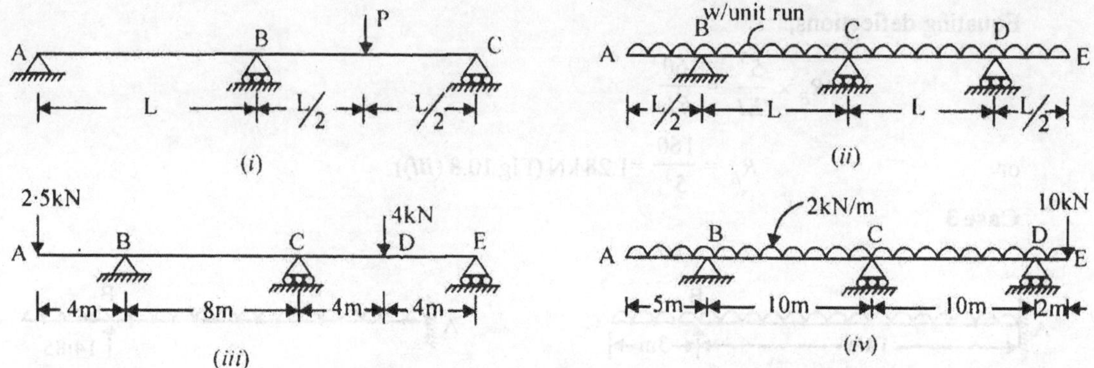

Fig 10.10. Example 10.2

Solution:

Case 1 (Fig 10.10 (*i*) and Fig. 10.11)

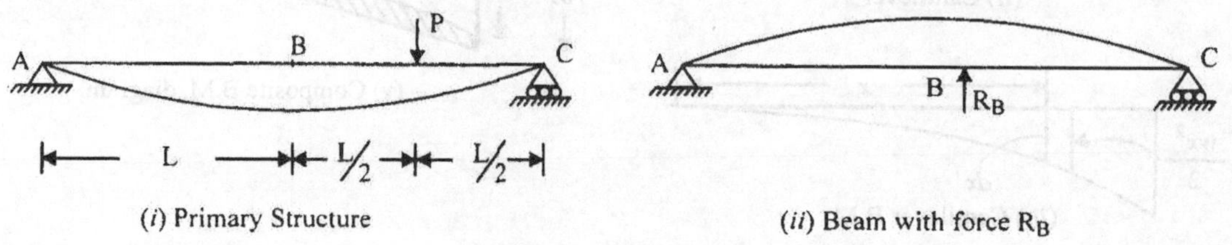

Fig. 10.11. (*i*) and (*ii*)

From Table 10.1

Deflection at B due to load P

$$= \frac{11P(2L)^3}{768EI} = \frac{11}{96}.\frac{PL^3}{EI}$$

$$\text{Deflection at } B \text{ due to } R_B = \frac{R_B(2L)^3}{48EI} = \frac{R_B.L^3}{6EI}$$

Equating Deflections

$$\frac{R_B.L^3}{6EI} = \frac{11}{96}.\frac{PL^3}{EI}$$

or $$R_B = \frac{11}{16}.P\uparrow$$

Taking moments about A

$$R_C \times 2L - P \times \frac{3L}{2} + \frac{11}{16}P.L = 0$$

or $$R_C = PL\left(\frac{3}{2} - \frac{11}{16}\right)\frac{1}{2L} = \frac{13}{32}P\uparrow$$

Taking moments about C

$$R_A \times 2L + R_B \times L - \frac{PL}{2} = 0$$

or $$R_A = \frac{1}{2L}\left(\frac{PL}{2} - R_B \times L\right)$$

$$= \frac{P}{4} - \frac{11}{32}.P = -\frac{3}{32}P$$

Check $\Sigma V = P + \frac{3}{32}.P - \frac{13}{32}.P - \frac{11}{16}.P = 0$

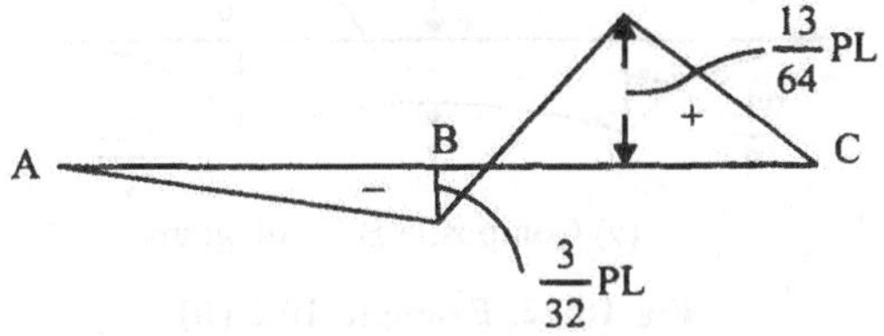

(*iii*) Composite B.M diagram.

Fig 10.11. (*iii*) Example 10.2 (*i*)

Case II (Fig 10.10 (*ii*)) and Fig. 10.12)

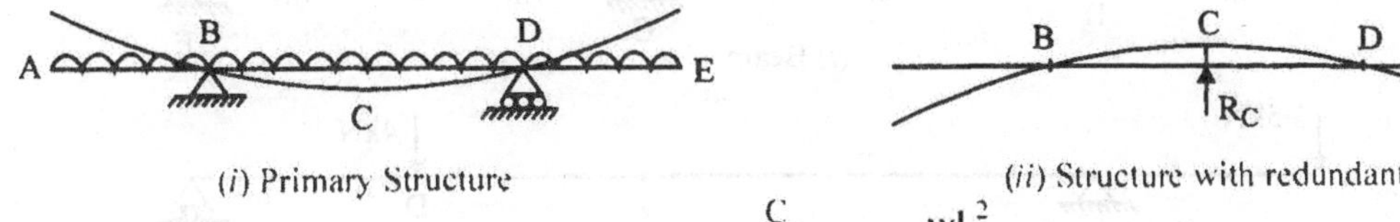

(*i*) Primary Structure (*ii*) Structure with redundant R_C

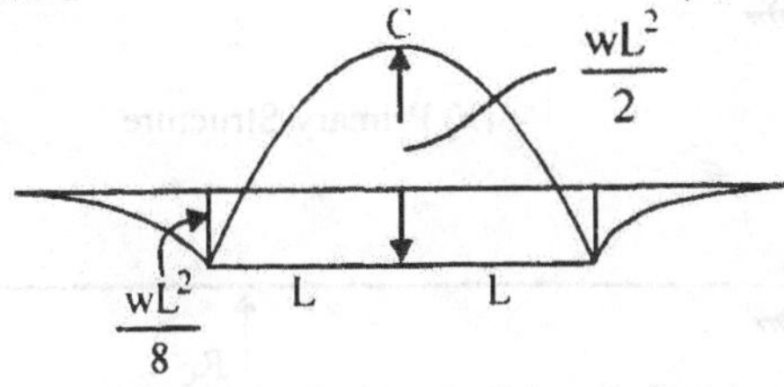

(*iii*) B.M. diagram on Primary Structure.

Fig. 10.12. (Contd)

Deflection of B with respect to tangent at C.

$$= \frac{2}{3} \times \frac{L}{EI} \times \frac{wL^2}{2} \times \frac{5}{8}.L - \frac{L}{2} \times \frac{wL^2}{8} \times \frac{L}{EI}$$

$$= \frac{wL^4}{EI}\left(\frac{5}{24} - \frac{1}{16}\right) = \frac{wL^4}{EI}.\frac{7}{48}$$

Deflection at centre due to $R_C = \frac{R_C \times (2L)^3}{48EI} = \frac{R_C.L^3}{6EI}$

Equating deflections,

$$\frac{R_C.L^3}{6EI} = \frac{wL^4}{EI}.\frac{7}{48}$$

or $R_C = \frac{7}{8}.wL$

$\frac{17}{16}$wL
$\frac{7}{8}$wL
$\frac{17}{16}$wL

(iv)

$\frac{wL^2}{16}$
$\frac{wL^2}{8}$
$\frac{wL^2}{8}$

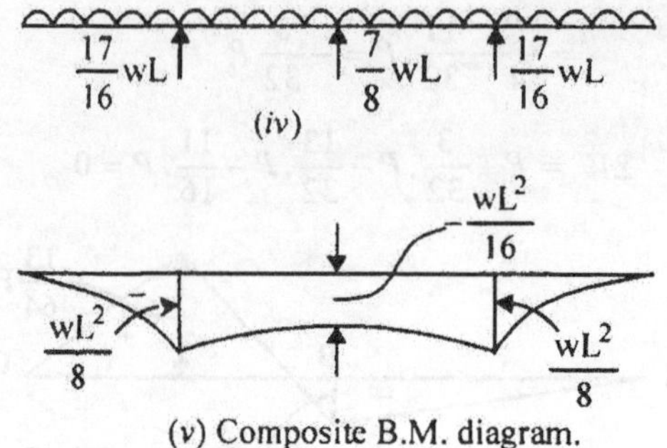

(v) Composite B.M. diagram.

Fig 10.12. Example 10.2 (*ii*)

Case III (Fig. 10.10 (*iii*) and Fig. 10.13)

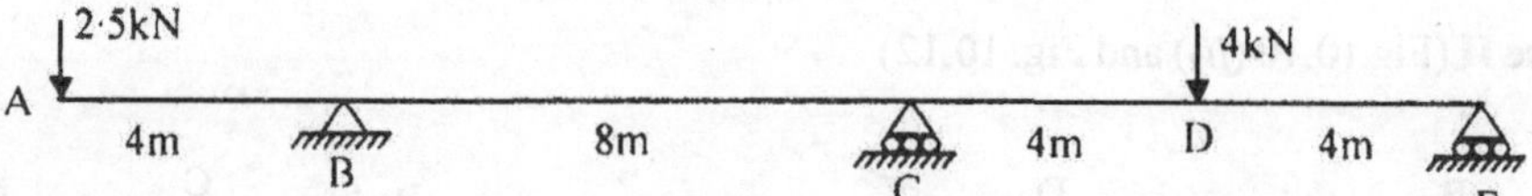

2·5kN
4kN
A
4m
B
8m
C
4m
D
4m
E

(*i*) Beam

2·5kN
4kN
A
B
C
D
E

(*ii*) Primary Structure

+

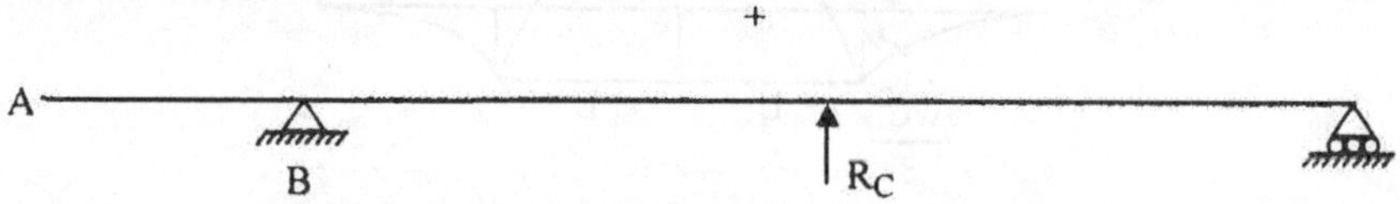

A
B
R_C

(*iii*) Redundant

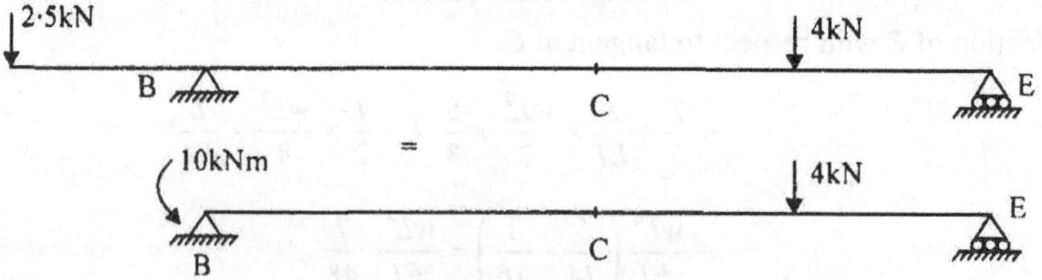

2·5kN
4kN
B
C
E
=
10kNm
4kN
E
B
C

(*iv*) Calculation of deflection at C.

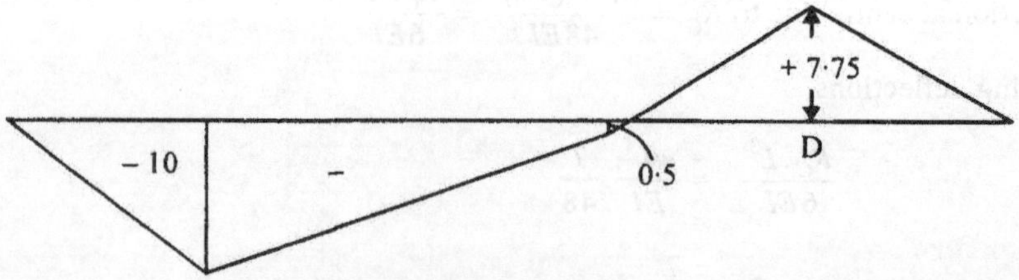

+ 7·75
– 10
–
0·5
D

(*v*) B.M. diagram.

Fig 10.13. Example 10.2 (*iii*)

Deflection at point C due to applied loads.

$$\text{Deflection due to 4 } kN = \frac{11PL^3}{768EI} = \frac{11\times4\times16^3}{768EI} = \frac{234.66}{EI}\downarrow$$

$$\text{Deflection due to moment of 10 } kNm = -\frac{ML^2}{16EI} = \frac{-10\times16^2}{16EI} = -\frac{160}{EI}$$

$$\text{Total deflection at } C \text{ due to applied load (fig 10.13(v))} = \frac{74.66}{EI}\downarrow$$

$$\text{Deflection at } C \text{ due to } R_C \text{ (Fig 10.13(iii))} = \frac{RC\times L^3}{48EI} = \frac{R_C\times16^3}{48EI}\uparrow$$

Equating deflections.

$$\frac{R_C\times16^3}{48EI} = \frac{74.66}{EI}$$

or

$$R_C = \frac{48\times74.66}{16^3} = \mathbf{0.8736}\ \boldsymbol{kN.}$$

Taking monents at E

$$-2.5\times20+R_B\times16+R_c\times8-4\times4=0$$

or

$$R_B = \frac{1}{16}[50+16-R_c\times8]$$

$$= \frac{1}{16}[50+16-.8736\times8] = \mathbf{3.688}\ \boldsymbol{kN.}$$

Taking Moments about B

$$-R_E\times16+4\times12-R_c\times8-2.5\times4=0$$

or

$$R_E = \frac{1}{16}[48-.8736\times8-10] = 1.938$$

Check $\Sigma V = 1.938+3.688+.8736-2.5-4.0 \approx 0.$

$M_B = -2.5\times4 = -10\ kNm$ $M_C = -2.5\times12+3.688\times8 = -0.5\ kNm.$

$M_D = 1.938\times4 = 7.75\ kNm$

Case (iv) (Fig. 10.10 (iv)) and Figs. 10.14 and 10.15)

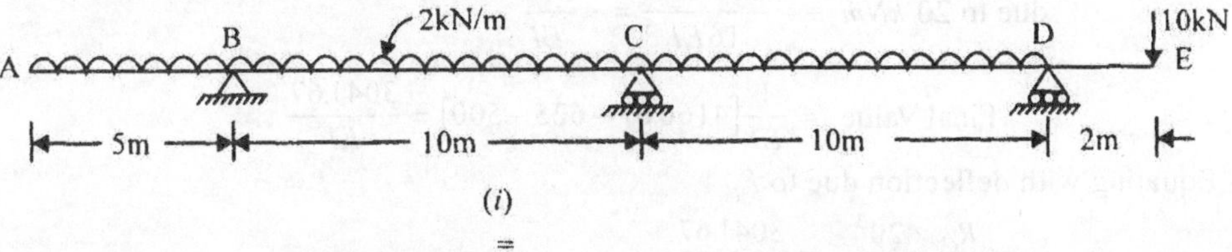

(*i*)

=

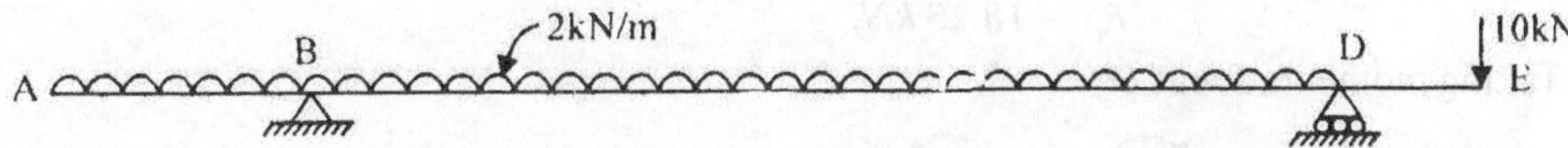

(*ii*)

+

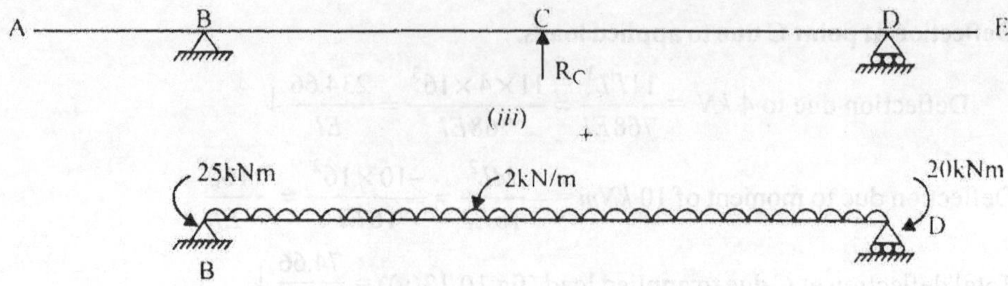

(iv) Calculation of Deflection at C.

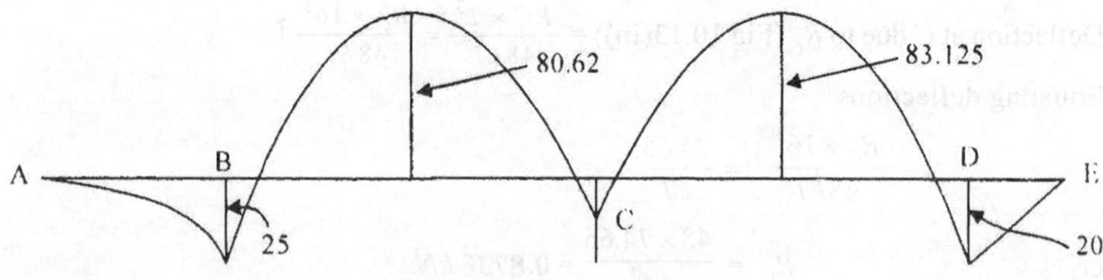

(v) B.M. diagram

Fig 10.14. Example 10.2 (iv)

Deflection at C due to applied load. Fig 10.14 (iii)

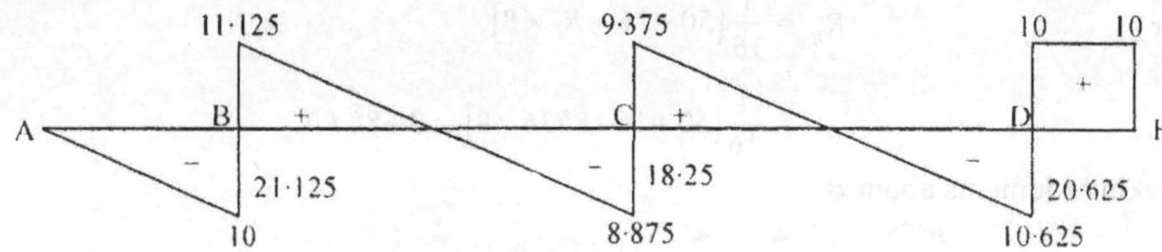

Fig. 10.15. S.F., Diagram.

Deflection at Centre due to $u.d.l.$ of 2 kN/m.

$$= \frac{5}{384} \cdot \frac{wL^4}{EI} = \frac{5}{384} \times \frac{2 \times 20^4}{EI} = \frac{4166.67}{EI}$$

$$\text{Due to 25 } kNm = -\frac{ML^2}{16EI} = -\frac{25 \times 20^2}{16EI} = -\frac{625}{EI}$$

$$\text{due to 20 } kNm = -\frac{20 \times 20^2}{16EI} = -\frac{500}{EI}$$

$$\text{Final Value} = \frac{1}{EI}[4166.67 - 625 - 500] = \frac{3041.67}{EI}$$

Equating with deflection due to R_C.

$$\frac{R_C \times 20^3}{48EI} = \frac{3041.67}{EI}$$

or R_C = **18.25 kN.**

Taking moments about D

$$R_B \times 20 - \frac{2 \times 25^2}{2} + 18.25 \times 10 + 10 \times 2 = 0$$

or $$R_B = \frac{1}{20}[625 - 182\cdot5 - 20] = \mathbf{21.125}\ kN.$$

Taking moments about B

$$R_D \times 20 - 10 \times 22 - \frac{2 \times 20^2}{2} + \frac{2 \times 5^2}{2} + 18.25 \times 10 = 0$$

or $$R_D = \frac{1}{20}[220 + 400 - 25 - 182.5] = 20.625$$

Total upward force $= 18.25 + 21.125 + 20.625$

$= 60.00 =$ Total downward force.

Support Moments

$$M_B = -\frac{2 \times 5^2}{2} = -25\ kNm.$$

$$M_C = -\frac{2 \times 15^2}{2} + 21.125 \times 10 = -13.75\ kNm.$$

$$M_D = -10 \times 2 = -20$$

B.M. diagram is shown is Fig 10.14 (v)

Example 10.3. Draw the B.M. and S.F. diagrams for the following portals (Fig 10.16)

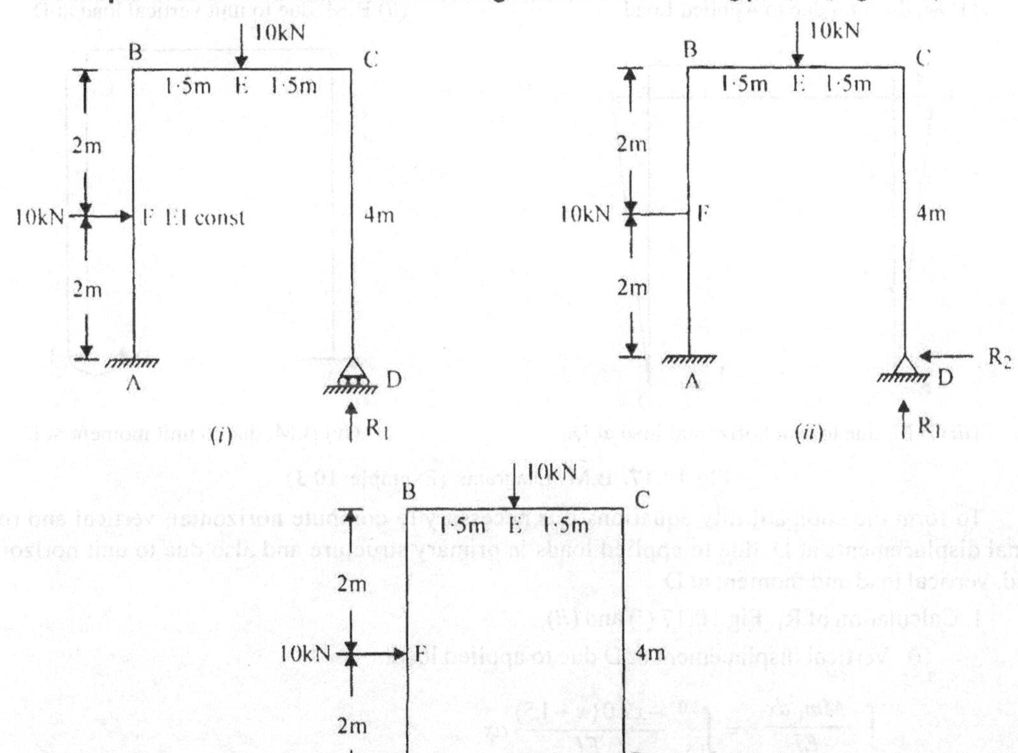

Fig 10.16. (Example 10.3)

Solution: The problems involve degree of indeterminacy of (*i*) one in Fig 10.16 (*i*), (*ii*) two in Fig 10.16 (*ii*) and (*iii*) three in Fig 10.16 (*iii*)

In Fig 10.16 (*i*) Redundant is R_1 and compatibility condition is no vertical displacement at D.

In Fig 10.16 (*ii*) redundants are R_1 and R_2 and compatibility conditions are no vertical and horizontal displacement at D.

In Fig 10.16 (*iii*) redundants are R_1, R_2 and R_3 and compatibility conditions are no horizontal, vertical and rotational displacement at D.

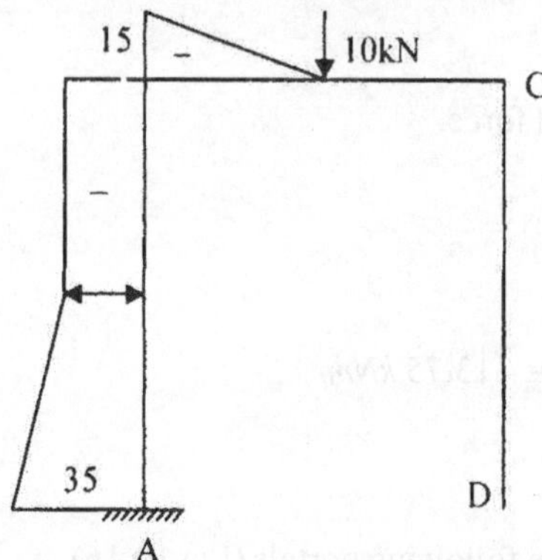

(*i*) B.M. diagram due to Applied Load.

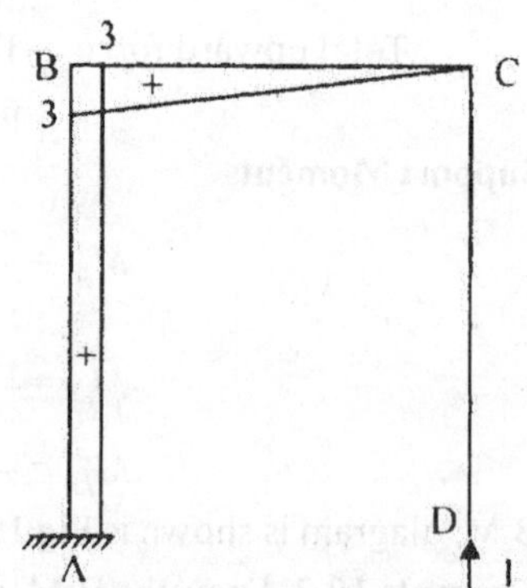

(*ii*) B.M. due to unit vertical load at D.

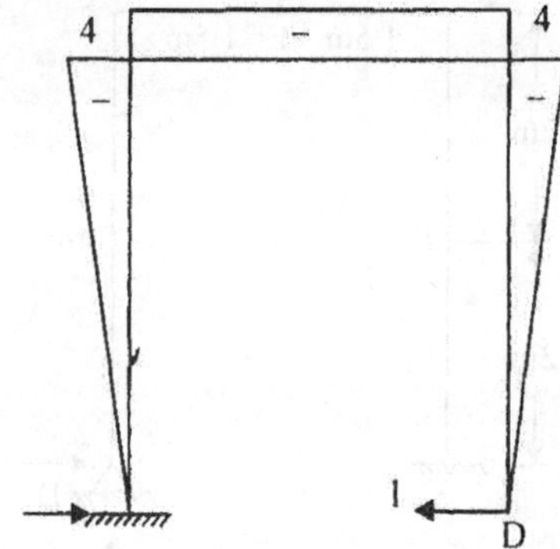

(*iii*) B.M. due to unit horizontal load at D.

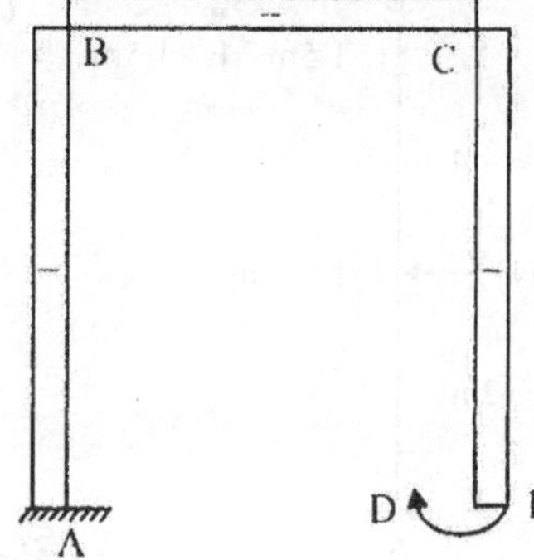

(*iv*) B.M. due to unit moment at D.

Fig 10.17. B.M. Diagrams (Example 10.3)

To form the compatibility equations it is necessary to compute horizontal, vertical and rotational displacements at D. due to applied loads in primary structure and also due to unit horizontal load, vertical load and moment at D.

1. Calculation of R_1. Fig 10.17 (*i*) and (*ii*)

(*i*) Vertical displacement at D due to applied load.

$$\int \frac{Mm_1\,dx}{EI} = \int_{1.5}^{3.0} \frac{-x.10(x-1.5)}{EI}dx$$

$$+\frac{1}{EI}\int_0^2 -3(15)\,dx + \frac{1}{EI}\int_0^2 (-3)(15+10x)\,dx$$

$$= \frac{1}{EI}\left[-10.\frac{x^3}{3}+\frac{15.x^2}{2}\right]_{1.5}^{3.0} - \frac{1}{EI}[+45x]_0^2 - \frac{1}{EI}\left[+45x+30\times\frac{x^2}{2}\right]_0^2$$

$$= \frac{1}{EI}\left[\frac{-10}{3} \times 23.625 + 7.5 \times 6.75\right] - \frac{1}{EI}[90] - \frac{1}{EI}[90 + 60]$$

$$= -\frac{1}{EI}[268.12] \text{ (downward)}$$

(*ii*) Vertical displacement at D due to unit vertical load

$$\int \frac{m_1^2\, dx}{EI} = \int_0^3 \frac{(+x)^2 . dx}{EI} + \int_0^4 \frac{9\, dx}{EI}$$

$$= \frac{1}{EI}\left[\frac{27}{3} + 36\right] = \frac{45}{EI} \text{ (upward)}$$

(*iii*) Compatibility Equation

$$-\frac{268.12}{EI} + R_1 \times \frac{45}{EI} = 0$$

or $$R_1 = \frac{268.12}{45} = \mathbf{5.958\ kN}$$

2. Calculation of R_1 and R_2 (Fig. 10.17 (*i*), (*ii*) and (*iii*))

(*i*) Horizontal displacement at D due to applied load.

$$\int \frac{M m_2 dx}{EI} = \int_0^{1.5} \frac{10.x.4.dx}{EI} + \int_0^2 \frac{x(35 - 10x)}{EI} \cdot dx + \int_2^4 \frac{x.15dx}{EI}$$

$$= \frac{1}{EI}[45 + 70 - 26.67 + 90]$$

$$= \leftarrow \frac{178.33}{EI}.$$

(*ii*) (*a*) Horizontal displacement at D due to unit vertical load.

$$\int \frac{m_1 m_2\, dx}{EI} = \int_0^3 \frac{-x(4).dx}{EI} + \int_0^4 \frac{-x(+3)\,dx}{EI}$$

$$= \frac{1}{EI}\left[-\frac{4 \times 9}{2} - \frac{3 \times 16}{2}\right]$$

$$= -\frac{42}{EI} \rightarrow$$

(b) Vertical displacement at D due to unit horizontal load $= -\frac{42}{EI} \downarrow$

(iii) Horizontal displacement at D due to unit horizontal load.

$$\int \frac{m_2^2\, dx}{EI} = 2\int_6^4 \frac{x^2\, dx}{EI} + \int_0^3 \frac{16dx}{EI}$$

$$= \frac{1}{EI}\left[\frac{2 \times 64}{3} + 48\right]$$

$$= \leftarrow \frac{90.66}{EI}$$

(*iv*) Compatibility conditions at D

Horizontal displacement at $D = O$

Vertical displacement at $D = O$

The compatibility equations are :

$$\text{Horizontal displacement } \Delta H = \underbrace{\frac{178.33}{EI}}_{\text{due to primary load}} + \underbrace{R_2\left(\frac{90.66}{EI}\right)}_{\text{due to } R_2} + \underbrace{R_1\left(-\frac{42}{EI}\right)}_{\text{due to } R_1} = 0$$

$$\text{Vertical displacement } \Delta V = \underbrace{-\frac{268.12}{EI}}_{\text{due to Primary Load}} + \underbrace{R_2\left(-\frac{42}{EI}\right)}_{\text{due to } R_2} + \underbrace{R_1\left(\frac{45}{EI}\right)}_{\text{due to } R_1} = 0$$

or $\quad -42\,R_1 + 90.66\,R_2 = -178.33,\ 45\,R_1 - 42\,R_2 = 268.12$

Solving $\quad R_1 = 7.2608$ kN

$R_2 = 1.3964$ kN.

3. Calculation of R_1, R_2 and R_3. Fig. 10.17 (*i*) to and (*iv*)

(*i*) Rotation at D due to applied load.

$$\int \frac{M\,m_3\,.dx}{EI} = \int_0^{1.5} \frac{10x\,.1.dx}{EI} + \int_0^{2} \frac{15.1.dx}{EI} + \int_0^{?} \frac{(15+10x).1.dx}{EI}$$

$$= \frac{1}{EI}[11.25 + 30 + 30 + 20]$$

$$= \frac{91.25}{EI}$$

(*ii*) Rotation at D due to unit vertical force

$$\int \frac{m_1 m_3 .dx}{EI} = \int_0^{3} -\frac{x.1.dx}{EI} + \int_0^{4} \frac{-3.1.dx}{EI}$$

$$= \frac{1}{EI}\left[-\frac{3^2}{2} - 3\times 4\right]$$

$$= -\frac{16.5}{EI}$$

(*iii*) Rotation at D due to unit horizontal force

$$\int \frac{m_2 m_3\, dx}{EI} = 2\int_0^{4} \frac{x.1.dx}{EI} + \int_0^{3} \frac{4.1.dx}{EI}$$

$$= \frac{1}{EI}\left[2.\frac{4^2}{2} + 4\times 3\right]$$

$$= \frac{28}{EI}$$

(*iv*) Vertical displacement at D due to unit moment

$$\int \frac{m_3\, m_1\, dx}{EI} = -\frac{16.5}{EI}$$

(v) Horizontal displacement at D due to unit moment.

$$\int \frac{m_3 m_2}{EI} . dx = \frac{28}{EI}$$

(vi) Rotation at D due to unit moment

$$\int \frac{m_3^2 . dx}{EI} = 2\int_0^4 \frac{(1)^2 . dx}{EI} + \int_0^3 \frac{(1)^2 . dx}{EI} = \frac{11}{EI}$$

(vii) **Compatibility conditions at D**

Horizontal displacement

$$= \underbrace{\frac{178.33}{EI}}_{\text{Due to applied load}} + \underbrace{R_1\left(\frac{-42}{EI}\right)}_{\text{due to } R_1} + \underbrace{R_2\left(\frac{90.66}{EI}\right)}_{\text{due to } R_2} + \underbrace{R_3\left(\frac{28}{EI}\right)}_{\text{due to moment } R_3} = 0$$

Vertical displacement

$$= \frac{-268.12}{EI} + R_1\left(\frac{45}{EI}\right) + R_2\left(\frac{-42}{EI}\right) + R_3\left(-\frac{16.5}{EI}\right) = 0.$$

Rotational displacement

$$\frac{91.25}{EI} + R_1\left(\frac{-16.5}{EI}\right) + R_2\left(\frac{+28}{EI}\right) + R_3\left(\frac{11}{EI}\right) = 0$$

or $\quad -R_1 + 2.159\,R_2 + 0.667\,R_3 = -4.246,\; R_1 - 0.933\,R_2 - 0.367\,R_3 = 5.958$

$-R_1 + 1.696\,R_2 + 0.667\,R_3 = -5.530.$

Solving $\quad R_1 = 6.48, \qquad R_2 = 2.7732, \qquad R_3 = -5.626.$

10.4 NOTES FOR DRAWING THE B.M. DIAGRAM FOR PORTALS. (FIG 10.18, 10.19 AND 10.20)

(i) The sign convention for beam bending moment (sagging + and Hogging-) is not suitable for vertical members. Hence for portals separate convention (outside compression is positive) and plotted on compression side is followed.

(2) To draw the B.M. diagram compute the B.M. at critical points as A,B.C,.D. These are Joints moments and should be plotted on the compression side.

(3) Over the Joint moments is superposed the span moment diagram. Care should be taken about the sign of the span moment.

Plotted on the compression side Compression outside +

$M_D = 0$

$M_C = 0 \; M_E = +8.937$

$M_B = 5.958 \times 3 - 10 \times 1.5$

$= +2.874$

$M_F = +2.874$

$M_A = 5.958 \times 3 - 10 \times 1.5 - 10 \times 2$

$= -17.126.$

Fig. 10.18 B.M. diagram with unit redundancy.
(Fig. 10.16 (i))

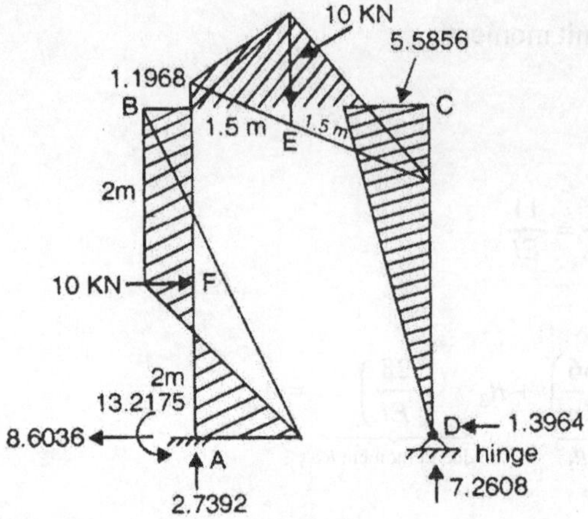

Fig. 10. 19. B.M. diagram with two redundancies (Fig. 10.16 (*ii*))

$$M_C = -1.3964 \times 4$$
$$= -5.5856.$$
$$M_E = 7.2608 \times 1.5 - 5.5856$$
$$= 5.3056$$
$$M_B = 7.2608 \times 3 - 1.3964 \times 4 - 10 \times 1.5$$
$$= 1.1968$$
$$M_F = 1.1968 + 1.3964 \times 2$$
$$= 3.9896$$
$$M_A = -13.2176$$
$$M_D = 0$$

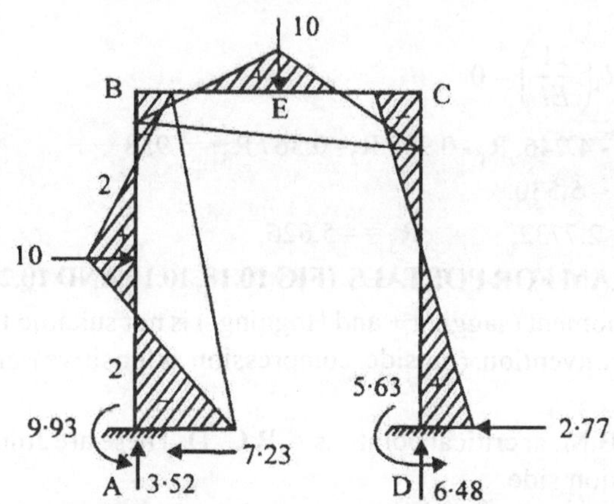

Fig. 10.20. B.M. diagram with three redundancies. (Fig. 10.16 (*iii*))

$$M_D = +5.63$$
$$M_C = +5.63 - 2.77 \times 4$$
$$= -5.45.$$
$$M_E = +5.63 - 2.77 \times 4 + 6.48 \times 1.5$$
$$= +4.27.$$
$$M_B = 5.63 + 6.48 \times 3 - 2.77 \times 4 - 15$$
$$= -1{\cdot}010.$$
$$M_F = 5.63 + 6.48 \times 3 - 2.77 \times 2 - 15$$
$$= 4.53$$
$$M_A = 5.63 + 6.48 \times 3 - 15 - 20$$
$$= -9.93.$$

Example 10.4. If in the previous example support D moves vertically downward by 1 cm, horizontally by 2 cm to the right and rotates by 0.001 radian clockwise what will be the reactions R_1, R_2 and R_3 ?

$EI = 20{,}000\ kNm^2$ when there is

(*i*) No Load in the structure

(*ii*) Applied load as given in Example 10.3.

Case 1 There is no load in the structure

(*i*) One degree indeterminate

$$R_1 \times \frac{45}{EI} = -.01\ m$$

or $$R_1 = \frac{EI}{45}.(-.01) = -4.444\ kN$$

(*ii*) Two degrees indeterminate

$$R_2\left(\frac{90.66}{EI}\right)+R_1\left(-\frac{42}{EI}\right) =-\cdot 02\ m$$

$$R_2\left(-\frac{42}{EI}\right)+R_1\left(\frac{45}{EI}\right) =-\cdot 01\ m.$$

or
$$90.66\,R_2-42\,R_1 =-400$$
$$-42\,R_2+45\,R_1 =-200$$

Solving
$$R_1 =-14.561$$
$$R_2 =-11.156.$$

(*iii*) Three degree indeterminate .

$$R_1\left(-\frac{42}{EI}\right)+R_2\left(\frac{90.66}{EI}\right)+R_3\left(\frac{28}{EI}\right)=-.02$$

$$R_1\left(\frac{45}{EI}\right)+R_2\left(-\frac{42}{EI}\right)+R_3\left(-\frac{16}{EI}\right)=-.01$$

$$R_1\left(-\frac{16.5}{EI}\right)+R_2\left(\frac{28}{EI}\right)+R_3\left(\frac{11}{EI}\right)=+.001$$

Substituting the value of $EI = 20{,}000\ kNm^2$ and solving

$$R_1 =-8.3957$$
$$R_2 =-23.2580$$
$$R_3 =48.4327$$

Case II

With applied load and support movements. The equations are modified only by support movements as specified in this example . Hence the equations are :

$$\frac{178.33}{EI}+R_1\left(-\frac{42}{EI}\right)+R_2\left(\frac{90.66}{EI}\right)+R_3\left(\frac{28}{EI}\right)=-.02$$

$$-\frac{268.12}{EI}+R_1\left(\frac{45}{EI}\right)+R_2\left(-\frac{42}{EI}\right)+R_3\left(-\frac{16.5}{EI}\right)=-.01$$

$$\frac{91.25}{EI}+R_1\left(-\frac{16.5}{EI}\right)+R_2\left(\frac{28}{EI}\right)+R_3\left(\frac{11}{EI}\right)=.001$$

or
$$178.33-42\,R_1+90.66\,R_2+28\,R_3 =-400$$
$$-268.12+45\,R_1-42\,R_2-16.5\,R_3 =-200$$
$$91.25-16.5\,R_1+28\,R_2+11\,R_3 =+20$$

Solving the above three equations.

$$R_1 =-1.913,\ R_2 =-20.412,\ R_3 =42.56$$

This solves the case when support D is fixed. When support D is hinged, there are only two unknowns R_1 and R_2.

The equations then redece to

$$\frac{178.33}{EI}+R_2\left(\frac{90.66}{EI}\right)+R_1\left(-\frac{42}{EI}\right)=-.02$$

$$-\frac{268.12}{EI}+R_2\left(-\frac{42}{EI}\right)+R_1\left(\frac{45}{EI}\right)=-.01$$

$$\text{or} \qquad 90.66\,R_2 - 42\,R_1 = -400 - 178.33 = -578.33$$

$$-42\,R_2 + 45\,R_1 = -200 + 268.12 = +68.12$$

Solving $\qquad R_1 = -7.813\ kN,\ R_2 = -9.996\ kN.$

When support D is roller, there is only one redundant R_1 and the equation becomes

$$-\frac{268.12}{EI} + \frac{R_1 \times 45}{EI} = -0.01$$

or $\qquad R_1 \times 45 = -200 + 268.12 = 68.12$

or $\qquad R_1 = +1.5130$

Table 10.2. Analysis of Results of Portal Frame.

		R_1	R_2	R_3
With Load only	One redundant	5.958	0	0
	Two redundants	7.268	1.3964	0
	Three redundants	6.48	2.7732	−5.626
With Support Settlements only	One redundant	−4.444	0	0
	Two redundants	−14.561	−11.156	0
	Three redundants	−8.3937	−23.258	+48.4327
With Load and Support Settlements	One redundant	+1.5130	0	0
	Two redundants	−7.813	−9.996	0
	Three redundants	−1.913	−20.412	+42.56

Example 10.5 Analyse the continuons beam show in Fig 10.21 to draw the B.M. diagram. Take internal moments as redundants.

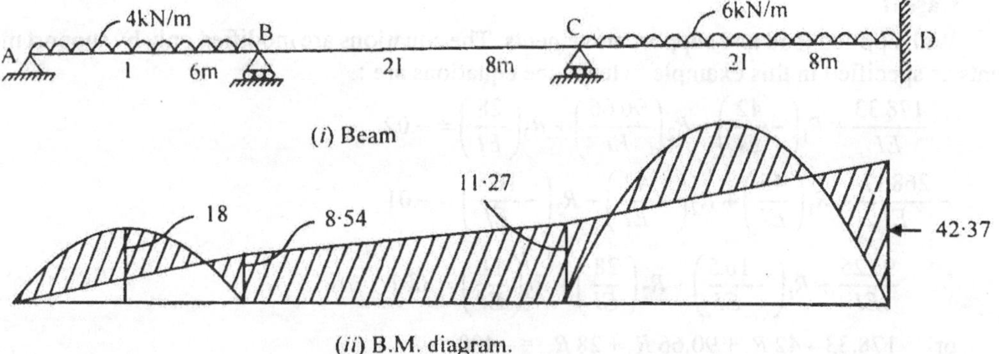

Fig 10.21. Example 10.5

Solution:

Here internal moments at B, C and D are taken as redundants. M_A is obviously zero. The unknown moments are taken as positive.

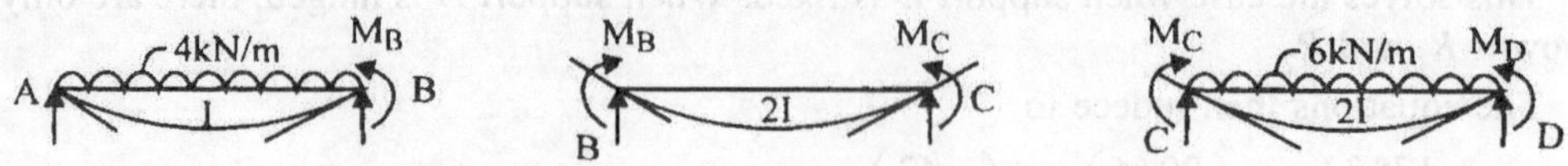

Fig 10.22. *FBDs* of Span *AB*, *BC* and *CD*.

By splitting the beam into three simply supported beams AB, BC and CD continuity of supports at A and B have been destroyed. As a result there will be discontinuous slopes at B, C and D. The internal moments M_B, M_C and M_D will restore continuity of the structure *i.e.* relative slope at B and C will be zero and slope at D will also be zero.

$$\text{For span } AB \text{ slope at } B = \frac{4\times 6^3}{24EI} + \frac{M_B.6}{3EI}$$

$$\text{For span } BC \text{ slope at } B = \frac{M_B.8}{6EI} + \frac{M_C.8}{12EI}$$

$$\text{For span } BC \text{ slope at } C = \frac{M_C\times 8}{6EI} + \frac{M_B\times 8}{12EI}$$

$$\text{For span } CD \text{ slope at } C = \frac{M_C\times 8}{6EI} + \frac{6\times 8^3}{48EI} + \frac{M_D\times 8}{12EI}$$

$$\text{For span } CD \text{ slope at } D = \frac{6\times 8^3}{48EI} + \frac{M_D\times 8}{6EI} + \frac{M_C\times 8}{12EI}$$

$$\text{Relative slope at } B = \frac{4\times 6^3}{24EI} + \frac{6M_B}{3EI} + \frac{8M_B}{6EI} + \frac{8M_C}{12EI} = 0$$

or $\quad 80\,M_B + 16\,M_C = -4\times 6^3.$

$$\text{Relative slope at } C = \frac{8M_C}{6EI} + \frac{8M_B}{12EI} + \frac{8M_C}{6EI} + \frac{8M_D}{12EI} + \frac{6\times 8^3}{48EI} = 0$$

or $\quad M_B + 4M_C + M_D = -96$

Slope at $D = 0$

$$\frac{8M_C}{12EI} + \frac{8M_D}{6EI} + \frac{6\times 8^3}{48EI} = 0$$

or $\quad M_C + 2M_D = -96$

Solving the three equations

$$M_B = -8.546\ kNm,\ M_C = -11.27\ \text{kNm}.$$

and $\quad M_D = \mathbf{-42.37\ kNm.}$

Example 10.6. The beam of Fig 10.23 is subjected to a concentrated load P at $L/4$ of the beam., where L is the span. Compute the redundant reaction R_C

(i) when there is no support settlement

(ii) when support settlement is Δ

(iii) when the support is a spring with spring stiffness K.

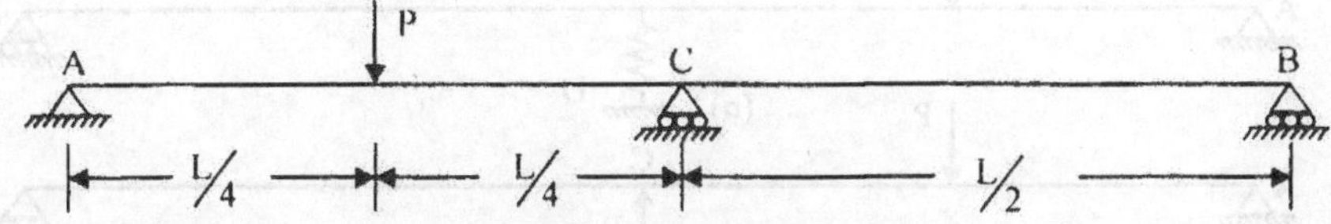

Fig 10.23 — Example 10.6

Solution:

The indeterminate beam can be split into two simply supported beams and compatibility conditions at C is then applied to compute the reaction R_C.

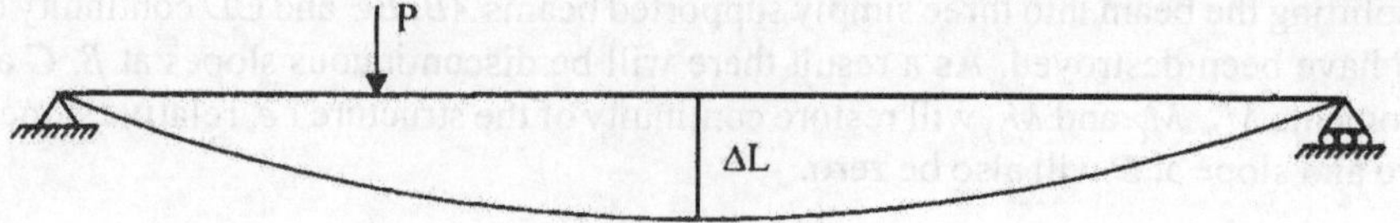

(*i*) Primary Structure with Applied load.

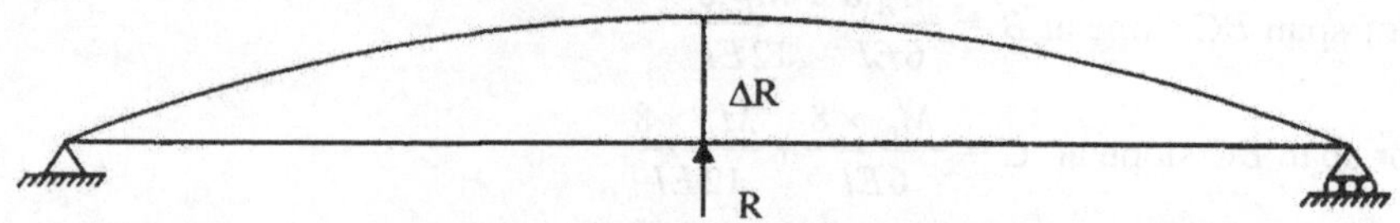

(*i*) Structure subjected to redundant force R.

Fig. 10.24. Consistent Deformation.

Dis placement at C due to load P at $L/4$

$$\Delta_L = \frac{11PL^3}{768EI} \quad [Fig. 10.24\,(i)]$$

Due to reaction in R $\quad \Delta_R = \frac{RL^3}{48EI} [Fig 10.24(ii)]$

If Δ_C is the settlement at C

$$-\frac{11PL^3}{768EI} + \frac{RL^3}{48EI} = -\Delta_C$$

Upward direction of R is taken positive. Hence displacement upward is taken positive, settlement of support which occurs in opposite duection to R is, therefore, taken as negative

Solving
$$R = \frac{48EI}{L^3}\left(\frac{11PL^3}{768} - \Delta_C\right)$$

$$= \frac{11}{16}.P - \frac{48EI}{L^3}.\Delta_C.$$

Notice that Δ_C tends to reduce the value of the redundant when Δ_C is zero i.e no support dis placement, $R = \frac{11}{16} P.$

The free body diagram of the beam and the spring is shown in Fig 10.25 (*b*) and (*c*). It can be seen that while the spring reaction acts upward on the beam, the spring deformation is downward i.e. in opposite direction to the redundant R.

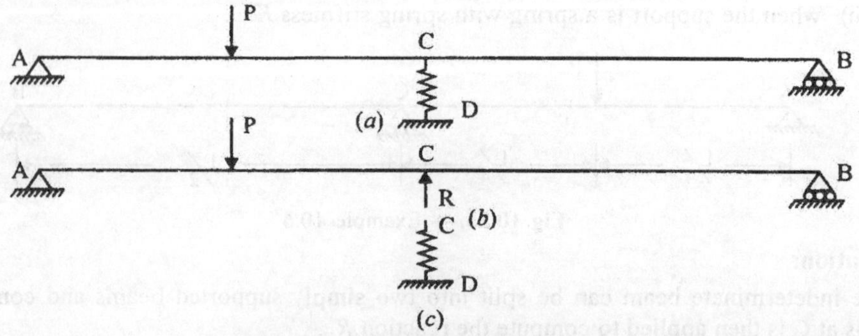

Fig. 10.25 — With Spring Support at C.

Hence the compatibility condition is again

$$\Delta_L + \Delta_R = -\Delta\, spring$$

$$-\frac{11PL^3}{768EI} + \frac{RL^3}{48EI} = -\frac{R}{K} \text{ where } K \text{ is the spring shiffness.}$$

or

$$R = \frac{\dfrac{11PL^3}{768EI}}{\dfrac{L^3}{48EI} + \dfrac{1}{K}}$$

Effect of spring stiffness on R can now be studied. When $K \to 0$ *i.e.* spring offers negligible resistance $\frac{1}{K} \to \infty$ and $R \to 0$ *i.e.*, it appears there is no support at R. On the other hand when $K \to \infty \frac{1}{K} \to 0$ and $R = \frac{11}{16}.P$ *i.e* support becomes rigid.

For any other value of K, there is finite deformation and R gets reduced.

10.5 SOLUTION OF STATICALLY INDETERMINATE TRUSSES

Statically indeterminate trusses may be

(*a*) Externally indeterminate : Support reactions are more than that can be solved by statics alone.

(*b*) Internally indeterminate : when the number of truss members are more than what can be determined statically.

(*c*) A combination of above.

Degree of indeterminacy in a truss $= m + r - 2n$

where m = no. of members

r = no. of reactions

n = no. of joints

Normally, it is applied load which causes forces in the numbers. But they may also be due to (*i*) support settlement (*ii*) environmental effects such as temperature rise or fall (*iii*) fabrication error i.e. any member too large or too small causing force fit.

Basic approach in solving statically indeterminate truss is the same as that followed in solving statically indeterminate beams or frames as already explained. In the case of externally indeterminate truss, the redundant support is removed and displacement at the support is computed. As there is no displacement at the support, the support reaction should be such that when applied as load the truss will have equal and opposite deflection.

In the case of internally indeterminate truss, the indeterminacy is removed by making a cut in the redundant member (s). As a result a gap occurs in the cut. By applying suitable force/ forces in the redundant member/members the cut/cuts are closed. These are explained by means of illustrative examples.

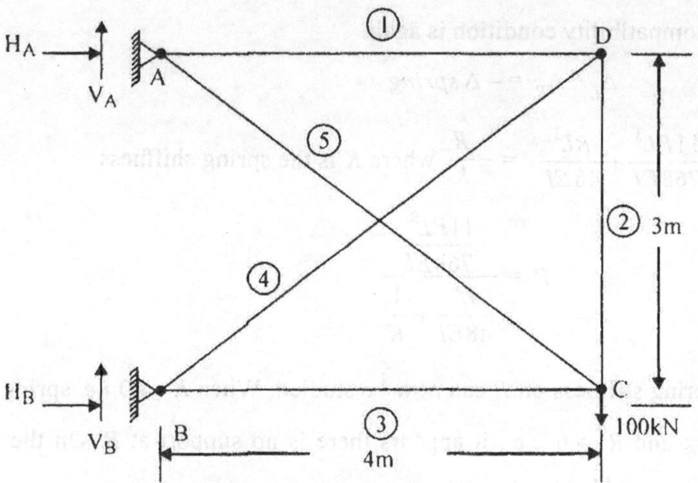

Fig. 10.26. Example 10.7

Example 10.7. Analyse the truss shown. All areas are the same. (Fig 10.26).

Solution : The truss is externally redundant as we have four unknowns H_A, V_A, H_B, V_B at supports with only three equations of statics. Once the reactions are known all the member forces can be computed.

From the formula D.O.I $= (m + r) - 2n$

$= (5 + 4) - 2 \times 4 = 1$

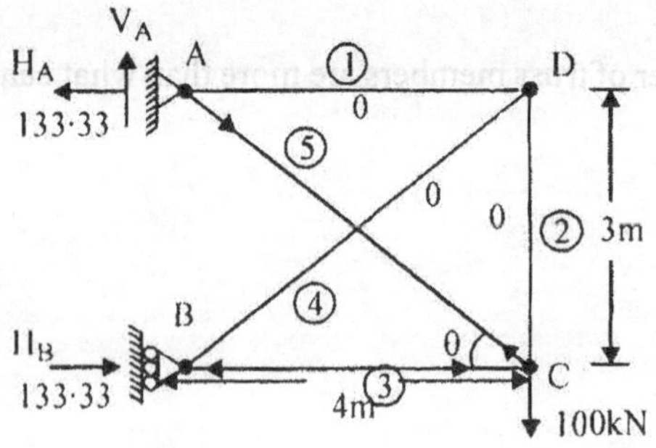

(*i*) Primary Structure

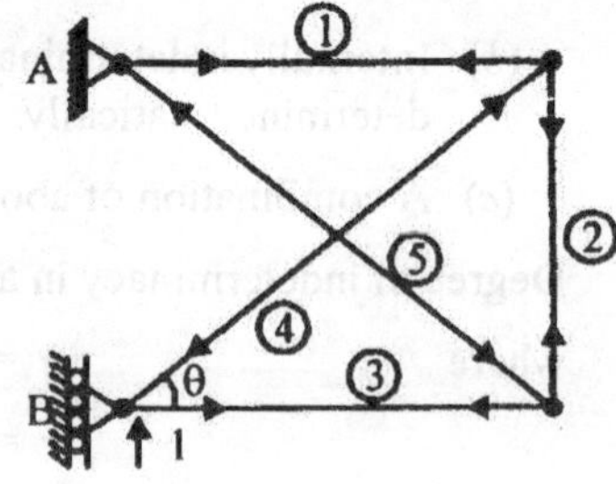

(*ii*) Truss subjected to unit vertical forces

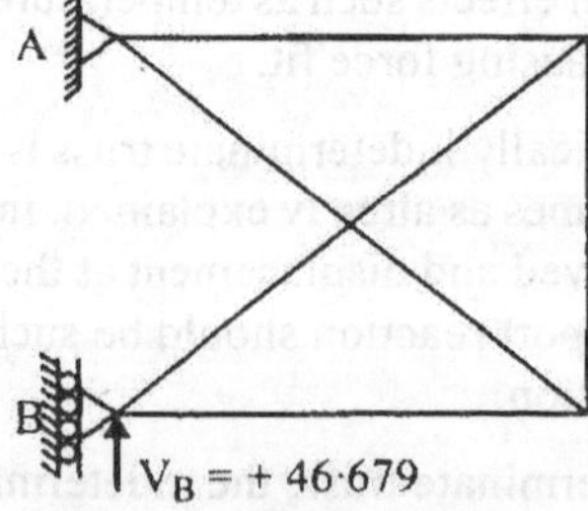

(*iii*) Value of forces

Fig. 10.27. Example 10.7

1. The redundancy is first removed by making the support B a roller. Forces in the statically determinate truss is then calculated.

$$100 \times 4 = H_B \times 3$$

or $H_B = \frac{400}{3} = 133.33\ kN$

$V_A - 100 = 0$

or $V_A = 100\ kN.$

Considering joint B

$F_3 = H_B = -133.33$

and $F_4 = 0$ as there is no vertical force at joint B.

Considering joint C

$F_5 \cos\theta = F_3 = 133.33\ kN$

$$F_5 = \frac{133.33 \times 5}{4} = \mathbf{+166.625\ kN}$$

$F_5 \sin\theta + F_2 = 100$

$$166.625 \times \frac{3}{5} + F_2 = 100$$

or $F_2 = 100 - 100 = 0$

2. Since B is a roller we next find the movement of the roller B. To do this a unit load is applied at B and force in all the member are found out.

Joint B $f_4 \sin\theta = 1$

$$f_4 = \frac{1}{\sin\theta} = \frac{1}{3/5} = \frac{5}{3} = 1.667 \text{ (comp.)}$$

$$f_2 = f_4 . \sin\theta = 1.667 \times \frac{3}{5} = +1$$

$$f_3 = f_4 . \cos\theta = 1.667 \times \frac{4}{5} = +1.333$$

$$f_1 = f_4 \cos\theta = +1.333$$

$$f_5 = -1.667$$

3. The third step is to compute vertical displacement of the joint B in the primary truss. This is given by the formula

$$l.\Delta = \Sigma \frac{Ff.L}{AE}$$

where F = Forces in the primary structure

f = forces in the member due to unit load.

The same sign convention tension + and compression – is followed. The result is computed in Table 10.3.

Table 10.3 Example 10.7

Member	F (kN)	L (m)	f	f^2	FfL	f^2L	fV_B	$F+fV_{B'}$
1.	0	4	+1.333	+1.776	0	+7.104	62.223	+62.223
2.	0	3	+1.000	+1.000	0	+3.000	46.679	+46.679
3.	–133.33	4	+1.333	+1.776	–710.91	+7.104	+62.223	–71.107
4.	0	5	–1.667	+2.778	0	+13.89	–77.814	–77.814
5.	166.66	5	–1.667	+2.778	–1389.11	+13.89	+77.814	+88.846

$V_B \times 44.988$ — Displacement in the direction of unit load (upward)

$-2100.02 = 0$ — Displacement in the opposite direction of unit load (downward) or $\quad V_B = 46.679$

Example 10.8 For all members AE = constant. Calculate the member forces and reactions.

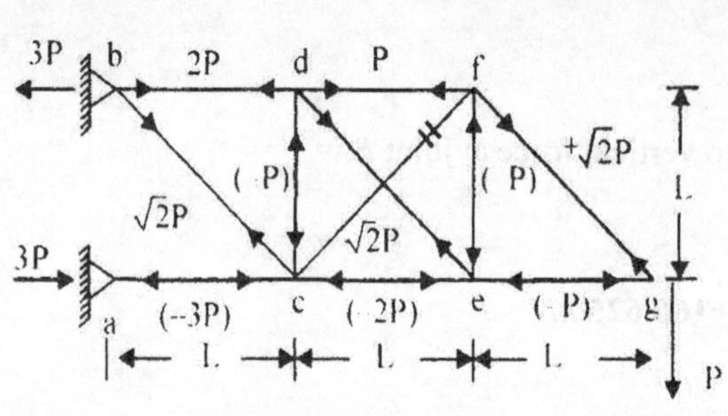

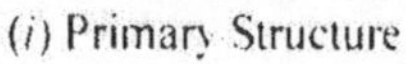

(i) Primary Structure

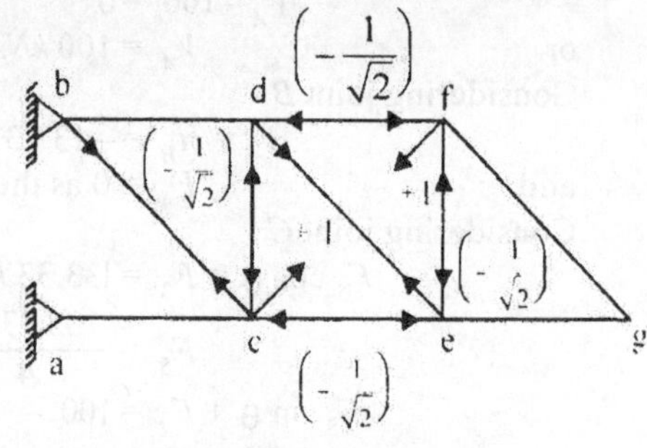

(ii) Forces due to unit load along cf.

Fig. 10.28. Example 10.8

Solution : Taking cf as redundant, forces in the determinate structure are

Joint g. $\quad F_{fg}\dfrac{1}{\sqrt{2}} = P$ or $F_{fg} = \sqrt{2}P$ (tension)

$$F_{fg}\cdot\frac{1}{\sqrt{2}} = F_{ge} = \frac{\sqrt{2}P}{\sqrt{2}} = P \text{ (Compression)}$$

Joint f. $\quad F_{fe} = \dfrac{1}{\sqrt{2}}F_{fg} = \dfrac{\sqrt{2}P}{\sqrt{2}} = P$ (Compression)

$F_{fd} = P$. (tension)

From Joint e, $\quad F_{ed} = \sqrt{2}P$ (tension)

$F_{ec} = 2P$ (compression)

From joint d. $\quad F_{bd} = 2P$ (tension)

$F_{dc} = P$ (compression)

From joint C $\quad F_{cb} = \sqrt{2}P \cdot F_{ca} = 3P$ (Compression)

Forces in the members due to unit load along fc.

f_{ge} and f_{gf} are both zero

$$f_{fd} = -\frac{1}{\sqrt{2}} \text{ (Comp)} \quad f_{fe} = -\frac{1}{\sqrt{2}} \text{ (Comp)}$$

$$f_{ed} = +1 \qquad f_{ec} = -\frac{1}{\sqrt{2}} \text{ (comp).}$$

$$f_{cd} = -\frac{1}{\sqrt{2}} \text{ (Comp)} \quad f_{bd} = f_{bc} = f_{ac} = 0$$

Tabel 10.4 Example 10.8

Member	Length	F	f	LFf	Lf^2	X	$F+X.f$
bd	L	$2P$	0	0	0		$2P$
df	L	P	$-\frac{1}{\sqrt{2}}$	$-\frac{PL}{\sqrt{2}}$	$\frac{L}{2}$		$1.6035P$
fg	$\sqrt{2}L$	$\sqrt{2}P$	0	0	0		$\sqrt{2}P$
ge	L	$-P$	0	0	0		$-P$
ec	L	$-2P$	$-\frac{1}{\sqrt{2}}$	$+\frac{2PL}{\sqrt{2}}$	$\frac{L}{2}$		$-1.396P$
ca	L	$-3P$	0	0	0		$-3P$
bc	$\sqrt{2}L$	$\sqrt{2}P$	0	0	0		$\sqrt{2}P$
cd	L	$-P$	$-\frac{1}{\sqrt{2}}$	$\frac{PL}{\sqrt{2}}$	$\frac{L}{2}$		$-0.3964P$
cf	$\sqrt{2}L$	0	$+1$	0	$\sqrt{2}L$		$-0.8535P$
de	$\sqrt{2}L$	$\sqrt{2P}$	$+1$	$2PL$	$\sqrt{2}L$		$0.5607P$
fe	L	$-P$	$-\frac{1}{\sqrt{2}}$	$\frac{PL}{\sqrt{2}}$	$\frac{L}{2}$		$-0.3964P$

$$\Sigma\, PL\left(\frac{3}{\sqrt{2}}+2\right)\left(2L+2\sqrt{2}L\right)$$

$$X = -\frac{PL\left(\frac{3}{\sqrt{2}}+2\right)}{L\left(2+2\sqrt{2}\right)} = -\mathbf{0.8535\ P.}$$

Example 10.9. Find the forces in all members of the truss shown in fig 10.29.

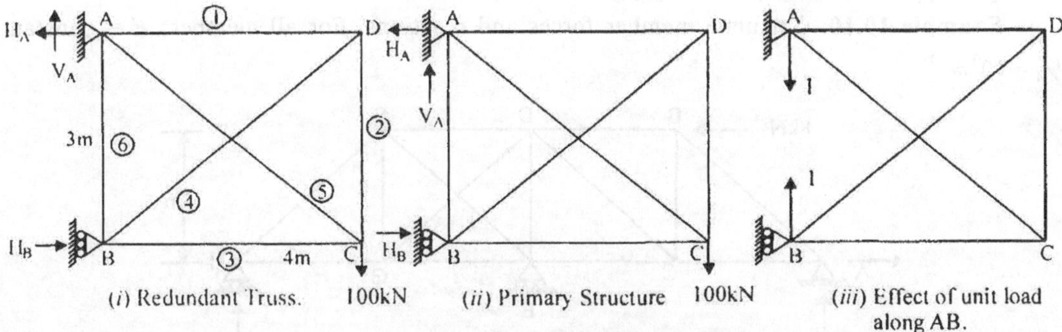

(*i*) Redundant Truss. (*ii*) Primary Structure (*iii*) Effect of unit load along AB.

Fig 10.29. Example 10.9.

Solution : This truss is externally determinate because of roller at B instead of hinge as given in Example 10.7 However it is internally redundant because of additional member 6.

$$\text{D.O.I.} = (m+r)-2n$$
$$= (6+3)-2\times 4 = 1$$

1. The truss is made internally determinate by making a cut in member 6. It is already externally dcterminate, hence forces is in all the membe^rs can be determined. This is already known from previous example.
2. Because of cut in member 6, there will be gap in member 6. This gap is first computed by applying unit load method. This is nothing but the movement of support B. (already computed)
3. Finally as there is no gap in the member the gap is closed by applying a force R along the member 6. The results are obtained by tabulation.

Table 10.5 (Example 10.9)

Member	$F\,(kN)$	$L\,(m)$	f	f^2	FfL	f^2L	$f.R$	Final force $F+fR$
1	0	4	+1.333	+1.776	0.0	7.104	58.332	58.332
2	0	3	+1.000	+1.000	0.0	3.000	43.76	43.76
3	−133.33	4	+1.333	+1.776	−710.91	7.104	58.332	−74.998
4	0	5	−1.667	+2.778	0	13.890	−72.947	−72.947
5	166.66	5	−1.667	+2.778	−1389.11	13.890	−72.947	+93.713
6	0	3	+1.000	+1.000	0.00	3.000	43.76	43.76
					Σ−2100.02	Σ47.988		

$$\frac{-2100.02}{EI} + R.47.988 = 0$$

displacement in the opposite direction to unit load   displacement in the direction of unit load

or $$R = \frac{2100.02}{47.988.} = \mathbf{43.76\ kN.}$$

This is less than $V_B = 46.679kN$ obtained in the previous example (why ?)

The reason is elastic member 6 allows some movement of the support B, hence the value is less than rigid support value. Final forces are given in Table 10.5.

Example 10.10. Calculate member forces and reactions. For all members E = constant, $L/A = 10^3\,m^{-1}$

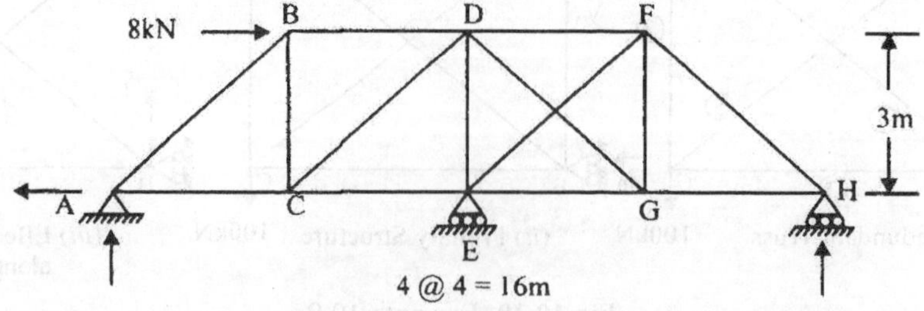

Fig 10.30. Example 10.10

Solution : The structure is statically indeterminate to 2nd degree. There are four external reactions. Hence, external indeterminacy = 4 – 3 = 1.

No of joints $n = 8$ no. of members = 14.

$$\text{D.O.I.} = (m + r) - 2n$$
$$= (14 + 4) - 2 \times 8 = 2.$$

D.O. Internal indeterminacy = 2 – 1 = 1

1. First step is to make the structure statically determinate and compute the bar forces in the statically determinate structure. This is done by

(*i*) removing the external support at *E*

and (*ii*) cutting the internal member *EF*.

This is shown in Fig 10.31 and Table 10.6.

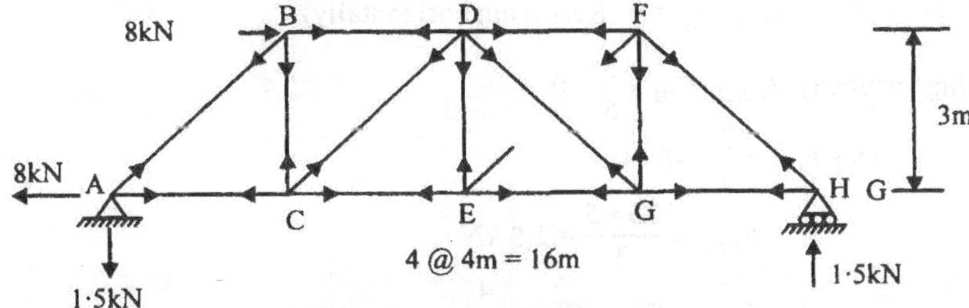

Fig 10.31. Primary Structure

2. Next step is to calculate

(*i*) Vertical deflection at *E*.

and (*ii*) the gap or discontinuity in member *EF*

This is computed by unit load method and tabulated in Table 10.6

3. The third step is to compute the redundant forces R_1 at support E and R_2 in the member EF such that the structure is restored to its original position ie there is

(*i*) no support displacement at *E*

and (*ii*) no discontinuity in member *EF*.

Forces in the determinate structure (Fig 10.31)

(*a*) Determination of reactions :

$$V_H \times 16 = 8 \times 3$$
$$V_H = \frac{24}{16} = 1.5\ kN \uparrow$$
$$V_A = 1.5\ kN \downarrow$$
$$H_A = 8\ kN \leftarrow$$

(*b*) Forces in the members due to applied load.

Joint A $F_{AB} \cdot \frac{4}{5} + F_{AC} - 8 = 0$ (Resolving horizontally)

Resolving vertically $F_{AB} \cdot \frac{3}{5} - 1.5 = 0$

or $$F_{AB} = \frac{5 \times 1.5}{3} = \mathbf{2.5\ kN.}$$

Substituting, $2.5 \times \frac{4}{5} + F_{AC} - 8 = 0$

$$F_{AC} = 8 - 2 = \mathbf{6\ kN.}$$

Joint B $F_{BA} \times \frac{3}{5} + F_{BC} = 0$ (Resolving vertically)

Resolving horizontally $F_{BA} \times \frac{4}{5} - 8 - F_{BD} = 0$

$$F_{BC} = -F_{BA} \times \frac{3}{5} = \frac{2.5 \times 3}{5} = -1.5\ kN \text{ (Comp.)}$$

$$F_{BD} = F_{BA} \times \frac{4}{5} - 8$$

$$= 2.5 \times \frac{4}{5} - 8 = -6\ kN.$$

Joint C $F_{CA} - F_{CE} - F_{CD} \times \frac{4}{5} = 0$ (Resolving horizontally)

Resolving vertically $F_{CB} + F_{CD} \times \frac{3}{5} = 0$

or $\quad -1.5 + F_{CD} \times \frac{3}{5} = 0$

or $\quad F_{CD} = \frac{1.5 \times 5}{3} = 2.5\ kN$

$$F_{CE} = F_{CA} - F_{CD} \times \frac{4}{5}$$

$$= 6 - \frac{2.5 \times 4}{5} = 4.$$

Joint E $\quad F_{EC} - F_{EG} = 0$ (Resolving horizontally)

$$F_{EG} = F_{EC} = 4$$

$$F_{ED} = 0$$

Joint D Resolving horizontally

$$F_{DB} + F_{DC} \times \frac{4}{5} - F_{DG} \times \frac{4}{5} - F_{DF} = 0$$

Resolving vertically $F_{DC} \times \frac{3}{5} + F_{DG} \times \frac{3}{5} = 0$

Substituting,

$$-6 + 2.5 \times \frac{4}{5} - F_{DG} \times \frac{4}{5} - F_{DF} = 0$$

or $\quad F_{DG} \times \frac{4}{5} + F_{DF} = -4$

$$2.5 \times \frac{3}{5} + F_{DG} \times \frac{3}{5} = 0$$

or $\quad F_{DG} = -1.5 \times \frac{5}{3} = -2.5\ kN.$

and $\quad F_{DF} = -4 - F_{DG} \times \frac{4}{5}$

$$= -4 + 2.5 \times \frac{4}{5} = -2$$

Joint F $\quad F_{FH} \times \frac{3}{5} + F_{FG} = 0$

$$F_{FD} - F_{FH} \times \frac{4}{5} = 0$$

or $\quad F_{FH} = F_{FD} \times \frac{5}{4} = -2 \times \frac{5}{4} = -2.5\ kN.$

$$F_{FG} = -F_{FH} \times \frac{3}{5} = \frac{3 \times 2.5}{5} = \mathbf{1.5\ kN.}$$

Jont G $F_{GD} \times \frac{3}{5} + F_{GF} = 0$

$$F_{GE} + F_{GD} \times \frac{4}{5} - F_{GH} = 0$$

$$F_{GF} = -F_{GD} \times \frac{3}{5} = +2.5 \times \frac{3}{5} = +1.5$$

$$F_{GH} = 4 + (-2.5) \times \frac{4}{5} = 2$$

$$F_{FH} \times \frac{3}{5} + 1.5 = -2.5 \times \frac{3}{5} + 1.5 = 0$$

$$F_{HG} + F_{HF} \times \frac{4}{5} = 2 + (-2.5) \times \frac{4}{5} = 0$$

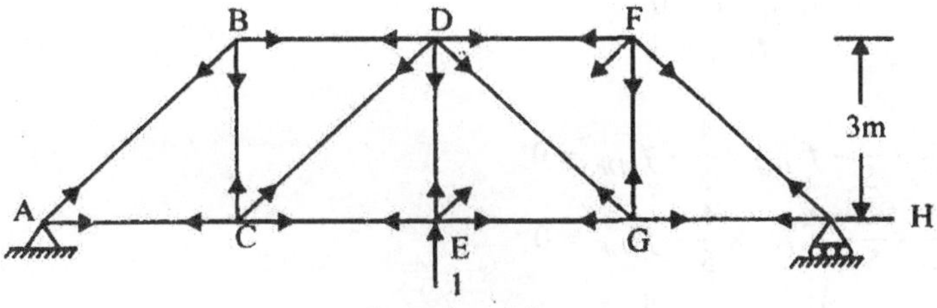

4 @ 4 = 16m.

Fig 10.32. Effect of Unit vertical load at *E*

(*c*) Forces in the members due to unit load at *E*

Joint A Resolving horizontally.

$$f_{1AB} \times \frac{3}{5} - \frac{1}{2} = 0$$

or $$f_{1AB} = \frac{1}{2} \times \frac{5}{3} = \frac{5}{6}$$

Resolving vertically.

$$f_{1AB} \times \frac{4}{5} + f_{1AC} = 0$$

or $$f_{1AC} = -f_{1AB} \times \frac{4}{5} = -\frac{5}{6} \times \frac{4}{5} = -\frac{2}{3}$$

Joint B $f_{1BA} \times \frac{3}{5} + f_{1BC} = 0$ **(Resolving vertically)**

$f_{1BA} \times \frac{4}{5} - f_{1BD} = 0$ **(Resolving Horizontally)**

or $$f_{1BD} = f_{1BA} \times \frac{4}{5} = \frac{5}{6} \times \frac{4}{5} = \frac{2}{3}$$

$$f_{1BC} = -f_{1BA} \times \frac{3}{5} = -\frac{5}{6} \times \frac{3}{5} = -\frac{1}{2}.$$

Joint C Resolving Horizontally

$$f_{1CA} - f_{1CD} \times \frac{4}{5} - f_{1CE} = 0$$

Resolving vertically

$$f_{1CB} + f_{1CD} \times \frac{3}{5} = 0$$

or $$-\frac{1}{2}+f_{1CD}\times\frac{3}{5}=0$$

or $$f_{1CD}=\frac{1}{2}\times\frac{5}{3}=\frac{5}{6}$$

$$f_{1CE}=-f_{1CD}\times\frac{4}{5}+f_{1CA}$$

$$=-\frac{5}{6}\times\frac{4}{5}+\left(-\frac{2}{3}\right)=-\frac{4}{3}.$$

Joint E

$$f_{1EC}-f_{1EG}=0$$

or $$-\frac{4}{3}-f_{1EG}=0$$

or $$f_{1EG}=-\frac{4}{3}$$

$$f_{1ED}+1=0$$

or $$f_{1ED}=-1$$

Joint D

$$f_{1DB}+f_{1DC}\times\frac{4}{5}-f_{1DG}\times\frac{4}{5}-f_{1DF}=0$$

$$\frac{2}{3}+\frac{5}{6}\times\frac{4}{5}-f_{1DG}\times\frac{4}{5}-f_{1DF}=0$$

or $$f_{1DG}\times\frac{4}{5}+f_{1DF}=\frac{4}{3}.$$

Resolving vertically

$$f_{1DC}\times\frac{3}{5}+f_{1DE}+f_{1DG}\times\frac{3}{5}=0$$

or $$f_{1DG}\times\frac{3}{5}-1+\frac{5}{6}\times\frac{3}{5}=0$$

or $$f_{1DG}=\frac{1}{2}\times\frac{5}{3}=\frac{5}{6}$$

$$f_{1DF}=\frac{4}{3}-\frac{5}{6}\times\frac{4}{5}=\frac{2}{3}$$

(*d*) Forces in the members due to unit load along *EF*

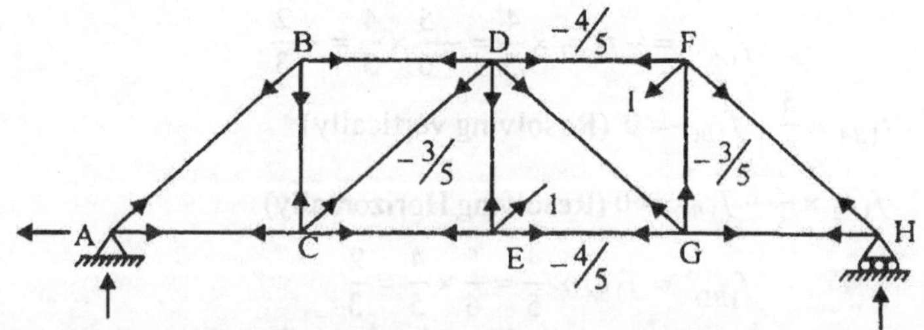

Fig 10.33 Effect of unit load along *EF*

Reactions at *A* and *H* are zero because forces are self balancing

Joint H Since vertical reaction at $H=0$

$$f_{2HF}=0 \text{ and } f_{2HG}=0.$$

Joint F $$f_{2FG}+1.\frac{3}{5}=0 \text{ (Resolving horizontally)}$$

or $$f_{2FG} = -\frac{3}{5}$$

Resolving vertically

$$f_{2FD} + 1 \times \frac{4}{5} = 0$$

or $$f_{2FD} = -\frac{4}{5}$$

Joint G Resolving Vertically

$$f_{2GD} \times \frac{3}{5} + f_{2GF} = 0$$

Resolving Horizontally,

$$f_{2GD} \times \frac{4}{5} + f_{2GE} - f_{2GH} = 0$$

$$f_{2GD} = -f_{2GF} \times \frac{5}{3}$$

$$= +\frac{3}{5} \times \frac{5}{3} = 1$$

$$f_{2GE} = f_{2GH} - f_{2GD} \times \frac{4}{5}$$

$$= 0 - 1 \times \frac{4}{5} = -\frac{4}{5}$$

Joint E $f_{2EC} - 1 \times \frac{4}{5} - f_{2EG} = 0$

or $$f_{2EC} = \frac{4}{5} + f_{2EG} = \frac{4}{5} - \frac{4}{5} = 0$$

$$f_{2ED} + 1 \times \frac{3}{5} = 0$$

$$f_{2ED} = -\frac{3}{5}$$

Joint D $f_{2DB} - f_{2DF} + f_{2DC} \times \frac{4}{5} - f_{2DG} \times \frac{4}{5} = 0$

$$f_{2DC} \times \frac{3}{5} + f_{2DG} \times \frac{3}{5} + f_{2DE} = 0$$

Substituting

$$f_{2DB} - \left(-\frac{4}{5}\right) + f_{2DC} \times \frac{4}{5} - 1 \times \frac{4}{5} = 0$$

or $$f_{2DB} + f_{2DC} \times \frac{4}{5} = 0$$

$$f_{2DC} \times \frac{3}{5} + 1 \times \frac{3}{5} + \left(-\frac{3}{5}\right) = 0$$

or $$f_{2DC} = 0 \text{ Hence } f_{2DB} = 0$$

$$f_{2AB} = f_{2AC} = f_{2CB} = 0.$$

As vertical deflection at $E = 0$

$$\underbrace{-21.332}_{\text{due to applied load}} + \underbrace{R_E \times 9.606}_{\text{due to } R_E} + \underbrace{2.226 \times F_{EF}}_{\text{Due to } F_{EF}} = 0$$

Table 10.6 Example 10.10

Member	F	f_1	f_2	Ff_1	Ff_2	f_1f_2	f_1^2	f_2^2	$f_1 \times 2.2227$	$f_2 \times -0.00922$	*Final force*
AB	2.5	5/6	0	+ 2.083	0.000	0.000	0.694	0	1.824	0	4.352
BD	– 6.0	2/3	0	– 4.000	0.000	0.000	0.444	0	1.482	0	– 4.518
DF	– 2.0	2/3	– 4/5	– 1.333	1.600	– 0.533	0.444.	0.64	1.482	+ 0.007	– 0.511
FH	– 2.5	5/6	0	– 2.083	0.000	0.000	0.694	0.00	1.852	0	– 0.648
HG	2.0	– 2/3	0	– 1.333	0.000	0.000	0.444	0.00	– 1.482	0	+ 0.518
GE	4.0	– 4/3	– 4/5	– 5.333	– 3.200	1.066	1.777	0.64	– 2.964	+0 .007	1.043
EC	4.0	– 4/3	0	– 5.333	0.000	0.000	1.777	0.00	– 2.964	0	1.036
CA	6.0	– 2/3	0	– 4.000	0.000	0.000	0.444	0.00	– 1.482	0	4.518
CB	– 1.5	– 1/2	0	+ 0.750	0.000	0.000	0.250	0.00	– 1.111	0	– 2.611
CD	2.5	5/6	0	+ 2.083	0.000	0.000	0.694	0.00	1.852	0	+ 4.352
ED	0.0	– 1	– 3/5	0.000	0.000	0.600	1.000	0.36	– 2.223	+ 0.0055	– 2.218
EF	0.0	0	1	0.000	0.000	0.000	0.000	1.00	0	– 0.0092	–0.00922
GD	– 2.5	5/6	1	– 2.083	– 2.500	0.833	0.694	1.00	1.852	–0.0092	– 0.657
GF	1.5	– 1/2	– 3/5	– 0.750	– 0.900	0.300	0.250	0.36	– 1.111	+ 0.0055	0.395
				Σ –21.332	Σ –5.00	Σ 2.266	Σ 9.606	Σ 4.00			

As there should be no discontinuity along EF

$$\underbrace{\boxed{-5.0}}_{\text{due to applied load}} + \underbrace{\boxed{R_E \times 2.26}}_{\text{due to } R_E} + \underbrace{\boxed{4.00 \times F_{EF}}}_{\text{due to } F_{EF}} = 0$$

Solving the two equations

$$F_{EF} = -.00922$$
$$R_E = 2.2227$$

Resultant value of the forces are given in Table 10.6.

10.6 INDETERMINATE STRUCTURES SUBJECTED TO SUPPORT SETTLEMENTS, TEMPERATURE VARIATION, FABRICATION ERROR ETC

The method of consistent deformation is very general and versatile and can, therefore, be used to compute effects due to all these causes :

If (*i*) The deflection is small.

and (*ii*) the structure is linear elastic

the effects can be superposed. Superposing deflections

$\Delta_i = \Delta_{is} + \Delta_{it} + \Delta_{ie} + \Delta_{io} + X_i \delta_{ii}$ ..... (10.1)

where Δ_i = resultant deflection of *i*th redundancy in actual structure .

Δ_{is} = deflection of the ith redundancy in primary structure due to support displacement.

Δ_{it} = Do due to temperature variation

Δ_{ie} = Do due to fabrication error.

Δ_{io} = Do due to load

X_i = redundant force

δ_{ii} = deflection in the direction of redundant force due to unit force at the redundant support or member.

Usually Δ_i is zero or a known value. For internal truss member it is zero as the imaginery cut in the member must be restored to continuity. If there are settlements at other supports, Δ_i is the net settlement at the actual structure and Δ_{is} is the net settlement for the primary structure .

The theorem of least work is limited to structures in which the displacements are caused by loads only since it is derived from Castigliano's Theorem part II, in which the strain energy is assumed to be equal to the complementary energy. As such it cannot be used to study effects caused by environmental variations.

Example 10.11 Calculate the reaction R_B of the beam if (*a*) the structure is loaded as shown (*b*) the loads are removed and support B settles 10 mm. $I = 150 \times 10^{-6}\ m^4$. (*c*) loads and support displacement act together $E = 200\ GPa$.

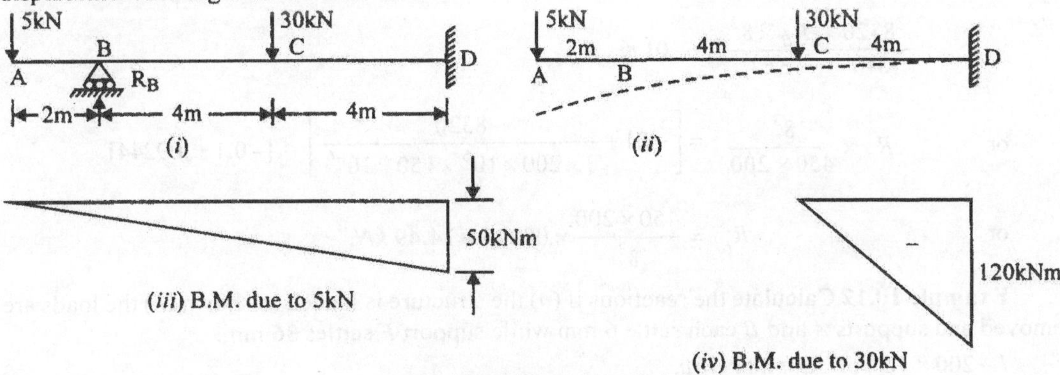

Fig. 10.34. Example 10.11

Deflection due to load.

Deflection of point B with respect to tangent at D

= Moment of the $M/_{EI}$ diagram between B and D about B.

$$= \frac{-10\times8\times4}{EI} - \frac{1}{2}\times\frac{40\times8\times16}{3EI} - \frac{1}{2}\times120\times4\left(4+\frac{2}{3}\times4\right)$$

$$= -\frac{1}{EI}\left[320+\frac{2560}{3}+\frac{4800}{3}\right]$$

$$= -\frac{1}{EI}\left[\frac{8320}{3}\right]$$

Deflection due to redundant $R_B = \frac{R_B\times8^3}{3EI}$

(*a*) Total deflection at B is zero

$$\frac{-8320}{3EI} + \frac{R_B\times8^3}{3EI} = 0$$

downward upward.

or $R_B = \frac{8320}{8^3} = 16.25\ kN$

(*b*) When support settles 10mm with no load in the beam.

$$\frac{R_B\times8^3}{3EI} = -.01\ m$$

$$R_B = \frac{-.01\times3EI}{8^3}$$

$$R_B = \frac{-.01\times3\times200\times10^6\times150\times10^{-6}}{8^3}$$

$$= -1.75\ kN.$$

(*c*) When both load and support act together

$$-\frac{8320}{3EI}+\frac{R_B\times8^3}{3EI} = -.01\ m$$

or $R_B\times\frac{8^3}{450\times200} = \left[-.01+\frac{8320}{3\times200\times10^6\times150\times10^{-6}}\right] = [-0.1+.09244]$

or $R_B = \frac{450\times200}{8^3}\times.08244 = 14.49\ kN.$

Example 10.12 Calculate the reactions if (*a*) the structure is loaded as shown (*b*) the loads are removed and supports A and B each settle 6 mm while support F settles 36 mm. $I = 200\times10^{-6}\ m^4$ $E = 200\ GPa$.

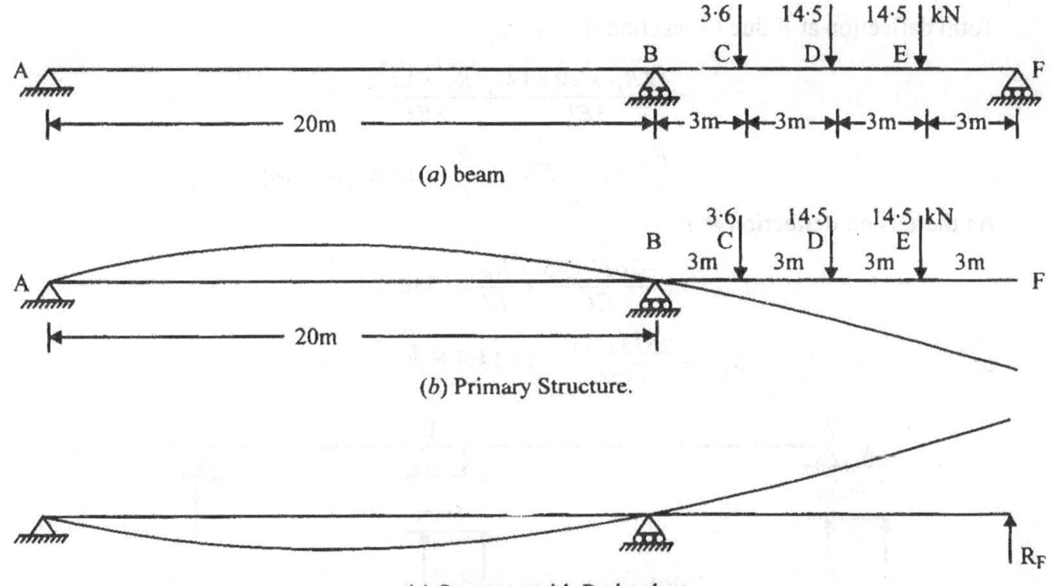

(a) beam

(b) Primary Structure.

(c) Structure with Redundant

Fig 10.35. Example 10.12

Solution : First step is to compute deflection at F for the primary structure. Deflection at F due to slope θ_B at $B = \theta_B \times 12$.

$$\theta_B = \frac{ML}{3EI} = \left(\frac{3.6\times3+14.5\times6+14.5\times9}{3EI}\right)\times20$$

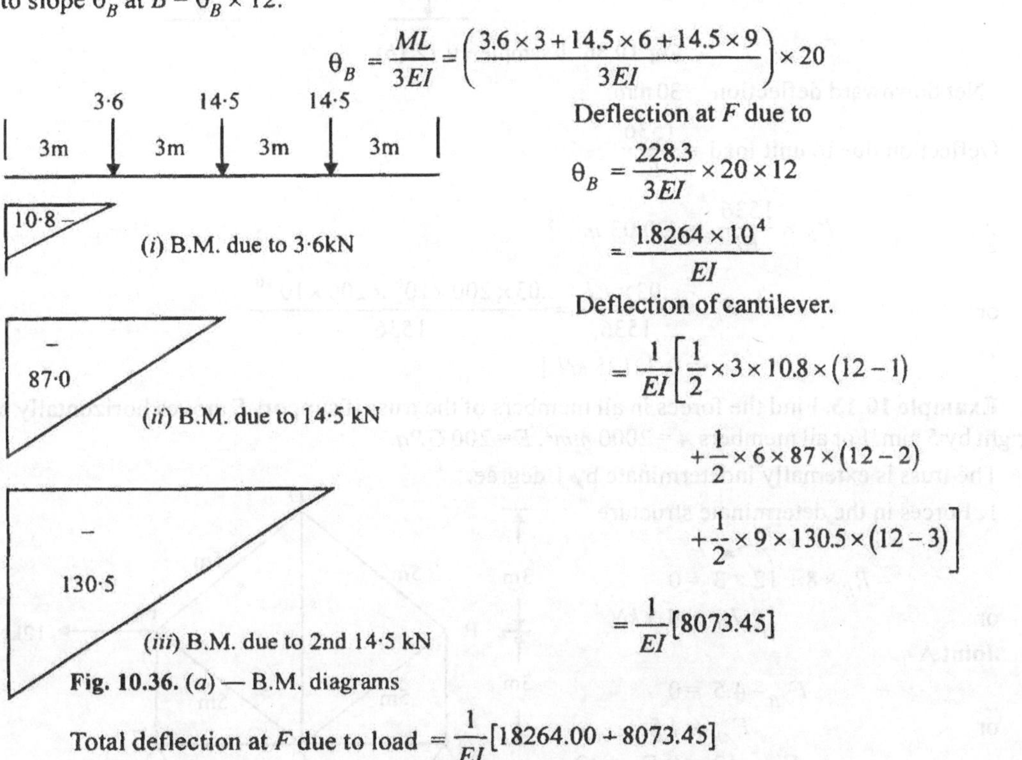

(i) B.M. due to 3·6kN

(ii) B.M. due to 14·5 kN

(iii) B.M. due to 2nd 14·5 kN

Fig. 10.36. (a) — B.M. diagrams

Deflection at F due to

$$\theta_B = \frac{228.3}{3EI}\times20\times12$$

$$= \frac{1.8264\times10^4}{EI}$$

Deflection of cantilever.

$$= \frac{1}{EI}\left[\frac{1}{2}\times3\times10.8\times(12-1) + \frac{1}{2}\times6\times87\times(12-2) + \frac{1}{2}\times9\times130.5\times(12-3)\right]$$

$$= \frac{1}{EI}[8073.45]$$

Total deflection at F due to load $= \dfrac{1}{EI}[18264.00+8073.45]$

$$= \frac{26337.45}{EI}\ (downward)$$

Total deflection at F due to reaction R_F

$$= \frac{12R_F \times 20 \times 12}{3EI} + \frac{R_F \times 12^3}{3EI}$$

$$= \frac{R_F}{EI}[960 + 576] = \frac{R_F}{EI} \times 1536 \text{ (upward)}$$

As there is no deflection at F

$$\frac{-26337.45}{EI} + \frac{R_F}{EI} \times 1536 = 0$$

or $$R_F = \frac{26337.45}{1536} = \mathbf{17.146kN} \uparrow$$

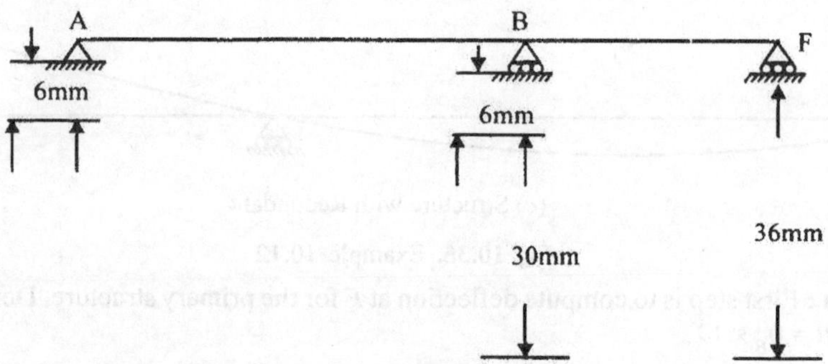

Fig 10.36. Example 10.12 (*b*)

Net downward deflection = 30 mm.

$$\text{Deflection due to unit load} = \frac{1536}{EI}$$

$$\therefore \quad R_F \times \frac{1536}{EI} = -0.03 \, m$$

or $$R_F = \frac{.03 \times EI}{1536} = \frac{.03 \times 200 \times 10^6 \times 200 \times 10^{-6}}{1536}$$

$$= 0.78125 \, kN \downarrow$$

Example 10.13. Find the forces in all members of the truss if support E moves horizontally to the right by 5 mm. For all members $A = 2000 \, mm^2$, $E = 200 \, GPa$.

The truss is externally indeterminate by 1 degree.

1. Forces in the determinate structure

$$-R_E \times 8 + 12 \times 3 = 0$$

or $$R_E = 4.5 \, kN$$

Joint A

$$F_{AB} - 4.5 = 0$$

or $$F_{AB} = 4.5$$

$$F_{AC} - 12 = 0 \; F_{AC} = 12$$

Joint B

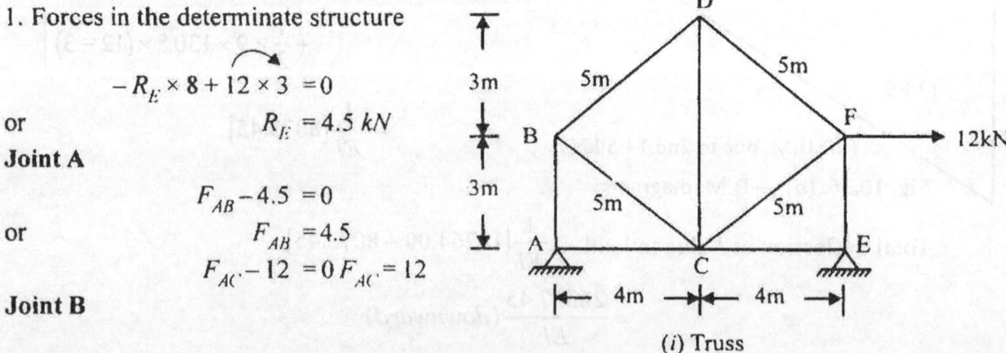

(*i*) Truss

$$F_{BA} + F_{BC} \times \frac{3}{5} - F_{BD} \times \frac{3}{5} = 0$$

$$F_{BD} \times \frac{4}{5} + F_{BC} \times \frac{4}{5} = 0$$

or $$F_{BC} \times \frac{3}{5} - F_{BD} \times \frac{3}{5} = -4.5$$

$$F_{BC} \times \frac{3}{5} + F_{BD} \times \frac{3}{5} = 0$$

Solving $$F_{BC} = -3.75\ kN$$
$$F_{BD} = 3.75\ kN$$

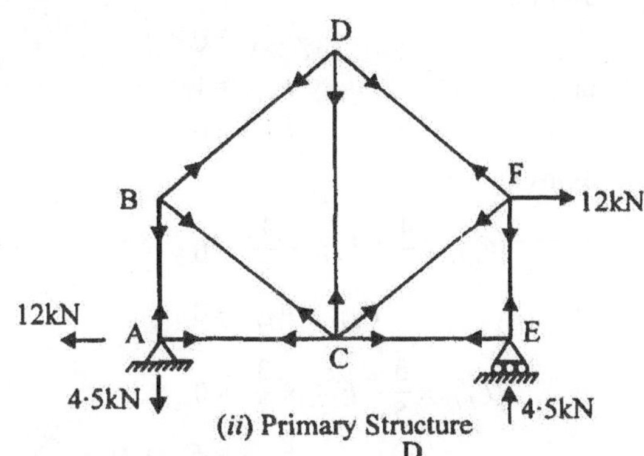

(*ii*) Primary Structure

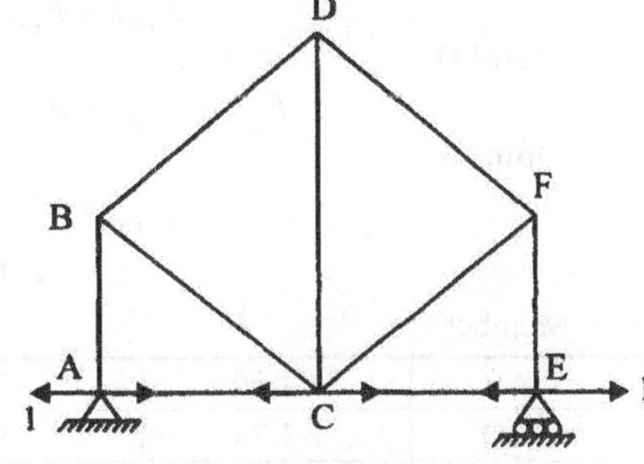

(*iii*) Effect of unit horizontal load at E.

Fig. 10. 37. Example 10.13.

Joint D

$$F_{DB} \times \frac{3}{5} + F_{DC} + F_{DF} \times \frac{3}{5} = 0$$

$$F_{DB} \times \frac{4}{5} - F_{DF} \times \frac{4}{5} = 0$$

or $$F_{DF} = F_{DB} = +3.75\ kN$$

$$F_{DC} = -\left(F_{DF} \times \frac{3}{5} + F_{DB} \times \frac{3}{5}\right)$$

$$= -\left(3.75 \times \frac{3}{5} + 3.75 \times \frac{3}{5}\right) = -4.5\ kN$$

Joint F

$$F_{FD} \times \frac{4}{5} + F_{FC} \times \frac{4}{5} - 12 = 0$$

$$F_{FD} \times \frac{3}{5} - F_{FC} \times \frac{3}{5} - F_{FE} = 0$$

or $$F_{FC} = \frac{5}{4}\left(12 - \frac{3.75 \times 4}{5}\right) = 11.25\ kN.$$

$$F_{FE} = \frac{3}{5} \times 3.75 - 11.25 \times \frac{3}{5} = -4.50$$

Joint E

$$F_{EF} + 4.5 = 0$$

or $$F_{EF} = -4.5$$
$$F_{EC} = 0$$

Joint C

$$F_{CA} + F_{CB} \times \frac{4}{5} - F_{CF} \times \frac{4}{5} - F_{CE}$$

$$= 12 + (-3.75) \times \frac{4}{5} - 11.25 \times \frac{4}{5} - 0$$

$$= 12 - 3 - 9 - 0 = 0 \text{ (Check).}$$

2. Forces in the members due to unit load at *E*.

Joint E

$$F_{EC} - 1 = 0$$

or
$$F_{EC} = 1$$
$$F_{EF} = 0$$

Joint F

$$F_{FD} \times \frac{4}{5} + F_{FC} \times \frac{4}{5} = 0$$

or
$$F_{FD} + F_{FC} = 0$$

$$F_{FD} \times \frac{3}{5} - F_{FC} \times \frac{3}{5} = 0$$

$$F_{FD} = F_{FC} = 0$$

Joint D

$$F_{DB} = F_{DC} = 0$$

Joint A

$$F_{AC} = 1$$

Table 10.7 Example 10.12

Member	F	L	f	$F.f.L$	f^2L	$F + Rf$
AB	+ 4.50	3	0	0		+ 4.50
BD	+ 3.75	5	0	0		+ 3.75
DF	+ 3.75	5	0	0		+ 3.75
FE	− 4.50	3	0	0		− 4.50
EC	0.00	4	1	0	4	− 244.0
CA	+ 12.00	4	1	48	4	− 232.0
BC	− 3.75	5	0	0		− 3.75
CF	+ 11.25	5	0	0		+ 11.25
DC	− 4.5	6	0	0		− 4.5
				$\Sigma 48$	$\Sigma 8$	

$$\frac{\Sigma FfL}{AE} - R\frac{f^2L}{AE} = .005\text{ m}$$

Movement due to load towards right. — Movement due to reaction at the support

$$\frac{48 \times 10^6}{2000 \times 200 \times 10^6} - R \cdot \frac{8 \times 10^6}{2000 \times 200 \times 10^6} = .005\text{m}$$

or
$$R = \frac{48}{4 \times 10^5} \times \frac{4 \times 10^5}{8} - \frac{.005 \times 4 \times 10^5}{8}$$

$$= 6 - 250 = -244\ kN$$

Hence force is towards right.

Example 10.14. Find forces in all members of the truss when

(*a*) The truss is loaded as shown in Fig 10.38 (*a*)

(*b*) The loads are removed and member *AB* increases in temperature by 15°C

$(\alpha = 1.2 \times 10^{-5}\, mm / mm.^{0} c.)$

(c) The loads are removed and member AB has fabrication error of $-10\ mm$.

Area of
$$AB = 30 \times 10^{-4}\ m^2$$
$$AC = 20 \times 10^{-4}\ m^2$$
$$AD = 15 \times 10^{-4}\ m^2$$
$$E = 200\ GPa$$

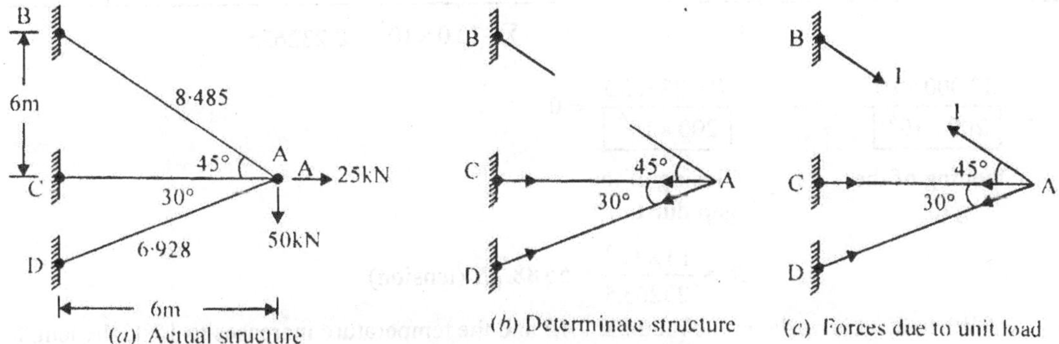

Fig 10.38. Example 10.14.

Solution: The structure is statically indeterminate of degree 1 Cutting member AB makes it determinate.

1. Solution of determinate structure

Joint A

$$F_{AC} + F_{AD} \cos 30° - 25 = 0$$
$$F_{AD} \sin 30° + 50 = 0$$

or
$$F_{AD} = -100$$
$$F_{AC} = 25 - F_{AD} \cos 30°$$
$$= 25 + 100 \frac{\sqrt{3}}{2} = \mathbf{111.60\ kN.}$$

2. Forces due to unit load.

$$f_{AC} + 1.\cos 45° + f_{AD}.\cos 30° = 0$$
$$f_{AD}.\sin 30° - 1 \times \sin 45° = 0$$

or
$$f_{AD} = \frac{\sin 45°}{\sin 30°} = \frac{1}{\sqrt{2}} \times 2 = \sqrt{2}$$
$$f_{AC} = -f_{AD}.\cos 30° - \cos 45°$$
$$= -\sqrt{2} \times \frac{\sqrt{3}}{2} - \frac{1}{\sqrt{2}}$$
$$= -\left(\frac{\sqrt{3}}{\sqrt{2}} - \frac{1}{\sqrt{2}}\right) = \mathbf{-1.932}$$

Table 10.8 Example 10.14

Member	F	L	A	f	$\frac{FL}{A}.f$	$\frac{f^2L}{A}$	$F+R_f$	Remarks
AB	0	8.485	30×10⁻⁴	1	0	2828.30	55.88	
AC	111.6	6.000	20×10⁻⁴	−1.932	-6.468×10⁵	11197.90	3.64	
AD	−100.00	6.928	15×10⁻⁴	$\sqrt{2}$	-6.532×10⁵	9237.3	− 20.97	
					$\Sigma -13.0\times10^5$	Σ 23263.5		

$$\frac{-13.000\times10^5}{200\times10^6} + \frac{R\times23263.5}{200\times10^6} = 0$$

Opening of the gap — Closing of the gap due to R

or $$R = \frac{13\times10^5}{23263.5} = 55.88\ kN\ \text{(tension)}$$

(b) When a cut is made in the reduntant AB and the temperature increases to 15°C the length increases to

$$l\alpha t = 8.485\times1.2\times10^{-5}\times1000\times15$$
$$= \underline{1.5273\ mm}$$

Force required to fit this length

$$\frac{R\times23263.5\times10^3}{200\times10^6} = -1.5273$$

or $$R = -\frac{200\times10^6\times1.5273}{23263.5\times10^3}$$
$$= -13.13\ kN\ \text{(Compression)}$$

Substituting value of R in the expression $F + R_f$, forces in the members can be obtained ($F = 0$ in this case).

(c) When there is a fabrication error of -10mm The member has to be force fit which will induce forces in other members of the truss. The gap up 10 mm is closed by extension of member AB and coming closer of joints A and B

$$R\times\frac{23263.5}{200\times10^6}\times10^3 = 10$$

(Coming closer of joints and extension of member)

or $$R = \frac{10\times200\times10^6}{23263.5\times10^3}$$
$$= \mathbf{85.97}\ kN\ \text{(tension)}.$$

Force in $AB = 85.97\ (t)$

$AC = -166.094\ (c)$

$AD = 121.56\ (t)$

Example 10.15. Find the forces in the members when the top chord members experience a rise of temperature of 50°C. $\alpha = 1.2\times10^{-5}\ mm/mm°C$. For all members $A = 1500\ mm^2$, $E = 200\ GPa$

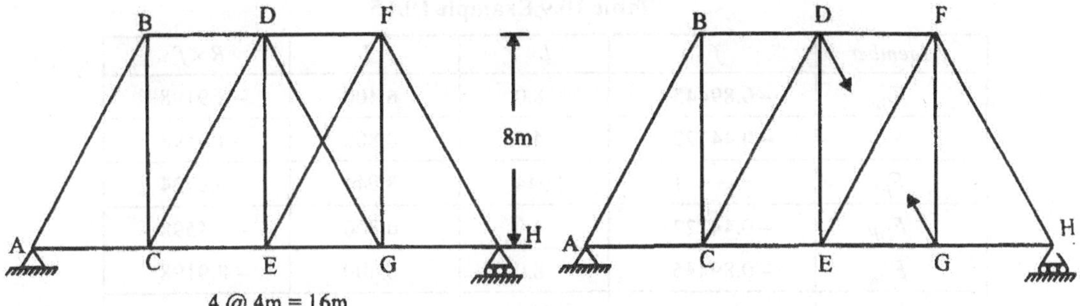

Fig. 10.39. Ex. 10.15 **Fig. 10.40.** Determinate structure.

Solution: The truss is made determinate by cutting the member *DG*.

Forces in the members due to unit load along *DG*

Joint H F_{FH} and $F_{HG} = 0$

Joint G $1 \times \frac{8}{8.944} + F_{GF} = 0$

or $F_{GF} = -\frac{8}{8.944} = -0.89445$

$$F_{GE} + 1 \times \frac{4}{8.944} = 0$$

$$F_{GE} = -\frac{4}{8.944}$$

Joint F

$$F_{FE} \times \frac{8}{8.944} + F_{FG} = 0$$

or $F_{FE} = -\frac{8.944}{8} \times F_{FG} = 1$

$$F_{FD} + F_{FE} \times \frac{4}{8.944} = 0$$

or $F_{FD} = -F_{FE} \times \frac{4}{8.944}$

$$= -\frac{4}{8.944}$$

Joint D

$$F_{DE} + 1 \times \frac{8}{8.944} = 0$$

or $F_{DE} = -\frac{8}{8.944}$.

It may be noted that as the forces are self balancing reaction at *A* is zero. Hence forces in members *AB*, *BD*, *AC*, *CE* and *BC* are zero

Displacement along *DG* due to rise in temperature in *BD* and *DF*.

1. Δ = Real displacement × virtual force

$$= L.\alpha.t \times f$$

$$= 4 \times 10^3 \times 1.2 \times 10^{-5} \times 50 \times 0 + 4 \times 10^3 \times 1.2 \times 10^{-5} \times 50 \times \left(-\frac{4}{8.944}\right)$$

$$= -\frac{4 \times 1.2 \times 50 \times 4}{8.944} \times 10^{-2} = -1.0733 mm.$$

Table 10.9 Example 10.15

Member	f	L	f^2L.	$R \times f$
F_{GF}	−0.89445	8.0	6.400	−8.9198
F_{GE}	−0.44722	4.0	0.800	−4.4598
F_{FE}	1	8.944	8.944	9.9724
F_{DF}	−0.44722	4.0	0.800	−4.4598
F_{DE}	−0.89445	8.0	6.400	−8.9198
F_{DG}	1	8.944	8.944	9.9724

$$\Sigma 32.288$$

$$\frac{R \times 32.288 \times 10^6}{1500 \times 200 \times 10^6} \times 1000 - 1.0733 = 0$$

or
$$R = \frac{1.0733 \times 1500 \times 200 \times 10^6}{32.288 \times 10^9}$$

$$= \mathbf{9.9724}\ \boldsymbol{kN} \text{ (tensile)}$$

10.7 APPLICATION OF ENERGY METHODS IN STATICALLY INDETERMINATE PROBLEMS

In section 8.13 important energy theorems were enunciated and explained. Then a number of examples were solved showing use of energy theorems in computing deflections of statically determinate problems. We concentrate now on the application of energy theorems in statically indeterminate problems. The following energy theorems can be used to determine reduntant forces in indeterminate structures.

1. Principle of Minimum Total potential Energy.
2. Casligliano's First Theorem.
3. Principle of Stationary Complementary potential Energy.
4. Casligliano's 2nd theorem and Theorem of Least work.
5. Engesser's theorem of compatibility.

All the above principles and theorems were discussed in Section 8·13. For linear elastic structures which from the major part of study in this book, Casligliano's 2nd Theorem and the Theorem of Least Work are very useful. As strain energy is used in these theorems instead of complementary energy, these two theorems are applicable only to linear elastic structures with no temperature effects or fabrication error. It means that these Theorems should be used for structures whose deflections are caused by loads only.

Symbolically, Casligliano's 2nd Theorem is

$$\frac{\partial U}{\partial P_i} = D_i$$

and Theorem of Least work is $\frac{\partial U}{\partial P_i} = 0$ It follows then that Theorem of Least work can be derived from Casligliano's Theorem with zero support displacement.

Of the five energy methods stated above we concentrate in this chapter on

(*a*) Casligliano's 2nd Theorem and theorem of least work

(*b*) Engesser's theorem of compatibility as they lead to force method of analysis which form the subject of this chapter. The other three theorems are used to solve statically indeterminate structures by displacement or stiffness methods of analysis and will be discussed in the next chapter.

Example 10.16. Find the forces in all members of the truss given in Fig 10.41 (*i*) The stress-strain curve is nonlinear and is shown in Fig 10.41 (*ii*)

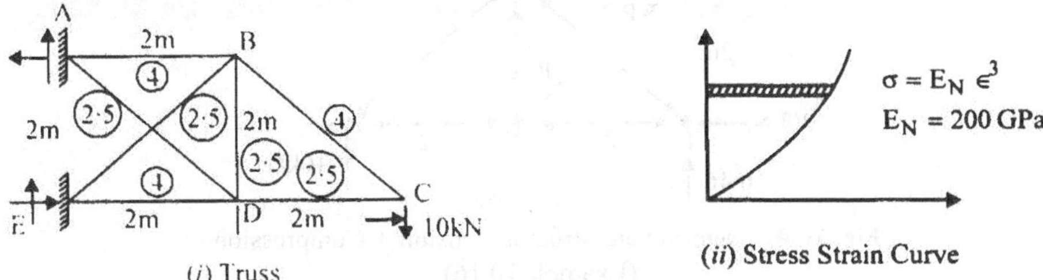

(*i*) Truss

(*ii*) Stress Strain Curve

Fig 10.41. Example 10.16.

Solution: No. of Joints = 5

No. of members = 7

No of reactions = 4

D.O.I = $m-(2n-4)$

$= 7-(10-4) = 1$

The truss is statically indeterminate of first degree. Since the stress strain curve is non linear Engesser's Theorem of compatiblity has to be used if energy method is to be applied. With only one redundancy R, Engesser's Theorem of compatibility becomes

$$\frac{\partial U_c}{\partial R} = 0$$

$$U_c = \int \epsilon.d\sigma.d_{vol.}$$

$$= \int \left(\frac{\sigma}{E_N}\right)^{\frac{1}{3}} d\sigma.d_{vol}$$

$$= \frac{3}{4}.\frac{1}{E_N^{\frac{1}{3}}}.\sigma^{\frac{4}{3}}.d_{vol}$$

For a truss of members with constant cross section this is equal to

$$\frac{3}{4}\frac{1}{E_N^{1/3}}\left(\frac{F}{A}\right)^{\frac{4}{3}}.AL. = \frac{3}{4}.\frac{1}{E_N^{1/3}}.\frac{F^{\frac{4}{3}}}{A^{\frac{1}{3}}}.L$$

$$\frac{\partial U_c}{\partial R} = \frac{3}{4}.\frac{1}{E_N^{\frac{1}{3}}}\frac{4}{3}.\frac{F^{\frac{1}{3}}}{A^{\frac{1}{3}}}.\frac{\partial F}{\partial R}.L$$

$$= \frac{1}{E_N^{\frac{1}{3}}}\left[\frac{F^{\frac{1}{3}}}{A^{\frac{1}{3}}}.\frac{\partial F}{\partial R}.L\right]$$

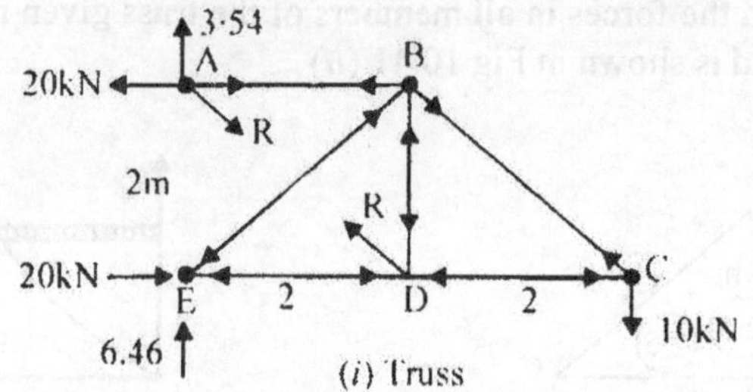

(i) Truss

Fig. 10.42. Determinate Structure tension + Compression–
(Example 10.16)

Table 10.10 Example 10.16

Member	F	$\frac{\partial F}{\partial R}$	$F^{1/3}$	L	$A^{1/3}$	Final value of forces
AB	$20-\frac{R}{\sqrt{2}}$	$-\frac{1}{\sqrt{2}}$	$\left(20-\frac{R}{\sqrt{2}}\right)^{1/3}$	2	$4^{1/3}$	16.46
BC	$10\sqrt{2}$	0	$(10\sqrt{2})^{1/3}$	$2\sqrt{2}$	$4^{1/3}$	14.14
CD	-10	0	$(-10)^{1/3}$	2	$2.5^{1/3}$	-10.00
DE	$-10-\frac{R}{\sqrt{2}}$	$-\frac{1}{\sqrt{2}}$	$\left(-10-\frac{R}{\sqrt{2}}\right)^{1/3}$	2	$4^{1/3}$	-13.53
BD	$-\frac{R}{\sqrt{2}}$	$-\frac{1}{\sqrt{2}}$	$\left(-\frac{R}{\sqrt{2}}\right)^{1/3}$	2	$2.5^{1/3}$	-3.54
BE	$-10\sqrt{2}+R$	1	$\left(-10\sqrt{2}+R\right)^{1/3}$	$2\sqrt{2}$	$2.5^{1/3}$	-9.14
AD	R	1	$R^{1/3}$	$2\sqrt{2}$	$2.5^{1/3}$	$+5.00$

The value of R can be computed by making

$$\frac{\partial U_C}{\partial R} = \frac{1}{E_N{}^{1/3}} \sum \left[\frac{F^{1/3}}{A^{1/3}} \cdot \frac{\partial F}{\partial R} \cdot L\right] = 0$$

The expression $\frac{F^{1/3}}{A^{1/3}} \cdot \frac{\partial F}{\partial R} \cdot L$ for

$$AB = -\frac{1}{\sqrt{2}} \cdot \frac{2}{4^{1/3}}\left(20-\frac{R}{\sqrt{2}}\right)^{1/3} = -0.8909\left(20-\frac{R}{\sqrt{2}}\right)^{1/3}$$

$$DE = -\frac{1}{\sqrt{2}} \cdot \frac{2}{4^{1/3}}\left(-10-\frac{R}{\sqrt{2}}\right)^{1/3} = -.8909\left(-10-\frac{R}{\sqrt{2}}\right)^{1/3}$$

$$BD = -\frac{1}{\sqrt{2}} \cdot \frac{2}{2.5^{1/3}}\left(-\frac{R}{\sqrt{2}}\right)^{1/3} = -1.042\left(-\frac{R}{\sqrt{2}}\right)^{1/3}$$

$$BE = \frac{1\times 2\sqrt{2}}{2.5^{1/3}}\left(-10\sqrt{2}+R\right) = 2.084\left(-10\sqrt{2}+R\right)^{1/3}$$

$$AD = 1\times\frac{2\sqrt{2}}{2.5^{1/3}}\left(R^{1/3}\right) = \frac{2.084\left(R^{1/3}\right)}{\Sigma = 0}$$

The summation equation is solved for R in a table by trial and error

Table 10.11 Example 10.16

	With $R=4$	With $R=5$	With $R=6$	With $R=7$
$\left(20-\frac{R}{\sqrt{2}}\right)^{1/3}$ (–0.8909)	–2.2984	–2.266	–2.2335	–2.1990
$\left(-10-\frac{R}{\sqrt{2}}\right)^{1/3}$ (–0.8909)	+2.0855	+2.123	2.1595	+2.1947
$\left(-\frac{R}{\sqrt{2}}\right)^{1/3}$ (–1.042)	1.4736	+1.5873	1.6869	+1.7758
$\left(-10-\frac{R}{\sqrt{2}}\right)^{1/3}$ (2.084)	–4.8785	–4.9665	–5.0515	–5.1338
$R^{1/3}$ (2.084)	+3.3081	+3.5636	+3.7869	+3.9865
	$\Sigma -0.3097$	$\Sigma 0.0414$	$\Sigma 0.3483$	$\Sigma 0.6269$

Hence $R \approx 5$ kN

Final values of forces are shown in Table 10.10

Example 10.17. Calculate the reactions using theorem of least work.

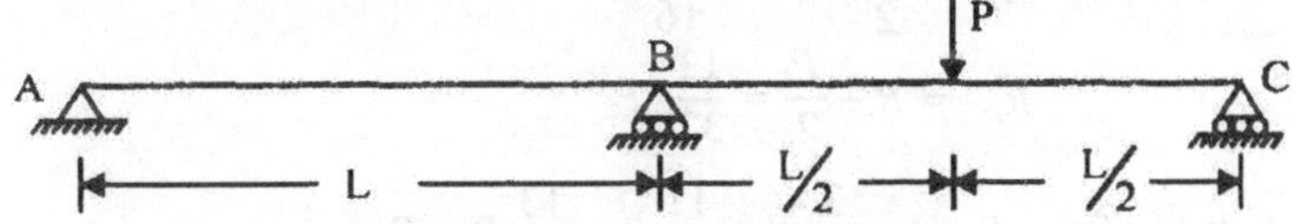

Fig. 10.43. Example 10.17

Let the redundant reaction at A be R.

Then $$R\times 2L + R_B\times L - P\times\frac{L}{2} = 0$$

or $$R_B = \frac{P}{2} - 2R$$

$$R_C = P - R - \frac{P}{2} + 2R = R + \frac{P}{2}$$

Span	Origin	M	$\frac{\partial M}{\partial R}$	Limit
AB	A	$R \cdot x$	x	$0-L$
CD	C	$\left(R+\frac{P}{2}\right)\cdot x$	x	$0-\frac{L}{2}$
DB	C	$\left(R+\frac{P}{2}\right)\cdot x - P\left(x-\frac{L}{2}\right)$	x	$\frac{L}{2}-L$

$$U = \int \frac{M^2 \cdot dx}{2\,EI}$$

$$\frac{\partial U}{\partial R} = \frac{1}{EI}\int M \cdot \frac{\partial M}{\partial R}\,dx$$

$$= \frac{1}{EI}\left[\int_0^L R\cdot x\cdot x\,dx + \int_0^{L/2}\left(R+\frac{P}{2}\right)\cdot x^2\,dx + \int_{L/2}^{L}\left[\left(R+\frac{P}{2}\right)\cdot x - P\left(x-\frac{L}{2}\right)\right] x\,dx\right]$$

$$= \frac{1}{EI}\left[R\cdot\frac{L^3}{3} + \left(R+\frac{P}{2}\right)\frac{L^3}{24} + \frac{\left(R+\frac{P}{2}\right)}{3}\left[L^3 - \left(\frac{L}{2}\right)^3\right]\right.$$

$$\left. - \frac{P}{3}\left[L^3 - \left(\frac{L}{2}\right)^3\right] + \frac{PL}{4}\left(L^2 - \left(\frac{L}{2}\right)^2\right)\right]$$

When the expression is equated to zero

$$R = -\frac{3P}{32}\cdot \downarrow$$

$$R_B = \frac{P}{2} - 2R = \frac{11}{16}\cdot P$$

$$R_C = R + \frac{P}{2} = \frac{13}{32}\cdot P$$

$$R + R_B + R_C - P = -\frac{3}{32}P + \frac{11}{16}P + \frac{13}{32}P - P$$

$$= 0 \text{ (check)}$$

Example 10.18. Calculate the reactions if (*a*) the structure is loaded as shown (*b*) the loads are removed and support *B* settles 10 mm. $I = 150 \times 10^{-6}\,m^4$ $E = 200G$ Pa.

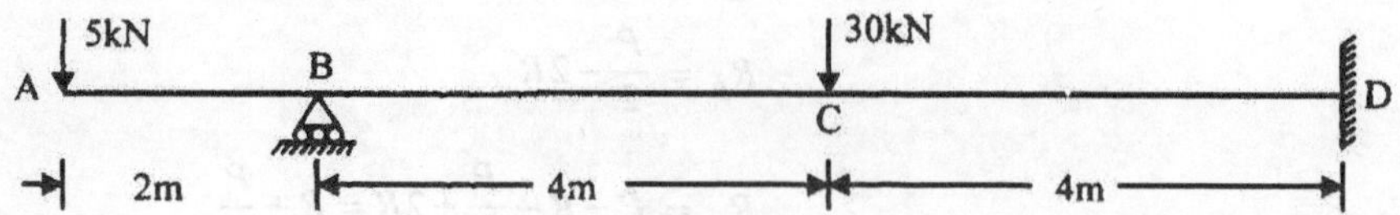

Fig. 10.44*a*. Example 10.18 (*a*)

Solution: Case (*a*) Let R_B be the redundant. The relevant expressions for B.M. with origin and limit are given below.

Member	M	$\dfrac{\partial M}{\partial R_B}$	Origin	Limit	Remarks
AB	$-5.x$	0	A	$0-2$	
BC	$-5x$ $+R_B(x-2)$	$x-2$	A	$2-6$	
CD	$-5x+R_B(x-2)$ $-30(x-6)$	$x-2$	A	$6-10$	

$$\frac{\partial U}{\partial R_B} = \frac{1}{EI}\int_0^2 -5\cdot x(0)\cdot dx + \frac{1}{EI}\int_2^6 [-5x+R_B(x-2)](x-2).dx$$

$$+\frac{1}{EI}\int_6^{10}[-5.x+R_B(x-2)-30(x-6)](x-2).dx.$$

$$\int_2^6 [-5x+R_B(x-2)](x-2).dx$$

$$=\left[-5\cdot\frac{x^3}{3}+\frac{10.x^2}{2}+\frac{R_B}{3}(x-2)^3\right]_2^6$$

$$=-346.666+160+21.33R_B$$

$$=-186.66+21.33R_B$$

$$\int \{-5x+R_B(x-2)-30(x-6)\}(x-2).dx$$

$$=\int[-5x^2+10x+R_B(x-2)^2-30(x^2-8x+12)]dx$$

$$=\left[-35\cdot\frac{x^3}{3}+250.\frac{x^2}{2}-360.x+\frac{R_B}{3}(x-2)^3\right]_6^{10}$$

$$=-9146.667+8000-1440+R_B\times 149.33$$

Adding the two results and equating to zero $-2773.327+170.66R_B=0$

or $\quad$ $\mathbf{R_B = 16.25\ kN}$

Case (*b*)

A B C D
2m 4m 4m

Fig. 10.4. (*b*) Example 10.18 (*b*)

Let R_B be the redundant

From A to B $\quad$ B.M. = 0

B to D $\quad$ B.M. $=R_B.x$

$$\frac{\partial M}{\partial R_B}=x$$

origin at B and limit $0-8$

$$\frac{\partial U}{\partial R_B} = \frac{1}{EI}\int M\cdot\frac{\partial M}{\partial R_B}\cdot dx$$

$$= \frac{1}{EI}\int_0^8 R_B\,.x.x\,dx$$

$$= R_B\cdot\frac{8^3}{3EI}$$. Support settlement at

$$B = 10\text{ mm} = \frac{10}{1000}\text{ m}$$

$$\therefore \quad R_B\,\frac{8^3}{3\,EI} = \frac{-10}{1000}\text{ m}$$

or $$R_B = \frac{-10}{1000}\times\frac{3\times150\times10^{-6}\times200\times10^6}{512}$$

$= -1.75$ kN Hence downward.

Example 10.19. Compute the reactions at A and B of the cantilever beam ABC loaded as shown in Fig. 10.44. The spring which approximates an elastic foundation has a stifness of 3500 kN/m. The beam has constant $EI = 4200$ kN m^2.

Solution: Applying Minimum Energy principle $\frac{\partial U}{\partial R} = 0$

Strain energy of the system

= Strain energy of the beam + Strain energy of the spring

$$\frac{1}{2}\int\frac{M^2\,dx}{EI} + \frac{R^2}{2\times3500}$$

$$\frac{\partial U}{\partial R} = \frac{1}{EI}\int\frac{MdM}{\partial R}dx + \frac{R}{3500}$$

Expressions of B.M. of different parts are given below.

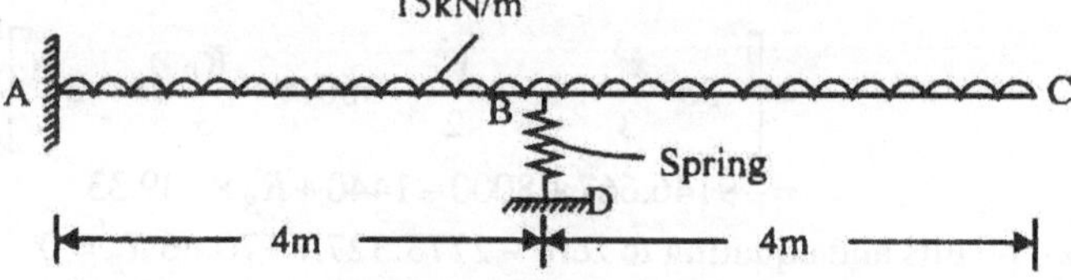

Fig. 10.45. Example 10.19

Origin at C

For CB $\quad M = -15.\frac{x^2}{2} \quad \frac{\partial M}{\partial R} = 0$

For BA $\quad M = -15\cdot\frac{x^2}{2} + R.(x-4)$

$$\frac{\partial M}{\partial R} = x - 4.$$

Limits of integration 4 to 8

$$\int_4^8 \frac{1}{4200}\left[-7.5x^2 + R(x-4)\right](x-4)\,dx + \frac{R}{3500} = 0$$

or $$\frac{1}{4200}\left[-7.5\left[\frac{x^4}{4}\right]\right]_4^8 + 30\left[\frac{x^3}{3}\right]_4^8 + R\left[\frac{(x-4)^3}{3}\right]_4^8 + \frac{R}{3500} = 0$$

or $\quad \dfrac{1}{4200}[-7200+4480+21.33R]+\dfrac{R}{3500}=0$

or $\quad R = \mathbf{120.82\ kN}$

Example 10.20. Find by theorem of Least work the redundants R_1, R_2 and R_3.

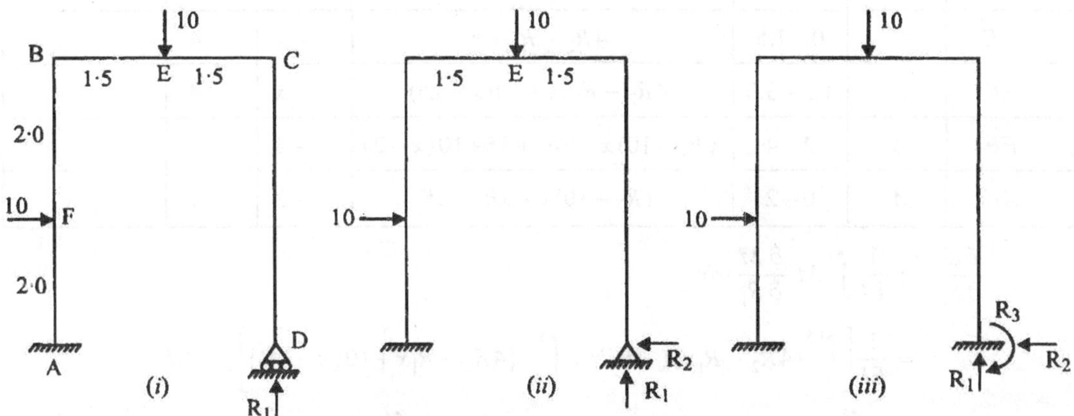

Fig. 10.46. Example 10.20.

Case (*i*) By Method of Least work

$\dfrac{1}{EI}\displaystyle\int M\cdot\dfrac{\partial M}{\partial R}\cdot dx=0$ taking only bending deformations into account.

Expressions for B.M. and other details are given below.

Span	Origin	M	$\dfrac{\partial M}{\partial R_1}$	Limit
CE	C	$R_1 . x$	x	$0-1.5$
EB	C	$R_1 . x-10\,(x-1.5)$	x	$1.5-3.0$
BF	B	$3\,R_1-10\times(1.5)$	3	$0-2$
FA	F	$3\,R_1-10\times(1.5)-10x$	3	$0-2$

$$\frac{\partial U}{\partial R}=\frac{1}{EI}\left[\int_0^{1.5}R_1x^2\,dx+\int_{1.5}^{3.0}[(R_1-10).x+15]\,x\,dx\right.$$

$$\left.+\int_0^2(9R_1-45)\,dx+\int_0^2(9R_1-45-30x)\,dx\right]$$

$$=\frac{1}{EI}\left[R_1\cdot\frac{1.5^2}{3}+(R_1-10)\times\frac{1}{3}\times[3^3-1.5^3]\right.$$

$$\left.+\frac{15}{2}[3^2-1.5^2]+9R_1\times2-45\times2+9R_1\times2-90-\frac{30\times4}{2}\right]$$

$$=\frac{1}{EI}[45R_1-268.12]=0$$

or $\quad R_1 = \mathbf{5.958\ kN}$

Case (*ii*)

Span	Origin	Limits	Expression for B.M.	$\frac{\partial M}{\partial R_1}$	$\frac{\partial M}{\partial R_2}$	Remarks
DC	*D*	0–4	$R_2 \times x$	0	x	
CE	*C*	0–1.5	$4R_2 - R_1 \times x$	$-x$	4	
EB	*C*	1.5–3.0	$4R_2 - R_1 . x + 10(x-1.5)$	$-x$	4	
FB	*A*	2–4	$(R_2 - 10)x - 3R_1 + 35 + 10(x-2)$	-3	x	
AF	*A*	0–2	$(R_2 - 10)x - 3R_1 + 35$	-3	x	

$$\frac{\partial U}{\partial R_1} = \frac{1}{EI}\int M \frac{\partial M}{\partial R_1}\cdot dx$$

$$= \frac{1}{EI}\left[\int_0^{1.5}(4R_2 - R_1 x)(-x)\,dx + \int_{1.5}^{3}[4R_2 - R_1 x + 10(x-1.5)]\right](-x)\,dx$$

$$+\int_2^4 [R_2 . x - 3R_1 + 35 + 10(x-2)](-3)\,dx + \int_0^2 [R_2 - 10)x - 3R_1 + 35](-3)\,dx$$

$$= \frac{1}{EI}[-4.5R_2 + 1.125R_1 - 13.5R_2 + 7.875R_1 - 78.75 + 50.625 - 18R_2 + 18R_1 - 210$$

$$+120 - 6R_2 + 60 + 18R_1 - 210]$$

$$= \frac{1}{EI}[45R_1 - 42R_2 - 268.125] = 0$$

$$\frac{\partial U}{\partial R_2} = \frac{1}{EI}\int M . \frac{\partial M}{\partial R_2}\cdot dx .$$

$$= \frac{1}{EI}\left[\int_0^4 R_2 .x.x\,dx + \int_0^{1.5}(4R_2 - R_1 x)(4)\,dx + \int_{1.5}^{3}[4R_2 - R_1 x + 10(x-1.5)]4\,dx\right.$$

$$\left.+\int_0^2 [(R_2 - 10)x - 3R_1 + 35]x\,dx + \int_2^4 [(R_2 - 10)x - 3R_1 + 35 + 10(x-2)x\,dx]\right]$$

$$= \frac{1}{EI}\left[R_2\frac{4^3}{3} + 16R_2(1.5) - 4R_1 . \frac{1.5^2}{2} + 16R_2(3-1.5)\right.$$

$$-4\frac{R_1}{2}\cdot(3^2 - 1.5^2) + \frac{40}{2}(3^2 - 1.5^2) - 60(3-1.5)$$

$$+R_2\frac{2^3}{3} - \frac{3R_1}{2}(2^2) + 35 . \frac{2^2}{2} - \frac{10\times 2^3}{3}$$

$$\left.+\frac{R_2}{3}(4^3 - 2^3) - 3R_1 . \frac{4^2 - 2^2}{2} + \frac{35}{2}(4^2 - 2^2) - \frac{20}{2}\cdot(4^2 - 2^2)\right]$$

$$= \frac{1}{EI}[90.67R_2 - 42R_1 + 178.33] = 0 .$$

The equations then are

$$45R_1 - 42R_2 - 268.125 = 0$$

$$-42R_1 + 90.67\,R_2 + 178.33 = 0$$

Solving the equation

$$R_1 = 7.2608\text{ kN.}$$
$$R_2 = 1.3964\text{ kN.}$$

Case (*iii*) The bending moment expressions are given below.

Member	Origin	Limit	M	$\frac{\partial M}{\partial R_1}$	$\frac{\partial M}{\partial R_2}$	$\frac{\partial M}{\partial R_3}$	
DC	D	0−4	$R_2x + R_3$	0	x	1	
CE	C	0−1.5	$4R_2 + R_3 - R_1x$	$-x$	4	1	
EB	C	1.5−3.0	$4R_2 + R_3 - R_1x + 10(x-1.5)$	$-x$	4	1	
FB	A	2−4	$(R_2-10)x - 3R_1 + 35 + R_3 + 10(x-2)$	−3	x	1	
AF	A	0−2	$(R_2-10)x - 3R_1 + 35 + R_3$	−3	x	1	

Using integrated values of example

$$\frac{\partial U}{\partial R_1} = \frac{1}{EI}[45R_1 - 42R_2 - 268.125]$$

$$+\frac{1}{EI}\left[\int_0^{1.5} R_3(-x)\,dx + \int_{1.5}^{3} R_3(-x)\,dx\right] + \int_2^4 -3R_3dx + \int_0^2 -3R_3dx$$

$$= \frac{1}{EI}[45R_1 - 42R_2 - 268.125 - (1.125 + 3.375 + 6 + 6)R_3]$$

$$= \frac{1}{EI}[45R_1 - 42R_2 - 16.5R_3 - 268.125] = 0$$

$$\frac{\partial U}{\partial R_2} = \frac{1}{EI}[-42R_1 + 90.67R_2 + 178.33]$$

$$+\frac{1}{EI}\left[\int_0^4 R_3x\,dx + \int_0^{1.5} 4R_3\,dx + \int_{1.5}^{3} 4R_3dx + \int_2^4 R_3x\,dx + \int_0^2 R_3.x\,dx\right]$$

$$= \frac{1}{EI}[-42R_1 + 90.67R_2 + 28R_3 + 178.33]$$

$$= 0$$

$$\frac{\partial U}{\partial R_3} = \frac{1}{EI}\left[\int_0^4 (R_2.x + R_3)\,dx + \int_0^{1.5} (4R_2 + R_3 - R_1x)\,dx\right]$$

$$+\int_{1.5}^{3.0} [4R_2 + R_3 - R_1.x + 10(x-1.5)]dx$$

$$+\int_2^4 [(R_2-10)x - 3R_1 + 35 + R_3 + 10(x-2)]dx$$

$$+\int_0^2 [(R_2-10)x - 3R_1 + 35 + R_3]dx$$

$$= \frac{1}{EI}[-16.5R_1 + 28R_2 + 11R_3 + 91.25] = 0.$$

The three equations then are:

$$45\,R_1 - 42\,R_2 - 16.5\,R_3 = 268.125$$
$$-42\,R_1 + 90.67\,R_2 + 28\,R_3 = -178.33$$
$$-16.5\,R_1 + 28\,R_2 + 11R_3 = -91.25$$

giving

$$R_1 = 6.48 \text{ kN}$$
$$R_2 = 2.77 \text{ kN}$$
$$R_3 = -5.63 \text{ kNm}$$

Example 10.21. Analyse the truss shown in Fig. 10.46. All areas are the same.

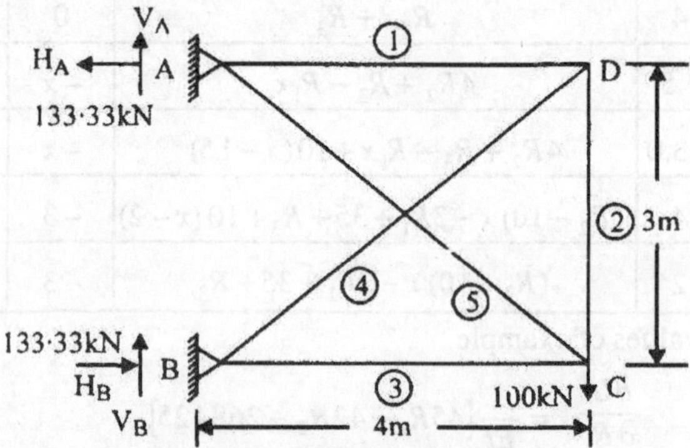

Fig. 10.47. Example 10.21.

Solution: Let V_B be the redundant, H_A and and H_B are 133.33 kN as shown in Fig. 10.47.

Joint B $\quad V_B + F_4 \cdot \frac{3}{5} = 0$ (Resolving vertically)

$$133.33 + F_3 + F_4 \times \frac{4}{5} = 0 \text{ (Resolving horizontally)}$$

or $\quad F_4 = -V_B \cdot \frac{5}{3}$

$$F_3 = -133.33 + V_B \times \frac{5}{3} \times \frac{4}{5}$$
$$= -133.33 + 1.33\,V_B.$$

Joint A $\quad F_1 - 133.33 + F_5 \times \frac{4}{5} = 0$ (Resolving horizontally)

$$V_A - F_5 \times \frac{3}{5} = 0 \text{ (Resolving vertically)}$$

or $\quad F_5 \times \frac{3}{5} = V_A = 100 - V_B$

or $\quad F_5 = \frac{5}{3} \times 100 - \frac{5}{3} \times V_B$

$$F_1 = 133.33 - F_5 \times \frac{4}{5}$$
$$= 133.33 - \frac{4}{5}\left(\frac{5}{3} \times 100 - \frac{5}{3} \times V_B\right)$$
$$= 133.33 - 133.33 + 1.333\,V_B$$
$$= +1.333\,V_B$$

Joint D Resolving vertically

$$F_2 + F_4 \frac{3}{5} = 0$$

or

$$F_2 = -F_4 \times \frac{3}{5}$$

$$= +V_B \times \frac{5}{3} \times \frac{3}{5}$$

$$= +V_B$$

$$U = \sum \frac{F^2 L}{2AE}$$

$$\frac{\partial U}{\partial V_B} = \sum \frac{F \cdot \frac{\partial F}{\partial V_B} . L}{AE}$$

Table 10.12 Example 10.21.

Member	F	$\frac{\partial F}{\partial V_B}$	L	$F \cdot \frac{\partial F}{\partial V_B} \cdot L$	Final Value of forces
1	$+1.333\ V_B$	+1.333	4	$7.1075\ V_B$	+62.223 kN
2	$+V_B$	+1	3	$3\ V_B$	+46.679 kN
3	-133.33 $+1.33\ V_B$	+1.333	4	-709.315 $+7.075\ V_B$	−71.246 kN
4	$-V_B \times \frac{5}{3}$	−1.667	5	$+13.894\ V_B$	−77.798 kN
5	$\frac{5}{3}(100 - V_B)$	−1.667	5	-1389.44 $+13.894\ V_B$	+88.868 kN

$$\Sigma\, F \cdot \frac{\partial F}{\partial V_B} \cdot L = (7.1075 + 3 + 7.075 + 13.894 + 13.894)\, V_B - (709.315 + 1389.44) = 0$$

or $\quad$ V_B **= 46.679 kN**

Example 10.22. Calculate the member forces and reactions. For all members E = constant and $\frac{L}{A} = 10^3\ \text{m}^{-1}$.

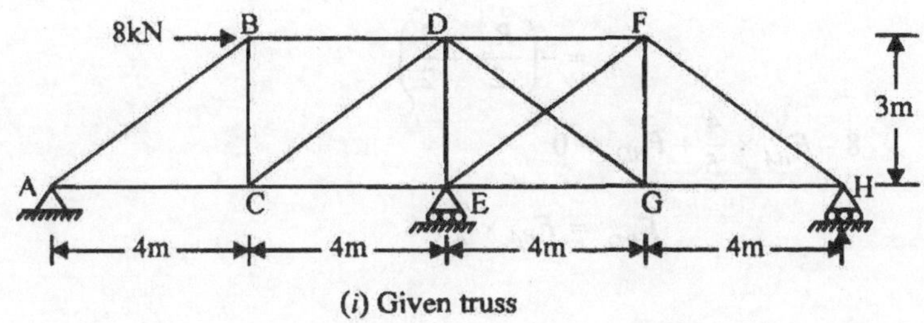

(i) Given truss

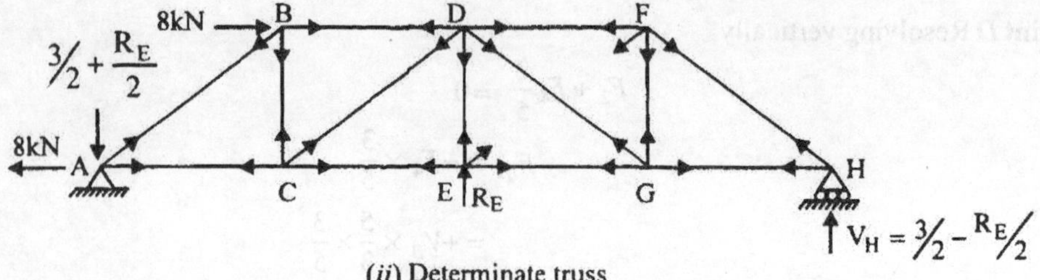

(*ii*) Determinate truss

Fig. 10.48. Example 10.22

Solution: The truss as already explained is statically in determinate to 2nd degree. It is made statically determinate by assuming support reaction at E as R_E and force in the member EF as F_{EF}.

Taking moments about A

$$8 \times 3 - R_E \times 8 - V_H \times 16 = 0$$

or
$$V_H = \frac{1}{16}[24 - 8R_E]$$

$$= \frac{3}{2} - \frac{R_E}{2}.$$

Forces are now resolved at different joints

Joint A Resolving horizontally

$$F_{AB} \times \frac{3}{5} - \frac{R_E}{2} - \frac{3}{2} = 0 \quad \text{or} \quad F_{AB} = \frac{5}{3}\left(\frac{R_E}{2} + \frac{3}{2}\right)$$

Resolving vertically

$$F_{AB} \times \frac{4}{5} + F_{AC} - 8 = 0.$$

or
$$\frac{5}{3} \times \frac{4}{5}\left(\frac{R_E}{2} + \frac{3}{2}\right) + F_{AC} - 8 = 0$$

or
$$F_{AC} = 8 - \frac{4}{3}\left(\frac{R_E}{2} + \frac{3}{2}\right)$$

$$= 6 - \frac{2}{3} \cdot R_E$$

Joint B
$$F_{BA} \times \frac{3}{5} + F_{BC} = 0$$

or
$$F_{BC} = -F_{BA} \times \frac{3}{5}$$

$$= -\frac{5}{3} \times \frac{3}{5}\left(\frac{R_E}{2} + \frac{3}{2}\right)$$

$$= -\left(\frac{R_E}{2} + \frac{3}{2}\right)$$

$$8 - F_{BA} \times \frac{4}{5} + F_{BD} = 0$$

or
$$F_{BD} = F_{BA} \times \frac{4}{5} - 8$$

$$= \frac{5}{3} \times \frac{4}{5}\left(\frac{R_E}{2} + \frac{3}{2}\right) - 8$$

$$= \frac{2}{3} R_E - 6$$

Joint *C* $\quad F_{CA} - F_{CD} \times \frac{4}{5} - F_{CE} = 0$

$$F_{CD} \times \frac{3}{5} + F_{CB} = 0$$

or $\quad 6 - \frac{2}{3} \cdot R_E - F_{CD} \cdot \frac{4}{5} - F_{CE} = 0$

$$F_{CD} \times \frac{3}{5} - \left(\frac{R_E}{2} + \frac{3}{2}\right) = 0$$

or $\quad F_{CD} = \frac{5}{3}\left(\frac{R_E}{2} + \frac{3}{2}\right)$

$$= \frac{5}{6} \cdot R_E + \frac{5}{2}$$

$$F_{CE} = 6 - \frac{2}{3} R_E - \frac{4}{5}\left(\frac{5}{6} R_E + \frac{5}{2}\right)$$

$$= 6 - \frac{4}{3} R_E - 2 = 4 - \frac{4}{3} \cdot R_E$$

Joint E $\quad F_{EC} - F_{EF} \times \frac{4}{5} - F_{EG} = 0$

$$F_{ED} + F_{EF} \times \frac{3}{5} + R_E = 0$$

or $\quad F_{EG} = F_{EC} - F_{EF} \times \frac{4}{5}$

$$= 4 - \frac{4}{3} R_E - F_{EF} \times \frac{4}{5}$$

$$F_{ED} = -\frac{3}{5} F_{EF} - R_E.$$

Joint *D*

$$F_{DB} + F_{DC} \cdot \frac{4}{5} - F_{DC} \times \frac{4}{5} - F_{DF} = 0$$

$$F_{DC} \times \frac{3}{5} + F_{DG} \times \frac{3}{5} + F_{DE} = 0$$

$$F_{DG} \times \frac{3}{5} = -F_{DE} - F_{DC} \cdot \frac{3}{5}$$

$$= \frac{3}{5} F_{EF} + R_E - \frac{3}{5}\left(\frac{5}{6} R_E + \frac{5}{2}\right)$$

$$F_{DG} = \frac{5}{3}\left(\frac{3}{5} F_{EF} + R_E - \frac{R_E}{2} - \frac{3}{2}\right)$$

$$= F_{EF} + \frac{5}{3} R_E - \frac{5}{6} R_E - 2.5$$

$$= F_{EF} + \frac{5}{6} R_E - 2.5.$$

$$F_{DF} = F_{DB} + F_{DC} \times \frac{4}{5} - F_{DG} \times \frac{4}{5}$$

$$= \frac{2}{3} R_E - 6 + \frac{4}{5}\left(\frac{5}{6} R_E + \frac{5}{2}\right) - \frac{4}{5}\left(F_{EF} + \frac{5}{6} R_E - 2.5\right).$$

$$= \frac{2}{3} R_E + \frac{2}{3} R_E - \frac{2}{3} R_E - 6 + 2 + 2 - \frac{4}{5} F_{EF}$$

$$= \frac{2}{3} R_E - 2 - \frac{4}{5} F_{EF}.$$

Joint F

$$F_{FD} + F_{EF} \times \frac{4}{5} - F_{FH} \times \frac{4}{5} = 0$$

$$F_{EF} \times \frac{3}{5} + F_{FG} + F_{FH} \times \frac{3}{5} = 0$$

or $$\frac{2}{3} R_E - 2 - \frac{4}{5} F_{EF} + F_{EF} \times \frac{4}{5} - F_{FH} \times \frac{4}{5} = 0$$

or $$F_{FH} = \frac{5}{4}\left(\frac{2}{3} R_E - 2\right)$$

$$= \frac{5}{6} R_E - \frac{5}{2}$$

$$F_{FG} = -\left(F_{EF} \times \frac{3}{5} + F_{FH} \times \frac{3}{5}\right)$$

$$= -\left[F_{EF} \times \frac{3}{5} + \frac{3}{5}\left(\frac{5}{6} R_E - \frac{5}{2}\right)\right]$$

$$= -\left[F_{EF} \times \frac{3}{5} + \frac{R_E}{2} - \frac{3}{2}\right].$$

Joint G

$$F_{GD} + \frac{4}{5} + F_{GE} - F_{GH} = 0$$

and $$F_{GF} + F_{GD} \times \frac{3}{5} = 0$$

or $$F_{GD} = -\frac{5}{3} F_{GF}$$

$$= \frac{5}{3}\left[F_{EF} \times \frac{3}{5} + \frac{R_E}{2} - \frac{3}{2}\right]$$

$$= F_{EF} + \frac{5}{6} R_E - 2.5$$

$$\frac{4}{5} \cdot \left(F_{EF} + \frac{5}{6} R_E - \frac{5}{2}\right) + 4 - \frac{4}{3} R_E - F_{EF} \times \frac{4}{5} = F_{GH}$$

or $$F_{GH} = \frac{4}{5} F_{EF} + \frac{2}{3} R_E - 2 + 4 - \frac{4}{3} R_E - F_{EF} \times \frac{4}{5}.$$

Table 10.14 Example 10.22

Member	Force	$\frac{\partial F}{\partial R_E}$	$\frac{\partial F}{\partial F_{EF}}$	$F\frac{\partial F}{\partial R_E}$	$F\frac{\partial F}{\partial F_E}$	Value of Forces
AB	$5/3(R_{E/2}+3/2)$	5/6	0	$25/18(R_{E/2}+3/2)$	0	4·35
BD	$\frac{2}{3}R_E-6$	2/3	0	$4/9R_E-4$	0	–4·52
DF	$\frac{2}{3}R_E-2-4/5F_{EF}$	2/3	–4/5	$4/9R_E-4/3-\frac{8}{15}F_{EF}$	$-\frac{8}{15}R_E+\frac{8}{5}+\frac{16}{25}F_{EF}$	–0·510
FH	$5/6R_E-5/2$	5/6	0	$25/36R_E-25/12$	0	–0·648
HG	$2-2/3R_E$	–2/3	0	$-4/3+4/9R_E$	0	0·518
GE	$4-4/3R_E-4/5F_{EF}$	–4/3	–4/5	$-16/3+\frac{16}{9}R_E+\frac{16}{15}F_{EF}$	$-16/5+16/15R_E+\frac{16}{25}\cdot F_{EF}$	1·043
EC	$4-4/3R_E$	–4/3	0	$-16/3+16/9R_E$	0	1·036
CA	$6-\frac{2}{3}R_E$	–2/3	0	$-4+4/9R_E$	0	4·52
CB	$-(R_E/2+3/2)$	–1/2	0	$R_E/4+3/4$	0	–2·61
CD	$5/6R_E+5/2$	5/6	0	$25/36R_E+25/12$	0	+4·35
ED	$-3/5F_{EF}-R_E$	–1	–3/5	$+3/5F_{EF}+R_E$	$+\frac{9}{25}F_{EF}+\frac{3}{5}R_E$	–2·217
EF	F_{EF}	0	1	0	F_{EF}	–·00922
GD	$F_{EF}+5/6R_E-5/2$	5/6	1	$F_{EF}\,5/6+\frac{25}{36}R_E-\frac{25}{12}$	$F_{EF}+5/6R_E-5/2$	–0·65697
GF	$-(F_{EF}\,3/5+R_{E/2}-3/2)$	–1/2	–3/5	$F_{EF}\cdot\frac{3}{10}+\frac{RE}{4}-3/4$	$F_{EF}\cdot\frac{9}{25}+\frac{3R_E}{10}-\frac{9}{10}$	0·394

$$\Sigma F\frac{\partial F}{\partial R_E} = 9\cdot611R_E+2\cdot266F_{EF}-21\cdot33=0$$

$$\Sigma F\frac{\partial F}{\partial F_E} = 2\cdot266R_E+4F_{EF}\quad -5\cdot00=0$$

Solving $R_E = 2\cdot2227$

$F_{EF} = -\cdot00922.$

Resultant values of the forces are given in the Table 10.14. (above)

$$= 2 - \frac{2}{3} R_E$$

Joint *H*

$$F_{HG} + F_{HF} \cdot \frac{4}{5} = 0$$

$$F_{HF} \cdot \frac{3}{5} + \frac{3}{2} - \frac{R_E}{2} = 0$$

$$F_{HF} \times \frac{4}{5} = -F_{HG}$$

$$= -2 + \frac{2}{3} R_E$$

$$F_{HF} = \frac{5}{4}\left(-2 + \frac{2}{3} R_E\right)$$

$$= -\frac{5}{2} + \frac{5}{6} R_E \quad \frac{3}{5}\left(-\frac{5}{2} + \frac{5}{6} \cdot R_E\right) + \frac{3}{2} - \frac{R_E}{2}$$

$$= -\frac{3}{2} + \frac{R_E}{2} + \frac{3}{2} - \frac{R_E}{2} = 0 \text{ (check).}$$

10.8. GENERALISED SOLUTION OF FIXED PORTAL FRAMES (Elastic Centre Method)

It has already been shown in Example 10.3 how a fixed portal frame which is statically indeterminate to the third degree can be solved by consistent deformation method. With three unknown forces and three compatibility conditions, three simultaneous equations were obtained, solution of which gave the unknown forces. However, these equations can be solved in simpler form if elastic centre method is used.

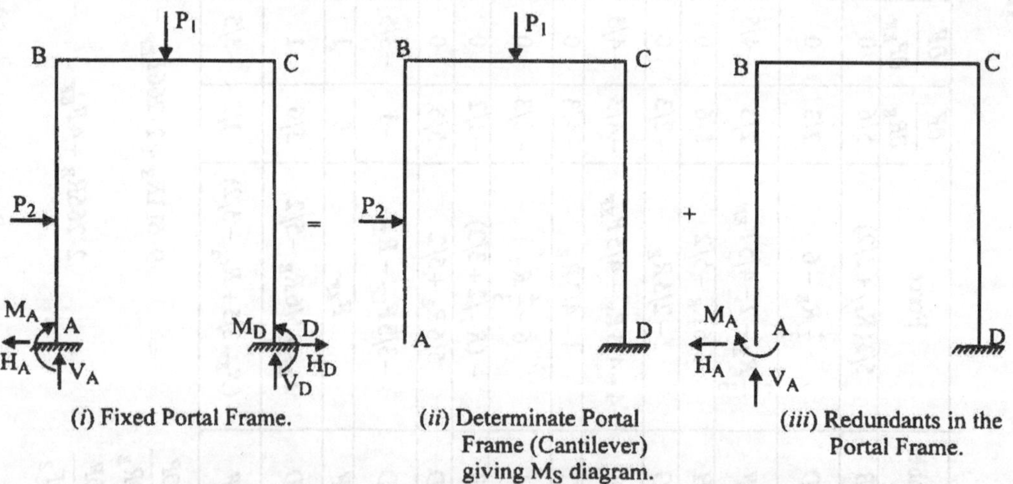

(*i*) Fixed Portal Frame. (*ii*) Determinate Portal Frame (Cantilever) giving M_S diagram. (*iii*) Redundants in the Portal Frame.

Fig. 10.49*a*. Solution of fixed portal by elastic centre method.

The fixed portal frame has been split into two determinate cantilever structures.

(*i*) With applied loads only as shown in Fig. 10.49 (*a*) (*ii*)

(*ii*) Subjected to redundants only as shown in Fig. 10.49 (*a*) (*iii*).

The direction of the redundants are so assumed that they produce positive bending moments which cause tension inside the frame. These directions of redundant reactions are to be regarded as positive.

The compatibility conditions are

$$\Delta_{VA} = \text{Vertical displacement at } A = 0$$

$$\Delta_{HA} = \text{Horizontal displacement at } A = 0$$

$$\theta_A = \text{Rotation at } A = 0$$

The displacements at A due to applied load and the redundants can be obtained by unit load method. The unit loads and the corresponding value of m are given in Fig. 10.49*b*. (*i*), (*ii*) and (*iii*).

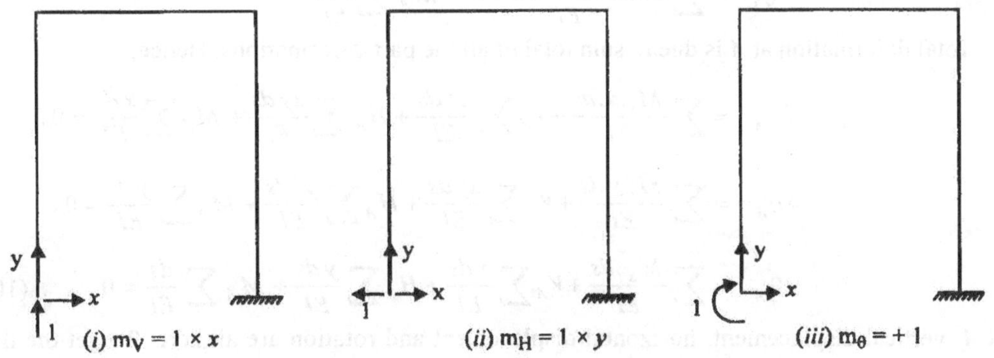

Fig. 10.49*b*. Application of unit forces

1. When M_s is the bending moment due to applied loads (Fig. 10.49*a* and 10.49*b*). displacements at A due to applied load are :

$$\text{Vertical displacement } \Delta_{V0} = \sum \frac{M_s . m_v . ds}{EI} = \sum \frac{M_s \, x ds}{EI}$$

$$\text{Horizontal displacement } \Delta_{H0} = \sum \frac{M_s m_H . ds}{EI} = \sum \frac{M_s . y ds}{EI}$$

$$\text{Rotation } \theta_0 = \sum \frac{M_s . m_\theta . ds}{EI} = \sum \frac{M_s . ds}{EI}$$

2. Deformation at A due to V_A acting alone (Fig. 10.49*a* (i))

$$\Delta_{V1} = \sum \frac{(m_v . V_A)(m_v).ds}{EI} = V_A \sum \frac{x^2 \, ds}{EI}$$

$$\Delta H_1 = \sum \frac{(m_v . V_A) m_H . ds}{EI} = V_A \sum \frac{x \, y ds}{EI}$$

$$\theta_1 = \sum \frac{(m_v . V_A) m_\theta . ds}{EI} = V_A \sum \frac{x ds}{EI}$$

3. Deformations at A due to H_A acting alone (Fig. 10.49*b* (ii))

$$\Delta V_2 = \sum \frac{(m_H . H_A)(m_v).ds}{EI} = H_A \sum \frac{x \, y ds}{EI}$$

$$\Delta H_2 = \sum \frac{(m_H . H_A)(m_H) ds}{EI} = H_A \sum \frac{y^2 \, ds}{EI}$$

$$\theta_2 = \sum \frac{(m_H . H_A).(m_\theta).ds}{EI} = H_A \sum \frac{y\,ds}{EI}$$

4. Deformations at A due to M_A acting alone Fig. 10.49*b* (iii)

$$\Delta_{V_3} = \sum \frac{(m_\theta . M_A)(m_v).ds}{EI} = M_A \sum \frac{x\,ds}{EI}$$

$$\Delta H_3 = \sum \frac{(m_\theta . M_A).(m_H).ds}{EI} = M_A \sum \frac{y\,ds}{EI}$$

$$\theta_3 = \sum \frac{(m_\theta . M_A)(m_\theta).ds}{EI} = M_A \sum \frac{ds}{EI}$$

Total deformation at A is due to sum total of all the part deformations. Hence,

$$\Delta_{V_A} = \sum \frac{M_s.x.ds}{EI} + V_A \sum \frac{x^2\,ds}{EI} + H_A \sum \frac{xy\,ds}{EI} + M_A \sum \frac{x\,ds}{EI} = 0.$$

$$\Delta_{H_A} = \sum \frac{M_s\,y\,ds}{EI} + V_A \sum \frac{xy\,ds}{EI} + H_A \sum \frac{y^2 ds}{EI} + M_A \sum \frac{y\,ds}{EI} = 0.$$

$$\theta_A = \sum \frac{M_s.ds}{EI} + V_A \sum \frac{x\,ds}{EI} + H_A \sum \frac{y\,ds}{EI} + M_A \sum \frac{ds}{EI} = 0. \quad ...(10.2)$$

as at A, vertical displacement, horizontal displacement and rotation are all zero. To get the three unknowns it is necesary to solve the above three equations simultaneously. Solving the above equations will be greatly simplified if we shift the origin from end A to a point which is known as elastic centre. This was first shown in 1886 by German Engineer Carl Culmann. The elastic centre may be defined as the centroid of elastic quantities $\frac{ds}{EI}$ of any structure where ds is a very small length of the axis of the structure. As the elastic centre is the centroid of the elastic quantities $\frac{ds}{EI}$

$$\sum \frac{x\,ds}{EI} = 0, \quad \sum \frac{y\,ds}{EI} = 0$$

If a rigid arm AO be connected between the points A and O, and if A remains fixed, O also wil remain fixed and the following modified equations will be obtained.

$$\sum \frac{M_s.x\,ds}{EI} + V_0 \sum \frac{x^2\,ds}{EI} + H_0 \sum \frac{xy\,ds}{EI} = 0$$

$$\sum \frac{M_s.y\,ds}{EI} + V_0 \sum \frac{xy\,ds}{EI} + H_0 \sum \frac{y^2\,ds}{EI} = 0$$

$$\sum \frac{M_s.ds}{EI} + M_0 \sum \frac{ds}{EI} = 0$$

...(10.3)

Fig. 10.50. Elastic Centre

with H_0, V_0 and M_0, moment at any point with coordinates x, y can be obtained as

$$M(x, y) = M_s + M_0 + H_0 y + V_0 \cdot x$$

with appropriate signs for x, y

Hence $$M_A = M_s + M_0 + H_0 y_A + V_0 x_A$$

with appropriate signs for x_A and y_A

when a portal is symmetrical $\sum \frac{xy\,ds}{EI} = 0$

with elastic centre as origin and Equations (10.3) further reduce to

$$\sum \frac{M_s . x\,ds}{EI} + V_0 \sum \frac{x^2\,ds}{EI} = 0$$

$$\sum \frac{M_s . y\,ds}{EI} + H_0 \sum \frac{y^2\,ds}{EI} = 0$$

$$\sum \frac{M_s . ds}{EI} + M_0 \sum \frac{ds}{EI} = 0$$

giving
$$M_0 = -\frac{\sum \frac{M_s ds}{EI}}{\sum \frac{ds}{EI}}$$

$$V_0 = -\frac{\sum \frac{M_s . x . ds}{EI}}{\sum \frac{x^2 ds}{EI}}$$

$$H_0 = -\frac{\sum \frac{M_s y\,ds}{EI}}{\sum \frac{y^2 ds}{EI}} \quad ...(10.5)$$

In the case of symmetrical portal frame, however, it is easier to cut the arch at the centre assuming that rigid arm is connected to the part of the frame on each side of the cut. This is shown in Fig. 10.51.

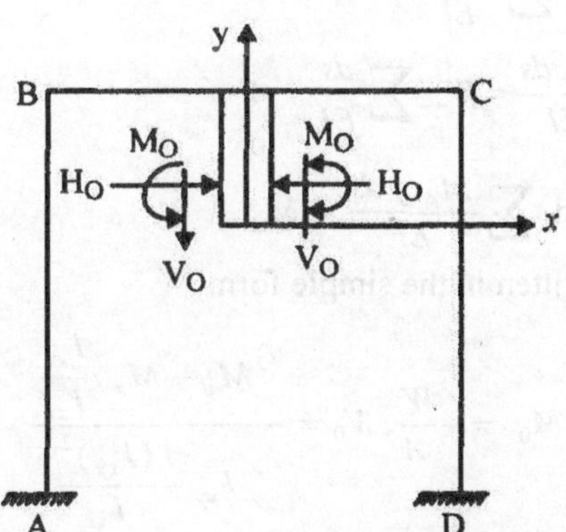

Fig. 10.51. Elastic Centre for Symmetrical Portal

The directions of H_0, V_0 and M_0 are shown in Fig. 10.51 again should conform to the condition that positive directions will all produce tension inside the frame. In the case of unsymmetrical portal frame the rigid arm connecting one end of the frame to the elastic centre is preferable.

For unsymmetrical frame $\sum \frac{xy\,ds}{EI} \neq 0$. Hence the results obtained for symmetrical arch is not valid. With $\sum \frac{x\,ds}{EI}$ and $\sum \frac{y\,ds}{EI}$ both equal to zero but $\sum \frac{xy\,ds}{EI} \neq 0$, the solution of the three equations give.

$$M_0 = -\frac{\sum \frac{M_s\,ds}{EI}}{\sum \frac{ds}{EI}}$$

$$V_0 = -\frac{\sum \frac{M_s\,.x\,ds}{EI} - \left(\sum \frac{M_s\,.y\,ds}{EI}\right)\left(\frac{\sum \frac{xy\,ds}{EI}}{\sum \frac{y^2\,ds}{EI}}\right)}{\sum \frac{x^2\,ds}{EI} - \frac{\left(\sum \frac{xy\,ds}{EI}\right)^2}{\sum \frac{y^2\,ds}{EI}}}$$

$$H_0 = -\frac{\sum \frac{M_s\,.y\,ds}{EI} - \left(\sum \frac{M_s\,.x\,ds}{EI}\right)\left(\frac{\sum \frac{xy\,ds}{EI}}{\sum \frac{x^2\,ds}{EI}}\right)}{\sum \frac{x^2\,ds}{EI} - \frac{\left(\sum \frac{xy\,ds}{EI}\right)^2}{\sum \frac{x^2\,ds}{EI}}} \qquad \text{...(10.6)}$$

Taking $\sum \frac{x^2\,ds}{EI} = Iyy,\ \sum \frac{y^2\,ds}{EI} = I_{xx}$

$$\sum \frac{xy\,ds}{EI} = I_{xy},\ \sum \frac{M_s\,ds}{EI} = W,\ \sum \frac{ds}{EI} = A$$

$$\sum \frac{M_s\,.x\,ds}{EI} = M_y \text{ and } \sum \frac{M_s\,.y\,ds}{EI} = M_x$$

the equations can be written in the simple form

$$M_0 = -\frac{W}{A},\ V_0 = -\frac{M_y - M_x \cdot \frac{I_{xy}}{I_{xx}}}{I_{yy} - \frac{(I_{xy})^2}{I_{xx}}}$$

and
$$H_0 = -\frac{M_x - M_y \cdot \frac{I_{xy}}{I_{yy}}}{I_{xx} - \frac{(I_{xy})^2}{I_{yy}}} \qquad \text{...(10.7)}$$

Example 10.23. Analyse the frame shown in Fig. 10.52 below by elastic centre method and draw the B.M. diagram. E is same for all members.

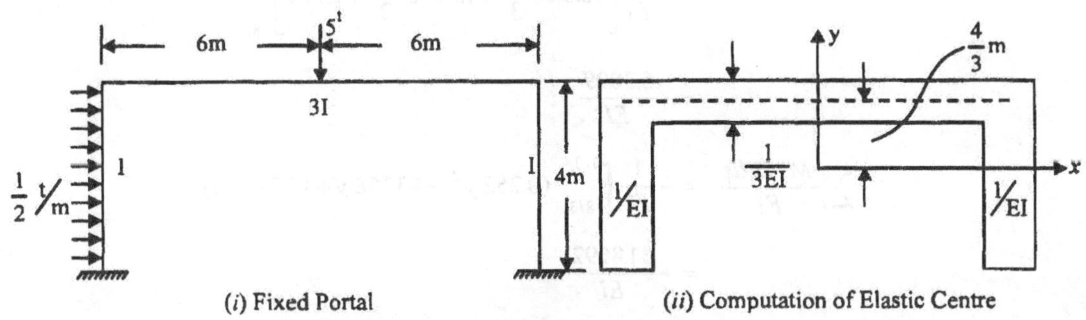

(i) Fixed Portal (ii) Computation of Elastic Centre

Fig. 10.52. Example 10.23

Solution: The figure is symmetrical, hence the elastic centre will lie on the symmetrical axis. The depth of the elastic centre from top

$$= \frac{2\times4\times\frac{1}{EI}\times2}{2\times4\times\frac{1}{EI}+12\times\frac{1}{3EI}}$$

$$= \frac{4}{3} \text{ m from top}$$

$$\sum\frac{x^2\,ds}{EI} = \frac{1}{12}\times\frac{1}{3\,EI}\times12^3+2\times4\times\frac{1}{EI}\times6^2$$

$$= \frac{1}{EI}[48+288] = \frac{336}{EI}$$

$$\sum\frac{y^2\,ds}{EI} = 2\times\frac{1}{EI}\times\frac{4^3}{12}+2\times4\times\frac{1}{EI}\times\left(\frac{2}{3}\right)^2+12\times\frac{1}{3\,EI}\times\left(\frac{4}{3}\right)^2$$

$$= \frac{1}{EI}[10.667+3.55+7.11]$$

$$= \frac{21.328}{EI}$$

$$\sum\frac{ds}{EI} = 2\times4\times\frac{1}{EI}+12\times\frac{1}{3EI} = \frac{12}{EI}$$

Equation of the curve

$$M_s = -0.252\,y^2 - 1.3358\,y - 1{\cdot}77,$$

at
$$y = -\frac{8}{3}, M_s = 0$$

At $y = \frac{4}{3},\ M_s = -4$

$y = 0,\ M_s = -1.77$

For AB

$$\sum M_s \frac{ds}{EI} = -\frac{1}{EI}\int_{-8/3}^{4/3} (0.252\,y^2 + 1.3358y + 1.77)\,dy$$

$$= -\frac{1}{EI}\left[0.252 \cdot \frac{y^3}{3} + 1.3358\frac{y^2}{2} + 1.77y\right]_{-8/3}^{4/3}$$

$$= -\frac{5.3099}{EI}.$$

$$\sum \frac{M_s.x\,ds}{EI} = -\frac{1}{EI}\int_{-8/3}^{4/3} (0.252\,y^2 + 1.3358\,y + 1.77).(-6)\,dy$$

$$= +\frac{31.8592}{EI}$$

$$\sum M_s \cdot \frac{y\,ds}{EI} = -\frac{1}{EI}\int_{-8/3}^{4/3} (0.252\,y^2 + 1.3358\,y + 1.77)\,y\,dy$$

$$= -\frac{1}{EI}\left[0.252\frac{y^4}{4} + 1.3358\frac{y^3}{3} + 1.77\frac{y^2}{2}\right]$$

$$= -\frac{1}{EI}[-2.9866 + 9.499 - 4.72]$$

$$= -\frac{1.7924}{EI}$$

For BC

$$\sum \frac{M_s\,ds}{EI} = \int_{-6}^{6} \frac{-4\,dx}{3\,EI} + \int_0^6 \frac{-5\,x\,dx}{3\,EI}$$

$$= \frac{1}{3\,EI}[-48 - 90] = \frac{-138}{3\,EI} = \frac{-46}{EI}$$

$$\sum \frac{M_s\,x\,ds}{EI} = \int_{-6}^{6} \frac{4x\,dx}{3\,EI} + \int_0^6 -\frac{5x.x\,dx}{3\,EI}$$

$$= \frac{-120}{EI}.$$

$$\sum \frac{M_s y\,ds}{EI} = \frac{1}{3\,EI}\left[\int_{-6}^{6} -4 \cdot \frac{4}{3}\,dx + \int_0^6 -5x \cdot \frac{4}{3}\,dx\right.$$

$$= \frac{1}{3\,EI}[-64 - 120] = \frac{-61 \cdot 333}{EI}$$

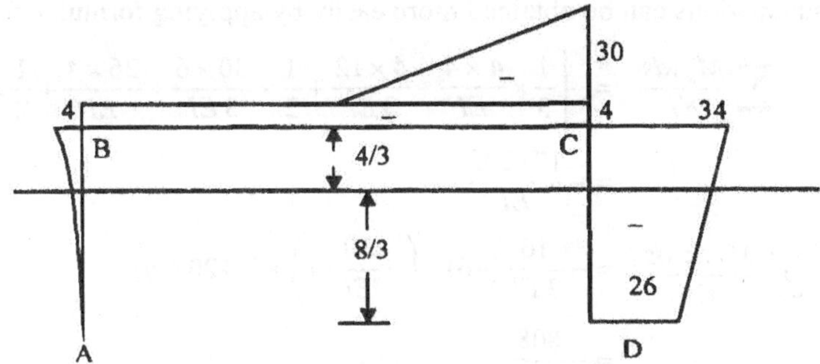

Fig. 10.53 M_s **diagram Examnple 10.23.**

For ***CD***

$$\sum \frac{M_s\,ds}{EI} = \frac{1}{EI}\int_{-8/3}^{4/3}\left(-2y-\frac{94}{3}\right)dy$$

$$= -\frac{1}{EI}\left(2\cdot\frac{y^2}{2}+\frac{94}{3}\cdot y\right)$$

$$= -\frac{1}{EI}\left[\left(\frac{4}{3}\right)^2-\left(-\frac{8}{3}\right)^2+\frac{94}{3}\left\{\frac{4}{3}-\left(-\frac{8}{3}\right)\right\}\right]$$

$$= -\frac{1}{EI}[120]$$

$$\sum \frac{M_s.x\,ds}{EI} = -\frac{1}{EI}(120)\times 6 = -\frac{720}{EI}$$

$$\sum \frac{M_s.y.ds}{EI} = \frac{1}{EI}\int_{-8/3}^{4/3}\left(-2y-\frac{94}{3}\right).y.dy$$

$$= -\frac{1}{EI}\int\left(+2y^2+\frac{94}{3}\cdot y\right)dy$$

$$= -\frac{1}{EI}\left[2\cdot\frac{y^3}{3}+\frac{94}{3}\cdot\frac{y^2}{2}\right]_{-8/3}^{4/3}$$

$$= -\frac{1}{EI}[14.222-83.555]$$

$$= +\frac{69.3335}{EI}$$

Total

$$\sum \frac{M_s\,ds}{EI} = -\frac{5.3099}{EI}-\frac{46}{EI}-\frac{120}{EI} = -\frac{1}{EI}\cdot 171.3099$$

$$\sum \frac{M_s.x\,ds}{EI} = \frac{1}{EI}[31.8592-120-720] = \frac{-808.1408}{EI}$$

$$\sum \frac{M_s.y.ds}{EI} = \frac{1}{EI}[-1.7924-61.333+69.335] = \frac{6.2096}{EI}$$

The above summations can be obtained more easily by applying formulae for areas and cgs.

$$\sum \frac{M_s.ds}{EI} = -\left[\frac{1}{3}\times\frac{4\times4}{EI}+\frac{4\times12}{3\,EI}+\frac{1}{2}\times\frac{30\times6}{3\,EI}+\frac{26\times4}{EI}+\frac{1}{2}\times\frac{8\times4}{EI}\right]$$

$$= -\frac{171.333}{EI}$$

$$\sum \frac{M_s.x.ds}{EI} = -\frac{16}{3\,EI}(-6)+\left(-\frac{30}{EI}\cdot4\right)+(-120\times6)$$

$$= -\frac{808}{EI}$$

$$\sum \frac{M_s.y.ds}{EI} = -\frac{16}{3\,EI}\left(\frac{1}{3}\right)-(16+30)\times\frac{4}{3}-26\times4\left(-\frac{2}{3}\right)-\left(\frac{1}{2}\times8\times4\times0\right)$$

$$= -\frac{16}{9}-\frac{48\times4}{3}+\frac{26\times8}{3}+0$$

$$= \frac{6.222}{EI}$$

Then,
$$M_0 = -\sum\frac{M_s.ds}{EI}\Big/\sum\frac{ds}{EI} = \frac{+171.3099}{12} = +14.2758$$

$$V_0 = -\sum\frac{M_s x ds}{EI}\Big/\sum\frac{x^2\,ds}{EI} = +\frac{808.1408}{336} = +2.452$$

$$H_0 = -\sum\frac{M_s\,y ds}{EI}\Big/\sum\frac{y^2\,ds}{EI} = -\frac{6.2096}{21.328} = -0.2911$$

as shown in Fig. 10.54.

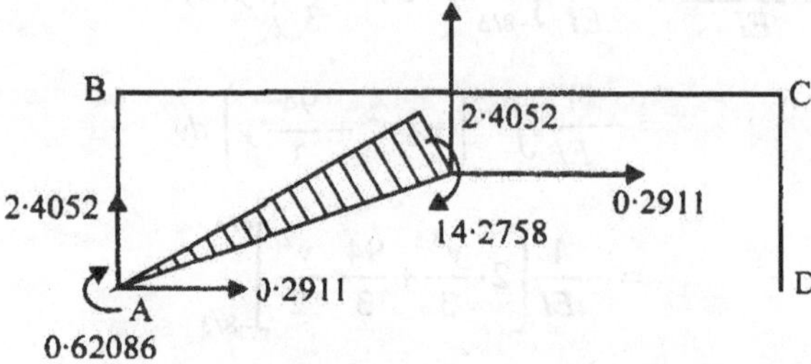

Fig. 10.54. Forces at elastic centre and at A

$$M_A = 14.2758-2.4052\times6+0.2911\times\frac{8}{3}$$
$$= 0.62086.$$

From fig 10.54

$$M_{AB} = 0.62086$$

$$M_{BA} = 0.62086-0.2911\times4-\frac{1}{2}\times4\times2$$
$$= -4.54354.$$

$$M_{CB} = 0.62086-0.2911\times4-\frac{1}{2}\times4\times2-5\times6+2.4052\times12$$
$$= \mathbf{-5.68114.}$$

$$M_{DC} = 0.62086 + \frac{1}{2} \times 4 \times 2 - 5 \times 6 + 2.4052 \times 12$$
$$= 3.48326.$$

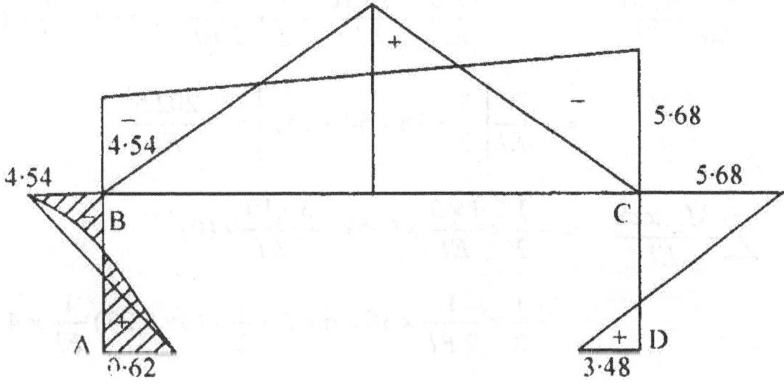

Fig. 10.55. B.M. diagram

Example 10.24. Analyse the portal frame shown in Fig. 10.56 by elastic centre method.

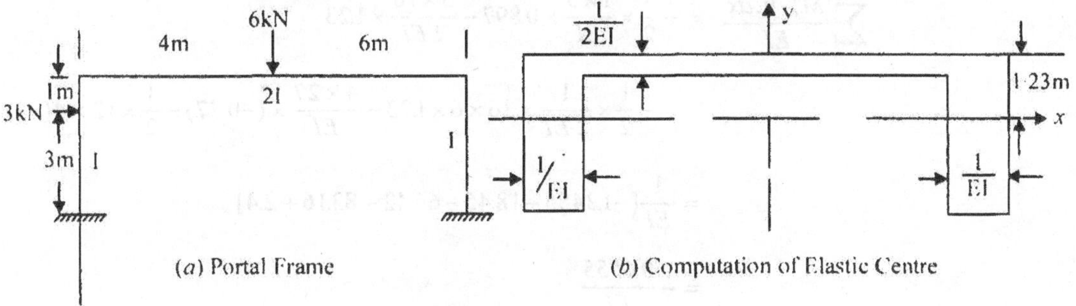

Fig. 10.56. Example 10.24.

Solution. The frame is symmetrical. Hence the elastic centre will lie on the axis of symmetry. The depth from the top $\bar{y}$

$$= \frac{2 \times 4 \times \frac{1}{EI} \times 2}{2 \times 4 \times \frac{1}{EI} + 10^5 \times \frac{1}{2EI}} = \frac{16}{13} = 1.23\,\text{m}$$

$$\sum \frac{x^2\,ds}{EI} = 5^2 \times \frac{4}{EI} + (-5)^2 \times \frac{4}{EI} + \frac{1}{2EI} \times \frac{10^3}{12}$$

$$= \frac{1}{EI}[100 + 100 + 41.667] = \frac{1}{EI} \cdot 241.667.$$

$$\sum \frac{y^2 ds}{EI} = \frac{1}{EI}\left(\frac{4^3}{12} + 4 \times 0.77^2\right) \times 2 + \frac{1}{2EI} \cdot 10 \cdot 1.23^2$$

$$= \frac{1}{EI}[15.40 + 7.56] = \frac{22.96}{EI}$$

$$\sum \frac{ds}{EI} = \frac{2\times 4}{EI} + \frac{10}{2\,EI} = \frac{13}{EI}$$

$$\sum \frac{M_s.ds}{EI} = -\frac{1}{2}\times\frac{1\times 3}{EI} - \frac{3\times 10}{2\,EI} - \frac{1}{2}\times\frac{1}{2\,EI}\times 36\times 6 - \frac{1}{2}\times(39+27)\frac{1}{EI}\times 4$$

$$= -\frac{1}{EI}\left[\frac{3}{2}+15+54+132\right] = -\frac{202.5}{EI}$$

$$\sum \frac{M_s.x\,dx}{EI} = -\frac{1}{2}\times\frac{1\times 3}{EI}\times(-5) - \frac{3\times 10}{2\,EI}\times(0)$$

$$-\frac{1}{2}\times\frac{1}{2\,EI}\times 36\times 6\times 3 - \frac{1}{2}\times(39+27)\frac{1}{EI}\times 4\times 5$$

$$= \frac{1}{EI}[7.5-0-162-660] = \frac{-814.50}{EI}$$

$$\sum \frac{M_s.y.dx}{EI} = -\frac{1}{2}\times\frac{1\times 3}{EI}\times 0.897 - \frac{3\times 10}{2\,EI}\times 1.23$$

$$-\frac{1}{2}\times\frac{1}{2\,EI}\times 36\times 6\times 1.23 - \frac{4\times 27}{EI}\times(-0.77) - \frac{1}{2}\times 12\times 4(-1).$$

$$= \frac{1}{EI}[-1.3455-18.45-66.42+83.16+2.4]$$

$$= -\frac{0.6555}{EI}$$

Then,

$$M_0 = -\frac{\sum \frac{M_s.ds}{EI}}{\sum \frac{ds}{EI}} = -\frac{\frac{-202.50}{EI}}{\frac{13}{EI}}$$

$$= +15.577 \text{ kNm}$$

$$V_0 = -\frac{\sum \frac{M_s.x.ds}{EI}}{\sum \frac{x^2\,ds}{EI}} = -\frac{-\frac{814.50}{EI}}{\frac{241.667}{EI}}$$

$$= +3.37 \text{ kN}$$

$$H_0 = -\frac{\sum M_s\,y\frac{ds}{EI}}{\sum \frac{y^2\,ds}{EI}} = -\frac{-\frac{0.6555}{EI}}{\frac{22.96}{EI}}$$

$$= +0.0285 \text{ kN}$$

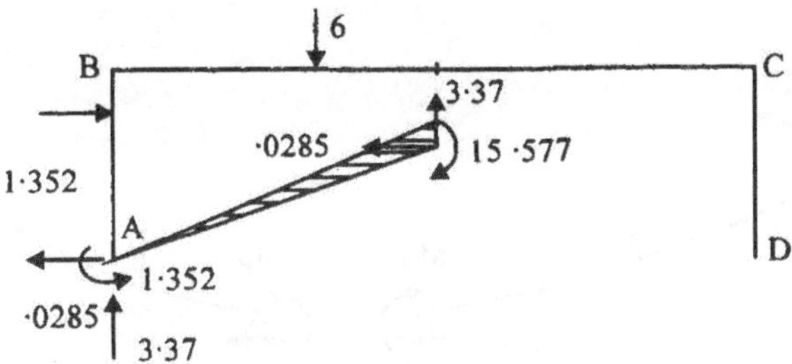

Fig. 10.57. Forces at Elastic Centre transfered to A

$$M_A = 15.577 - .0285 \times 2.77 - 3.37 \times 5$$
$$= -1.35 \text{ kNm (anticlockwise)}$$
$$M_B = -1.352 - 3 \times 1 + .0285 \times 4 = -4.2379 \text{ kNm}$$
$$M_C = -1.352 - 3 \times 1 + .0285 \times 4 + 3.37 \times 10 - 6 \times 6 = -6.538 \text{ kNm}$$
$$M_D = -1.352 + 3 \times 3 + 3.37 \times 10 - 6 \times 6$$
$$= +5.348.$$

10.9. THEOREM OF THREE MOMENTS

The three moment equation expresses the relation between the bending moments at three successive supports of a continuous beam subjected to loading on the various spans with or without settlement of supports. Here moments at the supports are taken as unknowns. It is a force method of analysis. A statically determinate solution is first assumed satisfying the equilibrium requirements. The compatibility condition is then imposed. In the case of continuous beam the compatibility condition is continuity of the beam over intermediate support, that is the slope of the tangent at the left end of the right span is equal to the slope of the tangent of the right end of the left span.

10.9.1. DERIVATION OF THE EQUATION

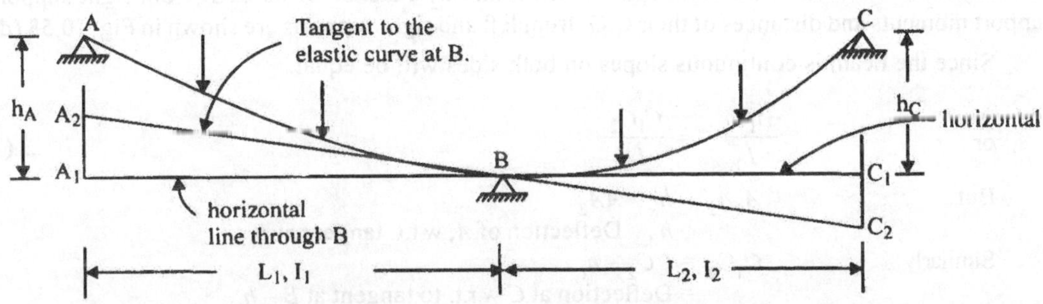

Fig. 10.58 (*a*) Continuous beam with support settlement.

Let *ABC* be a part of the continuous beam with unequal settlement of supports. Let the middle support *B* be at the lowest point and let the height of *A* be h_A and that of *C* be h_C above *B*. The effect of continuity at *A, B* and *C* is to introduce continuity moment at the support. We can then draw the free body diagram of each span *AB* and *BC* separately provided we put in continuity moments at the supports. Let them be M_A, M_B and M_C.

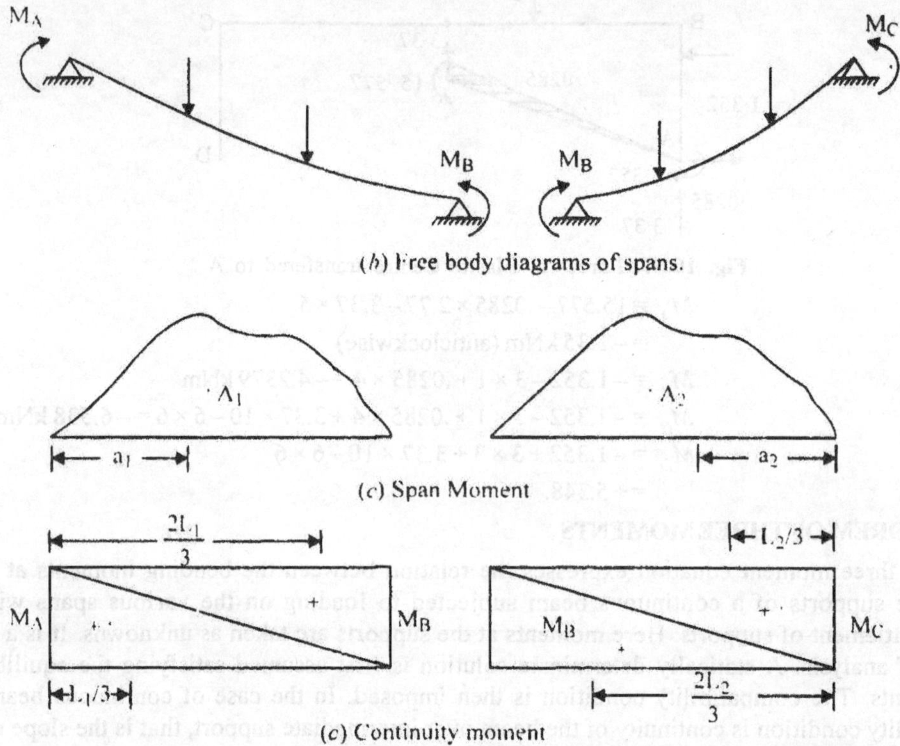

Fig. 10.58. (*b*), (*c*) and (*d*) — Derivation of Three Moment Equation.

Fig. 10.58 (b) shows the free body diagrams of the spans. Fig. 10.58 (c) gives the span moments. Let A_1 be the area of the left span moment and a_1, distance of its C.G. from the left support. Similarly A_2 is the area of the right span moment and a_2 distance of its C.G. from right support. Support moments and distances of their C.G. from left and right supports are shown in Fig. 10.58 (d).

Since the beam is continuous slopes on both sides will be equal.

or $$\frac{A_1A_2}{L_1} = \frac{C_1C_2}{L_2} \qquad \text{.....(1)}$$

But $A_1A_2 = h_A - AA_2$

$= h_A -$ Deflection of A, w.r.t. tangent at B.

Similarly $C_1C_2 = CC_2 - h_C$

= Deflection at C w.r.t. to tangent at $B - h_C$.

But deflection of A w.r.t. tangent at B = Moment of the $\frac{M}{EI}$ diagram between A and B about A. which is equal to

$$= \frac{1}{EI_1}\left[A_1a_1 + \frac{M_A.L_1}{2}\left(\frac{L_1}{3}\right) + \frac{M_B.L_1}{2}\left(\frac{2L_1}{3}\right)\right]$$

Similarly deflection of C w.r.t. tangent at B.

$$= \frac{1}{EI_2}\left[A_2a_2 + \frac{M_C.L_2}{2}\left(\frac{L_2}{3}\right) + \frac{M_B.L_2}{2}\left(\frac{2L_2}{3}\right)\right]$$

Therefore $$A_1A_2 = h_A - \frac{1}{EI_1}\left[A_1a_1 + \frac{M_A.L_1}{2}\left(\frac{L_1}{3}\right) + \frac{M_B.L_1}{2}\left(\frac{2L_1}{3}\right)\right] \quad ....(2)$$

Similarly $$C_1C_2 = \frac{1}{EI_2}\left[Aa_2a_2 + \frac{M_C.L_2}{2}\left(\frac{L_2}{3}\right) + \frac{M_B.L_2}{2}\left(\frac{2L_2}{3}\right)\right] - h_C \quad ....(3)$$

Substituting Eqns. (2) and (3) in (1) we get

$$\frac{h_A}{L_1} - \frac{1}{EI_1L_1}\left[A_1a_1 + \frac{M_A.L_1}{2}\left(\frac{L_1}{3}\right) + \frac{M_B.L_1}{2}\left(\frac{2L_1}{3}\right)\right]$$

$$= \frac{1}{EI_2L_2}\left[A_2a_2 + \frac{M_C.L_2}{2}\left(\frac{L_2}{3}\right) + \frac{M_B.L_2}{2}\left(\frac{2L_2}{3}\right)\right] - \frac{h_C}{L_2} \quad ....(4)$$

Multiply both sides by $6E$ and rearranging terms :

$$M_A.\left(\frac{L_1}{I_1}\right) + 2M_B\left(\frac{L_1}{I_1} + \frac{L_2}{I_2}\right) + M_C\left(\frac{L_2}{I_2}\right) = -\frac{6A_1a_1}{I_1L_1} - \frac{6A_2a_2}{I_2L_2} + \frac{6Eh_A}{L_1} + \frac{6Eh_C}{L_2} \quad .....(10.8)$$

This is known as Three Moment Equation as it gives a relation between three moments M_A, M_B and M_C. It is the work of three French investigators, Berlot (1855), Clapeyron (1857) and Bresse (1862). When I is constant for the whole beam and there is no support settlement $I_2 = I_1 = I$ and $h_A = h_C = 0$. The equation then reduces to.

$$M_AL_1 + 2M_B(L_1 + L_2) + M_CL_2 = -\left(\frac{6A_1a_1}{L_1} + \frac{6A_2a_2}{L_2}\right) \quad .....(10.8\,a)$$

Initially all moments are assumed to be sagging and positive.

Example 10.25. Use three moment equation to determine the support moments for the continuous beam shown in Fig. 10.59. Draw the B.M. and S.F. diagrams.

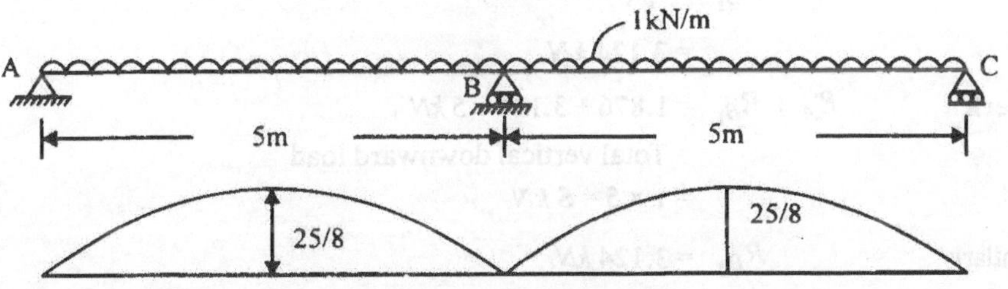

Fig. 10.59. Example 10.25

Solution: As the supports A and C are simple supports $M_A = M_C = 0$

Applying three moment equations to spans AB and BC.

$$M_AL_1 + 2M_B(L_1 + L_2) + M_CL_2 = -\frac{6A_1a_1}{L_1} - \frac{6A_2a_2}{L_2}$$

$$A_1 = \frac{2}{3} \times \frac{wL^2}{8} \times L = \frac{2}{3} \times \frac{1 \times 25 \times 5}{8} = \frac{50 \times 5}{24}$$

$a_1 = 2.5\ m$. Similarly $\quad A_2 = \dfrac{250}{24}, a_2 = 2.5$

Hence $\quad 2M_B(5+5) = -\dfrac{6}{5}\left(\dfrac{2 \times 2.5 \times 250}{24}\right)$

or $\quad M_B = -\dfrac{6 \times 2 \times 2.5 \times 250}{20 \times 24 \times 5}$

$= -3.12\ kNm$

Calculation of reactions

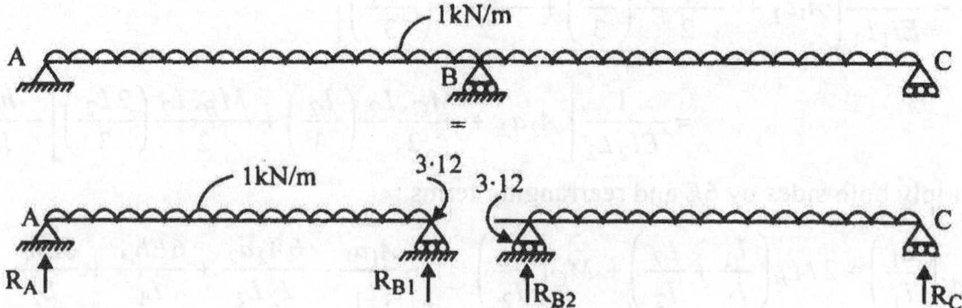

Fig. 10.60. Calculations of Reactions.

$R_A \times 5 - \dfrac{1 \times 5^2}{2} + 3.12 = 0$

or $\quad R_A = \dfrac{1}{5}(-3.12 + 12.50)$

$= 1.876\ kN.$

$R_{B_1} \times 5 - \dfrac{1 \times 5^2}{2} - 3.12 = 0$

or $\quad R_{B_1} = \dfrac{1}{5}(12.5 + 3.12)$

$= 3.124\ kN.$

Check $\quad R_A + R_{B_1} = 1.876 + 3.124 = 5\ kN \uparrow$

= Total vertical downward load

$= 1 \times 5 = 5\ kN$

Similarly $\quad R_{B_2} = 3.124\ kN.$

$R_C = 1.876\ kN.$

Total reaction at $B = R_{B_1} + R_{B_2}$

$= 6.248\ kN$

(a) Beam with reactions

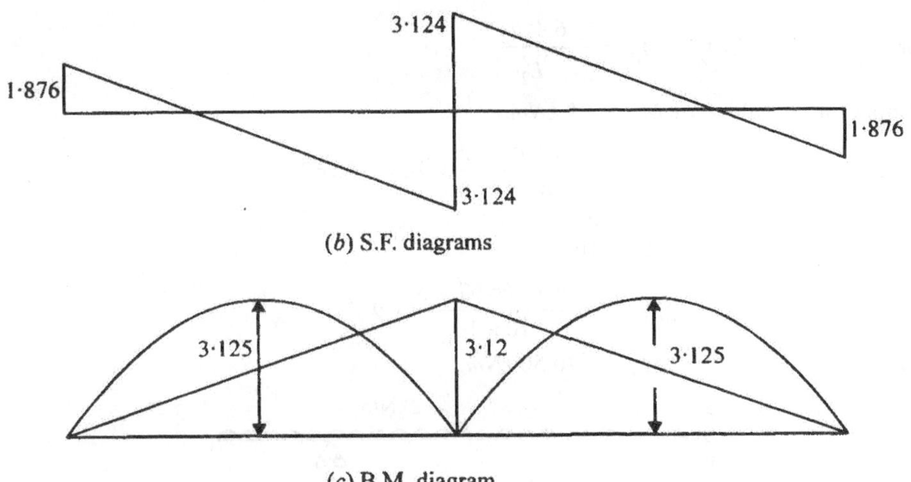

(*b*) S.F. diagrams

(*c*) B.M. diagram.

Fig. 10.61. S.F. and B.M. diagrams.

Example 10.26. Use three moment equation to find M_A and M_B. Draw B.M. and S.F. diagrams.

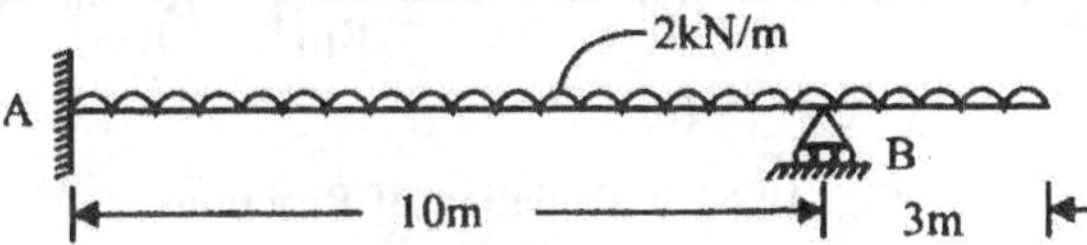

Fig. 10.62. Example 10.26

Solution: Here M_B is known and is equal $-\frac{2 \times 3^2}{2} = -9\ kNm$

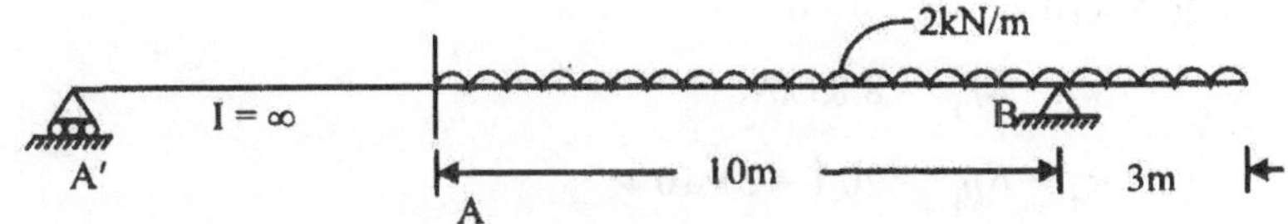

Fig. 10.63

To compute M_{A}, assume a span AA' to the left of A with $I = \infty$.

The general equation of three moment method without support settlement is

$$M_A\left(\frac{L_1}{I_1}\right) + 2M_B\left(\frac{L_1}{I_1} + \frac{L_2}{I_2}\right) + M_C\left(\frac{L_2}{I_2}\right) = -\frac{6A_1a_1}{I_1L_1} - \frac{6A_2a_2}{I_2L_2} \quad \text{....(10.8)}$$

With spans $A'A$ and AB, the equation becomes

$$M_A'\left(\frac{L_1}{I_1}\right) + 2M_A\left(\frac{L_1}{I_1} + \frac{L_2}{I_2}\right) + M_B\left(\frac{L_2}{I_2}\right) = -\frac{6A_1a_1}{I_1L_1} - \frac{6A_2a_2}{I_2L_2}$$

with $I_1 = \infty$ and $\frac{L_1}{I_1} = 0$ the above equation becomes

$$2M_A\left(\frac{L_2}{I_2}\right) + M_B\left(\frac{L_2}{I_2}\right) = -\frac{6A_2a_2}{I_2L_2}$$

or $$2M_A + M_B = -\frac{6A_2a_2}{L_2^2}$$

Substituting values of $M_B = -9\ kNm$.

$$A_2 = \frac{2 \times 10^2}{8} \times \frac{2}{3} \times 10 = 166.67$$

$$a_2 = 5$$

$$L_2 = 10.$$

$$2M_A = -\frac{6 \times 166.67 \times 5}{10 \times 10} + 9$$

or $$M_A = -20.50\ kNm.$$

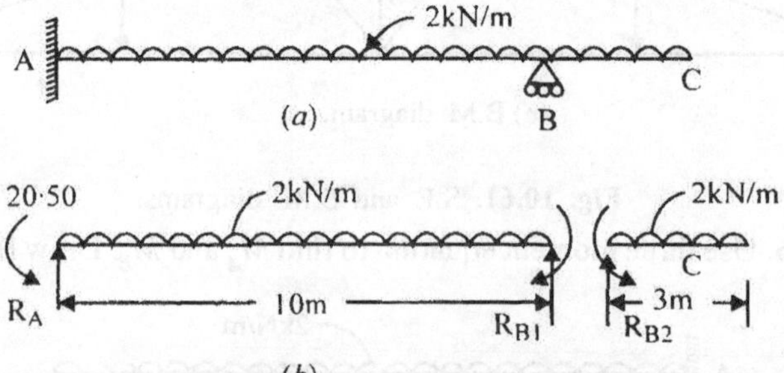

(a)

(b)

Fig. 10.64. Calculation of Reactions.

$$R_A \times 10 - 20 \times 50 - 10 \times 2 \times 5 + 9 = 0$$

or $$R_A = 11.15\ kN.$$

$$R_{B_1} \times 10 + 20.5 - 10 \times 2 \times 5 - 9 = 0$$

or $$R_{B_1} = 8.85\ kN$$

Check $$R_A + R_{B_1} = 20 \uparrow = 2 \times 10 \downarrow$$

$$R_{B_2} \times 3 - 9 - \frac{2 \times 3^2}{2} = 0$$

$$R_{B_2} = 6kN$$

$$R_{B_1} + R_{B_2} = 14.85\ kN.$$

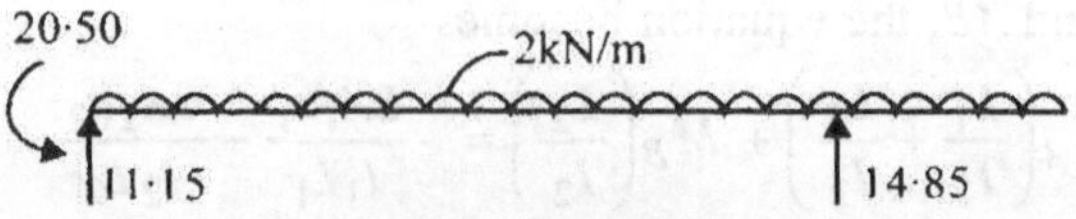

(a) Beam with reactions

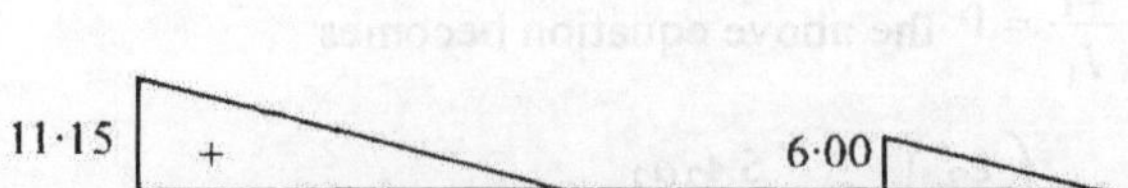

(b) S.F. diagram

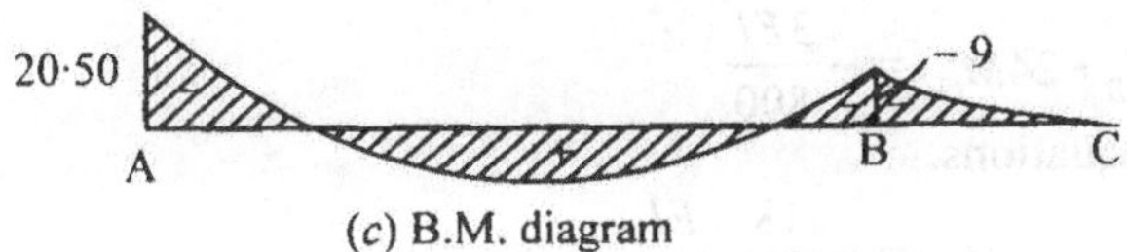

(c) B.M. diagram

Fig. 10.65. S.F., B.M. Diagrams (Example 10.26)

Example 10.27. Use Three Moment Equation to find the support moments for the beam showr below (Fig. 10.66) when support B settles by $\frac{1}{2}$ cm down. $E = 2000' / cm^2$. $I = 40{,}000\ cm^4$.

Solution:

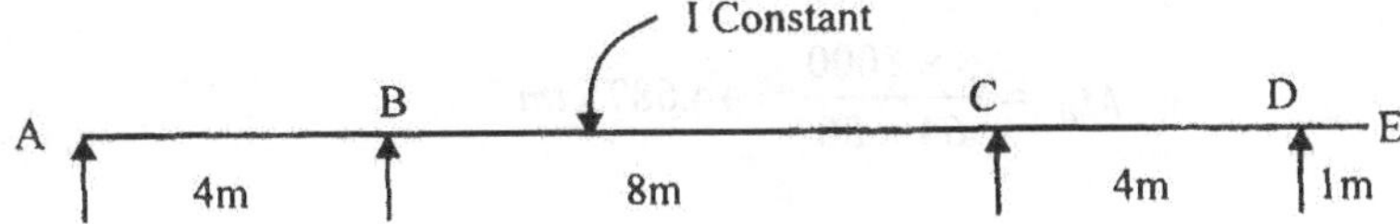

Fig. 10.66. Example 10.27

1. General Three Moment Equation with constant I

$$M_A(L_1) + 2M_B(L_1 + L_2) + M_C(L_2) = -\frac{6A_1a_1}{L_1} - \frac{6A_2a_2}{L_2} + \frac{6Eh_A.I}{L_1} + \frac{6Eh_C.I}{L_2} \qquad (10.8\text{ b})$$

Here $M_A = 0$, $M_D = 0$ and with noload A_1 and A_2 are zero.
Hence for spans AB and BC Eqn. (10.8 *b*) reduces to

$$2M_B(L_1 + L_2) + M_C(L_2) = \frac{6Eh_A.I}{L_1} + \frac{6Eh_C.I}{L_2}$$

with middle support B, $\frac{1}{2}$ *cm* down

$$h_A = h_C = +\frac{1}{200}m$$

Substituting values for span AB and BC

$$2M_B(4+8) + M_C(8) = 6EI\left(\frac{1}{200 \times 4} + \frac{1}{200 \times 8}\right)$$

or
$$24M_B + 8M_C = \frac{9}{800}.EI$$

Similarly for spans BC and CD.

$$M_B.L_1 + 2M_C(L_1 + L_2) + M_D(L_2) = +\frac{6EIh_B}{L_1} + \frac{6EIh_D}{L_2}$$

Here $L_1 = 8\,m$ $\quad h_B = -\frac{1}{200}m$ (as B is below C)

$L_2 = 4\,m$ $\quad h_D = 0$

$M_D = 0$

The equation then reduces to

$$8M_B + 2M_C(12) = -6EI.\frac{1}{200}.\frac{1}{8}$$

$$= -EI.\frac{6}{1600}.$$

or $\quad 8M_B + 24M_C = -\dfrac{3EI}{800}$

Solving the two equations.

$$M_C = -\frac{18}{800}\cdot\frac{EI}{64}$$

$$M_B = \frac{3}{64}\cdot\frac{EI}{80}$$

$$EI = 2,000 \times 40,000 = 8,000\ tm^2$$

Hence $\quad M_C = -\dfrac{18 \times 8000}{800 \times 64} = -2.8125\ tm$

$$M_B = \frac{3 \times 8000}{64 \times 80} = +4.6875\ tm$$

Check.

$$\text{Slope at } B \text{ left side} = \frac{1}{800} - \frac{4.6875 \times 4}{3 \times 8000} = +0.00046875$$

$$\text{Slope at } B \text{ right side} = \frac{1}{1600} - \frac{4.6875 \times 8}{3 \times 8000} + \frac{2.8125 \times 8}{6 \times 8000} = -0.00046875$$

Example 10.28. A continuous beam ACB of uniform section has two spans $AC = L_1$ and $CB = L_2$. It is freely supported at the ends A and B and carries a uniformly distributed load w_1/ unit length over span AC and w_2 / unit length over the span CB. If the support C is δ_1 below A and δ_2 below B, show that the reaction at C is given by

$$R_C = \frac{(w_1 L_1 + w_2 L_2)}{2} + \frac{(w_1 L_1^2 + w_2 L_2^2)}{8 L_1 L_2} - \frac{3EI}{L_1 L_2}\left(\frac{\delta_1}{L_1} + \frac{\delta_2}{L_2}\right)$$

[I.A.S. 1958]

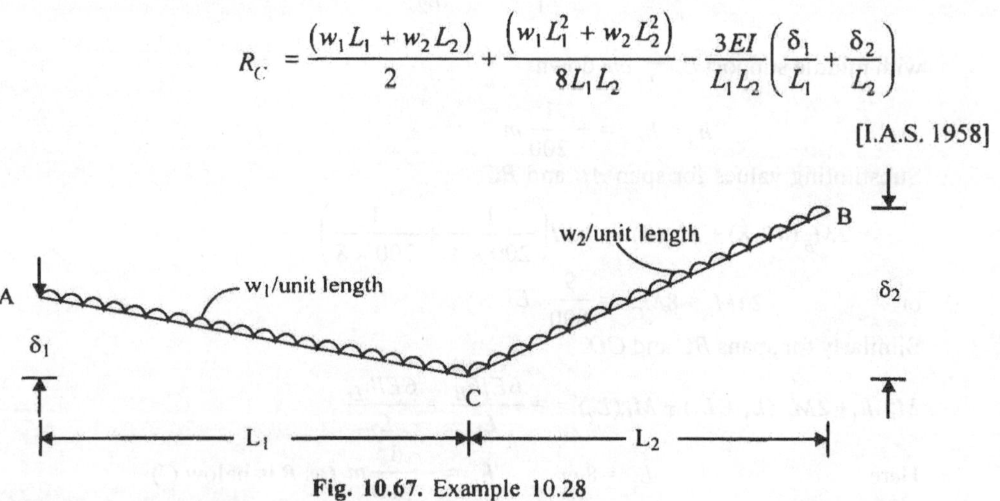

Fig. 10.67. Example 10.28

Solution: $\quad M_A = 0$ and $M_B = 0$

Applying Theorem of three moments.

$$2M_C(L_1 + L_2) = -\frac{6A_1a_1}{L_1} - \frac{6A_2a_2}{L_2} + \frac{6EI\delta_1}{L_1} + \frac{6EI\delta_2}{L_2}$$

$$= -6 \times \frac{2}{3} \times \frac{w_1 L_1^2}{8} \times L_1 \times \frac{L_1}{2} \times \frac{1}{L_1}$$

$$-6 \times \frac{2}{3} \times \frac{w_2 L_2^2}{8} \times L_2 \times \frac{L_2}{2} \times \frac{1}{L_2} + \frac{6E\delta_1}{L_1} \cdot I + \frac{6E\delta_2}{L_2} \cdot I$$

$$= -\frac{w_1 L_1^3}{4} - \frac{w_2 L_2^3}{4} + \frac{6E\delta_1}{L_1} . I + \frac{6E\delta_2}{L_2} . I$$

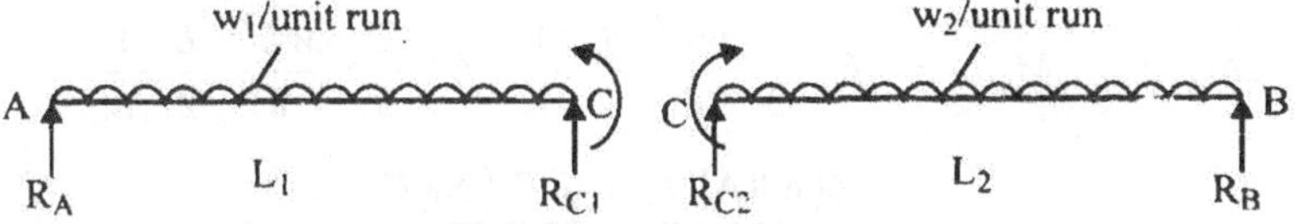

Fig. 10.68. Example 10.28

Taking moments about A.

$$R_{C_1} \times L_1 + M_C - \frac{w_1 L_1^2}{2} = 0$$

or $$R_{C_1} = \frac{1}{L_1}\left(\frac{w_1 L_1^2}{2} - M_C\right)$$

Similarly $$R_{C_2} = \frac{1}{L_2}\left(\frac{w_2 L_2^2}{2} - M_C\right)$$

$$R_{C_1} + R_{C_2} = \frac{w_1 L_1}{2} - \frac{M_C}{L_1} + \frac{w_2 L_2}{2} - \frac{M_C}{L_2}$$

$$= \frac{w_1 L_1 + w_2 L_2}{2} - \left(\frac{1}{L_1} + \frac{1}{L_2}\right) . M_C$$

$$= \left(\frac{w_1 L_1 + w_2 L_2}{2}\right) - \left(\frac{L_1 + L_2}{L_1 L_2}\right) . \frac{1}{2(L_1 + L_2)}$$

$$\left[-\frac{w_1 L_1^3}{4} - \frac{w_2 L_2^3}{4} + \frac{6EI\delta_1}{L_1} + \frac{6EI\delta_2}{L_2}\right]$$

$$= \left(\frac{w_1 L_1 + w_2 L_2}{2}\right) + \left(\frac{w_1 L_1^3 + w_2 L_2^3}{8 L_1 L_2}\right) - \frac{3EI}{L_1 L_2}\left(\frac{\delta_1}{L_1} + \frac{\delta_2}{L_2}\right)$$

$= R_C$ (Proved).

Example 10.29. Find the bending moments at the supports of a continuous beam on elastic (spring) supports. The beam has a constant flexural rigidity EI and the stiffness of the elastic supports is $K = 20\ EI/L^3$.

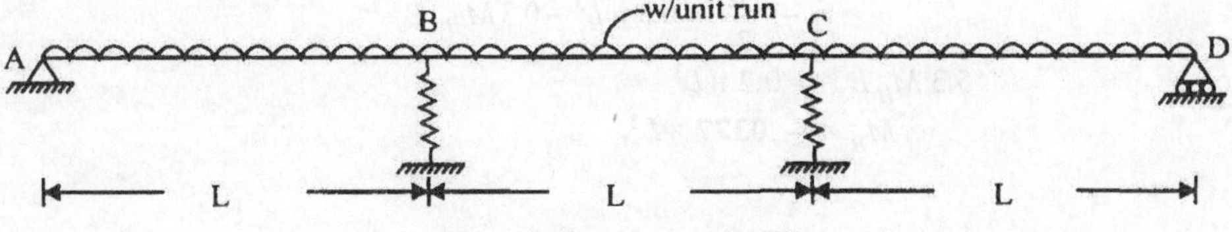

Fig. 10.69. Example 10.29

Solution: $M_A = M_D = 0$

Because of symmetry let the support settlement be Δ at B and C. $M_B = M_C$ from symmetry.

For span *ABC*.

$$M_A(L_1) + 2M_B(L_1 + L_2) + M_C(L_2) = -\frac{6A_1 a_1}{L_1} - \frac{6A_2 a_2}{L_2} + \frac{6Eh_A . I}{L_1} + \frac{6Eh_C . I}{L_2}$$

$$2M_B(2L) + M_B . L = -6 \times \frac{2}{3} . \frac{wL^3}{8} . \frac{L}{2} . \frac{1}{L} - 6 \times \frac{2}{3} \times \frac{wL^3}{8} . \frac{L}{2} . \frac{1}{L}$$

$$+ \frac{6(E)(\Delta)(I)}{L} + \frac{6(E)(\Delta)(I)}{L}$$

or
$$5M_B . L = -\frac{wL^3}{2} + \frac{6EI\Delta}{L}$$

Calculation of Reactions

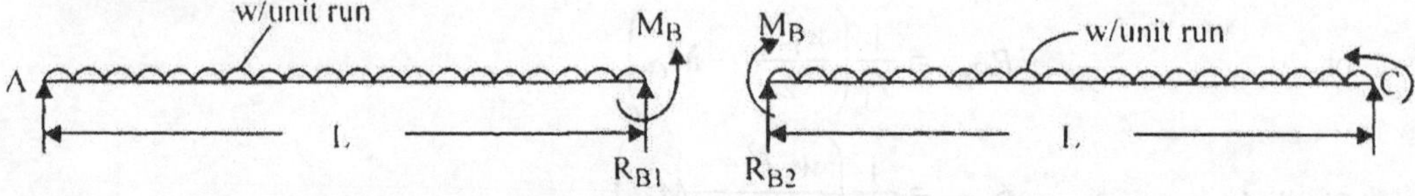

Fig. 10.70. Example 10.29

$$R_{B_1} \times L + M_B = \frac{wL^2}{2}$$

or
$$R_{B_1} = \frac{wL}{2} - \frac{M_B}{L}$$

$$R_{B_2} \times L + M_B - M_C - \frac{wL^2}{2} = 0$$

or
$$R_{B_2} = \frac{wL}{2}$$

$$R_{B_1} + R_{B_2} = R_B = \left(wL - \frac{M_B}{L}\right) = \frac{20EI\Delta}{L^3}$$

or
$$\Delta = \frac{L^3}{20EI}\left(wL - \frac{M_B}{L}\right)$$

Substituting

$$5M_B L = -\frac{wL^3}{2} + \frac{6EI}{L} \times \frac{L^3}{20EI} \times \left(wL - \frac{M_B}{L}\right)$$

$$= -\frac{wL^3}{2} + 0.3\,wL^3 - 0.3\,M_B . L$$

or
$$5.3\,M_B L = -0.2\,wL^3$$

$$M_B = -.0377\,wL^2.$$

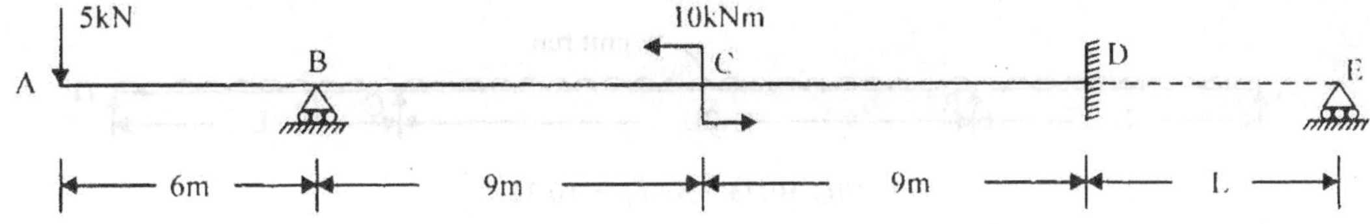

Fig. 10.71. Example 10.30.

Example 10.30. Compute the B.M. at the supports for the beam given below.

Solution: $M_B = -30\ kNm.$

Assume a span DE with infinite I consider span BD and DE.

$$M_B\left(\frac{18}{I}\right) + 2M_D\left(\frac{18}{I} + \frac{L}{\infty}\right) + M_E \cdot \frac{L}{\infty} = -6\frac{\left(\frac{1}{2}\times 5\times 9\times 6 - \frac{1}{2}\times 5\times 9\times 12\right)}{I\times 18}$$

10kNm
B
D
10kNm
R_B
R_D

(*i*) Simply supported span BC.

5kNm
+
5kNm
–

(*ii*) Free B.M. diagram for span BD.

Fig. 10.72. Free B.M. diagram for span *BD.*

$$-30(18) + 2M_D \times 18 = 45$$

$$36\,M_D = 45 + 540 = 585$$

$$M_D = \frac{585}{36} = 16.25\ kNm.$$

Example 10.31. *ABCD* is a straight uniform beam of length $4L$. It is freely supported at its ends A and D and two intermediate supports B and C distant L from either end. The supports at A and D are rigid but those at B and C are such that they deflect by an amount λ for each unit of load which is placed upon them. The beam carries a uniformly distributed load w per unit length along its entire length.

Show that the reactions at the supports are :

$$\frac{wL}{8}\left[\frac{7L^3 + 48EI\lambda}{4L^3 + 3EI\lambda}\right] \text{ and } \frac{3wL}{8}\left[\frac{19L^3}{4L^3 + 3EI\lambda}\right]$$ (Cambridge)

Solution:

Three Moment Equation

$$M_A \cdot L + 2M_B(L + 2L) + M_C \cdot 2L = -\frac{6A_1a_1}{L} - \frac{6A_2a_2}{2L} + 6EhI\left(\frac{1}{L} + 0\right).$$

with $M_A = 0$ and $M_B = M_C$

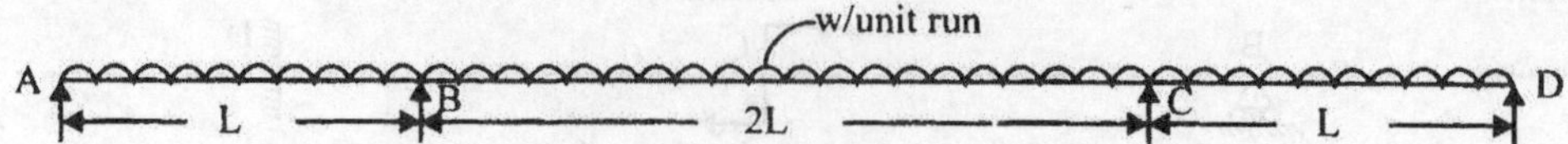

Fig. 10.73. Example 10.31

$$8\,M_B L = -\frac{6}{L}\times\frac{2}{3}\times L\times\frac{wL^2}{8}\times\frac{L}{2} - 6\times\frac{2}{3}\times 2L\times w\frac{(2L)^2}{8}\times\frac{L}{2L}+\frac{6EI\Delta}{L}$$

$$= -\frac{wL^3}{4} - 2wL^3 + \frac{6EI\Delta}{L}$$

$$= -\frac{9}{4}wL^3 + \frac{6EI\Delta}{L}$$

$$M_B = -\frac{9}{32}wL^2 + \frac{3}{4}.\frac{EI\Delta}{L^2}$$

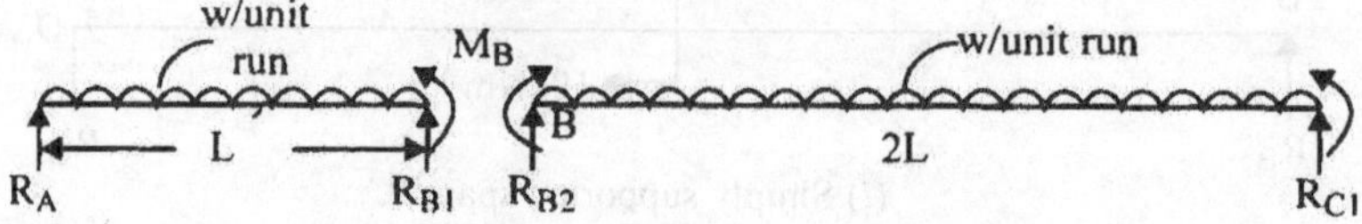

Fig. 10.74. Calculation of Reactions.

$$R_A.L - \frac{wL^2}{2} - M_B = 0$$

$$R_A = \frac{1}{L}\left[\frac{wL^2}{2} - M_B\right]$$

$$R_{B_1}\times L - \frac{wL^2}{2} + M_B = 0$$

$$R_{B_1} = \frac{1}{L}\left[\frac{wL^2}{2} - M_B\right]$$

$$= \frac{wL}{2} - \frac{M_B}{L}$$

$$R_{B_2} = \frac{w\times 2L}{2} = wL$$

$$R_B = \frac{wL}{2} - \frac{M_B}{L} + wL = \frac{3wL}{2} - \frac{M_B}{L}$$

As per data given, $\left(\frac{3wL}{2} - \frac{M_B}{L}\right).\lambda = \Delta$

$$M_B = -\frac{9}{32}wL^2 + \frac{3}{4}.\frac{EI}{L^2}\left(\frac{3wL}{2} - \frac{M_B}{L}\right).\lambda$$

or $M_B\left(1+\frac{3}{4}.\frac{EI}{L^2}.\frac{1}{L}.\lambda\right) = -\frac{9}{32}wL^2 + \frac{9}{8}.\frac{EI}{L}.w.\lambda$

or
$$M_B = \frac{-\dfrac{9}{32}wL^2 + \dfrac{9}{8}.\dfrac{EIw}{L}.\lambda}{1+\dfrac{3}{4}.\dfrac{EI\lambda}{L^3}}$$

$$R_A = \frac{wL}{2} + \frac{M_B}{L}$$

$$= \frac{wL}{2} + \frac{-\dfrac{9}{32}wL^2 + \dfrac{9}{8}.\dfrac{EIw}{L}.\lambda}{4L^3 + 3EI\lambda} \times 4L^2$$

$$= \frac{2wL^4 + \dfrac{3}{2}wL.EI\lambda - \dfrac{9}{8}wL^4 + \dfrac{9}{2}EIw\lambda L}{4L^3 + 3EI\lambda}$$

$$= \frac{\dfrac{7}{8}wL^4 + 6EI\lambda wL}{4L^3 + 3EI\lambda}$$

$$= \frac{wL}{8}\left[\frac{7L^3 + 48EI\lambda}{4L^3 + 3EI\lambda}\right]$$

$$R_B = \frac{3wL}{2} - \frac{M_B}{L}$$

$$= \frac{3wL}{2} - \frac{-\dfrac{9}{32}wL^2 + \dfrac{9}{8}.\dfrac{EIw}{L}.\lambda}{4L^3 + 3EI\lambda} \times 4L^2$$

$$= \frac{3wL\left(4L^3 + 3EI\lambda\right) + \dfrac{9}{4}wL^4 - 9EIw\lambda L}{2\left(4L^3 + 3EI\lambda\right)}$$

$$= \frac{12wL^4 + 9wLEI\lambda + \dfrac{9}{4}wL^4 - 9EIw\lambda L}{2\left(4L^3 + 3EI\lambda\right)}$$

$$= \frac{57wL^4}{8\left(4L^3 + 3EI\lambda\right)} = \frac{3}{8}wL\left[\frac{19L^3}{4L^3 + 3EI\lambda}\right].$$

10.10 TEMPERATURE CHANGE AND FABRICATION ERROR

Temperature difference in the top and bottom of a beam will produce curvature and if the beam is indeterminate it will produce bending moments. Similarly fabrication error will generate moment. These are self straining problems as moments are generated without application of external loads.

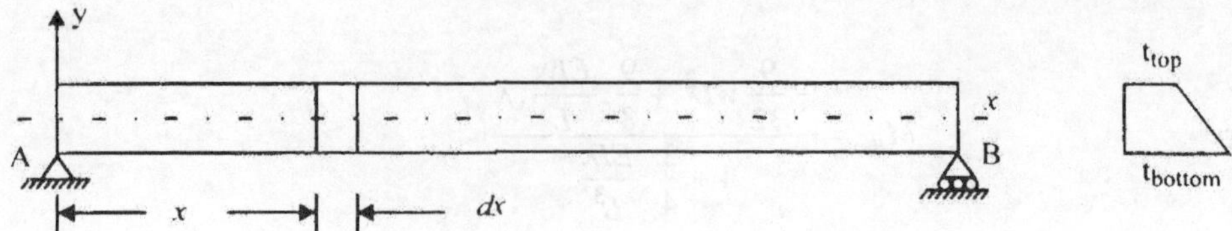

Fig. 10.75. Effect of Temperature Change.

Consider a beam AB subjected to a temperature difference of ΔT over a depth h of beam. If an element of length dx is considered at a distance x from origin the expansion at top $= \alpha.dx.t_{top}$

$$\text{at bottom} = \alpha.dx.t_{bottom}$$

$$\text{Difference in expansion} = \alpha.dx.\Delta T$$

$$\text{and slope of the line} = \frac{\alpha.dx.dT}{h} = d\theta.$$

or

$$\frac{d\theta}{dx} = \frac{\alpha.dT}{h}$$

Now $\frac{d\theta}{dx}$ = curvature and is constant through out the length of the beam. Hence the curvature diagram is rectangular as shown in Fig. 10.76.

Fig. 10.76. Curvature Diagram.

Now area of the curvature diagram gives the slope at

$$A \text{ and } B = \frac{\alpha.dT}{h}.\frac{L}{2}$$

$$= \frac{L\alpha.dT}{2h}$$

Example 10.32. Compute the moment developed in the beam given below due to a temperature gradient of 10°C applied to both spans.

Fig. 10.77. Example 10.32.

Take $\alpha_t = 0.000012\,/\,°C$, $h = 40$ mm and

$EI = 500{,}000$ kN m^2.

Solution: The three moment equation can be rederived from continuity of slope at the mid-point when there is no support settlement.

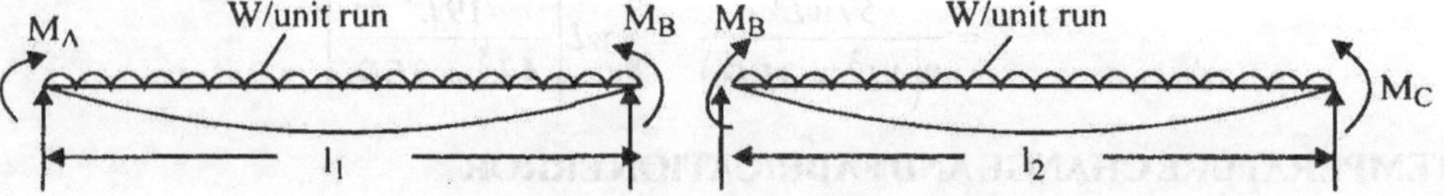

Fig. 10.78. Derivation of three moment Equation.

$$\theta_B \text{ left} = \frac{M_A\,l_1}{6\,EI} + \frac{A_1 a_1}{EI\,l_1} + \frac{M_B l_1}{3\,EI} + \frac{\alpha\,dT\,l_1}{2h}$$

$$\theta_B \text{ right} = \frac{M_B . l_2}{3EI} + \frac{A_2 a_2}{EI l_2} + \frac{M_C l_2}{6EI} + \frac{\alpha dT l_2}{2h}$$

Since rotation at B should be the same on both sides of B (either clockwise or anticlockwise)

$$\frac{M_A l_1}{6EI} + \frac{A_1 a_1}{EI l_1} + \frac{M_B l_1}{3EI} + \alpha \cdot \frac{dT}{h} \cdot \frac{l_1}{2} = -\left(\frac{M_B . l_2}{3EI} + \frac{A_2 a_2}{EI l_2} + \frac{M_C l_2}{6EI} + \frac{\alpha\, dT}{h} \cdot \frac{l_2}{2} \right)$$

With transformation and multiplying by 6 EI

$$M_A l_1 + 2 M_B (l_1 + l_2) + M_C l_2 = -\frac{6 A_1 a_1}{l_1} - \frac{6 A_2 a_2}{l_2} - \frac{\alpha dT l_1}{2h} . 6EI - \frac{\alpha . dT}{2h} l_2 . 6EI$$

As there is no load $A_1 = A_2 = 0$ and $M_A = M_C = 0$ and $l_1 = l_2 = 10$
The above equation becomes.

$$2 M_B (10 + 10) = -\frac{6 EI \alpha . dT}{h} \cdot l$$

$$= \frac{-6 \times (500{,}000).(.000012) \times 10 \times 10}{.04}$$

or $\qquad M_B = \mathbf{2250\ kNm.}$

SUMMARY

1. Indeterminate structures can be solved by two methods.
   (*a*) Force or flexibility or compatibility method
   (*b*) Displacement, stiffness or equilibrium method.
   In the force or flexibility method force is taken as unknown and the equations are obtained considering compatibility.
2. Classical force methods are:
   1. Consistent deformation.
   2. Method of Leastwork.
   3. Theorem of three moments.
3. A fixed portal frame is statically indeterminate to the third degree. Consistent deformation method can be applied to find the three unknowns by solving the three consistent deformation equations obtained from the condition that rotation, horizontal and vertical displacement are all zero at the redundant support. The equations are, however, coupled. Elastic centre method can be applied to get the unknowns more easily and in a more straight forward manner.
4. Three moment Equation expresses the relation between the bending moments at three successive supports of a continuous beam subjected to an arbitrary loading with or without support settlements. Here moments at the supports are taken as unknowns.

PROBLEM

A. Theoretical.

10.1.(*i*) Write the theorem of three moments considering support settlement, explaining meaning of notations used. [AMIE 'Summer' 95

(*ii*) Derive the theorm of three moments.

(*iii*) Explain the elastic centre method of structural analysis.

(*iv*) Explain "force method" and "displacement method" by simple examples.

B. Objective, Tick the correct answer.

10.2.(*i*) The three moment equation written for an intermediate support of a continuous beam expresses the condition of

(*a*) slope compatibility at that point

(*b*) moment equilibrium at that point

(*c*) zero deflection at the support point

(*d*) structural stability **[AMIE Winter '98]**

(*ii*) Three moment equation is applicable when

(*a*) the beam is prismatic.

(*b*) there is no settlement of support.

(*c*) there is no discontinuity within the span.

(*d*) the spans are equal. **[AMIE Winter '97]**

(*iii*) The theorem of three moments expresses the condition of

(*a*) equilibrium of forces.

(*b*) slope compatibility

(*c*) Maxwell's reciprocal relation **[AMIE Summer' 99]**

C. Mathematical

10.3. Calculate the slope at B and vertical deflection at D in the continuous beam shown in Fig. P. 10.1 Take EI = constant **[AMIE Winter 97]**

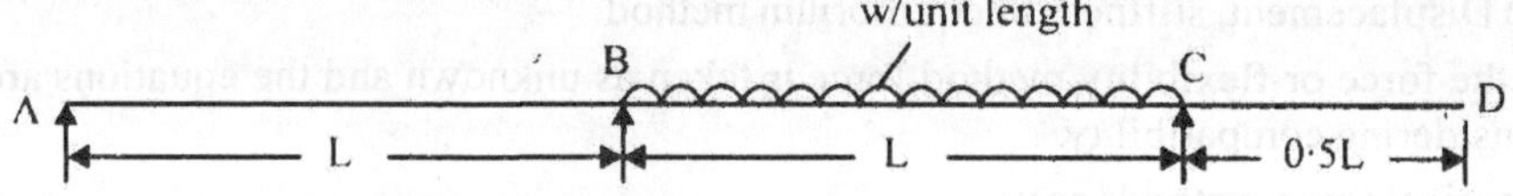

Fig. P. 10.1. (Problem 10.3)

10.4 Find the load in the member CF shown in Fig. P 10.2 below in the plane pin jointed frame loaded as shown. Vertical and horizontal members each have a cross sectional area 'a' while diagonal members are each of area = $2a$. All members are of the same material.

[AMIE Winter 1978]

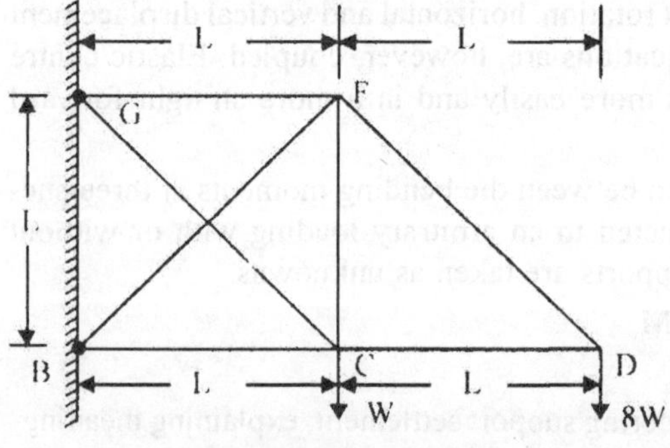

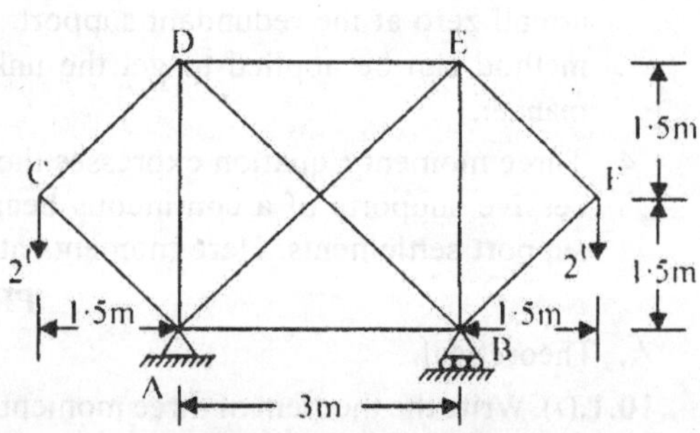

Fig. P 10.2. (Problem 10.4) **Fig. P 10.3.** (Problem 10.5)

10.5 The pin jointed truss shown in Fig. P 10.3 below is simply supported at A and B and carries load as shown. All members have a cross-sectional area of 10 cm^2 and the members are initially unstressed. Calculate vertical deflection of joint C if E = 1,900 t/cm^2.

[AMIE Summer 1979]

10.6 In the plane braced frame shown in Fig. P 10.4 below, all the members have same cross-sectional area of 8 cm^2 and are made of the same material. The member *AC* in the frame is initially short by 0.25 cm. Determine the force in each member. Also determine the final forces in each member when member *AC* is subjected to a rise of temperature of 20°C. Take

$$E = 2 \times 10^6 \text{ kg/cm}^2 \quad \alpha = 1.1 \times 10^{-5} / °C$$

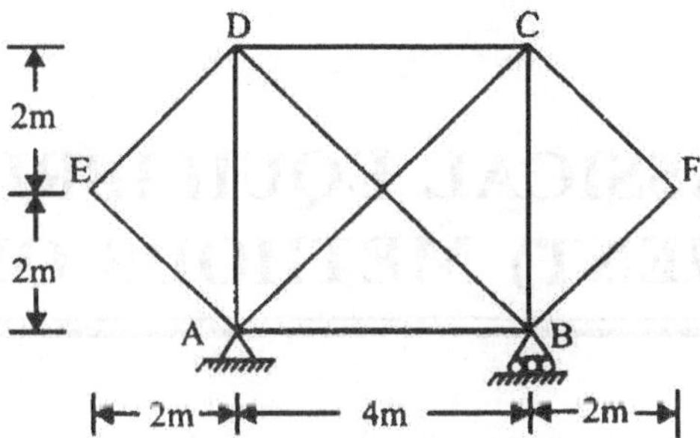

Fig. P. 10.4. (Problem 10.6)

10.7 Calculate the slope at *B* and vertical deflection at *D* in the continuous beam. shown in Fig. P. 10.5. Take *EI* as constant.

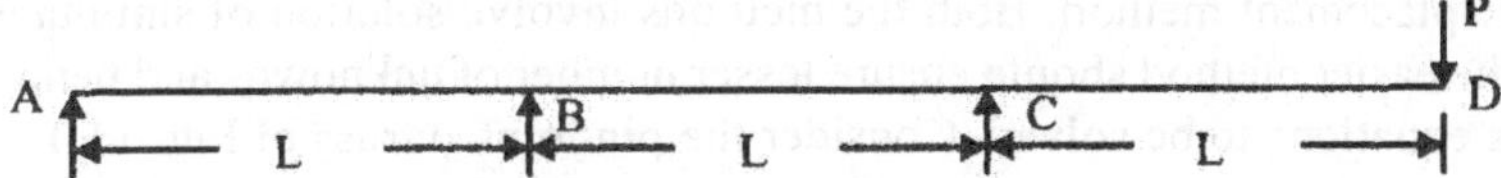

Fig. P. 10.5. Problem 10.7

10.8 *A* beam *ABC* of uniform *EI* rests on three simple supports *A*, *B* and *C* initially at the same level. The spans *AB* and *BC* are each of length *L*. The span *AB* is loaded with a load *W* concentrated at mid-section and the span *BC* carries the load *W* uniformly distributed over the entire span. During loading the middle support *B* sinks by an amount $\frac{5WL^3}{96\,EI}$. Using three moment equation analyse the beam and compute the support reactions. Hence draw the S.F. and B.M. diagrams.

(AMIE Summer '95)

11

CLASSICAL EQUILIBRIUM (DISPLACEMENT) METHODS OF ANALYSIS

11.1 INTRODUCTION

In the last chapter force method was explained. In this chapter we examine displacement method. Here displacement of joint is taken as unknown and equilibrium equations are formed to solve for the unknown displacements. An indeterminate structure can be solved either by force method or by displacement method. Both the methods involve solution of simultaneous equations and obviously the easier method should ensure lesser number of unknowns and hence lesser number of simultaneous equations to be solved. Consider the pin jointed truss of Fig. 11.1.

The structure has only one redundancy. Hence if force method is used there will be only one equation to be solved involving the unknown force. However, if displacement method is used, there will be 17 unknown displacements to be solved. (two in each joint of B, C, D, E, G, H, I, J and one in joint F). This will require 17 simultaneous equations in the displacement method of analysis. Obviously, therefore, force method is easier in solving the truss problem. Let us now consider the frame of Fig. 11.2. In the force method of analysis we have three unknown forces at joint C to solve.

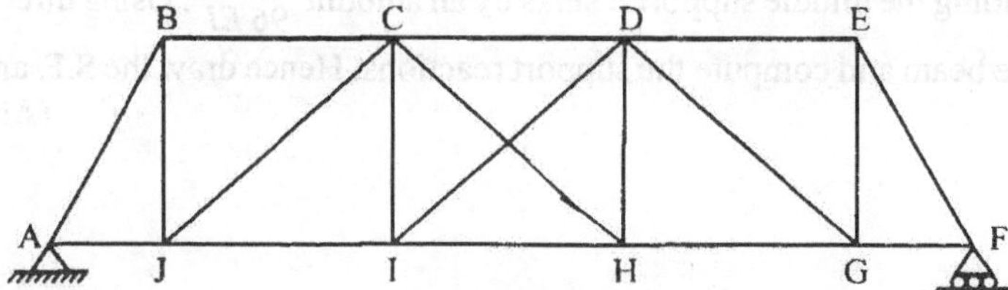

Fig. 11.1. Pin jointed Truss.

In the displacement method of analysis if axial compression is neglected there is only one unknown displacement *i.e.*, rotation at B to be solved. Hence displacement method is obviously advantageous in this case. Many classical methods in structural analysis are based on equilibrium approach through they are not always obvious. The following classical methods based on equilibrium approach are widely used in structural analysis.

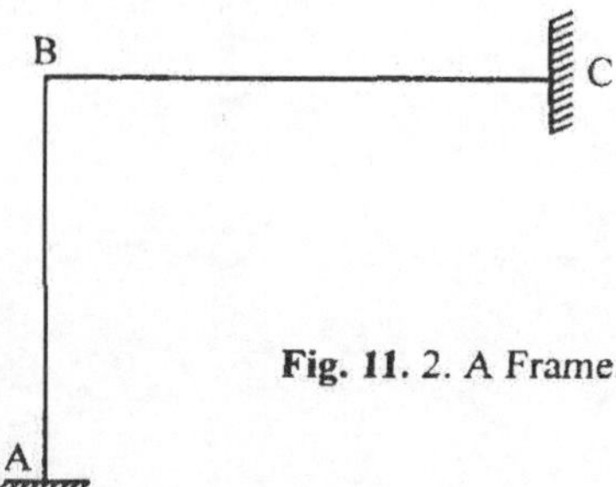

Fig. 11. 2. A Frame.

1. Slope deflection method
2. Moment distribution method

3. Kani's method.

The following energy theorems can also be used to derive equilibrium equations.

1. Principle of Minimum Total Potential Energy
2. Castigliano's First Theorem.
3. Principle of Stationary Complementary Potential Energy.

Of all the above methods slope deflection method was very popular before the advent of computers. Moment distribution and Kani's method are iterative solutions of the simultaneous equations of the slope deflection method. The energy principles above leading to equilibrium equations are of theoretical interest as the first three methods appear simpler to apply compared to the energy principles.

11.2 SLOPE DEFLECTION METHOD

The method was developed by Axel Bendexon in Germany in 1914. It was presented in greater detail by Prof. G.A. Maney of University of Minnesota in 1915. In the slope deflection method all joints are considered rigid *i.e.*, angles between members at the joints are considered not to change with application of loads. Thus the joints at the interior supports of statically indeterminate beams can be considered as 180° rigid joints and joints in the right angled rigid frames as 90° rigid joints. When beams or rigid frames are deformed, the rigid joints are considered to rotate as a whole which means that the angles between tangents to the various parts of the elastic curve meeting at a joint remain the same as those in the original underformed structure. In this way compatibility condition of the structure is satisfied. Deflection due to shear and axial stresses are not considered as the effects are small in case of beams and rigid frames.

Consider again the frame of Fig. 11. 3(*a*). The only joint displacement (degree of freedom) possible in the frame is rotation θ at joint *B*. The joint *B* and the members *BA* and *BC* are isolated in Fig. 11.3 (*b*). Now the equilibrium equation when applied to the joint *B* with direction of moments as shown gives :

$M_{BA} + M_{BC} = 0$. M_{BA} means moment acting on the joint *B* from member *BA*, M_{BC} means moment acting on joint *B* from member *BC*. Notice that equal and opposite moments act on the member *BA* and *BC*. In the slope deflection method M_{BA} and M_{BC} are expressed interms of θ and the member characterstics of *BA* and *BC* respectively. Thus all three requirements of structural analysis namely,

1. Compatibility
2. Equilibrium
3. Member characteristics are satisfied in slope deflection equations

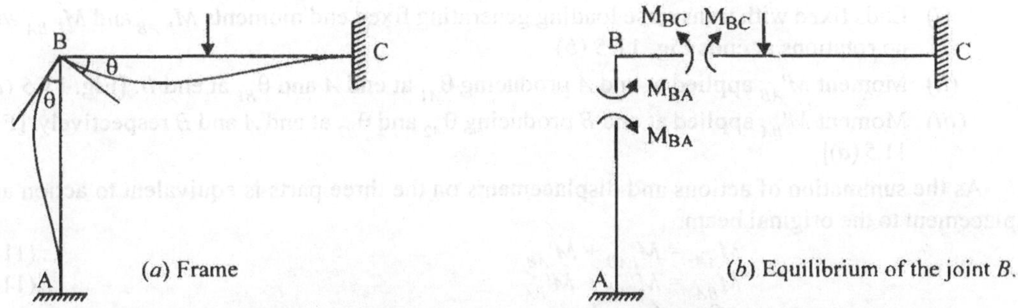

Fig. 11.3. Displacement method of analysis.

The bending moment sign convention (sagging + ve and hogging –ve) is not suitable for slope deflection equations. Instead the following sign convention is adopted.

"Clockwise moments and rotations at the ends of a member and upward displacement of right end relative to the left end producing clockwise fixed end moments are all considered positive." These are shown in Fig. 11.4.

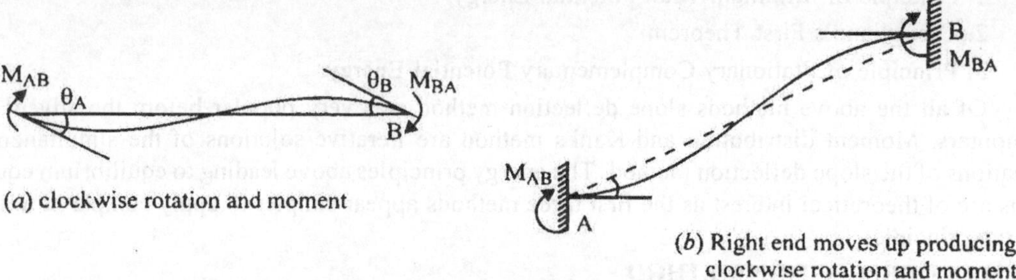

Fig. 11.4. Sign Convention (Positive).

After getting all the unknown moments with proper signs (clockwise +*ve* and anticlockwise –*ve*) bending moment diagrams can be drawn following the usual bending moment sign convention (sagging +*ve* and hogging –*ve* quantity). It should be clearly understood that different sign conventions are applicable for different purposes and they should not be clubbed together.

11.3 DERIVATION OF SLOPE DEFLECTION EQUATIONS

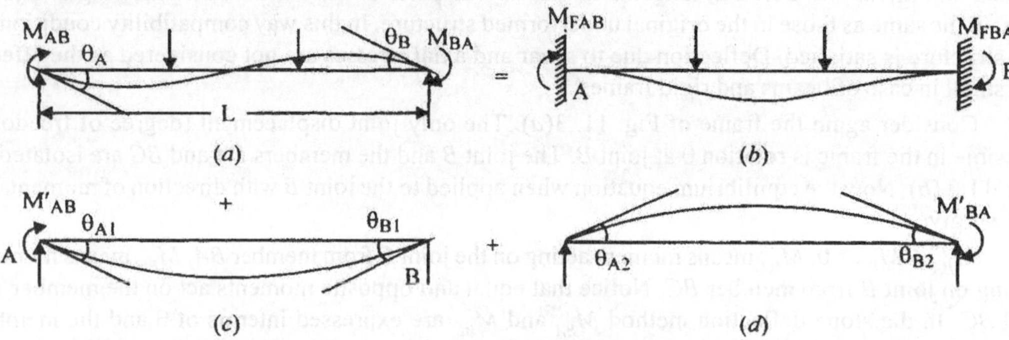

Fig. 11.5. Derivation of Slope Deflection Equations.

Fig. 11.5 shows the free body diagram of a beam subjected to arbitrary transverse loading, end moments M_{AB} and M_{BA} (clockwise) and end rotations θ_A and θ_B (clockwise). The beam can be split into the three parts :

(*i*) Ends fixed with transverse loading generating fixed end moments $M_{F\,AB}$ and $M_{F\,BA}$ with no rotations at ends Fig. 11.5 (*b*).

(*ii*) Moment M'_{AB} applied at end *A* producing θ_{A1} at end *A* and θ_{B1} at end *B*. [Fig. 11.5 (*c*)]

(*iii*) Moment M'_{BA} applied at end *B* producing θ_{A2} and θ_{B2} at end *A* and *B* respectively. [Fig. 11.5 (*d*)]

As the summation of actions and displacements on the three parts is equivalent to action and displacement to the original beam.

$$M_{AB} = M_{FAB} + M'_{AB} \quad \text{....(11.1)}$$

$$M_{BA} = M_{FBA} + M'_{BA} \quad \text{....(11.2)}$$

$$\theta_A = \theta_{A1} - \theta_{A2}$$

$$\theta_B = \theta_{B2} - \theta_{B1}$$

From conjugate beam method,

$$\theta_{A1} = \frac{M'_{AB}\,L}{3EI} \qquad \theta_{B1} = \frac{M'_{AB}\,L}{6EI}$$

$$\theta_{A2} = \frac{M'_{BA}\,L}{6EI} \qquad \theta_{B2} = \frac{M'_{BA}\,L}{3EI}$$

Substituting these values in Eqns. (11.2)

$$\theta_A = \frac{M'_{AB}\,L}{3EI} - \frac{M'_{BA}\,L}{6EI}$$

$$\theta_B = \frac{M'_{BA}\,L}{3EI} - \frac{M'_{AB}\,L}{6EI}$$

Solving for M'_{AB} and M'_{BA}

$$M'_{AB} = \frac{2EI}{L}(2\theta_A + \theta_B)$$

$$M'_{BA} = \frac{2EI}{L}(2\theta_B + \theta_A)$$

Hence with substitution in Eqns. (11.1)

$$M_{AB} = M_{FAB} + \frac{2EI}{L}(2\theta_A + \theta_B) \quad \text{....(11.3)}$$

$$M_{BA} = M_{FBA} + \frac{2EI}{L}(2\theta_B + \theta_A) \quad \text{....(11.4)}$$

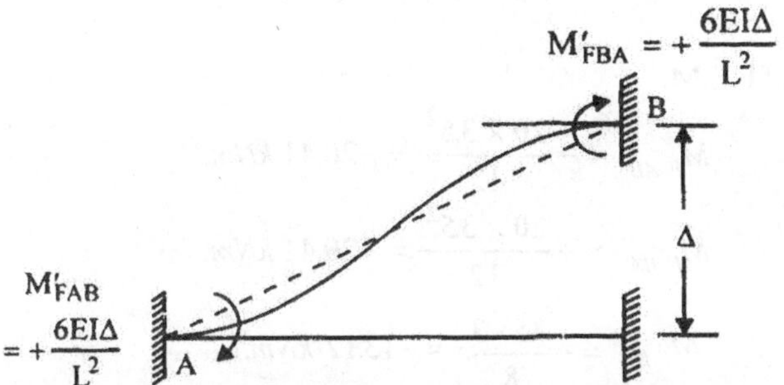

Fig. 11.6. F.E.M. due to support movement.

Fig. 11.6 shows the deflected shape and moment generated when the right support goes up with respect to the left by an amount Δ. From the figure it can be seen that additional clockwise moments $M'_{F\,AB} = M'_{F\,BA} = +\dfrac{6EI\,\Delta}{L^2}$ are generated as a result of this displacement. This is already derived in Chapter 7 (Ex. 7.24). With support movements the final moments, therefore, becomes

$$M_{AB} = M_{FAB} + \frac{2EI}{L}\left(2\theta_A + \theta_B + 3.\frac{\Delta}{L}\right) \quad \text{....(11.5)}$$

$$M_{BA} = M_{FBA} + \frac{2EI}{L}\left(2\theta_B + \theta_A + 3.\frac{\Delta}{L}\right) \quad \text{....(11.6)}$$

In solving the slope deflection equations it is desirable to use a relative stiffness $\bar{K}$ as $\dfrac{2EI}{L}$ is usually a large number. The coeffecient $\dfrac{2EI}{L}$ before the parenthesis is different for each span. If the $\dfrac{2EI}{L}$ values for all spans are made N time smaller, the effect will be only to make all θ values N times larger while the products of the expressions $\frac{2EI}{L}$ and $(2\theta_A + \theta_B)$ or the values of the end moments remain unchanged. To obtain actual rotations, however, the calculated values should be divided by N. The slope deflection equation can be expressed in matrix form as :

$$\begin{bmatrix} M_{AB} \\ M_{BA} \end{bmatrix} = \begin{bmatrix} M_{FAB} \\ M_{FBA} \end{bmatrix} + \begin{bmatrix} \frac{4EI}{L} & \frac{2EI}{L} \\ \frac{2EI}{L} & \frac{4EI}{L} \end{bmatrix} \begin{bmatrix} \theta_A \\ \theta_B \end{bmatrix} \quad \text{....(11.7)}$$

It can now be summarised that the slope deflection method is a displacement (stiffness) method that employs the equations of equilibrium in the direction of the degrees of freedom of the structure interms of the joint rotations and displacements perpendicular to each member. This method is the forerunner of the matrix stiffness method but is not easily programmed and hence has limited use. For small structures it can serve as an efficient hand method.

Example 11.1 Find the moments at the critical points of the beam shown in Fig. 11.7

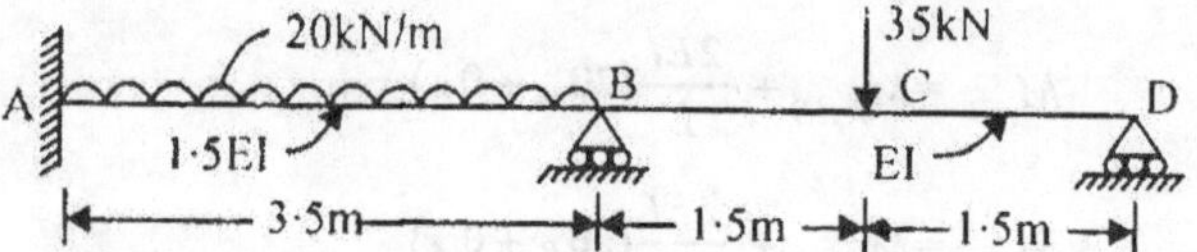

Fig. 11.7. Example 11.1.

Solution:

(*a*) Calculation of F.E.M.

$$M_{F\,AB} = -\frac{20 \times 3.5^2}{12} = -20.41\ kNm.$$

$$M_{F\,BA} = +\frac{20 \times 3.5^2}{12} = +20.41\ kNm.$$

$$M_{F\,BD} = -\frac{35 \times 3}{8} = -13.12\ kNm.$$

$$M_{F\,DB} = +\frac{35 \times 3}{8} = +13.12\ kNm.$$

(*b*) Calculation of stiffnesses

$$K_{AB} = \frac{2 \times 1.5 \times EI}{3.5} = \frac{3}{3.5} EI$$

$$K_{BD} = \frac{2 \times 1 \times EI}{3.0} = \frac{2}{3} EI$$

Multiplying both by $\frac{3 \times 3.5}{EI}$

$$\overline{K}_{AB} = 9$$

$$\overline{K}_{BD} = 7$$

(*c*) Slope deflection equations.

$$M_{AB} = -20.41 + 9\,(2\theta_A + \theta_B).$$
$$M_{BA} = +20.41 + 9\,(2\theta_B + \theta_A).$$
$$M_{BD} = 13.12 + 7\,(2\theta_B + \theta_D)$$
$$M_{DB} = +13.12 + 7\,(2\theta_D + \theta_B).$$

(*d*) Boundary conditions

$$\theta_A = 0 \qquad M_{DB} = 0$$

(*e*) Equilibrium Equations

$$M_{BA} + M_{BD} = 0$$

With given boundary conditions and applying Equilibrium equations, the slope deflection equations become.

$$M_{AB} = -20.41 + 9\theta_B.$$
$$M_{BA} = +20.41 + 18\theta_B.$$
$$M_{BD} = -13.12 + 14\theta_B + 7\theta_D.$$
$$M_{DB} = +13.12 + 14\theta_D + 7\theta_B = 0$$
$$M_{BA} + M_{BD} = 7.29 + 32\theta_B + 7\theta_D = 0.$$

We now have five unknows θ_B, θ_D, M_{AB}, M_{BA} and M_{BD}. The last two equations are sufficient to solve for the five unknowns.

Solution gives

$$M_{AB} = -20.64\ kNm.$$
$$M_{BA} = +19.96\ kNm.$$
$$M_{BD} = -19.96\ kNm.$$
$$\theta_B = -0.025$$
$$\theta_D = -0.9246$$

Fig. 11.8 gives the B.M. diagram.

Check. $\qquad M_{DB} = +13.12 + 14(-0.9246) + 7(-0.025) = 0$

Since we have used relative stiffnesses actual rotation

$$\overline{\theta_B} = -0.025 \times \frac{3 \times 3.5}{EI} = -\frac{0.2625}{EI}$$

$$\overline{\theta_D} = -0.9246 \times \frac{3 \times 3.5}{EI} = -\frac{9.7183}{EI}$$

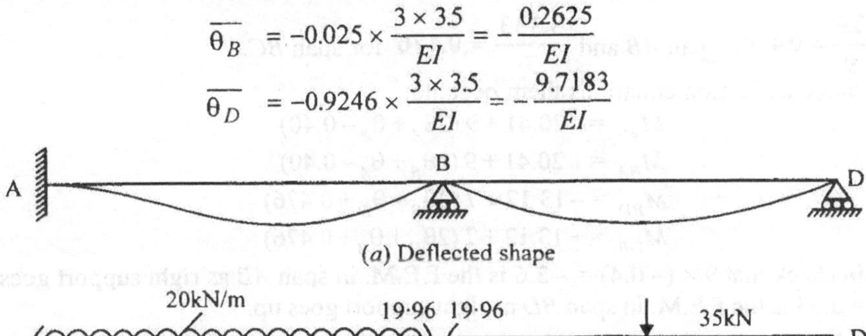

(*a*) Deflected shape

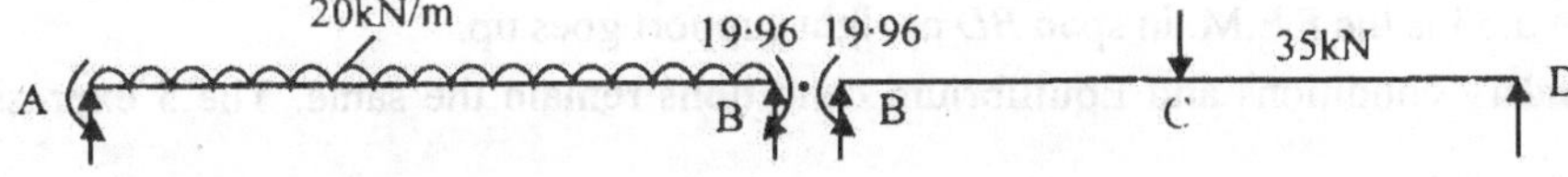

(*b*) F.B. diagrams of beams

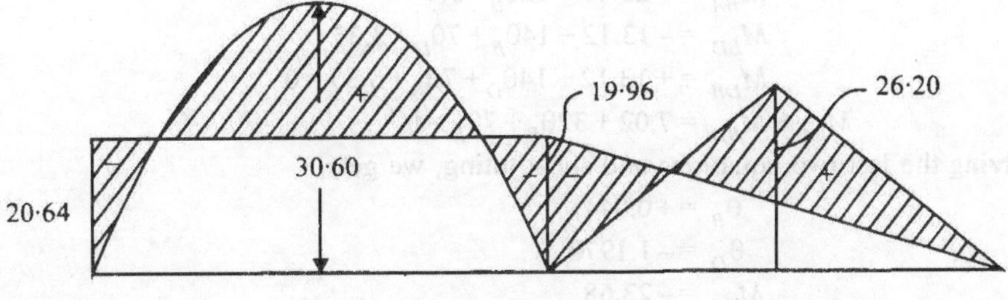

(*c*) B.M. diagram.

Fig. 11.8. B.M. Diagram. (Ex. 11.1).

Example 11.2 Same problem when the support B settles 1 cm. $EI = 500\ kNm^2$. (Fig. 11.9)

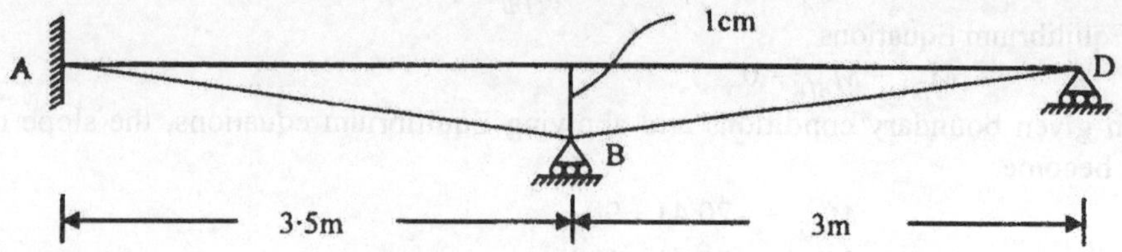

Fig. 11.9. Example 11.2.

Solution:

$$\frac{\Delta}{L} \text{ for } AB = -\frac{0.01}{3.5} = -0.0028$$

$$\frac{\Delta}{L} \text{ for } BD = +\frac{0.01}{3} = +0.0033$$

F.E.M. due to support settlement for AB

$$= -\frac{6EI\Delta}{L^2} = -\frac{6 \times 1.5 \times 500 \times 0.0028}{3.5} = -3.6$$

F.E.M. due to support settlement for BC

$$= +\frac{6 \times 500 \times 0.0033}{3} = +3.333$$

Since absolute values of F.E.M. must be used, with relative stiffness 9, the value of $\frac{3\Delta}{L}$ becomes $\frac{3.6}{9} = 0.4$ for span AB and $\frac{3.333}{7} = 0.476$ for span BC.

The slope deflection equations them become

$$M_{AB} = -20.41 + 9\,(2\theta_A + \theta_B - 0.40)$$
$$M_{BA} = +20.41 + 9\,(2\theta_B + \theta_A - 0.40)$$
$$M_{BD} = -13.12 + 7\,(2\theta_B + \theta_D + 0.476)$$
$$M_{DB} = +13.12 + 7\,(2\theta_D + \theta_B + 0.476)$$

Again check that $9 \times (-0.4) = -3.6$ is the F.E.M. in span AB as right support goes down and $7 \times 0.476 = 3.33$ is the F.E.M. in span BD as right support goes up.

Boundary conditions and Equilibrium conditions remain the same. The 5 expressions then become

$$M_{AB} = -20.41 + 9\theta_B - 3.6$$
$$M_{BA} = +20.41 + 18\theta_B - 3.6$$
$$M_{BD} = -13.12 + 14\theta_B + 7\theta_D + 3.33$$
$$M_{DB} = +13.12 + 14\theta_D + 7\theta_B + 3.33 = 0$$
$$M_{BA} + M_{BD} = 7.02 + 32\theta_B + 7\theta_D = 0$$

Solving the last two equations and substituting, we get.

$$\theta_B = +0.0441$$
$$\theta_D = -1.1970$$
$$M_{AB} = -23.68$$
$$M_{BA} = +17.542$$

$$M_{BD} = -17.542$$
$$M_{DB} = 0.00$$

Fig. 11.12 shows the S.F. and B.M. diagrams.

Since we have used relative stiffness actual rotations are

$$\theta_B = +\frac{0.0441 \times 3 \times 3.5}{500} = 0.0009261$$

$$\theta_D = -\frac{1.197 \times 3 \times 3.5}{500} = -0.02513.$$

Computations can be checked by computing the slope from the free body diagram of span *BD*. (Fig. 11.10)

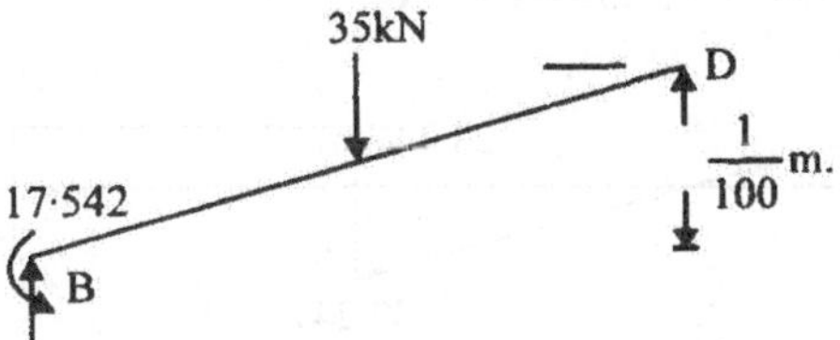

Fig. 11.10. Free body diagram of span BD.

$$\theta_{DB} = +\frac{17.542 \times 3}{6 \times 500} - \frac{1}{100 \times 3} - \frac{35 \times 3 \times 3}{16 \times 500}$$
$$= -0.02516$$

$$\theta_{BD} = -\frac{1}{100 \times 3} - \frac{17.542 \times 3}{3 \times 500} + \frac{35 \times 3^2}{16 \times 500}$$
$$= 0.000958$$

$$\theta_{BA} = \frac{17.542 \times 3.5}{3 \times 1.5 \times 500} + \frac{23.68 \times 3.5}{6 \times 1.5 \times 500} + 0.0028 - \frac{20 - \times 3.5^3}{24 \times 1.5 \times 500}$$
$$= 0.000924$$

$$\theta_{AB} = -\frac{23.68 \times 3.5}{3 \times 1.5 \times 500} + \frac{20 \times 3.5^3}{24 \times 1.5 \times 500} - \frac{17.542 \times 3.5}{6 \times 1.5 \times 500} + \frac{1}{3.5 \times 100}$$
$$= 0.00001 \approx 0$$

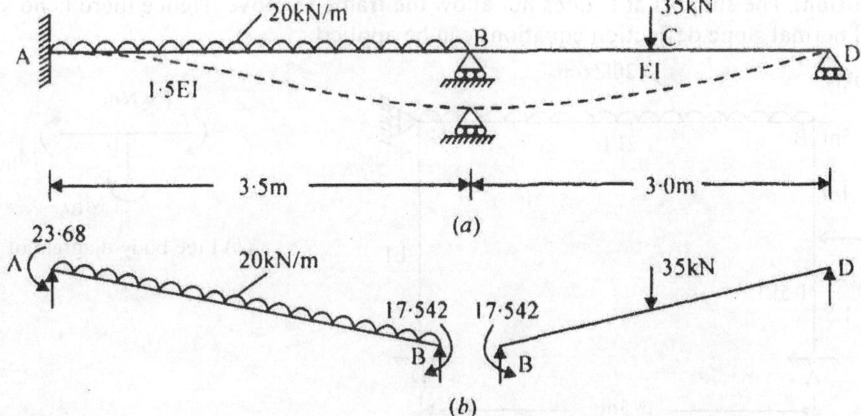

Fig. 11.11. Calculation of Reactions.

Calculation of Reactions.

$$R_A = \frac{23.68}{3.5}\uparrow - \frac{17.542}{3.5}\downarrow + \frac{20\times 3.5}{2}\uparrow = 36.753\uparrow$$

$$R_B = \frac{17.542}{3.5}\uparrow - \frac{23.68}{3.5}\uparrow + \frac{17.542}{3}\uparrow + \frac{35}{2}\uparrow + 35\uparrow = 56.594\uparrow$$

$$R_D = -\frac{17.542}{3.5} + \frac{35}{2} = 11.653\uparrow$$

Check.

$$36.753 + 56.594 + 11.653 - 105 = 0$$

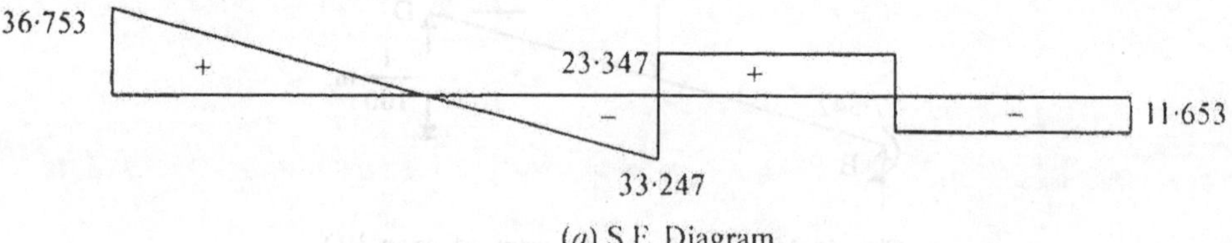

(*a*) S.F. Diagram.

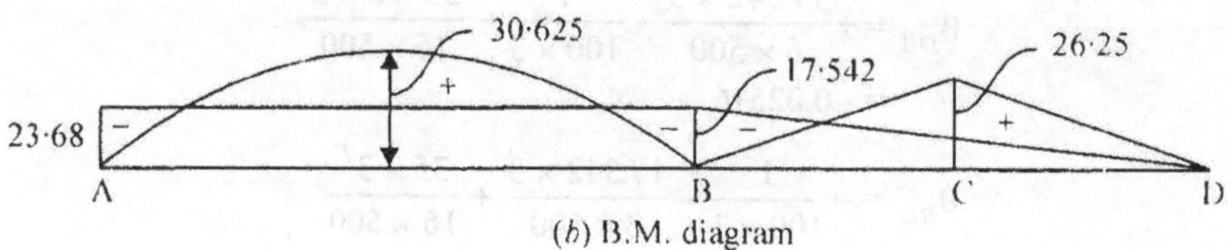

(*b*) B.M. diagram

Fig. 11.12. S.F. and B.M. diagrams.

11.4 SLOPE DEFLECTION EQUATIONS FOR FRAMES

When there is no sway or known sway the frame can be treated like a beam and original beam equations can be applied. When there is unknown side sway Δ, additional unknown (side sway) occurs. So another equation is required. This is obtained by applying $\Sigma H = 0$ to the whole frame. This is known as shear equation or bent equation.

Example 11.3 Find the moment and rotations at the critical sections of the frame shown in Fig. 11.13.

Solution: The support at C does not allow the frame to move. Hence there is no sway in the frame and normal slope deflection equations can be applied.

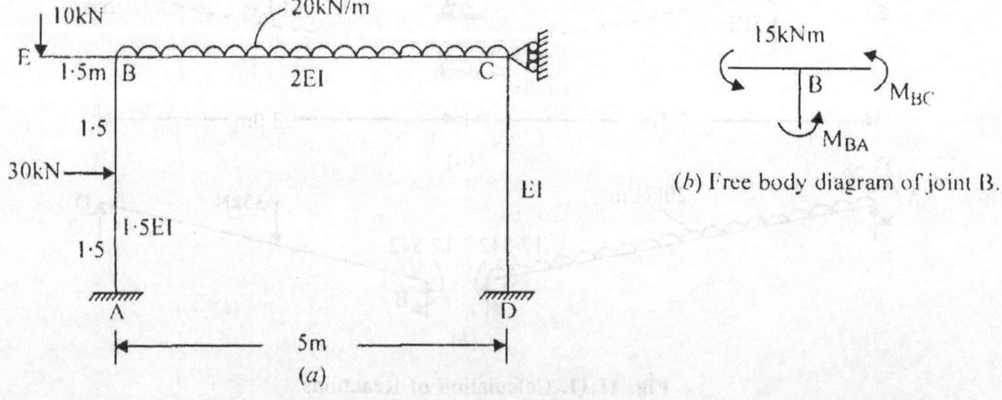

(*a*)

(*b*) Free body diagram of joint B.

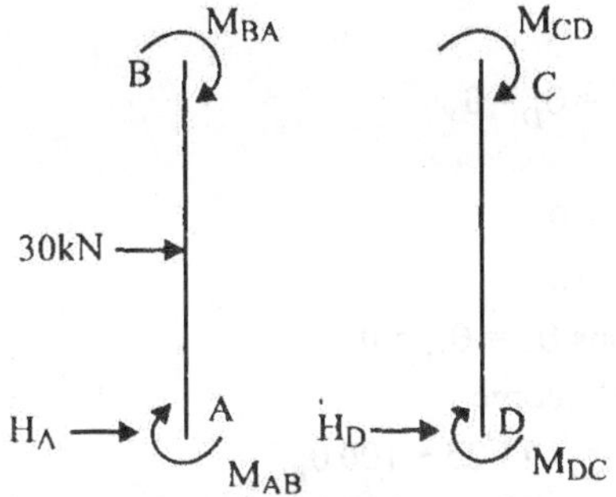

(c) Free body diagrams of AB and DC

Fig. 11.13. Example 11.3

Fixed end Moments.

$$M_{FAB} = -\frac{30 \times 3}{8} = -11.25$$

$$M_{FBA} = +\frac{30 \times 3}{8} = +11.25$$

$$M_{FBC} = -\frac{20 \times 5^2}{12} = -41.66$$

$$M_{FCB} = +\frac{20 \times 5^2}{12} = +41.66$$

Stiffnesses.

$$k_{AB} = \frac{2 \times 1.5\,EI}{3} = EI$$

$$k_{BC} = \frac{2 \times 2\,EI}{5} = \frac{80\,EI}{100}$$

$$k_{CD} = \frac{2 \times EI}{3} = 0.67\,EI = \frac{67}{100}\,EI$$

Multiplying by $\frac{100}{EI}$

$$\overline{K}_{AB} = 100$$
$$\overline{K}_{BC} = 80$$
$$\overline{K}_{CD} = 67$$

Slope deflection Equations

$$M_{AB} = -11.25 + 100\,(2\theta_A + \theta_B)$$
$$M_{BA} = +11.25 + 100\,(2\theta_B + \theta_A)$$
$$M_{BC} = -41.66 + 80\,(2\theta_B + \theta_C)$$
$$M_{CB} = +41.66 + 80\,(2\theta_C + \theta_B)$$
$$M_{CD} = 0 + 67\,(2\theta_C + \theta_D)$$
$$M_{DC} = 0 + 67\,(2\theta_D + \theta_C)$$

Boundary Conditions

$$\theta_A = \theta_D = 0$$

Continuity Conditions

$$M_{BC} + M_{BA} + 15\ kNm = 0$$
$$M_{CB} + M_{CD} = 0$$

with the boundary conditions $\theta_A = \theta_D = 0$

Slope deflection equations become.

$$M_{AB} = -11.25 + 100\,\theta_B$$
$$M_{BA} = +11.25 + 200\,\theta_B$$
$$M_{BC} = -41.66 + 160\,\theta_B + 80\,\theta_C$$
$$M_{CB} = +41.66 + 160\,\theta_C + 80\,\theta_B$$
$$M_{CD} = 134\,\theta_C$$
$$M_{DC} = 67\,\theta_C$$

Substituting the values in the continuity equations and solving

$$\theta_B = 0.07862$$
$$\theta_C = -0.16289$$

Substituting

$$M_{AB} = -11.25 + 100 \times 0.07862$$
$$= -3.388\ kNm$$
$$M_{BA} = +11.25 + 200 \times 0.07862$$
$$= 26.974\ kNm$$
$$M_{BC} = -41.66 + 160 \times 0.07862 + 80\,(-0.16289)$$
$$= -42.112.$$
$$M_{CB} = +41.66 + 160\,(-0.16289) + 80\,(0.07862)$$
$$= +21.8872.$$
$$M_{CD} = 134\,(-0.16289) = -21.827$$
$$M_{DC} = 67\,(-0.16289) = -10.913.$$

Check. $M_{CB} + M_{CD} = 21.8872 - 21.827 \approx 0$

$$M_{BC} + M_{BA} + 15 = -42.112 + 26.974 + 15.00 \approx 0$$

Example 11.4 When the support at C is removed, that is, sway is allowed, find the B.Ms at critical points.

Solution :

Let the sway to the left be Δ slope deflection equations becomes :

$$M_{AB} = -11.25 + 100\left(2\theta_A + \theta_B + 3.\frac{\Delta}{3}\right)$$

$$M_{BA} = +11.25 + 100\left(2\theta_B + \theta_A + 3.\frac{\Delta}{3}\right)$$

$$M_{BC} = -41.66 + 80\,(2\theta_B + \theta_C)$$
$$M_{CB} = +41.66 + 80\,(2\theta_C + \theta_B)$$

$$M_{CD} = 0 + 67\left(2\theta_C + \theta_D + 3.\frac{\Delta}{3}\right)$$

$$M_{DC} = 0 + 67\left(2\theta_D + \theta_C + 3.\frac{\Delta}{3}\right)$$

To tackle unknown displacement Δ we require another condition known as shear condition. Considering equilibrium of the frame in the horizontal direction (when there is no support at C) shear condition is

$$H_A + H_D + 30\ kN = 0$$

From free body diagrams other equilibrium equations are,

$$H_A \times 3 - M_{AB} - M_{BA} + 30 \times 1.5 = 0.$$

$$H_D \times 3 - M_{DC} - M_{CD} = 0$$

Substituting in shear condition above

$$\frac{1}{3}(-30 \times 1.5 + M_{AB} + M_{BA}) + \left(\frac{M_{DC} + M_{CD}}{3}\right) + 30 = 0$$

or, $$-45 + M_{AB} + M_{BA} + M_{DC} + M_{CD} + 90 = 0$$

or, $$M_{AB} + M_{BA} + M_{CD} + M_{DC} = -45$$

With boundary conditions θ_A and θ_D zero, the slope deflection equations become.

$$M_{AB} = -11.25 + 100\,\theta_B + 100\,\Delta$$

$$M_{BA} = +11.25 + 200\,\theta_B + 100\,\Delta$$

$$M_{BC} = -41.66 + 160\,\theta_B + 80\,\theta_C$$

$$M_{CB} = +41.66 + 160\,\theta_C + 80\,\theta_B$$

$$M_{CD} = 134\,\theta_C + 67\theta_D + 67\Delta$$

$$M_{DC} = 134\,\theta_D + 67\theta_C + 67\Delta.$$

Substituting the values in the equilibrium equation $M_{BA} + M_{BC} + 15 = 0$

we get, $(11.25 - 41.66) + 360\,\theta_B + 80\,\theta_C + 100\,\Delta + 15 = 0$

From the equation, $M_{CB} + M_{CD} = 0$

we get, $41.66 + 294\,\theta_C + 80\,\theta_B + 67\,\Delta = 0$

From the relation

$$M_{AB} + M_{BA} + M_{CD} + M_{DC} = -45$$

$$-100\,\theta_B - 67\,\theta_C - 111.334 - 15 = 0$$

Arranging sequentially

$$360\,\theta_B + 80\,\theta_C + 100\,\Delta - 15.41 = 0$$

$$80\,\theta_B + 294\,\theta_C + 67\,\Delta - 41.66 = 0$$

$$+100\,\theta_B + 67\,\theta_C + 111.33 + 15.00 = 0$$

Solving $\theta_C = -0.13739$

$\theta_B = 0.11643$

$\Delta = -0.15513$

indicating sway is towards right

giving $M_{AB} = 15.12$, $M_{BA} = 19.023$

$M_{BC} = -34.022$, $M_{CB} = 28.992$

$M_{CD} = -28.80$, $M_{DC} = -19.59$

Check. $M_{CB} + M_{CD} = 0$

$$M_{AB} + M_{BA} + M_{CD} + M_{DC} + 45 = 0.$$

In the simultaneous equations not only are the coeffecients on the diagonal predominates in their own equations but the other coefficients are symmetrical with respect to this diagonal. It is important to arrange the unknowns in the order of θ_A, θ_B, θ_C, and θ_D along the horizontal direction and the joint conditions in the order of joints A, B, C and D in the vertical direction.

Example 11.5 A fixed beam ACB of length 2 L is fixed at A and B and $AC = CB = L$. While AC has a moment of inertia = $2I$, CB has a moment of inertia I. Find the fixing end moments at A and B due to uniformly distributed load of w per unit length over the entire span AB. Use slope deflection method.

[AMIE Summer' 97]

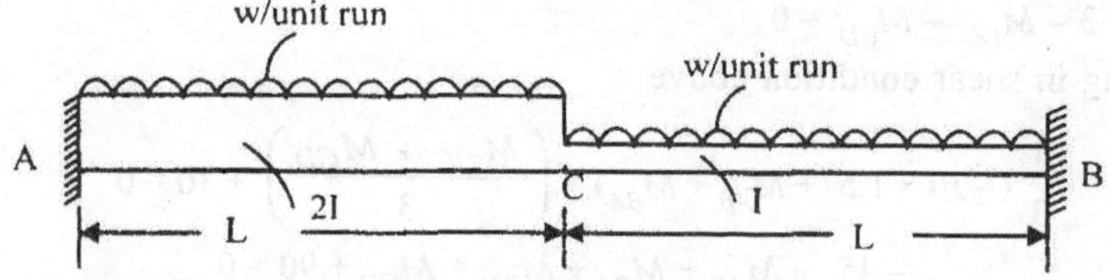

Fig. 11.14. Example 11.5

Solution: As the moment of inertia changes at C it is treated as a node/joint. Thus the beam is composed of two spans AC and CB.

F.E.M. For uniformly distributed load the FEMs are :

$$M_{F\,AC} = -\frac{wL^2}{12} \qquad M_{F\,CA} = +\frac{wL^2}{12}$$

$$M_{F\,CB} = -\frac{wL^2}{12} \qquad M_{F\,BC} = +\frac{wL^2}{12}.$$

Slope Deflection Equations.

Let the displacement at C be δ downward in the vertical direction. δ is negative for member AC and positive for member CB.

$$M_{AC} = M_{F\,AC} + \frac{2EI_{AC}}{L}\left[2\theta_A + \theta_C + \frac{3(-\delta_{AC})}{L}\right]$$

$$= -\frac{wL^2}{12} + \frac{2E(2I)}{L}\left[0 + \theta_C - 3.\frac{\delta}{L}\right]$$

$$= -\frac{wL^2}{12} + \frac{4EI}{L}\left(\theta_C - 3.\frac{\delta}{L}\right)$$

$$M_{CA} = +\frac{wL^2}{12} + \frac{4EI}{L}\left(2\theta_C + 0 - 3.\frac{\delta}{L}\right)$$

$$= +\frac{wL^2}{12} + \frac{4EI}{L}\left(2\theta_C - 3.\frac{\delta}{L}\right)$$

$$M_{CB} = M_{F\,CB} + \frac{2EI_{CB}}{L_{CB}}\left[2\theta_C + \theta_B + 3.\frac{\delta_{CB}}{L_{CB}}\right]$$

$$= -\frac{wL^2}{12} + \frac{2EI}{L}\left(2\theta_C + \frac{3\delta}{L}\right)$$

$$M_{BC} = +\frac{wL^2}{12} + \frac{2EI}{L}\left(\theta_C + \frac{3\delta}{L}\right)$$

Equilibrium Equations.

At joint C $\qquad \Sigma M_C = M_{CA} + M_{CB} = 0$

Hence, $+\frac{wL^2}{12} + \frac{4EI}{L}\left(2\theta_C - \frac{3\delta}{L}\right) - \frac{wL^2}{12} + \frac{2EI}{L}\left(2\theta_C + \frac{3\delta}{L}\right) = 0$

or $$\frac{8EI}{L}.\theta_C - \frac{12EI\,\delta}{L^2} + \frac{4EI}{L}.\theta_C + \frac{6EI\,\delta}{L^2} = 0$$

or $$\frac{12EI.\theta_C}{L} - \frac{6EI\,\delta}{L^2} = 0$$

Shear Equation $\qquad \Sigma V = V_{CA} + V_{CB} = 0$

$$V_{CA} = \frac{wL}{2} + \frac{1}{L}\left[\frac{wL^2}{12} + \frac{4EI}{L}\left(2\theta_C - \frac{3\delta}{L}\right)\right] + \left[-\frac{wL^2}{12} + \frac{4EI}{L}\left(\theta_C - \frac{3\delta}{L}\right)\right]$$

$$= \frac{wL}{2} + \frac{12EI}{L^2}.\theta_C - \frac{24EI\,\delta}{L^3}$$

$$V_{CB} = \frac{wL}{2} - \frac{1}{L}\left[-\frac{wL^2}{12} + \frac{2EI}{L}\left(\theta_C + \frac{3\delta}{L}\right) + \frac{wL^2}{12} + \frac{2EI}{L}\left(2\theta_C + \frac{3\delta}{L}\right)\right]$$

$$= \frac{wL}{2} - \frac{4EI\,\theta_C}{L^2} - \frac{6EI\,\delta}{L^3} - \frac{2EI\,\theta_C}{L^2} - \frac{6EI\,\delta}{L^3}$$

$$= \frac{wL}{2} - \frac{6EI}{L^2}.\theta_C - \frac{12EI\,\delta}{L^3}$$

$$\Sigma V = wL + \frac{6EI}{L^2}.\theta_C - \frac{36EI\,\delta}{L^3} = 0$$

From the relation

$$\frac{12EI\,\theta_C}{L} - \frac{6EI\,\delta}{L^2} = 0$$

or $$\theta_C = \frac{1}{2}.\frac{\delta}{L}$$

Substituting in the relation $\Sigma V = 0$

$$wL + \frac{6EI}{L^2}.\frac{1}{2}.\frac{\delta}{L} - \frac{36EI\,\delta}{L^3} = 0$$

or $$wL = \frac{33EI\,\delta}{L^3}$$

or $$\delta = \frac{wL^4}{33EI}$$

$$\theta_C = \frac{1}{2L}\cdot\frac{wL^4}{33EI} = \frac{wL^3}{66EI}$$

$$M_{AC} = -\frac{wL^2}{12} + \frac{4EI}{L}\left(\frac{wL^3}{66EI} - \frac{3wL^4}{33EI}\cdot\frac{1}{L}\right)$$

$$= -\frac{51}{132}wL^2$$

$$M_{BC} = +\frac{wL^2}{12} + \frac{2EI}{L}\left(\frac{wL^3}{66EI} + \frac{3wL^4}{33EI}\cdot\frac{1}{L}\right)$$

$$= +\frac{39}{132}wL^2$$

Example 11.6 Determine the moments for the frame as shown in Fig. 11.15 and draw the B.M. diagram.

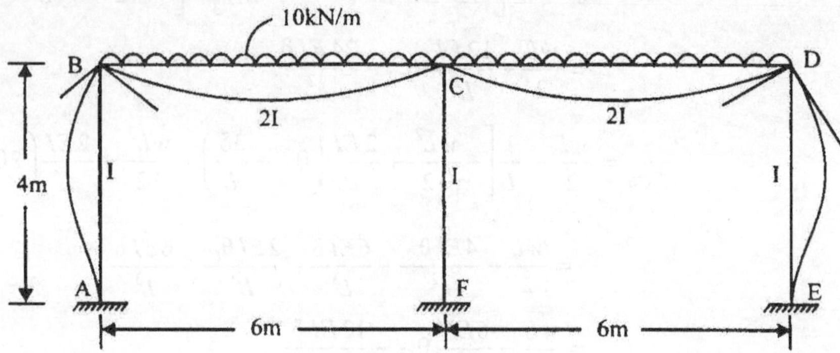

Fig. 11.15. Example 11.6.

Solution: Because of symmetry, there will be no sway in the frame. $\theta_B = -\theta_D$, $\theta_C = 0$

From fixity $\theta_A = \theta_F = \theta_E = 0$

$$M_{AB} = M_{F\,AB} + \frac{2EI}{4}\left(2\theta_A + \theta_B - \frac{3\Delta}{L}\right)$$

$$= 0 + \frac{2EI}{4}.(0 + \theta_B) = \frac{EI\,\theta_B}{2}$$

$$M_{BA} = M_{F\,BA} + \frac{2EI}{4}(2\theta_B + \theta_A)$$

$$= 0 + \frac{2EI}{4}.2\theta_B = EI.\theta_B$$

$$M_{BC} = M_{F\,BC} + \frac{4EI}{6}.(2\theta_B + \theta_C)$$

$$= -\frac{10 \times 6^2}{12} + \frac{4EI}{6}.2.\theta_B$$

$$= -30 + \frac{4}{3}.EI.\theta_B$$

From equilibrium conditions

$$M_{BA} + M_{BC} = EI\theta_B - 30 + \frac{4}{3}EI\theta_B = 0$$

or $\frac{7}{3}.EI\theta_B = 30$

$$\theta_B = \frac{90}{7EI}$$

$$M_{AB} = \frac{EI}{2}.\frac{90}{7EI} = 6.43 \ kNm.$$

$$M_{BA} = EI.\frac{90}{7EI} = 12.85 \ kNm.$$

$$M_{BC} = -30 + \frac{4EI}{L}.(2\theta_B)$$

$$= -30 + \frac{8EI}{6}.\frac{90}{7EI} = -12.857$$

$$M_{CB} = +30 + \frac{4EI}{L}.(\theta_B)$$

$$= +30 + \frac{4EI}{6}.\frac{90}{7EI} = 30 + 8.57 = 38.57$$

$$M_{CD} = -30 + \frac{4EI}{6}.(0 + \theta_D)$$

$$= -30 + \frac{4EI}{6}.\left(-\frac{90}{7EI}\right) = -38.57$$

Fig. 11.16 shows the B.M. diagram.

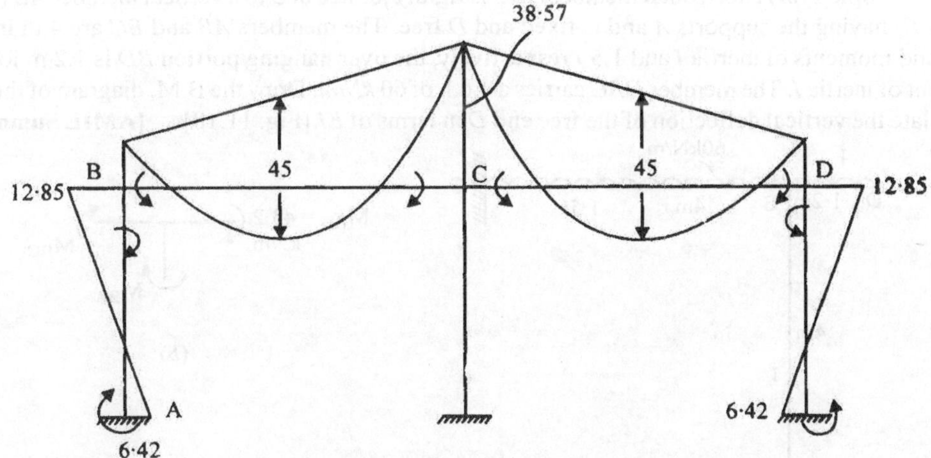

Fig. 11.16. B.M. diagram.

Example 11.7 A prismatic member *AB* of length 9 m and sectional moment of inertia 18,000 cm^4 has partially fixed ends. A point load of 90 *kN* acts at a distance of 3 m from *A* and produces the following displacements.

At *A*, rotation = 0.01 radian clockwise and settlement 1.5 cm.

At *B*, rotation = 0.0075 radian anticlockwise and settlement = 1 cm.

Calculate the support moments. Take $E = 20{,}000\ kN/cm^2$ [AMIE Winter' 94]

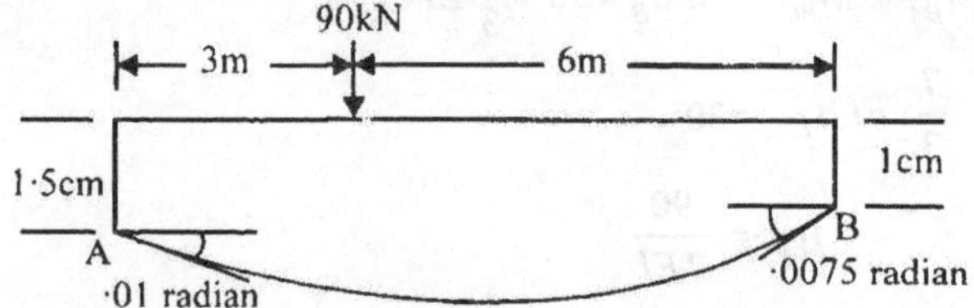

Fig. 11.17. Example 11.7

Solution:

$$M_{F\,AB} = -\frac{Pab^2}{L^2} = -\frac{90 \times 3 \times 6^2}{9^2} = -120\ kNm.$$

$$M_{F\,BA} = +\frac{Pa^2b}{L^2} = +\frac{90 \times 9 \times 6}{9^2} = +60\ kNm.$$

$$M_{AB} = -120 + \frac{2EI}{9}\left[0.01 \times 2 + (-0.0075) + \frac{3(+0.5)}{900}\right]$$

$$= -120 + \frac{2 \times 18{,}000 \times 20{,}000}{9 \times 10^4}(0.0141666)$$

$$= -120 + 113.33 = -6.666\ kNm.$$

$$M_{BA} = +60 + \frac{2 \times 18{,}000 \times 20{,}000}{9 \times 10^4} \times \left[0.01 + 2 \times (-0.0075) + \frac{3 \times 0.005}{9}\right]$$

$$= +60 + \frac{72{,}000}{9} \times (-0.00333)$$

$$= +60 - 26.66 = \mathbf{+33.34\ kNm.}$$

Example 11.8 A horizontal member *DBC* is rigidly jointed at *B* to a vertical member *AB* (*A* being below *B*) having the supports *A* and *C* fixed and *D* free. The members *AB* and *BC* are 4 m in length each and moments of inertia *I* and 1.5 *I* respectively, the over hanging portion *BD* is 1.2 m long with moment of inertia *I*. The member *DBC* carries a u.d.l. of 60 *kN/m*. Draw the B.M. diagram of the frame. Calculate the vertical deflection of the free end *D* in terms of *EI*. [Fig. 11.18] **[AMIE Summer' 94]**

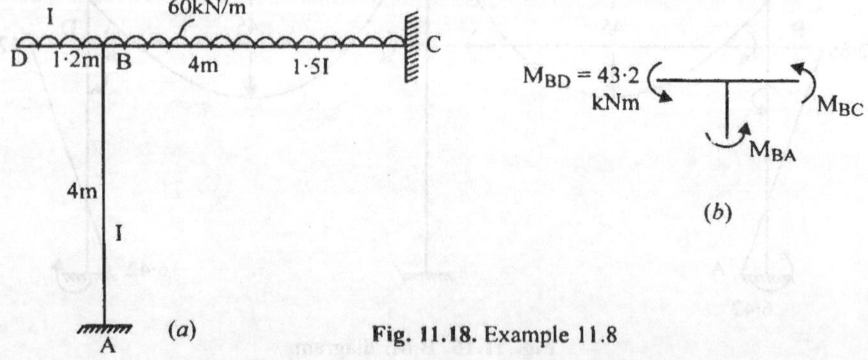

Fig. 11.18. Example 11.8

Solution:

FEM's

$$M_{F\,BC} = -\frac{wL^2}{12} = -\frac{60 \times 4^2}{12} = -80\ kNm.$$

$$M_{F\,CB} = +80\ kNm. \quad M_{BA} = M_{AB} = 0$$

$$M_{BD} = +\frac{60 \times 1.2^2}{2} = +43.2\ kNm.$$

Slope deflection Equations.

As C and A are fixed, there is no sway in the frame.

$$M_{BC} = M_{F\,BC} + \frac{2E \times 1.5I}{L}(2\theta_B + \theta_C)$$

$$= -80 + \frac{2E \times 1.5I}{2}(2\theta_B) \qquad [\theta_C = 0]$$

$$= -80 + 1.5\,EI\theta_B$$

$$M_{CB} = +80 + \frac{3EI}{2}.\theta_B$$

$$M_{BA} = 0 + \frac{2EI}{4}.(2\theta_B) = EI.\theta_B$$

$$M_{AB} = \frac{2EI}{4}.\theta_B = \frac{EI.\theta_B}{2}$$

Equilibrium conditions.

$$\Sigma M_B = 0$$

or $\quad M_{BD} + M_{BA} + M_{BC} = 0$

or $\quad 43.2 + EI\theta_B + (-80 + 1.5\,EI\theta_B) = 0$

or $\quad EI\,\theta_B = \frac{36.8}{2.5} = 14.72$ (clockwise)

Final Moments.

$$M_{BC} = -80 + 1{\cdot}5.\,EI\theta_B$$

$$= -80 + 1.5 \times 14.72 = -57.92\ kNm.$$

$$M_{CB} = +80 + \frac{3}{4}\,EI.\theta_B$$

$$= +80 + \frac{3}{4} \times 14.72 = 91.04\ kNm.$$

$$M_{BA} = 14.72$$

$$M_{AB} = \frac{1}{2}(14.72) = 7.36\ kNm.$$

Fig. 11.19 shows the B.M. diagram.

Check. $\quad 43.2 + 14.72 - 57.92 = 0$

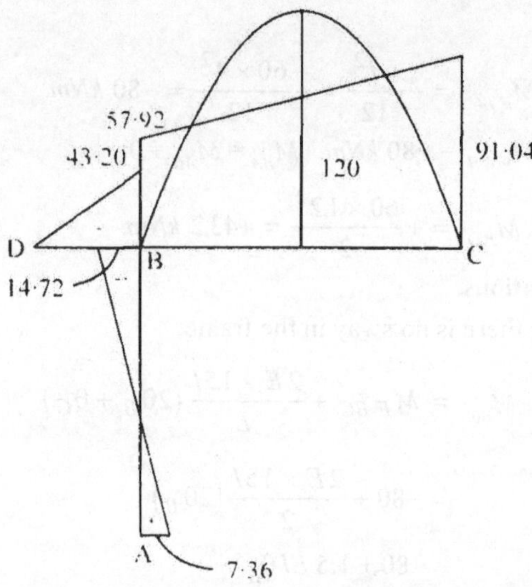

Fig. 11.19. B.M. diagram.

Deflection at *D*.

$$= 1.2\theta_B - \frac{wL^4}{8EI}$$

$$= \frac{1.2 \times 14.72}{EI} - \frac{60 \times 1.2^4}{8EI}$$

$$= \frac{2.112}{EI} \text{ (upward)}$$

Example 11.9 A beam *ABCD* is fixed at end *A* and simply supported at intermediate points *B* and *C*. The other end *D* being kept free. The lengths *AB*, *BC* and *CD* are 6 m, 8 m and 2 m respectively. The beam carries a u.d.l. of 3 t/m acting on the span *BC* and the over hang *CD*. The flexural rigidity is *EI* for span *AB* and is 2*EI* for the rest of the beam. Using slope deflection method determine the member end moments for all members of the beam. Hence draw the *BMD* and *SFD*. [Fig. 11.20]

[AMIE Summer' 95]

Solution:

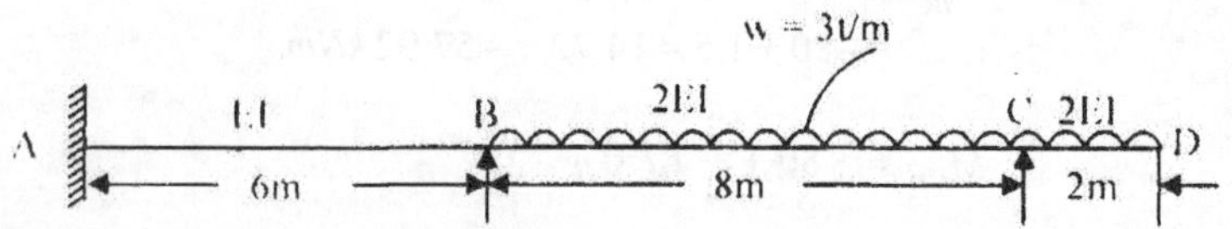

Fig. 11.20. Example 11.9.

FEMs

$$M_{F\,BC} = -\frac{wL^2}{12} = -\frac{3 \times 64}{12} = -16\ t.m$$

$$M_{F\,CB} = +16\ t.m$$

$$M_{F\,AB} = M_{F\,BA} = 0$$

$$M_{F\,CD} = -\frac{3 \times 2^2}{2} = -6\ t.m.$$

Slope deflection Equations.

$\theta_A = 0$. Unknown displacements are θ_B and θ_C.

$$M_{AB} = M_{F\,AB} + \frac{2EI}{6}.\left(\frac{2\theta}{A} + \theta_B\right)$$

$$= 0 + \frac{EI}{3}.\theta_B = \frac{EI}{3}.\theta_B$$

$$M_{BA} = 0 + \frac{2EI}{6}.\left(2\theta_B + \frac{\theta}{A}\right) = \frac{2EI}{3}.\theta_B$$

$$M_{BC} = -16 + \frac{2 \times 2EI}{8}.(2\theta_B + \theta_C)$$

$$= -16 + EI.\theta_B + \frac{EI}{2}.\theta_C$$

$$M_{CB} = +16 + \frac{4EI}{8}.(2\theta_C + \theta_B)$$

$$= +16 + EI.\theta_C + \frac{EI.\theta_B}{2}$$

$$M_{CD} = -6\,t.m$$

Equilibrium Equations.

$$\Sigma M_B = M_{BA} + M_{BC} = 0$$

$$\frac{2}{3}.EI.\theta_B + EI.\theta_B + \frac{EI.\theta_C}{2} - 16 = 0$$

or $$\frac{5}{3}.EI.\theta_B + \frac{EI.\theta_C}{2} - 16 = 0$$

or $$10\,EI.\theta_B + 3\,EI.\theta_C - 96 = 0$$

Similarly from $$\Sigma M_C = 0$$

$$M_{CB} + M_{CD} = 0$$

$$16 + EI.\theta_C + \frac{EI.\theta_B}{2} - 6 = 0$$

or $$2\,EI\theta_C + EI.\theta_B - 20 = 0$$

Solving $$EI.\theta_C = -17.41$$

and $$EI.\theta_B = -20 - 2EI.\theta_C$$

$$= -20 + 17.41 \times 2$$

$$= +14.82$$

$$M_{AB} = \frac{EI.\theta_B}{3} = +\frac{14.82}{3} = +4.94\,t.m$$

$$M_{BA} = \frac{2}{3}EI.\theta_B = +9.88\,t.m$$

$$M_{BC} = -16 + EI.\theta_B + \frac{EI.\theta_C}{2}$$

$$= -16 + (14.82) + \left(-\frac{17.41}{2} \right)$$

$$= \mathbf{-9.88\ t.m}$$

$$M_{CB} = 16 + EI.\theta_C + \frac{EI.\theta_B}{2}$$

$$= 16 + (-17.41) + 7.41 = +6.00\ t.m$$

Fig. 11.21 shows the S.F. and B.M. diagrams

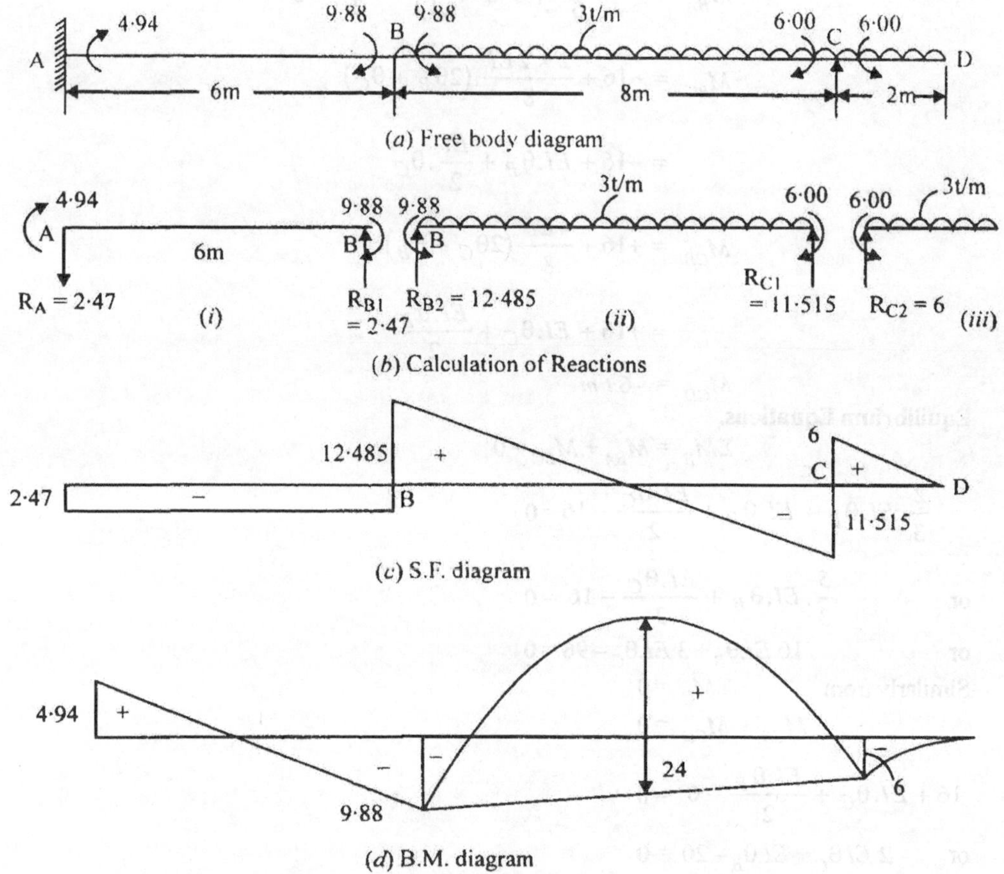

Fig. 11.21. Reactions, S.F. and B.M. diagrams (Example 11.9).

$$\curvearrowright \quad R_A \times 6 + 4.94 + 9.88 = 0 \qquad \text{[Fig. 11.21 (b)(i)]}$$

$$R_A = -\frac{1}{6}(4.94 + 9.88) = -2.47\ t$$

$$R_{B1} = 2.47\ t$$

$$\curvearrowright \quad R_{B2} \times 8 - 9.88 - 3 \times 8 \times 4 + 6{\cdot}00 = 0 \qquad \text{[Fig. 11.21 (b)(ii)]}$$

$$R_{B2} = \frac{1}{8}[9.88+90] = 12.485$$

$$R_{C1}\times 8 - 6.00 - 3 \times 8\times 4 + 9.88 = 0 \qquad \text{[Fig. 11.21 (b)(iii)]}$$

or
$$R_{C1} = \frac{1}{8}[6+96-9.88]$$
$$= \mathbf{11.515}$$
$$R_{C1}+R_{B2} = 12.485+11.515 = 24.0 \text{ Hence Ok}$$

$$R_{C2}\times 2 - 6.00 - 2\times 3 = 0$$
$$R_{C2} = \frac{12}{2} = 6.$$

Example 11.10 The continuous steel beam shown in Fig. 11.22 below has a constant M.I. of 15,000 cm^4. The supports are initially at the same level. If the support B sinks by 1 cm after loading, draw the *SFD* and *BMD* indicating all salient values. Take $E = 2000$ t/cm^2 **[AMIE Winter 1979]**

Solution:

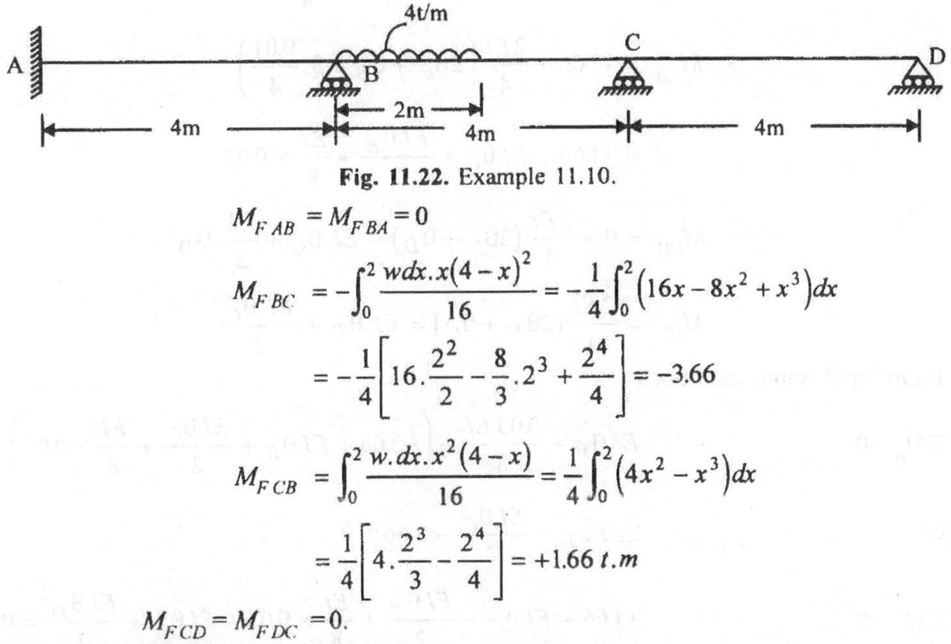

Fig. 11.22. Example 11.10.

$$M_{F\,AB} = M_{F\,BA} = 0$$

$$M_{F\,BC} = -\int_0^2 \frac{wdx.x(4-x)^2}{16} = -\frac{1}{4}\int_0^2 (16x - 8x^2 + x^3)dx$$

$$= -\frac{1}{4}\left[16.\frac{2^2}{2} - \frac{8}{3}.2^3 + \frac{2^4}{4}\right] = -3.66$$

$$M_{F\,CB} = \int_0^2 \frac{w.dx.x^2(4-x)}{16} = \frac{1}{4}\int_0^2 (4x^2 - x^3)dx$$

$$= \frac{1}{4}\left[4.\frac{2^3}{3} - \frac{2^4}{4}\right] = +1.66\ t.m$$

$$M_{F\,CD} = M_{F\,DC} = 0.$$

Support conditions.

$$\theta_A = 0$$
$$M_{DC} = 0$$

Equilibrium conditions.

$$\Sigma M_B = 0$$
or
$$M_{BA} + M_{BC} = 0$$
$$\Sigma M_C = 0$$
or
$$M_{CB} + M_{CD} = 0$$

Slope Deflection Equations.

$$M_{AB} = M_{F\,AB} + \frac{2EI}{L}\left(2\theta_A + \theta_B + 3.\frac{\Delta}{L}\right)$$

$$= 0 + \frac{2EI}{4}\left(0 + \theta_B + 3.\frac{-(0.01)}{4}\right)$$

$$= \frac{EI.\theta_B}{2} - \frac{EI}{8} \times 0.03$$

$$M_{BA} = M_{FBA} + \frac{2EI}{4}\left(2\theta_B + \theta_A + 3.\frac{-(0.01)}{4}\right)$$

$$= EI.\theta_B - \frac{EI}{8} \times 0.03$$

$$M_{BC} = M_{F\,BC} + \frac{2EI}{4}\left(2\theta_B + \theta_C + 3.\frac{0.01}{4}\right)$$

$$= -3.66 + EI.\theta_B + \frac{EI\,\theta_C}{2} + \frac{EI}{8} \times 0.03$$

$$M_{CB} = +1.66 + \frac{2EI}{4}\left(2\theta_C + \theta_B + 3.\frac{0.01}{4}\right)$$

$$= +1.66 + EI\,\theta_C + \frac{EI\,\theta_B}{2} + \frac{EI}{8} \times 0.03$$

$$M_{CD} = 0 + \frac{2EI}{4}(2\theta_C + \theta_D) = EI.\theta_C + \frac{EI}{2}.\theta_D$$

$$M_{DC} = \frac{2EI}{4}(2\theta_D + \theta_C) = EI.\theta_D + \frac{EI.\theta_C}{2}.$$

From Equilibrium conditions

$\Sigma M_B = 0$ $$EI\,\theta_B - \frac{0.03\,EI}{8} + \left(-3.66 + EI\,\theta_B + \frac{EI\,\theta_C}{2} + \frac{EI}{8} \times 0.03\right)$$

or $$2EI\,\theta_B + \frac{EI\,\theta_C}{2} - 3.66 = 0$$

$\Sigma M_C = 0$ $$+1.66 + EI\,\theta_C + \frac{EI\,\theta_B}{2} + \frac{EI}{8} \times 0.03 + EI\,\theta_C + \frac{EI\,\theta_D}{2} = 0$$

or $$2EI\,\theta_C + \frac{EI\,\theta_B}{2} + \frac{EI\,\theta_D}{2} + \frac{EI \times 0.03}{8} + 1.66 = 0$$

$M_{DC} = 0$ $$EI\,\theta_D + \frac{EI\,\theta_C}{2}$$

or $$\theta_D = -\frac{\theta_C}{2}$$

Substituting in the equation $\Sigma M_C = 0$

$$EI.\frac{\theta_B}{2} + 1.75\,EI\,\theta_C = -\left(\frac{EI \times 0.03}{8} + 1.66\right)$$

$$EI = \frac{2{,}000 \times 15{,}000}{10^4} = 3 \times 10^3$$

Substituting the equation becomes

$$EI.\frac{\theta_B}{2} + 1.75\,EI\,\theta_C = -\left(\frac{3 \times 10^3 \times 0.03}{8} + 1.66\right)$$

$$= -12.91$$

Solving this equn. along with the equation

$$2EI\,\theta_B + \frac{EI\,\theta_C}{2} - 3.66 = 0$$

gives $\theta_B = 0.0013189$

$\theta_C = -0.0028358$

Substituting these values.

$$M_{AB} = \frac{EI\,\theta_B}{2} - \frac{EI \times 0.03}{8}$$

$$= \frac{3 \times 10^3 \times 0.0013189}{2} - \frac{3 \times 10^3 \times 0.03}{8}$$

$$= -9.2716\ t.m$$

$$M_{BA} = EI\,\theta_B - \frac{EI}{8} \times 0.03$$

$$= 3 \times 10^3 \times 0.0013189 - \frac{3 \times 10^3}{8} \times 0.03$$

$$= -7.2933$$

$$M_{BC} = -3.66 + EI\,\theta_B + \frac{EI\,\theta_C}{2} + \frac{EI}{8} \times 0.03$$

$$= -3.66 + 3.9567 - 4.2537 + 11.25$$

$$= +7.2930$$

$$M_{CB} = +1.66 + EI\,\theta_C + \frac{EI\,\theta_B}{2} + \frac{EI}{8} \times 0.03$$

$$= +1.66 + 3 \times 10^3\left(-2.8358 \times 10^{-3} + \frac{0.0013189}{2} + \frac{0.03}{8}\right)$$

$$= +6.381$$

$$M_{CD} = EI\,\theta_C + \frac{EI}{2}.\theta_D$$

$$= EI\left(\theta_C + \frac{\theta_D}{2}\right) = 3 \times 10^3\left[\frac{3}{4}.\left(-2.8358 \times 10^{-3}\right)\right]$$

$$= -6.381$$

Fig. 11.23 shows the S.F. and B.M. diagrams.

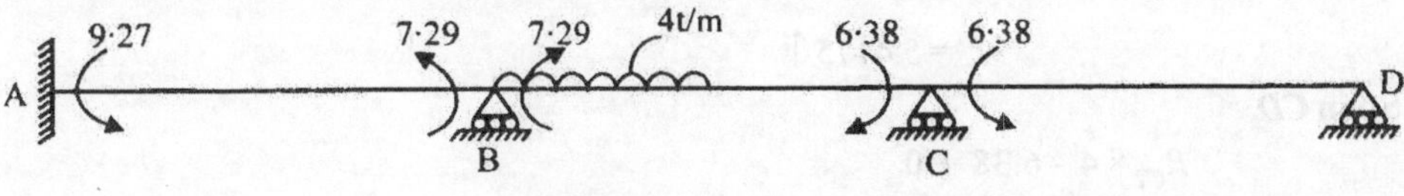

(*a*) beam

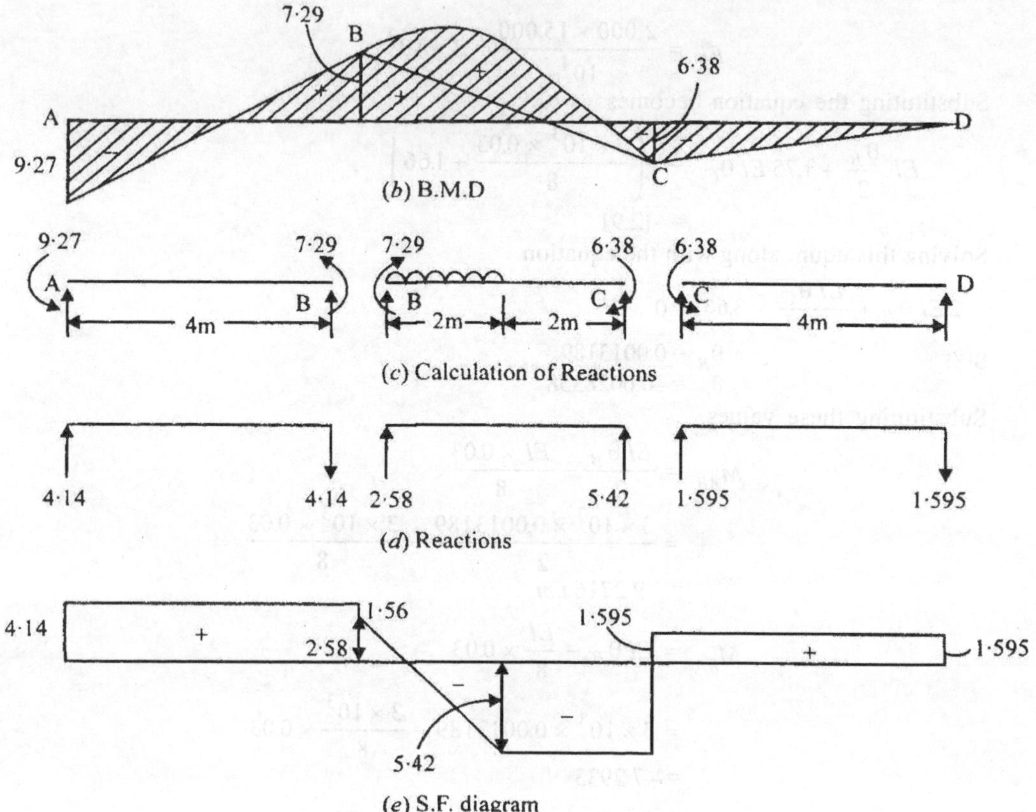

Fig. 11.23. S.F. and B.M. diagrams.

Span *AB*.

$$R_A \times 4 - 9.27 - 7.29 = 0$$

$$R_A = \frac{1}{4}(9.27 + 7.29) = 4.14\ kN \uparrow$$

$$R_{B1} = 4.14 \downarrow$$

Span *BC*.

$$R_{B2} \times 4 + 7.29 - 4 \times 2 \times 3 + 6.38 = 0$$

OR

$$R_{B2} = \frac{1}{4}(24 - 7.29 - 6.38)$$

$$= 2.5825 \uparrow$$

$$R_{B1} + R_{B2} = 4.14 - 2.58 = 1.56 \downarrow$$

$$R_{C1} \times 4 - 6.38 - 4 \times 2 \times 1 - 7.29 = 0$$

or

$$R_{C1} = \frac{1}{4}(6.38 + 8 + 7.29)$$

$$= 5.4175 \uparrow$$

Span *CD*.

$$R_{C2} \times 4 - 6.38 = 0$$

$$\text{or} \qquad R_{C2} = \frac{6.38}{4} = 1.595 \uparrow$$

$$R_D = 1.595 \downarrow$$

$$R_{C1} + R_{C2} = 5.4175 + 1.5950$$

$$= 7.0125^t$$

Example 11.11 A box culvert *ABCD* 6 m wide and 3 m deep has the horizontal members *BC* and *AD* each 6 m long and the verticals *AB* and *DC* each 3 m high. M.I. of the horizontal members is twice that of the verticals. The box culvert is loaded as shown in Fig. 11.24. Analyse the box culvert by slope deflection method and draw the complete bending moment diagram. **[AMIE Winter 1980]**

Solution:

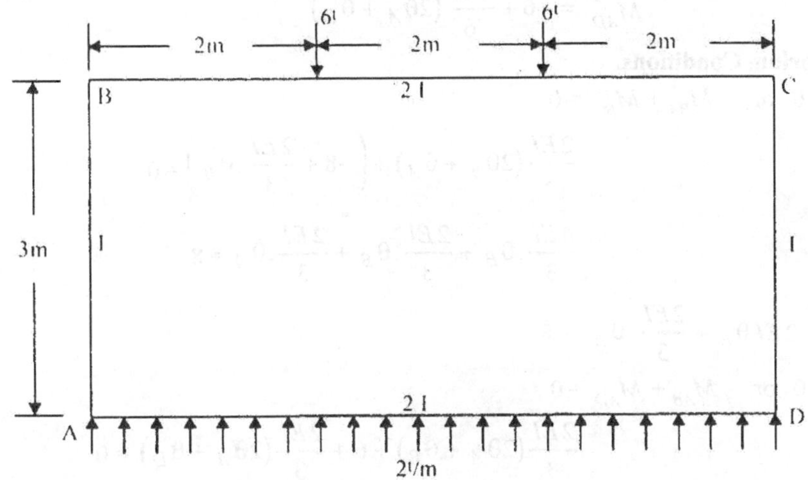

Fig. 11.24. Example 11.11.

From Symmerty $\theta_B = -\theta_C. \qquad M_{BC} = -M_{CB}$

$\theta_A = -\theta_D. \qquad M_{AD} = -M_{DA}$

FEMs

$$M_{FBC} = -\left(\frac{6 \times 4^2 \times 2}{6} + \frac{6 \times 4 \times 2^2}{6^2}\right)$$

$$= -8\ t.m$$

$$M_{FCB} = +8\ t.m$$

$$M_{FBA} = M_{FAB} = M_{FCD} = M_{FDC} = 0$$

$$M_{FAD} = +\frac{2 \times 6^2}{12} = +6$$

$$M_{FDA} = -6.$$

Slope deflection Equations.

$$M_{BC} = -8 + \frac{4EI}{6}(2\theta_B + \theta_C)$$

$$= -8 + \frac{4EI}{6}(2\theta_B - \theta_B) = -8 + \frac{2EI}{3}.\theta_B$$

$$M_{CB} = +8 + \frac{4EI}{6}(2\theta_C + \theta_B)$$

$$= +8 + \frac{2EI}{3}.\theta_C$$

$$M_{BA} = 0 + \frac{2EI}{3}(2\theta_B + \theta_A)$$

$$M_{AB} = 0 + \frac{2EI}{3}(2\theta_A + \theta_B)$$

$$M_{AD} = +6 + \frac{4EI}{6}(2\theta_A + \theta_D)$$

Equilibrium Conditions.

$\Sigma M_B = 0$ or $M_{BA} + M_{BC} = 0$

or
$$\frac{2EI}{3}(2\theta_B + \theta_A) + \left(-8 + \frac{2EI}{3}.\theta_B\right) = 0$$

or
$$\frac{4EI}{3}.\theta_B + \frac{2EI}{3}.\theta_B + \frac{2EI}{3}.\theta_A = 8$$

or
$$2EI\theta_B + \frac{2EI}{3}.\theta_A = 8$$

$\Sigma M_A = 0$ or $M_{AB} + M_{AD} = 0$

or
$$\frac{2EI}{3}(2\theta_A + \theta_B) + 6 + \frac{2EI}{3}(2\theta_A + \theta_D) = 0$$

or
$$2EI\theta_A + \frac{2}{3}EI\theta_B = -6$$

Solving the two equations

$$EI\theta_B = 5.625$$
$$EI\theta_A = -4.875$$

Substituting in slope deflection equations

$$M_{BC} = -8 + \frac{2}{3}EI.\theta_B = -8 + \frac{2}{3} \times 5.625$$
$$= -4.25$$

$$M_{CB} = +8 + \frac{2}{3}EI\theta_C = +8 - \frac{2}{3} \times 5.625 = +4.25$$

$$M_{BA} = \frac{2EI}{3}(2\theta_B + \theta_A) = \frac{4}{3} \times 5.625 + \frac{2}{3}(-4.875) = +4.25$$

$$M_{AB} = \frac{2EI}{3}(2\theta_A + \theta_B) = \frac{4}{3}(-4.875) + \frac{2}{3}.5.625 = -2.75$$

$$M_{AD} = 6 + \frac{4EI}{6}(2\theta_A + \theta_B) = 6 + \frac{4}{3}(-4.875) + \frac{2}{3}.(4.875) = +2.75\,t.m$$

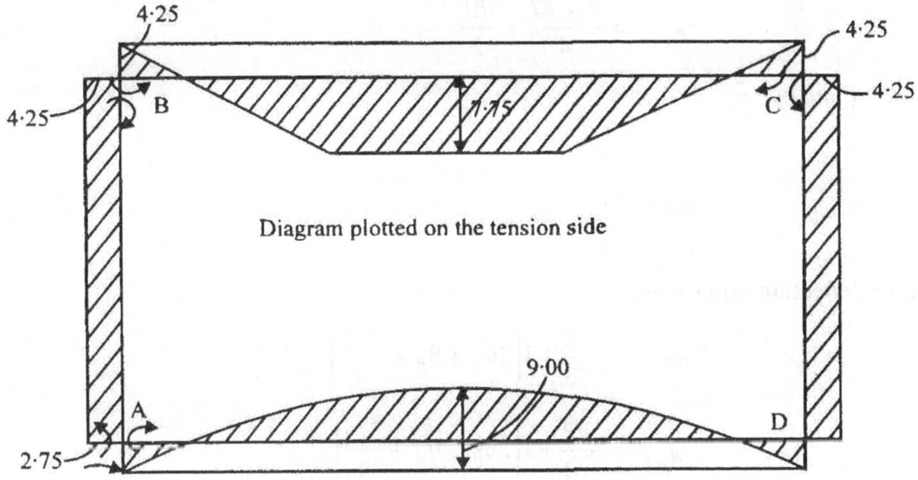

Fig. 11.25. B.M. diagram.

Example 11.12 Analyse the frame shown in Fig. 11.26 below by slope deflection method.

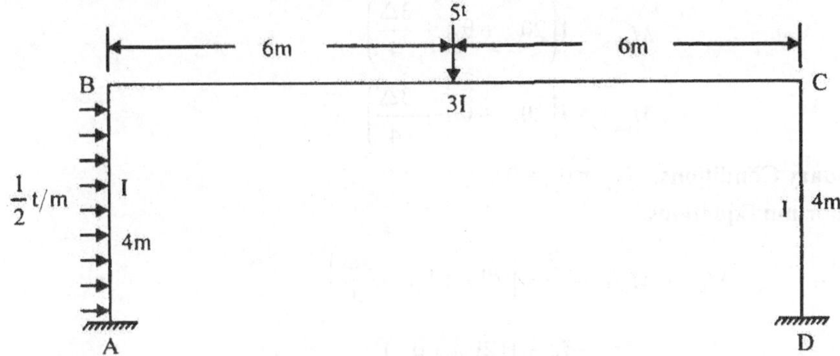

Fig. 11.26 Example 11.12.

Solution: There are three degrees of freedom θ_B, θ_C and sway Δ to the left (assumed).

FEMs

$$M_{F\,AB} = -\frac{1}{2} \times \frac{4^2}{12} = -\frac{2}{3}\,t.m$$

$$M_{F\,BA} = +\frac{2}{3}\,t.m$$

$$M_{F\,BC} = -\frac{5 \times 12}{8} = -7.5\,t.m$$

$$M_{F\,CB} = +7.5\,t.m$$

$$M_{F\,CD} = 0$$

$$M_{F\,DC} = 0$$

Stiffness

$$K_{AB} = \frac{2EI}{4} = \frac{EI}{2}$$

$$K_{BC} = 2 \times \frac{3EI}{12} = \frac{EI}{2}$$

$$K_{CD} = \frac{2 \times EI}{4} = \frac{EI}{2}$$

Multiplying by $\frac{2}{EI}$,

$$\overline{K}_{AB} = 1$$

$$\overline{K}_{BC} = 1$$

$$\overline{K}_{CD} = 1$$

Solpe deflection Equations.

$$M_{AB} = -\frac{2}{3} + 1\left(2\theta_A + \theta_B + \frac{3.\Delta}{4}\right)$$

$$M_{BA} = +\frac{2}{3} + 1\left(2\theta_B + \theta_A + \frac{3\Delta}{4}\right)$$

$$M_{BC} = -7.5 + 1\,(2\theta_B + \theta_C)$$

$$M_{CB} = +7.5 + 1\,(2\theta_C + \theta_B)$$

$$M_{CD} = 1\left(2\theta_C + \theta_D + \frac{3\Delta}{4}\right)$$

$$M_{DC} = 1\left(2\theta_D + \theta_C + \frac{3\Delta}{4}\right)$$

Boundary Conditions. $\theta_A = \theta_D = 0.$

Equilibrium Equations.

$$M_{BA} + M_{BC} = \frac{2}{3} + \left(2\theta_B + \theta_A + \frac{3\Delta}{4}\right)$$

$$-7.5 + 1(2\theta_B + \theta_C)$$

$$= -6.833 + 4\,\theta_B + \theta_C + \frac{3\Delta}{4} = 0$$

$$M_{CB} + M_{CD} = +7.5 + 1\,(2\theta_C + \theta_B) + 2\,\theta_C + \theta_D + \frac{3\Delta}{4}$$

$$= 7.5 + 4\theta_C + \theta_B + \frac{3\Delta}{4} = 0$$

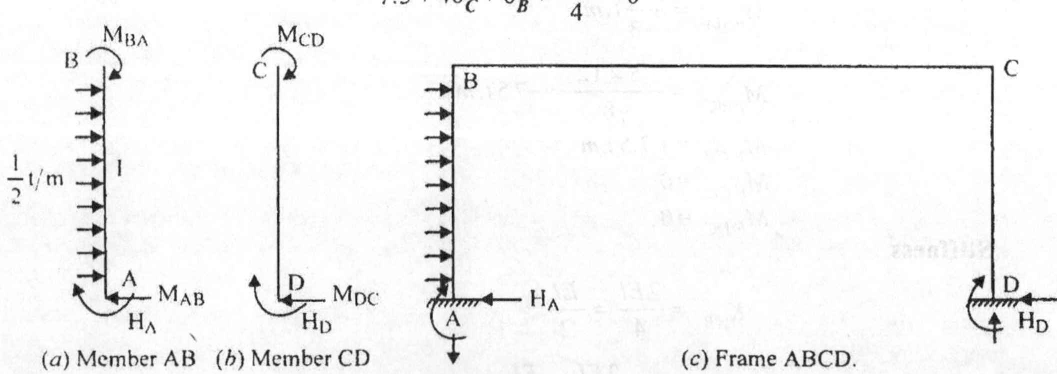

(a) Member AB (b) Member CD (c) Frame ABCD.

Fig. 11.27. Freebody Diagrams

Considering equilibrium of the frame as a whole

$$H_A + H_D - \frac{1}{2} \times 4 = 0$$

or $\qquad H_A + H_D = 2$

From Equilibrium of member AB

$$H_A \times 4 + M_{AB} + M_{BA} - \frac{1}{2} \times 4 \times 2 = 0$$

or $\qquad H_A = \frac{1}{4}(4 - M_{AB} - M_{BA})$

Similarly, $\qquad H_D = -\frac{1}{4}(M_{DC} - M_{CD})$

Substituting $\quad \frac{1}{4}(4 - M_{AB} - M_{BA} - M_{DC} - M_{CD}) - 2$

or $\qquad M_{AB} + M_{BA} + M_{DC} + M_{CD} = -4$

or $\quad -\frac{2}{3} + \theta_B + \frac{3\Delta}{4} + \frac{2}{3} + 2\theta_B + \frac{3\Delta}{4} + 2\theta_C + \frac{3\Delta}{4} + \theta_C + \frac{3\Delta}{4} = 4.$

or $\qquad 3\theta_B + 3\theta_C + 3\Delta = -4.$

or $\quad \frac{3}{4}\theta_B + \frac{3}{4}\theta_C + \frac{3}{4}\Delta = -1$

Arranging the equations sequentially

$$4\theta_B + \theta_C + \frac{3\Delta}{4} = 6.833$$

$$\theta_B + 4\theta_C + \frac{3\Delta}{4} = -7.500$$

$$\frac{3}{4}.\theta_B + \frac{3}{4}.\theta_C + \frac{3}{4}.\Delta = -1.00$$

Solving $\qquad \theta_C = -2.1984, \theta_B = 2.57926$

and $\qquad \Delta = -1.71418$

$$M_{AB} = -\frac{2}{3} + 2.57926 + \frac{3}{4}(-1.71418)$$

$$= 0.62695$$

$$M_{BA} = +\frac{2}{3} + 2\theta_B + \frac{3\Delta}{4}$$

$$= +\frac{2}{3} + 2 \times 2.57926 + \frac{3}{4}(-1.71418)$$

$$= 4.5395$$

$$M_{BC} = -7.5 + 2\theta_B + \theta_C$$

$$= -7.5 + 2 \times 2.57926 + (-2.1984)$$

$$= -4.5398$$

$$M_{CB} = +7.5 + 2\theta_C + \theta_B$$

$$= +7.5 + 2 \times (-2.1984) + 2.57926$$

$$= 5.68246$$

$$M_{CD} = 2\theta_C + \frac{3\Delta}{4}$$

$$= 2(-2.1984) + \frac{3}{4}(-1.71418)$$

$$= -5.68243$$

$$M_{DC} = \theta_C + \frac{3\Delta}{4}$$

$$= -2.1984 + \frac{3}{4}(-1.71418)$$

$$= \mathbf{-3.48403.}$$

Example 11.13 Find moments at critical points of the continuous beam given in Fig. 11.28.

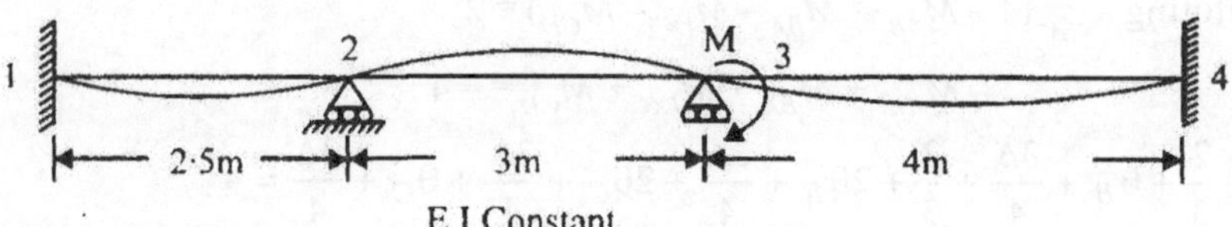

Fig. 11.28. Example 11.13.

Solution: Unknown displacements are θ_2 and θ_3. All FEMs are zero as there is no load in the spans. $\theta_1 = \theta_4 = 0$

Slope deflection Equations.

$$M_{12} = 0 + \frac{2EI}{2.5}(2\theta_1 + \theta_2) = \frac{2EI}{2.5}.\theta_2$$

$$M_{21} = \frac{2EI}{2.5}.(2\theta_2 + \theta_1) = \frac{4EI}{2.5}.\theta_2$$

$$M_{23} = \frac{2EI}{3.0}.(2\theta_2 + \theta_3)$$

$$M_{32} = \frac{2EI}{3.0}.(2\theta_3 + \theta_2)$$

$$M_{34} = \frac{2EI}{4}.(2\theta_3 + \theta_4)$$

$$M_{43} = \frac{2EI}{4}.(2\theta_4 + \theta_3)$$

Equilibrium Equations

$$M_{32} + M_{34} - M = 0$$

or

$$M = M_{32} + M_{34}$$

$$= \frac{2EI}{3.0}(2\theta_3 + \theta_2) + \frac{2EI}{4}(2\theta_3)$$

$$= \frac{EI}{3}(2\theta_2 + 7\theta_3)$$

or $$2\theta_2 + 7\theta_3 = \frac{3M}{EI}$$

$$M_{21} + M_{23} = 0$$

or $$\frac{4EI}{2.5}.\theta_2 + \frac{2EI}{3}(2\theta_2 + \theta_3) = 0$$

or $$\frac{EI}{7.5}(22\theta_2 + 5\theta_3) = 0$$

or $$22\theta_2 + 5\theta_3 = 0$$

Solving the two equilibrium equations

$$\theta_2 = -\frac{M}{EI}(0.1042)$$

$$\theta_3 = +\frac{M}{EI}(0.4583)$$

giving
$$M_{12} = -0.08336\,M$$
$$M_{21} = -0.16672\,M$$
$$M_{23} = 0.1666\,M$$
$$M_{32} = 0.5416\,M$$
$$M_{34} = 0.4583\,M$$
$$M_{43} = 0.2292\,M$$

Check. $$M_{32} + M_{34} = (0.5416 + 0.4583)\,M = 0.9999\,M \approx M.$$

11.5. ITERATIVE SOLUTION OF SLOPE DEFLECTION EQUATIONS

In the last example (11.13) we solved simultaneously the two equilibrium equations to get exact values of θ_2 and θ_3. The equations can also be solved by iterative method as shown below.

1st Cycle. Assume θ_3 is allowed to occur but θ_2 is temporarily prevented from rotation *i.e.*, $\theta_2 = 0$.

From the 1st equation.

$$2\theta_2 + 7\theta_3 = \frac{3M}{EI}$$

$$\theta_3 = \frac{3M}{7EI} \text{ as } \theta_2 \text{ is assumed to be zero.}$$

Substituting this value of θ_3 in the 2nd equation

$$22\,\theta_2 + 5\,\theta_3 = 0$$

$$\theta_2 = -0.0974\frac{M}{EI}$$

2nd Cycle. Substituting this value of θ_2 in 1st equation

$$\theta_3 = 0.4564\frac{M}{EI}$$

From 2nd equation $\theta_2 = -0.1037\dfrac{M}{EI}$

3rd Cycle. Substituting this value of θ_2 in 1st equation

$$\theta_3 = 0.4582\frac{M}{EI}$$

and from 2nd euqation $\theta_3 = -0.1041\dfrac{M}{EI}$

4th Cycle. From 1st equation with

$$\theta_2 = -0.1041\frac{M}{EI}$$

$$\theta_3 = 0.4583\frac{M}{EI}$$

Substituting this value of θ_3 in 2nd equation

$$\theta_2 = -0.1042\frac{M}{EI}$$

These are the exact solutions of the simultaneous equations.

The graphical presentation of this iterative process is given in Fig. 11.29. It can be seen that after 4 cycles of iteration we get very close to the exact values. The convergence on the exact answers is called a monotonic convergence which means that successive values always fall between the last value computed and the convergence value.

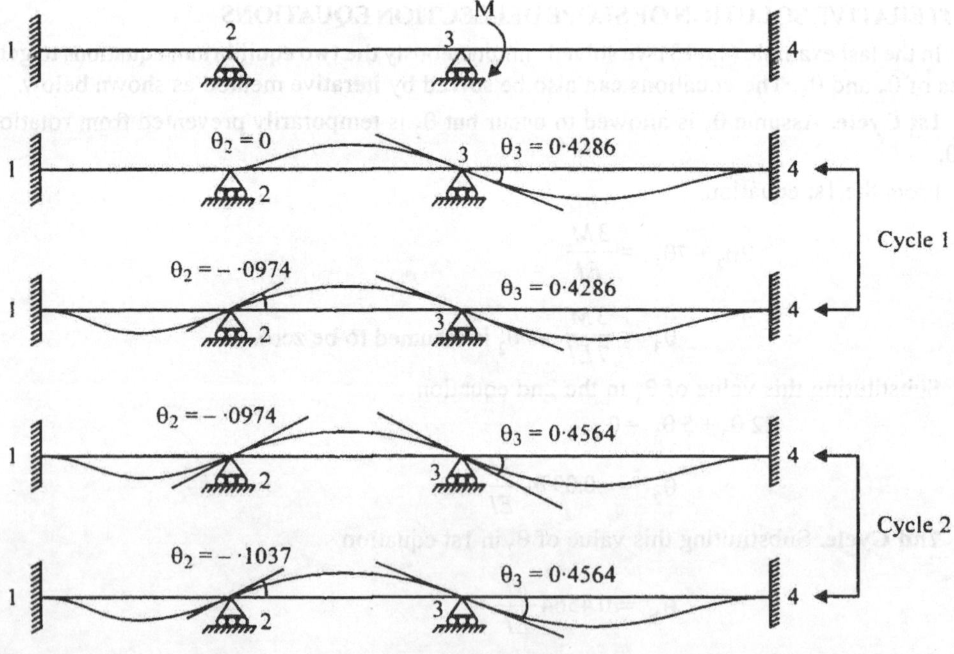

Fig. 11.29 (Contd)

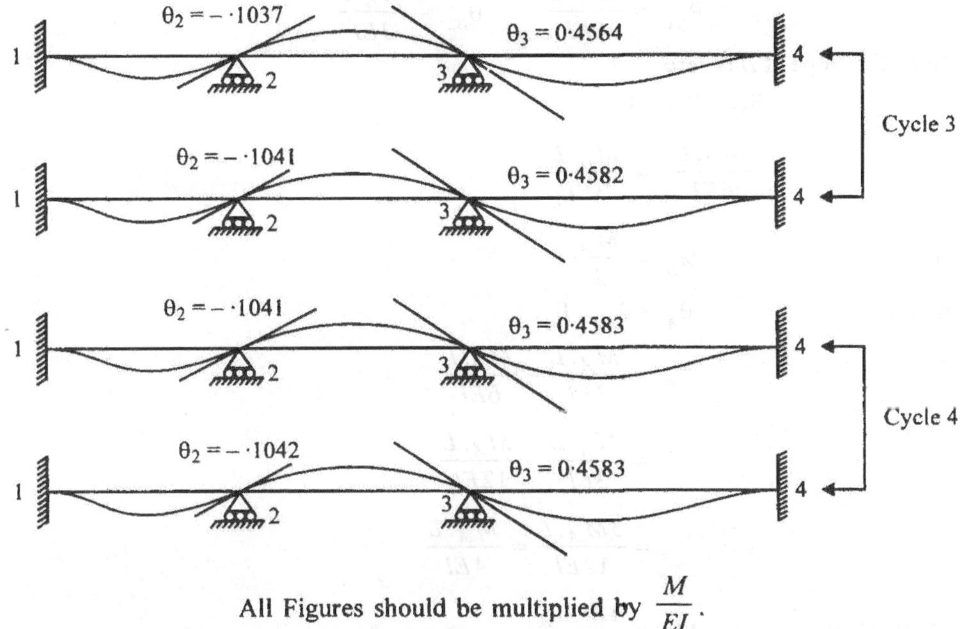

All Figures should be multiplied by $\frac{M}{EI}$.

Fig. 11.29. Visual presentation of iterative process.

11.6 MOMENT DISTRIBUTION METHOD

Like slope deflection method, moment distribution is a displacement method of analysis of indeterminate structure. In fact essentially it consists of solving the simultaneous equations of the slope deflection method by an iterative technique as shown in the last section. It is not an approximate method but a method of successive approximation where any degree of accuracy can be obtained by repeated iteration. The moment distribution is not always recognised as a deformable variable method because the computations are carried out in terms of moments produced by the changes in the deformable variable. This is all the more advantageous because in structural problems we are interested in finding the bending moments rather than slopes and deflections. Before the advent of computers, this method developed by Hardy Cross (1930) was very popular. Even in this age of computer, the method will be useful in preliminary designs and in checking of computer results.

11.7 PRELIMINARY IDEAS

1. Moment and rotation of a propped cantilever (Fig. 11.30)— If a clockwise moment M_A is applied at the end A of a propped cantilever of prismatic section, the value of θ_A and the moment M_B can be obtained from the method of consistent deformation.

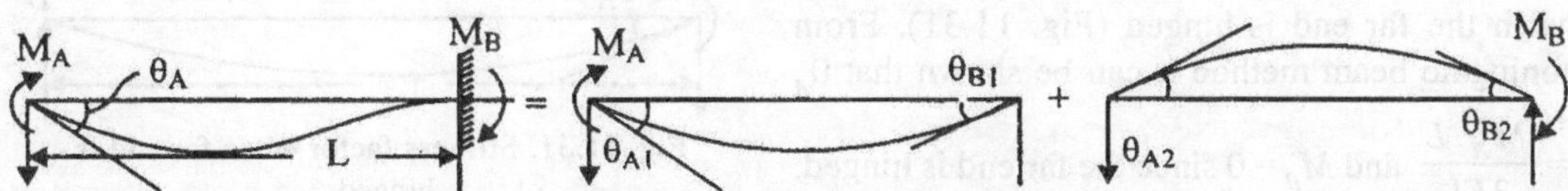

Fig. 11.30. Analysis of propped cantilever.

Slope $\theta_{A1} = \frac{M_A . L}{3EI}$ $\theta_{A2} = \frac{M_B . L}{6EI}$

$$\theta_{B1} = \frac{M_A.L}{6EI} \qquad \theta_{B2} = \frac{M_B.L}{3EI}$$

Since the slope at B is zero.

$$-\theta_{B1} + \theta_{B2} = 0$$

or
$$\frac{M_A.L}{6EI} = \frac{M_B.L}{3EI}$$

or
$$M_B = \frac{M_A}{2}$$

Again
$$\theta_A = \theta_{A1} - \theta_{A2}$$
$$= \frac{M_A.L}{3EI} - \frac{M_B.L}{6EI}$$
$$= \frac{M_A.L}{3EI} - \frac{M_A.L}{12EI}$$
$$= \frac{3M_A.L}{12EI} = \frac{M_A.L}{4EI}$$

which gives
$$M_A = \frac{4EI.\theta_A}{L}$$

The above derivation shows that when θ_A is unity, M_A is equal to $\frac{4EI}{L}$ which is known as stiffness factor. The stiffness factor is therefore, the moment required for unit rotation of the near end A of a prismatic section when end A is simply supported while the far end is fixed. Since $M_B = \frac{M_A}{2}$ it can be deduced that when moment M_A is applied at the near end A, the moment induced at the far end is $\frac{M_A}{2}$, both M_A and M_B are of same direction, clockwise or anti-clockwise. The factor $\frac{1}{2}$ is known as carryover factor which means the ratio of the moment carried to the far end to the moment applied at the near end. In moment distribution method clockwise moment at the end of a member is taken as positive and anticlockwise moment negative as in slope deflection method. This sign convention is very helpful in moment distribution method as it makes C.O.F. positive. B.M. sign convention would have made the C.O.F., $-\frac{1}{2}$ which is cumbersome.

2. Stiffness factor at the near end of a member when the far end is hinged (Fig. 11.31). From conjugate beam method it can be shown that $\theta_A = \frac{M_A.L}{3EI}$ and $M_B = 0$ since the far end is hinged.

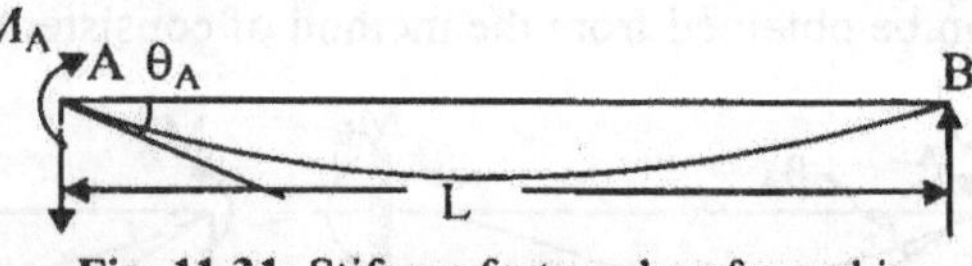

Fig. 11.31. Stifness factor when far end is hinged.

When θ_A is unity $M_A = \frac{3EI}{L}$. Thus the stiffness factor at the near end when the far end is hinged is three fourth of that when the far end is fixed. This is known as modified thickness.

3. Similarly θ_A, when equal moments are applied at both ends of a simply supported beam is given by (Fig. 11.32)

$$\theta_A = = \frac{M_A L}{3EI} + \frac{M_A L}{6EI}$$

$$= \frac{M_A L}{2EI}$$

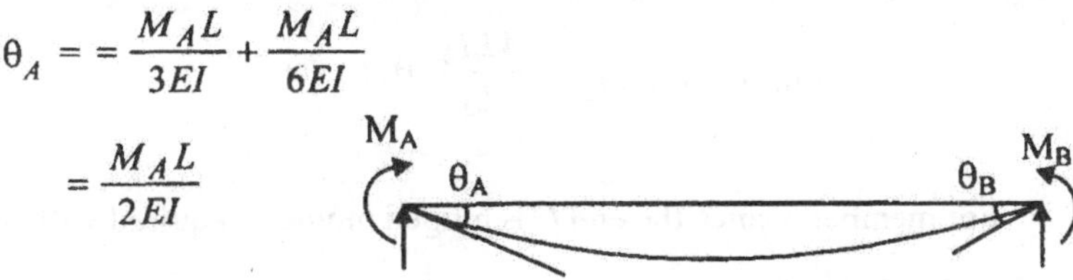

Fig. 11.32. Stiffness factor with equal moments at both ends.

giving modified stiffness as $\frac{2EI}{L}$ for θ_A = unity.

For antisymmetrical case (Fig. 11.33).

$$\theta_A = \frac{M_A.L}{3EI} - \frac{M_A.L}{6EI}$$

$$= \frac{M_A.L}{6EI}$$

giving modified stiffness as $\frac{6EI}{L}$ when θ_A = unity.

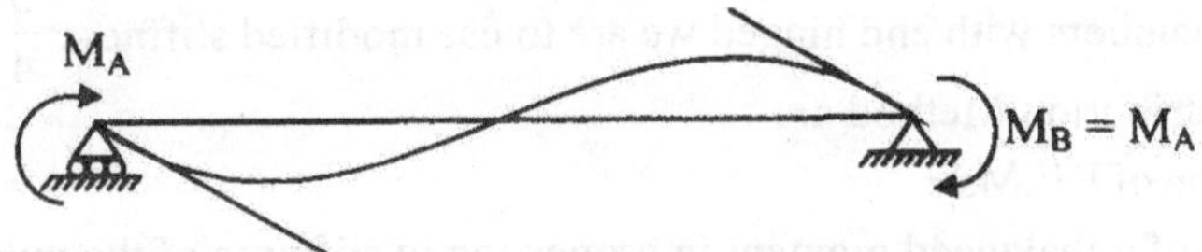

Fig. 11.33 Stiffness factor with antisymmetrical moments

4. Distribution of moment in several members meeting at a joint.

In Fig. 11.34 is shown members 1 to 4 meeting at joint *E*. Ends *A*, *B* and *D* are fixed while end *C* is hinged. Suppose an external moment *M* is applied at *E*. Since the joint *E* is rigid, all the members will rotate through the same angle θ when the joint rotates through an angle θ.

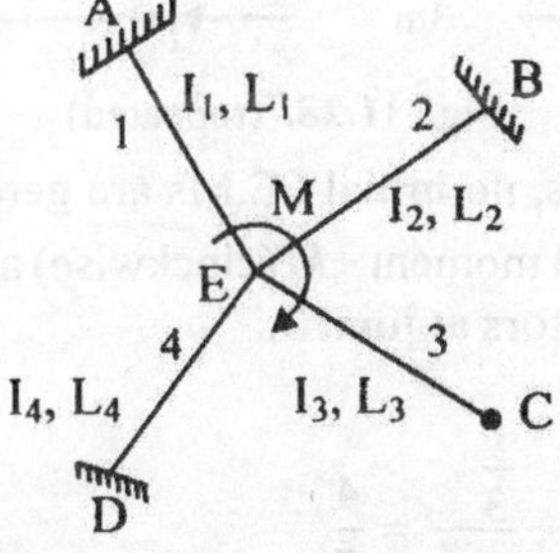

Fig. 11.34. Distribution of moments at a joint.

Moment required to rotate member

$$1 \text{ through } \theta, M_1 = \frac{4EI_1}{L_1}.\theta.$$

$$2 \text{ through } \theta, M_2 = \frac{4EI_2}{L_2}.\theta.$$

$$4 \text{ through } \theta, M_4 = \frac{4EI_4}{L_4}.\theta.$$

For member 3 since the end C is hinged moment required will be $\frac{3EI_3}{L_3}.\theta.$

Hence $M_1 : M_2 : M_3 : M_4 = \frac{I_1}{L_1} : \frac{I_2}{L_2} : \frac{3}{4}\frac{I_3}{L_3} : \frac{I_4}{L_4}$

Also $M_1 + M_2 + M_3 + M_4 = M.$

Therefore
$$M_1 = \frac{\frac{I_1}{L_1}}{\frac{I_1}{L_1} + \frac{I_2}{L_2} + \frac{3}{4}.\frac{I_3}{L_3} + \frac{I_4}{L_4}}.M$$

$$= d_1 M.$$

d_1 is known as distribution factor for member 1. As can be seen from above, distribution factor for a member $= \dfrac{\text{Stiffness of the member}}{\text{Sum of stiffness of all members meeting at the joint}}$.

Note that for members with end hinged we are to use modified stiffness $\frac{3}{4}.\frac{I}{L}$. The important steps in Moment Distribution Method are :

1. Determination of F.E.Ms.
2. Distribution of unbalanced moment in proportion to stiffness of the members at a joint.
3. To carry over half the moment to the far end of the member assuming it to be fixed.

11.8 APPLICATION OF PRELIMINARY IDEAS

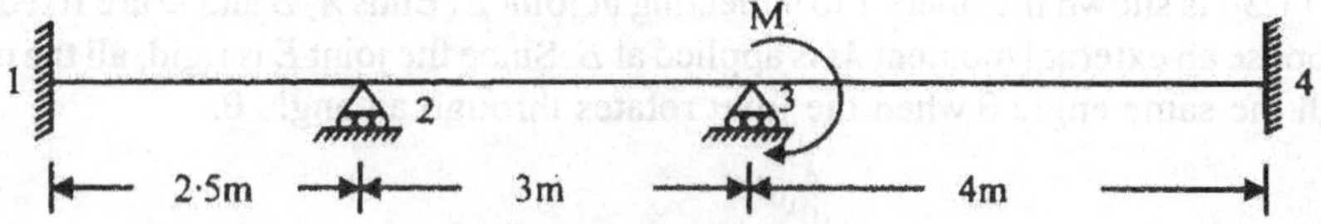

Fig. 11.28. (repeated)

As there is no load in the spans, no initial FE.Ms are generated.

1st Cycle. To distribute external moment $+M$ (clockwise) among the two members 32 and 34. To distribute we require distribution factors at joint 3.

For member 3 – 2 $\quad d_{32} = \dfrac{\frac{I}{3}}{\frac{I}{3} + \frac{I}{4}} = \dfrac{4}{7}$

For member 3 – 4 $\quad d_{34} = \dfrac{3}{7}$

Hence $$M_{32} = \frac{4}{7} M$$

$$M_{34} = \frac{3}{7} M.$$

As joint 2 is assumed to be fixed when this distribution is done, moment carried or generated at far end 2 is $\frac{2}{7} M$ and moment transferred to joint 4 is $\frac{3}{14} M$.

2nd Cycle. We have assumed joint 2 to be fixed (locked) which is not correct. If we now unlock the joint the carried over moment $+\frac{2}{7} M$ gets distributed amongst members 21 and 23 in proportion to their stiffnesses till the joint is balanced.

Distribution factors for joint 2

$$d_{21} = \frac{\frac{I}{2.5}}{\frac{I}{2.5} + \frac{I}{3}} = \frac{3}{5.5}$$

$$d_{23} = \frac{\frac{I}{3}}{\frac{I}{2.5} + \frac{I}{3}} = \frac{2.5}{5.5}$$

Hence moment generated in member.

2 – 1 $$\frac{3}{5.5} \times \frac{2}{7} M = 0.1558\, M.$$

2 – 3 $$\frac{2.5}{5.5} \times \frac{2}{7} M = 0.1299\, M.$$

In order that the joint 2 is balanced *i.e.* moment in members 21 and 23 are equalised the distributed moments should have opposite sign to the moment to be distributed (also called unbalanced or unlocked moment). Hence the distributed moment will be, for member

$$2-1 = -0.1558\, M$$

$$2-3 = -0.1299\, M.$$

$\frac{1}{2}$ of $(-0.1558\, M)$ *i.e.*, $-0.0779\, M$ will be transferred to joint 1. $\frac{1}{2}$ of $(-0.1299\, M)$ *i.e.*, $-0.0649\, M$ will be transferred to joint 3 as joint 3 is assumed fixed when distribution is done at joint 2. Notice that joint 1 and joint 4 are fixed ends and as such no distribution is required at these joints. These ends can absorb any amount of moment as they are fixed.

3rd Cycle. While distributing moments at joint 2 we assumed joint 3 to be fixed which however, is not correct. We therefore, unlock joint 3 and distribute the unbalanced moment there *i.e.*, $-0.0649\, M$ among members 32 and 34 in proportion to their stiffness.

$$M_{32} = +0.0649\, M \times \frac{4}{7} = +0.0371\, M$$

$$M_{34} = +0.0649\,M \times \frac{3}{7} = +0.0278\,M.$$

The moment should be positive in accordance with the reasoning that it should be opposite to the unbalanced moment.

Half of + 0.0371 *M i.e.*, + 0.0185 *M* will be transferred to joint 2. Similarly $\frac{1}{2}$ of + 0.0278 *M i.e.*, 0.0139 *M* will be transferred to joint 4.

4th Cycle. As before when moments are carried over to far ends, they are assumed fixed. Hence joint 2 is to be unlocked and unbalanced moments are to be distributed amongst 21 and 23 in proportion to their stiffness.

$$M_{21} = -\frac{0.0371\,M}{2} \times \frac{3}{5.5} = -0.0101\,M$$

$$M_{23} = -\frac{0.0371\,M}{2} \times \frac{2.5}{5.5} = -0.0084\,M$$

Carried over moment to joint 1 = $\frac{1}{2}$ (– 0.0101 *M*) = – 0.00505 *M*.

Carried over moment to joint 3 = $\frac{1}{2}$ (– 0.0084 *M*) = – 0.0042 *M*.

The whole process of moment distribution is explained in Fig. 11.35. As the explanations and figures for M.D. process are both long and cumbersome the whole process is presented in the form of Table 11.1 as shown below.

Table 11.1 Tabulation of M.D. data

Jt	1	2		3		4	$\theta_2 \times \frac{M}{EI}$	$\theta_3 \times \frac{M}{EI}$
Mem	12	21	23	32	34	43		
D.F	0	$\frac{3}{5\cdot5}$	$\frac{2\cdot5}{5\cdot5}$	$\frac{4}{7}$	$\frac{3}{7}$	0		
C.O.F	1/2		1/2	1/2		1/2		
F. E.M.	0	0	0	0	0	0		
Bal				0·5714M	0·4286M		0	0·4286
C.O			0·2857M			0·2143M	0	0·4286
Bal		−0·1558*M*	−0·1299M				−·0974	0·4286
C.O	−0·0779*M*			0·0649M			−·0974	0·4286
Bal				+0·0371M	+0·0278M		−0974	0·4564
C.O			+0·0185M			0·0134M	−·0974	0·4564
Bal		−·0101M	−·0084M				−·1037	0·4564
C.O	−·0050M			−·0042M			−·1037	0·4564
Bal				+·0024M	+·0018M		−·1037	0.4581
C.O						+·0009M		
F.M	−·0829M	−·1659M	+·1659M	+·5418M	+·4582M	+·2285M		

Jt. — Joint , Mem. — Member.

D.F. — Distribution Factor, C.O.F. — Carry Over Factor.

F.E.M. — Fixed End Moment, Bal. — Balance, C.O. — Carry Over.

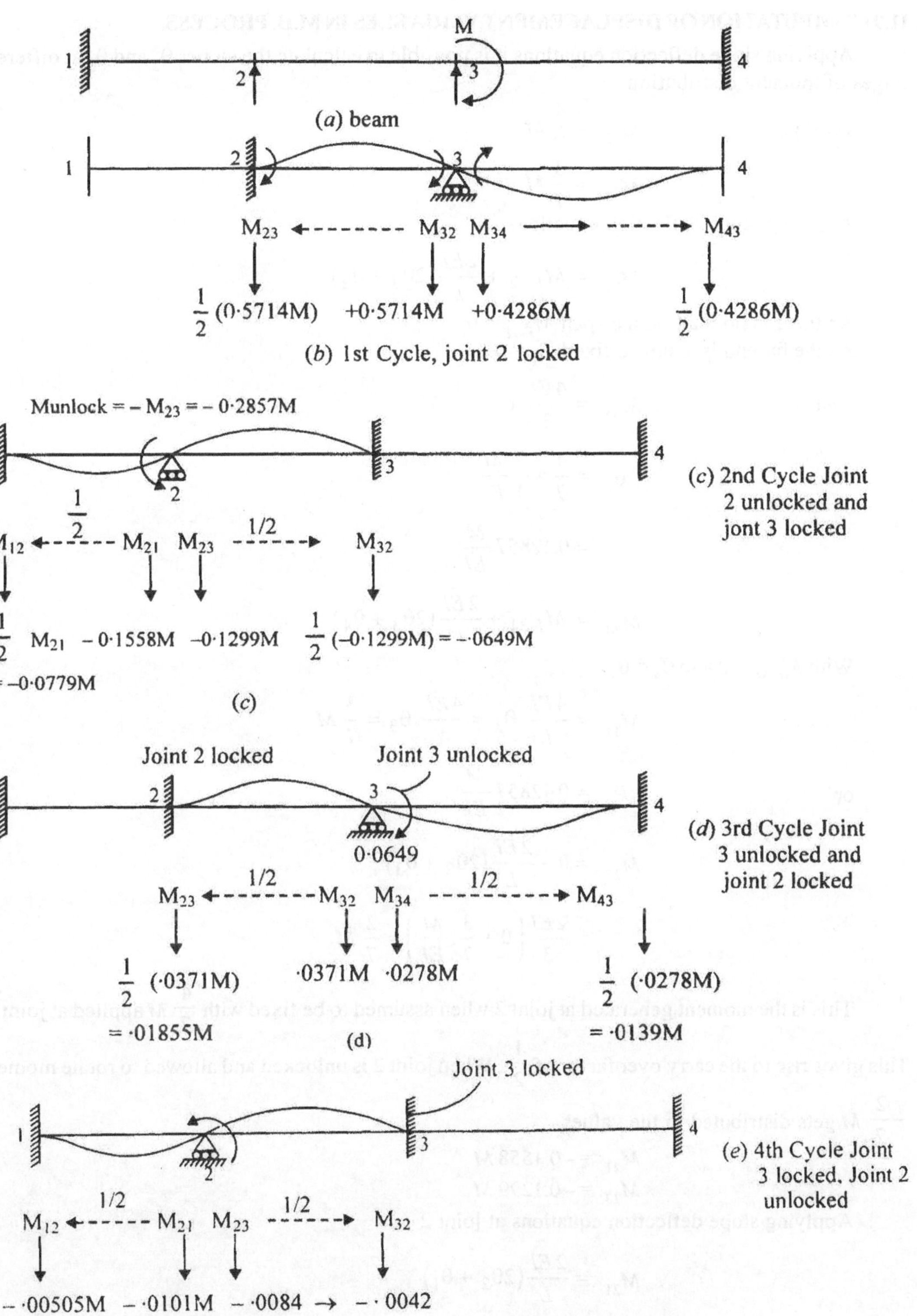

Fig. 11.35. Graphical presentation of Moment distribution.

11.9 COMPUTATION OF DISPLACEMENT VARIABLES IN M.D. PROCESS.

Applying slope deflection equations it is possible to calculate the slopes θ_2 and θ_3 at different stages of moment distribution.

1st Cycle. $M_{32} = \frac{4}{7} M$

$$M_{34} = \frac{3}{7} M$$

Applying slope deflection equations.

$$M_{32} = M_{F32} + \frac{2EI}{L}(2\theta_3 + \theta_2)$$

As there is no load on the span $M_{F32} = 0$

As the far end is assumed fixed. $\theta_2 = 0$.

Hence $M_{32} = \frac{4EI}{3}.\theta_3 = \frac{4}{7} M$

or $\theta_3 = \frac{4}{7} \times \frac{3}{4} \frac{M}{EI}$

$$= 0.42857 \frac{M}{EI}$$

$$M_{34} = M_{F34} + \frac{2EI}{L}(2\theta_3 + \theta_4)$$

With $M_{F34} = 0$ and $\theta_4 = 0$.

$$M_{34} = \frac{4EI}{L}.\theta_3 = \frac{4EI}{4}.\theta_3 = \frac{3}{7} M$$

or $\theta_3 = 0.42857 \frac{M}{EI}$

$$M_{23} = 0 + \frac{2EI}{L}(2\theta_2 + \theta_3)$$

$$= \frac{2EI}{3}\left(0 + \frac{3}{7}.\frac{M}{EI}\right) = \frac{2}{7} M.$$

This is the moment generated at joint 2 when assumed to be fixed with $\frac{4}{7}$ M applied at joint 3.

This gives rise to the carry over factor of $\frac{1}{2}$. When joint 2 is unlocked and allowed to rotate moment $-\frac{2}{7}$ M gets distributed in the values.

$$M_{21} = -0.1558\, M$$

$$M_{23} = -0.1299\, M$$

Applying slope deflection equations at joint 2

$$M_{21} = \frac{2EI}{L}(2\theta_2 + \theta_1)$$

$$= \frac{4EI\theta_2}{L} \qquad \text{as } \theta_1 = 0$$

$$= \frac{4EI}{2.5}.\theta_2$$

or $$\theta_2 = \frac{2.5}{4EI}(-0.1558\ M)$$

$$= -0.09737\frac{M}{EI}$$

Similarly, $$M_{23} = \frac{2EI}{L}(2\theta_2 - \theta_3)$$

$$= \frac{4EI\theta_2}{3} + \frac{2EI}{3}.\theta_3$$

As θ_3 is kept at its original value and not allowed to rotate further at does not contribute to M_{23}

or $$\frac{4EI\theta_2}{3} = M_{23} = -0.1299\,M$$

or $$\theta_2 = -\frac{0.1299\ M \times 3}{4EI}$$

$$= -0.09737\frac{M}{EI}.$$

These are values of the rotations at the end of the 1st Cycle.

2nd Cycle. When joint 3 is again released $M_{32} = 0.0371\ M = \frac{4EI}{3}.\theta'_3$ which is the additional rotation at joint 3 due 0.0371 M.

$$\theta'_3 = \frac{0.0371\ M \times 3}{4EI} = 0.02775\frac{M}{EI}$$

Hence new $$\theta_3 = (0.02775 + 0.42857)\frac{M}{EI}$$

$$= 0.45632\frac{M}{EI}$$

Similarly from $$M_{34} = 0.0278\,M.$$

$$\theta'_3 = .02780\frac{M}{EI}$$

$$\theta_3 = (0.02780 + 0.42857)\frac{M}{EI}$$

$$= 0.45637\frac{M}{EI}.$$

3rd Cycle. $$M_{21} = -0.0101\,M$$

$$M_{23} = -0.0084\,M$$

$$\theta_2' = \frac{2.5}{4EI}(-0.0101\,M)$$

$$= -0.0063\frac{M}{EI}$$

θ_2' for $\qquad M_{23} = \frac{3}{4EI}(-0.0084\,M)$

$$= -0.0063\frac{M}{EI}$$

Final value of $\qquad \theta_2 = [-0.0974 + (-0.0063)]\frac{M}{EI}$

$$= -0.1037\frac{M}{EI}$$

4th Cycle. When joint 3 is again released,

$$M_{32} = +0.0024\,M = \frac{4EI}{3}.\theta''_3$$

which is the additional rotation at joint 3 due to 0.0024 M.

$$\theta''_3 = \frac{0.0024 \times 3}{4}\frac{M}{EI} = 0.0018\frac{M}{EI}$$

Hence New $\qquad \theta_3 = (0{\cdot}45632 + 0018)\,M/EI$
$= 0.45812.$

These values of θ_2 and θ_3 are shown in the moment distribution table (Table 11.1). These values exactly tally with the iterative solution of the slope deflection eqautions as shown in Fig. 11.29. They are also very close to the exact solution of the simultaneous equations of the slope deflection method. The final moments obtained by moment distribution method are very close to the exact values obtained by slope deflection method. It can, therefore, be concluded that though Prof. Hardy Cross derived the method intuitively it is based on very sound principles and reasoning. Hardy Cross based the method on physical conception of locking and unlocking of joints, distribution of moments, and carry over of moments to far ends. His sound engineering judgment endabled him to derive a method based on sound theoretical concepts.

Example 11.14 Calculate the end moments by moment distribution method of (Fig. 11.36).

(*a*) all member lengths are equal and I is the same for all members.

(*b*) all member lengths are equal and $I_{ab} = I_{ac} = I_{ae} = I$ with $I_{ad} = 0.5\,I$.

(*c*) all member lengths are equal, $I_{ab} = I_{ac} = I_{ae} = I$ and $I_{ad} = 2I$.

(*d*) I is the same for all members and $L_{ab} = L_{ac} = L_{ae} = L$ with $L_{ad} = 0.5\,L$.

(*e*) I is the same for all members and $L_{ab} = L_{ac} = L_{ae} = L$ with $L_{ad} = 2\,L$ [Sack].

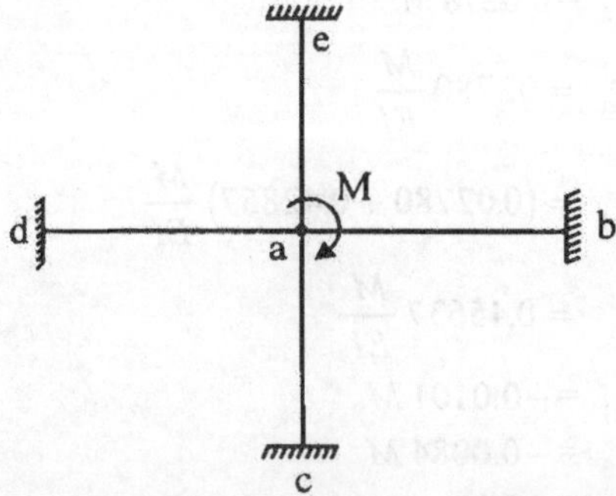

Fig. 11.36. Example 11.14

Solution:

(*a*) $$M_{ab} = M_{ac} = M_{ad} = M_{ae} = \frac{M.\frac{I}{L}}{4\frac{I}{L}} = \frac{M}{4}.$$

(*b*) $$M_{ab} = M_{ac} = M_{ae} = \frac{M.\frac{I}{L}}{3\frac{I}{L} + 0.5\frac{I}{L}} = \frac{M}{3.5}$$

$$M_{ad} = \frac{M.0.5\frac{I}{L}}{3.5\frac{I}{L}} = \frac{M}{7}$$

(*c*) $$M_{ab} = M_{ac} = M_{ae} = \frac{M.\frac{I}{L}}{3\frac{I}{L} + 2\frac{I}{L}} = \frac{M}{5}$$

$$M_{ad} = \frac{M.2\frac{I}{L}}{5\frac{I}{L}} = \frac{2M}{5}$$

(*d*) $$M_{ab} = M_{ac} = M_{ae} = \frac{M.\frac{I}{L}}{3\frac{I}{L} + \frac{I}{0.5\,L}} = \frac{M}{5}$$

$$M_{ad} = \frac{M.\frac{I}{0.5\,L}}{3\frac{I}{L} + 2\frac{I}{L}} = \frac{M}{2.5}$$

(*e*) $$M_{ab} = M_{ac} = M_{ae} = \frac{M.\frac{I}{L}}{3\frac{I}{L} + \frac{I}{2\,L}} = \frac{M}{3.5}$$

$$M_{ad} = \frac{M.\frac{I}{2L}}{3.5\frac{I}{L}} = \frac{M}{7}$$

11.10. SYSTEMATIC STEPS OF MOMENT DISTRIBUTION METHOD

Before starting solving examples we recapitulate the following steps of moment distribution method :

1. Calculate stiffness factors, distribution factors and carry over factors for all members.
2. Lock joints against rotation.
3. Calculate fixed end moments for all members.
4. Select joint to be unlocked first. Calculate unbalanced moment at this joint.
5. Calculate carry over moments to the far ends of the members.
6. Relock joints and select next joint to be unlocked and continue steps 4, 5, 6 until all joints have been balanced once and one cycle is completed.

In practice, however, all joints may be unlocked at the same time and balancing and carryover done simultaneously.

11.11 DISTRIBUTION FACTOR AT A PINNED END AND AT A FIXED END

At a pinned end we can inagine that is fictitious member outside the structure exists but it has $I = 0$. Therefore, the distribution factors at a pinned joint are 1 for the real member side and 0 for the otherside. At a fixed joint we can think of the wall as a member with infinite stiffness. As shown in Fig. 11.37 the distribution factors yield a value of zero for the member and 1 for the wall.

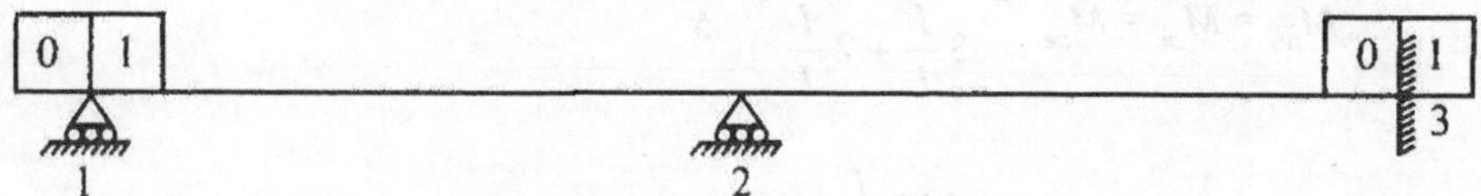

Fig. 11.37. Distribtution factors at pinned end and at fixed end.

Example 11.15. Draw the B.M. diagram for the beam. (Fig. 11.38)

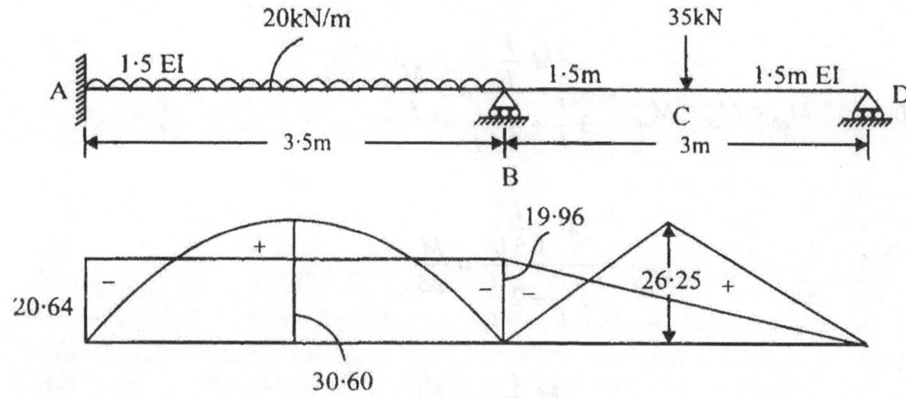

Fig. 11.38. Example 11.15.

Solution: FEMs

$$M_{F\,AB} = -\frac{wL^2}{12} = -\frac{20 \times 3.5^2}{12}$$

$$= -20.41 \; kNm$$

$$M_{F\,BA} = +20.41$$

$$M_{F\,BD} = -\frac{35 \times 3}{8} = -13.12$$

$$M_{F\,DB} = +13.12\ kNm$$

Stiffness factors.

$$K_{BA} = \frac{1.5\,EI}{3.5} = 0.43\,EI$$

$$K_{BD} == \frac{3}{4} \cdot \frac{EI}{3.0} = 0.25\,EI$$

Distribution factors.

$$\text{for } BA = \frac{0.43}{0.43 + 0.25} = 0.63$$

$$\text{for } BD = \frac{0.25}{0.68} = \frac{0.37}{1.00} \text{ (Check).}$$

The moment distribution is shown in Table 11.2.

1. 1st step is to fix all joints and compute FEMs.

2. Joint D is then released as there is no moment at D. While releasing joint D, $\frac{1}{2}$ the moment *i.e.*, $\frac{1}{2}$ (13.12) is transferred to joint B.

3. Joint D is now kept permanently hinged. Therefore while computing distribution factors at B, $\frac{3}{4}$ th EI should be taken for BD.

4. Next joint B is released and unbalanced moment distributed between BA and BD.

5. $\frac{1}{2}$ the moment at B for BA *i.e.*, $\frac{1}{2}$ (– 0.46) = – 0.23 is carried to A. As the end D is permanently hinged, there is no carry over to D.

It is also possible to lock, unlock, distribute and carry over from all joints simultaneously. This is shown in Table 11.3.

Table 11.2 Example 11.15

Joint	A	B		D	Remarks
Member	AB	BA	BD	DB	
D.F.	0	0.63	0.37	1	
C.O.F.	½ ←		½ ←	→ 0	
F.E.M.	– 20.41	+ 20.41	– 13.12	+ 13.12	All joints fixed
Balancing				– 13.12	Joint D is being released.
C.O.			– 6.56 ↙		
Balancing		– 0.46	– 0.27		Joint B is being released.
C.O.	– 0.23 ↙				
Final Moments	– 20.64	+ 19.95	– 19.95	0.00	

Table 11.3 Example 11.15

Joint	A	B		D	Remarks
Member	AB	BA	BD	DB	
FEM	−20.41	+20.41	−13.12	+13.12	
Balance		−4.59	−2.70	−13.12	Joints *B* and *D* being unlocked simultaneously
C.O.	−2.29 ↙		−6.56 ↙		C.O. from *B* to *A* and *D* to *B* No C.O. from *B* to *D*. *B* & *A* locked.
Balance		+4.13	+2.43		Joint *B* again unlocked.
C.O.	+2.06				C.O. from *B* to *A*. No carry over from *B* to *D*.
F. M.	−20.64	+19.95	−19.95	0.00	Final Moment

Example 11.16. Find the moments when support B settles 1cm EI = 500kNm². (Fig. 11.38)

Solution : We assume all the joints locked and allow support B to settle 1cm.

FEM generated at A and B of AB due to settlement of B.

$$= -\frac{6EI\Delta}{L^2} = -\frac{6 \times 1\cdot 5 \times 500 \times 1}{100 \times 3\cdot 5^2} = \mathbf{-3\cdot 67kNm}$$

FEM generated at B and D in BD due to settlement of B $= \dfrac{+6EI\Delta}{L^2} = +\dfrac{6 \times 500 \times 1}{100 \times 3^2}$

$$= +\,3\cdot 33\text{kNm}$$

The moment distribution is shown in Table 11.4.

Table 11.4 Example 11.16.

Joint	A	B		D	Remaks
Member	*AB*	*BA*	*BD*	*DB*	
D.F.	1	0·63	0·37	0	
C.O.	1/2 ←	→ 1/2	1/2 ←	→ 0	
FEM due to load	−20·41	+20·41	−13·12	+13·12	
FEM due to support settlement	−3·67	−3·67	+3·33	+3·33	
Total FEM	−24·08	+16·74	−9·79	+16·45	
Balance		−4·38	−2·57	−16·45	Joint B and D unlocked
C.O.	−2·19		−8·23		Joint A and B fixed
Balance		+5·19	+3·04		Joint B unlocked
C.O.	+2·59				Joint A fixed
Final Moments	−23·68	+17·55	−17·55	0·00	Final Moment

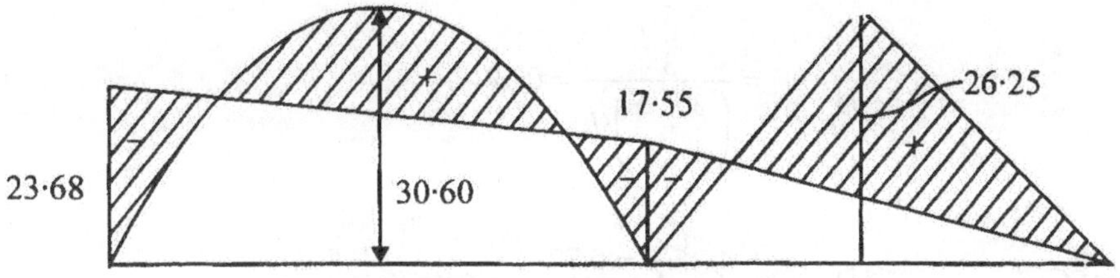

Fig. 11.39. B.M. diagram. (Example 11.16).

Example 11.17 Analyse the continuous beam shown in Fig. 11.40 by moment distribution method. Draw the S.F. and B.M.D. **[AMIE Summer' 98]**

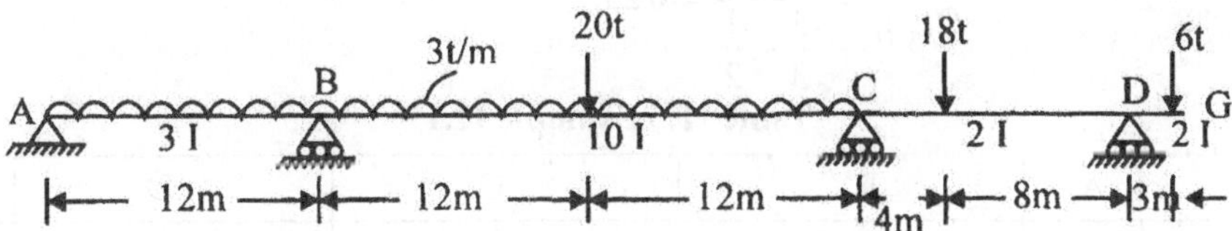

Fig. 11.40. Example 11.17

Solution:

FEMs

$$M_{F\,AB} = -\frac{3 \times 12^2}{12} = -36$$

$$M_{F\,BA} = +\frac{3 \times 12^2}{12} = +36$$

$$M_{F\,BC} = -\frac{3 \times 24^2}{12} + \left(-\frac{20 \times 24}{8}\right)$$

$$= -144 + (-60)$$

$$= -204 \text{ kNm.}$$

$$M_{F\,CB} = +204 \text{ kNm.}$$

$$M_{F\,CD} = -\frac{18 \times 4 \times 8^2}{12^2} = -32$$

$$M_{F\,DC} = +\frac{18 \times 8 \times 4^2}{12^2} = +16$$

$$M_{F\,DG} = -6 \times 3 = -18$$

Stiffnesses.

$$K_{BA} = 3 \times \frac{3I}{12} = \frac{3}{4}.I$$

$$K_{BC} = \frac{4 \times 10I}{24} = \frac{5}{3}.I$$

$$K_{CB} = \frac{4 \times 10I}{24} = \frac{5}{3}.I$$

$$K_{CD} = 3 \times \frac{2I}{12} = \frac{I}{2}$$

$$K_{DG} = 0$$

Distribution Factors.

Joint *B*

$$D_{F\,BA} = \frac{\frac{3}{4}I}{\left(\frac{3}{4} + \frac{5}{3}\right)I} = 0.31$$

$$D_{FBC} = \frac{\frac{5}{3}I}{\left(\frac{3}{4}+\frac{5}{3}\right)I} = 0.69$$

Joint C

$$D_{FCB} = \frac{\frac{5}{3}}{\frac{5}{3}+\frac{1}{2}} = 0.77$$

$$D_{FCD} = \frac{\frac{1}{2}}{\frac{5}{3}+\frac{1}{2}} = 0.23$$

Table 11.5 Example 11.17

Joint	A	B		C		D		Remarks
Members	AB	BA	BC	CB	CD	DC	DG	
D.F.		0.31	0.69	0.77	0.23	1	0	
C.O.	0 ←	→ 1/2	1/2 ←	→ 1/2	1/2 ←	→ 0		
F.E.M.	−36	+36	−204	+204	−32.0	+16	−18	
Balance	+36	+52	+116.00	−132.44	−39.56	+02		
C.O.		+18	−66.22	+58.00	+1.00			
Balance		+14.95	+33.27	−45.43	−13.57			
C.O.			−22.72	+16.64				
Balance		+7.04	+15.68	−12.81	−3.83			
C.O.			−6.41	+7.84				
Balance		+1.99	+4.42	−6.04	−1.80			
C.O.			−3.02	+2.21				
Balance		+0.94	+2.08	−1.70	−0.51			
C.O.			−0.85	+1.04				
Balance		+0.26	+0.59	−0.80	−0.24			
C.O.			−0.40	+0.30				
Balance		+0.12	+0.28	−0.23	−0.07			
C.O.		+0.0	−0.12	0.14				
Balance		+0.04	+0.08	−0.11	−0.03			
Final moment	0.00	+131.34	−131.34	+90.61	−90.61	+18.00	−18.00	

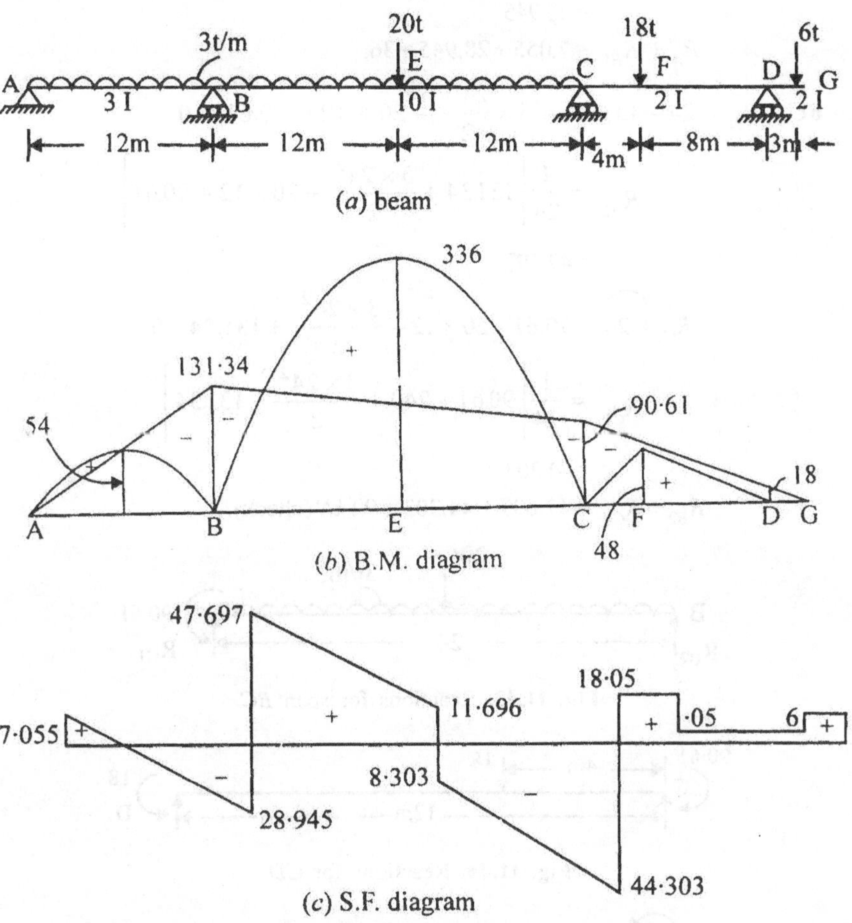

Fig. 11.41. S.F.and B.M. diagrams (Example 11.17).

S.F. Calculations.

Fig. 11.42. Reactions for span *AB*.

Span *AB*.

$$R_A \times 12 - 3 \times \frac{12^2}{2} + 131.34 = 0$$

$$R_A = \frac{1}{12}\left[\frac{3 \times 12^2}{2} - 131.34\right]$$

$$= 7.055$$

$$R_{B1} \times 12 - 131.34 - 3 \times 12 \times 6 = 0$$

$$R_{B1} = \frac{1}{12}[131.34 + 3 \times 12 \times 6]$$

$$=28.945$$

$$R_A + R_{B1} = 7.055 + 28.945 = 36.$$

Span *BC*. $R_{B2} \times 24 - 131.34 - 3 \times \frac{24^2}{2} - 20 \times 12 + 90.61 = 0$

or
$$R_{B2} = \frac{1}{24}\left[131.34 + \frac{3 \times 24^2}{2} + 20 \times 12 - 90.61\right]$$

$$=47.697$$

$$R_{C1} \times 24 - 90.61 - 20 \times 12 - \frac{3 \times 24^2}{2} + 131.34 = 0$$

or
$$R_{C1} = \frac{1}{24}\left[90.61 + 240 + \frac{3 \times 24^2}{2} - 131.34\right]$$

$$=44.303$$

$$R_{B2} + R_{C1} = 47.697 + 44.303 = 92\ kN \text{(check)}.$$

Fig. **11.43.** Reactions for span *BC*

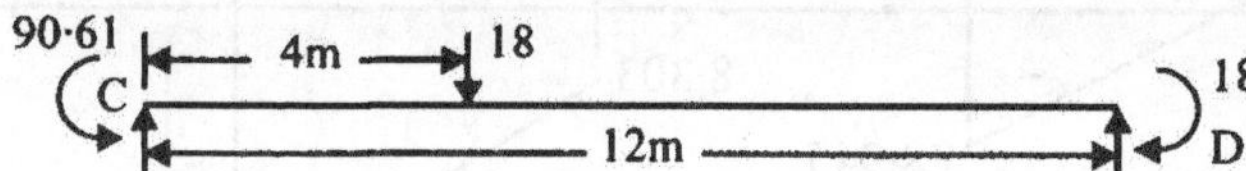

Fig. **11.44.** Reactions for *CD*

Span *CD*. $R_{C2} \times 12 - 90.61 - 18 \times 8 + 18 = 0$

or
$$R_{C2} = \frac{1}{12}[90.61 + 18 \times 8 - 18]$$

$$= 18.05$$

$$R_C = 44.303 + 18.05 = 62.353.$$

$$R_{D1} \times 12 - 18 - 18 \times 4 + 90.61 = 0$$

$$R_{D1} = \frac{1}{12}[18 + 72 - 90.61]$$

$$= -0.05$$

$$18.05 - 0.05 = 18 \text{ (check)}$$

$$R_{D2} = 6 \uparrow$$

$$R_D = 6 - 0.05 = 5.95 \uparrow$$

Fig. **11.45.** Span *DE*

Checking of M.D. results.

$$\theta_{BL} = \frac{131.34 \times 12}{3 \times 3\,EI} - \frac{3 \times 12^3}{24 \times 3\,EI}$$

$$= 175.12 - 72.00$$

$$= \frac{103.12}{EI}$$

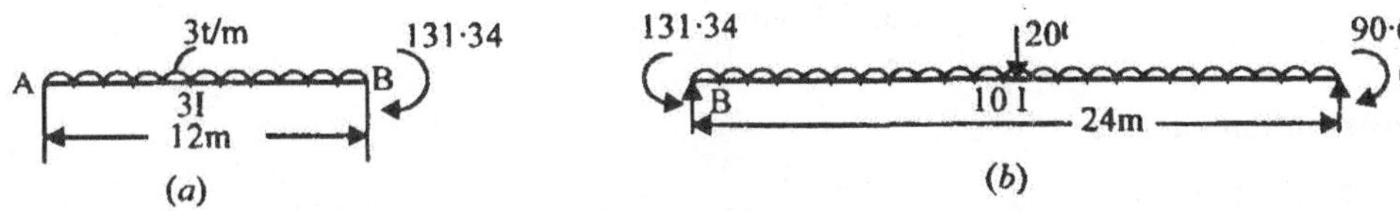

Fig. 11.46. Checking of M.D. results by continuity of slope.

$$\theta_{BR} = \frac{131.34 \times 24}{3 \times 10\,EI} - \frac{3 \times 24^3}{24 \times 10\,EI} - \frac{20 \times 24^2}{16 \times 10\,EI} + \frac{90.61 \times 24}{6 \times 10\,EI}$$

$$= \frac{1}{EI}(105.072 - 172.80 - 72.00 + 36.244)$$

$$= -\frac{103.484}{EI}.$$

Example 11.18 Find the support moments in the continuous beam shown in Fig. 11.47 below if the support at A settles by 10 mm. $E = 2200$ t/cm^2 $I = 5{,}000$ cm^4. (Fig. 11.47)

EI Constant

Fig. 11.47. Example 11.18.

Solution:

FEMs

$$M_{FAB} = -\frac{1.5 \times 1^2 \times 2}{3^2} = -0.333$$

$$M_{FBA} = +\frac{1.5 \times 2^2 \times 1}{3^2} = +0.646$$

$$M_{FBC} = -\frac{1.5 \times 1.5 \times 0.9^2}{2.4^2} = -0.316$$

$$M_{FCB} = +\frac{1.5 \times 0.9 \times 1.5^2}{2.4^2} = +0.527$$

$$M_{FCD} = -\frac{2 \times 1.9 \times 1.1^2}{3^2} = -0.511$$

$$M_{FDC} = +\frac{2 \times 1.1 \times 1.9^2}{3^2} = +0.882$$

$$M_{FDE} = -1 \times 1.2 = -1.2$$

FEM due to support settlement at A

$$M_{FAB} = M_{FBA} = +\frac{6\,EI\,\delta}{L^2}$$

$$= +\frac{6 \times 2200 \times 5000 \times 1}{(300)^2}$$

$$= +733.33\ t\,\text{cm}$$

$$= 7.33\ t.m$$

Distribution Factors

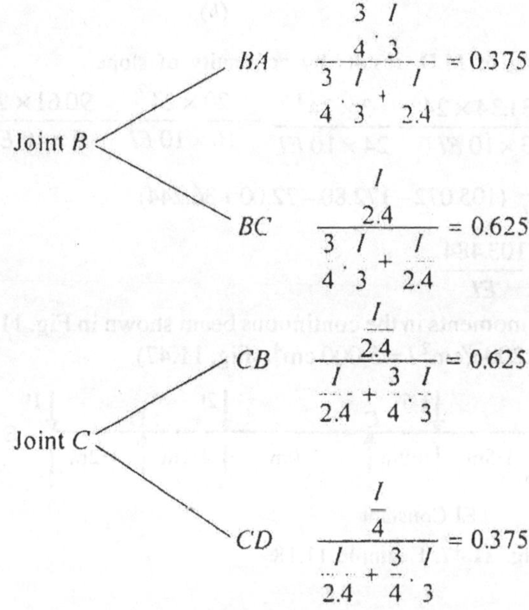

Table 11.6 (Example 11.18)

Joint	A	B		C		D		Remarks
Member	AB	BA	BC	CB	CD	DC	DE	
D.F.		0.375	0.625	0.625	0.375			
C.O.F.	0 ←	→ 1/2	1/2 ←	→ 1/2	1/2 ←	→ 0		
F.E.M. due to load	−0.333	+0.666	−0.316	+0.527	−0.511	+0.882	−1.2	
F.E.M. due to support settlement	+7.333	+7.333						
Total FEM	+7.000	+7.999	−0.316	+0.527	−0.511	+0.882	−1.200	
Balance	−7.000	−2.881	−4.802	−0.010	−0.006	+0.318		

C.O.		−3.500	−0.005	−2.401	+0.159			
Balance		+1.314	+2.191	+1.401	+0.841			
C.O.			+0.700	+1.095				
Balance		−0.263	−0.437	−0.684	−0.411			
C.O.			−0.342	−0.218				
Balance		+0.128	+0.214	+0.136	+0.082			
C.O.			+0.068	+0.107				
Balance		−0.026	−0.042	−0.067	−0.040			
Final moment	0	+2.771	−2.771	−0.114	+0.114	+1.20	−1.20	

Check. Slope at *C*. (taking EI = unity)

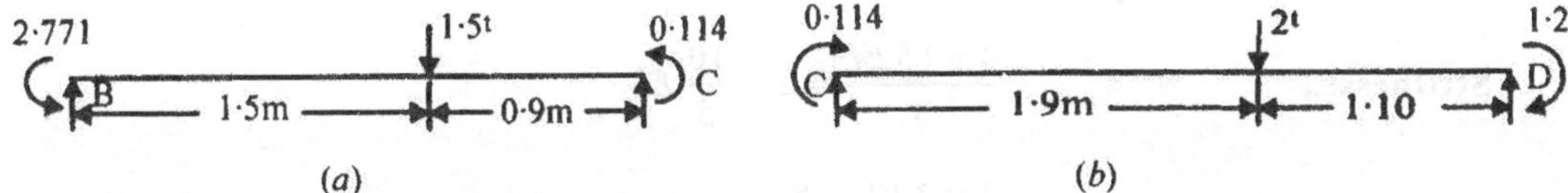

Fig. 11.48. Checking of M.D. results.

Slope at C from BC

$$\frac{2.771 \times 2.4}{6} - \frac{0.114 \times 2.4}{3} - \frac{1.5 \times 1.5 \times 0.9}{2.4 \times 2.4} \times \left[\frac{1}{2} \times 1.5 \times 1.00 + \frac{1}{2} \times 0.9 \times 1.8\right]$$

$$= \mathbf{0.4687625}$$

from CD

$$\frac{0.114 \times 3}{3} - \frac{1.2 \times 3}{6} + \frac{2 \times 1.9 \times 1.1}{3 \times 3} \times \left[\frac{1}{2} \times 1.9 \times \left(1.1 + \frac{1.9}{3}\right) + \frac{1}{2} \times 1.1 \times \frac{2}{3} \times 1.1\right]$$

$$= \mathbf{0.46596404.}$$

Example 11.19 Find the moments at the critical sections. Draw the B.M. diagram. (Fig. 11.49)

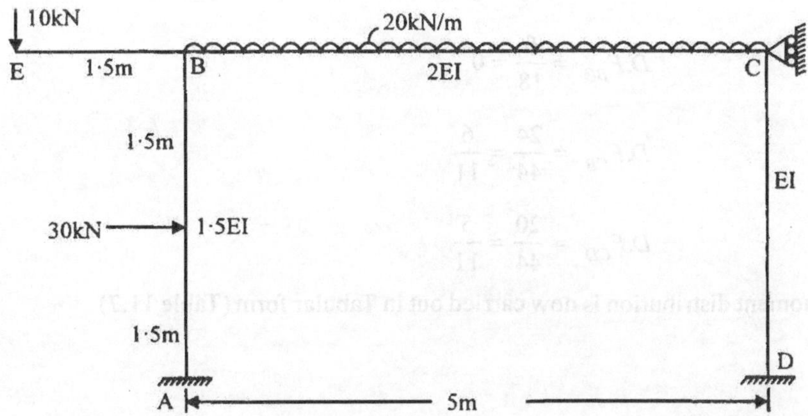

Fig. 11.49. Example 11.19.

Solution: As there is a support at C, lateral movement of the frame is prevented. Hence there will be no sway of the frame. In this case normal moment distribution can be done.

FEMs

$$M_{F\,AB} = -\frac{30 \times 3}{8} = -11.25\ kNm.$$

$$M_{F\,BA} = +\frac{30 \times 3}{8} = +11.25\ kNm.$$

$$M_{F\,BC} = -\frac{20 \times 5^2}{12} = -41.66\ kNm.$$

$$M_{F\,CB} = +\frac{20 \times 5^2}{12} = +41.66\ kNm.$$

$$M_{F\,BE} = +15\ kNm.$$

Stiffnesses

$$K_{BA} = \frac{4 \times 1.5\,EI}{3} = 2 = \frac{10}{5}EI$$

$$K_{BC} = \frac{4 \times 2\,EI}{5} = \frac{8}{5}EI$$

$$K_{BE} = 0$$

$$K_{CB} = \frac{4 \times 2\,EI}{5} = \frac{8}{5}EI = 24\,EI$$

$$K_{CD} = \frac{4 \times EI}{3} = \frac{4}{3}EI = 20\,EI$$

D.F.

$$D.F_{BA} = \frac{10}{18} = \frac{5}{9}$$

$$D.F_{BC} = \frac{8}{18} = \frac{4}{9}$$

$$D.F_{BE} = \frac{0}{18} = 0$$

$$D.F_{CB} = \frac{24}{44} = \frac{6}{11}$$

$$D.F_{CD} = \frac{20}{44} = \frac{5}{11}$$

The moment distribution is now carried out in Tabular form (Table 11.7)

Table 11.7 M.D. Table (Example 11.19)

Joint	A	B			C		D	Remarks
Member	AB	BE	BA	BC	CB	CD	DC	Stiffness of canti-lever zero
D.F.		0	$\frac{5}{9}$	$\frac{4}{9}$	$\frac{6}{11}$	$\frac{5}{11}$		
C.O.			1/2 ←	1/2 ←	→ 1/2	1/2 ←	→ 1/2	
F.E.M.	11.25	+ 15	+ 11.25	41.66	+ 41.66	0	0	
Balance			+ 8.56	+ 6.85	22.72	18.94		
C.O.	+ 4.28			11.36	+ 3.43		9.47	
Balance			+ 6.31	+ 5.05	1.87	1.56		
C.O.	+ 3.16			0.94	+ 2.52		0.78	
Balance			+ 0.52	+ 0.42	− 1.37	− 1.15		
C.O.	+ 0.26			− 0.68	+ 0.21		− 0.58	
Balance			+ 0.37	+ 0.30	− 0.11	0.10		
C.O.	+ 0.18			− 0.05	+ 0.15		− 0.05	
Balance			+ 0.03	+ 0.02	− 0.08	− 0.07		
Final moment	− 3.27	+ 15	+ 27.04	− 42.04	+ 21.82	− 21.82	− 10.88	

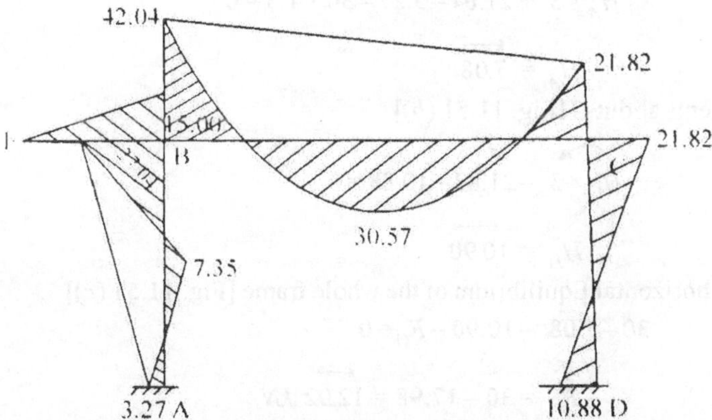

Fig. 11.50. B.M. diagram

Example 11.20 Same problem with sway *i.e.*, when there is no support at *C*.

In this case moment distribution is to be carried out in two steps.

1. Without sway *i.e.*, with support at *C*.
2. Then superposing the effect of sway.

The moments generated with support at *C* (non sway moment) are already known.

$$M_{AB} = -3.27 \qquad M_{BA} = +27.04$$
$$M_{BC} = -42.04 \qquad M_{CB} = +21.82$$
$$M_{CD} = -21.82 \qquad M_{DC} = -10.88$$

Fig. 11.51 (*a*) and (*b*) show free body diagrams of members *AB* and *DC*.

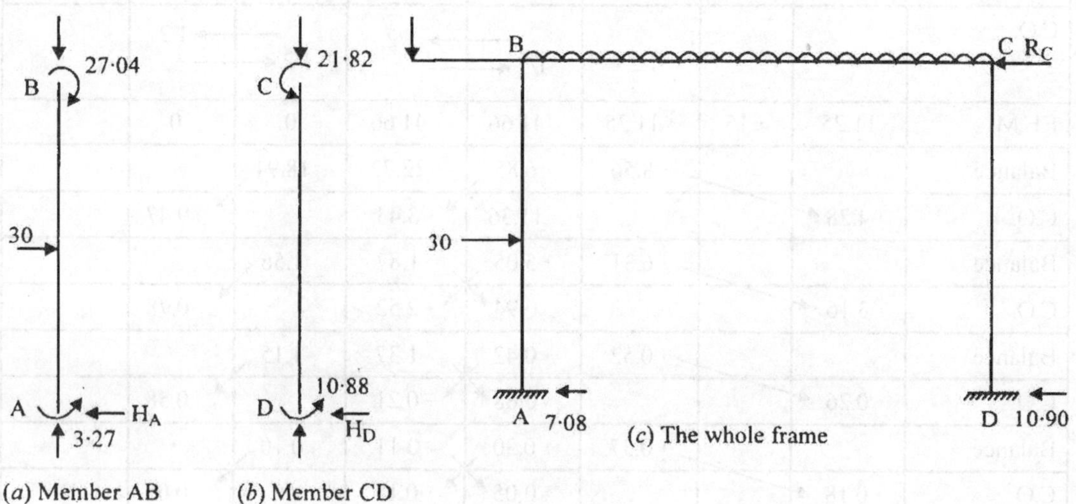

(*a*) Member AB (*b*) Member CD

Fig. 11.51. Computing R_C

Taking moments about *B* [Fig. 11.51 (*a*)]

$$H_A \times 3 + 27.04 - 3.27 - 30 \times 1.5 = 0.$$

or $$\overleftarrow{H_A} = 7.08$$

Taking moments about *C* [Fig. 11.51 (*b*)]

$$H_D \times 3 - 21.82 - 10.88 = 0$$

or $$\overleftarrow{H_D} = 10.90$$

Considering horizontal Equilibrium of the whole frame [Fig. 11.51 (*c*)]

$$30 - 7.08 - 10.90 - R_C = 0$$

or $$\overleftarrow{R_C} = 30 - 17.98 = 12.02 \text{ kN}$$

In the actual frame there is no such support at *C*. Hence we should neutralise its effect by putting a force of 12.02 *kN* at *C* towards the right. The bending moment due to this load which is known as sway moment is to be superposed on the non-sway moment.

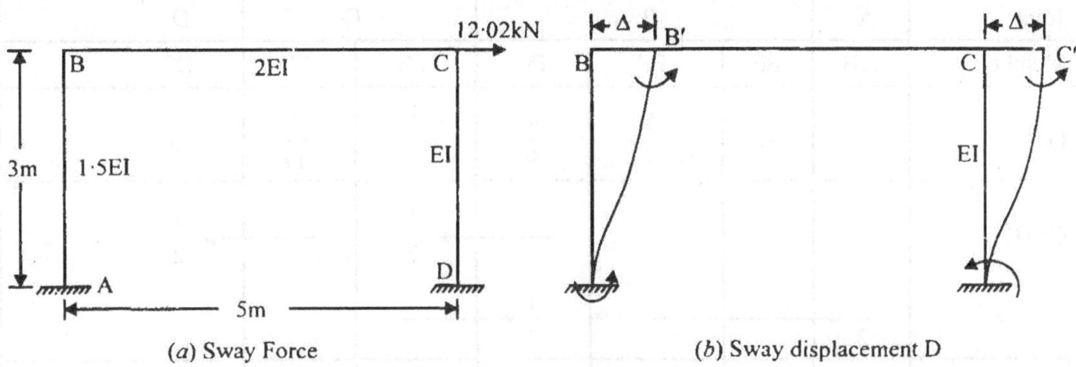

Fig. 11.52. Sway moment

Since the amount of sway is not known before hand, the problem is to be solved in an indirect way. Let the frame sway Δ towards the right. Considering the joints *B* and *C* locked, sway moments generated at *A, B, C* and *D* are : [Fig. 11.52 (*b*)]

$$M_{AB} = -\frac{6(1.5\,EI)\Delta}{3^2}$$

$$M_{BA} = -\frac{6(1.5\,EI)\Delta}{3^2}$$

$$M_{CD} = -\frac{6\,EI.\Delta}{3^2}$$

$$M_{DC} = -\frac{6\,EI.\Delta}{3^2}$$

Here axial strain in *BC* has been ignored as we are considering bending deformation only. Hence sway of *B* is equal to sway of *C*.

Therefore $M_{AB} : M_{BA} : M_{CD} : M_{DC} = 3 : 3 : 2 : 2$

Hence arbitrary moments

$$M_{BA} = M_{AB} = -30\ kNm.$$

$$M_{CD} = M_{DC} = -20\ kNm.$$

are assumed and moment distribution carried out (Table 11.8) .

Table 11.8 Distribution of arbitrary sway.

Joint	A		B		C		D
Member	AB	BE	BA	BC	CB	CD	DC
D.F.	1	0	$\frac{5}{9}$	$\frac{4}{9}$	$\frac{6}{11}$	$\frac{5}{11}$	0
C.O.F.	$\frac{1}{2}$ ←			→ $\frac{1}{2}$ ←	$\frac{1}{2}$	→	$\frac{1}{2}$
F.E.M.	−30		−30			−20	−20
Balance			+16.66	+13.34	+10.90	+9.10	
C.O.	+8.33			+5.45	+6.67		+4.55
Balance			−3.02	−2.43	−3.64	−3.02	
C.O.	−1.51			−1.82	−1.22		−1.52
Balance			+1.01	+0.81	+0.66	+0.56	
C.O.	+0.50			+0.33	+0.40		+0.28
Balance			−0.19	−0.14	−0.22	−0.18	
C.O.	−0.09			−0.11	−0.07		−0.09
Balance			+0.06	+0.05	+0.04	+0.03	
C.O.	+0.03			+0.02	+0.02		+0.02
Final Moment	−22.74		−15.48	+15.50	+13.54	−13.54	−16.76

Since the sway moments have been arbitrarily chosen sway will not correspond to a force of 12.02 *kN*. It is therefore, necessary, to calculate the force at *C* which corresponds to this arbitrary sway moment.

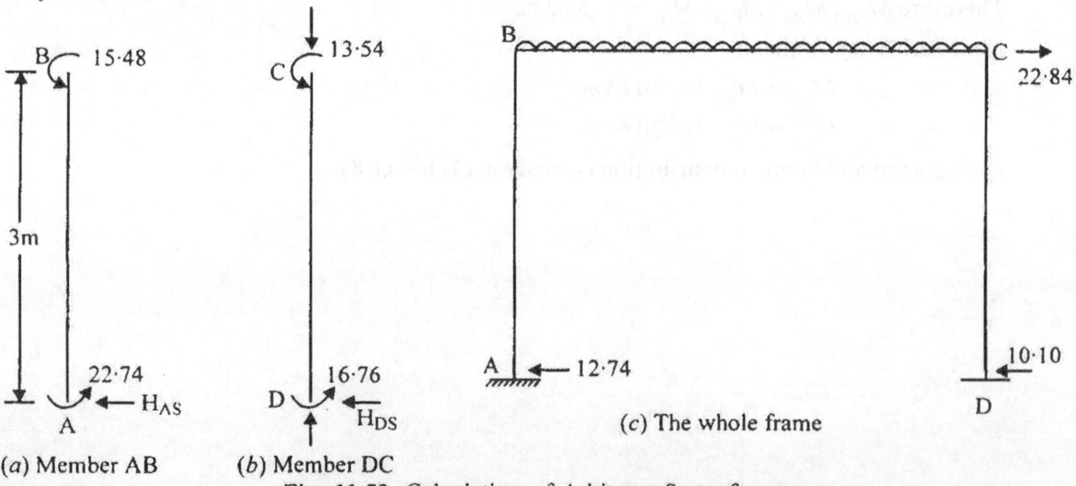

(*a*) Member AB (*b*) Member DC (*c*) The whole frame

Fig. 11.53. Calculation of Arbitrary Sway force.

From Fig. 11.53 (*a*)

$$H_{AS} \times 3 - 22.74 - 15.48 = 0$$

or

$$H_{AS} = \frac{15.48 + 22.74}{3} = 12.74\ kN$$

From Fig. 11.53 (*b*)

$$H_{DS} = \frac{13.54 + 16.76}{3} = 10.10\ kN$$

Considering horizontal equilibrium of the frame as a whole from $\Sigma H = 0$. (Fig. 11.53 c)

Force at $C = 12.74 + 10.10 = 22.84\ kN$

However the actual force is 12.02 *kN*

Hence the sway moments have to be reduced in the ratio $\frac{12.02}{22.84} = 0.5262$.

The final moments can be obtained in a Tabular form as follows : (Table 11.9)

Table 11.9 Computation of Final Results.

Joint	A	B			C	D	Remarks
Member	AB	BE	BA	BC	CB	CD	DC	
Arbitrary Sway Moment	− 22.74	0	− 15.48	+ 15.50	+ 13.54	− 13.54	− 16.76	(1)
Actual Sway Moment	− 11.97	0	− 8.15	+ 8.15	+ 7.12	− 7.12	− 8.82	(2) = (1) × 0.5262
Non Sway Moment	− 3.27	+ 15.00	+ 27.04	− 42.04	+ 21.82	− 21.82	− 10.88	(3)
Final Moment	− 15.24	+ 15.00	+ 18.89	− 33.89	− 28.94	− 28.94	− 19.70	(2) + (3)

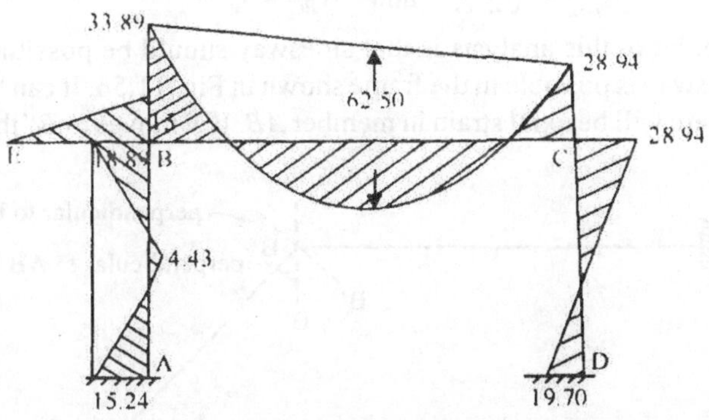

Fig. 11.54. B.M. diagram (Example 11.20)
Plotted on the tension side.

11.12 MOMENT DISTRIBUTION FOR FRAMES WITH SLOPING MEMBERS

For frames with sloping members deflected shape of the frame is of great importance as sway moments generated will depend on the deflected shape. We examine first the deflected shape of a frame with only one inclined member and one independent degree of freeedom sidesway for easy understanding.

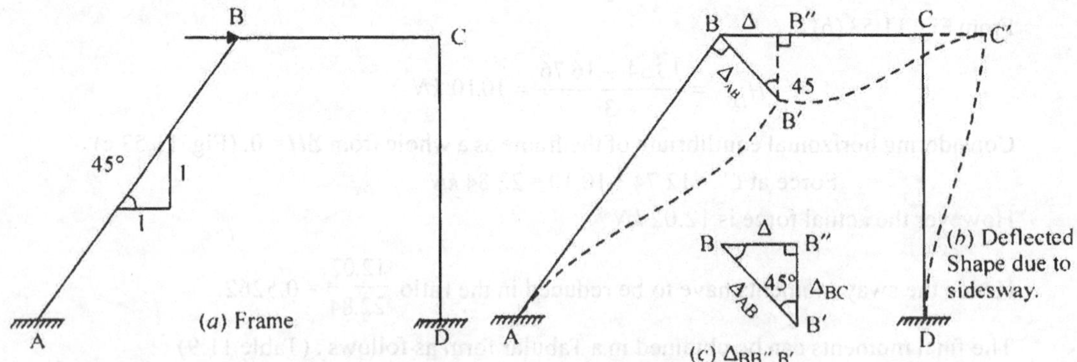

Fig. 11.55. Sway for sloping members.

Fig. 11.55 shows such a frame and its deflected shape. While sketching the deflected shape the following assumptions are made.

1. The joints are assumed fixed and not allowed to rotate.
2. The angle between members at a rigid joint remains unchanged.
3. The members do not elongate or shorten due to axial strain.
4. The distance between member ends does not decrease due to the curvature of the member.
5. The member ends can only translate in a direction perpendicular to the member.

These rules follow from the small displacement theory. Rule 5 follows directly from rules 3 and 4. However, it has been explicitly stated to facilitate drawing of the deflected shape.

From the relation, angle between two lines is angle between their perpendicular $\angle BB'B'' = 45°$

Hence $\qquad \Delta_{AB} \cdot \cos 45° = \Delta$

or $\qquad \Delta_{AB} = \sqrt{2}.\Delta \quad \text{and} \quad \Delta_{BC} = \Delta.$

The important point in this analysis is that sidesway should be possible without axial strain. From this view no sidesway is possible in the frame shown in Fig. 11.56. It can be seen from the figure that if B moves to B' there will be axial strain in member AB. If B moves to B'' there will be axial strian in member BC.

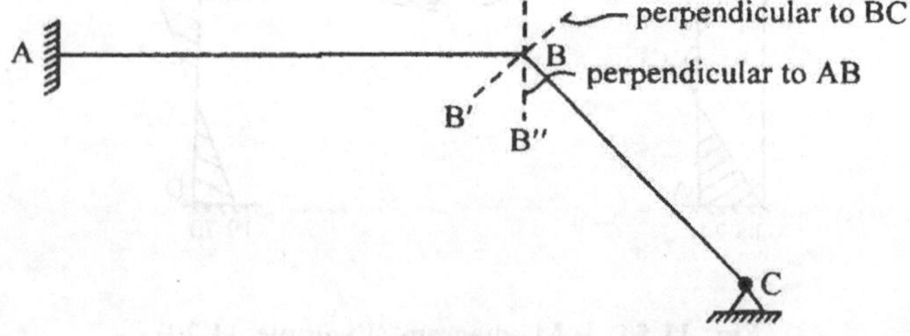

Fig. 11.56. Inclined Frame (No sway possible)

Sway without axial strain is, however, possible in Fig. 11.57 as joint C is a roller. The member AB first moves to AB' with vertical movement Δ downward. Joint C'' then moves to C'. So that C' is on horizontal line of the possible movement of the roller.

$C''C'$ = perpendicular distance between two ends B' and C' the member BC.

From geometry $C'C'' \cos\theta = \Delta$

$$C'C'' = \frac{\Delta}{\cos\theta} = \frac{\Delta \times 5}{4}.$$

We now derive relative displacements of the ends of members of a non-orthogonal frame.

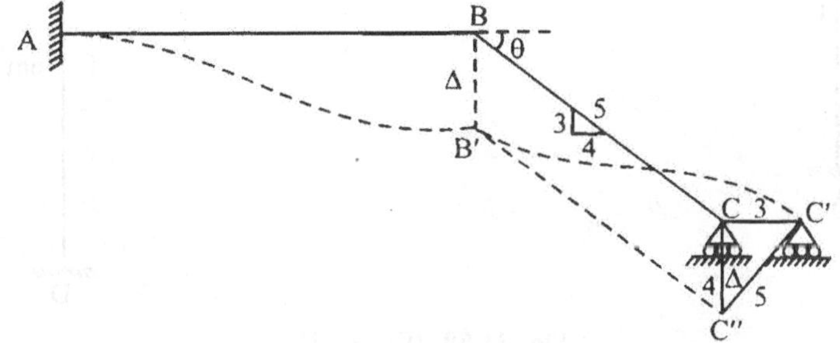

Fig. 11.57. Inclined Frame (Sway Possible)

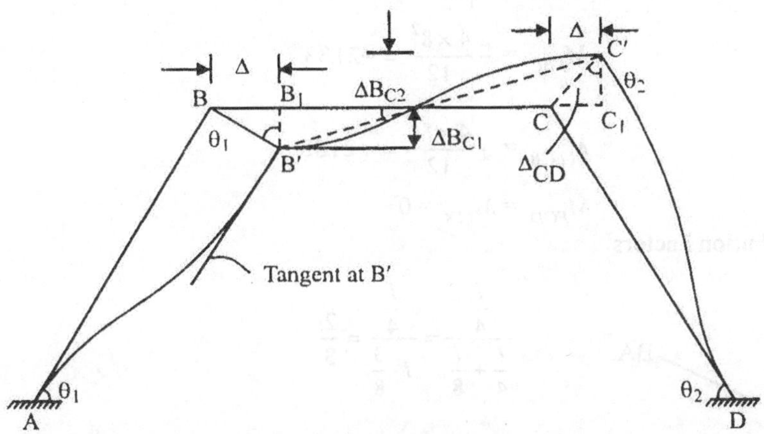

Fig. 11.58. Sway of non orthogonal frame.

$$\Delta_{AB} = BB' = \Delta \operatorname{cosec} \theta_1$$

$$\Delta_{CD} = CC' = \Delta \operatorname{cosec} \theta_2$$

$$\Delta_{BC} = \Delta_{BC_1} + \Delta_{BC_2}$$

$$B_1 B' + C_1 C'$$

$$= \Delta \cot\theta_1 + \Delta \cot\theta_2$$

$$= \Delta (\cot\theta_1 + \cot\theta_2)$$

Example 11.21. A rigid portal *ABCD* has two unequal vertical legs. $AB = 4$ m, and $CD = 6$ m with member *BC* horizontal and 8 m long. The bases *A* and *D* are fixed and all the members have same moment of inertia *I*. The member *BC* carries a *u.d.l* of 4*t*/*m* and a horizontal force *H* acts at *C*. Find the magnitude and sense of *H* such that the portal does not sway under the given loading. Draw the B.M. diagram for the portal under the same loading condition. Use the moment distribution method.

[AMIE Summer 1985]

Solution:

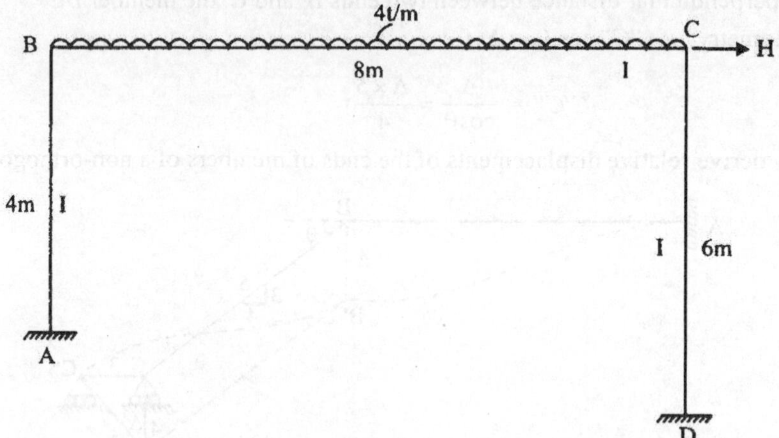

Fig. 11.59. (Ex. 11.21)

1. F.E.M.

$$M_{FAB} = M_{FBA} = 0$$

$$M_{FBC} = -\frac{4\times 8^2}{12} = -21.33$$

$$M_{FCB} = +\frac{4\times 8^2}{12} = +21.33$$

$$M_{FCD} = M_{FDC} = 0$$

2. Distribution Factors

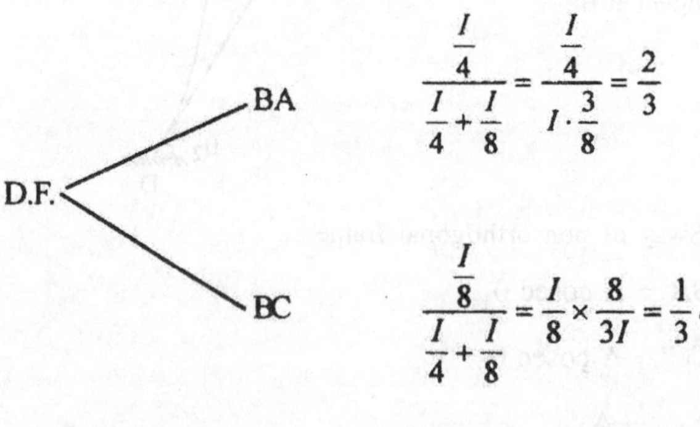

$$\frac{\frac{I}{4}}{\frac{I}{4}+\frac{I}{8}} = \frac{\frac{I}{4}}{I\cdot\frac{3}{8}} = \frac{2}{3}$$

$$\frac{\frac{I}{8}}{\frac{I}{4}+\frac{I}{8}} = \frac{I}{8}\times\frac{8}{3I} = \frac{1}{3}.$$

D.F. — CB

$$\frac{\frac{I}{8}}{\frac{I}{8}+\frac{I}{6}} = \frac{I}{8}\times\frac{24}{7I} = \frac{3}{7}.$$

D.F. — CD

$$\frac{\frac{I}{6}}{\frac{I}{8}+\frac{I}{6}} = \frac{I}{6}\times\frac{24}{7I} = \frac{4}{7}.$$

3. Distribution of Non Sway moment. *i.e.* assuming that there is support at C which provides a force H to prevent sway.

Table 11.10 Distribution of Non Sway moment

Joint	A	B		C		D
Member	AB	BA	BC	CB	CD	DC
D.F.		$\frac{2}{3}$	$\frac{1}{3}$	$\frac{3}{7}$	$\frac{4}{7}$	
C.D.	$\frac{1}{2}$ ←		→ $\frac{1}{2}$	$\frac{1}{2}$ ←	→	$\frac{1}{2}$
F.E.M.			−21.33	+21.33		
Balance		+14.22	+7.11	−9.14	−12.19	
C.O.	+7.11		−4.57	+3.55		−6.09
Balance		+3.05	+1.52	−1.52	−2.03	
C.O.	+1.2		−0.76	+0.76		−1.02
Balance		+0.51	+0.25	−0.33	−0.43	
C.O.	+0.25		−0.16	+0.12		−0.22
Balance		+0.10	+0.06	−0.05	−0.07	
Final Moment	+8.88	+17.88	−17.88	+14.72	−14.72	−7.33

4. Computation of required H.

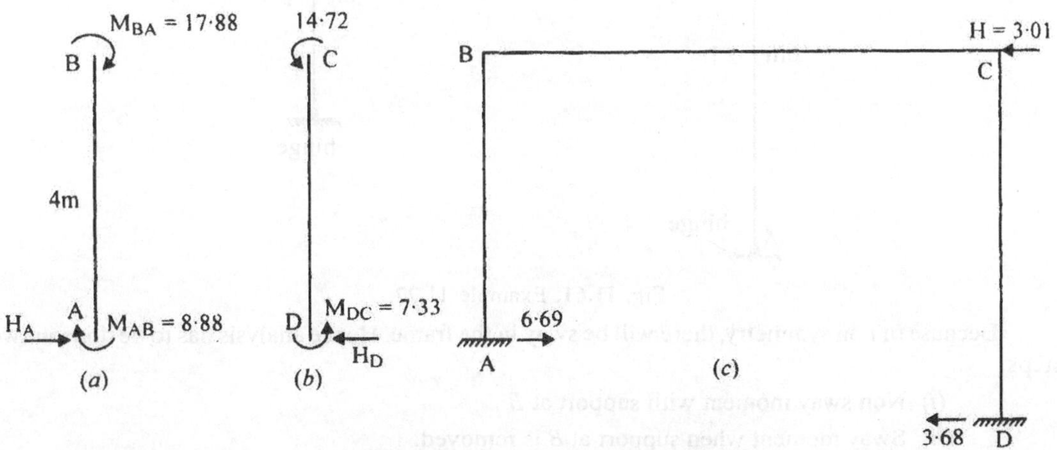

Fig. 11.60. Computation of Required H

Equilibrium of AB (Fig. 11.60a)

$$H_A \times 4 - 17.88 - 8.88 = 0$$

or

$$H_A = \frac{1}{4}[17.88 + 8.88] = \overrightarrow{6.69}$$

Equilibrium of *CD*. (Fig. 11.60*b*)

$$H_D \times 6 - 7.33 - 14.72 = 0$$

$$H_D = \frac{1}{6}[7.33 + 14.72]$$

$$= \overleftarrow{3.68}$$

From the equilibrium of the frame as a whole (Fig. 11.60*c*) force required at *C* for no sway = 3.01′←.

Example 11.22. A rigid jointed portal frame *ABCD* has the horizontal beam member *BC* of length 6 m and moment of inertia 2*I*, the left vertical column *AB* of height 6 *m* and moment of inertia 2*I* and the right vertical column *DC* of length 4 m and moment of inertia 1.5 *I*. The bases *A* and *D* are hinged.

A uniformly distributed load of 10 kN per metre acts on the full length of the beam *BC*. Using moment distribution method, analyse the portal and draw the bending moment diagram. Make use of the following information. The analysis of the frame for a sway load of 1 kN at *B* in the direction of $\overrightarrow{BC}$ gives the end moments M_{BC} = +2 kN m and M_{CB} = + 2.66 kNm both clockwise.

[AMIE Summer '97]

Solution:

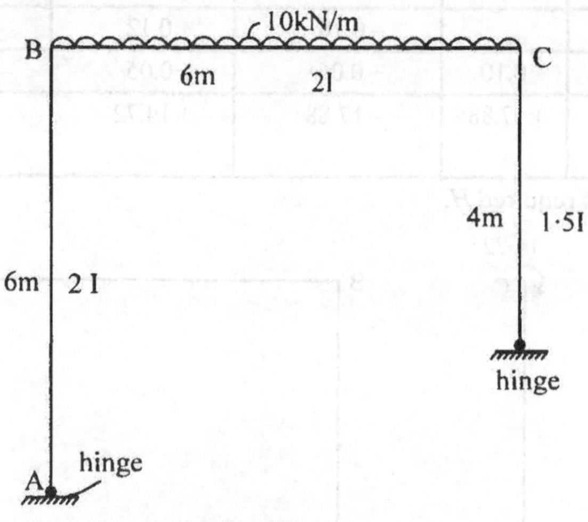

Fig. 11.61. Example 11.22.

Because of non symmetry, there will be sway in the frame. Hence analysis has to be done in two steps.

(*i*) Non sway moment with support at *B*

(*ii*) Sway moment when support at *B* is removed.

Non Sway moment

1. F.E.M. $$M_{BC} = -\frac{10 \times 6^2}{12} = -30$$

$$M_{CB} = +\frac{10 \times 6^2}{12} = +30$$

2. *Stiffness*

$$k_{BA} = \frac{3 \times 2EI}{6} = EI$$

$$k_{BC} = \frac{4 \times 2EI}{6} = \frac{4}{3}EI$$

$$k_{CD} = \frac{3 \times 1.5EI}{4} = \frac{9}{8}EI$$

3. *Distribution factors*

Joint *B*

$$BA = \frac{1}{1+\frac{4}{3}} = 0.429$$

$$BC = \frac{\frac{4}{3}}{1+\frac{4}{3}} = 0.571$$

Total: 1.00

Joint *C*

$$CB = \frac{\frac{4}{3}}{\frac{4}{3}+\frac{9}{8}} = 0.542$$

$$CD = \frac{\frac{9}{8}}{\frac{4}{3}+\frac{9}{8}} = 0.458$$

Total: 1.00

Table 11.11. Example 11.21.

Joint	*A*	*B*		*C*		*D*
Member	*AB*	*BA*	*BC*	*CB*	*CD*	*DC*
Stiffness		*EI*	$\frac{4}{3}EI$	$\frac{4}{3}EI$	$\frac{9}{8}EI$	
C.O.F.	0 ←		→ ½ ½ ←			→ 0
D.F.		0.429	0.571	0.542	0.458	
FEM	0	–	−30	+30	–	0
Balance		+12.87	+17.13	−16.26	−13.74	
C.O.			−8.13	+8.56	–	
Balance		+3.49	+4.64	−4.64	−3.92	
C.O.			−2.32	+2.32		
Balance		+1.00	+1.32	−1.26	−1.06	
C.O.			−0.63	+0.66		
Balance		+0.27	+0.36	−0.36	−0·30	
C.O.			−0.18	+0.18		
Balance		+.08	−0.10	−0.10	−0.08	
C.O.			−0.05	+0.05		
Balance		+0.02	+0.03	−0.03	−0.02	
Final Moment	0.00	+17.73	−17.76	+19.12	−19.12	0.00

Check:

Slope at B for $BA = \frac{17.73 \times 6}{6\,EI} = \frac{17.73}{EI}$

Slope at B for $BC = \frac{17.73 \times 6}{6\,EI} - \frac{10 \times 6^3}{24 \times 2\,EI} + \frac{19.12 \times 6}{12\,EI} = \frac{-17.71}{EI}$

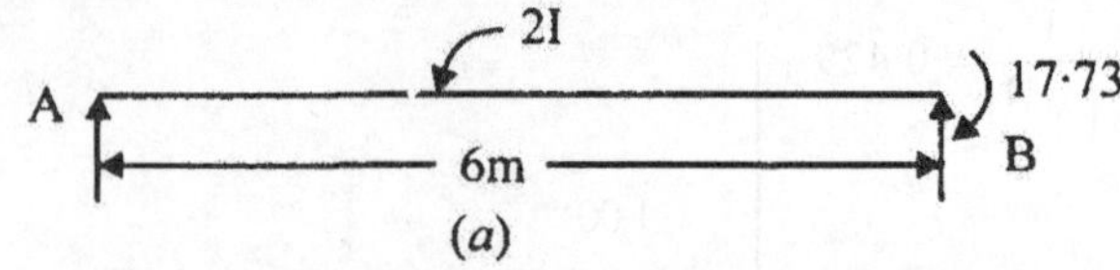

Fig. 11.62. Checking of Moment Distribution.

Hence moment distribution is O.K.

Calculation of horizontal forces at the bases.

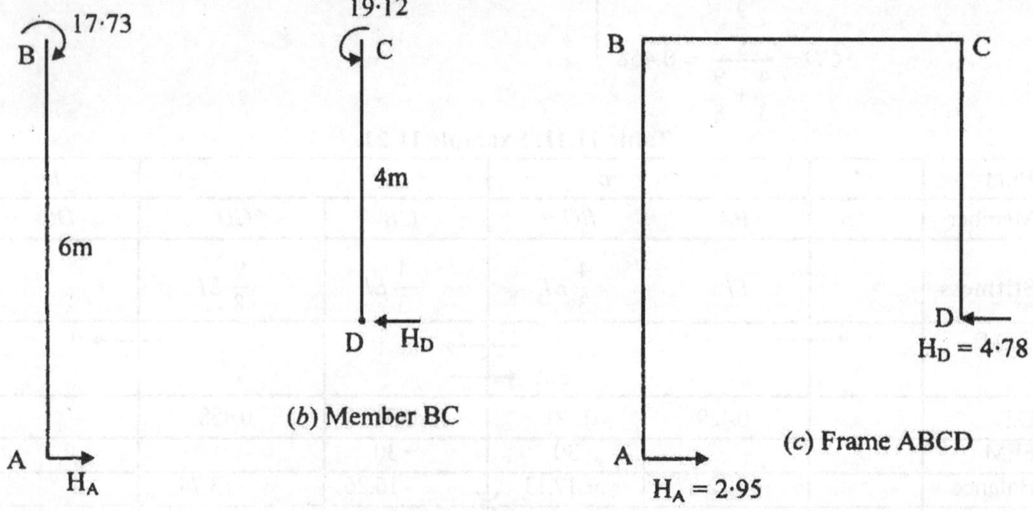

Fig. 11.63. Computing base horizontal forces.

$$H_A \times 6 - 17.73 = 0$$

or $$H_A = \frac{17.73}{6} = 2.95 \rightarrow$$

$$H_D = \frac{19.12}{4} = 4.78 \leftarrow$$

From equilibrium of the frame as a whole

$$R_B + 2.95 - 4.78 = 0$$

$$R_B = 1.83 \text{ kN} \rightarrow$$

To neutralise the effect of R_B a force equal to 1.83 kN should act at B from right to left.
From the information given 1 kN from left to right gives moments.

$M_{BC} = +2$ kNm and $M_{CB} = +2.66$ kN both clockwise.

Hence 1.83 kN from right to left will produce.

$$M_{BC} = -2 \times 1.83 = -3.66 \text{ kNm}$$

and

$$M_{CB} = -2.66 \times 1.83 = -4.87 \text{ kNm}$$

Hence final moments.

$M_{AB} = 0$

$$M_{BA} = +17.73 + 3.66 = 21.39$$
$$M_{BC} = -17.76 - 3.66 = -21.42$$
$$M_{CB} = +19.12 - 4.87 = +14.25$$
$$M_{CD} = -19.12 + 4.87 = -14.25$$
$$M_{DC} = 0$$

Fig. 11.64. Final B.M. Diagram

Example 11.23. A rigid jointed portal frame *ABCD* is loaded, dimensioned and supported as shown in Fig. 11.65. Determine the member end moments of all members of the frame by using moment distribution method (Compute for four cycles). Hence draw the B.M.D. on the tension side of the frame. Also calculate the support reaction components. **[AMIE Summer '95]**

Solution:

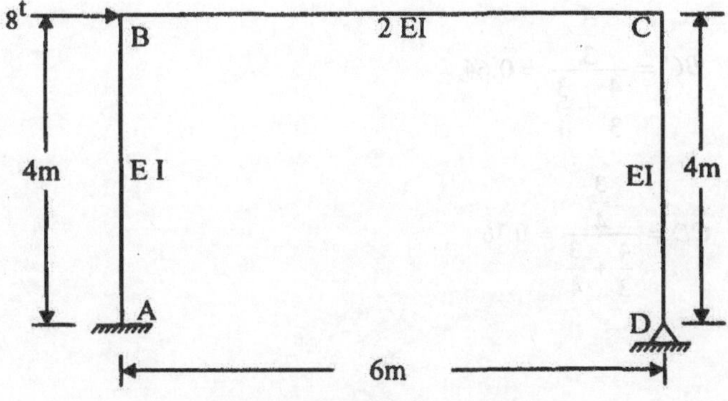

Fig. 11.65. Example 11.23.

As there is no transverse load in the frame and it is subjected to a horizontal load at B, there is only sway moment in the frame towards right. As the frame sways through an amount Δ, fixed end moment developed.

$$M_{FBA} = -\frac{6EI\Delta}{16}$$

$$M_{FAB} = -\frac{6EI\Delta}{16}$$

As the end D is hinged,

$$M_{FCD} = -\frac{3EI\Delta}{16}$$

$$M_{FDC} = 0$$

Hence $M_{FBA} : M_{FAB} : M_{FCD} : M_{FDC}$

$$= 2 : 2 : 1 : 0$$

Hence assume the following moments for indirect method of calculating the moments:

$$M_{FBA} = -20,\ M_{FAB} = -20,\ M_{FCB} = -10,\ M_{FDC} = 0$$

Stiffness Stiffness of $BA = \frac{4EI}{4} = EI$

Stiffness of $BC = \frac{8EI}{6} = \frac{4}{3}EI$

Stiffness of $CD = \frac{3EI}{4} = \frac{3}{4}EI$

Distribution Factors

Joint B

$$BA = \frac{1}{1+\frac{4}{3}} = 0.43$$

$$BC = \frac{\frac{4}{3}}{1+\frac{4}{3}} = 0.57$$

Joint C

$$BC = \frac{\frac{4}{3}}{\frac{4}{3}+\frac{3}{4}} = 0.64$$

$$CD = \frac{\frac{3}{4}}{\frac{4}{3}+\frac{3}{4}} = 0.36$$

Table 11.12. Sway Moment Distribution.
(Example 11.23)

Joint	A	B		C	D	
Member	*AB*	*BA*	*BC*	*CB*	*CD*	*DC*
Stiffness		*EI*	$\frac{4}{3}EI$	$\frac{4}{3}EI$	$\frac{3}{4}EI$	
D.F.		0.43	0.57	0.64	0.36	
C.O.F.	$\frac{1}{2}$ ←		← $\frac{1}{2}$	→ $\frac{1}{2}$		→ 0
F.E.M.	−20	−20			−10	0
Balance		+8.6	+11.40	+6.4	+3.60	
C.O.	+4.30		+3.20	+5.7		
Balance		−1.38	+1.82	−3.65	−2.05	
C.O.	−0.69		−1.83	−0.91		
Balance		+0.79	+1.04	+0.58	+0.33	
C.O.	+0.40		+0.29	+0.52		
Balance		−0.12	+0.17	+0.33	−0.19	
Final Moment	−15.99	−12.11	+12·11	+8.31	−8.31	0.00

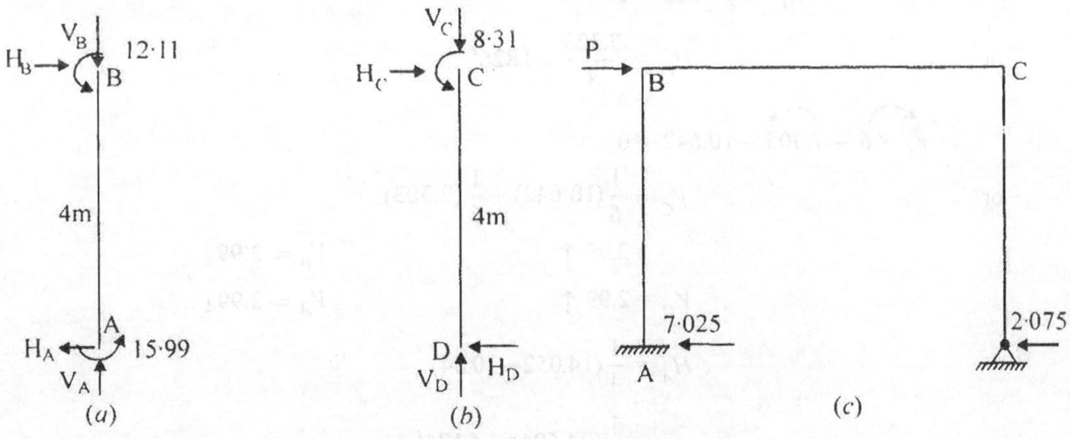

Fig. 11.66. Calculation of *P*.

$$H_A = \frac{12.11 + 15.99}{4} = \frac{28.10}{4} = 7.025' \text{ (Fig. 11.66}a)$$

$$H_D = \frac{8.31}{4} = 2.0775' \text{ (Fig. 11.66}b\text{)}$$

Total $H = 9.1025$

But actual force at $B = 8'$

Hence the reducing factor $= \dfrac{8}{9.1025} = 0.8788$

Actual moments

$$M_{AB} = -15.99 \times 0.8788 = -14.052$$

$$M_{BA} = -M_{BC} = -12.11 \times 0.8788 = -10.642$$

$$M_{CB} = -M_{CD} = 8.31 \times 0.8788 = +7.303$$

$$M_{DC} = 0$$

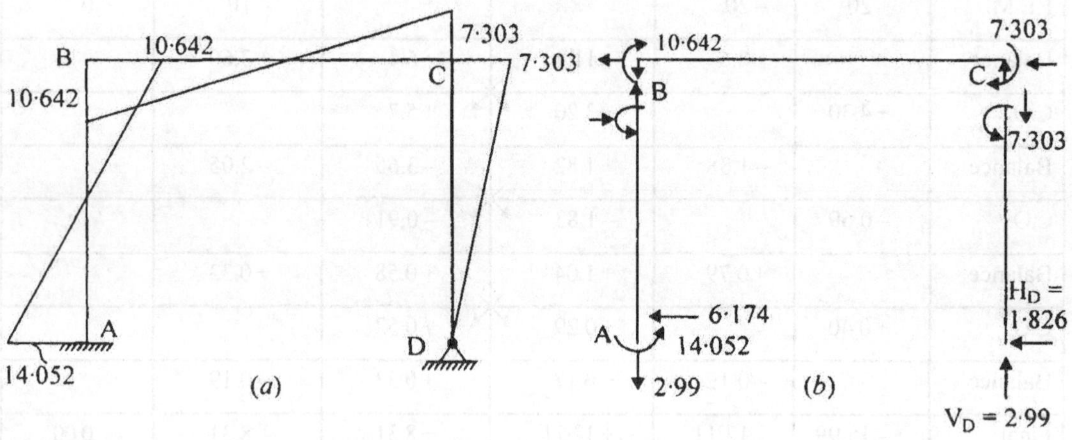

Actual Reactions

$$H_D \times 4 - 7.303 = 0$$

or $$H_D = \frac{7.303}{4} = 1.826'$$

$$V_C \times 6 - 7.303 - 10.642 = 0$$

or $$V_C = \frac{1}{6}(10.642) + \frac{1}{6}(7.303)$$

$$= 2.99 \uparrow \qquad V_B = 2.99 \downarrow$$

$$V_D = 2.99 \uparrow \qquad V_A = 2.99 \downarrow$$

$$H_A = \frac{1}{4}(14.052 + 10.64)$$

$$= \frac{1}{4}(24.694) = 6.174' \leftarrow$$

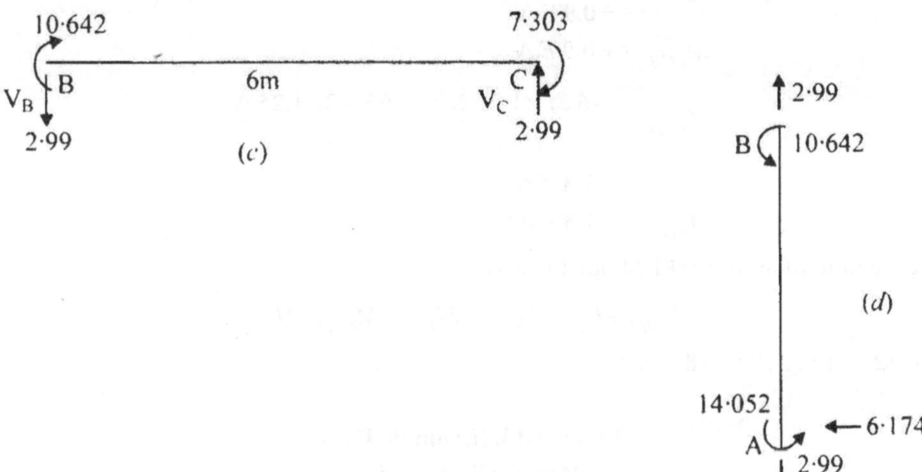

Fig. 11.67. Calculation of Reactions.

Example 11.24. Determine the moments in the rigid frame shown in Fig. 11.68.

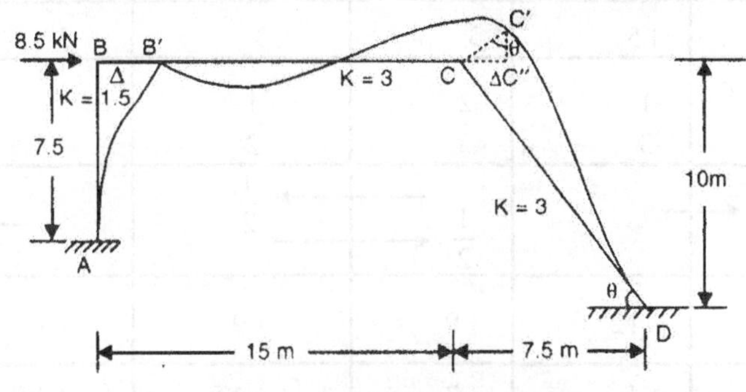

Fig. 11.68. Example 11.24

Solution: The deflected shape of the frame is shown in Fig. 11.68. From geometry it is clear that when B moves to B' through a distance Δ, $CC' \sin\theta = \Delta$ as there is no axial strain in member BC.

Hence, $$CC' = \frac{\Delta}{\sin\theta} = \frac{\Delta}{\dfrac{10}{\sqrt{10^2 + 7.5^2}}} = 1.25\Delta$$

and $$C'C'' = 1 \cdot 25\Delta \cdot \frac{7 \cdot 5}{12 \cdot 5} = 0 \cdot 75\,\Delta$$

Therefore, FEMs generated due to arbitrary sway Δ

$$M_{FAB} = -\frac{6\,EI\,\Delta}{L^2} = -\frac{6E\,.1.5.\Delta}{7.5} = -1.2\,E\,\Delta$$

$$M_{FBA} = -1.2\,E\,\Delta$$

$$M_{FBC} = +\frac{6\,EI\,.0.75\,\Delta}{L^2} = +\frac{6E \times 3 \times 0.75\,\Delta}{15}$$

$$= +0.9\,E\,\Delta$$

$$M_{FCB} = +0.9\,E\,\Delta$$

$$M_{FCD} = \frac{-6\,EI \times 1.25\,\Delta}{L^2} = -\frac{6E \times 3 \times 1.25\,\Delta}{12.5}$$

$$= +1.8\,E\,\Delta$$

$$M_{FDC} = -1.8\,E\,\Delta$$

Hence ratio of arbitrary FEM due to sway

$$M_{FAB} : M_{FBA} : M_{FBC} : M_{FCB} : M_{FCD} : M_{FDC}$$

$$= -12 : -12 : 9 : 9 : -18 : -18$$

Table 11.13. (Example 11.24)
Moment Distribution

Joint	*A*	*B*		*C*		*D*
Member	*AB*	*BA*	*BC*	*CB*	*CD*	*DC*
Stiffness		1.5	3	3	3	
D.F.		$\frac{1}{3}$	$\frac{2}{3}$	$\frac{1}{2}$		$\frac{1}{2}$
C.O.F.	$\frac{1}{2}$ ←		$\frac{1}{2}$ ←	→ $\frac{1}{2}$	→	$\frac{1}{2}$
F.E.M.	−12	−12	+9	+9	−18	−18
Balance		+1	+2	+4.5	+4·5	
C.O.	+0.5		+2·25	+1·0		2·25
Balance		−0.75	−1.50	−0.5	−0.5	
C.O.	−0.37		−0.25	−0.75		−0.25
Balance		+0.08	+0.16	+0.37	+0.37	
C.O.	+0.04		+0.19	+0.08		+0.19
Balance		−0.06	−0.12	−0.04	−0.04	
Moments	−11.83	−11.73	+11.73	+13.66	−13.67	−15.81
× 1.15	−13.71	−13.60	+13.60	+15.83	−15.83	−18.83

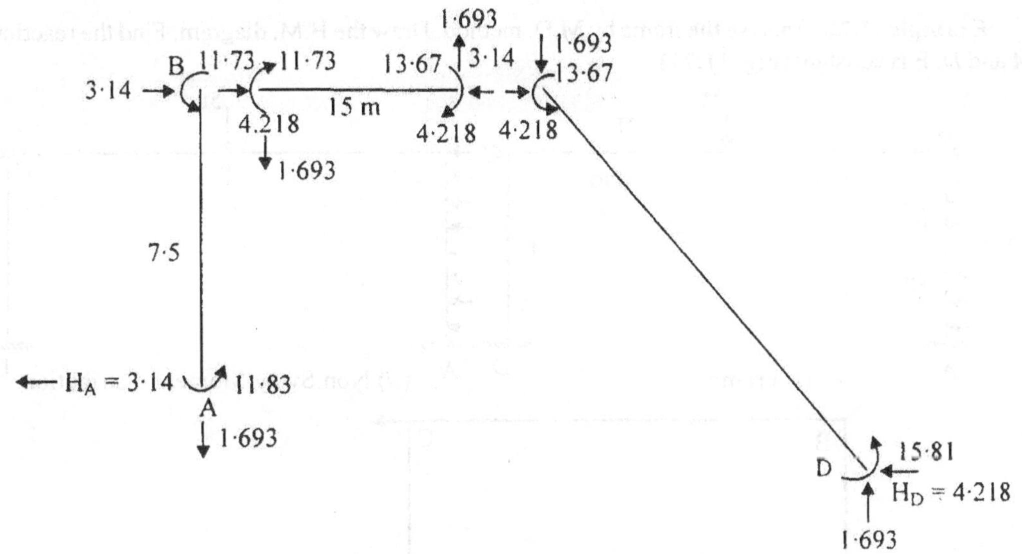

Fig. 11.69. Free body diagrams of the members of the frame.

$$H_A \times 7.5 = 11.83 + 11.73 = 23.56$$

$$H_A = \frac{23.56}{7.5} = 3.14$$

$$V_B \times 15 + 11.73 + 13.67 = 0$$

$$V_B = -\frac{25.40}{15} = -1.693$$

$$H_D \times 10 - 13.67 - 15.81 - 1.693 \times 7.5 = 0$$

$$H_D = 4.22$$

Total $\qquad H = 3.14 + 4.22 = 7.36$

Load applied = 8.5 kN. ratio = $\frac{8.5}{7.36}$

$= 1.15.$

Final moments are given in the last row of Table 11.13 B.M. diagram is given in fig. 11.70.

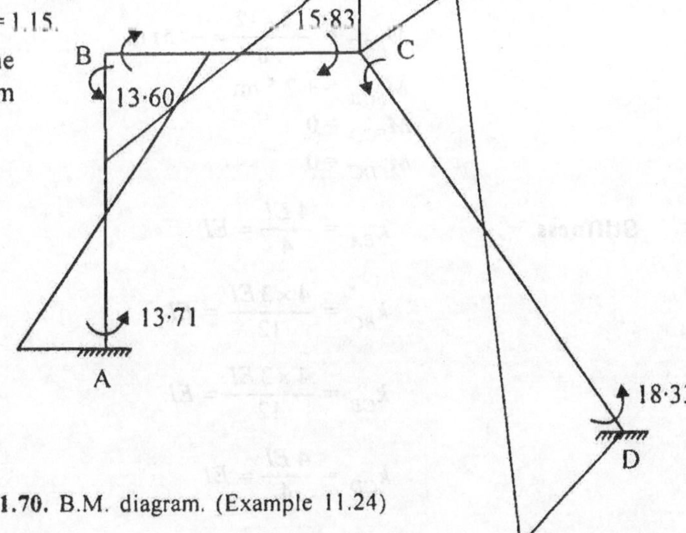

Fig. 11.70. B.M. diagram. (Example 11.24)

Example 11.25. Analyse the frame by M.D. method. Draw the B.M. diagram. Find the reactions at *A* and *D*. E is constant (Fig. 11.71)

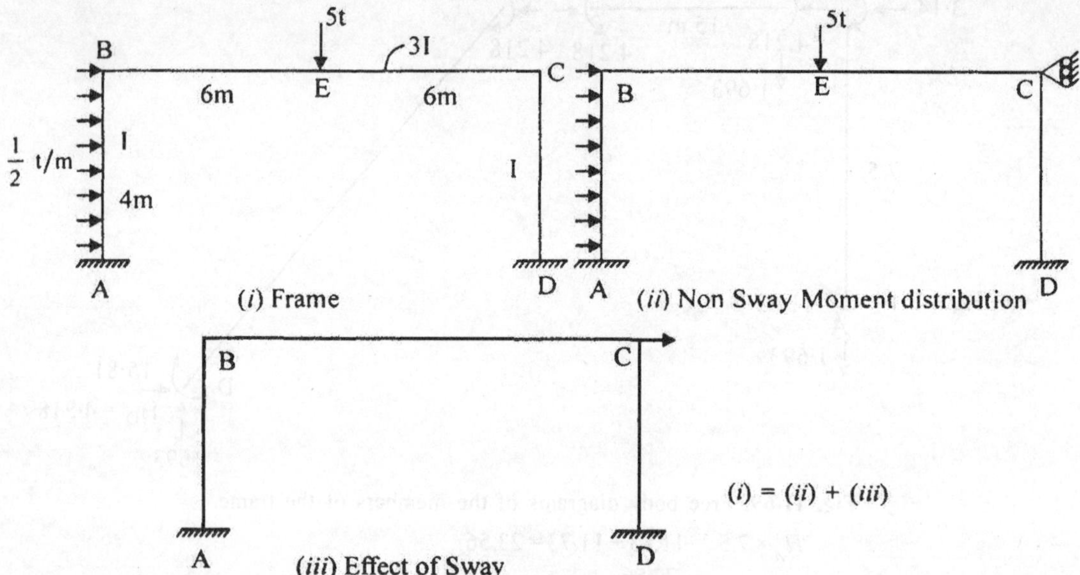

Fig. 11.71. Example 11.25

Solution: The moment distribution will be carried out in two parts.

(*i*) Non Sway moment distribution with support at *C*.

(*ii*) Sway moment distribution with support at *C* removed.

FEMs

$$M_{FAB} = \frac{-\frac{1}{2}\times 4^2}{12} = -\frac{2}{3}\,tm$$

$$M_{FBA} = +\frac{2}{3}\,tm$$

$$M_{FBC} = -\frac{5\times 12}{8} = -7.5\,t\,m$$

$$M_{FCB} = +7.5\,tm$$

$$M_{FCD} = 0$$

$$M_{FDC} = 0$$

Stiffness

$$k_{BA} = \frac{4\,EI}{4} = EI$$

$$k_{BC} = \frac{4\times 3\,EI}{12} = EI$$

$$k_{CB} = \frac{4\times 3\,EI}{12} = EI$$

$$k_{CD} = \frac{4\,EI}{4} = EI$$

Distribution Factors. Joint B $BA = \frac{1}{2}$

$BC = \frac{1}{2}$

Joint C $CB = \frac{1}{2}$

$CD = \frac{1}{2}$

Table 11.14 M.D. (Example 11.25)

Joint	A	B		C		D
Member	AB	BA	BC	CB	CD	DC
D.F.		$\frac{1}{2}$	$\frac{1}{2}$	$\frac{1}{2}$	$\frac{1}{2}$	
C.O.	$\frac{1}{2}$ ←		→ $\frac{1}{2}$ ←	$\frac{1}{2}$	→	$\frac{1}{2}$
F.E.M.	−0.667	+0.667	−7.50	+7.50	0.00	0.00
Balance		+3.420	+3.420	−3.750	−3.750	
C.O.	+1.710		−1.875	+1.710		−1.875
Balance		+0.937	+0.937	−0.855	−0.855	
C.O.	+0.468		−0.427	+0.468		−0.427
Balance		+0.213	+0.213	−0.234	−0.234	
C.O.	+0.106		−0.117	+0.106		−0.117
Balance		+0.058	+0.058	−0.053	−0.053	
C.O.	+0.029		−0.027	+0.029		−0.027
Balance		+0.013	+0.013	−.015	−.015	
Final Moment	+1.646	+5.308	−5.305	+4.906	−4.907	−2.446

Sway Moments

$$M_{AB} : M_{BA} : M_{CD} : M_{DC} = 1 : 1 : 1 : 1$$

Hence trial sway moments

$$M_{AB} = -10 \text{ tm}$$
$$M_{BA} = -10 \text{ tm}$$
$$M_{CD} = -10 \text{ tm}$$
$$M_{DC} = -10 \text{ tm}$$

Table 11.15 shows moment distribution of arbitrary sway moments.

Sway Force

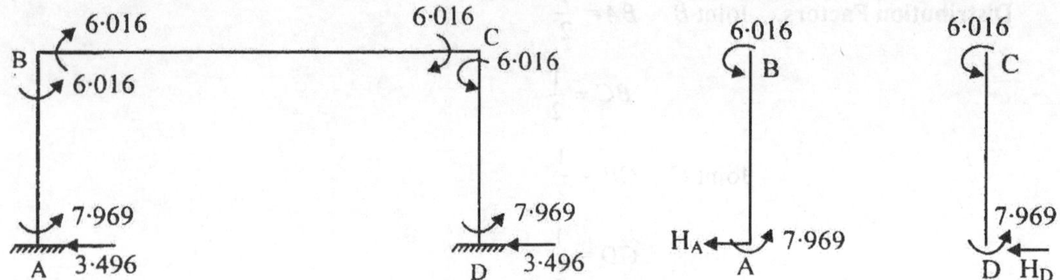

Fig. 11.72. Calculation of Sway Force.

$$H_A = H_D = \frac{7.969 + 6.016}{4} = \frac{13.985}{4} = 3.496'$$

Total sway force = 3.496 × 2 = **6.992′**

Table 11.15 Sway M.D. (Example 11.25)

Joint	A	B		C		D
Member	AB	BA	BC	CB	CD	DC
D.F.		$\frac{1}{2}$	$\frac{1}{2}$	$\frac{1}{2}$	$\frac{1}{2}$	
C.O.	←	$\frac{1}{2}$	$\frac{1}{2}$ ⇄	$\frac{1}{2}$	→	$\frac{1}{2}$
F.E.M.	− 10.00	− 10.00			− 10.00	− 10.00
Balance		+ 5.00	+ 5.00	+ 5.00	+ 5.00	
C.O.	2.50		+ 2.50	+ 2.50		+ 2.50
Balance		− 1.25	− 1.25	− 1.25	− 1.25	
C.O.	−0.625		−0.625	−0.625		−0.625
Balance		+0.312	+0.312	+0.312	+0.312	
C.O.	+0.156		+0.156	+0.156		+0.156
Balance		−0.078	−0.078	−0.078	−0.078	
Final Moment	−7.969	−6.016	+6.015	+6.015	−6.016	−7.969

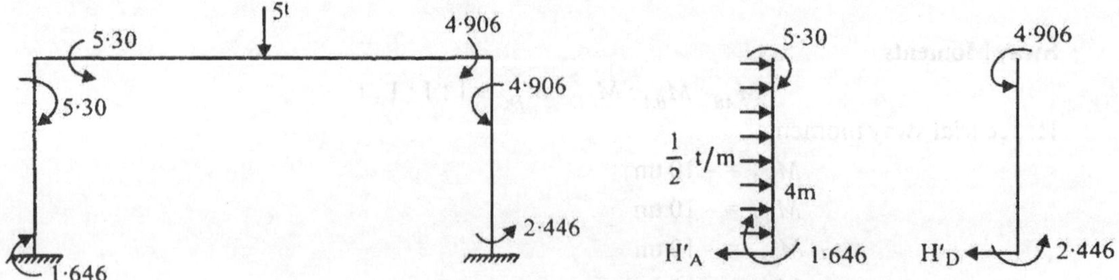

Fig. 11.73. Calculation of Non Sway Force. (Example 11.25)

$$H'_A \times 4 + 1.65 + 5.301 - \frac{1}{2} \times 4 \times 2 = 0$$

$$H'_A \times 4 = 4 - 1.65 - 5.301$$

$$H'_A = \frac{-2.951}{4} = -0.738 \rightarrow$$

$$H'_D = \frac{1}{4}(4.906 + 2.446) = 1.838 \leftarrow$$

Net $H' = 2 + 0.738 - 1.838$

$= 0.9$

Hence Sway Force Provided by

$$R_C = 0.9 \leftarrow$$

But as there is no support at C a sway force should be applied at C of magnitude 0.897 $\rightarrow$ in the opposite direction.

$$\text{Ratio} = \frac{0.9}{6.992} = 0.12872$$

Final moments are computed as follows.

Table 11.16 : Calculation of Final Moments

Joint	*A*	*B*		*C*		*D*
(1) Arbitrary S. Moment	−7.9690	−6.016	+6.016	+6.016	−6.016	−7.969
Sway Moment	−1.0223	−0.7718	+0.7718	+0.7718	−0.7718	−1.0223
Non Sway Moment	+1.6430	+5.3010	−5.3010	+4.9060	−4.9060	−2.446
Final Moment	0.6277	+4.5292	−4.5292	+5.6778	−5.6778	−3.4683

Sway Moment = (1) × (0.128)

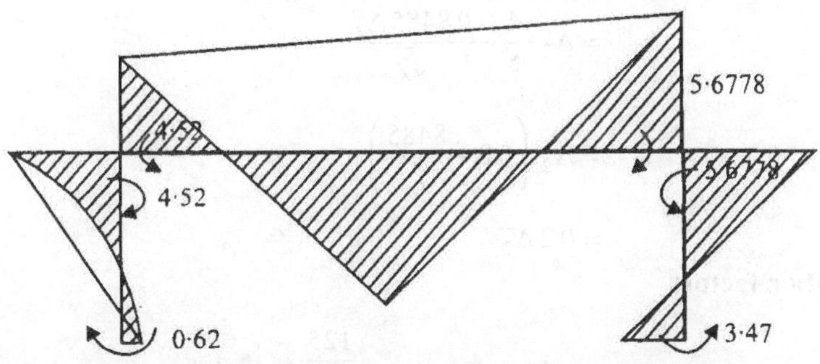

Fig. 11.74. B.M. Diagram (Plotted on the tension side).

Example 11.26. The rigid frame has $I_{AB} = 125 \times 10^{-6}$ m^4 $I_{BC} = 100 \times 10^{-6}$ m^4 and $I_{CD} = 141 \times 10^{-6}$ m^4. Calculate the end moments if (*a*) the structure is loaded as shown and (*b*) the load illustrated is removed and 100 kN force is placed 1 m to the left of joint C acting downward.

Solution

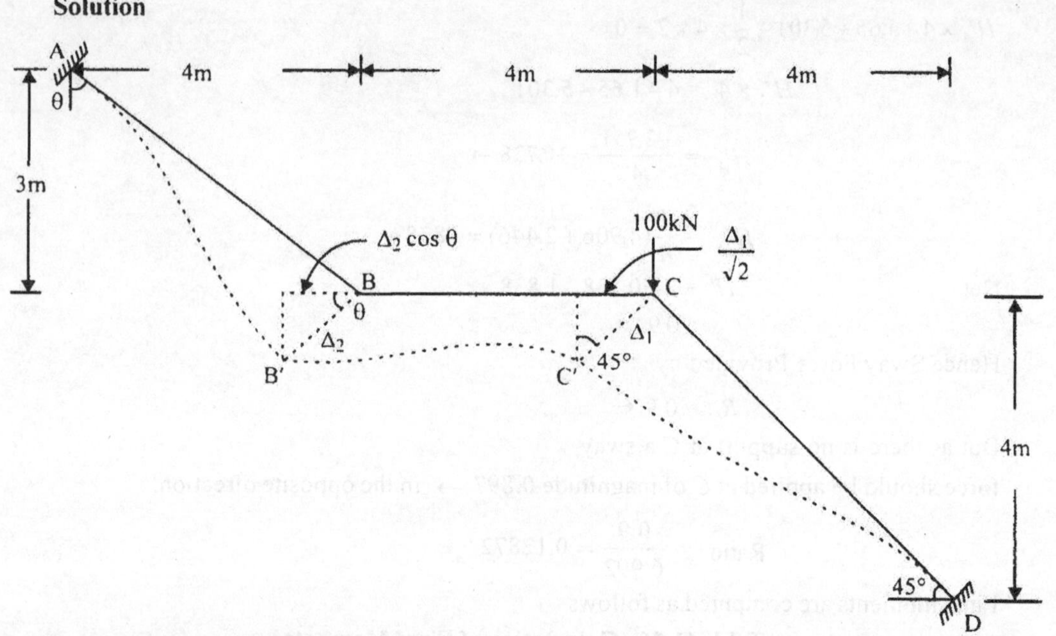

Fig. 11.75. Ex. 11.26.

1. Relative displacement of Joints

For no axial strain

$$0.6\,\Delta_2 = \frac{\Delta_1}{1.414} \quad \text{or} \quad \Delta_1 = 0.6\sqrt{2}\,\Delta_2.$$

Displacement between A and $B = \Delta_2$

$$C' \text{ and } D' = 0.6\sqrt{2}\,\Delta_2$$

$$B' \text{ and } C' = \Delta_2 \sin\theta - \frac{\Delta_1}{\sqrt{2}}$$

$$= \Delta_2 \cdot \frac{4}{5} - \frac{0.8485\,\Delta_2}{\sqrt{2}}$$

$$= \Delta_2 \left(0.8 - \frac{8485}{\sqrt{2}}\right)$$

$$= 0.2\,\Delta_2$$

2. Distribution factors

$$\text{Joint } B \quad BA \quad \frac{I_{AB}/5}{I_{AB}/5 + \frac{I_{BC}}{4}} = \frac{\frac{125}{5}}{\frac{125}{5} + \frac{100}{4}} = \frac{1}{2}$$

$$BC \quad \frac{I_{BC}/4}{I_{AB}/5 + \frac{I_{BC}}{4}} = \frac{1}{2}$$

Joint C

$$CB \qquad \frac{I_{CB}/4}{I_{CB}/4 + I_{CD}/4\sqrt{2}} = \frac{\frac{100}{4}}{\frac{100}{4} + \frac{141}{4\sqrt{2}}} = \frac{1}{2}$$

$$CD \qquad \frac{I/4\sqrt{2}}{I_{CB}/4 + I_{CD}/4\sqrt{2}} = \frac{1}{2}$$

3. F.E.M. due to sway

$$M_{FAB} = M_{FBA} = \frac{6EI\Delta_2}{5^2} = -\frac{6 \times 125}{25} E \cdot \Delta_2 \times 10^{-6}$$

$$M_{FBC} = M_{FCB} = \frac{6E\, I_{BC}\, 0.2\Delta}{16} = +\frac{12 \times 100 \times 10^{-6}}{16} E\Delta_2$$

$$M_{FCD} = M_{FDC} = \frac{6EI_{CD}.0.8485\Delta_2}{32} = \frac{6 \times 141 \times 10^{-6}}{32} E \times 0.8485\Delta_2$$

$$M_{FAB} = M_{FBA} : M_{FBC} : M_{FCB} : M_{FCD} : M_{FDC}$$
$$= -30 : -30 : 7.5 : 7.5 : 22.5 : 22.5$$

Table 11.17 Moments generated due to arbitrary sway.

Joint	A	B		C		D
Member	AB	BA	BC	CB	CD	DC
D.F.	–	0.5	0.5	0.5	0.5	
C.O.	←	$\frac{1}{2}$	→ $\frac{1}{2}$ ←	$\frac{1}{2}$	→	$\frac{1}{2}$
F.E.M.	–30	–30	+7.5	+7.5	+22.50	+22.50
Balance		+11.25	+11.25	–15.00	–15.00	
C.O.	5.625		–7.50	+5.625		–7.50
Balance		+3.75	+3.75	–2.812	–2.812	
C.O.	+1.875		–1.40	+1.875		–1.406
Balance		+0.70	+0.70	–0.937	–0.937	
C.O.	+0.35		–0.46	+0.35		–0.47
Balance		+0.23	+0.23	–0.175	–0.175	
C.O.	+0.12		–0.09	+0.12		–0.087
Balance		+0.05	+0.05	+0.06	+0.06	
Final Moment	–22.03	–14.02	+14.03	–3.51	+3.51	+13.04

Table 11.18. Final Moments for case (a)

Joint	A	B		C		D	Remarks
Member	AB	BA	BC	CB	CD	DC	
	–22.03	–14.02	+14.02	–3.51	+3.51	+13.04	Load at C 17.04
$\times \frac{100}{17.04}$	–129.28	–82.28	+82.28	–20.60	+20.60	+76.53	Load at C 100

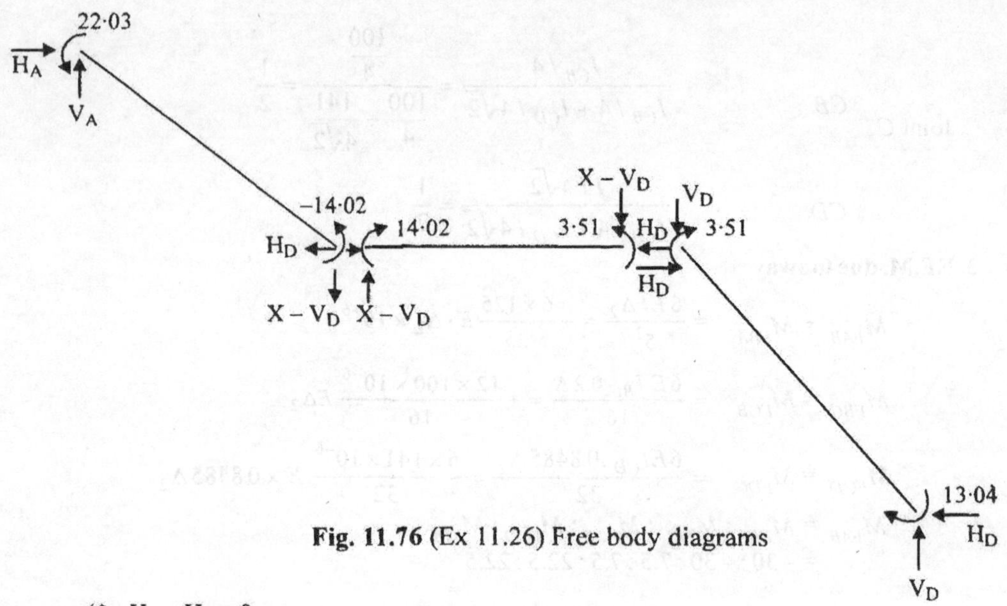

Fig. 11.76 (Ex 11.26) Free body diagrams

(*i*) $H_A - H_D = 0$

(*ii*) $V_A + V_D = X$

(*iii*) $H_D \times 4 - V_D \times 4 + 3.51 + 13.04 = 0$

(*iv*) $(X - V_D) \times 4 + 14.02 - 3.51 = 0$

(*v*) $(X - V_D) \times 4 + H_D \times 3 - 22.03 - 14.02 = 0$

From (*v*) $(X - V_D) \times 4 = 22.03 + 14.02 - H_D \times 3$

From (*iv*) $(X - V_D) \times 4 = -10.51$

or $-10.51 = 22.03 + 14.02 + H_D \times 3$

or $H_D \times 3 = 46.56$

$H_D = 15.52$

From (*iii*) $V_D \times 4 = H_D \times 4 + 3.51 + 13.04$

$= 15.52 \times 4 + 3.51 + 13.04$

or $V_D = \frac{1}{4}\,[78.66]$

$= 19.66$

$(X - V_D) \times 4 = -10.51$

or $X = \frac{-10.51}{4} + V_D$

$X = \frac{-10.51}{4} + 19.66$

$= 17.04$ kN.

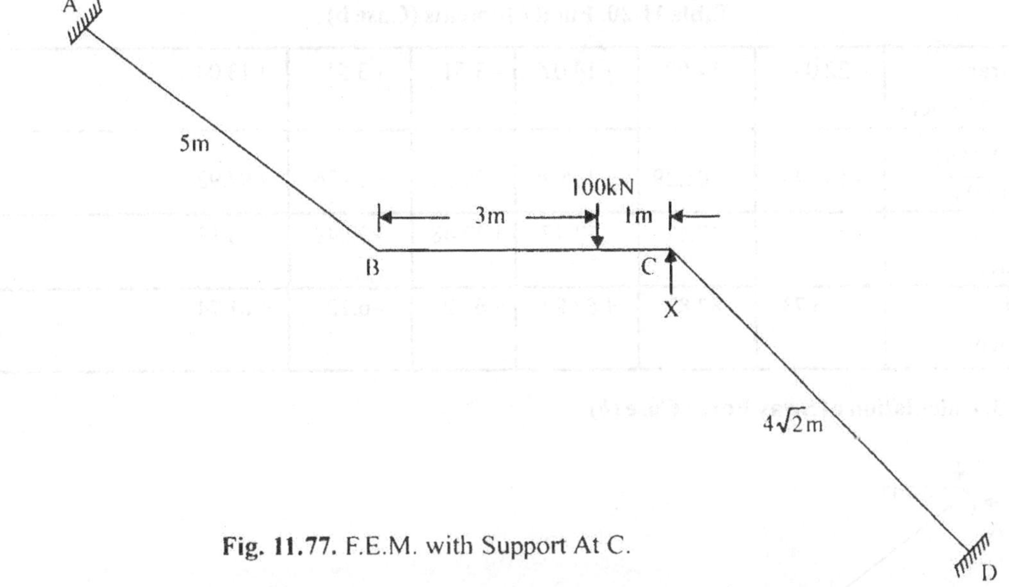

Fig. 11.77. F.E.M. with Support At C.

Table 11.19. Non Sway Moment Distribution (Case *b*).

Joint	*A*	*B*		*C*		*D*	Remarks
Member	*AB*	*BA*	*BC*	*CB*	*CD*	*DC*	
D.F.		0.5	0.5	0.5	0.5		
C.O.F.	← $\frac{1}{2}$		$\frac{1}{2}$ ←	→ $\frac{1}{2}$		→ $\frac{1}{2}$	
F.E.M.			− 18.75	+ 56.25			
Balance		+9.375	+9.375	−28.125	−28.125		
C.O.	+4.688		− 14.062	+4.688		−14.062	
Balance		+7.031	+7.031	−2.344	−2.344		
C.O.	3.516		− 1.172	+3.516		− 1.172	
Balance		+0.586	+0.586	− 1.758	− 1.758		
C.O.	0.283		−0.879	+0.283		−0.879	
Balance		+0.439	+0.439	−0.141	−0.141		
C.O.	+0.220		−0.070	+0.220		−0.070	
Balance		+0.035	+0.035	−0.110	−0.110		
Non Sway Final Moment	+ 8.707	+ 17.466	− 17.467	+32.479	−32.478	− 16.183	

Table 11.20. Final Moments (Case b)

Arbitrary Sway Moment	−22.03	−14.02	+14.02	−3.51	+3.51	+13.04
$\times \frac{127.9655}{17.04}$	−165.44	−105.29	+105.29	−26.36	+26.36	+97.92
Non Sway Moment	+8.71	+17.47	−17.47	+32.48	−32.48	−16.18
Final Moment	−156.73	−87.82	+87.82	+6.12	−6.12	+81.74

5. Calculation of Sway Force Case (*b*)

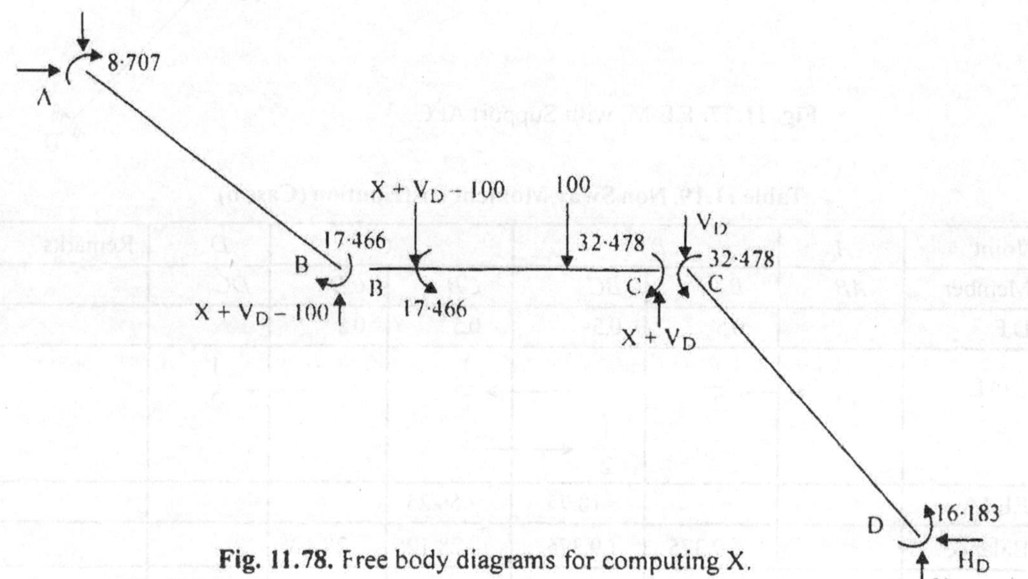

Fig. 11.78. Free body diagrams for computing X.

Equations available

(*i*) $V_D \times 4 + 16.183 + 32.478 - H_D \times 4 = 0.$

(*ii*) $(X + V_D) \times 4 - 32.478 + 17.466 - 100 \times 3 = 0.$

(*iii*) $(X + V_D - 100) \times 4 - 17.466 - 8.707 - H_D \times 3 = 0.$

or $(X + V_D) \times 4 = 315.013$ (from (*ii*))

or $X + V_D = 78.75$

$H_D \times 3 = -17.466 - 8.707 + (78.75 - 100) \times 4 = -111.16$ (from (*iii*))

or $H_D = \frac{1}{3}[-111.16] = -37.05$

$V_D = \frac{1}{4}[-37.05 \times 4 - 16.183 - 32.478]$

$= -49.22$

or $X = 78.75 + 49.22$

$= 127.97$

Final moments are calculated in Table 11.20.

11.13. KANI'S METHOD

Like moment distribution Kani's method is an iterative solution of the slope deflection equations. The iteration follows Gauss Seidel iteration technique. However, there is one difference. In the moment distribution method, increments to the member end moments are iterated. But Kani's method iterates the member end moments themselves. The method was developed by Dr Gasper Kani in 1947. The following are the advantages of Kani's method

(*i*) The method is very rapid.

(*ii*) The method is error eliminating. Even if an error is committed in the calculation it gets automatically corrected in the course of further work.

(*iii*) Sway calculations and non sway calculations are carried out in a single continuous calculation operation which results in considerable saving of time and effort.

(*iv*) Even if flexural rigidities of some members are altered after some steps in calculation (as is the practice in design office) current values of rotation contributions and displacement contributions can be taken as initial values with new rigidities for further calculations. This is not possible in any other method where a completely new analysis has to be made with new flexural rigidities.

(*v*) Checking of results in Kani's method is easy as checking can be confined to final values of rotation contributions and displacement contributions. All rounds of calculations need not be checked.

11.13.1. Derivation of Rotation Contribution Method

The general slope deflection equations derived in section 11.3 are

$$M_{AB} = M_{FAB} + \frac{2EI}{L}\left(2\theta_A + \theta_B + \frac{3\Delta}{L}\right) \quad ...(11.5)$$

and

$$M_{BA} = M_{FBA} + \frac{2EI}{L}\left(2\theta_B + \theta_A + \frac{3\Delta}{L}\right) \quad ...(11.6)$$

The sign conventions adopted are:

(1) Clockwise moment and clockwise rotations at the ends of members are positive.

(2) Anti-clockwise moment and anticlockwise rotations are negative.

(3) Displacement of right support upward is taken as positive.

Removing the brackets

$$M_{AB} = M_{FAB} + 2\left(\frac{2EI}{L}\cdot\theta_A\right) + \frac{2EI}{L}\cdot\theta_B + \frac{6EI\Delta}{L^2} \quad ...(11.5a)$$

and a similar expression for M_{BA}.

In Kani's method the terms are defined as

M_{FAB} = Fixed End Moment at end A.

$\frac{2EI}{L}\cdot\theta_A$ = rotation contribution at end A for member AB and denoted by M'_{AB}.

$\frac{2EI}{L}\cdot\theta_B$ = rotation contribution at end B and denoted by M'_{BA}.

$\frac{6EI\Delta}{L^2}$ = displacement contribution at end A and denoted by M''_{AB}.

E, I, L correspond to member AB

In the member AB, A is the near end and B is the far end. The equation for M_{AB} may now be written as

$$M_{AB} = M_{FAB} + 2M'_{AB} + M'_{BA} + M''_{AB} \quad \text{...(11.8)}$$

In words, the moment at the near end of a member is given by the algebraic sum of

(*i*) The fixed end moment at the near end A corresponding to the actual loading on AB.

(*ii*) Twice the rotation contribution at near end.

(*iii*) The rotation contribution at far end.

(*iv*) The displacement contribution.

If a number of members meet at joint A from equilibrium of joint.

$$\Sigma M_{AB} = 0$$

Substituting the value of M_{AB} from equation 11.8

$$\Sigma M_{FAB} + 2\,\Sigma M'_{AB} + \Sigma M'_{BA} + \Sigma M''_{AB} = 0 \quad \text{...(11.10)}$$

or $$\Sigma M'_{AB} = -\frac{1}{2}(\Sigma M_{FAB} + \Sigma M'_{BA} + \Sigma M''_{AB})$$

where $\Sigma M'_{AB}$ = algebraic sum of rotation contributions at A for all the members meeting at A.

ΣM_{FAB} = algebraic sum of fixed end moments at A for all the member meeting at A.

$\Sigma M'_{BA}$ = algebraic sum of rotation contributions at all the joints which are "far" from the "near" joint A.

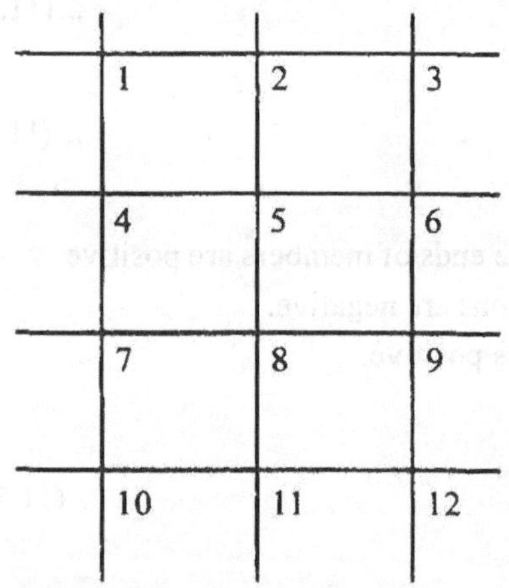

Fig. 11.79. Equilibrium of a joint

Suppose we are considering joint 5 (Fig. 11.79). The members meeting at this joint are

5 – 2, 5 – 6, 5 – 8 and 5 – 4.

Joint 5 is the near end and 2, 4, 8, 6 are far ends for members meeting at joint 5.

$\Sigma M'_{AB}$ is the algebraic sum of M'_{52}, M'_{56}, M'_{58} and M'_{54}.

ΣM_{FAB} is the algebraic sum of fixed end moments M_{F52}, M_{F56}, M_{F58} and M_{F54}.

$\Sigma M'_{BA}$ is the algebraic sum of M'_{25}, M'_{65}, M'_{85} and M'_{45}

Equation 11.10 gives the sum of rotation contributions at near end. From this sum $(\Sigma M'_{AB})$, the individual rotation contributions of the members at A are determined as follows:

By definition

$$M'_{AB} = \frac{2\,EI}{L}.\theta_A$$

As the members are rigidly connected, all members connected at A rotate through the same angle. As E is assumed to be the same for all the members, M'_{AB} is proportional to $\frac{I}{L} = k$.

Hence $$\frac{M'_{AB}}{\Sigma M'_{AB}} = \frac{k}{\Sigma k}$$

or $$M'_{AB} = \frac{k}{\Sigma k,} \cdot \Sigma M'_{AB}$$

$$= \frac{k}{\Sigma k}\left(-\frac{1}{2}\right)(\Sigma M_{FAB} + \Sigma M'_{BA} + \Sigma M''_{AB}).$$

The factor $\frac{k}{\Sigma k} \cdot \left(-\frac{1}{2}\right)$ is termed as rotation factor (R.F.). Thus rotation factor is obtained by distributing $\left(-\frac{1}{2}\right)$ among the members in proportion to their k values.

The sum of the rotation factors of members meeting at a joint is equal to $-\frac{1}{2}$. [This can be compared to distribution factors in moment distribution method which is also proportional to stiffness of the members $\frac{I}{L}$ but their sum is equal to 1.]

The above result for the individual rotation contribution may be re-written as.

$$M'_{AB} = R.F.(\Sigma M_{FAB} + \Sigma M'_{BA} + \Sigma M''_{AB}) \quad ...(11.11)$$

Trial values are initially assumed for the rotation contributions at far ends and also for displacement contributions and equation 11.11. is applied to "near" joint one at a time.

11.13.2. End Conditions

(1) When one end is fixed, the end rotation is zero so that rotation contribution $\left(\frac{2EI}{L} \cdot \theta\right)$ at the end is zero.

(2) When an end is pinned or hinged to a support : In such a case like moment distribution modified stiffness has to be used. Rotation at end B with far end A hinged $= \frac{3EI}{L} = 3E.K.$

with far end fixed, rotation $= \frac{4EI}{L} = 4EK'$

For the replacement to be justified.

$$4EK' = 3EK$$

or $$K' = \frac{3}{4}K$$

This is shown graphically in 11.80.

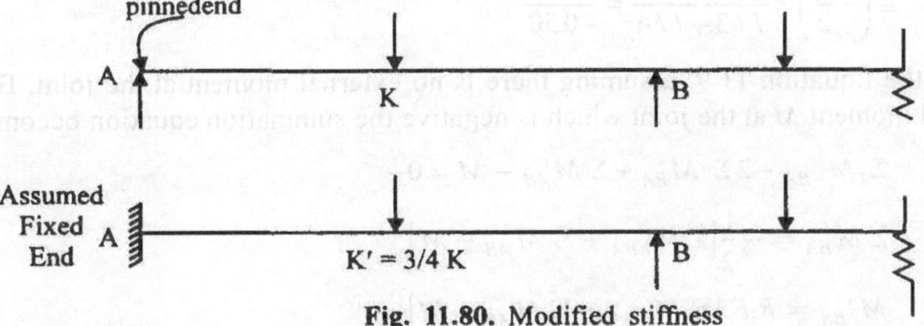

Fig. 11.80. Modified stiffness

The end A which is actually pinned will have zero moment. Hence no calculation is required at this joint. As k has been replaced by 3/4 k for the computation for the rest of the beam, end A will be assumed to be fixed and hence rotation contribution will be taken as zero.

3. When there is an over hang, stiffness k will be taken as zero. The over hang may be considered as supported at infinite distance ie $L = \infty$. Hence $\frac{I}{L} = 0$.

Example 11.27 — Find moment at supports of the continuous beam shown in Fig. 11.81.

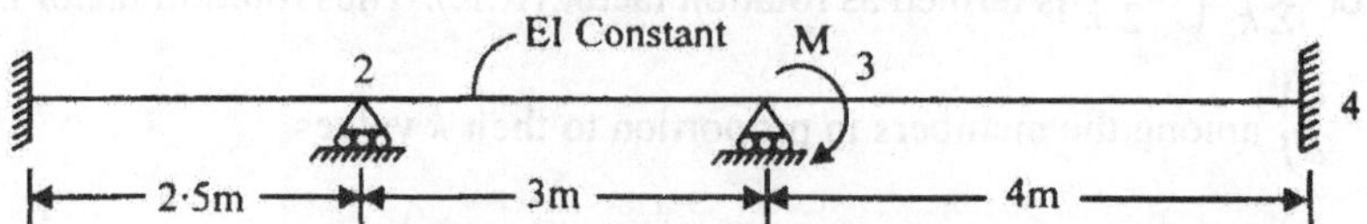

Fig. 11.81. Example 11.27.

Solution: As there is no load in the spans FEM'S are zero. The rotation factors at joints 2 and 3 only are to be determined as rotations are possible only at these joints.

For member 2 – 1 stiffness $= \frac{I}{2.5}$

2 – 3 stiffness $= \frac{I}{3}$

3 – 4 stiffness $= \frac{I}{4}$

Rotation factors at 2.

$$\text{R.F. of } 2-1 = \left(-\frac{1}{2}\right) \times \frac{I/2.5}{I/2.5 + I/3}$$

$$= -0.27$$

$$\text{R.F. of } 2-3 = \left(-\frac{1}{2}\right) \times \frac{I/3}{I/2.5 + I/3}$$

$$= -0.23$$

$$-0.27 + (-0.23) = -0.5 \text{ (check)}$$

Rotation factor at 3

$$\text{R.F. of } 3-2 = \left(-\frac{1}{2}\right) \times \frac{I/3}{I/3 + I/4} = -0.29$$

$$\text{of } 3-4 = \left(-\frac{1}{2}\right) \times \frac{I/4}{I/3 + I/4} = \frac{-0.21}{-0.50}$$

We derived the Equation 11.9. assuming there is no external moment at the joint. But with clockwise external moment M at the joint which is negative the summation equation becomes

$$\Sigma M_{FBA} + 2\Sigma M'_{BA} + \Sigma M'_{AB} - M = 0$$

or
$$\Sigma M'_{BA} = -\frac{1}{2}\left[\Sigma M_{FBA} + \Sigma M'_{AB} - M\right]$$

or
$$M'_{BA} = R.F.\left[\Sigma M_{FBA} + \Sigma M'_{AB} - M\right]$$

1st Cycle

Joint 3

Sum of *FEM'S* = 0 [as there is no transverse loads in 32 and 24]

External clockwise moment = – *M*

Sum of rotation contributions at far ends

0 at end 4 (known)

0 at end 2 (assumed)

$\Sigma = -M$

Hence $M'_{32} = -0.29 \times (-M) = +0.29\ M.$

$M'_{34} = -0.21 \times (-M) = +0.21\ M.$

Now we consider **joint 2**

Sum of *FEM'S* = 0

rotation contribution of joint 1 = 0 (known)

r.c. at joint 3 = 0.29 M

$\Sigma = 0.29\ M$

M'_{21} = R.F. [+ 0.29 *M*]

$= -0.27\ (+0.29\ M)$

$= -0.08\ M$

$M'_{23} = -0.23\ (+0.29\ M)$

$= -0.07\ M$

This completes 1st cycle.

2nd cycle

Joint 3

Sum of *FEM'S* = 0

External Moment = – *M*

rotation contribution

at end 2 = – .07 *M*

at end 4 = 0.00

$\Sigma = -1.07\ M$

$M'_{32} = -0.29\ (-1.07\ M) = +0.31\ M$

$M'_{34} = -0.21\ (-1.07\ M) = +0.22\ M$

Joint 2

Sum of *FEM'S* = 0

rotation contribution

at joint 1 = 0

at joint 3 = 0.31 *M*

$\Sigma = 0.31\ M$

$M'_{21} = -0.27\ (0.31\ M)$

$= -0.08\ M$

$M'_{23} = -0.23\ (0.31\ M)$

$= -0.071\ M$

The rotation contributions are shown in Fig. 11.82 (*a*) Final Moments are calculated in Fig. 11.82 (*b*).

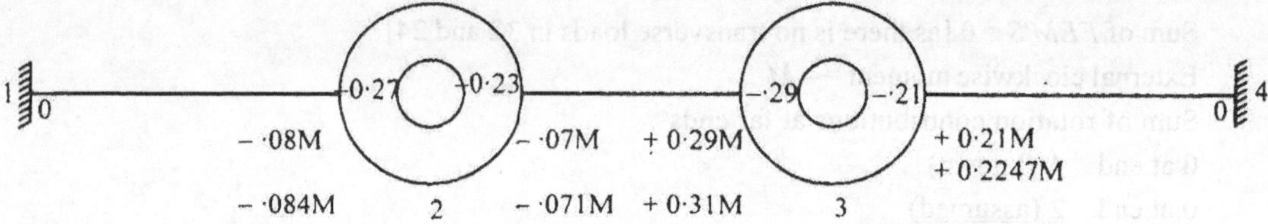

(*a*) Calculation of rotation contributions.

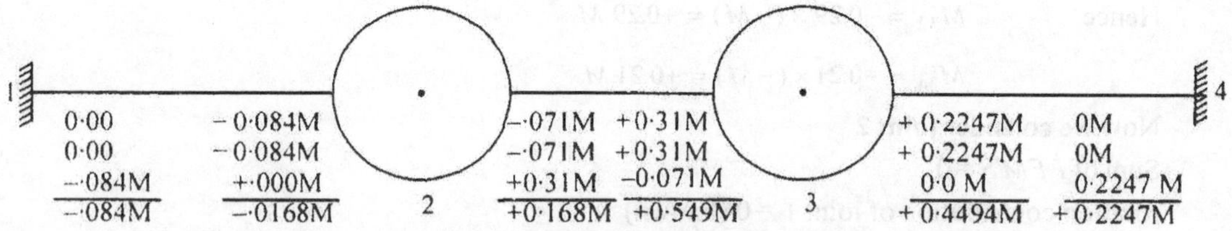

(*b*) Computations of final moments

Fig. 11.82. Rotation contributions and Final Moments

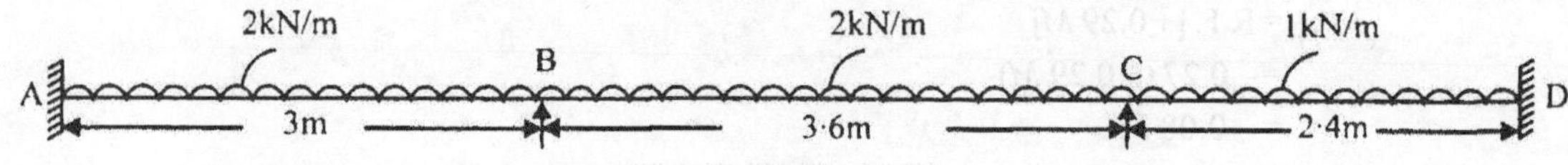

Fig. 11.83. Ex 11.28

Ex 11.28. Determine the support moments in the continuous beam shown in fig 11.83.

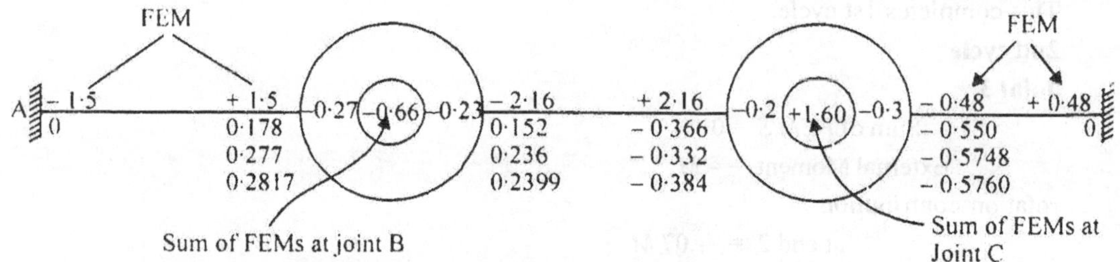

Fig. 11.84. Calculation of Rotation Contributions.

Step 1. Determine the fixed end moments

$$M_{FAB} = -\frac{2 \times 3^2}{12} = -1.5 \text{ kNm}$$

$$M_{FBA} = +\frac{2 \times 3^2}{12} = 1.5 \text{ kNm}$$

$$M_{FBC} = \frac{-2 \times 3.6^2}{12} = -2.16 \text{ kNm}$$

$$M_{FCB} = +\frac{2 \times 3 \cdot 6^2}{12} = +2 \cdot 16 kNm$$

$$M_{FCD} = -\frac{1 \times 2.4^2}{12} = -0.48 \text{ kNm}$$

$$M_{FDC} = +\frac{1 \times 2.4^2}{12} = + 0.48 \text{ kNm}$$

Step 2. Calculate rotation factors at joint B and C.

$$k_{AB} = \frac{I}{3} \quad k_{BC} = \frac{I}{3.6} \quad k_{CD} = \frac{I}{2.4}$$

Rotation factors at B

$$\text{R.F. for } BA = -\frac{1}{2} \times \frac{I/3}{I/3 + I/3.6} = -0.27$$

$$\text{for } BC = -\frac{1}{2} \times \frac{I/3.6}{I/3 + I/3.6} = \frac{-0.23}{-0.50}$$

Rotation factor at C

$$\text{R.F. for } BC = -\frac{1}{2} \times \frac{I/3.6}{I/3.6 + I/2.4} = -0.20$$

$$CD = -\frac{1}{2} \times \frac{I/2.4}{I/3.6 + I/2.4} = \frac{-0.30}{-0.50}$$

Step 3. These are written in schematic form for facilitation of repeated use.

Main features are:

(*i*) Each of the joints B and C is represented by two circles.

(*ii*) All the fixed end moments are entered just above the horizontal line (and at the respective joints) outside the circles.

(*iii*) Rotation contributions will be entered below the horizontal line.

(*iv*) As we require sum of fixed end moments at a joint for every calculation of rotation contribution it is entered in the central small circle

(*v*) Similarly rotation factors are entered between the circles for easy reference.

This is illustrated for Example 11.28 in Fig. 11.84

Step 4. Apply equation 11.11 to joint B.

At B Sum of F.E.M = −0.66

rotation contribution at far ends.

At A (known) 0.00

At C (assumed) 0.00

−0.66

Rotation contribution

$$\text{for } BA = -0.27 \times (-0.66) = +0.178$$

$$\text{for } BC = -0.23 \times (-0.66) = +0.152$$

These rotation contributions are entered at B below the line BA and BC respectively.

Apply equation 11.11 to joint C

Sum of F.E.M. = 1.68

Rotation contribution

at D (known) = 0.00

at B (just calculated) = 0.152

+1.832

Rotation contribution

$$\text{for } CB = (-0.2) \times (+1.832) = -0.366$$

$$\text{for } CD = (-0.3) \times (1.832) = -0.550$$

These rotation contributions are entered below the line under appropriate joint and member. This completes the 1st cycle. The process is then repeated to the level of accuracy desired.

2nd Cycle

Joint *B* Sum of FEM = – 0.660

Sum of rotation contribution at far ends

At A (known) = 0.00

At C from last cycle = $\underline{-0.366}$

– 1.026

R.C. for BA = (– 1.026) × (–0.27)

= + 0.277.

for BC = (– 1.026) × (–0.23)

= + 0.236

These are entered below B at BA and BC respectively cancelling the previous values.

Joint *C* Sum of F.E.Ms = + 1.680

Rotation contribution

at D (known) = 0.000

at B (just calculated) = $\underline{0.236}$

+ 1.916

Rotation for CB = (–0.2) × (1.916) = – 0.3832

for CD = (–0.3) × (1.916) = – 0.5748

These are entered below C at CB and CD respectively cancelling the previous values.

3rd Cycle

Joint B Sum of F.E.M. = – 0.660

Rotation contribution

at A (known) = 0.0000

at C (from last cycle) = $\underline{-0.3832}$

– 1.0432

Rotation contribution

for BA = (–0.27) (– 1.0432) = + 0.2817

BC = (–0.23) (– 1.0432) = + 0.2399.

Joint C Sum of F.E.M. = 1.680

RC at D (known) = 0.000

B (from last cycle) = $\underline{0.2399}$

1.9199.

Rotation contribution

for CB = – 0.2 × (1.9199) = – 0.384

CD = – 0.3 × (1.9199) = – 0.576

Step 4. The final end moments are now calculated from the formula.

$$M_{AB} = M_{FAB} + 2M'_{AB} + M'_{BA}$$

For example at A

$$M_{FAB} = -1.5$$

$$2M'_{AB} = 0.00 \text{ [As rotation contribution at } A = 0]$$

$$M'_{BA} = \frac{+0.2817}{-1.2183}$$

It is however, more convenient to calculate moments in a separate scheme as shown in Fig. 11.85.

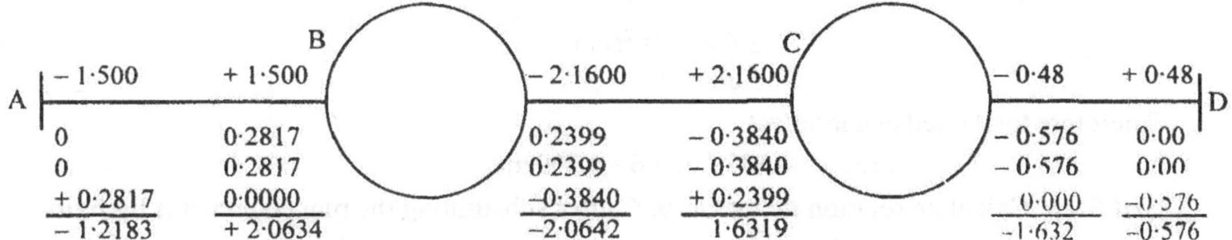

Fig. 11.85. Calculation of final moments (All figure in kNm units)

Example 11.29. Determine the support moment in the continuous beam shown in Fig. 11.86. below if the support at A settles by 10 mm. $E = 2200\ t/\text{cm}^2$ $I = 500\text{cm}^4$

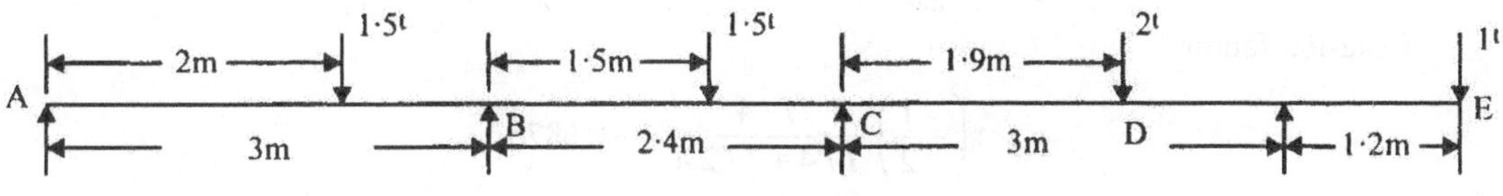

Fig. 11.86. Example 11.29

Solution: This example demonstrates the method of solution for beam with (*i*) hinged end (*ii*) over hang (*iii*) Support settlement.

Step 1. Calculate the fixed end moments for the actual end conditions.

$$M_{FAB} = -\frac{1.5 \times 2 \times 1^2}{3^2} = -.333 \text{ tm}$$

$$M_{FBA} = +\frac{1.5 \times 1 \times 2^2}{3^2} = 0.666 \text{ tm}$$

As support A is hinged moment at A is zero and

$$M_{FBA} = 0.666 + \frac{1}{2} \times 0.333$$

$$= +0.833 \text{ tm}$$

$$M_{FBC} = -\frac{1.5 \times 1.5 \times 0.9^2}{2.4^2} = -0.316 \text{ tm}$$

$$M_{FCB} = +\frac{1.5 \times 0.9 \times 1.5^2}{2.4^2} = +0.527 \text{ tm}$$

$$M_{FCD} = -\frac{2 \times 1.9 \times 1.1^2}{3^2} = +0.511 \text{ tm}$$

$$M_{FDC} = +\frac{2 \times 1 \cdot 9^2 \times 1 \cdot 1}{3^2} = +0 \cdot 882 \text{ tm}$$

$$M_{FDE} = -1.2 \text{ tm}$$

2nd Step. Calculate moment at B due to support settlement at A. M_{FBA} due to

Support settlement

$$= +\frac{3\,EI\,\delta}{L^2} \text{ as end } A \text{ is hinged.}$$

$$= +\frac{3 \times 2200 \times 5000 \times 1}{(300)^2}$$

$$= +3.666 \times 10^2 t\,\text{cm}$$

$$= +3.666\ \text{tm}.$$

Therefore total fixed end moment.

$$M_{FBA} = 0.833 + 3.666 = 4.499\ \text{tm}.$$

3rd Step. Calculate rotation factors at B, C after substituting the pinned end at A by fixity.

Joint B $\quad k_{BA} = \frac{3}{4}\frac{I}{3} = \frac{I}{4}$

$$k_{BC} = \frac{I}{2.4}$$

Rotation factor

$$BA : \left(-\frac{1}{2}\right)\frac{I/4}{I/4 + I/2.4} = -0.1875$$

$$BC : \left(-\frac{1}{2}\right)\frac{I/2.4}{I/4 + I/2.4} = -0.3125$$

Joint C

$$k_{CB} = \frac{I}{2.4} \quad R.F_{CB} = \left(-\frac{1}{2}\right)\frac{I/2.4}{I/2.4 + I/.3} = -0.278$$

$$k_{CD} = \frac{I}{3} \quad R.F_{CD} = \left(-\frac{1}{2}\right)\frac{I/3}{I/2.4 + I/3} = -0.222.$$

Joint D Joint stiffness of cantilever is zero.

Hence R.F for

$DC = 0.5$

$DE = 0.$

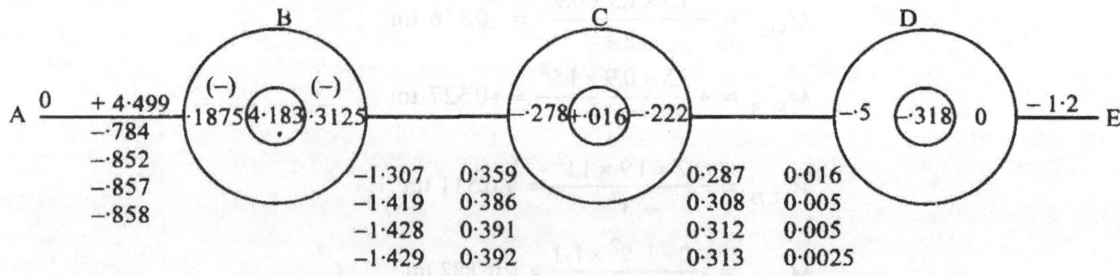

(a) Calculation of rotation contributions

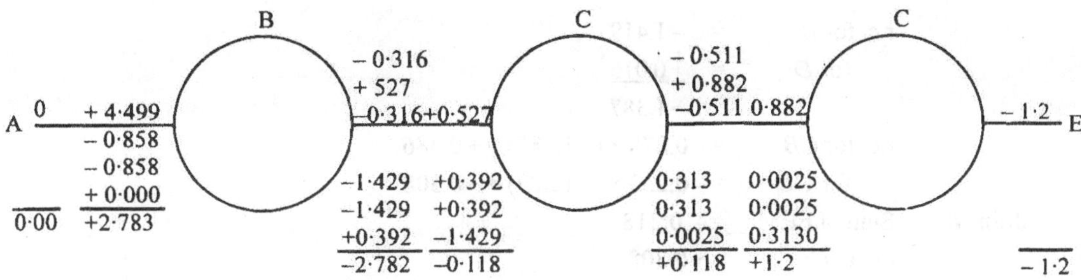

(*b*) Calculation of Final Moment

Fig. 11.87. Rotation Contributions and Final Moments. (Example 11.29)

4th Step. Start the iteration process

Joint B Sum of F.E.M. = +4.183

rotation contribution

at A substituted = 0.000

at C assumed = 0.000

+4.183

Rotation contribution for BA = −0.1875 × (4.183)

= −0.7843.

for BC = −0.3125 (4.183) = −1.3072

Joint C Sum of F.E.M. = +0.016

r.c. for B = −1.307

r.c. for D = 0.000

(assumed) −1.291

r.c. for CB = −1.291 × (−0.278) = 0.359.

for CD = −1.291 × (0.222) = 0.287.

Joint D Sum of F.E.M. = −0.318

r.c. at C = +0.287

r.c at E = 0.000

−0.031

r.c. at D for DC = $-\frac{1}{2} \times (-0.031)$

= +0.016.

2nd Cycle.

Joint B Sum of FEM = +4.183

r.c at A = 0.000

r.c. at C = 0.359

4.542

r.c. at B for BA = −0.1875 × 4.542

= −0.852

for BC = −0.3125 × 4.542

= −1.419.

Joint C Sum of FEM = +0.016

r.c. for B = -1.419

for D = $+0.016$

-1.387

r.c. for CB $= -0.278 \times (-1.387) = +0.386$

for CD $= -0.222 \times (-1.387) = +0.308$

Joint D Sum of FEM $= -0.318$

r.c. at C $= +0.308$

r.c. at E $= 0.000$

-0.010

r.c. for DC $= -\dfrac{1}{2} \times (-0.01)$

$= +0.005$

3rd Cycle

Joint B Sum of F.E.M. $= +4.183$

r.c. at A $= 0.000$

at C $= 0.386$

4.569

r.c. for $BA = -0.1875 \times (4.569) = -0.857$

for $BC = -0.3125 \times (4.569) = -1.428$

Joint C Sum of F.E.M. $= +0.016$

r.c for B $= -1.428$

r.c for D $= +0.005$

-1.407

r.c. for $CB = (-0.278)(-1.407) = 0.391$

for $CD = (-0.222)(-1.407) = 0.312$

Joint D Sum of F.E.M. $= -0.318$

r.c. at C $= +0.312$

-0.006

r.c. for DC $= -\dfrac{1}{2} \times (-0.006)$

$= 0.003.$

4th Cycle

Joint B Sum of F.E.M. $= +4.183$

r.c at A $= 0.000$

r.c at C $= 0.391$

$+4.574$

r.c. for BA $= -0.1875 \times (4.574) = -0.858$

$-0.3125 \times (4.574) = -1.429$

Joint C Sum of F.E.M. $= +0.016$

r.c for B $= -1.429$

r.c for D $= +0.003$

1.410

r.c. for CB $=(-0.278)(1.410)=0.392$

for $CD=(-0.222)(1.410)=0.313.$

Joint *D* Sum of F.E.Ms $= -0.318$

r.c. at C $= \underline{+0.313}$

$-\cdot 005$

r.c. for DC $= -\frac{1}{2}\times(-0.005)$

$=+0.0025.$

Alternative Procedure

Alternative shorter procedure involves use of subtitute stiffness of CD and known moment of 1.2 tm at D.

Fixed End Moments at CD.

$$M_{FCD} = -0.511 \text{ tm}$$
$$M_{FDC} = +0.882 \text{ tm}$$

With known moment of + 1.2 tm at D

$$M_{FCD} = -0.511 + \frac{1}{2}(1.2 - 0.882)$$
$$= -0.511 + 0.519$$
$$= -\mathbf{0.352}$$

Modified rotation factors

Joint B R.F. for $BA = -0.1875$

for $BC = -0.3125$

Joint C

$$k_{CB} = \frac{I}{2.4}$$

$$k_{CD} = \frac{3}{4}\times\frac{I}{3} = \frac{I}{4}$$

$$R.F. \text{ for } CB = \left(-\frac{1}{2}\right)\cdot\frac{I/2.4}{I/2.4 + I/4} = -0.3125$$

$$\text{for } CD = \left(-\frac{1}{2}\right)\cdot\frac{I/4}{I/2.4 + I/4} = -0.1875.$$

With this modifications joints A and D are assumed to be permanently fixed and as such these two joints do not contribute to the iteration process and can be ignored.

A	BA	Joint B	BC	CB	Joint C	CD	D
	+4·499	(−) ·1875 / +4·183 / (−) ·3125	−0·316	+0·527	(−) ·3125 / (+) 0·175 / (−) ·1875	−0·352	
	−0·7843		−1·3072	+0·3538		+0·2123	
	−0·8507		−1·4178	+0·3883		+0·2330	
	−0·8571		−1·4285	+0·3917		+0·2350	

(*a*) Rotation Contributions

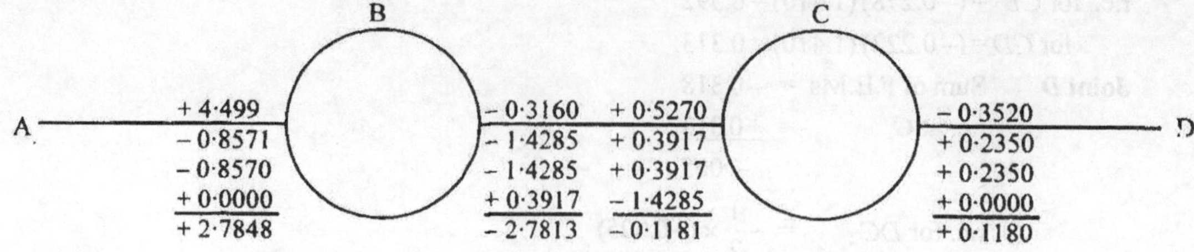

(*b*) Final Moments

Fig. 11.88. (Example 11.29 Alternate procedure)

1st Cycle

Joint B Sum of F.E.M. = 4·183

r.c at A (assumed permanently fixed) = 0.000

r.c. at C (assumed) = 0.000

+4.183

r.c for BA = −0.1875 × 4.183 = −0.7843

for BC = −0.3125 × 4.183 = −1.3072

Joint C Sum of F.E.M. = 0.1750

r.c. at B = −1.3072

r.c. at D (assumed permanently fixed) = 0.0000

−1.1322

r.c. for CB = (−0.3125) × (−1.1322)

= 0.3538

for CD = (−0.1875) × (−1.1322)

= 0.2123.

2nd Cycle

Joint *B* Sum of F.E.M. = +4.1830

r.c at A = 0.0000

r.c at C = +0.3538

4.5368

r.c. for BA = −0.1875 × (4.5368) = −0.8507

for BC = −0.3125 × (4.5368) = −1.4178

Joint *C* Sum of F.E.M. = + 0.1750

r.c. at B = −1.4178

r.c. at D = 0.000

−1.2428

r.c. for CB = (−0.3125) × (−1.2428) = 0.3883.

for CD = (−0.1875) − (−1.2428) = 0.2330.

3rd Cycle

Joint B Sum of F. E. M $= +4.1830$

r.c at A $= 0.0000.$

r.c at C $= \underline{+0.3883}$

$+.4.5713$

r.c for BA $= -0.1875 \times (4.5713)$

$= -0.8571$

for BC $= -0.3125 \times (4.5713)$

$= -1.4285.$

Joint C Sum of F.E.M $= +0.1750$

r.c at B $= -1{\cdot}4285$

r.c. at D $= \underline{0.0000}$

-1.2535

r.c. for CB $= (-1.2535) \times (-0.3125) = +0.3917$

for CD $= (-0.1875) \times (-1.2535) = +0.2350$

The iteration for moment and final calculated values are shown in Fig. 11.88 (a) and (b).

11.13.3 Symmetrical Frames with out Lateral Sway

Symmetrical frames with symmetrical loading can be of two types

(*i*) Frames with even number of bays. (Symmetry axis passing through columns)

(*ii*) Frames with odd number of bays (Summetry axis passing through centres of beams).

Fig. 11.89 shows a symmetrical frame with even number of bays and its equivalent substitute frame for analytical purposes when loading is symmetrical.

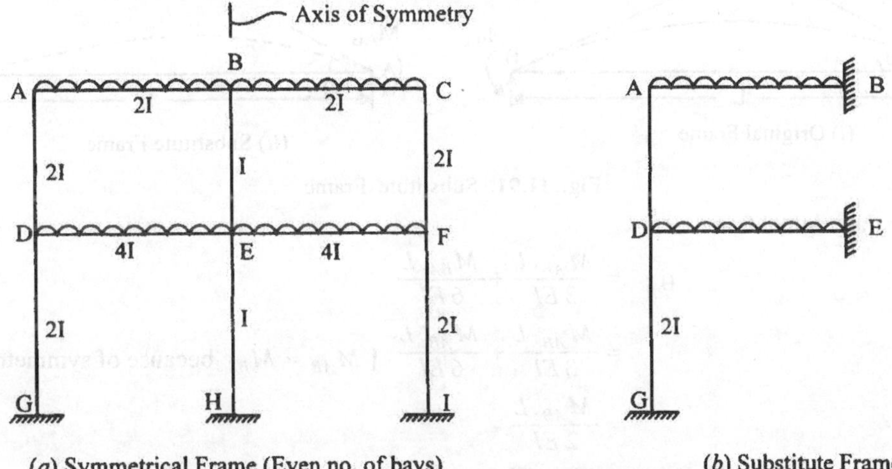

(*a*) Symmetrical Frame (Even no. of bays)

(*b*) Substitute Frame

Fig. 11.89. Symmetrical Frame (Even Number of bays).

Fig. 11.90 shows a symmetrical frame with odd number of bays (Axis of symmetry passing through centres of beams).

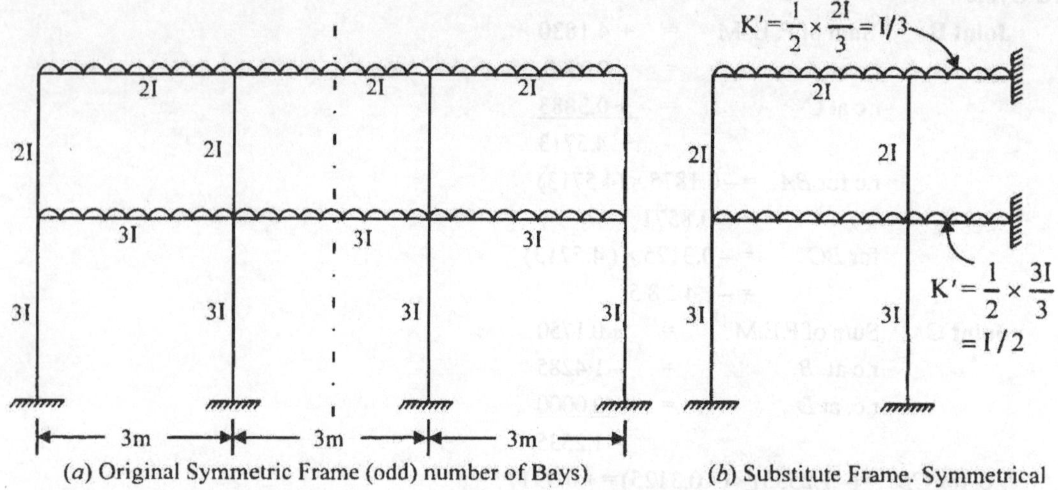

(*a*) Original Symmetric Frame (odd) number of Bays)

(*b*) Substitute Frame. Symmetrical Frame (odd number of bays)

Fig. 11.90. Symmetrical Frames (Odd number of bays).

In the first case because of symmetry joint at *B* will not rotate and hence can be taken as fixed. Solution can, therefore, be obtained by considering only half of the frame.

In case (*ii*) there will be no rotation at the centre of the beam through which the symmetrical axis passes. Hence the centre of beams can be considered as fixed as shown in Fig. 11.90 (*b*).

To consider equivalence of the original frame and the substitute frame, θ_A is computed for both cases as shown in Fig. 11.91.

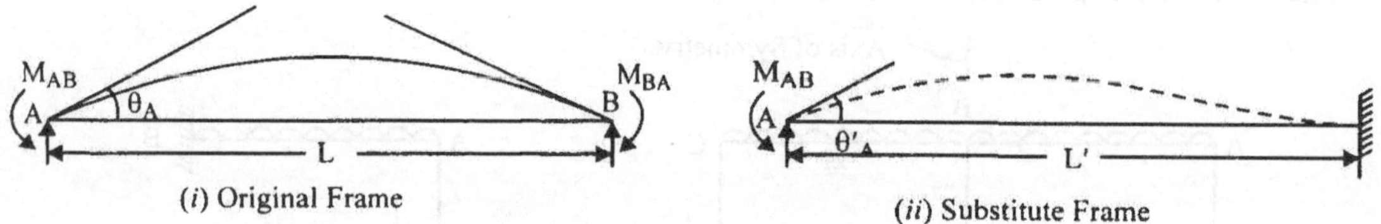

(*i*) Original Frame

(*ii*) Substitute Frame

Fig. 11.91. Substitute Frame

For the original frame

$$\theta_A = \frac{M_{AB}.L}{3\,EI} + \frac{M_{BA}.L}{6\,EI}$$

$$= \frac{M_{AB}.L}{3\,EI} + \frac{M_{AB}.L}{6\,EI} \quad [\, M_{AB} = M_{BA} \text{ because of symmetry)}$$

$$= \frac{M_{AB}.L}{2\,EI}$$

$$\theta'_A = \frac{M_{AB}.L'}{4\,EI'}$$

For equivalence $\theta_A = \theta_A{}'$

or $$\frac{M_{AB}.L}{2\,EI} = \frac{M_{AB}.L'}{4\,EI'}$$

or $$\frac{I'}{L'} = \frac{I}{2L} \text{ or } k' = \frac{k}{2}$$

This is shown in Fig. 11.90(*b*).

11.13.4. Frames with Lateral Sway, Vertical Loading Only

Because of Sway, sway moments will be developed. This is an additional unknown and an additional equation is required to solve this unknown. This is obtained by considering shear condition at each storey.

Consider the free body diagram of a column of the frame. (Fig. 11.92).

Taking moments of all forces about *P*

$$M_{QP} + M_{PQ} + H.h = 0$$

or $$H = -\left(\frac{M_{PQ} + M_{QP}}{h}\right)$$

Considering all columns in one storey say *r*th storey

$$\Sigma H = -\frac{\left(\sum_r M_{PQ} + \sum_r M_{QP}\right)}{h_r}$$

= storey shear in the rth storey

$= S_r$

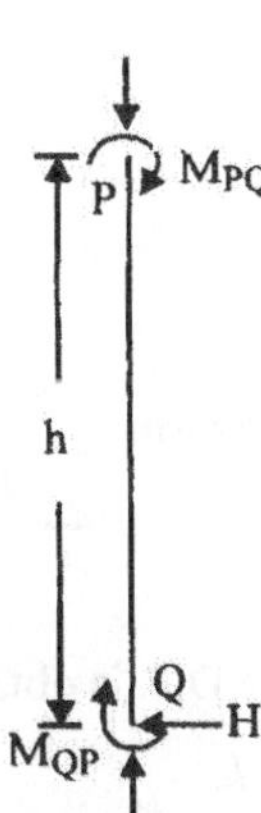

Fig. 11.92. Free body diagram of column

But with vertical loading only and no horizontal force S_r must be equal to zero.

Therefore $$\sum_r M_{PQ} + \sum_r M_{QP} = 0 \qquad ...(11.12)$$

But it is already derived

$$M_{PQ} = M_{FPQ} + 2M'_{PQ} + M'_{QP} + M''_{PQ}$$

and $$M_{QP} = M_{FQP} + 2M'_{QP} + M'_{PQ} + M''_{PQ}$$

(M''_{PQ} is displacement contribution at each end).

When there is no transverse load

$$M_{FPQ} = M_{FQP} = 0$$

Therefore $$M_{PQ} + M_{QP} = 3\,M'_{PQ} + 3\,M'_{QP} + 2\,M''_{PQ}$$

Substituting in Eq. 11.12

$$3\sum_r M'_{PQ} + 3\sum_r M'_{QP} + 2\sum_r M''_{PQ} = 0$$

or $$\sum M''_{PQ} = -\frac{3}{2}\left(\sum_r M'_{PQ} + \sum_r M'_{QP}\right) \qquad ...(11.13)$$

Equation 11.13, shows the relation between displacement contributions and rotation contributions. The term $\sum_r M''_{PQ}$ is the sum of the dispacement contributions of the columns of the rth storey. It is now necessary to compute individual displacement contribution. Individual displacement contribution is given by $\frac{6EI\delta}{L^2}$. In a framed structure usually only I varies in columns in a particular storey. Lateral sway δ is the same for all columns in a storey. Displacement contribution is, therefore, proportional to I or $\frac{I}{L} = k$ as L is constant. Hence,

$$M''_{PQ} = \frac{k}{\sum k} \sum_r M''_{PQ}$$

$$= \frac{k}{\sum_r k} \times \left(-\frac{3}{2}\right) \cdot \left(\sum_r M'_{PQ} + \sum_r M'_{QP}\right) \quad \text{...(11.14)}$$

The term $\frac{k}{\sum_r k} \times \left(-\frac{3}{2}\right)$ is known as displacement factor D.F. (Compare with rotation factor R.F.). The D.F. is obtained by distributing $\left(-\frac{3}{2}\right)$ among the columns of every storey in proportion to stiffness k.

Therefore $$M''_{PQ} = D.F. \left(\sum_r M'_{PQ} + \sum_r M'_{QP}\right) \quad \text{...(11.14a)}$$

Rotation contributions are obtained by applying equation 11.11 repeatedly at the joints.

Displacement contributions are obtained by applying equation 11.14 repeatedly to the columns of a particular storey.

Example 11.30. Apply Kani's method to compute moments at the joints of the portal frame shown in Fig. 11.93.

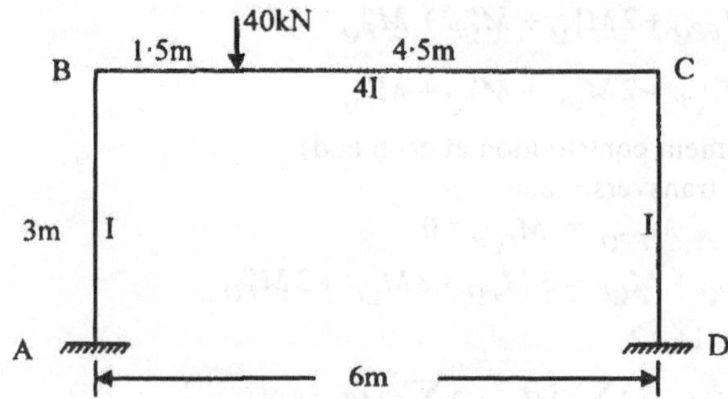

Fig. 11.93. Example 11.30.

Solution: Compute

(*i*) FEMs

$$M_{FBC} = \frac{-40 \times 1 \cdot 5 \times 4 \cdot 5^2}{6^2} = -33 \cdot 75 \text{ kNm}$$

$$M_{FCB} = \frac{+40 \times 4.5 \times 1.5^2}{6^2} = +11 \cdot 25 \text{kNm}$$

(*ii*) *K*-values

$$BA = \frac{I}{3}$$

$$BC = \frac{4I}{6} = \frac{2I}{3}$$

$$CD = \frac{I}{3}.$$

proportion 1 : 2 : 1

(*iii*) **Rotation Factors**

Joint *B* $\quad BA \quad \frac{1}{3} \times \left(-\frac{1}{2}\right) = -\frac{1}{6}$

$\quad BC \quad \frac{2}{3} \times \left(-\frac{1}{2}\right) = -\frac{1}{3}$

Joint *C* $\quad CB \quad \frac{2}{3} \times \left(-\frac{1}{2}\right) = -\frac{1}{3}$

$\quad CD \quad \frac{1}{3} \times \left(-\frac{1}{2}\right) = -\frac{1}{6}$

(*iv*) Displacement factors

$$AB = \frac{1}{2} \times \left(-\frac{3}{2}\right) = -\frac{3}{4}$$

$$DC = \frac{1}{2} \times \left(-\frac{3}{2}\right) = -\frac{3}{4}.$$

The starting scheme is shown in Fig. 11.94

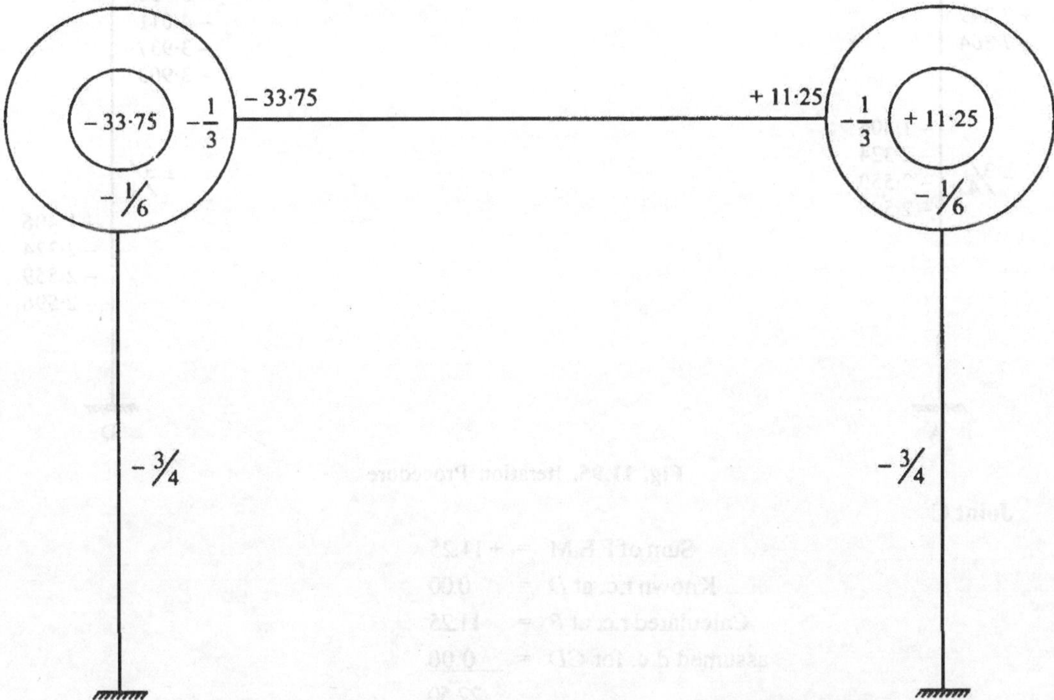

Fig. 11.94. Rotation factors, Distribution factors and fixed end moments.

Calculation. Calculation is carried out in the following order.

Joint *B*, Joint *C* Member *BA*, Member *CD*. This completes one cycle. Repeat the same order for successive cycles.

1st Cycle

Joint B

Sum of FEM'S at *B*	=	−33.75
Known rotation contribution at *A*	=	0.00
assumed r.c. at *C*	=	0.00
assumed d.c. from *BA*	=	0.00
		−33.75

r.c. for $BC = -\frac{1}{3} \times (-33.75) = 11.25$ kNm

for $BA = -\frac{1}{6} \times (-33.75) = 5.625$ kNm

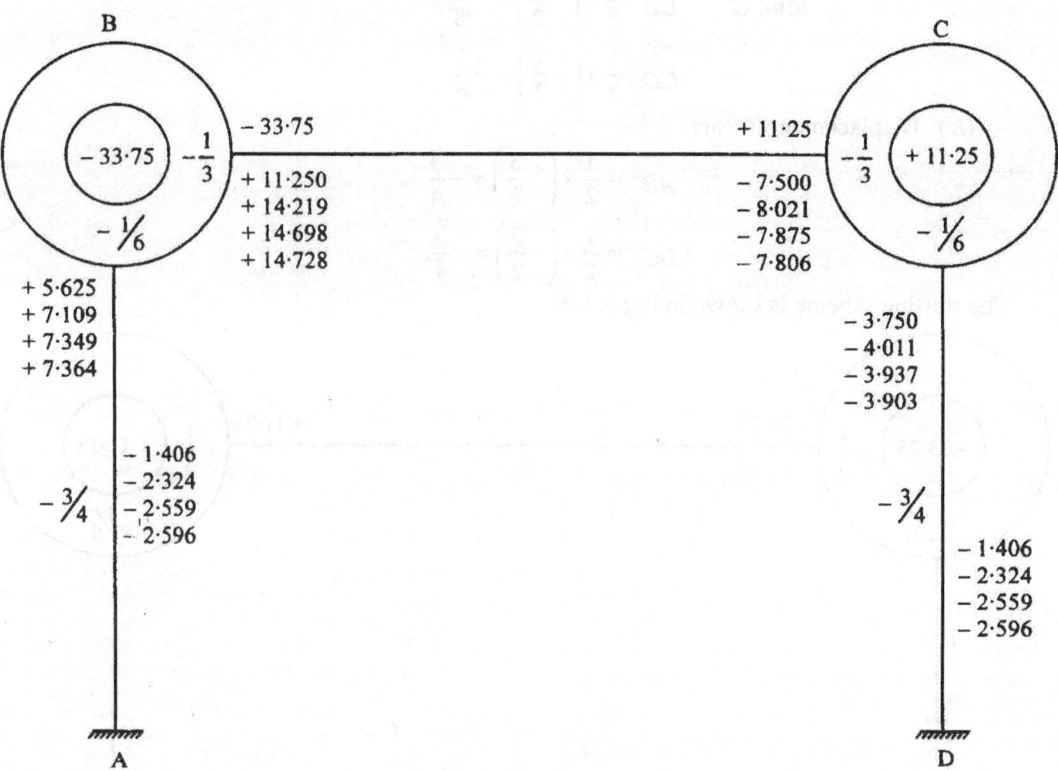

Fig. 11.95. Iteration Procedure

Joint C

Sum of F.E.M	=	+11.25
Known r.c. at *D*	=	0.00
Calculated r.c. at *B*	=	11.25
assumed d.c. for *CD*	=	0.00
		22.50

r.c. for $CB = -\frac{1}{3} \times (22.50) = -7.50$ kNm

r.c. for $CD = -\frac{1}{6} \times (22.50) = -3.75$ kNm.

Storey Calculation

r.c. for BA = 5.625
r.c. for CD = −3.750
r.c. at A = 0.000
r.c. at D = 0.000
1.875

displacement contribution for

$$BA = -\frac{3}{4} \times 1.875 = -1.406$$

$$CD = -\frac{3}{4} \times 1.875 = -1.406$$

This completes 1st Cycle

2nd Cycle

Joint B

Sum of *FEMs* at B = −33.750
Known r.c. at A = 0.000
Calculate r.c. at C = −7.500
d.c. for BA = −1.406
−42.656

$$\text{r.c. for } BC = -\frac{1}{3}(-42.656) = 14.219$$

$$\text{r.c. for } BA = -\frac{1}{6}(-42.656) = 7.109$$

Joint C

Sum of *FEMS* at C = + 11·250
Calculated r.c. at B = + 14.219
Known r.c. at D = 0.000
d.c. for C = −1.406
+24.063

$$\text{r.c. for } CB = -\frac{1}{3} \times 24.063 = -8.021$$

$$\text{for } CD = -\frac{1}{6} \times 24.063 = -4.011$$

Storey Calculation

r.c. for BA = 7.109
r.c. for CD = −4.011
r.c. for A = 0.000
for D = 0.000
3.098

$$\text{d.c. for } BA = -\frac{3}{4} \times 3.098 = -2.324$$

$$\text{for } CD = -\frac{3}{4} \times 3.098 = -2.324.$$

3rd Cycle

Joint B

Sum of *FEMs* at B = −33.750
Known r.c. at A = 0.000
Calculated r.c. at C = −8.021
d.c. at B = −2.324
−44.095

$$\text{r.c. for } BC = -\frac{1}{3}(-44.095) = +14.698$$

$$BA = -\frac{1}{6}(-44.095) = +7.349$$

Joint C

Sum of *FEMs* at C = +11.250
r.c. at B = +14.698
r.c. at D = 0.000
d.c. at C = −2·324
23.624

$$\text{r.c. for } CB = -\frac{1}{3}(23.624) = -7.875$$

$$\text{for } CD = -\frac{1}{6}(23.624) = 3.937$$

Storey Calculation

r.c. for BA = 7.349
for CD = −3.937
r.c. at A and D = 0.000
3.412

$$\text{d.c. for } BA = -\frac{3}{4}(3.412) = -2.559.$$

$$\text{for } CD = -\frac{3}{4}(3.412) = -2.559$$

4th Cycle

Joint B

Sum of *FEMs* = −33.750
r.c. at A = 0.000
at C = −7.875
d.c. for BA = −2.559
−44.184

$$\text{r.c. for } BC = -\frac{1}{3}(-44.184) = +14.728$$

$$BA = -\frac{1}{6}(-44.184) = 7.364.$$

Joint C

$$\text{Sum of } FEMs = +11.250$$
$$\text{r.c. at } B = +14.728$$
$$\text{r.c. at } D = 0.000$$
$$\text{d.c. at } C = \underline{-2.559}$$
$$23.419$$

$$\text{r.c. for } CB = -\frac{1}{3}(23.419) = -7.806$$

$$\text{for } CD = -\frac{1}{6}(23.419) = -3.903$$

Storey Calculation

$$\text{r.c. for } BA = +7.364$$
$$\text{r.c. for } CD = -3.903$$
$$\text{r.c. at } A \text{ and } D = \underline{0.000}$$
$$+3.461$$

$$\text{d.c. for } BA = \frac{-3}{4}(3.461) = -2.596$$

$$\text{for } CD = \frac{-3}{4}(3.461) = -2.596$$

Calculation of end moments

M_{AB}

$$FEM \text{ at } A = 0.000$$
$$\text{Twice rotation contribution at } A = 0.000$$
$$\text{r.c. at } B = 7.364$$
$$\text{d.c. for } BA = \underline{-2.596}$$
$$4.768$$

M_{BA}

$$FEM \text{ at } B = 0.000$$
$$\text{Twice r.c. at } B = 7.364$$
$$7.364.$$
$$\text{r.c. at } A = 0.000$$
$$\text{d.c. for } AB = \underline{-2.596}$$
$$12.132$$

M_{BC}

$$F.E.M. \text{ at } B = -33.750$$
$$\text{Twice r.c. at } B = 14.728$$
$$14.728$$
$$\text{r.c. at } C = \underline{-7.806}$$
$$12.100$$

M_{CB}

$$FEM \text{ at } C = +11.250$$
$$\text{Twice r.c. at } C = -7.806$$
$$-7.806$$
$$\text{r.c. at } B = \underline{14.728}$$
$$10.366$$

M_{CD}

$$FEM \text{ at } C = 0.000$$
$$\text{Twice r.c. at } C = -3.903$$

		=	−3.903
	r.c. at D	=	0.000
	d.c. for CD	=	<u>−2.596</u>
			−10.402
M_{DC}	FEM at D	=	0.000
	Twice r.c. at D	=	0.000
	r.c. at C	=	−3.903
	d.c. for DC	=	<u>−2.596</u>
			−6.499

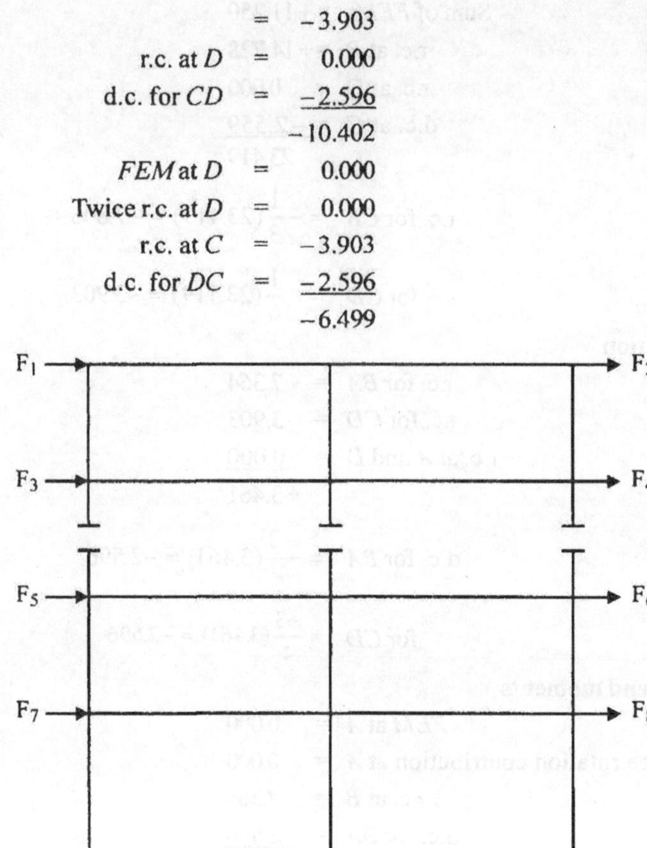

Fig. 11.96. Frames with Lateral Sway.

11.13.5. Frames with Lateral Sway

Horizontal loading at nodal points and vertical loading:

As there are horizontal loading at nodal points storey shear at the rth storey is not equal to zero. Hence Equation 11.12 modifies to

$$\sum_r M_{PQ} + \sum_r M_{QP} + S_r . h_r = 0 \qquad ...(11.15)$$

Where S_r is the storey shear in the rth storey and h_r is the height of the rth storey. Storey shear S_r. is the algebriae sum of horizontal loads above the rth storey. Form Fig. 11.96

$$S_r = F_1 + F_2 + F_3 + F_4$$

Since clockwise moments are positive according to our sign convention S_r should be taken as positive if the algebraic sum of horizontal forces above the storey is towards the right.

As already derived in Eq. 11.8

$$M_{PQ} = M_{FPQ} + 2M'_{PQ} + M'_{QP} + M''_{PQ}$$

$$M_{QP} = M_{FQP} + 2M'_{QP} + M'_{PQ} + M''_{PQ}$$

Since PQ is a vertical member with only nodal load

(There is no load between P and Q)

$$M_{FPQ} = M_{FQP} = 0$$

$$\therefore \quad M_{PQ} + M_{QP} = 3M'_{PQ} + 3M'_{QP} + 2M''_{PQ}$$

Substituting eq. 11.15.

$$3\sum_r M'_{PQ} + 3\sum_r M'_{QP} + 2\sum_r M''_{PQ} + S_r.h_r = 0$$

or
$$\sum M''_{PQ} = -\frac{3}{2}\left(\frac{S_r h_r}{3} + \sum M'_{PQ} + \sum M'_{QP}\right)$$

and
$$M''_{PQ} = D.F.\left(\frac{S_r.h_r}{3} + \sum_r M'_{PQ} + \sum_r M'_{QP}\right) \qquad ...(11.16)$$

Example 11.31. Analyse the portal frame shown in Fig. 11.97 below by Kani's method and sketch the bending moment diagram giving all salient features.

[AMIE Advanced Structures, Summer, 1988]

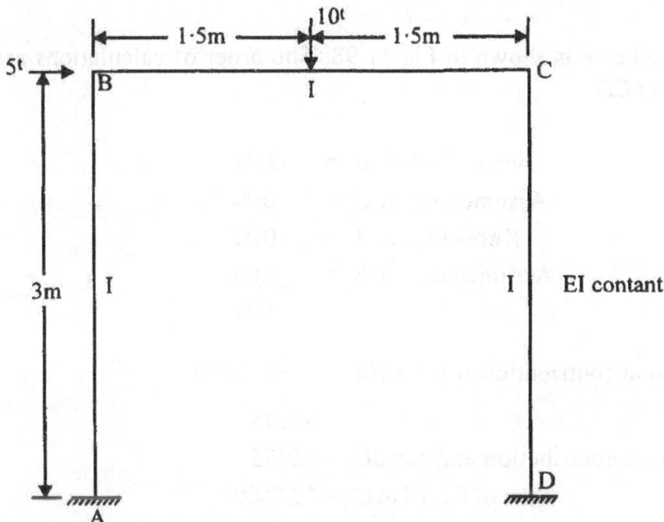

Fig. 11.97. (Example 11.31)

Solution:

(*i*) Fixed End Moments

$$M_{FBC} = -\frac{10 \times 1.5 \times 1.5^2}{3^2} = -3.75$$

$$M_{FCB} = +\frac{10 \times 1.5 \times 1.5^2}{3^2} = +3.75$$

(*ii*) K values

$$AB = \frac{I}{3}, \quad BC = \frac{I}{3}$$

$$CD = \frac{I}{3}$$

ratio 1 : 1 : 1

(*iii*) Rotation factors

Joint B

$BA \quad \frac{1}{2}\left(-\frac{1}{2}\right)=-\frac{1}{4}$

$BC \quad \frac{1}{2}\left(-\frac{1}{2}\right)=-\frac{1}{4}$

Joint C

$CB \quad =-\frac{1}{4}$

$CD \quad =-\frac{1}{4}$

(*iv*) Displacement factors

$AB: \quad \frac{1}{2}\times\left(-\frac{3}{2}\right)=\frac{-3}{4}.$

$CD: \quad \frac{1}{2}\times\left(-\frac{3}{2}\right)=\frac{-3}{4}$

The starting scheme is shown in Fig 11.98. The order of calculations are : Joint B, Joint C Member BA, Member CD.

1st Cycle

Joint B

Sum of FEM at B = −3.75
Assumed r.c. at C = 0.00
Known r.c. at A = 0.00
Assumed d.c. at B = 0.00
−3.75

Rotation contribution at B for BA $= -\frac{1}{4}(-3.75)$
$=0.9375$

Rotation contribution at B for BC $=0.9375$

Joint C

Sum of F.E.M at C = +3.7500
Known r.c. at D = 0.0000
Calculated r.c. at B = 0.9375
Assumed d.c. at C = 0.0000
4.6875

R.C. at C for CB $= -\frac{1}{4}(4.6875)$
$=-1.1719$

R.C. at C for CD $=-1.1719$

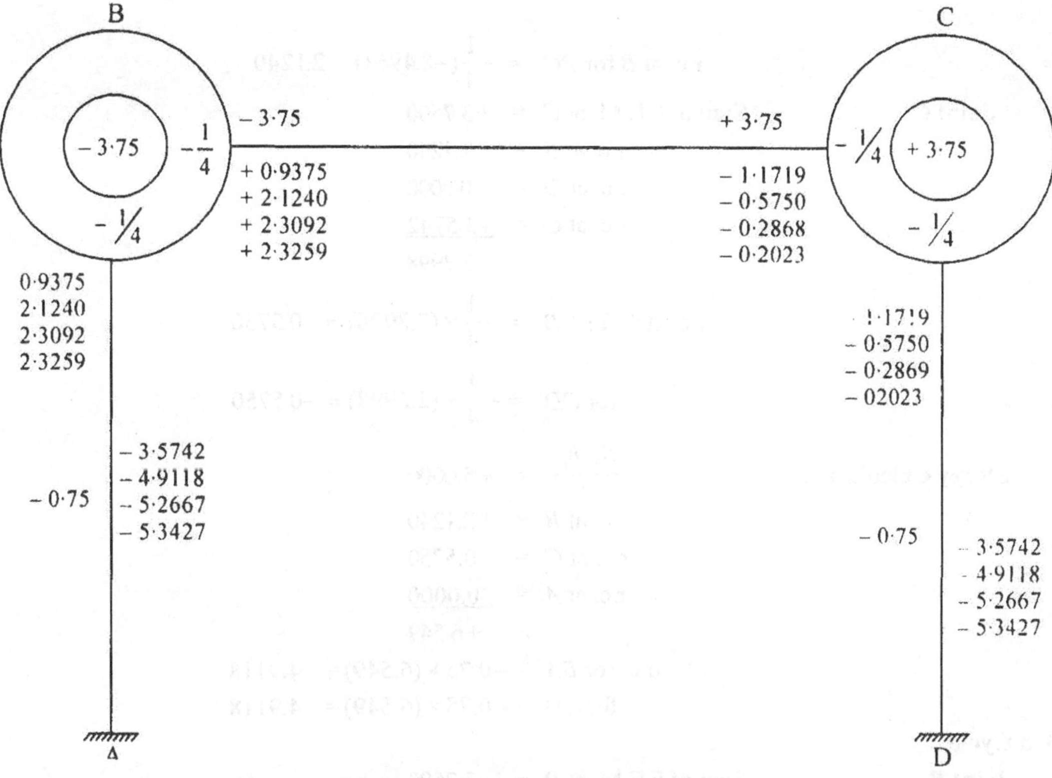

Fig. 11.98. Iteration Process (Ex. 11.31)

Storey Calculation

$$\frac{S_r\ h_r}{3} = +5.0000$$

r.c. at B = +0.9375

r.c. at B = −1.1719

r.c. at A = 0.0000

at D = 0.0000

4.7656

Displacement contribution for BA = −0.75 × (4.7656)

= −3.5742

for CD = −3.5742

This completes 1st Cycle

2nd Cycle

Joint *B* Sum of F.E.M. at B = −3.7500

r.c. at C = −1.1719

r.c. at A = 0.0000

d.c. at B = −3.5742

−8.4961

$$\text{r.c. at } B \text{ for } BA = -\frac{1}{4}(-8.4961) = 2.1240$$

r.c. at B for BC $= -\frac{1}{4}(-8.4961) = 2.1240$

Joint C

Sum of F.E.M. at C = +3.7500
r.c. at B = 2.1240
r.c. at D = 0.0000
r.c. at C = –3.5742
+2.2998

r.c. at C for CB $= -\frac{1}{4} \times (2.2998) = -0.5750$

for CD $= -\frac{1}{4} \times (2.2998) = -0.5750$

Storey Calculation

$\frac{S_r h_r}{3}$ = +5.0000
r.c. at B = +2.1240
r.c. at C = –0.5750
r.c. at A = 0.0000
+6.549

d.c. for BA $= -0.75 \times (6.549) = -4.9118$

for CD $= -0.75 \times (6.549) = -4.9118$

3rd Cycle

Joint B

Sum of F.E.M. at B = +3.7500
r.c. at C = 0.5750
r.c. at A = 0.0000
r.c. at B = –4.9118
–9.2368

r.c. at B for BA $= \left(-\frac{1}{4}\right)(-9.2368) = 2.3092$

for BC $= -\left(\frac{1}{4}\right)(-9.2368) = 2.3092$

Joint C

Sum of F.E.M. at C = +3.7500
r.c. at B = +2.3092
r.c. at D = 0.0000
r.c. at C = –4.9118
+1.1474

r.c at C for CB $= \left(-\frac{1}{4}\right)(1.1474) = -0.2869$

for CD $= \left(-\frac{1}{4}\right)(1.1474) = -0.2869$

Storey Calculation

$\frac{S_r \cdot h_r}{3}$ = +5.0000

$$
\begin{aligned}
\text{r.c. at } B &= +2.3092 \\
\text{at } C &= -0.2869 \\
\text{r.c. at } A &= 0.0000 \\
\text{at } D &= \underline{0.0000} \\
& \quad 7.0223
\end{aligned}
$$

$$
\begin{aligned}
\text{d.c. for } AB &= (-0.75)(7.0223) = -5.2667 \\
CD &= (-0.75)(7.0223) = -5.2667
\end{aligned}
$$

4th Cycle

Joint *B*

$$
\begin{aligned}
\text{Sum of F.E.M. at } B &= -3.7500 \\
\text{r.c. at } C &= -0.2869 \\
\text{r.c. at } A &= 0.0000 \\
\text{r.c. at } B &= \underline{-5.2667} \\
& \quad -9.3036
\end{aligned}
$$

$$\text{r.c. at } B \text{ for } BA = \left(-\frac{1}{4}\right) \times (-9.3036) = 2.3259$$

$$\text{for } BC = = \left(-\frac{1}{4}\right) \times (-9.3036) = 2.3259$$

Joint *C*

$$
\begin{aligned}
\text{Sum of F.E.M.S} &= +3.7500 \\
\text{r.c. at } B &= +2.3259 \\
\text{r.c. at } AD &= 0.0000 \\
\text{d.e. at } C &= \underline{-5.2667} \\
& \quad 0.8092
\end{aligned}
$$

$$\text{r.c. at } C \text{ for } CB = \left(-\frac{1}{4}\right)(0.8092) = -0.2023$$

$$\text{for } CD = \left(-\frac{1}{4}\right)(0.8092) = -0.2023$$

Storey Calculation

$$
\begin{aligned}
\frac{S_r \cdot h_r}{3} &= 5.0000 \\
\text{r.c. at } B &= 2.3259 \\
\text{r.c. at } C &= -0.2023 \\
\text{r.c. at } D &= 0.0000 \\
\text{at } A &= \underline{0.0000} \\
& \quad 7.1236
\end{aligned}
$$

$$\text{d.c. for } AB = -\frac{3}{4}(7.1236) = -5.3427$$

$$DC = \frac{-3}{4}(7.1236) = -5.3427$$

Calculation of moments

M_{AB}	F.E.M. at A	=	0.0000
	2 times r.c. at A	=	0.0000
	r.c. at B	=	2.3259
	d.c. at A	=	<u>−5.3427</u>
			−3.0168
M_{BA}	F.E.M. at B	=	0.0000
	2 times r.c. at B	=	2.3259
			2.3259.
	r.c. at A	=	0.0000
	d.c. at B	=	<u>−5.3427</u>
			0.6909
M_{BC}	F.E.M. at B	=	−3.7500
	2 times r.c. at B	=	+2.3259
			+2.3259
	r.c. at C	=	<u>−0.2023</u>
			−0.6995
M_{CB}	F.E.M. at C	=	+3.7500
	2 times r.c. at C	=	−0.2023
			−0.2023
	r.c. at B	=	<u>+2.3259</u>
			+5.6713
M_{CD}	F.E.M. at C	=	0.0000
	2 times r.c. at C	=	−0.2023
			−0.2023
	r.c at D	=	0.0000
	d.c. at C	=	<u>−5.3427</u>
			− 5.7473
M_{DC}	F.E.M. at D	=	0.0000
	2 times r.c. at D	=	0.0000
	r.c at C	=	−0.2023
	d.c. at D	=	<u>−5.3427</u>
			−5.5450

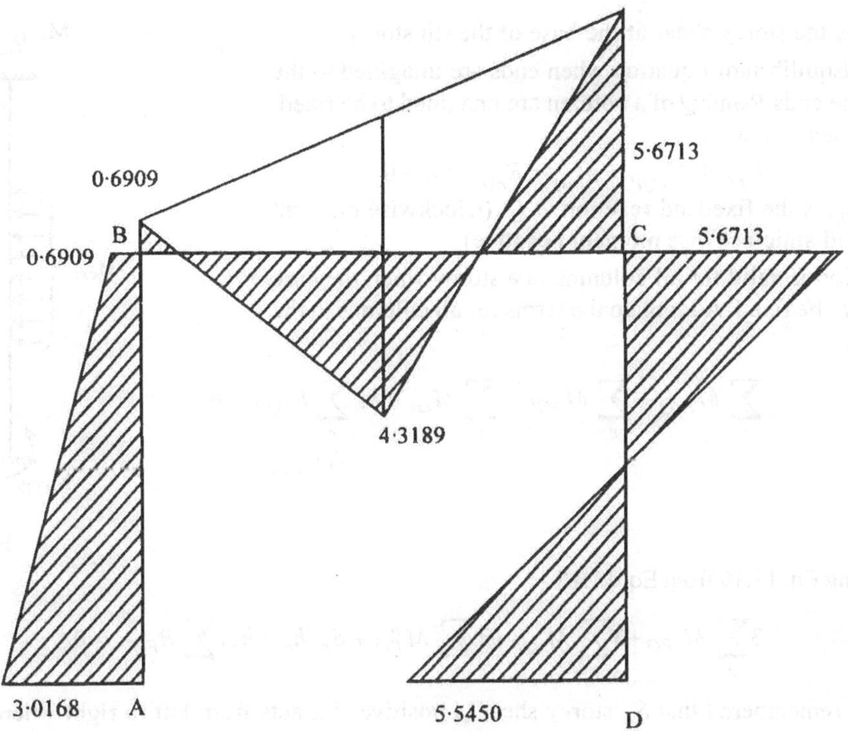

Fig. 11.99. B.M. Diagram (Plotted on the tension side)

11.13.6. Frames with lateral sway

General Case of loading: Here any type of loading is possible.

(*i*) General Equilibrium Equation.

Consider the free body diagram of a column in the frame.

Taking moment of all forces about *P*,

$$M_{PQ} + M_{QP} + H.h + M_{ext} = 0 \quad ...(11.17)$$

where M_{ext} denote the moment of the force H_{ext} about *P*. M_{ext} will be positive if moment is clockwise and negative if anti-clockwise.

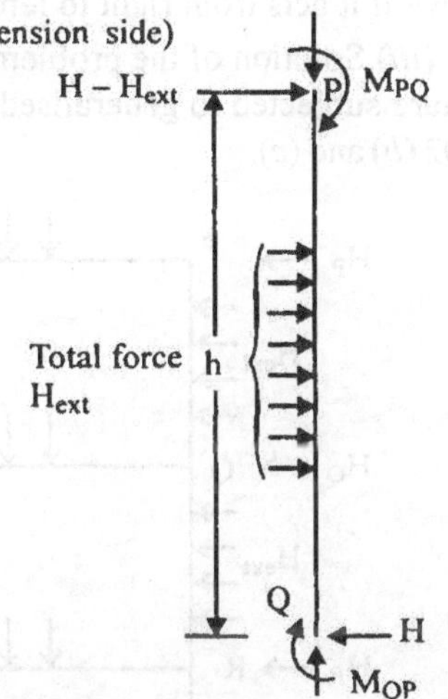

Fig. 11.100. Derivation of General Equilibrium Equation.

Again

$$M_{PQ} = M_{FPQ} + 2M'_{PQ} + M'_{QP} + M''_{PQ}$$
$$M_{QP} = M_{FQP} + 2M'_{QP} + M'_{QP} + M''_{PQ}$$

($M''_{PQ} = M''_{QP}$ = displacement contribution of member *PQ*).

Substituting these values in Eq. 11.17

$$M_{FPQ} + M_{FQP} + 3M'_{PQ} + 3M'_{QP} + 2M''_{PQ} + M_{ext} + H.h = 0$$

Now this relation holds good for every column in the frame. Adding up the terms of such equations for all the columns of the rth storey

$$\sum_r M_{FPQ} + \sum_r M_{FQP} + 3\sum_r M'_{PQ} + 3\sum_r M'_{QP} + 2\sum_r M''_{PQ} + \sum_r M_{ext} + S_r.h_r = 0 \quad (11.18)$$

where S_r is the storey shear at the base of the rth storey.

(*ii*) Equilibrium Equation when ends are imagined to the fixed. If the ends *P* and *Q* of a column are imagined to be fixed, taking moments about *P*,

$$M_{FPQ} + M_{FQP} + M_{ext} + R_{FQP} \times h = 0$$

where R_{FQP} is the fixedend reaction at *Q*. (Clockwise moment positive and anticlockwise moment negative)

This relation is valid for all columns in a storey when they are imagined to be fixed. Adding up the terms for all columns in the rth storey.

$$\sum_r M_{FPQ} + \sum_r M_{FQP} + \sum_r M_{ext} + h_r \sum_r R_{FQP} = 0 \quad ...(11.19)$$

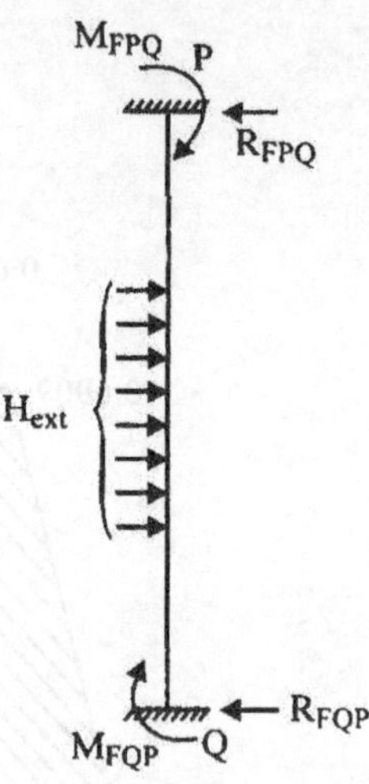

Fig. 11.101.

Substracting Eq. 11.19 from Eq. 11.18

$$3\sum_r M'_{PQ} + 3\sum_r M'_{QP} + 2\sum_r M''_{PQ} + S_r.h_r - h_r.\sum_r R_{FQP} = 0 \quad ...(11.20)$$

It is to be remembered that S_r, storey shear is positive if it acts from left to right whereas R_{FQP} is positive if it acts from right to left since in that case it causes clockwise moment about *P*.

(*iii*) Solution of the problem in two stages: Fig. 11.102 (*a*) shows a typical multistorey frame structure subjected to generalised loading. The problem is analysed in two stages as shown in Fig. 11.102 (*b*) and (*c*).

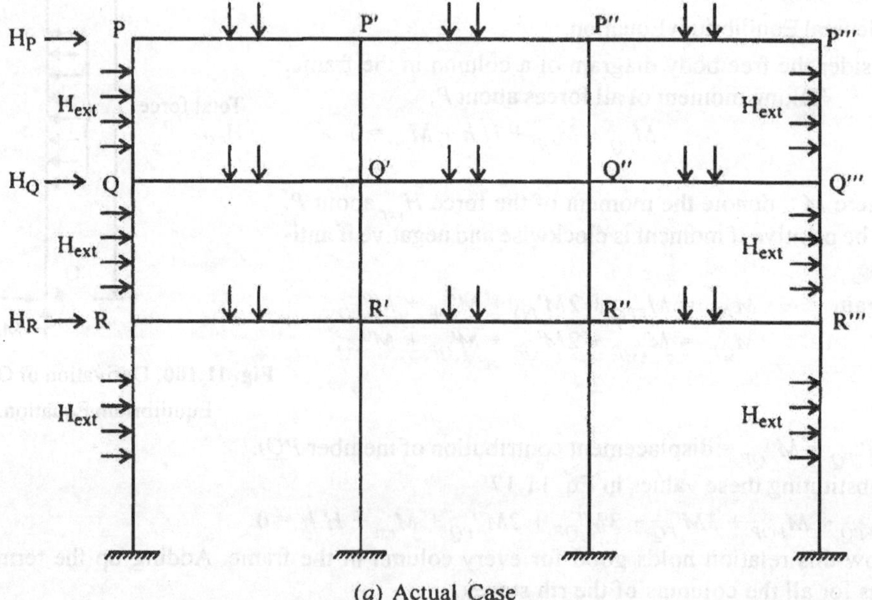

(*a*) Actual Case

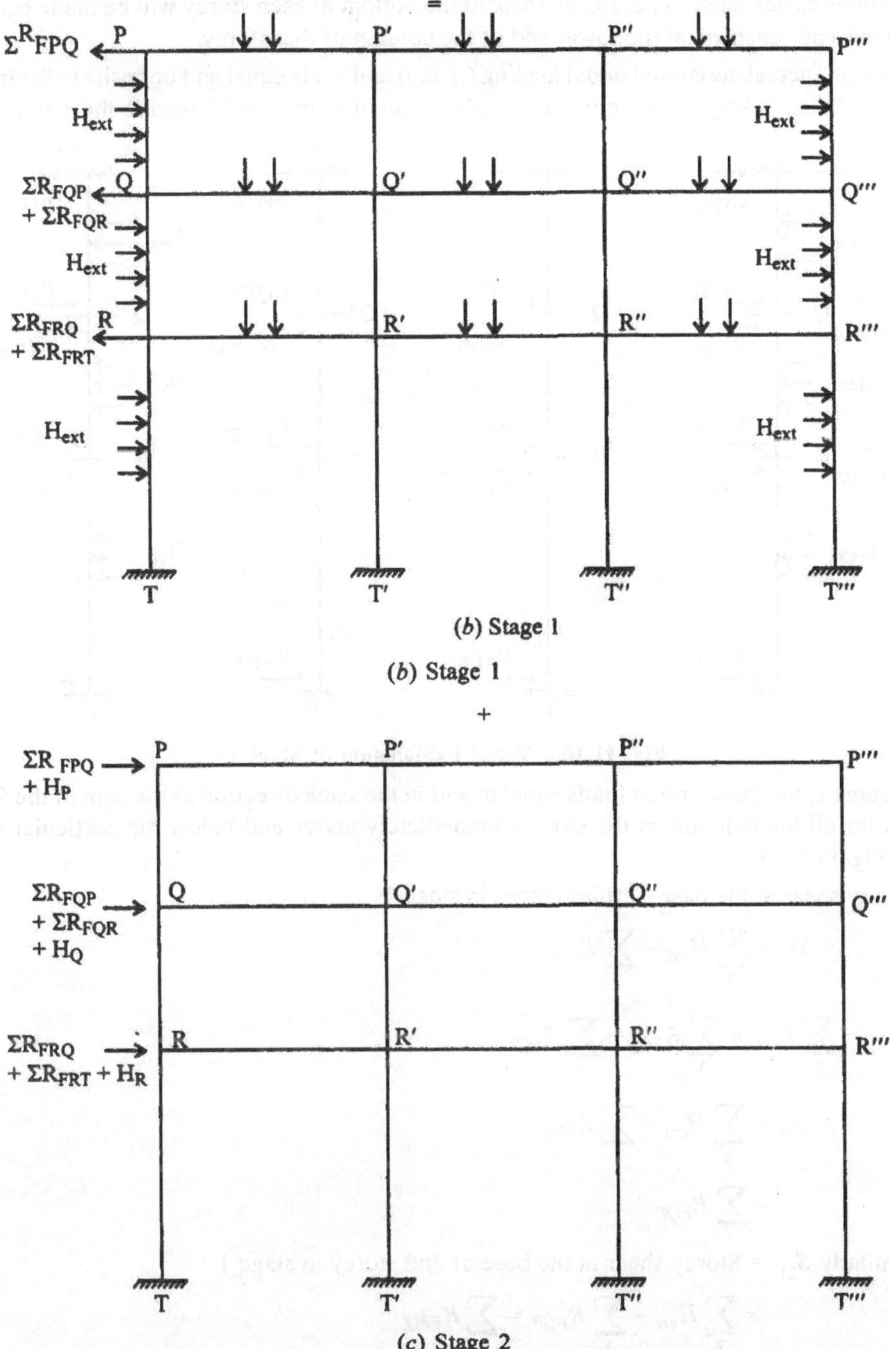

(b) Stage 1

(b) Stage 1

(c) Stage 2

Fig. 11.102. Frame with general loading

In stage 1. (Fig. 11.102 (b)) only the non nodal horizontal loading and the vertical loading are taken into account. In addition certain imaginery horizontal point loads are applied at the various nodes of the frame such that

$S_{r1} = \sum_r R_{FQP}$ at every storey (where the index 1 after S_r indicates that the values correspond to calculations as per stage 1) *i.e.* storey shear at the bottom of each storey will be made equal to the sum of fixed-end reactions at the lower end of the column of that storey.

In stage 2 actual horizontal nodal loading (if any) and loads equal and opposite to the imaginery loads (introduced in stage 1) are applied. Combination of stage 1 and 2 lead to the actual case.

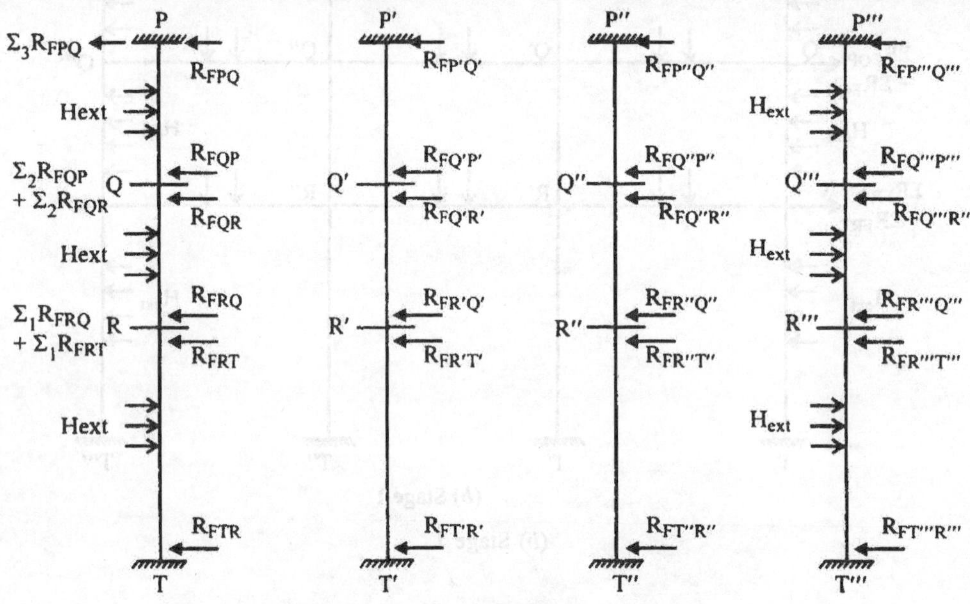

Fig. 11.103. Visual Explanation of Stage 1

In stage 1, imaginery point loads equal to and in the same direction as the sum of the fixed end reactions for all the columns in the storeys immediately above and below the particular node are applied. (Fig. 11.103)

Storey shear at the base of third storey in stage 1

$$= S_{31} = \sum_3 H_{ext} - \sum_3 R_{FPQ}$$

But $\sum_3 H_{ext} = \sum_3 R_{FPQ} + \sum_3 R_{FQP}$

$$\therefore S_{31} = \sum_3 H_{ext} - \sum R_{FPQ}$$

$$= \sum_3 R_{FQP}$$

Similarly S_{21} = Storey shear at the base of 2nd storey in stage 1

$$= \sum_2 H_{ext} - \sum_2 R_{FQR} = \sum_2 R_{FRQ}$$

S_{11} = Storey shear at the base of 1st storey in stage 1

$$= \sum_1 H_{ext} - \sum_1 R_{FRT} = \sum R_{FTR}$$

In stage 1 $Sr_1 = \sum_r R_{FQP}$ at every storey.

On substitution, the equation

$$3\sum_r M'_{PQ} + 3\sum_r M'_{QP} + 2\sum_r M''_{PQ} + S_r \cdot h_r - h_r \sum_r R_{FQP} = 0 \quad \text{...(11.20)}$$

becomes $$3\sum_r M'_{PQ} + 3\sum_r M'_{QP} + 2\sum_r M''_{PQ} = 0 \quad \text{...(11.21)}$$

In stage 2 with only nodal loads

$$\sum_r R_{FQP} = 0$$

and the equation 11.19 reduces to

$$\sum_r M'_{PQ2} + 3\sum_r M'_{QP2} + 2\sum M''_{PQ2} + S_r h_{r2} = 0 \quad \text{...(11.22)}$$

(*iv*) The displacement contribution equation for the general case of loading:

Adding Equations 11.20 and 11.21 and considering that the addition of quantities for stage 1 and stage 2 gives the quantities for actual case (which are expressed without any index) we get,

$$3\sum_r M'_{PQ} + 3\sum_r M'_{QP} + 2\sum_r M''_{PQ} + S_{r2} h_r = 0$$

or` $$\sum_r M''_{PQ} = -\frac{3}{2}\left(\frac{S_{r2}.h_r}{3} + \sum_r M'_{PQ} + \sum_r M'_{QP}\right) \quad \text{....(11.23)}$$

or $$M''_{PQ} = D.F.\left(\frac{Sr_2 h_r}{3} + \sum_r M'_{PQ} + \sum_r M'_{QP}\right) \quad \text{...(11.24)}$$

where D.F. is the distribution factor as already defined.

It should be clearly understood that S_{r2} is the storey shear for stage 2 only and not the actual storey shear.

Example 11.32. Find the moments at the critical points of the frame given below. (Fig. 11.104)

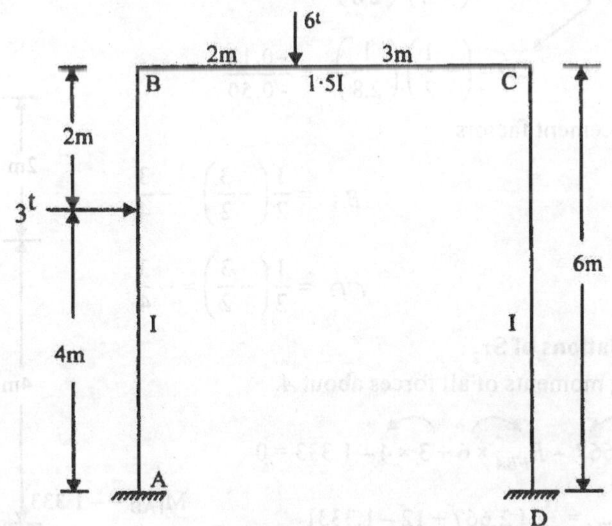

Fig. 11.104. Ex. 11.32.

Solution:

(*i*) FEMS

$$M_{FAB} = \frac{-3 \times 4 \times 2^2}{6^2} = -1.333 \text{ tm}$$

$$M_{FBA} = \frac{+3 \times 2 \times 4^2}{6^2} = 2.667$$

$$M_{FBC} = \frac{-6 \times 2 \times 3^2}{5^2} = -4.32$$

$$M_{FBC} = \frac{+6 \times 3 \times 2^2}{5^2} = 2.88$$

$$M_{FCD} = M_{FDC} = 0$$

(*ii*) *k* values

$$AB = \frac{I}{6} = \frac{I}{6}$$

$$BC = \frac{1.5I}{5} = 0.3I$$

$$CD = \frac{I}{6} = \frac{I}{6}$$

ratio of $k_{AB} : k_{BC} : k_{CD} = 1 : 1.8 : 1$

(*iii*) **Rotation factors**

Joint *B*

$$BA = \left(-\frac{1}{2}\right) \times \frac{1}{2.8} = -0.18$$

$$BC = \left(-\frac{1}{2}\right) \times \left(\frac{1.8}{2.8}\right) = \frac{-0.32}{-0.50}$$

Joint *C*

$$CB = \left(-\frac{1}{2}\right)\left(\frac{1.8}{2.8}\right) = -0.32$$

$$CD = \left(-\frac{1}{2}\right)\left(\frac{1}{2.8}\right) = \frac{-0.18}{-0.50}$$

(*iv*) Displacement factors

$$BA = \frac{1}{2}\left(-\frac{3}{2}\right) = -\frac{3}{4}$$

$$CD = \frac{1}{2}\left(-\frac{3}{2}\right) = -\frac{3}{4}$$

(*v*) **Calculations of Sr_2**

Taking moments of all forces about *A*.

$$2.667 - R_{FBA} \times 6 + 3 \times 4 - 1.333 = 0$$

or $$R_{FBA} = \frac{1}{6}[2.667 + 12 - 1.333]$$

$$= \overleftarrow{2.222}$$

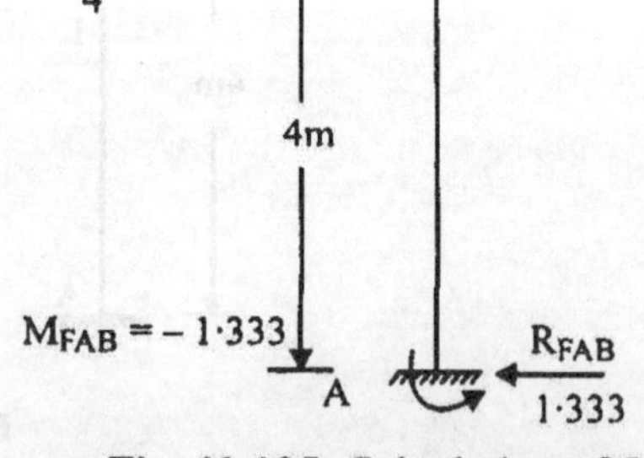

Fig. 11.105. Calculation of Sr_2.

Stage 1, therefore, consists of

(*i*) Actual loading i e. vertical loading of 6^t and horizontal load $3t$ and

(*ii*) an imaginery restraining force of $= \overleftarrow{2.222^t}$ at *B* towards left.

Stage 2 consists of a horizontal force of 2.222 → at *B*

Sr_2 is then + 2.222 (Storey shear at the base of the storey for stage 2 loading)

$$\frac{S_{r2}\,h_r}{3} = \frac{2.222 \times 6}{3} = 4.444\,t$$

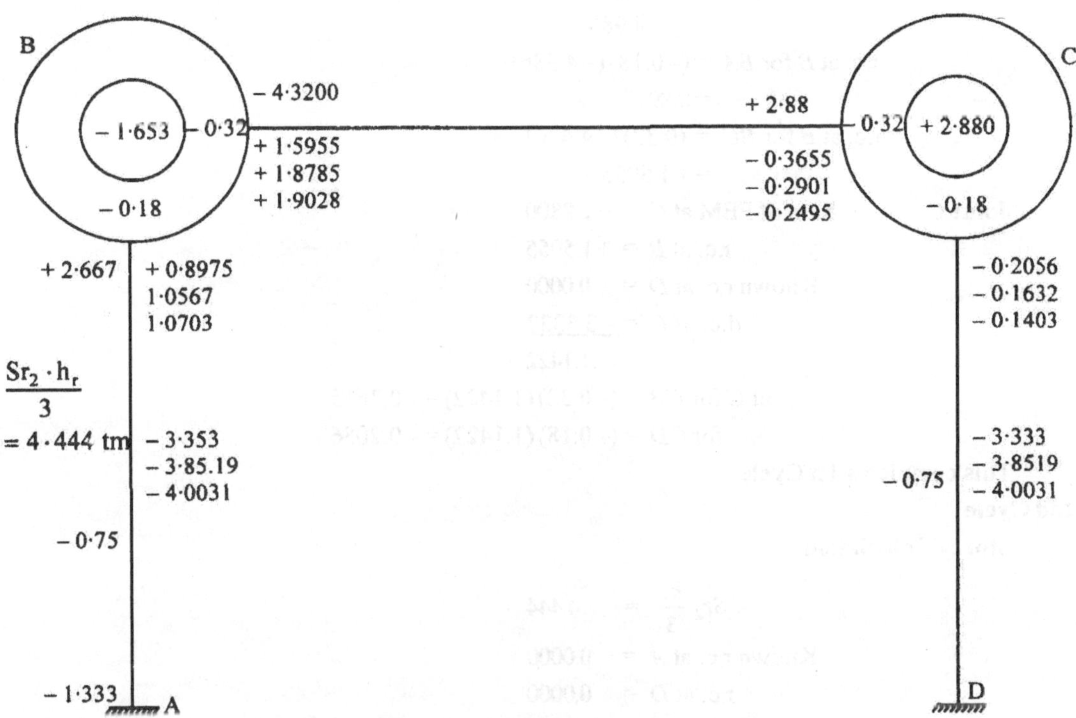

Fig. 11.106. Iteration Process.

The order in which these calculations have been carried out are:

Members *BA*, *CD* (Storey)

B, *C* (Joints)

1st Cycle

Storey Calculation

$$\frac{S_{r2}h_r}{3} = 4.444$$

Known r.c. at *A* = 0.000

at *D* = 0.000

Assumed r.c. at *B* = 0.000

Assumed r.c. at *B* = 0.000

4.444

displacement contribution.

$$BA = -\frac{3}{4} \times 4.444 = -3.333$$

$$CD = -\frac{3}{4} \times 4.444 = -3.333$$

Joint B Sum of FEM = -1.653

Known r.c. at A = 0.000

assumed r.c. at C = 0.000

d.c. at B = $\underline{-3.333}$

-4.986

r.c. at B for BA = $(-0.18)(-4.986)$

= 0.8975

r.c. at B for BC = $(0.32)(-4.436)$

= +1.5955

Joint C Sum of FEM at C = +2.8800

r.c. at B = +1.5955

Known r.c. at D = 0.0000

d.c. at C = $\underline{-3.3333}$

1.1422

r.c. at C for CB = $(-0.32)(1.1422) = -0.3655$

for CD = $(-0.18)(1.1422) = -0.2056$

This completes 1st Cycle

2nd Cycle

Storey Calculation

$$Sr_2 \frac{h_r}{3} = 4.444$$

Known r.c. at A = 0.0000

r.c. at D = 0.0000

r.c. at B = 0.8975

r.c. at C = $\underline{-0.2056}$

5.1359

displacement contribution for BA $= -\frac{3}{4}(5.1359) = -3.8519$

CD $= -\frac{3}{4}(5.1359) = -3.8519$

Joint B Sum of F.E.M. = -1.6530

r.c. at A = 0.0000

r.c. at C = -0.3655

d.c. at B = $\underline{-3.8519}$

-5.8704

r.c. for BA = $(-0.18)(-5.8704)$

$= 1.0567$

for BC $= (-0.32)(-5.8704)$

$=$ **1.8785**

Joint C

Sum of F.E.M. $= +2.8800$

r.c. at B $= +1.8785$

r.c. at D $= 0.0000$

d.c. at C $= \underline{-3.8519}$

$+0.9066$

r.c. for CB $= (-0.32)(+0.9066)$

$=$ **– 0.2901**

for CD $= (-0.18)(+0.9066)$

$=$ **– 0.1632**

3rd Cycle

Storey Calculation

$$Sr_2 \cdot \frac{h_r}{3} = 4.4440$$

Known r.c. at A $= 0.0000$

r.c. at D $= 0.0000$

r.c. at B $= 1.0567$

r.c. at C $= \underline{-0.1632}$

5.3375

displacement contribution

for AB $= (-0.75)(5.3375)$

$= -4.0031$

for CD $= (2 0.75)(5.3375)$

$=$ **– 4.0031**

Joint *B*

Sum of F.E.Ms $= 1.6530$

r.c. at A $= 0.0000$

r.c. at C $= -0.2901$

d.c. at B $=- \underline{-4.0031}$

-5.9462

r.c. at B for BA $= (-0.18)(-5.9462)$

$= +1.0703$

for BC $= (-0.32)(-5.9462)$

$=$ **+ 1.9028**

Joint *C*

Sum of F.E.M $= +2.880$

r.c. at B $= 1.9028$

r.c. at D $= 0.0000$

d.c. for CD $= \underline{-4.0031}$

0.7797

r.c. for CB $= -0.32 \times 0.7797$

$= -0.2495$

$$\text{for } CD = -0.18 \times 0.7797$$
$$= -0.1403$$

Final Moments

M_{AB} =

$$\begin{aligned} \text{F.E.M.} &= -1.3330 \\ \text{Twice r.c. at } A &= 0.0000 \\ \text{r.c. at } B &= 1.0703 \\ \text{d.c. at } A &= \underline{-4.0031} \\ & -4.2661 \end{aligned}$$

M_{BA} =

$$\begin{aligned} \text{F.E.M.} &= +2.6670 \\ \text{Twice r.c. at } B &= 1.0703 \\ & 1.0703 \\ \text{r.c. at } A &= 0.0000 \\ \text{d.c. at } B &= \underline{-4.0031} \\ & +0.8045 \end{aligned}$$

M_{BC} =

$$\begin{aligned} \text{F.E.M.} &= -4.3200 \\ \text{Twice r.c. at } B &= +1.9028 \\ & +1.9028 \\ \text{r.c. at } C &= \underline{-0.2495} \\ & -0.7639 \end{aligned}$$

M_{CB} =

$$\begin{aligned} \text{F.E.M.} &= +2.8800 \\ \text{Twice r.c. at } B &= -0.2495 \\ & -0.2495 \\ \text{r.c. at } B &= \underline{1.9028} \\ & +4.2838 \end{aligned}$$

M_{BD} =

$$\begin{aligned} \text{F.E.M.} &= 0.0000 \\ \text{Twice r.c. at } C &= -0.1403 \\ & -0.1403 \\ \text{r.c. at } D &= 0.000. \\ \text{d.c. at } C &= \underline{-4.0031} \\ & -4.2837 \end{aligned}$$

M_{DC} =

$$\begin{aligned} \text{F.E.M.} &= 0.0000 \\ \text{Twice r.c. at } D &= 0.0000 \\ \text{r.c. at } C &= -0.1403 \\ \text{d.c. at } D &= \underline{-4.0031} \\ & -4.1434 \end{aligned}$$

Example 11.33. Analyse the frame shown in Fig. 11.107 below by Kanis method.

[AMIE Winter 1987]

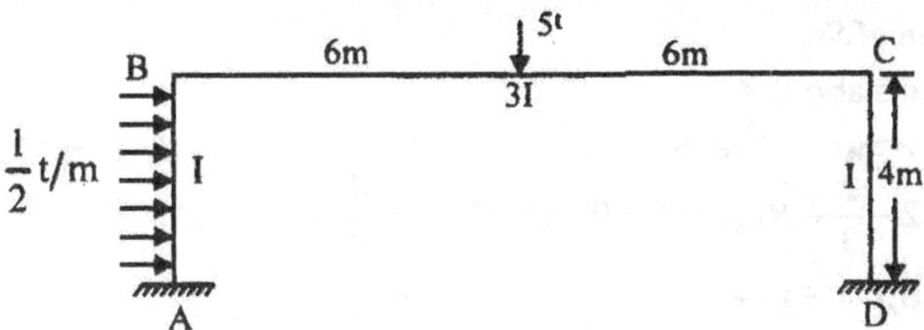

Fig. 11.107. Ex. 11.33

Solution: (*i*). F.E.M.

$$M_{FAB} = -\frac{wL^2}{12} = -\frac{1 \times 16}{2 \times 12}$$

$$= -\frac{2}{3} \text{ tm}$$

$$M_{FBA} = +\frac{wL^2}{12} = \frac{1}{2} \times \frac{16}{12} = +\frac{2}{3} \text{ tm}$$

$$M_{FBC} = -\frac{PL}{8} = \frac{-5 \times 12}{8} = -7.5 \text{tm}$$

$$M_{FCB} = +\frac{PL}{8} = +7.5 \text{tm}$$

(*ii*) Stiffness

$$k_{BA} = \frac{I}{4}$$

$$k_{BC} = \frac{3I}{12} = \frac{I}{4}$$

$$k_{CD} = \frac{I}{4}$$

Ratio of stiffnesses = 1 : 1 : 1.

(*iii*) Rotation factors

Joint *B* $\quad BA \quad \left(-\frac{1}{2}\right)\left(\frac{1}{2}\right) = -0.25$

$\quad BC \quad \left(-\frac{1}{2}\right)\left(\frac{1}{2}\right) = -0.25$

Joint *C* $\quad CB \quad \left(-\frac{1}{2}\right)\left(\frac{1}{2}\right) = -0.25$

$\quad CD \quad \left(-\frac{1}{2}\right)\left(\frac{1}{2}\right) = -0.25$

(*iv*) Displacement factors

$$AB = \frac{1}{2}\left(-\frac{3}{2}\right) = -\frac{3}{4}$$

$$CD = \frac{1}{2}\left(-\frac{3}{2}\right) = -\frac{3}{4}$$

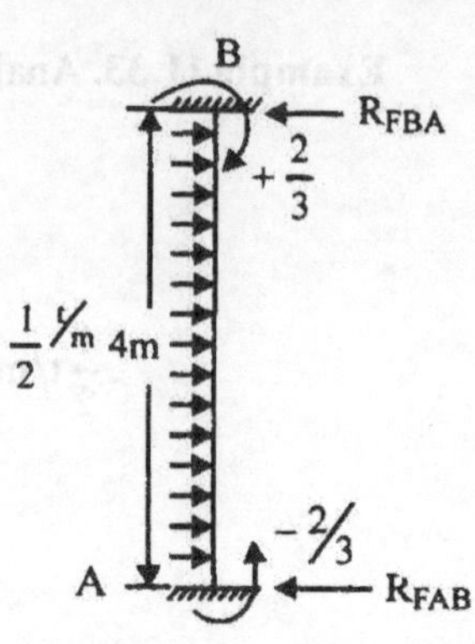

Fig. 11.108. Calculation of Sr_2

(v) Calculation of Sr_2

Taking moments about A

$$+\frac{2}{3}+\frac{1}{2}\times 4\times 2-\frac{2}{3}-R_{FBA}\times 4=0 \quad \text{or} \quad R_{FBA}=\frac{4}{4}=1$$

Storey shear $S_{r2} = +1 \rightarrow$

$$\frac{S_{r2}\times h_r}{3} = \frac{1\times 4}{3} = 1.333 \rightarrow$$

(vi) Iteration

Ist Cycle **Storey Calculation**

$\frac{S_r \cdot h_r}{3}$ = 1.3333

Known r.c. at A and D = 0.0000

Assumed r.c. at B = 0.0000

at C = 0.0000

1.3333

d.c. for BA = −0.75 (1.3333) = −1.000

CD = −0.75 (1.3333) = −1.000

Joint B

Sum of F.E.M = −6.8333

Known r.c. at A = 0.0000

Assumed r.c. at C = 0.0000

d.c. at B = −1.0000

−7.8333

r.c. for BA = (−0.25) (−7.8333) = 1.9583

for BC = (−0.25) (−7.8333) = 1.9583

Joint C

Sum of F.E.M. = +7.5000

Known r.c. at D = 0.0000

Known r.c. at B = +1.9583

d.c. at C = −1.0000

+8.4583

r.c. for CB = (−0.25) (8.4583) = −2.1146

CD = (−0.25) (8.4583) = −2.1146

2nd Cycle **Storey Calculation**

$\frac{S_r . h_r}{3}$ = +1.3333

Known r.c. at A and D = 0.0000

r.c. at B = +1.9583

at C = −2.1146

1.1770

d.c. for BA and CD = (−0.75) (1.1770) = −0.8827

Joint B

Sum of FEM = −6.8333

Known r.c. at A = 0.0000

Known r.c. at C = −2.1146

d.c. at B = −0.8827

−9.8306

r.c. for BA and BC = (−0.25) (−9.8306)

= +2.4577.

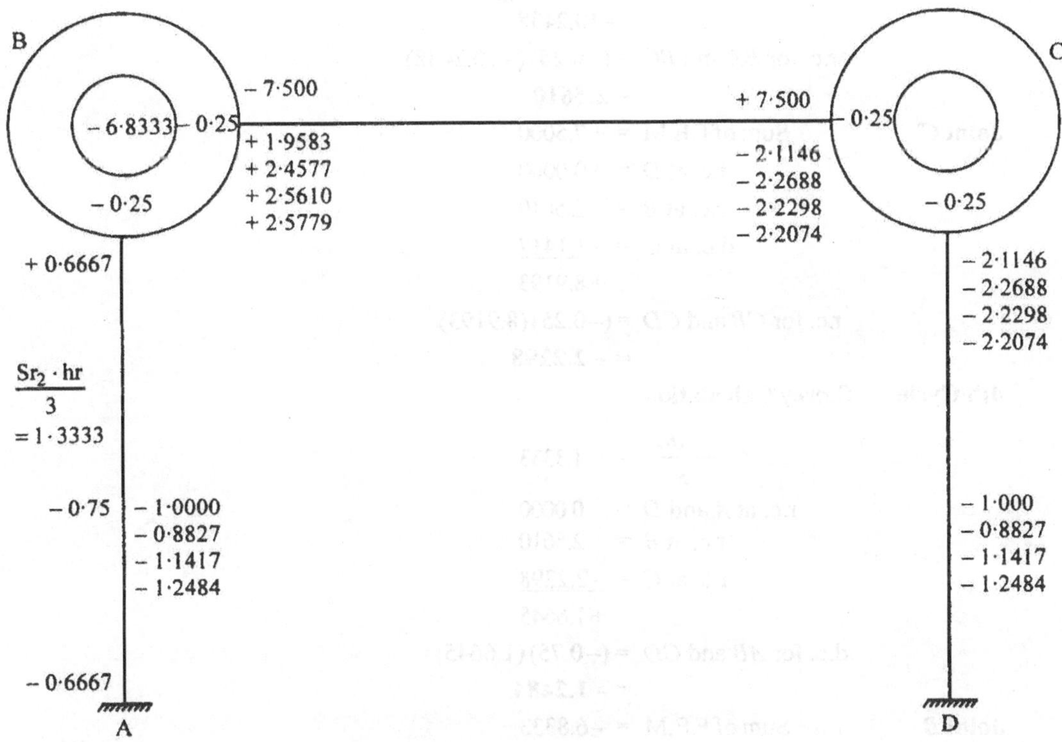

Fig. 11.109. Iteration Process Ex. 11.33

Joint C

Sum of F.E.Ms = +7.5000

r.c. at D = +0.0000

r.c. at B = +2.4577

d.c. at C = −0.8827

+9.0750

r.c. for CB and CD = (−0.25) (9.075)

= −2.2688

3rd Cycle **Storey Calculation**

$$\frac{Sr.hr}{3} = 1.3333$$

r.c. at A & D = 0.0000

r.c. at B = 2.4577

at C = $\underline{-2.2688}$

1.222

d.c. for AB and CD = (−0.75) (1.5222)

= −1.1417

Joint *B* Sum of F.E.Ms = −6.8333

r.c. at A = +0.0000

r.c. at C = −2.2688

d.c. at B = $\underline{-1.1417}$

−10.2438

r.c. for BA and BC = (−0.25) (−10.2438)

= **2.5610**

Joint *C* Sum of F.E.M = +7.5000

r.c. at D = +0.0000

r.c. at B = 2.5610

d.c. at C = $\underline{-1.1417}$

+8.9193

r.c. for CB and CD = (−0.25) (8.9193)

= **− 2.2298**

4th Cycle **Storey Calculation**

$$\frac{S_r \cdot h_r}{3} = 1.3333$$

r.c. at A and D = 0.0000

r.c. at B = 2.5610

r.c. at C = $\underline{-2.2298}$

+1.6645

d.c. for AB and CD = (−0.75) (1.6645)

= **− 1.2484**

Joint *B* Sum of F.E.M = −6.8333

r.c. at A = 0.0000

r.c. at C = −2.2298

d.c. at B = $\underline{-1.2484}$

−10.3115

r.c. for BA and BC = (−0.25) (−10.3115)

= **− 2.5779**

Joint *C* Sum of F.E.M = +7.5000

r.c. at D = 0.0000

r.c. at B = +2.5779

d.c. at C = $\underline{-1.2484}$

+8.3295

r.c. for CB = (−0.25) (8.8295)

and CD = −2.2074.

Calculation of Final Moments

M_{AB}

F.E.M =	−0.6667
Twice r.c. at A =	0.0000
r.c. at B =	+2.5779
d.c. at A =	−1.2484
	+0.6628

M_{BA}

F.E.M =	+0.6667
Twice r.c. at A =	+2.5779
	+2.5779
r.c. at A =	0.0000
d.c. at B =	−1.2484
	4.5741

M_{BC}

F.E.M =	−7.5000
Twice r.c. at B =	+2.5779
	+2.5779
r.c. at C =	−2.2074
	−4.5516

M_{CB}

FEM =	+7.5000
Twice r.c. at C =	−2.2074
	−2.2074
r.c. at B =	+2.5779
	+5.6631

M_{CD}

F.E.M =	0.0000
Twice r.c. at C =	−2.2074
	−2.2074
r.c. at D =	0.0000
d.c. at C =	−1.2484
	−5.6632.

M_{DC}

FEM =	0.0000
Twice r.c. at D =	0.0000
r.c. at C =	−2.2074
d.c. at D =	−1.2484
	−3.4558.

11.13.7 Derivation of Special Cases

(*i*) When columns are not of same height in a storey.

Equation of equilibrium for any column is already derived as,

$$M_{PQ} + M_{QP} + M_{ext.} + H.h = 0 \qquad \text{...(11.17)}$$

This can be rewritten as

$$\frac{M_{PQ}}{h} + \frac{M_{QP}}{h} + \frac{M_{ext.}}{h} + H = 0 \qquad \text{.....(11.17 a)}$$

Substituting the values of M_{PQ} and M_{QP} in terms of rotation contribution and displacement contribution and then adding the terms of such equations for all columns of the rth storey the equation transforms to

$$\sum_r\left(\frac{M_{F\,PQ}}{h}\right)+\sum_r\left(\frac{M_{F\,QP}}{h}\right)+3\sum_r\left(\frac{M'_{PQ}}{h}\right)+3\sum_r\left(\frac{M'_{QP}}{h}\right)$$

$$+2\sum_r\left(\frac{M''_{PQ}}{h}\right)+\sum_r\left(\frac{M_{ext.}}{h}\right)+S_r=0 \qquad \text{....(11.25)}$$

Similarly considering the column to be fixed at ends and adding up the terms of equilibrium equations for all such columns of the rth storey we get,

$$\sum_r\left(\frac{M_{F\,PQ}}{h}\right)+\sum_r\left(\frac{M_{F\,QP}}{h}\right)+\sum_r\left(\frac{M_{ext.}}{h}\right)+\sum_r R_{F\,QP}=0 \qquad \text{....(11.26)}$$

Subtracting eqn. 11.26 from equation 11.25 we get the following equation,

$$3\sum_r\left(\frac{M'_{PQ}}{h}\right)+3\sum_r\left(\frac{M'_{QP}}{h}\right)+2\sum_r\left(\frac{M''_{PQ}}{h}\right)+Sr-\sum_r R_{F\,QP}=0 \qquad \text{....(11.27)}$$

As already derived, the frame problem is visualised in two stages.

In stage 1 $S_{r1}=\sum_r R_{F\,QP}$ and equation 11.27 reduces to

$$3\sum_r\left(\frac{M'_{PQ1}}{h}\right)+3\sum_r\left(\frac{M'_{QP1}}{h}\right)+2\sum_r\left(\frac{M''_{PQ1}}{h}\right)=0 \qquad \text{....(11.28)}$$

In stage 2, there are no fixed end moments or fixed end reactions and Equation 11.27 then becomes

$$3\sum_r\left(\frac{M'_{PQ2}}{h}\right)+3\sum_r\left(\frac{M'_{QP2}}{h}\right)+2\sum_r\left(\frac{M''_{PQ2}}{h}\right)+S_{r2}=0$$

Adding the equations for stage 1 and stage 2

$$3\sum_r\left(\frac{M'_{PQ}}{h}\right)+3\sum_r\left(\frac{M'_{QP}}{h}\right)+2\sum_r\left(\frac{M''_{PQ}}{h}\right)+S_{r2}=0$$

i.e.

$$\sum_r\left(\frac{M'_{PQ}}{h}\right)=-\frac{3}{2}\left\{\frac{S_{r2}}{3}+\sum_r\left(\frac{M'_{PQ}}{h}+\frac{M'_{QP}}{h}\right)\right\}$$

Multiplying the equation 11.29 by h_r which is the height of the most of the column of the rth storey

$$\sum_r\left(\frac{M''_{PQ}}{h}\times h_r\right)=-\frac{3}{2}\left\{\frac{S_{r2}.h_r}{3}+\sum_r\left(\frac{M'_{PQ}\,h_r}{h}\times\frac{M'_{QP}\,h_r}{h}\right)\right\}$$

Putting $\frac{h_r}{h}=C_{PQ}$ = height modification factor (h, m) the equation reduces to

$$\sum_r C_{PQ}.M''_{PQ}=-\frac{3}{2}\left[\frac{S_{r2}.h_r}{3}+\sum_r\left\{C_{PQ}\left(M'_{PQ}+M'_{QP}\right)\right\}\right] \qquad \text{...(11.30)}$$

To get individual displacement components we start with

$$M''_{PQ} = \frac{6EI\delta}{h^2}$$

$$= \frac{AK}{h} = B \times K \times C_{PQ}$$

where $6E\delta = A$ = constant and $\frac{I}{h} = K$

$$B = \frac{A}{h_r} \text{ and } C_{PQ} = \frac{h_r}{h}.$$

Now

$$\frac{M''_{PQ}}{\sum_r \left(C_{PQ} \times M''_{PQ}\right)} = \frac{B \times K \times C_{PQ}}{\sum\left(C_{PQ} \times B \times K \times C_{PQ}\right)}$$

$$= \frac{K \times C_{PQ}}{\sum\left(K \times C_{PQ}^2\right)}$$

It should be observed that B being a constant for a particular storey can be cancelled from the numerator and denominator

$$\therefore \quad M''_{PQ} = \frac{K \times C_{PQ}}{\sum_r\left(K \times C_{PQ}^2\right)} . \sum_r \left(C_{PQ} \times M''_{PQ}\right)$$

$$= \frac{K \times C_{PQ}}{\sum_r\left(K \times C_{PQ}^2\right)} \times -\frac{3}{2}\left\{\frac{S_{r2}.h_r}{3} + \sum_r C_{PQ}\left(M'_{PQ} + M'_{QP}\right)\right\} \quad \text{....(11.31)}$$

or

$$M''_{PQ} = D.F.\left\{\frac{S_{r2}.h_r}{3} + \sum_r C_{PQ}\left(M'_{PQ} + M'_{QP}\right)\right\} \quad \text{....(11.31}a\text{)}$$

where D.F. = displacement factor and is modified to $\frac{K \times C_{PQ}}{\sum_r\left(K \times C_{PQ}^2\right)} \times \left(-\frac{3}{2}\right)$

when $h_r = h$ *i.e.*, all columns are of same height $C_{PQ} = \frac{h_r}{h} = 1$ and D.F. $= -\frac{3}{2}$ as already derived.

Equation 11.30 is valid for general loading. If loading is only at nodal points $S_{r2} = S_r$ and for purely vertical loading $S_{r2} = 0$.

The general, procedure, therefore involves

(*i*) Rotation contributions at upper and lower ends of columns

(*ii*) height modification factor C_{PQ}.

(*iii*) S_{r2} — storey shear for stage 2.

(*iv*) Displacement factor.

(*ii*) Column with hinged end : To replace a column with a hinged end by a substitute column with fixed end, the following two conditions must be statisfied.

(*i*) a moment *M* applied at the top of the column must give rise to equal rotations at the top in both cases.

(*ii*) a lateral sway δ at the top of a column must give rise to equal fixed end moments at the top in both cases.

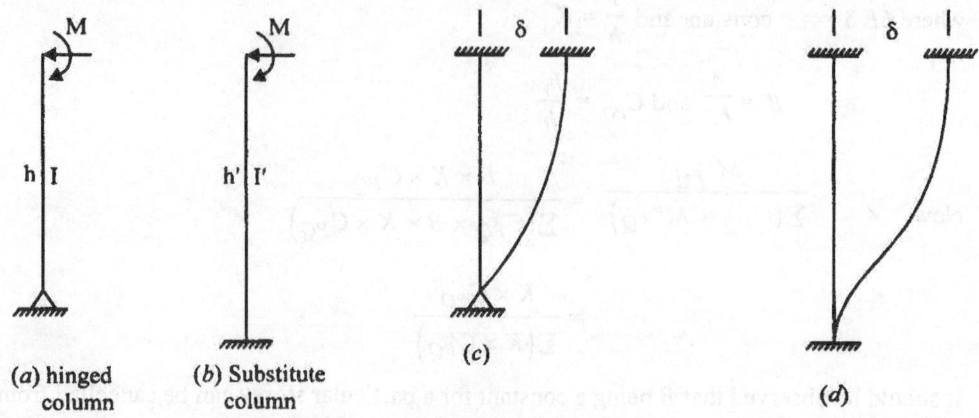

Fig. 11.110. Substitute Column.

Consider a column with hinged base and a substitute column with fixed base [Fig. 11.110 (*a*) and (*b*)]

Due to applied moment *M*.

$$\text{rotation of hinged column} = \frac{Mh}{3EI}$$

$$\text{rotation with fixed base} = \frac{Mh'}{4EI'}$$

To give same rotation.

$$\frac{Mh}{3EI} = \frac{Mh'}{4EI'}$$

or $$\frac{I'}{h'} = \frac{3}{4}\frac{I}{h}$$

or $$K' = \frac{3}{4}.K$$

Similarly for same lateral sway δ [Fig. 11.110 (*c*) and (*d*)]

$$\text{Fixed end moment at top with hinged base} = \frac{3EI\delta}{h^2}$$

$$\text{Fixed end moment at top with fixed base} = \frac{6EI'\delta}{h'^2}$$

Equalising :

$$\frac{3EI\delta}{h^2} = \frac{6EI'\delta}{h'^2}$$

or $$\frac{I}{h^2} = \frac{2I'}{h'^2}$$

or $$\frac{K}{h} = \frac{2K'}{h'}$$

or $$\frac{h}{h'} = \frac{K}{2K'} = \frac{4}{3} \times \frac{1}{2} = \frac{2}{3}$$

or $$h' = \frac{3}{2}.h.$$

Hence a column with hinged base can be replaced by a column with fixed base.

provided K' is taken as $\frac{3}{4}.K$

and h' is taken as $\frac{3}{2}.h$.

(*iii*) Modification to the displacement contribution equation— The displacement contribution equation must be modified when hinged column is replaced by fixed column with substitute height.

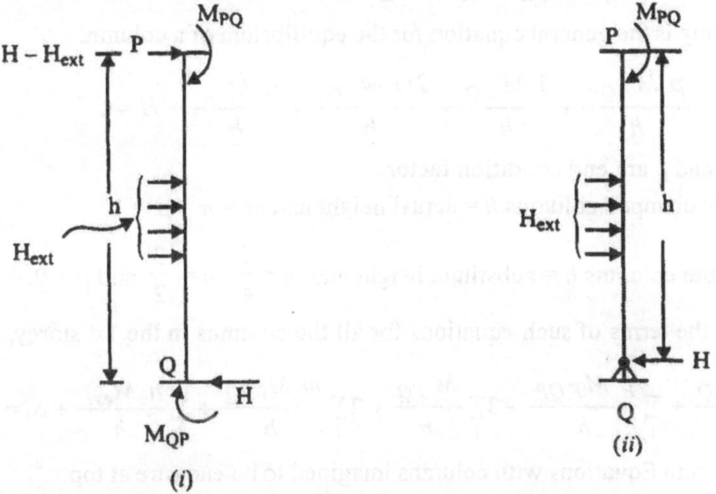

Fig. 11.111. Modification of displacement contribution equation.

(*a*) Equilibrium equation with actual and substitute columns.

In Fig. 11.111 (*i*) the column which is part of a frame has bottom end fixed.

In Fig. 11.111 (*ii*) the bottom end is hinged.

From Fig. (*i*) when bottom end is fixed taking moments about P.

$$M_{PQ} + M_{Qp} + M_{ext.} + H.h = 0$$

but $$M_{PQ} = M_{FPQ} + 2M'_{PQ} + M''_{PQ}. \qquad [M'_{QP} = 0]$$

and $$M_{QP} = M_{FQP} + M'_{PQ} + M''_{QP}. \qquad [M'_{QP} = 0]$$

$\therefore$ $$M_{FPQ} + M_{FQP} + 3M'_{PQ} + 2M''_{QP} + M_{ext.} + H.h = 0$$

i.e., $$\frac{M_{FPQ}}{h} + \frac{M_{FQP}}{h} + \frac{3M'_{PQ}}{h} + \frac{2M''_{QP}}{h} + \frac{M_{ext.}}{h} + H = 0 \qquad ....(11.32)$$

From Fig. (*ii*) when bottom end is hinged taking moments about P

$$M_{PQ} + M_{ext.} + H.h = 0$$

where h is the actual height of the hinged column.

If the hinged column is now substituted by an encastre column, then

$$M_{PQ} = M_{FPQ} + 2M'_{PQ} + M''_{PQ}$$

(M'_{QP} is zero for the substitute column since Q is encastre) when M_{FPQ} is the fixed end moment calculated for the actual end condition (hinged) while M''_{PQ} is the rotation contribution for the substitute column.

$$\therefore \quad M_{FPQ} + 2M'_{PQ} + M''_{PQ} + M_{ext.} + H.h = 0$$

i.e., $$\frac{M_{FPQ}}{h} + \frac{2M'_{FPQ}}{h} + \frac{M''_{PQ}}{h} + \frac{M_{ext.}}{h} + H = 0.$$

Now h' = height of substitute column

$$= \frac{3}{2}.h.$$

$$\therefore \frac{3}{2}.\frac{M_{FPQ}}{h'} + \frac{3M'_{PQ}}{h'} + \frac{3}{2}.\frac{M''_{PQ}}{h'} + \frac{3}{2}\frac{M_{ext.}}{h'} + H = 0 \quad ....(11.33)$$

The following is the general equation for the equilibrium of a column.

$$\frac{n\,M_{FPQ}}{h} + \frac{p\,M_{FQP}}{h} + \frac{3\,M'_{PQ}}{h} + \frac{2m\,M''_{PQ}}{h} + \frac{n.M_{ext.}}{h} + H = 0 \quad ....(11.34)$$

when m, n and p are end condition factors.

For actually clamped columns h = actual height and $m = n = p = 1$

For substitute columns h = substitute height and $m = \frac{3}{4}$, $n = \frac{3}{2}$ and $p = 0$.

Adding up the terms of such equations for all the columns in the 1st storey, we get

$$\sum_1 \frac{n\,M_{FPQ}}{h} + \sum_1 \frac{p\,M_{FQP}}{h} + 3\sum_1 \frac{M'_{PQ}}{h} + 2\sum_1 \frac{m\,M''_{PQ}}{h} + \sum_1 \frac{n.M_{ext.}}{h} + S_1 = 0 \quad ..(11.35)$$

(*b*) Equilibrium Equations with columns imagined to be encastre at top.

Case 1 Fig. 11.112 (*a*) **Column fixed at base.**

Taking moments about P

$$M_{FPQ} + M_{FQP} + M_{ext.} + R_{FQP}.h = 0$$

or $$\frac{M_{FPQ}}{h} + \frac{M_{FQP}}{h} + \frac{M_{ext.}}{h} + R_{FQP} = 0 \quad ....(11.36)$$

Case II. Column hinged at base (Fig. 11.112 (*b*))

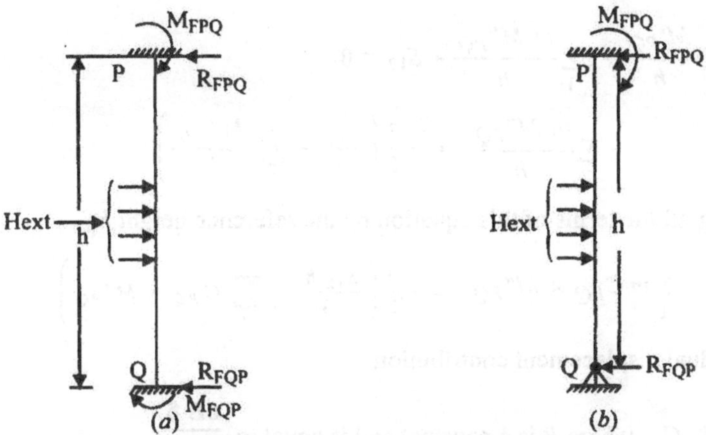

Fig. 11.112. Column hinged at base.

Taking moments about P

$$M_{FPQ} + M_{ext.} + R_{FQP} \cdot h = 0$$

or
$$\frac{M_{FPQ}}{h} + \frac{M_{ext.}}{h} + R_{FQP} = 0$$

Now h' = height of substitute column

$$= \frac{3}{2} h$$

or
$$\frac{3}{2}\frac{M_{FPQ}}{h'} + \frac{3}{2}\frac{M_{ext.}}{h'} + R_{FQP} = 0 \quad \text{....(11.37)}$$

The generalised equation in corporating the two end conditions is

$$\frac{n\, M_{FPQ}}{h} + \frac{p\, M_{FQP}}{h} + \frac{n.\, M_{ext.}}{h} + R_{FQP} = 0 \quad \text{....(11.38)}$$

where for clamped condition $m = n = p = 1$

for hinged condition $m = \frac{3}{4}$, $n = \frac{3}{2}$ and $p = 0$

Adding up the terms of such equations for all the columns of the first storey, we get,

$$\sum_{1} \frac{n\, M_{FPQ}}{h} + \sum_{1} \frac{p\, M_{FQP}}{h} + \sum_{1} \frac{n.\, M_{ext.}}{h} + \sum_{1} R_{FQP} = 0 \quad \text{....(11.39)}$$

Subtracting Equation (11.39) from Equation 11.35

we get
$$3\sum \frac{M'_{PQ}}{h} + 2\sum_{1} \frac{m\, M''_{PQ}}{h} + S_1 - \sum_{1} R_{FQP} = 0$$

Visualising the frame problem in two stages (stage 1 and stage 2) for stage 1

$$S_1 = \sum_{1} R_{FQP_1}$$

and stage 2
$$\sum_{1} R_{FQP_2} = 0 \text{ as usual.}$$

So that $3\sum_{1}\frac{M'_{PQ}}{h}+2\sum_{1}\frac{m\,M''_{PQ}}{h}+S_{12}=0$

i.e.,
$$\sum_{1}\frac{m\,M''_{PQ}}{h}=-\frac{3}{2}\left(\frac{S_{12}}{3}+\sum_{1}\frac{M'_{PQ}}{h}\right) \qquad \text{...(11.40)}$$

Multiplying all the terms of this equation by the reference height h_1

$$\sum_{1} m\,C_{PQ}\times M''_{PQ}=-\frac{3}{2}\left(\frac{S_{12}\,h_1}{3}+\sum_{1}C_{PQ}\times M'_{PQ}\right) \qquad \text{...(11.41)}$$

The individual displacement contribution

$M''_{PQ}=B.K.\,C_{PQ}$ where B is a constant and is equal to $\frac{6E\,\delta}{h_1}$

$$\therefore \qquad \frac{M''_{PQ}}{\sum_{1} m C_{PQ}.\,M''_{PQ}}=\frac{B.K.C_{PQ}}{\sum_{1} m C_{PQ}.\,B.K.C_{PQ}}=\frac{K.C_{PQ}}{\sum m\,K C_{PQ}^2}$$

$$\therefore \qquad M''_{PQ}=\frac{K.C_{PQ}}{\sum m\,K C_{PQ}^2}.\sum_{1} m C_{PQ}.\,M''_{PQ}$$

Substituting the value of $\sum m\,C_{PQ}\times M''_{PQ}$ in the above relation, we get,

$$M''_{PQ}=\frac{K.C_{PQ}}{\sum m K C_{PQ}^2}\times\frac{-3}{2}\left(\frac{S_{12}h_1}{3}+\sum_{1}C_{PQ}\times M'_{PQ}\right) \qquad \text{....(11.42)}$$

where h_1 = arbitrary chosen reference height for the first storey.

C_{PQ} = reduction factor $\frac{h_1}{h}$

where h is the actual height for columns really encastre' and is the substitute height for substitute columns. S_{12} shall correspond to the actual end conditions of the columns of the first storey.

m = end condition factor taken as 1 for fixed end and $\frac{3}{4}$ for substitute column.

$K=\frac{I}{L}$ (actual value in the case of really clamped column and $\frac{3}{4}\frac{I}{L}$ value for hinged columns)

Example 11.34 Analyse the rigid frame shown in Fig. 11.113.

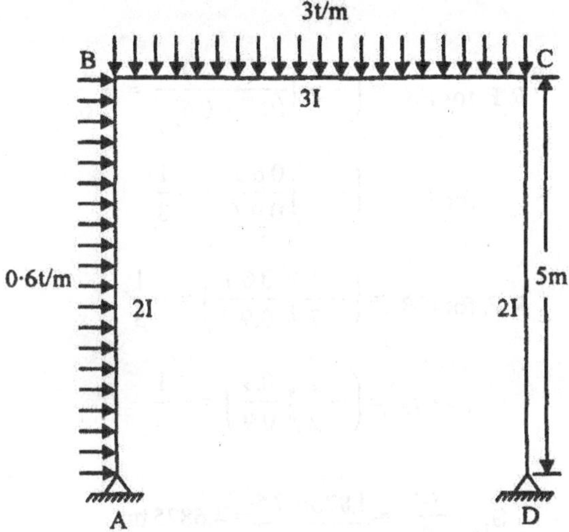

Fig. 11.113. Example 11.34

Solution:

1. Calculate FEM for actual column end conditions.

$$M_{F\,BA} = +\frac{0.6 \times 5^2}{8} = +1.875 \text{ tm}$$

$$M_{F\,BC} = -\frac{3 \times 5^2}{12} = -6.25 \text{ tm}$$

$$M_{F\,CB} = +\frac{3 \times 5^2}{12} = +6.25 \text{ tm}$$

2. Calculate $R_{F\,BA}$. Taking moment about A.

$$R_{F\,BA} \times 5 - 1.875 - 0.6 \times 5 \times 2.5 = 0$$

or $$R_{F\,BA} = \frac{1}{5}(1.875 + 7.5)$$

$$= 1.875 \leftarrow$$

$\therefore$ $$S_{12} = 1.875 \rightarrow$$

Fig. 11.114. Calculation of $R_{F\,BA}$

3. The hinged columns BA and CD are now replaced by encastre' columns with substitute heights.

$$\text{For } BA \text{ and } CD \text{ length } = \frac{3}{2} \times 5 = 7.5 \text{ m}$$

4. Modified K values.

For BA and CD

$$K' = \frac{3}{4}.K = \frac{3}{4}.\frac{2I}{5} = 0.3\,I$$

$$k \text{ for } BC = \frac{3\,I}{5} = 0.6\,I$$

Joint *B*. R.F. for *BA* $= \left(-\frac{1}{2}\right)\frac{(0.3\,I)}{(0.3+0.6)\,I} = -\frac{1}{6}$

for *BC* $= \left(-\frac{1}{2}\right)\frac{0.6\,I}{0.9\,I} = -\frac{1}{3}$

Joint *C* R.F. for *CB* $= \left(-\frac{1}{2}\right)\left(\frac{0.6\,I}{0.9\,I}\right) = -\frac{1}{3}$

for *CD* $= \left(-\frac{1}{2}\right)\left(\frac{0.3}{0.9}\right) = -\frac{1}{6}$

5. $$S_{12} \times \frac{7.5}{3} = \frac{\overrightarrow{1.875 \times 7.5}}{3} = 4.6875 \text{ tm}$$

6. Calculations of displacement factors

For $BA, +\frac{1}{2}\left(-\frac{3}{2}\right) = -\frac{3}{4}$

For $CD = -\frac{3}{4}$

Ist Cycle. Storey Calculation

$$\frac{S_{12} \times h_1}{3} = 4.6875$$

known r.c. at *A* & *D* = 0.0000

assumed r.c. at *B* & *D* = 0.0000

4.6875

d.c. for *BA* $= \left(-\frac{3}{4}\right)(4.6875) = -3.5156$

for *CD* $= \left(\frac{3}{4}\right)(4.6875) = -3.5156$

Joint *B* Sum of F.E.M = –4.3750.

r.c. at *A* = 0.0000

assumed r.c. at *C* = 0.0000

d.c. at *B* = –3.5156

–7.8906

r.c. for *BA* $= -\frac{1}{6}(-7.8906) = +1.3151$

for *BC* $= -\frac{1}{3}(-7.8906) = +2.6302.$

Joint *C* Sum of FEM = +6.2500

r.c at B = 2.6302

r.c. at D = 0.0000

r.c. at C = $\underline{-3.5156}$

5.3646

$$\text{r.c. for } CB = -\frac{1}{3}(5.3646) = 1.7882$$

$$CD = -\frac{1}{6}(5.3646) = 0.8941.$$

Fig. 11.115. Iteration process.

2nd Cycle. Storey Calculation

$$S_{12} \times \frac{h_1}{3} = 4.6875$$

r.c. at A and D = 0.0000

r.c. at B = +1.3151

r.c. at C = $\underline{-0.8941}$

5.1085

$$\text{d.c. for } BA \text{ and } CD = -\frac{3}{4} \times 5.1085 = -3.8314$$

Joint *B* Sum of FEM = –4.3750

r.c. at A = 0.0000

r.c. at C = −1.7882
d.c. at B = −3.8314
−9.9946

r.c. for BA $= -\frac{1}{6} \times (-9.9946) = 1.6658$

for BC $= -\frac{1}{3} \times (-9.9946) = 3.3315$

Joint C

Sum of FEM = +6.2500
r.c. at D = 0.0000
r.c. at B = +3.3315
d.c. at C = −3.8314
5.7501

r.c. for CB $= -\frac{1}{3} \times (5.7501) = -1.9167$

for CD $= -\frac{1}{6} \times (5.7501) = -0.9584.$

3rd Cycle Storey Calculation

$S_{12} \times \frac{h}{3}$ = 4.6875
r.c. at A & D = 0.0000
r.c. at B = 1.6658
r.c. at C = −0.9584
+5.3949

d.c. for BA $= -0.75 \times 5.3949 = -4.0462$
CD = −4.0462.

Joint B

Sum of FEM = −4.3750
r.c. at A = 0.0000
r.c. at C = 1.9167
d.c. at B = −4.0462
−10.3379

r.c. for BA $= -\frac{1}{6} \times (-10.2600) = +1.7230$

BC $= -\frac{1}{3} \times (10.2600) = +3.4460.$

Joint C

Sum of F.E.M = +6.2500
r.c. at B = +3.4460
r.c. at D = 0.0000
d.c. at C = −4.0462
+5.6498

r.c. for CB $= -\frac{1}{3} \times (+5.6498) = -1.8833$

for CD = −0.9416.

4th Cycle Storey calculation

$$\frac{S_{12} \times h_1}{3} = 4.6875$$

r.c. at A and D = 0.0000

r.c. at B = +1.7230

r.c. at C = $\underline{-0.9416}$

+5.4689

d.c. for BA and CD = (−0.75)(5.4689) = −4.1017.

Joint *B* Sum of FEM = −4.3750

r.c. at A = 0.0000

r.c. at C = −1.8833

d.c. at B = $\underline{-4.1017}$

−10.3600

$$\text{r.c. for } BA = -\frac{1}{6} \times (-10.36) = +1.7267$$

$$\text{for } BC = -\frac{1}{3} \times (-10.36) = +3.4533.$$

Joint *C* Sum of FEM = +6.2500

r.c. at B = +3.4533

r.c. at D = 0.0000

d.c. at C = $\underline{-4.1017}$

+5.6016

$$\text{r.c. for } CB = -\frac{1}{3} \times (5.6016) = -1.8672.$$

$$CD = -\frac{1}{6} \times (5.6016) = -0.9336.$$

Calculation of moments

$M_{AB} = 0$

M_{BA} FEM at B = + 1.8750

Twice r.c. at B = 1.7267

1.7267

r.c. at A = 0.0000

d.c. at B = $\underline{-4.1017}$

1.2267.

M_{BC} FEM at B = −6.2500

Twice r.c. at B = +3.4533

+3.4533

d.c. at C = $\underline{-1.8672}$

1.2106.

M_{CB} FEM at C = +6.2500

$$2\text{r.c. at } C = -1.8672$$
$$-1.8672$$
$$\text{r.c. at } B = \underline{+3.4533}$$
$$+5.9689.$$

M_{CD}

$$\text{FEM at } C = 0.0000$$
$$2\text{ r.c. at } C = -0.9336$$
$$-0.9336$$
$$\text{d.c. at } C = \underline{-4.1017}$$
$$-5.9689.$$

Example 11.35 Compute moments FEMs at critical sections of the portal given in Fig. 11.116.

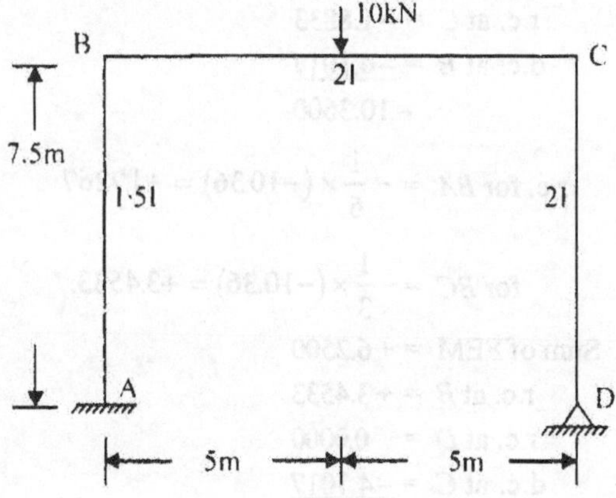

Fig. 11.116. (Example 11.35)

Solution:

(i) FEM

$$M_{FAB} = M_{FBA} = 0$$

$$M_{FBC} = \frac{-10 \times 10}{8} = -12.5 \; kNm.$$

$$M_{FCB} = +12.5 \; \boldsymbol{kNm}$$

(ii) Rotation factors. $U = -\dfrac{1}{2}\dfrac{K_{e\,rel}}{\Sigma_{Ke\,rel}}$

Joint	Member	K	Modified Stiffness	$K_{e\,rel}$	U	Check
B	BA	$\frac{1.5I}{7.5}$	$\frac{I}{5}$	1	$-\frac{1}{4}$	
	BC	$\frac{2I}{10}$	$\frac{I}{5}$	1	$-\frac{1}{4}$	$\Sigma U = -\frac{1}{2}$
C	CB	$\frac{2I}{10}$	$\frac{I}{5}$	1	$-\frac{1}{4}$	$\Sigma U = -\frac{1}{2}$
	CD	$\frac{2I}{7.5}$	$\frac{2I}{7.5} \times \frac{3}{4} = \frac{I}{5}$	1	$-\frac{1}{4}$	

***(iii)* Correction factor**

$$le(AB) = 7.5$$

$$le(CD) = \frac{7.5 \times 3}{2} = 11.25$$

Taking reference height of storey as 11.25

$$C_{AB} = \frac{lR_1}{le(AB)} = \frac{11.25}{7.5} = 1.5$$

$$C_{AD} = \frac{lR_1}{le(CD)} = 1 = 1.0$$

***(iv)* Displacement factors**

$$V = -\frac{3}{2} \frac{K_{e\,rel} . C}{\sum_r K_{e\,rel} C^2}$$

Storey	Columns	Stiffness	$K_{(e)}$	K_{erel}	C	$K_{erel} \times C$	m	$m K_{erel} \times C^2$	V
1	AB	$\frac{1 \cdot 5I}{7 \cdot 5}$	I/5	1	1.5	1.5	1	2.25	−0.75
	CD	$\frac{2I}{7 \cdot 5}$	$\frac{2I}{7 \cdot 5} \times 3/4 = I/5$	1	1	1	3/4	0.75	−0.5

$$\Sigma m\, K_{e\,rel}\, c^2 = 3.00$$

check $\Sigma mCV = 1 \times 1.5 \times (-0.75) + 0.75 \times 1 \times (-0.5)$
$= -1.5$

(*v*) Storey shear = 0 Hence $\frac{Q_{12} h_1}{3} = 0$

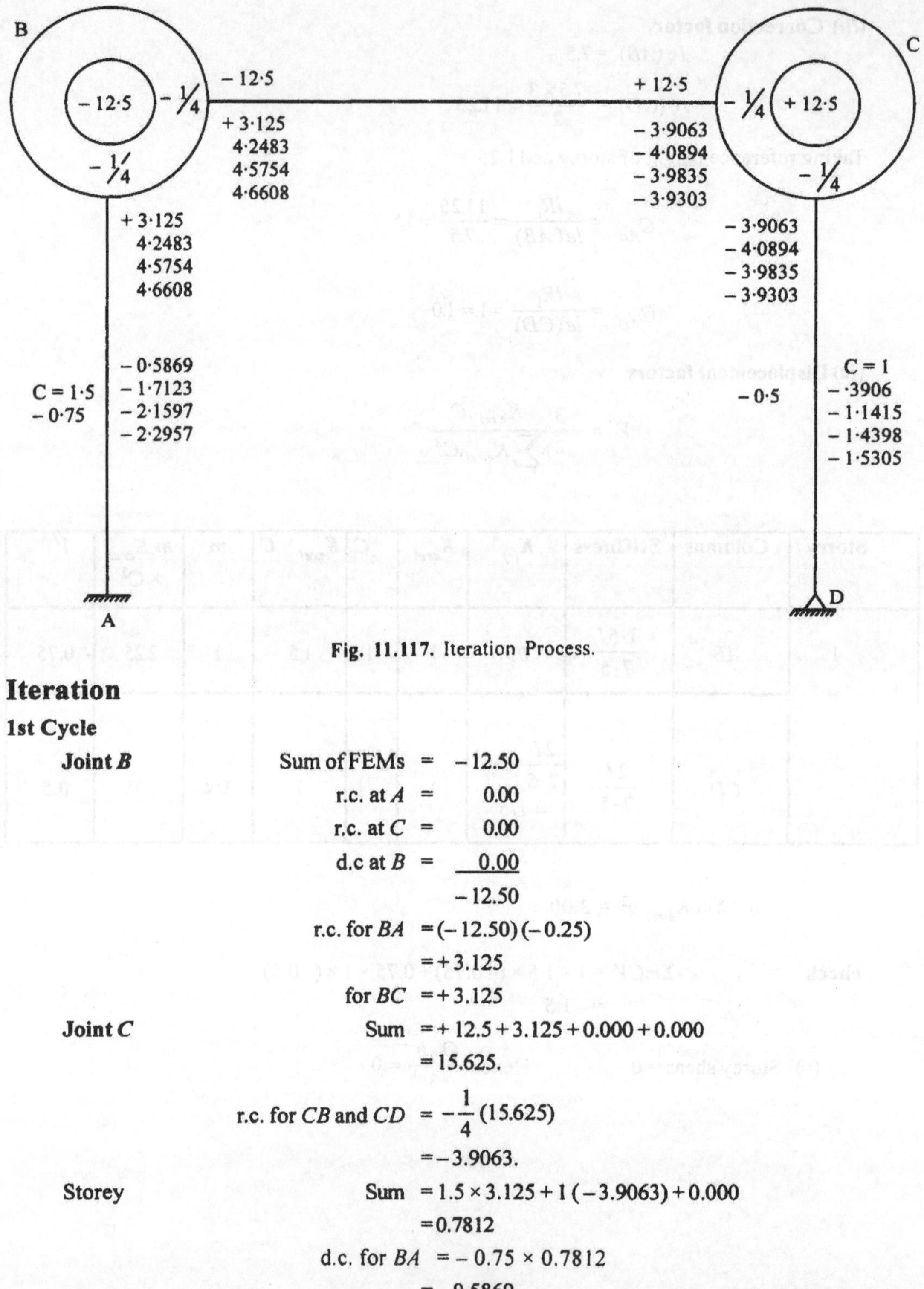

Fig. 11.117. Iteration Process.

Iteration

1st Cycle

Joint *B*

$$\begin{aligned}
\text{Sum of FEMs} &= -12.50\\
\text{r.c. at } A &= 0.00\\
\text{r.c. at } C &= 0.00\\
\text{d.c at } B &= \underline{0.00}\\
&\quad -12.50\\
\text{r.c. for } BA &= (-12.50)(-0.25)\\
&= +3.125\\
\text{for } BC &= +3.125
\end{aligned}$$

Joint *C*

$$\begin{aligned}
\text{Sum} &= +12.5 + 3.125 + 0.000 + 0.000\\
&= 15.625.\\
\text{r.c. for } CB \text{ and } CD &= -\frac{1}{4}(15.625)\\
&= -3.9063.
\end{aligned}$$

Storey

$$\begin{aligned}
\text{Sum} &= 1.5 \times 3.125 + 1(-3.9063) + 0.000\\
&= 0.7812\\
\text{d.c. for } BA &= -0.75 \times 0.7812\\
&= -0.5869
\end{aligned}$$

for CD $= -0.5 \times 0.7812$

$= -0.3906.$

2nd Cycle Jont B

Sum $= -12.50 + 0.00 + (-3.9063) + (-0.5869)$

$= -16.9932$

r.c. for BA $= (-1/4) \times (-16.9932) = +4{\cdot}2483$

$= +4.2483$

Joint C

Sum $= 12.50 + 0.00 + 4.2483 + (-0.3906)$

$= +16.3577$

r.c. for CB and CD $= (-0.25)(+16.3577)$

$= -4.0894$

Storey

Sum $= 1.5 \times 4.2483 + (-4.0894) + 0.00 + 0.00$

$= 2.28305$

d.c. for BA $= -0.75 \times (2.2831) = -1.7123$

for CD $= -0.50 \times (2.2831) = -1.1415.$

3rd Cycle Joint B

Sum $= -12.50 + 0.00 + (-4.0894) - 1.7123$

$= -$ **18.3017**

r.c. for BA and BC $= 4.5754.$

Joint C

Sum $= +12.50 + 0.000 + (+4.5754) + (-1.1415)$

$= 15.9339$

r.c. for CB and CD $= (-0.25)(15.9339)$

$= -3.9835$

Storey

Sum $= 1.5 \times 4.5754 + (-3.9835) + 0.00 + 0.00$

$= 2.8796$

d.c. for BA $= (-0.75)(2.87967)$

$= -2.1597$

for CD $= (-0.5)(2.8796) = -$ **1.4398**

4th Cycle

Joint B

$= -1250 + 0.00 + (-3.9835) - 2.1597$

$= -18.643$

r.c. for BA and BC $= (-0.25)(-18.643)$

$= 4.6608$

Joint C

Sum $= 12.50 + 4.6608 + 0 + (-1.4398)$

$= 15.721$

r.c. for CB $= (-0.25)(15.721)$

and CD $= -3.9303.$

Storey

Sum $= 1.5 \times 4.6608 + (-3.9303) + 0.00 + 0.00$

$= 3.0609.$

d.c. for BA $= (-0.75)(3.0609)$

$$= -2.2957$$

$$\text{for } BC = (-0.5)(3.0609)$$

$$= \mathbf{1.9305}$$

$$M_{AB} = 4.6608 - 2.2957 = 2.3651$$

$$M_{BA} = 0.0000 + 2 \times 4.6608 + 0.0000 - 2.2957$$

$$= 7.0259 \text{ kNm}$$

$$M_{BC} = -12.5000 + 2 \times (4.6608) + (-3.9303)$$

$$= -7.1087$$

$$M_{CB} = +12.5000 + 2 \times (-3.9303) + 4.6608$$

$$= 9.3002$$

$$M_{CD} = 2 \times (-3.9303) - 1.5305$$

$$= 9.3909$$

$$M_{DC} = 0$$

Example 11.36. Analyse the frame shown in Fig. 11.118 by Kani's Method.

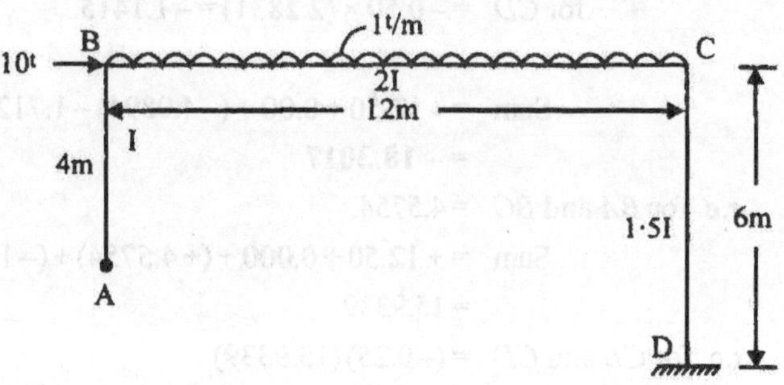

Fig. 11.118. (Example 11.36)

Solution : (*i*) Fixed end moments

$$M_{FAB} = M_{FBA} = 0$$

$$M_{FBC} = -\frac{wL^2}{12} = \frac{-1 \times 144}{12} = -12 \text{ tm}$$

$$M_{FCB} = +12 \text{ tm}$$

(*ii*) Rotation factors U

Joint	Member	K	Modified Stiffness Ke	Ke (rel)	U	Check
B	BA	$\frac{I}{4}$	$\frac{3}{4} \cdot \frac{I}{4} = \frac{3I}{16}$	9	−0.265	
	BC	$\frac{2I}{12}$	$\frac{2I}{12} = \frac{I}{6}$	8	−0.235	Σ = −0.50
C	CB	$\frac{I}{12}$	$\frac{I}{12}$	1	−0.125	
	CD	$\frac{1.5I}{6}$	$\frac{I}{4}$	3	−0.375	Σ = −0.50

(iii) **Correction Factors :**

$$l_e(AB) = \frac{3}{2} \times 4 = 6\,\text{m}.$$

$$l_e(CD) = 6\,\text{m}.$$

Taking reference height = 6 m

$$C_{AB} = 1,\ C_{CD} = 1.$$

(iv) **Displacement Factors**

Storey	Columns	Stiffness	$K_{(e)}$	$K_{e\,rel}$	C	$K_{e\,rel} \times C$	m	$m\,K_{erel}\,C^2$	V
1	AB	$\frac{I}{4}$	$\frac{3}{4}\cdot\frac{I}{4} = \frac{3}{16}$	3	1	3	$\frac{3}{4}$	$\frac{3}{4}.3.1 = \frac{9}{4}$	−0.72
1	CD	$\frac{15I}{6}$	$\frac{I}{4}$	4	1	4	1	$1 \times 4 \times 1 = 4$	−0.96

Check $= \Sigma m.C.V. = \frac{3}{4}\cdot 1\cdot(-0.72) + 1 \times 1(-0.96)$

$= -1.5$ (O.K)

(v) Storey Shear = 10$'$

$$\frac{S_{12} \times h_1}{3} = \frac{10 \times 6}{3} = +20\,\text{tm}$$

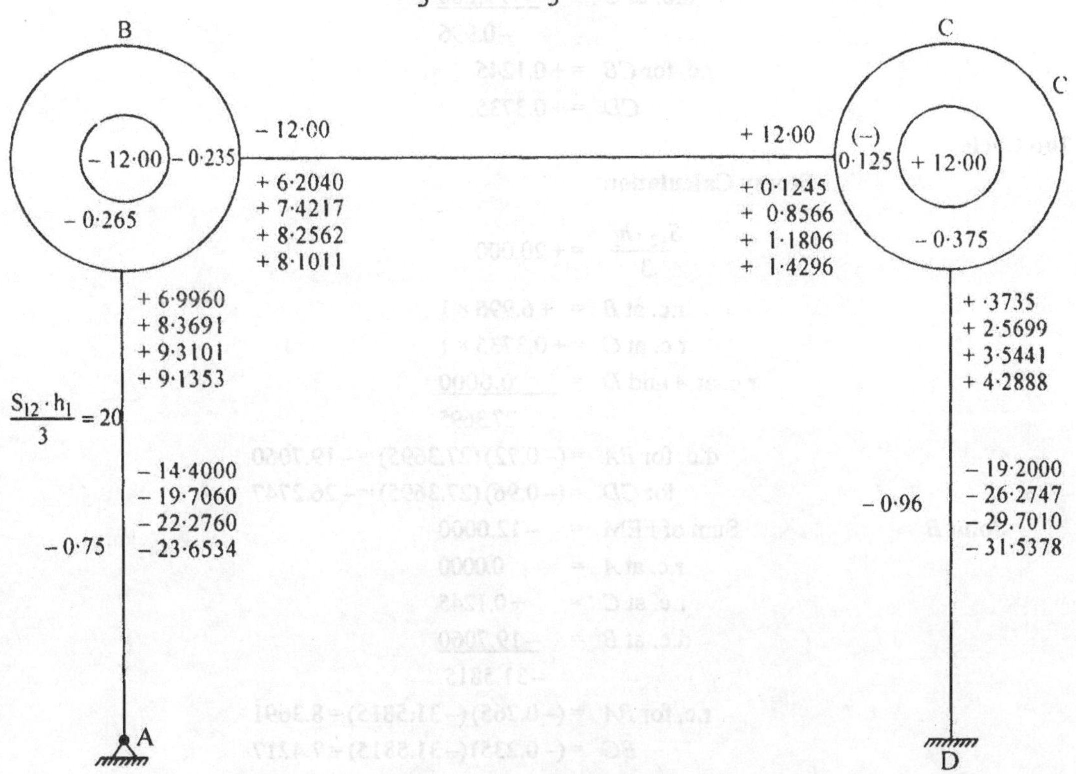

Fig. 11.119. Iteration Process

1st Cycle **Storey Calculation**

$$\frac{S_{12} \cdot h_1}{3} = +20.000$$

r.c. at B = 0.000

r.c. at C = 0.000

r.c. at A and D = 0.000

+20.000

d.c. for BA $= -0.72 \times 20 = -14.40$

CD $= -0.96 \times 20 = -19.20$

Joint B Sum of FEM = −12.000

r.c. at C = 0.000

r.c. at A = 0.000

d.c. at B = −14.40

−26.40

r.c. for BA $= (-0.265)(-26.40) = +6.996$

for BC $= (-0.235)(-26.40) = +6.204$

Joint C Sum of FEM = +12.000

r.c. at B = +6.204

r.c. at D = 0.000

d.c. at C = −19.200

−0.996

r.c. for CB = +0.1245

CD = +0.3735.

2nd Cycle

Storey Calculation

$$\frac{S_{12} \cdot h_1}{3} = +20.000$$

r.c. at B $= +6.996 \times 1$

r.c. at C $= +0.3735 \times 1$

r.c. at A and D = 0.0000

27.3695

d.c. for BA $= (-0.72)(27.3695) = -19.7060$

for CD $= (-0.96)(27.3695) = -26.2747$

Joint B Sum of FEM = −12.0000

r.c. at A = 0.0000

r.c. at C = +0.1245

d.c. at B = −19.7060

−31.5815.

r.c. for BA $= (-0.265)(-31.5815) = 8.3691$

BC $= (-0.235)(-31.5815) = 7.4217$

Joint C Sum of FEM = +12.0000

$$\text{r.c. at } B = +7.4217$$
$$\text{r.c. at } D = +0.0000$$
$$\text{d.c. at } C = \underline{-26.2747}$$
$$-6.8930$$
$$\text{r.c. for } CB = (-0.125)(-6.853) = +0.8566$$
$$\text{for } CD = (-0.375)(-6.853) = +2.5699$$

3rd Cycle

$$\frac{S_{12}h_1}{3} = +20.000$$
$$\text{r.c. at } A \text{ and } B = 0.0000$$
$$\text{r.c. at } B = 8.3691$$
$$\text{r.c. at } C = \underline{+2.5699}$$
$$= 30.9390$$

$$\text{d.c. for } BA = -0.72 \times (30.9390) = 22.276$$
$$\text{for } CD = -0.96 \times (30.9390) = -29.701.$$

Joint *B*

$$\text{Sum of FEM} = -12.0000$$
$$\text{r.c. at } A = 0.0000$$
$$\text{r.c. at } C = 0.8566$$
$$\text{d.c. at } B = \underline{-22.2760}$$
$$-35.1326$$
$$\text{r.c. for } BA = (-0.265)(-35.1326) = 9.3101.$$
$$\text{for } BC = (-0.235)(-35.1326) = 8.2562.$$

Joint *C*

$$\text{Sum of FEM} = -12.0000$$
$$\text{r.c. at } D = 0.0000$$
$$\text{r.c. at } B = 8.2562$$
$$\text{d.c. at } C = \underline{-29.7010}$$
$$-9.4448$$
$$\text{r.c. for } CB = (-0.125)(-9.4448) = 1.1806$$
$$\text{for } CD = (-0.375)(-9.4448) = 3.5418$$

4th Cycle

Storey Calculation

$$\frac{S_{12} \cdot h_1}{3} = +20.000.$$
$$A \text{ and } D = 0.0000$$
$$\text{r.c. at } B = 9.3101$$
$$\text{r.c. at } C = \underline{3.5418}$$
$$32.8519.$$
$$\text{d.c. for } BA = -0.72\,72 \times (32.8519) = -23.6534$$
$$\text{for } CD = -0.96 \times (32.8119) = -31.5378.$$

Joint *B*

$$\text{Sum of FEM} = -12.0000$$
$$\text{r.c. at } A = 0.0000$$

$$\begin{aligned}
\text{r.c. at } C &= 1.1806 \\
\text{d.c. at } B &= \underline{-23.6534} \\
&\quad -34.4728 \\
\text{r.c. for } BA &= -0.265 \times (-34.4728) \\
&= +\mathbf{9.135.} \\
\text{for } BC &= (-0.235) \times (-34.4728) \\
&= +\mathbf{8.1011}
\end{aligned}$$

Joint C

$$\begin{aligned}
\text{Sum of FEM} &= 12.0000 \\
\text{r.c. at } D &= 0.0000 \\
\text{r.c. at } B &= 8.1011 \\
\text{r.c. at } C &= \underline{-31.5378} \\
&\quad -11.4367 \\
\text{r.c. for } CB &= (-0.125)(-11.4367) = +1.4296 \\
CD &= (-0.375)(-11.4367) = +4.2888
\end{aligned}$$

Calculation of Moments

$$\begin{aligned}
M_{AB} &= 0 \\
M_{BA} &= 2 \times 9.1353 + 0.000 - 23.6534 \\
&= -5.3828. \\
M_{BC} &= -12.00 + 2 \times 8.1011 + 1.4296 \\
&= +5.6318. \\
M_{CB} &= +12.00 + 2 \times 1.4296 + 8.1011 \\
&= 22.9603. \\
M_{CD} &= 2 \times 4.288 - 31.5378 + 0.0000 \\
&= -22.9618. \\
M_{DC} &= 4.2888 - 31.5378 \\
&= -27.249.
\end{aligned}$$

Example 11.37. Use Kani's method to determine the joint moments of the frame given in Fig. 11.120. Assume all members to have identical cross-section. Hence draw the bending moment diagram.

[AMIE Summer 1987]

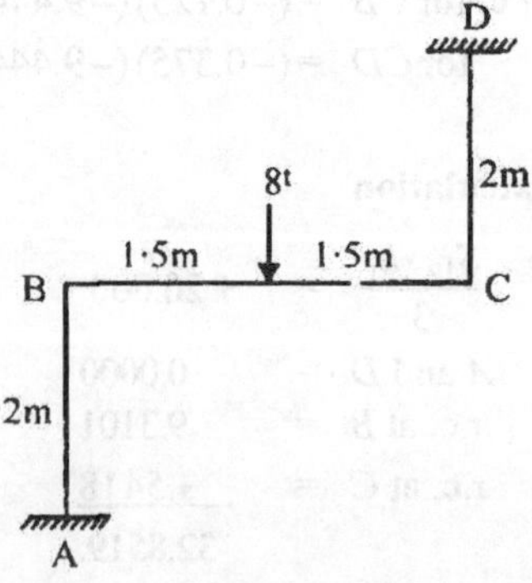

Fig. 11.120. Example 11.37

Solution: Because of central loading, there will be no sway along BC.

(*i*) FEM $\quad M_{FBC} = -\dfrac{8 \times 3}{8} = -3\,\text{tm}$

$M_{FCB} = +3$ tm

(*ii*) Stiffness $K_{BA} = \frac{I}{2}:3$

$K_{BC} = \frac{I}{3}:2$

$K_{CD} = \frac{I}{2}:3$

Rotation factors

$$U_{BA} = \left(-\frac{1}{2}\right)\frac{3}{5} = -0.3$$

$$U_{BC} = \left(-\frac{1}{2}\right)\left(\frac{2}{5}\right) = -0.2$$

$$U_{CB} = -0.2$$

$$U_{CD} = -0.3$$

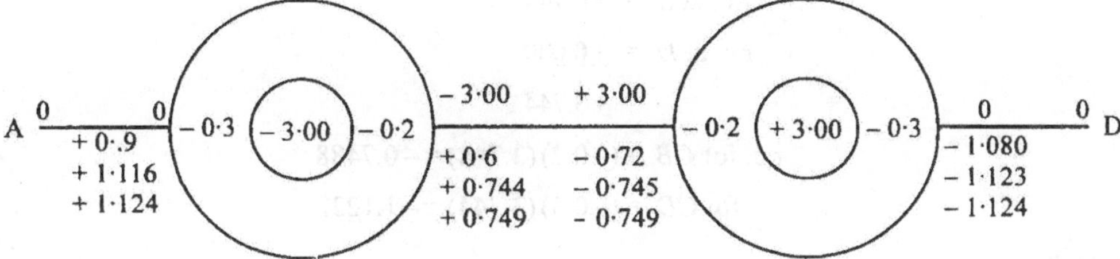

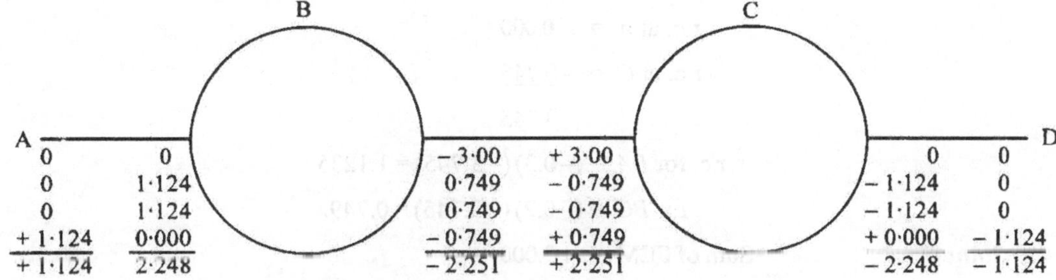

Fig. 11.121. Iteration Process and calculation of B.M.

1st Cycle

Joint *B*

Sum of FEM = −3.00
r.c. at A (known) = 0.00
r.c. at *C* (assumed) = 0.00
−3.00

r.c. at *B* for *BA* = −0.3 × (−3.00) = +0.9
for *BC* = −0.2 × (−3.00) = +0.6

Joint *C*

Sum of FEM = +3.00
r.c. at *B* = +0.60
r.c. at *D* = 0.00
+3.60

r.c. at *C* for *CB* = (−0.2) (3.60) = −0.72

$CD = (-0.3)(3.60) = -1.08$

2nd Cycle

Joint *B*

Sum of FEM = −3.00

r.c. at A = 0.00

r.c. at C = $\underline{-0.72}$

−3.72

r.c. for $BA = (-0.3)(-3.72)$

$= +1.116$

for $BC = (-0.2)(-3.72)$

$= 0.744$

Joint *C*

Sum of FEM = +3.000

r.c. at B = +0.744

r.c. at D = $\underline{+0.000}$

+3.744

r.c. for $CB = (-0.2)(3.744) = -0.7488$

for $CD = (-0.3)(3.744) = -1.1232.$

3rd Cycle

Joint B

Sum of FEM = −3.000

r.c. at A = 0.000

r.c. at C = $\underline{-0.745}$

−3.745

r.c. for $BA = (-0.3)(-3.745) = 1.1235$

for $BC = (-0.2)(-3.745) = 0.749.$

Joint *C*

Sum of FEM = +3.000

r.c. at B = 0.749

r.c. at D = $\underline{+0.000}$

3.749

r.c. for $CB = -0.2 \times 3.749$

$= -0.749$

for $CD = -0.3 \times 3.749$

$= -1.124$

Fig. 11.121 shows the iteration process and the calculation of B.M. Fig. 11.122 shows the B.M. diagram.

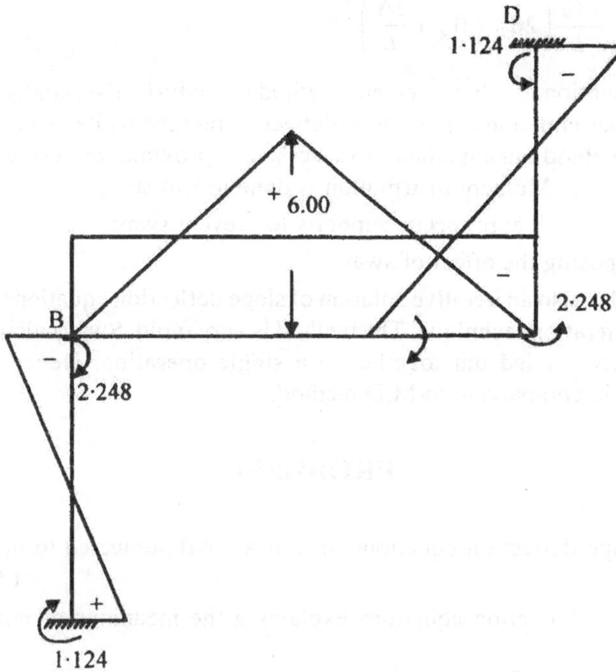

Fig. 11.122 B.M. diagram

SUMMARY

1. In displacement method of analysis displacement is taken as unknown and equilibrium equations are formed to solve for the unknown displacements. The following classical methods based on equilibrium approach are widely used in structural analysis .
   1. Slope deflection method
   2. Moment Distribution method
   3. Kani's method.

   The following energy theorems can also be used to derive equilibrium equations
   1. Principle of Minimum Total Potential Energy.
   2. Castigliano's first theorem
   3. Principle of Stationary Complementary Potential Energy.

   The energy principle above leading to equilibrium equations are of theoretical interest as the first three methods appear simpler to apply compared to the energy principles.
2. In slope deflection method all joints are considered rigid i.e angles between members at the joints are considered not to change with the application of loads. In deriving slope deflection equations only bending deformation is considered. Shear and axial force are neglected. The three basic requirements of structural analysis namely

   1. Compatibility
   2. Equilibrium
   3. Member characteristics are satisfieu ... ..., e deflection equations which are :

$$M_{AB} = M_{FAB} + \frac{2EI}{L}\left(2\theta_A + \theta_B + \frac{3\Delta}{L}\right)$$

$$M_{BA} = M_{FBA} + \frac{2EI}{L}\left(2\theta_B + \theta_A + \frac{3\Delta}{L}\right)$$

3. Moment distribution is a displacement method of analysis. Essentially it consists of solving the simultaneous equations of the slope deflection method by iterative technique. It is not an approximate method but a method of successive approximation. For unsymmetrical Portals where sway occurs Moment distribution is done in two steps:
   1. Without sway with appropriate supports to prevent sway.
   2. Then superposing the effect of sway.
4. Kani's method is also an iterative solution of slope deflection equations, the iterations follow Gauss Seidel iteration technique. The method is very rapid. Sway calculations and non sway calculations are carried out together in a single operation. Hence the method is more advantageous in comparison to M.D method.

PROBLEM

A. Theoretical.

11.1 (*i*) Derive the slope deflection equations for a beam AB subjected to moments and shears at ends. **(AMIE Summer' 94)**

(*ii*) Write the slope deflection equations explaining the meanings of notations used. (AMIE summer, 95)

(*iii*) Explain the terms: (*i*) stiffness (*ii*) distribution factor (*iii*) carry over factor. **(AMIE Summer, 1978)**

(*iv*) What is flexural stiffness of a structural member ? Show that the members rigidly connected to a joint and all lying in one plane share the moment applied to the joint in the same plane in proportion to flexural stiffness. **(AMIE Summer 1990)**

(*v*) Explain the terms: absolute stiffness, relative stiffness, carry over factor, shear equations, sway correction. **(AMIE Winter 1989)**

(*vi*) Define stiffness and carry over factors for a prismatic beam member and derive the stiffness of such a beam when the other end is hinged.

(*vii*) Derive Kani's first iteration equation required to compute rotation contribution. **(AMIE Summer' 95)**

(*viii*) What is the distinct advantage of Kani's method over Hardy Cross method of moment distribution.

(*ix*) Define the following terms:

(*i*) rotation contribution (*ii*) displacement contribution (*iii*) rotation factor (*iv*) displacement factor as used in Kanis method. **(AMIE Summer 87)**

(*x*) Explain the terms :

Rotation factor; linear displacement factor, storey moment, restraint moment as used in Kani's method. **(AMIE Winter 1988)**

B. Objective.

11.2 Choose the correct answer.

(*i*) A rectangular portal will have horizontal sway

(*a*) only if it is subjected to horizontal load

(*b*) only if its geomentry is nonsymmetric.

(*c*) only if its loading is unsymmetric.

(*d*) None of the above. **(AMIE Winter '98).**

(*ii*) In a rigid jointed structure at any joint

(*a*) The members under go the same rotation

(*b*) The members have zero bending moment.

(*c*) There is no deflecltion

(*d*) The bending moments in the members are indeterminate **(AMIE Winter '98)**

(*iii*) the slope deflection method of structural analysis is a (*a*) displacement method

(*b*) force method (*c*) hybrid method **(AMIE Summer 99).**

(*iv*) Rotation at end B of a simply supported beam AB of length L and uniform EI subjected to a moment M acting at end A is

(*a*) $\dfrac{ML}{3EI}$ (*b*) $\dfrac{ML}{4EI}$ (*c*) $\dfrac{ML}{6EI}$ (*d*) $\dfrac{ML}{12EI}$

(*v*) The modified stiffness of a members which crosses the axis of symmetry

(*a*) $\dfrac{6EI}{L}$ (*b*) $\dfrac{4EI}{L}$ (*c*) $\dfrac{3EI}{L}$ (*d*) $\dfrac{2EI}{L}$

(AMIE Summer '95)

(*vi*) The carry over factor in a prismatic member whose far end is hinged is

(*a*) 0 (*b*) 2 (*c*) 0.5 (*d*) 1

(*vii*) Distribution factor for an overhanging beam for a cantilever portion is

(*a*) 0.5 (*b*) 1 (*c*) 0 (*d*) none of the above

(AMIE Winter, 97,

(*viii*) The rotational stiffness of a cantilever beam at its free end is

(a) $\dfrac{EI}{L}$ (b) $\dfrac{2EI}{L}$ (c) $\dfrac{3EI}{L}$ (d) $\dfrac{4EI}{L}$ (AMIE 3,97).

(*ix*) A horizontal prismatic member AB of length 2.5 m fixed at B and pinned at A is subjected to a rotation of 0.01 rad at A and the magnitude of the corresponding BM at the fixed end is 0.5 KNm. If the same member is used as a vertical column with pinned ends, critical Eulerload will be

(*a*) $10\pi^2$ kN (*b*) $100\pi^2$ kN (*c*) 50π kN (*d*) None of the above

(AMIE Winter 1998).

(*x*) The rotational stiffness of a prismatic element is

(*a*) directly proportional to its area of cross section.

(*b*) inversely proportional to the square of its length

(*c*) directly proportional to its length

(*d*) independent of its sectional polar moment of inertia **(AMIE Winter 98)**

(*xi*) In a portal frame with columns of same height and fixed base, the sum of displacement factors of column in Kanis method is

(*a*) – 2.0 (*b*) – 1.5 (*c*) – 1.0 (*d*) – 0.5

[AMIE Summer '97]

(*xii*) What is the value of moment developed at *A*? **[AMIE S '97] (Fig. P. 11.1)**

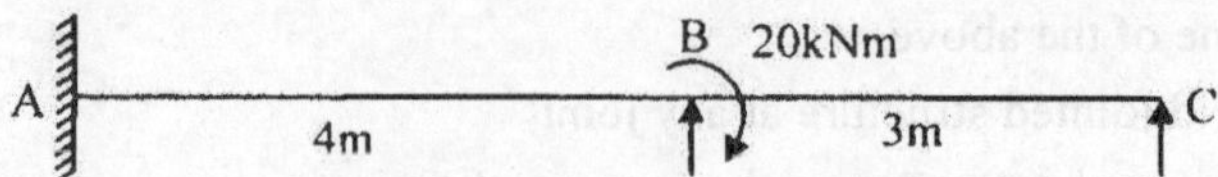

Fig. P. 11.1. Prob. 11.2 (*xii*)

(*xiii*) Find the magnitude of the couple required to be applied at the middle support of a two span continuous beam having both the span of length l and flexural rigidity EI to cause unit rotation of the joints. Take end supports to be hinged. **[AMIE Winter 1981]**

C. Mathematical.

11.3. By using slope deflection method analyse the beam shown in Fig. P. 11.2 and draw the B.M. diagram. EI is constant. **[AMIE Winter 1987]**

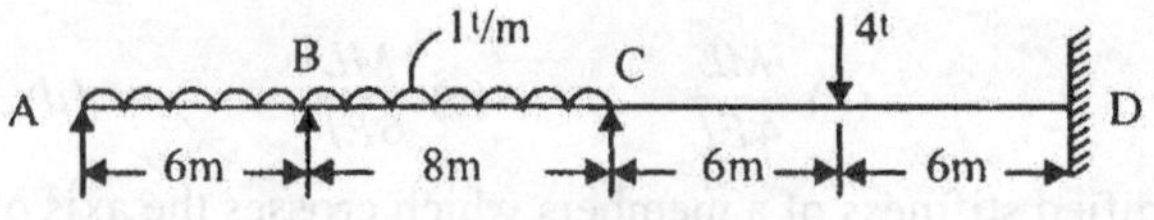

Fig. P. 11.2 Prob 11.3

11.4 By using the slope deflection method draw B.M. and S.F. diagrams for the continuous beam shown in Fig P. 11.3 below giving values at salient points. Take EI uniform throughout.

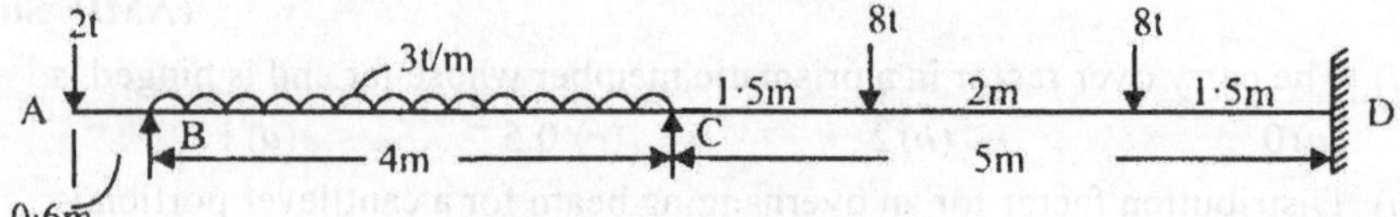

Fig. P. 11.3 Prob. 11.4

11.5 A portal $ABCD$ has a span of 6 m and height 4 m. The vertical members AB and CD are fixed at their ends A and D and the horizontal member BC carries a uniformly distributed load of 2t/m throughout its length. The moment of inertia of the vertical members is I and that of horizontal member 2I. Using the slope deflection method, determine the moments at A, B, C and D and draw the B.M. diagram. **[AMIE Winter 1988]**

11.6 A horizontal beam ABC has support A hinged, B rigid jointed to a vertical column BD and the over hang BC is free at C. The bottom of the column BD is fixed at D. The Span AB is 6.4 m and BC is 2.0 m long. The height of the column BD is 4.2 m. The beam carries a vertical load of 60.0 kN at the free end C of the overhang BC. Analyse the frame by slope Deflection Method and draw the Bending Moment Diagram. Assume uniform flexural rigidity EI = 9600 kN/m^2, for the beam and the column. Calculate the slope and deflection at point C. Sketch the deflected shape indicating the points of contraflexure, if any. **[AMIE Summer '99]**

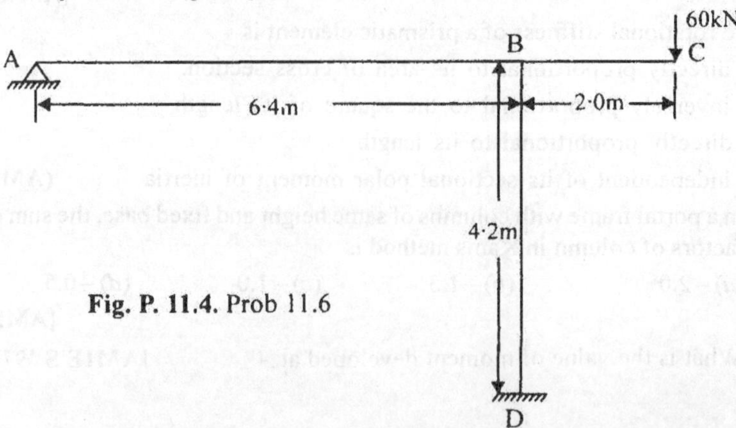

Fig. P. 11.4. Prob 11.6

11.7 Determine the movement of the roller C of the frame shown in Fig. P 11.5 by slope deflection method. **[AMIE Summer' 91]**

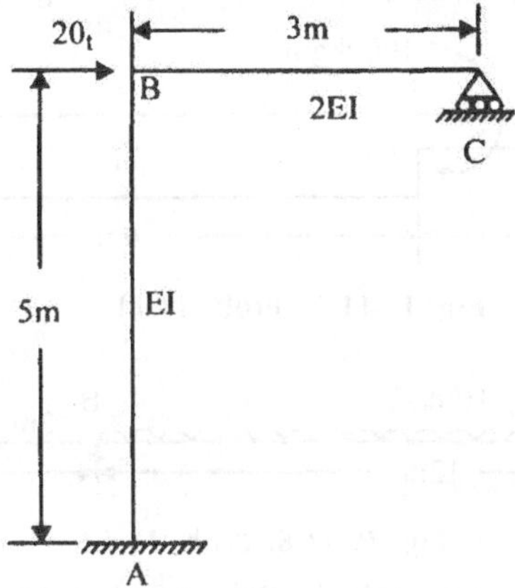

Fig. P. 11.5 Prob. 11.7

11.8 A rectangular rigid frame $ABCD$ has its base A and D fixed. The vertical members $AB = DC$ = 12 m and the horizontal girder BC = 10 m. The EI values of all three members are equal. A concentrated horizontal load of 5 tonnes acts at the joint B from left to right. Calculate the joint moments by slope deflection method. **[AMIE Winter 1985]**

11.9 Four beams OA, OB, OC and OD are connected at O, the joint O being rigid and pinned in position. All the beams are in one plane. The ends A, B and C are fixed while D is pinned in position. A moment of 500 tm acting in the plane of the beams is applied at O. Calculate the moments at the supports A, B, C and D. The lengths are:

OA = 2 m, OB = 4 m, OC = 4 m and OD = 3 m.

Assume the appropriate flexural rigidity same for all the beams.

(Solve from the fundamentals without using any formula). **[AMIE Summer 1991]**

11.10 A continuous beam ABC fixed at A and hinged at C is supported by a column BD which is rigidly connected to the beam ABC at B. It is loaded as shown in Fig. P 11.6 below. Analyse the frame by the method of moment distribution and draw the B.M. diagram. Take moment of inertia of $AB = 2I$ and that of BD and BC as I. (Axial deformations may be neglected). **[AMIE Summer 1978]**

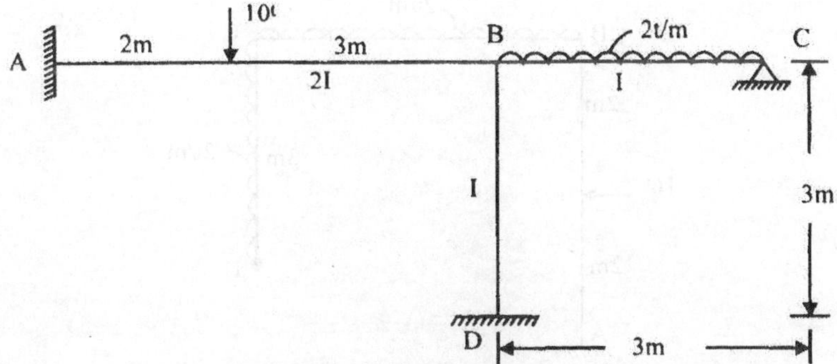

Fig. P. 11.6. Prob. 11.10

11.11 Analyse the fixed beam shown in Fig. P. 11.7 below by moment distribution method. **[AMIE Advanced Structures, Winter 1979]**

11.12 A continuous beam ABC shown in Fig. P. 11.8 below is fixed at A and C, simply supported at B. The support B sinks by $10mm$ relative to A and C. Calculate the support moments using moment distribution method and sketch the moment diagram $EI = 6000$ tm^2

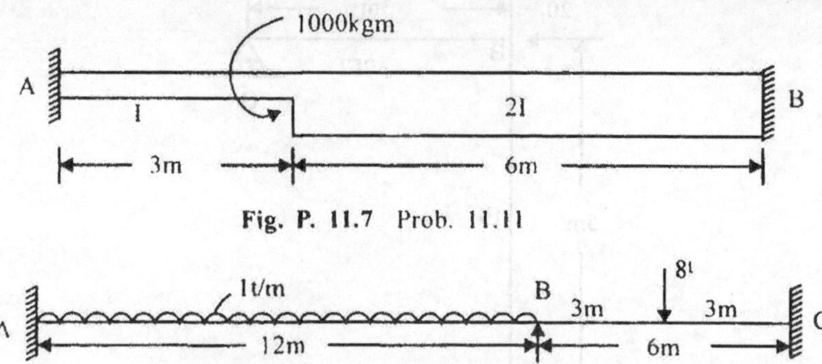

Fig. P. 11.7 Prob. 11.11

Fig. P. 11.8. Prob. 11.12

11.13 A fixed base rectangular portal in which all the members have the same moment of inertia has a span of 6 m and height 4.5 m. Analyse the frame by moment distribution method when it is subjected to a horizontal sway load of $10t$ acting at the beam level from left to right and draw the bending moment diagram. Also compute the reaction components at each base.

[AMIE Summer 1981]

11.14 A rigid jointed portal frame $ABCD$ has the horizontal beam member BC of length 6 m and moment of inertia $2I$, the left vertical column AB of height 6 m and moment of inertia $2I$ and the right vertical column DC of length 4 m and moment of inertia $1.5I$. The bases A and D are hinged. *A u.d.l* of 2 t/m acts on the full length of the member BC. By using Moment Distribution Method analyse the portal and draw the bending moment diagram. Make use of the following information.

The analysis of the frame for a sway load of $1t$ acting at B in the direction of $\overrightarrow{BC}$ gives the end moments.

$M_{BC} = +2$ tm and $M_{CB} = +2.66$ tm both clockwise. **[AMIE Winter 1981]**

11.15 Analyse the portal frame shown in Fig. P. 11.9 below by the moment distribution method. (i) Compute the reaction components at the base (ii) draw the B.M. diagram. The support A to be built in and support D to be hinged. **[AMIE Summer 1982]**

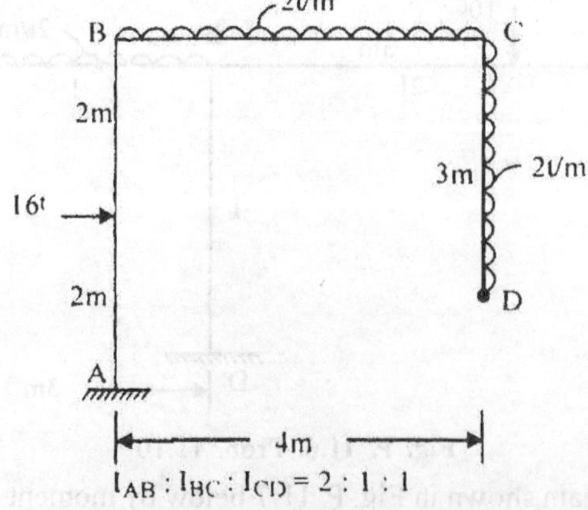

Fig. P 11.9 Prob. 11.15

11.16 Analyse the frame shown in Fig. P. 11.10 by moment distribution method. Draw the B.M. diagram.

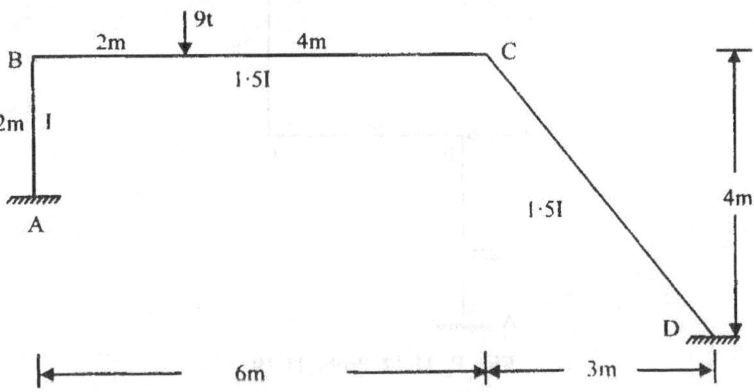

Fig. P. 11.10. Prob. 11.16

11.17 A box culvert *ABCD* has 12 m span and 6 m height with uniform section for all the four members. The top horizontal member *BC* is subjected to two concentrated loads each of magnitude 10*t* separated by 4m, symmetrically placed about the central axis. The vertical members *AB* and *DC* are subjected to triangular earth pressure having zero intensity at the top ends *B* and *C* and an intensity of 5 *t/m* at the bottom ends *A* and *D*. The bottom member *AD* is subjected to uniform soil reaction

(*i*) Analyse the structure either by slope deflection or moment distribution method.

(*ii*) Draw the complete B.M. diagram mentioning values of B.M. at the salient points.

[AMIE Summer, 1984]

11.18 In the loaded structure shown in Fig. P. 11.11 ends *A* and *E* are fixed and end *D* is hinged and the joints *B* and *C* are rigid. *EI* values for various members are as shown in figure.

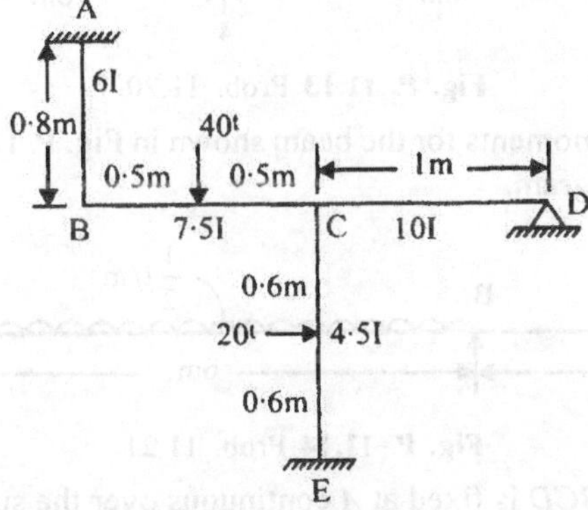

P. 11.11 Prob. 11.18

Using the moment distribution method

(*a*) Obtain the final moments at the joints

(*b*) Draw the bending moment diagram of the loaded structure stating values at the salient points.

(*c*) Find the component of the reaction at *A* along *BA*. **[AMIE Summer 1987]**

11.19 By using moment distribution method analyse the frame shown in Fig. P. 11.12 and determine the end moment. Take *EI* constant through out. **[AMIE Summer 1988]**

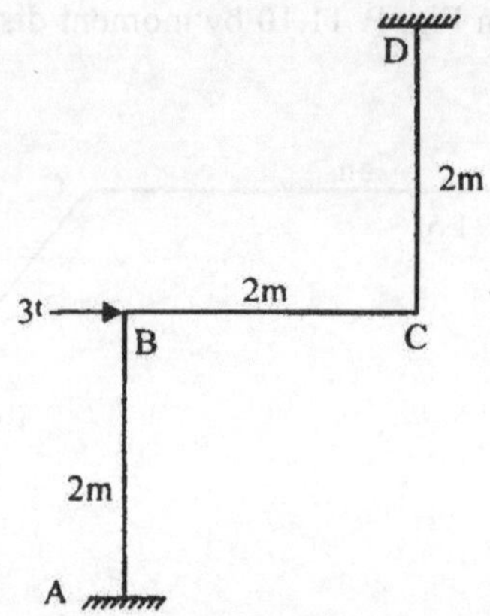

Fig. P. 11.12 Prob. 11.19.

11.20 **By using Moment distribution Method analyse the frame shown in Fig. P. 11.13 Draw bending moment and shear force diagrams and determine the end reactions. [AMIE Winter 1988]**

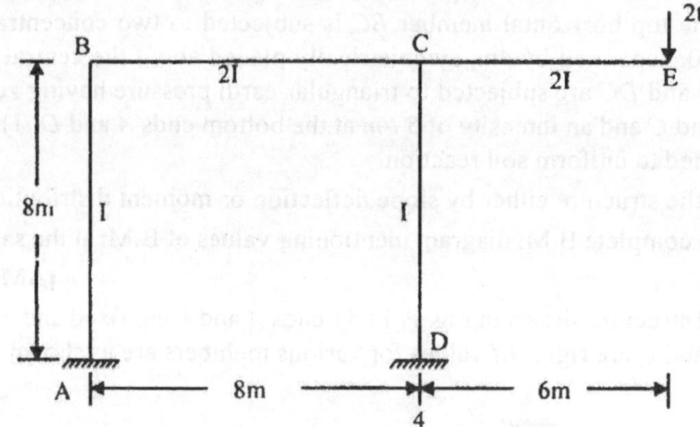

Fig. P. 11.13 Prob. 11.20.

11.21 **Calculate the support moments for the beam shown in Fig. P. 11.13 using Kani's method and sketch the moment diagram. [AMIE Summer 1980]**

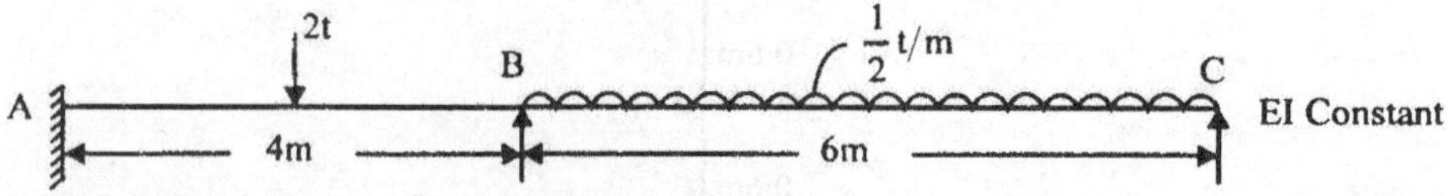

Fig. P. 11.14 Prob. 11.21

11.22 **A continuous beam *ABCD* is fixed at *A* continuous over the supports *B* and *C* and free at *D* $AB = 4$ m, $BC = 3$ m and $CD = 2$ m. $I_{AB} = I_0$, and $I_{BC} = 1.5\,I_0$. Both the spans *AB* and *BC* carry a *u.d.l* of 5 *t/m* and a concentrated load of 2.5ᵗ acts vertically down ward at the frame end *D*. Analyse the beam by Kani's method and draw the bending moment diagram.**

[AMIE Summer 1981]

11.23 **A rigid jointed rectangular portal frame *ABCD* has two equal length $AB = CD = 3$m and a horizontal member $BC = 6$ m. The bases *A* and *D* are fixed. The moments of inertia of the sections of the members are in the proportion $AB : BC : CD = 1 : 4 : 1$ The frame is subjected to a vertical load of 8*t* acting at a distance of 1.5 m from *B*. Analyse the frame by moment distribution or Kani's method and draw the bending moment diagram. [AMIE Winter 1986]**

11.24 Analyse the portal frame shown in Fig P. 11.15 by Kanis method and sketch the bending moment diagram giving all the salient values. **[AMIE Winter '88]**

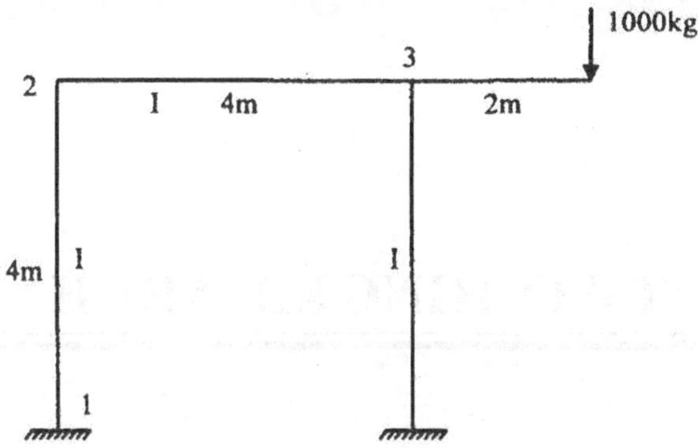

Fig. P. 11.15. Prob. 11.24

11.25 Using Kanis method analyse the portal frame shown in Fig. P. 11.16 and draw the B.M. diagram. Moment of inertia of all members is the same. **[AMIE Winter 1989]**

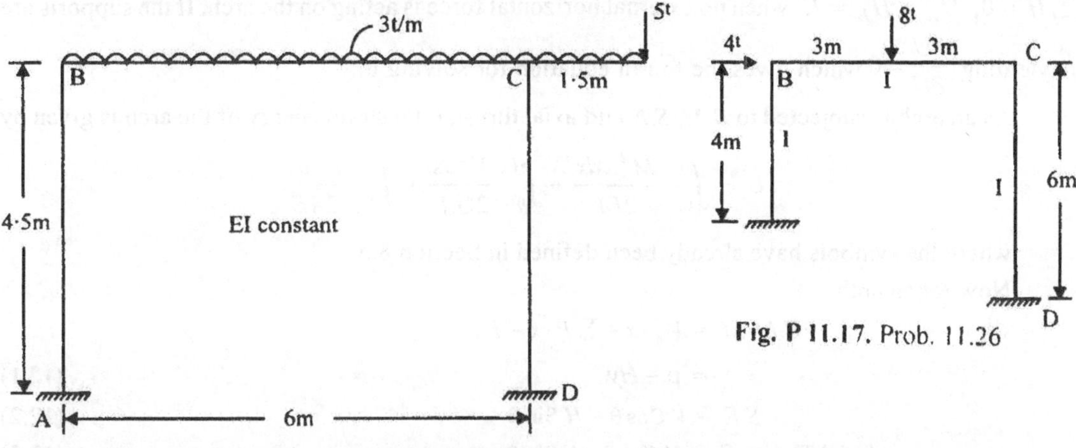

Fig. P. 11.16 Prob. 11.25

Fig. P 11.17. Prob. 11.26

11.26 Analyse the portal frame shown in Fig. P. 11.17 by Kani's method and sketch the bending moment diagram giving all salient values. **[AMIE Winter 1989]**

11.27 Analyse the portal frame shown in Fig. P. 11.18 by Kani's method.

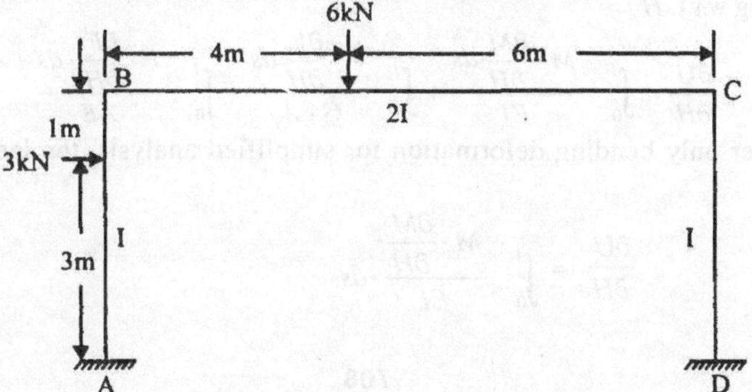

Fig. P. 11.18. Prob. 11.27

12

TWO HINGED ARCH

12.1. INTRODUCTION

Two hinged arch, as explained in Sec 6.8, is statically indeterminate to the 1st degree. We have four unknowns, V_A, H_A, V_B and H_B but only three equations of Equilibrium. From $\Sigma H = 0$, $H_A = H_B = H$ when no external horizontal force is acting on the arch. If the supports are unyielding $\frac{\partial U}{\partial H} = 0$ which gives the fourth equation for solving it.

As an arch is subjected to *B.M, S.F,* and axial thrust, total strain energy of the arch is given by

$$U = \int_0^l \frac{M^2 \cdot dx}{2EI} + \int_0^l \frac{V^2 ds}{2GA_s} + \int_0^l \frac{F^2 ds}{2AE}$$

where the symbols have already been defined in Section 8.5

Now for an arch

$$B.M, M = V_A \cdot x + \Sigma P \cdot x - Hy$$

$$= \mu - Hy \quad \ldots(12.1)$$

$$S.F = V \cos\theta - H \sin\theta \quad \ldots(12.2)$$

$$\text{Axial Thrust } F = H \cos\theta + V \sin\theta \quad \ldots(12.3)$$

where μ is beam bending moment

V is beam shear at a section

and θ is angle between the tangent at a section and the horizontal.

Differentiating w.r.t. H

$$\frac{\partial U}{\partial H} = \int_0^l \frac{M \cdot \frac{\partial M}{\partial H} ds}{EI} + \int_0^l \frac{V \cdot \frac{\partial V}{\partial H} ds}{G \cdot A_s} + \int_0^l \frac{F \cdot \frac{\partial F}{\partial H} \cdot ds}{AE}$$

If we consider only bending deformation for simplified analysis, the last two terms can be neglected.

Hence
$$\frac{\partial U}{\partial H} = \int_0^l \frac{M \cdot \frac{\partial M}{\partial H}}{EI} \cdot ds.$$

$$M = \mu - H.y$$

$$\frac{\partial M}{\partial H} = -y$$

Then $$\frac{\partial U}{\partial H} = \int_0^l \frac{(\mu - Hy)\ (-y)}{EI}$$

$= 0$ as there is no movement along H, supports being unyielding.

Therefore, $$H = \frac{\int_0^l \frac{\mu y ds}{EI}}{\int_0^l \frac{y^2 ds}{EI}} \quad \ldots(12.4)$$

For a parabolic arch the value of I at the springings is greater than that at the crown and it is assumed that the cross section varies in such a way that

$$I = I_c \text{ Sec } \theta$$

where I_c = moment of inertia at the crown.

Now $$\text{Sec } \theta = \frac{ds}{dx} \text{ or } ds = dx \sec \theta$$

Then $$H = \frac{\int_0^l \frac{\mu y\ dx \text{ Sec } \theta}{EI_0 \text{Sec } \theta}}{\int_0^l \frac{y^2\ dx \text{ Sec } \theta}{EI_0 \text{Sec } \theta}} = \int_0^l \frac{\mu y\ dx}{\int_0^l y^2 dx} \quad \ldots(12.5)$$

12.2. STANDARD CASES OF TWO HINGED SEMICIRCULAR ARCH OF CONSTANT FLEXURAL RIGIDITY *EI*.

(*i*) Concentrated load W acting at any point. Let the load W act at D and let OD make an angle of α with OA. To ensure symmetry of load we introduce another load W at E where OE makes an angle α with OB. This load will introduce additional horizontal reaction H so that the total reaction will be $2H$ and we get

$$2H = \frac{\int \mu y ds}{\int y^2 ds} = \frac{N}{D}$$

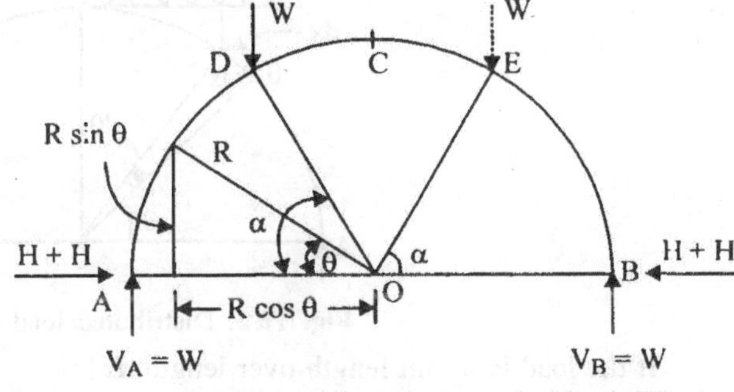

(*a*) Arch with two symmetrical loads W.

(*b*) μ diagram.

Fig. 12.1. Concentrated load on arch.

Now for AD,
$$\mu = WR(1 - \text{Cos } \theta)$$
$$y = R \sin \theta$$
$$ds = R.d\theta$$

for DC
$$\mu = WR(1 - \text{Cos } \alpha)$$

$$N = 2\int_0^{\alpha} WR(1 - \text{Cos } \theta)(R \sin \theta) R\, d\theta$$
$$+ 2\int_{\alpha}^{\pi/2} WR(1 - \text{Cos } \alpha)(R \sin \theta) R d\theta$$

$$= 2 \cdot \frac{WR^3}{2} . (1 - \text{Cos } \alpha)^2 + 2WR^3 \text{ Cos } \alpha \ (1 - \text{Cos } \alpha)$$

$$= WR^3 (1 - 2 \text{ Cos } \alpha + \text{Cos}^2 \alpha + 2\text{Cos } \alpha - 2\text{Cos}^2 \alpha)$$

$$= WR^3 (1 - \text{Cos}^2 \alpha) = WR^3 \text{ Sin}^2 \alpha$$

$$D = \int y^2 ds = 2 \int_0^{\pi/2} (R \text{ Sin } \theta)^2 . R.d\theta$$

$$= \frac{\pi R^3}{2}$$

Now $$2H = \frac{N}{D} = \frac{WR^3 \text{ Sin}^2 \alpha}{\pi R^3} \times 2$$

$$= \frac{2W}{\pi} \text{ Sin}^2 \alpha .$$

or $$H = \frac{W}{\pi} \text{ Sin}^2 \alpha .$$

(ii) When load W acts at the crown

$$\alpha = \pi/2, \text{ Sin } \alpha = 1$$

and $$H = \frac{W}{\pi}.$$

(iii) When a uniformly distributed load acts on half the span (Fig. 12.2)

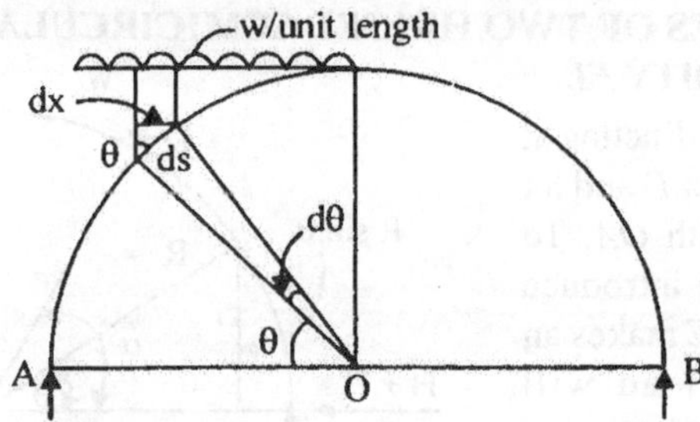

Fig. 12.2. Distributed load over half span.

If the load is w/unit length over length dx

$$W = w.dx. \text{ but } dx = ds \text{ Sin } \theta$$

Hence $$W = w \text{ Sin } \theta .ds.$$

Applying the result of concentrated load W

$$dH = \frac{W}{\pi} \cdot \sin^2\theta$$

$$= \frac{w.\sin^3\theta \cdot ds}{\pi} = \frac{Rw \cdot \text{Sin}^3\theta \cdot d\theta}{\pi}$$

But for the given load θ varies from 0 to $\pi/2$.

Hence $$H = \frac{wR}{\pi} \int_0^{\pi/2} \sin^3\theta d\theta.$$

$$= \frac{2}{3} \cdot \frac{wR}{\pi}.$$

(*iv*) When uniformly distributed load is spread over the whole span.

$$H = 2\times\frac{2}{3}\frac{wR}{\pi} = \frac{4}{3}\cdot\frac{wR}{\pi}.$$

(*v*) When the load is triangular in nature.

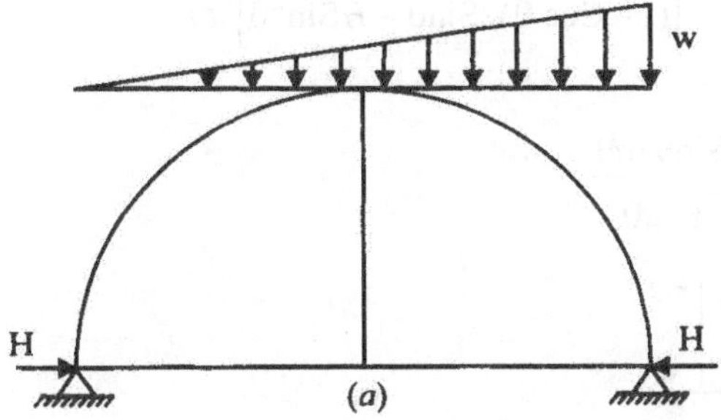

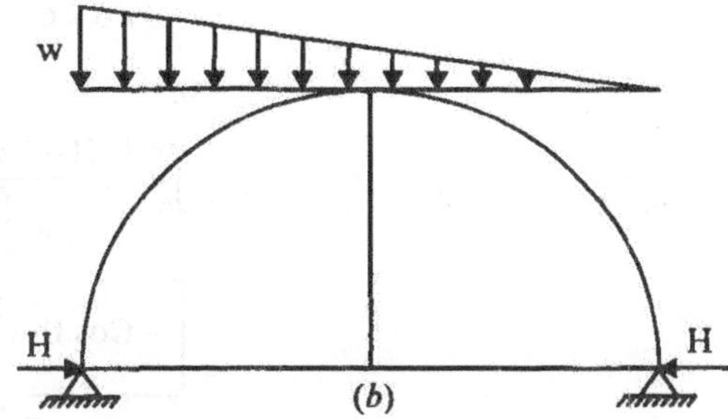

Fig. 12.3. Triangular Load.

When the two triangular loads are added together the load becomes uniformly distributed load *w*/unit length and horizontal reaction adds to 2*H*.

Hence $$2H = \frac{4}{3}\cdot\frac{wR}{\pi}$$

and $$H = \frac{2}{3}\cdot\frac{wR}{\pi}.$$

Example 12.1 A semi circular two hinged arch has a span of 12 metres with the hinged supports at the same horizontal level. The rise of the arch is 6 m. A concentrated load (downward) of 2 tonnes acts at the central crown point. Find the horizontal thrust developed at the hinged supports by using principle of minimum strain energy. Also draw the B.M. diagram. The arch is of uniform cross section. Its own weight may be neglected . **[AMIE Winter 1985]**

Solution.

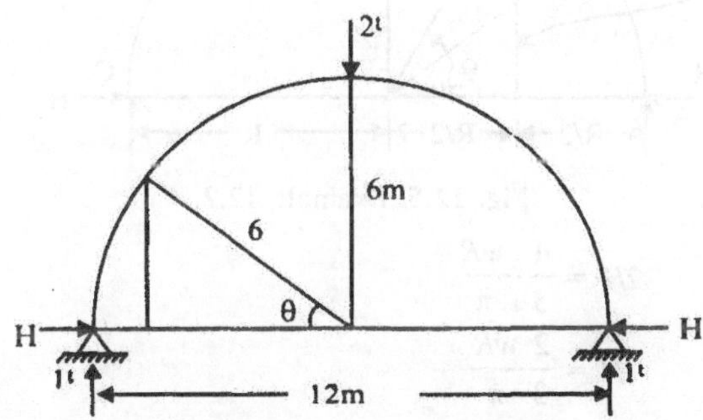

Fig. 12.4. Example 12.1.

$$U = \int\frac{M^2 ds}{2EI}$$

$$\frac{\partial U}{\partial H} = M\cdot\frac{\partial M}{dH}\cdot\frac{ds}{EI}$$

$$M = 1\times(6-6\,\text{Cos}\theta) - H.6\,\text{Sin}\theta$$

$$\frac{\partial M}{\partial H} = -6\,\text{Sin}\,\theta$$

$$ds = R \cdot d\theta.$$

$$\frac{\partial U}{\partial H} = 2\int_0^{\pi/2} (6(1-\text{Cos }\theta) - 6H\text{ Sin }\theta)\cdot(-6\text{ Sin }\theta)\cdot 6\cdot d\theta$$

$$= -2\times 6\times 6\times 6\int_0^{\pi/2}\left[(1-\text{Cos }\theta)\cdot\text{Sin}\theta - H\text{Sin}^2\theta\right]d\theta$$

$$= 0.$$

or
$$H = \frac{\int_0^{\pi/2} (1-\text{Cos }\theta)\text{ Sin}\theta\cdot d\theta}{\int_0^{\pi/2} \text{Sin}^2\theta\cdot d\theta}$$

$$= \frac{\left[-\text{Cos }\theta - \frac{\text{Sin}^2\theta}{2}\right]_0^{\pi/2}}{\pi/4}$$

$$= \frac{+1-\frac{1}{2}\cdot 1}{\pi/4} = \frac{1}{2}\cdot\frac{4}{\pi} = \frac{2}{\pi}$$

Example 12.2. A two hinged semi circular arch of uniform *EI* and of radius *R* supported at the same level, carries a distributed vertical load linearly varying from zero at the left end to *w* per unit run at the right end. Obtain the reaction components at the right end. Hence compute the bending moment, the normal thrust and the radial shear at left quarter span section. **[AMIE Summer, 95]**

Solution:

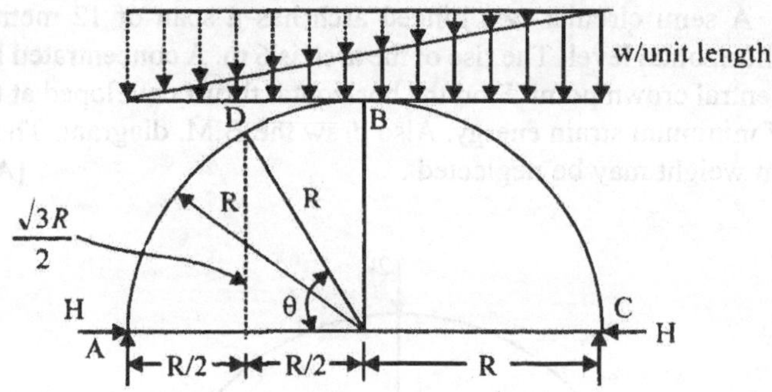

Fig. 12.5. Example 12.2.

$$2H = \frac{4}{3}\cdot\frac{wR}{\pi}$$

or
$$H = \frac{2}{3}\frac{wR}{\pi}$$

$$V_A\times 2R - \frac{1}{2}\times w\times 2R\times\frac{2R}{3}$$

$$= 0$$

or
$$V_A = \frac{wR}{3}$$

$$V_C = \frac{2wR}{3}$$

B. M at D

$$V_A(R/2) - H \cdot R\frac{\sqrt{3}}{2} - \frac{1}{2}\cdot\frac{w}{4}\cdot\frac{R}{2}\cdot\frac{1}{3}\cdot\frac{R}{2}$$

$$= \frac{wR}{3}\cdot\frac{R}{2} - \frac{2}{3}\frac{wR}{\pi}\cdot R\cdot\frac{\sqrt{3}}{2} - \frac{wR^2}{96}$$

$$= wR^2\left(\frac{1}{6} - \frac{1}{96} - \frac{1}{\sqrt{3}\pi}\right)$$

$$= -0.0275\, wR^2.$$

$$\theta = \tan^{-1}\frac{\sqrt{3}R}{2}\frac{2}{R} = \tan^{-1}\sqrt{3} = 60°$$

Normal Thrust.

$$\left(V_A - \frac{1}{2}\cdot\frac{w}{4}\cdot\frac{R}{2}\right)\text{Cos}\,\theta + H\,\text{Sin}\,\theta$$

$$= \left(\frac{wR}{3} - \frac{wR}{16}\right)\text{Cos}\,60° + \frac{2}{3}\cdot\frac{wR}{\pi}\cdot\text{Sin}\,60°$$

$$= \frac{wR\cdot 13}{96} + \frac{2}{3}\cdot\frac{wR}{\pi}\cdot\frac{\sqrt{3}}{2}$$

$$= wR\left(\frac{13}{96} + \frac{2}{3}\cdot\frac{1}{\pi}\cdot\frac{\sqrt{3}}{2}\right)$$

$$= \mathbf{0\cdot 3192\ wR.}$$

Shear Force.

$$\left(V_A - \frac{1}{2}\cdot\frac{w}{4}\cdot\frac{R}{2}\right)Sin\ 60° - \frac{2}{3}\cdot\frac{wR}{\pi}\ \text{Cos}\ 60°$$

$$= \left(\frac{wR}{3} - \frac{wR}{16}\right)\cdot\frac{\sqrt{3}}{2} - \frac{2}{3}\cdot\frac{wR}{\pi}\cdot\frac{1}{2}$$

$$= wR\left(\frac{13}{96}\cdot\sqrt{3} - \frac{1}{3\pi}\right)$$

$$= \mathbf{0\cdot 1284\ wR.}$$

12.3. STANDARD CASES OF TWO HINGED PARABOLIC ARCH

(*a*) Uniformly distributed load over whole span.

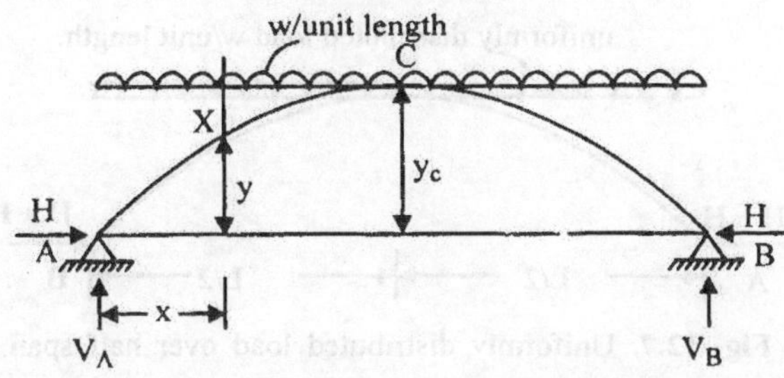

Fig. 12.6. Uniformly distributed load over whole span of parabolic arch.

$$H = \frac{\int_0^L \mu\, y dx}{\int_0^L y^2 \cdot dx} = \frac{N}{D}$$

$$V_A = V_B = \frac{wL}{2}.$$

Hence B.M. at any section x

$$= \frac{wL}{2} \cdot x - \frac{wx^2}{2} = \mu$$

Equation of a parabola with A as origin

$$y = \frac{4y_c}{L^2}\left(Lx - x^2\right)$$

$$N = \int_0^L \left(\frac{wLx}{2} - \frac{wx^2}{2}\right) \cdot \frac{4y_c}{L^2}\left(Lx - x^2\right) dx$$

$$= \frac{2y_c}{L^2} \int_0^L \left(wLx - wx^2\right)\left(Lx - x^2\right) dx$$

$$= \frac{2wy_c}{L^2} \int_0^L \left(Lx - x^2\right) \cdot dx$$

$$= \frac{wL^3}{15} \cdot y_c \cdot$$

$$D = \int_0^L y^2 \cdot dx = \int_0^L \frac{16yc^2}{L^4} \ (Lx - x^2) \ dx \cdot$$

$$= \frac{16yc^2}{L^4} \cdot \frac{L^5}{30} = \frac{8}{15} \cdot yc^2 \cdot L$$

$$\therefore \quad H = \frac{\frac{wL^3}{15} \cdot y_c}{\frac{8}{15} \cdot y_c{}^2 \cdot L} = \frac{wL^2}{8y_c}$$

$$= \frac{\text{B M at centre of simple beam}}{\text{Rise at centre}}$$

(*b*) Uniformly distributed load over half the span.

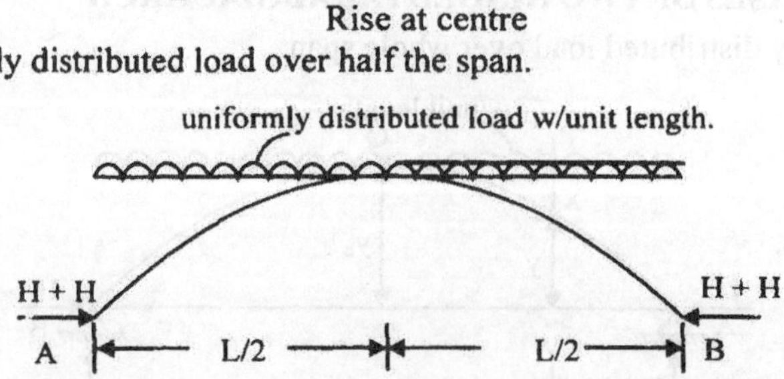

Fig. 12.7. Uniformly distributed load over half span.

As in the case of two hinged circular arch if the right half is also loaded with the uniformly distributed load, the horizontal reaction will be doubled *i.e.* it will become $2H$. Hence

$$2H = \frac{wL^2}{8y_c} \text{ or } H = \frac{wL^2}{16y_c}$$

$$= \frac{\text{B M at Centre of simple beam AB}}{\text{Rise in arch}}$$

(c) Concentrated load

$$V_A = \frac{W}{L}(L - \alpha L)$$

$$= W(1-\alpha)$$

$$V_B = W.\alpha.$$

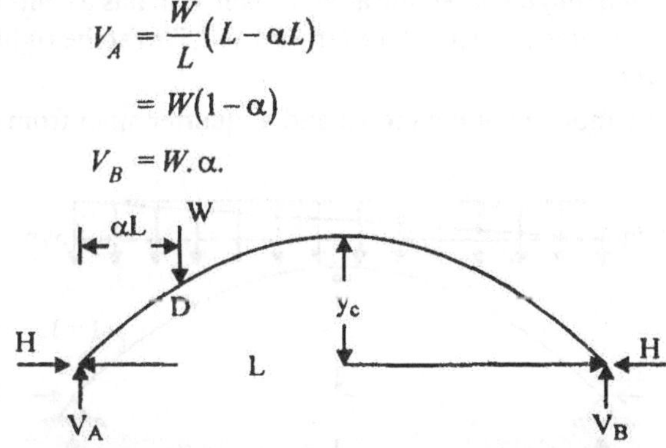

Fig. 12.8. Concentrated load on parabolic arch.

Value of μ

(*i*) Between *A* and *D* with origin at '*A*'

$$V_A \cdot x = W(1-\alpha)\cdot x$$

Between D and B with origin at B

$$V_B \cdot x = W \cdot \alpha \cdot x \cdot$$

$$N = \int_0^L \mu \, y dx = \int_0^{\alpha L} W \, (1-\alpha)\cdot x \cdot \frac{4y_c}{L^2} \, (Lx - x^2) \, dx$$

$$+ \int_0^{L-\alpha L} W \cdot \alpha \cdot x \cdot \frac{4y_c}{L^2} \, (Lx - x^2) \, dx$$

$$= \frac{4W(1-\alpha)\cdot yc}{L^2}\int_0^{\alpha L} (Lx^2 - x^3)\, dx + \frac{4Wy_c.\alpha}{L^2}\int_0^{L-\alpha L} (Lx^2 - x^3)\, dx$$

$$= \frac{4W(1-\alpha)\cdot Y_c}{L^2}\left[L\cdot\frac{x^3}{3} - \frac{x^4}{4}\right]_0^{\alpha L} + \frac{4Wy_c \cdot d}{L^2}\left[\frac{Lx^3}{3} - \frac{x^4}{4}\right]_0^{L-\alpha L}$$

$$= \frac{4W(1-\alpha)\cdot y_c}{L^2}\left(\frac{L^4 \cdot \alpha^3}{3} - \frac{\alpha^4 L^4}{4}\right)$$

$$+ \frac{4W.\alpha y_c}{L^2}\left[\frac{L}{3}\cdot(L-\alpha L)^3 - \frac{(L-\alpha L)^4}{4}\right]$$

$$= \frac{W.y_c.\alpha L^2}{3}(\alpha^3 - 2\alpha^2 + 1)$$

$$D = \int_0^L y^2 dx = \frac{8}{15}\cdot y_c^2 \cdot L$$

$$H = \frac{N}{D} = \frac{\dfrac{Wy_c.\alpha L^2}{3}(\alpha^3 - 2\alpha^2 + 1)}{\dfrac{8}{15}\cdot Y_c^2 \cdot L}$$

$$= \frac{5}{8}\frac{W\alpha L}{y_c}(1 - 2\alpha^2 + \alpha^3).$$

Example 12.3. (*a*) A two hinged parabolic arch of span 30m has a central rise of 4m. The arch is loaded by a triangular load of intensity zero at the left end and 5^t/m at the right support. Determine the horizontal thrust on the arch.

(*b*) Find the bending moment at the crown and at quarter span from the left hand support.

[AMIE summer 1978]

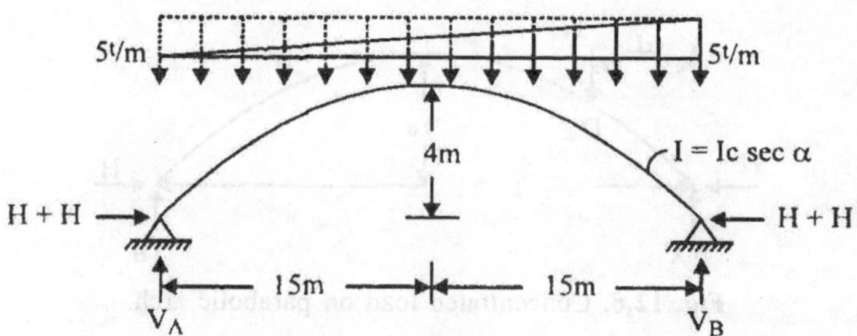

Fig. 12.9. Example 12·3

Solution :

If over the given triangular load an additional same triangular load varying from 5t/m from left support to zero at right support is superposed, the load will become uniformly distributed load of 5tm throughout and the horizontal reaction will be doubled.

Hence $\qquad 2H = \dfrac{wL^2}{8y_c} = \dfrac{5 \times 30^2}{8 \times 4}$

or $\qquad H = \dfrac{5 \times 30^2}{64} = 70 \cdot 3125^t$

Moment at B $\qquad V_A \times 30 - \dfrac{1}{2} \times 5 \times 30 \times 10 = 0$

or $\qquad V_A = \dfrac{5 \times 30 \times 10}{2 \times 30} = 25^t$

$$B.M. \text{ at crown } = V_A \times 15 - H \times 4 - \frac{1}{2} \times 2.5 \times 15 \times 5$$

$$= 25 \times 15 - 70{\cdot}3125 \times 4 - \frac{1}{2} \times 2.5 \times 15 \times 5$$

$$= 0$$

B.M. at quarter span.

At $x = 7 \cdot 5$, $\qquad y = \dfrac{4Y_c}{L^2}(Lx - x^2)$

$$= \frac{4 \times 4}{30^2}(30 \times 7.5 - 7.5^2)$$

$$= 3\text{m}$$

$$B.M. = V_A \times 7\cdot5 - H \times 3 - \frac{1}{2} \times 7\cdot5 \times 1\cdot25 \times \frac{7\cdot5}{3}$$

$$= 25 \times 7\cdot5 - 70\cdot3125 \times 3 - \frac{1}{2} \times \frac{7\cdot5 \times 1\cdot25 \times 7\cdot5}{3}$$

$$= -35\cdot156^{t}m$$

Example 12.4 A symmetrical two pinned parabolic arch spans 15m and has a central rise of 3m. Assuming $I = I_c \sec\theta$, find the horizontal thrust at the supports if the span carries a central load of 10^t. **[AMIE Summer, 1980]**

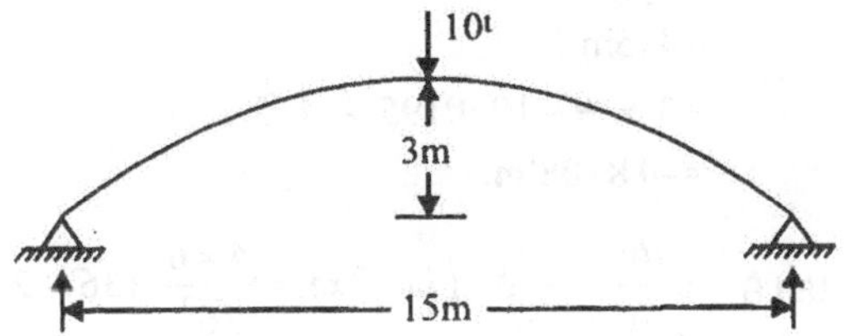

Fig. 12.10. Example 12.4

Solution : It is already derived

$$H = \frac{5}{8}\frac{W\cdot\alpha L}{y_c}(1-2\alpha^2+\alpha^3)$$

Here $\alpha = 0\cdot5$

Putting numerical values

$$H = \frac{5}{8} \times \frac{10 \times 0\cdot5 \times 15}{3}\left(1-2\times(0\cdot5)^2+(0\cdot5)^3\right)$$

$$= \mathbf{9\cdot765625^{t}}$$

Example 12.5 A two hinged symmetrical parabolic arch has a span of 36m and a rise of 6m. The moment of inertia of its section at any point varies as the secant of the slope of the rib axis at that point. The arch carries a vertical load of 12 tonnes at the left quarter span. Find due to this load the horizontal thrust at the supports and the values of the bending moment and normal thrust at the right quarter span. [AMIE Summer 1981]

Solution: The horizontal thrust due to a concentrated load W.

$$= \frac{5}{8}\frac{W\cdot\alpha L}{y_c}[1-2\alpha^2+\alpha^3]$$

[If necessary, this formula should be derived]

Substituting the numerical values

$$H = \frac{5}{8} \times \frac{12 \times 0\cdot25 \times 36}{6}\left[1-2\times(0\cdot25)^2+(0\cdot25)^3\right]$$

$$= 10\cdot0195^{t}$$

Fig. 12.11. Example 12.5

$$V_B \times 36 - 12 \times 9 = 0$$

or $$V_B = \frac{12 \times 9}{36} = 3^t$$

Equation of a parabola

$$y = \frac{4y_c}{L^2}(Lx - x^2)$$

$$= \frac{4 \times 6}{36^2}(36 \times 9 - 9^2)$$

$$= 4.5\text{m}$$

B.M. at D $$= 3 \times 9 - 10.0195 \times 4.5$$

$$= \mathbf{-18.08^t m.}$$

$$\tan\theta = \frac{dy}{dx} = \frac{4y_c}{L^2}(L - 2x) = \frac{4 \times 6}{36^2}(36 - 2 \times 9)$$

$$= 0.333$$

$$\theta = 18.4349^\circ \quad \sin\theta = 0.3162 \quad \cos\theta = 0.94868$$

Axial Thrust:

$$= 10.0195 \cos\theta + 3 \sin\theta$$

$$= 10.0195 \times 0.94868 + 3 \times 0.3162$$

$$= 10.4539^t$$

Example 12.6 A two hinged parabolic arch of span 20m and rise 4m carries a uniformly distributed load of 5^t per meter on the left half of span as shown in Fig. 12.12. The moment of mertia I of the arch section at any point is given by $I = I_o \sec\theta$ where θ = inclination of the tangent at the point with the horizontal and I_o is the moment of inertia at the crown. Find (*i*) the reactions at the supports (*ii*) the position and (*iii*) the value of the maximum bending moment in the arch.

[AMIE Summer 1982]

Solution:

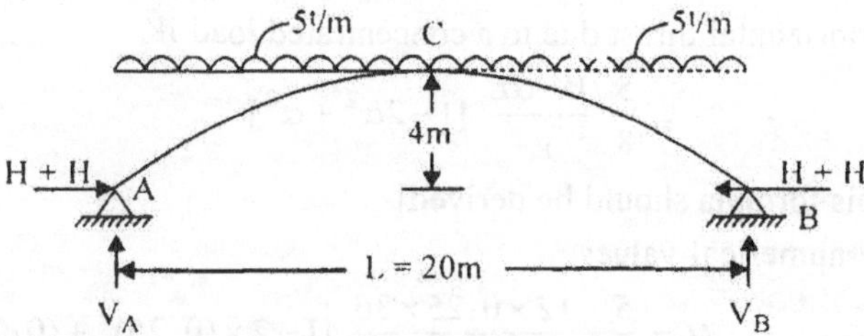

Fig. 12.12. Example 12.6

If the right part is also loaded with a uniformly distributed load of 5^t/m, the horizontal reaction will be doubled.

Hence $$2H = \frac{wL^2}{8y_c}$$

or $$H = \frac{wL^2}{16y_c} = \frac{5 \times 20^2}{16 \times 4} = \mathbf{31.25^t}$$

$$V_A \times 20 - 5 \times 10 \times 15 = 0$$

or
$$V_A = \frac{5 \times 10 \times 15}{20} = 37 \cdot 5^t$$

$$V_B = 12 \cdot 5^t$$

B.M.

In *AC*

B.M. at any section $= V_A.x - H.y - \dfrac{5x^2}{2}$

Again, $y = \dfrac{4y}{L^2}(Lx - x^2)$

$$= \frac{4 \times 4}{400}(20x - x^2)$$

$$= 0 \cdot 8x - \cdot 04x^2$$

B.M then becomes

$$37 \cdot 5x - 31 \cdot 25\,(0 \cdot 8x - \cdot 04x^2) - 2 \cdot 5x^2$$

$$= 12 \cdot 5x - 1 \cdot 25x^2$$

$$\frac{dM}{dx} = 12 \cdot 5 - 2 \cdot 50x = 0 \text{ for maximum B.M.}$$

or $x = 5\text{m}$

$$M = 12 \cdot 5 \times 5 - 1 \cdot 25 \times 5^2$$

$$= +31 \cdot 25 \text{ (sagging)}$$

For CB $B.M = V_B \times x - H.y$

$$= 12 \cdot 5x - 31 \cdot 25 \times \frac{4y_c}{L^2}(Lx - x^2)$$

$$= 12 \cdot 5x - \frac{31 \cdot 25 \times 4 \times 4}{20^2}(20x - x^2)$$

$$= 1 \cdot 25x^2 - 12 \cdot 5x$$

$$\frac{dM}{dx} = 2 \cdot 5x - 12 \cdot 5 = 0$$

$$M = 1 \cdot 25 \times 25 - 12 \cdot 5 \times 5$$

$$= -31 \cdot 25\text{tm}$$

Example 12.7 (*a*) A two hinged parabolic arch of span l and rise h carries a concentrated load W at a distance 'a' from the left end. Making the secant assumption of the variation of the *M.I.* at any cross-section show that the horizontal thrust is given by

$$H = \frac{5}{8}\frac{W}{h\,l^3} \cdot a\,(l - a)\,(l^2 + la - a^2)$$

(*b*) If the above arch carries a uniformly distributed load w per unit run for a horizontal distance of 's' from the left end instead of the concentrated load, show that the horizontal thrust is given by

$$H = \frac{1}{16}\frac{ws^2}{hl^3}\left[5l^3 - 5ls^2 + 2s^3\right]$$

[AMIE Summer 1983]

Solution: We have already derived

$$H = \frac{5}{8}\cdot\frac{W}{h}\cdot \alpha l \ (1-2\alpha^2+\alpha^3)$$

Here $\alpha \ = \frac{a}{l}$

$$H = \frac{5}{8}\cdot\frac{W}{h}\cdot\frac{a}{l}\cdot l \left(1-2\left(\frac{a}{l}\right)^2+\left(\frac{a}{l}\right)^3\right)$$

$$= \frac{5}{8}\cdot\frac{W}{h}\cdot\frac{a}{l^3}\ (l^3-2a^2l+a^3)$$

$$= \frac{5}{8}\cdot\frac{W}{h}\cdot\frac{a}{l^3}\ (l^3+l^2a-la^2-al^2-a^2l+a^3)$$

$$= \frac{5}{8}\cdot\frac{W}{h}\cdot\frac{a}{l^3}\ (l-a)\ (l^2+la-a^2)$$

When there is a distributed load from 0 to s α varies from 0 to $\frac{s}{L}$

$$\alpha = \frac{s}{l} \qquad d\alpha = \frac{ds}{l}.$$

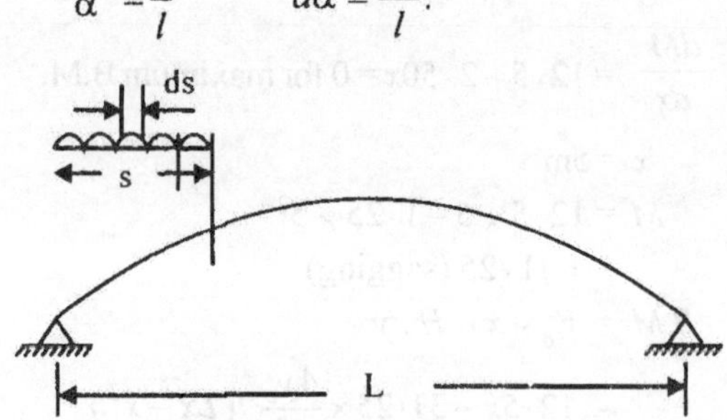

Fig. 12.13. Example 12.7

$$H = \int_0^{\frac{s}{l}} \frac{5}{8}\frac{wl\ d\alpha\cdot\alpha l}{h}\ (1-2\alpha^2+\alpha^3)$$

$$= \frac{5}{8}\ \frac{wl^2}{h}\int_o^{s/l}(\alpha-2\alpha^3+\alpha^4)\ d\alpha$$

$$= \frac{5}{8}\ \frac{wl^2}{h}\left[\frac{\alpha^2}{2}-\frac{2\alpha^4}{4}+\frac{\alpha^5}{5}\right]_o^{s/l}$$

$$= \frac{5wL^2}{8h}\left[\frac{s^2}{2l^2}-\frac{2\cdot s^4}{4l^4}+\frac{s^5}{5l^5}\right]$$

$$= \frac{5}{8h}\cdot wl^2\cdot\frac{s^2}{10l^5}\ [5l^3-5ls^2+2s^3]$$

$$= \frac{1}{16}\cdot\frac{ws^2}{hl^3}\ \left[5l^3-5ls^2+2s^3\right].$$

12.4. EFFECT OF TEMPERATURE CHANGE AND YIELDING OF SUPPORTS ON HORIZONTAL THRUST.

If there is a rise in temperature there will be an increase of curved length of elementary length δs by $\delta s \cdot \alpha \cdot t$ where α is the coefficient of linear expansion of the material of the rib.

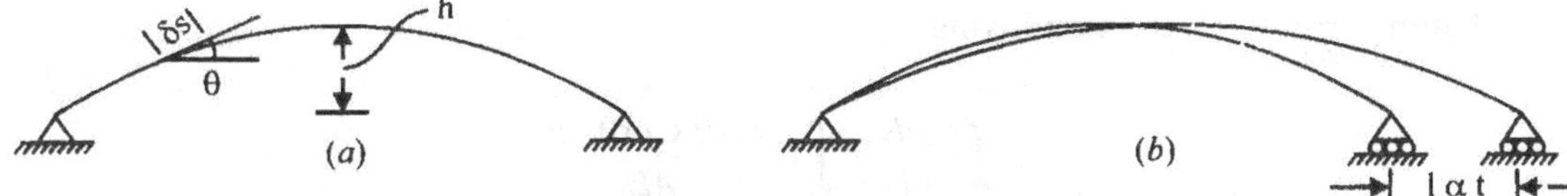

Fig. 12.14. Effect of temperature change.

The increase in the horizontal direction

$$= \delta s \cdot \alpha \cdot t \cdot \text{Cos}\,\theta$$

$$\text{Total increase} = \int_o^l \alpha \cdot t \cdot \delta s\ \text{Cos}\,\theta$$

$$= \int_o^l \alpha \cdot t \cdot dx = \alpha t \cdot l \cdot$$

If in addition there is load

$$\text{increase in length} = \int_o^l \frac{\mu\, yds}{EI}$$

$$\text{Hence total extension} = \alpha t \cdot l + \int_o^l \frac{\mu\, yds}{EI}$$

If the supports are rigid, horizontal reaction will produce equal displacement in the opposite direction. If the supports yield by δ the displacement will be reduced by δ.

Hence
$$H = \frac{\alpha t l + \int_o^l \frac{\mu\, yds}{EI} - \delta}{\int \frac{y^2 ds}{EI}}, \qquad \ldots(12.6)$$

In the numerator appropriate terms should be used for computing H.

12.5 DERIVATION OF HORIZONTAL THRUST WITHOUT NEGLECTING AXIAL FORCE :

The formulae derived so far were obtained by considering bending only and effect of shear force, and axial thrust on strain energy were neglected. We now derive expressions where effect of axial force is included but that of shear is again neglected as its effect is very small.

By differentiating the general energy equation

$$\frac{\partial U}{\partial H} = \int_o^l \frac{M \cdot \frac{\partial M}{\partial H} \cdot ds}{EI} + \int_o^l \frac{V \cdot \frac{\partial V}{\partial H} \cdot ds}{GAs} + \int_o^l \frac{F \cdot \frac{\partial F}{\partial H} \cdot ds}{AE}$$

$$= \int_o^l \frac{(\mu - Hy)(-y) \cdot ds}{EI} + \int_o^l \frac{(\text{H Cos}\theta + \text{V Sin}\theta)\text{Cos}\theta \cdot \text{ds}}{AE}$$

$$= -\int_o^l \frac{\mu yds}{EI} + H\int_o^l \frac{y^2 ds}{EI} + H\int \frac{Cos^2\theta ds}{AE} + \int_o^l \frac{V \cdot Sin\theta \cdot Cos\theta ds}{AE}$$

Now $\frac{\partial U}{\partial H} = 0$ when there is no support displacement.

$= -\delta$ when support yields by an amount δ

Putting $\frac{\partial U}{\partial H} = -\delta$ and simplifying

$$H = \frac{\int_o^l \frac{\mu y ds}{EI} - \int_o^l V \frac{Sin\theta\, Cos\theta \cdot ds}{AE} - \delta}{\int_o^l \frac{y^2 ds}{EI} + \int_o^l \frac{Cos^2\theta \cdot ds}{AE}}$$

Neglecting the 2nd term in the numerator whose effect on the result is negligible

$$H = \frac{\int_o^l \frac{\mu y ds}{EI} - \delta}{\int_o^l \frac{y^2 ds}{EI} + \int_o^l \frac{Cos^2\theta \cdot ds}{AE}} \qquad \text{......(12.7)}$$

where l = curved length of the arch. For parabolic arch with $I = I_c\, Sec\,\theta$ equation reduces to

$$H = \frac{\int_o^l \frac{\mu \cdot y dx}{EI_c} - \delta}{\int_o^l \frac{y^2 \cdot dx}{EI_c} + \int_o^l \frac{Cos\theta \cdot dx}{AE}}$$

where l = horizontal span length

For an arch of small rise $Cos\,\theta = 1$

and putting $A = A_c$ for all areas of the arch the equation can be further simplified to

$$H = \frac{\int_o^l \frac{\mu y dx}{EI_c} - \delta}{\int_o^l \frac{y^2 dx}{EI_c} + \int_o^l \frac{dx}{A_c E}} \qquad \text{.....(12.7a)}$$

$$= \frac{\int_o^l \frac{\mu y dx}{EI_c} - \delta}{\int_o^l \frac{y^2 dx}{EI_c} + \frac{l}{EA_c}} \qquad \text{.....(12.7b)}$$

If the support yields by an amount K for unit value of horizontal thrust $\delta = KH$.

Hence

$$H = \frac{\int_o^l \frac{\mu y dx}{EI_c} - KH}{\int_o^l \frac{y^2 dx}{EI_c} + \frac{l}{EA_c}}$$

$$= \frac{\int \frac{\mu y dx}{EI_c}}{\int_o^l \frac{y^2 dx}{EI_c} + \frac{l}{EA_c} + K} \qquad \text{....(12.7c)}$$

For a tied arch where a tie rod connects the two supports

δ = extension of the tie rod

= yielding of the support

$= \frac{H.l.}{A_t . E_t}$ where A_t and E_t

refer to tie rod and the expression reduces to

$$H = \frac{\int_o^l \frac{\mu y dx}{EI_c}}{\int_o^l \frac{y^2 dx}{EI_c} + \frac{l}{EA_c} + \frac{l}{E_t . A_t}} \qquad \text{.... (12.7d)}$$

Example 12.8. A bow string girder (i.e. tied two hinged parabolic arch) of span 50m and rise 8m is subjected to two loads each of 20ᵗ acting at a distance of 15m from either support. Area of arch rib=360cm² Area of tie = 50cm². I of arch rib at any point = 54,000 Sec α cm⁴. where α = slope of arch axis at P. Earch = 160tf/cm² $E_{(tie)}$ = 2000 tf/cm². Compute the B.M. and axial thrust at the crown.

[AMIE Summer 1985]

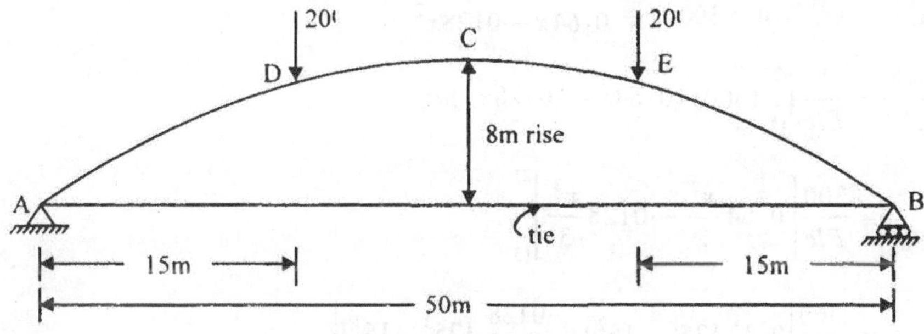

Fig. 12.15. Example 12.8

Solution:

(*i*) Applicable formula.

$$H = \frac{\int_o^l \frac{\mu y dx}{EIc}}{\int_o^l \frac{y^2 dx}{EIc} + \frac{l}{EAc} + \frac{l}{E_t . A_t}}$$

(*ii*) $$\int_o^l \frac{y^2 dx}{EIc} = \frac{8}{15} \cdot \frac{h^2 l}{EIc} = \frac{8}{15} \times \frac{8^2 \times 50 \times 10^6}{160 \times 54{,}000}$$

$$= 0 \cdot 19753 \times 10^3$$
$$= 197 \cdot 53 \text{ cm} = 1 \cdot 9753 \text{ m.}$$

(*iii*) $\int_o^l \frac{\mu y dx}{EIc}$

For AD, $\mu = 20x.$

$$y = \frac{4y_c}{L^2}(Lx - x^2)$$

$$= \frac{4 \times 8}{50 \times 50}(50x - x^2)$$

$$= .0128\,(50x - x^2)$$

$$= 0.64x - .0128x^2$$

$$\int_0^{15} \frac{\mu y.dx.}{EIc} = \frac{1}{EIc}\int_0^{15}(20x)(0.64x - 0128x^2)dx.$$

$$= \frac{20}{EIc}\int_0^{15}(0.64x^2 - .0128x^3)dx$$

$$= \frac{20}{EIc}\left[0.64\frac{x^3}{3} - .0128\frac{x^4}{4}\right]_0^{15}$$

$$= \frac{11160}{EIc}$$

For DC $\mu = 300$ $y = 0.64x - .0128x^2$

$$\int_{15}^{25} \frac{\mu y dx}{EIc} = \frac{1}{EIc}\int_{15}^{25}(300)\,(0.64x - \;.0128x^2)dx$$

$$= \frac{300}{EIc}\left[0.64\frac{x^2}{2} - .0128\frac{x^3}{3}\right]_{15}^{25}$$

$$= \frac{300}{EIc}\left[0.32\,[25^2 - 15^2] - \frac{.0128}{3}\,[25^3 - 15^3]\right]$$

$$= \frac{300}{EIc}[128 - 52.26] = \frac{22720}{EIc}$$

$$\text{Total value} = \frac{2}{EIc}[11160 + 22720] = \frac{67760}{EIc}$$

$$= \frac{67760 \times 10^4}{160 \times 54,000} = 78.426 \text{ m}$$

(*iv*) Horizontal Thrust : To avoid numerical error ton and m are used throughout

$$\frac{l}{EAc} = \frac{50}{360 \times 160} = .000868 \text{ m/t}$$

$$\frac{l}{E_t A_t} = \frac{50}{2000 \times 50} = \cdot 0005 \text{ m/t}$$

$$H = \frac{78 \cdot 426 \text{ m}}{(1 \cdot 9753 + \cdot 000868 + \cdot 0005) \text{ m/t}}$$

$$= 39 \cdot 675^{t}$$

(v) Bending Moment at Crown.

$$M_c = 20 \times 25 - 20 \times 10 - 39 \cdot 675 \times 8$$

$$= \mathbf{-17 \cdot 4tm}$$

Example 12.9. Two parabolic arches have a common hinge at B that forms part of a roller bearing there and are hinged to the springings at A and C. If EIc is the same for both arches and I=Ic Sec θ, determine the normal thrust and bending moment at the crown of the left hand arch under the given load.

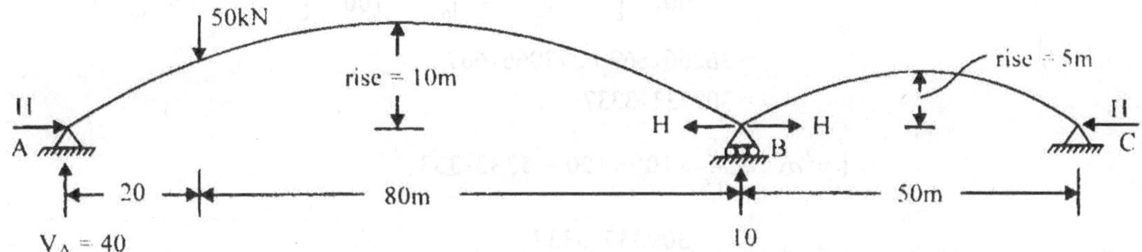

Fig. 12.16. Example 12.9.

Solution: If δ is the movement of the roller B towards right

$$H = \frac{\int_o^l \frac{\mu y dx}{EI_c} - \delta}{\int_o^l \frac{y^2 dx}{EIc}}$$

From the 2nd arch

$$H \int_o^l \frac{y^2 dx}{EIc} = \delta$$

$$\text{or } \delta = H \cdot \frac{8}{15} \frac{y_c^2 l}{EIc}$$

$$= H \cdot \frac{8}{15} \cdot \frac{5^2 \times 50}{EIc}$$

$$= 666 \cdot 67 \frac{H}{EIc}$$

$$H = \frac{\int_o^l \frac{\mu y dx}{EI_c} - \frac{666 \cdot 67 H}{EI_c}}{\int_o^l \frac{y^2 dx}{EIc}}$$

$$= \frac{\int_0^l \mu y dx}{\int_0^l y^2 dx + 666 \cdot 67}$$

$$\int_0^l \mu y dx = \int_0^{20} (40x) \left(\frac{4y_c}{l^2}\right)(lx - x^2)dx$$

$$+ \int_0^{80} (10x) \left(\frac{4y_c}{l^2}\right)(lx - x^2)dx$$

$$= \frac{160y_c}{l^2}\int_0^{20} (100x^2 - x^3)dx + \frac{40y_c}{l^2}\int_0^{80} (100x^2 - x^3)dx$$

$$= \frac{160 \times 10}{100^2}\left[100\frac{x^3}{3} - \frac{x^4}{4}\right]_0^{20} + \frac{40 \times 10}{100^2}\left[\frac{100x^3}{3} - \frac{x^4}{4}\right]_0^{80}$$

$$= 36266 \cdot 667 + 273066 \cdot 667$$

$$= 309333 \cdot 3337$$

$$\int y^2 dx = \frac{8}{15} \times 10^2 \times 100 = 5333 \cdot 333$$

$$H = \frac{309333 \cdot 3337}{5333 \cdot 333 + 666 \cdot 67}$$

$$= \mathbf{51 \cdot 555\ kN}$$

Normal thrust at the crown

$$= H = 51 \cdot 555 \text{ kN.}$$

B.M. at the crown$= 40 \times 50 - 51 \cdot 555 \times 10 - 50 \times 30$

$$= -15 \cdot 55 \text{ kNm}$$

Summary

1. Two hinged arch is statically indeterminate to 1st degree. There are four unknowns: V_A, H_A, V_B and H_B but only three equations of statical equilibrium. From $\Sigma H = 0$, $H_A = H_B = H$ when no external horizontal force is acting on the arch. If the supports are unyielding $\frac{\partial U}{\partial H} = 0$ which gives the fourth equation for solving H.

2. If only bending deformation is considered H is given by

$$H = \frac{\int_0^l \frac{\mu y ds}{EI}}{\int \frac{y^2 ds}{EI}}$$

For a parabolic arch the value of I at the springings is greater than that at the crown and it is assumed that the section varies in such a way that

$$I = I_c \cdot \text{Sec}\,\theta$$

and $H = \dfrac{\int_o^l \mu y dx}{\int_o^l y^2 dx}$.

The value of H gets modified when there is temperature change or when support yields. H then becomes

$$H = \frac{\alpha t l + \int_o^l \frac{\mu y ds}{EI} - \delta}{\int_o^l \frac{y^2 ds}{EI}}.$$

If axial force is taken into account (but shear still remains reglected)

$$H = \frac{\alpha t l + \int_o^l \frac{\mu y ds}{EI} - \delta}{\int_o^l \frac{y^2 ds}{EI} + \int_o^l \frac{Cos^2\theta ds}{AE}}.$$

PROBLEMS

A. Theoretical

12.1 (*i*) Derive an expression for determination of horizontal thrust of a two hinged arch considering

(*i*) Bending deformation only

(*ii*) Bending and axial thrust

Why shear is usually not considered in analysis of an arch?

(*ii*) Considering bending deformation only derive expressions for horizontal thrust in a semi circular arch with constant flexural rigidity EI when it is subjected to:

(*i*) Concentrate load W at any point

(*ii*) Uniformly distributed load on half the span.

(*iii*) Uniformly distributed load over the whole span.

(*iii*) For a parabolic arch with $I = Ic$ Sec θ derive expressions for H. when as before it is subjected to:

(*i*) Concentrated load W at any point

(*ii*) Uniformly distributed load on half the span

(*iii*) Uniformly distributed load over whole span.

B. Objective. Choose the correct answer.

12.2 (*i*) A parabolic two hinged arch subjected to uniform distributed loading per unit horizontal loading length over the entire span has (*a*) Zero B.M. at all sections (*b*) maximum normal thrust at the crown (*c*) has varying radial shear over its length (*d*) a parabolic variation of BM over the span. **[AMIE Winter'98]**

(*ii*) For a two hinged arch if one of the supports settles down vertically, then the horizontal thrust (*a*) is increased (*b*) is decreased (*c*) remains unchanged.

(*iii*) Bending moment in a two hinged parabolic arch carrying u.d.l. through out (*a*) varies linearly (*b*) is maximum at the crown (*c*) is zero everywhere **[AMIE Summer, 99]**

C. Mathematical

12.3. (*a*) A two hinged parabolic arch of span 20m has a central rise of 4m. The arch is loaded by a triangular load of intensity zero at the left abutment to 5t/m at the right abutment.

Determine the horizontal thrust on the arch.

(*b*) Find the bending moment at the crown and at quarter span from the left hand support **[AMIE Summer, 1978]**

12.4. (*a*) A two hinged parabolic arch of span 20m and rise 4m carries a u.d.l of 10t/m on the left half of the span. The arch rib has moment of inertia of its x section at any point varying as the recant of the slope of the rib axis at that point. Find the reactions at the supports.

(*b*) Determine the three - stress resultants (B.M., normal Thrust and radial shear) at the left quarter span **[AMIE Winter 1984]**

12.5. A two hinged parabolic arch of span L with both supports at same level and central rise '*h*' is subjected to a uniformly distributed load of an intensity *w* per meter covering the left half of the span and another of an intensity 2*w* per meter over the right half. Determine over the right half of the span. Determine from the first principles the horizontal thrust developed and draw the bending. Moment diagram. Assume that the moment of inertia of any cross-section of the arch $I = Io \sec\theta$ where Io is the M.I. at the crown and θ is the inclination to the horizontal of the tangent to the arch centre line of concerned cross section. **[AMIE Summer 1987]**

12.6. A two hinged parabolic arch of span L and rise $L/4$ is subjected to a concentrated load *W* at its quarter span. Show that the value of the horizontal thrust at the ends is $\frac{285}{512} \cdot W$. Take $I = I_o \sec\theta$. **[AMIE Winter 1989]**

12.7. A two hinged parabolic arch of span 42m and rise 14m carries two vertical concentrated loads of value 160 kN each acting symmetrically at a distance of 14m from the nearest support. The moment of inertia of the cross section of the arch rib varies as the secant of the slope of the rib axis. Calculate the support reactions. Draw the B.M. diagram showing values of maximum B.M. and location of points of contra flexure. Also find the shear force and normal thrust at the load points. How the B.M. under the load will change if the supports yield horizontally. **[AMIE Summer'99]**

13

COLUMN ANALOGY

13.1 INTRODUCTION

Column analogy method was developed by Prof. Hardy Cross in 1930. It is used for solving statically indeterminate structures like indeterminate beams and fixed arches. It is also useful in calculating stiffness and carry over factors for beams particularly with variable section. General solution of a statically indeterminate structure of 3rd degree involves solution of three simultaneous equations which is long and tedious. Column analogy method transforms the problem to routine calculation of pressures in a short column under biaxial eccentric load.

13.2 GENERAL EQUATION OF STRESSES IN A COLUMN

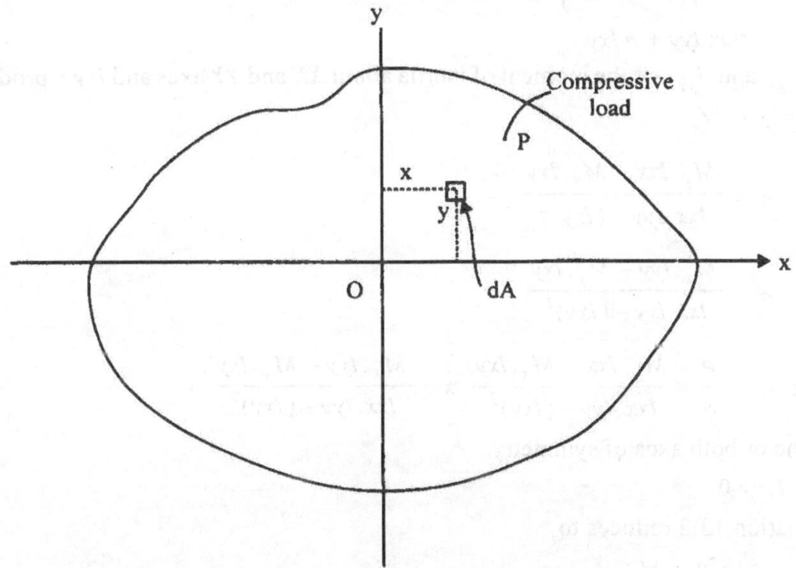

Fig. 13.1. Column under eccentric load.

Fig. 13.1 shows cross section of a column with two reference axes in the x and y directions subjected to compressive load P. From the assumption of planar distribution of pressure over the cross-section, the stress σ at any point is given by.

$$\sigma(x, y) = c + mx + ny \qquad \text{... (13.1)}$$

The constants c, m and n can be determined by three equations of equilibrium of statics.

$$P = \int \sigma dA, \; M_x = \int \sigma . y \cdot dA$$

$$M_y = \int \sigma . x \cdot dA$$

These can be rewritten as

$$P = c\int dA + m\int xdA + n\int ydA$$

$$M_x = c\int ydA + m\int xydA + n\int y^2 dA$$

$$M_y = c\int xdA + m\int x^2 dA + n\int xydA.$$

If reference axes pass through the centroid of the section, $\int xdA = 0$ and $\int ydA = 0$ and the equations reduce to

$$P = cA$$

$$M_x = m\int xydA + n\int y^2 dA$$

$$= m . Ixy + n . Ixx$$

$$M_y = m\int x^2 dA + n\int xydA$$

$$= m \, Iyy + n \, Ixy.$$

where I_{xx} and I_{yy} are the moment of inertia about XX and YY axes and Ixy = product of inertia

Solving $c = P/A$

$$m = \frac{M_y . Ixx - M_x . Ixy}{Ixx . Iyy - (Ixy)^2}$$

$$n = \frac{M_x . Iyy - M_y . Ixy}{Ixx . Iyy - (Ixy)^2}$$

whence $\sigma = \dfrac{P}{A} + \dfrac{M_y . Ixx - M_x . Ixy}{Ixx . Iyy - (Ixy)^2} \cdot x + \dfrac{M_x . Iyy - M_y . Ixy}{Ixx . Iyy - (Ixy)^2} \cdot y$ ...(13.2)

With one or both axes of symmetry

$$I_{xy} = 0$$

and equation 13.2 reduces to

$\sigma = \dfrac{P}{A} + \dfrac{M_y}{Iyy} \cdot x + \dfrac{M_x}{Ixx} \cdot y$ ...(13.2*a*)

13.3. DERIVATION OF COLUMN ANALOGY METHOD

In section 10.8 generalised solution of fixed portal frame by elastic centre method was derived. Moment at any point (x, y) is given by

$M_{(x,y)} = M_s + M_o + H_o y + V_o x$ (with appropriate signs for x and y)

Where $$M_o = -\frac{\Sigma \frac{M_s\, ds}{EI}}{\Sigma \frac{ds}{EI}}$$

$$V_o = -\frac{\frac{\Sigma M_s\, xds}{EI} - \left(\Sigma \frac{M_s yds}{EI}\right)\left(\frac{\Sigma \frac{xyds}{EI}}{\Sigma \frac{y^2 ds}{EI}}\right)}{\Sigma \frac{x^2\, ds}{EI} - \frac{(\Sigma \frac{xyds}{EI})^2}{\Sigma \frac{y^2 ds}{EI}}} \quad \text{...(10.6)}$$

$$H_o = -\frac{\frac{\Sigma M_s\, y \cdot ds}{EI} - \left(\Sigma \frac{M_s \cdot xds}{EI}\right)\frac{\Sigma \frac{xyds}{EI}}{\Sigma \frac{x^2 ds}{EI}}}{\Sigma \frac{y^2 ds}{EI} - \frac{(\Sigma \frac{xyds}{EI})^2}{\Sigma \frac{x^2 ds}{EI}}}$$

Taking $\Sigma \frac{x^2 ds}{EI} = Iyy,\ \Sigma \frac{y^2 ds}{EI} = Ixx$

$\Sigma \frac{xyds}{EI} = Ixy,\ \Sigma \frac{M_s \cdot ds}{EI} = W,\ \Sigma \frac{ds}{EI} = A$

$\Sigma \frac{M_s \cdot xds}{EI} = M_y$ and $\Sigma \frac{M_s \cdot yds}{EI} = M_x$.

The equations can be written in the simple form

$$M_o = -W/A,\ V_o = -\frac{M_y - M_x \cdot \frac{Ixy}{Ixx}}{Iyy - \frac{(Ixy)^2}{Ixx}}$$

and $$H_o = -\frac{M_x - M_y \frac{Ixy}{Iyy}}{Ixx - \frac{(Ixy)^2}{Iyy}} \quad \text{...(10.7)}$$

and $M(x,y) \quad = M_s + M_o + V_o x + H_o y$

$$= M_s - \left[\frac{W}{A} + \frac{M_y - M_x \cdot \frac{Ixy}{Ixx}}{Iyy - \frac{(Ixy)^2}{Ixx}} \cdot x + \frac{M_x - M_y \cdot \frac{Ixy}{Iyy}}{Ixx - \frac{(Ixy)^2}{Iyy}} \cdot y \right]$$

$$= M_s - M_i$$

Where $M_i = \frac{W}{A} + \left(\frac{M_y . Ixx - M_x . Ixy}{Iyy . Ixx - (Ixy)^2} \right) \cdot x$

$$+ \left(\frac{M_x . Iyy - M_y . Ixy}{Ixx . Iyy - (Ixy)^2} \right) \cdot y \qquad \text{... (13.3)}$$

But from the column flexure formula

$$\sigma = \frac{P}{A} + \left(\frac{M_y . Ixx - M_x Ixy}{Iyy . Ixx - (Ixy)^2} \right) \cdot x$$

$$+ \left(\frac{M_x . Iyy - M_y Ixy}{Ixx . Iyy - (Ixy)^2} \right) \cdot y \qquad \text{...(13.2)}$$

The right hand side of equation 13.3 and 13.2 are identical in pattern though equation 13.3 calculates indeterminate moment M_i and equation 13.2 calculate column stress under biaxial eccentricity. Hence column flexure formula can be applied in computing M_i provided we take

$$W = \Sigma \frac{M_s . ds}{EI} \text{ as } P.$$

and $\Sigma \frac{ds}{EI} = A \qquad \Sigma \frac{x^2 ds}{EI} = Iyy$

$$\Sigma \frac{M_s . xds}{EI} = M_y \quad \Sigma \frac{y^2 ds}{EI} = Ixx$$

$$\Sigma \frac{M_s . yds}{EI} = M_x \quad \Sigma \frac{xyds}{EI} = Ixy$$

in the column flexure formula.

Hence the portal frame with fixed end can be visualised as a column of the same shape with width equal to $\frac{1}{EI}$. This is known as analogous column as shown in Fig. 13.2 and the method as column analogy method

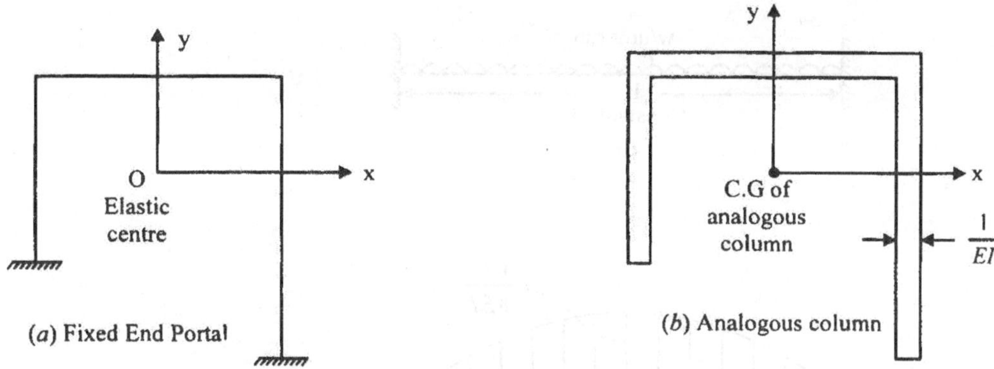

Fig. 13.2. Fixed Portal and Analogous Column.

If the portal is symmetrical Ixy for the portal is zero. Hence.

$$M_i = \frac{W}{A} + \frac{M_y}{Iyy} \cdot x + \frac{M_x}{Ixx} \cdot y \quad \text{...(13.3a)}$$

The corresponding column formula when the analogous column is symmetrical with $I_{xy} = 0$

$$\sigma = \frac{P}{A} + \frac{M_y}{Iyy} \cdot x + \frac{M_x}{Ixx} \cdot y \text{ as expected} \quad \text{... (13.2a)}$$

For a beam

$$M_i = \frac{W}{A} + \frac{M_y}{Iyy} \cdot x \quad \text{... (13.3b)}$$

and

$$\sigma = \frac{P}{A} + \frac{M_y}{Iyy} \cdot x \quad \text{.. (13.2b)}$$

and the analogy holds.

Sign Convention. The following sign conventions are applied (Fig. 13.3)

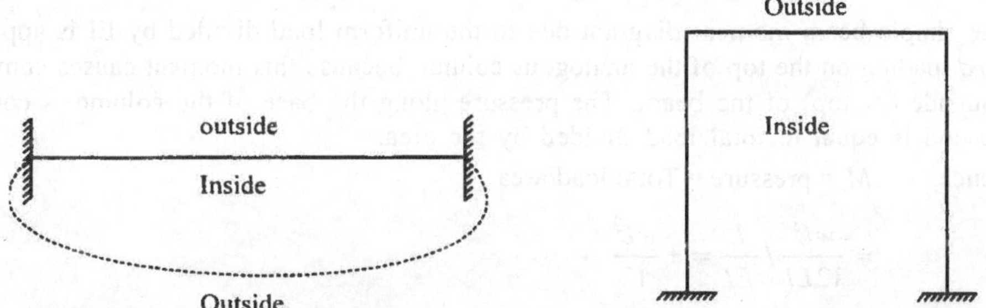

Fig. 13.3. Sign Convention

1. Loading on top of column is downward for positive M_x which causes compression ou' side.
2. Upward pressure on the bottom of the column is positive.
3. Resultant moment at any point is positive if it causes compression on the out side.

Example 13.1. Determine the fixed end moments for the fixed end beam shown in Fig. 13.4 under uniformly distributed load w/unit length.

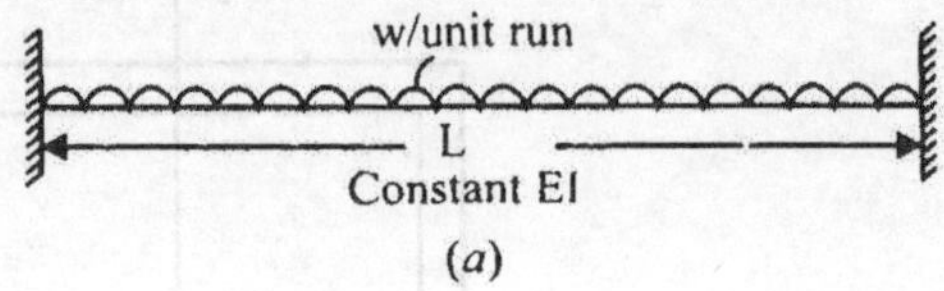

(*a*)

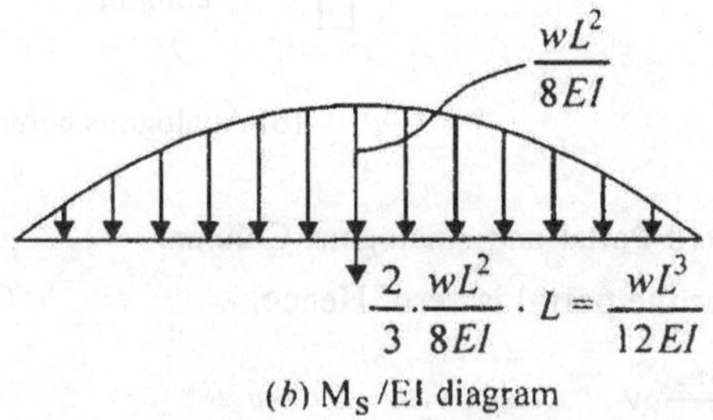

(*b*) M_S/EI diagram

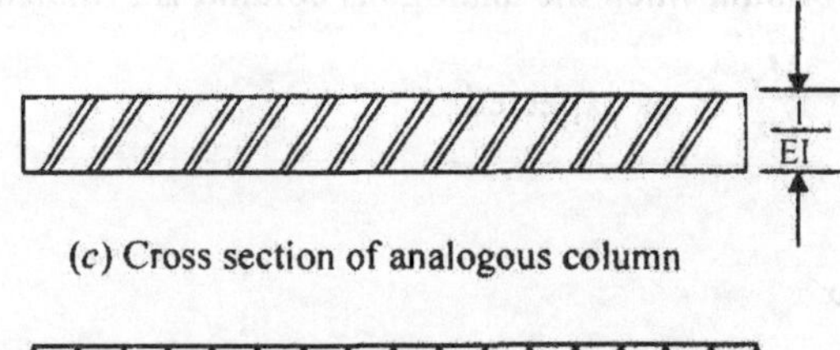

(*c*) Cross section of analogous column

(*d*) M_i diagram (pressure on bottom of analogous column)

Fig. 13.4. Example 13.1

The simple beam moment diagram due to the uniform load divided by EI is applied as a *downward* loading on the top of the analogous column because this moment causes compression on the outside (or top) of the beam. The pressure along the base of the column is constant in this case and is equal to total load divided by the area.

Hence $\quad M_i$ = pressure = Total load/area

$$= \frac{wL^3}{12EI} / \frac{L}{EI} = +\frac{wL^2}{12}$$

$$\text{Moment at } A = Ms - Mi = O - \frac{wL^2}{12} = -\frac{wL^2}{12}$$

The minus sign for M_A (or M_B) indicates that M_A or M_B is in such a direction as to cause compression on the inside (or bottom) of the beam.

Example 13.2. Find the fixed end moments for the beam shown in Fig. 13.5. EI is constant.

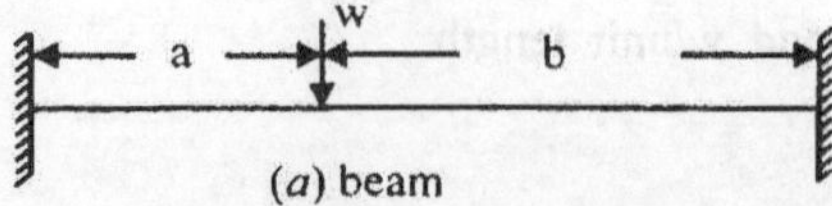

(*a*) beam

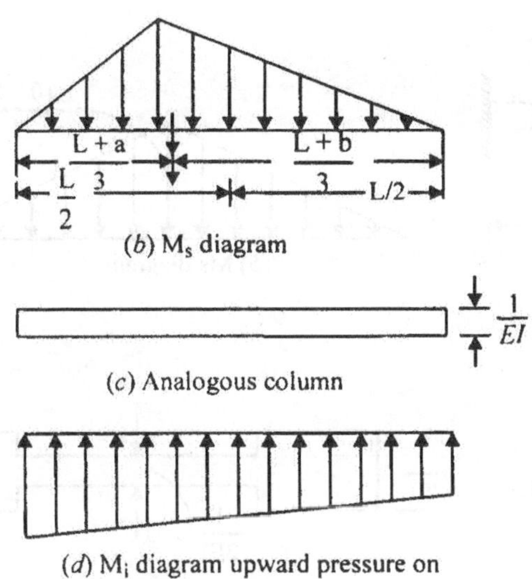

(*b*) M_s diagram

(*c*) Analogous column

(*d*) M_i diagram upward pressure on analogous column

Fig. 13.5. Example 13.2

Total downward load on column.

$$= \frac{1}{2} \cdot \frac{wab}{L} \cdot \frac{L}{EI} = \frac{wab}{2EI}$$

acting at an eccentricity of $\frac{L+b}{3} - \frac{L}{2} = \frac{2b-L}{6}$

$$\sigma = \frac{P}{A} \pm \frac{P.e.L/2}{1 \times \frac{L^3}{12}} = \frac{wab}{2EI} \cdot \frac{EI}{L} \pm \frac{wab}{2EI} \times \frac{2b-L}{L^2} \cdot EI$$

$$= \frac{wab}{2L} + \frac{wab}{2L^2}(2b-L) \quad \text{or} \quad = \frac{wab}{2L} - \frac{wab}{2L^2}(2b-L)$$

$$= \frac{wab(L+2b-L)}{2L^2} \qquad = \frac{wab(L-2b+L)}{2L^2}$$

$$= \frac{wab^2}{L^2} \qquad = \frac{wab}{2L^2} \cdot (2L-2b) \cdot$$

$$= \frac{wab}{2L^2} \cdot 2a = \frac{wa^2b}{L^2}$$

$$M_A = M_x - M_i = O - \frac{wab^2}{L^2} = -\frac{wab^2}{L^2}$$

$$M_B = M_x - M_i = O - \frac{wa^2b}{L^2} = -\frac{wa^2b}{L^2}$$

Example 13.3. Find the fixed end moments for the beam shown in Fig. 13.6

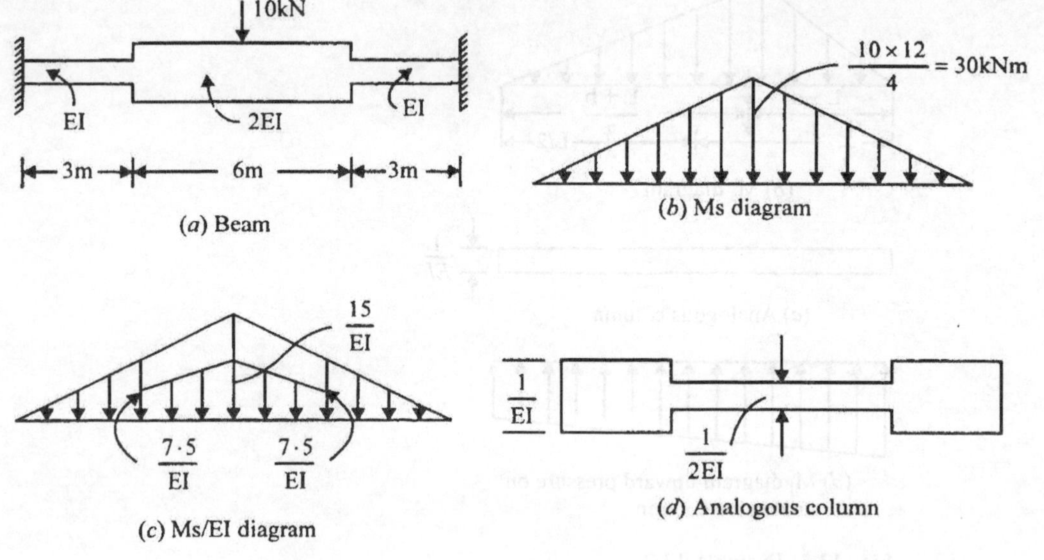

Fig. 13.6. Example 13.3.

M_i = σ over analogous column

$$= \frac{\text{Load}}{\text{Area}} \text{ as the load is central over symmetrical analogous column}$$

$$= \frac{\frac{1}{2} \times \frac{15 \times 6}{EI} + \frac{2}{2} \times \frac{(7 \cdot 5 + 15)}{EI} \times 3}{\frac{3}{EI} + \frac{6}{2EI} + \frac{3}{EI}}$$

$= 12 \cdot 5$ kNm.

M_A = $Ms - Mi$

$= 0 - 12 \cdot 5$

$= -$ **12·5 kNm.**

13.4. INDETERMINATE AND CORRESPONDING DETERMINATE STRUCTURES

Fig. 13.7 shows indeterminate and corresponding determinate structures.

(*i*)

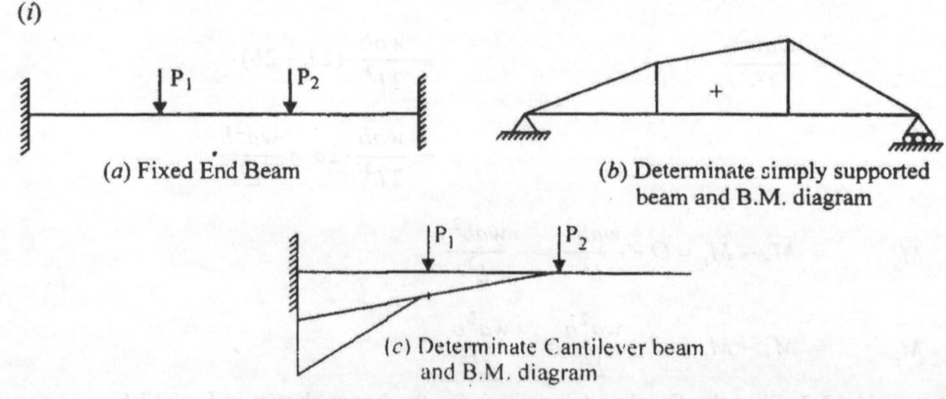

(*a*) Fixed End Beam

(*b*) Determinate simply supported beam and B.M. diagram

(*c*) Determinate Cantilever beam and B.M. diagram

(*ii*)

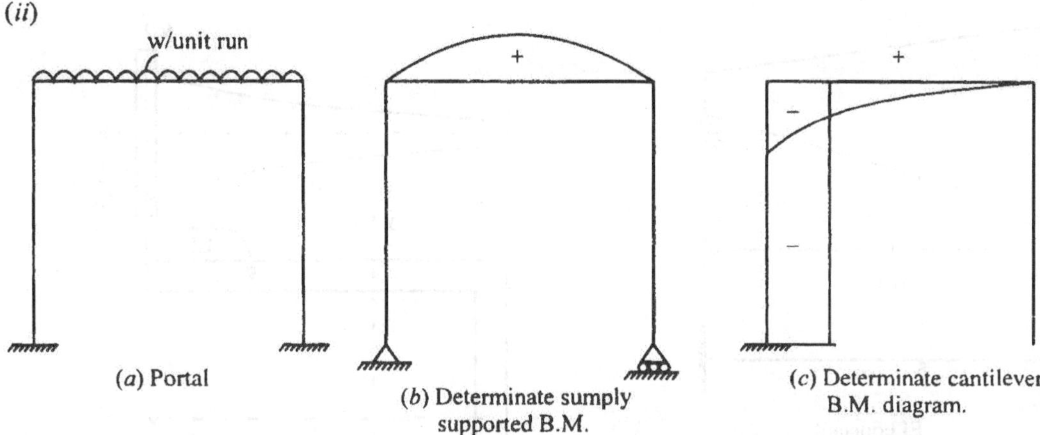

(*a*) Portal (*b*) Determinate sumply supported B.M. (*c*) Determinate cantilever B.M. diagram.

Fig. 13.7. Indeterminate and corresponding determinate structure.

13.5. ANALOGOUS COLUMN – THE MAIN CHARACTERISTIC OF ANALOGOUS COLUMN ARE :

(*i*) Width $= \dfrac{1}{EI}$

(*ii*) At hinge point flexural rigidity is zero and hence $\dfrac{1}{EI}$ is infinite. Thus the analogous column has an infinite area. The centroid is therefore at the hinge and the two principal axes pass through the hinge but their directions are unknown. (They can be determined by the equation $\int \dfrac{xyds}{EI} = 0$).

(*iii*) For fixed support flexural rigidity is infinite and $\dfrac{1}{EI}$ is zero.

(*iv*) If the bent has one roller support the support may be replaced by a short straight pin jointed member. Axis perpendicular to the surface of the roller will be a principal axis (say, y axis) of the analogous column.

(*v*) If the frame is composed of a bent with extending parts such as overhangs in which the internal forces are statically determinate, the analogous column is derived from the frame without the overhangs. The effect of the force on the extending parts is taken into account by applying to the frame appropriate internal forces at the cross section where the extending part meets the frame.

Some typical structures, corresponding analogous columns and rectangular reference axes are shown in Fig. 13.8.

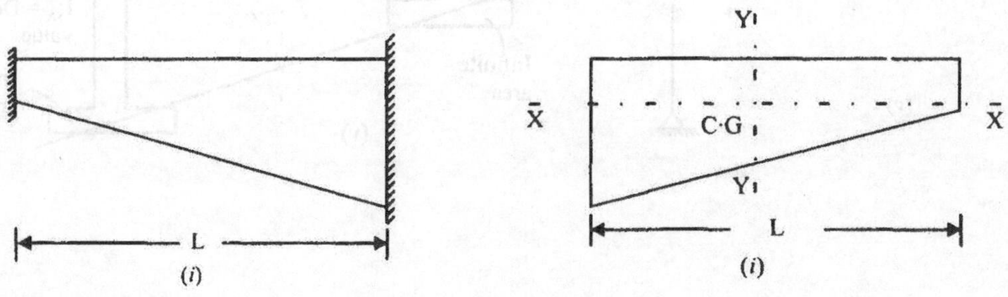

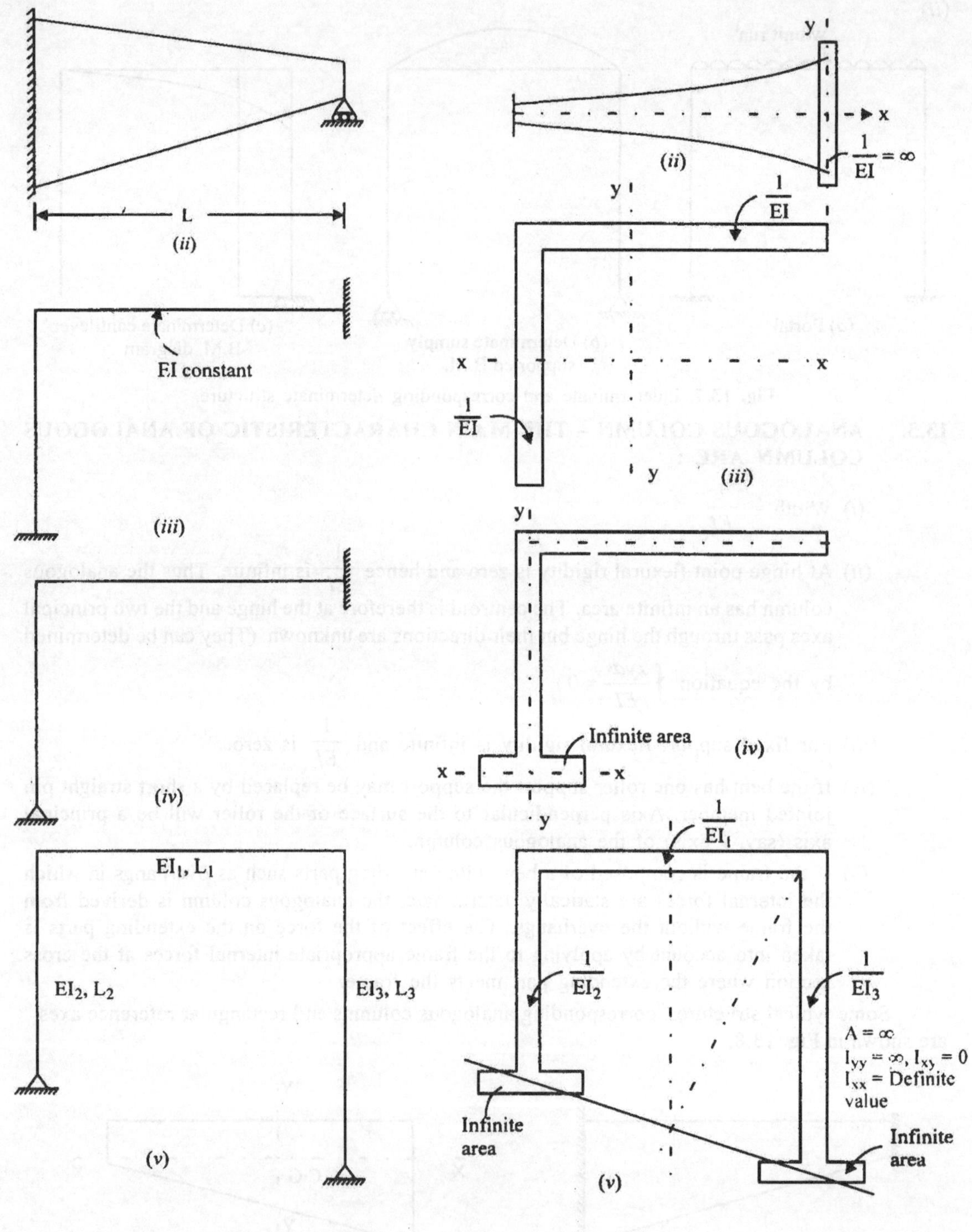
L
(ii)
y
x
(ii)
$\frac{1}{EI} = \infty$
y
$\frac{1}{EI}$
EI constant
x
x
$\frac{1}{EI}$
y
(iii)
(iii)
y
Infinite area
x
x
(iv)
y
(iv)
$\frac{1}{EI_1}$
EI_1, L_1
EI_2, L_2
EI_3, L_3
$\frac{1}{EI_2}$
$\frac{1}{EI_3}$
$A = \infty$
$I_{yy} = \infty, I_{xy} = 0$
I_{xx} = Definite value
Infinite area
Infinite area
(v)
(v)

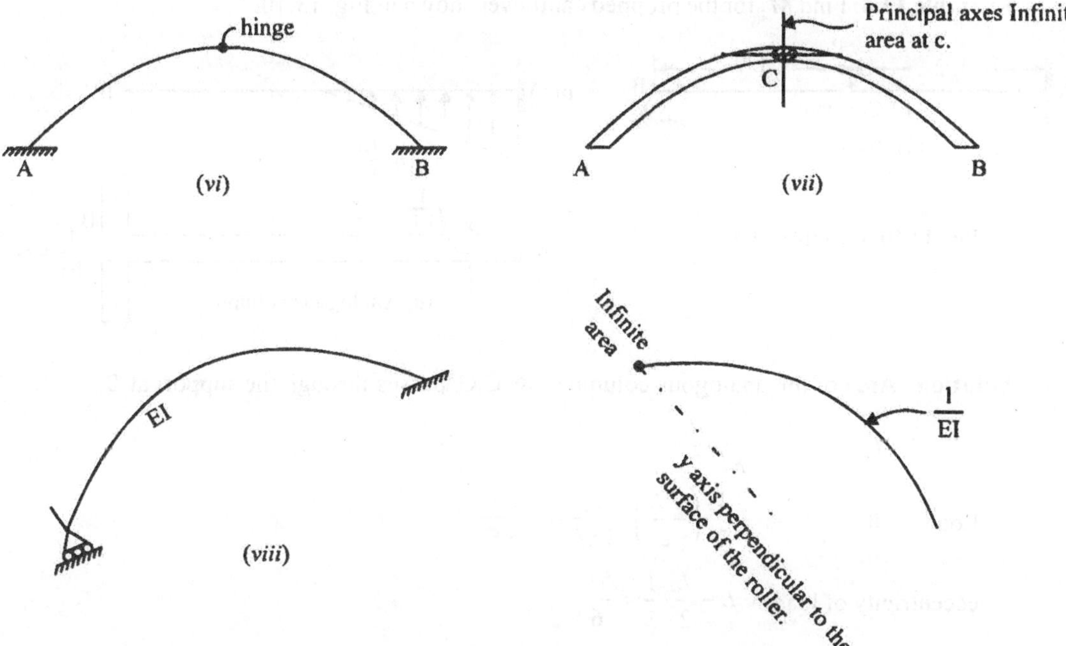

Fig. 13.8. Structures and corresponding analogous column

13.6. PROPERTIES OF ANALOGOUS COLUMN SECTIONS (FIG. 13.9)

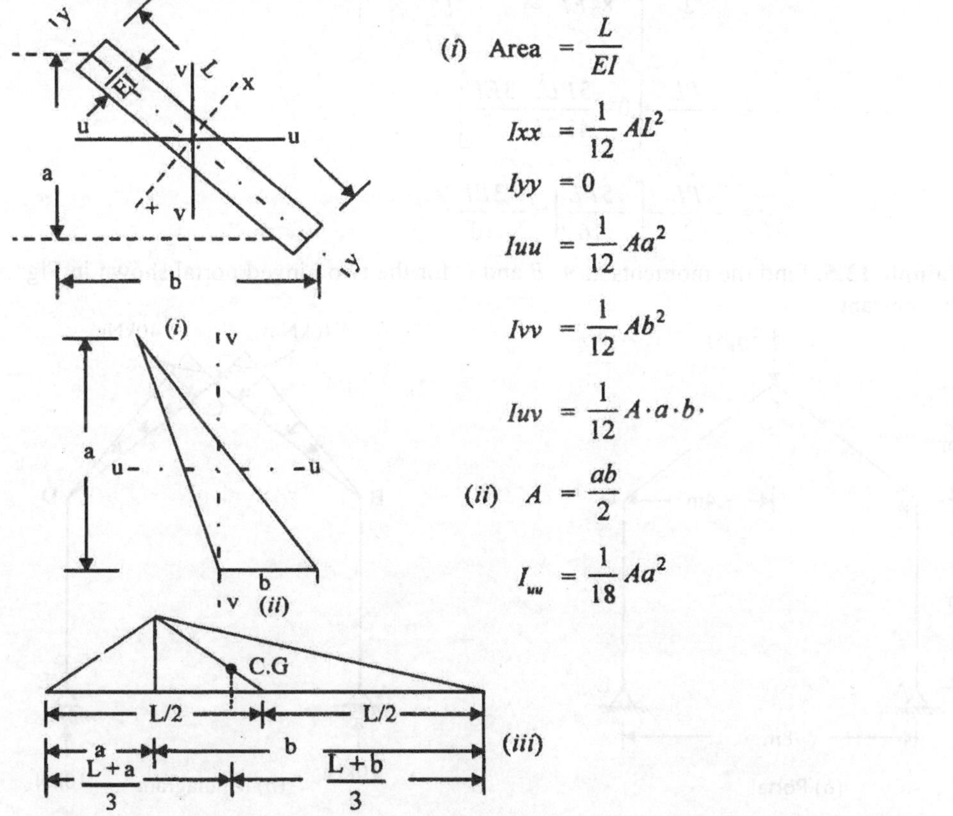

(*i*) Area $= \dfrac{L}{EI}$

$Ixx = \dfrac{1}{12} AL^2$

$Iyy = 0$

$Iuu = \dfrac{1}{12} Aa^2$

$Ivv = \dfrac{1}{12} Ab^2$

$Iuv = \dfrac{1}{12} A \cdot a \cdot b \cdot$

(*ii*) $A = \dfrac{ab}{2}$

$I_{uu} = \dfrac{1}{18} Aa^2$

Fig. 13.9. Properties of Analogous column Section.

Example 13.4. Find M_A for the propped cantilever shown in Fig. 13.10.

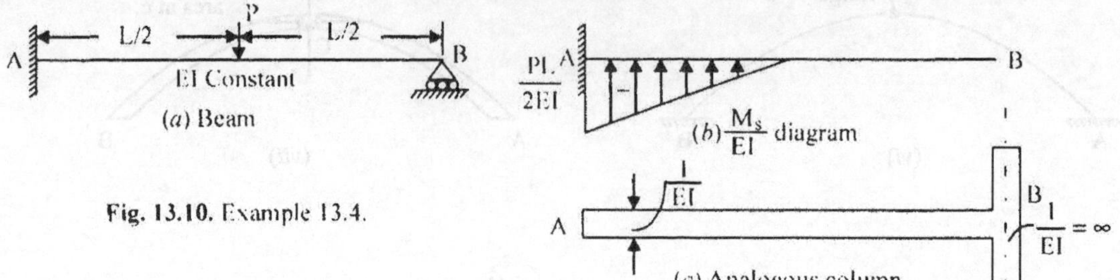

Fig. 13.10. Example 13.4.

Solution. Area of the analogous column = ∞ C.G. passes through the support at *B*

$$I_{yy} = \frac{1}{EI}\cdot\frac{L^3}{3}$$

$$\text{Load} \quad W = -\frac{1}{2}\cdot\left(\frac{PL}{2}\right)\cdot\frac{L}{2EI} = -\frac{PL^2}{8EI}$$

$$\text{eccentricity of load} = L - \frac{L}{2}\cdot\frac{1}{3} = \frac{5L}{6}$$

$$M_A = Ms\text{--}Mi = -\frac{PL}{2} - \left| \frac{PL^2}{8\cdot EI}\cdot\frac{1}{\infty} - \frac{\frac{PL^2}{8EI}\cdot\frac{5L}{6}}{\frac{L^3}{3EI}} L \right|$$

$$= -\frac{PL}{2} - \left[0 - \frac{5PL^3}{48EI}\cdot\frac{3EI}{L^2}\right]$$

$$= -\frac{PL}{2} - \left[-\frac{5PL}{16}\right] = -\frac{3PL}{16}.$$

Example 13.5. Find the moments at *A*, *B* and *C* for the two hinged portal shown in Fig. 13.11. EI Constant.

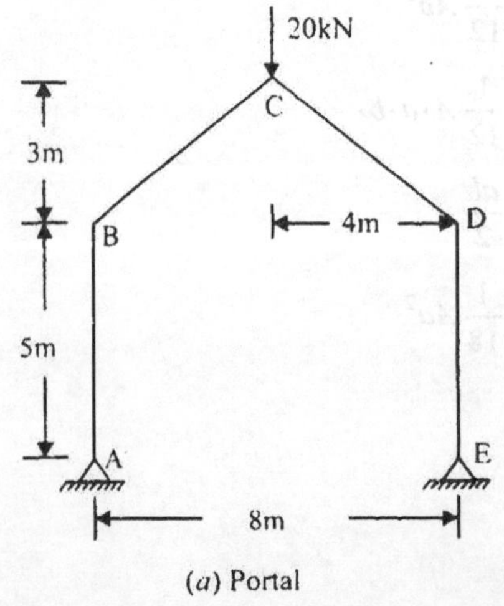

(*a*) Portal

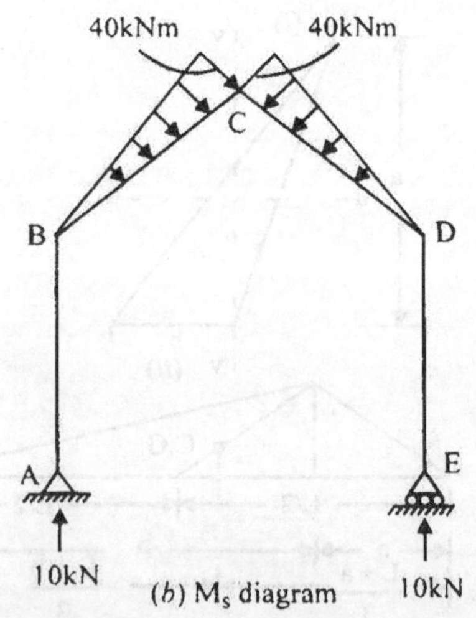

(*b*) M_s diagram

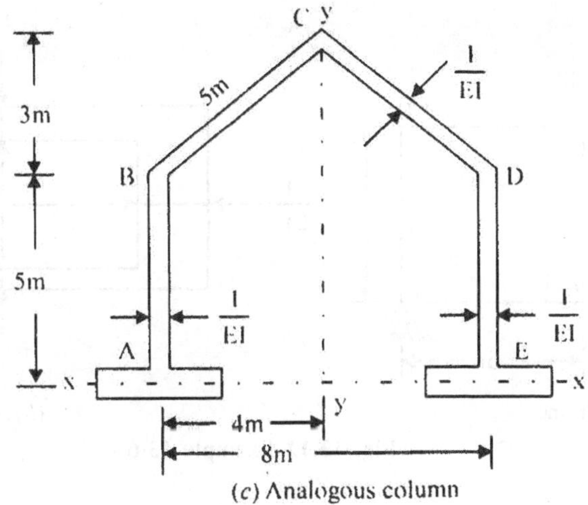

(c) Analogous column

Fig. 13.11. Example 13.5

Solution:

$$Ixx = \frac{1}{EI}\cdot\frac{5^3 \times 2}{3} + \frac{1}{12}\times\frac{5}{EI}\times 3^2 \times 2 + \frac{1}{EI}\times 5 \times 6\cdot 5^2 \times 2$$

$$= \frac{513\cdot 32}{EI}$$

$$\text{Total load} = \frac{1}{2}\times 40 \times \frac{5}{EI}\times 2$$

$$= \frac{200}{EI} \text{ acting at 7m from AE}$$

Area of analogous column

$$= 2\times\left(\frac{1}{EI}\times 5 + \frac{1}{EI}\times 5\right) + \infty$$

$$= \infty$$

$$M_i \text{ at } A = \frac{200}{EI}\cdot\frac{1}{\infty} = 0$$

$$M_i \text{ at } B = \frac{200}{EI}\times\frac{7\times 5}{513\cdot 32}\times EI = +13\cdot 63 \text{ kNm}$$

$$M_i \text{ at } C = \frac{200\times 7}{513\cdot 32}\times 8 = 21\cdot 82 \text{ k Nm}$$

$$M \text{ at } B = Ms - Mi = 0 - (+13\cdot 63) = -13\cdot 62 \text{ kNm.}$$

$$M \text{ at } C = Ms - Mi = 40 - (21\cdot 82) = 18\cdot 18 \text{ kNm.}$$

Example 13.6. Find the moments at the critical points of the closed frame shown in Fig. 13.12.

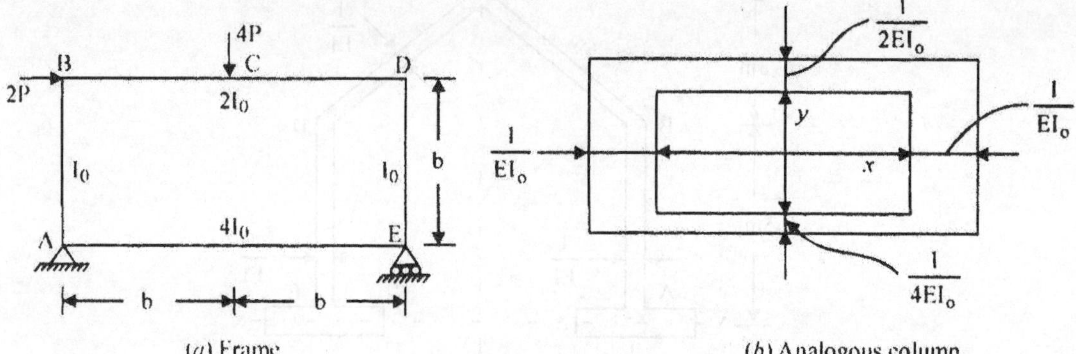

Fig. 13.12 Example 13.6

Solution :

(*i*) Area of analogous column$= \frac{1}{EIo}\left(\frac{1}{2}\times 2b + \frac{1}{4}\cdot 2b + 2b\cdot 1\right)$

$$= \frac{3\cdot 5b}{EIo}$$

(*ii*) C.G. of the analogous column

Taking moments about BD.

$$\frac{1}{E}\left(\frac{1}{4Io}\times 2b\times b + \frac{1}{Io}\times b\times\frac{b}{2}\times 2\right) = \frac{1\cdot 5b^2}{EIo}$$

distance of C.G. from BD.

$$= \frac{1\cdot 5b^2}{EIo}\times\frac{EIo}{3\cdot 5b} = 0\cdot 428b$$

As the frame is symmetrical C.G. lies on the axis of symmetry.

(*iii*) Calculation of *Ixx* and *Iyy*. (*Ixy*=0).

$$Ixx = \left[\frac{1}{2EIo}\cdot 2b\cdot(0\cdot 428b)^2 + \frac{1}{4EIo}\times 2b(0\cdot 572b)^2\right]$$

$$+ 2\left[\frac{(0\cdot 428b)^3}{3} + \frac{(0\cdot 572b)^3}{3}\right]\frac{1}{EIo}$$

$$= \frac{0\cdot 524b^3}{EIo}$$

$$Iyy = \frac{1}{2EIo}\cdot\frac{(2b)^3}{12} + \frac{1}{4EIo}\frac{(2b)^3}{12} + \frac{1}{EIo}\cdot 2b\cdot b^2$$

$$= \frac{2\cdot 5b^3}{EIo}.$$

(*iv*) Ms diagram – Ms diagram is obtained by cutting the frame just to the right of *C* and making it determinate. (Fig. 13.13). *Mx* and *My* are computed in Tabular form. (Table 13.1).

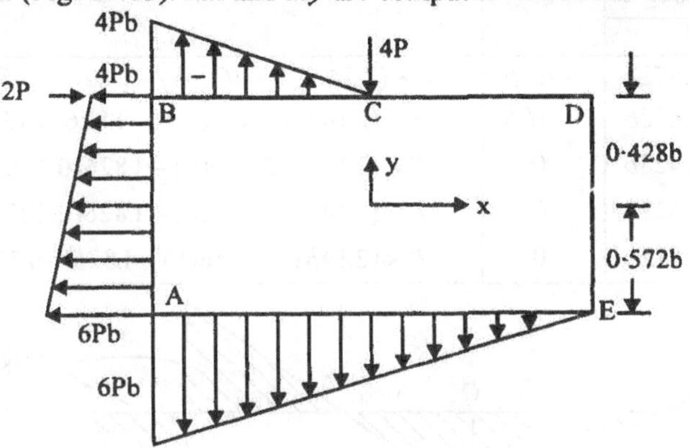

Fig. 13.13. Ms diagram

Table 13.1 (Ex. 13.6)

No	Load	x	y	Moment about X axis	Moment about Y axis
1.	$-\frac{1}{2}\times\frac{4Pb\times b\cdot}{2EIo}$	–0.667b.	+0.428b	$-0.428\frac{Pb^3}{EIo}$	$+0.667\frac{Pb^3}{EIo}$
2.	$-\frac{1}{2}\times\frac{4Pb\times b\cdot}{EIo}$	–1.000b	+0.095b.	$-0.19\frac{Pb^3}{EIo}$	$+2.00\frac{Pb^3}{EIo}$
3.	$-\frac{1}{2}\times\frac{6Pb\times b\cdot}{EIo}$	–1.000b	–0.239b	$+0.717\frac{Pb^3}{EIo}$	$+3.00\frac{Pb^3}{EIo}$
4.	$-\frac{1}{2}\times\frac{6Pb\times 2b\cdot}{4EIo}$	–0.333b.	–0.572b	$+0.858\frac{Pb^3}{EIo}$	$+.499\frac{Pb^3}{EIo}$

$\Sigma-7.5\frac{Pb^2}{EIo}$ $\qquad$ $\Sigma+0.957\frac{Pb^3}{EIo}$ $\qquad$ $\Sigma+6.164\frac{Pb^3}{EIo}$

$$M_i = \frac{W}{A}+\frac{My}{Iyy}\cdot x+\frac{Mx}{Ixx}\cdot y$$

$$= -\frac{7.5b^2}{EIo}\cdot\frac{EIo}{3.5b}+\left(\frac{6.164\,Pb^3}{EIo}\right)\times\frac{EIo}{2.5b^3}\cdot x$$

$$+\left(\frac{0.957.\,Pb^3}{EIo}\times\frac{EIo}{0.524b^3}\right)\cdot y$$

$$= (-2.14b+2.466x+1.826y)P.$$

$$\text{or } -M_i = (2.14b-2.466x-1.826y)P.$$

Moments at different points are computed in Table 13.2

Table 13.2 (Example 13.6)

Point	Coordinates		M_s	$-M_i$	$M_s - M_i$
	x	y			
A	$-b$	$-0.572b$	$-6Pb$	$P \times (2.14b - 2.446(-b) - 1.826(-0.572b))$	$-0.369Pb$
B	$-b$	$+0.428b$	$-4Pb$	$P \times (2.14b - 2.446(-b) - 1.826(0.428b))$	$-0.195Pb$
C	o	$+0.428b$	0	$P \times (2.14b - 2.446(0) - 1.826(0.428b))$	$+1.358Pb$
D	b	$+0.428b$	0	$P \times (2.14b - 2.446(b) - 1.826(0.428b))$	$-1.0875Pb$
E	b	$-0.572b$	0	$P \times (2.14b) - 2.446(b) - 1.826(-0.572b)$	$+0.738Pb$

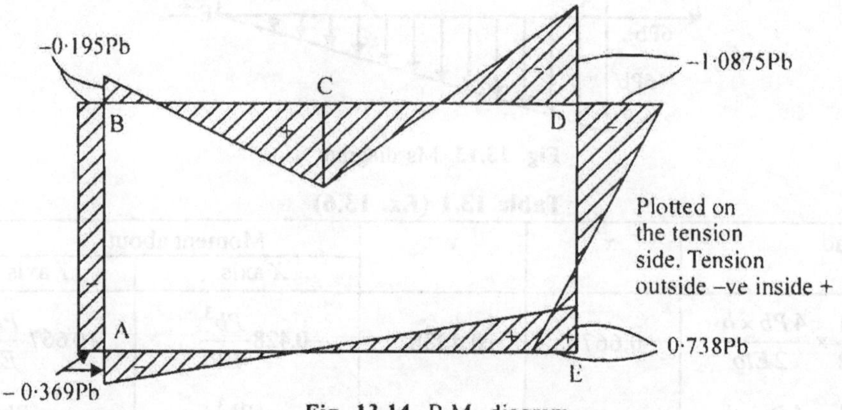

Fig. 13.14. B.M. diagram

Example 13.7. Find the bending moment in the unsymmetrical frame shown in Fig. 13.15 a. All members of the frame have constant flexural rigidity EI.

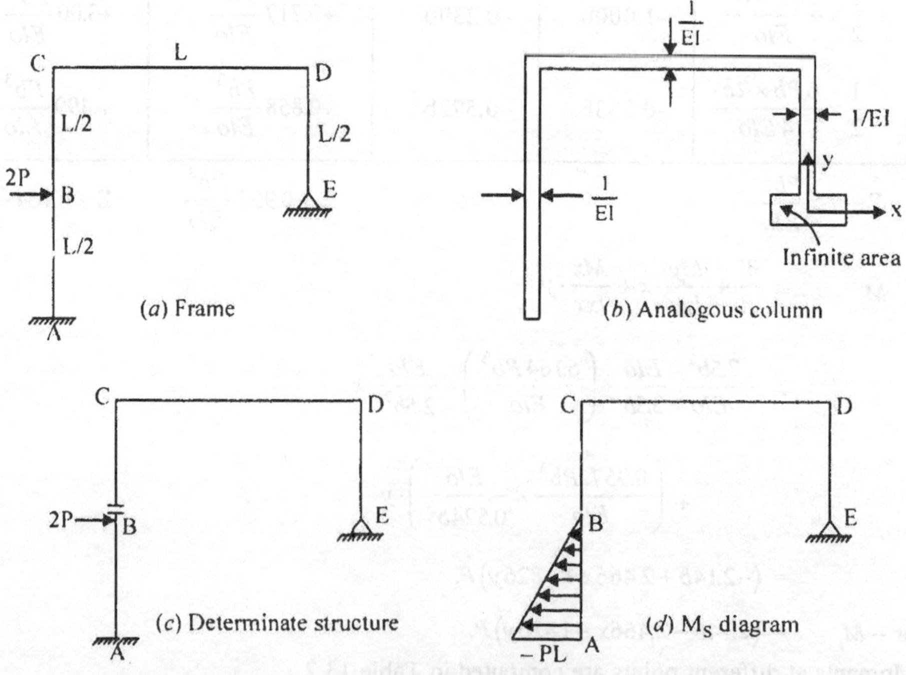

Fig. 13.15. (Example 13.7)

The analogous column is shown in Fig. 13.15(*b*). The column has an infinite area at *E* where we have the C.G. The rectangular axes *X* and *Y* pass through the C.G. but as they are not the principal axes, *Ixx, Iyy* and *Ixy* are to be calculated.

$$Ixx = L\times\frac{1}{EI}\times\left(\frac{L}{2}\right)^2+\frac{1}{EI}\times\frac{L^3}{12}+\frac{1}{EI}\left(\frac{L}{2}\right)^3\frac{1}{3}$$

$$= \frac{3L^3}{8EI}$$

$$Iyy = L\times\frac{1}{EI}\times L^2+\frac{L^3}{3EI}+0=\frac{L^3}{EI}+\frac{L^3}{3EI}=\frac{4L^3}{3EI}$$

$$Ixy = \frac{1}{EI}\times L\times O\times L+\frac{1}{EI}\times L\left(\frac{L}{2}\times(-1/2)\right)$$

$$+\frac{1}{EI}\times\frac{L}{2}\times O\times\frac{L}{4}.$$

$$= -\frac{L^3}{4EI}.$$

The released structure is shown in Fig. 13.15 (*c*) and the corresponding B.M. diagram is shown in Fig. 13.15(*d*). The loads in the analogous column and *Mx* and *My* are calculated by means of Table 13.3.

Table 13.3 (Example 13.7)

Load	x	y	Moment about	
			X axis	Y axis
$W=-\frac{1}{2}\cdot\frac{PL}{EI}\cdot\frac{L}{2}$ $=\frac{-PL^2}{4EI}$	$-L$	$-\frac{L}{2}\cdot\frac{2}{3}$ $=-\frac{L}{3}$	$\left(-\frac{PL^2}{4EI}\right)\left(-\frac{L}{3}\right)$ $=\frac{PL^3}{12EI}$	$+\frac{PL^3}{4EI}$

Substituting in column flexure formula

$$\sigma = +\frac{W}{A}+\left(\frac{My\,Ixx-Mx\,Ixy}{Ixx\,Iyy-(Ixy)^2}\right)x$$

$$+\left(\frac{Mx\,Iyy-My\,Ixy}{Ixx\,Iyy-(Ixy)^2}\right)\cdot y$$

$$= -\frac{PL^2}{4EI}\cdot\frac{1}{\infty}+\left(\frac{\frac{PL^3}{4EI}\cdot\frac{3L^3}{8EI}+\frac{PL^3}{12EI}\cdot\frac{L^3}{4EI}}{\frac{3L^3}{8EI}\cdot\frac{4L^3}{3EI}-\left(\frac{L^3}{4EI}\right)^2}\right)\cdot x$$

$$+\frac{\frac{PL^3}{12EI}\cdot\frac{4L^3}{3EI}+\frac{PL^3}{4EI}\cdot\frac{L^3}{4EI}}{\frac{3L^3}{8EI}\cdot\frac{4L^3}{3EI}-\left(\frac{L^3}{4EI}\right)^2}\cdot y$$

$$= 0 + P\left(\frac{11}{96}\cdot\frac{16}{7}\right)x + P\left(\frac{25}{144}\cdot\frac{16}{7}\right)y$$

$$= P\left(\frac{11}{42}\cdot x+\frac{25}{63}\cdot y\right)$$

or $-Mi$ $$= -P\left(\frac{11}{42}\cdot x+\frac{25}{63}\cdot y\right)$$

Table 13.4 gives the values of Moments at different points.

Table 13.4 – Moments at points (Ex. 13.7)

Point	Coordinates		M_s	$-M_i$	$M_s - M_i$
	x	y			
A	$-L$	$-L/2$	$-PL$	$+P(0.46L)$	$-0.54PL$
B	$-L$	0	0	$-P\left(\frac{11}{42}\cdot(-L)\right)$	$+0.26PL$
C	$-L$	$+L/2$	0	$-P\left(\frac{11}{42}(-L)+\frac{25}{63}\cdot\frac{L}{2}\right)$	$0.063PL$
D	0	$L/2$	0	$-P\left(\frac{25}{63}\cdot\frac{L}{2}\right)$	$-.198PL$
E	0	0	0	0	0

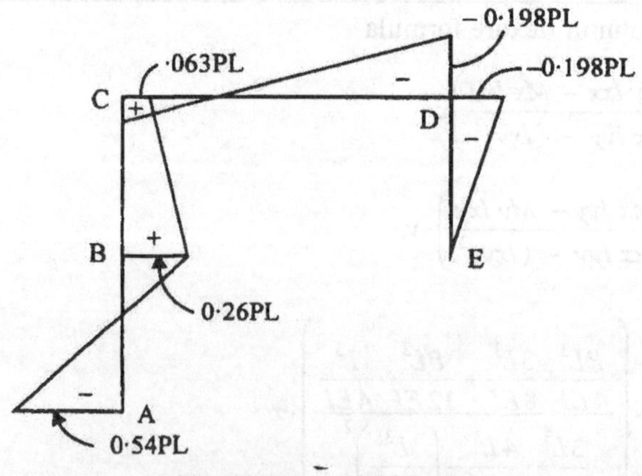

Fig. 13.16. B.M diagram

Example 13.8. Find the moments at the critical sections of the frame by column analogy method.

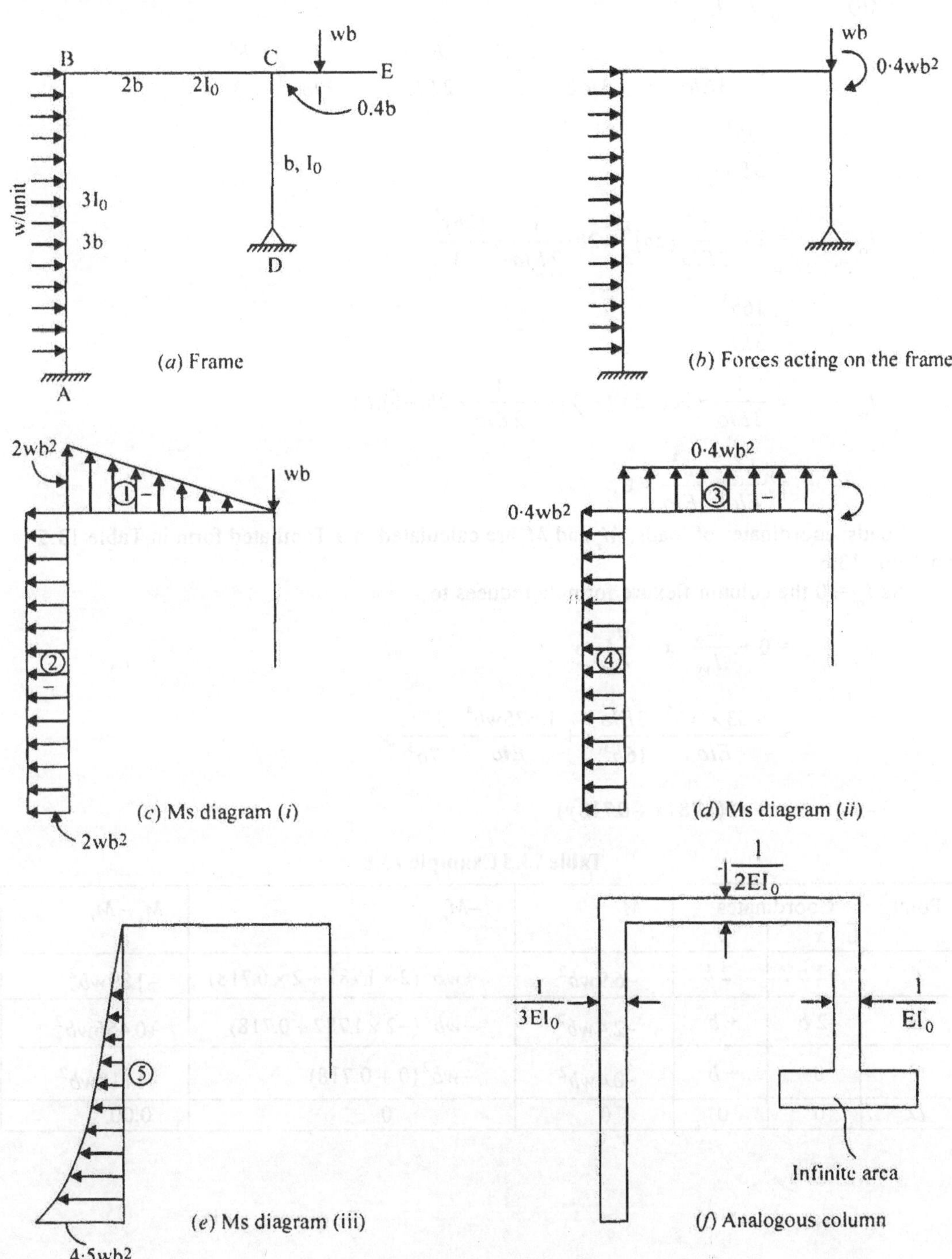

Fig. 13.17. Example 13.8

Solution: Origin at D

(*i*) Area of the analogous column = ∞

(*ii*) I_{yy}, I_{xx}, I_{xy}

$$I_{xx} = b \times \frac{1}{3EIo} \cdot \frac{b^2}{3} + \frac{2b}{3IoE} \cdot \frac{(2b)^2}{3} + \frac{2b \times 1}{2EIo} \cdot b^2 + \frac{1}{EIo} \cdot b \cdot \frac{b^2}{3}$$

$$= \frac{7b^3}{3EIo}$$

$$I_{yy} = 3b \cdot \frac{1}{3EIo}(2b)^2 + 2b \cdot \frac{1}{2EIo} \times \frac{(2b)^2}{3}$$

$$= \frac{16b^3}{3EIo}$$

$$I_{xy} = \frac{1}{3EIo} \times 3b(-2b)(-b/2) + \frac{1}{2EIo} \times 2b(-b)(b)$$

$$= \frac{b^3}{EIo} - \frac{b^3}{EIo} = 0.$$

Loads, coordinates of loads, M_x and M_y are calculated in a Tabulated form in Table 13.5 and Table 13.6

As $I_{xy} = 0$ the column flexure formula reduces to

$$\sigma = 0 + \frac{M_y}{I_{yy}} \cdot x + \frac{M_x}{I_{xx}} \cdot y$$

$$= \frac{9.53 \times wb^4}{EIo} \cdot \frac{3EIo}{16b^3} x + \frac{1.675wb^4}{EIo} \cdot \frac{3EIo}{7b^3} y$$

$$-M_i = -wb(1.787x + 0.718y)$$

Table 13.5 Example 13.8

Point	Coordinates		M_s	$-M_i$	$M_s - M_i$
	x	y			
A	−2 b	−2 b	$-6.9wb^2$	$+wb^2(2 \times 1.787 + 2 \times 0.718)$	$-1.89wb^2$
B	−2 b	+ b	$-2.4wb^2$	$-wb^2(-2 \times 1.787 + 0.718)$	$+0.456wb^2$
C	0	+ b	$-0.4wb^2$	$-wb^2(0 + 0.718)$	$-1.118wb^2$
D	0	0	0	0	0.00

Table 13.6 (Example 13.8)

Load	x	y	Moment about X axis M_x	Moment about Y axis M_y
$W_1 = -\frac{1}{2}\cdot\frac{2wb^2 \times 2b}{2EIo} = -\frac{wb^3}{EIo}$	$\frac{-4}{3}\cdot b$	$+b$	$-\frac{wb^4}{EIo}$	$+\frac{4wb^4}{3EIo}$
$W_2 = -\frac{2wb^2 \cdot 2b}{3EIo} = -\frac{2wb^3}{EIo}$	$-2b$	$-b/2$	$+\frac{wb^4}{EIo}$	$+\frac{4wb^4}{EIo}$
$W_3 = -\frac{0.4wb^2 \times 2b}{2EIo} = -\frac{0.4wb^3}{EIo}$	$-b$	$+b$	$\frac{-0\cdot 4wb^4}{EIo}$	$+\frac{0.4wb^4}{EIo}$
$W_4 = -\frac{0.4wb^2 \cdot 3b}{3EIo} = -\frac{0.4wb^3}{EIo}$	$-2b$	$-b/2$	$+\frac{0.2wb^4}{EIo}$	$+\frac{0.8wb^4}{EIo}$
$W_5 = -\frac{4.5wb^2 \times 3b}{3 \times 3EIo} = -\frac{1.5wb^3}{EIo}$	$-2b$	$-\frac{5b}{4}$	$+\frac{1.875wb^4}{EIo}$	$+\frac{3wb^4}{EIo}$
			$\Sigma + \frac{1.675wb^4}{EIo}$	$\Sigma 9.53\frac{wb^4}{EIo}$.

13.7 DETERMINATION OF STIFFNESS AND CARRY OVER FACTORS BY COLUMN ANALOGY METHOD:

Column analogy method is very useful in computing stiffness and carry over factors for beams. Let us consider a beam AB simply supported at A and fixed at B subjected to moment M_A at end A. Stiffness K_A is the moment required to rotate the beam through unit angle at A. Hence if θ_A is the rotation at A due to M_A. Then

$$M_A = K_A \cdot \theta_A$$

When M_A is applied at A, the moment generated at B is M_B. The ratio $M_B/M_A = C_{AB}$ which means carry over factor from A to B.

When $\theta_A = 1$ $M_A = K_A$ and $M_B = C_{AB} \cdot M_A$

Now applying Moment Area Method, change of slope between A and B with $\theta_B = 0$, that is,

$$\theta_A = \int \frac{Mdx}{EI}. \quad \text{Now from Column formula}$$

$$\overset{\downarrow}{P} = \int \overset{\downarrow\,\downarrow}{\sigma\, dA}$$

From the above two formulae, the following analogies can be deduced.

(*i*) Rotation θ_A is analogous to load P

(*ii*) Bending moment M is analogous to stress σ.

(*iii*) $\frac{dx}{EI}$ is analogous to dA.

Now since $\int dA = A$ and $\int_o^l \frac{dx}{EI} = \frac{l}{EI}$, the area A of the analogous column is given by a column of length l and width $\frac{1}{EI}$. θ_A acting at end A is chosen as load for the analogous column. Since M_A is positive moment $\theta_A = 1$, the column load should act at the downward direction.

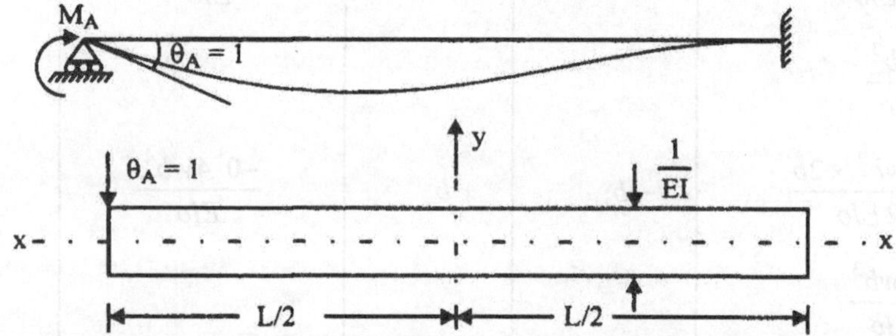

Fig. 13.18. Stiffness and carry over factors for a propped cantilever.

Stiffness & C.O.F. for a propped cantilever

$$\sigma_A = \frac{P}{A} + \frac{M_{yy}}{I_{xx}} \cdot x$$

$$= \frac{1}{\left(\frac{1}{EI} \times L\right)} + \frac{\frac{1 \times L/2}{1 \times L^3} \times \frac{L}{2}}{EI \times 12}$$

$$= \frac{EI}{L} + \frac{6EI}{L^2} \cdot L/2$$

$$= \frac{4EI}{L}.$$

But $\sigma_A = M_A$ (sagging) $= \frac{4EI}{L}$ = stiffness at A

Similarly $\sigma_B = M_B = -\frac{2EI}{L}$ (hogging)

C.O.F. $= -\frac{2EI/L}{4EI/L} = -1/2$ when B.M. sign convention is followed

$= \frac{1}{2}$ when M.D. sign convention is followed.

Example 13.9. Find the stiffness and C.O. Fs of the beam with variable cross section as shown in Fig. 13.19.

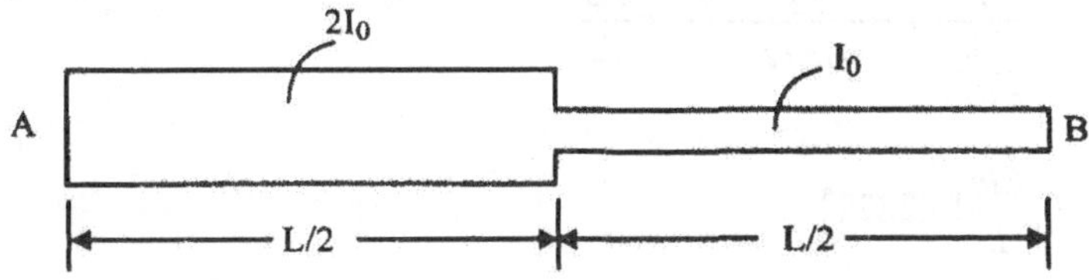

Fig. 13.19 Example 13.9

Solution : Rotational stiffness at *A* and C.O. from *A* to *B*.

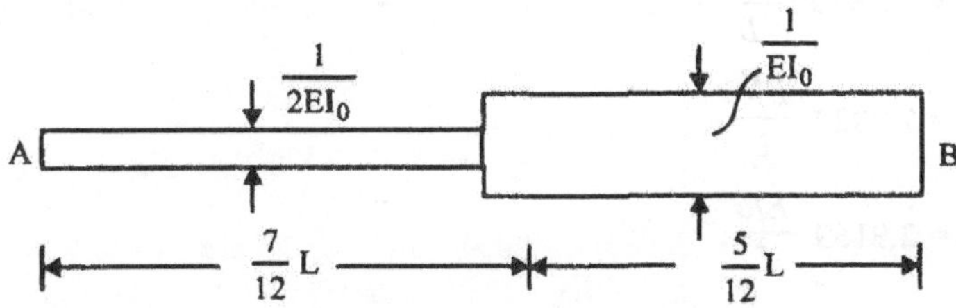

Fig. 13.20 C.G. of Section

C.G. of the section :

Taking moments about *A*.

$$\bar{x} = \frac{\frac{1}{2EIo}\times\frac{L}{2}\times\frac{L}{4}+\frac{1}{EIo}\times\frac{L}{2}\times\frac{3L}{4}}{\frac{1}{2EIo}\times\frac{L}{2}+\frac{1}{EIo}\times\frac{L}{2}}$$

$$= \frac{7}{12}\cdot L.$$

M.I about YY axis :

$$= \frac{1}{EIo}\times\frac{5L}{12}\cdot\frac{\left(\frac{5L}{12}\right)^2}{3}+\frac{1}{2EIo}\times\left(\frac{L}{2}\right)^3\times\frac{1}{12}$$

$$+\frac{1}{2EIo}\cdot\frac{L}{2}\times\left(\frac{7}{12}\cdot L-\frac{L}{4}\right)^2+\frac{1}{EIo}\times\left(\frac{L}{12}\right)^3\cdot\frac{1}{3}$$

$$= 0.0572\frac{L^3}{EIo}.$$

Area of analogous column :

$$= \frac{L}{2}\times\frac{1}{2EIo}+\frac{L}{2}\times\frac{1}{EIo}$$

$$= \frac{3L}{4EIo}.$$

When unit load is applied at *A* :

$$\text{Stress at } A = \frac{1}{\dfrac{3L}{4EIo}} + \frac{1\times\dfrac{7L}{12}\times EIo}{.0572L^3}\times\frac{7L}{12}$$

$$= \frac{EIo}{L}[7.2822]$$

$$\text{Stress at } B = \frac{EIo}{L}\left[1.333 - \frac{7}{12}\times\frac{5}{12}\times\frac{1}{.0572}\right]$$

$$= -2.9159\frac{EIo}{L}$$

$$M_A = 7.2822\ \frac{EIo}{L}$$

$$M_B = 2.9159\ \frac{EIo}{L}$$

$$\text{C.O.F.} = \frac{M_B}{M_A} = \frac{2.9159}{7.2822} = 0.40.$$

(*ii*) When unit load is placed at *B*.

$$\text{Stress at } B = \frac{1}{\dfrac{3L}{4EIo}} + \frac{\dfrac{1\times 5L}{12}\times EIo}{.0572L^3}\times\frac{5L}{12}$$

$$= \frac{EIo}{L}\left[\frac{4}{3} + \frac{5\times 5}{12\times 12\times .0572}\right]$$

$$= 4.368\ \frac{EIo}{L} = M_{B'}$$

$$\text{Stress at } A = \frac{4EIo}{3L} - \frac{5\times 7}{12\times 12\times .0572}\cdot\frac{EIo}{L}$$

$$= \frac{EIo}{L}\left[\frac{4}{3} - \frac{5\times 7}{12\times 12\times .0572}\right]$$

$$= 2.916\ \frac{EIo}{L}$$

$$\text{C.O.F.} = \frac{2.916}{4.368} = \mathbf{0.6676.}$$

Example 13.10. Using the moment distribution method determine the moments developed at all the joints and sketch the B.M. diagram for the portal frame shown in Fig. 13.21. The beam is non prismatic and its middle $\frac{1}{3}$rd has M.I. equal to 2I while the remaining length has moment of inertia I.

[AMIE Advanced Structures Summer 1985]

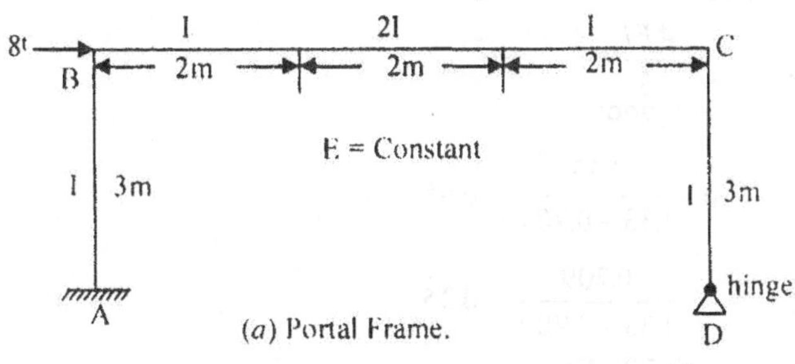

(a) Portal Frame.

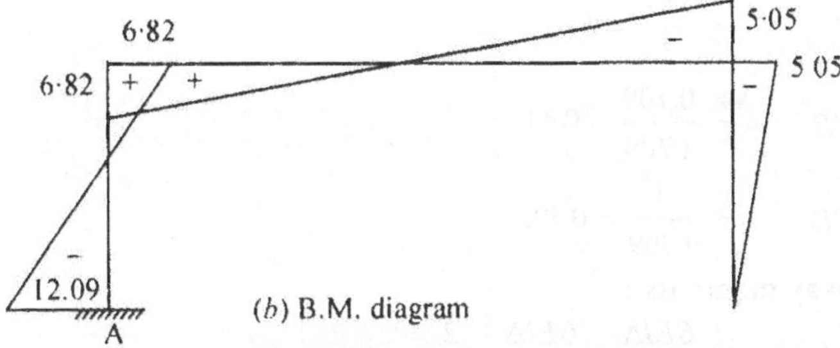

(b) B.M. diagram

Fig. 13.21. (Example 13.10)

Solution : The first step is computation of stiffness and carry over factors for the portion BC (Fig. 13.22)

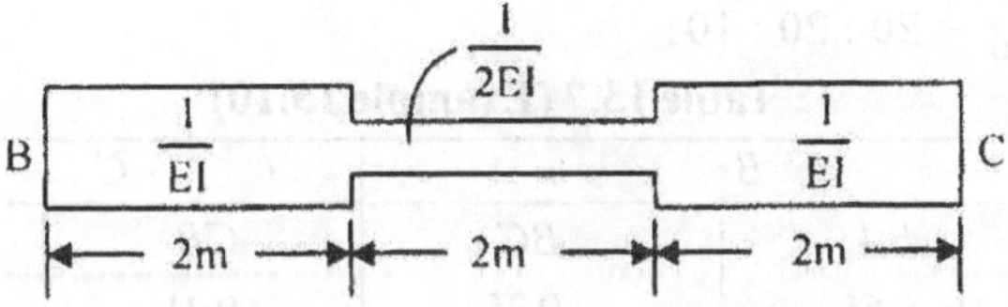

Fig. 13.22. Analogous column for BC.

M.I. about YY axis :

$$\frac{1}{EI}\cdot\frac{6^3}{12} - \frac{1}{2EI}\cdot\frac{2^3}{12}$$
$$= 17.667/EI$$

$$\text{Area of analogous column} = \frac{1}{EI}\cdot 4 + \frac{1}{2EI}\cdot 2$$
$$= 5/EI$$

$$M_B = \sigma_B = \frac{1}{5/EI} + \frac{1\times 3\times 3}{17.667/EI}$$
$$= 0.709\ EI$$

$$M_C = \sigma_C = -0.309\ EI$$

Stiffness at $B = 0.709$ EI

C.O.F. from B to $C = \dfrac{0.309}{0.709} = 0.4358$

Since the section is symmetrical stiffness at C and C.O.F. from C to B are the same as those at B.

Distribution factors.

Joint B $\quad K_{BA} = \dfrac{4EI}{3}$

$K_{BC} = 0.709\text{EI}$

D.F. $\quad BA = \dfrac{1.33}{1.33+0.709} = 0.65$

$BC = \dfrac{0.709}{1.33+0.709} = 0.35$

Joint C $\quad K_{CB} = 0.709\text{ EI}$

$K_{CD} = \dfrac{3EI}{3} = EI$

D.F. $\quad CB = \dfrac{0.709}{1.709} = 0.41$

$CD = \dfrac{1}{1.709} = 0.59.$

Arbitrary Sway moments :

$M_{FAB} : M_{FBA} = \dfrac{6EI\Delta}{L^2} = \dfrac{6EI\Delta}{9} = \dfrac{2}{3}$

$M_{FCD} = \dfrac{3EI\Delta}{L^2} = \dfrac{3EI\Delta}{9} = \dfrac{1}{3}$

$M_{FAB} : M_{FBA} : M_{FCD} = 20 : 20 : 10$

Table 13.7 (Example 13.10)

Joint	A	B		C		D
Member	AB	BA	BC	CB	CD	DC
D.F.		0.65	0.35	0.41	0.59	
C.O.F.	½ ←		0.44 ←	→ 0.44		→ 0
FEM	− 20	− 20			− 10	
Balance		+ 13	+ 07	+ 4.1	+ 5.9	
C.O	+ 6.50		+ 1.80	+ 3.08		
Balance		− 1.17	− 0.63	− 1.26	− 1.82	
C.O.	− 0.59		− 0.55	− 0.28		
Balance		+ 0.36	+ 0.19	+ 0.11	+ 0.17	
C.O.	+ 0.18		+ 0.05	+ 0.08		
Balance		− 0.03	− 0.02	− 0.03	− 0.05	
Arbitrary Moment	− 13.91	− 7.84	+ 7.84	+ 5.80	− 5.80	
Final Moment	− 12.09	− 6.82	+ 6.82	+ 5.05	− 5.05	0.00

Calculation of Sway Force

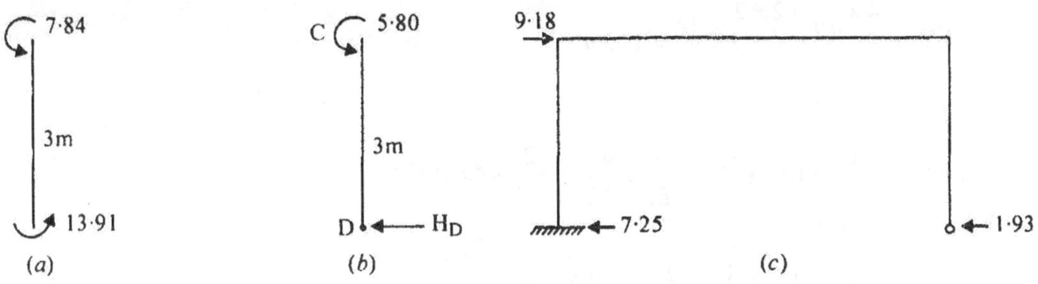

Fig. 13.23. Calculation of Sway force.

$$H_A = \frac{13.91+7.84}{3} = 7.25 \qquad H_D = \frac{5.80}{3} = 1.93$$

$$\text{Total H} = 7.25 + 1.93 = 9.18$$

$$\text{ratio} = \frac{8}{9.18} = 0.87.$$

Example 13.11. Analyse the portal frame shown in Fig. 13.34

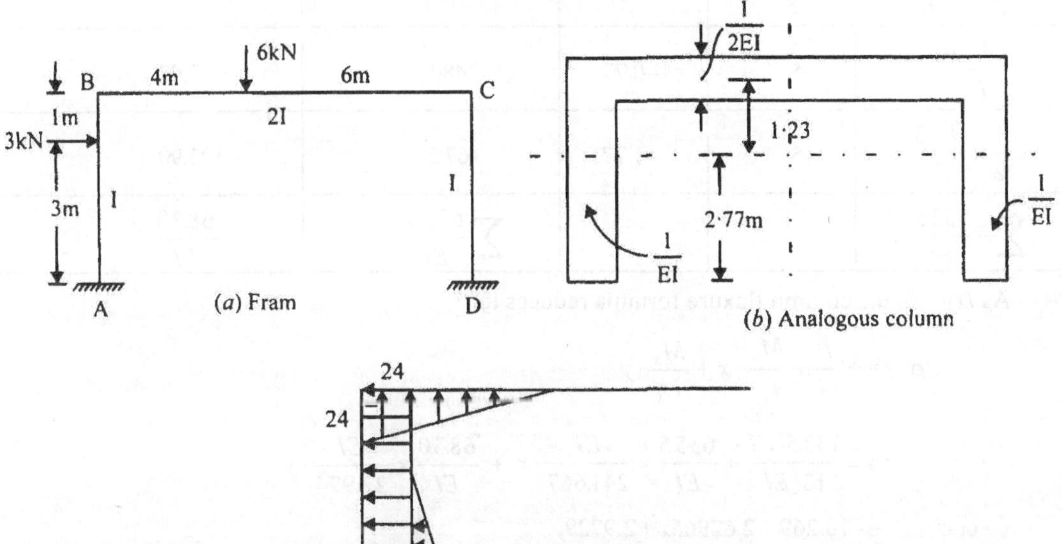

Fig. 13.24. Example 13.11

Area of Analogous column

$$= \frac{1}{2EI} \times 10 + \frac{1}{EI} \times 4 \times 2$$

$$= \frac{13}{EI}.$$

Y axis is the axis of symmetry.

Taking moments of all areas about the top

$$\bar{y} = \frac{4 \times \frac{1}{EI} \times 2 \times 2}{\frac{13}{EI}} = \frac{16}{3} = 1.23m.$$

$$\text{M.I } I_{xx} = \frac{1}{2EI} \times 10 \times 1.23^2 + \frac{2}{EI} \times \frac{4^3}{12} + \frac{2}{EI} \times 4 \times 0.77^2$$

$$= 22.974$$

$$I_{yy} = 10^3 \times \frac{1}{2EI} \times \frac{1}{12} + \frac{2}{EI} \times 4 \times 5^2$$

$$= \frac{241.667}{EI}$$

Table 13.8 (Example 13.11)

Load	x (m)	y (m)	Moment about	
			y axis M_y EI	x axis M_x/EI
$-\frac{1}{2} \times \frac{24 \times 4}{2EI}$	$-\frac{11}{3}$	+1.23	+88	–29.52
$-\frac{24 \times 4}{EI}$	–5	–0.77	+480	+73.92
$-\frac{1}{2} \times \frac{9 \times 3}{EI}$	–5	–1.77	+67.5	+23.90
$\sum -\frac{133.5}{EI}$			$\sum \frac{635.5}{EI}$	$+\frac{68.30}{EI}$

As $Ixy = 0$, the column flexure formula reduces to.

$$\sigma = +\frac{P}{A} + \frac{M_y}{I_{yy}} \cdot x + \frac{M_x}{I_{xx}} \cdot y$$

$$= -\frac{133.5/EI}{13/EI} + \frac{635.5}{EI} \times \frac{EI}{241.667} \cdot x + \frac{68.30}{EI} \times \frac{EI}{22.974} \cdot y$$

$$\text{or } \sigma = -10.269 + 2.62965x + 2.9729y.$$

Table 13.9 (Example 13.11)

Computation of moments at different points

Point	Coordinates		M_s	$-M_i$	$M_s - M_i$
	x	y			
A	–5	–2.77	–33	31.652	–1.3478
B	–5	+1.23	–24	19.7605	–4.2394
C	+5	+1.23	00	–6.5359	–6.5359
D	+5	–2.77	00	5.3556	+5.3556

Example 13.12. Using column analogy method draw the bending moment diagram for the portal frame shown in Fig.13.25

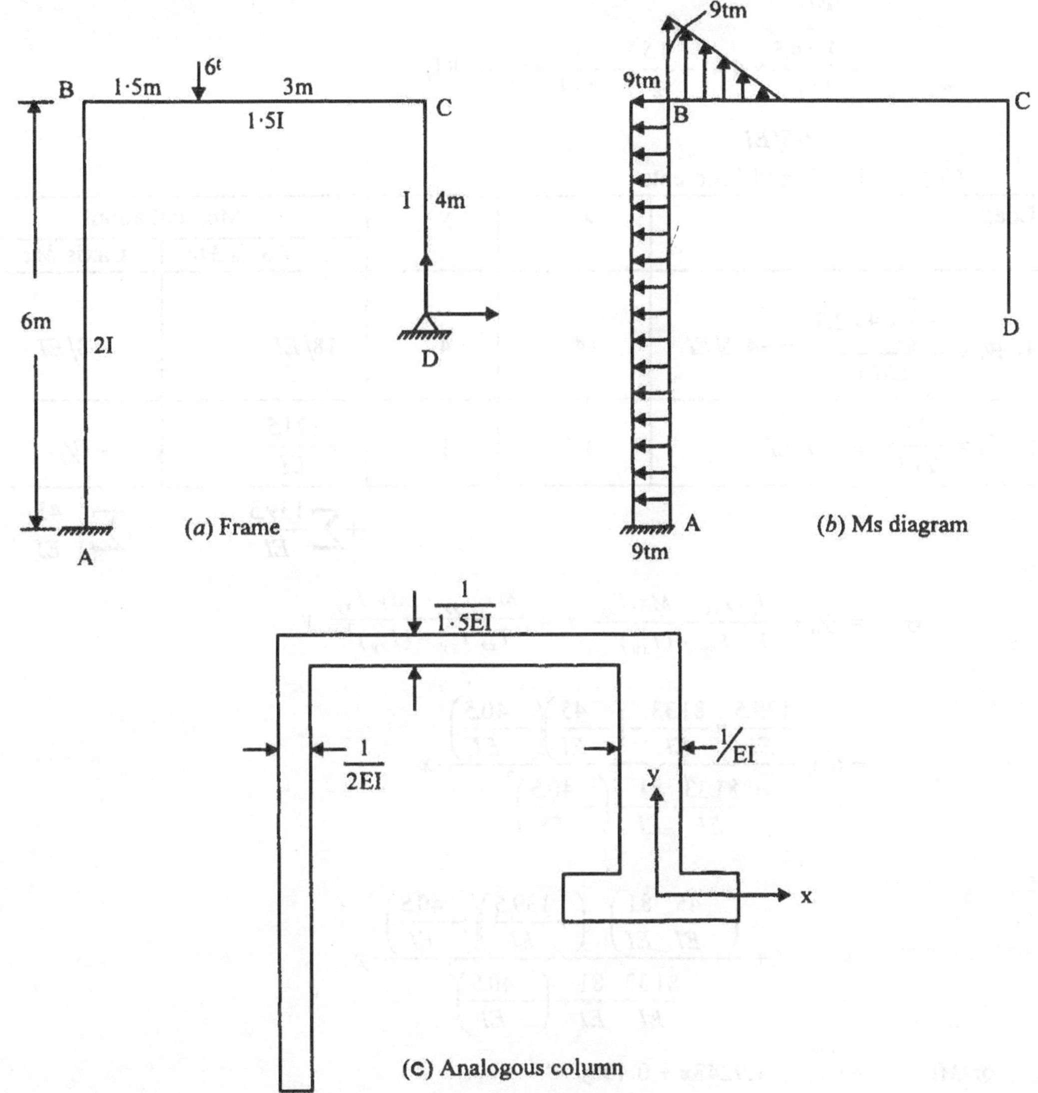

Fig. 13.25 Ex. 13.12

(*i*) Area of analogous column = ∞

(*ii*) I_{xx}, I_{xy}, I_{yy}.

$$I_{xx} = \frac{1}{EI}\cdot\frac{4^3}{3} + \frac{1}{1.5EI}\times 4.5\times 4^2 + \frac{1}{2EI}\times\frac{6^3}{12} + \frac{1}{2EI}\times 6\times 1^2$$

$$= \frac{81.33}{EI}$$

$$I_{yy} = \frac{1}{1.5EI}\times\frac{4.5^3}{3} + \frac{1}{2EI}\times 6\times 4.5^2$$

$$= \frac{81}{EI}$$

$$I_{xy} = \frac{1\times 4.5}{1.5EI}\times 4\times\left(-\frac{4.5}{2}\right)+\frac{1}{2EI}\times 6(-4.5)(1)$$

$$= -40.5/EI$$

(*iii*) Loads and Moments

Load	x	y	Moment about	
			Y axis My	X axis Mx
1. $W_1 = \frac{-\frac{1}{2}\times 9\times 1.5}{1.5EI} = -4\cdot 5/EI$	–4	4	$18/EI$	$-18/EI$
2. $W_2 = \frac{-9\times 6}{2EI} = -27/EI$	–4.5	1	$+\frac{121.5}{EI}$	$-27/_{EI}$

$$+\sum\frac{139.5}{EI} \qquad \sum -\frac{45}{EI}$$

$$\sigma = P/_A + \frac{My.I_{xx} - Mx.I_{xy}}{I_{xx}\,I_{yy}-(I_{xy})^2}\cdot x + \frac{Mx.I_{yy} - My.I_{xy}}{I_{xx}\,I_{yy}-(I_{xy})^2}\cdot y$$

$$= 0 + \frac{\frac{139.5}{EI}\times\frac{81.33}{EI} - \left(-\frac{45}{EI}\right)\left(-\frac{40.5}{EI}\right)}{\frac{81.33}{EI}\cdot\frac{81}{EI} - \left(-\frac{40.5}{EI}\right)^2}\cdot x$$

$$+ \frac{\left(-\frac{45}{EI}\cdot\frac{81}{EI}\right) - \left(+\frac{139.5}{EI}\right)\left(-\frac{40.5}{EI}\right)}{\frac{81.33}{EI}\cdot\frac{81}{EI} - \left(-\frac{40.5}{EI}\right)^2}\cdot y$$

or $Mi = 1.9248x + 0.405y$

$-Mi = -1.9248x - 0.405y$

Point	Coordinates		M_s	$-M_i$	$M_s - M_i$
	x	y			
A	– 4.5	– 2	– 9	9.4716	+ 0.4716
B	– 4.5	+ 4	– 9	7.0416	–1.9584
C	0	+ 4	0	– 1.62	–1.62
D	0	0	0	0	0

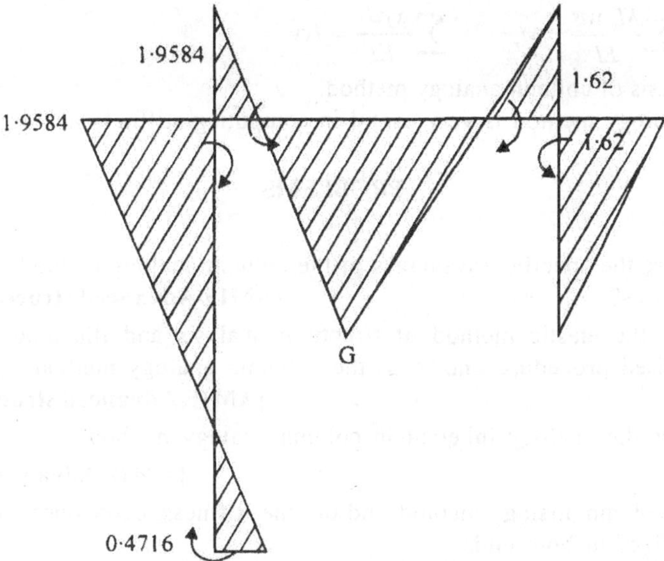

Fig. 13.26. B.M.diagram

SUMMARY

1. Column analogy method is used for solving statically indeterminate structures like indeterminate beams and fixed arches. It is also useful in calculating stiffness and carry over factors for beams particularly with variable section. General solution of a statically indeterminate structure of 3rd degree involves solution of three simultaneous equations which is long and tedious. Column analogy method transforms the problem to routine calculation of pressures in a short column under biaxial eccentric load.
2. By elastic centre method moment at any point (x,y) of a fixed portal is given by

$$M_{(x,y)} = Ms - Mi$$

where
$$M_i = \frac{W}{A} + \left(\frac{M_y . I_{xx} - M_x . I_{xy}}{I_{yy} . I_{xx} - (I_{xy})^2}\right) \cdot x + \left(\frac{M_x . I_{yy} - M_y . I_{xy}}{I_{xx} . I_{yy} - (I_{xy})^2}\right) \cdot y$$

But the column – flexure formula is

$$\sigma = \frac{P}{A} + \left(\frac{M_y . I_{xx} - M_x . I_{xy}}{I_{xx} . I_{yy} - (I_{xy})^2}\right) \cdot x + \left(\frac{M_x . I_{yy} - M_y . I_{xy}}{I_{xx} . I_{yy} - (I_{xy})^2}\right) \cdot y$$

These two equations appear identical and column flexure formula can be applied in computing Mi provided we take

$$W = \sum \frac{M_s . ds}{EI} \text{ as } P.$$

and
$$\sum \frac{ds}{EI} = A \qquad \sum \frac{x^2 ds}{EI} = Iyy$$

$$\sum \frac{M_s . xds}{EI} = M_y \qquad \sum \frac{y^2 ds}{EI} = Ixx$$

$$\sum \frac{M_s.yds}{EI} = M_x \qquad \sum \frac{xyds}{EI} = Ixy$$

This is the basis of column analogy method.

3. Column analogy method is very useful in computing stiffness and carry over factors for beams.

PROBLEMS

A. Theoretical

13.1 (*i*) What are the specific advantages of the column analogy method over other methods of analysis? **[AMIE Advanced structures Winter 1978]**

(*ii*) Explain the elastic method of structural analysis and illustrate how it leads to a generalised procedure known as the "column analogy method" **[AMIE Advanced structure Winter 1985]**

(*iii*) What is the analogy inherent in column analogy method? **[AMIE Advanced structures 1991]**

(*iv*) Using column analogy method find out the stiffness, carry over factors for a uniform beam fixed at both ends.

B. Objective

13.2 Choose the correct answer :

The width of analogous column in the method of column analogy is

(*a*) $\frac{1}{EI}$ (*b*) $2/EI$ (*c*) $3/EI$ (*d*) $\frac{4}{EI}$ **[AMIE Winter '97]**

C. Mathematical

13.3. A beam AB of 10m span is fixed at its both ends. It carries a point load of 5ᵗ at its mid point. Determine the fixed end moments by using column analogy method. Assume that M.I. of the section is 2I for the left half of the beam and I for the right half

13.4 Using column analogy method plot B.M.D. for the portal shown in Fig. P. 13.1

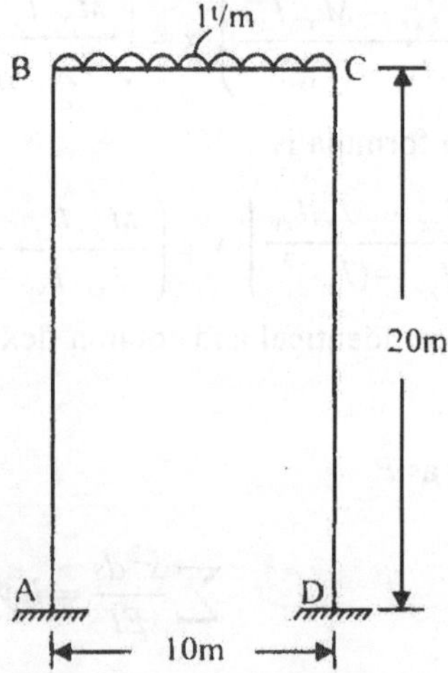

Fig. P. 13.1 (Prob. 13.4)

13.5 Using the column analogy method draw the bending moment diagram for the portal frame shown in Fig. P. 13.2

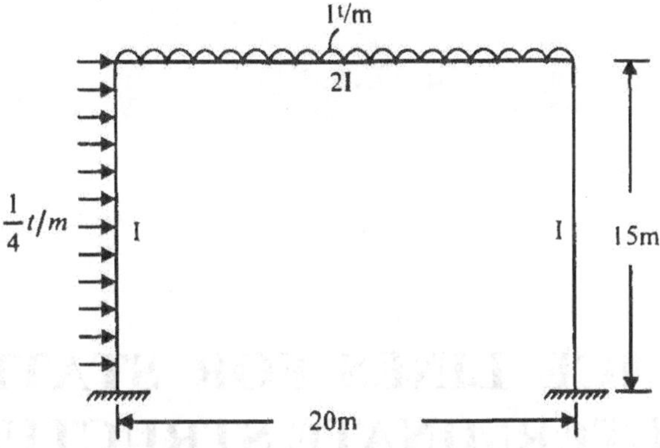

Fig. P 13.2 Problem 13.5

13.6 A beam of constant width (= unity) ABC is fixed at two ends A and C and carries a concentrated load of 8'f at its midpoint B. Portion AB is of 2.4m length and 0.2m indepth and the portion BC is of 2.4m in length and 0.4m in depth. The entire beam ABC is of same material ($E = 1.2 \times 10^7$ tf/m²).

Using column analogy method find.

(*a*) Fixed end moments at *A* and *C*.

(*b*) Absolute and relative stiffnesses and also the carry over factors for this stepped beam.

[AMIE Advanced Structures Winter 1987]

13.7 Using column analogy method analyse the frame shown in Fig. P 13.3. The frame is hinged at both the supports viz. A and E. **[AMIE Advanced structures Winter 1979]**

13.8 **A fixed beam of span *l* carries a uniformly distributed load *w* per unit run over the whole length. The M.I. of the beam is I_1 for the middle half and I_2 for the remaining portion. Ignoring the weight of the beam, find the ratio of I_2 to I_1 so that the magnitude of the bending moment at centre is one third that of fixing moments at ends**

[AMIE Advanced structures Summer 1986]

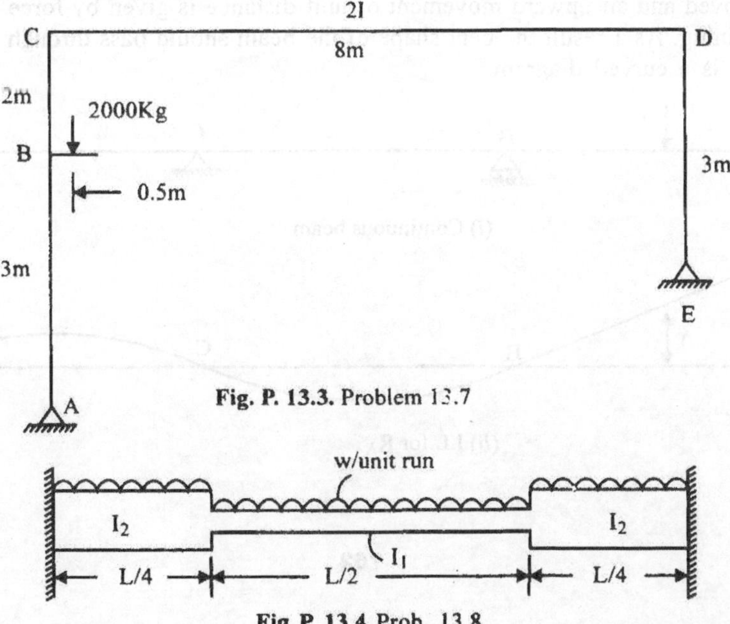

Fig. P. 13.3. Problem 13.7

Fig P 13.4 Prob. 13.8

14

INFLUENCE LINES FOR STATICALLY INDETERMINATE STRUCTURES

14.1 INTRODUCTION

In Chapter 9 influence line for staticlly determinate structures was discussed. It is now proposed to study influence lines for indeterminate structures. The influence line for statically determinate structures is obtained by using only the equations of equilibrium whereas to obtain influence lines for statically indeterminate structures equations of static equilibrium must be augmented with deflection equations to yield a tractable solution method for indeterminate structures. Hence drawing influence lines for statically indeterminate structures is a little more complicated than that of statically determinate structures. Müler Breslau's principle can be used to determine the influence line for indeterminate structures. Its use in the case of determinate structure was explained in section 9.5. It is restated as "The ordinates of the influence line for any action in a structure are equal to those of the deflection curve obtained by releasing the constraint corresponding to this action and giving a unit displacement in the direction of the action". This principle is now used to draw the influence line of various parameters for a three span continuous beam. Fig. 14.1 (*i*) shows the influence line for R_A. Here resistance against movement of the beam upward is removed and an upward movement of unit distance is given by force F with all other restraints remaining. As a result the bent shape of the beam should pass through B, C and D and resultant shape is a curved diagram.

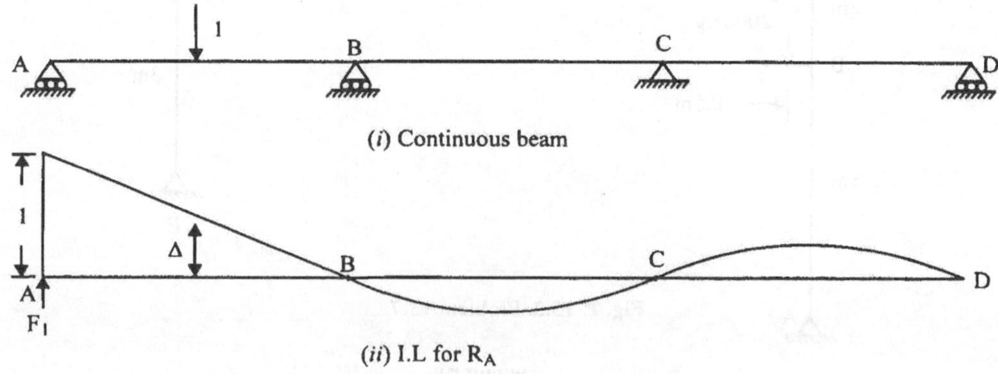

(*i*) Continuous beam

(*ii*) I.L for R_A

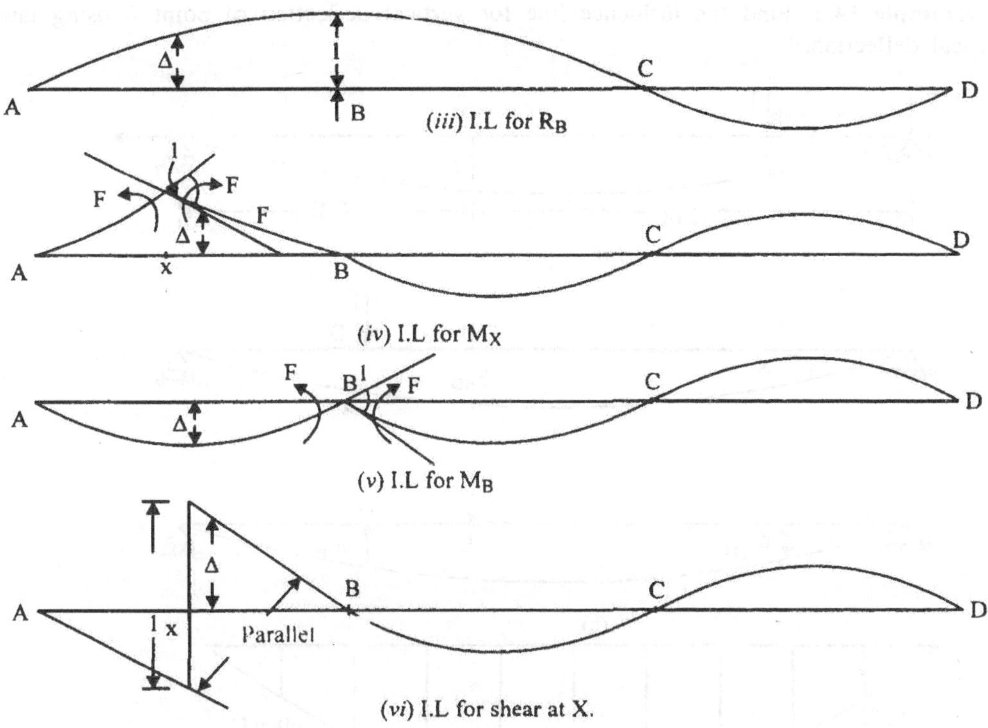

Fig. 14.1. Influence lines for different functions of a continuous beam.

Δ Corresponds to unit load

If Betti's law is now applied to Fig. 14.1 (*i*) and 14.1 (*ii*)

Ist system of forces –1, R_A, R_B, R_C and R_D

2nd system of corresponding displacement Δ, 1, 0 0 and 0

2nd system of forces F_1, R'_B, R'_C, R'_D.

Ist system of corresponding displacement 0 0 0 0

Therefore $-1 \times \Delta + R_A \times 1 = 0$ or $R_A = \Delta$

Hence the deflected shape gives the influence line for R_A. Similarly for R_B.

To get the influence line for B.M. at *X* or at support B, the bending resistance at the point is removed by putting a hinge there and unit rotation is given by applying suitable moment with all other restraints remaining. The resultant curved shape is the influence line diagram. To construct the influence line for shear at *X*, the beam is cut at *X* and a relative displacement of unity is imposed between the two parts without causing relative rotation. The tangents at the two ends must be parallel as shown in Fig. 14.1 (*vi*). It has already been repeatedly said that the influence lines for statically indeterminate structure unlike those for statically determinate structures are curved lines. The reason for this is that when a restraint for a determinate structure is removed it becomes a mechanism and there is rigid body motion which are all straight lines. In the case of statically indeterminate structure the removal of a restraint reduces the degree of indeterminacy by one. The remaining structure is still stable and responds to the inducement of a unit displacement by a curved deflected shape.

In drawing the influence line diagram by Müller Breslau's principle it is necessary to compute deflections at different points with unit deflection or rotation at the point desired. This can be done by using conjugate beam method. Moment distribution method may also be used for computing influence line ordinates.

Example 14.1. Find the influence line for vertical deflection of point B using law of reciprocal deflection.

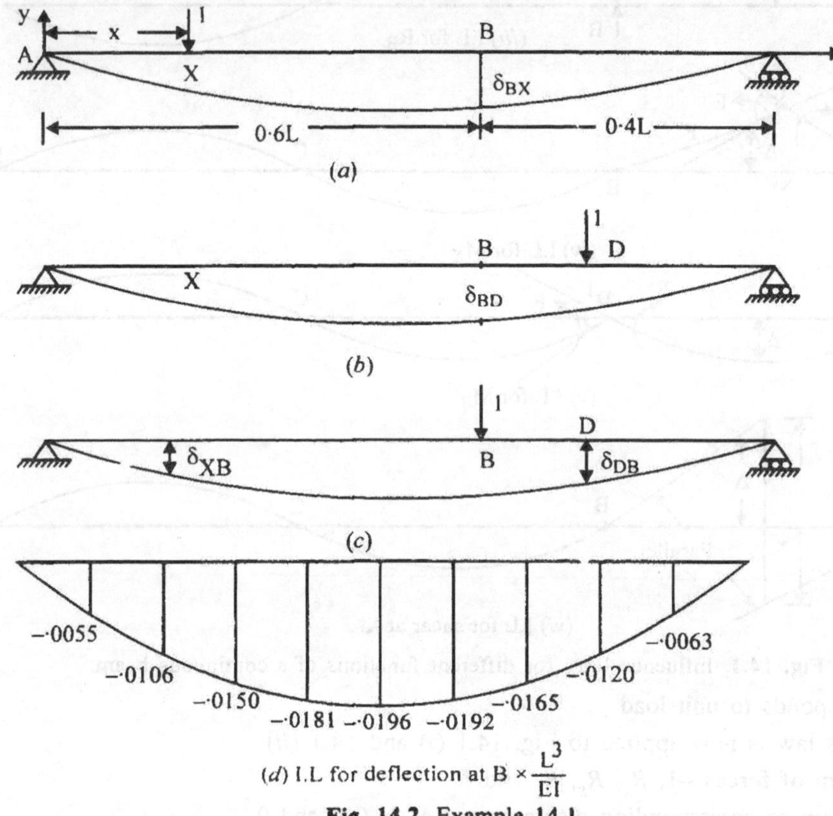

(d) I.L for deflection at B $\times \frac{L^3}{EI}$

Fig. 14.2. Example 14.1

Solution : It is necessary to find the deflection at B when the unit load is placed at X where x is the variable. This involves calculation of δ_{BX} with x taking different values. But from the reciprocal theorem.

$\delta_{BX} = \delta_{XB}, \delta_{BD} = \delta_{DB}$ and so on. This means the total influence line for the deflection at point B can be obtained by placing a unit concentrated load at point B and calculating the deflection at all points of the beam. The expression for deflection at any point X with unit load at B is

$$y = \frac{0.4L}{6EIL}\left[x^3 - \left(L^2 - (0.4L)^2\right)x - \frac{L}{0.4L}\langle x - 0.6L\rangle^3\right]$$

$$= \frac{1}{15EI}\left[x^3 - 0.84L^2x - 2.5\langle x - 0.6L\rangle^3\right]$$

The results are tabulated below (Table 14.1).

Table 14.1 (Example 14.1) and plotted in Fig 14.2 (d).

x	y
0·1L	– ·0055
0·2L	– ·0106
0·3L	– ·0150
0·4L	– ·0181

0·5L	– ·0196
0·6L	– ·0192
0·7L	– ·0165
0·8L	– ·0120
0·9L	– ·0063

Example 14.2. Compute the ordinates at intervals of $L/4$ of the influence line for R_A for the two span continuous beam of Fig. 14.3 for which EI is constant.

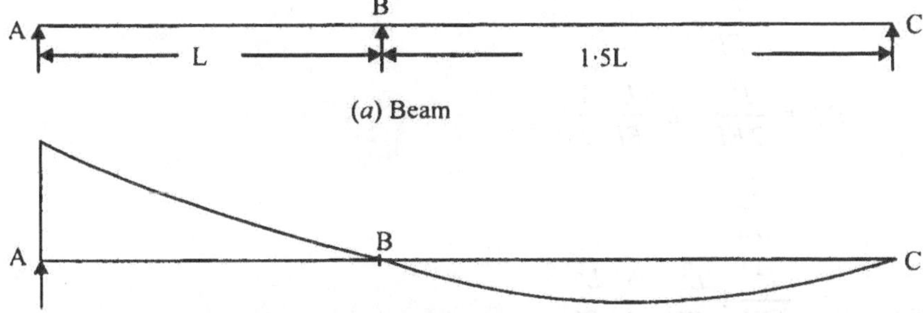

(*a*) Beam

(*b*) Deflected shape of the beam with unit load at A.

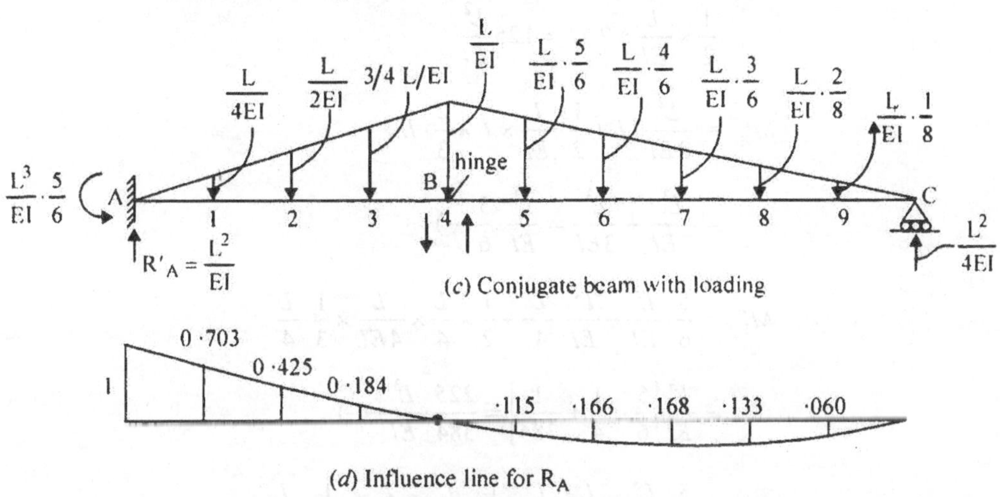

(*c*) Conjugate beam with loading

(*d*) Influence line for R_A

Fig. 14.3. Example 14.2

Solution : The redundant reaction at *A* is removed and a unit vertical force is supplied at *A*. The conjugate beam for computing deflections at various points is shown in Fig. 14.3 (*c*). The B.M. diagram will be the loading diagram. B.M. at different points of the conjugate beam are computed to get deflections at corresponding points. By making deflection at *A* unity, the influence line for R_A can be obtained. Results are given in Table 14.2

Sagging moment, downward load and deflection downward positive

Taking moments about B.

$$R'_c \times 1.5L - \frac{1}{2} \times \frac{L}{EI} \times 1.5L \times \frac{1}{3} \times 1.5L = 0$$

$$\text{or } R'_c = \frac{L^2}{4EI}$$

Total downward force $= \frac{1}{2} \times \frac{L}{EI} \times 1.5L$

$$= \frac{3}{4} \cdot \frac{L^2}{EI}$$

$$R_B' = \left(\frac{3}{4} - \frac{1}{4}\right) \frac{L^2}{EI}$$

$$= \frac{L^2}{2EI}$$

$$R_A' = \frac{L^2}{2EI} + L \cdot \frac{L}{EI} \cdot \frac{1}{2}$$

$$= \frac{L^2}{EI}$$

Check : $\frac{L^2}{4EI} + \frac{L^2}{EI} = \frac{5}{4} \cdot \frac{L^2}{EI}$

$$\frac{1}{2} \times \frac{L}{EI} \times 2.5L = 1.25 \frac{L^2}{EI}$$

$$M'_A = \frac{L^2}{2EI} \cdot L + \frac{1}{2} \cdot \frac{L}{EI} \times L \times \frac{2}{3} \cdot L$$

$$= \frac{L^3}{2EI} + \frac{L^3}{3EI} = \frac{L^3}{EI} \cdot \frac{5}{6} (-)$$

$$M_1' = \frac{5}{6} \cdot \frac{L^3}{EI} - \frac{L^2}{EI} \cdot \frac{L}{4} \cdot + \frac{1}{2} \times \frac{L}{4} \times \frac{L}{4EI} \times \frac{1}{3} \cdot \frac{L}{4}$$

$$= \frac{L^3}{EI} \left[\frac{5}{6} - \frac{1}{4} + \frac{1}{384} \right] = \frac{225}{384} \cdot \frac{L^3}{EI} \cdot (-)$$

$$M_2' = \frac{5}{6} \cdot \frac{L^3}{EI} - \frac{L^2}{EI} \cdot \frac{L}{2} + \frac{1}{2} \times \frac{L}{2} \times \frac{L}{2EI} \times \frac{1}{3} \times \frac{L}{2}$$

$$= \frac{L^3}{EI} \frac{17}{48} (-)$$

$$M_3' = \frac{5}{6} \cdot \frac{L^3}{EI} - \frac{L^2}{EI} \cdot \frac{3L}{4} + \frac{1}{2} \times \frac{3L}{4} \times \frac{3L}{4EI} \times \frac{1}{3} \times \frac{3L}{4}$$

$$= \frac{L^3}{EI} \cdot \frac{59}{384} \cdot (-)$$

$$M_4' = \frac{5}{6} \times \frac{L^3}{EI} - \frac{L^2}{EI} \cdot L + \frac{1}{2} \times \frac{L}{EI} \times L \times \frac{1}{3} \cdot L$$

$$= 0$$

$$M_9' = \frac{L^2}{4EI}\cdot\frac{L}{4} - \frac{1}{2}\times\frac{L}{4}\times\frac{L}{EI}\times\frac{L}{4}\times\frac{2}{3L}\times\frac{L}{4}\times\frac{1}{3}$$

$$= \frac{L^3}{EI}\left[\frac{35}{16\times 36}\right].$$

$$M_8' = \frac{L^2}{4EI}\cdot\frac{L}{2} - \frac{1}{2}\times\frac{L}{2}\times\frac{L}{EI}\times\frac{L}{2}\times\frac{2}{3L}\times\frac{L}{2}\times\frac{1}{3}$$

$$= \frac{L^3}{EI}\cdot\frac{1}{9}.$$

$$M_7' = \frac{L^2}{4EI}\cdot\frac{3L}{4} - \frac{1}{2}\times\frac{3L}{4}\times\frac{L}{EI}\times\frac{3L}{4}\times\frac{2}{3L}\times\frac{3L}{4}\times\frac{1}{3}.$$

$$= \frac{L^3}{EI}\left[\frac{3}{16} - \frac{9}{192}\right] = \frac{L^3}{EI}\times\frac{9}{64}$$

$$M_6' = \frac{L^2}{4EI}\cdot L - \frac{1}{2}\times L\times\frac{L}{EI}\times L\times\frac{2}{3L}\times L\times\frac{1}{3}$$

$$= \frac{L^3}{EI}\left[\frac{1}{4} - \frac{1}{9}\right] = \frac{L^3}{EI}\cdot\frac{5}{36}.$$

$$M_5' = \frac{L^2}{4EI}\cdot\frac{5L}{4} - \frac{1}{2}\times\frac{5L}{4}\times\frac{L}{EI}\times\frac{5L}{4}\times\frac{2}{3L}\times\frac{5L}{4}\times\frac{1}{3}$$

$$= \frac{L^3}{EI}\left[\frac{5}{16} - \frac{125}{64\times 9}\right]$$

$$= \frac{L^3}{EI}\cdot\frac{55}{64\times 9}.$$

Table 14.2 (Example 14.2) upward deflection negative

	M'_A	M'_1	M'_2	M'_3	M'_4	M'_5	M'_6	M'_7	M'_8	M'_9
Value $\times \frac{L^3}{EI}$	$5/6$	$\frac{225}{384}$	$\frac{17}{48}$	$\frac{59}{384}$	0	$\frac{55}{64\times 9}$	$\frac{5}{36}$	$\frac{9}{64}$	$\frac{1}{9}$	$\frac{35}{36\times 16}$
$/ M'_A$	−1	−0·703	−0·425	−0·184	0	0·115	0·166	0·168	0·133	0·060

Examples 14.3. Draw I.L. for R_B, R_C, M_B, B.M at midpoint of AB and S.F. at mid point of AB for the continuous beam ABC (Fig. 14.4).

(*a*) Beam

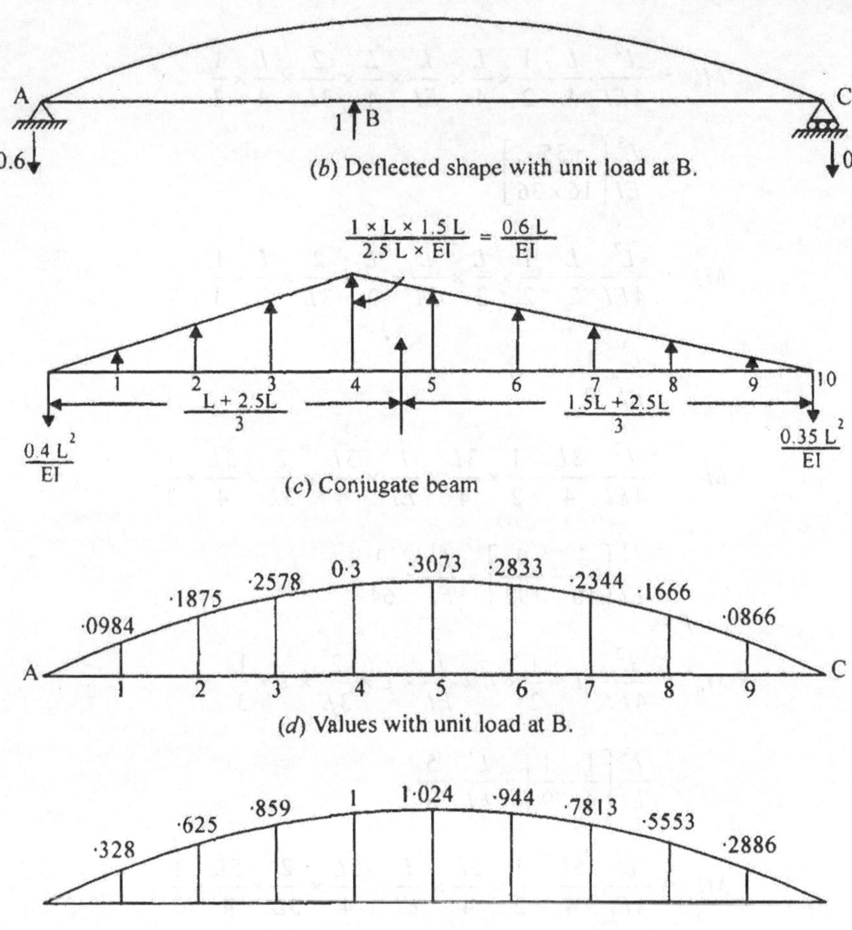

(e) I.L for R_B.

Fig. 14.4. I.L. for R_B Example 14.3

Solution : Fig. 14.4 (*a*), (*b*) and (*c*) show the beam, deflected shape and the corresponding conjugate beam.

B.M in conjugate beam at $L/4$ intervals are calculated now

$$M'_A = 0$$

$$M'_1 = \frac{0.4L^2}{EI}\cdot\frac{L}{4} - \frac{1}{2}\times\frac{0.6L}{EI}\times\frac{L}{4}\times\frac{1}{L}\times\frac{L}{4}\times\frac{1}{3}\cdot\frac{L}{4}$$

$$= \frac{L^3}{EI}\left[\frac{0.4}{4} - \frac{0.6}{8\times 48}\right] = \frac{L^3}{EI}\times 0.0984$$

$$M'_2 = \frac{0.4L^2}{EI}\cdot\frac{L}{2} - \frac{1}{2}\times\frac{0.6L}{EI}\times\frac{L}{2}\times\frac{1}{L}\times\frac{L}{2}\times\frac{1}{3}\times\frac{L}{2}$$

$$= \frac{L^3}{EI}\left[\frac{0.4}{2} - \frac{0.6}{48}\right] = \frac{L^3}{EI}\times 0.1875.$$

$$M'_3 = \frac{0.4L^2}{EI} \cdot \frac{3}{4} \cdot L - \frac{1}{2} \times \frac{0.6L}{EI} \times \frac{3}{4}L \times \frac{1}{L} \times \frac{3}{4} \cdot L \times \frac{1}{3} \times \frac{3}{4} \cdot L$$

$$= \frac{L^3}{EI}\left[\frac{1.2}{4} - \frac{0.6 \times 27}{32 \times 12}\right] = 0.2578\frac{L^3}{EI}$$

$$M'_4 = \frac{0.4L^2}{EI} \times L - \frac{1}{2} \times \frac{0.6L}{EI} \times L \times \frac{L}{3}$$

$$= \frac{L^3}{EI}\left[0.4 - \frac{0.6}{6}\right] = \frac{0.3L^3}{EI}$$

$$M'_{10} = 0$$

$$M'_9 = \frac{0.35L^2}{EI} \times \frac{L}{4} - \frac{1}{2} \times \frac{0.6L}{EI} \times \frac{L}{4 \times 1.5L} \times \frac{L}{4} \times \frac{1}{3} \times \frac{L}{4}$$

$$= \frac{L^3}{EI}\left[\frac{.35}{4} - \frac{.6}{4 \times 144}\right] = 0.0866\frac{L^3}{EI}$$

$$M'_8 = \frac{0.35L^2}{EI} \times \frac{L}{2} - \frac{1}{2} \times \frac{0.6L}{EI} \times \frac{L}{2 \times 1.5L} \times \frac{L}{2} \times \frac{1}{3} \times \frac{L}{2}$$

$$= \frac{L^3}{EI}\left[\frac{.35}{2} - \frac{0.6}{12 \times 6}\right] = 0.1666\frac{L^3}{EI}$$

$$M'_7 = \frac{0.35L^2}{EI} \times \frac{3L}{4} - \frac{1}{2} \times \frac{0.6L}{EI} \times \frac{3L}{4 \times 1.5L} \times \frac{3L}{4} \times \frac{1}{3} \times \frac{3L}{4}$$

$$= \frac{L^3}{EI}\left[\frac{3 \times .35}{4} - \frac{0.6 \times 9}{12 \times 16}\right] = 0.2344\frac{L^3}{EI}$$

$$M'_6 = \frac{.35L^2}{EI} \times L - \frac{1}{2} \times \frac{0.6L}{EI} \times \frac{L}{1.5L} \times L \times \frac{L}{3}$$

$$= \frac{L^3}{EI}\left[0.35 - \frac{0.6}{9}\right] = 0.2833.$$

$$M'_5 = \frac{0.35L^2}{EI} \times 1.25L - \frac{1}{2} \times \frac{0.6L}{EI} \times \frac{1.25L}{1.5L} \times 1.25L \times \frac{1}{3} \times 1.25L$$

$$= \frac{L^3}{EI}\left[1.25 \times 0.35 - \frac{1.25^3 \times 0.6}{9}\right] = 0.3073\frac{L^3}{EI}$$

$$M'_4 = \frac{0.35L^2}{EI} \times 1.5L - \frac{1}{2} \times \frac{0.6L}{EI} \times \frac{1.5L}{1.5L} \times 1.5L \times \frac{1}{3} \times 1.5L$$

$$= \frac{L^3}{EI}\left[.35 \times 1.5 - \frac{0.6 \times 1.5^3}{9}\right] = 0.3\frac{L^3}{EI}$$

These values are plotted in Fig. 14.4 (*d*) I.L. for R_B is plotted in Fig. 14.4(*e*) by dividing the ordinates of Fig. 14.4 (d) by 0.3.

I.L. for R_c Fig. 14.5 (*a*), (*b*) and (*c*) show the beam, deflected shape and the conjugate beam loading when restraint at *C* is removed and unit force applied.

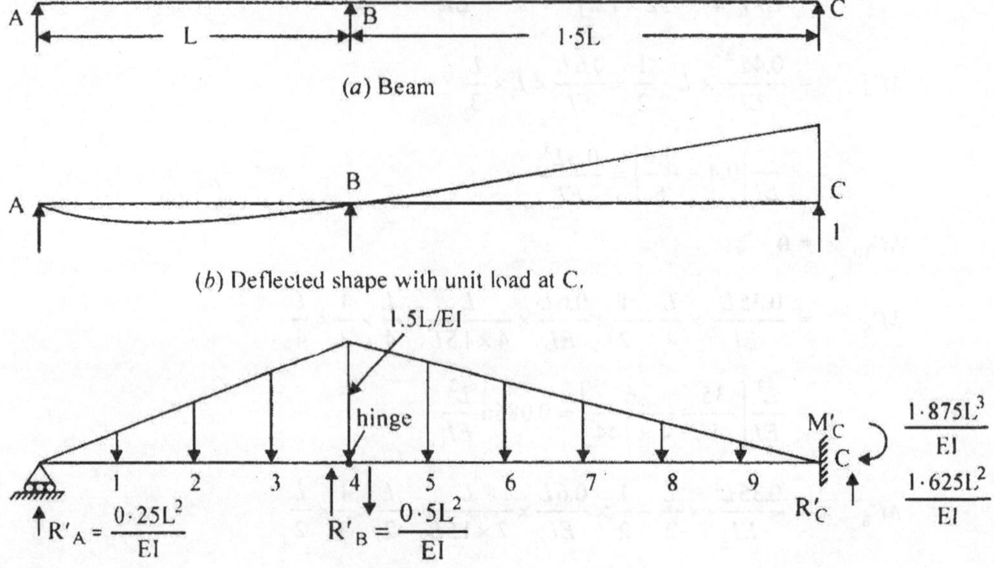

Fig. 14.5. Example 14.3

Calculation of deflection from conjugate beam.

$$R'_A \times L - \frac{1}{2} \times \frac{1.5L}{EI} \times L \times \frac{L}{3} = 0$$

or $$R'_A = \frac{0.25L^2}{EI}$$

$$R'_B = \frac{1}{2} \times \frac{1.5L}{EI} \times L - \frac{0.25L^2}{EI}$$

$$= \frac{0.5L^2}{EI}.$$

$$R'_C = \frac{1}{2} \times \frac{1.5L}{EI} \times 1.5L + \frac{0.5L^2}{EI}$$

$$= (1.125 + 0.5)\frac{L^2}{EI} = \frac{1.625L^2}{EI}$$

$$M'_1 = \frac{0.25L^2}{EI} \times \frac{L}{4} - \frac{1}{2} \times \frac{L}{4} \times \frac{1.5L}{EI} \times \frac{L}{4} \times \frac{1}{L} \times \frac{1}{3} \times \frac{L}{4}$$

$$= \frac{L^3}{EI}\left[\frac{.25}{4} - \frac{1.5}{32 \times 12}\right] = 0.0586 \cdot \frac{L^3}{EI}$$

$$M'_2 = \frac{0.25L^2}{EI} \times \frac{L}{2} - \frac{1}{2} \times \frac{L}{2} \times \frac{1.5L}{EI} \times \frac{L}{2} \times \frac{1}{L} \times \frac{1}{3} \times \frac{L}{2}$$

$$= \frac{L^3}{EI}\left[\frac{0.25}{2} - \frac{1.5}{48}\right] = 0.09375\frac{L^3}{EI}.$$

$$M'_3 = \frac{0.25L^2}{EI} \times \frac{3}{4} \cdot L - \frac{1}{2} \times \frac{3}{4} \cdot L \times \frac{1.5L}{EI} \times \frac{3}{4} \cdot L \times \frac{1}{L} \times \frac{1}{3} \times \frac{3}{4} \cdot L$$

$$= \frac{L^3}{EI}\left[\frac{0.75}{4} - \frac{1.5 \times 27}{32 \times 12}\right] = 0.0820\frac{L^3}{EI}$$

$$M'_4 = \frac{0.25L^2}{EI} \times L - \frac{1}{2} \times L \times \frac{1.5L}{EI} \times L \times \frac{1}{L} \times \frac{1}{3} \times L$$

$$= \frac{L^3}{EI}[0.25 - .25] = 0$$

$$M'_5 = \frac{1.875L^3}{EI} - \frac{1.625L^2}{EI} \times 1.25L + \frac{1}{2} \times \frac{1.25L}{EI} \times 1.25L \times \frac{1}{3} \times 1.25L$$

$$= \frac{L^3}{EI}\left|1.875 - 1.625 \times 1.25 + \frac{1.25^3}{6}\right| = 0.1693 \cdot \frac{L^3}{EI}(-)$$

$$M'_6 = \frac{1.875L^3}{EI} - \frac{1.625L^2}{EI} \times L - \frac{1}{2} \times \frac{L}{EI} \times L \times \frac{1}{3} \cdot L$$

$$= \frac{L^3}{EI}\left|1.875 - 1.625 + \frac{1}{6}\right| = 0.4166 \cdot \frac{L^3}{EI}(-)$$

$$M'_7 = \frac{1.875L^3}{EI} - \frac{1.625L^2}{EI} \times \frac{3L}{4} + \frac{1}{2} \times \frac{3L}{4EI} \times \frac{3L}{4} \times \frac{1}{3} \times \frac{3L}{4}$$

$$= \frac{L^3}{EI}\left|1.875 - 1.625 \times \frac{3}{4} + \frac{27}{32 \times 12}\right| = 0.7265\frac{L^3}{EI}(-)$$

$$M'_8 = \frac{1.875L^3}{EI} - \frac{1.625L^2}{EI} \times \frac{L}{2} + \frac{1}{2} \times \frac{L}{2EI} \times \frac{L}{2} \times \frac{1}{3} \times \frac{L}{2}$$

$$= \frac{L^3}{EI}\left|1.875 - \frac{1.625}{2} + \frac{1}{48}\right| = 1.0833 \cdot \frac{L^3}{EI}(-)$$

$$M'_9 = \frac{1.875L^3}{EI} - \frac{1.625L^2}{EI} \times \frac{L}{4} + \frac{1}{2} \times \frac{1.5L}{EI} \times \frac{L}{4} \times \frac{1}{1.5L} \times \frac{L}{4} \times \frac{1}{3} \times \frac{L}{4}$$

$$= \frac{L^3}{EI}\left|1.875 - \frac{1.625}{4} + \frac{1}{32 \times 12}\right| = 1.4714 \cdot \frac{L^3}{EI}(-)$$

Table 14.3 gives the B.M ordinates and coeffecients for influence line for R_C.

Table 14.3 I.L. for R_C (upward negative)

	M'_1	M'_2	M'_3	M'_4	M'_5	M'_6	M'_7	M'_8	M'_9
$\times L^3/EI$	·0586	·0938	·0823	0	−0·1693	−0·4166	−0·7265	−1·0833	−1·4714
$/M'_C$	·0313	0·0500	·0439	0	−·0903	−0·2222	−0·3875	−0·5778	−0·7847

$$M'_C = \frac{1.125L^2}{EI} \times \frac{2}{3} \times 1.5L + \frac{0.5L^2}{EI} \times 1.5L = \frac{L^3}{EI} \times 1.875$$

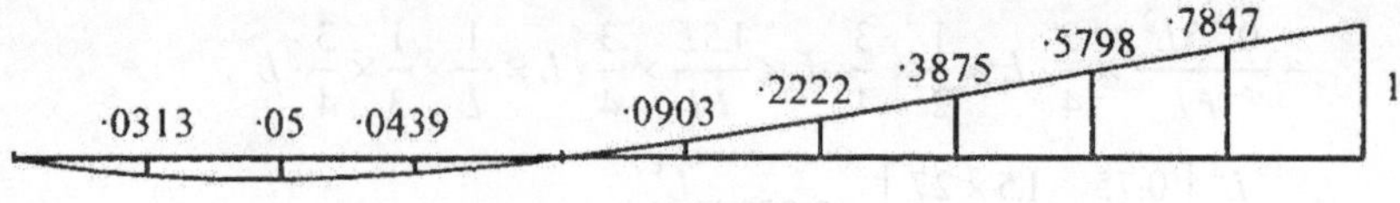

(*d*) Influence line for R_C.

Fig. 14.6. I.L for R_C (Example 14.3)

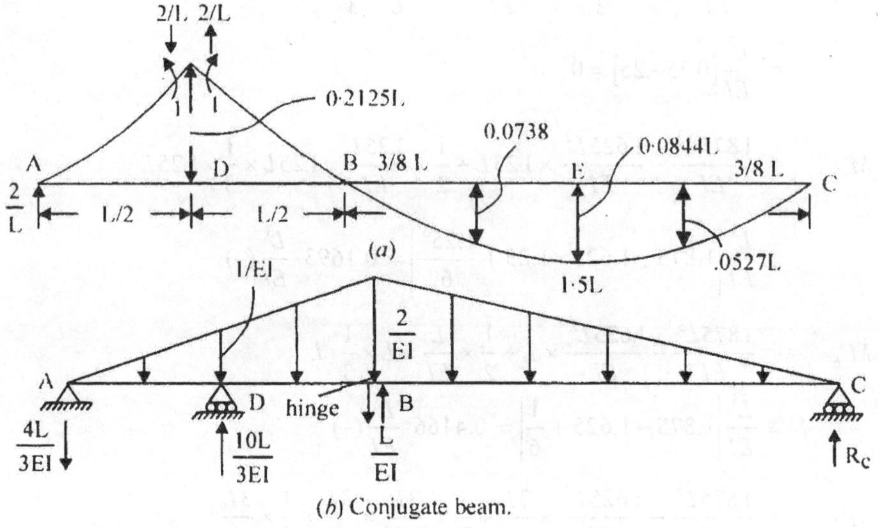

(*b*) Conjugate beam.

Fig. 14.7. Conjugate beam for I.L. for M_D.

I.L for M_D.

Solution : Fig. 14.7 (*b*) gives the conjugate beam for I.L. for M_D shown in Fig. 14.7 (*a*)

Calculation of deflection from conjugate beam

Taking moments about B

$$R_C \times 1.5L - \frac{1}{2} \times 1.5L \times \frac{2}{EI} \times \frac{L}{2} = 0$$

or
$$R_C = \frac{1.5L \times L}{2EI \times 1.5L} = \frac{L}{2EI}$$

$$R_B + R_C = \frac{1}{2} \times \frac{2}{EI} \times 1.5L = \frac{1.5L}{EI}$$

$$\therefore R_B = \frac{3L}{2EL} - \frac{L}{2EI} = \frac{L}{EI}$$

Taking moments about *A*

$$\frac{L}{EI} \times L + \frac{1}{2} \times \frac{2}{EI} \times L \times \frac{2}{3}L - R_D \times \frac{L}{2} = 0$$

$$\text{or } R_D \times \frac{1}{2} = \frac{L^2}{EI}\left[1 + \frac{2}{3}\right]$$

$$\text{or } R_D = \frac{L^2}{EI} \cdot \frac{2}{L} \cdot \frac{5}{3} = \frac{10L}{3EI}$$

$$R_A + \frac{10L}{3EI} - \frac{L}{EI} - \frac{1}{2} \times L \times \frac{2}{EI} = 0$$

$$\text{or } R_A = \frac{2L}{EI} - \frac{10L}{3EI} = -\frac{4L}{3EI}$$

$$M_D = \frac{4L}{3EI} \cdot \frac{L}{2} + \frac{1}{2} \times \frac{1}{EI} \times \frac{L}{2} \times \frac{1}{3} \times \frac{L}{2}$$

$$= \frac{2}{3} \cdot \frac{L^2}{EI} + \frac{L^2}{24EI} = \frac{17}{24} \cdot \frac{L^2}{EI}$$

$$M_E = \frac{L}{2EI} \times \frac{3}{4} L - \frac{1}{2} \times \frac{1}{EI} \times \frac{3}{4} \cdot L \times \frac{1}{3} \times \frac{3}{4} \cdot L$$

$$= \frac{3}{8} \cdot \frac{L^2}{EI} - \frac{3}{32} \frac{L^2}{EI} = \frac{9}{32} \cdot \frac{L^2}{EI}$$

$$\text{Change in slope} = R_D = \frac{10L}{3EI}$$

$$M'_D = \frac{17}{24} \cdot \frac{L^2}{EI} \times \frac{3EI}{10L} = \frac{51}{240} \times L = 0.2125L$$

$$M'_E = \frac{9}{32} \times \frac{L^2}{EI} \times \frac{3EI}{10L} = \frac{27}{320} \cdot L = 0.0844L$$

The values are shown in Fig. 14.7(a)

I.L. for M_B

Fig. 14.8(*b*) gives the conjugate beam for drawing the I.L. for M_B shown in Fig. 14.8 (*a*)

Calculation of deflection by conjugate beam (EI = unity)

$$R_A = 1 \times L \times \frac{1}{2} \times \frac{1}{3} = \frac{L}{6}$$

$$R_C = 1 \times \frac{3L}{2} \times \frac{1}{2} \times \frac{1}{3} = \frac{L}{4}$$

$$R_B = \frac{1}{2} \times 1 \times \frac{5}{2} \cdot L - \frac{L}{6} - \frac{L}{4}$$

$$= \frac{5}{6} L \cdot$$

Rotation at B = $\frac{5}{6} L$

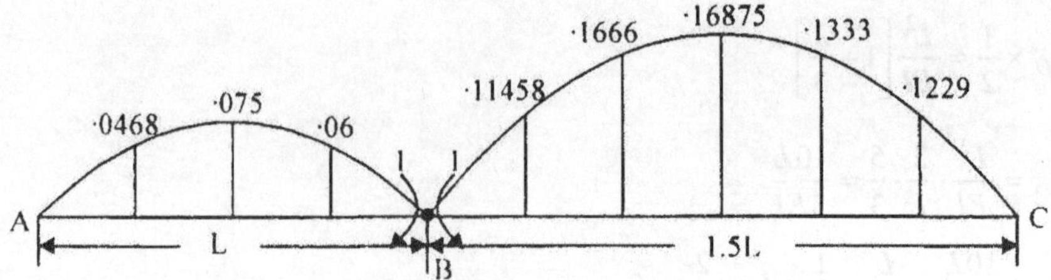

(*a*) Influence line (ordinates to be multiplied by L)

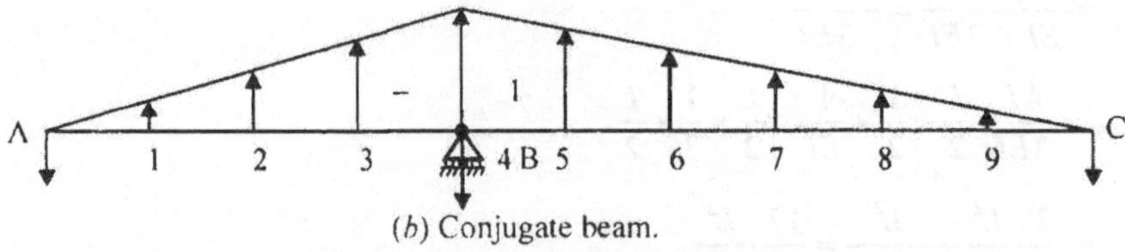

(*b*) Conjugate beam.

Fig. 14.8. I.L. for M_B (Example 14.3)

$$M_1 = \frac{L}{6} \times \frac{L}{4} - \frac{1}{2} \times \frac{1}{4} \times \frac{L}{4} \times \frac{L}{12} = .0390625 L^2$$

Dividing by $\frac{5}{6} \cdot L$, ratio = ·046875*L*.

$$M_2 = \frac{L}{6} \times \frac{L}{2} - \frac{1}{2} \times \frac{1}{2} \times \frac{L}{2} \times \frac{1}{3} \times \frac{L}{2} = .0625 L^2 \text{ ratio} = .075\text{L}$$

$$M_3 = \frac{L}{6} \times \frac{3}{4} \cdot L - \frac{1}{2} \times 3/4 \times \frac{3}{4} \cdot L \times \frac{1}{3} \times \frac{3}{4} \cdot L = .0546875 L^2 \text{ ratio} = .065625\text{L}$$

$$M_4 = \frac{L}{6} \times L - \frac{1}{2} \times 1 \times L \times \frac{1}{3} \times L = 0 \text{ ratio} = 0$$

$$M_9 = \frac{L}{4} \times \frac{L}{4} - \frac{1}{2} \times \frac{1}{1.5L} \times \frac{L}{4} \times \frac{L}{4} \times \frac{1}{3} \times \frac{L}{4} = 0.102438 L^2 \text{ ratio} = 0.1229.\text{L}$$

$$M_8 = \frac{L}{4} \times \frac{L}{2} - \frac{1}{2} \times \frac{1}{1.5L} \times \frac{L}{2} \times \frac{L}{2} \times \frac{1}{3} \times \frac{L}{2} = 0.1111 L^2 \text{ ratio} = 0.13333\text{L}$$

$$M_7 = \frac{L}{4} \times \frac{3L}{4} - \frac{1}{2} \times \frac{1}{1.5L} \times \frac{3L}{4} \times \frac{1}{3} \times \frac{3L}{4} = 0.1406259 L^2 \text{ ratio} = 0.16875\text{L}$$

$$M_6 = \frac{L}{4} \times L - \frac{1}{2} \times \frac{L}{1.5L} \times L \times \frac{1}{3} \times L = 0.13888 L^2 \text{ ratio} = 0.1666\text{L}$$

$$M_5 = \frac{L}{4} \times \frac{5L}{4} - \frac{1}{2} \times \frac{1}{1.5L} \times \frac{5L}{4} \times \frac{5L}{4} \times \frac{1}{3} \times \frac{5L}{4} = 0.09548 L^2 \text{ ratio} = 0.11458\text{L}$$

The values of the ordinates are shown in Fig. 14.8(*a*). All values are negative.

Influence line for S. F at D. To construct the influence diagram for the shear at *D*, the mid span of AB, the beam is first cut at *D* and a sliding device is introduced as shown in Fig. 14.9(*c*). Next we apply a pair of equal and opposite unit forces to separate the ends vertically

Δ_D with out causing relative rotation as shown in Fig. 14.9(*a*). The deflected shape and the induced reactions and moments required by equilibrium are also shown in Fig. 14.9(*a*).

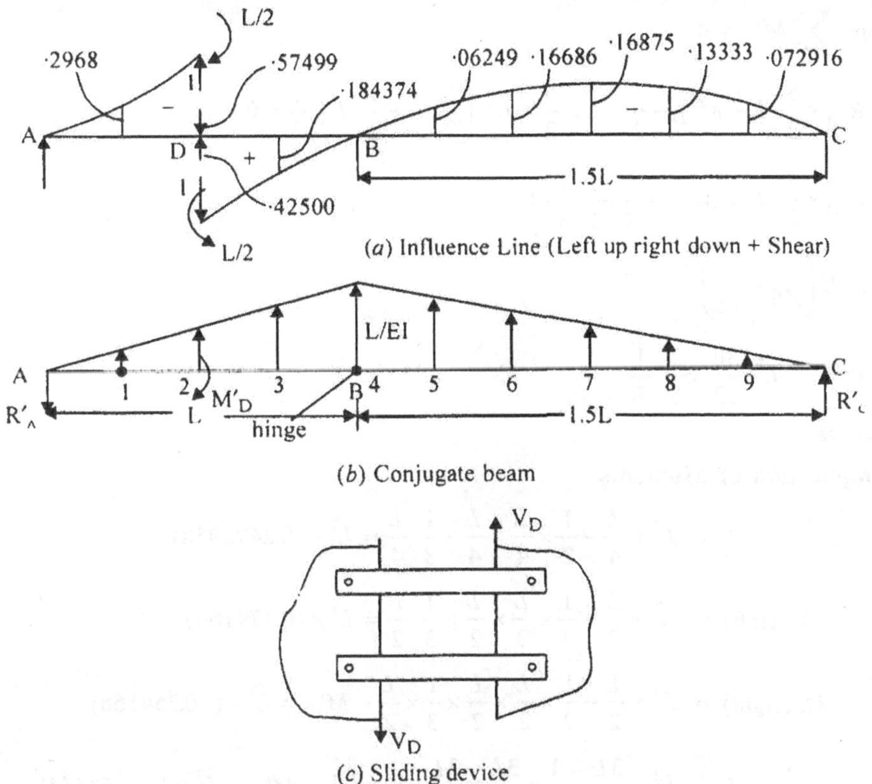

Fig. 14.9. Influence line for shear at *D* (Example 14.3)

The conjugate beam and its loading are shown in Fig. 14.9(*b*). Since we imposed a relative movement of the two cut ends equal to Δ_D without relative rotation a moment M_D' must be applied at *D'* of the conjugate beam. Such a moment satisfies the requirement that there must be a moment difference (with out a shear difference) between the two cut ends. The reactions R_A' and R_C' and the moment M_D' are obtained from the equilibrium conditions and the condition equaton that the bending moment at the internal hinge *B* is zero.

Calculation of deflection from conjugate beam (with $EI = 1$)

Taking moment about B

$$R'_C \times \frac{3}{2} \cdot L - \frac{1}{2} \times \frac{3}{2} L \times L \times \frac{1}{3} \times \frac{3}{2} \cdot L = 0$$

or $R'_C = L^2/_4 \cdot \downarrow$

From $\sum M'_A = 0$

$$R'_C \times \frac{5}{2} \cdot L - \frac{1}{2} \times L \times \frac{5}{2} \cdot L \times (L + {}^5/_2 L) \cdot \frac{1}{3} + M'_D = 0$$

$$\text{or } M'_D = \frac{5}{4} L^2 \cdot \frac{7}{2} \cdot L \cdot \frac{1}{3} - \frac{5}{2} \cdot L \times \frac{L^2}{4}.$$

$$= L^3 \cdot \frac{5}{6}$$

From $\sum M'_C = 0$

$$R'_A \times \frac{5}{2} \cdot L - M'_D - \frac{1}{2} \times L \times \frac{5}{2} \times L \times \left(\frac{5}{2} \cdot L + \frac{3}{2} \cdot L\right) \cdot \frac{1}{3} = 0$$

$$\text{or } R'_A \times \frac{5}{2} \cdot L = M'_D + \frac{5}{4} \cdot \frac{8}{2} \cdot \frac{1}{3} \cdot L^3$$

$$= L^3 \left(5/6 + \frac{20}{12} \right)$$

$$\text{or } R'_A = L^3 \cdot \frac{30}{12} \times \frac{2}{5} \cdot \frac{1}{L}$$

$$= L^2$$

Computation of Moments

$$M_1 = L^2 \times \frac{L}{4} - \frac{1}{2} \times \frac{L}{4} \times \frac{L}{4} \times \frac{1}{3} \cdot \frac{L}{4} = L^3 \times (0.2473958)$$

$$M_2 \text{ (left)} = L^2 \times \frac{L}{2} - \frac{1}{2} \times \frac{L}{2} \times \frac{L}{2} \times \frac{1}{3} \cdot \frac{L}{2} = L^3 \times (0.479166)$$

$$M_2 \text{ (right)} = L^2 \times \frac{L}{2} - \frac{1}{2} \times \frac{L}{2} \times \frac{L}{2} \times \frac{1}{3} \times \frac{L}{2} - M'_D = L^3 \times (-0.354166).$$

$$M_3 = L^2 \times \frac{3L}{4} - \frac{1}{2} \times \frac{3L}{4} \times \frac{3L}{4} \times \frac{1}{3} \times \frac{3L}{4} - M'_D = L^3 \times (-0.153645)$$

$$M_4 = L^2 \cdot L - \frac{1}{2} \times L \times L \times \frac{1}{3} L - M'_D = 0.$$

$$M_5 = \frac{L^2}{4} \times 1.25L - \frac{1}{2} \times \frac{1}{1.5L} \times 1.25L \times 1.25L \times \frac{1}{3} \times 1.25L = L^3 (.05208)$$

$$M_6 = \frac{L^2}{4} \times L - \frac{1}{2} \times \frac{L}{1.5L} \times L \times L \times \frac{1}{3} \times L = L^3 \times (0.13888)$$

$$M_7 = \frac{L^2}{4} \times \frac{3L}{4} - \frac{1}{2} \times \frac{L}{1.5L} \times \frac{3L}{4} \times \frac{3L}{4} \times \frac{1}{3} \times \frac{3L}{4} = L^3 \times (0.140625)$$

$$M_8 = \frac{L^2}{4} \times \frac{L}{2} - \frac{1}{2} \times \frac{L}{1.5L} \times \frac{L}{2} \times \frac{L}{2} \times \frac{1}{3} \times \frac{L}{2} = L^3 \times (0.11111)$$

$$M_9 = \frac{L^2}{4} \times \frac{L}{4} - \frac{1}{2} \times \frac{L \times L}{1.5L \times 4} \times \frac{L}{4} \times \frac{1}{3} \times \frac{L}{4} = L^3 \times (0.06076)$$

Now M_2 (left) + M_2 (right) = L^3 (0·479166 + 0.354166)

(When sign is neglected)

= L^3 (·833332)

All the moment values are divided by

L^3 (0·833332) to make M_2 (left) + (right) *i.e.* deflection at 2 equal to unity.

The results are as follows :

$M'_1 = 0.2968$ $\qquad$ $M'_5 = 0.06249$

$$\left.\begin{aligned} M'_2\,(left) &= 0.57499 \\ M'_2\,(right) &= -0.42500 \end{aligned}\right] 1.000$$ $\qquad$ $M'_6 = 0.16666$

$M'_3 = -0.184374$ $\qquad$ $M'_7 = 0.16875$

$M'_4 = 0$ $\qquad$ $M'_8 = 0.13333$

$M'_9 = 0.072916$

The data are plotted in Fig. 14.9(*a*). Positive and negative sign indicated deflection upward and down respectively. In Fig. 14.9 (a) this is converted to S.F. sign convention.

Example 14.4. Obtain an influence line at $L/4$ intervals for the moment at the support B shown in Fig. 14.10.

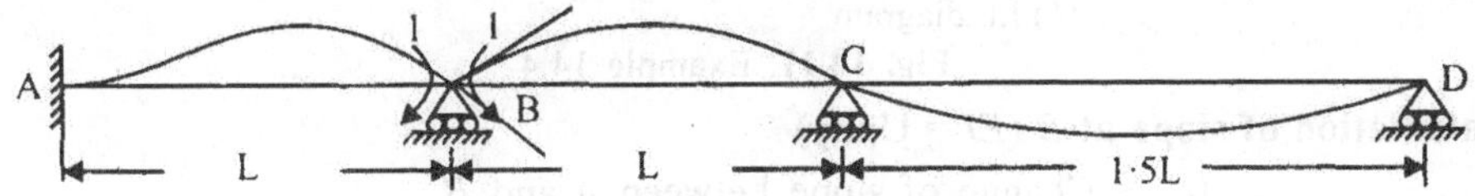

Fig. 14.10. Example 14.4

Solution : With hinge at B

$$\text{Stiffness of } CB = 3EI/L = 3\text{ K}$$

$$\text{of } CD = \frac{3EI}{1.5L} = 2\text{ K}$$

$$\text{D.F for } CB = 3/5 = 0.6$$

$$\text{for } CD = 2/5 = \frac{0.4}{1.00}$$

Table 14.4. Moment Distribution

Joint	A	B		C		D
Member	*AB*	*BA*	*BC*	*CB*	*CD*	*DC*
D.F.				0.6	0.4	
C.O.F.	½ ←			→ ½		← ½
F.E.M.		1.00	– 1.00			
C.O.	0.50			–0.50		
				+030	+0.20	
Final Moment	+0.50	+ 1.00	– 1.00	–0.20	+0.20	0.00

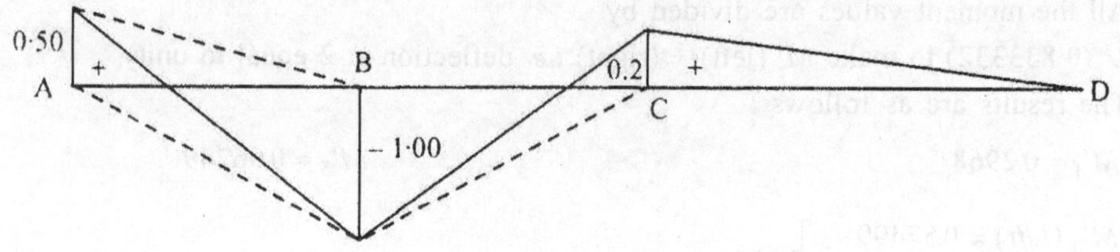

(*a*) B.M. diagram

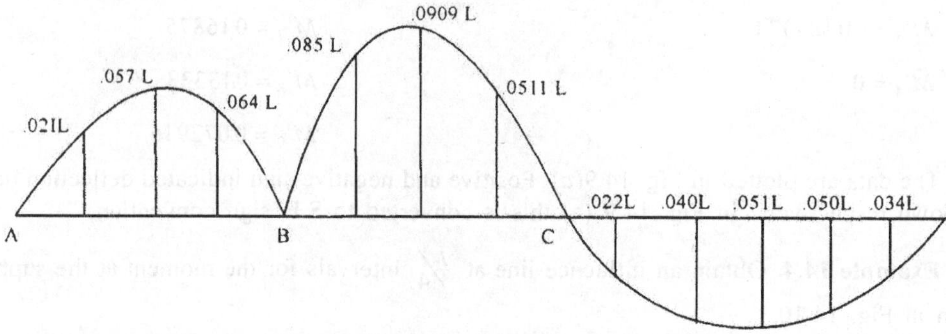

(*b*) I.L diagram

Fig. 14.11. Example 14.4

Calculation of slope at *B* (*EI* = Unity)

$$\theta_{BA} = \text{Change of slope between } A \text{ and } B$$

$$= \text{area of the } M/_{EI} \text{ diagram between } A \text{ and } B$$

$$= +\frac{1}{2} \times 0.50 \times L - \frac{1}{2} \times 1.00 \times L$$

$$= -0.25L \text{ (clockwise)}$$

$$\theta_{BC} = \frac{1 \times L}{3EI} - \frac{0.2 \times L}{6EI} = \frac{1.8L}{6EI} = 0.3L \text{ (anticlockwise) } EI = 1$$

Change in slope at $B = -0.25L - (0.3L) = -0.55L$

Deflection for span *AB* at a distance *x*.

$$= \frac{1}{2} \times \frac{(-1)}{L} \times x \times x \times \frac{x}{3} + \frac{1}{2} \times \frac{0.50}{L} \times (L-x) \times x \times \frac{x}{3} + \frac{1}{2} \times 0.50 \times x \times \frac{2x}{3}$$

$$= -\frac{x^3}{6L} + \frac{1}{4L} \cdot \frac{x^2}{3}(L-x) + \frac{1}{2} \cdot \frac{x^2}{3}$$

$$= -\frac{x^3}{6L} + \frac{x^2}{12} - \frac{x^3}{12L} + \frac{x^2}{6}$$

$$= \frac{x^2}{4}\left(1 - \frac{x}{L}\right)$$

Fig. 14.12 (*a*) Deflection of *AB*

Deflection for unit rotation

For unit rotation expression becomes

$$y = \frac{x^2}{4 \times (-0.55)} \cdot \frac{1}{L}\left(1 - \frac{x}{L}\right)$$

$$= \frac{x^2}{-2.2L} \cdot \left(1 - \frac{x}{L}\right)$$

Values at different points

At $x = 0$ deflection $= 0$

$x = L/4$ $\quad y = \dfrac{L^2}{16 \times (-2.2) \cdot L}\left(1 - \dfrac{1}{4}\right)$

$= \mathbf{0.021L}$ **(neglecting - sign).**

$x = L/2$ $\quad y = \dfrac{L^2}{4(-2.2) \cdot L} \cdot \dfrac{1}{2}$

$= \mathbf{0.057L}$

$x = \dfrac{3L}{4}$ $\quad y = \dfrac{9L^2}{16(-2.2) \cdot L} \cdot \left(\dfrac{1}{4}\right)$

$= \mathbf{.064L}$

$x = L$ $\quad y = 0$

Deflection for span BC For unit moment at B

$$y = \frac{1.(L-x)}{6EIL} \cdot \left(L^2 - (L-x)^2\right)$$

$$= \frac{L-x}{6EIL}\left(2Lx - x^2\right)$$

with $EI = 1$

$$y = -\frac{L-x}{6L}(2Lx - x^2).$$

Fig. 14.12 (*b*) Deflection of BC

For 0.20 moment at C $\quad y = \dfrac{(0.20)x(L^2 - x^2)}{6L}$

Resultant deflection

$$y = \frac{1.8L^2x - 3Lx^2 + 1.2x^3}{6L}$$

For unit rotation resultant deflection

$$= \frac{1.8L^2x - 3Lx^2 + 1.2x^3}{6 \times 0.55L^2}$$

Values at different points

At $x = L/4$ $\quad y = 0.08522L$

$x = L/2$ $\quad y = 0.0909L$

$x = \dfrac{3L}{4}$ $\quad y = 0.051136L$

Deflection for *CD*

$$y = \frac{Mo.x}{6EIL \times 1.5} \cdot \left((1.5L)^2 - x^2\right)$$

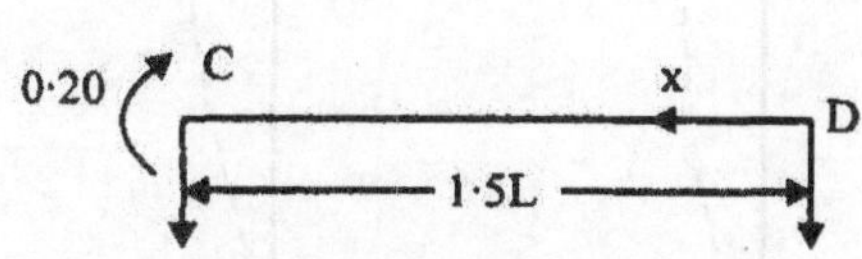

Fig. 14.12 (*c*) Deflection of CD

$$= \frac{0.2.x}{6EI \times 1.5L} \cdot \left((1.5L)^2 - x^2\right)$$

with $EI = 1$ $\qquad y = \frac{0.2}{6L \times 1.5}\left((1.5L)^2 x - x^3\right)$

For unit rotation, $\qquad y = \frac{0.2(2.25L^2 x - x^3)}{9L \times 0.55L}$

Table 14.5. Deflection of CD

At	$x = L/4$	$y = .022L$
	$= L/2$	$y = .040L$
	$= 3L/4$	$y = .051L$
	= L	$y = .050L$
	$= 5L/4$	$y = .034L$
	= 1.5L	$y = 0$

14.2 INFLUENCE LINE DIAGRAMS FOR PLANE FRAMES

Müller Breslau's principle can be utilised to compute the influence line ordinates of a plane frame. Consider a fixed portal ABCD (Fig. 14.13 (*a*))

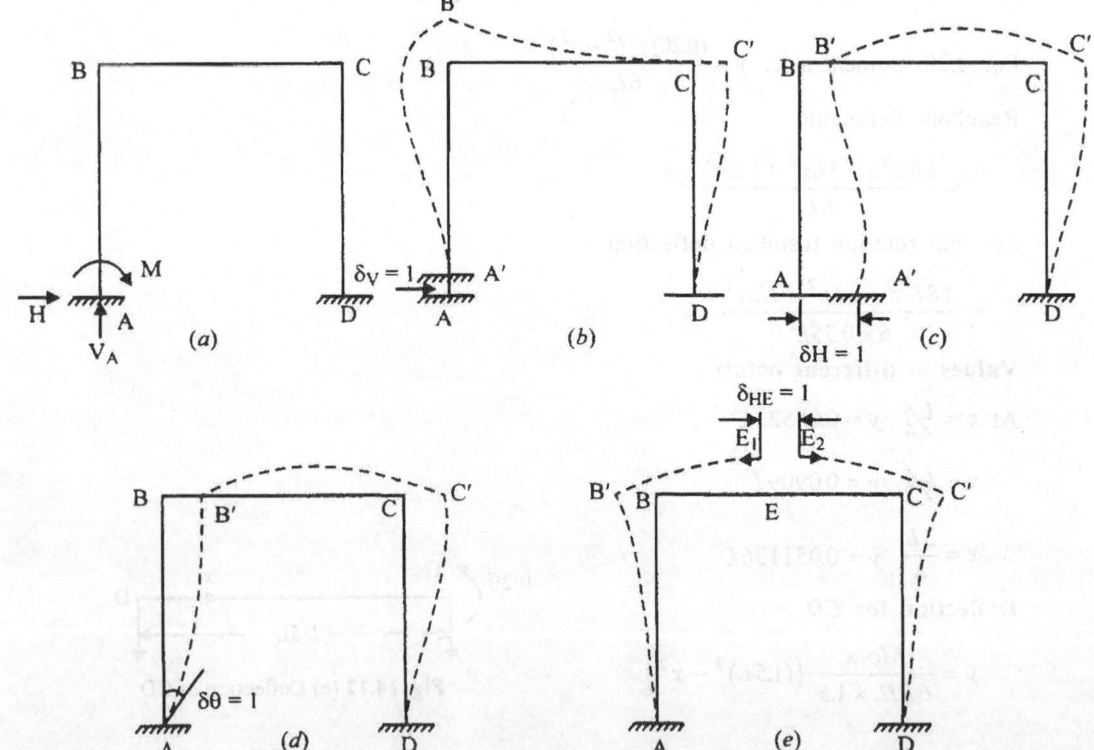

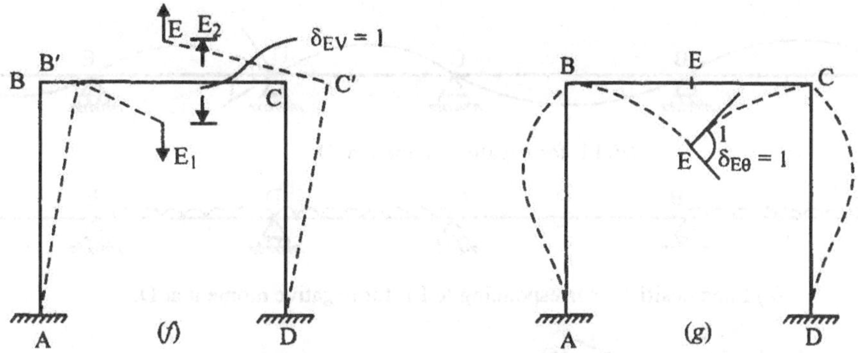

Fig. 14.13. Influence lines for a rigid frame by the Müller Breslau's principle

Influence line for V_A is obtained by moving joint A vertically, by unit distance, keeping end A fixed against rotation and against horizontal displacement. The resultant shape is shown by dotted line in Fig. 14.13 (*b*).

Similarly influence line for horizontal reaction is obtained by making horizontal movement of joint A by unity Fig. 14.13 (*c*)

Similarly for rotation joint A should be rotated through unit angle. The resultant shape is given in Fig. 14.13(*d*) by dotted lines.

To find the influence line for internal forces same principle is to be followed. For example for

(*i*) axial force E — unit horizontal displacement is to be given at E. Fig. 14.13 (*e*)

(*ii*) Shear force at E — unit shear displacement is to be given Fig. 14.13 (*f*) at *E*.

and (*iii*) moment at E — unit rotation at the joint E will be necessary Fig. 14.13 (*g*).

It is important to remember that only one displacement in the direction of only one redundant quantity can be allowed at a time since it is essential to prevent other forces from doing work. When horizontal displacement is allowed at *E*, rotation as well as vertical movement of the two cut ends must be equal so that zero work is done by shear and moment at the cut ends.

14.3 QUALITATIVE INFLUENCE LINE:

Qualitative influence lines are drawn to know usually where the live loads should be placed for producing the maximum effect of a function say, B.M, S.F., Reaction or Axial force. Fig. 14.14 shows a five span continuous beam with qualitative influence line for various functions. From the qualitative influence line it is possible to find out the spans to be loaded for the maximum effects.

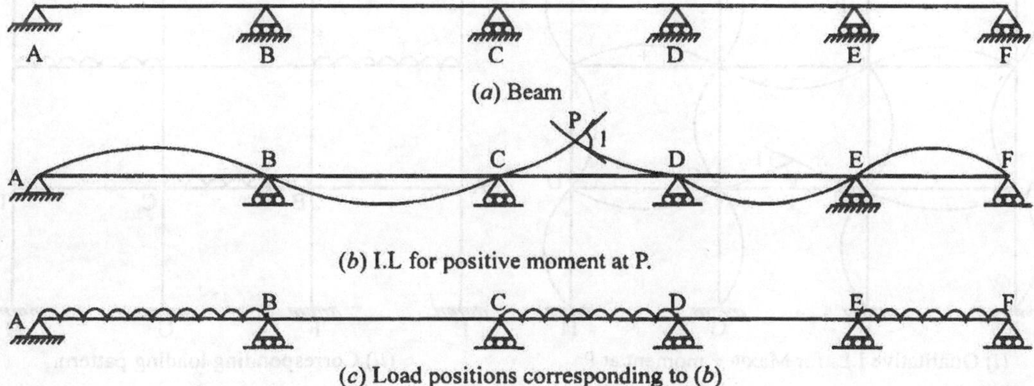

(*a*) Beam

(*b*) I.L for positive moment at P.

(*c*) Load positions corresponding to (*b*)

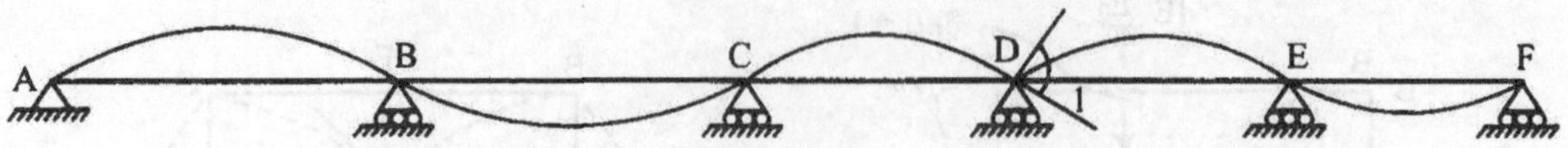

(*d*) I.L for negative moment at D.

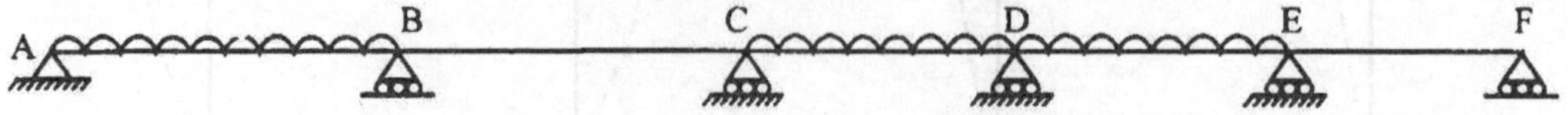

(*e*) Load positi[illegible] corresponding to I.L for negative moment at D.

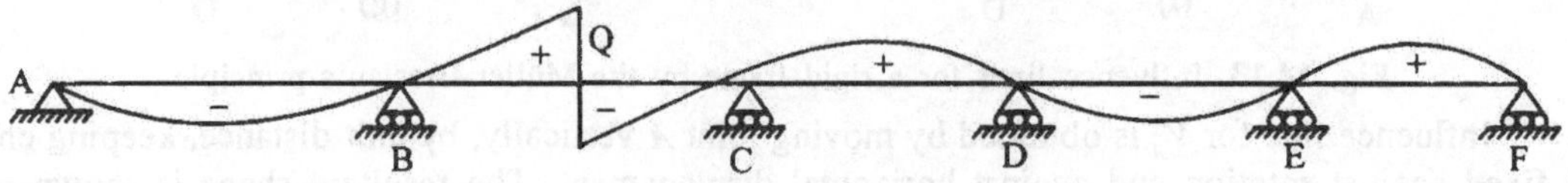

(*f*) I.L for shear at Q.

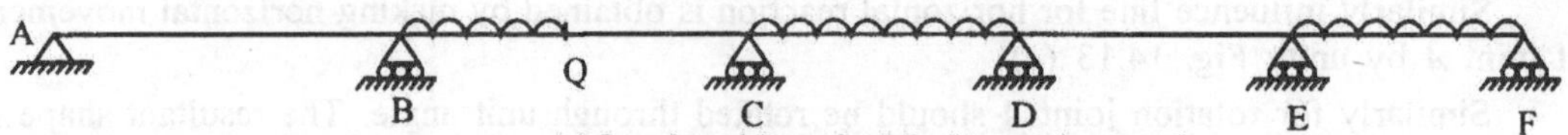

(*g*) Load positions for Maximum shear at Q.

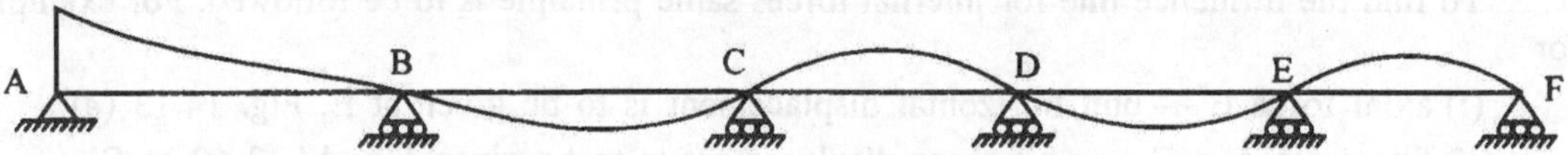

(*h*) I.L for R_A

(*i*) Load positions for maximum value of R_A.

Fig. 14.14. qualitative Influence Line.

In the analysis of multistorey building frames the qualitative influence line method is very useful in determining the spans which should be loaded for maximum effect.

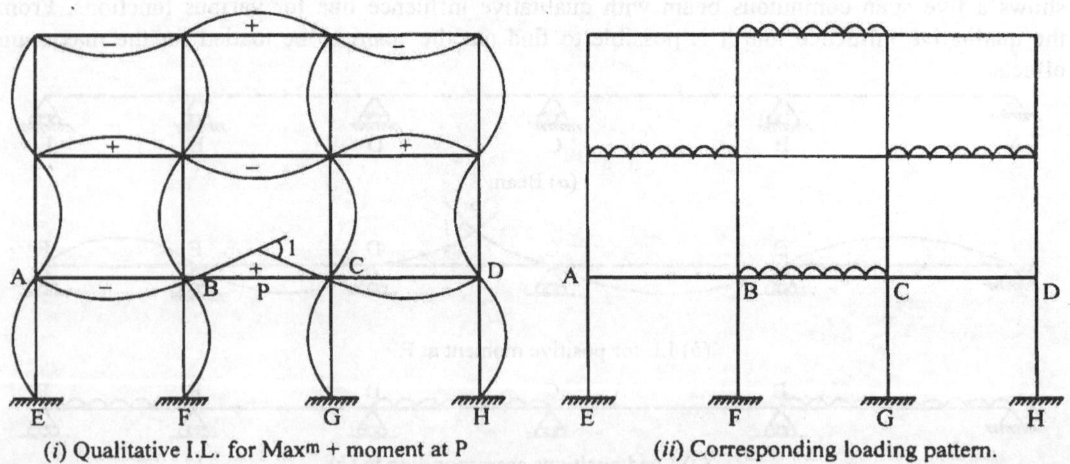

(*i*) Qualitative I.L. for Max^m + moment at P (*ii*) Corresponding loading pattern.

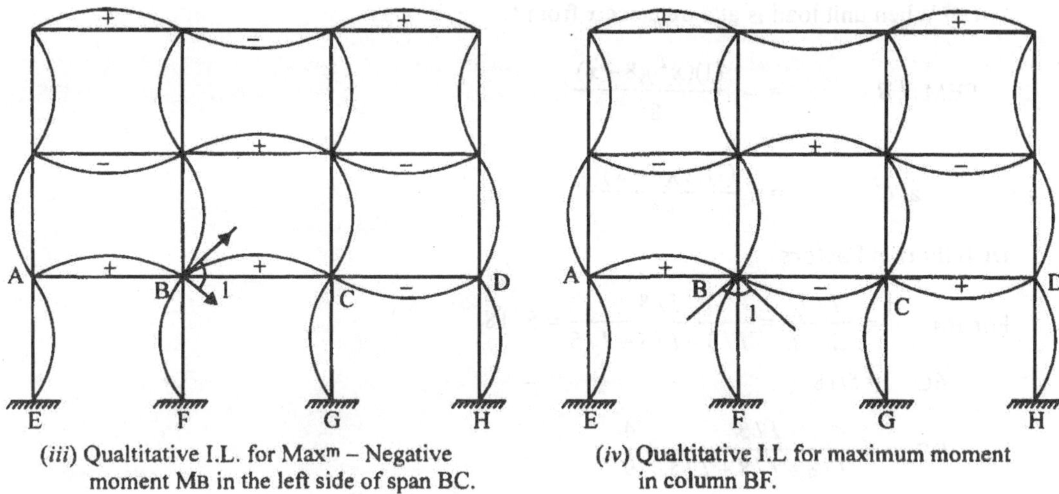

(*iii*) Qualtitative I.L. for Max^m – Negative moment M_B in the left side of span BC.

(*iv*) Qualtitative I.L for maximum moment in column BF.

Fig. 14.15. Qualitative Influence line for building frame.

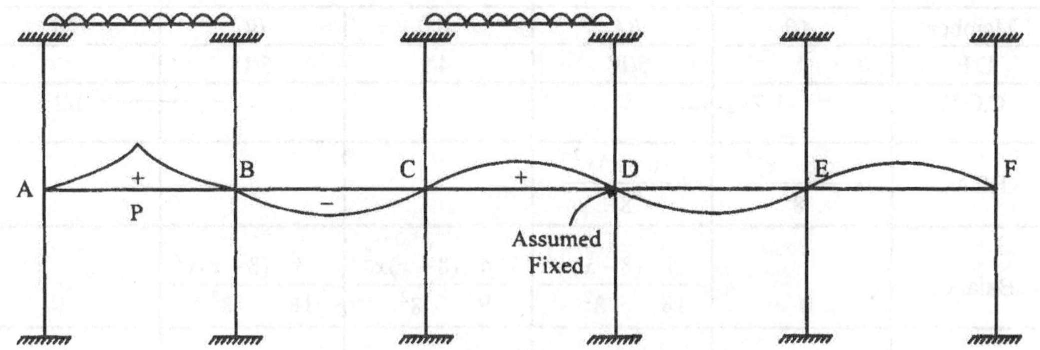

Fig. 14.16. Substitute frame for Multistorey building

Example 14.5 : The members of a rigid jointed frame shown in Fig. 14.17 are of uniform cross section and material through out. Find the influence line for bending moment in the beam at *B*. Use this influence line to find the approximate maximum value of the bending momont at *B* due to the passage from *A* to *C* of two axle loads of 2kN at 2m apart.

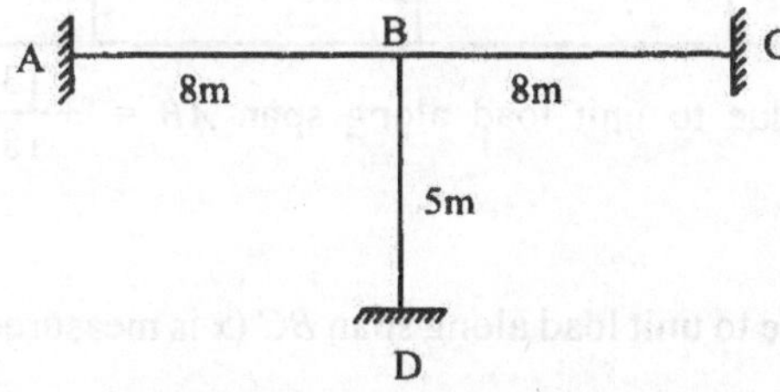

Fig. 14.17. Example 14.5

Solution : (*i*) When unit load is at a distance *x* from *A*

$$\text{FEM at } A = -\frac{(1)(x)(8-x)^2}{8^2}$$

$$\text{FEM at } b = +\frac{(1)(x^2)(8-x)}{8^2}$$

(*ii*) When unit load is at a distance x from C

$$\text{FEM at } B = -\frac{(1)(x^2)(8-x)}{8^2}$$

$$\text{at } C = +\frac{(1)(x)(8-x)^2}{8^2}$$

Distribution Factors

$$\text{For } BA = \frac{I/L}{\Sigma I/L} = \frac{I/8}{I/8+I/8+I/5} = 5/18$$

$$BC = 5/18$$

$$BD = \frac{I/5}{I/8+I/8+I/5} = \frac{4}{9}.$$

Table 14.6 (Moment distribution Example 14.5)

Joint	A	B			C
Member	AB	BA	BD	BC	CB
D.F.		5/18	4/9	5/18	
C.O.F.	1/2 ←				→ 1/2
F.E.M.	$-\frac{x(8-x)^2}{8^2}$	$+\frac{(8-x)x^2}{8^2}$			
Balance		$-\frac{5}{18}\cdot\frac{(8-x)x^2}{8^2}$	$-\frac{4}{9}\cdot\frac{(8-x)x^2}{8^2}$	$-\frac{5}{18}\cdot\frac{(8-x)x^2}{8^2}$	
C.O.	$-\frac{5}{36}\cdot\frac{(8-x)x^2}{8^2}$				$-\frac{5}{36}\cdot\frac{(8-x)x^2}{8^2}$
Final Moment		$+\frac{13}{18}\cdot\frac{(8-x)x^2}{8^2}$		$-\frac{5}{18}\cdot\frac{(8-x)x^2}{8^2}$	

I.L ordinate for M_{BA} due to unit load along span AB = $+\frac{13}{18}\cdot\frac{(8-x)x^2}{8^2}$ (x is measured from A)

I.L. ordinate for M_{BA} due to unit load along span BC (x is measured from C) = $-\frac{5}{18}\cdot\frac{(8-x)x^2}{8^2}$

Let the live load move from left to right

If the forward axle is xm from A in the span AB, the near axle will be $(x-2)$ m from A in span AB.

$\therefore$ M_{BA} due to two axle loads of 2kN each

$$(2)\left[\frac{13}{18}\cdot\frac{(8-x)\cdot x^2}{8^2}\right]+(2)\left[\frac{13}{18}\frac{[8-(x-2)](x-2)^2}{8^2}\right]$$

$$= 2\times\frac{13}{18\times 8^2}\left[(8-x)x^2+[8-(x-2)](x-2)^2\right]$$

Max^m value of M_{BA} occurs when

$$\frac{\partial M_{BA}}{\partial x} = 0$$

or $2x(8-x)+x^2(-1)$

$+\ [8-(x-2)]2(x-2)$

$+\ (x-2)^2(-1) = 0$

or $16x-2x^2-x^2+16(x-2)-2(x-2)^2-(x-2)^2 = 0$

or $-6x^2+44x-44 = 0$ Solving $x = 6\cdot 138$m or $1\cdot 236$m

Smaller value is not admissible as then the near axle will not be in the span.

Value of Max^m B.M.

$$= \frac{2\times 13}{18\times 8^2}\left[(8-6\cdot 138)(6\cdot 138)^2+(10-6\cdot 138)(4.138)^2\right]$$

$= \mathbf{3\cdot 075\ kNm.}$

Example 14.6. For the propped cantilever beam shown in Fig. 14.18(*a*), (*i*) determine the equation for the influence line for reaction at *B* (*ii*) draw the influence line for shear force at section *C* (*iii*) draw the influence line diagram for bending moment at section *C* (*iv*) determine the maximum reaction at B due to the passage from *A* to *B* of a triangular load shown in Fig. 14.18(*b*).

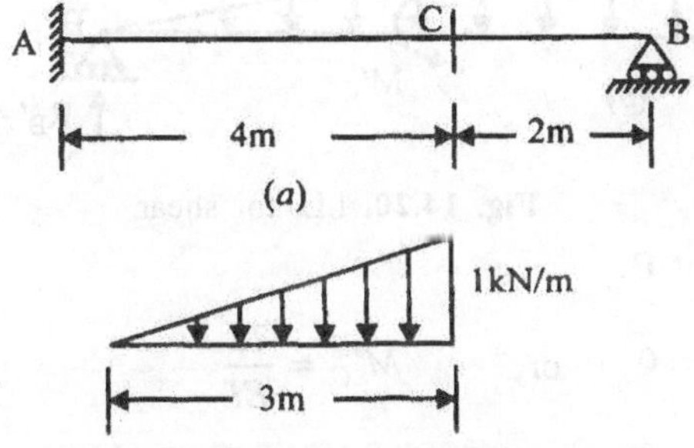

Fig. 14.18. Example 14.6.

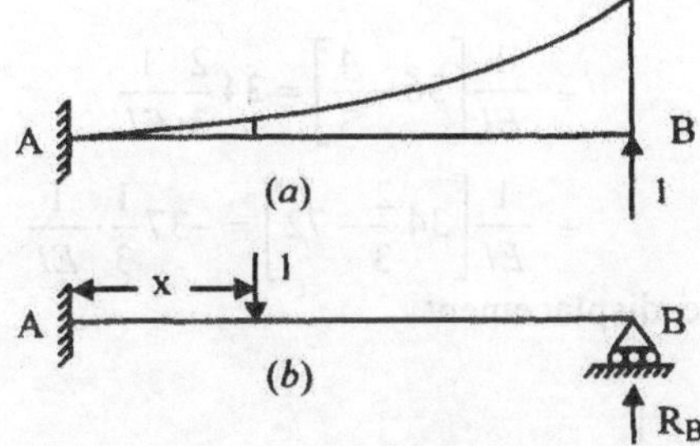

Fig. 14.19. I.L. for R_B.

Solution: (*a*) **I.L. for R_B.** Equation of elastic curve due to unit load at a distance x (Fig. 14.19)

$$y = \frac{(3L-x)\cdot x^2}{6EI}$$

At B, $x = L$ and $\Delta_B = \dfrac{L^3}{3EI}$

Applying Maxwell and Bettis' law

$$1 \times \left(-\frac{(3L-x)x^2}{6EI}\right) + R_B \times \frac{L^3}{3EI} = 0$$

or $$R_B = \frac{(3L-x)x^2}{2L^3}$$

when $L = 6$ $R_B = \dfrac{(18-x)\cdot x^2}{432}$

(*b*) **I.L. ordinate for shear**

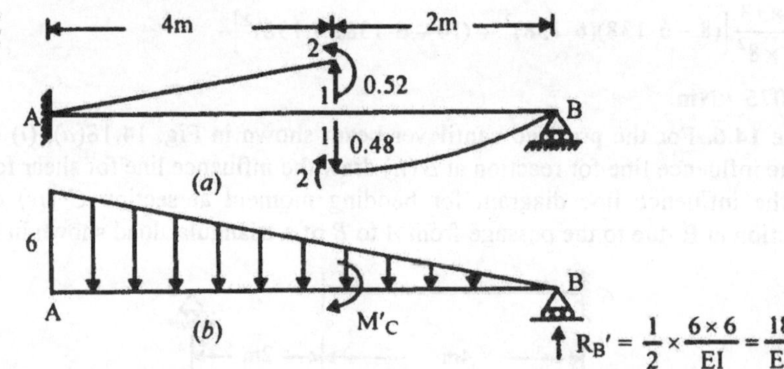

Fig. 14.20. I.L. for shear.

Taking moments about B.

$$\frac{1}{2} \times 6 \times \frac{6}{EI} \times 4 - M'_C = 0 \qquad \text{or} \qquad M'_C = \frac{72}{EI}$$

Moment to the right of C $= \dfrac{18}{EI} \times 2 - \dfrac{1}{2} \times 2 \times \dfrac{2 \times 2}{3EI}$

$$= \frac{1}{EI}\left[36 - \frac{4}{3}\right] = 34\frac{2}{3}\frac{1}{EI}$$

To the left of C $= \dfrac{1}{EI}\left[34\dfrac{2}{3} - 72\right] = -37\dfrac{1}{3} \cdot \dfrac{1}{EI}$

Ratio when the relative displacement at C = unity

$$\frac{34\frac{2}{3}}{72} = 0\cdot48, \qquad \frac{-37\frac{1}{3}}{72} = 0\cdot52.$$

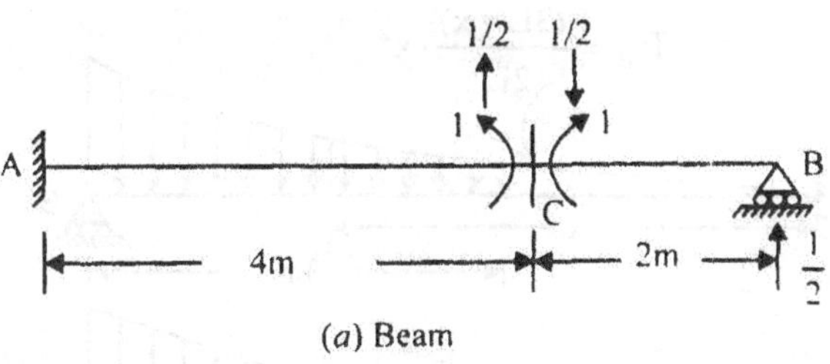

(a) Beam

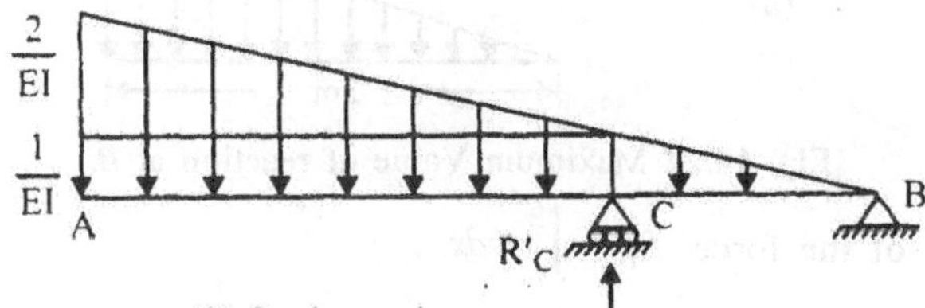

(b) Conjugate beam

(c) Influence Line for moment at C.

Fig. 14.21. I.L for Moment at C.

Solution : $R'_C \times 2 = \left[1\times 4\times(2+2)+\frac{1}{2}\times 2\times 4\times\left(2+\frac{8}{3}\right)+\frac{1}{2}\times 1\times 2\times\frac{4}{3}\right]\frac{1}{EI}$

$$= \frac{1}{EI}\left[16+\frac{56}{3}+\frac{4}{3}\right]=\frac{108}{3EL}$$

$R'_C = 18/\text{EI}$

Moment at C $= \left[1\times 4\times 2+\frac{1}{2}\times 2\times 4\times\frac{8}{3}\right]\frac{1}{EI}$

$$= \frac{1}{EI}\left[8+\frac{32}{3}\right]=\frac{56}{3EI}=\frac{18\cdot 66}{EI}$$

Ratio $= \dfrac{18\cdot 667}{18} = 1\cdot 037$

Moment at

1m from left $\dfrac{1}{EI}+\dfrac{5}{2}\times\dfrac{1}{2}+\dfrac{1}{EI}\times\dfrac{1}{2}\times\dfrac{1}{2}\times\dfrac{2}{3} = \dfrac{17}{12EI}$

Ratio $= 0.0787$

2m from left $= \dfrac{1}{EI}\times 2\times 2+\dfrac{1}{2EI}\times\dfrac{2\times 2\times 2}{3} = \dfrac{16}{3EI}$

Ratio $= 0.296$

(*d*)

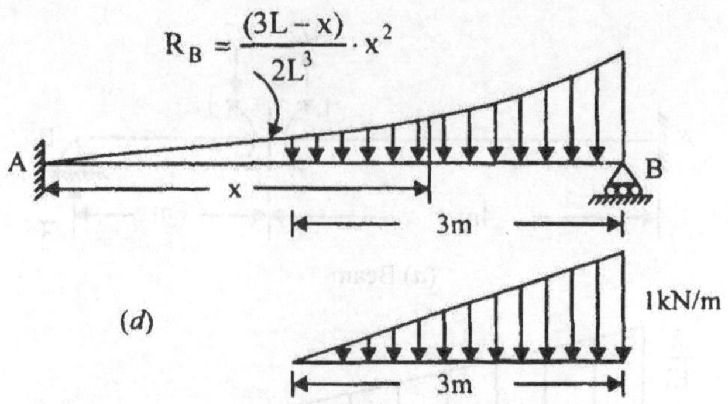

Fig. 14.22 Maximum Value of reaction at *B*.

Maximum value of the force $R_B = \int_3^6 ywdx.$

$$= \int_3^6 \frac{(3L-x)x^2}{2L^3}\cdot\frac{1}{3}(x-3)dx$$

with $L = 6$

$$= \frac{1}{6L^3}\int_3^6 x^2(18-x)(x-3)dx\cdot$$

$$= \frac{1}{6\times 216}\int_3^6 (21x^3 - x^4 - 54x^2)dx$$

$$= \frac{1}{6\times 216}\left[21\cdot\frac{x^4}{4} - \frac{x^5}{5} - 54\cdot\frac{x^3}{3}\right]_3^6$$

$$= \frac{1}{6\times 216}[6378\cdot 75 - 1506\cdot 6 - 3402]$$

$= \mathbf{1\cdot 1344\ kN.}$

14.4 INFLUENCE LINE FOR INDETERMINATE TRUSS

Influence lines for statically indeterminate trusses are obtained in the usual manner by systematically placing the unit load at each panel point. The truss analysis required for each position of the unit load can be done by (*i*) consistent deformation method or (*ii*) energy method. Two examples are solved, one on internal indeterminacy and the other on external indeterminacy.

Example 14.7. Construct the influence lines for members *cf*, *dc* and *df* if vertical loads are applied at the bottom chord. AE = constant for all members.

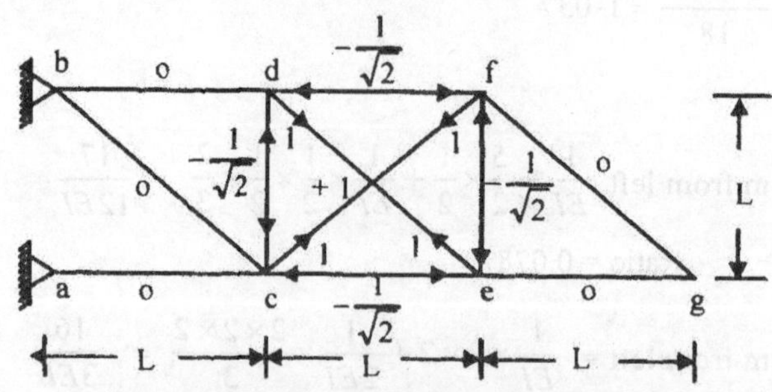

Fig. 14.23. F_Q forces (Example 14.7)

Solution : Take *cf* as redundant. With unit force along *cf*, forces F_Q in all members are determined. F_{fg} and $F_{eg} = 0$.

$F_{fe} = -\frac{1}{\sqrt{2}}$, $F_{fd} = -\frac{1}{\sqrt{2}}$. $F_{ac} = 0$ $F_{bd} = F_{bc} = 0$ $F_{de} = 1$ $F_{dc} = -\frac{1}{\sqrt{2}}$ $F_{ce} = -\frac{1}{\sqrt{2}}$ The forces are shown in Fig. 14.23.

Unit Vertical load at *C*

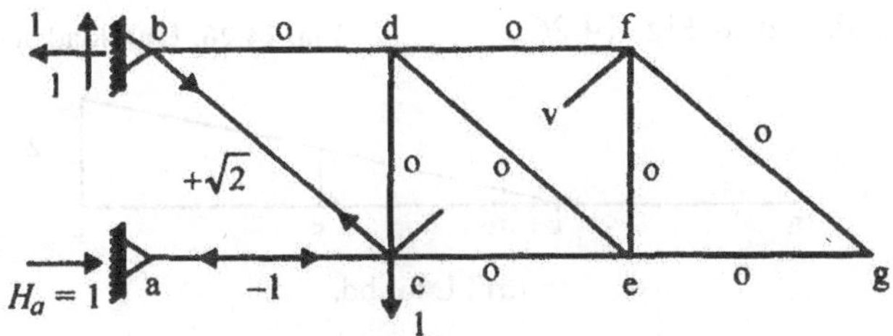

Fig. 14.24. Unit vertical load at C.

Taking moments about *b*, $1 \times L - H_a \times L = 0$

or $H_a = 1$

$$F_{bc} \cdot \frac{1}{\sqrt{2}} = 1$$

$$F_{bc} = +\sqrt{2}\,.$$

$$F_{bc} \cdot \frac{1}{\sqrt{2}} - 1 + F_{bd} = 0$$

or $F_{bd} = 0$

Similarly $F_{cd} = 0$

and $F_{de} = F_{df} = F_{fg} = 0$

And also $F_{ce} = F_{eg} = 0$

F_p forces are given in Fig. 14.24.

Unit vertical load at *e*.

Taking moments about b,

$$1 \times 2L - H_a \times L = 0$$

$$H_a = 2$$

The forces in members due to this load are shown in Fig. 14.25

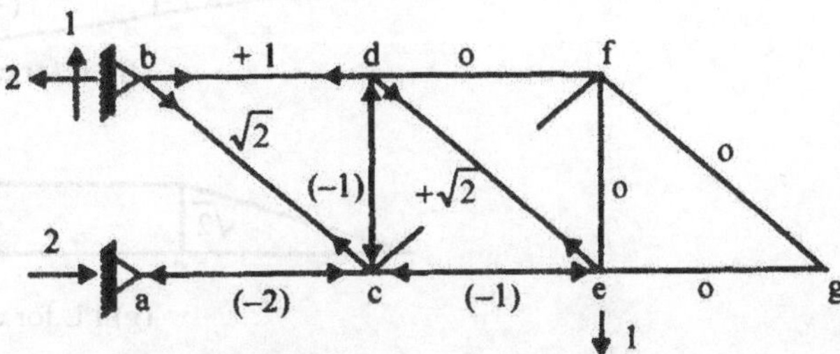

Fig. 14.25. Unit vertical load at *e*.

Unit vertical load at *g*.

Taking moments about *b*,

$$1 \times 3L = H_a \times L$$

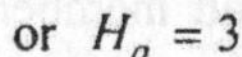

or $H_a = 3$

Resolving vertically

$$F_{bc} \cdot \frac{1}{\sqrt{2}} = 1$$

$$F_{bc} = \sqrt{2}$$

Other forces are shown in Fig. 14.26

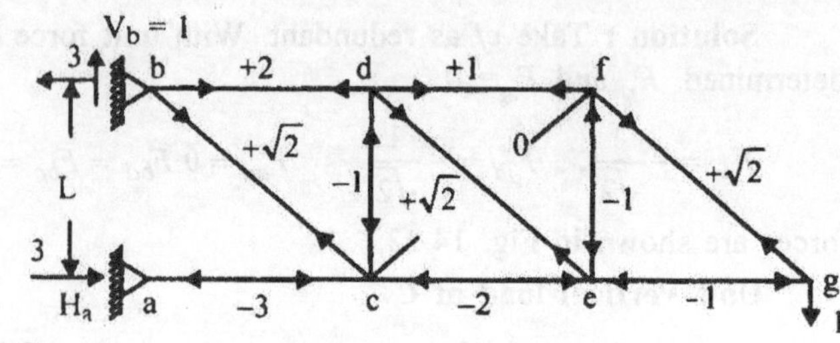

Fig. 14.26. Unit load at *g*

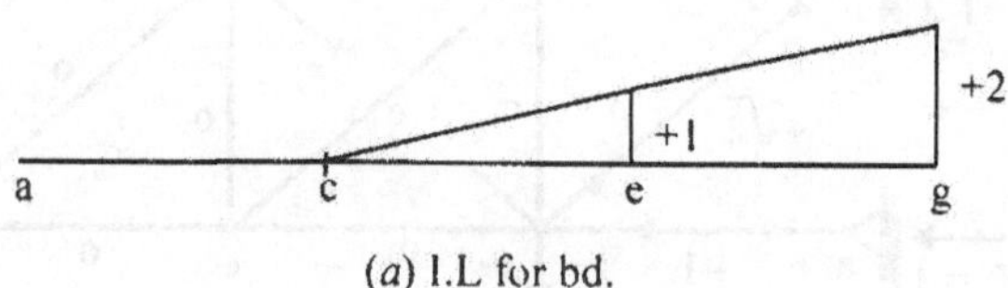

(*a*) I.L for bd.

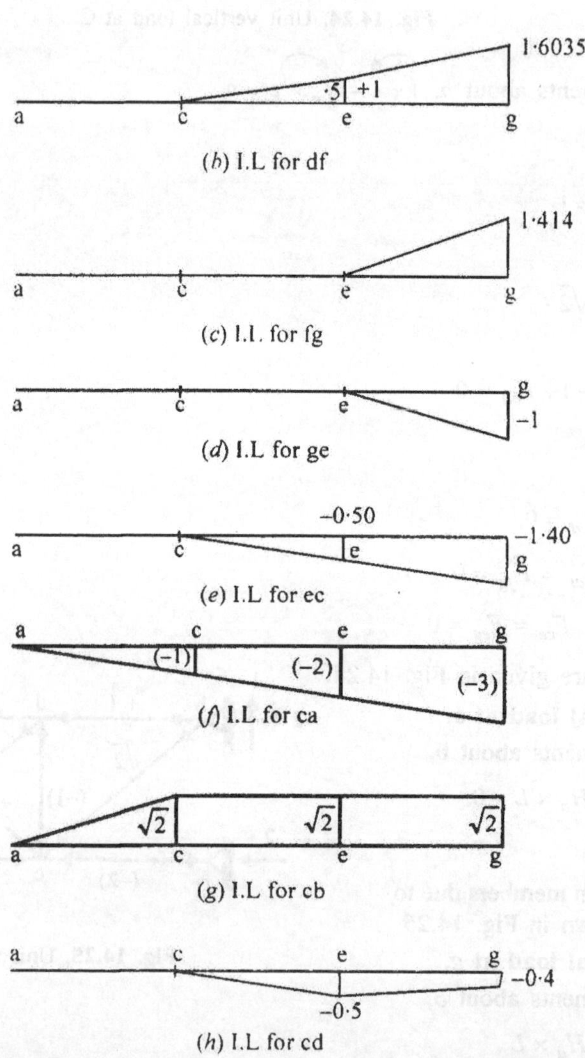

(*b*) I.L for df

(*c*) I.L. for fg

(*d*) I.L for ge

(*e*) I.L for ec

(*f*) I.L for ca

(*g*) I.L for cb

(*h*) I.L for cd

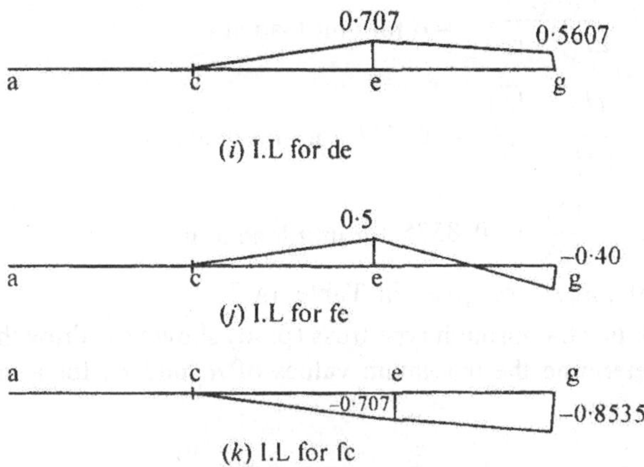

Fig. 14.27. Influence line diagrams (Example 14.7)

Table 14.7

Member	L	F_Q	$F_Q^2.L$	F_P Unit	Load	at	F_PF_QL Unit	Load	at	$F_P + XF_Q$ Unit	load	at
				c	e	g	c	e	g	c	e	g
bd	L	0	0	0	+1	+2	0	0	0	0	+1	+2
df	L	$-\frac{1}{\sqrt{2}}$	$\frac{L}{2}$	0	0	+1	0	0	$-\frac{L}{\sqrt{2}}$	0	+0·499	+1·6035
fg	$\sqrt{2}L$	0	0	0	0	$+\sqrt{2}$	0	0	0	0	0	+1·414
ge	L	0	0	0	0	–1	0	0	0	0	0	–1
ec	L	$-\frac{1}{\sqrt{2}}$	L/2	0	–1	–2	0	$\frac{L}{\sqrt{2}}$	$\sqrt{2}L$	0	–0·499	–1·396
ca	L	0	0	–1	–2	–3	0	0	0	–1	–2	–3
bc	$\sqrt{2}L$	0	0	$+\sqrt{2}$	$+\sqrt{2}$	$+\sqrt{2}$	0	0	0	$+\sqrt{2}$	$+\sqrt{2}$	1·414
dc	L	$-\frac{1}{\sqrt{2}}$	L/2	0	–1	–1	0	$+\frac{L}{\sqrt{2}}$	$\frac{L}{\sqrt{2}}$	0	–0·499	–3964
de	$\sqrt{2}L$	+1	$\sqrt{2}L$	0	$+\sqrt{2}$	$+\sqrt{2}$	0	2L	2L	0	+0·707	0·5607
fe	L	$-\frac{1}{\sqrt{2}}$	L/2	0	0	–1	0	0	$\frac{L}{\sqrt{2}}$	0	+0·499	–·3964
fc	$\sqrt{2}L$	+1	$\sqrt{2}L$	0	0	0	0	0	0	0	–0·707	–0·8535

$\Sigma 2L(1+\sqrt{2})$ $\quad\quad$ $\Sigma 0 \quad \Sigma L(2+\sqrt{2}) \cdot \quad \Sigma 4 \cdot 121L$

$$\Delta_{xo} = \frac{1}{AE}\Sigma F_p\, F_Q . L$$

$$\delta_{xx} = \frac{1}{AE}\Sigma F_Q^{\,2} \cdot L$$

$$X = -\frac{\Sigma F_P F_Q . L}{\Sigma F_Q^2 \cdot L} = 0 \text{ for unit load at C}$$

$$= -\frac{L(2+\sqrt{2})}{2L(1+\sqrt{2})} = -0 \cdot 707 \text{ for unit load at e}$$

$$= -\frac{4 \cdot 121}{4 \cdot 828} = -0 \cdot 8535 \text{ for unit load at g.}$$

Final values of forces are given in Table 14.7

Example 14.8. For the through type truss (pratt) shown (*a*) draw the influence line diagrams for R_B and R_A (*b*) determine the maximum values of R_B and R_A for a moving trailor load shown in Fig. 14.28 (*b*).

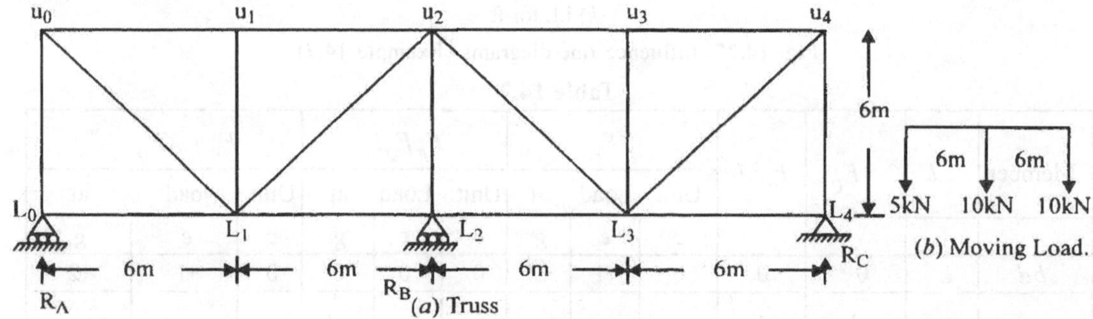

Fig. 14.28. Example 14.8

Solution: The truss is indeterminate to the 1st degree. Let the reaction R_B be the redundant quantity, we remove it and introduce a unit force in the line of R_B as shown in Fig. 14.28(*a*). To find the influence line for R_B we require the deflections of joints L_1 or L_3 and L_2. These forces are found by the unit load applied at L_2 and L_1. These are given in Table 14.8 Fig. 14.29(*a*) gives the forces in members due to unit load at L_2(K_1 forces) Fig. 14.29(*b*) gives K_2 forces *i.e.* forces in members due to unit load at L_1. Fig. 14.29(*c*) gives the deflection diagram when unit load is placed at L_2. Calculations for computation of deflection are shown in Table 14.8. Influence line for R_B is obtained by dividing the ordinates of Fig. 14.29(*c*) by the deflection at *B i.e.* by making the ordinate at *B* equal to unity. This is shown in Fig. 14.29(*d*). Influence line for R_A can be obtained by utilising the results of influence line for R_B.

Influence line for R_A

(*i*) When unit load is at *A*, $R_A = 1$

(*ii*) When unit load is at L_1, taking moments about L_4.

$$R_A \times 24 - 1 \times 18 + R_B \times 12 = 0$$

or $$R_A = \frac{1}{24}[18 - R_B \times 12]$$

with $$R_B = 0 \cdot 534$$

$$R_A = \frac{1}{24}[18 - 0 \cdot 534 \times 12]$$

$$= 0 \cdot 483.$$

(*iii*) Unit load at R_B, $R_A = 0$

(iv) Unit load at L_3

$$R_A \times 24 + R_B \times 12 - 1 \times 6 = 0$$

or $$R_A \times 24 + 0{\cdot}534 \times 12 - 1 \times 6 = 0$$

or $$R_A = \frac{1}{24}[6 - 6{\cdot}408]$$

$$= -\frac{{\cdot}408}{24} = -{\cdot}017$$

From Maxwell's Theorem

deflection at L_1 when unit load is at L_2

= deflection at L_2 when unit load is at L_1

$$= \frac{\sum K_1 K_2 L}{AE} = \frac{23{\cdot}485}{AE}$$

$$\frac{\sum K_1^2 L}{AE} = \frac{43{\cdot}97}{AE}.$$

(b) Maximum values are obtained

(i) for R_B when central load is placed over L_2.

Value $= 5 \times {\cdot}534 + 10 \times 1{\cdot}00 + 10 \times {\cdot}534$

$= 17{\cdot}97$ upward

(ii) for R_A when two 10kN inside span

value $= 10 \times 1 + 10 \times 0{\cdot}483 = 14{\cdot}83$ kN ↑

Table 14.8

Member	K_1 forces	K_2 forces	L	LK_1K_2	LK_1^2
$u_0\ u_1$	−1/2	−3/4	6	$+\frac{3}{8} \times 6$	$+\frac{1}{4} \times 6$
$u_1\ u_2$	−1/2	−3/4	6	$+\frac{3}{8} \times 6$	$+\frac{1}{4} \times 6$
$u_2\ u_3$	−1/2	−1/4	6	$+\frac{1}{8} \times 6$	$+\frac{1}{4} \times 6$
$u_3\ u_4$	−1/2	−1/4	6	$+\frac{1}{8} \times 6$	$+\frac{1}{4} \times 6$
$L_o L_1$	0	0	6	0×6	0
$L_1 L_2$	+1	1/2	6	$\frac{1}{2} \times 6$	$+1 \times 6$
$L_2 L_3$	+1	1/2	6	$\frac{1}{2} \times 6$	$+1 \times 6$
$L_3 L_4$	+0	0	6	0×6	$+0 \times 6$
$u_0\ L_0$	−1/2	−3/4	6	$3/8 \times 6$	$+1/4 \times 6$

$u_0 L_1$	$\frac{1}{\sqrt{2}}$	$3\sqrt{2}/4$	$6\sqrt{2}$	$3/4 \times 6\sqrt{2}$	$+\frac{1}{2} \times 6\sqrt{2}$
$u_1 L_1$	0	0	6	0×6	0
$L_1 u_2$	$-1/\sqrt{2}$	$+\sqrt{2}/4$	$6\sqrt{2}$	$-1/4 \times 6\sqrt{2}$	$\frac{1}{2} \times 6\sqrt{2}$
$u_2 L_2$	+1	0	6	0×6	$+1 \times 6$
$u_2 L_3$	$-\frac{1}{\sqrt{2}}$	$-\sqrt{2}/4$	$6\sqrt{2}$	$+1/4 \times 6\sqrt{2}$	$+\frac{1}{2} \times 6\sqrt{2}$
$u_3 L_3$	0	0	6	0×6	0×6
$L_3 u_4$	$\frac{1}{\sqrt{2}}$	$\sqrt{2}/4$	$6\sqrt{2}$	$\frac{1}{4} \times 6\sqrt{2}$	$\frac{1}{2} \times 6\sqrt{2}$
$u_4 L_4$	$-1/2$	$-1/4$	6	$+\frac{1}{8} \times 6$	$1/4 \times 6$
				$\sum 23 \cdot 485$	$\sum 43 \cdot 97$

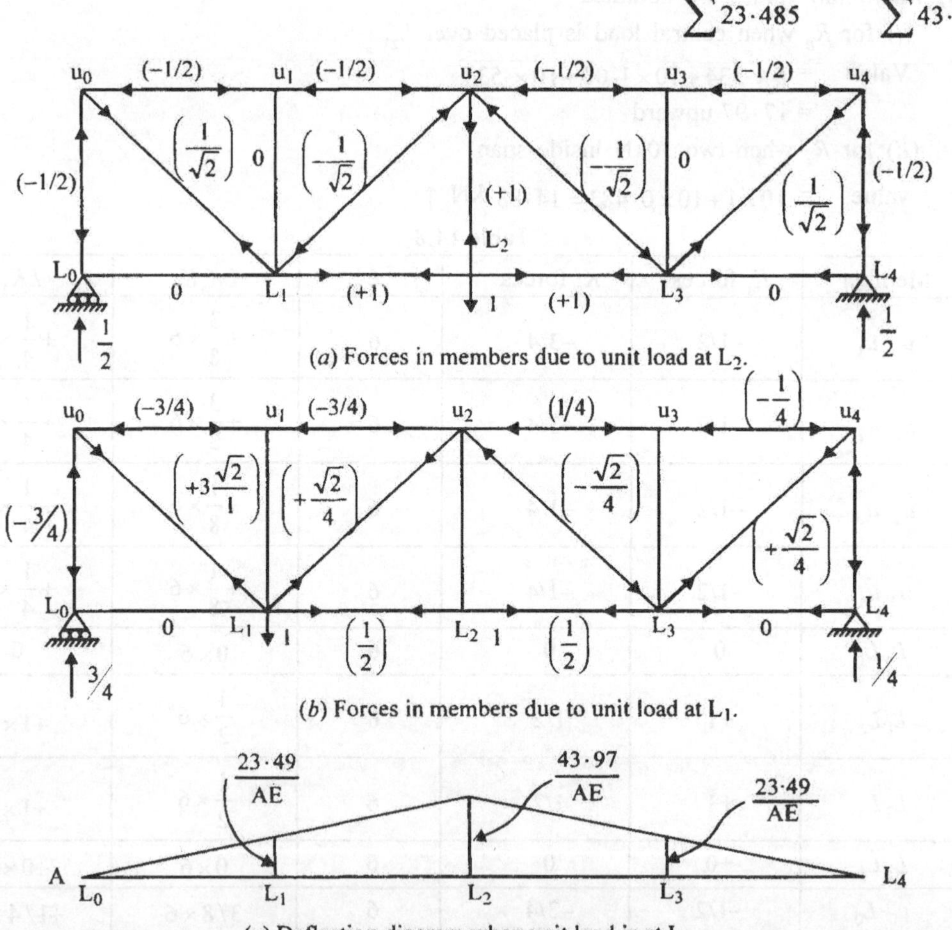

(a) Forces in members due to unit load at L_2.

(b) Forces in members due to unit load at L_1.

(c) Deflection diagram when unit load is at L_2.

1·00
0·534
0·534
L0
L1
L2
L3
L4

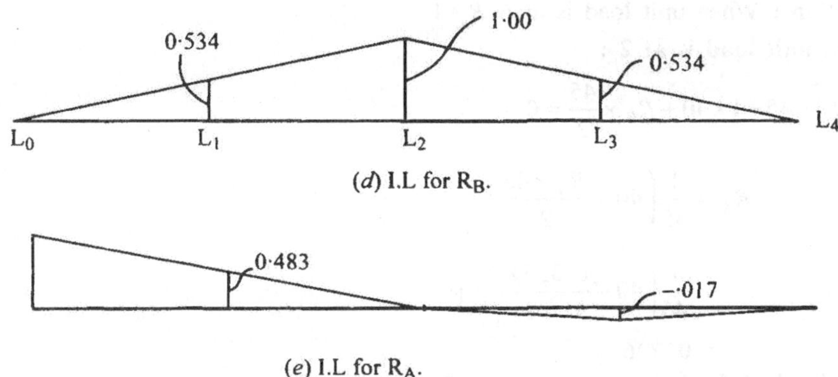

(*d*) I.L for R_B.

0·483
–·017

(*e*) I.L for R_A.

Fig. 14.29. Example 14.8 – Influence Lines for R_A, R_B.

Example 14.9. Construct the influence line diagram for member 5-7 of the two span continuous highway bridge truss shown in Fig. 14.30(*a*). The influence line for the reaction at *B* is given by Fig. 14.30(*b*). Also determine the maximum reaction at B and the maximum force in member 5-7 for the two axle loading shown in Fig. 14.30(*d*).

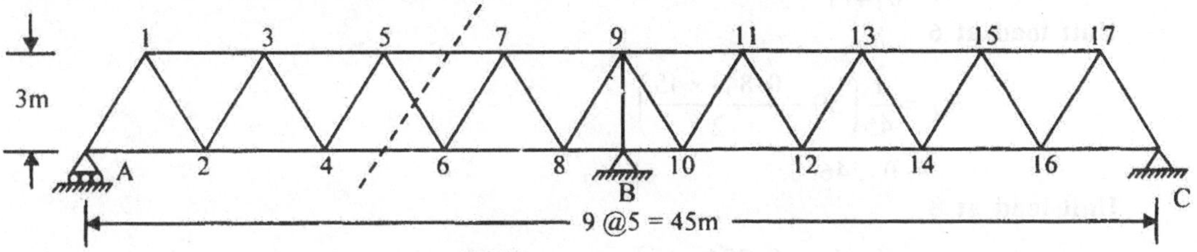

1
3
5
7
9
11
13
15
17
3m
A
2
4
6
8
10
12
14
16
B
C
9 @5 = 45m

(*a*) Truss

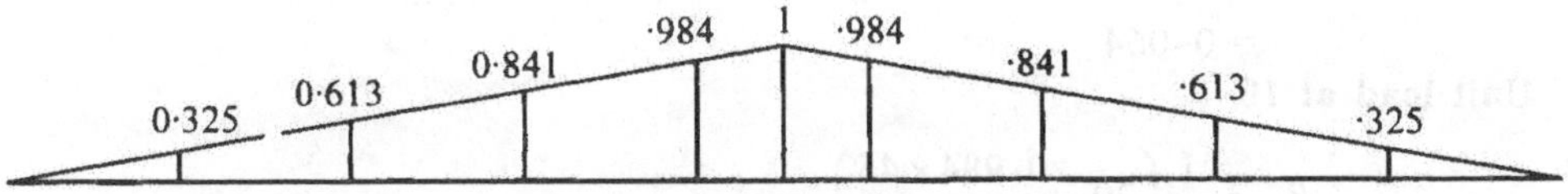

0·325
0·613
0·841
·984
1
·984
·841
·613
·325

(*b*) I.L for R_B.

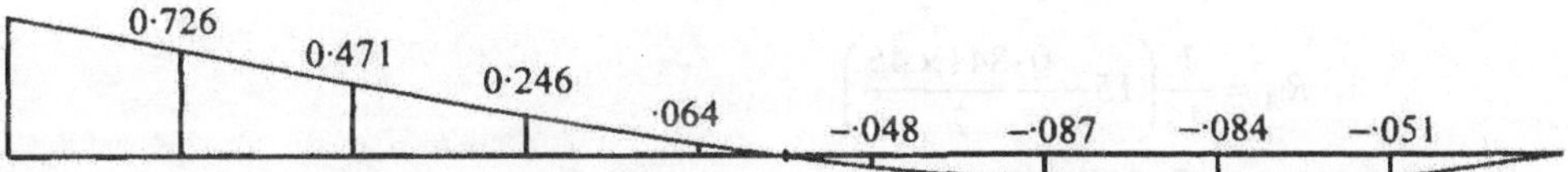

0·726
0·471
0·246
·064
–·048
–·087
–·084
–·051

(*c*) I.L for R_A.

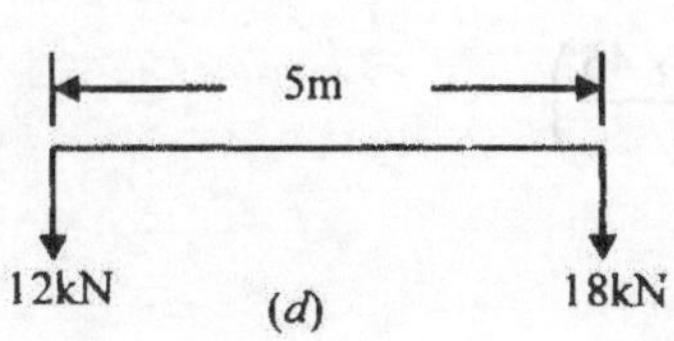

5m
12kN
18kN

(*d*)

Fig. 14.30. Example 14.9

Solution : When unit load is at A $R_A = 1$

When unit load is at 2 :

$$R_A \times 45 - 1 \times 40 + R_B \times \frac{45}{2} = 0$$

or
$$R_A = \frac{1}{45}\left(40 - \frac{R_B \times 45}{2}\right)$$

$$= \frac{1}{45}\left(40 - \frac{0.325 \times 45}{2}\right)$$

$$= 0.726$$

Unit load at 4

$$R_A \times 45 - 1 \times 35 + R_B \times \frac{45}{2} = 0$$

or
$$R_A = \frac{1}{45}\left(35 - \frac{0.613 \times 45}{2}\right)$$

$$= 0.471.$$

Unit load at 6

$$R_A = \frac{1}{45}\left(30 - \frac{0.841 \times 45}{2}\right)$$

$$= 0.246.$$

Unit load at 8

$$R_A = \frac{1}{45}\left(25 - \frac{0.984 \times 45}{2}\right)$$

$$= 0.064$$

Unit load at 10

$$R_A = \frac{1}{45}\left(20 - \frac{0.984 \times 45}{2}\right)$$

$$= 0.048.$$

Unit load at 12

$$R_A = \frac{1}{45}\left(15 - \frac{0.841 \times 45}{2}\right)$$

$$= -0.087.$$

Unit load at 14

$$R_A = \frac{1}{45}\left(10 - \frac{0.613 \times 45}{2}\right)$$

$$= -0.084.$$

Unit load at 16

$$R_A = \frac{1}{45}\left(5 - \frac{0.325 \times 45}{2}\right)$$

$$= -0.051.$$

Unit load at C $\quad R_A = 0$

Influence line for member 5-7

Force in member 5-7 is obtained by taking moments of all forces about opposite joint 6.

When unit load is at 2

$$R_A \times 15 + F_{57} \times 3 - 1 \times 10 = 0$$

$$F_{57} = \frac{1}{3}(10 - R_A \times 15)$$

$$= \frac{1}{3}(10 - 0{\cdot}726 \times 15)$$

$$= -0{\cdot}296.$$

Unit load at 4

$$R_A \times 15 + F_{57} \times 3 - 1 \times 5 = 0$$

or $$F_{57} = \frac{1}{3}(5 - R_A \times 15)$$

$$F_{57} = \frac{1}{3}(5 - 0{\cdot}471 \times 15)$$

$$= -\,0{\cdot}688$$

Unit load at 6

$$F_{57} \times 3 + R_B \times 7{\cdot}5 + R_C \times 30 = 0$$

or $$F_{57} = -\frac{1}{3}[R_B \times 7{\cdot}5 + R_C \times 30]$$

$$= -\frac{1}{3}[0{\cdot}841 \times 7{\cdot}5 + (-0{\cdot}087) \times 30]$$

$$= -1{\cdot}23.$$

Unit load at 8

$$F_{57} \times 3 - 1 \times 5 + R_B \times 7{\cdot}5 + R_C \times 30 = 0$$

$$F_{57} = \frac{1}{3}[5 - R_B \times 7{\cdot}5 - R_C \times 30]$$

$$= \frac{1}{3}[5 - 0{\cdot}984 \times 7{\cdot}5 - (-{\cdot}048) \times 30]$$

$$= -0{\cdot}3133.$$

Load at B $\quad F_{57} = 0.$

Unit load at 10

$$F_{57} \times 3 - 1 \times 10 + R_B \times 7{\cdot}5 + R_C \times 30 = 0$$

$$F_{57} = \frac{1}{3}[10 - R_B \times 7{\cdot}5 - R_C \times 30]$$

$$= \frac{1}{3}[10 - 0{\cdot}984 \times 7{\cdot}5 - ({\cdot}064) \times 30]$$

$$= 0{\cdot}233.$$

Unit load at 12

$$F_{57} = \frac{1}{3}[15 - R_B \times 7 \cdot 5 - R_C \times 30]$$

$$= \frac{1}{3}[15 - 0 \cdot 841 \times 7 \cdot 5(-0 \cdot 246) \times 30]$$

$$= 0 \cdot 438$$

Unit load at 14

$$F_{57} = \frac{1}{3}[20 - R_B \times 7 \cdot 5 - R_C \times 30]$$

$$= \frac{1}{3}[20 - 0 \cdot 613 \times 7 \cdot 5 - 0 \cdot 471 \times 30]$$

$$= \frac{1}{3}[1 \cdot 2725] = 0 \cdot 424.$$

Unit load at 16

$$F_{57} = \frac{1}{3}[25 - R_B \times 7 \cdot 5 - R_C \times 30]$$

$$= \frac{1}{3}[25 - 0 \cdot 325 \times 7 \cdot 5 - 0 \cdot 726 \times 30]$$

$$= 0 \cdot 261.$$

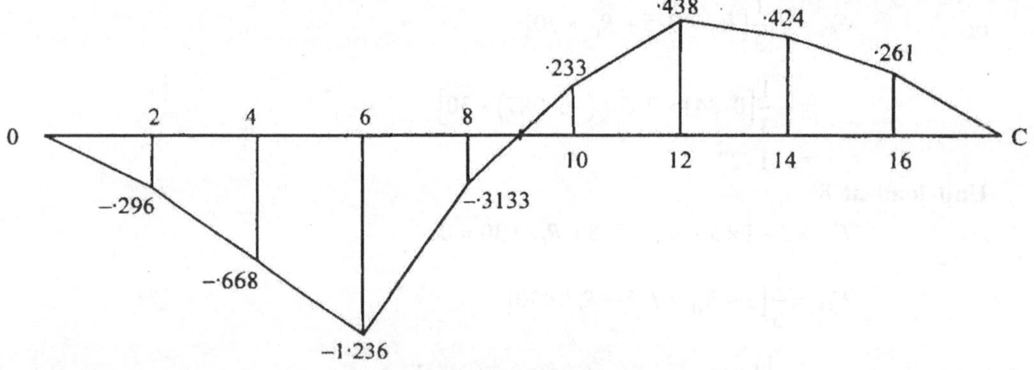

Fig. 14.31. I.L. for member 5-7

Maximum Value

R_B = (12+18) 0·984

= $30 \times 0 \cdot 984 = 29 \cdot 52$ kN.

F_{57} = $-18 \times 1 \cdot 236 - 12 \times \cdot 668$ = **– 30·268 kN.**

14.5 INFLUENCE LINE FOR TWO HINGED ARCH

Fig. 14.32 shows a two hinged arch. As already explained it is statically indeterminate to 1st degree. It is required to draw the influence line for

(*i*) Horizontal reaction

(*ii*) B.M., shear Force and Axial thrust for any point *P* (*x,y*)

(*i*) Horizontal reaction – *H* is given by the expression H = $\frac{5}{8} \cdot \frac{\alpha L}{y_c}\left(1 - L\alpha^2 + \alpha^3\right)$

where $\alpha = a/l$

The influence line can be plotted by putting $\alpha = 0{\cdot}1,\ 0{\cdot}2,\ 0{\cdot}3,\ 0{\cdot}4$, and $0{\cdot}5$. The corresponding values of H are

$0{\cdot}063\frac{L}{y_c}$, $0{\cdot}116\frac{L}{y_c}$, $0{\cdot}159\frac{L}{y_c}$, $0{\cdot}187\frac{L}{y_c}$, and $0{\cdot}195\frac{L}{y_c}$. This is shown in Fig. 14.33(*a*).

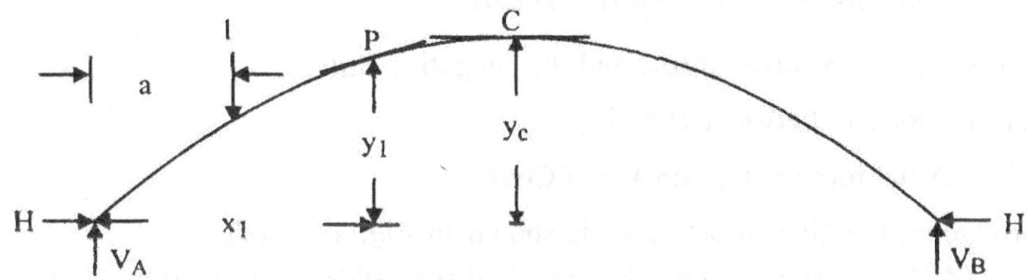

Fig. 14.32. Two hinged arch.

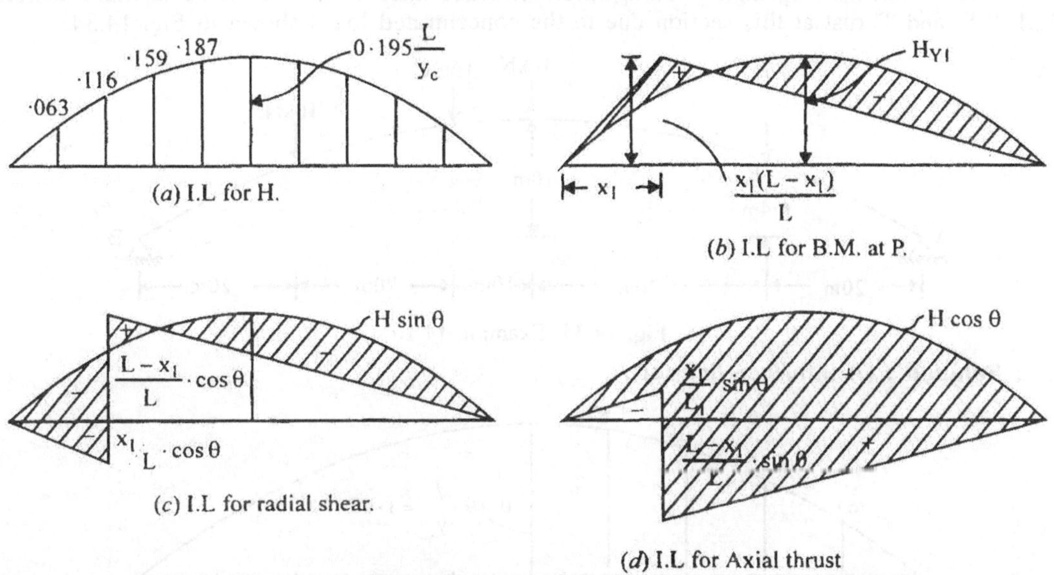

Fig. 14.33. Influence Line for two hinged arch

(*ii*) B.M. at P – B.M. at P is given by

$$M_P = \mu - Hy_1$$

Hence influence line for B.M. can be obtained by subtracting y_1 times influence line for H from the simple beam B.M. influence line at P. This is shown in Fig. 14.33(*b*).

(*iii*) Influence line for Radial shear.

When the unit load is between *A* and *P*.

$$\text{radial shear} = H \text{ Sin } \theta + V_B \text{ Cos } \theta$$

where θ is the angle between horizontal and tangent at *P*. Both the forces tend to move right hand of the section upward and hence negative. Similarly when unit load is between P and B.

$$\text{radial shear} = V_A \cos\theta - H \sin\theta$$

$V_A \cos\theta$ is positive. This is shown in Fig. 14.33(c)

(*iv*) Influence line for axial thrust.

(*a*) When unit load is between A and P

$$\text{Axial force} = -V_B \sin\theta + H \cos\theta$$

$H \cos\theta$ produces positive thrust and V_B negative thrust.

(*b*) When unit load is between P and B.

$$\text{Axial force} = V_A \sin\theta + H \cos\theta .$$

both producing positive thrust. This is shown in Fig. 14.33(*d*).

Example 14.10. A two hinged parabolic arch of span 100m a rise of 10m. Draw the influence line diagrams for Horizontal reaction bending moment, shear force and axial thrust at section F, 20m from the left hand springing. Using these influence lines determine the horizontal reaction, B.M. S.F. and Thrust at this section due to the concentrated loads shown in Fig. 14.34.

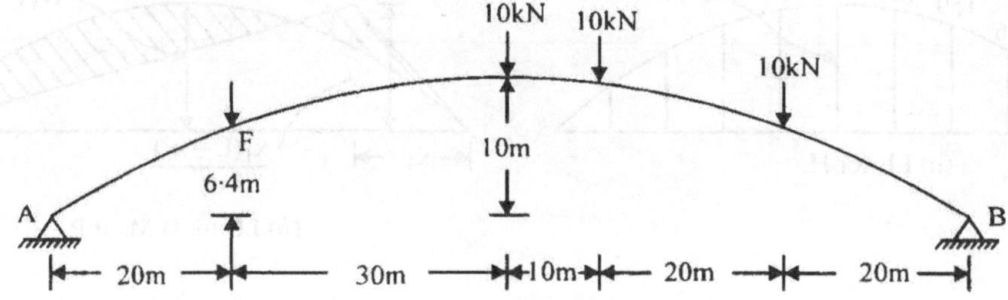

Fig. 14.34. Example 14.10

Solution : (*i*) Influence line for H.

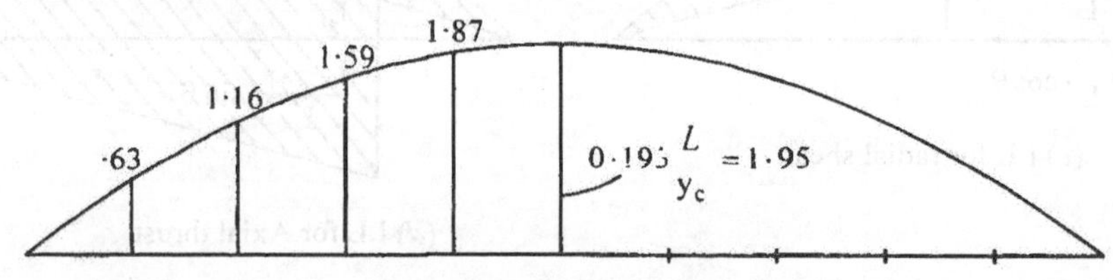

Fig. 14.35 I.L. for H

Total horizontal thrust = $10 \times 1{\cdot}95 + 1{\cdot}87 \times 10 + 10 \times 1{\cdot}16$

= **49.8 kN.**

(*ii*) **Influence line for B.M.**

$$y_1 = \frac{4y_c}{L^2}\left(Lx_1 - x_1^2\right)$$

$$= \frac{4 \times 10}{100^2}\left(100 \times 20 - 20^2\right)$$

$$= 6 \cdot 4.$$

$$\tan\theta = \frac{dy}{dx} = \frac{4y_c}{L^2}(L-2x)$$

$$= \frac{4\times10}{100^2}(100-40)$$

$$= 0\cdot24$$

$$\cos\theta = 0\cdot9723$$

$$\sin\theta = 0\cdot2333$$

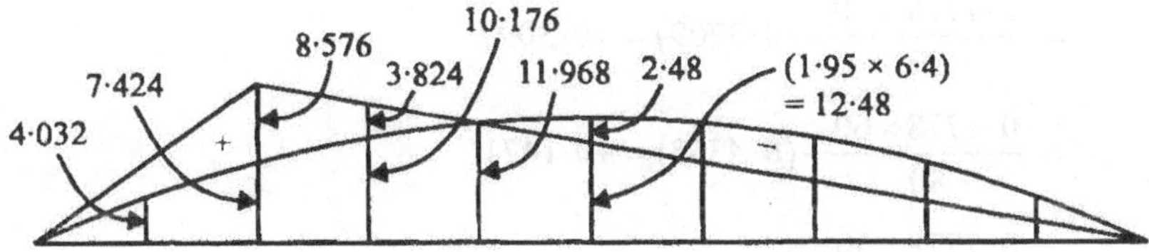

Fig. 14.36. I.L. for B.M. diagram

Value for the given load

$$\frac{x_1\times(L-x_1)}{L} = \frac{20\times80}{100} = 16$$

$$\text{B.M} = 10\times\left(-12\cdot48+\frac{16}{80}\times50\right)+10\times\left(\frac{16}{80}\times40-11\cdot968\right)$$

$$+\ 10\times\left(\frac{16}{80}\times20-7\cdot424\right) = -\ 98\cdot72$$

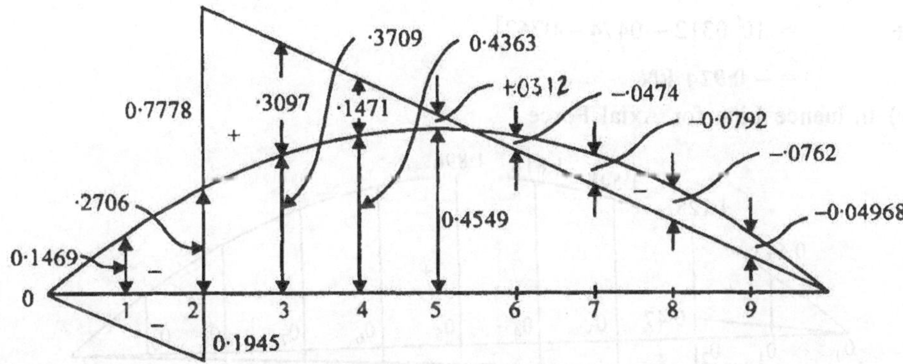

Fig. 14.37. I.L. for S.F.

***(iii)* I.L. for shearing force**

$$\frac{x}{L}\cdot\ \cos\theta = \frac{20}{100}\times\cdot9723$$

$$= 0\cdot1945$$

$$\frac{(L-x)\cdot Cos\theta}{L} = \frac{80\times\cdot9723}{100}$$

$$= 0\cdot 7778$$

$$0_0 \quad = 0$$

$$0_1 \quad = -0\cdot 1469 + \left(-\frac{0\cdot 1945}{2}\right) = -0\cdot 2442$$

$$0_2 \text{ (left)} = -0\cdot 1945 + (-0\cdot 2706) = -0\cdot 4651$$

$$0_2 \text{ (right)} = 0\cdot 7778 - (0\cdot 2706) = +0\cdot 5072$$

$$0_3 \quad = \frac{0\cdot 7778 \times 70}{80} - (0\cdot 3709) = +0\cdot 3097$$

$$0_4 \quad = \frac{0\cdot 7778 \times 60}{80} - (0\cdot 4363) = +0\cdot 1471$$

$$0_5 \quad = \frac{0\cdot 7778 \times 50}{80} - (0\cdot 4549) \; +0\cdot 0312$$

$$0_6 \quad = \frac{0\cdot 7778 \times 40}{80} - (0\cdot 4363) = -0\cdot 0474$$

$$0_7 \quad = \frac{0\cdot 7778 \times 30}{80} - (0\cdot 3709) = -0\cdot 0792$$

$$0_8 \quad = \frac{0\cdot 7778 \times 20}{80} - 0\cdot 2706 = -0\cdot 0762$$

$$0_9 \quad = \frac{0\cdot 7778 \times 10}{80} - 0\cdot 1469 = -0\cdot 04968$$

$$\text{S.F.} \quad = 10[\cdot 0312 - \cdot 0474 - \cdot 0762]$$

$$= -\,\mathbf{0{\cdot}924\ kN.}$$

(*iv*) Influence Line for Axial Force

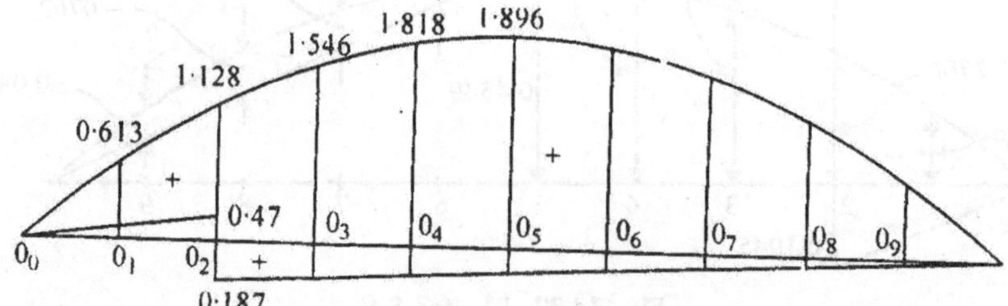

Fig. 14.38. I.L. for Axial Force

$$0_1 \quad = 0\cdot 613 - \frac{\cdot 047}{2} = 0\cdot 5895$$

$$0_2 \text{ (left)} = 1\cdot 128 - \cdot 047 = 1\cdot 081$$

$$0_2 \text{ (right)} = 1\cdot 128 + 0\cdot 187 = 1\cdot 3150$$

$$0_3 \quad = 1\cdot 546 + \frac{0\cdot 187 \times 70}{80} = 1\cdot 7096$$

$0_4 \quad = 1\cdot818 + \dfrac{0\cdot187\times60}{80} = 1\cdot9582$

$0_5 \quad = 1\cdot896 + \dfrac{0\cdot187\times50}{80} = 2\cdot0129$

$0_6 \quad = 1\cdot818 + \dfrac{0\cdot187\times40}{80} = 1\cdot9115$

$0_7 \quad = 1\cdot546 + \dfrac{0\cdot187\times30}{80} = 1\cdot6161$

$0_8 \quad = 1\cdot128 + \dfrac{0\cdot187\times20}{80} = 1\cdot1748$

$0_9 \quad = 0\cdot613 + \dfrac{0\cdot187\times10}{80} = \cdot6364.$

Value of Thrust $= 10\,[2\cdot0129 + 1\cdot9115 + 1\cdot1748]$

$= \mathbf{50\cdot992\ kN.}$

Example 14.11. A two hinged parabolic arch of span l and rise h carries a concentrated unit load at a distance 'a' from the left support the second moment of area varies as the slope of the rib axis with respect to horizontal. Starting from fundamental show that the horizontal thrust is given by.

$$H = 5/8\frac{a}{hl^3}(l-a)\left(l^2 + la - a^2\right).$$

Hence draw the influence line for H for an arch with $h = l/5$ at $0\cdot1l$ intervals.

Use the influence line to find the horizontal thrust in an arch of above proportions (l = 50m) for two concentrated loads of 40 kN amd 80 kN at 20m and 40m from the left end support respectively **[AMIE Summer'97]**

Solution: The first part is already proved in section 12.3 (c).

$$H = \frac{5}{8}\times\frac{a}{hl^3}(l-a)\left(l^2 + la - a^2\right) \quad \text{with} \quad h = \frac{l}{5},$$

$$H = \frac{5}{8}\times\frac{5}{l}\frac{a\times(l-a)\left(l^2 + la - a^2\right)}{\times l^3}$$

$$= \frac{25}{8}\cdot\frac{a}{l^4}(l-a)\left(l^2 + la - a^2\right).$$

a	H	value
$0 \cdot 1l$	$25/8 \times \frac{1}{l^4} \times 0 \cdot 1l(0 \cdot 9l)\left(l^2 + 0 \cdot 1l^2 - \cdot 01l^2\right)$	0·3065
$0 \cdot 2l$	$25/8 \times \frac{1}{l^4} \times 0 \cdot 2l(0 \cdot 8l)\left(l^2 + 0 \cdot 2l^2 - \cdot 04l^2\right)$	0·5800
$0 \cdot 3l$	$25/8 \times \frac{1}{l^4} \times 0 \cdot 3l(0 \cdot 7l)\left(l^2 + 0 \cdot 3l^2 - \cdot 09l^2\right)$	0·7900
$0 \cdot 4l$	$25/8 \times \frac{1}{l^4} \times 0 \cdot 4l(0 \cdot 6l)\left(l^2 + 0 \cdot 4l^2 - \cdot 16l^2\right)$	0·9300
$0 \cdot 5l$	$25/8 \times \frac{1}{l^4} \times 0 \cdot 5l(0 \cdot 5l)\left(l^2 + 0 \cdot 5l^2 - \cdot 25l^2\right)$	0·9765
$0 \cdot 6l$	$25/8 \times \frac{1}{l^4} \times 0 \cdot 6l(0 \cdot 6l)\left(l^2 + 0 \cdot 6l^2 - \cdot 36l\right)$	0·9300
$0 \cdot 7l$	$25/8 \times \frac{1}{l^4} \times 0 \cdot 7l(0 \cdot 7l)\left(l^2 + 0 \cdot 7l - 0 \cdot 49l^2\right)$	0·7900
$0 \cdot 8l$	$25/8 \times \frac{1}{l^4} \times 0 \cdot 8l(0 \cdot 8l)\left(l^2 + 0 \cdot 8l - \cdot 64l^2\right)$	0·5800
$0 \cdot 9l$	$25/8 \times \frac{1}{l^4} \times 0 \cdot 9l(0 \cdot 9l)\left(l^2 + 0 \cdot 9l^2 - 0 \cdot 81l^2\right)$	0·3065

The influence line is plotted in Fig. 14.40

Fig. 14.39. Influence Line for H.

Position of loads

40 kN at $\frac{20}{50} = 0 \cdot 4l$ Coefficient = 0·93

80 kN at $\frac{40}{50} = 0 \cdot 8l$ Coeffiecient = 0·58

$$H = 40 \times 0 \cdot 93 + 80 \times 0 \cdot 58$$

$$= 37 \cdot 2 + 46 \cdot 4 = \mathbf{83 \cdot 6\ kN.}$$

SUMMARY

1. To get influence lines for statically indeterminate structures equations of static equilibrium must be augmented with deflection equations. Müller Breslau's principles can be used to determine the influence lines for indeterminate structures like beam or frame. All these influence lines are curved unlike influence lines for determinate structures which are straight lines.
2. Qualitative influence lines are drawn to know visually where live load should be placed to produce maximum effect of a function, say, B.M, S.F, Reaction or axial force. In the analysis of multistorey building frames the qualitative influence line method is very useful in determining the spans which should be loaded for maximum effect.
3. Influence lines for indeterminate truss are obtained in the usual manner by systematically placing the unit load at each panel point using consistent displacement method or energy method.
4. Influence line for (*i*) Horizontal reaction (*ii*) B.M. (*iii*) S.F. and (*iv*) Axial Force for two hinged arch can be obtained by usual method by putting unit load at different positions and utilising already derived results.

   I.L. for H is obtained from the formula

$$H = \frac{5}{8} \cdot \frac{\alpha L}{y_c}\left(1 - 2\alpha^2 + \alpha^3\right)$$

   I.L. for B.M. is obtained by
   Superposing beam B.M. diagram on Hy diagram with proper signs.

   Similarly I.L. for S.F. and A.F. is obtained by superposing I.L. of Beam shear on $H \operatorname{Sin}\theta$ or $H \operatorname{Cos}\theta$ as required.

PROBLEMS

A. Theoretical

14.1. (*i*) State the Müller Breslau's principle and explain its use for obtaining the influence lines for statically indeterminate structures. **[AMIE Summer 1978]**

(*ii*) State the Müller Breslau's and explain how it is related to Maxwell's theorem. Using the principle explain the method of drawing the influence line for the prop reaction for a uniform propped cantilever **[AMIE Advanced structures Summer 1979]**

(*iii*) Explain Betti's law and Maxwell's law as applied to structures **[AMIE Summer 1983]**

B. Objective. Tick the correct answer.

14.2. (*i*) Müller Breslau's principle is used to obtain the

(*a*) deflected shape (*b*) influence line
(*c*) B.M. diagram (*d*) S.F. at any section.

(*ii*) Müller Breslau's principle can be used to

(*a*) determine the shape of the influence line
(*b*) Calculate ordinates of the influence line
(*c*) All of the above. **[AMIE Winter '97]**

(*iii*) Müller Breslau's principle is applicable.

(*a*) to only statically determinate structure
(*b*) to only beams

(*c*) to only statically indetermiante structe

(*d*) when principle of superposition is valid **[AMIE Winter '99]**

(*iv*) Sketch the influence line for reaction at *A* in the continuous beam ABC (Fig. P 14.1)

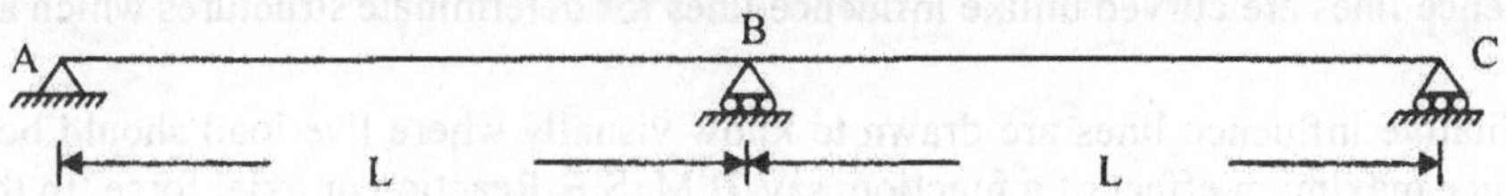

Fig. P 14.1 Prob 14.2 (*iv*)

C. Mathematical

14.3 Compute the ordinates at intervals of 2m of the influence line for the bending moment at the midpoint of the span BC for the continuous beam shown in Fig. P14.2.EI is constant. B is the hinged support while A,C and D are roller supports **[AMIE Summer 1978]**

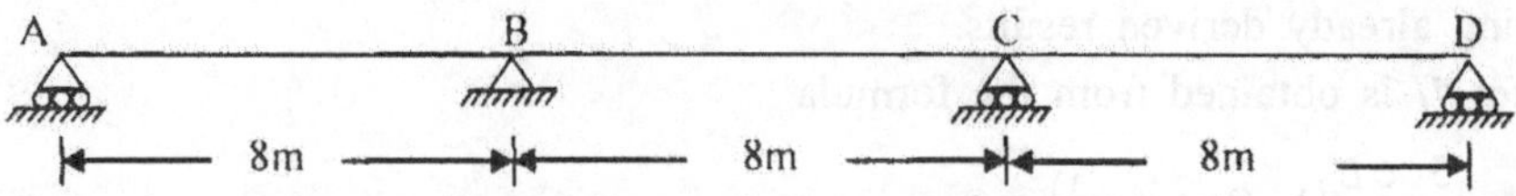

Fig. P 14.2 Problem 14.3

14.4 Compute the ordinates at intervals of 2.5m of the influence line for the reaction at B for the continuous beam shown in Fig. P 14.3 EI is contant. **[AMIE Advanced Structures Winter 1970]**

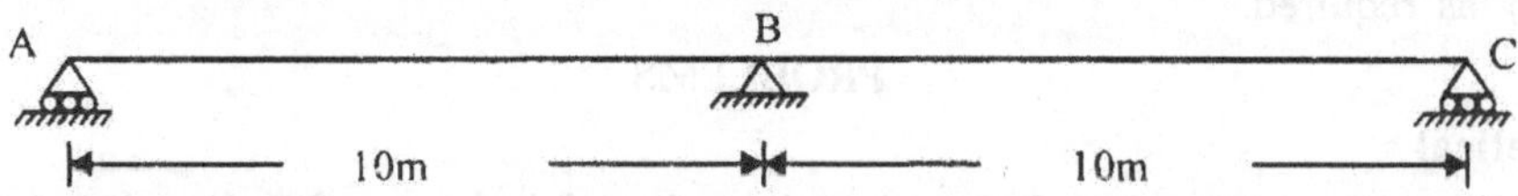

Fig. P. 14.3 Problem 14.4

14.5 ABC is a beam which is continuous over the support B and supply supported over ends *A* and *C*. span AB = 4m and BC = 2m. Using Müller Breslau's principle, draw the influence line for support reaction at *A*. Adopt 1m interval for obtaining the values of R_A **[AMIE Structures Summer, 1983]**

14.6 Construct the influence line for shearing force at the section m of the uniform continuous beam shown in Fig. P14.4

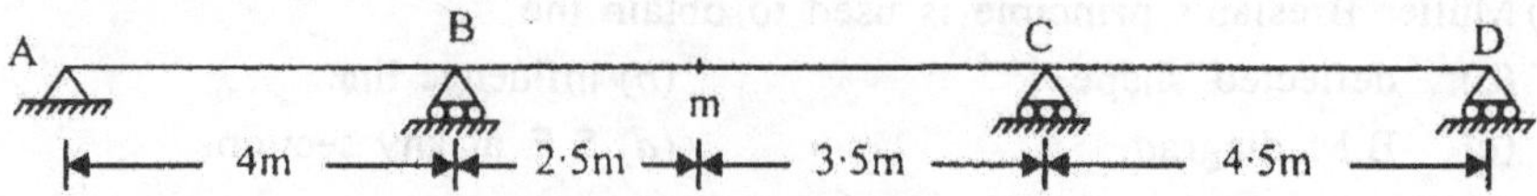

Fig. P14.4 Problem 14.6

15

APPROXIMATE METHODS OF ANALYSIS OF INDETERMINATE STRUCTURES

15.1 INTRODUCTION

It is to be clearly understood that no analytical method, however, refined it may be, gives us 'exact' answer. The first "approximation" is made when a physical problem is converted to a mathematical model. Examples of these approximations are:

1. Rivetted, bolted or welded joints are *assumed* to be frictionless pins and so called exact analytical tools are then applied.

2. A concentrated load is assumed to act through a knife edge when it is distributed over a small area.

3. For easy analysis, engineering materials are usually assumed to be homogeneous and isotropic.

Inspite of these limitations available analytical tools give reasonable good results not far from experimental values.

However, in some cases all available analytical tools may not be used in solving a problem. This may be due to

(*i*) Prohibitive cost (use of computer programs in solving a highly redundant structure).

(*ii*) Insufficient time available for a refined analysis.

(*iii*) Structure too complicated for exact analysis.

In the case of an indeterminate structure it is necessary to estimate the proportions of its members in order to know their relative stiffness upon which the analysis depends.

These dimensions maybe obtained on the basis of approximate analysis.

Also, even with the availability of computers, most engineers find it desirable to make a rough check of results using approximate means to detect gross errors.

In practice approximate analysis converts a highly indeterminate structure to a statically determinate one by making suitable intelligent educated assumptions. The assumptions are often based on observations of actual behaviour of structures.

15.2 APPROXIMATE ANALYSIS OF STATICALLY INDETERMINATE TRUSSES

Indeterminate trusses can be approximately analysed by considering equilibrium equaions only by

1. Assuming that the diagonals are resistant to tension only. This is acceptable as diagonals

are very slender. In such a case the shear will be resisted by one of the two diagonals in a given panel which will be under tension.

2. Assuming that the diagonals in a panel each resist half the shear. This is possible if the diagonals in any panel are stocky (i.e. not slender) and with equal cross sectional area.

3. Assuming that all the vertical members are removed and the shear at a cross section is resisted as though the truss functioned as two independent component statically determinate trusses. For the method to be applicable it should be possible to split the given truss into a number of component trusses equal to be degree of indeterminacy plus one. Further, it should be possible to apportion the external loads on the redundant truss to the component trusses.

All these ideas are further clarified by working a problem by all these methods.

Example 15.1. Calculate the forces is all members of the truss (Fig. 15.1) by approximate method assuming

(*i*) diagonals can resist tension only.

(*ii*) diagonals share shear equally.

(*iii*) that all the intermediate vertical member are removed and the shear at a cross-section is resisted as through the truss functioned as two independent component statically determinate trusses.

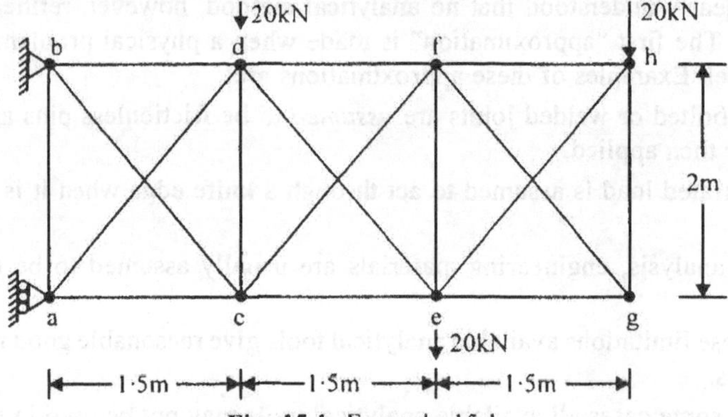

Fig. 15.1. Example 15.1.

Solution: The structure is statically indeterminate. Degree of indeterminacy.

$$= (m+3) - 2n$$

$$= (16+3) - 2 \times 8 = 3$$

(*i*) By first assumption (diagonals can resist tension only) (Fig. 15.2)

From inspection it can be seen that members *eh*, *cf* and *ad* are in compression and *if*, *ed* and *cb* are in tension. If diagonals can resist tension only as assumed member *eh*, *cf* and *ad* are ineffective or forces in them are zero. The truss then becomes statically determinate and can be analysed as such.

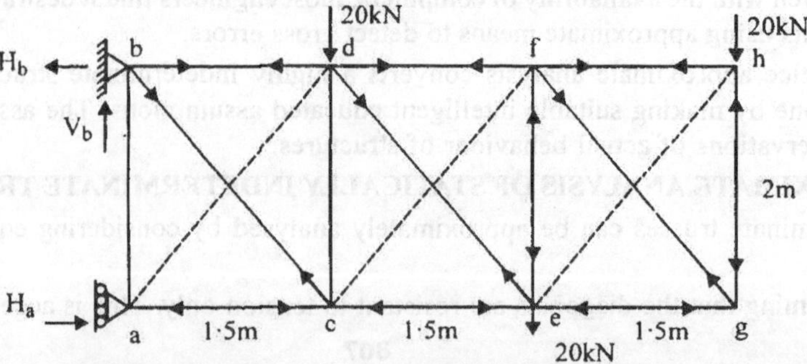

Fig. 15.2. Analysis of truss based on 1st assumption.

Taking moments about B

$$H_a \times 2 = 20 \times 1.5 + 20 \times 3 + 20 \times 4.5$$
$$= 180$$

oı

$$H_A = 90 \text{ kN}$$
$$V_B = 60 \text{ kN}$$

Calculation of forces

Joint *h* $F_{hg} = +20$ (compression)

Joint *g* $F_{gf} \cdot \frac{2}{2.5} = 20$

$$F_{gf} = +25 \text{ (tension)}$$

$$F_{ge} = F_{gf} \cdot \frac{1.5}{2.5}$$

$$= 25 \times \frac{1.5}{2.5} = 15 \text{ (compresion)}$$

$$F_{fd} = F_{fg} \times \frac{1.5}{2.5} = \frac{25 \times 1.5}{2.5} = 15 \text{ (tension)}$$

Joint *e* $F_{ed} \times \frac{2.0}{2.5} = 40$

$$F_{ed} = \frac{2.5 \times 40}{2.0} = 50 \text{ (tension)}$$

$$F_{ed} \times \frac{1.5}{2.5} - F_{ce} + 15 = 0$$

or $F_{ce} = 45$ (compression)

Check $F_{ce} \times 2 - 90 \times 2 + 60 \times 1.5 = 0$

Joint *a*

$$F_{ac} = 90 \text{ (compression)}$$
$$F_{ab} = 0$$

$$F_{bc} \times \frac{2}{2.5} = 60$$

$$F_{bc} = 75 \text{ (tension)}$$

$$F_{bc} \times \frac{15}{2.5} + F_{bd} - H_b = 0$$

or $45 + F_{bd} - 90 = 0$

or $F_{bd} = 45$ kN (tension)

$$F_{cd} = 60 \text{ (compression)}$$
$$F_{ab} = 0.$$

(*ii*) By second assumption (diagonals share shear equally). (Fig. 15.3)

This assumption enables one to find forces in each diagonal in each panel and the structure becomes statically determinate.

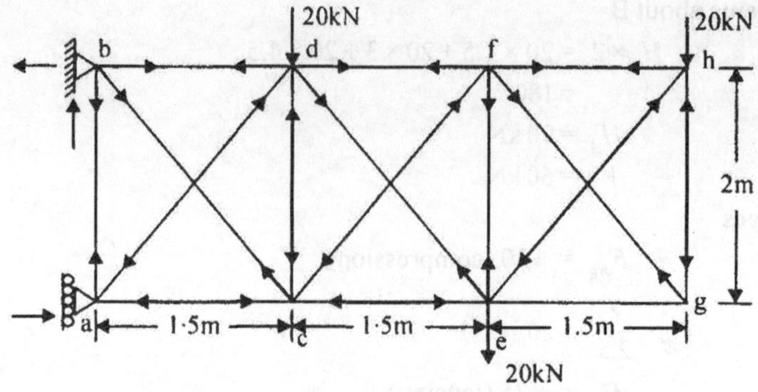

Fig. 15.3. Analysis based on 2nd assumption.

Panel *efgh*

Shear $= 20\text{ kN}$

$$F_{eh} \cdot \frac{2}{2.5} = 10 \quad \text{or} \quad F_{eh} = 12.5\text{ kN}$$

$$F_{gf} \cdot \frac{2}{2.5} = 10 \quad \text{or} \quad F_{gf} = 12.5\text{ kN}$$

$$F_{hf} = F_{eh} \cdot \frac{1.5}{2.5} \quad \text{or} \quad F_{hf} = 7.5\text{ kN}$$

$$F_{gh} = 20 - F_{ch} \cdot \frac{2}{2.5} = 20 - \frac{12.5 \times 2}{2.5}$$

$$= 10\text{ kN}$$

$$F_{ge} = F_{gf} \cdot \frac{1.5}{2.5} = \frac{12.5 \times 1.5}{2.5} = 7.5\text{ kN}.$$

Panel cdfe

Shear $= 40\text{kN}.$

$$F_{fc} \times \frac{2}{2.5} = \frac{40}{2}$$

$$F_{fc} = 25\text{kN}$$

$$F_{ed} = 25\text{kN}.$$

Joint *f*

$$F_{fd} - F_{fc} \cdot \frac{1.5}{2.5} - F_{fg} \cdot \frac{1.5}{2.5} - F_{fh} = 0$$

$$F_{fd} = \frac{25 \times 1.5}{2.5} + \frac{12.5 \times 1.5}{2.5} + 7.5$$

$$= \mathbf{30\ kN}$$

$$F_{fc} \times \frac{2}{2.5} - F_{fe} - F_{fg} \cdot \frac{2}{2.5} = 0$$

$$F_{fe} = F_{fc} \times \frac{2}{2.5} - F_{fg} \cdot \frac{2}{2.5}$$

$$= \frac{25 \times 2}{2.5} - \frac{12.5}{2.5}$$

$$= 20 - 10 = \mathbf{10\ kN}$$

Joint *e* Resolving vertically

$$F_{ed}\cdot\frac{2}{2.5}-F_{eh}\cdot\frac{2}{2.5}+F_{ef}-20$$
$$=\frac{25\times2}{2.5}-\frac{12.5\times2}{2.5}+10-20$$
$$=20-10+10-20=0 \text{ (check)}$$

Resolving horizontally

$$F_{ec}-Fed\cdot\frac{1.5}{2.5}-F_{eh}\cdot\frac{1.5}{2.5}-F_{eg}=0$$

$$F_{ec}=F_{ed}\times\frac{1.5}{2.5}+F_{eh}\cdot\frac{1.5}{2.5}+F_{eg}$$
$$=\frac{25\times1.5}{2.5}+\frac{12.5\times1.5}{2.5}+7.5$$
$$=30 \text{ kN.}$$

Panel *abdc*

Panel shear $=60$ kN

Shear carried by each diagonal $=\frac{60}{2}=30$ kN

$$F_{cb}=F_{da}=\frac{30\times2.5}{2}=37.5 \text{ kN.}$$

Joint *a* Resolving horizontally

$$90-F_{ac}-F_{ad}\cdot\frac{1.5}{2.5}=0$$

or
$$F_{ac}=90-\frac{37.5\times1.5}{2.5}=90-22.5$$
$$=67.5 \text{ kN.}$$

Resolving vertically,

$$F_{ab}-F_{ad}\cdot\frac{2}{2.5}=0$$

or
$$F_{ab}=F_{ab}\cdot\frac{2}{2.5}$$
$$=\frac{37.5\times2}{2.5}=\mathbf{30\ kN.}$$

Joint *b* Resolving vertically,

$$F_{bc}\cdot\frac{2}{2.5}+F_{ba}-60=\frac{37.5\times2}{2.5}+30-60=0 \textit{ (check)}$$

Resolving horizontally

$$90-F_{bc}\cdot\frac{1.5}{2.5}-F_{bd}=0$$

or
$$F_{bd}=90-\frac{37.5\times1.5}{2.5}$$
$$=\mathbf{67.5\ kN}$$

Joint *d* Resolving horizontally,

$$67.5\times F_{da}\times\frac{1.5}{2.5}-F_{de}\times\frac{1.5}{2.5}-F_{df}=67.5-\frac{37.5\times1.5}{2.5}-\frac{25\times1.5}{2.5}-30$$
$$=67.5-22.5-15-30=0 \text{ (check)}$$

Resolving vertically

$$F_{dc} \downarrow + 20 \downarrow + F_{de} \cdot \frac{2}{2.5} \downarrow - F_{da} \times \frac{2}{2.5} \uparrow = 0$$

or

$$F_{dc} = F_{da} \cdot \frac{2}{2.5} - F_{dc} \cdot \frac{2}{2.5} - 20$$

$$= \frac{37.5 \times 2}{2.5} - \frac{25 \times 2}{2.5} - 20 = -10 \uparrow$$

Joint *c*

$$10 - F_{cb} \times \frac{2}{2.5} + F_{cf} \times \frac{2}{2.5} = 10 - \frac{37.5 \times 2}{2.5} + \frac{25 \times 2}{2.5}$$

$$= 10 - 30 + 20 = 0 \text{ (check)}$$

(*iii*) By third assumption (Fig. 15.4 (*a*) and (*b*))

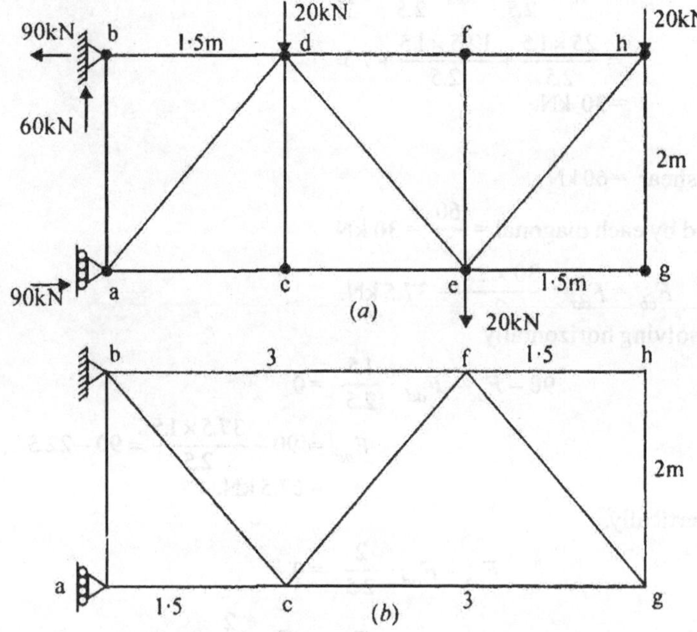

Fig. 15.4. Analysis based on 3rd assumption.

$$F_{bd} = 90 \qquad F_{ba} = 60 \qquad F_{ad} \times \frac{2}{2.5} = 60$$

$$F_{ad} = 75$$

$$F_{ae} + F_{ad} \times \frac{1.5}{2.5} - 90 = 0$$

$$F_{ae} = 90 - \frac{75 \times 1{\cdot}5}{2{\cdot}5} = \mathbf{45kN.}$$

$$F_{ge} = F_{gh} = 0$$

$$F_{he} \times \frac{2}{2.5} = 20 \qquad F_{he} = 25$$

$$F_{hd} = F_{he} \times \frac{1.5}{2.5} = \frac{2.5 \times 1.5}{2.5} = 15 = F_{df} = F_{fh}.$$

$$F_{ad} \times \frac{2}{2.5} - 20 - F_{de} \times \frac{2}{2.5} = 0$$

or $$0.8\,F_{de} = \frac{75 \times 2}{2.5} - 20 = 60 - 20 = 40 \text{ kN}.$$
$$F_{de} = 50 \text{ kN}.$$

The truss is statically indeterminate to 3rd degree with two vertical members removed, the indeterminate reduces to 1. Hence the truss should be split to (1 + 1) *i.e.* 2 independently loaded trusses as shown. With no load in 2nd truss, the forces are all zero. Hence the final force in members is the force in the 1st truss only. Table 15.1. shows the values of the forces in different members by different method.

Table 15.1 (Example 15.1)

Member	by Assump. 1	By Assump. 2	By Assump. 3
ab	00	30.00 *t*	60 *t*
bd	45 *t*	67.50 *t*	90 *t*
dc	60 *c*	10.00 *c*	00
ca	90 *c*	67.50 *c*	45 *c*
bc	75 *t*	37.50 *t* ⌉	00 ⌉
ad	00	37.50 *c* ⌋	75 *c* ⌋
df	15 *t*	30.00 *c*	15 *t*
fe	20 *c*	10.00 *t*	00
ec	45 *c*	30.00 *c*	45 *c*
de	50 *t* ⌉	25.00 *t* ⌉	50 *t* ⌉
cf	00 ⌋	25.00 *c* ⌋	00 ⌋
fh	00	7.50 *t*	15 *t*
hg	20 *c*	10.00 *c*	00
ge	1.5 *c*	7.50 *c*	00
fg	25 *t* ⌉	12.50 *t* ⌉	00 ⌉
eh	00 ⌋	12.50 *c* ⌋	25 *c* ⌋

15.3. APPROXIMATE ANALYSIS OF BEAMS AND FRAMES

The indeterminate beams and frames can be reduced to determinate forms by making assumptions about (*i*) the location of points of zero moment (inflection points) and (*ii*) the distribution of forces among several members.

The location of infection points is studied first. A simply supported beam, a fixed beam and continuous beam of same span subjected to uniformly distributed load are considered.

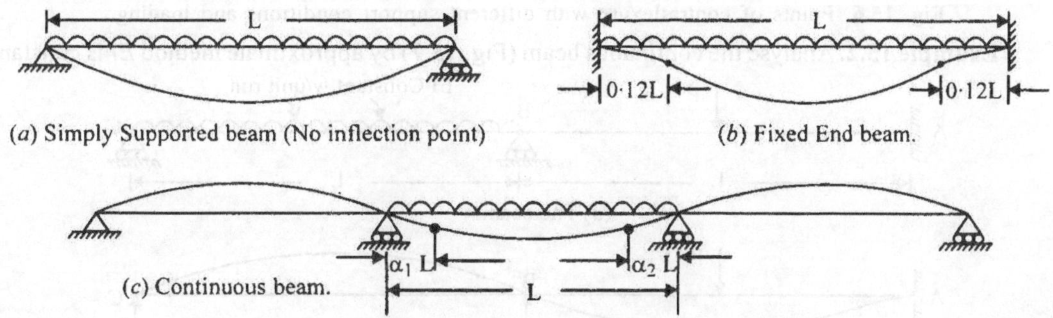

(*a*) Simply Supported beam (No inflection point)

(*b*) Fixed End beam.

(*c*) Continuous beam.

Fig. 15.5. Deflected shapes for different types of beams.

Simply supported beam and fixed end beam represent two extreme positions. Simply supported beam is subjected only to positive moment, hence there is no point of contraflexure. (Fig. 15.5(a)). In a fixed end beam subjected to uniformly distributed load the inflection points are 0.21 L from the supports (Fig. 15.5. (*b*)). The intermediate span of continuous beam lies in between a simply supported beam and a fixed end beam. Hence α_1 and α_2 will lie between 0 and 0.21 L, a reasonable assumption will be $\alpha_1 = \alpha_2 = 0.1$ L provided the two end conditions are the same. When end conditions are not the same α_1 and α_2 will differ (Fig. 15.5 (*c*)).

Fig. 15.6 shows the values of α_1 and α_2 for some typical loading with different end conditions. These results help us in making "educated" guesses about approximate location of inflection points of continuous beams and frames.

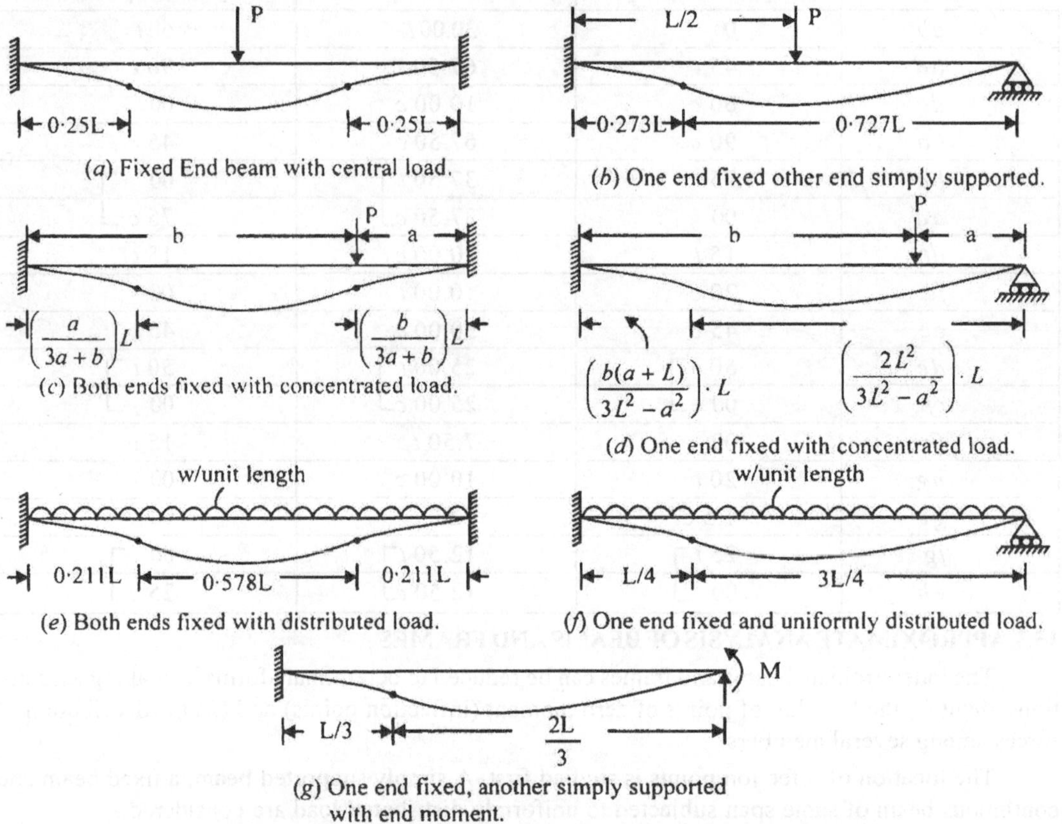

Fig. 15.6. Points of contraflexure with different support conditions and loading.

Example 15.2. Analyse the continuous beam (Fig. 15.7) by approximate method *EI* is constant

P
A
B
C
EI Constant w/unit run.
L
L

(*a*) The beam.

=

P
A
B
C
+

(*b*) With concentrated load only

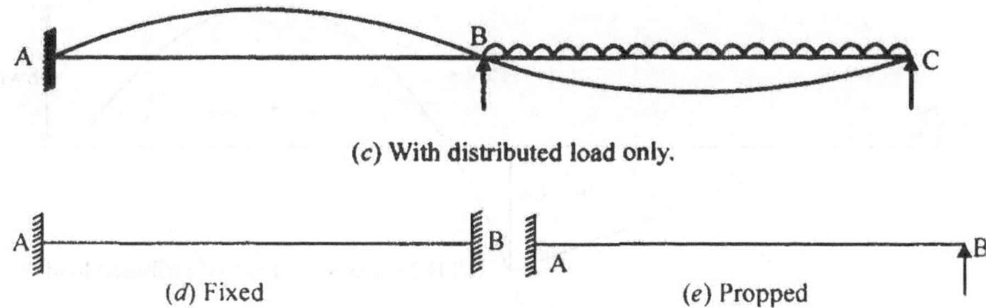

(c) With distributed load only.

(d) Fixed (e) Propped

Fig. 15.7. Example 15.2

Solution: Since the beam has two loads, each load is analysed separately and the values superposed.

The end B lies between support conditions Fig. 15.7 (d) and (e) as end A is fixed but end B is neither fixed nor free but there is some negative moment.

Hence, the point of inflection from A lies between 0.25 L and 0.27 L say 0.26 L Inflection point from end B will lie between 0 and 0.27 L say 0.15 L.

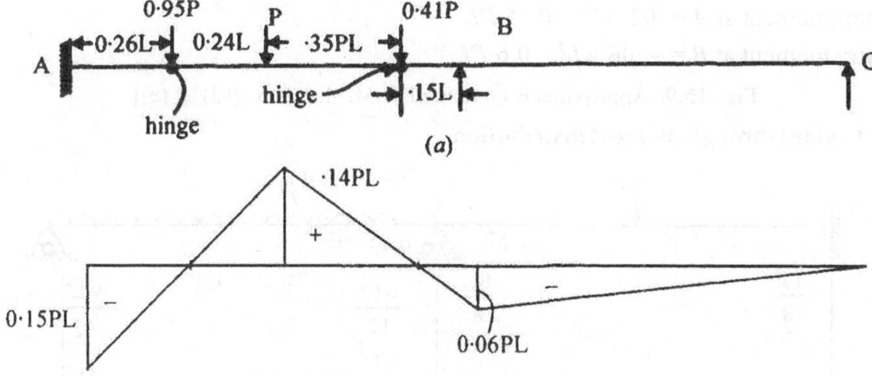

(a)

(b) B.M. due to concentrated load.

Fig. 15.8. B.M. due to concentrated load (Ex. 15.2).

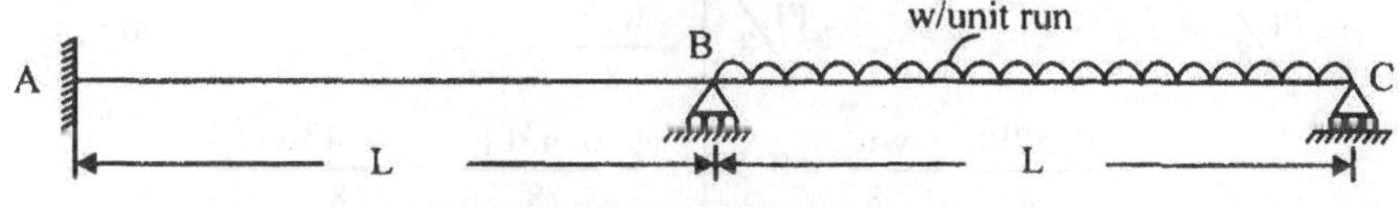

(a) Beam under distributed load only.

horizontal

A B C

·26L 0·74L ·15L ·85L

(b) Deflected shape.

·09wL 0·42wL w/unit run

A B C

·09wL 0·66wL 0·42wL 0·42wL

$$\frac{0\cdot 42\times 0\cdot 89\text{wL}}{0\cdot 74}=0\cdot 5\text{wL}$$

$$+\frac{0\cdot 15\text{wL}\times 81\text{L}}{0\cdot 74\text{L}}=\frac{0\cdot 16\text{wL}}{0\cdot 66\text{wL}}$$

(c) Free body diagrams.

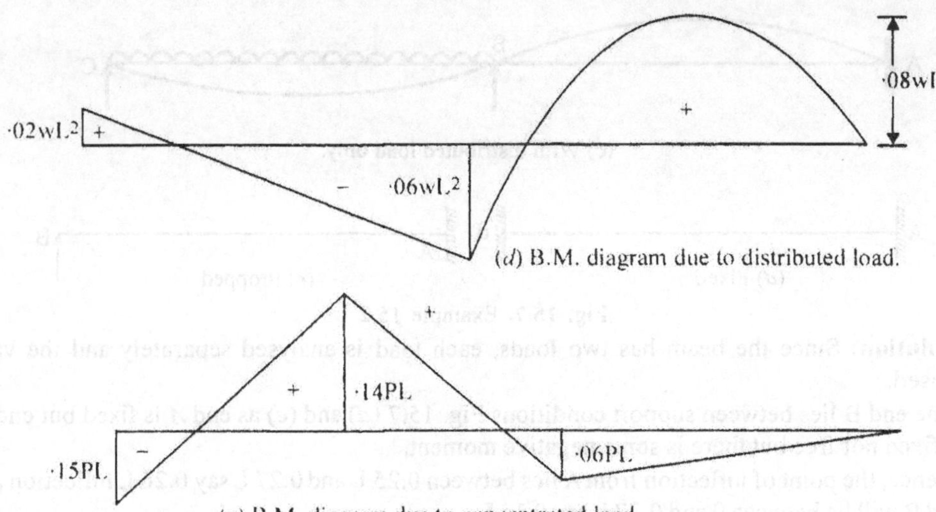

(e) B.M. diagram due to concentrated load.

Support moment at $A = \cdot 02\ wL^2 - 0.15\ PL$

Support moment at $B = -\cdot 06\ wL^2 - 0.6\ PL$

Fig. 15.9. Approximate combined B.M. diagram. [(d) + (e)]

Exact values through moment distribution

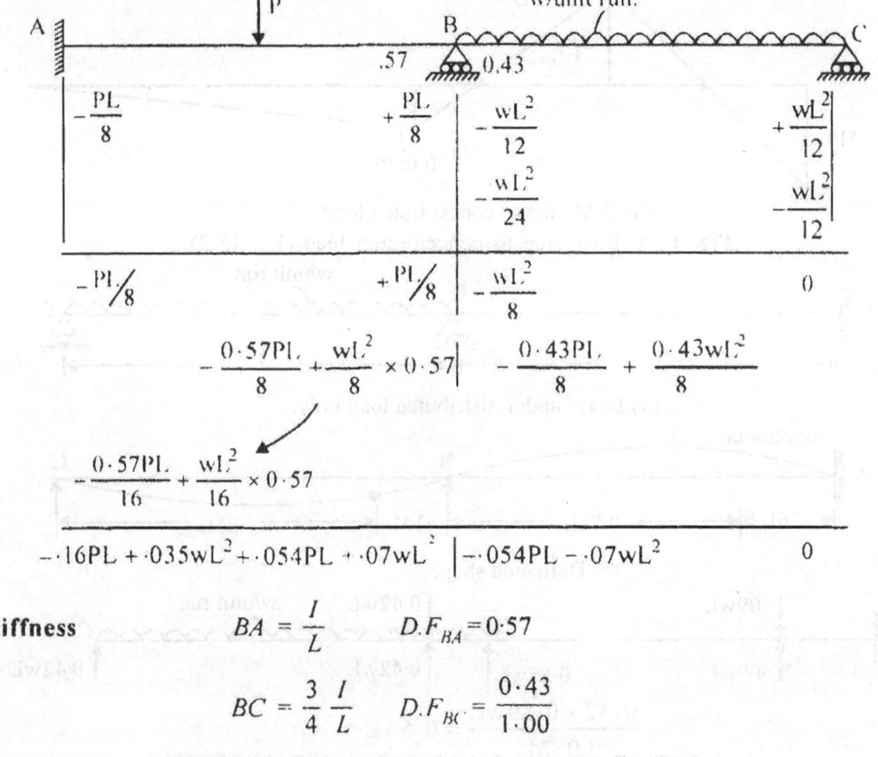

Stiffness $BA = \frac{I}{L}$ $D.F_{BA} = 0{\cdot}57$

$BC = \frac{3}{4}\frac{I}{L}$ $D.F_{BC} = \frac{0{\cdot}43}{1{\cdot}00}$

Fig. 15.10. Exact values through moment distribution.

15.4. PORTALS AND MILLS BENTS SUBJECTED TO LATERAL FORCES.

Portals are efficient in transmitting lateral forces and as such they are incorporated in bridge and other structures. to transmit wind loads etc into the foundation Fig. 15.11. shows a typical portal frame subjected to a lateral load at the top. The resulting bending moment diagram and the deflected shape are also shown.

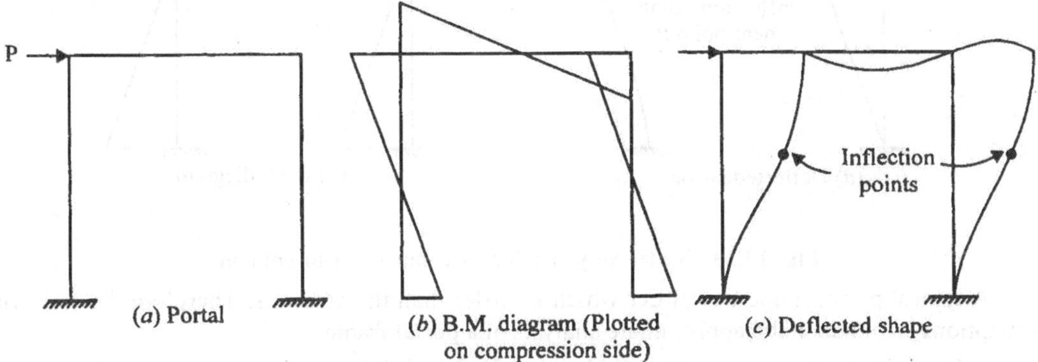

Fig. 15.11. Deflection of Portal

The deflected shape will obviously depend on the relative stiffness of beam and columns. Two possibilities may arise.

(*i*) Girder is very stiff $\left(\frac{I}{L} = \infty\right)$ in comparison with columns. In such a case the bending of the grider is very small, the bending moment at the base of each column will be almost equal to that at the top. Consequently, the inflection points are at the mid height of columns. (Fig. 15.12)

(*ii*) Girder is very flexible in comparison to columns. The point of contraflexures lie near the top of column the deflected shape and the bending moment diagrams are shown in Fig. 15.13.

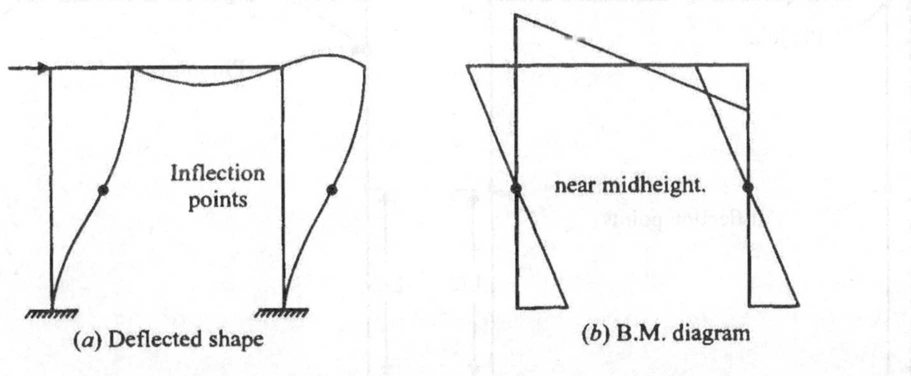

Fig. 15.12. Girder very stiff in comparison to column.

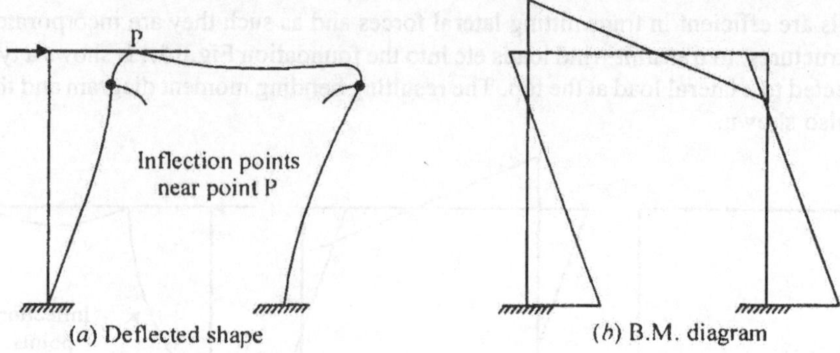

Fig. 15.13. Girder very flexible in comparison to column.

A typical portal frame has girder which is stiffer than the columns. Therefore the following assumptions are made in the approximate analysis of a portal frame.

1. Inflection points occur at the mid height of left and right columns.
2. The shear in each column is equal to half the applied lateral load.

With these three assumptions a portal frame which is indeterminate to the third degree becomes statically determinate.

Fig. 15.14. shows a trussed portal and a mill bent. Here instead of a beam there is a truss which is very stiff and hence behaves like a rigid girder. The truss is attached to the column through pins. In the case of mill bent Knee bases are incorporated to lend lateral stability to the configuration.

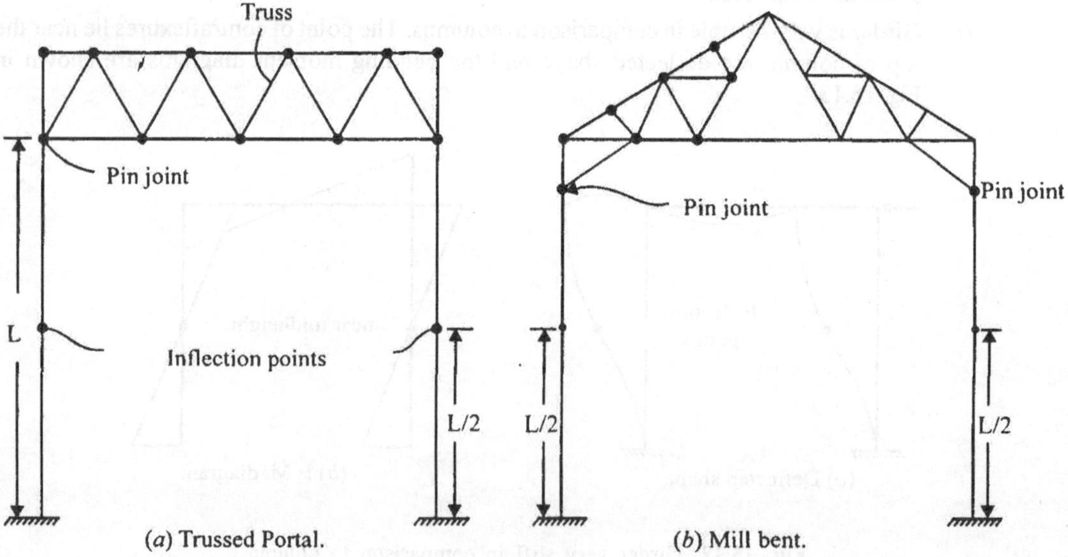

Fig. 15.14. Trussed Portal and Mill Bent

15.5 BUILDING FRAMES SUBJECTED TO VERTICAL FORCES ONLY

Building frames are highly statically indeterminate. For example the building frame as shown in Fig. 15.15 is statically indeterminate to the 24th degree.

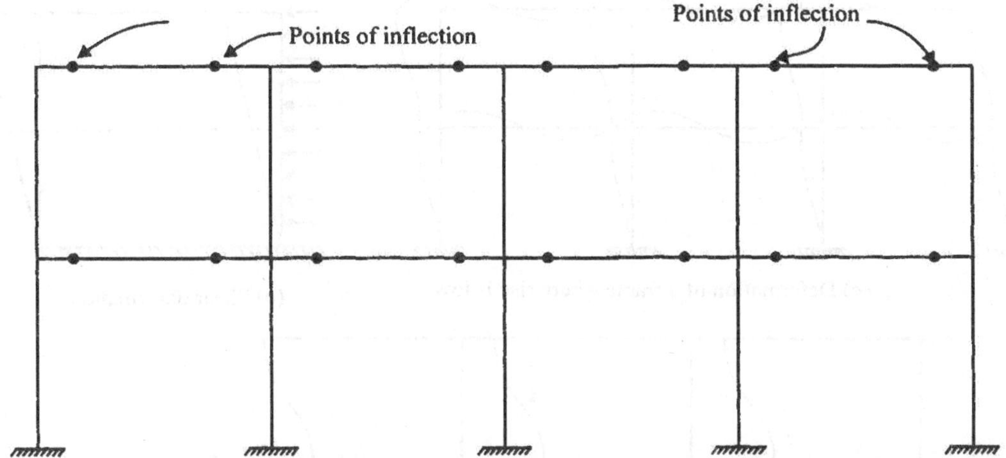

Fig. 15.15. Building frames subjected to vertical loads only.

This can be derived as follows:

If b is the number of members, r the number of independent generalised reaction forces, total number of unknowns $= 3b + r$.

If n is the number of joints, no. of independent equations of equilibrium available $= 3\,n$.

Hence, degree of indeterminacy $N = 3b + r - 3n$.

Here (Fig. 15.5) $b = 18,\ r = 15,\ n = 15$

Hence, $N = 3 \times 18 + 15 - 45 = 24$

The behaviour of the girders of a building frame is like that of a continuous beam with elastic end restraints provided by the stiffness of the members connected at each joint. Like continuous beams there will be two inflection points (zero moment) near the two ends. Good correlation is obtained if it is assumed that inflection points occur at 0.1 L from the ends. In addition it is observed that when the structure is subjected to vertical loads only, the girders will have virtually no axial force. Hence, a rigid frame under *vertical loads* only can be approximately analysed by statics only with the following assumptions.

1. Inflection points will occur at 0.1 L from two ends of the girder.
2. There is no axial force in the girder.

For the frame given in Fig. 15.15 1st assumption gives 16 additional conditions while 2nd assumption gives another eight. The frame then becomes statically determinate. The last assumption, however, is violated if we want to calculate the maximum axial force in an exterior column as girder moment when distributed in a joint will introduce column moment which will lead to column shear. Column shear will introduce axial load into the girder which violates the 2nd assumption used in the analysis.

15.6 BUILDING FRAMES SUBJECTED TO LATERAL FORCES

Environmental effects such as wind and earthquake introduce lateral forces in frames. There forces are assumed to act at the joints. Two methods are available for analysis of frames:

1. Portal method
2. Cantilever method

The methods are based on the behaviour of the frames under lateral loads. The behaviour

again depends on the height width ratio of the frames. When the width of a multistorey frame is greater than or at least equal to the total height, the deformation most nearly resembles shearing. This is shown in Fig. 15.16 (*a*) and (*b*).

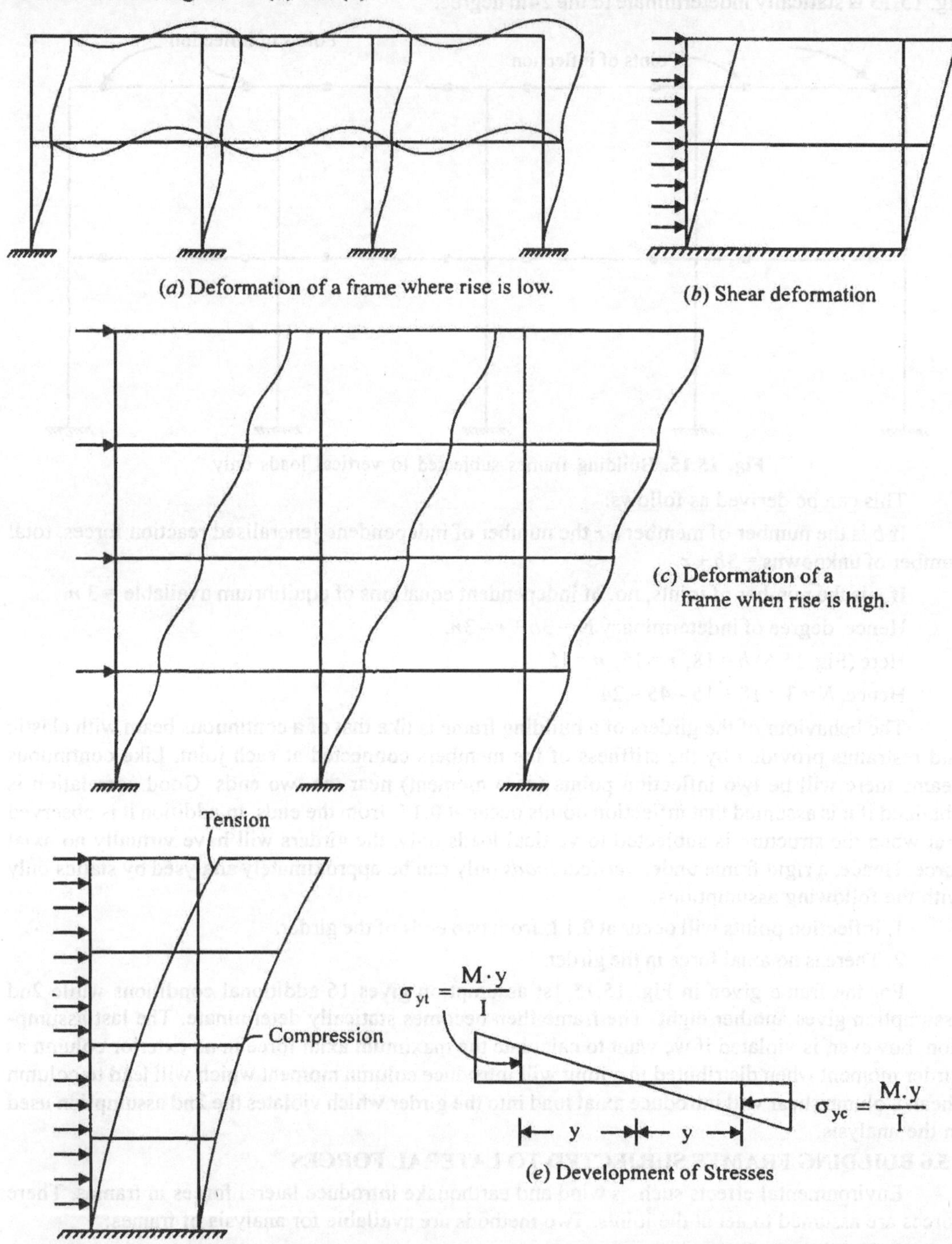

(*a*) Deformation of a frame where rise is low.

(*b*) Shear deformation

(*c*) Deformation of a frame when rise is high.

(*d*) Bending deformation

(*e*) Development of Stresses

Fig. 15.16. Physical basis of approximate analysis of multistorey frames.

Portal method is based on this concept. Lateral shear in each storey is distributed equal in each panel. Since the interior columns connect two panels, it will carry twice the shear compared to exterior column. The distribution can also be visualised by dividing a frame into a number of portals (from which originates the name) and assuming that each leg carries equal shear.

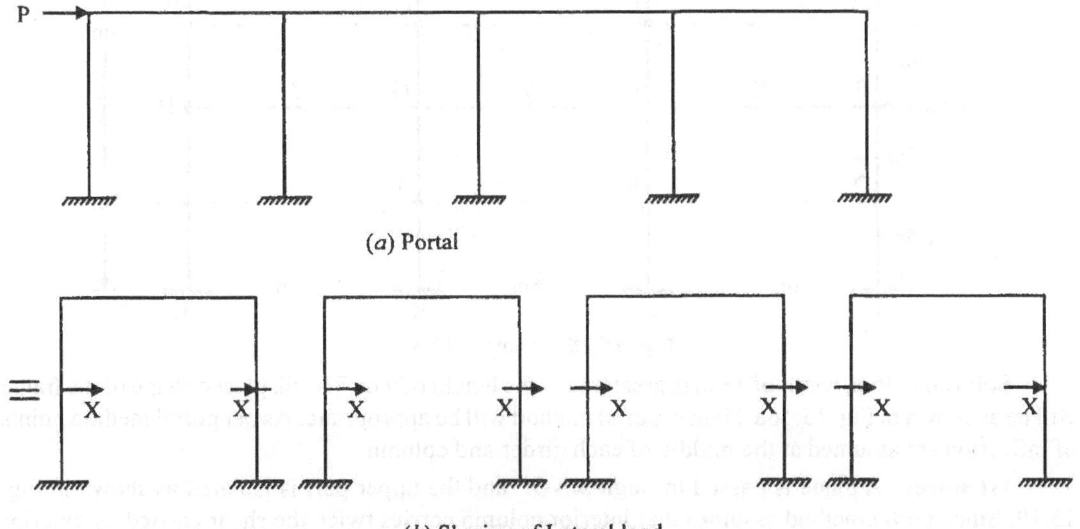

(*a*) Portal

(*b*) Sharing of Lateral load.

Fig. 15.17. Derivation of Portal Method

This is same as assuming that each interior column takes $2X$ where X is the force in exterior column as shown in Fig. 15.17. P, then is equal to $8X$ or $X = \frac{P}{8}$. From the displaced shape of the frame it is also clear that there are inflection points in each column and girder. Portal method, therefore, is based on the following assumptions.

1. There is a point of inflection at the centre of each girder.
2. There is a point of inflection at the centre of each rigidly connected column. This does not apply to columns with pinned bases where the moment is zero.
3. The shear resisted by an exterior column of any storey is equal to $\frac{1}{2n}$ of the total storey shear and the shear resisted by each of the $(n-2)$ adjacent interior columns is $\frac{1}{n}$ where n is the number of bays in any storey. In Fig. 15.17 $n = 4$.

Hence, shear resisted by exterior column $= \frac{P}{2 \times 4} = \frac{P}{8}$ and interior column $= 2 \times \frac{P}{8} = \frac{P}{4}$.

In the portal method the shear in column is determined first and then other forces and moments.

The cantilever method uses the concept of a vertical cantilever subjected to horizontal load. The basic deformation is bending and not shear. Hence the frame can be analysed using the elastic bending theory and using the relation $\sigma = \frac{M.y.}{I}$ which means stresses in columns will be proportional to its distance from the centroid of the column areas. The other assumptions about location of points of inflection at the middle of girders and columns remains valid in this method also. In this method axial stresses and axial forces are determined first and then other forces and moments. The two methods are explained with two examples.

Example 15.3. Analyse the frame by approximate method. Draw the final moment diagram.

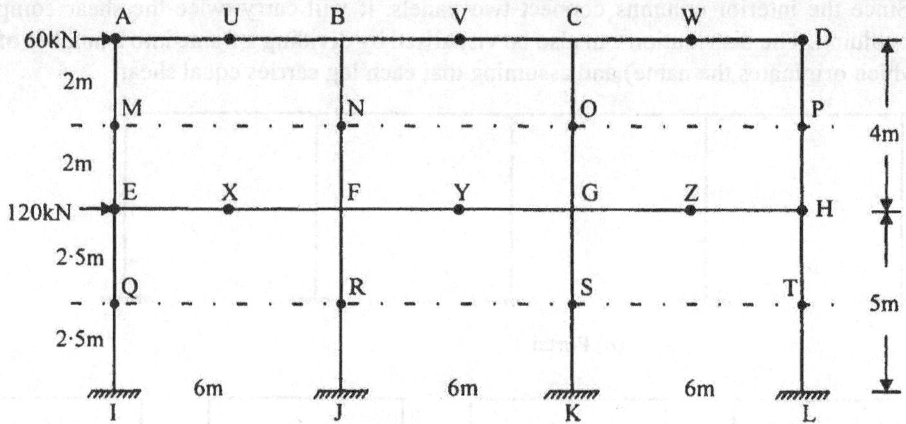

Fig. 15.18. Example 15.3.

Solution: Here, width of 18 m is greater than the height of 9 m. The displaced shape of the frame will be as shown in Fig. 15.16a. Hence, portal method will be appropriate. As per portal method points of inflection are assumed at the middle of each girder and column.

1st storey. A plane is passed through *MNOP* and the upper part is isolated as shown in Fig. 15.19. Since portal method assumes that interior column carries twice the shear carried by exterior column.

From $\Sigma H = 0$

$$60 = V + 2V + 2V + V$$
$$= 6V$$

or $V = 10\,\text{kN}$

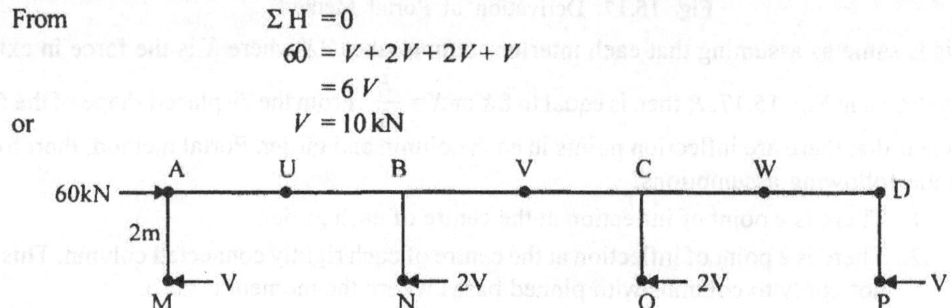

Fig. 15.19 Part of the frame.

(*i*) Frame MAU (Fig. 15.20)

From $\Sigma H = 0$

$$F_{UA} = 50$$

Taking moments about *A*

$$10 \times 2 = V_{uA} \times 3$$

or $V_{UA} = \dfrac{20}{3} = 6.67\,\text{kN}$

Fig. 15.20

(*ii*) Frame UBVN (Fig. 15.21)

From $\Sigma H = 0$

$$Fu_B - 20 - Fv_B = 0$$

or $F_{VB} = 30$

Fig. 15.21

Taking moments about B.

$$6.67 \times 3 - 20 \times 2 + V_{UB} \times 3 = 0$$

or $$V_{UB} = \frac{1}{3}(40 - 20) = \frac{20}{3}$$

$$= 6.67 \text{ kN}$$

From $$\Sigma V = 0$$

$$V_{NB} = 0$$

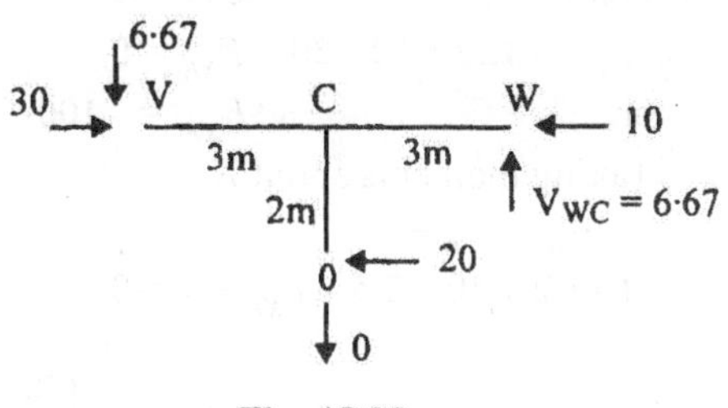

Fig. 15.22

Frame VCOW (Fig 15.22)

Taking moments about C

$$6.67 \times 3 - 20 \times 2 + V_{WC} \times 3 = 0$$

or $$V_{WC} = \frac{1}{3}(40 - 20) = 6.67 \text{ kN} \uparrow$$

$$V_{OC} = 6.67 - 6.67 = 0$$

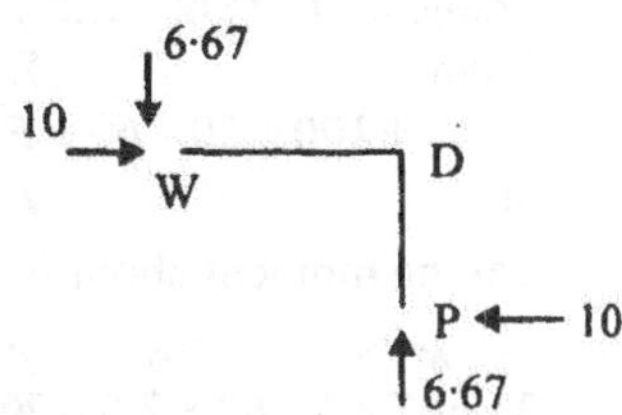

Fig. 15.23

From $\Sigma H = 0$, $F_{PD} = 6.67$

From $\Sigma H = 0$, $V_{PD} = 6.67$ kN.

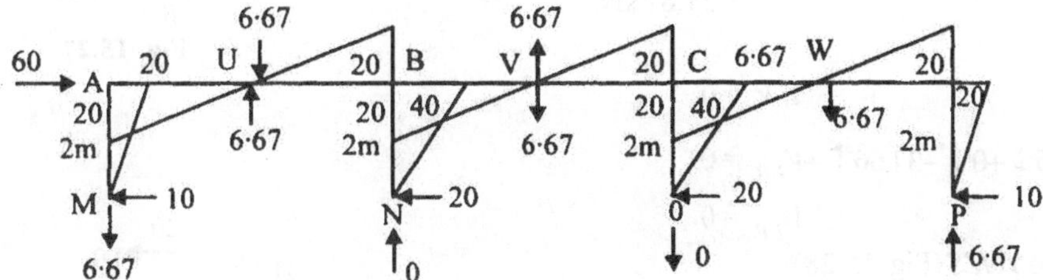

Fig. 15.24. B.M. diagram.

Ground Floor (Fig. 15.25)

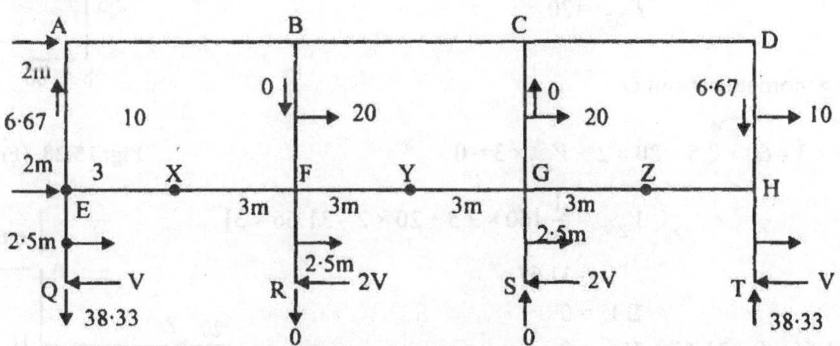

Fig. 15.25. Frame QADT.

Passing a plane through $QRST$ and from $\Sigma H = 0$

$$6V = 60 + 120$$

$$= 180$$

$$\mathbf{V = 30 \text{ kN}}$$

Frame MEQX (Fig. 15.26)

From $\Sigma H = 0$

$$120 + 10 - 30 + F_{XE} = 0$$

or $F_{XE} = -100$

Taking moments about E

$$10 \times 2 + 30 \times 2.5 - V_{XE} \times 3 = 0$$

or $V_{XE} = \frac{1}{3}[20 + 75]$

$= 31.667$ kN

Fig. 15.26

Frame $XNYR$ (Fig. 15.27)

From $\Sigma H = 0$

$$+100 + 20 - 60 + F_{YE} = 0$$

or $F_{YE} = -60$

Taking moment about F

$$31.667 \times 3 - 60 \times 2.5 - 20 \times 2 + V_{YF} \times 3 = 0$$

or $V_{YF} = \frac{1}{3}[95 + 150 + 40]$

$= 31.67$ kN.

Fig. 15.27

From $\Sigma V = 0$

$$31.66\downarrow + 0\downarrow - 31.66\uparrow + V_{FR} = 0$$

or $V_{FR} = 0$

Frame $YOZS$ (Fig. 15.28)

From $\Sigma H = 0$

$$60 + 20 - 60 - F_{ZG} = 0$$

or $F_{ZG} = 20$

Taking moments about G

$$31.66 \times 3 - 60 \times 2.5 - 20 \times 2 + V_{ZG} \times 3 = 0$$

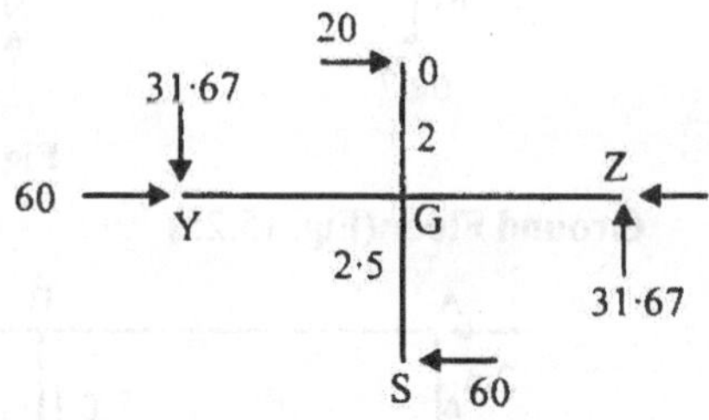

Fig. 15.28. (Frame YOZS)

or $V_{ZG} = \frac{1}{3}[60 \times 2.5 + 20 \times 2 - 31.66 \times 3]$

$= 31.67$

From $\Sigma V = 0$

$$31.66 - 0 - 31.67 + V_{SG} = 0$$

or $V_{SG} = 0$

$31.67 \times 3 - 10 \times 2 - 30 \times 2.5 = 0$ (check)

Frame $ZHPT$ (Fig. 15.29)

From $\Sigma V = 0$

$$31.67 + 6.67 - V_{TH} = 0$$

or $V_{TH} = $ **38.34 kN**

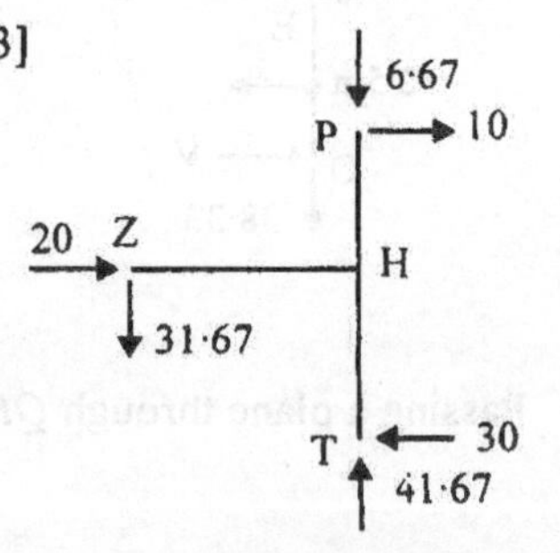

Fig. 15.29. (Frame $ZHPT$)

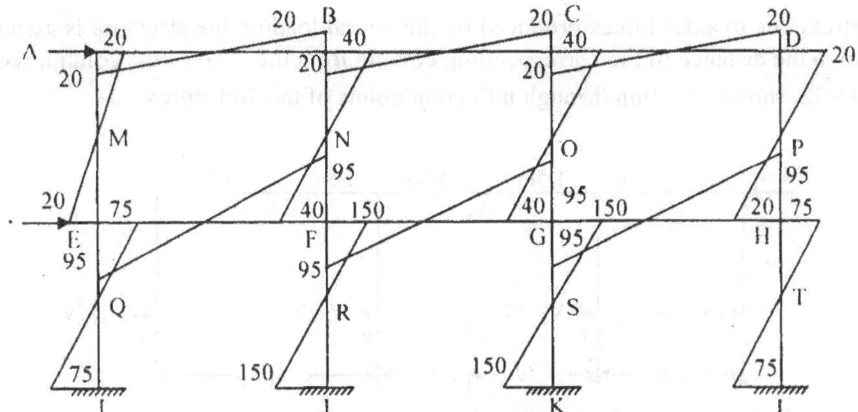

Fig. 15.30. B.M. diagram.

Example 15.4. Find the forces in the building frame by cantilever method.

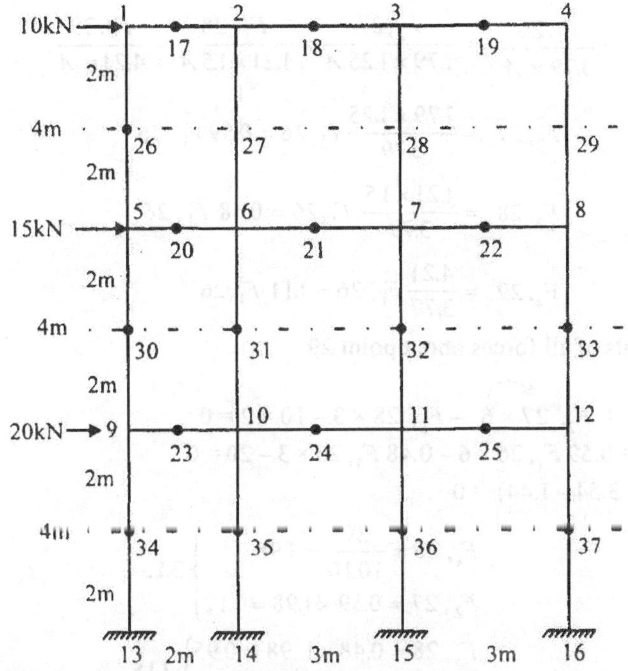

Fig. 15.31. Example 15.4

Solution : Since the height of the frame is greater than its width the displaced shape will be more like 15.16 (b) than 15.16 (a). Hence, cantilever method will be more appropriate. As already explained the inflection points are at the middle of both columns and beams. Area of columns is A, 1.25 A, 1.5 A and A in 1st, 2nd, 3rd and 4th row respectively.

Centre of gravity of the frame [assuming all columns in a row are of same area] from the line 1 – 5 – 9 – 13,

$$\bar{x} = \frac{A \times 0 + 1.25\,A \times 2 + 1.5\,A \times 5 + A \times 8}{A + 1.25\,A + 1.5\,A + A}$$

$$= \mathbf{3.79\ m}$$

The stress due to axial forces produced by the lateral load on the structure is assumed to be proportional to the distance of the corresponding column from the C.G. of the column sections.

Fig. 15.32. shows a section through inflection points of the 2nd storey.

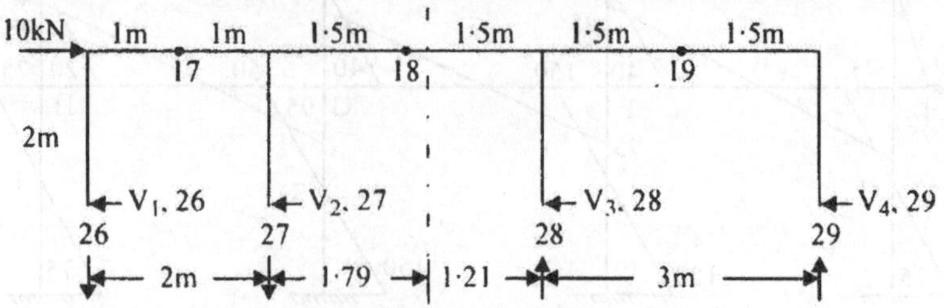

Fig. 15.32. Section through 2nd storey.

The column axial forces have the following relationship.

$$\frac{F_1,26}{3.79 \times A} = \frac{F_2,27}{1.79 \times 1.25\,A} = \frac{F_3,28}{1.21 \times 1.5\,A} = \frac{F_4,29}{4.21 \times A}$$

or
$$F_2,27 = \frac{1.79 \times 1.25}{3.79} F_1,26 = 0.59\,F_1,26$$

$$F_3,28 = \frac{1.21 \times 1.5}{3.79} F_1,26 = 0.48\,F_1,26$$

$$F_4,29 = \frac{4.21}{3.79} F_1,26 = 1.11\,F_1,26$$

Taking moments of all forces about point 29

$$F_1,26 \times 8 + F_2,27 \times 6 - F_3,28 \times 3 - 10 \times 2 = 0$$

or $F_1,26 \times 8 + 0.59\,F_1,26 \times 6 - 0.48\,F_1,26 \times 3 - 20 = 0$

or $F_1,26\,(8 + 3.54 - 1.44) = 0$

or
$$\left.\begin{aligned} F_1,26 &= \frac{20}{10.10} = 1.98 \\ F_2,27 &= 0.59 \times 1.98 = 1.17 \end{aligned}\right\} 3.15$$

$$\left.\begin{aligned} F_3,28 &= 0.48 \times 1..98 = 0.95 \\ F_4,29 &= 1.11 \times 1.98 = 2.20 \end{aligned}\right\} 3.15$$

Take moments of forces about Point 17

$$V_1,26 \times 2 = F_1,26 \times 1$$

or
$$V_1,26 = F_1,26 \times \frac{1}{2}$$

$$= 1.98 \times \frac{1}{2} = 0.99\text{ kN}$$

$$M_1,26 = V_1,26 \times 2 = 0.99 \times 2 = 1.98\text{ kNm}$$

Taking moments about hinge 18 for all forces to its left.

$$V_1, 26 \times 2 - F_1, 26 \times 3.5 + V_2, 27 \times 2 - F_2, 27 \times 1.5 = 0$$

or $\quad 0.99 \times 2 - 1.98 \times 3.5 + V_2, 27 \times 2 - 1.17 \times 1.5 = 0$

or $\quad V_2, 27 \times 2 = 1.98 \times 3.5 + 1.17 \times 1.5 - 0.99 \times 2$

$$= 6.705$$

or $\quad V_2, 27 = 3.35$

and $\quad M_2, 27 = 3.35 \times 2 = 6.70$

Taking moments of all forces about point 19

$$V_1, 26 \times 2 - F_1, 26 \times 6.5 + V_2, 27 \times 2 - F_2, 27 \times 4.5 + V_3, 28 \times 2 + F_3, 28 \times 1.5 = 0$$

or $\quad 0.99 \times 2 - 1.98 \times 6.5 + 3.35 \times 2 - 1.17 \times 4.5 + V_3, 28 \times 2 + 0.95 \times 1.5 = 0$

or $\quad V_3, 28 = \frac{1}{2}\ [1.98 \times 6.5 + 1.17 \times 4.5 - 3.35 \times 2 - 0.99 \times 2 - 0.95 \times 1.5)$

$= \mathbf{4.015}$

Considering right part of point 19

$$V_4, 29 \times 2 = F_4, 29 \times 1.5$$

$$= 2.20 \times 1.5$$

or $\quad V_4, 29 = \frac{2.20 \times 1.5}{2} = 1.65$

Check $V_1, 26 + V_2, 27 + V_3, 28,\ V_4, 29$

$= 0.99 + 3.35 + 4.015 + 1.65$

$= 10.006$

$\approx \mathbf{10\ kN}$

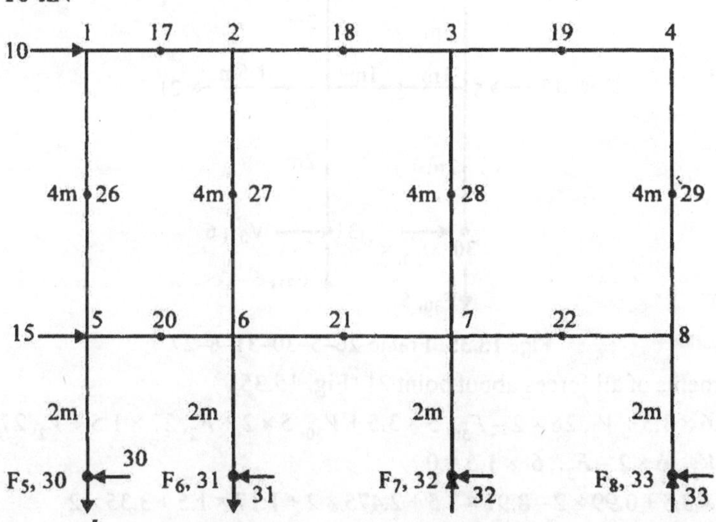

Fig. 15.33. Section through 1st Storey.

Section is passed through 30 – 31 – 32 – 33 to isolate the upper part. The stresses in the column will be proportional to their distance from the centre of the columns. Hence

$$F_6, 31 = 0.59\ F_5, 30$$

$$F_7, 32 = 0.48\ F_5, 30$$

$$F_8, 33 = 1.11\ F_5, 30$$

Taking moments of all forces, about 33 (Fig. 15.33)

$$10 \times 6 + 15 \times 2 - F_5, 30 \times 8 - F_6, 31 \times 6 + F_7, 32 \times 3 = 0$$

or $60 + 30 - F_5, 30 \times 8 - 0.59.F_{5,30} \times 6 + 0.48, F_5, 30 \times 3 = 0$

or $F_5, 30\,(8 + 6 \times 0.59 - 0.48 \times 3) = 90$

or
$$\left.\begin{aligned} F_5, 30 &= \frac{90}{10.1} = 8.91 \\ F_6, 31 &= 5.25 \end{aligned}\right\} 14.17$$
$$\left.\begin{aligned} F_7, 32 &= 4.38 \\ F_8, 33 &= 9.89 \end{aligned}\right\} 14.17$$

Taking moments of all forces about point 20 (Fig. 15.34)

$$F_1, 26 \times 1 + V_1, 26 \times 2 + V_5, 30 \times 2 \times F_5, 30 \times 1 = 0$$

or $1.98 \times 1 + 0.99 \times 2 + V_5, 30 \times 2 - 8.91 \times 1 = 0$

or $V_5, 30 = \frac{1}{2}[8.91 - 1.98 \times 2]$

$= 2.475$

$M_5, 30 = \uparrow 2.475 \times 2 = \mathbf{4.95 kNm.}$

Fig. 15.34. Frame 26-5-30-20.

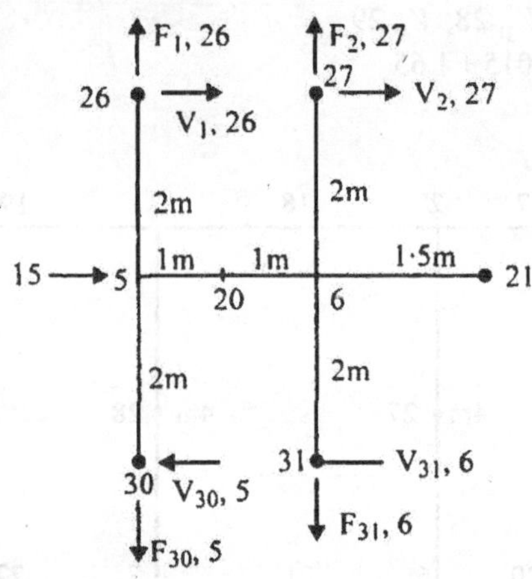

Fig. 15.35. Frame 26–5–30–31–6–27

Taking moments of all forces about point 21 (Fig. 15.35)

$$F_1, 26 \times 3.5 + V_1, 26 \times 2 - F_{30}, 5 \times 3.5 + V_{30}, 5 \times 2 + F_2, 27 \times 1.5 + V_2, 27 \times 2$$
$$+ V_{31}, 6 \times 2 - F_{31}, 6 \times 1.5 = 0$$

or
$$1.98 \times 3.5 + 0.99 \times 2 - 8.91 \times 3.5 + 2.475 \times 2 + 1.17 \times 1.5 + 3.35 \times 2$$
$$+ V_{31}, 6 \times 2 - 5.26 \times 1.5 = 0$$

or $V_{31}, 6 = \frac{1}{2}[5.26 \times 1.5 - 3.35 \times 2 - 1.17 \times 1.5 - 2.475 \times 2 + 8.91 \times 3.5 - 1.98 - 1.98 \times 3.5]$

$= \frac{1}{2}[16.76] = = \mathbf{8.38 kN}$

$M_6, 31 = 8.38 \times 2 = \mathbf{16.76 kN}$

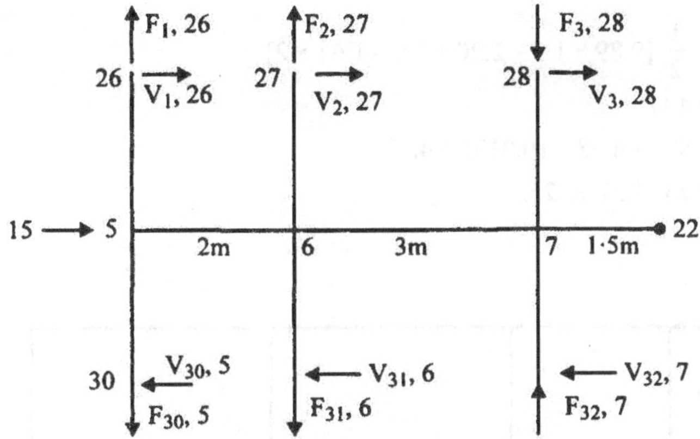

Fig. 15.36. Frame 26–5–30–32–7–28

Taking moments of all forces about point 22 (Fig. 15.36)

$$F_1, 26 \times 6.5 + F_2, 27 \times 4.5 - F_3, 28 \times 1.5 + (V_1, 26 + V_2, 27 + V_3, 28)$$

$$2 + (V_{30}, 5 + V_{31}, 6 + + V_{32}, 7) - 2\,F_{30}, 5 \times 6.5 - F_{31}, 6 \times 4.5 + F_{32}, 7 \times 1.5 = 0$$

or
$$V_{32}, 7 \times 2 = [F_{30}, 5 \times 6.5 + F_{31}, 6 \times 4.5 + F_3, 28 \times 1.5$$

$$- F_1, 26 \times 6.5 - F_2, 27 \times 4.5 - 2\,(V_1, 26 + V_2, 27 + V_3, 28)$$

$$- (V_{30}, 5 + V_{31}, 6) \times 2 - F_{32}, 7 \times 1.5]$$

$$V_{32}, 7 = \frac{1}{2}\;[(8.91 \times 6.5) + (5.26 \times 4.5) + (0.95\; \times 1.5 - 1.98\; \times 6.5)$$

$$-(1.17\; \times 4.5) - 2\,(0.99 + 3.35 + 4.015) - (2.475 + 8.38) \times 2 - (4.28 \times 1.5)]$$

$$= \frac{1}{2}\;[20.035] = 10.0175$$

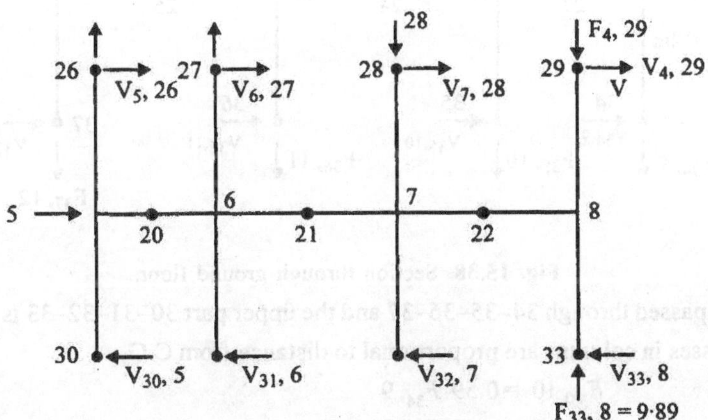

Fig. 15.37. Frame 26–30–33–29.

Taking moments of forces in the frame portion 22, 29, 8,33 about 22 (Fig. 15.37)

$$F_4, 29 \times 1.5 + V_4, 29 \times 2 + V_{33}, 8 \times 2 - F_{33}, 8 \times 1.5 = 0$$

or
$$V_{33}, 8 = \frac{1}{2}\,[F_{33}, 8 \times 1.5 - F_4, 29 \times 1.5 - V_4, 29 \times 2]$$

$$= \frac{1}{2}\,[9.89 \times 1.5 - 2.20 \times 1.5 - 1.65 \times 2]$$

$$= 4.12$$

Check: $2.475 + 4.38 + 10.0175 + 4.12$

$$= 24.9925 \approx 25.$$

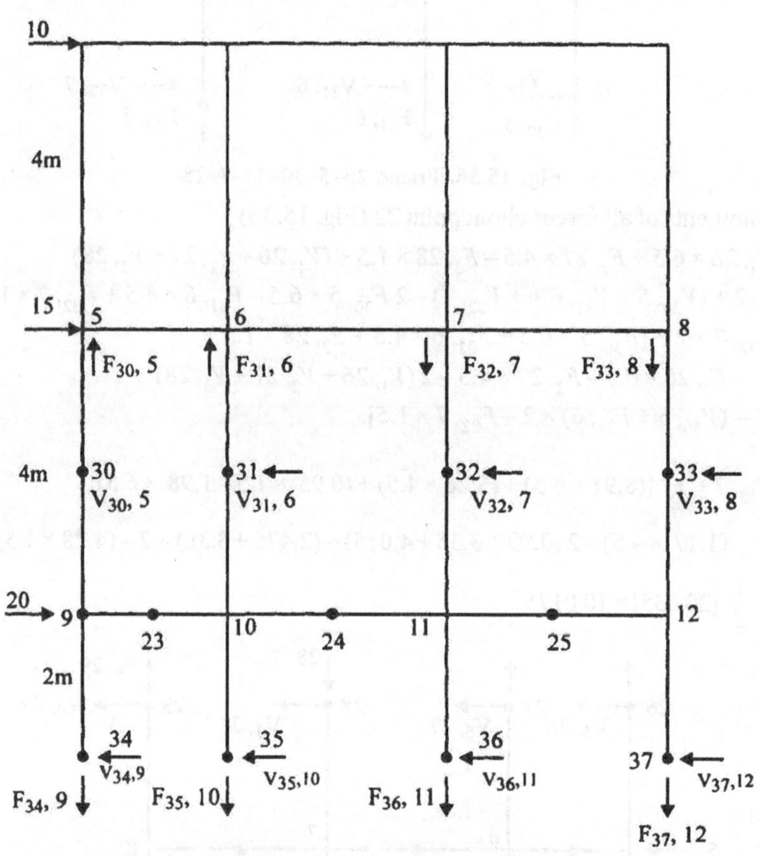

Fig. 15.38. Section through ground floor.

Section is passed through 34–35–36–37 and the upper part 30–31–32–33 is considered. Since stresses in columns are proportional to distance from C.G.

$$F_{35}, 10 = 0.59\, F_{34}, 9$$

$$F_{36}, 11 = 0.48\, F_{34}, 9$$

$$F_{37}, 12 = 1.11\, F_{34}, 9$$

Taking moments of all forces about 37.

$$F_{34}, 9 \times 8 + F_{35}, 10 \times 5 - F_{36}, 11 \times 3 - 20 \times 2 - 15 \times 6 - 10 \times 10 = 0$$

$$F_{34}, 9\,(8 + 0.59 \times 6 - 0.48 \times 3) = 40 + 90 + 100$$

or $\quad F_{34}, 9 = \dfrac{230}{10.1} = 22.77$

36.20

$F_{35}, 10 = 0.59 \times 22.77 = 13.43$

$F_{36}, 11 = 0.48 \times 22.77 = 10.90$

36.17

$F_{37}, 12 = 1.11 \times 22.77 = 25.27$

Taking moments of all forces about 23 (Fig. 15.39)

$$\overset{\curvearrowright}{F_{30}, 5 \times 1} + V_{30}, 5 \times 2 - V_{34}, 9 \times 2 - F_{34}, 9 \times 1 = 0$$

$$\overset{\curvearrowright}{8.91 \times 1} + 2.475 \times 2 - \overset{\curvearrowright}{22.77 \times 1} + V_{34}, 9 \times 2 = 0$$

or $\quad V_{34}, 9 \times 2 = 22.77 - 8.91 - 4.950 = 8.91$

$V_{34}, 9 = 4.455$

Fig. 15.39. Frame 30, 23-34.

Taking moments of all forces about 24 (Fig. 15.40)

$$V_{30}, 5 \times 2 + V_{31}, 6 \times 2 + F_{35}, 10 \times 2 + V_{34}, 9 \times 2 + V_{30}, 5 \times 3.5 + V_{31}, 6 \times 1.5$$
$$- F_{35}, 10 \times 1.5 - V_{34}, 9 \times 3.5 = 0$$

or $\quad 2.475 \times 2 + 8.38 \times 2 + V_{35}, 10 \times 2 + 4.455 \times 2 + 8.91 \times 3.5 + 5.26 \times 1.5 - 13.43 \times 1.5$
$-22.77 \times 3.5 = 0$

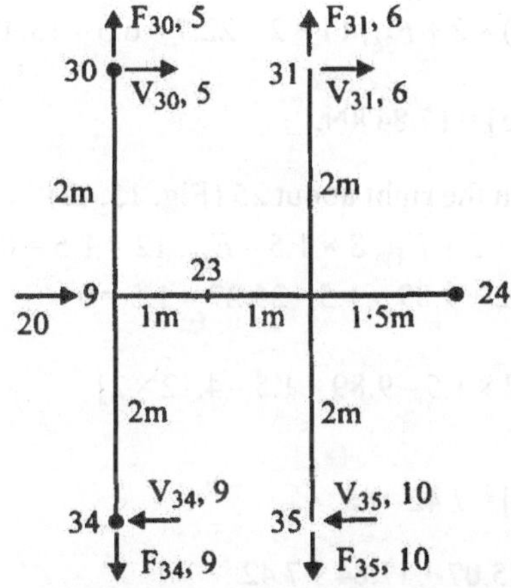

Fig. 15.40. Frame 30–34–35–31

or $\quad V_{35}, 10 = \dfrac{1}{2}\,[22.77 \times 3.5 + 13.43 \times 1.5 - 2.475 \times 2 - 8.38 \times 2 - 4.455 \times 2$
$- 8.91 \times 3.5 - 5.26 \times 1.5]$

$= \dfrac{1}{2}\,[30.145] = \mathbf{15.07.}$

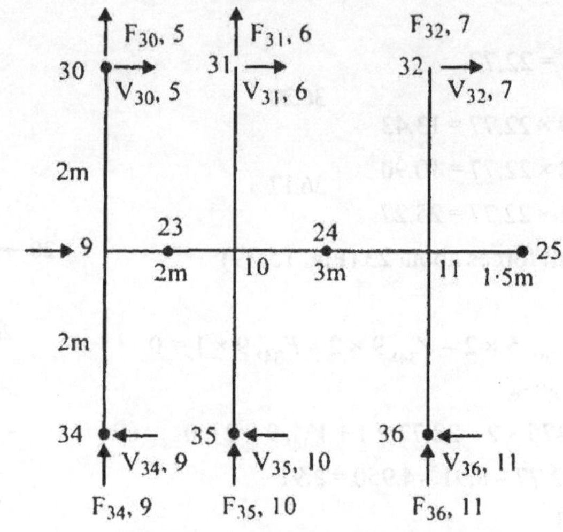

Fig. 15.41. Frame 30–34–36–32

Taking moments of all forces about hinge 25 (Fig. 15.41)

$$F_{30},5\times 6.5+F_{31},6\times 4.5-F_{32},7\times 1.5$$
$$+(V_{30},5+V_{31},6+F_{32},7)\times 2-(V_{34},9+V_{35},10)\times 2$$
$$+V_{36},11\times 2-F_{34},9\times 6.5-F_{35},10\times 4.5+F_{36},11\times 1.5=0$$

or $$8.91\times 6.5+5.26\times 4.5-4.28\times 1.5+(2.475+8.38+10.0175)\times 2$$
$$+(4.455+15.07)\times 2+F_{36},11\times 2-22.7\times 6.5-13.43\times 4.5+10.90\times 1.5=0$$

or $$V_{36},11=\frac{1}{2}[35.675]=17.84\text{ kN.}$$

Taking moment of forces on the right about 25 (Fig. 15.42)

$$V_{33},8\times 2+V_{37},12\times 2+F_{33},8\times 1.5-F_{37},12\times 1.5=0$$
$$4.12\times 2+V_{37},12\times 2+9.89\times 1.5-25.27\times 1.5=0$$

or $$V_{37},12=\frac{1}{2}[25.27\times 1.5-9.89\times 1.5-4.12\times 2]$$
$$=\frac{1}{2}[14.83]=7.42$$
$$\Sigma=4.455+15.07+17.84+7.42$$
$$=44.785\approx 45.$$

Fig. 15.42. Frame 25-33-37.

Summary

1. No analytical method, however refined it may be, is exact. A method will be called approximate when most refined available tool is not used for its analysis. This may be due to (*i*) prohibitive cost (*ii*) insufficient time (*iii*) complicated analysis. Before the advent of computers approximate analysis was very useful in analysis of a multi storey building frame as it converted a highly indeterminate structure to a statically determinate one.
2. In determinate trusses can be approximately analysed by equilibrium equations only after making any one of the following assumptions.

   1. Diagonals can resist tension only

2. Diagonals in a panel each resist half the shear.

3. All the vertical members are removed and the shear at a cross-section is resisted as though the truss functioned as two independent component statically determinate trusses.

3. The indeterminate beams and frames can be reduced to determinate forms by assuming location of inflection points and distribution of forces among several members.

4. A typical portal frame can be analysed approximately by making the following assumptions.
   (*a*) Inflection points occur at the mid height of left and right columns.
   (*b*) The shear in each column is equal to half the applied lateral load.

5. Building frames under vertical load only can be approximately analysed by statics only with the following assumptions.
   1. Inflection points will occur at 0.1 L from two ends of the girder.
   2. There is no axial force in the girder.

6. Building frames subjected to lateral forces can be analysed by two approximate methods
   1. Portal method
   2. Cantilever method.

7. Portal method is based on the following assumptions:
   (*a*) There is a point of inflection at the centre of each girder.
   (*b*) There is a point of inflection at the centre of each rigidly connected column. This does not apply to columns with pinned bases where the moment is zero.
   (*c*) The shear resisted by an exterior column of any storey is equal to $\frac{1}{2n}$ of the total storey shear and the shear resisted by each of the $(n-2)$ adjacent interior columns is $\frac{1}{n}$ where n is the number of bays in any storey.

   In the portal method the shear in column is determined first and then other forces and moments.

8. In the cantilever method the basic deformation is bending not shear. Stresses in columns are assumed to be proportional to its distance from the centroid of the column areas. The other assumptions about location of points of inflection at the middle of girders and columns remain valid in this method also. In this method axial stresses and axial forces are determined first and then other forces and moments.

16

MATRIX METHOD OF STRUCTURAL ANALYSIS

16.1 GENERAL INTRODUCTION

The developments in the methods of structural analysis which have been taking place during the last decades have been due almost entirely to the impact of the digital computer. The easy availability of high speed digital computers, revolutionary growth in the fieled of numerical techniques have made it possible to investigate and analyze large complicated structures with even thousands of unknown. Although the principles of equilibrium and compatibility, and the relationship between forces and displacement, remain basic in all the methods, their interpretations within various numerical processes have received much attention. An outstanding example of this development is the matrix method that has gained popularity for analyzing skeletal i.e. framed structure in recent years. The elegant finite element method is nothing but the extension of this method. The matrix method of structural analysis were developed and initially formulated for truss and beam. At early developments compatibility is enforced at discrete points throughout the structure, yielding a set of equations in which the forces are unknown and the coefficients in the equations are flexibility quantities. Hence, the method is termed as force method or flexibility method. Almost, simultaneously, work was being carried out on parallel method, in which equilibrium is enforced at discrete points, giving a set of equations with displacements as the unknowns and the coefficient as the stiffness quantities. The method is known as displacement method or the stiffness method. In both the methods, the governing equaions are constituted in matrix form. Matrix relation is a compact and convenient method for generating systems of equations,and the arrays represent the efficient format for writing systems of equations and storing the information, in computer. Most structural engineering offices now have access to computers that permits the use of the stiffness method for analyzing traditional civil engineering structures.

This chapter is intended to provide a reasonably comprehensive treatment of matrix method in structural analysis suitable for digital computer. The basic concepts of the stiffness method are explained. Then various basic structural elements i.e. truss; beam, frame etc. are discussed covering theoretical formulation, numerical examples and programming aspects. It is intended to give engineering students and practicing professionals the fundamentals of the background theory necessary in commercial frame analysis program. Also an attempt has been made to give an idea about the step by step procedure of core program development for the structural analysis software. This is of course not intended for the expert analysis, but it should serve as an introductory text to more advanced studies as the methodology and notations have been chosen to make easy the transition to finite element theory.

16.1.1 Classification of Framed / Skeletal structure

Framed structures are structures that can be satisfactorily idealized using line elements. Such structures are often referred to as skeletal structures. Usually the members of the structures are assumed to be connected either by friction-less pins or by rigid joints. Framed structure are usually idealized as skeletal structures as shown in Figure 16.1

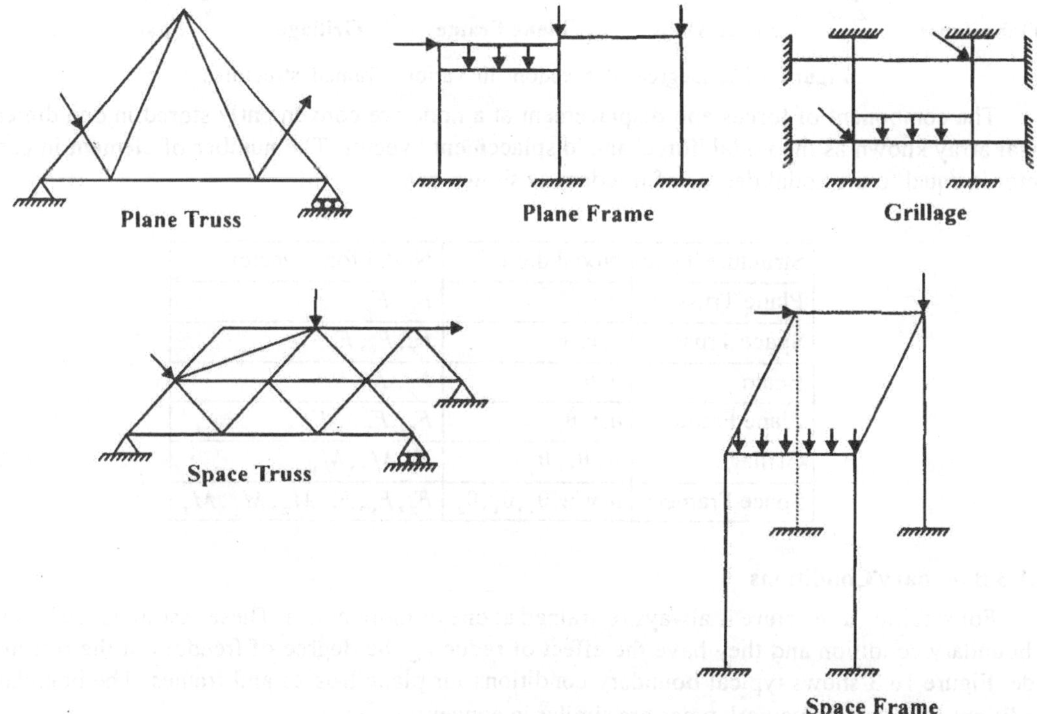

Figure 16.1 Various Framed structures.

Trusses are usually loaded at their joints and as the joints can not transmit bending moment; trusses must be triangular to avoid mechanism formation. Plane trusses are loaded in their own plane, whereas the joints of a space truss can be loaded from any direction. Rigid jointed frames are often loaded along their members as well as at the joints. Plane frame, like plane tursses are loaded only in their own plane. In contrast, grillages are always loaded normal to the structure. Space frames can, of course, be loaded in any plane.

16.1.2 Degree of Freedom

In the context of the stiffness method, the degree of freedom of a structure is the number of displacement components to be found during the analysis. Finding the displacements by stiffness method involves the solution of a system of linear equation relating the known applied forces to the degree of freedom of the structure. In general elastic structures have an infinite number of independent displacement components as displacements vary continuously through out the structure. Obviously, to solve an infinite number of simultaneous equations it will take an unacceptable long time. The matrix method renders the anlysis tractable by considering displacements and forces at selected points, called 'nodes'. When the matrix method is applied to structural form-works a node is usually located at each joint of the structure. The analysis determines the nodal displacements caused by applied loading. The number of possible dispalcement components at each joint is known as the 'degree of freedom'. Figure 16.2 illustrate the nodal degree of freedoms for different types of skeletal structure.

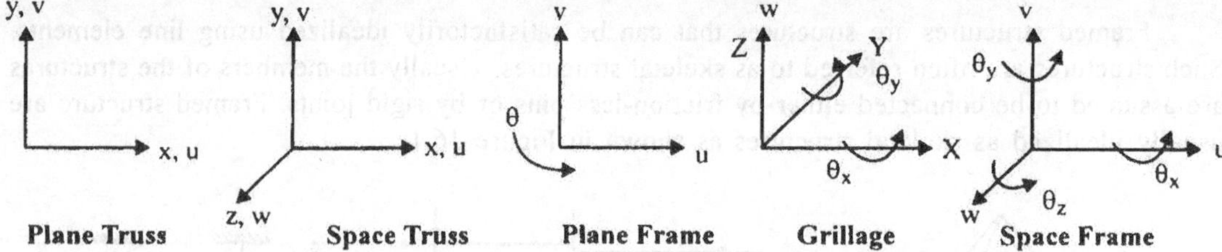

Figure 16.2. Degree of Freedom in various framed structures.

The component of forces and displacement at a node are conveniently stored in one dimensional array known as the nodal 'force' and 'displacement ' vector. The number of element in each vector is equal to the nodal degree of freedom as shown below:

Structure Type	Nodal d.o.f.	Nodal force vector
Plane Truss	u, v	$F_x,\ F_y$
Space Truss	u, v, w	F_x, F_y, F_z
Beam	$v,\ \theta_z$	$F_y,\ M_z$
Plane Frame	u, v, θ_z	$F_x,\ F_y,\ M_z$
Grillage	$w,\ \theta_x,\ \theta_y$	F_z, M_x, M_y
Space Frame	$u, v, w, \theta_x, \theta_y, \theta_z$	$F_x, F_y, F_z, M_x, M_y, M_z$

16.1.3 Boundary Conditions

For stability, a structure is always restrained at one or more nodes. These restraints are known as boundary condition and they have the effect of reducing the degree of freedom of the restraint node. Figure 16.3 shows typical boundary conditions for plane trusses and frames. The boundary conditions for other framework types are similar in concept.

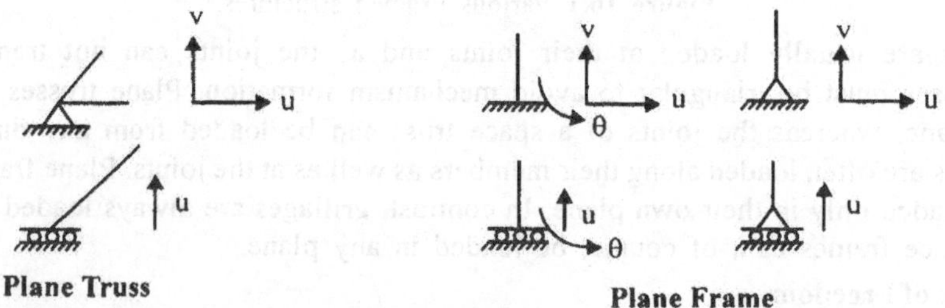

Figure 16.3. Various Boundary Conditions for plane frame and truss.

16.1.4 Force Displacement Relationships

The purpose of structural analysis is to determine the state of deformation or stress of the structure based on known force-displacement characteristics of the component parts, and satisfying all the conditions of compatibility and equilibrium. These relationships are known for standard structural elements, and in other cases can be derived by established procedures depending on the known geometry of the element and the elastic properties of the material.

Two simple structural elements are shown in figure 16.4 along with their force vectors and corresponding desplacement vectors. If force vector is denoted by $\{f\}$ and the displacement vector by $\{d\}$ then in general, there will be a relationship,

$$\{f\} = \{K\}\{d\} \quad 16.1$$

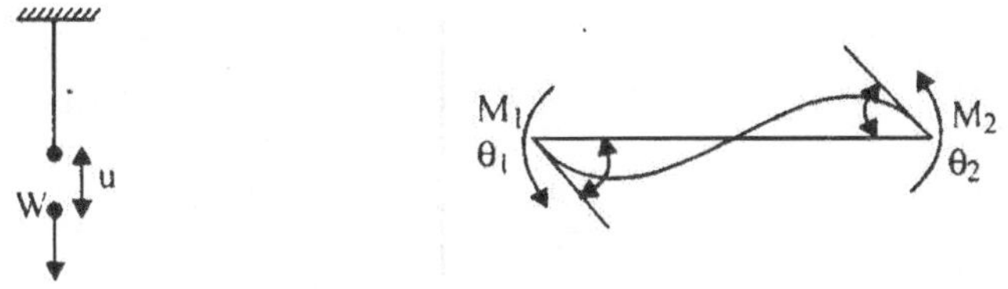

$$\{f\} = \begin{Bmatrix} M_1 \\ M_2 \end{Bmatrix} \text{ and } d = \begin{Bmatrix} \theta_1 \\ \theta_2 \end{Bmatrix}$$

$\{f\} = \{W\}$ and $d = \{u\}$

(a) Axial Bar **(b) Beam Segment**

Figure 16.4. Force Displacement relations.

Where, the matrix $[K]$ contains the coefficient of displacements in the force displacement relationship, the force vector $\{f\}$ may contain a number of different types of forces i.e. bending moment, twisting moment, shear forces, thrust etc. and the displacement vector $\{d\}$will contain the corresponding displacements. If the relationships are assumed linear (majority of structural problems involve linear relationships)the force displacement relationship for figure 16.4a can be written as

$$\{w\} = \frac{AE}{L}u \quad 16.2$$

so $[K]= \frac{AE}{L}$ and for figure 16.4b it can be written as (recalling slope deflection equations)

$$M_1 = \frac{2EI}{L}[2\theta_1 + \theta_2]$$

$$M_2 = \frac{2EL}{L}[2\theta_2 + \theta_1]$$

$$\text{i.e.} \quad \begin{Bmatrix} M_1 \\ M_2 \end{Bmatrix} = \frac{2EI}{L}\begin{bmatrix} 2 & 1 \\ 1 & 2 \end{bmatrix}\begin{Bmatrix} \theta_1 \\ \theta_2 \end{Bmatrix} \quad 16.3$$

Thus, $$[k] = \frac{2EI}{L}\begin{bmatrix} 2 & 1 \\ 1 & 2 \end{bmatrix}$$

16.1.5 Fundamental Concept of Matrix Method of Structural Analysis

The response of any complex structure can be cast as a set of algebraic equations that relate the forces and displacements functionally. If a structures is in a state of equilibrium and small displacement theory is valid then there is a unique relationship between the deformation of the structure and the load system applied to it.This dependency can be described either by the flexiblity or stiffness coefficient. Figure 16.5 shows a rectangular portal frame. The deformed shape is completely defined by the degree of freedoms as shown in the figure.

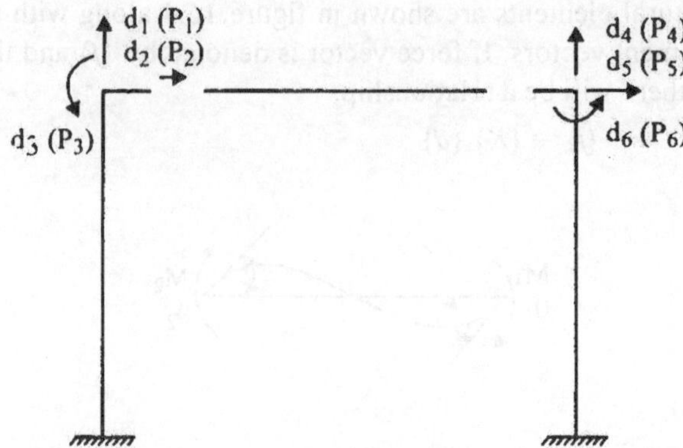

Figure 16.5. Unrestrained nodal displacements of Rectangular Frame.

If the loads in the directions of displacements are P_1, P_2,..., P_6 and the displacements are d_1, d_2, ..., d_6, then we can write either,

$$\begin{aligned} d_1 &= \phi_1 (P_1, P_2, ..., P_6) \\ d_2 &= \phi_2 (P_1, P_2, ..., P_6) \\ d_6 &= \phi_6 (P_1, P_2, ..., P_6) \end{aligned} \tag{16.4}$$

or

$$\begin{aligned} P_1 &= \psi_1 (d_1, d_2, ... d_6) \\ P_2 &= \psi_2 (d_1, d_2, ... d_6) \\ P_6 &= \psi_6 (d_1, d_2, ... d_6) \end{aligned} \tag{16.5}$$

where, ϕ_1, ϕ_2, ... ϕ_6 are some functions relating the unknown nodal displacements with applied forces or ψ_1, ψ_2 ...... ψ_6 are some other functions which relate the applied load with the nodal displacements.

Now if the structure is assumed to be linearly elastic then there is a linear relationship between the loads and the resulting displacements. These relationships can be expressed with either the loads or the displacements as the independent variables. Thus equation 16.4 can be written as,

$$\begin{aligned} d_1 &= f_{11} P_1 + ... + f_{16} P_6 \\ d_6 &= f_{61} P_1 + ... + f_{66} P_6 \end{aligned} \tag{16.6}$$

or

$$\begin{aligned} P_1 &= k_{11} d_1 + ... + k_{16} d_6 \\ P_6 &= k_{61} d_1 + ... + k_{66} d_6 \end{aligned} \tag{16.7}$$

Rewriting the above equations in matrix form as

$$\begin{Bmatrix} d_1 \\ d_2 \\ d_3 \\ d_4 \\ d_5 \\ d_6 \end{Bmatrix} = \begin{bmatrix} f_{11} & f_{12} & f_{13} & f_{14} & f_{15} & f_{16} \\ f_{21} & f_{22} & f_{23} & f_{24} & f_{25} & f_{26} \\ f_{31} & f_{32} & f_{33} & f_{34} & f_{35} & f_{36} \\ f_{41} & f_{42} & f_{43} & f_{44} & f_{45} & f_{46} \\ f_{51} & f_{52} & f_{53} & f_{54} & f_{55} & f_{56} \\ f_{61} & f_{62} & f_{63} & f_{64} & f_{65} & f_{66} \end{bmatrix} \begin{Bmatrix} P_1 \\ P_2 \\ P_3 \\ P_4 \\ P_5 \\ P_6 \end{Bmatrix} \text{ or } \{d\} = [f]\,[P] \tag{16.8}$$

Where $\{d\}$ and $\{P\}$ contains all the components of displacement and force respectively and the square matrix $[f]$ is the flexibility matrix. Note that f_{ij} represents the displacement at coordinate i due to a unit force at coordinate j. Since $f_{ij} = f_{ji}$ according to Maxwell's reciprocal theorem, it is evident that the flexibility matrix is symmetric.

Now equation (16.5) can be rewritten as

$$\begin{Bmatrix} P_1 \\ P_2 \\ P_3 \\ P_4 \\ P_5 \\ P_6 \end{Bmatrix} = \begin{Bmatrix} k_{11} & k_{12} & k_{13} & k_{14} & k_{15} & k_{16} \\ k_{21} & k_{22} & k_{23} & k_{24} & k_{25} & k_{26} \\ k_{31} & k_{32} & k_{33} & k_{34} & k_{35} & k_{36} \\ k_{41} & k_{42} & k_{43} & k_{44} & k_{45} & k_{46} \\ k_{51} & k_{52} & k_{53} & k_{54} & k_{55} & k_{56} \\ k_{61} & k_{62} & k_{63} & k_{64} & k_{65} & k_{66} \end{Bmatrix} \begin{Bmatrix} d_1 \\ d_2 \\ d_3 \\ d_4 \\ d_5 \\ d_6 \end{Bmatrix} \text{ or } \{f\} = [k]\{d\} \tag{16.9}$$

k_{ij} is the force at coordinate i due to a unit force at coordinate j. $k_{ij} = k_{ji}$ also .

Equation (16.8) and (16.9) form the basis of the stiffness method of structural analysis. For most structural syatems many manipulations are required to obtain the governing equations in the form of (16.8) or (16.9). It is desirable to have these equations generated systematically so that it can be programmed for the computer. The selection of the released structure forms an essential part of the flexibility method. On the other hand, the restrained structure used in the stiffness method is unique. A special feature of the restrained structure is that a displacement given at any one of the joints deforms only those members which are connected directly to the joint. The localized nature of the distrubance caused by a displacement at a joint is responsible for making most of the coefficient of the stiffness matrix equal to zero. Consequently, the development of the stiffness matrix involves lesser computations. On the contrary, a force applied at one of the joints of the released structure generally produces displacements at most of or all the joints. Thus, absence of localized effect causes most of the terms in the flexibility matrix non zero. Consequently, the development of flexibilty matrix involves greater computations. The computations of displacements in force method are generally more cumbersome as compared to the computations of forces in the stiffness method. The stiffness method is used almost exclusively and the equations can be easily generated with the computer, consequently, the flexiblity method will be abandoned at this point in this text.

16.2 MATRIX METHOD OF ANALYSIS OF PLANE TRUSSES

16.2.1 Introduction

A truss is a structure composed of straight bars connected at their points of intersection by means of momentless joints called pinned or hinged. All loading are assumed to be applied at those points of intersection. Thus each straight bar is subjected to axial load only without any bending or twist. The independent displacements at a joint of such structure are translations along a convenient set of orthogonal coordinate axes. The plane truss is the most important of all structural forms. Examples of application can be found in bridge trusses, roof trusses, truss beams and girders, truss walls etc. Three-dimensional or space trusses are also an important structural form. The examples include roof trusses for antenna, tower for supporting power line, truss structure for space satellite etc. In present chapter formation of basic stiffness matrix, assembly for the total structure and soloution procedure will be discussed.

16.2.2 Element Stiffness for Plane Truss in Axial Co-ordinate

The first step in formulating the analysis method of truss is to obtain the relationships between the force and displacements for an individual member. The simplest truss member can be thought of as a two noded bar element. The forces and displacements are defined in axial co-ordinate system. Displacements can be systematically imposed at the points and the forces are found so that the system maintained in equilibrium under prescribed displacement configuration.

Let us refer to figure 16.6 which consists of a truss member extending between node 1and 2. The member is assumed to have a constant cross sectional area A, modulus of elasticity E and length L.

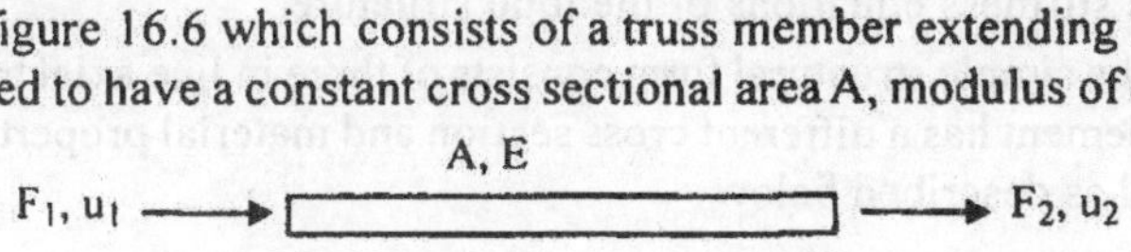

Fig. 16.6. Two noded simple bar.

Displacement Case 1:

If node 2 is fixed and force F_1 is applied at node 1 *i.e* $u_1 = u_1$ and $u_2 = 0$, then the force deformation relation may be written as

$$u_1 = \frac{F_1 L}{AE} \text{ or } F_1 = \left[\frac{AE}{L}\right] u_1 \tag{16.10}$$

Since node 2 is fixed, F_2 becomes a reaction. Imposing equilibrium on the forces we can write

$$F_2 = -F_1 \text{ i.e. } F_2 = -\left[\frac{AE}{L}\right] u_1 \tag{16.11}$$

Displacement Case 2:

If node 1 is fixed and force F_2 is applied at node 2 *i.e* $u_1 = 0$ and $u_2 = u_2$ then the force deformation relation may be written as

$$u_2 = \frac{F_2 L}{AE} \text{ or } F_2 = \left[\frac{AE}{L}\right] u_2 \tag{16.12}$$

Since node 1 is fixed, F_1 becomes a reaction. Imposing equilibrium on the forces we can write

$$F_1 = -F_2 \text{ i.e. } F_1 = -\left[\frac{AE}{L}\right] u_2 \tag{16.13}$$

General Displacement Case:

If both displacement components are simultaneously given, the force conforming to Hooke's law maintaining overall equilibrium of the member can be written as (combining equation 16.10 and 16.11, 16.12 and 16.13)

$$F_1 = F_1 \textit{ for case } 1 + F_1 \textit{ for case } 2 = \frac{AE}{L} u_1 - \frac{AE}{L} u_2$$

$$F_2 = F_2 \textit{ for case } 1 + F_2 \textit{ for case } 2 = -\frac{AE}{L} u_1 - \frac{AE}{L} u_2$$

which can be written in a matrix form as,

$$\begin{Bmatrix} F_1 \\ F_2 \end{Bmatrix} = \frac{AE}{L} \begin{bmatrix} 1 & -1 \\ -1 & 1 \end{bmatrix} \begin{Bmatrix} u_1 \\ u_2 \end{Bmatrix} \tag{16.14}$$

This equation defines the stiffness matrix for the truss member. However equation (16.14) is the simplest form of the stiffness equation for the truss member defined in local coordinates, that is, with the axis lying along the member.

16.2.3 System Stiffness Matrix

In the earlier section we have obtained the force displacement relationship of an isolated truss member. Thus the stiffness matrix for a particular truss member is obtained. However, a truss structure is combinations of more than one such member. Thus we must know the final force displacement relation for the total structure consists of all the members which will give the assembled stiffness matrix for the total structure. In present section, we will illustrate the basic principles of obtaining the assembled form of the stiffness equations of the total structure.

Let us consider the simple structural form consists of three in line axial truss member as shown in Figure 16.7. Each element has a different cross section and material property, the system stiffness matrix can be obtained as described below

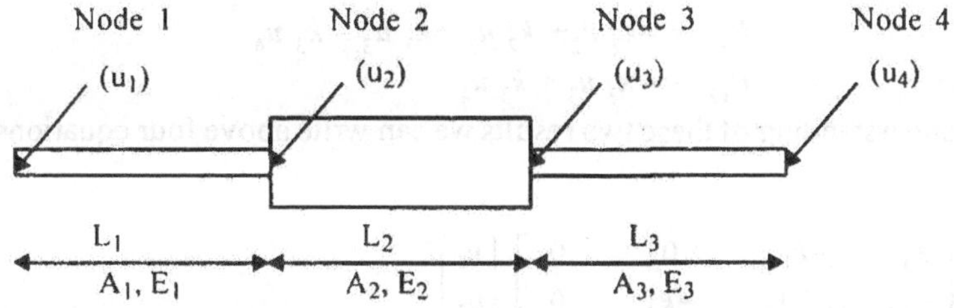

Fig. 16.7. Three member System.

The element level equilibrium equation for member 1 (between node 1 and 2)

$$\begin{Bmatrix} F_{x_1} \\ F_{x_2} \end{Bmatrix}_1 = \begin{bmatrix} k_1 & -k_1 \\ -k_1 & k_1 \end{bmatrix} \begin{Bmatrix} u_1 \\ u_2 \end{Bmatrix}, \text{ thus, } [k]_{1-2} = \begin{bmatrix} k_1 & -k_1 \\ -k_1 & k_1 \end{bmatrix} \quad (16.15)$$

The element level equilibrium equation for member 2(between node 2 and 3)

$$\begin{Bmatrix} F_{x_1} \\ F_{x_2} \end{Bmatrix}_2 = \begin{bmatrix} k_2 & -k_2 \\ -k_2 & k_2 \end{bmatrix} \begin{Bmatrix} u_2 \\ u_3 \end{Bmatrix}, \text{ thus, } [k]_{2-3} = \begin{bmatrix} k_2 & -k_2 \\ -k_2 & k_2 \end{bmatrix} \quad (16.16)$$

The element level equilibrium equation for member 3 (between node 3 and 4)

$$\begin{Bmatrix} F_{x_1} \\ F_{x_2} \end{Bmatrix}_3 = \begin{bmatrix} k_3 & -k_3 \\ -k_3 & k_3 \end{bmatrix} \begin{Bmatrix} u_3 \\ u_4 \end{Bmatrix}, \text{ thus, } [k]_{3-4} = \begin{bmatrix} k_3 & -k_3 \\ -k_3 & k_3 \end{bmatrix} \quad (16.17)$$

Where,

$$k_1 = \frac{A_1 E_1}{L_1},\ k_2 = \frac{A_2 E_2}{L_2},\ k_3 = \frac{A_3 E_3}{L_3} \quad (16.18)$$

Now let us consider that the external load applied at nodes are P_{x1}, P_{x2}, P_{x3} and P_{x4}. Internal forces acting are F_{x1}, and F_{x2} at each node for each element as shown in Figure 16.8

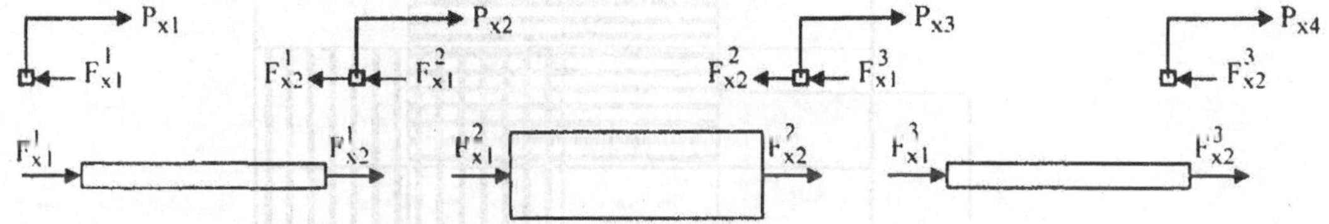

Fig. 16.8. Free Body Diagram showing internal forces.

We must bear in mind that the sum of the internal forces in x direction at any nodal point is equal to the external load applied at the nodal point in the same direction. Thus

$$\begin{aligned} P_{x1} &= F_{x1} \text{ of element 1} \\ P_{x2} &= F_{x2} \text{ of element 1} + F_{x1} \text{ of element 2} \\ P_{x3} &= F_{x2} \text{ of element 2} + F_{x1} \text{ of element 3} \\ P_{x4} &= F_{x2} \text{ of element 3} \end{aligned} \quad (16.19)$$

We must also bear in mind that at each nodal point the displacements at the element level are the same as those after assemblage. Above equation sets can be rewritten using equation 16.15 to 16.18 in 16.19 i.e.

$$P_{x1} = k_1 u_1 - k_1 u_2$$

$$P_{x2} = -k_1 u_1 + k_1 u_2 + k_2 u_3 - k_2 u_3$$

$$P_{x3} = -k_2 u_2 + k_2 u_3 + k_3 u_3 - k_3 u_4 \qquad (16.20)$$
$$P_{x2} = -k_3 u_3 + k_3 u_4$$

With the understanding of these two results we can write above four equations in a matrix form as,

$$\begin{Bmatrix} P_{x_1} \\ P_{x_2} \\ P_{x_3} \\ P_{x_4} \end{Bmatrix} = \begin{bmatrix} k_1 & -k_1 & 0 & 0 \\ -k_1 & k_1 + k_2 & -k_2 & 0 \\ 0 & -k_2 & k_2 + k_3 & -k_3 \\ 0 & 0 & -k_3 & k_3 \end{bmatrix} \begin{Bmatrix} u_1 \\ u_2 \\ u_3 \\ u_4 \end{Bmatrix} \text{ i.e. } \{P\} = [K]\{u\} \qquad (16.21)$$

Above matrix [K] is the assembled matrix. Thus the system stiffness matrix $[K]$ can be obtained by forming a matrix of pigeon hole, each pigeon hole corresponding to the displacements u_1, u_2 u_3 u_4. The next process is to add together the components of stiffness from each element stiffness matrix with respect to the nodal displacements into the appropriate pigeon hole as explained in figure 16.9

u_1	u_2	u_3	u_4	
k_1	$-k_1$	0	0	u_1
$-k_1$	$-k_1+k_2$	$-k_2$	0	u_2
0	$-k_2$	k_2+k_3	$-k_3$	u_3
0	0	$-k_3$	k_3	u_4

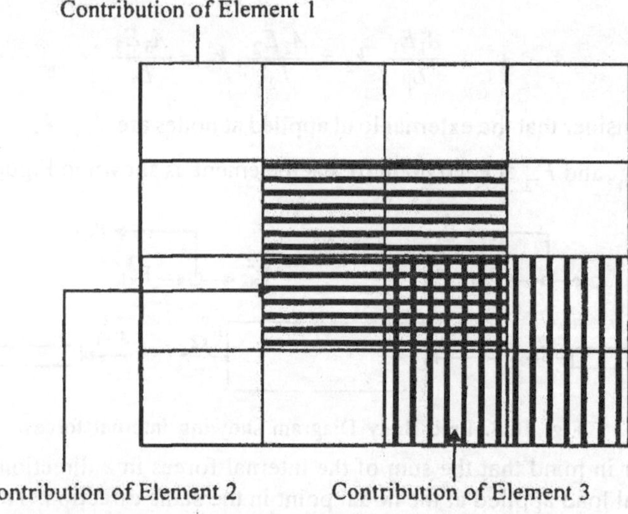

Fig. 16.9. Assembly of Element Stiffness matrix.

16.2.4 Stiffness Matrix in Global Coordinate System

Since truss is composed of many members, it will be convenient to introduce both local and reference coordinates. The local coordinates are always chosen to represent the individual member conveniently. Reference coordinates i.e. global coordinates, on the contrary are chosen to be convenient for the overall structure. Figure 16.10 shows a truss member 1– 2 and two coordinate system. The local coordinates are shown as $\bar{x}$ and $\bar{y}$ and reference or global coordinates are shown as x and y. Note that the member is arbitrarily located with respect to the global axes system. Throughout this text displacements, force and stiffness written with respect to local coordinates, are designated by $\bar{u}$, $\bar{F}$ and $\bar{k}$ respectively and the bar is omitted when described with respect to global axis.

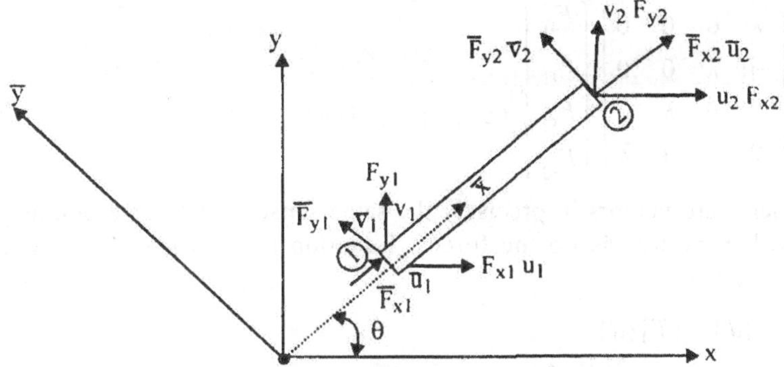

Fig. 16.10. The truss member oriented arbitrary.

In Figure 16.10 forces are shown at node *1* and *2*. In local coordinate system we have forces $\bar{F}_{x1}$ and $\bar{F}_{x2}$. To describe in reference coordinate the force needs to replaced by two components F_x *and* F_y. Note that $\bar{F}_{y1}$ and $\bar{F}_{y2}$ does not exist since the truss member can not sustain a normal components of force. The important point emerging here is the following: whereas a total of two components of forces appear for the member in terms of local coordinate, a total of four components arises when reference coordinates are used. Therefore, prior to transformation of equation (16.14) to reference coordinates, it must be expanded to the order which is consistent with the use of reference coordinates. This is simple to do. We introduce $\bar{F}_{y1}$ and $\bar{F}_{y2}$ as well as $\bar{v}_1$ and $\bar{v}_2$ into equation 16.14. This is done by adding corresponding rows and columns of zeros in $\bar{k}$. In other words, we rewrite equation (16.14) as

$$\begin{Bmatrix} \bar{F}_{x1} \\ \bar{F}_{y1} \\ \bar{F}_{x2} \\ \bar{F}_{y2} \end{Bmatrix} = \frac{AE}{L} \begin{bmatrix} 1 & 0 & -1 & 0 \\ 0 & 0 & 0 & 0 \\ -1 & 0 & 1 & 0 \\ 0 & 0 & 0 & 0 \end{bmatrix} \begin{Bmatrix} \bar{u}_1 \\ \bar{v}_1 \\ \bar{u}_2 \\ \bar{v}_2 \end{Bmatrix} \qquad (16.22)$$

16.2.5 Transformation of the Stiffness Matrix

An operation of considerable importance is the transformation of a stiffness matrix from one set of coordinates to another. This is often used in solving problems particularly on the computer; it is also of great value in deriving the general form of stiffness matrices. Referring to figure 16.10, considering the component of forces along the member axis it can be written as,

$$\bar{F}_{x1} = F_{x1}\cos\theta + F_{y1}\sin\theta$$

$$\bar{F}_{y1} = -F_{x1}\sin\theta + F_{y1}\cos\theta \qquad (16.23)$$

Similar equations hold at node 2 i.e.

$$\bar{F}_{x2} = F_{x2}\cos\theta + F_{y2}\sin\theta$$

$$\bar{F}_{y2} = -F_{x2}\sin\theta + F_{y2}\cos\theta$$

If we put $\lambda = \cos\theta$ and $\mu = \sin\theta$ the above set of equation may be written in matrix form as

$$\begin{Bmatrix} \bar{F}_{x1} \\ \bar{F}_{y1} \\ \bar{F}_{x2} \\ \bar{F}_{y2} \end{Bmatrix} = \begin{bmatrix} \lambda & \mu & 0 & 0 \\ -\mu & \lambda & 0 & 0 \\ 0 & 0 & \lambda & \mu \\ 0 & 0 & -\mu & \lambda \end{bmatrix} \begin{Bmatrix} F_{x1} \\ F_{y1} \\ F_{x2} \\ F_{y2} \end{Bmatrix} \quad \text{i.e. } \{\bar{F}\} = \{T\}\{F\} \tag{16.24}$$

Displacements are vectors in precisely the same sense as the corresponding forces. Consequently, they will transform as do the forces. Equation (16.24) may be written for the nodal displacements as well *i.e.*

$$\{\bar{d}\} = [T]\{d\} \tag{16.25}$$

16.2.6. Stiffness matrix in reference coordinate system

Substituting equation (16.2.16) and (16.2.17) in equation (16.2.13),

$$[T]\{F\} = [\bar{k}][T]\{d\}$$

i.e. $$\{F\} = [T]^{-1}[\bar{k}][T]\{d\} \tag{16.26}$$

As the transformation matrix $[T]$ is an orthogonal matrix, then $[T]^{-1} = [T]^T$. So, equation 16.26 can be written as

$$\{F\} = [T]^T[\bar{k}][T]\{d\}$$

i.e. $$\{F\} = [K]\{d\}, \text{ where } [k] = [T]^T[\bar{k}][T] \tag{16.27}$$

Hence, the stiffness matrix to an arbitrarily oriented reference system can be found as

$$[K] = [T]^T[\bar{k}][T] = \begin{bmatrix} \lambda & -\mu & 0 & 0 \\ \mu & \lambda & 0 & 0 \\ 0 & 0 & \lambda & -\mu \\ 0 & 0 & \mu & \lambda \end{bmatrix} \frac{AE}{L} \begin{bmatrix} 1 & 0 & -1 & 0 \\ 0 & 0 & 0 & 0 \\ -1 & 0 & 1 & 0 \\ 0 & 0 & 0 & 0 \end{bmatrix} \begin{bmatrix} \lambda & \mu & 0 & 0 \\ -\mu & \lambda & 0 & 0 \\ 0 & 0 & \lambda & \mu \\ 0 & 0 & -\mu & \lambda \end{bmatrix}$$

$$= \frac{AE}{L} \begin{bmatrix} \lambda^2 & \lambda\mu & -\lambda^2 & -\lambda\mu \\ \lambda\mu & \mu^2 & -\lambda\mu & -\mu^2 \\ -\lambda^2 & -\lambda\mu & \lambda^2 & \lambda\mu \\ -\lambda\mu & -\mu^2 & \lambda\mu & \mu^2 \end{bmatrix} \tag{16.28}$$

Thus, the equilibrium equation at element level can be written as

$$\begin{Bmatrix} F_{x1} \\ F_{y1} \\ F_{x2} \\ F_{y2} \end{Bmatrix} = \frac{AE}{L} \begin{bmatrix} \lambda^2 & \lambda\mu & -\lambda^2 & -\lambda\mu \\ \lambda\mu & \mu^2 & -\lambda\mu & -\mu^2 \\ -\lambda^2 & -\lambda\mu & \lambda^2 & \lambda\mu \\ -\lambda\mu & -\mu^2 & \lambda\mu & \mu^2 \end{bmatrix} \begin{Bmatrix} u_1 \\ v_1 \\ u_2 \\ v_2 \end{Bmatrix} \tag{16.29}$$

16.2.7 Properties of Stiffness Matrix

Let us consider the two dimensional bar element as a free body shown in figure 16.11.

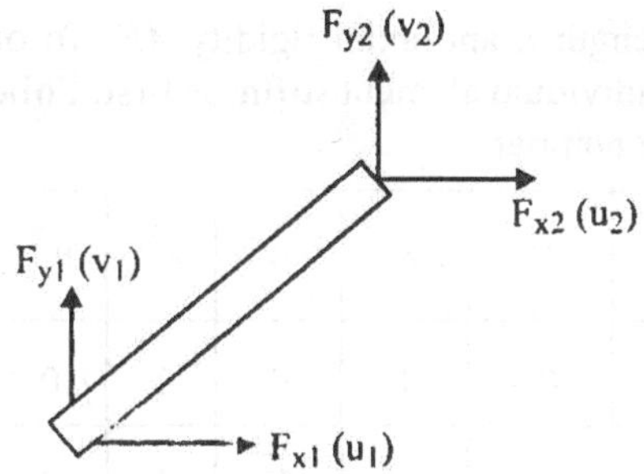

Fig. 16.11. Free body diagram of the truss element.

We see in the stiffness matrix as described in equation (16.29) that the coefficient in the first row are the same but opposite in sign as the coefficient in the third row. The same relation holds between the second and fourth ro'vs. If we multiply the matrix equation, we find that,

$$F_{x1} = \lambda^2 u_1 + \lambda\mu v_1 - \lambda^2 u_2 - \lambda\mu v_2$$

$$F_{x2} = -\lambda^2 u_1 - \lambda\mu v_1 + \lambda^2 u_2 + \lambda\mu v$$

So, $F_{x2} = -F_{x1}$, Similarly we can show $F_{y2} = -F_{y2}$. So the equilibrium in both x and y directions for the free body is satisfied. Now we take the moment about nodal point 1.

$$\sum M = \frac{AE}{L}[-\lambda^2 u_1 - \lambda\mu v_1 + \lambda^2 u_2 + \lambda\mu v_2] L\mu - [-\lambda\mu m_1 - \mu^2 v_1 - \lambda\mu u_2 - \mu^2 v_2]\ L\lambda = 0$$

which satisfy the moment equilibrium conditions.

In the expression of $[k]$ we note that the first and third rows are same but opposite in sign,as are the second and fourth rows. Thus $[k]$ is singular matrix. It cannot be inverted and solution to stiffness equation (16.29) is not possible. Physically, singular stiffness matrix means that the element without any support is an unstable free body. The element becomes stable and the stiffness matrix can be nonsingular if it is properly supported.

Also we see that the diagonal element k_{ii} for all i is positive. This implies a force F_i directed towards right will not produce displacement towards left. For example, if all degree of freedom is zero except u_i, then from stiffness equilibrium equation

$$F_i = k_{ii} u_i$$

which shows that F_i and u_i have the same sign as k_{ii} is positive.

16.2.8 Assembly of Element Matrix in Global Coordinate System

To demonstrate the method of assemblage, an example of three bar system as shown in figure 16.12 is chosen.

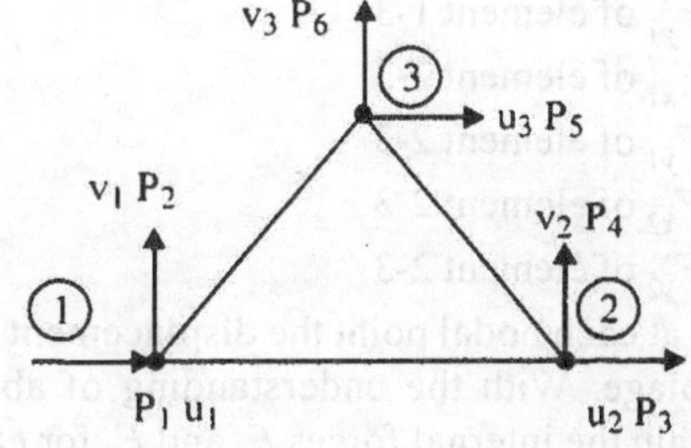

Fig. 16.12. Three bar truss.

All bars have the same length L and axial rigidity AE. To obtain the total system stiffness matrix, we must formulate the individual element stiffness first. Following table shows the direction consines and sines for the three number.

Element	θ	λ	μ	λ^2	μ^2	λμ
1 – 2	0°	1	0	1	0	0
2 – 3	120°	$-\frac{1}{2}$	$\frac{\sqrt{3}}{2}$	$\frac{1}{4}$	$\frac{3}{4}$	$-\frac{\sqrt{3}}{4}$
1 – 3	60°	$\frac{1}{2}$	$\frac{\sqrt{3}}{2}$	$\frac{1}{4}$	$\frac{3}{4}$	$\frac{\sqrt{3}}{4}$

For element 1-2 using equation (16.29) and the value of λ and μ from the above table.

$$\begin{Bmatrix} F_{x1} \\ F_{y1} \\ F_{x2} \\ F_{y2} \end{Bmatrix} = \frac{AE}{L}\begin{bmatrix} 1 & 0 & -1 & 0 \\ 0 & 0 & 0 & 0 \\ -1 & 0 & 1 & 0 \\ 0 & 0 & 0 & 0 \end{bmatrix}\begin{Bmatrix} u_1 \\ v_1 \\ u_2 \\ v_2 \end{Bmatrix} \qquad (16.30)$$

For Element 2-3,

$$\begin{Bmatrix} F_{x2} \\ F_{y2} \\ F_{x3} \\ F_{y3} \end{Bmatrix} = \frac{AE}{4L}\begin{bmatrix} 1 & -\sqrt{3} & -1 & \sqrt{3} \\ -\sqrt{3} & 3 & \sqrt{3} & -3 \\ -1 & \sqrt{3} & 1 & -\sqrt{3} \\ \sqrt{3} & -3 & -\sqrt{3} & 3 \end{bmatrix}\begin{Bmatrix} u_2 \\ v_2 \\ u_3 \\ v_3 \end{Bmatrix} \qquad (16.31)$$

For Element 1-3

$$\begin{Bmatrix} F_{x_1} \\ F_{y_1} \\ F_{x_3} \\ F_{y_3} \end{Bmatrix} = \frac{AE}{4L}\begin{bmatrix} 1 & \sqrt{3} & -1 & -\sqrt{3} \\ \sqrt{3} & 3 & -\sqrt{3} & -3 \\ -1 & -\sqrt{3} & 1 & \sqrt{3} \\ -\sqrt{3} & -3 & \sqrt{3} & 3 \end{bmatrix}\begin{Bmatrix} u_1 \\ v_1 \\ u_3 \\ v_3 \end{Bmatrix} \qquad (16.32)$$

We must bear in mind that F_x's and F_y's are the internal nodal forces, when the three sets of element stiffness equations are assembled, the sum of such internal forces in either x or y direction at each nodal point must be equal to the external load P applied in the same direction Thus,

$P_1 = F_{x1}$ of element 1-2 + F_{x1} of element 1-3
$P_2 = F_{y1}$ of element 1-2 + F_{y1} of element 1-3
$P_3 = F_{x2}$ of element 1-2 + F_{x1} of element 2-3
$P_4 = F_{y2}$ of element 1-2 + F_{y1} of element 2-3
$P_5 = F_{x2}$ of element 1-3 + F_{x2} of element 2-3
$P_6 = F_{y2}$ of element 1-3 + F_{y2} of element 2-3 (16.33)

We must bear in mind that at each nodal point the displacement u and v at the element level are the same as those after assemblage. With the understanding of above two rules, the method of assemblage is simply first to obtain the internal forces F_x and F_y for each element and then use those values in the above static equilibrium equation. Thus

$$P_1 = \frac{AE}{L}\left[(4+1)u_1 + \sqrt{3}v_1 - 4u_2 - u_3 - \sqrt{3}v_3\right]$$

$$P_2 = \frac{AE}{L}\left[\sqrt{3}u_1 + 3v_1 - \sqrt{3}u_3 - 3v_3\right]$$

$$P_3 = \frac{AE}{L}\left[-4u_1 + (4+1)u_2 - \sqrt{3}v_2 - u_3 + \sqrt{3}v_3\right]$$

$$P_4 = \frac{AE}{L}\left[-\sqrt{3}u_2 + 3v_2 + \sqrt{3}u_3 - 3v_3\right]$$

$$P_5 = \frac{AE}{L}\left[-u_1 + \sqrt{3}v_1 - u_2 + \sqrt{3}v_2 + (1+1)u_3 + (\sqrt{3} - \sqrt{3})v_3\right]$$

$$P_6 = \frac{AE}{L}\left[-\sqrt{3}u_1 - 3v_1 + \sqrt{3}u_2 - 3v_2 + \left(\sqrt{3} - \sqrt{3}\right)u_3 + (3+3)v_3\right]$$

we can write these equation in matrix form

$$\begin{Bmatrix} P_1 \\ P_2 \\ P_3 \\ P_4 \\ P_5 \\ P_6 \end{Bmatrix} = \frac{AE}{4L}\begin{bmatrix} (4+1) & \sqrt{3} & -4 & 0 & -1 & -\sqrt{3} \\ \sqrt{3} & 3 & 0 & 0 & -\sqrt{3} & -3 \\ -4 & 0 & (4+1) & -\sqrt{3} & -1 & \sqrt{3} \\ 0 & 0 & -\sqrt{3} & 3 & \sqrt{3} & -3 \\ -1 & -\sqrt{3} & -1 & \sqrt{3} & (1+1) & \left(\sqrt{3}-\sqrt{3}\right) \\ -\sqrt{3} & -3 & \sqrt{3} & -3 & \left(\sqrt{3}-\sqrt{3}\right) & (3+3) \end{bmatrix}\begin{Bmatrix} u_1 \\ v_1 \\ u_2 \\ v_2 \\ u_3 \\ v_3 \end{Bmatrix} \quad (16.34)$$

The foregoing assemblage is equivalent to move 4 × 4 element stiffness matrix into a 6 × 6 total stiffness matrix as illustrated in figure 16.13. The position where each element matrix block moves depends on the related degree of freedom number associated with that element.

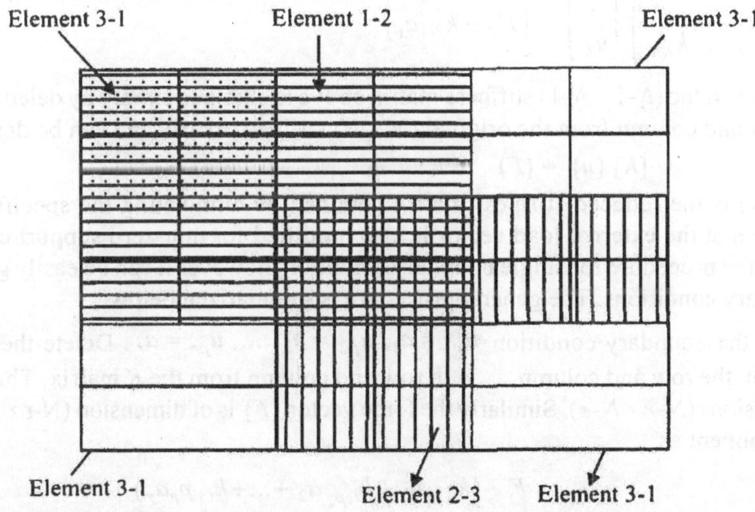

Fig. 16.13. Assmebly of element stiffness matrix.

16.2.9. Application of Boundary Conditions

In the structural stiffness equation $[K]\{u\} = \{F\}$, matrix $[K]$ is singular and an unique solution for displacement vector $\{u\}$ is possible if the structure is supported. Some d.o.f. in $\{u\}$ must be prescribed to enable a solution. In this section we consider procedure for imposing prescribed values of one or more d.o.f. u_i's. It should be emphasized that improper specification of boundry conditions can lead to erroneous result. Boundary conditions eliminate the possibilities of the structures moving as a rigid body. Further, boundary conditions should properly model the physical system. Two approaches will now be discussed for handling specified displacement boundary conditions.

16.2.9a Elimination Approach

The structural stiffness equation $[K]\{u\} = \{F\}$ a basically a set of n simultaneous linear equations which can be written as

$$\begin{aligned} k_{11}u_1 + k_{12}u_2 + k_{13}u_3 + \ldots\ldots + k_{1N}u_N &= F_1 \\ k_{21}u_1 + k_{22}u_2 + k_{23}u_3 + \ldots\ldots + k_{2N}u_N &= F_2 \\ k_{31}u_1 + k_{32}u_2 + k_{33}u_3 + \ldots\ldots + k_{3N}u_N &= F_3 \\ k_{N1}u_1 + k_{N2}u_2 + k_{N3}u_3 + \ldots\ldots + k_{NN}u_N &= F_N \end{aligned} \tag{16.35}$$

To illustrate the basic idea, consider the single boundary condition $u_1 = a_1$. Note that we have to determine N-1 unknown u value, as u_1 is already known. Thus, it is sufficient to solve N-1 equations for obtaining N-1 unknown d.o.f. Excluding first of the above set of equations, we can rewrite by substituting the known value of $u_1 = a_1$, Then the above equation can be written in matrix form as

$$\begin{aligned} k_{22}u_2 + k_{23}u_3 + \ldots\ldots + k_{2N}u_N &= F_2 - k_{21}a_1 \\ k_{32}u_2 + k_{33}u_3 + \ldots\ldots + k_{3N}u_N &= F_3 - k_{31}a_1 \\ k_{N2}u_2 + k_{N3}u_3 + \ldots\ldots + k_{NN}u_N &= F_N - k_{N1}a_1 \end{aligned} \tag{16.36}$$

i.e. in matrix form

$$\begin{bmatrix} k_{22} & k_{23} & . & k_{2N} \\ k_{32} & k_{33} & . & k_{3N} \\ . & . & . & . \\ k_{2N} & k_{3N} & . & k_{NN} \end{bmatrix} \begin{Bmatrix} u_2 \\ u_3 \\ . \\ u_N \end{Bmatrix} = \begin{Bmatrix} F_2 - k_{21}a_1 \\ F_3 - k_{31}a_1 \\ F_N - k_{N1}a_1 \end{Bmatrix} \tag{16.37}$$

we now observe that the (N-1 × N-1) stiffness matrix above is obtained simply by deleting or eliminating the first row and column from the original (N × N) stiffness matrix and can be denoted as

$$[K]\{u\} = \{F\} \tag{16.38}$$

Where $[K]$ is the reduced stiffness matrix obtained by eliminating the specified or support d.o.f. Also note that the external load vector is also modified for non-zero support conditions. We have explained the procedure for single boundary condition, however it can be easily generalized for multiple boundary conditions. The general procedure is summarized below:

Consider the boundary condition $u_{p1} = a_1, u_{p2} = a_2, \ldots, u_{pr} = a_r$. Delete the p_1th row and column, the p_2th the row and column, ..., p_rth row and column from the K matrix. The resulting [K] matrix of dimensions (N–r × N–r). Similarly the force vector $\{F\}$ is of dimension (N-r × 1). Modify the each load component as

$$F_i = F_i - (k_{i,p1}a_1 + k_{i,p2}a_2 + \ldots + k_i, p_r a_r)$$

for each d.o.f.i. that is not a support. Now solve $[K]\{u\} = \{F\}$ for the displacement vector $\{u\}$

16.2.9b Penalty Method

A second approach to incorporate boundry conditions will now be discussed. This approach is easy to implement in a computer program. To illustrate the basic idea, consider the single boundary condition $u_1 = a_1$. A spring with large suffiness k_s is added to model the corresponding support. Now add this large stiffness to the corresponding diagonal i.e. with k_{11} and apply the large force $k_s \times a_1$ and discard the load F_1. The corresponding matrix equation becomes.

$$(k_{11}+k_s)u_1 + k_{12}u_2 + k_{13}u_{32} + ...+ k_{1N}Nu_N = F_1 + k_s a_1$$

$$k_{21}u_1 + k_{22}u_2 + k_{23}u_3 + ...+ k_{2N}u_N = F_2 \tag{16.39}$$

$$k_{31}u_1 + k_{32}u_2 + k_{33}u_3 + ...+ k_{3N}u_N = F_3$$

$$k_{N1}u_1 + k_{N2}u_2 + k_{N3}u_3 + ...+ k_{NN}u_N = F_N$$

Now the first of above set of equation can be written as

$$u_1 = \frac{F_1 + k_s a_1 - (k_{12}u_2 + k_{13}u_3 + ...+ k_{1N}u_N)}{k_{11}+k_s} \approx a_1$$

as k_s is very large compare to other quantities in the right hand side. From above we see that the only modifications to handle $u_1 = a_1$ are that a large number k_s gets added to the first diagonal element of $[K]$, and that $k_s a_1$ gets added to F. Solution of equation (16.39) yields the displacement vector $\{u\}$. The above procedure can be easily generalized for multiple boundary conditions. The general procedure is summarized below

Consider the boundry condition $u_{p1} = a_1, u_{p2} = a_2, ..., u_{pr} = a_r$. Modify the structural stiffness matrix $[K]$ by adding a large number k_s to each p_1th, p_2th, ... p_rth diagonal elements of K. Also, modify the global load vector $\{F\}$ by adding $k_s a_1$ to $F_{p_1}, k_s a_2$ to $F_{p_2}, k_s a_r$ to F_{pr}. Now solve $[K]\{u\} = \{F\}$ for the displacement vector $\{u\}$.

Example

Consider the truss bar system discussed in 16.12. Let it be assumed that the truss is supported as shown in Figure 16.14. The supporting conditions are $u_1 = v_2 = u_3 = v_3 = 0$. Only u_2 and v_1 remain to be solved. The reduced stiffness matrix is obtained by deleting the rows and columns of corresponding dof to be fixed. In this problem 2nd, 3rd, 5th & 6th d.o.f. is restrained due to support conditions. Thus, those rows and columns in equation 16.34 need to be deleted and the reduced stiffness matrix becomes as:

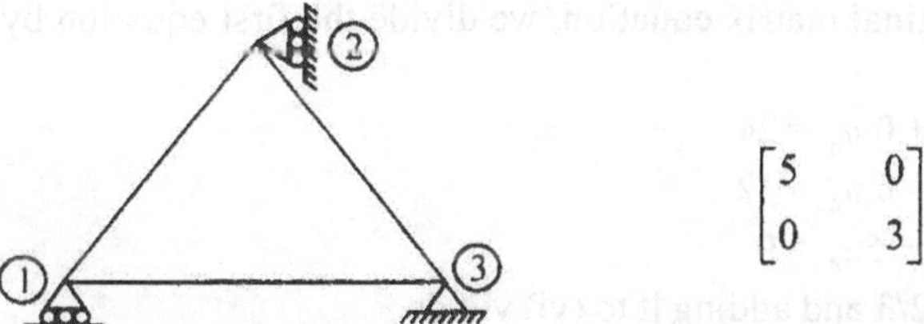

$$\begin{bmatrix} 5 & 0 \\ 0 & 3 \end{bmatrix}$$

Fig. 16.14. Three member truss with support.

16.2.10. Solutions

Structural equation $[K]\{d\} = \{P\}$ can be solved by a direct method or an indirect (iterative) method. In either case there are many algorithms to choose from. Computational aspects of equations solving may found in various numerical analysis book. However, in present section, we will discuss the Gauss elimination method which is used in various program in this book for solving simultaneous equations.

Gauss Elimination solution

We will first present the method by means of an example, followed by general solution and algorithm. Consider the example shown in Figure 16.15.

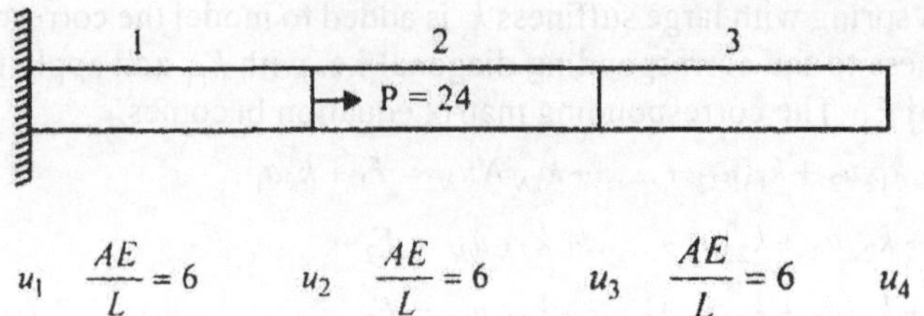

$u_1 \quad \frac{AE}{L} = 6 \quad u_2 \quad \frac{AE}{L} = 6 \quad u_3 \quad \frac{AE}{L} = 6 \quad u_4$

Fig. 16.15. Three element axial bar.

The stiffness matrix for all the elements is same as below.

$$k = \frac{AE}{L}\begin{bmatrix} 1 & -1 \\ -1 & 1 \end{bmatrix}$$

and the global stiffness matrix is as below

$$K = \frac{AE}{L}\begin{bmatrix} 1 & -1 & 0 & 0 \\ -1 & 1+1 & -1 & 0 \\ 0 & -1 & 1 & -1 \\ 0 & 0 & -1 & 1 \end{bmatrix} = 6\begin{bmatrix} 1 & -1 & 0 & 0 \\ -1 & 2 & -1 & 0 \\ 0 & -1 & 2 & -1 \\ 0 & 0 & -1 & 1 \end{bmatrix}$$

After application of the boundary conditions using penalty method, the equilibrium equation becomes

$$\begin{bmatrix} 12 & -6 & 0 \\ -6 & 12 & -6 \\ 0 & -6 & 6 \end{bmatrix}\begin{Bmatrix} u_2 \\ u_3 \\ u_4 \end{Bmatrix} = \begin{Bmatrix} 24 \\ 0 \\ 0 \end{Bmatrix}$$

This matrix equation is nothing but a set of simultaneous equation

$$12u_2 - 6u_3 + 0\,u_4 = 24 \qquad (i)$$

$$-6\,u_2 + 12\,u_3 - 6\,u_4 = 0 \qquad (ii)$$

$$0\,u_2 - 6u_3 + 6u_4 = 0 \qquad (iii)$$

Starting with the original matrix equation, we divide the first equation by 2 and add it to the second equation yielding.

$$12u_2 - 6u_3 + 0\,u_4 = 24 \qquad (iv)$$

$$0\,u_2 + 9\,u_3 - 6\,u_4 = 12 \qquad (v)$$

$$0\,u_2 - 6u_3 + 6u_4 = 0 \qquad (vi)$$

Now multiple (v) by 2/3 and adding it to (vi) yields

$$12u_2 - 6\,u_3 + 0\,u_4 = 24 \qquad (vii)$$

$$0\,u_2 + 9\,u_3 - 6\,u_4 = 12 \qquad (viii)$$

$$0\,u_2 + 0u_3 + 2u_4 = 8 \qquad (ix)$$

The solution is now virtually complete, since the last equation yields $u_4 = \frac{8}{2} = 4$, which upon substitution on (*vii*) gives $u_3 = 4$ and u_2 from equation (*viii*), $u_2 = 4$.

$$\begin{bmatrix} 12 & -6 & 0 & : 24 \\ -6 & 12 & -6 & : 0 \\ 0 & -6 & 6 & : 0 \end{bmatrix} \rightarrow \begin{bmatrix} 12 & -6 & 0 & : 24 \\ 0 & 9 & -6 & : 12 \\ 0 & -6 & 6 & : 0 \end{bmatrix} \rightarrow \begin{bmatrix} 12 & -6 & 0 & : 24 \\ 0 & 9 & -6 & : 12 \\ 0 & 0 & 2 & : 8 \end{bmatrix}$$

Which upon back substitution yields

$$u_4 = 4.,\ u_3 = 3.,\ u_2 = 4$$

General Algorithm

We have discussed the Gaussian elimination process above by means of an example. This process will be now stated in general suitable for computer programming. It is a procedure applied to the general system of equations.

$$\begin{aligned} a_{11}x_1 + a_{12}x_2 + \ldots\ldots + a_{1n}x_n &= b_1 \\ a_{21}x_1 + a_{22}x_2 + \ldots\ldots a_{2n}x_n &= b_2 \\ &\vdots \\ a_{n1}x_1 + a_{n2}x_2 + \ldots\ldots + a_{nn}x_n &= b_n \end{aligned} \qquad (16.40)$$

i.e. In matrix form,

$$\begin{bmatrix} a_{11}\ a_{12} \ldots\ldots a_{1n} \\ a_{21}\ a_{22} \ldots\ldots a_{2n} \\ \cdot \\ a_{n1}\ a_{n2} \ldots\ldots a_{nn} \end{bmatrix} \begin{Bmatrix} x_1 \\ x_2 \\ \cdot \\ \cdot \\ \cdot \\ x_n \end{Bmatrix} = \begin{Bmatrix} b_1 \\ b_2 \\ \cdot \\ \cdot \\ \cdot \\ b_n \end{Bmatrix} \quad \text{i.e. } [A]\{x\} = \{b\}$$

to reduce it to the form

$$\begin{bmatrix} a_{11}\ a_{12} \ldots\ldots a_{1n} \\ 0\ a_{22} \ldots\ldots a_{2n} \\ \cdot\ \ a_3 \ldots\ldots a_{3n} \\ \cdot \\ 00.0 \ldots\ldots a_{nn} \end{bmatrix} \begin{Bmatrix} x_1 \\ x_2 \\ \cdot \\ x_n \end{Bmatrix} = \begin{Bmatrix} b_1 \\ b_2 \\ \cdot \\ b_n \end{Bmatrix} \qquad (16.41)$$

where prime indicates new values in position a_{ij} and b_i. This set is obtained by scaling and adding rows to obtain the zero values below each diagonal. The process of obtaining this upper triangular form is called forward substitution. Once the system is in the form, the values of x_n is easily obtained from the last row as $x_n = b_n / a_{nn}$. Moving back to the row above x_{n-1} can be now easily found, since x_n has been previously determined. This process of successive solving for x starting with the last value is called back substitution.

The simplest way to carry out forward elimination is to start with row 1. Divide every element of this row (including b_1) by the diagonal term a_{11}. The term a_{11} is now 1. We can now obtain zeros in first column by multiplying row 1 by a_{21} and subtracting the results from row 2; i.e. $a_{21} = a_{21} - a_{21}\ (1) = 0$. The same may be done for b values. The zero in row 3 is then obtain by mutiplying row 1 by a_{31} and subtracting the factored row 1 from row 3, and so on. The reduction process at step 1 is as.

$$a_{ij} = a_{ij} - \frac{a_{i1}}{a_{11}} \times a_{1j}$$

$$b_i = b_i - \frac{a_{i1}}{a_{11}} \times b_1$$

when the first column is done we move to second column. Divide row 2 by the present value of a_{22} and proceed as before by multiplying row 2 by a_{32} and subtracting from row 3 and so on. Thus, at step k of the reduction procedure

$$a_{ij}^{(k)} = a_{ij}^{(k-1)} - \frac{a_{ik}^{(k-1)}}{a_{kk}^{(k-1)}} a_{kj}^{(k-1)} \qquad i, j = k+1, \ldots\ldots, N$$

$$b_i^{(k)} = b_i^{(k-1)} - \frac{a_{ik}^{(k-1)}}{a_{kk}^{(k-1)}} b_k^{(k-1)} \qquad i = k+1, \ldots\ldots, N$$

After n–1 steps we get the upper triangular matrix. The back substitution process is given by

$$x_n = \frac{b_n'}{a_{mn}'}$$

$$x_i = \frac{b_i' - \sum_{j=i+1}^{n} a_{ij}' x_j}{a_{ii}'}, \; i = N-1, N-2, \ldots\ldots, 1$$

16.2.11 Internal Forces

To complete the total analysis it is now necessary to determine the unknown internal forces in the various elements of the structure. For the truss type structure these are simply the tensile or compressive forces in each member. Figure 16.16 shows the Free body diagram of element *i-j*.

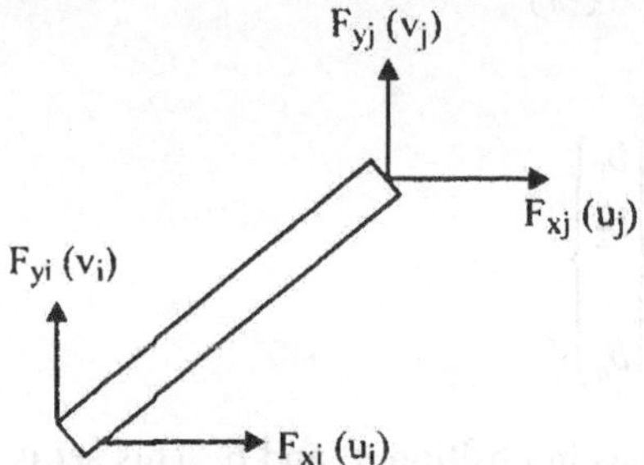

Fig. 16.16. Free body diagram of *i-j*th truss element.

Now the static equilibrium equation for the (i–j)th element

$$\begin{Bmatrix} F_{xi} \\ F_{yi} \\ F_{xj} \\ F_{yj} \end{Bmatrix} = \left(\frac{AE}{L}\right)_{ij} \begin{bmatrix} \lambda^2 & \lambda\mu & -\lambda^2 & -\lambda\mu \\ \lambda\mu & \mu^2 & -\lambda\mu & -\mu^2 \\ -\lambda^2 & -\lambda\mu & \lambda^2 & \lambda\mu \\ -\lambda\mu & -\mu^2 & \lambda\mu & \mu^2 \end{bmatrix} \begin{Bmatrix} u_i \\ v_i \\ u_j \\ v_j \end{Bmatrix} \tag{16.42}$$

Thus, $F_{xi} = \left(\frac{AE}{L}\right)_{i-j} \left[\lambda^2 u_i + \lambda\mu v_i - \lambda^2 u_j - \lambda\mu v_j\right]$ and so on for other forces. Now the axial forces $S_{i\,j}$ can be obtained in terms of F_i and F_j's as

$$S_{i\text{-}j} = F_{xi}\lambda + F_{yi}\mu = F_{xj}\lambda + F_{yj}\mu$$

16.2.12 Example Problem

Determine the displacements and internal forces for the truss shown in figure 16.17. Use the following data: Modulus of Elasticity E = 2 × 10^{11} N/m^2, Cross-section area of Member 1-2: 4 × 10^{-4} m^2, Member 2-3: 4 × 10^{-4} m^2, Member 2-4: 6 × 10^{-4} m^2

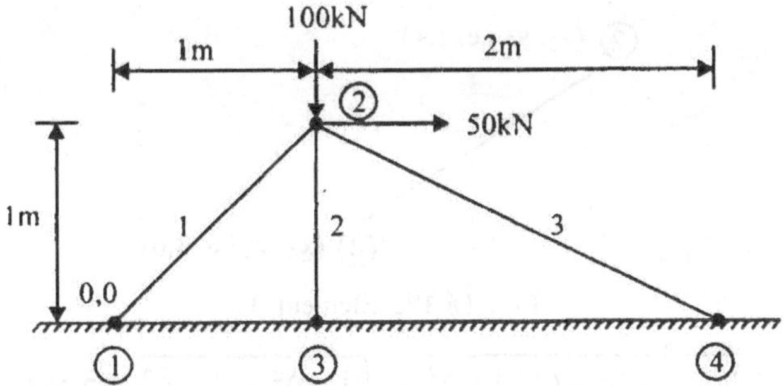

Fig. 16.17. Three member truss.

Element 1(1-2):

By defining the element as 1-2 it means that the element points from 1 to 2, as shown below.

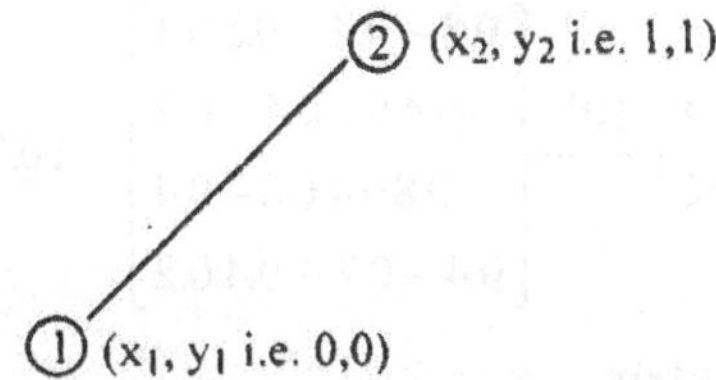

Fig. 16.17*a* Element 1.

Now, $L_{1-2} = \sqrt{(x_2 - x_1)^2 + (y_2 - y_1)^2} = \sqrt{(1-0)^2 + (1-0)^2} = 1.414$

$$\lambda = \cos\alpha = \frac{x_2 - x_1}{L_{1-2}} = \frac{1-0}{1.414} = 0.707,\ \mu = \sin\alpha = \frac{y_2 - y_1}{L_{1-2}} = \frac{1-0}{1.414} = 0.707$$

$$\text{Then, } [k_{1-2}]^e = \frac{4\times10^{-4}\times2\times10^{11}}{1.414}\begin{bmatrix} 0.5 & 0.5 & -0.5 & -0.5 \\ 0.5 & 0.5 & -0.5 & -0.5 \\ -0.5 & -0.5 & 0.5 & 0.5 \\ -0.5 & -0.5 & 0.5 & 0.5 \end{bmatrix} = 28.284\times10^6\begin{bmatrix} 1 & 1 & -1 & -1 \\ 1 & 1 & -1 & -1 \\ -1 & -1 & 1 & 1 \\ -1 & -1 & 1 & 1 \end{bmatrix}$$

Element 2-3:

2 (x_2, y_2 i.e. 1,1)

3 (x_3, y_3 i.e. 1,0)

Fig. 16.17*b* Element 2

$$L_{2-3} = \sqrt{(x_3 - x_2)^2 + (y_3 - y_2)^2} = \sqrt{(1-1)^2 + (0-1)^2} = 1.0$$

$$\lambda = \cos\alpha = \frac{x_3 - x_2}{L_{2-3}} = \frac{1-1}{1.0} = 0.0,\ \mu = \sin\alpha = \frac{y_3 - y_2}{L_{2-3}} = \frac{0-1}{1.0} = -1.0$$

$$\text{Then, } [k_{2-3}]^e = \frac{4\times10^{-4}\times2\times10^{11}}{1.0}\begin{bmatrix} 0 & 0 & 0 & 0 \\ 0 & 1 & 0 & -1 \\ 0 & 0 & 0 & 0 \\ 0 & -1 & 0 & 1 \end{bmatrix} = 80.0\times10^6\begin{bmatrix} 0 & 0 & 0 & 0 \\ 0 & 1 & 0 & -1 \\ 0 & 0 & 0 & 0 \\ 0 & -1 & 0 & 1 \end{bmatrix}$$

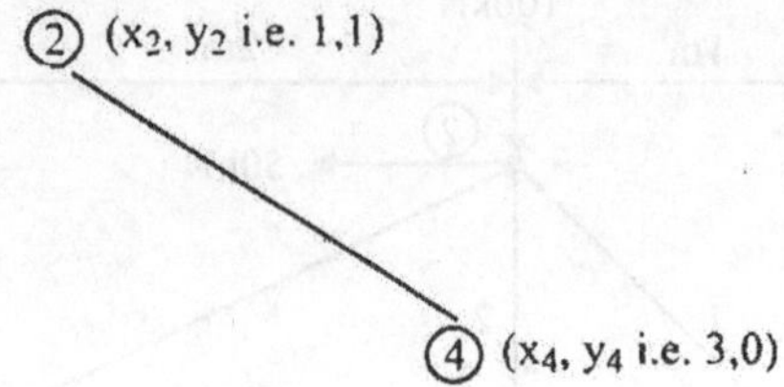

Fig. 16.17*c* Element 3.

$$L_{4-2} = \sqrt{(x_4 - x_2)^2 + (y_4 - y_2)^2} = \sqrt{(1-3)^2 + (1-0)^2} = 2.236$$

$$\lambda = \cos\alpha = \frac{x_4 - x_2}{L_{4-2}} = \frac{3-1}{2.236} = 0.894, \mu = \sin\alpha = \frac{y_4 - y_2}{L_{4-2}} = \frac{0-1}{2.236} = -0.447$$

$$\text{Then, } [k_{2-4}]^e = \frac{6 \times 10^{-4} \times 2 \times 10^{11}}{2.236} \begin{bmatrix} 0.8 - 0.4 - 0.8\,0.4 \\ -0.4\,0.2\,0.4 - 0.2 \\ -0.8\,0.4\,0.8 - 0.4 \\ 0.4 - 0.2 - 0.4\,0.2 \end{bmatrix} = 10.733 \times 10^6 \begin{bmatrix} 4 & -2 & -4 & 2 \\ -2 & 1 & 2 & -1 \\ -4 & 2 & 4 & -2 \\ 2 & -1 & -2 & 1 \end{bmatrix}$$

Assembly of Element Stiffness matrix

The truss has four nodal points and each node has 2 d.o.f. Thus the total d.o.f for the truss is eight, hence the size of the global stiffness matrix is 8 by 8. The system stiffness matrix can be obtained by forming a pigeon hole as described earlier for axial problem. Following table shows the positions of each 4 × 4 element stiffness matrix move during assemblage.

u_1	v_1	u_2	v_2	u_3	v_3	u_4	v_4	
28.29	28.29	−28.29	−28.29					u_1
28.29	28.29	−28.29	−28.29					v_1
−28.29	−28.29	28.29 + 0 + 49.26	28.29 + 0 − 21.46	0	0	− 49.26	21.46	u_2
−28.29	−28.29	28.29 + 0 − 21.46	28.29 + 80 + 10.73	0	−80	21.46	−10.73	v_2
		0	0	0	0			u_3
		0	− 80	0	80			v_3
		− 49.26	21.46			49.26	−21.46	u_4
		21.46	−10.73			− 21.46	10.73	v_4

Thus, the assemblage global stiffness matrix is

$$[K] = 10^6 \begin{bmatrix} 28.29* & 28.29 & -28.29 & -28.29 & 0 & 0 & 0 & 0 \\ 28.29 & 28.29* & -28.29 & -28.29 & 0 & 0 & 0 & 0 \\ -28.29 & -28.29 & 77.55 & 6.83 & 0 & 0 & -49.26 & 21.46 \\ -28.29 & -28.29 & 6.83 & 139.02 & 0 & 0 & 21.46 & -10.73 \\ 0 & 0 & 0 & 0* & 0 & 0 & 0 & 0 \\ 0 & 0 & 0 & -80 & 0 & 80* & 0 & 0 \\ 0 & 0 & -49.26 & 21.46 & 0 & 0 & 49.26* & -21.46 \\ 0 & 0 & 21.96 & -10.73 & 0 & 0 & -21.46 & 10.73* \end{bmatrix}$$

Application of Boundary Conditions:

The supporting conditions for the present truss are $u_{x1} = 0, u_{y1} = 0, u_{x3} = 0, u_{y3} = 0, u_{x4} = 0, u_{y4} = 0$. Thus, we apply the boundary conditions to solve the matrix equation by multiplying the corresponding diagonal as shown by asterisk marks by large number say 1.0×10^{-20}.

Solution: Now the external laod vector can be form readily from the external load acting on various joints and the final equilibrium equation will be

$$10^6 \begin{bmatrix} 28.29\times10^{20} & 28.29 & -28.29 & -28.29 & 0 & 0 & 0 & 0 \\ 28.29 & 28.29\times10^{20} & -28.29 & -28.29 & 0 & 0 & 0 & 0 \\ -28.29 & -28.29 & 77.55 & 6.83 & 0 & 0 & -49.26 & 21.46 \\ -28.29 & -28.29 & 6.83 & 139.02 & 0 & 0 & 21.46 & -10.73 \\ 0 & 0 & 0 & 0 & 0\times10^{20} & 0 & 0 & 0 \\ 0 & 0 & 0 & -80 & 0 & 0\times10^{20} & 0 & 0 \\ 0 & 0 & -49.26 & 21.46 & 0 & 0 & 49.26\times10^{20} & -21.46 \\ 0 & 0 & 21.96 & -10.73 & 0 & 0 & -21.46 & 10.73\times10^{20} \end{bmatrix}$$

$$\begin{Bmatrix} u_1 \\ v_1 \\ u_2 \\ v_2 \\ u_3 \\ v_3 \\ u_4 \\ v_4 \end{Bmatrix} = 10^3 \begin{Bmatrix} P_1 \\ P_2 \\ P_3 = 50.0 \\ P_4 = -100.0 \\ P_5 \\ P_6 \\ P_7 \\ P_8 \end{Bmatrix}$$

Above matrix equation can be solved by any numerical method i.e. by Gauss elemination or decomposition algorithm.

However here only unknown displacements are u_2 and v_2. For finding the above displacements, the matrix equation to be solved can be immediately determined from the original stiffness matrix. It is merely necessary to cross out columns and corresponding rows in the total stiffness matrix for which zero displacements have been prescribed. Thus the reduced stiffness equation will be

$$10^6 \begin{bmatrix} 77.56 & 6.83 \\ 6.83 & 119.02 \end{bmatrix} \begin{Bmatrix} u_2 \\ v_2 \end{Bmatrix} = 10^3 \begin{Bmatrix} 50 \\ -100 \end{Bmatrix}$$

$$\begin{Bmatrix} u_2 \\ v_2 \end{Bmatrix} = 10^{-6} \begin{bmatrix} 0.012 & -7.43\times10^{-4} \\ -7.43\times10^{-4} & 0.008426 \end{bmatrix} 10^3 \begin{Bmatrix} 50 \\ -100 \end{Bmatrix} = 10^{-3} \begin{Bmatrix} 0.674 \\ -0.881 \end{Bmatrix}$$

Internal Forces:

The internal forces can be obtained by substituting the known quantities i.e. nodal displacements of the individual element equilibrium equation.

$$\text{Member 1 (1–2):} \begin{Bmatrix} F_{x1}^1 \\ F_{y1}^1 \\ F_{x2}^2 \\ F_{y2}^2 \end{Bmatrix} = 28.29\times10^6 \begin{bmatrix} 1 & 1 & -1 & -1 \\ 1 & 1 & -1 & -1 \\ -1 & -1 & 1 & 1 \\ -1 & -1 & 1 & 1 \end{bmatrix} \begin{Bmatrix} 0 \\ 0 \\ 0.675 \\ -0.881 \end{Bmatrix} 10^{-3} = \begin{Bmatrix} 2.784 \\ 2.784 \\ -2.784 \\ -2.784 \end{Bmatrix}$$

$$\text{Member 2 (2–3):}\ \begin{Bmatrix} F_{x1}^2 \\ F_{y1}^2 \\ F_{x2}^2 \\ F_{y2}^2 \end{Bmatrix} = 80.0 \times 10^6 \begin{bmatrix} 0 & 0 & 0 & 0 \\ 0 & 1 & 0 & -1 \\ 0 & 0 & 0 & 0 \\ 0 & -1 & 0 & 1 \end{bmatrix} 10^{-3} \begin{Bmatrix} .674 \\ -.881 \\ 0 \\ 0 \end{Bmatrix} = \begin{Bmatrix} 0 \\ -70 \\ 0 \\ -2.784 \end{Bmatrix}$$

$$\text{Member 3 (2–4):}\ \begin{Bmatrix} F_{x1}^3 \\ F_{y1}^3 \\ F_{x2}^3 \\ F_{y2}^3 \end{Bmatrix} = 10.73 \begin{bmatrix} 4 & -2 & -4 & 2 \\ -2 & 1 & 2 & -1 \\ -4 & 2 & 4 & -2 \\ 2 & -1 & -2 & 1 \end{bmatrix} 10^{-3} \begin{Bmatrix} 0.674 \\ -0.881 \\ 0 \\ 0 \end{Bmatrix} = \begin{Bmatrix} 52.784 \\ -26.393 \\ -52.784 \\ 26.394 \end{Bmatrix}$$

16.2.13 Step by Step Procedure for Analysis of Pin Jointed Structure

The steps to be followed in the analysis of 2-D and 3-D pin jointed structures will be explained with reference to the 2D structures shown in Figure 16.18.

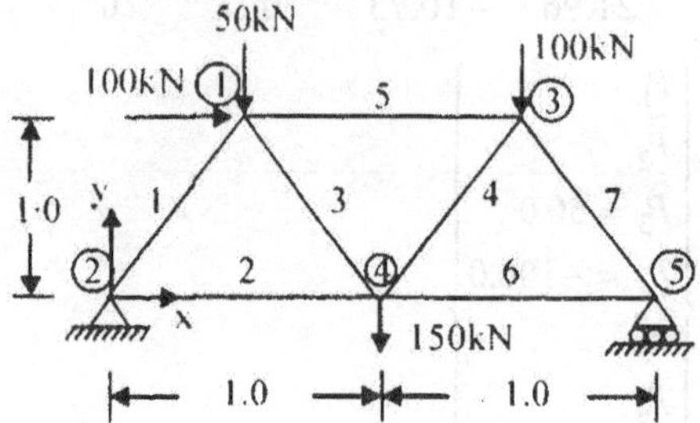

Fig. 16.18. Plane truss.

(1) Choose a convenient set of global co-ordinate axes to represent the structure. Origin is at node 2 with positive x and y axes as shown in figure 16.18.

(2) Next step is to identify the elements or members of the structure and their nodes and feed the co-ordinates of each joint. Thus, number the joints and determine their co-ordinate. For the present example

Node	1	2	3	4	5
X	0.5	0.0	1.5	1.0	2.0
Y	1.0	0.0	1.0	0.0	0.0

(3) Number the elements and for each elements identify the nodal connectivity, i.e., which joint is end 1 and which is end 2. This information is presented in Table. Note that once this information is known we can identify the global degree of freedom which are connected with a particular element. It should be appreciated that it is perfectly arbitrary as to which end is 1 and which is end 2. However, it is important to ensure that the positive direction of the displacements coincide with the positive direction of the co-ordinate axes.

Element Number	1	2	3	4	5	6	7
End 1	2	2	4	4	1	4	3
End 2	1	4	1	3	3	5	5

(4) For each element obtain the cross section area, elasticity etc. compute the length, and direction cosines.

$$L_{ij} = \sqrt{(x_j - x_i)^2 + (y_j - y_i)^2}$$

For *i-j* the element,

$$\lambda_{ij} = \frac{x_j - x_i}{L_{ij}}, \ \mu_{ij} = \frac{y_j - y_i}{L_{ij}}$$

Various quantities form the present problem is shown in the following table

Element Number	Length (m)	λ	μ	Cross - section Area (mm^2)	Modulus of Elasticity (kN/mm^2)
1	1.12	0.45	0.89	1200	200
2	1.0	1.0	0.0	1000	200
3	1.12	−0.45	0.89	1200	200
4	1.12	0.45	0.89	1200	200
5	1.0	1.0	0.0	1000	200
6	1.0	1.0	0.0	1000	200
7	1.12	0.45	0.89	1200	200

(5) For each element form the stiffness matrix (equation 16.28). For example the element stiffness matrix for element 5 is

$$[k]_5 = \frac{1000 \times 200}{1000.0}\begin{bmatrix} 1 & 0 & -1 & 0 \\ 0 & 0 & 0 & 0 \\ -1 & 0 & 1 & 0 \\ 0 & 0 & 0 & 0 \end{bmatrix} = 200\begin{bmatrix} 1 & 0 & -1 & 0 \\ 0 & 0 & 0 & 0 \\ -1 & 0 & 1 & 0 \\ 0 & 0 & 0 & 0 \end{bmatrix}$$

(6) From the element stiffness matrix of each individual truss member, form the overall stiffness matrix by adding up the contribution from all the member in corresponding positions (identified by - connectivity number) as described in step 2. For the present problem the size of stiffness matrix -will be 10 by 10 as there is total 5 node.

(7) Now apply the boundary conditions. Depending on the supporting node number and associated degree of freedom multiply the corresponding diagonal element of the overall stiffness matrix $[K]$ by a large number. For the present problem node 2 and 5 has supports. Node 2 is pinned i.e. both u_x and v_y is restrained and corresponding degree of freedom in global system is 3 and 4. Thus multiply K_{33} and K_{44} by large number and for node 5 which is a roller support, only u_y is restrained and corresponding global d.o.f. is 10. Thus multiply K_{1010} by a large number.

(8) Form the external load vector. Here the load vector is $\{F\}=[100 - 50\,0\,0\,0 - 100\,0 - 150\,0]^T$.

(9) Now solve the matrix equation $[K]\{U\}=\{F\}$ by Gauss elimination method as explained in 16.19 to obtain the displacement vector $\{U\}$.

(10) Compute the internal force quantities using equation 16.42 as described in 16.20.

16.2.14 COMPUTER Programming

It is highly interesting in the learning process to apply newly learned concepts to practical problems. Example problem is discussed in 16.21 to elucidate the theoretical ideas. However, it is seen that the volumes of arithmetic generated by even with simple problems are likely to dull the enthusiasm. To circumvent this difficulty it is essential to quickly move from hand calculations to computer program for solving the problems. Unfortunately standard structural analysis program tend to appear as 'black boxes' to their users with input going in one end and solution appearing magically at other end. Keeping this view in mind, the present section explain the various aspects of developing computer program for analysing the structure by matrix method. The intention is to give the students and practising engineers the background knowledge necessary to make safe and efficient use of frame analysis program. First various program parts needed for step by step solution of a 2-D truss structure as explained in 16.22 are presented in module form. These program parts from various modules are then connected together to form the final programs. The programs are written in FORTRAN and are designed to run on any computer from desktop microcomputer to mainframe machine. However, the programs are written with sole aim of clarity and it is to mention here that efficient and compactness is not the considerations. The flow diagram for a typical matrix analysis program is shown in Figure 16.19.

INPUT and OUTPUT of the program

The input to the computer is through the input file(for example **trusel.in**) and output obtained is written in a file (for example named as **truse·out)**. For this purpose, in the main program the input file and output file will be opened first by the following Fortran statements:

```
open(10,file='trusel.in)
open(12,file='trusel.out)
```

Thus all the input quantities to be typed in the **trusel.in** file in the sequence as in the input subroutine. Following three i.e.

(*i*) Total number of node:	ntnd
(*ii*) Total number of element	nelm
(*iii*) Total number of d.o.f. to restrained	nfix

which are read in the main program directly from screen as these are used as variable in the input module.

Module 1: Input data

The computations carried out by the programs are essentially same as those explained in step by step procedure. It can be also seen from flow diagarm that once the element numbering, node numbering are done, it is necessary to feed the nodal connectivity and global co-ordinates of the nodes. Then read the cross section area and modulus of elasticity of each element and compute the element length, λ, μ for each element. The subroutine program named **'input'** reads the input in the following sequence:

(*i*) The global co-ordinate $x(i)$ and $y(i)$ in turn for each node. Thus size of array x ( ) and y( ) is equal to ntnd (number of nodes)

(*ii*) It then loops for each element and reads the following data:

A(*i*) Cross section area of the each element

E(*i*) Modulus of elasticity of each element.

Thus, Size of array A( ) and E( ) is equal to nelm

Node c(*i,j*) is the nodal connectivity number ;i varies from 1 to total number of element (nelm). j varies from 1 to 2 (1st position for node number of end 1 and 2nd position for node number of end 2).

Thus size of array nodec ( ) is nelm × 2

(*iii*) Code number for boundary conditions.

It reads: (*a*) Total number of d.o.f. to be restrained i.e nfix (*b*) The node at which restrained to be applied. It is an array nnf( ) of size nfix.

(*c*) The associated d.o.f. to be restrained at corresponding nnf( ) array. It is also an array ndl( ) of same size as nnf( ) array. If u is restrained then 1 and v is restrained then type 2. (*iv*) Finally the externally applied load vector is read. It is an array of size equals to the total d.o.f. i.e. ndof(==2*ntnd, for plane truss). Horizontal load first and then the vertical load.

Output: After reading all the input from **trusel.in** file, this module also compute the element length, λ, μ for each element and produced output from this mudule is L(*i*), λ(*i*) i.e. μl(*i*), A(*i*), E(*i*). Various input information needed for a truss problems are explained w.r.t. the example problem as discussed in 16.21.

First one needs to read the total number of nodes, element and fixity interactively in the screen.

The required Fortran statement are to be written in main program(see the final complete program).

Thus for this problem

ntnd = 4

nelm = 3

nfix = 6

Once these values are given in screen, it calls the input subroutine and loops for each a(i) and e(*i*) and nodec (*i,j*). So in the input file one have to type following:

```
1   410e-4    210e+11   1   2
2   410e-4    210e+11   2   3
3   410e-4    210e+11   2   4
```

Then the *x* and *y* co-ordinate of each node are typed as follows:

```
1   0.0   0.0
2   1.0   1.0
3   1.0   0.0
4   2.0   0.0
```

Next is the boundary conditions. In present problem, as node 1,2 and 4 are restrained both for u and v, thus nfix = 3 × 2 = 6 already known from screen input. Thus input will be as below:

1 1 3 3 4 4 (the array nnf i.e. the node which is restrained , repeated if both u and v are restrained)

1 2 1 2 1 2 (the array ndl i.e. the associated d.o.f. which is retrained)

Finally the load vector is type in the input file as below:

```
0.0    0.0
50.0   -100.0
0.0    0.0
0.0    0.0
```

First type the load in horizontal direction then for the vertical direction sequentially for each node. Also note the sign convention with the theoretical formulation.Note that the unit of cross section area, nodal co-ordinate, modulus of elasticity and load vector should be consistent

The computer printout of the input file 'trusel.in' needed for the problem is shown below:

```
1 410e-4   210e+11   1  2
2 410e-4   210e+11   2  3
3 410e-4   210e+11   2  4
1 0.0   0.0
2 1.0   1.0
3 1.0   0.0
4 2.0   0.0
1 2 1 2 1 2
1 2 1 2 1 2
0.0   0.0
50.0  -100.0
0.0   0.0
0.0   0.0
```

Modeule 2: Element stifiness matrix generation subroutine

Once the properties of each element is known from module 1, the element stiffness matrix can be obtained for each element using equation (16.28). To obtain the element stiffness matrix for i-th element the following subroutine will be used:

subroutine elstif(ielm,a,e,nelm,cx,cy,estif)

Note that input parameters for this subroutine is

ielm :the element number for which element stiffness is required to compute

a :cross section area

cx,cy :direction cosine of the element

el :length of element

nodec :nodal connectivity

These input parameters are known already from the called subroutine input. Output from this segment of the program will be the element stiffness matrix for i-th element stored in estif matrix of size 4 × 4. Note that this subroutine can be used repeatedly for computing the element stiffness matrix for all the member of the truss structure.

Module 3: Assembley subroutine to obtain overall stifiness matrix

After computation of element stiffness matrix from module 2 it has to be placed in the global stiffness matrix in corresponding position. This is done by following procedure taking the element number, corresponding element stiffness matrix, connectivity number of the element as known quantity

subroutine asemb(ielm,nodec,estif,gstil)

In actual programming above two procedures are performed separately in two subroutine. The global stiffness matrix is first initialise by zeroed in all the element of gstif matrix of size ndof,ndof(ndof is the total d.o.f. of the structure) as below:

```
ndof=2*ntnd
do i=l,ndof
= do j = l, ndof
gstif(i,j)=0.0
end do
end do
```

Then a loop is performed as explained in the following:

```
      do l04 ielm=l,nelm
      call elstif(ielm,a,e,nelm,cx,cy,estif)
      call asemb(ielm,nodec,estif,gstif)
104   continue
```

After completion of the above loop, the global stiffness matrix is obtained in [gstif].

Module 4: Application of boundary conditions

After assembling the stiffness matrix, it is necessary to feed boundary conditions. This is done by multiplying the associated diagonal coefficient of the global stiffness matrix i.e. gstif(i,i) is multiplied by a large number if u(*i*) is zero(see 16.18). The input information needed for applying the boundary condition on global stiffness matrix gstif are : nfix, nnf( ), ndl( ) which are already known from input module and feed as input parameter to the boundary subroutine as can be seen below:

subroutine boundary (ndof,gstif,nfix,nnf,ndl)

Module 5 : Solution of unknown displacements

Once the boundary condition is applied on the global stiffness matrix gstif, the solution vector i.e. nodal displacements can be obtained simply by calling the following gauss elimination program:

subroutine gausselim(a,b,n)

Here in the calling process the input parameter will be the gstif matrix and the load vector f. Note that the solution i.e. displacement vector will be stored in the load vector.

Module 6 : Computation of internal forces

Once the displacement vector is known, the internal force in each element can be obtain very easily by multiplying the corresponding 4×4 element stiffness with 4×1 displacement vector. The following subroutine compute the force at i-th element by taking the estif for i -th element as input parameter:

subroutine force(ielim,nodec,estif,cx,cy,estif,u,f)

Note that above subroutine takes estif as input parameter for i-th element to get the force in each member. The following loop is to be performed which in turn calls the subroutine elstif, force and gets the internal forces:

```
      do 101 ielm=l,nelm
      call elstif(ielm,a,e,nnel,cx,cy,estif)
      call force(ielm,nodec,estif,cx,cy,u,f)
101   continue
```

Flow Chart

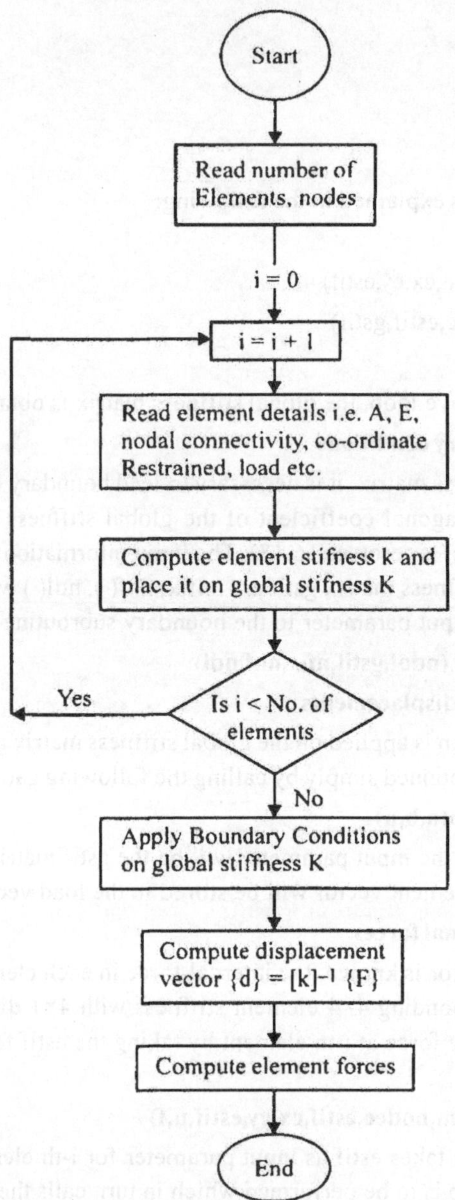

Fig. 16.19. Flow chart

Program Truss2D. for

```
cxxxxxxxxxxxxxxxxxxxxxxxxxxxxxxxxxxxxxxxxxxxxxxxxxxxxxxxxxxxxxxxxxxxxxxxxxxxxxxxxxxxxxxxxxx
c The program is written for maximum of 40 elements assuming
c maxium 40 nodes. However it can be used for otherwise
c simply by modification on parameter statements to suit the
c actual size of the structure.
cxxxxxxxxxxxxxxxxxxxxxxxxxxxxxxxxxxxxxxxxxxxxxxxxxxxxxxxxxxxxxxxxxxxxxxxxxxxxxxxxxxxxxxxxxx
   parameter (no=80,ne=40,nn=40,nfi=20)
   dimension nodec(ne,2),x(ne),y(ne),estif(4,4),gstif(no,no)
   dimension eload(no),a(ne),e(ne),el(ne),cx(ne),cy(ne) dimension
   (4),nnf(nfi),ndl(nfi),u(no)
c*******************************************************************************
    character file 1*10,file2*10
    write(*,'(a\)') ' Input file name? '
    read(*;(a)') file 1
   write(*;(a\)') ' Output file name ? '
    read(*;(a)') file2
    open( 10,file= file 1)
    open( 12,file= file 2)
c*******************************************************************************
  write(*,*)'Give number of members in the truss:'
      read(*,*)nelm
   write(*,*)'Give number of joints in the truss:'
      read(*,*)ntnd
      write(*,*) 'number of fixity'
      read(*,*) nf
      ndof=2*ntnd
c---->Following statement calls the input data
  call input(ntnd,ntnd,nelm,nelm,ndof,a,e,nodec,x,y,el,cx,cy,
  1 eload,nf,nf,nnf,ndl)
 c----> Initialize the global structural stiffness matrix
    do 14i=l,ndof
    do 14 j==l,ndof
    gstif(i,j)=0.0
 14 continue
 c---->The following loop compute the elemement and global stiffness matrix.
    do 15 ielm=l,nelm
   call elstif(ielm,nelm,a,e,el,cx,cy,estif)
   call asemb(ielm,nodec,estif,gstif,ndof,nelm)
   write(12,*)'Element Number ',ielm
  write(12,*)'   Length ';'  cos ','  sin '
  write(12,*)el(ielm),cx(ielm),cy(ielm)
  write(12,*)'Element Stiffness Matrix is'
  write(12,l)((estif(i,j),j=l,4),i=l,4)
  c  write(12,*)'Global Stiffness Matrix '
  c  write(12,2)((gstif(i,j),i=l,ndof),j=l,ndof)
```

```
   15  continue
   1  format(4f18.4/)
c----> Once above loop is over gstif contains the globalstiffness matrix
2  format(8e10.2/)
          write(12,*)'Load Vector'
          write(12,*)(eload(i),i=l,ndof)
c----> Following statement apply the boundary conditions on gstif
     call boundary (ndof,ndof,gstif,nf,nf,nnf,ndl)
c ----> Following subrouirie solve matrix equlibrium equation
      call gausselim(gstif,eload,ndof,ndof)
c---->Displacement vector is stored in the vector u
     dc i==l,ndof
     u(i)=eload(i)
      end do
     write(12,*)'DISPLACEMENT VECTOR'
     write(12,3)(u(i),i= 1 ,ndof)
3  format(4el8.6/)
c---->This loop compute the internal forces for each elements in turn
          do 101 ielm=l,nelm
     alm=cx(ielm)
     amu=cy(ielm)
    call elstif(ielm,nelm,a,e,el,cx,cy,estif)
   call force(ielm,nodec,estif,alm,amu,nelm,ndof,u,f)
 101  continue
     stop
     end
c************************************************************************
     subroutine input (nn,ntnd,ne,nelm,no,a,e,nodecxi,y,el,cx,cy,
   I eload,nf,nfix,nnf,ndl)
    dimension a(ne),e(ne),nodec(ne,2),xr(nn),y(nn),cx(ne),
   I cy(ne),el(ne) ,eload(no),nnf(nf),ndl(nf)
c---->Reads the sectional area,elasticity and nodal connectivity of elements
       ndf=ntnd*2
       do nl=ndm
      read(10,*)i,e(i),(nodec(i,j),j=l,2)
        end do
c----> Read the nodal coordinate of each node
     do n=l,ntnd
     read(10,*)i,x(i),y(i)
       end do
c----> Compute length and direction cosines of each element
          do 103 i=l,nelm
     nl =nodec(i,l)
     n2=nodec(i,2)
     el(i)=sqrt((x(n2)-x(nl))**2+(y(n2)-y(nl))**2)
     cx(i)=(x(n2)-x(nl))/el(i)
     cy(i)=(y(n2)-y(nl))/el(i)
 103   continue
c----> Read Boundary Conditions
```

```
      read(10,*)(nnf(i),i=l,nfix)
      read(10,*)(ndl(i),i=l,nfix)
c----> Read the external load vector
          read(10,*)(eload(i),i=l,ndf)
     return
     end
c******************************************************************************
      subroutine elstif(ielm,ne,a,e,el,cx,cy,estif)
       dimension a(ne),estif(4,4)
       dimension cx(ne),cy(ne),el(ne), e(ne)
c******************************************************************************

     do 18 i=l,4
      do 18 j=l,4
     estif(i,j)=0.0
18  continue
     ael=a(ielm)*e(ielm)/el(ielm.)
c   element Stiffness Matrix
     estif(l,l)=cx(ielm)*cx(ielm.)*ael
     estif(1,2) = cx(ielm)*cy(ielm)*ael
     stif(l,3)=-estif(l,l)
      estif(l,4)=-estif(l,2)
     estif(2,l)=estif(l,2)
     es1if(2,2)=cy(ielm)*cy(ielm)*ael
     estif(2,3)=estif(l,4)
     estif(2,4)=-estif(2,2)
     estif(3,l)=esttf(l,3)
     estif(3,2)=estif(23)
     estif(3,3)=estif(l,l)
     estif(3,4)=estif(l,2)
     estif(4,3)=estif(3,4)
     estif(4,l)=estif(l,4)
     cstif(4,2)=estif(2,4)
     cstif(4,4)=cstif(2,2)
     return
     end
c******************************************************************************
     subroutine asemb(ielm,nodec,estif,gstif,no,ne)
     dimension estif(4,4),gstif(no,no),nodec(ne,2),kk(4)
c ******************************************************************************
          kk(2)=2*nodec(ielm,l)
          kk(l)=kk(2)-l
          kk(4)=2*nodec(ielm,2)
          kk(3)=kk(4)-l
       do 301 i=l,4
      ki=kk(i)
     do 301 j=l,4
      kj=kk(j)
         gstif(ki,kj)=gstif(ki,kj)+estif(i,j)
```

```
301  continue
     return
     end
c*******************************************************************
      subroutine boundary (no,ndof,gstif,nf,nfix,nnf,ndl)
      dimension gstif(no,no),nnf(nf),ndl(nf)
        cnst=1.0
        do i=l,ndof
     if (cnst.le.gstif(i,i))then
     cnst=gstif(i,i)
     end if
        end do
      cnst=(l.+cnst)*1.0e+20
      do 401 i=l,nfix
      il=2*nnf(i)-l
      i2=il+l
      if(ndl(i).eq.l)go to 402
      if(ndl(i).eq.2)go to 403
402  gstif(il,il)=cnst+gstif(il,il)
       go to 401
403  gstif(i2,i2)=cnst+gstif(i2,i2)
401  continue
     return
     end
c*******************************************************************
      subroutine gausselim(a,b,no,n)
     dimension a(no,no),b(no)
     nl = n - 1
     do 130 k=l, nl
     kl = k + 1
     do 130 i = kl, n
        c = a(i, k) / a(k, k)

       do 120 j = kl, n
 120  a(i, j) = ai, j)-c * a(k, j)
 130 b(i)=b(i) - c* b(k)
 C   *** back substitution ***
    b(n)=b(n)/a(n,n)
    do 140 ii = I, nl
     i = n- ii
     ii = i + I
     c=l/a(i,i)
     b(i) = c * b(i)
     do 140 k = il, n
 140 b(i) = b(i) - c * a(i, k) * b(k)
c   print *, 'solution vector'
c  write (*;(i4,el2.4)')(i, b(i),i = I, n)
    return
    end
```

```
c
******************************************************************************
      subroutine force(ielm,nodec,estif,alm,amu,ne,ndf,u,f)
      dimension nodec(ne,2),k(4),estif(4,4),f(4),u(ndf)
      do 45 i 1,4
45  f(i) 0.0
      nl nodec(ielm,l)
      n2 = nodec (ielm,2)
      k(2)=2*nl
      k(1)=k(2)-l
      k(4)=2*n2
      k(3)=k(4)-l
      do 50 i=l,4
      do 50 j=l,4
50   f(i)=f(i)+estif(i,j)*u(k(j))
c------>Calculation of Axial Forces
      faxl = f(l)*alm+f(2)*amu
      fax2=-f(3)*alm+f(4)*amu
      write(12,*)'Member Number', ielm
      write(12,*)'forces at first node',f(l); ',f(2)
       write(12,*)'forces at second node',f(3); ',f(4)
      write(12,*)'axial force', faxl,fax2
      write(12,*)
      return
      end
******************************************************************************
```

Example 1. The numerical example already worked out by hand to explain the procedure is taken for analyzing again by computer. The joint and member numbering are same as earlier. Typical input file (trusel.in) and the detailed output file (truse l .out) are given below.

Input File (truse l.in)

```
1 4.0e-4    2.0e+11    1    2
2 4.0e-4    2.0e+11    2    3
3 6.0e-4    2.0e+11    2    4
1 0.0       0.0
2 1.0       1.0
3 1.0       0.0
4 3.0       0.0
1   1   3   3   4   4
1   2   1   2   1   2
0.0      0.0
50.0     -100.0
0.0      0.0
0.0      0.0
```

Output

Element Number 1 Length   cos   sin

1.414214 7.071068E-01 7.071068E-01

Element Stiffness Matrix is

28284270.0000 28284270.0000 -28284270.0000 -28284270.0000

28284270.0000 28284270.0000 -28284270.0000 -28284270.0000

```
 -28284270.0000 -28284270.0000 28284270.0000 28284270.0000
 -28284270.0000 -28284270.0000 28284270.0000 28284270.0000
Global Stiffness Matrix (after placing contribution of element 1)
-28E+08 .28E+08 -.28E+08 -28E+08 .00E+00.00E+00 .00E+00 -00E+00
-28E+08 .28E+08 -.28E+08 -.28E+08 .00E+00.00E+00 .00E+00 .00E+00
-.28E+08-.28E+08 .28E+08 .28E+08 .00E+00.00E+00 .00E+00 .00E+00
-.28E+08-.28E+08 .28E+08 .28E+08 .00E+00.00E+00 .00E+00 .00E+00
.00E+00 .00E+00 .00E+00 .00E+00 .00E+00.00E+00 .00E+00 .00E+00
.00E+00 .00E+00 .00E+00 .00E+00 .00E+00.00E+00 .00E+00 .00E+00
.00E+00 .00E+00 .00E+00 .00E+00 .00E+00.00E+00 .00E+00 .00E+00
.00E+00 .00E+00 .00E+00 .00E+00 .00E+00.00E+00 .00E+00 .00E+00
Element Number 2 Length cos sin
 1.000000 0.000000E+00 -1.000000
Element Stiffness Matrix is
 .0000 .0000 .0000 .0000
 .0000 80000000.0000 .0000 -80000000.0000
 .0000 .0000 .0000 .0000
 .0000 -80000000.0000 .0000 80000000.0000
Global Stiffness Matrix (after placing contribution of element I and 2)
 .28E+08 .28E+08 -.28E+08 -.28E+08 .00E+00.00E+00 .00E+00.00E+00
 .28E+08 .28E+08 -.28E+08 -.28E+08 .00E+00.00E+00 .00E+00 .00E+00
-.28E+08 -.28E+08 .28E+08 .28E+08 .00E+00.00E+00 .00E+00 .00E+00
-.28E+08 -.28E+08 .28E+08 .11E+09 .00E+00-.80E+08.00E+00 .00E+00
.00E+00 .00E+00 .00E+00 .00E+00 .00E+00 .00E+00.00E+00 .00E+00
.00E+00 .00E+00 .00E+00 -.80E+08 .00E+00 .80E+08.00E+00 .00E+00
.00E+00 .00E+00 .00E+00 .00E+00 .00E+00 .00E+00 .00E+00 .00E+00
.00E+00 .00E+00 .00E+00 .00E+00 .00E+00 .00E+00.00E+00 .00E+00
Element Number 3 Length cos sin
 2.236068 8.944272E-01 -4.472136E-01
Element Stiffness Matrix is
 42932500.0000 -21466250.0000 -42932500.0000 21466250.0000
 -21466250.0000 10733130.0000 21466250.0000 -10733130.0000
 -42932500.0000 21466250.0000 42932500.0000 -21466250.0000
 21466250.0000 -10733130.0000 -21466250.0000 10733130.0000
Final Global Stiffness Matrix (contribution of all elements i.e-1,2 and 3)
.28E+08 .28E+08 -.28E+08 -.28E+08 .00E+00 .00E+00 .00E+00 .00E+00
.28E+08 .28E+08 -.28E+08 -.28E+08 .00E+00 .00E+00 .00E+00 .00E+00
-.28E+08 -.28E+08 .71E+08 .68E+07 .00E+00 .00E+00-.43E+08 .21E+08
-.28E+08 -.28E+08 .68E+07 .12E+09.00E+00 -.80E+08 .21E+08 -11E+08
.00E+00 .00E+00 .00E+00 .00E+00.00E+00 .00E+00 .00E+00 .00E+00
.00E+00 .00E+00 .00E+00 -.80E+08.00E+00 .80E+08 .00E+00 .00E+00
.00E+00 .00E+00 -.43E+08 .21E+08 .00E+00 .00E+00 .43E+08 -.21E+08
.00E+00 .00E+00 .21E+08 -.11E+08 .00E+00 .00E+00-.21E+08 .11E+08
Load Vector
0.00.0 50.0-100.0 0.0 0.0 0.0 0.00.0
DISPLACEMENT VECTOR
 -.233969E-17 -.233969E-17 .786836E-06 -.885288E-06
 .000000E+00 -.595064E-16 .443504E-16 -.221752E-16
Member Number 1
```

forces at first node 2.784642 2.784642
forces at second node -2.784642 -2.784642
axial force 3.938079 -3.938079
Member Number 2
forces at first node 0.000000E+00 -70.823030
forces at second node 0.000000E+00 70.823030
axial force 70.823030 -70.823030
Member Number 3
forces at first node 52.784640 -26.392320
forces at second node -52.784640 26.392320
axial force 59.015020 -59.015020
****

Example 2. Analyze the truss as in the figure, the node and member numbering is also shown in the figure. Input

(truse2.in)
1 450.0 210.0 1 2
2 450.0 210.0 1 3
3 450.0 210.0 2 3
4 450.0 210.0 2 4
5 450.0 210.0 3 4
6 450.0 210.0 3 5
7 450.0 210.0 4 5
1 0.0 0.0
2. 1500.0 2598.0
3. 3000.0 0.0
4 4500.0 2598.0
4 5000.0 0.0
1 1 5
1 2 2
0.0 0.0
0.0 100.0
0.0 0.0
50.0 0.0
0.0 0.0

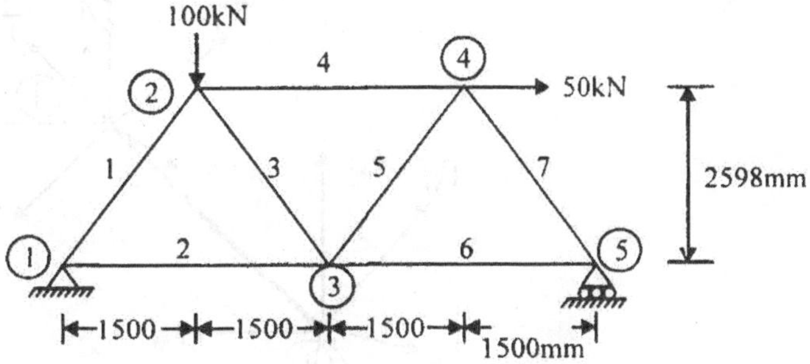

100kN
4
50kN
2598mm
1 3 5 7
2 6
−1500−−1500−−1500−
1500mm

A = 450mm^2 (for all member)
E = 210 kN/mm^2

16.20. Plane Truss Example

Output(truse2.out, only final results)
DISPLACEMENT VECTOR
.626951E-20 -.668957E -20 .378623E+01 -.444423E+01 .256516E+01 .317465E+01
.366343E+01 -.183416E+01.342021E+01 -.584945E-20
Forces at Frist Node at second node
Member Number 1
axial force 61.603730 -61.603730
Member Number 2
axial force -80.802540 80.802540
Member Number 3
axial force 53.867180 -53.867180
Member Number 4
axial force 3.868358 -3.868358
Member Number 5
axial force -53.867180 53.867180

Member Number 6
axial force 53.867180 -53.867180

16.2.15 Space Truss

The most difficult truss to analyze are those which form a network in space. In proceeding with space trusses, the first step is to obtain the stiffness matrix for the member. The bar element in space has three forces and three displacements at each node(i.e F_x, F_y, F_z and u, v, w) in the global co-ordinate system; hence the order of element stiffness matrix will be 6 by 6. The truss member in three-dimensional space is shown in Figure 16.21. It will be assumed to have length L and uniform axial rigidity AE. The element in local co-ordinate system has only one axial force F and one axial d.o.f. u at each nodal point. Let us first augment the 2 × 2 stiffness matrix equation in local co-ordinate system to 6 × 6 stiffness matrix equations by creating two more non-existence forces $\overline{F}_y$ and $\overline{F}_z$ and displacements d.o.f. $\overline{v}$ and $\overline{w}$ in the local co-ordinate system $\overline{y}$ and $\overline{z}$.

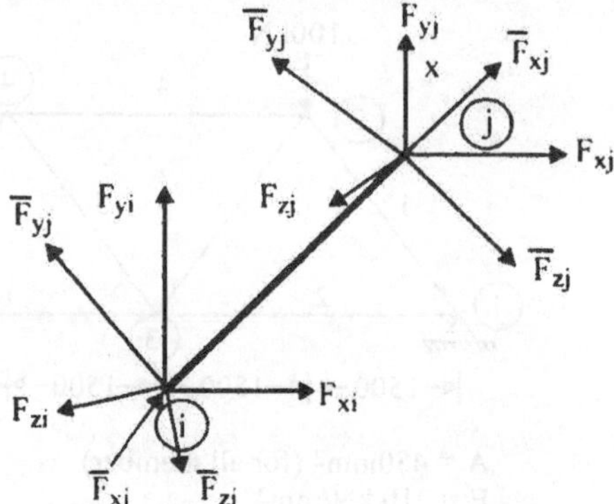

Fig. 16.21. Space truss element.

$$\begin{Bmatrix} \overline{F}_{xi} \\ \overline{F}_{yi} \\ \overline{F}_{zi} \\ \overline{F}_{xj} \\ \overline{F}_{yj} \\ \overline{F}_{zj} \end{Bmatrix} = \frac{AE}{L} \begin{Bmatrix} \overline{u}_i \\ \overline{v}_i \\ \overline{w}_i \\ \overline{u}_j \\ \overline{v}_j \\ \overline{w}_j \end{Bmatrix} \begin{bmatrix} 1 & 0 & 0 & -1 & 0 & 0 \\ 0 & 0 & 0 & 0 & 0 & 0 \\ 0 & 0 & 0 & 0 & 0 & 0 \\ -1 & 0 & 0 & 1 & 0 & 0 \\ 0 & 0 & 0 & 0 & 0 & 0 \\ 0 & 0 & 0 & 0 & 0 & 0 \end{bmatrix} \text{i.e.} \{\overline{f}\} = [\overline{k}]\{d\} \tag{16.43}$$

Considering equilibrium of forces at node i i.e.

$$\begin{aligned} \overline{F}_{xi} &= F_{xi}\cos(x,\overline{x}) + F_{yi}\cos(y,\overline{x}) + F_{zi}\cos(z,\overline{x}) = F_{xi}\lambda_{x\overline{x}} + F_{yi}\lambda_{y\overline{x}} + F_{zi}\lambda_{z\overline{x}} \\ \overline{F}_{yi} &= F_{xi}\cos(x,\overline{y}) + F_{yi}\cos(y,\overline{y}) + F_{zi}\cos(z,\overline{y}) = F_{xi}\lambda_{x\overline{y}} + F_{yi}\lambda_{y\overline{y}} + F_{zi}\lambda_{z\overline{y}} \\ \overline{F}_{zi} &= F_{xi}\cos(x,\overline{z}) + F_{yi}\cos(y,\overline{z}) + F_{zi}\cos(z,\overline{z}) = F_{xi}\lambda_{x\overline{z}} + F_{yi}\lambda_{y\overline{z}} + F_{zi}\lambda_{z\overline{z}} \end{aligned} \tag{16.44}$$

Above relations can be written in matrix form as

$$\begin{Bmatrix} \overline{F}_{xi} \\ \overline{F}_{yi} \\ \overline{F}_{zi} \end{Bmatrix} = \begin{bmatrix} \lambda_{x\overline{x}} & \lambda_{y\overline{x}} & \lambda_{z\overline{x}} \\ \lambda_{x\overline{y}} & \lambda_{y\overline{y}} & \lambda_{z\overline{y}} \\ \lambda_{x\overline{z}} & \lambda_{y\overline{z}} & \lambda_{z\overline{z}} \end{bmatrix} \begin{Bmatrix} F_{xi} \\ F_{yi} \\ F_{zi} \end{Bmatrix} \tag{16.45}$$

i.e. $\{\overline{F}_i\} = [t]\{F_i\}$

Similarly for node j

$$\{\overline{F}_j\} = [t]\{F_j\}$$

Now above two matrix equation we can couple in a single matrix equation as follows,

$$\begin{Bmatrix} \overline{F}_i \\ \overline{F}_j \end{Bmatrix} = \begin{bmatrix} [t] & [0] \\ [0] & [t] \end{bmatrix} \begin{Bmatrix} F_i \\ F_j \end{Bmatrix} \text{ i.e. } \{\overline{\mathrm{F}}\} = [\mathrm{T}]\{\mathrm{F}\} \qquad (16.46)$$

Where $[T]$ is the 6 × 6 transformation matrix. Similarly we can write the relationship between local and global displacement vector i.e.

$$\{\overline{d}\} = [T]\{d\} \qquad (16.47)$$

Now at element level

$$[\overline{k}]\{\overline{d}\} = \{\overline{F}\}$$

$$[\overline{k}][T]\{d\} = [T]\{F\}$$

$$\therefore \quad [T]^{-1}[\overline{k}][T]\{d\} = \{F\}$$

i.e. $$[k]\{d\} = \{F\} \qquad (16.48)$$

where, $[k] = [T]^{-1}[\overline{k}][T]$ is the element stiffness matrix in the global co-ordinate system. Now it can be observed that the transformation matrix $[T]$ is orthogonal in nature. Thus, $[T]^{-1} = [T]^{\mathrm{T}}$. Note that the direction in which the angle is measured in immaterial as cos (360° - θ) = cos θ. Thus, $\lambda_{x\bar{y}} = \lambda_{y\bar{x}}$, $\lambda_{z\bar{x}} = \lambda_{x\bar{z}}$ $\lambda_{z\bar{y}} = \lambda_{y\bar{z}}$ and $\lambda_{x\bar{x}} = \lambda_{y\bar{y}} = \lambda_{z\bar{z}}$ and denoting $\lambda_{x,\bar{x}} = \lambda$, $\lambda_{y,\bar{x}} = \mu$, $\lambda_{z,\bar{x}} = \nu$, the element stiffness matrix can be written as

$$\begin{bmatrix} \lambda^2 & \lambda\mu & \lambda\upsilon & -\lambda^2 & -\lambda\mu & -\lambda\upsilon \\ \mu\lambda & \mu^2 & \mu\upsilon & -\mu\lambda & -\mu^2 & -\mu\upsilon \\ \upsilon\lambda & \upsilon\mu & \upsilon^2 & -\upsilon\lambda & -\upsilon\mu & -\upsilon^2 \\ -\lambda^2 & -\lambda\mu & -\lambda\upsilon & \lambda^2 & \lambda\mu & \lambda\upsilon \\ \mu\lambda & -\mu^2 & -\mu\upsilon & \mu\lambda & \mu^2 & \mu\upsilon \\ -\upsilon\lambda & -\upsilon\mu & -\upsilon^2 & \upsilon\lambda & \upsilon\mu & \upsilon^2 \end{bmatrix} \qquad (16.49)$$

where, $\lambda = \dfrac{x_j - x_i}{L_{ij}}, \mu = \dfrac{y_j - y_i}{L_{ij}}, \nu = \dfrac{z_j - z_i}{L_{ij}}$

Program Truss3D. for

```
cxxxxxxxxxxxxxxxxxxxxxxxxxxxxxxxxxxxxxxxxxxxxxxxxxxxxxxxxxxxxxxxxxxxxxxxxxxxxxxxxxx
c The program is written for maximum of 40 elements assuming
c maxium 40 nodes. However it can be used for otherwise by
e simply by modification on parameter statements to suit the
c  actual size of the structure.
cxxxxxxxxxxxxxxxxxxxxxxxxxxxxxxxxxxxxxxxxxxxxxxxxxxxxxxxxxxxxxxxxxxxxxxxxxxxxxxxxxx
      parameter (no=120,ne=40,nn=40,nfi=20)
      dimension nodec(ne,2),x(nn)y(nn),z(nn),estif(6,6),gstif(no,no)
      dimension eload(no),a(ne),e(ne),el(ne),cx(ne),cy(ne),cz(ne)
      dimension f(6),nnf(nfi),ndl(nfi),u(no)
c************************************************************************
        character filel*10,file2*10
        write(*,'(a\)) ' Input file name? '
```

```
read(*,'(a)3 filel
write(*;(a\)') ' Output file name ?'
read(*,l(a)file2
open( 10,file=filel)
open( 12,File=file2)
write(*,*)'Give number of members in the truss:'
read(*,*)nelm
write(*,*)'Give number of joints in the truss:'
read(*,*)ntnd
write (*,*) 'number of fixity'
read(*,*) nif
ndof=3*ntnd
call input(ntnd,ntnd,nelm,nelm,ndof,a,e,nodec,x,y,z,el,cx,cy,
1 cz,eload,nf,nf,nnf,ndl)
do 14 i=l,ndof
do 14 j=l,ndof
gstif(i,j)=0.0
14 continue
c - >The following loop compute the element and global stiffness matrix.
do 15 ielm=l,nelm
call elstif(ielm,nelm,a,e,el,cx,cy,cz,estif)
call asemb(ielm,nodec,estif,gstif,ndof,nelm)
c write(12,*) 'Element Stiffness Matrix is'
c write(12,l)((estif(i,j),j=l,6),i=l,6)
15 continue
1 format(6f12.4/)
2 fomat(8el0.2/)
write(12,*)'Load Vector'
write(12,*)(eload(i)i=l,ndof)
c-> Following statement apply the boundary conditions on
c gstiff matrix
call boundary (ndof,ndof,gstif,nf,nf,nnf,ndl)
c write(12,*)'Global Stiffness Matrix (after applying B. C.)'
c write(12,2)((gstif(i,j),j=l,ndof),i=l,ndof)
c.> Following subroutine solve matrix equilibrium equation and
call gausselim(gstif,eload,ndof,ndof)
c - >Displacement vector is stored in the vector u
do i=l,ndof
u(i)=eload(i)
end do
write(12,*)'node', ',' x-displacement y-displacement
1 z-displacement'
```

```
      in=0
      do i=l,ndof,3
      in=in+l
      write(12,*)in,u(i), u(i+l), u(i+2)
      end do
3   fomat(4el8.6/)
        write (12,*)'Member Number ',' "axial force'
     write(l2,*)'       ', 'at node 1 at node 2'
      do 101 ielm=l,nelm
      alm=cx(ielm)
      amu=cy(ielm)
      anu=cz(ielm)
     call elstif(ielm,nelm,a,e,el,cx,cy,cz,estif)
     call force(ielm,nodec,estif,alm,amu anu, nelm,ndof,u,f)
 101  continue
       stop
      end
**************************************************************************
     subroutine input (nn,ntnd,ne,nelm,no,a,e,nodec,x,y,z,el,cx,cy,
    l cz,eload,nf,nfix,nnf,ndl)
      dimension a(ne),e(ne),nodec(ne,2),x(nn),y(nn),z(nn),
    l cx(ne),cy(ne),cz(ne),el(ne),eload(no),nnf(nf),ndl(nf)
c---->Read the cross section area,modulus of elasticity,nodal
c   connectivity of each element
        ndf=ntnd*3
do n=l,nelm
       read(10,*)i,a(i),e(i),(nodec(i,j),i=l,2)
        end do
c----> Read the nodal coordinate of each node
      do n=l,ntnd
     read(10,*)i,x(i),y(i)
       end do
c-----> Compute length and direction cosines of each element
         do 103 i=l,nelm
nl=nodec(i,l)
      n2=nodec(i,2)
     el(i)=sqrt((x(n2)-x(nl))**2+(y(n2)-y(nl))**2+(z(n2)-z(nl))**2)
      cx(i)=(x(n2)-x(nl))/el(i)
      cy(i)=(y(n2)-4(n1)/el(i)
      cz(i)=(z(nz)-2(nl))/el(i)
103    continue
 c----> Read Boundary Conditions
```

```
        read(10,*)(nnf(i),i=l,nfix)
      read(10,*)(ndl(i),i=l,nfix)
c----> Read the external load vector
          read(10,*)(eload(i),i=l,ndf)
      return
      end
c*************************************************************************
      subroutine elstif(ielm,ne,a,c,el,cx,cy,c2,estif)
       dimension estif(6,6)
       dimension cx(ne),cy(ne),cz(ne),a(ne),el(ne), e(ne)
       do l8 i=l,6
       do 18 j=1,6
      estif(i,j)=0.0
 18   continue
      ae=a(ielm)*e(ielm)/el(ielm)
      estif(l,l)=(cx(ielm)**2)*ae
       estif(l,2)=cx(ielm)*cy(ielm)*ae
       estif(l,3)=cx(ielm)*cz(ielm)*ae
       estif(l,4)=-estif(1,1)
       estif(l,5)=-estif(1,2)
       estif(l,6)=-estif(1,3)
        estif(2,l)=estif(1,2)
       estif(2,2)=(cy(ielm)**2)*ae
       estif(2,3)=cy(ielm)*cz(ielm)*ae
       estif(2,4)=-estif(2,1)
       estif(2,5)=-estif(2,2)
       estif(2,6)=-estif(2,3)
       estif(3,l)=cz(ielm)*cx(ielm)*ae
       estif(3,2)=cz(ielm)*cy(ielm)*ae
       estif(3,3)= (cz(ielm)**2)*ae
       estif(3,4)=-estif(3,l)
       estif(3,5)=-estif(3,2)
       estif(3,6)=-estif(3,3)
       do i=4,6
       do j=l,3
       estif(i,j)=estif(j-i)
       estif(i,j+3)=estif(i-3,j)
       end do
       end do
       return
       end
c*************************************************************************
```

```
      subroutine asemb(ielm,nodec,estif,gstif,no,ne)
      dimension estif(6,6),gstif(no,no),nodec(ne,2),kk(6)
      kk(3)=3*nodec(ielm,1)
      kk(2)=kk(3)-1
      kk(1)=kk(3)-2
      kk(6)=3*nodec(ielm,2)
      kk(5)=kk(6)-1
      kk(4)=kk(6)-2
       do 301 i=1,6
        ki=kk(i)
      do 301 i=1,6
        kj=kk(j)
       gstif(ki,kj)=gstif(ki,kj)+estif(i,j)
301   continue
     return
     end
c*******************************************************************************
      subroutine boundary (no,ndof,gstif,nf,nfix,nnf,ndl)
      dimension gstif(no,no),nnf(nf),ndl(nf)
          cnst=1.0
       do i=l,ndof
     if (cnst.le.gstif(i,i))then
     cnst=gstif(i,i)
     end if
       end do
       cnst=(l.+cnst)*1.0e+20
        do 401 i=l,nfix
      il=3*nnf(i)-2
      i2=il+1
      i3=il+2
      if(ndl(i).eq.l)go to 402
      if(ndl(i).eq.2)go to 403
      if(ndl(i).eq.3)goto 404
402   gstif(il il)=cnst+gstif(i1,i1)
        go to 401
403   gstif(i2,i2)=cnst+gstif(i2,i2)
        go to 401
404   gstif(i3,i3)=cnst+gstif(i3,i3)
401   continue
     return
     end
c*******************************************************************************
```

```
      subroutine gausselim(a,b,no,n)
    dimension a(no,no),b(no)
          n1 = n - 1
    do 130 k = 1, n1
    k1 = k+1
    do 130 i=k1,n.
       c=a(i,k) / a(k,k)
       do 120 j = k1, n
 120  a(i, j) = a(i, j) - c * a(k, j)
 130 b(i)=b(i)-c*b(k)
C *** back substitution ***
    b(n) = b(n) / a(n, n)
    do 140 ii = 1, n1
    i = n- ii
    i1 = i + 1
    c=l/a(i, i)
    b(i)=c*b(i)
    do 140 k = i1, n
 140 b(i) = b(i) - c * a(i, k) * b(k)
c   print *, 'solution vector'
c   write (*;(i4,el2.4)')(i, b(i),i = 1, n)
    return
    end
c*****************************************************************************
      subroutine force(ielm,nodec,estif,alm,amu,anu,ne,ndf,u,f)
      dimiension nodec(ne,2),k(6),estif(6,6),f(6),u(ndf)
          do 45 i=l,6
45  f(i)=0.0
       nl =nodec(ielm,1)
      n2=nodec(ielm,2)
          k(3)=3*n1
      k(2)=k(3)-1
      k(l)=k(3)-2
      k(6)=3*n2
      k(5)=k(6)-l
      k(4)=k(6)-2
       do 50i=1,6
      do 50 j=1,6
50      f(i)=f(i)+estif(i,j)* u(k(j))
c-----> Calculation of Axial Forces
      fax1=f(1)*alm+f(2)*amu+f(3)*anu
      fax2=f(4)*alm+f(5)*amu+f(6)*amu
```

```
c      write(12,*) 'forces at first node', f(1), 'f(2)', 'f(3)'
c      write(12,*) 'forces at second node', f(4), 'f(5)', 'f(6)'
       write (12,*)ielm, fax1, fax2
       write (12,*)
       return
       end
```

****

Example 1. Analyse the truss as shown in the figure, the node and member numbering is also shown in the figure. Each member has cross-section area of 5000 mm² and E = 200 kN/mm².

Input (trus3d1.in)

```
1 5000.0 200.0 1 2
2 5000.0 200.0 1 3
3 5000.0 200.0 1 4
4 5000.0 200.0 1 5
5 5000.0 200.0 1 6
1    8000.0    4000.0    8000.0
2    8000.0    4000.0    0.0
3    8000.0    0.0       8000.0
4    0.0       4000.0    2000.0
5    0.0       4000 .0   8000.0
6    0.0       4000.0    14000.0

2 2 2 3 3 3 4 4 4 5 5 5 6 6 6
1 2 3 1 2 3 1 2 3 1 2 3 1 2 3
500.0 800.0 0.0
0.0 0.0 0.0
0.0 0.0 0.0
0.0 0.0 0.0
0.0 0.0 0.0
0.0 0.0 0.0
```

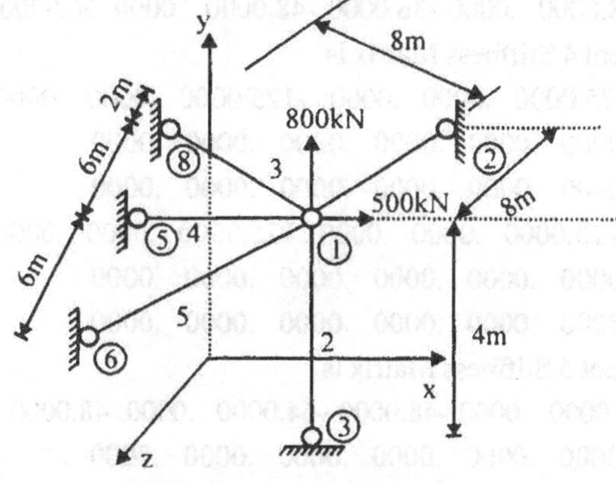

Fig. 16.22. 3D Truss example 1.

Output

Element 1 Stiffness Matrix is

```
.0000  .0000   .0000     .0000  .0000   .0000
.0000  .0000   .0000     .0000  .0000   .0000
.0000  .0000   125.0000  .0000  .0000   –125.0000
.0000  .0000   .0000     .0000  .0000   .0000
.0000  .0000   .0000     .0000  .0000   .0000
.0000  .0000   –125.0000 .0000  .0000   125.0000
```

Element 2 Stiffness Matrix is

```
.0000  .0000     .0000  .0000  .0000     .0000
.0000  250.0000  .0000  .0000  -250.0000 .0000
.0000  .0000     .0000  .0000  .0000     .0000
```

.0000 .0000 .0000 .0000 .0000 .0000
.0000 -250.0000 .0000 .0000 250.0000 .0000
.0000 .0000 .0000 .0000 .0000 .0000

Element 3 Stiffness Matrix is

64.0000 .0000 48.0000 -64.0000 .0000 -48.0000
.0000 .0000 .0000 .0000 .0000 .0000
-48.0000 .0000 36.-0000 -48.0000 .0000 -36.0000
-64.0000 .0000 48.0000 64.0000 .0000 48.0000
.0000 .0000 .0000 .0000 .0000 .0000
-48.0000 .0000 -36.0000 -48.0000 .0000 36.0000

Element 4 Stiffness Matrix is

125.0000 .0000 .0000 -125.0000 .0000 .0000
.0000 .0000 .0000 .0000 .0000 .0000
.0000 .0000 .0000 .0000 .0000 .0000
-125.0000 .0000 .0000 125.0000 .0000 .0000
.0000 .0000 .0000 .0000 .0000 .0000
.0000 .0000 .0000 .0000 .0000 .0000

Element 5 Stiffness Matrix is

64.0000 .0000 -48.0000 -64.0000 .0000 48.0000
.0000 .0000 .0000 .0000 .0000 .0000
-48.0000 .0000 36.-0000 48.0000 .0000 -36.0000
-64.0000 .0000 48.0000 64.0000 .0000 -48.0000
.0000 .0000 .0000 .0000 .0000 .0000
48.0000 .0000 -36.0000 -48.0000 .0000 36.0000

Load Vector

500.00 800.000.00 0.00 0.00 0.00 0.00 0.000.00 0.000.000.00 0.00 0.000.00 0.00 0.00 0.00

Global Stiffness Matrix (after applying B. C.)

.25E+03 -00E+00 -.36E+02 .00E+00 -00E+00 .00E+00 -00E+00 -00E+00 -00E+00 .00E+00 .00E+00 0E+00
.00E+00 .00E+00 .00E+00 .00E+00 .00E+00 .00E+00 .00E+00 .00E+00 .00E+00 .00E+00 .00E+00 0E+00 .00E+00
.00E+00 .00E+00 .00E+00 .00E+00 .00E+00 .00E+.00 .00E+00 .00E+00 .00E+00 .00E+00 0E+00 .00E+00 .
.00E+00 .00E+00 .00E+00 .00E+00 .00E+00 .00E+00 .00EP00 .00E+00 .005+00 .00E+00 OE+00 .00E+00 .00E+00
.00E+00 .00E+00 .00E+00 .00E-fc00 .00E+eO .00E+00 .00E+.00 .00E+00 .00E+00 OE+00 .00E+00 .00E+00 .00E+00
.00E+00 .00E+00 .00E+00 .00E+-00 .00E+&0 .00E+00 .00E+00 .00E+00 .00E+00 .00E+00 .00E+00 .00E.+00 .00E+00
.25E+23 I00E+00 .00E+00 .00E+00 .00E+00 .00E+00 .00E+00 .00E+00 00E+00 .00E+00 .00E+00 .00E+00 .00E+00
.00E+00 .00E+00 .00E+00 .00E+00 .00E,+00 .00E+00 .00E+00 0E+00 .00E+00 .00E+00 .00E+00 .00E+00
.00E+00 .00E+QO .00E+00 .00E+00 .00E+00 -OQE+00 .00E+00 OE+00 .00E+00 .00E+00 .00E+00 .48E+02
.00E+00 .00E+,00 .00E+JDO .00E+00 .00E+00 .00E+00 .00E+.Q00E+00 .00E+-00 .00E+00 .00E+00 .00E+OP .00E+00
.00E+00 .00E+00 .00E+00 .48E+02 .00E+00 .00E+00 OE+00 .00E+00 .00E+00 .00E+00 .00E+00 :00E+00
.00E+00 .00E+00 .00E+00 .00E+00 .00E+00 .00E+00 OE+O .00E+00 .00E+00 .00E+00 -.36E-?-02 .00E+00 .00E+00
.00E+00.00E+00 .00E+00,.00E+00.00E+00QE+00 .00E+00.00E+00 .00E+00 .00E+00 .00E+00 .00E+00-

.64E+02.00EtPO25Et-23.00E+00.oo-t-ooOE+0.00E+00.00E+00.00E+00.00E+00.00E+00.00E+00
.00E+00.00E+00.00E+00-00E+0.0.00E+00.00E+,00.00E+00.00E+00.00E+00.00E+00.00E+00
.00E+00.00E+00.00E+00.00E+00.00E+00.00E+00.00E+00.00E+00'.00E+00.00E+00.00E+00
.00E+00.00E+00-.48E+02.00E+00.00E+00.00E+00.00E+00.00E+00.00E+00.00E+00.00E+00
.00E+O.OE+00.00E+00.00E+00.00E+00.00E+00.00E+00.00E+00.00E+00-.13E+03.00E+00.00E+O
.OE+00.00E+00'.00E+00.00E+00.00E+00.00E+00.00E+00.00E+00.00E+00.00E+00.00E+00
.00E+00.00E+00.00E+00.00E+00.00E+00.00E+00.00E+00.00E+00.00E+00.00E+00.00E+00
.00E+00.00E+00.00E+00.00E+00.00E+00.00E+00OE+00.00E+00.00E+00.00E+00.00E+00
.00E+00.25E+.23.00E+00.00E+00-00E+.00.00E+00-00E+09.00E+QO.00E+00.00E+00.00E+00
.00E+00.00E+00.00E+00.00E+00.00E+00.00E+00.00E+00-.64E+02.00E+00.00E+00.00E+00
.00E+00.00E+00.00B+00-00E+.00.00E+00.00E+00.00E+00.QOE+00.00E+00.00E+00.00E+00
.00E+00.00E+00.00E+00.00E+00.00E+00-.13E+03.00E+00.00E+00.00E+00.00E+00.00E+00
.00E+00.00E+00.00E+00.00E+00.00E+00.00E+00.00E+00.00E+00.00E+00.00E+00.48E+02
.00E+00.00E+00'.00E+00.00E+00.00E+00.00E+00.00E+00.00E+00.00E+00.00E+00.00E+00
.00E+00.00E+00.00E+00.00E+00.00E+00.00E+00

NODE X-DISPLACEMENT Y-DISPLACEMENT Z-DISPLACEMENT

1 1.976285 3.200000 1.259299E-28
2 0.000000E+00 0.000000E+00 0.000000E+00
3 0.000000E+00 3.149606E-20 0.000000E+00
4 4.979615E-21 0.000000E+00 3.734712E-21
5 9.725810E-21 0.000000E+00 0.000000E+00
6 4.979615E-21 0.000000E+00 -3.734712E-21

Member Number 'axial force

	at node 1	at node 2
1	-1.574123E-26	1.574123E-26
2	-800.000000	800.000000
3	-158.102800	158.102800
4	-217.035600	217.035600
5	-158.102800	158.102800

****

Example 2. Analyze the truss as in the figure, the node and member numbering is also shown in the figure. Input (trus3d2.in)

1 80.0 210.0 1 5
2 80.0 210.0 4 8
3 100.0 210.0 2 6
4 100.0 210.0 3 7
5 30.0 210.0 5 6
6 30.0 210.0 7 8
7 30.0 210.0 5 7
8 30.0 210.0 6 8
9 50.0 210.0 1 6
10 50.0 210.0 2 8

11 50.0 210.0 3 5
12 50.0 210.0 4 7
1 0.0 0.0 0.0
2 3000.0 0.0 0.0
3 0.0 3000.0 0.0
4 3000.0 3000.0 0.0
5 0.0 0.0 4000.0
6 3000 0.0 4000.0
7 0.0 3000.0 4000.0
8. 3000.0 3000.0 4000.0
1 1 1 2 2 2 3 3 3 4 4 4
1 2 3 1 2 3 1 2 3 1 2 3
0.0 0.0 0.0
0.0 0.0 0.0
0.0 0.0 0.0
0.0 0.0 0.0
0.0 0.0 -10.0
2.0 0.0 -10.0
0.0 0.0 -10.0
2.0 0.0 -10.0

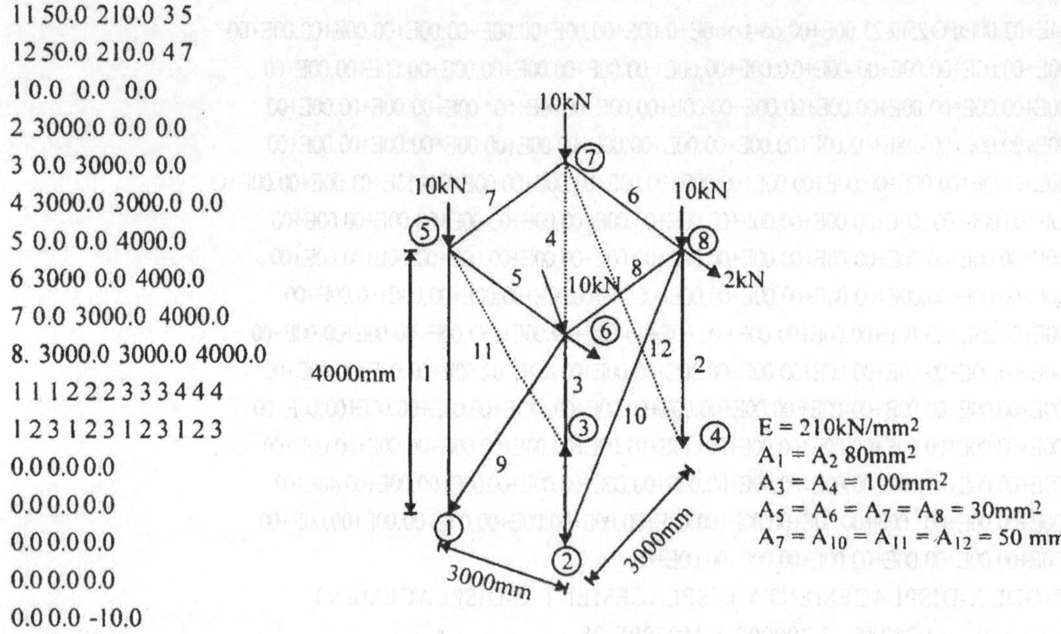

Fig. 16.23. 3D Truss Example 2.

Output (only final results)

NODE	X-DISPLACEMENT	Y-DISPLACEMENT	Z-DISPLACEMENT
1	2.633658E-21	0.000000E+00	-9.656748E-21
2	0.000000E+00	1.071270E-28	-1.667484E-20
3	0.000000E+00	-3.090754E-28	-9.656748E-21
4	2.633658E-21	0.000000E+00	-1.667484E-20
5	5.862433	-3.174604	-2.380953
6	5.862433	.174603	-2.412698
7	7.830685E-01	-3.174604	-1.396825
8	1.735449	.174603	-2.380953

Member Number 'axial force

	at node 1	at node 2
1	10.000000	-10.000000
2	10.000000	-10.000000
3	12.666670	-12.666670
4	7.333333	-7.333333
5	-1.08468E-07	1.08468E-07
6	-2.000000	2.000000
7	5.078693E-07	-5.078693E-07
8	1.665148E-08	-1665148E-08
9	-3.333333	3.333333
10	-7.81047E-08	7.811047E-08
11	-4.118965E-7	4.118965E-7
12	3.333333	-3.333333

16.3. MATRIX METHOD OF ANALYSIS FOR BEAM STRUCTURES

16.3.1 Introduction

A continuous beam system represents the most simple skeletal system used for supporting transverse loading. Long horizontal members used in buildings and bridges shafts supported to bearings are some common examples of beams. The concept of matrix stiffness method as described and used to analyze truss problems in earlier sections can be readily extend for beam problems. However, a unique problem emerges for the beam structure is that it is not always apparent where the nodal points are located. The nodes are well defined for trusses, but with various factors obscuring the location of the nodes for beams (e.g. presence of load between support points). Thus the nodal points must be located considering the behavior properties of the beam. This process known as 'idealization' must be done carefully to obtain solutions that closely approximate the behavior of the actual beam. In this section the stiffness method is extended to include problems in beam bending. The stiffness matrix for the basic beam element is derived from known available beam theory.

16.3.2 Beam element stiffness matrix

A typical beam element of uniform cross section is shown in Figure 16.24. The longitudinal axis of the element lies along the *x*-axis. The element has constant moment of inertia *I*, modulus of elasticity *E* and length *L* the element is assumed to have two *d.o.f.* at each end (nodal point) a transverse deflection *v* and an angle of rotation (slope) θ. Corresponding to the two *d.o.f.*, *v* and θ, a transverse shear force F_y and a bending moment *M* respectively acts at each nodal point.

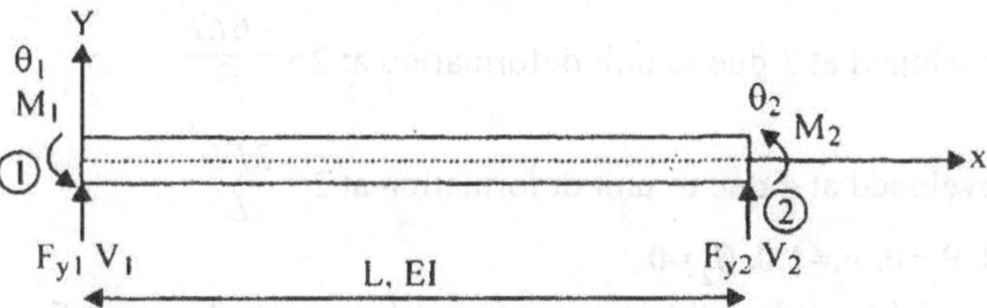

Fig. 16.24. Straight Beam with uniform cross section.

The stiffness matrix relating the nodal forces and their corresponding displacement is obtained by deforming the beam element applying four distinct degree of freedom. Various stiffness coefficient are derived by considering separate displacement cases for which force displacement relationships are known. Now the final force displacement relationships are obtained by superposing the results of four individual cases on the basis of linear structural behavior.

Case 1: $v_1=1.0, \theta_1=\theta_2=v_2=0$

Under the imposed boundary the beam deformed shape as well various internal forces developed to such deformation is shown in figure 16.25*a*. Not that this corresponds to a beam that is fixed against rotation at both ends and given an upward translation at the left end. Now from the fundamental definition of stiffness coefficient we can easily write

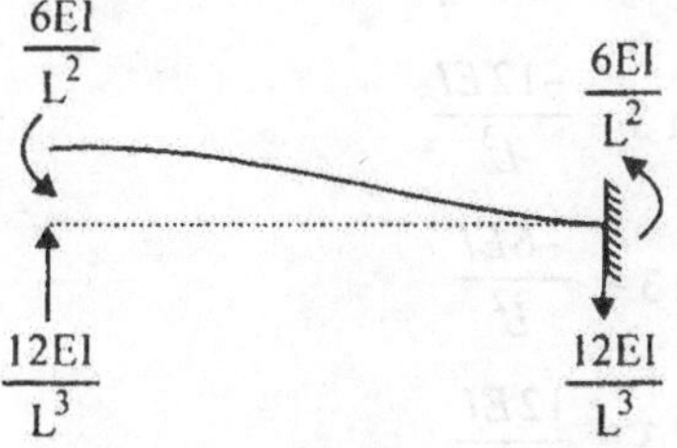

Fig. 16.25*a*. Displacement case 1

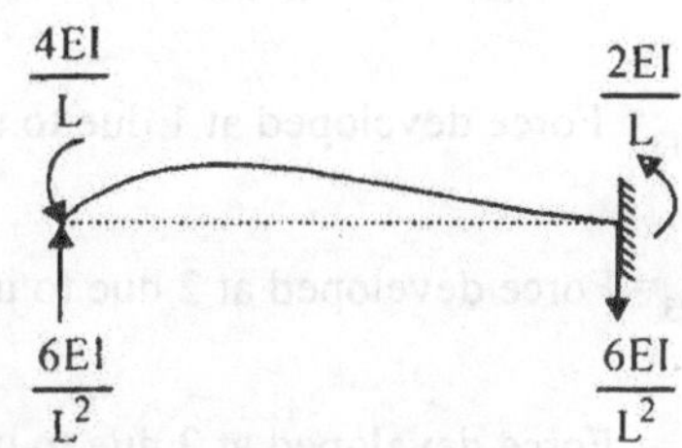

Fig. 16.25*b*. Displacement case 2

k_{11}= Force developed at 1 due to unit deformation at $1 = \dfrac{12EI}{L^3}$

k_{21}= Force developed at 2 due to unit deformation at $1 = \dfrac{6EI}{L^2}$

k_{31}=Force developed at 4 due to unit deformation at $1 = \dfrac{-12EI}{L^3}$

k_{41}= Force developed at 4 due to unit deformation at $1 = \dfrac{6EI}{L^2}$ (16.50)

Case 2: $v_1 = 0, \theta_1 = 1.0, \theta_2 = v_2 = 0$

Under the imposed boundary the beam deformed shape is shown in figure 16.25*b*. Note that this corresponds to a beam that is simply supported at node 1 and fixed at end 2. Various internal forces developed to such deformation is also shown in the same figure. Now from the fundamental definition of stiffness coefficient we can easily write

k_{12}= Force developed at 1 due to unit deformation at $2 = \dfrac{6EI}{L^2}$

k_{22}= Force developed at 2 due to unit deformation at $2 = \dfrac{4EI}{L}$

k_{32}= Force developed at 3 due to unit deformation at $2 = \dfrac{-6EI}{L^2}$ (16.51)

k_{42}= Force developed at 4 due to unit deformation at $2 = \dfrac{2EI}{L}$

Case 3: $v_1 = 0, \theta_1 = 0, v_2 = 1.0, \theta_2 = 0$

Under the imposed boundary the beam deformed shape is shown in figure 16.25c. Note that this corresponds to a beam that is fixed against a rotation at both ends and given an upward translation at the right end. Various internal forces developed to such deformation is also shown in the same figure. Now from the fundamental definition of stiffness coefficient we can easily write

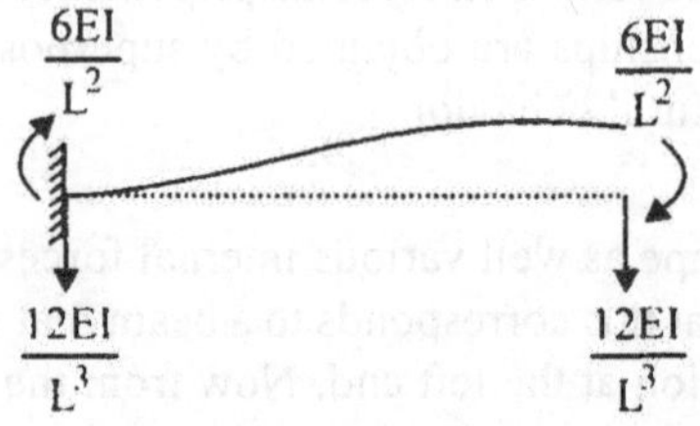

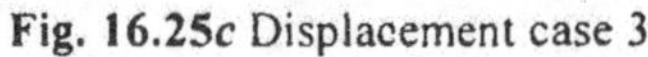

Fig. 16.25*c* Displacement case 3

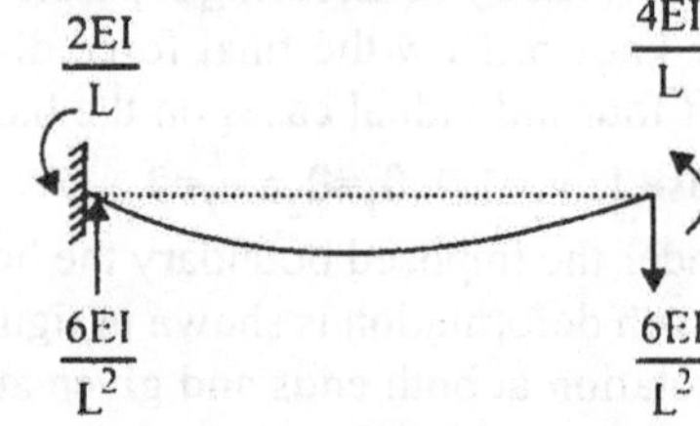

Fig. 16.25*d* Displacement case 4

k_{13}= Force developed at 1 due to unit deformation at $3 = \dfrac{-12EI}{L^3}$

k_{23}= Force developed at 2 due to unit deformation at $3 = \dfrac{-6EI}{L^2}$ (16.52)

k_{33}= Force developed at 3 due to unit deformation at $3 = \dfrac{12EI}{L^3}$

k_{43}= Force developed at 4 due to unit deformation at 3 = $\frac{-6EI}{L^2}$

Case 4: v_1=0, θ_1= v_2 =0, θ_2 =1.0

Under the imposed boundary the beam deformed shape is shown in figure 16.25d. Note that this corresponds to a beam that is supported at node 2 on a roller and fixed at end 1. Various internal forces developed to such deformation is also shown in the same figure. Now from the fundamental definition of stiffness coefficient we can easily write

$$k_{14} = \text{Force developed at 1 due to unit deformation at 4} = \frac{6EI}{L^2}$$

$$k_{24} = \text{Force developed at 2 due to unit deformation at 4} = \frac{2EI}{L}$$

$$k_{34} = \text{Force developed at 3 due to unit deformation at 4} = \frac{-6EI}{L^2} \quad (16.53)$$

$$k_{44} = \text{Force developed at 4 due to unit deformation at 4} = \frac{4EI}{L}$$

General Displacement Case:

If each of the generalized displacements at both node of the beam member is simultaneously given nonzero values, the forces that maintain equilibrium for a linear elastic structure can be obtain by superposing the four individual displacement cases. Thus adding the forces defined in equation (16.50) to (16.51) gives the general force deformation relationships for beam element when all four dispalcements have non zero values *i.e.* having value v_1, θ_1 at node 1 and v_2, θ_2 at node 2. Then we can write,

$$F_{y1} = k_{11}v_1 + k_{12}\theta_1 + k_{13}v_2 + k_{14}\theta_2$$

$$M_1 = k_{21}v_1 + k_{22}\theta_1 + k_{23}v_2 + k_{24}\theta_2$$

$$F_{y2} = k_{31}v_1 + k_{32}\theta_1 + k_{33}v_2 + k_{34}\theta_2$$

$$M_2 = k_{41}v_1 + k_{42}\theta_1 + k_{43}v_2 + k_{44}\theta_2$$

i.e.
$$F_{y1} = \frac{12\,EI}{L^3}v_1 + \frac{6\,EI}{L^2}\theta_1 - \frac{12\,EI}{L^3}v_2 + \frac{6\,EI}{L^2}\theta_2$$

$$M_1 = \frac{6\,EI}{L^2}v_1 + \frac{4\,EI}{L}\theta_1 - \frac{6\,EI}{L^2}v_2 + \frac{2\,EI}{L}\theta_2$$

$$F_{y2} = \frac{-12\,EI}{L^3}v_1 - \frac{6\,EI}{L^2}\theta_1 + \frac{12\,EI}{L^3}v_2 - \frac{6\,EI}{L^2}\theta_2$$

$$M_2 = \frac{6\,EI}{L^2}v_1 + \frac{2\,EI}{L}\theta_1 - \frac{6\,EI}{L^2}v_2 + \frac{4\,EI}{L}\theta_2$$

These forces displacement relationships can be cast into matrix form as

$$\begin{Bmatrix} F_{y1} \\ M_1 \\ F_{y2} \\ M_2 \end{Bmatrix} = EI \begin{bmatrix} \frac{12}{L^3} & \frac{6}{L^2} & -\frac{12}{L^3} & \frac{6}{L^2} \\ \frac{6}{L^2} & \frac{4}{L} & -\frac{6}{L^2} & \frac{2}{L} \\ -\frac{12}{L^3} & -\frac{6}{L^2} & \frac{12}{L^3} & -\frac{6}{L^2} \\ \frac{6}{L^2} & \frac{2}{L} & -\frac{6}{L^2} & \frac{4}{L} \end{bmatrix} \begin{Bmatrix} v_1 \\ \theta_1 \\ v_2 \\ \theta_2 \end{Bmatrix} \; i.e. \{f\} = [k]\{d\} \tag{16.54}$$

Note that for beam it is not necessary to transform quantities from one co-ordinate system to another since the member and structural co-ordinate coincide.

16.3.3 Properties of Beam Stiffness Matrix

If we consider the beam element as shown in figure 16.24 as free body, it must be in equilibrium under four nodal forces which are related to four nodal *d.o.f.* by the stiffness equation in (16.54). It is seen in the stiffness matrix $[k]$ that the first row and the third row are exactly identical but with opposite signs. This means that forces F_{y1} and F_{y2} are equal in magnitude but opposite in direction. So the equilibrium in the transverse direction is satisfied. Now taking the moment about nodal point 2 *i.e.*

$$\Sigma M_2 = F_{y1} L - M_1 - M_2$$

$$= EI\left[\frac{12}{L^3} v_1 \times L + \frac{6}{L^2}\theta_1 \times L - \frac{12}{L^3} v_2 \times L + \frac{6}{L^2}\theta_2 \times L - \frac{6}{L^2} v_1 - \frac{4}{L}\theta_1 + \frac{6}{L^2} v_2 - \frac{2}{L}\theta_2 - \frac{6}{L^2} v_1 - \frac{2}{L}\theta_1 + \frac{6}{L^2} v_2 - \frac{4}{L}\theta_2\right]$$

$$= EI\left[\left(\frac{12}{L^2} - \frac{6}{L^2} - \frac{6}{L^2}\right)v_1 + \left(\frac{6}{L} - \frac{4}{L} - \frac{2}{L}\right)\theta_1 + \left(\frac{6}{L^2} + \frac{6}{L^2} - \frac{12}{L^2}\right)v_2 + \left(\frac{6}{L} - \frac{2}{L} - \frac{4}{L}\right)\theta_2\right] = 0$$

which again satisfy the moment equilibrium condition. Also note that the element stiffness matrix [k] is singular in nature.

16.3.4 Application of Beam Element

In order to apply the stiffness method to beams, one must sub-divide the beam into its component elements. In general, each element must be free from load and have a prismatic cross-section. For this, the nodes should be located at a support points, where an external load is applied, cross-section area suddenly changes or where the displacement at a point to be determined. The method of assemblage of element stiffness matrix to form the overall structural stiffness matrix, application of boundary conditions and procedure for solution are same as described for truss problem. However, the complete procedure is explained with the aid of illustrative examples.

Example Using two beam elements to model the structure as shown in Figure 16.26, find deflection and internal force quantities.

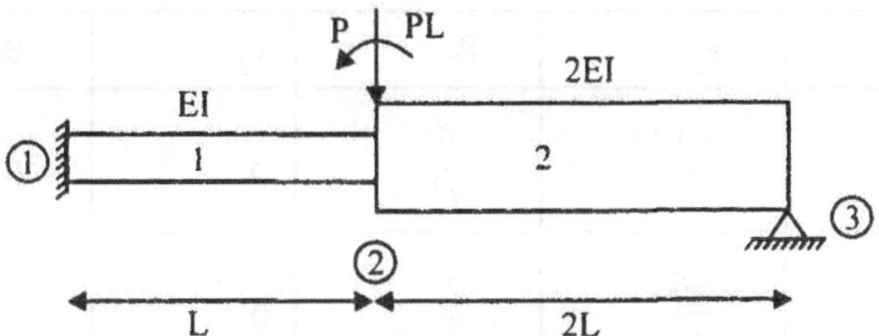

Fig. 16.26. Two member beam structure.

For element 1-2; The stiffness equilibrium equation is as below

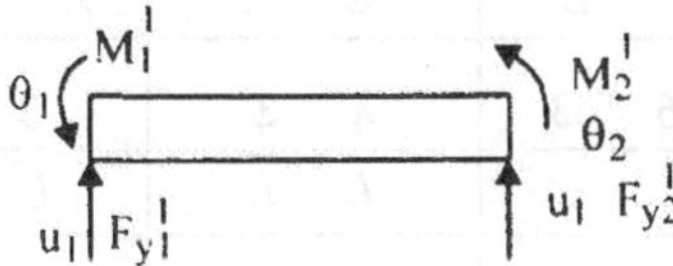

Fig. 16.27*a*. Free body of element 1-2

$$\begin{Bmatrix} F_{y_1}^1 \\ M_1^1 \\ F_{y2}^1 \\ M_2^1 \end{Bmatrix} = EI \begin{bmatrix} \frac{12}{L^3} & \frac{6}{L^2} & -\frac{12}{L^3} & \frac{6}{L^2} \\ \frac{6}{L^2} & \frac{4}{L} & -\frac{6}{L^2} & \frac{2}{L} \\ -\frac{12}{L^3} & -\frac{6}{L^2} & \frac{12}{L^3} & -\frac{6}{L^2} \\ \frac{6}{L^2} & \frac{2}{L} & \frac{6}{L^2} & \frac{4}{L} \end{bmatrix} \begin{Bmatrix} v_1 \\ \theta_1 \\ v_2 \\ \theta_2 \end{Bmatrix}$$

For element 2.3: The stiffness eqiluibrium equation is as

$$\begin{Bmatrix} F_{y_1}^2 \\ M_1^2 \\ F_{y2}^2 \\ M_2^2 \end{Bmatrix} = EI \begin{bmatrix} \frac{3}{L^3} & \frac{3}{L^2} & \frac{3}{L^3} & \frac{3}{L^2} \\ \frac{3}{L^2} & \frac{4}{L} & -\frac{3}{L^2} & \frac{2}{L} \\ -\frac{3}{L^3} & -\frac{3}{L^2} & \frac{3}{L^3} & -\frac{3}{L^2} \\ \frac{2}{L^2} & \frac{3}{L} & -\frac{3}{L^2} & \frac{4}{L} \end{bmatrix} \begin{Bmatrix} v_2 \\ \theta_2 \\ v_3 \\ \theta_3 \end{Bmatrix}$$

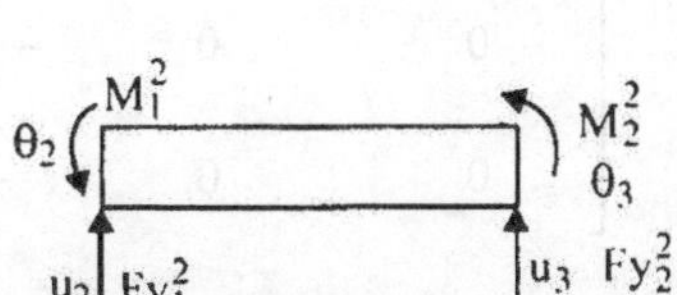

Fig. 16.27*b*. Free body of element 2-3

The beam has three nodal points and eight *d.o.f.* Thus the global stiffness matrix has size of 8.The system stiffness matrix can be obtain similarly as done for the truss problems. The final global stiffness matrix is now obtained as

v_1	θ_1	v_2	θ_2	v_3	θ_3		
$\frac{12}{L^3}$	$\frac{6}{L^2}$	$\frac{-12}{L^3}$	$\frac{6}{L^2}$	0	0	v_1	
$\frac{6}{L^2}$	$\frac{4}{L}$	$\frac{-6}{L^2}$	$\frac{2}{L}$	0	0	θ_1	
$\frac{-12}{L^3}$	$\frac{-6}{L^2}$	$\frac{12}{L^3}+\frac{3}{L^3}$	$\frac{-6}{L^2}+\frac{3}{L^2}$	$\frac{-3}{L^3}$	$\frac{3}{L^2}$	u_1	× EI
$\frac{6}{L^2}$	$\frac{2}{L}$	$\frac{-6}{L^2}+\frac{3}{L^2}$	$\frac{4}{L}+\frac{4}{L}$	$-\frac{3}{L^2}$	$\frac{2}{L}$	θ_2	
0	0	$\frac{-3}{L^3}$	$\frac{-3}{L^2}$	$+\frac{3}{L^3}$	$\frac{-3}{L^2}$	v_3	
0	0	$\frac{2}{L^2}$	$\frac{3}{L}$	$\frac{-3}{L^2}$	$\frac{4}{L}$	θ_3	

Load : $P_3 = -P$, $P_4 = PL$

Application of Boundary Conditions:

we have three boundary condtions $v_1 = \theta_1 = v_2 = 0$. Thus we can apply boundary condition to solve the matrix equation by multiplying the corresponding diagonal by large quantities *i.e.* 10×10^{20}

$$\frac{EI}{L}\begin{bmatrix} \frac{12}{L^2}\times 10^{20} & \frac{6}{L} & -\frac{12}{L^2} & \frac{6}{L} & 0 & 0 \\ \frac{6}{L} & 4\times 10^{20} & -\frac{6}{L} & 2 & 0 & 0 \\ -\frac{12}{L^2} & -\frac{6}{L} & \frac{15}{L^2} & \frac{3}{L} & -\frac{3}{L^2} & \frac{3}{L} \\ \frac{6}{L} & 2 & \frac{3}{L} & 8 & -\frac{3}{L} & 2 \\ 0 & 0 & -\frac{3}{L^2} & -\frac{3}{L} & \frac{3}{L^2}\times 10^{20} & -\frac{3}{L} \\ 0 & 0 & \frac{3}{L} & 2 & -\frac{3}{L} & 4 \end{bmatrix}\begin{Bmatrix} v_1 \\ \theta_1 \\ v_2 \\ \theta_2 \\ v_3 \\ \theta_3 \end{Bmatrix} = \begin{Bmatrix} P_1 \\ P_2 \\ P_3 \\ P_4 \\ P_5 \\ P_6 \end{Bmatrix}$$

Solution : Above matrix equation can be soved by Gauss elimination method to obtain the unknown displacement vector. However for solving the above problem by hand (as number of unknowns are only three) we can obtain the reduce matrix by cross out corresponding column and rows as done for truss problem as below,

$$\frac{EI}{L}\begin{bmatrix} \frac{15}{L^2} & -\frac{3}{L} & \frac{3}{L} \\ -\frac{3}{L} & 8 & 2 \\ \frac{3}{L} & 2 & 4 \end{bmatrix}\begin{Bmatrix} v_2 \\ \theta_2 \\ \theta_3 \end{Bmatrix} = \begin{Bmatrix} P_3 = -P \\ P_4 = PL \\ P_5 = 0 \end{Bmatrix}$$

then,
$$\begin{Bmatrix} v_2 \\ \theta_2 \\ \theta_3 \end{Bmatrix} = \frac{L^3}{276\,EI}\begin{bmatrix} 28 & \frac{18}{L} & \frac{30}{L} \\ \frac{18}{L} & \frac{51}{L^2} & -\frac{39}{L^2} \\ -\frac{30}{L} & -\frac{39}{L^2} & \frac{111}{L^2} \end{bmatrix}\begin{Bmatrix} -P \\ -PL \\ 0 \end{Bmatrix} = \frac{PL^3}{276\,EI}\begin{Bmatrix} -10 \\ 33/L \\ -9/L \end{Bmatrix}$$

Internal forces:

The internal shear forces and bending moments are obtained by substituting the known quantities of nodal displacements into each individual element stiffness matrix. For Element 1

$$\begin{Bmatrix} F'_{y1} \\ M_1^1 \\ F'_{12} \\ M_2^1 \end{Bmatrix} = EI\begin{bmatrix} \frac{12}{L^3} & \frac{6}{L^2} & -\frac{12}{L^3} & \frac{6}{L^2} \\ \frac{6}{L^2} & \frac{4}{L} & -\frac{6}{L^2} & \frac{2}{L} \\ -\frac{12}{L^3} & -\frac{6}{L^2} & \frac{12}{L^3} & -\frac{6}{L^2} \\ \frac{6}{L^2} & \frac{2}{L} & -\frac{6}{L^2} & \frac{4}{L} \end{bmatrix}\frac{PL^3}{276\,EI}\begin{Bmatrix} 0 \\ 0 \\ -10 \\ \frac{33}{L} \end{Bmatrix} = \begin{Bmatrix} \frac{53}{46}P \\ \frac{21}{46}PL \\ -\frac{53}{46}P \\ \frac{16}{23}PL \end{Bmatrix}$$

For Element 2
$$\begin{Bmatrix} F_{y_1}^2 \\ M_1^2 \\ F_{y2}^2 \\ M_2^2 \end{Bmatrix} = EI\begin{bmatrix} \frac{3}{L^3} & \frac{3}{L^2} & \frac{3}{L^3} & \frac{3}{L^2} \\ \frac{3}{L^2} & \frac{4}{L} & \frac{3}{L^2} & \frac{2}{L} \\ -\frac{3}{L^3} & \frac{3}{L^2} & \frac{3}{L^3} & -\frac{3}{L^2} \\ \frac{2}{L^2} & \frac{3}{L} & \frac{3}{L^2} & \frac{4}{L} \end{bmatrix}\frac{PL^3}{276\,EI}\begin{Bmatrix} -10 \\ \frac{33}{L} \\ 0 \\ -\frac{9}{L} \end{Bmatrix} = \begin{Bmatrix} \frac{7}{46}P \\ \frac{7}{23}PL \\ -\frac{7}{46}P \\ 0 \end{Bmatrix}$$

16.3.5 Beam Under Distributed Loading :

In the illustrated examples, the forces applied are all concentrated at the nodal points. As a matter of fact, the basic beam stifiness matrix expressed the relationship between the nodal degree of freedom and force quantities. In practical engineering application, the loads are not always concentrated at nodal points, the members of beam and frame structures are often loaded along their length. If the member loads consists of point loads only, then by locating a node under such load the problem can be solved using the earlier procedure. However, the introduction of additional nodes increases the problem size and hence the cost of the analysis. For such cases, special treatment is required to transfer the distributed loads to concentrated loads so that we can apply them at nodal points.

Equivalent Static Loads :

Simplest procedure to perform this task is the statically equivalent load method. In this procedure, we need to find only the concentrated loads that are in statical equilibrium with the distributed loads. This is done by simply assigning the total distributed loads over one half of the element to one nodal point and those over the other half of the element to the other nodal point. Although this approximate method appears to be crude, it is convenient and provides results with good accuracy when sufficient number of elements are used.

The procedure of generating the load vector is now explained with reference to figure consists of a beam model having 3 element.

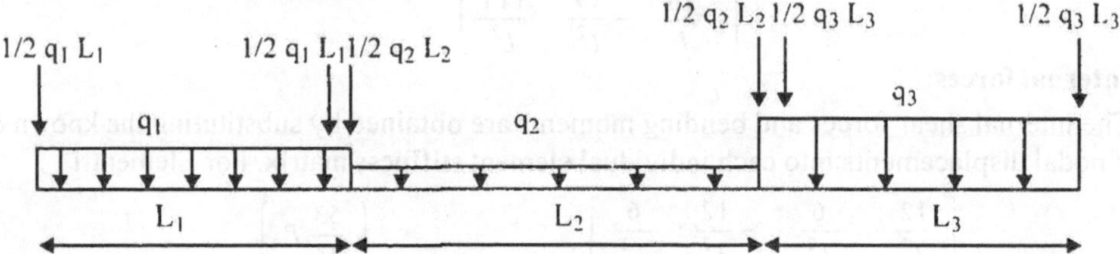

Fig. 16.28. Three element Beam.

Here the load vector will be

$$\left[P_1 = \frac{q_1 L_1}{2} \; P_2 = 0 \; P_3 = \frac{q_1 L_1}{2} + \frac{q_2 L_2}{2} \; P_4 = 0 \; P_5 = \frac{q_2 L_2}{2} + \frac{q_3 L_3}{2} \; P_6 = 0 \; P_7 = \frac{q_3 L_3}{2} \; P_8 = 0 \right]$$

Consistent Load vector:

As mentioned above distributed loads cannot be satisfactorily catered by the introduction of additional nodes as more number of elements is required for complex load. The use of equivalent joint load should be used whenever member loads are present. Such forces can be dealt with using the concepts of fixed end force, equivalent joint force, and superposition, as a illustrated in Figure 16.29

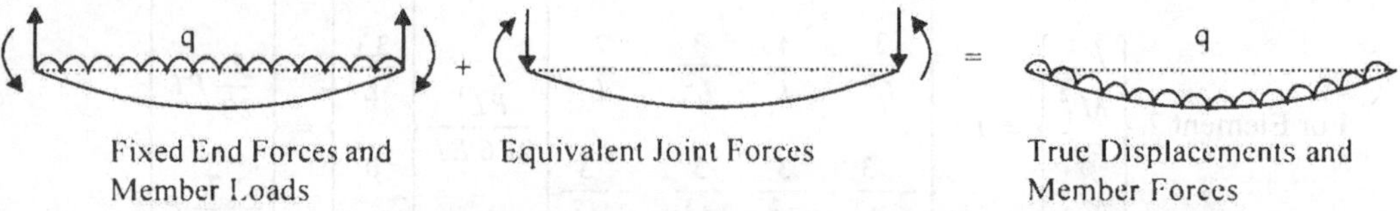

Fig. 16.29. Superposition of the Fixed End Forces and The Equivalent Joint forces.

The fixed end forces are an artificial system of applied loads that serve to hold the nodal displacements at zero. To return to the true structural behavior it is necessary to superimpose the negative of these forces. The procedure of obtaining the equivalent joint forces are as follows:

1. Lock all joints against displacements.
2. Apply member loads and calculate the forces developed at the locked ends of the member *i.e.* fixed end forces.

3. Take the negative of the fixed end forces as nodal loads and analyze the structure.
4. The final member forces are obtained by adding the fixed end forces with the nodal force vector as obtained from equation 16.54.

Consistent load vector for the beam model as shown in figure 16.30 is derived below

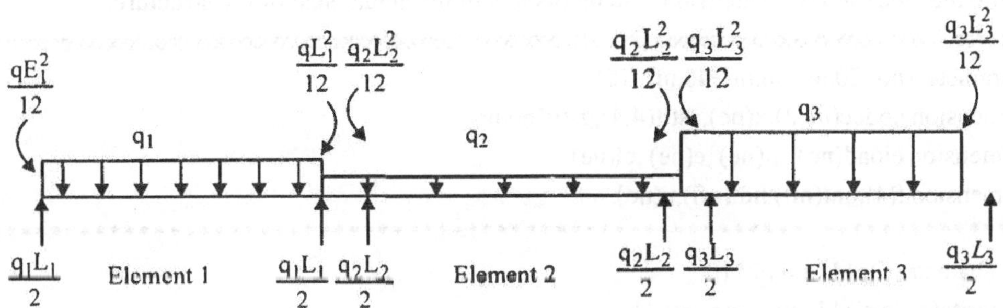

Fig. 16.30. Fixed End Forces

Fixed end force vector and Equivalent joint force vector for element 1, 2 and 3 element 1, 2 and 3

$$\begin{Bmatrix} \frac{q_1L_1}{2} \\ \frac{q_1L_1^2}{12} \\ \frac{q_1L_1}{2} \\ -\frac{q_1L_1^2}{12} \end{Bmatrix} \begin{Bmatrix} \frac{q_2L_2}{2} \\ \frac{q_2L_2^2}{12} \\ \frac{q_2L_2}{2} \\ -\frac{q_2L_2^2}{12} \end{Bmatrix} \begin{Bmatrix} \frac{q_3L_3}{2} \\ \frac{q_3L_3^2}{12} \\ \frac{q_3L_3}{2} \\ -\frac{q_3L_3^2}{12} \end{Bmatrix} \qquad \begin{Bmatrix} -\frac{q_1L_1}{2} \\ -\frac{q_1L_1^2}{12} \\ -\frac{q_1L_1}{2} \\ \frac{q_1L_1^2}{12} \end{Bmatrix} \begin{Bmatrix} -\frac{q_2L_2}{2} \\ -\frac{q_2L_2^2}{12} \\ -\frac{q_2L_2}{2} \\ \frac{q_2L_2^2}{12} \end{Bmatrix} \begin{Bmatrix} -\frac{q_3L_3}{2} \\ -\frac{q_3L_3^2}{12} \\ -\frac{q_3L_3}{2} \\ \frac{q_3L_3^2}{12} \end{Bmatrix}$$

Thus, Final load vector will be

$$\left[P_1 = -\frac{q_1L_1}{2} \;\middle|\; P_2 = -\frac{q_1L_1^2}{12} \;\middle|\; P_3 = \frac{-q_1L_1}{2} - \frac{q_2L_2}{2} \;\middle|\; P_4 = \frac{-q_1L_1^2}{12} - \frac{q_2L_2^2}{12} \;\middle|\; P_5 = \frac{-q_2L_2}{2} - \frac{q_3L_3}{2} \right.$$

$$\left. P_6 = \frac{-q_2L_2^2}{12} - \frac{q_3L_3^2}{12} \;\middle|\; P_7 = \frac{-q_3L_3}{2} \;\middle|\; P_8 = \frac{-q_3L_3^2}{12} \right]$$

Program Beam. for

```
cxxxxxxxxxxxxxxxxxxxxxxxxxxxxxxxxxxxxxxxxxxxxxxxxxxxxxxxxxxxxxxxxxxxxxxxxxxxxxxxxxxxxx
c The program is written for maximum of 40 element. However it can be used for any beam by"
c simple modification on dimension statements to suit the actual size of the structure.
cxxxxxxxxxxxxxxxxxxxxxxxxxxxxxxxxxxxxxxxxxxxxxxxxxxxxxxxxxxxxxxxxxxxxxxxxxxxxxxxxxxxxxxxx
   parameter (no=80,ne=40,nn=40,nfi=15)
   dimension nodec(ne,2),x(ne),estif(4,4),gstif(no,no)
   dimension eload(no),ai(ne),e(ne),el(ne)
   dimension f(4),nnf(nfi),ndl(nfi),u(no)
c ******************************************************************************
     character file1*10,file2*10
     write(*,'(a\)') ' Input file name? '
     read(*,'(a) file1
     'write(*,'(a\)') ' Output file name ?'
     read("',(a)3 file2
     open( 10,file=file1)
     open( 12,file=file2)
c ******************************************************************************
    write(*,*)'max number of element:'
       read(*,*)nelm
       wiite(*,*)'max number of node:'
       read(*,*)ntnd
       write*,*) 'number of fixity'
       read(*,*) nf
      ndof=2*ntnd
c---- >Following statement calls the input data
   call input(ntnd,ntnd,nelm,nelm,ndof,ai,e,nodec,x,el,
    1 eload, nf, nf, nnf, ndl
c----> Initialize the global structural stifness matrix
     do 14i=l,ndof
     do 14 j=l,ndof
    gstif(i,j)=0.0
  14 continue
c---->The following loop compute the element stiffness matrix
c  in turn and then place it in the corresponding position of the
c global stifness matrix.
     do 15 ielm=l,nelm
     call elstif(ielm,nodec,ai,e,el,estif,
     callasemb(ielm,nodec,estif,gstif,ndof;,nelm)
     write(12,*)'Element Number ',ielm
```

```
     write(12,*)'Length=',el(ielm)
      write(12,*)'Element Stiffness Matrix is'
       write(12,l)((estif(i,j),j=l,4),i=l,4)
 15  continue
 1  format(4fl8.4/)
c----> Once above loop is over gstif matrix contains the global
c   stiffness matrix of the entire structure
c-----> Following statement apply the boundary conditions on
c  gstiff matrix
       call boundary (ndof,ndof,gstif,nf,nf,nnf,ndl)
        write(12,2)'Global Stiffness Matrix is'
       write(12,*)((gstif(i,j)),i=l,ndof),j=l,ndof)
 2  format(6el0.3/)
c----> Following subroutine solve matrix equilbrium equation and
c   stored the displacements in eload vector
         call gausselim(gstif,eloadndof,ndof)
 c---->Displacement vector is stored in the vector u
        do i=l,ndof
     u(i)=eload(i)
       end do
     write(12,*)'DISPLACEMENT VECTOR'
     write(12,3) u(i),i=1,ndof)
3  format(4el8.6/)
c---->This loop call the force subroutine for each elements in turn and
c  compute the internal forces
       do 101 ielm=l nelm
             call elstif(ielm,nodec,ai,e,el,elestif)
    call force(ielm,f nodec,estif,nelm,ndof,u,f)
  101  continue
        stop
       end
c**********************************************************************
        subroutine input (nn,ntnd,ne,nelms no,ai,e,nodec,x,el
      l  eload,nf,nfix,nnf,ndl)
        dimension ai(ne),e(ne),nodec(ne,2),x(nn),el(ne),el(no)
     I nnf(nf),ndl(nf)
                 ndf=ntnd*2
c----> Read the cross section area,modulus of elasticity,nodal
c  connectivity of each element
         do n=l,nelm
          read(10,*)i,ai(i),e(i),(nodec(i,j),j=l,2)
           end do
```

```
c----> Read the nodal coordinate of each node
      do n=l,ntnd
      read(l0,*)i,x(i)
        end do
c----> Compute element length
            do 103 i=l,nelm
        nl=nodec(i,l)
        n2=nodec(i,2)
        el(i)=x(n2)-x(nl)
103     continue
c----> Read Boundary Conditions
        read(10,*)(nnf(i),i=l,nfix)
      read(10,*)(ndl(i),i=l,nfix)
c----> Read the external load vector
          read(10,*)(eload(i),i=l,ndf)
      return
      end
************************************************************************
      subroutine elstif(ielm,ne,ai,e,el,estif)
       dimension estif(4,4)
       dimension e(ne),ai(ne),el(ne)
c***********************************************************************
       do 18 i=l,4
      do l8 j=l,4
      estif(i,j)=0.0
 18   continue
      c=e(ielm)*ai(ielm)/(el(ielm)**3)
       estif(1,1)=12.0*c
       estif(2,1)=c*6.0*el(ielm)
       estif(3,1)=-estif(1,1)
       estif(4,1)=estif(2,1)
       estif(l,2)=estif(2,1)
       estif(2,2)=c*4.0*(el(ielm)**2)
       estif(3,2)=-estif(2,1)
       estif(4,2)=c*2.0*(el(ielm)**2)
       estif(1,3)=estif(3,l)
     estif(2,3)=estif(3,2)
     estif(3,3)=estif(l,l)
     esrif(4,3)=estif(3,2)
     esti?(l,4)=estif(4,l)
     estif(2,4)=estif(4,2)
     estif(3,4)=estif(4,3)
```

```
      esrif(4,4)=estif(2,2)
      return
      end
c***********************************************************************
      subroutine asemb(ielm,nodec,estif,gstif,no,ne)
      dimension estif(4,4),gstif(no,no),nodec(ne,2),kk(4)
c***********************************************************************
          kk(2) =2*nodec(ielm,1)
         kk(l)=kk(2)-1
         kk(4)=2*nodec(ielm,2)
         kk(3)=kk(4)-1
       do 301 i=1,4
      ki=k(i)
     do 301 j=l,4
      kj=kk(j)
         gstif(ki,kj) =gstif(ki,kj) +estif(i,j)
301  continue
     return
     end
c***********************************************************************
      subroutine boundary (no,ndof,gstif,nf,nfix,nnf,ndl)
      dimension gstif(no,no),nnf(nf),ndl(nf)
        cnst=1.0
        do i=1,ndof
     if (cnst.le.gstif(i,i.))then
     cnst=gstif(i,i)
     end if
        end do
          cnst=(1.+cnst)*1.0e+25
          do 401 i=1nfix
      i1=2*nnf(i)-1
      i2=i1+1
      if(ndl(i).eq.1)go to 402
      if(ndl(i).eq.2)go to 403
402 gstif(il,il)=cnst+cnst*gstif(il,il)
          go to 401
403  gstif(i2,j2)=cnst+cnst*gstif(i2,i2)
401 contitine
    return
    end
c***********************************************************************
      subroutine gausselim(a,b,no,n)
```

```
   dimension a(no,no),b(no)
      nl = n-1
   do 130 k = 1, nl
   kl = k+1
   do 130 i = kl, n
      c = a(i,k)/a(k,k)
      do 120 ) = kl, n
 120 a(i,j))=a(i,j)-c*a(k,j)
 130 b(i)=b(i)-c*b(k)
 C *** back substitution ***
   b(n)=b(n) / a(n,n)
      do 140 ii = 1, nl
   i = n - ii
   il = i + I
   c=l/a(i,i)
   b(i)=c*b(i)
   do 140 k = il, n
 140 b(i) = b(i) - c * a(i, k) * b(k)
 c  print *, 'solution vector'
 c  write *(i4,el2.4)')(i, b(i),i = l, n)
   return
   end
c*************************************************************************
      subroutine force(ielm,nodec,estif,ne,ndf,u,f)
      dimension nodec(ne,2),k(4),estif(4,4),f(4),u(ndf)
            do 45 i=l,4
 45  f(i)=0.0
      nl=nodec(ielm,l)
      n2=nodec(ielm)
       k(2)=2*nl
      k(l)-k(2)-l
      k(4)=2*n2
      k(3)=k(4)-l
            do 50 i=l,4
      do 50j=l,4
 50  f(i)=f(i)+estif(i,j)*u(k(j))
     write(12,*)'Member Number', ielm
      write(12,*)f(l); ',f(2); ',f(3); ',f(4)
       write(12,*)
      return
      end
*************************************************************************
```

Example Problem 1: The numerical example already worked out by hand to explain the procedure is taken for analyzing again by computer. The joint and member numbering are same as earlier. Typical input file (beam l in) and the detailed output file (beam l.out) are given below.

(E=1.0, I=1.0, P=1.0, L=1.0)

Input File (Beam 1 .in)

```
1  1·0 1·0 1 2
2  2·0 1·0 2 3
1      0.0
2      1.0
3      3.0

113

121
0.0 0.0

-1.0 1.0
0.0 0.0
```

Output file (beam1 .out)

```
Element Number 1
 Length =  1.000000
 Element Stiffness Matrix is
 12.0000   6.0000      -12.0000  6.0000
 6.0000    4.0000      -6.0000   2.0000
 -12.0000  -6.0000     12.0000   -6.0000
 6.0000    2.0000      -6 0000   4.0000
 Element Number 2
 Lengdi=  2.000000
Element Stiffhess Matrix is
 3.0000   3.0000     -3.0000      3.0000
 3.0000   4.0000     -3.0000      2.0000
 -3.0000  -3.0000    3.0000       -3.0000
 3.0000   2.0000     -3.0000      4.0000
 Global Stiffness Matrix is
 .208E+28 .600E+01 -.120E+02 .600E+01 .000E+00 .000E+00
 .600E+01 .800E+27-.600E+01 .200E+01 .000E+00 .000E+00
-.120E+02-.600E+01 .150E+02-.300E+01 -300E+01 .300E+01
 .600E+01 .200E+01-.300E+01 .800E+01-.300E+01 .200E+01
 .000E+00 .000E+00 .300E+01 -.300E+01 .640E+27 -.300E+01
 .000E+00 .000E+00 .300E+01 .200E+01 .300E+01 .400E+01
DISPLACEMENT VECTOR
```

Node	Deflection	Slope
1	-.553930E-27	-.570652E-27
2	-.362319E-01	.119565E+00
3	.237772E-27	-.326087E-01

Internal Forces

Member Number 1

1.152174 4.565217E-01 -1.152174 6.956522E-01

Member Number 2

1.521739E-01 3.043478E-01 -1.521739E-01 1.490116E-08

****

Example Problem 2 Analyze the continuous beam as shown in the figure. "The node and member numbering is shown in the figure. As distributed load is acting on the structure, we will use equivalent load utilizing fixed end actions which are also shown in the figure. Note that final load at various nodes will be as

$$P_1=-40.0, P_2=26.67, P_3=-130.0, P_4=126.67-90=-63.33$$

$$P_5=-90.0-62.5=-152.5, P_6=90-52.08=37.92, P_7=-62.5 \text{ and } P_8=52.080$$

Typical input and output files are given below.

Input File (Beam2.in)

```
1 23.48e-6 210.e+6 1 2
2 55.44e-6 120.e+6 2 3
3 34.08e-6 210.e+6 3 4
1 0.0
2. 4.0
3. 10.0
4. 15.0
1 2 3 4
1 1 1 1
-40.0 -26.67
-130.0 -63.33
-152.5 37.92
-62.5.52.08
```

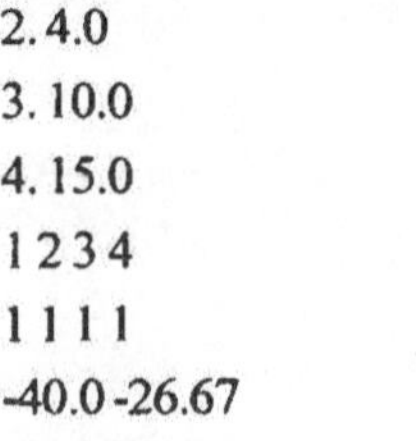

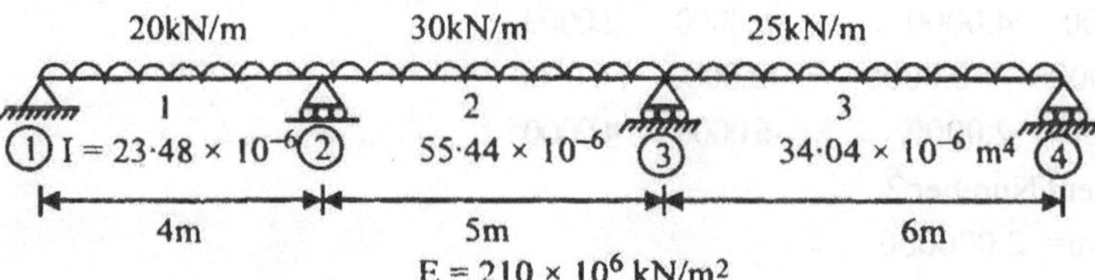

Fig. 16.30. Example Problem 2.

Output

Element Number 1

Length= 4.000000

Element Stifiness Matrix is

```
 924.5250   1849.0500   -924.5250   1849.0500
1849.0500   4930.8000  -1849.0500   2465.400C
-924.5250  -1849.0500    924.5250  -1849.0500
1849.0500   2465.4000  -1849.050C   4930 8000
```

Element Number 2

Length= 6.000000

Element Stifiness Matrix is

646.8000	1940.4000	-646.8000	1940.4000
1940.4000	7761.6000	-1940.4000	3880.8000
-646.8000	-1940.4000	646.8000	-1940.4000
1940.4000	3880.8000	-1940.4000	7761.6000

Element Number 3

Length= 5.000000

Element Stiffness Matrix is

686.2464	1715.6160	-686.2464	1715.6160
1715.6160	5718.7200	-1715.6160	2859.3600
-686.2464	-1715.6160	686.2464	-1715.6160
1715.6160	2859.3600	-1715.6160	57187200

Global Stiffness Matrix is

.125E+33 .185E+04-.925E+03 .185E+04 .000E+00 .000E+00 .000E+00 .000E+00

.185E+04 .493E+04-.185E+04 .247E+04 .000E+00 .000E+00 .000E+00 .000E+00

-.925E+03-.185E+04 .212E+33 .913E+02 -.647E+03 .194E+04 .000E+00 .000E+00

.185E+04 .247E+04 .913E+02 .127E+05-.194E+04 .388E+04 .000E+00 .000E+00

.000E+00 .000E+00 -.647E+03 -.194E+04 .180E+33 -.225E+03 -686E+03 .172E+04

.000E+00 .000E+00 .194E+04 .388E+04-.225E+03 .135E+05-.172E+04 .286E+04

.000E+00 .000E+00 .000E+00 .000E+00-.686E+03-.172E+04 .926E+32-.172E+04

.000E+00 .000E+00 .000E+00 .000E+00 .172E+04 .286E+04-.172E+04 .572E+04

DISPLACEMENT VECTOR

Node	Deflection	Slope
1	-.201367E-30	-.0.2773125E-2
2	-.659777E-30	-527143E-02
3	-.975538E-30	.268345E-02
4	-.481103E-30	.776521E-02

Member Number 1

-14.874810 -26.670000 14.874810 -32.829260

Member Number 2

-5.021699 -30.500750 5.021699 3.705527E-01

Member Number 3

17.925890 37.549440 -17.925890 52.080000

Now the final force vector are obtained by adding the fixed end actions with above forces and will be

Member Number 1

-14.874810	-26.670000	14.874810	-32.829260
40.0	26.67	40.0	-26.67

=25.126	0.0	54.8748	-59.5

Member Number 2

-5.021699	-30.500750	5.021699	3.705527E-01
90.0	90.0	90.0	-90.0
=84.98	59.5	95.0216	-90

Member Number 3

17.925890	37.549440	-17.925890	52.080000
+ 62.5	52.08	62.5	-52.08
= 80.42	89.67	44.57	0

16.4 MATRIX METHOD OF ANALYSIS OF FRAME STRUCTURE

16.4.1 Introduction

Two dimensional plane frames are extensively used in civil engineering. Examples of most common such structures are all types of building frame works etc. Plain frames constitute one of the most important structural entities. In most of the situations, a real structural system may be viewed as an assembly of beam grids and plane frames, wherein, the former consists load transferring and the later a load bearing entity. The main feature of these structures is that the members meeting at a joint are rigidly connected together by welding or bolt as in steel structure or by monolithic construction as reinforced and prestressed concrete structures. Because of the rigid joint large moments are transferred at the joints and such members are subjected to axial force, shear and bending moment as shown in Figure 16.32.

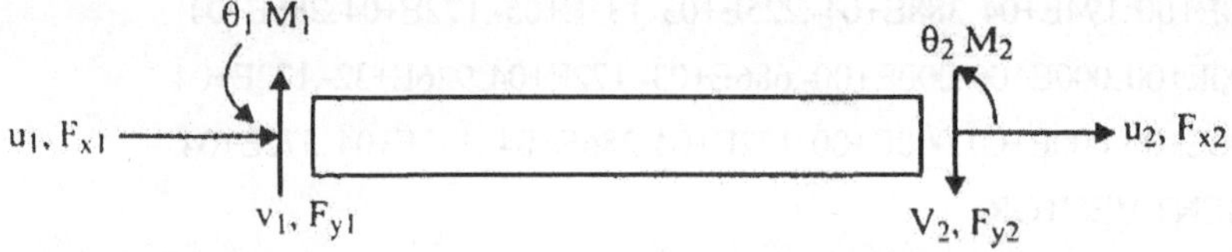

Fig. 16.32. Two noded Plane Frame Element.

A plane rigid frame can be envisioned as being composed of individual beam members with axial effect i.e. axial forces and deformations are present, oriented in various directions and connected together at their ends. To cope with the plane frame problem, an element must possess three d.o.f. at each nodal points; two displacement component u and v in the x and y directions respectively and one angle of rotation θ. Corresponding to the d.o.f. u,v and θ there are two forces F_x and F_y and a bending moment M at each nodal point. Such an element is shown in Figure 16.33 oriented arbitrarily, having modulus of elasticity E, cross section area A, moment of inertia I and length L.

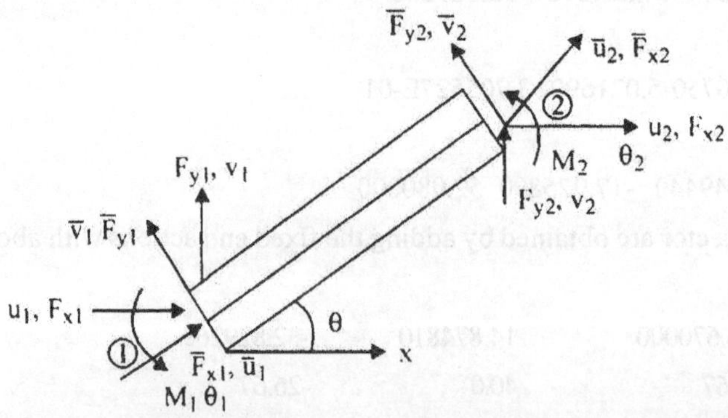

Fig 16.33. Two noded Plane Frame Element Oriented θ° with x axis.

It is clear that the element stiffness matrix for above element will be of size 6×6 relating the six d.o.f. with the corresponding 6 internal force quantities. The stiffness matrix can be thought of as the combination of an axial 2×2 truss bar and 4×4 stiffness of a beam element in member axis system and then transform the above combined 6x6 stiffness matrix to the global coordinate system ' using the similiar procedure described in 16.2 for truss members.

16.4.2 Axial-Flexural Beam Element in Local Coordinate System:

When a truss bar element and a beam element are combined, we obtain the axial flexural beam-element as shown in figure 16.34, the stiffness equation for which are

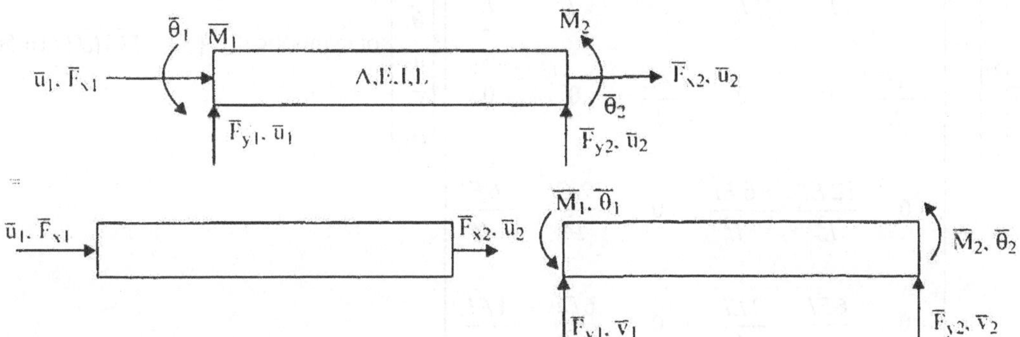

Fig. 16.34. Axial-Flexural Beam Element.

$$\begin{Bmatrix} \overline{F}_{x1} \\ \overline{F}_{x2} \\ \overline{F}_{y1} \\ \overline{F}_{y2} \\ \overline{M}_1 \\ \overline{M}_2 \end{Bmatrix} = \begin{bmatrix} \frac{AE}{L} & -\frac{AE}{L} & 0 & 0 & 0 & 0 \\ -\frac{AE}{L} & \frac{AE}{L} & 0 & 0 & 0 & 0 \\ 0 & 0 & \frac{12EI}{L^3} & \frac{6EI}{L^2} & -\frac{12EI}{L^3} & \frac{6EI}{L^2} \\ 0 & 0 & \frac{6EI}{L^2} & \frac{4EI}{L} & -\frac{6EI}{L^2} & \frac{2EI}{L} \\ 0 & 0 & \frac{12EI}{L^3} & -\frac{6EI}{L^2} & \frac{12EI}{L^3} & -\frac{6EI}{L^2} \\ 0 & 0 & \frac{6EI}{L^2} & \frac{2EI}{L} & -\frac{6EI}{L^2} & \frac{4EI}{L} \end{bmatrix} \begin{Bmatrix} \overline{u}_1 \\ \overline{u}_2 \\ \overline{v}_1 \\ \overline{\theta}_1 \\ \overline{v}_2 \\ \overline{\theta}_2 \end{Bmatrix} \qquad (16.55)$$

Equation 16.55 indicates that the axial and flexural stiffness are uncoupled. Thus the solution for axial and bending effect can be carried out separately. For the convenience of aseemblage, it is desirable to number all the d.o.f. at each nodal point in a certain sequence. For this purpose, we rearrange the above equation.

$$\begin{Bmatrix} \bar{F}_{x1} \\ \bar{F}_{y1} \\ \bar{M}_1 \\ \bar{F}_{x2} \\ \bar{F}_{y2} \\ \bar{M}_2 \end{Bmatrix} = \begin{bmatrix} \frac{AE}{L} & 0 & 0 & -\frac{AE}{L} & 0 & 0 \\ 0 & \frac{12EI}{L^3} & \frac{6EI}{L^2} & 0 & \frac{12EI}{L^3} & \frac{6EI}{L^2} \\ 0 & \frac{6EI}{L^2} & \frac{4EI}{L} & 0 & -\frac{6EI}{L^2} & \frac{2EI}{L} \\ -\frac{AE}{L} & 0 & 0 & \frac{AE}{L} & 0 & 0 \\ 0 & \frac{12EI}{L^3} & -\frac{6EI}{L^2} & 0 & \frac{12EI}{L^3} & -\frac{6EI}{L^2} \\ 0 & \frac{6EI}{L^2} & \frac{2EI}{L} & 0 & \frac{6EI}{L^2} & \frac{4EI}{L} \end{bmatrix} \begin{Bmatrix} \bar{u}_1 \\ \bar{v}_1 \\ \bar{\theta}_1 \\ \bar{u}_2 \\ \bar{v}_2 \\ \bar{\theta}_2 \end{Bmatrix}$$

or symbolically $\{\bar{f}\} = [\bar{k}]\{\bar{d}\}$ (16.56)

16.4.3. Coordinate Transformation:

An axial flexural beam element with its longitudinal axis lying along the $\bar{x}$ -axis is shown in figure 16.33. The axes $(\bar{x}, \bar{y})$ are the local coordinate axis and (x, y) are the global or reference axis. The $\bar{x}$ axis is ordinated at an angle θ with the x-axis. Nodal forces and d.o.f. in local axis (bars are added) and global axis systems are shown in Figure 16.33. Now considering the equilibrium of nodal forces at node (1), we can write

$$\bar{F}_{x1} = F_{x1}\cos\theta + F_{y1}\sin\theta$$

$$\bar{F}_{y1} = -F_{x1}\sin\theta + F_{y1}\cos\theta$$

$$\bar{M}_1 = M_1 \qquad (16.57)$$

i.e. in matrix form

$$\begin{Bmatrix} \bar{F}_{x1} \\ \bar{F}_{y1} \\ \bar{M}_1 \end{Bmatrix} = \begin{Bmatrix} \cos\theta & \sin\theta & 0 \\ -\sin\theta & \cos\theta & 0 \\ 0 & 0 & 1 \end{Bmatrix} \begin{Bmatrix} F_{x1} \\ F_{y1} \\ M_1 \end{Bmatrix}$$

Similar relation can be written at node 2 i.e. (16.58)

$$\begin{Bmatrix} \bar{F}_{x2} \\ \bar{F}_{y2} \\ \bar{M}_2 \end{Bmatrix} = \begin{Bmatrix} \cos\theta & \sin\theta & 0 \\ -\sin\theta & \cos\theta & 0 \\ 0 & 0 & 1 \end{Bmatrix} \begin{Bmatrix} F_{x2} \\ F_{y2} \\ M_2 \end{Bmatrix}$$

Above two equations can be written in a single matrix equation as

$$\begin{Bmatrix} \bar{F}_{x1} \\ \bar{F}_{y1} \\ \bar{M}_1 \\ \bar{F}_{x2} \\ \bar{F}_{y2} \\ \bar{M}_2 \end{Bmatrix} = \begin{bmatrix} \lambda & \mu & 0 & 0 & 0 & 0 \\ -\mu & \lambda & 0 & 0 & 0 & 0 \\ 0 & 0 & 1 & 0 & 0 & 0 \\ 0 & 0 & 0 & \lambda & \mu & 0 \\ 0 & 0 & 0 & -\mu & \lambda & 0 \\ 0 & 0 & 0 & 0 & 0 & 1 \end{bmatrix} \begin{Bmatrix} F_{x1} \\ F_{y1} \\ M_1 \\ F_{x2} \\ F_{y2} \\ M_2 \end{Bmatrix} \text{ or symbolical } \{\bar{F}\} = [T]\{F\} \qquad (16.59)$$

where $[T]$ is the transformation matrix and $\lambda = \cos\theta$, $\mu = \sin\theta$. It can be shown the above $[T]$ matrix is orthogonal. Then we can also write, $[T]^{-1} = [T]^T$. In the mean time, we can also transform the six nodal d.o.f. in the local coordinate system into those in global coordinate system i.e.

$$\{\bar{d}\} = [T]\{d\} \qquad (16.60)$$

Now substituting equation (16.59) for the vector $\{\bar{F}\}$ and equation (16.60) for the vector $\{\bar{d}\}$ into the stiffness equation (16.56) and making use of orthogonal properties of $[T]$, we obtain

$$[T] = \{f\} = [\bar{k}][T]\{d\}$$

$$\{f\} = [T]^T[\bar{K}][T]\{d\} \qquad (16.61)$$

i.e. $\quad \{f\} = [k]\{d\}$

The product $[T]^T[\bar{k}][T]$ is the final element stiffness matrix in reference i.e. member axis system which can be obtained as

$$\begin{bmatrix} \lambda & -\mu & 0 & 0 & 0 & 0 \\ \mu & \lambda & 0 & 0 & 0 & 0 \\ 0 & 0 & 1 & 0 & 0 & 0 \\ 0 & 0 & 0 & \lambda & -\mu & 0 \\ 0 & 0 & 0 & \mu & \lambda & 0 \\ 0 & 0 & 0 & 0 & 0 & 1 \end{bmatrix} \begin{bmatrix} \frac{AE}{L} & 0 & 0 & -\frac{AE}{L} & 0 & 0 \\ 0 & \frac{12EI}{L^3} & \frac{6EI}{L^2} & 0 & -\frac{12EI}{L^3} & \frac{6EI}{L^2} \\ 0 & \frac{6EI}{L^2} & \frac{4EI}{L} & 0 & -\frac{6EI}{L^2} & \frac{2EI}{L} \\ -\frac{AE}{L} & 0 & 0 & \frac{AE}{L} & 0 & 0 \\ 0 & -\frac{12EI}{L^3} & -\frac{6EI}{L^2} & 0 & \frac{12EI}{L^3} & -\frac{6EI}{L^2} \\ 0 & \frac{6EI}{L^2} & \frac{2EI}{L} & 0 & \frac{6EI}{L^2} & \frac{4EI}{L} \end{bmatrix} \begin{bmatrix} \lambda & \mu & 0 & 0 & 0 & 0 \\ -\mu & \lambda & 0 & 0 & 0 & 0 \\ 0 & 0 & 1 & 0 & 0 & 0 \\ 0 & 0 & 0 & \lambda & \mu & 0 \\ 0 & 0 & 0 & -\mu & \lambda & 0 \\ 0 & 0 & 0 & 0 & 0 & 1 \end{bmatrix}$$

$$= \begin{bmatrix} c_1 & & & & & \\ c_2 & c_3 & Symm. & & & \\ -c_4 & c_5 & c_6 & & & \\ -c_1 & -c_2 & c_4 & c_1 & & \\ -c_2 & -c_3 & -c_5 & c_2 & c_3 & \\ -c_4 & c_5 & c_7 & c_4 & -c_5 & c_6 \end{bmatrix} \tag{16.62}$$

$$c_1 = \lambda^2 \frac{AE}{L} + \mu^2 \frac{12\,EI}{L^2}, c_2 = \lambda\mu\left(\frac{AE}{L} - \frac{12\,EL}{L^3}\right)$$

where, $c_3 = \mu^2 \frac{AE}{L} + \lambda^2 \frac{12\,EI}{L^3}$, $c_4 = \frac{6\,EI}{L^2}\mu$, $c_5 = \frac{6\,EI}{L^2}\lambda$, $c_6 = \frac{4\,EI}{L}$, $c_7 = \frac{2\,EI}{L}$

Note that when the element undergoes a co-ordinate transformation the stiffness matrix no longer remains uncoupled. Now assembly of the element stiffness matrix to form global stiffness matrix, application of boundary conditions, solution procedure etc. follow the same as earlier for truss and beam problems.

16.4.4. Application of Frame Element.

The two member frame shown in figure 16.35. is constructed from steel beam sections having the following properties: $I=0.25 \times 10^9$ mm^4. $A = 9000$ mm^2. $E = 200$ kN/mm^2

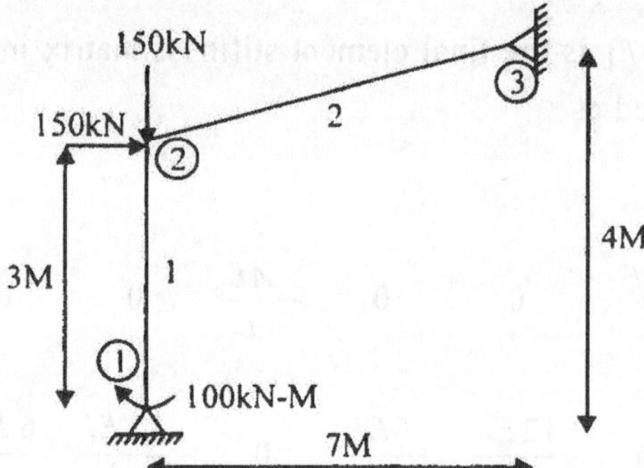

Fig. 16.35. Two member frame structure.

The structure has three node. So total d.o.f. will be 9. These are numbered as $[d_1, d_2, ..., d_9]$. Note that for element 1 $d_1 = u_1$, $d_2 = v_1$, $d_3 = \theta_1$ i.e. d.o.f at node 1. Similar numbering for other node is done.

Element 1-2:

$\theta = 90°$, then $\lambda = \cos 90° = 0$, $\mu = \sin 90° = 1.0$, $L = 3000$ MM.

$$c_1 = 0 \times \frac{AE}{L} + 1 \times 12 \times (50 \times 10^9)/(3000)^3 = 22.22,\ c_2 = 0,\ c_3 = 60,$$

$$c_4 = 33.33 \times 10^3,\ c_5 = 0,\ c_6 = 66.66 \times 10^6,\ c_7 = 33.33 \times 10^6$$

Thus,

$$[k]_{1-2} = \begin{bmatrix} 22.2 & 0 & -33.33e+3 & -22.2 & 0 & -33.33e+3 \\ 0 & 60 & 0 & 0 & -60 & 0 \\ -33.33e+3 & 0 & 66.7e+6 & 33.33e+3 & 0 & 33.3e+6 \\ -22.2 & 0 & 33.33e+3 & 22.2 & 0 & 33.33e+3 \\ 0 & -60 & 0 & 0 & 60 & 0 \\ -33.3e+3 & 0 & 33.3e+6 & 33.3e+3 & 0 & 66.6e+6 \end{bmatrix}$$

Element 2-3:

Similarly λ, μ and c's can obtained to compute the following stiffness matrix

$$[k]_{2-3} = \begin{bmatrix} 249.5 & 35.4 & -848.5 & -249.5 & -35.4 & -848.5 \\ 35.4 & 6.754 & 5940.0 & -35.4 & -6.754 & 5940 \\ -848.5 & 5940 & 28.3e+6 & 848.5 & -5940 & 14.1e+6 \\ -249.5 & -35.4 & +848.5 & 249.5 & 35.4 & 848.5 \\ -35.4 & -6.754 & -5940.0 & 35.4 & 6.754 & -5940 \\ -848.5 & 5940 & 14.1e+6 & 848.5 & -5940 & 28.28e+6 \end{bmatrix}$$

As in for truss examples, initially the global stiffness matrix zeroed and then the element stiffness matrix is added for each element in terms to obtain the global stiffness matrix as per the global d.o.f. number of each node. The final equilibrium equation will be as

$$\begin{bmatrix} 22.2 & 0 & 3.33\times10^3 & -22.2 & 0 & -3.33\times10^3 & 0 & 0 & 0 \\ 0 & 60 & 0 & 0 & -60 & 0 & 0 & 0 & 0 \\ 3.33\times10^3 & 0 & 66.76\times10^6 & 3.33\times10^3 & 0 & 3.33\times10^6 & 0 & 0 & 0 \\ -22.2 & 0 & 3.33\times10^6 & 271.7 & 35.4 & 32485 & -249.4 & -35.4 & -848.5 \\ 0 & -60 & 35.4 & 35.4 & 60.68 & 5940 & -35.4 & -6.75 & -5940 \\ -3.33\times10^6 & 0 & 3.33\times10^6 & 32485 & 5940 & 95.0\times10^6 & 848.5 & -5940 & 14.1\times10^6 \\ 0 & 0 & 0 & -249.4 & -35.4 & 848.5 & 249.5 & 35.4 & 848.5 \\ 0 & 0 & 0 & -35.4 & -6.75 & -5940 & 35.4 & 6.754 & -5940 \\ 0 & 0 & 0 & 848,5 & 5940 & 14.1\times10^6 & 848.5 & -5940 & 28.3\times10^6 \end{bmatrix}$$

$$\begin{Bmatrix} d_1 \\ d_2 \\ d_3 \\ d_4 \\ d_5 \\ d_6 \\ d_7 \\ d_8 \\ d_9 \end{Bmatrix} = \begin{Bmatrix} P_1 \\ P_2 \\ P_3 \\ P_4 \\ P_5 \\ P_6 \\ P_7 \\ P_8 \\ P_9 \end{Bmatrix} = \begin{Bmatrix} 0 \\ 0 \\ -1.0\times10^5 \\ 150 \\ -150 \\ 0 \\ 0 \\ 0 \\ 0 \end{Bmatrix}$$

Now to apply the boundary condition for solving the equilibrium equation to obtain the unknown displacements the diagonal element of the global stiffness matrix corresponding to zero displacements are multiplied by large number i.e. K_{11}, K_{22}, K_{88}, K_{77} are multiplied by large number so that d_1, d_2, d_8, d_7 becomes zero. Finally, the solution vector is obtained as

$$[d_1 d_2 d_3 d_4 d_5 d_6 d_7 d_8 d_9] = [0\ 0 - 2.16\times10^{-3}\ \ 0.792 - 0.297\ 0.532\times10^{-3}\ \ 0\ 0 - 0.179\times10^{-3}]^T$$

Internal Forces:

Member 1-2

$$\begin{Bmatrix} X_1^1 \\ Y_1^1 \\ M_1 \\ X_2^1 \\ Y_2^1 \\ M_2 \end{Bmatrix} = \begin{bmatrix} 22.2 & 0 & -33.3e+3 & -22.2 & 0 & -33.3e+3 \\ 0 & 60 & 0 & 0 & -60 & 0 \\ -33.33e+3 & 0 & 66.7e+3 & 33.3e+3 & 0 & 33.3e+6 \\ -22.2 & 0 & 33.e+3 & 22.2 & 0 & 33.3e+3 \\ 0 & -60 & 0 & 0 & 60 & 0 \\ -33.3e+3 & 0 & 33.3e+6 & 33.3e+3 & 0 & 66.6e+6 \end{bmatrix} \begin{Bmatrix} 0 \\ -2.16\times10^{-3} \\ 0.79 \\ 2 \\ -0.297 \\ 0.532\times10^{-3} \\ 0 \end{Bmatrix} \begin{Bmatrix} 178 \\ -36.67 \\ -99879 \\ -178 \\ 36.67 \\ 10113 \end{Bmatrix}$$

Member 2-3:

$$\begin{Bmatrix} X_1^2 \\ Y_1^2 \\ M_1^2 \\ X_2^2 \\ Y_2^2 \\ M_2^2 \end{Bmatrix} = \begin{bmatrix} 249.5 & 35.4 & -848.5 & -249.5 & -35.4 & -848.5 \\ 35.4 & 6.754 & 5940.0 & -35.4 & -6.754 & 5940 \\ -848.5 & 5940 & 28.3e+6 & 848.5 & -5940 & 14.1e+6 \\ -249.5 & -35.4 & +848.5 & 249.5 & 35.4 & 848.5 \\ -35.4 & -6.754 & -5940.0 & 35.4 & 6.754 & -5940 \\ -848.5 & 5940 & 14.1e+6 & 848.5 & -5940 & 28.28e+6 \end{bmatrix} \begin{Bmatrix} 0.792 \\ -0.297 \\ 0.532\times10^{-3} \\ 0 \\ 0 \\ -0.179\times10^{-3} \end{Bmatrix} \begin{Bmatrix} 189 \\ 1.43 \\ 10097 \\ -189 \\ -1.43 \\ 24.59 \end{Bmatrix}$$

Note that the forces obtained for each element in global axis system. However, one can easily transformed it to local axis system utilizing equation 16.57 and 16.58.

Program plane. for

```
cxxxxxxxxxxxxxxxxxxxxxxxxxxxxxxxxxxxxxxxxxxxxxxxxxxxxxxxxxxxxxxxxxxxxxxxxxxxxxxxxxxxxxxxxxx
c The program follows the same procedure and notations as earlier. It can be used
c for any plane frame by simple modification on dimension statements to suit the
c actual size of the structure.
c*************************************************************************
        parameter (no=18,ne=6,nn=6, nfi=6)
        dimension nodec (ne, 2), x (nn), y(nn), estif (6, 6), gstif (no, no)
        dimension eload(no) . a (ne), ai (ne), e (ne), el (ne), cx ( ne), cy(ne)
        dimension f (6) ,nnf (nfi) ,ndl (nfi), u (no)
c*************************************************************************
            character filel*10,file2*10
            write (*,'(a\)') 'Input file name?'
            read(*, '(a)') file1
            write (*,'(a\') 'Output file name ?'
            read(*, '(a)') file2
            open( 10,file=file l)
            open( 12,file=file2)
        write (*,*)' Give number of members in the frame:'
              read(*, *-)nelm
        write (*,*)' Give number of joints in the frames:'
              read(*, *)ntnd
              write (*,*) 'number of fixity'
              read(*,*) nf
              ndof=3*ntnd
c---->Following statement calls the input data
        call input (ntnd, ntnd, nelm, nelm, ndof, a, e, ai, nodec, x, y, el, cx, cy,
      I eload,nf,nf,nnf,ndl)
c----> Initialize the global structural stiffness matrix
          do 14 i=l,ndof
          do 14 j=l,ndof
        gstif(i,j)=0.0
   14   continue
c---- >The following loop compute the element stiffness matrix
c    in turn and then place it in the corresponding position of the
c    global stiffness matrix.
          do 15 ielm=l,nelm
          call elstif(ielm,nelm,a,e,ai,el,cx,cy,estif)
          call asemb (ielm,nodec, estif, gstif, ndof, nelm)
  15    continue
```

```
1     format(6el2.2/)
2     format(9e8.2/)
c----> Following statement apply the boundary conditions on gstif matrix
          call boundary (ndof, ndof, gstif, nf,nf,nnf,ndl)
          write(12,*) 'Global Stiffness Matrix (after applying B. C.)'
          write (12, 2) ( (gstif (i,j) ,j=l,ndof) ,i=l,ndof)
c----> Following subroutine solve matrix equilibrium equation
              call gausselim (gstif, eload, ndof, ndof)
c ---->Displacement vector is stored in vector u
            do i=l,ndof
            u (i) =eload (i)
              end do
            write (12,*) 'DISPLACEMENT VECTOR'
            write(12,3) (u(i),i=l,ndof)
 3    format(3el8.6/)
c---->This loop call the force subroutine for each elements in turn and
c     compute the internal forces
          write (12,*)'                    ','at 1st node',' ','at 2nd node'
            do 101 ielm=l,nelm
            call elstif (ielm, nelm, a, e, ai, el, ex, cy, estif)
              alm==cx (ielm)
              amu=cy(ielm)
          call force (ielm,nodec, estif, nelm,ndof,u, alm, amu, f)
          write (12,*)
          write (12, *) 'member' , ielm
          write(12,*)'Axial Force(Fx)=',f(l)' , f(4)
          write(12,*)'shearForce(Fy)=',f(2)' , f(5)
          write (12,*) 'Moment      =',f (3)' , f(6)
      101   continue
                stop
              end
c*************************************************************************
          subroutine input (nn,ntnd,ne, nelm, no, a, e, ai, nodec,x,y, el, cx, cy,
        I eload, nf,nfix,nnf,ndl)
            dimension a(ne),ai(ne),e(ne),nodec(ne,2),x(nn),y(nn),
        I cx (ne), cy (ne), el (ne), eload (no) ,nnf (nf) , ndl (nf)
c---->Read the cross section area,modulus of elasticity, nodal
  c     connectivity of each element
              do n=l,nelm
                  read (10, *)i, a (i) ,ai (i), e (i), (nodec (i,j) ,j=l,2)
                    end do
c----> Read the nodal coordinate of each node
```

```
            do  n==l,ntnd
        read (10, *)i,x (i), y (i)
                end do
c----> Compute length and direction cosines of each element
                        do 103 i=l,nelm
                nl=nodec(i,l)
                n2=nodec(i,2)
        el(i)=sqrt( (x(n2)-x(nl) )**2+(y(n2)-y(nl))**2)
                cx(i)=(x(n2)-x(nl))/el(i)
                cy(i)=(y(n2)-y(nl))/el(i)
   103        continue
c---->Read Boundary Conditions
        read(10),*) (nnf(i),i=l,nfix)
        read(10,*)(ndl(i),i=l,nfix)
c----> Read the external load vector
            ndf=ntnd*3
                read(10,*)(eload(i),i=l,ndf)
        return
        end
c*********************************************************************
        subroutine elstif (ielm,ne, aa, ae, ai, al, cx, cy, estif)
         dimension estif (6,6)
         dimension cx(ne) , cy(ne) , aa (ne) , ai (ne) , al (ne) , ae (ne)
c*********************************************************************
  do 200 i=l,6
  do 200 j=l,6
  estif(i,j)=0
200 continue
 dl=ai (ielm) *ae (ielm) / (al (ielm) **3)
 d2=ai (ielm) *ae (ielm) / (al (ielm) **2)
 d3=ai (ielm) *ae (ielm) /al (ielm)
  c=aa (ielm) *ae (ielm) /al (ielm)
  alm=cx (ielm)
  amu=cy (ielm)
  cl=12.* (amu**2) *dl+c*alm**2
  c2=amu*alm* (-12.0*dl+c)
  c3= (amu**2) *c+12 . * (alm**2) *dl
  c4=6.0*d2*amu
  c5=6.0*d2*alm
  c6=4.0*d3
  c7=2.0*d3
```

```
  estif(1,1)=c1
  estif(2,1)=c2
  estif(3,1)=-c4
  estif(4,1)=-c1
  estif(5,1)=-c2
  estif(6,1)=-c4
  estif(2,2)=c3
  estif(3,2)=c5
  estif(4,2)=-c2
  estif(5,2)=-c3
  estif(6,2)=c5
  estif(3,3)=c6
  estif(4,3)=c4
  estif(5,3)=-c5
  estif(6,3)=c7
  estif(4,4)=c1
  estif(5,4)=c2
  estif(6,4)=c4
  estif(5,5)=c3
  estif(6,5)=-c5
  estif(6,6)=c6
  do i=1,6
  do j=1,6
      if(i.lt.j)then
      estif (i, j ) =estif (j , i)
          else
      end if
  end do
  end do
    return
     end
c*****************************************************************************
        subroutine asemb (ielm,nodec, estif, gstif, no,ne)
        dimension estif (6, 6) , gstif (no, no) ,nodec (ne, 2) , kk(6)
c*****************************************************************************
        kk (3) =3*nodec (ielm, 1)
        kk(2)=kk(3)-1
        kk(1)=kk(3)-2
        kk(6)=3*nodec(ielm,2)
        kk(5)=kk(6)-1
        kk(4)=kk(6)-2
```

```
        do 301 i=l,6
          ki=kk(i)
        do 301 j=l,6
          kj=kk(j)
        gstif(ki, kj ) =gstif(ki, kj ) +estif(i,j )
301   continue
      return
      end
c*********************************************************************
      subroutine boundary (no,ndof,gstif,nf,nfix,nnf,ndl)
      dimension gstif (no, no) ,nnf (nf) ,ndl (nf)
        cnst=1.0
        do i=l,ndof
      if (cnst.le.gstif (i,i))then
      cnst=gstif (i, i)
      end if
        end do
        cnst=(l.+cnst)*1.0e+20
      do 401 i=l,nfix
      il=3*nnf(i)-2
      i2=il+l
      i3=il+2
      if(ndl(i).eq.l)go to 402
      if(ndl(i).eq.2)go to 403
      if(ndl(i).eq.3)go to 404
402   gstif(il,il)=cnst*gstif(il,il)
        go to 401
403   gstif (i2,i2)=cnst*gstif(i2,i2)
        go to 401
404   gstif(i3, i3)=cnst*gstif(i3,i3)
401   continue
      return
      end
c*********************************************************************
      subroutine gausselim(a,b,no,n)
      dimension a (no, no) ,b (no)
       nl = n-1
      do 130 k = 1, nl
      kl = k + 1
      do 130 1 = kl, n
        c = a(i, k) / a(k, k)
        do 120 j = kl, n
```

```
  120  a(i,j)=a(i,j)-c* a(k,j)
  130 b(i)== b(i) - c * b(k)
C   *** back substitution ***
        b(n) = b(n) / a(n, n)
        do 140 ii = I, nl
        i = n- ii
        il = i+l
        c = l/ a(i, i)
        b(i) = c * b(i)
        do 140 k.= il, n
  140 b(i) = b(i) - c * a(i, k) * b(k)
  return
  end
c*********************************************************************
  subroutine force (ielm, nodec, estif, ne,ndf, u, alm, amu, fl)
  dimension nodec (ne, 2) , k(6), estif (6,6),f(6),fl (6),u (ndf)
        do 45 i=l,6
45    f(i)=0.0
      nl=nodec (ielm, 1)
      n2=nodec ( ielm, 2 )
      k(3)==3*nl
      k(2)=k(3)-l
      k(l)=k(3)-2
      k(6)=3*n2
      k(5)=k(6)-l
      k(4)=k(6)-2
      do 50 i=l,6
      do 50 j=l,6
50    f(i)=f(i)+estif(i,j)*u(k(j))
C     To transform the load to local axis system
      fl(l)=f(l)*alm+f(2)*amu
      fl(2)=-f(l)*alm+f(2)*amu
      fl(3)=f(3)
      fl(4)=f(4)*alm+f(5)*amu
      fl(5)=-f(4)*amu+f(5)*alm
      fl(6)=f(6)
      return
      end
*********************************************************************
```

Example Problem 1 : The numerical example already worked out by hand to explain the procedure is taken for analyzing again by computer. The joint and member numbering are same as earlier. Typical input file (frame1.in) and the detailed output file (frame1.out) are given below.

Input File (Frame1.in)

```
1 9000.0   2.5e+8   200.0   1 2
2 9000.0   2.5e+8   200.0   2 3
1 0.0 0.0
2 0.0 3000.0
3 7000.0 4000.0
1 1 3 3
1 2 1 2
0.0 0.0   -100000.0
150.0     -150.0   0.0
0.0       0.0    0.0
```

Output

Element Number 1

3000.000000 9000.000000 2.500000E+08 0.000000E+00 1.000000

Element Stiffness Matrix is

```
0.22E+02   .00E+00  -0.33E+05   -.22E+02   .00E+00    -.33E+05
.00E+00    .60E+03  .00E+00     .00E+00    -.60E+03   .00E+00
-.33E+05   .00E+00  .67E+08     .33E+00    .00E+00    .33E+08
-.22E+02   .00E+00  .33E+05     .22E+02    .00E+00    .33E+05
.00E+00    -.60E+03 .00E+00     .00E+00    .60E+03    .00E+00
```

Element Number 2

7071.068000   9000.000000   2.5000000E+08   9.899495E-01

1.414213E-01

Element Stiffness Matrix is

```
.25E+03    .35E+02  -.85E+03    -.25E+03   -.35E+02   -85E+03
.35E+02    .68E+01  .59E+04     -.35E+02   -.68E+01   .59E+04
-.85E+03   .59E+04  .28E+08     .85E+03    -.59E+04   .14E+08
-.25E+03   -.35E+02 .85E+03     .25E+03    .35E+02    .85E+03
-.35E+02   -.68E+01 -.59E+04    .35E+02    .68E+01    -.59E+04
-.85E + 03  .59E + 0.4  .14E + 08  .85E + 03  -.59E + 04  .28E + 08
```

Load Vector

0.0E+00 0.0E+00 -1.0e+05 150.0 -150.0 0.0E+00 0.0E+00 0.0E+0 0.0E+0

Global Stiffness Matrix (after applying B. C.)

.21E+30 0.0E+00-.33E+05-.22E+02 .00E+00-.33E+05 .00E+00 .00E+ .00E+00

.00E+00 .57E+31 .00E+00 .00E+00-.60E+03 .00E+00 .00E+00 .00E+0 .00E+00

-.33E+05 .00E+00 .67E+08 .33E+05 .00E+00 .33E+08 .00E+00 .00E+0 .00E+00

-.22E+02 .0E+0 .33E+5 .27E+03 .35E+02 .32E+5 -.25E+03-.35E+0 -2.85E+03

-00E+00-.60E+03 0.0E+0 .35E+02 .61E+03 .59E+04 -35E+02 -.68E+01 .59E+04

-.33E+05 .00E+0.33E+08 .32E+05 .59E+04 .95E+08 .85E+03 -.59E+04 .14E+08

.00E+00 .00E+0 .00E+00-.25E+03-.35E+02 .85E+03 .24E+31 .35E+02 .85E+03

.00E+00 .00E+0 . 00E+00-.35E+02-.68E+01-.59E+04 .35E+02 .64E+29 -.59E+04

.00E+00 .00E+O .00E+00-.85E+03 .59E+04 .14E+08 .85E+03 -.59E+04 .28E+08

NODE DISPLACEMENT VECTOR

-.173946E-27 -.312646E-28 -.216278E-2 .791626 -.29686 .5333942E-03 .788097E-28 .438405E-27 -.180882E-03

Internal Forces

Element Number: 1

At first node global node= 1

Axial Force= 36.703050 shear= 178.116000 Moment= -99999.990000

At second node axial Force=

Axial Force= 36.703050 shear= -178.116000 Moment= -10109.130000

Element Number 2

At first node global node= 2

Axial Force= 186.703100 shear= 28.116020 Moment = 10109.130000

At second node Global node 3

Axial Force= –186.703100 shear= –28.116020 Moment = 1.159527E-03

[Note that forces in the members are shown in local axis. Earlier manually it was computed in global direction but was not transformed to local axis.]

****

Example Problem 2 Analyze the frame as shown in the figure. The node and member numbering is also shown in the figure. As distributed load is acting on the structure, we will use equivalent load utilizing fixed end actions. Final load at various nodes will be as

P_1=20.0 (direct concentrated load acting), P_2 =–90.0, P_3=–90.0, P_4=0.0, P_5=-90., P_6=90,
P_7=40.0 (direct concentrated load), P_8=-180.0, P_9=-180.0, P_{10}=0.0, P_{11}=-180.0, P_{12}=180.0
P_{13}=0.0, P_{14} = 0.0, P_{15} = 0.0, P_{16} = 0.0, P_{17}=0.0, P_{18} = 0.0

Typical input and output files are given below (frame2.in)

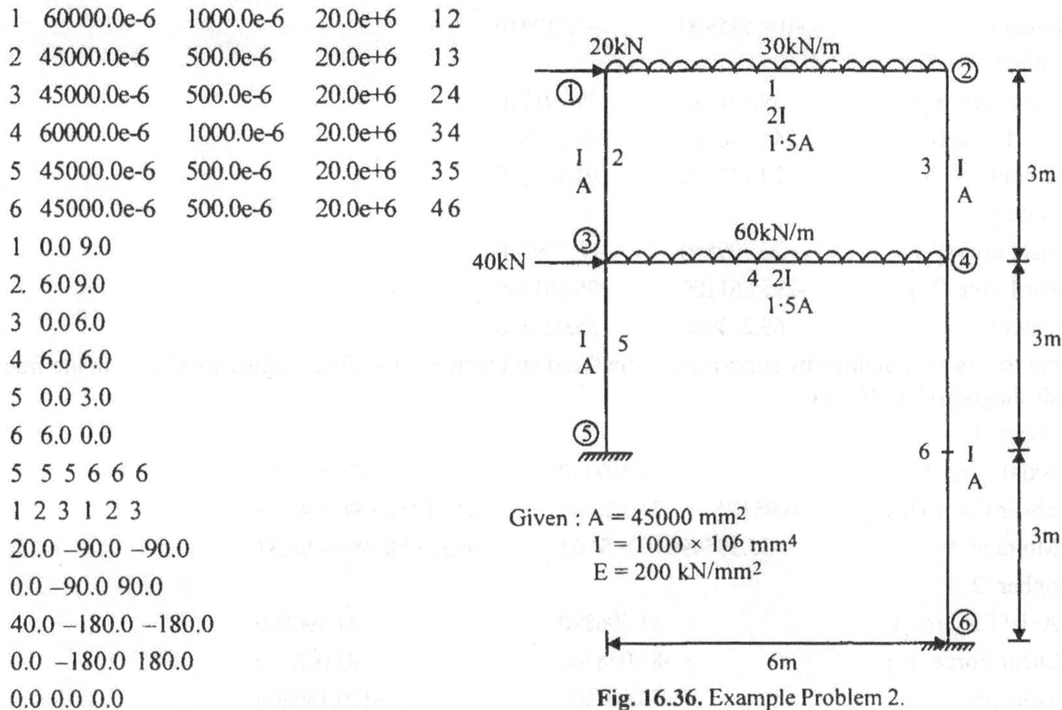

```
1  60000.0e-6   1000.0e-6   20.0e+6   1 2
2  45000.0e-6   500.0e-6    20.0e+6   1 3
3  45000.0e-6   500.0e-6    20.0e+6   2 4
4  60000.0e-6   1000.0e-6   20.0e+6   3 4
5  45000.0e-6   500.0e-6    20.0e+6   3 5
6  45000.0e-6   500.0e-6    20.0e+6   4 6
1  0.0 9.0
2. 6.0 9.0
3  0.0 6.0
4  6.0 6.0
5  0.0 3.0
6  6.0 0.0
5  5 5 6 6 6
1 2 3 1 2 3
20.0 –90.0 –90.0
0.0 –90.0 90.0
40.0 –180.0 –180.0
0.0 –180.0 180.0
0.0 0.0 0.0
0.0 0.0 0.0
```


Fig. 16.36. Example Problem 2.

output frame2.out (Only displacements and force vectors are shown)

DISPLACEMENT VECTOR

.33153E-01 -.10912E-02 -.34244E-02 .32795E-01 -.229424E-02 .93807E-03

.24233E-01 -.81446E-03 -.10261E-01 .24487E-01 -.19710E-02 .42631E-02

.14707E-27 -.13549E-28 -.11621E-27 .62013E-27 -.32790E-28 -.13730E-27

Internal Forces

	at 1st node	at 2nd node
member 1		
Axial Force (Fx)=	71.597190	-71.597190
Shear Force (Fy)=	-6.951256	6.951256
Moment =	-35.395490	-6.312048
member 2		
F_x =	-51.596890	51.596890
F_y =	-83.048740	83.048740
Moment =	-54.604500	-100.186200
member 3		
Axial Force (F_x)=	71.597150	-71.597150
shear Force (F_y)=	-96.951260	96.951260
Moment =	96.312030	118.479400
member 4		
Axial Force (F_x)=	-50.887950	50.887950

shear Force (F_y)=	-18.709880	18.709880
Moment =	-104.545900	-7.713410
member 5		
Axial Force (F_x)=	39.291810	-39.291810
Shear Force (F_y)=	-244.338800	244.338800
Moment =	24.732070	93.143330
member 6		
Axial Force (F_x)=	20.709550	-20.709550
shear Force (F_y)=	-295.661100	295.661100
Moment =	69.233950	55.023330

Final forces are obtained by superposing the fixed end actions. The final values are shown in the free body diagram (Fig. 16.37).

member 1		
Axial Force (F_x) =	71.597190	-71.597190
Shear Force (F_y) =	-6.951256 + 90 = 83.049	6.951256 + 90 = 96.95
Moment =	-35.395490+90=54.61	-6.312048-90=-.96.31
member 2		
Axial Force (F_x) =	-51.596890	51.596890
Shear Force (F_y) =	-83.048740	83.048740
Moment =	-54.604500	-100.186200
moment 3		
Axial Force (F_x) =	71.597150	-71.597150
Shear Force (F_y) =	-96.951260	96.951260
Moment =	96.312030	118.479400
member 4		
Axial Force (F_x) =	-50.887950	50.887950
Shear Force (F_y) =	-18.709880 + 180 = 161.29	18.709880+180
Moment =	- 104.545900 + 180=75.46	-7.713410-180
member 5		
Axial Force (F_x) =	39.291810	-39.291810
Shear Force (F_y) =	-244.338800	244.338800
Moment =	24.732070	93.143330
member 6		
Axial Force (F_x)	20.709550	-20.709550
Shear Force (F_y) =	-295.661100	295.661100
Moment =	69.233950	55.023330

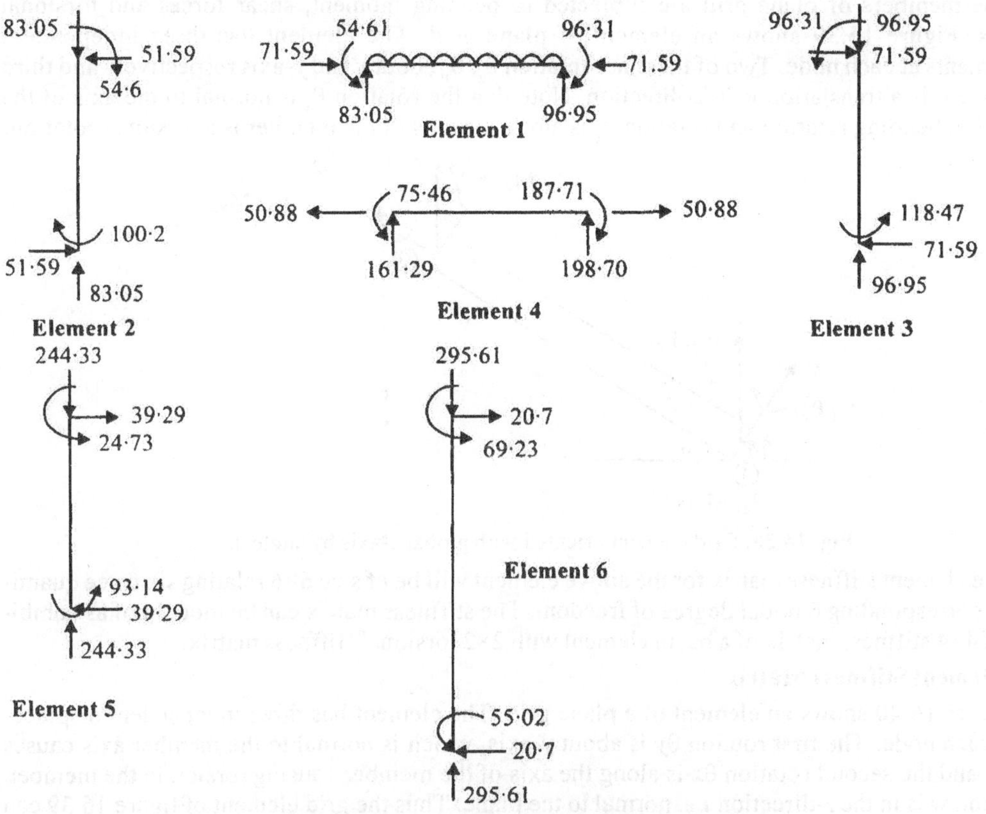

83·05
51·59
54·6
54·61
96·31
71·59
71·59
83·05
96·95
Element 1
96·31
96·95
71·59
75·46
187·71
50·88
50·88
100·2
161·29
198·70
118·47
71·59
51·59
83·05
96·95
Element 4
Element 2
Element 3
244·33
295·61
39·29
24·73
20·7
69·23
Element 6
93·14
39·29
244·33
Element 5
55·02
20·7
295·61

Fig. 16.37. Example Problem 2 : Free body diagram

16.5 MATRIX METHOD OF ANALYSIS OF GRILLAGES

16.5.1 Introduction

Plane grid *i.e.* grillage structure basically consists of beam in two or more directions and rigidly connected to each other at joints. The main difference between the grillage and two-dimensional rigid jointed frame structures are only in which they are loaded. Plane frames are loaded in the plane of the structures whereas grillages are loaded only perpendicular to the plane of the structure. Various reinforced or steel skeletons with their members confined to horizontal plane *(x, y)* of the global *(x, y, z)*space constitute beam grids. It plays a very important role in offering facilities to support a large floor or roof space without having any internal supports. They could be employed in the structural constitution of wide bridge decks. Furthermore, often a complex structural system can be transformed into equivalent beam grid for the purpose of at least preliminary calculations.

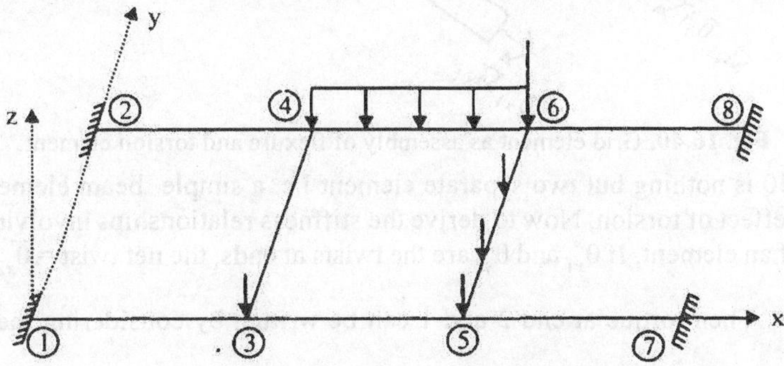

Fig. 16.38. A typical grillage structure.

The members of plane grid are subjected to bending moment, shear forces and torsional moments. Figure 16.39 shows an element of plane grid. The element has three independent displacements at each node. Two of these are rotation θ_x, θ_y about x and y-axis respectively and third displacement is a translation w in z-direction. Note that the rotation θ_y is normal to the axis of the member is a bending rotation and rotation θ_x is along the axis of the member is a torsional rotation.

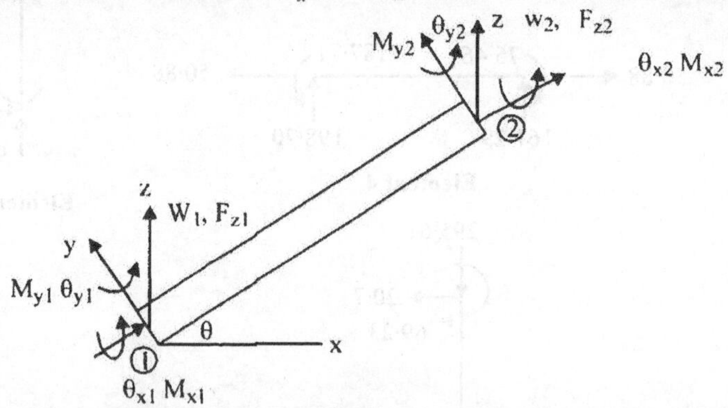

Fig. 16.38. Grid element oriented with global x-axis by angle θ.

The element stiffness matrix for the above element will be of size 6×6 relating six force quantities with corresponding 6 nodal degree of freedom. The stiffness matrix can be thought of as combination of 4×4 stiffness matrix of a beam element with 2×2 torsional stiffness matrix.

16.5.2 Element Stiffness Matrix:

Figure 16.40 shows an element of a plane grid. The element has three independent displacement at each node. The first rotaion θy is about y-axis, which is normal to the member axis causes bending, and the second rotation θx is along the axis of the member causing torsion in the member. Translation w is in the z-direction *i.e.* normal to the plane. Thus the grid element of figure 16.39 can be thought of as the assemly of two separate elements as shown in figure 16.40

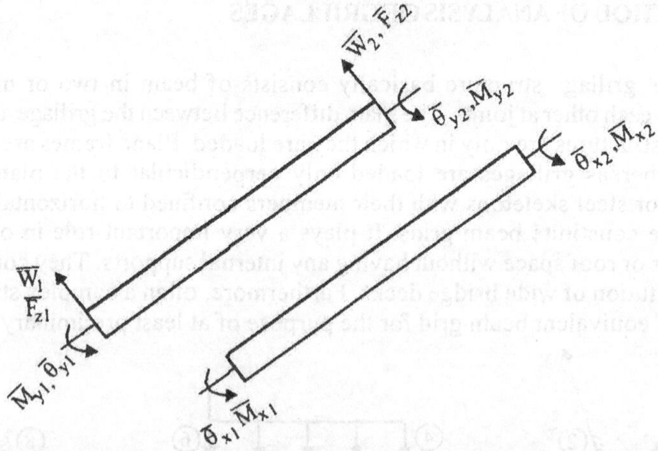

Fig. 16.40. Grid element as assembly of flexure and torsion element.

Figure 16.40 is nothing but two separate element i.e. a simple beam element and like a bar element with the effect of torsion. Now to derive the stiffness relationships involving the torque and twist at the end of an element. If θ_{x1} and θ_{x2} are the twists at ends, the net twist $= \bar{\theta}_{x2} - \bar{\theta}_{x1}$ and rate of twist $= \dfrac{\bar{\theta}_{x2} - \bar{\theta}_{x1}}{L}$. Then torque at end 2 and 1 can be written by considering the equilibrium *i.e.* $\bar{M}_{x1} + \bar{M}_{x2} = 0$ as

$$\bar{M}_{x1} = \frac{GJ}{L}(\bar{\theta}_{x1} - \bar{\theta}_{x2}) \tag{16.63}$$

$$\bar{M}_{x2} = -\frac{GJ}{L}(\bar{\theta}_{x2} - \bar{\theta}_{x1})$$

above two equations can be written in a matrix form as

$$\begin{Bmatrix} \bar{M}_{x1} \\ \bar{M}_{x2} \end{Bmatrix} = \frac{GJ}{L}\begin{bmatrix} 1 & -1 \\ -1 & 1 \end{bmatrix}\begin{Bmatrix} \bar{\theta}_{x1} \\ \bar{\theta}_{x2} \end{Bmatrix} \tag{16.64}$$

Then the force displacement relation for the combined system can be written as

$$\begin{Bmatrix} \bar{F}_{z1} \\ \bar{M}_{y1} \\ \bar{F}_{z2} \\ \bar{M}_{y2} \\ \bar{M}_{x1} \\ \bar{M}_{x2} \end{Bmatrix} = \begin{bmatrix} \frac{12EI}{L^3} & \frac{6EI}{L^2} & -\frac{12EI}{L^3} & \frac{6EI}{L^2} & 0 & 0 \\ \frac{6EI}{L^2} & \frac{4EI}{L} & -\frac{6EI}{L^2} & \frac{2EI}{L} & 0 & 0 \\ -\frac{12EI}{L^3} & \frac{6EI}{L^2} & \frac{12EI}{L^3} & -\frac{6EI}{L^2} & 0 & 0 \\ \frac{6EI}{L^2} & \frac{2EI}{L} & -\frac{6EI}{L^2} & \frac{4EI}{L} & 0 & 0 \\ 0 & 0 & 0 & 0 & \frac{GJ}{L} & -\frac{GJ}{L} \\ 0 & 0 & 0 & 0 & -\frac{GJ}{L} & \frac{GJ}{L} \end{bmatrix}\begin{Bmatrix} \bar{w}_1 \\ \bar{\theta}_{y1} \\ \bar{w}_2 \\ \bar{\theta}_{y2} \\ \bar{\theta}_{x1} \\ \bar{\theta}_{x2} \end{Bmatrix} \tag{16.65}$$

Now rearranging the above equation (as done in case of plane frame), we can write,

$$\begin{Bmatrix} \bar{F}_{z1} \\ \bar{M}_{x1} \\ \bar{M}_{y1} \\ \bar{F}_{z2} \\ \bar{M}_{x2} \\ \bar{M}_{y2} \end{Bmatrix} = \begin{bmatrix} \frac{12EI}{L^3} & 0 & \frac{6EI}{L^2} & -\frac{12EI}{L^3} & 0 & \frac{6EI}{L^2} \\ 0 & \frac{GJ}{L} & 0 & 0 & -\frac{GJ}{L} & 0 \\ \frac{6EI}{L^2} & 0 & \frac{4EI}{L} & \frac{6EI}{L^2} & 0 & \frac{2EI}{L} \\ -\frac{12EI}{L^3} & 0 & -\frac{6EI}{L^2} & \frac{12EI}{L^3} & 0 & -\frac{6EI}{L^2} \\ 0 & -\frac{GJ}{L} & 0 & 0 & \frac{GJ}{L} & 0 \\ \frac{6EI}{L^2} & 0 & \frac{2EI}{L} & -\frac{6EI}{L^2} & 0 & \frac{4EI}{L} \end{bmatrix}\begin{Bmatrix} \bar{w}_1 \\ \bar{\theta}_{x1} \\ \bar{\theta}_{y1} \\ \bar{w}_2 \\ \bar{\theta}_{x2} \\ \bar{\theta}_{y2} \end{Bmatrix} \text{ i.e.} \{\bar{f}\} = [\bar{k}]\{\bar{d}\} \tag{16.66}$$

16.5.3. Coordinate Transformation:

The member axis system adopted here demands that the member axis '$\bar{z}$' coincide with the global 'z' axis. Thus the forces and displacements are same in the z-direction for both the axis system. hence at nodel,

$$\bar{F}_{z1} = F_{z1} \tag{16.67}$$

However, the transformation of rotation and moments between the member '$\bar{x}$' and '$\bar{y}$' axis and the reference 'x' and 'y' axis is necessary. Bending and twisting at the ends of the member is shown in figure 16.41

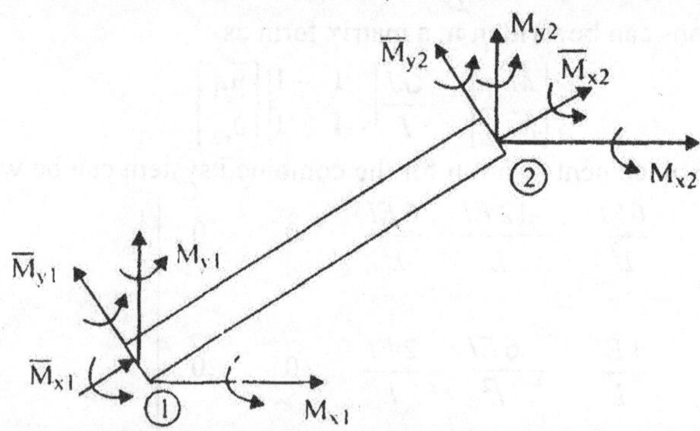

Fig. 16.41. Bending and twisting force on a two noded Grid element.

Resolving the moments and twist along the member axis '$\bar{x}$' and '$\bar{y}$', we get

$$\overline{M}_{x1} = M_{x1}\cos\theta + M_{y1}\sin\theta$$

$$\overline{M}_{y1} = -M_{x1}\sin\theta + M_{y1}\cos\theta \tag{16.68}$$

Then at node 1 the transformation of nodal force vector from global to local axis system takes the following form

$$\begin{Bmatrix} \overline{F}_{z1} \\ \overline{M}_{x1} \\ \overline{M}_{y1} \end{Bmatrix} = \begin{bmatrix} 1 & 0 & 0 \\ 0 & \lambda & \mu \\ 0 & -\mu & \lambda \end{bmatrix} \begin{Bmatrix} F_{z1} \\ M_{x1} \\ M_{y1} \end{Bmatrix} \tag{16.69}$$

Similar equation can be written for node 2

$$\begin{Bmatrix} \overline{F}_{z2} \\ \overline{M}_{x2} \\ \overline{M}_{y2} \end{Bmatrix} = \begin{bmatrix} 1 & 0 & 0 \\ 0 & \lambda & \mu \\ 0 & -\mu & \lambda \end{bmatrix} \begin{Bmatrix} F_{z2} \\ M_{x2} \\ M_{y2} \end{Bmatrix} \tag{16.70}$$

Above two equation can be written in a single matrix form as

$$\begin{Bmatrix} \overline{F}_{z1} \\ \overline{M}_{x1} \\ \overline{M}_{y1} \\ \overline{F}_{z2} \\ \overline{M}_{x2} \\ \overline{M}_{y2} \end{Bmatrix} = \begin{bmatrix} 1 & 0 & 0 & 0 & 0 & 0 \\ 0 & \lambda & \mu & 0 & 0 & 0 \\ 0 & -\mu & \lambda & 0 & 0 & 0 \\ 0 & 0 & 0 & 1 & 0 & 0 \\ 0 & 0 & 0 & 0 & \lambda & \mu \\ 0 & 0 & 0 & 0 & -\mu & \lambda \end{bmatrix} \begin{Bmatrix} F_{z1} \\ M_{x1} \\ M_{y1} \\ F_{z2} \\ M_{x2} \\ M_{y2} \end{Bmatrix} \quad \text{or symbolically } \{\overline{F}\} = [T]\{F\} \tag{16.71}$$

Where, $[T]$ is the transformation matrix. Similar expression can be obtain for the displacement also. Finally the stiffness martix in global coordinate system *i.e*, member axis system can be obtain as

$$[T]^T[\bar{K}][T] =$$

$$\begin{bmatrix} 1 & 0 & 0 & 0 & 0 & 0 \\ 0 & \lambda & -\mu & 0 & 0 & 0 \\ 0 & \mu & \lambda & 0 & 0 & 0 \\ 0 & 0 & 0 & 1 & 0 & 0 \\ 0 & 0 & 0 & 0 & \lambda & -\mu \\ 0 & 0 & 0 & 0 & \mu & \lambda \end{bmatrix} \begin{bmatrix} \frac{12EI}{L^3} & 0 & \frac{6EI}{L^2} & -\frac{12EI}{L^3} & 0 & \frac{6EI}{L^2} \\ 0 & \frac{GJ}{L} & 0 & 0 & -\frac{GJ}{L} & 0 \\ \frac{6EI}{L^2} & 0 & \frac{4EI}{L} & -\frac{6EI}{L^2} & 0 & \frac{2EI}{L} \\ -\frac{12EI}{L^3} & 0 & -\frac{6EI}{L^2} & \frac{12EI}{L^3} & 0 & -\frac{6EI}{L^2} \\ 0 & -\frac{GJ}{L} & 0 & 0 & \frac{GJ}{L} & 0 \\ \frac{6EI}{L^2} & 0 & \frac{2EI}{L} & -\frac{6EI}{L^2} & 0 & \frac{4EI}{L} \end{bmatrix} \begin{bmatrix} 1 & 0 & 0 & 0 & 0 & 0 \\ 0 & \lambda & \mu & 0 & 0 & 0 \\ 0 & -\mu & \lambda & 0 & 0 & 0 \\ 0 & 0 & 0 & 1 & 0 & 0 \\ 0 & 0 & 0 & 0 & \lambda & \mu \\ 0 & 0 & 0 & 0 & -\mu & \lambda \end{bmatrix}$$

$$= \frac{EL}{L} \begin{bmatrix} c_1 & -c_2 & c_3 & -c_1 & -c_2 & c_3 \\ & c_4 & -c_5 & c_2 & c_6 & -c_7 \\ & & c_8 & -c_3 & -c_9 & c_{10} \\ & & & c_1 & -c_2 & -c_3 \\ & & & & c_4 & -c_5 \\ & & & & & c_8 \end{bmatrix} \qquad (16.72)$$

where, $\alpha = \frac{GJ}{EI}$ and

$$c_1 = \frac{12}{L^2}, c_2 = \frac{6\mu}{L}, c_3 = \frac{6\lambda}{L}, c_4 = \lambda^2\alpha + 4\mu^2, c_5 = \lambda\mu(4-\alpha), c_6 = -\lambda^2\alpha + 2\mu^2$$

$$c_1 = 2\lambda\mu + \alpha\lambda\mu,\ c_8 = 4\lambda^2 + \alpha\mu^2,\ c_9 = \lambda\mu(2+\alpha),\ c_{10} = 2\lambda^2 - \alpha\mu^2$$

16.5.4 Application Problem:

For the purpose of illustraion, we shall consider the problem of a three-member beam grid system as shown in figure 16.42. Each member has a span of 6.0 m, E = 25491 N/mm^2, v = 0.15, I = 0.54 × 10^4 mm^4 and torsional constant J = 0.372 × 10^4 mm^4, udl = 4.5 N/mm.

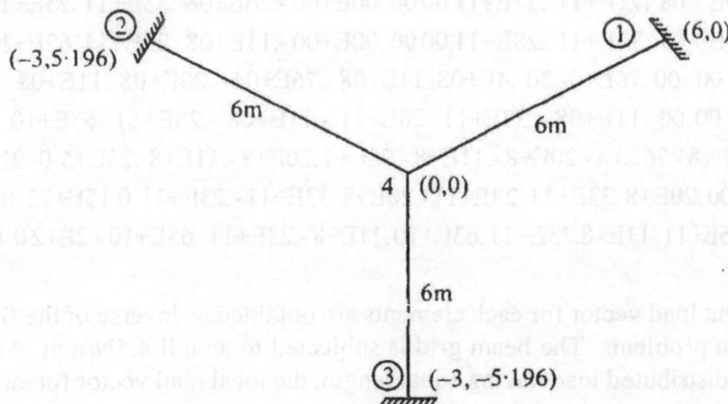

Fig. 16.42. Three member beam grid system.

For Element 1: (4-1)

Nodal co-ordinate : $x_4 = 0, y_4 = 0$ and $x_1 = 6000.0\ y_1 = 0.0$. Thus $L = 6.0$ m, $\lambda = 1.0$ and $\mu = 0.0$.

$$k_{4-1} = \begin{bmatrix} .76E+04 & .00E+00 & .23E+08 & -.76+04 & .00E+00 & .23E+08 \\ .00E+00 & .69E+10 & .00E+00 & .00E+00 & -.69E+10 & .00E+00 \\ .23E+08 & .00E+00 & .92E+11 & -.23E+08 & .00E+00 & .46E+11 \\ -.76E+04 & .00E+00 & -.23E+08 & .76E+04 & .00E+00 & -.23E+08 \\ .00E+00 & -.69E+10 & .00E+00 & .00E+00 & .69E+10 & .00E+00 \\ .23E+08 & .00E+00 & .46E+11 & -.23E+08 & .00E+00 & .92E+11 \end{bmatrix}$$

Element 2: (4-2)

$$k_{4-2} = \begin{bmatrix} .76E+04 & -.20E+08 & -.11E+08 & -.76E+04 & -.20E+08 & -11E+08 \\ -.20E+08 & .71E+11 & .37E+11 & 20E+08 & .33E+11 & 23E+11 \\ -.11E+08 & .37E+11 & .28E+11 & .11E+08 & .23E+11 & 63E+10 \\ -.76E+04 & .20E+08 & .11E+08 & .76E+04 & .20E+08 & 11E+08 \\ -.20E+08 & .33E+11 & .23E+11 & .20E+08 & .71E+11 & 37E+11 \\ -.11E+08 & .23E+11 & .63E+10 & .11E+08 & .37E+11 & 28E+11 \end{bmatrix}$$

For Element 3 : (4-3)

$$k_{4-3} = \begin{bmatrix} .76E+04 & .20E+08 & -.11E+08 & -.76E+04 & .20E+08 & -11E+08 \\ .20E+08 & .71E+11 & -.37E+11 & -.20E+08 & .33E+11 & -23E+11 \\ -.11E+08 & -.37E+11 & .28E+11 & .11E+08 & -.23E+11 & 63E+10 \\ -.76E+04 & -.20E+08 & .11E+08 & .76E+04 & -.20E+08 & 11E+08 \\ .20E+08 & .33E+11 & -.23E+11 & -.20E+08 & .71E+11 & -37E+11 \\ -.11E+08 & -.23E+11 & .63E+10 & .11E+08 & -.37E+11 & 28E+11 \end{bmatrix}$$

Then the global stiffness matrix

.76E+04. 00E+00 -.23E+08 00 00 00 00 .00E+00 .00E+00 -.76E+04 .00E+00 -.23E+08

.0 0.69E+10 .0 .0 .0 0 .0E+9 .0E+0 .0E+0 .0E+0 -.69E+10 .00E+00

-.23E+08 .00E+00 .92E+11 0 0 0 0 0 .23E+08.00E+00 .46E+11

0 0 0 .76E+04 .20E+08 .11E+08 00 .00E+00 .00E+00 -.76E+|04 .20E+08 .11E+08

00 00 00 .20E+08 .71E+11 .37E+11 00 00 .00E+00 -.20E+08 .33E+11 .23E+11

00 00 00 .11E+08 .37E+11 .28E+11 00 00 .00E+00 -.11E+08 .23E+11 .63E+10

00 00 00 00 00 .00 .76E+0-.20 .4E+08 .11E+08 -.76E+04 -.20E+08 .11E+08

00 00 00 00 00 00 .11E+08 -.37E+11 .28E+11 -.11E+08 -.23E+11 .63E+10

-.76E+4 0.23E+8 -.76E+4 -.20E+8 -.11E+8 -.76E+4 .20E+8 -.11E+8 .23E+5 0-.2E+2

00 -.69E+10 00 .20E+8 .33E+11 .23E+11 -.20E+8 .33E+11 -.23E+11 0.15E+12 .00

-.23E+08 0.46E+11. 11E+8.23E+11 .63E+10 .11E+8 -.23E+11 .63E+10 -.2E+2 0.15E+12

Load Vector:

The consistent load vector for each elements are obtained as inverse of the fixed end action as discussed for beam problems. The beam grid is subjected to an udl 4.5N/mm. As each member is subjected to same distributed load having equal length, the local load vector for each will be reverse of the fixed end actions as below

i.e. $\{f\}_{4-1} = \{f\}_{4-2} = \{f\}_{4-3} = [-0.135e{+}5\ 0\ -0.135e{+}8\ -0.135e{+}5\ 0.0\ 0.135e{+}8]^T$

Note that these element load vectors are computed in local axis system. However, the final loads will be in global directions. Thus each element load vectors need to transform to global axis system, then add to get final load vector in global axis system. The element load vector in global axis system for each element can be easily obtained from equation 16.5.9 i.e.

$$\{F\} = [T]^{-1}\{\bar{F}\} = [T]^T\{\bar{F}\} \text{ thus, } \begin{bmatrix} F_{z1} \\ M_{x1} \\ M_{y1} \\ F_{z2} \\ M_{x2} \\ M_{y2} \end{bmatrix} = \begin{bmatrix} 1 & 0 & 0 & 0 & 0 & 0 \\ 0 & \lambda & -\mu & 0 & 0 & 0 \\ 0 & \mu & \lambda & 0 & 0 & 0 \\ 0 & 0 & 0 & 1 & 0 & 0 \\ 0 & 0 & 0 & 0 & \lambda & -\mu \\ 0 & 0 & 0 & 0 & \mu & \lambda \end{bmatrix} \begin{Bmatrix} \bar{F}z_1 \\ \bar{M}_{x1} \\ \bar{M}_{y1} \\ \bar{F}_{z2} \\ \bar{M}_{x2} \\ \bar{M}_{y2} \end{Bmatrix}$$

For element 1: λ=1 and μ=0, so,

$$\{f\}_{4-1} = [-0.134e+5\ \ 0.0\ \ -0.135e+8\ \ -0.135e+5\ \ 0.0\ \ 0.135e+8]^T$$

For element 2: λ=0.5 and μ=0.866 thus,

$$\{f\}_{4-2} = [-0.135e+5\ \ .1169e+8\ \ .0675e+8\ \ -0.135e+5\ \ -.1169e+8\ -.0675e+8]^T$$

For element 3: λ=0.5 and μ=0.866 thus,

$$\{f\}_{4-3} = [-0.135e+5\ -.1169e+8\ .0675e+8\ -0.135e+5\ -.1169e+8\ -.0675e+8]^T$$

And the final global load vector is obtained by adding these three element load vectors in global axis system as below

-0.135e+5 0.0 0.135e+8 -0.135e+5 -0.117e+8 -0.0675e+8

-0.135e+5 0.117e+8 -0.0675e+8 -0.405e+5 0.0 0.0

Solution: Unknown displacement vector can be obtained by solving the equilibrium equation as earlier and the displacement vector is obtained as

$[0.0\ 0.0\ 0.0\ 0.0\ 0.0\ 0.0\ 0.0\ 0.0\ 0.0\ 1.76\ 0.0\ 0.0]^T$

Internal Forces

The internal force quantities at each member can be obtained by multiplying the element stiffness matrix of each element with corresponding nodal degree of freedoms and then by adding the corresponding fixed end force vector. The various internal force quantities obtained by multiplying the stiffness matrix with associated displacements for each elemment are shown below

element number : 1

-135000.0 -4.896164E-28 -4.050000E+07 13500.0 4.896164E-28 -4.049999E+07

element number :2

-13500.0 -2.573834 -4.049999E+07 13500.00 -1.109734 -4.049999E+07

element number: 3

-13500.0 2.573834 -4.049999E+07 13500.0 1.109734 -4.049999E+07

The final internal load vector will be obtained by adding the corresponding fixed end force to this load

vectors as below:

element number: 1

-13500.0 +0.135e+5 =0.0

-4.896164E-28 +0 =0.0

```
-4.050000E+07 +.135e+8          =-0.27e+8
13500.0 +.135e+5                =0.27e+5
4.896164E-28 +0                 =0
-4.049999E+07 +(-.135e+8)       =-0.539e+8
element number:    2
-13500.0 +0.135e+5              =0.0
-2.573834 +0                    =-2.573834~00
-4.049999E+07 +.135e+8          =-0.27e+8
13500.0 +.135e+5                =0.27e+5
-1.109734  +0                   =0
-4.049999E+07 +(-.135e+8)       =-0.539e+8
element number:    3
-13500.0 +0.135e+5              =0.0
2.573834 +0                     =2.573834≈0
-4.049999E+07  +.135e+8         =-0.27e+8
13500.000000   +.135e+5         =0.27e+5
1.109734 +0                     =0
```

Program grid. for

```
xxxxxxxxxxxxxxxxxxxxxxxxxxxxxxxxxxxxxxxxxxxxxxxxxxxxxxxxxxxxxxxxxxxxxxxxxxxxxxxxxx
c The program follows the same procedure and notations as earlier. It can
c be used for any grid structure by simple modification on dimension statements c to
suit the actual size of the structure.
xxxxxxxxxxxxxxxxxxxxxxxxxxxxxxxxxxxxxxxxxxxxxxxxxxxxxxxxxxxxxxxxxxxxxxxxxxxxxxxxxx
parameter (no=18,ne=3, nn=4, nfi=9)
dimension nodec (ne, 2) ,x(nn) , y(nn) , estif (6, 6) , gstif (no, no)
dimension dmu (ne) , aj (ne) , ai (ne) , e (ne) , el(ne) , ex (ne) , cy (ne)
dimension eload(no) , f(6) ,nnf (nfi) ,ndl (nfi) ,u (no)
c*************************************************************************
character file1*10,file2*10
write (*, '(a\)') ' Input file name? '
read(*, '(a)') file1
write ( *, '(a\)') ' Output file name ?
read(*, '(a)') file2
open( 10,file=file1)
open( 12,file=file2)
c*************************************************************************
        write (*,*) 'Give number of members :'
                read(*, *)nelm
        write (*,*)' Give number of joints:'
```

```
              read(*, *) ntnd
              write (*,*) 'number of fixity'
              read(*,*) nf
              ndof=3*ntnd
c---->Following statement calls the input data
        call input (ntnd, ntnd, nelm,nelm, ndof,aj , e, ai, dmu,nodec, x, y, el,
     I cx,cy,eload,nf,nf,nnf,ndl)
c----> Initialize the global structural stiffness matrix
          do 14 i=l,ndof
          do 14 j=l,ndof
          gstif(i,j)=0.0
 14    continue
c---->The following loop compute the elememnt stiffness matrix
c    in turn and then place it in the corresponding position of the
c   global stiffness matrix.
           do 15 ielm=l,nelm
           call elstif(ielm,nelm, aj ,e, ai,dmu, el, ex, cy,estif)
           call asemb (ielm, nodec, estif, gstif, ndof,nelm)
          write (12,*)' Element Number ',ielm
c    write (12,*)' Length and direction cosines'
          write (12,*) el (ielm), aj (ielm), ai (ielm), cx (ielm) ,cy (ielm) .
          write(12,*) 'Element Stiffness Matrix is'
           write (12, 1) ( (estif (i,j),j=l, 6), i=l, 6)
15     continue
 I   format(6el2.2/)
2   format(9e8.2/)
c----> Once above loop is over gstif matrix contains the global
c     stiffness matrix of the entire structure
                    write (12,*)' Load Vector'
                    write (12,*) (eload(i),i=l,ndof)
          write(12,*) 'Global Stiffness Matrix(Before applying B. C.)'
          write(12,2) ( (gstif (i,j) ,j=l,ndof) ,i=l,ndof)
c----> Following statement apply the boundary conditions on
c   gstif matrix
          call boundary (ndof,ndof, gstif, nf,nf,nnf,ndl)
   write(12,*) 'Global Stiffness Matrix (after applying B. C.)'
          write(12,2) ( (gstif (i,j) ,j=l,ndof) ,i=l,ndof)
c----> Following subroutine solve matrix equiibrium equation and
              call gausselim (gstif, eload, ndof,ndof)
c---->Displacement vector is stored in the vector u
          do i=l,ndof
```

```
         u(i)=eload(i)
          end do
         write(12,*)'DISPLACEMENT VECTOR'
         write(12,3)(u(i),i=l,ndof)
3    format(4el8.6/)
c---->This loop call the force subroutine for each elements in tern and
c    compute the internal forces
        write(12,*)'Shear=', 'Moment mxl=', 'Moment mx2='
          do 101 ielm=l,nelm
          call elstif(ielm,nelm, aj, e, ai,dmu, el, cx, cy, estif)
          alm=cx(ielm)
          amu=cy(ielm)
          call force(ielm, nodec, estif, alm, amu, nelm, ndof, u, f)
        write(12,*)
        write(12,*)
        write(12,*)' element number:', ielm
        write(12,*)f(l),f(2),f(3)
        write(12,*)f(4),f(5),f(6)
   101    continue
          stop
          end
c*****************************************************************************
        subroutine input(nn,ntnd,ne, nelm, no, aj, e, ai,dmu, nodec, x,y, el,
      l cx, cy,eload,nf,nfix,nnf,ndl)
         dimension aj(ne),ai(ne), e(ne),dmu(ne), nodec(ne, 2),x(nn), y(nn),
      l ex(ne), cy(ne), el(ne), eload(no), nnf(nf),ndl(nf)
 c----->Read the cross section area,modulus of elasticity,nodal
 c    connectivity of each element
              do n=l,nelm
              read(10,*) i,aj(i),ai(i),e(i),dmu(i),(nodec(i,j),j=l,2)
                end do
 c----> Read the nodal coordinate of each node
          do  n=l,ntnd
        read(10,*) i,x(i),y(i)
             end do
c----> Compute length and direction cosines of each element
                    do 103 i=l,nelm
          nl=nodec(i,1)
          n2=nodec(i,2)
      el(i)=sqrt((x(n2)-x(nl))**2+(y(n2)-y(nl))**2)
          cx(i)=(x(n2)-x(nl))/el(i)
          cy(i)=(y(n2)-y(nl))/el(i)
```

```
 103        continue
c----> Read Boundary Conditions
           read (10,*) (nnf(i),i=l,nfix)
            read (10,*) (ndl(i),i=l,nfix)
c----> Read the external load vector
              ndf=ntnd*3
                  read (10,*) (eload(i),i=l,ndf)
           return
           end
c*****************************************************************************
 subroutine elstif (ielm, ne, aj , ae, ai, dmu, al, cx, cy, estif)
 dimension estif (6,6)
 dimension cx (ne) , cy(ne) ,aj (ne) , ai (ne) , al (ne) , ae (ne) , dmu(ne)
c*****************************************************************************
            do 200 i=l,6
            do 200 j=l,6
            estif(i,j )=0
200    continue
          g=ae (ielm) / (2.* (l+dmu(ielm) ) )
           q=g*aj (ielm) / (ae (ielm) *ai (ielm)
             alm=cx (ielm)
             amu=cy (ielm)
          cl=12./(al (ielm) **2)
          c2=amu*6.0/al (ielm)
          c3=6. *alm/al (ielm)
          c4=alm**2*q+4.*amu**2
          c5=alm*amu* ( 4 . -q)
          c6=-alm**2*q+2. *amu**2
          c7=alm*amu* (2.+q)
          c8=4. *alm**2+q*amu**2
          c9=alm*amu* ( 2 . +q)
          cl0=2. *alm**2-q*amu**2
          estif(l,l)=cl
          estif(2,l)=-c2
          estif(3,l)=c3
          estif(4,l)=-cl
          estif(5,l)=-c2
          estif(6,l)=c3
          estif(2,2)=c4
          estif(3,2)=-c5
          estif(4,2)=c2
```

```
      estif(5,2)=c6
      estif(6,2)=-c7
      estif(3,3)=c8
      estif(4,3)==-c3
      estif(5,3)=-c9
      estif(6,3)=cl0
      estif(4,4)=cl
      estif(5,4)==c2
      estif(6,4)=-c3
      estif(5,5)=c4
      estif(6,5)=-c5
      estif(6,6)=c8
      do i=l,6
      do j=l,6
            if(i.lt.j )then
            estif (i,j )=estif(j, i)

        else
        end if
        estif(i, j )=estif (i, j ) * (ae (ielm) *ai (ielm) ) /al (ielm
end do
end do
  return
    end

c*****************************************************************************
   subroutine asemb (ielm, nodec,, estif, gstif, no,ne)
   dimension estif (6, 6) , gstif (no, no) , nodec (ne, 2) , kk(6)
c*****************************************************************************
            kk(3)=3*nodec(ielm,l)
          kk(2)=kk(3)-l
          kk(l)=kk(3)-2
          kk ( 6) =3*nodec (ielm, 2)
          kk(5)=kk(6)-l
          kk(4)=kk(6)-2
        do 301 i=l,6
             ki=kk(i)
          do 301 j=l,6
             kj=kk(j)
           gstif (ki, kj )=gstif (ki, kj ) +estif (i, j )
301    continue
       return
```

```
      end
c*******************************************************************************
        subroutine boundary (no, ndof, gstif, nf,nfix,nnf,ndl)
        dimension gstif (no, no) , nnf (nf) , ndl (nf)
          cnst=1.0
            do i=l,ndof
        if (cnst.le.gstif (i,i))then
        cnst=gstif (i, i)
        end if
            end do
        cnst=(l.+cnst)*1.0e+15
        do 401 i=l,nfix
          il=3*nnf(i)-2
          i2=il+l
          i3=il+2
          if(ndl(i).eq.l)go to 402
          if(ndl(i).eq.2)go to 403
          if(ndl(i).eq.3)go to 404
 402  gstif (il,il)=cnst*gstif(il,il)
            go to 401
 403   gstif (i2,i2)=cnst*gstif(i2,i2)
            go to 401
 404   gstif (i3,i3)=cnst*gstif(i3,i3)
 401    continue
        return
        end
c*******************************************************************************
        subroutine gausselim(a,b,no,n)
      dimension a (-no, no) ,b (no)
        nl = n-l
      do 130 k = 1, nl
      kl = k+l
      do 130 I = kl, n
          c = a(i, k) / a (k, k)
          do 120 j = kl, n
 120    a(i, j) = a(i, j)-c * a(k, j)
 130 b(i) = b(i) - c * b(k)
 C   *** back substitution ****
        b(n) = b(n) / a(n, n)
        do 140 ii = 1, nl
        i = n- ii
    il = 1+1
    c= 1 / a(i, i)
    b(i) = c *'b(i)
```

```
      do 140 k= i1, n
140 b(i) = b(i) - c * a(i, k) * b(k)
      return
      end
c*************************************************************************
          subroutine force (ielm,nodec, estif, alm, amu,ne,ndf,u, fl)
            dimension nodec (ne, 2) , k(6) , estif (6, 6),f(6) ,u(ndf) , fl (6)
            do 45 i=1,6
45    f(i)=0.0
            nl=nodec ( ielm, 1 )
            n2=nodec ( ielm, 2 )
            k(3)=3*nl
            k(2)=k(3)-l
            k(l)=k(3)-2
            k(6)=3*n2
            k(5)=k(6)-l
            k(4)=k(6)-2
            do 50 i=l,6
            do 50 j=l,6
50    f(i)=f(i)+estif(i,j)*u(k(j))
c----> Following few lines convert the forces to local axis system
        fl(l)=f(l)
        fl(4)=f(4)
            fl(2)=f(2)*alm+f(3)*amu
            fl(3)=-f(2)*amu+f(3)*alm
            fl(5)=f(5)*alm+f(6)*amu
            fl(6)=-f(5)*amu+f(6)*alm
            return
            end
**************************************************************************
```

Example Problem 1 : Analyze the grid frame as shown in the figure. The node and member numbering is also shown in the figure.

The input file for the program is :

```
1 1821 .0e-6 2600.0e-6 14.0e+6  0.1 1 2
2 1821 .0e-6 2600.0e-6 14.0e+6  0.1 2 3
1 0.0 4.0
2 4.0 4.0
3 4.0 0.0
1 1 1 3 3 3
1 2 3 1 2 3
0.0 0.0 0.0
-55.0 -25.0 20.0
0.0 0.0 0.0
```

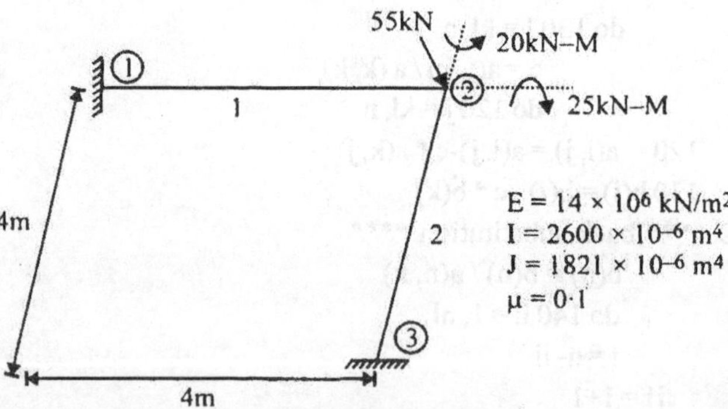

Fig. 16.43. Example Problem 1

Output from the program are :

Element Number 1

4 .000000 1.821000E-03 2.6000.00E-03 1.000000 0. 00000E+00

Element Stiffness Matrix: is

.68E+04	.00E+00	.14E+05	-.68E+04	.00E+00	.14E+05
.00E+00	.29E+04	.00E+00	.00E+00	-.29E+04	.00E+00
.14E+05	.00E+00	.36E+05	-.14E+05	.00E+00	.18E+05
-.68E+04	.00E+00	-.14E+05	.68E+04	.00E+00	-.14E+05
.00E+00	-.29E+04	.00E+00	.00E+00	.29E+Q4	.00E+00
.14E+05	.00E+00	.18E+05	-.14E+05	.00E+00	.36E+05

Element Number 2

4 .0000001.821000E-032.600000E-03. 0.000000E+00 -1.000000 Element stiffness Matrix

Element Sti	.68E+04	.14E+05,.	.00E+00	.68E+04	.14E+05	.00E+00
	.14E+05	.36E+05	.00E+00	-.14E+05	.18E+05	.00E+00
	.00E+00	.00E+00	.29E+04	.00E+00	.00E+00	-.29E+04
	-.68E+04	-.14E+05"	.00E+00	.68E+04	-.14E+05	.00E+00
	.14E+05	.18E+05	00E+00	-.14E+05	.36E+05	.00E+00
	.00E+00	.00E+00	-.29E+04	.00E+00	.00E+00	.29E+04

Load Vector

0.000000E+00 0.00000E+00 0.000000E+00 -55.000000 -25.0000 20.000000

0.000000E+00 0.0000E+00 0.000000E+00

Global Stiffness Matrix (after applying B. C.)

.27E+24 .14E+05 .00E+00 .00E+00 .00E+00 .00E+00 .00E+00 .00E+00 .00E+00

.00E+00 .00E+00-.29E+04 .00E+00 .00E+00 .00E+00 .00E+00 .00E+00 .00E+00

.14E+05 .14E+25 .00E+00 .00E+00 .00E+00 .00E+00 .00E+00 .00E+00 .00E+00

-.68E+04-.14E+05 .14E+05-.68E+04 .00E+00 .00E+00 .00E+00 .ooe+oo .00E+00

.00E+00 .00E+00 .39E+05-.14E+05 .00E+00 .00E+00 .00E+00 .00E+00 .00E+00

.14E+05 .18E+05 .00E+00 .00E+00-.29E+04 .00E+00 .00E+00 .00E+00 .00E+00

.00E+00 .00E+00-.14E+05 .27E+24 .00E+00 .00E+00 .00E+00 .00E+00 .00E+00

.00E+00 .00E+00 .18E+05-.14E+05 .00E+00 .00E+00 .00E+00 .00E+00 .00E+00

.00E+00 .00E+00 .00E+00 .00E+00 .11E+24 .00E+00 .00E+00 .00E+00 .00E+00

DISPLACEMENT VECTOR

-.9929E-22 .673159E-22 -.5487E-22 -.9447E-02 .2645E-02 -.2772E-02

.1057E-21 .564926E-22 -.7055E-22

Internal Forces

Shear	Moment mxl	=Moment mx2=
Element number: 1		
26.631620	-7.663792	78.494060
-26.631620	7.663792	28.032410
Element number: 2		
-28.368390	8.032400	-32.6638
28.368390	-8.032400	-80.809750

****

Example Problem 2: The numerical example already worked out by hand is also analyzing again by computer. The joint and member numbering are same as earlier. The nodal load vector is already computed. Typical input file (frame 1.in) and the detailed output file (frame 1.out) are given below.

Input File (Frame1.in)

```
1 3.72e+9 5.4e+9 25491.0 0.15 4 b 1
2 3.72e+9 5.4e+9 25491.0 0.15 4 b 2
3 3.72e+9 5.4e+9 25491.0 0.15 4 b 3
1 6000.0 0.0
2 -3000 5196.15
3 -3000.0 -5196.15
4 0.0 0.0
111222333
123123123
-0.135e+5 0.0 0.135e+8
-0.135e+5 -0.117e+8 -0.0675e+8
-0.135e+5 0.117e+8 -0.0675e+8
-0.405e+5 0.0 0.0
```

Output

Element Number 1

6000.000000 3.720000E+09 5.400000E+09 1.000000 0.000000E+00

Element Stiffness Matrix is

.76E+04	.00E+00	.23E+08	-.76E+04	.00E+00	.23E+08
.00E+00	.69E+10	.00E+00	.00E+00	-.69E+10	.00E+00
.23E+08	.00E+00	.92E+11	-.23E+08	.00E+00	.46E+11
-.76E+04	.00E+00	-.23E+08	.76E+04	.00E+00	-.23E+08
.00E+00	-.69E+10	.00E+00	.00E+00	.69E+10	.00E+00
.23E+08	.00E+00	.46E+11	-.23E+08	.00E+00	.92E+11

Element Number 2

5999.998000 3.720000E+09 5.400000E+09 -5.000002E-01 8.660253E-01

Element Stiffness Matrix is

.76E+04	-.20E+08	-.11E+08	-.76E+04	-.20E+08	-.11E+08
-.20E+08	.71E+11	.37E+11	.20E+08	.33E+11	.23E+11
-.11E+08	.37E+11	.28E+11	.11E+08	.23E+11	.63E+110
-.76E+04	.20E+08	.11E+08	.76E+04	.20E+08	.11E+05
-.20E+08	.33E+11	.23E+11	.20E+08	.71E+11	.37E+11
-.11E+08	.23E+11	.63E+10	.11E+08	.37E+11	.28E+11

Element Number 3

15999.998000 3.720000E+09 5.400000E+09 -5.000002E-01 -8.660253E-01

Element Stiffness Matrix is 3

.76E+04-	.20E+08	-.11E+08	-.76E+04	.20E+08	-.11E+08
.20E+08	.71E+11	-.37E+11	-.20E+08	.33E+11	-.23E+11
-.11E+08	-.37E+11	.28E+11	.11E+08	-.23E+11	.63E+10

-.76E+04	-.20E+08	.11E+08	.76E+04	-.20E+08	.11E+08
.20E+08	.33E+11	-.23E+11	-.20E+08	.71E+11	-.37E+11
-.11E+08	-.23E+11	.63E+10	.11E+08	-.37E+11	.28E+11

Load Vector

-13500.0 0.0E+00 1.35E+07 -13500.0 -1.17E+07 -6750000.0

-13500.0 1.17E+07 -675.0000 -40500.0 0.00E+00 0.000E+00

Global Stiffness Matrix

76E+04 .00E+00-.23E+08 00 00 00 00 .00E+00 .00E+00-.76E+04 .00E+00-.23E+08

.0 .69E+10 .0 .0 .0 0 .0E+0 .0E+0 .0E+0 .0E+0 -.69E+10 .00E+00 -.23E+08 .00E+00

.92E+11 000000 .23E+08 .00E+00 .46E+11

000 .76E+04 .20E+08 .11E+08 00 .00E+00 .00E+00-.76E+04 .20E+08

.11E+08 00 00 00 .20E+08 .71E+11 .37E+11 00 00 .00E+00-.20E+08 .33E+11 .23E+11

00 00 00 .11E+08 .37E+11 .28E+11 00 00 .00E+00-.11E+08 .23E+11 .63E+10

00 00 00 00 00 .00 .76E+04 -.20E+08 .11E+08 -.76E+04 -.20E+08 .11E+08

00 00 00 00 00 00 .11E+08 -.37E+11.28E+11 -.11E+08 -.23E+11 .63E+10

-.76E+4 0 .23E+8 -.76E+4 -.20E+8-.11E+8-.76E+4 .20E+8-.11E+8 -.23E+50 -.2E+2

00 -.69E+10 00 .20E+8 .33E+11 .23E+11 -.20E + 8 .33E+11 -.23E+11 0.15E+12 .00

-.23E+08 0.46E+11 .11E+8 .23E+11 .63E+10 .11E+8 -.23E+11 .63E+10 -.2E+2 0.15E+12

Internal Forces

	Shear	Moment mx1	Moment mx2
element number: 1			
	-13500.00000	-4.896164E-28	-4.050000E+07
	13500.00000	4.896164E-28	4.049999E+07
element number: 2			
	-13500.00000	-2.574834	-4.049999E+07
	13500.00000	-1.109734	-4.049999E+07
element number: 3			
	-13500.00000	2.573834	-4.049999E+07
	13500.00000	1.109734	-4.049999E+07

Now the final internal forces can be obtained by adding the fixed end actions as earlier.

16.6 SPACE FRAME

16.6.1 Introduction

Three-dimensional frames also called space frames are frequently encountered in the analysis of multistoried buildings. They are also found to be used in the modeling of aircraft, ships, car body, bicycle and many other mechanical structural systems. The space frames are the most complex of all skeletal structures met in practice. In fact, most of the structures are three-dimensional in behavior. Transformation of space frame into assemblage of components such as beam grids, plane frames, etc is an attempt to simplify the problem. A typical 3-D frame is shown in figure 16.44. Each node has 6 d.o.f. *i.e.*, translation (u_x, u_y, u_z) and rotations $(\theta_x, \theta_y, \theta_z)$ about x, y and z axis, making a total 12 d.o.f. for each member.

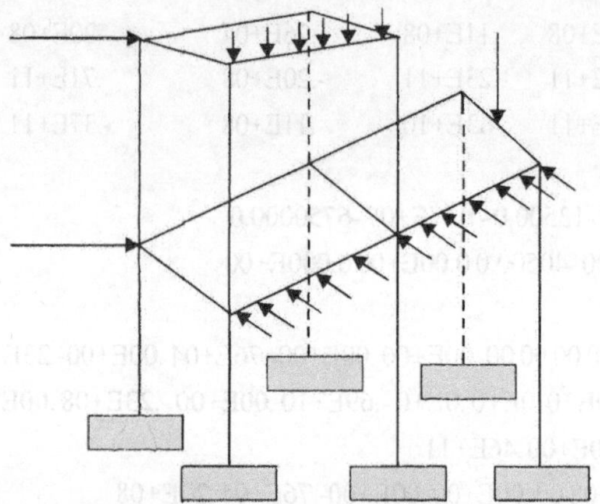

Fig. 16.44. A typical space frame.

16.6.2 Element Stiffness in Member Principal Axis

Consider a rigid jointed three-dimensional element as shown in figure 16.45. Let $\overline{x}$, $\overline{y}$, $\overline{z}$ represents the three orthogonal axes coinciding with the three principal axes of the member.

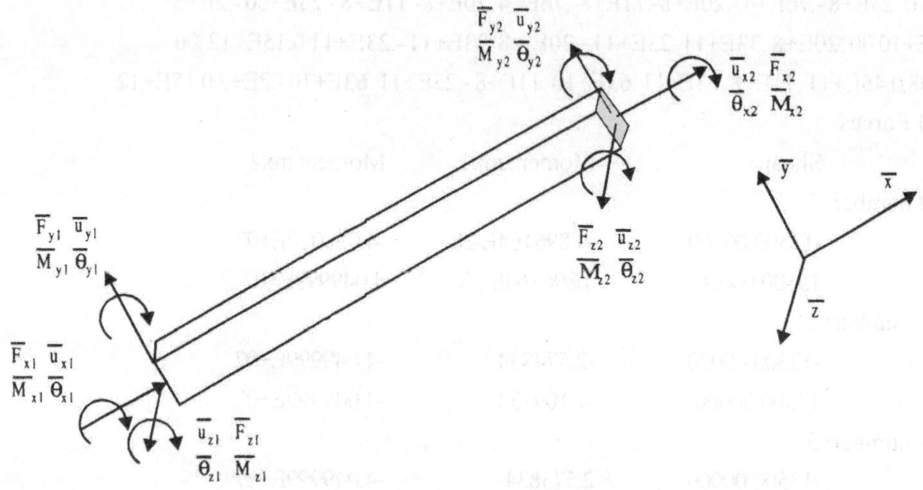

Fig. 16.45. Space frame element in principle axis system.

For three-dimensional behavior of this member, at each node we must consider six generalized displacement $\overline{u}_x, \overline{u}_y, \overline{u}_z, \overline{\theta}_x, \overline{\theta}_y, \overline{\theta}_z$ and six corresponding force quantities are $\overline{F}_x$, $\overline{F}_y$, $\overline{F}_z$, $\overline{M}_x$, $\overline{M}_y$, $\overline{M}_z$ as shown in figure 16.45. The linear elastic member can be envisioned to have four distinct response modes :

(*i*) Tension-Compression in *x*-direction

(*ii*) Flexure in *x*-*y* plane

(*iii*) Flexure in *x*-*z* plane

(*iv*) Torsion about *x*-axis

It may be noted that each stiffness coefficient associated with a space frame element has already been encountered when dealing with truss, beam, frame and grillage element. Thus, in accordance with that, the stiffness equilibrium equation for each four difference response modes can be described as :

(*i*) Axial Effect :

$$\begin{Bmatrix} \bar{F}_{x1} \\ \bar{F}_{x2} \end{Bmatrix} = \frac{AE}{L}\begin{bmatrix} 1 & -1 \\ 1 & 1 \end{bmatrix}\begin{Bmatrix} \bar{u}_{x1} \\ \bar{u}_{x2} \end{Bmatrix} \tag{16.73}$$

(*ii*) Bending about x-y plane

$$\begin{Bmatrix} \bar{F}_{y1} \\ \bar{M}_{z1} \\ \bar{F}_{y2} \\ \bar{M}_{z2} \end{Bmatrix} = EI_z\begin{bmatrix} \frac{12}{L^3} & \frac{6}{L^2} & -\frac{12}{L^3} & \frac{6}{L^2} \\ \frac{6}{L^2} & \frac{4}{L} & -\frac{6}{L^2} & \frac{2}{L} \\ -\frac{12}{L^3} & -\frac{6}{L^2} & \frac{12}{L^3} & -\frac{6}{L^2} \\ \frac{6}{L^2} & \frac{2}{L} & -\frac{6}{L^2} & \frac{4}{L} \end{bmatrix}\begin{Bmatrix} \bar{u}_{y1} \\ \bar{\theta}_{z1} \\ \bar{u}_{y2} \\ \bar{\theta}_{z2} \end{Bmatrix}, \text{ where } I_z = \int_A y^2\, da \tag{16.74}$$

(*iii*)Bending about x-z plane

$$\begin{Bmatrix} \bar{F}_{z1} \\ \bar{M}_{y1} \\ \bar{F}_{z2} \\ \bar{M}_{y2} \end{Bmatrix} = EI_y\begin{bmatrix} \frac{12}{L^3} & \frac{6}{L^2} & -\frac{12}{L^3} & \frac{6}{L^2} \\ \frac{6}{L^2} & \frac{4}{L} & -\frac{6}{L^2} & \frac{2}{L} \\ -\frac{12}{L^3} & -\frac{6}{L^2} & \frac{12}{L^3} & -\frac{6}{L^2} \\ \frac{6}{L^2} & \frac{2}{L} & -\frac{6}{L^2} & \frac{4}{L} \end{bmatrix}\begin{Bmatrix} \bar{u}_{z1} \\ \bar{\theta}_{y1} \\ \bar{u}_{z2} \\ \bar{\theta}_{y2} \end{Bmatrix}, \text{ where } I_y = \int_A x^2\, dA \tag{16.75}$$

(*iv*) Torsion about x-axis

$$\begin{Bmatrix} \bar{M}_{x1} \\ \bar{M}_{x2} \end{Bmatrix} = \frac{GJ}{L}\begin{bmatrix} 1 & -1 \\ -1 & 1 \end{bmatrix}\begin{Bmatrix} \bar{\theta}_{x1} \\ \bar{\theta}_{x2} \end{Bmatrix} \tag{16.76}$$

It is assumed that the shear centre and centroid of the cross-sections coincides *i.e.*, the cross-section of the members are doubly symmetric like I-sections. Torsion is assumed uniform on the member *i.e.*, members are warping free. For small displacement, these four individual phenomenon are uncoupled *i.e.*, generalized forces from each individual response mode produce neither force nor displacements in other modes. Thus, for linear behaviour, the general 12 × 12 stiffness matrix for the element can be obtained by combining the force displacement relationships for the four individual response mode. And combining the above four cases, we get the [k] matrix in local coordinate for the space frame *i.e.*,

$$\begin{Bmatrix} \bar{F}_{x1} \\ \bar{F}_{y1} \\ \bar{F}_{z1} \\ \bar{M}_{x1} \\ \bar{M}_{y1} \\ \bar{M}_{z1} \\ \bar{F}_{x2} \\ \bar{F}_{y2} \\ \bar{F}_{z2} \\ \bar{M}_{x2} \\ \bar{M}_{y2} \\ \bar{M}_{z2} \end{Bmatrix} = \frac{E}{L} \begin{bmatrix} A & 0 & 0 & 0 & 0 & 0 & -A & 0 & 0 & 0 & 0 & 0 \\ 0 & \frac{12I_z}{L^2} & 0 & 0 & 0 & \frac{6I_z}{L} & 0 & -\frac{12I_z}{L^2} & 0 & 0 & 0 & \frac{6I_z}{L} \\ 0 & 0 & \frac{12I_y}{L^2} & 0 & -\frac{6I_y}{L} & 0 & 0 & 0 & -\frac{12I_y}{L^2} & 0 & -\frac{6I_y}{L} & 0 \\ 0 & 0 & 0 & \frac{J}{2(1+\mu)} & 0 & 0 & 0 & 0 & 0 & -\frac{J}{2(1+\mu)} & 0 & 0 \\ 0 & 0 & -\frac{6I_y}{L} & 0 & 4I_y & 0 & 0 & 0 & \frac{6I_y}{L} & 0 & 2I_y & 0 \\ 0 & \frac{6I_z}{L} & 0 & 0 & 0 & 4I_z & 0 & -\frac{6I_z}{L} & 0 & 0 & 0 & 2I_z \\ -A & 0 & 0 & 0 & 0 & 0 & A & 0 & 0 & 0 & 0 & 0 \\ 0 & -\frac{12I_z}{L^2} & 0 & 0 & 0 & -\frac{6I_z}{L} & 0 & \frac{12I_z}{L^2} & 0 & 0 & 0 & \frac{6I_z}{L} \\ 0 & 0 & -\frac{12I_y}{L^2} & 0 & -\frac{6I_y}{L} & 0 & 0 & 0 & \frac{12I_y}{L^2} & 0 & -\frac{6I_y}{L} & 0 \\ 0 & 0 & 0 & -\frac{J}{2(1+\mu)} & 0 & 0 & 0 & 0 & 0 & \frac{J}{2(1+\mu)} & 0 & 0 \\ 0 & 0 & -\frac{6I_y}{L} & 0 & 2I_y & 0 & 0 & 0 & \frac{6I_y}{L} & 0 & 4I_y & 0 \\ 0 & \frac{6I_z}{L} & 0 & 0 & 0 & 2I_z & 0 & -\frac{6I_z}{L} & 0 & 0 & 0 & 4I_z \end{bmatrix} \begin{Bmatrix} \bar{u}_{x1} \\ \bar{u}_{y1} \\ \bar{u}_{z1} \\ \bar{\theta}_{x1} \\ \bar{\theta}_{y1} \\ \bar{\theta}_{z1} \\ \bar{u}_{x2} \\ \bar{u}_{y2} \\ \bar{u}_{z2} \\ \bar{\theta}_{x2} \\ \bar{\theta}_{y2} \\ \bar{\theta}_{z2} \end{Bmatrix} \quad (16.77)$$

16.6.3 Coordinate Transformation

The three-dimensional element in the member coordinate system as shown in figure 16.45 is now consider to orient in the structural *i.e.*, global axis system in as shown in figure 16.46.

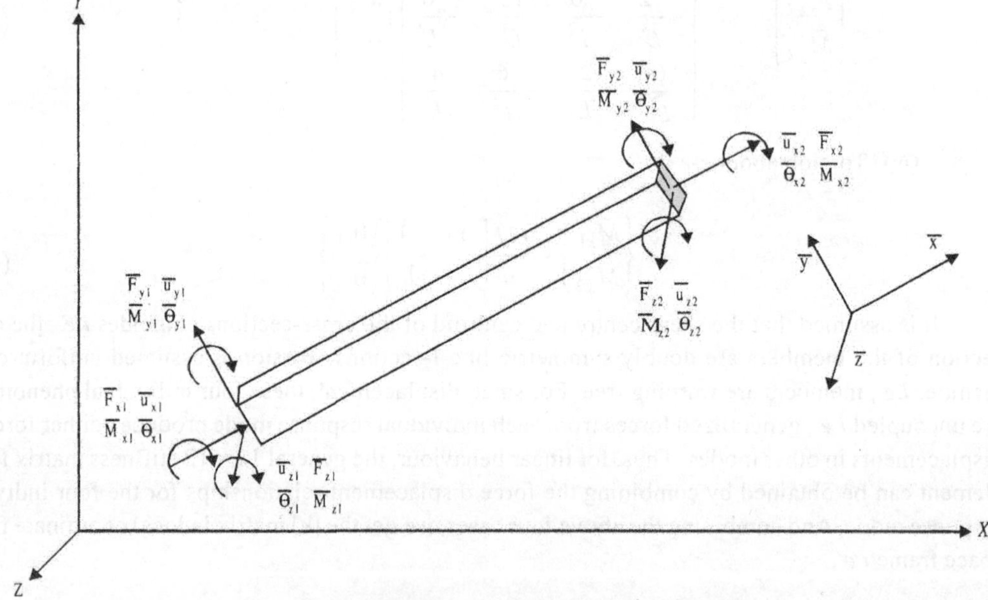

Fig. 16.46. Space frame element oriented arbitrarily.

It can be seen that 12×12 stiffness matrix in equation (16.5) are defined in member axis system. The transformation from local to global axis system in accomplished as,

$$K = [T]^T [\bar{K}][T] \qquad (16.78)$$

Where, $$[T]_{12\times12} = \begin{bmatrix} [\lambda] & [0] & [0] & [0] \\ [0] & [\lambda] & [0] & [0] \\ [0] & [0] & [0] & [\lambda] \end{bmatrix}$$

and $$[\lambda] = \begin{bmatrix} \lambda_{x\bar{x}} & \lambda_{y\bar{x}} & \lambda_{z\bar{x}} \\ \lambda_{x\bar{y}} & \lambda_{y\bar{y}} & \lambda_{z\bar{y}} \\ \lambda_{x\bar{z}} & \lambda_{y\bar{z}} & \lambda_{z\bar{z}} \end{bmatrix}, [0] = \begin{bmatrix} 0 & 0 & 0 \\ 0 & 0 & 0 \\ 0 & 0 & 0 \end{bmatrix} \qquad (16.79)$$

In above, $\lambda_{x\bar{x}}, \lambda_{y\bar{x}}, \lambda_{z\bar{x}}$ are the direction cosines of the $\bar{x}$ axis; $\lambda_{x\bar{y}}, \lambda_{y\bar{y}}, \lambda_{z\bar{y}}$ are the direction cosines of $\bar{y}$ axis and $\lambda_{x\bar{z}}, \lambda_{y\bar{z}}, \lambda_{z\bar{z}}$ are the direction cosines of $\bar{z}$ axis with respect to the global x, y, z axis. Finding direction cosines of $\bar{x}$ axis is straight forward *i.e.*,

$$\lambda_{z\bar{x}} = \frac{z_2 - z_1}{L}, \lambda_{x\bar{x}} = \frac{x_2 - x_1}{L}, \lambda_{y\bar{x}} = \frac{y_2 - y_1}{L} \qquad (16.80)$$

It may be noted that $\lambda_{y\bar{x}}, \lambda_{x\bar{y}}$ are not necessarily equal. The $\bar{y}$ axis is selected to be perpendicular to the $\bar{x}$ axis and z axis in such a way that the cross-product of global z with the $\bar{x}$ axis, as shown in figure 16.47 is $z \times \bar{x} = \bar{y} = \frac{1}{a}\begin{vmatrix} \bar{i} & \bar{j} & \bar{k} \\ 0 & 0 & 1 \\ \lambda_{x\bar{x}} & \lambda_{y\bar{x}} & \lambda_{z\bar{x}} \end{vmatrix} = -\frac{\lambda_{y\bar{x}}}{a}\bar{i} + \frac{\lambda_{x\bar{x}}}{a}\bar{j}, a = \sqrt{\lambda_{x\bar{x}}^2 + \lambda_{y\bar{x}}^2}$. The $\bar{z}$ axis will be determined by the orthogonality condition, $\bar{z} = \bar{x} \times \bar{y} = -\frac{\lambda_{x\bar{x}}}{a}\bar{i} + \frac{\lambda_{y\bar{x}}\lambda_{z\bar{x}}}{a}\bar{j} + a\bar{k}$. Thus,

$$[\lambda] = \begin{bmatrix} \lambda_{x\bar{x}} & \lambda_{y\bar{x}} & \lambda_{z\bar{x}} \\ -\frac{\lambda_{y\bar{x}}}{a} & \frac{\lambda_{x\bar{x}}}{a} & 0 \\ -\frac{\lambda_{x\bar{x}}\lambda_{z\bar{x}}}{a} & -\frac{\lambda_{y\bar{x}}\lambda_{z\bar{x}}}{a} & a \end{bmatrix} \qquad (16.81)$$

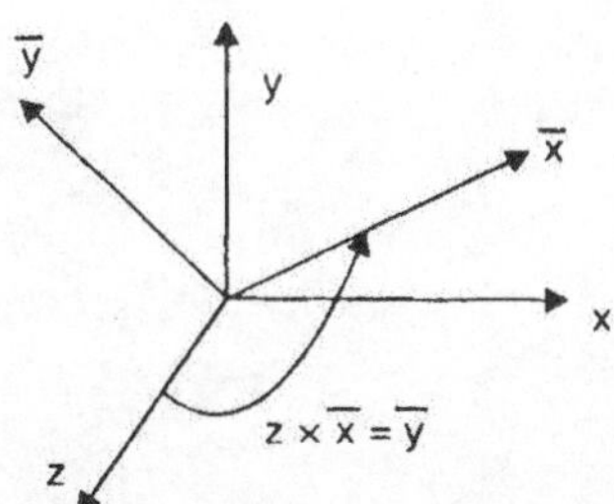

Fig. 16.47a Direction cosine

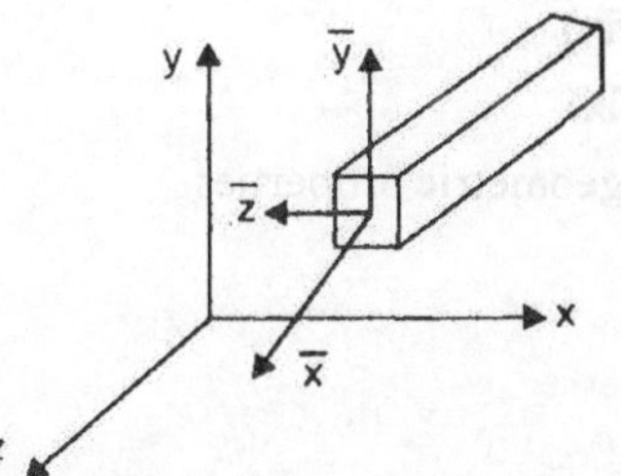

Fig. 16.47b $\bar{x}$ in same direction as z

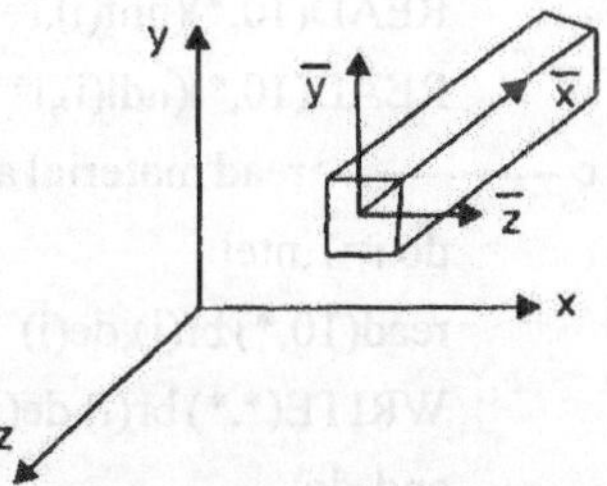

Fig. 16.47c $\bar{x}$ in opposite direction as z

Note that, if the local $\bar{x}$ axis coincides with global z axis (figure 16.47b) *i.e.*, the member parallel to global z axis, $\bar{y}$ axis becomes uncertain. In that case, the local $\bar{y}$ axis is selected as the global y-axis and transformation matrix will be $[\lambda]=\begin{bmatrix}0&0&1\\0&1&0\\-1&0&0\end{bmatrix}$. For positive $\bar{x}$ axis opposite the global z axis (figure 16.47c), transformation matrix will be $[\lambda]=\begin{bmatrix}0&0&-1\\0&1&0\\1&0&0\end{bmatrix}$.

Computer Implementation

A simple computer program is given to implement the space frame analysis as described above.

```
C-----------------------------------------------------------------------------------------------
parameter(ne=3, nd=4, no=24, nf=18)
implicit double precision(a-h,o-z)
dimension node(ne,2),stif(12,12),stif1(12,12),gstif(no,no)
dimension x(nd),y(nd),z(nd),br(ne),de(ne),dc(12,12),dct(12,12)
1, eload(no),u(no),ndl(nf),nnf(nf),al(no,no)au(no,no),bs(no)
c-----------------------------------------------------------------------------------------------
      open(10,file='ssr.in')
      open(12,file='ssr.out')
c--------------->Input data
      read(10,*)nfree,nel,ntel,ntnd
      read(10,*)e,g
      read(10,*)((node(i,j),j=1,2),i=1,ntel)
      do 11 i=1,ntnd
      >read(10,*) x(i),y(i),z(i)
11 continue
c enter the total no. of fixed nodes and total fixed degree of freedom
      read(10,*)nfix
      READ(10,*)(nnf(i),i=1,nfix)
      READ(10,*)(ndl(i),i=1,nfix)
c -------------->read material and geometric properties
      do i=1,ntel
      read(10,*) br(i),de(i)
      WRITE(*,*) br(i),de(i)
      end do
c -----------------------------------------------------------------------------------------------
      nfof=nfree*ntnd
      do i=1,ndof
      eload(i)=0.0
```

```
        do j=1,ndof
        gstif(i,j)=0.0
        end do
        end do
c ---------------------------------------------------------------------------------------------------
        do 20 ii=1,ntel
        eiy=e*br(ii)*(de(ii)**3)/12
        eiz=e*de(ii)*(br(ii)**3)/12
        gj=g*0.229*(de(ii)*br(ii)**3)
        ea=e*br(ii)*de(ii)
        n1=node(ii,1)
        n2=node(ii,2)
        call transform(ntnd,n1,n2,x,y,z,alij,dc,dct)
        call stifness(gj,eiy,eiz,ea,alij,stif)
        call mult(dct,stif,stif1,12,12,12)
        call mult(stif1, dc, stif, 12, 12, 12)
        call assem(n1,n2,nfree,gstif,stif,ndof)
c ---------------------------------------------------------------------------------------------------
20 continue
c---------------------->LOAD
        eload(20)=-10.0
c---------------------->structural boundary conditions
c----------------------> Following statement apply the boundary conditions on gstiff matrix
        call boundary (ndof,ndof,gstif,nfix,nfix,nnf,ndl)
        WRITE(12,*) (gstif(i,i),i=1,ndof)
c ----------------------> Following subrouine solve matrix equilibrium equation
        call decomlu(gstif,al,au,ndof)
        call fsub(al,eload,bs,ndof)
        call bsub(au,bs,u,ndof)
        write(12,*)'DISPLACEMENT VECTOR'
        write(12,3)(u(i),i=1,ndof)
        write(*,*)'DISPLACEMENT VECTOR'
        write(*,3)(u(i),i=1,ndof)
3 format(6e13.4/)
        stop
        end
C TRANSFORMATION MATRIX
        subroutine transform(nn,n1,n2,x,y,z,alij,dc,dct)
        implicit doule precision(a-h,o-z)
        dimension x(nn),y(nn),z(nn),dc(12,12),dct(12,12),al(3,3)
```

```
c ------------------------------------------------------------------------------------------
        do i=1,12
        do j=1,12
        dc(i,j)=0.0
        dct(i,j)=0.0
        end do
        end do
        alij=dsqrt((x(n2) - x(n1))**2+(y(n2)-y(nl))**2+(z(n2)-z(n1))**2)
        al(1,1)=(x(n2)-x(nl))/alij
        al(1,2)=(y(n2)-y(n1))/alij
        al(1,3)=(z(n2)-z(n1))/alij
        if(al(1,3).eq.1.) then
        al(2,2)=-1.
        al(3,1)=1
        else
        a=dsqrt((al(1,1))**2+(al(1,2))**2)
        al(2,1)=-al(1,2)/a
        al(2,2)=al(1,1)/a
        al(2,3)=0.0
        al(3,1)=-al(2,2)*al(1,3)
        al(3,2)=-al(2,1)*al(1,3)
        al(3,3)=a
        endif
        do 10 k=1,4
        do 10 i=1,3
        do 10 j=1,3
        m=3*(k-1)+j
        n=3*(k-1)+j
        dc(m,n)=al(i,j)
        dct(n,m)=al(i,j)
10 continue
        return
        end
c --------STIFNESS CALCULATION
subroutine stifness(gj,eiy,eiz,ea,alij,stif)
implicit double precision(a-h,o-z)
        dimension stif(12,12)
        do 200 i=1,12
        do 200 j=1,12
        stif(i,j)=0
```

```
200 continue
      cl=ea/alij
      c2=(12.0*eiz)/(alij**3)
      c3=(6.0*eiz)/(alij**2)
      c4=(12.0*eiy)/(alij**3)
      c5=(6.0*eiy)/alij**2)
      c6=gj/alij
      c7=(4.0*eiy)/alij
      c8=(4.0*eiz)/alij
      stif(1,1)=cl
      stif(1,7)=-cl
      stif(2,2)=c2
      stif(2,8)=-c2
      stif(2,6)=c3
      stif(2,12)=c3
      stif(3,3)=c4
      stif(3,9)=-c4
      stif(3,5)=-c5
      stif(3,11)=-c5
      stif(4,4)=c6
      stif(4,10)=-c6
      stif(5,5)=c7
      stif(5,11)=c7/2
      stif(5,9)=c5
      stif(6,6)=c8
      stif(6,8)=-c3
      stif(6,12)=c8/2
      stif(7,7) = cl
      stif (8, 8) = c2
      stif(8,12)=-c3
      stif(9,9)= c4
      stif (9, 11) = c5
      stif(10,10)=c6
      stif(11,11)=c7
      stif(12,12)=c8
      n=12
      call trans(stif,n,n)
      return
      end
c ------------------------------------------------------------------------
      subroutine trans(a,n,nn)
      implicit double precision(a-h,o-z)
      dimension a(nn,nn)
```

```
      do 10 i=2,n
      do 11 j=1,i-1
      a(i,j)=a(j,i)
11    continue
10    continue
      return
      end
c-----------------------------------------------------------------------
c  subroutine to assemble
      subroutine assem(i,j,nnf,a,b,norder)
      implicit double precision(a-h,o-z)
      dimension a(norder,norder),b(12,12)
      i1=(i-1)*nnf
      j1=(j-1)*nnf
      do 35 ii=1,nnf
      do 36 jj=1,nnf
      m1=i1+ii
      m2=j1+ii
      m3=i1+jj
      m4=j1+jj
      a(m1,m3)=a(m1,m3)+b(ii,jj)
      a(m1,m4)=a(m1,m4)+b(ii,jj+nnf)
      a(m2,m3)=a(m2,m3)+b(ii+nnf,jj)
      a(m2,m4)=a(m2,m4)+b(ii+nnf,jj+nnf)
36    continue
35    continue
      return
      end
c -----------------------------------------------------------------------
      subroutine mult(a,b,c,m,k,n)
      implicit double precision(a-h,o-z)
      dimension a(m,k),b(k,n),c(m,n)
      do 101 i=1,m
      do 101 j=1,n
      c(i,j)=0.0
      do 101 index=1,k
      c(i,j)=c(i,j)+a(i,index)*b(index,j)
101   continue
      return
      end
```

```
c -----------------------------------------------------------------------------------------------------------
      subroutine boundary (no,ndof,gstif,nf,nfix,nnf,ndl)
      implicit doule precision(a-h,o-z)
      dimension gstif(no,no),nnf(nf),ndl(nf)
      cnst=1.0
      doi=1,ndof
      if (cnst.le.gstif(i,i))then
      cnst=gstif(i,i)
      end if
      end do
      cnst=(1.+cnst)*1.0e+15
      do 400 i=1,nfix
      i1=6*nnf(i)-5
      i2=i1+1
      i3=i1+2
      i4=i1+3
      i5=i1+4
      i6=i1+5
      if(ndl(i).eq.1)go to 401
      if(ndl(i).eq.2)go to 402
      if(ndl(i).eq.3)go to 403
      if(ndl(i).eq.4)go to 404
      if(ndl(i).eq.5)go to 405
      if(ndl(i).eq.6)go to 406
401   gstif(i1,i1)=cnst*gstif(i1,i1)
      go to 400
402   gstif(i2,i2)=cnst*gstif(i2,i2)
      go to 400
403   gstif(i3,i3)=cnst*gstif(i3,i3)
      go to 400
404   gstif(i4,i4)=cnst*gstif(i4,i4)
      go to 400
405   gstif(15,i5)=cnst*gstif(i5,i5)
      go to 400
406   gstif(i6,i6)=cnst*gstif(i6,i6)
400   continue
      return
      end
```

```
c--------------> Subroutine of LU decomposition
      subroutine decomlu(a,al,au,n)
      implicit double precision (a-h,o-z)
      dimension a(n,n),al(n,n),au(n,n)
      do 50 i=1,n
      do 50 j=1,n
      al(i,j)=0.0
      au(i,j)=0.0
50    continue
      do 60 i=1,n
      au(i,i)=1.0
      al(i,1)=a(i,1)
60    au(1,i)=a(1,i)/al(1,1)
      do 10 j=2,n
      do 20 i=j,n
      sum=0.0
      do 30 k=1,j-1
30    sum=sum+al(i,k)*au(k,j)
20    al(i,j)=a(i,j)-sum
      if(j.eq.n)go to 10
      do 40 jj=j+1,n
      sum=0.0
      do 70 kk=1,j-1
70    sum=sum+al(j,kk)*au(kk,jj)
40    au(j,jj)=(a(j,jj)-sum)/al(j,j)
10    continue
       return
       end
c------ >Forward substitution to solve LZ=b
      subroutine fsub(al,b,z,n)
      implicit double precision(a-h,o-z)
      dimension al(n,n),b(n),z(n)
      z(1)=b(1)/al(1,1)
      do 80 i=2,n
      sum=0.0
      do 90 k=1,i-1
90    sum=sum+al(i,k)*z(k)
80    z(i)=(b(i)-sum)/al(i,i)
      return
      end
```

```
c----- >backward substitution to solve UX=Z
      subroutine bsub(au,z,x,n)
      implicit double precision(a-h,o-z)
      dimension au(n,n),z(n),x(n)
      x(n)=z(n)
c     do 100i=2,n-1
      do 100 i=2,n
      l=n-i+1
      sum=0.0
      do 110k = l + 1, n
110   sum=sum+au(l,k)*x(k)
100   x(l)=z(l)-sum
      return
      end
```

Example Problem : The space frame as shown in figure 16.48 is taken up to apply the above space frame program. Following data are assumed: Length of each member: 0.2m, cross-section area is 0.1m × 0.2 m, E = 2.0 × 10^6 N/cm², G = 0.8 × 10^6 N/cm². Joint 1, 2 and 3 are completely restrained against any rotation. The load acting on the frame is 10N vertically downward at joint 4. The input file:

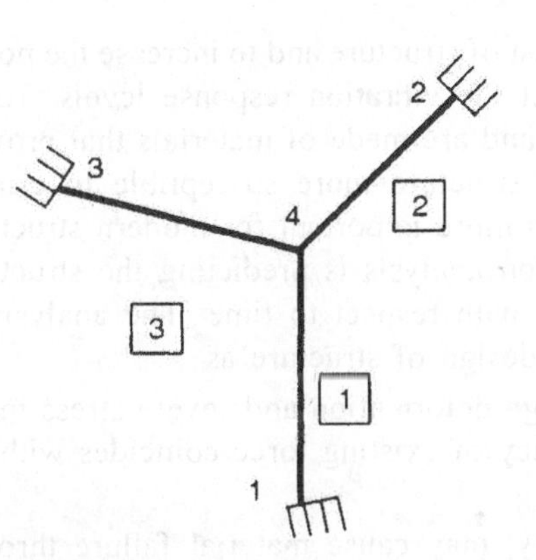

Fig. 16.48. The Space Frame

```
6234
2.0e+6 0.8e+6
14
34
24
20.0 20.0 0.0
20.0 0.0 20.0
0.0 20.0 20.0
20.0 20.0 20.0
18
111111222222333333
123456123456123456
1.0 2.0
1.0 2.0
1.0 2.0
```

The output results i.e., the displacements at various nodes of the frame are:

```
0.9273E-28   -0.7738E-25   0.1268E-27   0.6126E-26   0.6471E-29   -0.2979E-26
-0.2972E-25  -0.9029E-25   -0.1278E-25  -0.6360E-27  -0.1936E-29  0.1484E-26
0.7337E-28   -0.6050E-25   0.1074E-27   0.1291E-26   -0.8539E-29  -0.5282E-26
0.4048E-07   -0.4981E-04   0.6993E-07   0.7122E-06   -0.1068E-08  -0.1644E-05
```

17

DYNAMIC ANALYSIS OF STRUCTURE

17.1 INTRODUCTION

The response of physical objects to time varying loading is an important area of study in engineering. The physical object whose response is sought may either be treated as rigid body or deformable. The rigid body dynamics treat the object as rigid that will undergo motion without deformation when subjected to dynamic loading for example, movement of machinery, flight of an aircraft or space vehicle etc. In many instances, however, dynamic response involving deformation rather than simple rigid body motion is of primary concern. This is particularly so in the design of structural frames, for example, buildings, bridges, aircrafts, ships, offshore platforms etc. All of these structures and their supporting systems are subjected to dynamic disturbances during their service life.

17.2 IMPORTANCE OF VIBRATION ANALYSIS

The continuing tendency to reduce the weight and cost of structure and to increase the power to weight ratio of structure enhanced the need to predict the vibration response levels. Today building and bridges, for example, lighter, more flexible and are made of materials that provide much lower energy dissipation, all of which made the structure more susceptible to critical vibration. Dynamic analysis of structure is therefore even more important for modern structure, and this trend is likely to continue. The basis of vibration analysis is predicting the structural response when the external load acting on it is varying with respect to time. The analysis of structural response is of considerable importance in the design of structure as

(*i*) Under certain situations vibration may cause large deformation and severe stress in the structure. This may happen particularly when the frequency of existing force coincides with the natural frequency of the structure.

(*ii*) Fluctuating stresses, even of moderate intensity, may cause material failure through fatigue if the number of repetitions is large enough.

(*iii*) Oscillating motion may at times cause wearing and malfunctioning.

(*iv*) Vibration induced by rotating or reciprocating machinery, for example, transmitted through the support structure to delicate instruments elsewhere on it, causing such instruments to malfunction.

(*v*) Finally when structure is designed for human use, vibratory motion may result in severe discomfort to the occupants.

17.3 NATURE OF DYNAMIC INFLUENCES

Many natural and man made influences may cause dynamic response in structure. The most common influences are due to Initial Conditions, Applied Actions and Support Motions. A crane suddenly drops its load and rebounds from an initial condition of static equilibrium. A truck backs into a loading platform with some initial velocity. Consequently the frame and the body of the truck respond dynamically. Applied actions are in the form of lateral loads on building, machinery force on floor, moving load on bridges etc. The dynamic forces induced due to support motions are mainly due to earthquake. Various types of dynamic forces acting on structure may be categorize according to sources of their origin such as rotating machinery, wind, blast, earthquake etc.

Dynamic Forces Caused by Rotating Machinery

Rotating machinery that is not fully balanced about the centre of rotation will give rise to exciting force which may vary with time. Dynamic forces arising from unbalance rotating machinery are quite common in mechanical systems and the supporting structure in such case is designed to withstand the resulting deformation and stress.

Wind Loads

Structures subjected to wind experiences aerodynamic forces, which may be drag forces parallel to the direction of wind, lift forces perpendicular to the wind. Both the forces vary with respect to time. The response of the structure for wind is thus a dynamic phenomenon and to obtain the displacement, stress etc induced in the structure by wind load, a dynamic analysis is essential.

Blast Loads

The blast or shock wave caused by the detonation of a conventional explosive such as TNT or a bomb. In all the cases, the explosion results in the rapid release of a large amount of energy, which is turn induces dynamic forces on the structure.

Seismic Forces

Ground motions resulting from earthquakes of sufficiently large magnitude are one of the most severe and disastrous dynamic disturbances that affect man made structures.

17.4 CLASSIFICATION OF VIBRATION PROBLEMS

Vibration problem can be broadly classified into following two cases :

***(i)* Free Vibration :** Vibration of structure is started by the application of external loads which is subsequently remove the resulting oscillations where no external load is present is called free vibration.

***(ii)* Force Vibration :** This is to and fro motion of the structure due to external excitations as discussed earlier.

Again, depending on the nature of loading it may be classified as :

Periodic and Non periodic.

Dynamic loads may vary in their magnitude, direction or position with time. A special type of dynamic load that varies in magnitude with time and repeats itself at regular intervals, is called a periodic load. The duration of load that is repeated is called a cycle and the time taken to complete the cycle is called the period of the load. A general type of periodic load is shown in Figure 17.1. The loads that do not show any periodicity are called non-periodic load.

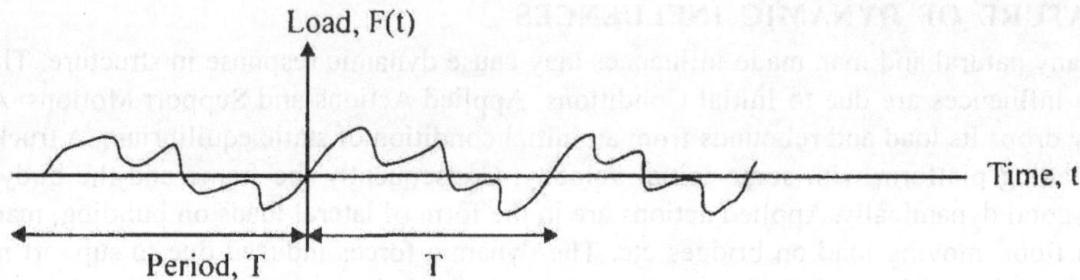

Fig. 17.1 General Periodic Load.

Deterministic and Stochastic Loads

Loads that can be specified as definite functions of time, irrespective of whether the time variation is regular or irregular is called deterministic load. Certain types of load cannot be specified as definite functions of time because of the inherent uncertainty in their magnitude and the form of variation with time. Such loads are described in a statistical sense only through certain statistical parameters such as mean, variances, power spectral density function etc. Such types of loads are known as stochastic or random loads. The analysis of a structure for such loads yield response values that are defined only in terms of some statistical parameters and is therefore known as random vibration analysis. Earthquake loads are random in nature because the magnitude and the frequency distribution of an acceleration record for a possible future earthquake can not be predicted with certainty but can be estimated only in a probabilistic sense. Similarly, wind loads, wave loads etc are also random in nature.

17.5 RESPONSE OF SINGLE DEGREE FREEDOM SYSTEM

17.5.1 Introduction

The displaced configuration of many structures subjected to dynamic loads can be completely described by specifying the time-varying displacement along only one coordinate direction. Such systems are designated as single degree of freedom (sdof) system. For preliminary or approximate analysis, simple structure may often be idealized as system with sdof system. Figure 17.2 shows the simplest of the sdof system and associated spring mass damper model

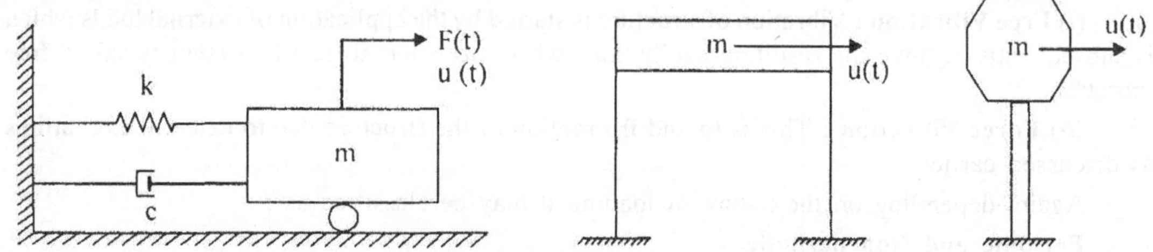

Fig. 17.2. SDOF system and spring mass model.

A sdof system has the following elements

(*a*) **Mass Element** : It represents a rigid body of mass *m* constrained to move along the x-direction in the plane of the paper. The mass is attached to a firm support by a spring and the interia force of magnitude $m\ddot{u}$ is developed due to acceleration $\ddot{u}$ of the concentrated mass.

(*b*) **Stiffness Element** : An elastic body undergoing deformation under the action of external forces sets up internal forces that resist the deformation. These forces of elastic constraints, alternatively known as spring forces occur, irrespective of whether the deformation is a result of static or dynamic forces. The simplest example of the forces of elastic constant is provided by the spring shown in figure 17.3*a*. When stretched by a force applied at its end, the spring resists the deformation by an internal forces, which is related to the magnitude of the deformation. The relationship between spring force and displacement is nonlinear and may take the form as shown in the figure. However, for small deformation, the force displacement relationship can be assumed as straight line. The slope of this straight line is denoted by the spring constant, *k*.

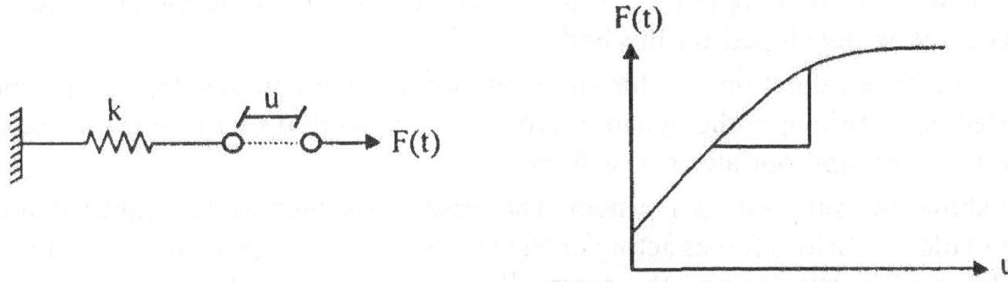

Fig. 17.3*a*. Force Displacement relation for spring.

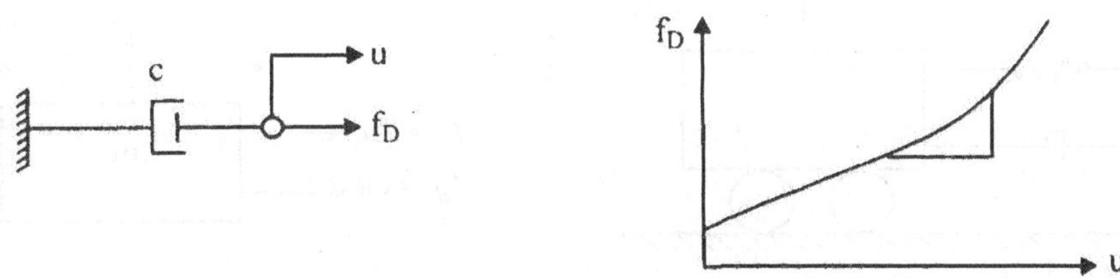

Fig. 17.3*b* Representation of Viscous Damping.

(*c*) Damping Element : In all practical cases due to resistance encountered during vibration the amplitude will not remain constant but will steadily diminish. Thus, in practice, the motion of the body is resisted by several kinds of damping forces. These forces are always opposite to the direction of motion and are small in magnitude compared to inertia or spring forces. Despite this, the damping force may significantly affect the response. A damping force may result from the resistance offered by air. When the velocity is small, the resistance offered by a fluid or gas is proportional to the velocity. A resisting force of this nature is called viscous damping given by

$$f_D = c\dot{u} \tag{17.1}$$

where c is a constant of proportionality called the damping coefficient. Forces resisting a motion may also arise from dry friction along a non lubricated surface. A resisting force of this nature is called the force of Coulomb friction. In addition to the forces of air resistance and external friction, damping forces may also arise because of imperfect elasticity of internal friction within the material. Resisting forces arising from internal friction are called forces of hysteretic or structural damping. It is observed that the damping forces are complex in nature and difficult to determine in general. Viscous damping forces are easiest to handle mathematically and are known to provide analytical results for the response of a system that conform reasonably well to experimental results. It is therefore usual to model the damping forces by viscous damping. The mechanism is indicated by a dashpot, as shown in fig. 17.3*b*. The general nature of damping force is represented by a nonlinear relationship between the force and the velocity. The relationship however, can be approximated by straight line, in which case the damping force is given by eqn. (17.1)

(*d*) Excitations : This is external actions *i.e.* time varying force acting on the structure due to various external cause as described earlier.

17.5.2 Dynamic Equilibrium Equations :

Depending on the magnitude of the forces F(t), the following situations can arise:

(*i*) Incipient motion : the magnitude of external force F(t) is such that the body is just on the verge of motion, which is nothing but in static equilibrium condition.

(*ii*) Moving with constant velocity : in this situation there is no acceleration on the body. Thus no inertia force is developed on the body.

(*iii*) Moving with acceleration : in this situation inertia force will develop. Under such cases following D' Alembert's principle the dynamic problem is converted to a problem of equilibrium of forces by introducing appropriate inertia forces.

Fig. 17.4 shows the simplest sdof system. The mass is assumed as concentrated at the mass centre of the rigid block. Various forces acting on the free body are the spring force (f_s), the damping force (f_d), the force of inertia (f_I) and the externally applied force F(t).

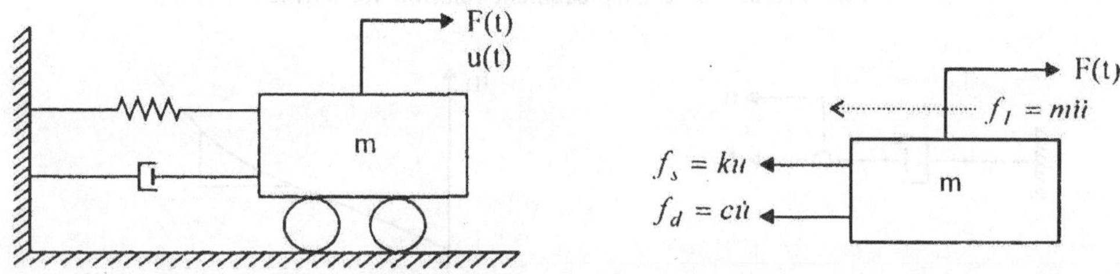

Fig. 17.4 Idealized SDOF System Free Body Diagram

The equation of motion is obtained directly by equating the sum of all the forces along x-direction *i.e.*

$$f_I + f_d + f_s = F(t) \tag{17.2}$$

If the spring is assumed to be linear with spring constant k and the damping is viscous with a damping coefficient c, equation (17.2) can be written as

$$m\ddot{u} + c\dot{u} + ku = F(t) \tag{17.3}$$

Above equation is linear as coefficient m, k and c are constant and power of u and its derivatives is no more than 1. Also note that it is a second order differential equation. Its solution will give u as a function of time.

17.5.3 Free Vibration Analysis :

The solution of second order differential equation as described in (17.3) not only depends on the nature of F(t) but also on original conditions from which the motion starts. In fact even when F(t) is zero, the system may undergo vibration, if it is given an initial displacement or velocity. Thus, the response of a system that is not subjected to external force but is excited by some initial disturbances is called free vibration. It is important to study such a response, not only because practical situations may arise when vibrations are excited by initial disturbances, but also because, when the response to an applied force is desired, the complete solution must include a free vibration component. This type of vibration may be in presence or absence of damping.

17.5.3*a* Undamped Free Vibration :

Although the systems without damping resistance do not exist in practices, the analysis of such response provides an insight into the nature of damped vibration. When damping is absent, the equation of motion becomes

$$m\ddot{u} + ku = 0 \tag{17.4}$$

This is a homogeneous second order differential equation. A possible solution of the above equation can be taken in the form of

$$u(t) = Ae^{\lambda t} \tag{17.5}$$

Substituting above in equation (17.4) yields

$$mA\lambda^2 e^{\lambda t} + kAe^{\lambda t} = 0 \qquad i.e. \qquad Ae^{\lambda t}\ (\lambda^2 m + k) = 0$$

Now, if the motion has to take place, 'A' must be non zero then $Ae^{\lambda t}$ can not be equal to zero, thus we obtain the characteristic equation as

$$\lambda^2 m + k = 0$$

$$i.e. \qquad \lambda = \pm\, i\omega, \text{ where } \omega^2 = \frac{k}{m}$$

The general solution then becomes,

$$u(t) = A_1 e^{i\omega t} + A_2 e^{-i\omega t} = A_1\ (\cos\omega t + i\sin\omega t) + A_2\ (\cos\omega t - i\sin\omega t) \tag{17.6}$$

$$= (A_1 + A_2)\cos\omega t + (A_1 - A_2)\, i\sin\omega t$$

$$i.e. \qquad u(t) = C\cos\omega t + D\sin\omega t$$

Where C and D are arbitrary constants. To determine these arbitrary constants, we use specific initial conditions of displacement and velocity at time $t = 0$ *i.e.* $u = u_0$ and $\dot{u} = \dot{u}_0$. Substitution of the initial conditions in (17.6) gives $C = u_0$ and $D = \dot{u}_0 / \omega$. Then the solution becomes as

$$u(t) = u_0\cos\omega t + \frac{\dot{u}_0}{\omega}\sin\omega t) \tag{17.7}$$

If we put $u_0 = \rho\sin\phi$ and $\dot{u}_0/\omega = \rho\cos\phi$, then equation (17.7) becomes

$$u(t) = \rho\sin\phi\cos\omega t + \rho\cos\phi\sin\omega t \;\; i.e.\;\; u(t) = \rho\sin(\omega t + \phi) \tag{17.8}$$

$$\text{where, } \rho = \sqrt{u_0^2 + \left(\frac{\dot{u}_0}{\omega}\right)^2}, \quad \tan\phi = u_0\omega / \dot{u}_0$$

Equation (17.7) or (17.8) is plotted in figure 17.5. It shows an oscillatory motion, which repeats itself.

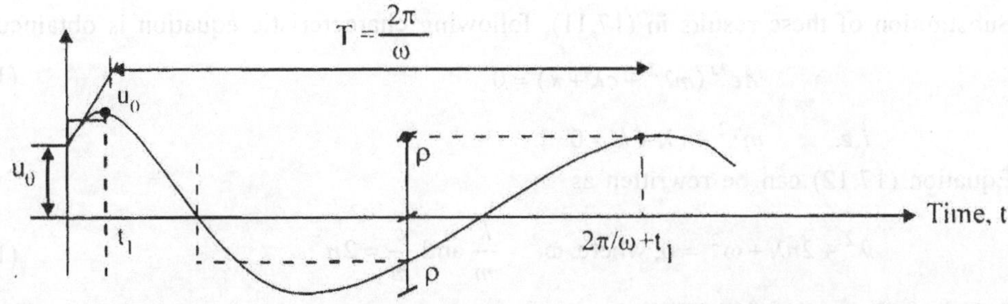

Fig. 17.5. Undamped Free vibration of SDOF System.

It can be easily verified that the response on two instance $2\pi/\omega$ apart, namely at t_1 and $\frac{2\pi}{\omega}+t_1$,

$$u(t_1) = \rho \sin(\omega t_1 + \phi)$$

$$u\left(\frac{2\pi}{\omega}+t_1\right) = \rho \sin\left[\left(\frac{2\pi}{\omega}+t_1\right)\omega+\phi\right] = \rho \sin(\omega t_1 + \phi) = u(t_1) \tag{17.9}$$

It can be also shown that the velocity is also same at two instance of time. This implies that the motion repeat itself after $T=\frac{2\pi}{\omega}$ seconds. This interval of time is called the period of motion. The system executes f cycles of motion in 1 seconds, where f is called the natural frequency given by,

$$f = \frac{1}{T} = \frac{\omega}{2\pi} \text{hertz(Hz)} \tag{17.10}$$

The maximum displacement of the system is equal to ρ and is called the amplitude of vibration and ϕ is the phase angle. The motion as shown in figure (17.5) represents a harmonic motion of the simplest type.

17.5.3*b* Damped Free Vibration

In previous discussion, we did not consider the effects of dissipative influences. Consequently, we found that amplitude of free vibration remains constant with time; but experiences show that the amplitude diminishes with time and the vibrations gradually damped out. Thus to bring our analytical discussion of vibration into better agreement with reality damping forces must be included.

Among all those sources of energy dissipation, the cases where the damping force is proportional to velocity is called viscous damping, the simplest one to deal mathematically. The equation governing the free vibration of a system with viscous damping is obtained, by setting $F(t) = 0$ in equation (17.3)

$$m\ddot{u} + c\dot{u} + ku = 0 \tag{17.11}$$

The solution of above equation can also be taken in the form

$$u(t) = Ae^{\lambda t}, \text{ then } \dot{u} = A\lambda e^{\lambda t}, \ddot{u} = A\lambda^2 e^{\lambda t}$$

Substitution of these results in (17.11), following characteristic equation is obtained

$$Ae^{\lambda t}(m\lambda^2 + c\lambda + k) = 0 \tag{17.12}$$

$$\text{i.e.} \quad m\lambda^2 + c\lambda + k = 0$$

Equation (17.12) can be rewritten as

$$\lambda^2 + 2n\lambda + \omega^2 = 0, \text{ where, } \omega^2 = \frac{k}{m} \text{ and } \frac{c}{m} = 2n \tag{17.13}$$

Solution of equation (17.13) is

$$\lambda_1 = -n + \sqrt{n^2 - \omega^2},\ \lambda_2 = -n - \sqrt{n^2 - \omega^2} \tag{17.14}$$

Now depending on the numerical value of n and ω, three distinct cases may arise which are discussed separately in the following sections.

Case I Critically damped System : n = w

A system, in which the magnitude of damping is such that the quantity in the root is zero, called a critically damped system. Hence c is denoted by c_r given by [as defined in eqn. (17.13)],

$$c = c_r = 2mn = 2m\omega = 2\sqrt{km} \tag{17.15}$$

The roots are equal in this situation *i.e.* $\lambda_1 = \lambda_2 = -n = -w$. Now for repeated roots, the general solution becomes

$$u = (A_1 + A_2 t)\, e^{-\omega t} \tag{17.16}$$

Where A_1 and A_2 are two constants to be determined from initial conditions. Substituting the initial conditions *i.e.* $u = u_0$ and $\dot{u} = \dot{u}_0$ at $t = 0$ in eqn. (17.16) A_1 and A_2 are obtained as, $A_1 = u_0$, $A_2 = \dot{u}_0 + \omega u_0$. The final solution then becomes,

$$u = [u_0 + (\dot{u}_0 + u_0\omega)t]e^{-\omega t} \tag{17.17}$$

Equation (17.17) is plotted in Fig. 17.6*a*. Note that the motion is not oscillatory. The system displacement decays exponentially with time and become very nearly zero after a while.

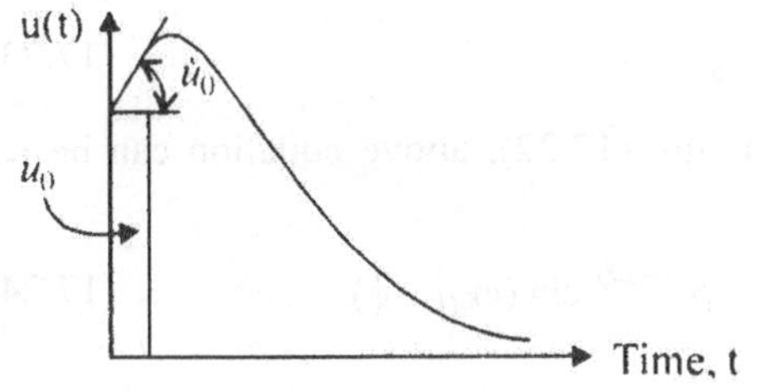

Fig. 17.6*a*. Free Vibration of Critically damped System.

Fig. 17.6*b*. Free Vibration of Overdamped System.

Case II Overdamped System : n > w

In this case the damping is greater than c_r, the system is said to be overdamped. For convenience we define damping ratio $\xi = \frac{n}{w} = \frac{c}{c_r}$. For overdamped system $\xi > 1.0$. Here the roots are negative and given by

$$\lambda_1 = -\omega\xi + \omega\sqrt{\xi^2 - 1}, \qquad \lambda_2 = -\omega\xi - \omega\sqrt{\xi^2 - 1} \tag{17.18}$$

If we denote $\omega\sqrt{\xi^2 - 1}$ as ωd, then the solutions become

$$u(t) = A_1 e^{\lambda_1 t} + A_2 e^{\lambda_2 t} = e^{-\omega\xi t}(A_1 e^{\omega d t} + A_2 e^{-\omega_d t})$$

$$= e^{-\omega\xi t}[A_1\{\cos h\omega_d t + \sin h\omega_d t\} + A_2\{\cos h\omega_d t - \sin h\omega_d t\}] \tag{17.19}$$

$$= e^{-\omega\xi t}[(A_1 + A_2)\cos h\omega_d t + (A_1 - A_2)\sin h\omega_d t]$$

Thus, equation (17.19) can be expressed as

$$u(t) = e^{-\omega_d \xi t}(A\cos\omega_d t + B\sin\omega_d t) \tag{17.20}$$

Where A and B are two arbitrary constants to be determine from initial conditions. Eqn. (17.20) has been plotted in Fig. 17.6*b*. In this case also, the motion is nonoscillatory. The displacement decays exponentially with time, although in comparison to a critically damped system, it takes longer time to return the system to zero displacement.

Case III Underdamped System : n < w

In all structural systems and majority of mechanical systems the damping is less than the critically damped system, known to be under damped system. The damping ratio ξ is again defined as c/c_c, and its value is less then 1.0. The solutions are now imaginary and given by

$$\lambda_1 = -\omega\xi + i\omega\sqrt{1-\xi^2},\ \lambda_2 = -\omega\xi - i\omega\sqrt{1-\xi^2} \tag{17.21}$$

Defining $\omega\sqrt{1-\xi^2}$ as ω_d the general solution can be written as,

$$u(t) = A_1 e^{(-\omega\xi + i\omega_d)t} + A_2 e^{(-\omega\xi - i\omega_d)t} = e^{-\omega\xi t}[A_1 e^{i\omega_d t} + A_2 e^{-i\omega d t}]$$

$$= e^{-\omega\xi t}\,[A_1\{\cos\omega_d t + i\sin\omega_d t\} + A_2\{\cos\omega_d t - i\sin\omega_d t\}] \tag{17.22}$$

$$i.e.\ u(t) = e^{-\omega\xi t}\,[A\cos\omega_d t + B\sin\omega_d t]$$

Where A and B are two arbitrary constants obtained by applying the initial conditions as earlier and given by $A = u_0,\ B = \dfrac{\dot{u}_0 + \omega\xi u_0}{\omega_d}$

Thus the final solution is as follows

$$u(t) = e^{-\omega\xi t}\,[u_0\cos\omega_d t + \frac{\dot{u}_0 + \omega\xi u_0}{\omega_d}\sin\omega_d t] \tag{17.23}$$

Again substituting $A = \rho\cos\phi$ and $B = \rho\sin\phi$ in eqn. (17.22), above equation can be re-written as

$$u(t) = e^{-\omega\xi t}\,[\rho\cos\omega_d t\cos\phi + \rho\sin\omega_d t\sin\phi] = \rho e^{-\omega\xi t}\sin(\omega_d t + \phi) \tag{17.24}$$

where,

$$\rho = \sqrt{A^2 + B^2} = \left[u_0^2 + \left(\frac{\dot{u}_0 + \xi\omega u_0}{\omega_d}\right)^2\right]^{\frac{1}{2}},\ \tan\phi = \frac{B}{A} = \frac{u_0\omega_d}{\dot{u}_0 + u_0\omega\xi} \tag{17.25}$$

Equation (17.24) has been plotted in Figure 17.6*c*. The motion is oscillatory or cyclic and repeat itself after a period $T_d = 2\pi/\omega_d$ referred to as damped period. However, the amplitude of displacement at the second instant of time is lower than at the first. This decay is caused due to the exponential terms in eqn. (17.24).

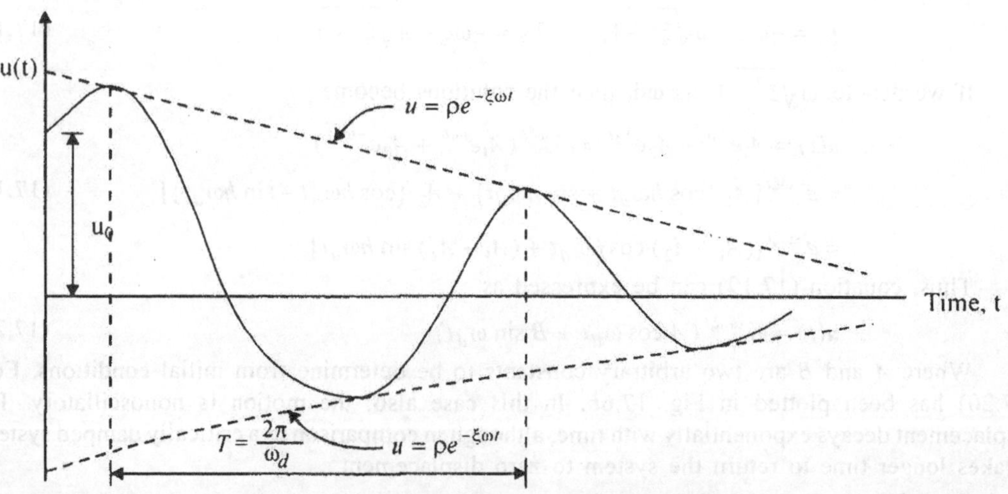

Fig. 17.6*c*. Free Vibration of Damped System.

Logarithmic decrement :

In the free vibration of an under-damped system, displacement amplitude decays exponentially with time. The rate of decrease depends on the damping ratio ξ. If we denote the displacement at any time t_1 by $u_1 = u(t_1)$, then

$$u(t_1) = \rho e^{-\xi\omega t_1} \sin(\omega_d t_1 + \phi)$$

The displacement at time $t_1 + 2\pi/\omega_d$ is given by

$$u(t_1 + \frac{2\pi}{\omega_d}) = \rho e^{-\xi\omega(t_1 + \frac{2\pi}{\omega_d})} \sin(\omega_d t_1 + \phi)$$

The ratio of $u(t_1)$ to $u(t_1 + 2\pi/\omega_d)$ provides a measure of the decrease in displacement over an cycle of motion. This ratio is constant and does not vary with time; its natural log is called logarithmic decrement and is denoted of δ. The value of δ is given by

$$\delta = \ln\left(\frac{e^{-\xi\omega t_1}}{e^{-\xi\omega\left(t_1 + \frac{2\pi}{\omega_d}\right)}}\right) = 2\pi\xi\frac{\omega}{\omega_d} = \frac{2\pi\xi}{\sqrt{1-\xi^2}} \tag{17.26}$$

For small value of ξ, $\delta = 2\pi\xi$.

17.5.4 Example Problems

Example 17.1. Determine the equivalent stiffness of a system comprises of two linear spring (constants K_1 and K_2) connected in (*a*) Parallel, (*b*) in series.

(*a*) Springs in Parallel : In this case both the springs undergo same displacement u and

$F_1 = K_1 u,\qquad F_2 = K_2 u\qquad$ thus, $\qquad F = F_1 + F_2 = (k_1 + k_2)u$

Thus, equivalent spring constant is given by

$$K_{eq} = \frac{F}{u} = K_1 + K_2$$

(*b*) Springs in Series : In this case, the forces in each spring should be the same and equal to the externally applied force F. The extension of the first spring $\Delta_1 = F/k_1$ and that of the second $\Delta_2 = F/k_2$, thus

$$\Delta = \Delta_1 + \Delta_2 = F\left(\frac{1}{k_1} + \frac{1}{k_2}\right) \text{ and, } \quad K_{eq} = \frac{F}{\Delta} = \frac{1}{\frac{1}{k_1} + \frac{1}{k_2}}$$

Example 17.2. Find the natural frequency of the simply supported beam system as shown in Figure 17.7 (*a*).

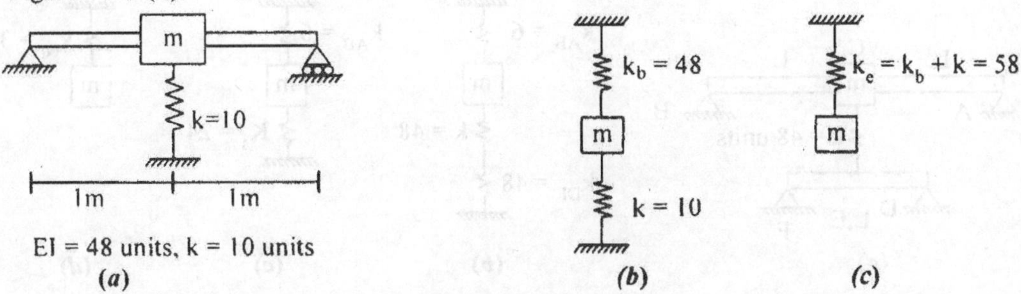

Fig. 17.7. Example 17.2

The system shown in Figure 17.7 (*a*) can be modeled as spring mass system as shown in Figure 17.7 (*b*). The simply supported beam can be considered as a linear spring having stiffness,

$$k_b = \frac{48EI}{L^3} = \frac{48 \times 8}{2^3} = 48 \text{ units}$$

In Figure 17.7 (*b*), the mass *m* connects the two spring in parallel. So equivalent stiffness,

$k_e = k_b + k = 48 + 10 = 58$. Thus the natural circular frequency, $\omega_n = \sqrt{\frac{k_e}{m}} = \frac{7.62}{\sqrt{m}}$.

Example 17.3. Find the natural frequency of the cantilever beam with attached mass system as shown in Figure. 17.8 (*a*) L = 1*m*, EI = unity, k = 3 units.

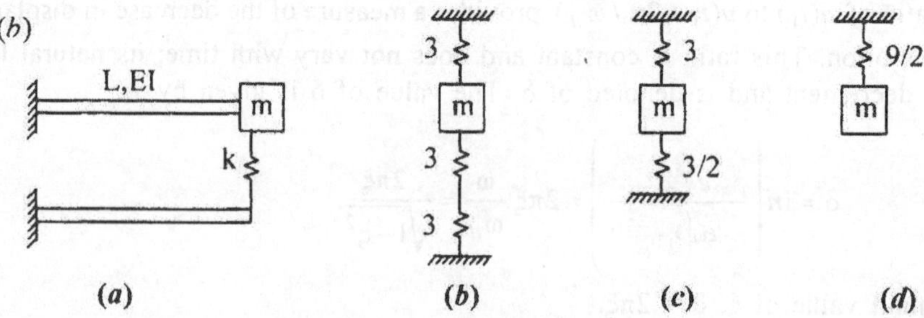

Fig. 17.8 Example 17.3

Sol. The spring stiffness for cantilever beam = $\frac{3EI}{l^3} = \frac{3 \times 1}{1^3} = 3$ units.

In Figure 17.8 (*b*), the bottom two springs are in series, thus $\frac{1}{k_s} = \frac{1}{k_1} + \frac{1}{k_h} = \frac{1}{3} + \frac{1}{3} = \frac{2}{3}$ *i.e.*

$$k_s = \frac{3}{2}$$

In Figure 17.8 (*c*) the mass connects the springs in parallel, thus equivalent spring factor for the system $k_e = k_s + k_2 = \frac{3}{2} + 3 = \frac{9}{2}$

Finally, the natural frequency of the equivalent sdof system (Figure 17.8 (*d*)) is

$$\omega_n = \sqrt{\frac{k_e}{m}} = \frac{2.12}{\sqrt{m}}$$

Example 17.4 Find natural frequency of the two simply supported beams with attached mass system as shown in Figure 17.9 L = 1*m*, EI = unity, K = 48 units.

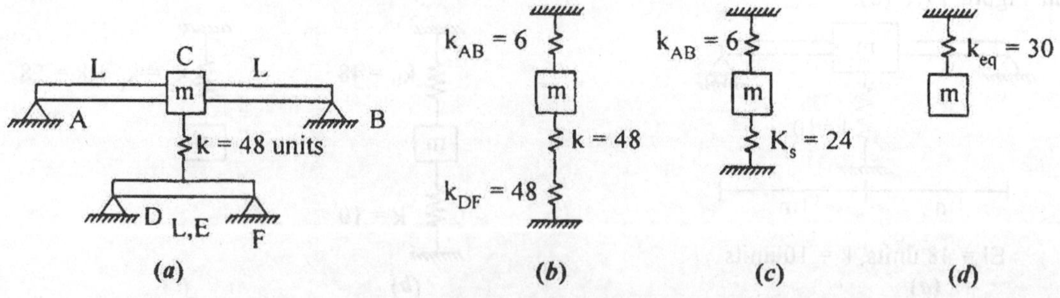

Fig. 17.9 (Example 17.4)

Stiffness of beam $AB = \frac{48 \times 1}{2^3} = 6\ units$ and Stiffness of beam $DF = \frac{48 \times 1}{1^3} = 48\ units$

In Figure 17.9 (*b*) the bottom two springs are in series, thus

$$\frac{1}{k_s} = \frac{1}{k_1} + \frac{1}{k_2} = \frac{1}{48} + \frac{1}{48} = \frac{1}{24} \quad i.e.\ k_s = 24$$

In the next figure 17.9 (*c*) the mass connects the springs in parallel, thus equivalent spring factor for the system $k_e = k_s + k_{AB} = 6 + 24 = 30$

Thus, the circular natural frequency of the system is $\omega_n = \sqrt{\frac{k_e}{m}} = \frac{5.477}{\sqrt{m}}$

Example 17.5. A single degree of freedom system (m = 9100 kg) with viscous damping is displaced from its position of rest by a distance of 30mm. If the maximum displacement on return swing is 20mm on 0.5 sec, determine (*i*) the spring constant k, (*ii*) the damping constant c.

Sol. Here two subsequent displacements are 30mm and 20mm on a return swing of 0.5 sec, thus $T = 0.5s$. Now using logarithmic decrement relations,

$$\delta = \ln\frac{u_1}{u_2} = \frac{2\pi\xi}{\sqrt{1-\xi^2}} \quad i.e.\ \ln\frac{30}{20} = \frac{2\pi\xi}{\sqrt{1-\xi^2}} \quad \text{on solving, } \xi = 0.0644$$

Now it is known that $T_d = 0.5s$, so

$$0.5 = \frac{2\pi}{\omega\sqrt{1-\xi^2}}$$

gives $\omega = 12.593$ rad / sec. So, $k = \omega^2 m = 1.443 \times 10^7\ N/m$

(*ii*) Critical damping constant

$$C_{cr} = 2m\omega = 2 \times 9100 \times 12.93 = 2.29 \times 10^6\ Ns/m$$

Finally, the damping constant $c = \xi C_{cr} = 147600\ Ns/m$

Example 17.6. The beam and attached mass system as shown in figure 17.10 modeled as single degree of freedom system with viscous damping. It is found that the logarithmic ratio of two consecutive displaced amplitude is 0.163. Determine the damped natural frequency of the system.

Sol.

Fig. 17.10 (Example 17.6)

$$\text{Stiffness of the beam} = \frac{48EI}{L^3} = \frac{48 \times 2 \times 10^6 \left(\frac{5x5^3}{12}\right)}{500^3} = 40kg/cm$$

The equivalent stifness of the system $= \frac{1}{k_q} = \frac{1}{k_b} + \frac{1}{k} = \frac{1}{40} + \frac{1}{50} \therefore k_{eq} = \frac{200}{9}$ ks/kms

Undamped natural frequency $\omega = \sqrt{\frac{k_e}{m}} = 0.3849$ sec

Now,

$$\delta = ln\frac{u_1}{u_2} = \frac{2\pi\xi}{\sqrt{1-\xi^2}} \text{ so } 0.163 = \frac{2\pi\xi}{\sqrt{1-\xi^2}} \text{ on solving, } \xi = 0.026$$

Hence, damped natural frequency $\omega_d = \omega\sqrt{1-\xi^2} = 0.385\sqrt{1-0.026^2} = 0.3848$ sec.

Example 17.7. A particle of mass 2g is making simple harmonic motion along x-axis. At distances 6cm and 10cm from the equilibrium position, the velocities of the particles are 5cm/s and 4cm/s respectively. Find the time period of vibration, the amplitude and maximum kinetic energy.

Son. The displacement and velocity can be written as,

$$u = \rho\sin(\omega t + \phi) \text{ and } v = \frac{du}{dt} = \rho\omega\cos(\omega t + \phi) = \omega\sqrt{\rho^2 - u^2}$$

then, $\quad v_1 = \omega\sqrt{\rho^2 - u_1^2}$ and $v_2 = \omega\sqrt{\rho^2 - u_2^2}$

thus, $\quad v_1^2 - v_2^2 = \omega^2(u_1^2 - u_2^2)$

then, $\quad \omega = \sqrt{\frac{v_1^2 - v_2^2}{u_1^2 - u_2^2}} = \sqrt{\frac{5^2 - 4^2}{10^2 - 6^2}} = \frac{3}{8}$ Thus period of vibration, $T = \frac{2\pi}{\omega} = 16.76$ sec.

and amplitude

$$\rho = \frac{v_1^2}{\omega^2} + u_1^2 = 14.62cm$$

$$\text{K.E.} = \frac{1}{2}mv_{\max}^2 = \frac{1}{2}m\omega^2\rho^2 = \frac{1}{2}2x10^{-3}\left(\frac{3}{8}\right)^2 x213.78x10^{-4} = 2.99 \times 10^{-6} J.$$

Example 17.8. A mass of 45 kg is dropped into the end of a cantilever beam with a velocity of 0.5 m/s. The mass sticks to the beam and vibrate with the beam. The I-beam made of steel (E = 210x10^9 N/m^2) is 3.5 m length, has a depth of 20 cm. I = 3 × 10^{-5} m^4. Find maximum stress developed in the beam as it vibrates, neglect damping.

Sol. The system is modeled as a mass of 45 kg hanging from a spring of stiffness

$$K_{eq} = \frac{3EI}{L^3} = \frac{3(210\times10^9 \times 3\times10^{-5}}{3.5^3} = 4.4\times10^5 N/m$$

Natural frequency of the vibrating system, $\omega_n = \sqrt{\frac{k}{m}} = \sqrt{\frac{4.4\times10^5}{45}} = 99.0$ rad

The beam is undeflected at $t = 0$, when the mass strike. If the systems were in static equilibrium, the tip of the beam would have a static deflection, thus u at $t = 0$

$$u_0 = -\delta_{st} = -\frac{mg}{k} = -\frac{g}{w_n^2} = -\frac{9.81}{99^2} = -1.0\times10^{-3} m$$

The initial velocity is the velocity at which the mass strikes the beam $i.e.$ $\dot{u}(0.) = 0.5m/s$. Then the response

$$u(t) = \rho \sin(99t + \phi),\ \phi = \tan^{-1}\left[\frac{99 \times (-1) \times 10^{-3}}{0.5}\right] = -11°,\ \rho = \sqrt{(-1x10^{-3})^2 + \left(\frac{0.5}{99}\right)^2} = 5.51x10^{-3}\, m$$

Maximum acceleration, $\ddot{u}_{max} = \omega_n^2\, \rho = 50.5 m/\sec^2$

Maximum moment developed at fixed end

$$= mgxL + m\ddot{u}_{max} L = 45[9.81 + 50.5] = 9.5 \times 10^3\, N-m \text{ thus, } \sigma_x = \frac{9.5 \times 10^3}{3x10^{-5}} = 3.2 \times 10^7\, N/m^2.$$

Example 17.9 An electric instrument having $m = 1$ kg, $k = 2400$ N/m, $c = 2$ N-S/m is displaced 20 mm from equilibrium and released. Find the vibration amplitude after 5 oscillations.

The natural frequency of the instrument, $\omega = \sqrt{\frac{k}{m}} = 49$ rad/s

The and damping ratio of the system, $\zeta = \frac{c}{2\sqrt{km}} = 0.0204$

Then, the damped frequency, $\omega_d = \omega\sqrt{1-\zeta^2} = 49\sqrt{1-.02^2} = 49$ rad/s.

The free vibration is defined by, $u(t) = u_0 e^{-\zeta\omega_n t}(\cos\omega_d t + \frac{\zeta}{\sqrt{1-\zeta^2}}\sin\omega_d t)$

Initial displacement, $u_0 = 20$ mm $= 0.02$ m, substituting in above.

$$u(t) \cong 0.02e^{-0.0249t}\cos 49t$$

$T = \frac{2\pi}{\omega_d} = 0.13$ sec. So, after 5 oscillation, $t = 5 \times T = 0.65$ sec. $\therefore$ $u(t = 0.65) = 0.0106$ m

Example 17.10 A *SDOF* viscously damped system has spring constant $k = 6000$ N/m, critical damping of 0.3 N-S/mm and a damping ratio of 0.3. If the system is given an initial velocity of 1 m/sec, determine the maximum displacment of the system.

The natural frequency, $\omega = \sqrt{\frac{k}{m}} = \sqrt{\frac{6000}{m}}$,

Critical Damping $c_c = 0.3\ Ns/mm = 300\ Ns/m = 2\sqrt{km}$ $\therefore$ $300 = 2\sqrt{6000 \times m}$ *i.e.* $m = 3.75$ kg

thus, $\omega = \sqrt{\frac{6000}{3.75}} = 40$ rad/sec

Damping ratio, $\zeta = \frac{c}{c_c} = 0.3$ $\therefore$ $c = \zeta \times c_c = 0.3 \times 0.3 = 90$ N-S/m.

$u_0 = 0$, $\dot{u}_0 = 1$m/s The free vibration equation gives,

$$u(t) = e^{-\xi\omega t}\frac{1.0}{\omega\sqrt{1-\xi^2}}\sin\sqrt{1-\xi^2} \times \omega t = e^{-0.3\times40\times t}\frac{1.0}{40\sqrt{1-0.3^2}}\sin\sqrt{1-0.3^2} \times 40t$$

For $u(t)$ to be maximum, $\sin\sqrt{1-\xi^2} \times \omega t$ which gives, $\omega t = \frac{\pi}{2}$

Thus, maximum amplitude $u_{max} = e^{-0.3 \times \frac{\pi}{2}} \dfrac{1.0}{40\sqrt{1-0.3^2}} = 0.01636\text{m}.$

Example 17.11. A *SDOF* viscously damped system makes five complete oscillations per second. The amplitude of vibrations reduces to 15% in 60 cycles. Find the damping ratio ξ.

Damped time period, $T_d = \dfrac{1}{5.0} = 0.2 \sec = \dfrac{2\pi}{\omega_d}$

∴ Damping Frequency, $\omega_d = 31.46\ rad/s$

If u_n is the amplitude of vibration in n cycles after amplitude of u_0, then

$$\frac{u_0}{u_n} = \frac{u_0}{u_1}\frac{u_1}{u_2}\cdots\cdots\frac{u_{n-1}}{u_n} = \left(\frac{u_i}{u_{i+1}}\right)^n$$

Thus, $\ln\dfrac{u_0}{u_n} = n\ln\dfrac{u_i}{u_{i+1}} = n.\delta.$ Then, $\delta = \dfrac{1}{n}\ln\dfrac{u_0}{u_n} = \dfrac{1}{60}\ln 15 = 0.0451$

$$\zeta = \frac{\delta}{\sqrt{(2\pi)^2 + \delta^2}} = \frac{0.0451}{\sqrt{(2\pi)^2 + (0.0451)^2}} = 0.007177.$$

17.6 FORCED VIBRATION

17.6.1 General

The equation of motion of a single degree of freedom system as described by Eqn. (17.3), represents a second order linear differential equation provided that the equation does not vary with time. Its solution can be expressed as the sum of two parts: a complementary function and a particular integral. The complementary solution is obtained with $F(t) = 0$ resulting in a homogeneous equation; its solution is already discussed in 17.5.3. The particular solution depends on the exciting force. Complementary solutions are transient in nature because they decay with time as a result of damping force. On the other hand, a particular solution persists as long as the exciting force does and represents the steady state part of the solution. When the exciting force is a simple mathematical function, the particular solution can be obtained by process of trial.

17.6.2 Undamped Harmonic Vibration

Neglecting damping and assuming the excitation is given by $F(t) = F_0 \sin \overline{\omega}t$, the equation of motion becomes

$$m\ddot{u} + ku = F_0 \sin\overline{\omega}t \tag{17.27}$$

The complementary solution of the equation is same as free vibration problem and for a particular solution, we try

$$u_p(t) = U_0 \sin\overline{\omega}t \tag{17.28}$$

where U_0 is a constant so chosen to satisfy equation (17.27). Substituting equation (17.28) in (17.27) gives

$$U_0(k - m\overline{\omega}^2)\sin\overline{\omega}t = F_0 \sin\overline{\omega}t,\ U_0 = \frac{F_0}{k - \overline{\omega}^2 m} = \frac{F_0}{k(1-\beta^2)} \tag{17.29}$$

where β is the frequency ratio defined as $\beta = \dfrac{\overline{\omega}}{\omega}$. The complete solution is now obtained by adding the complementary and particular solutions *i.e.*

$$u(t) = u_c(t) + u_p(t) \tag{17.30}$$

If the initial displacement is u_0 and velocity is v_0, the solution becomes

$$u(t) = u_0 \cos\omega t + \frac{v_0}{\omega}\sin\omega t + \frac{F_0}{k(1-\beta^2)}, (\sin\overline{\omega}t) \tag{17.31}$$

The static deflection under a load F_0 equal to the amplitude of harmonic excitation is obtained from

$$\delta_{st} = \frac{F_0}{k} \tag{17.32}$$

We now define the ratio of dynamic deflection to the static deflection as the dynamic load factor *D*. Thus, for steady state response of an undamped system subjected to harmonic excitation, the dynamic load factor is

$$D(t) = \frac{u(t)}{\delta_{st}} = \frac{1}{1-\beta^2}\sin\overline{\omega}t \tag{17.33}$$

The amplitude of dynamic load factor $A_D = \dfrac{1}{1-\beta^2}$ has been plotted in Figure 17.11 as a function of the frequency ratio, β. For small value of β, $A_D \approx 1.0$ and the maximum displacement is nearly equal to that produced under a static load of the same amplitude. The applied load in this case varies so gradually that the response is nearly equal to that produced in static case. When the frequency of the applied load is very high *i.e.* when β is very large, the displacement approaches zero. In this case the load is varying so rapidly that the system has no time to respond and remains at rest.

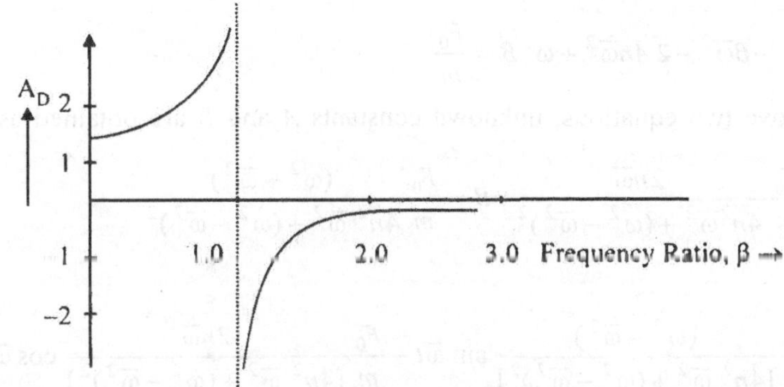

Fig. 17.11 Variation of Amplitude of dynamic load factor with frequency ratio.

For β = 1 *i.e.* the frequency of forcing function and natural frequency of the structure is same, the response is infinitely large. The state is known as resonance. It is clear that resulting situation may be catastrophic and should be avoided in practical design.

Also note that for β < 1, A_D is positive and the system is said to be vibrating in phase with the applied force. For β > 1, A_D is negative and the system is vibrating out of phase with the excitation.

17.6.3 Damped Harmonic Motion

When viscous damping is included, the equation of motion of a system subjected to $F(t) = F_0 \sin\overline{\omega}t$

$$m\ddot{u} + c\dot{u} + ku = F(t) = F_0 \sin \overline{\omega} t \tag{17.34}$$

which can be rewritten as

$$\ddot{u} + 2n\dot{u} + \omega^2 u = \frac{F_0}{m} \sin \overline{\omega} t \tag{17.35}$$

where, $\frac{c}{m} = 2n$ and $\omega^2 = k/m$. Now the total solution has two parts, the complementary function and the particular integral. The complementary solution can be readily obtained as in eqn. (17.23) i.e.

$$u_c = e^{-\xi\omega t}[u_0 \cos\omega_d t + \frac{\dot{u}_0 + \xi\omega u_0}{\omega_d} \sin \omega_d t] \tag{17.36}$$

$\omega_d = \sqrt{\omega^2 - n^2}$. And for the particular solution, we try the following form

$$u_p = A \cos \overline{\omega} t + B \sin \overline{\omega} t \tag{17.37}$$

where, A and B are as yet unknown constants. Substitutions of u and its derivative in equation (17.35) yields

$$-A\overline{\omega}^2 \cos\overline{\omega} t - B\overline{\omega}^2 \sin \overline{\omega} t + 2n(-A\overline{\omega} \sin \overline{\omega} t + B\overline{\omega} \cos \overline{\omega} t)$$

$$+ \omega^2 (A \cos \overline{\omega} t + B \sin \overline{\omega} t) = \frac{F_0}{m} \sin \overline{\omega} t \tag{17.38}$$

If equation (17.37) is to be a valid solution, eqn. (17.38) should hold for any values of t. This requires

$$-A\overline{\omega}^2 + 2n B\overline{\omega} + A\overline{\omega} + A\overline{\omega}^2 = 0$$

$$-B\overline{\omega}^2 - 2An\overline{\omega}^2 + \omega^2 B = \frac{F_0}{m} \tag{17.39}$$

Solving above two equations, unknown constants A and B are obtained as

$$A = -\frac{F_0}{m} \frac{2n\overline{\omega}}{4n^2\overline{\omega}^2 + (\omega^2 - \overline{\omega}^2)^2}, \; B = \frac{F_0}{m} \frac{(\omega^2 - \overline{\omega}^2)}{4n^2\overline{\omega}^2 + (\omega^2 - \overline{\omega}^2)^2} \tag{17.40}$$

Thus,

$$u_p = \frac{F_0}{m} \frac{(\omega^2 - \overline{\omega}^2)}{[4n^2\overline{\omega}^2 + (\omega^2 - \overline{\omega}^2)^2]} \sin \overline{\omega} t - \frac{F_0}{m} \frac{2n\overline{\omega}}{[4n^2\overline{\omega}^2 + (\omega^2 - \overline{\omega}^2)^2]} \cos \overline{\omega} t \tag{17.41}$$

Let, $(\omega^2 - \overline{\omega}^2)\lambda = \cos\phi$ and $2n\overline{\omega}\lambda = \sin\phi$

Then $\lambda = \frac{1}{\sqrt{(\omega^2 - \overline{\omega}^2)^2 + (2n\overline{\omega}^2)^2}}$, $\tan\phi = \frac{2n\overline{\omega}}{\omega^2 - \overline{\omega}^2}$

Then, particular integral can be written as

$$u_p = \frac{F_0}{m} \frac{1}{\sqrt{(\omega^2 - \overline{\omega}^2)^2 + (2n\overline{\omega})^2}} \sin(\overline{\omega} t - \phi) \tag{17.42}$$

Finally the complete solution becomes,

$$u(t) = u_c + u_p = e^{-\xi\omega t}[u_0 \cos \omega_d t + \frac{\dot{u}_0 + nu_0}{\omega_d} \sin \omega_d t]$$

$$+ \frac{F_0}{m} \frac{1}{\sqrt{(\omega^2 - \overline{\omega}^2)^2 + (2n\overline{\omega})^2}} \sin(\overline{\omega}t - \phi) \tag{17.43}$$

Above equation represents the response of SDOF system subjected to harmonic loading. It can be noted that the solution consists of two parts: The first part is called the transient response, presence of the exponential factor e^{-nt} will cause the component to die out. The second part is steady state response and continue as long as the harmonic force presents in the system.

The steady state response becomes

$$u = \frac{F_0}{m} \frac{1}{\sqrt{(\omega^2 - \overline{\omega})^2 + (2n\overline{\omega})^2}} \sin(\overline{\omega}t - \phi)$$

$$= \frac{F_0}{m\omega^2} \cdot \frac{1}{\sqrt{\left[1 - \left(\frac{\overline{\omega}}{\omega}\right)^2\right]^2 + \left(\frac{2n}{\omega}\right)^2 \left(\frac{\overline{\omega}^2}{\omega}\right)}} \sin(\overline{\omega}t - \phi) \tag{17.44}$$

$$= \frac{F_0}{k} \frac{1}{\sqrt{(1-\beta^2)^2 + (2\xi\beta)^2}} \sin(\overline{\omega}t - \phi) = \frac{\delta_{st}}{\sqrt{(1-\beta^2)^2 + (2\xi\beta)^2}} \sin(\overline{\omega}t - \phi)$$

Where, $\delta_{st} = \frac{F_0}{k}$ is the static displacement, $\beta = \frac{\overline{\omega}}{\omega}$ is the frequency ratio and $\xi = \frac{n}{n_c}$ is the critical damping ratio. As in the case of an undamped system, dynamic load factor $D(t)$ is defined as the ratio of dynamic deflection to static deflection under F_0. For steady state response, $D(t)$ is given by

$$D(t) = \frac{1}{\sqrt{(1-\beta^2)^2 + (2\xi\beta)^2}} \sin(\overline{\omega}t - \phi) = A_D \sin(\overline{\omega}t - \phi) \tag{17.45}$$

The amplitude of A_D is plotted in Figure 17.12 as a function of β for different value of ξ. The value of β for which A_D is maximum can be obtained by

$$\frac{d A_D}{d\beta} = 0, \text{ gives } \beta = \sqrt{1 - 2\xi^2}, \text{ thus, } A_{D_{max}} = \frac{1}{2\xi\sqrt{1-\xi^2}} \tag{17.46}$$

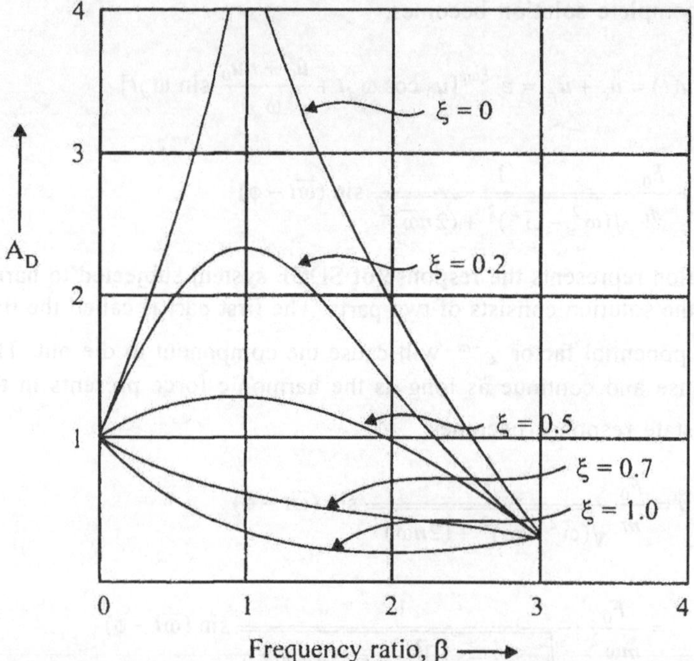

Fig. 17.12. Variation of dynamic load factor with frequency ratio and damping ratio

As in the case of an undamped system, the resonance is said to occur when the exciting frequency is equal to the natural frequency, that is when $\beta = 1$. Unlike the undamped case, however, maximum amplitude does not occur at resonance but at value of β given by eqn. (17.46). This value of β is slightly less than 1.

17.7 SUPPORT MOVEMENT

A structure attached to a moving support or a system mounted on such a support will have some of the support motion transmitted to it. For design of such system it is desirable that transmitted motion will be minimum. It is also extremely important in some case to estimate the support motion from measurement of the transmitted motion. In either case, a relationship must be established between the two motions.

Consider the simple sdof system shown in Fig. 17.13. The system consists of a mass m attached to a support through a dashpot and spring. The support movement is assumed as

$$u_g = u_0 \sin \overline{\omega} t \tag{17.47}$$

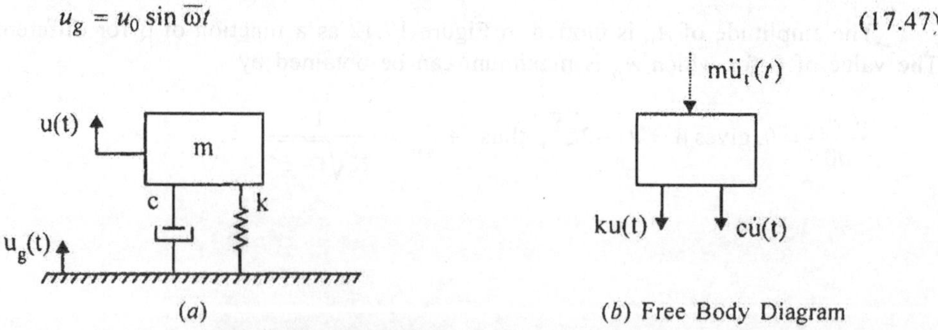

Fig. 17.13. Harmonic support motion

Let u is the relative displacement of the mass m with respect to support, while $u_t = u + u_g$. Considering the free body diagram, the equation of motion can be written as

$$m\ddot{u}_t + c\dot{u} + ku = 0 \text{ i.e. } m\ddot{u} + c\dot{u} + ku = -m\ddot{u}_g = mu_0\overline{\omega}^2 \sin \overline{\omega}t \tag{17.48}$$

On comparing this with equation (17.34), the steady state response of the mass is given simply with F_0 replaced by $mu_0\overline{\omega}^2$ in eqn. (17.44) i.e.

$$u = \frac{mu_0\overline{\omega}^2}{k} \frac{1}{\sqrt{(1-\beta^2)^2 + (2\xi\beta)^2}} \sin(\overline{\omega}t - \phi) = u_0 \frac{\beta^2}{\sqrt{(1-\beta^2)^2 + (2\xi\beta)^2}} \sin(\overline{\omega}t - \phi) \tag{17.49}$$

Now the total displacement

$$u_t = u_g + u = u_0 \sin \overline{\omega}t + u_0 \frac{\beta^2}{\sqrt{(1-\beta^2)^2 + (2\xi\beta)^2}} \sin(\overline{\omega}t - \phi) \tag{17.50}$$

$$= u_0 \sin \omega t + u_0 \frac{\beta^2}{\sqrt{(1-\beta^2)^2 + (2\xi\beta)^2}} [\sin \overline{\omega}t \cos \phi - \cos \overline{\omega}t \sin \phi])$$

Introducing $\rho \sin \phi = 2\xi\beta$ and $\rho \cos \phi = (1 - \beta^2)$ and substituting the value of $\sin \phi$ and $\cos \phi$

$$u_t = u_0 \sin \overline{\omega}t + u_0 \frac{\beta^2}{\sqrt{(1-\beta^2)^2 + (2\xi\beta)^2}} [\sin \overline{\omega}t \frac{1-\beta^2}{\sqrt{(1-\beta^2)^2 + (2\xi\beta)^2}}$$

$$- \cos \overline{\omega}t \frac{2\xi\beta}{\sqrt{(1-\beta^2)^2 + (2\xi\beta)^2}}] = \frac{u_0}{(1-\beta^2)^2 + (2\xi\beta)^2} [\sin \overline{\omega}t \{(1-\beta^2)^2$$

$$+ (2\xi\beta)^2 + (1-\beta^2)\beta^2\} - 2\xi\beta^3 \cos \overline{\omega}t]$$

$$= \frac{u_0}{(1-\beta^2)^2 + (2\xi\beta)^2} [\sin \overline{\omega}t \{(1-\beta^2) + 4\xi^2\beta^2\} - 2\xi\beta^3 \cos \overline{\omega}t]$$

Let $\quad \dfrac{u_0}{(1-\beta^2)^2 + (2\xi\beta)^2} \{(1-\beta^2) + 4\xi^2\beta^2\} = \chi \sin \gamma$

and $\quad \dfrac{u_0}{(1-\beta^2)^2 + (2\xi\beta)^2} 2\xi\beta^3 \cos \overline{\omega}t = \chi \cos \gamma,$

then $\quad u_t = \chi \sin(\overline{\omega}t - \gamma)$, where, $\chi = u_0 \sqrt{\dfrac{1 + (2\xi\beta)^2}{(1-\beta^2)^2 + (2\xi\beta)^2}}$,

$$\tan \gamma = \frac{2\xi\beta^3}{(1-\beta)^2 + 4\xi^2\beta^2} \tag{17.51}$$

17.8. VIBRATION ISOLATION AND TRANSMISSIBILITY :

In practical design it is required that the dynamic forces transmitted by a machine to its surrounding be minimized. Vibration isolation of this type is referred to as force isolation and is usually achieved by suitable insertion of spring and damping devices between the machine and its foundation. An inverse problem of isolation exists when it is required that vibrations from the surroundings are not transmitted to support systems.

Force Isolation :

Fig. 17.14 shows a machine of mass m mounted on a foundation through a set of springs and dampers, the total spring stiffness being k, and the damping coefficient c.

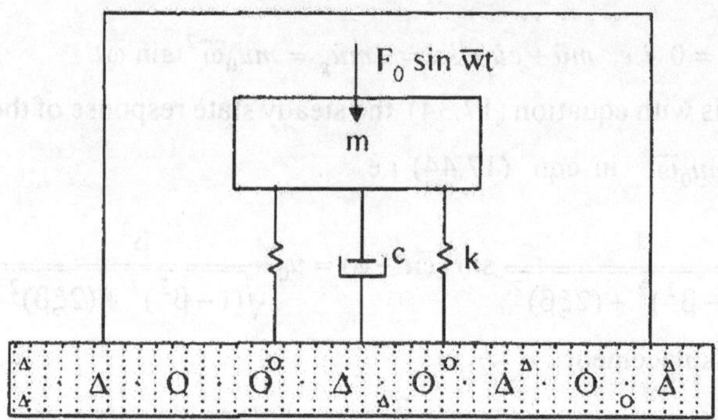

Fig. 17.14. Force transmission under harmonic excitation.

Let the dynamic force acting is $F_0 \sin \overline{w}t$. Now the steady state response is given by eq (17.44). Force is transmitted to the foundation through the spring and the damper. The total transmitted force F is given by

$$F = f_s + f_D = k.u + c\dot{u} = k\,\rho \sin(\overline{\omega}t - \phi) + c\overline{\omega}\rho \cos(\overline{\omega}t - \phi) = \chi \sin(\overline{\omega}t - \phi + \eta)$$

$$\text{Where, } \chi = \sqrt{(k\rho)^2 + (c\overline{\omega}p)^2},\ \rho = \frac{F_0}{k}\frac{1}{\sqrt{(1-\beta^2)^2 + (2\xi\beta)^2}} \text{ and } \phi = \frac{c\omega}{k} \qquad (17.52)$$

On substituting the value of ρ from equation (17.44) and noting that $c/k = 2\xi/\omega$, above equation reduces to

$$\chi = F_0 \left[\frac{1 + (2\xi\beta)^2}{(1-\beta^2)^2 + (2\xi\beta)^2}\right]^{1/2}, \quad \tan\phi = 2\xi\beta \qquad (17.53)$$

The ratio χ / F_0 is called the transmission ratio or transmissibility and denoted by TR, provides a measure of force transmitted to the foundation. In Fig. 17.15, TR is plotted as a function of β for several values of ξ. It is evident from the figure that if the force is to be smaller than the applied force, the natural frequency should be selected in such a way that β is greater than $\sqrt{2}$.

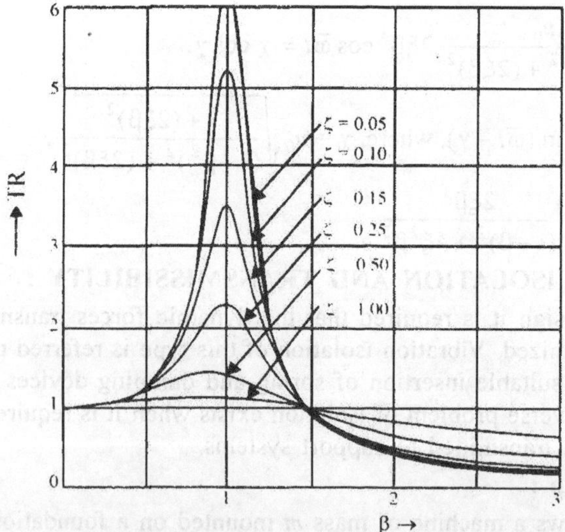

Fig. 17.15. Transmissibility Ratio.

17.9 ROTATING UNBALANCED FORCE :

Imbalance in rotating machinery is a common source of harmonic excitation. Let us consider the rotation of a motor with constant angular speed of $\overline{\omega}$, having an imbalance mass m_0 attached to it at radial distance 'e' from center as shown in figure 17.16.

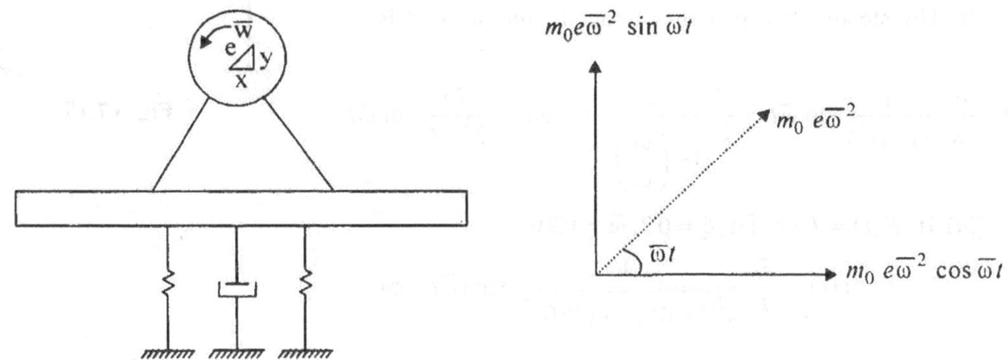

Fig. 17.16. Rotating Imbalance.

The coordinate of imbalance mass can be written as

$$x = e \cos \theta, y = e \sin \theta, \text{ where, } \theta = \overline{\omega}t \tag{17.54}$$

Now, the acceleration of the mass in x and y direction can be written as

$$\ddot{x} = -e\,\overline{\omega}^2 \cos \overline{\omega} t$$

$$\ddot{y} = -e\,\overline{\omega}^2 \sin \overline{\omega} t$$

These accelerations produce inertia forces in horizontal and vertical direction as shown in Figure 17.16 If the system is free to vibrate in the vertical direction, its steady state motion will be

$$m\ddot{u} + c\dot{u} + ku = m_0\, e\overline{\omega}^2 \sin \overline{\omega} t \tag{17.55}$$

where, m_0 is the total mass, k is the spring stiffness and c is the damping of the system. The steady state response is given by

$$u = \rho \sin(\overline{\omega} t - \phi)$$

$$\text{where,} \quad \rho = \frac{m_0\, e\overline{\omega}^2}{k\sqrt{(1-\beta^2)^2 + (2\xi\beta)^2}} = \frac{\dfrac{m_0\, e\beta^2}{m}}{\sqrt{(1-\beta^2)^2 + (2\xi\beta)^2}}, \; \tan\phi = \frac{2\xi\beta}{1-\beta^2} \tag{17.56}$$

17.10 EXAMPLE PROBLEM

Example 17.12. The rectangular plane frame (Fig. 17.17) has a very stiff girder of mass 'm' and flexible column of constant bending rigidity EI. (*i*) Find the natural frequency (ω) of the frame. (*ii*) The frame is subjected to harmonic force $F(t) = P \sin \overline{\omega} t$ as shown. Calculate the steady state response of vibration if $\overline{\omega} = \dfrac{5\omega}{6}$ (*iii*) If $F(t) = P \cos \overline{\omega} t$, $\xi = 0.2$, $\overline{\omega} = 0.9w$, calculate the steady state response (*iv*) The same frame is subjected to harmonic ground acceleration $\ddot{u}_g = a \sin \overline{\omega} t$. Find the steady state response for $\overline{\omega} = 3\omega$

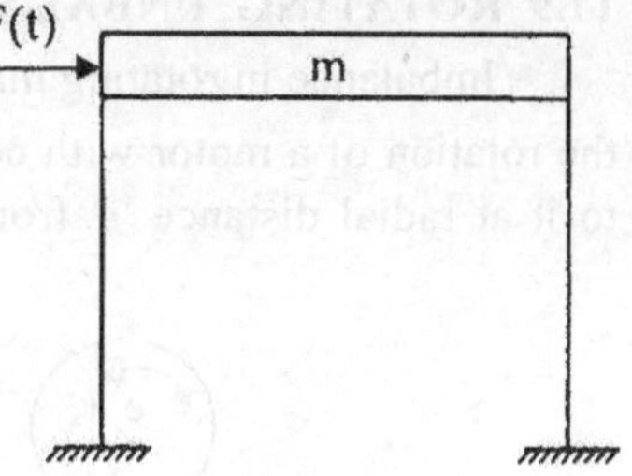

Fig. 17.17

Sol. (*i*) Stiffness coefficient $K = 2 \times 12EI/L^3 = 24EI/L^3$

Then, the natural frequency of the frame,

$$\omega = \sqrt{\frac{k}{m}} = \frac{2}{L}\sqrt{\frac{6EI}{mL}}$$

(*ii*) The steady state response under harmonic load is

$$u(t) = \frac{F_0}{k}\frac{1}{(1-\beta^2)}\sin\overline{\omega}t = \frac{P}{k}\frac{1}{1-\left(\frac{\overline{\omega}}{\omega}\right)^2}\sin\overline{\omega}t = \frac{3\,PL^3}{22EI}\sin\overline{\omega}t$$

(*iii*) If $F(t) = P\cos\overline{\omega}t, \xi = 0.2, \overline{\omega} = 0.9w$

$$u(t) = \frac{F_0}{k}\frac{1}{\sqrt{(1-\beta^2)^2+(2\xi\beta)^2}}\sin(\overline{\omega}t-\phi)$$

$$\frac{1}{[(1-0.9^2)^2+(2\times0.2\times0.9)^2]^{\frac{1}{2}}}\frac{P}{\frac{48EI}{L^3}\times 2}\cos(\overline{\omega}t-\theta) = 0.215\frac{PL^3}{EI}\cos(\overline{\omega}t-\theta)$$

(*iv*) Here the effective harmonic load acting on the frame due to ground acceleration is, $F(t) = -ma\sin\overline{\omega}t$. Thus the steady state response under harmonic ground acceleration is

$$u(t) = \frac{1}{1-\left(\frac{\overline{\omega}}{\omega}\right)^2}\frac{ma}{k}\sin\overline{\omega}t = \frac{ma\,L^3}{192EI}\sin\overline{\omega}t$$

Example 17.13. A machine of mass 25 Kg is placed on an elastic foundation. A sinusoidal force of magnitude 25N is applied to the machine. A frequency sweep reveals that the maximum steady state amplitude of 1.3 mm occurs when the period of response is 0.22 sec. Determine the equivalent stiffness and damping ratio of the foundation.

Sol. For a linear system, the frequency of response is same as the frequency of forcing function. Then, maximum response of 1.3mm occurs for a frequency of ω

$$\text{So, } 2\pi/\omega = 0.22s, \text{ i.e. } \omega = 28.6\ rad/s.$$

$$A_{D\max} = \frac{1}{2\xi\sqrt{1-\xi^2}} \text{ and } \beta = \sqrt{1-\xi^2} \text{ thus, } \omega_n = \frac{\overline{\omega}}{\sqrt{1-2\xi^2}}$$

$$u_{\max} = \delta_{st}\,A_{D\max} = \frac{F_0}{m\omega_n^2\,2\xi\sqrt{1-\xi^2}}$$

$$0.0013 = \frac{25}{25\frac{28.6^2}{1-2\xi^2}}\frac{1}{2\xi\sqrt{1-\xi^2}}$$

$$\xi^4 - \xi^2 + 0.117 \text{ i.e. } \xi = 0.368, 0.930$$

Larger value is discarded because a frequency of sweep would yield a maximum only $\xi \le \frac{1}{\sqrt{2}}$

Thus, $\xi = 0.368$.

$$\omega_n = \frac{28.6}{\sqrt{1-2\times 0.368^2}} = 33.5 \quad K = m\omega_n^2 = 2.81\times 10^4 \ N/m$$

Example 17.14. A steel frame fixed at base consisting of four legs each being a steel flanged section ($I = 25.8 \times 10^6$ mm^4, $Z = 249 \times 10^3$mm^3, Length = 3.0m), supports a rigid steel table. A rotating motor with an unbalance mass of 200 Kg at an eccentricity of 50mm is mounted on the table. If the total mass of the table and the motor is 2500 Kg, find the range of speed of the motor over which the maximum flexural stress in the leg will exceeds 100Mpa.

Sol. The lateral stiffness of the frame

$$k = \frac{4x12EI}{L^3} = \frac{4\times 12\times 200000\times 25.8\times 10^6}{3000^3} = 9173\times 10^3 \ N/m$$

Natural frequency of the system

$$\omega = \sqrt{\frac{k}{m}} = \sqrt{\frac{9173\times 10^3}{2500}} = 60.57 \ rad/s = 9.64 \ Hz$$

If the maximum deflection of the table is Δ, then moment developed at base

$$M = \frac{6EI\Delta}{L^3} \text{ and the flexural stress is } \sigma = \frac{M}{Z} = \frac{6EI\Delta}{L^2 Z}$$

Now for $\sigma = 100$, $100 = \dfrac{6\times 200{,}000\times 25.8\times 10^6\times\Delta}{3000^2\times 249\times 10^3}$

i.e. $\Delta = 7.24$mm

As there is no damping, $\rho = \dfrac{em_0}{m}\dfrac{\beta^2}{\sqrt{(1-\beta^2)^2}}$

For $\beta < 1$ $\rho = \dfrac{em_0}{m}\dfrac{\beta^2}{(1-\beta^2)} > 7.24$ *i.e.* $\dfrac{50\times 200}{2500}\dfrac{\beta^2}{(1-\beta^2)} > 7.24$ *i.e.* $\beta > 0.802$

For $\beta > 1$ $\rho = \dfrac{em_0}{m}\dfrac{\beta^2}{(\beta^2-)} > 7.24$ *i.e.* $\dfrac{50\times 200}{2500}\dfrac{\beta^2}{(1-\beta^2)} > 7.24$ *i.e.* $\beta < 1.495$

Thus the lower limit of the motor speed is $\beta\omega = 0.802 \times 9.64$ Hz = 464 rpm.

The upper limit is $\beta\omega = 1.495 \times 9.64$Hz = 865 rpm.

Thus in the range of 464 to 865 rpm, the deflection will be larger than 7.24 mm and the flexural stress will exceed 100MPa in the column sections.

Example 17.15. A centrifugal fan weighs 45Kg and has a rotating unbalance of 20Kg-cm. When dampers having damping factor $\xi = 0.2$ are used, specify the springs for mounting such that only 10% of the unbalance force is transmitted to the floor. The fan is running at constant speed of 1000rpm.

Sol. Transmissibility TR is

$$TR = \frac{\sqrt{1+(2\xi\beta)^2}}{\sqrt{(1-\beta^2)^2+(2\xi\beta)^2}}$$

Then, $0.1 = \frac{\sqrt{1+(0.16\beta)^2}}{\sqrt{(1-\beta^2)^2+(0.16\beta)^2}}$ *i.e.* $\beta^4 - 1.84\beta^2 - 16\beta - 99 = 0$ gives, $\beta = 3.7$

Then $\overline{\omega} = \frac{\omega}{\beta} = \frac{105}{3.7} = 28.4\ rad/\sec$, $k = m\omega^2 = 45 \times 28.4^2 = 3.6 \times 10^4\ N-cm$

Example 17.16. A machine of 100Kg mass is supported on spring of total stiffness 700 KN/n and has an unbalanced rotating element, which results in a disturbing force of 350N at a speed of 300 rev/min. $\xi = 0.2$. Determine (*a*) its amplitude of motion due to unbalance, (*b*) Transmissibility and transmitted force.

Sol. Natural frequency of the system, $\omega = \sqrt{\frac{700 \times 10^3}{100}} = 83.67\ rad/\sec$ and frequency of the disturbing force, $\overline{\omega} = \frac{2\pi \times 300}{60} = 31.42\ rad/\sec$, thus $\beta = \frac{31.42}{83.67} = 0.375$

(*a*) Amplitude of vibration

$$X = \frac{\frac{F_0}{K}}{\sqrt{(1-\beta^2)^2+(2\xi\beta)^2}} = \frac{\frac{350}{700\times10^3}}{\sqrt{[1-.375^2]^2+(2\times0.2\times.375)^2}} = 3.79 \times 10^{-5} = 0.0379mm$$

(*b*) The transmissibility

$$TR = \frac{\sqrt{1+(2\times0.2\times.375)^2}}{\sqrt{[1-(.375)^2]^2+(2\times0.2\times.375)^2}} = 0.137$$

transmitted force, $F_{TR} = 350 \times 0.137 = 47.89N$

Example 17.16. An automobile is travelling along a multispan elevated roadway supported every 12.2m. The deflected road profile (due to long term creep) can be approximated as sinusoidal with an amplitude of 3.05 cm and period of 12.2m The sdof system shown in Fig. Ex. 17.16 is an simple idealization of an automobile, appropriate for a 'first approximation' study of the ride quality of the vehicle. The weight of the vehicle is 1814Kg. The effective stiffness of the automobile suspension is 223.45 Kg/cm. and its viscus damping is such that the damping ratio of the system is 45%. Determine (*i*) the maximum vertical movement of the car when it is travelling at 45 mph. (20.11m/s) (*ii*) the maximum vertical movement of the car if damping is zero.) Determine the speed at which resonance will occur, if $\xi = 0.4$

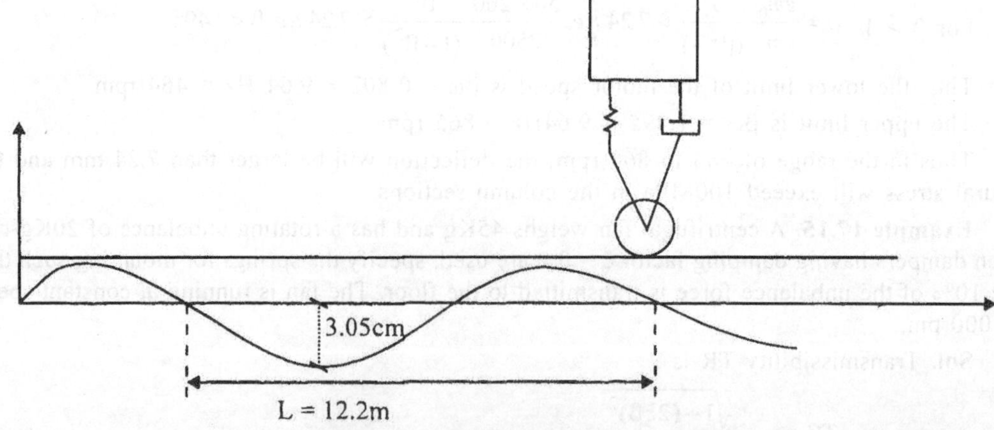

Fig. 17.18

Sol. Assuming that the tires are remaining in contact with the road, it can be modelled as shown in Fig. The vertical displacement of tires is

$$u_g(t) = u_{g0} \sin \overline{\omega}t, \text{ where, } u_{g0} = 3.05\,cm$$

The forcing period $T = \dfrac{L}{\text{velocity}}$ = 12.2/20.11 = 0.606

Period of the structure *i.e.* vehicle,

$$T = \frac{2\pi}{\omega} = 2\pi\sqrt{\frac{m}{k}} = 2\pi\sqrt{\frac{1814}{223.4 \times 980}} = 0.572 \text{ sec}$$

Then, $$\beta = \frac{\overline{\omega}}{\omega} = \frac{T}{\overline{T}} = \frac{0.572}{0.606} = 0.944$$

ξ = 0.45, then transmissibility, $TR = \sqrt{\dfrac{1+(2\xi\beta)^2}{(1-\beta^2)^2+(2\xi\beta)^2}} = 1.642$

Then maximum vertical movement of the car

$$U_0(t) = TR_{ugo} = 1.642 \times 3.05 = 5.008cm$$

If their is no damping *i.e.* ξ = 0, then

$$TR = \frac{1}{1-\beta^2} = 10.9 \text{ Hence, } u_0(t) = 10.9 \times 3.05 = 27.7cm$$

(*iii*) If ξ were small, resonance will occur approximately at β = 1. In this case, ξ = 0.4, and for such large damping the resonance frequency is significantly different from ω. By definition, resonance will occur when *TR* is maximum. Substituting ξ = 0.4,

$$TR^2 = \frac{1+0.64\beta^2}{(1-\beta^2)^2+0.64\beta^2} \therefore \frac{d(TR^2)}{d\beta} = 0 \text{ gives } \beta = 0.893$$

$$\text{so, } \overline{\omega} = 0.893\omega = 9.81 rad/\text{sec}$$

Thus, resonance occurs at forcing frequency of 9.81 rad/sec which implies a vehicle speed of $v = \dfrac{\overline{\omega}L}{2\pi}$ = 19.05m/s.

Example 17.18 A cantilever beam with a tip mass of 5 kg is subjected to a harmonic load as shown in figure. If the vibration amplitude is 0.5mm, then find the frequency of excitation. Assume, I_{beam} = 10^{-8}m^4, E = 200 × 10^9 N/m^2, F_0 = 150N.

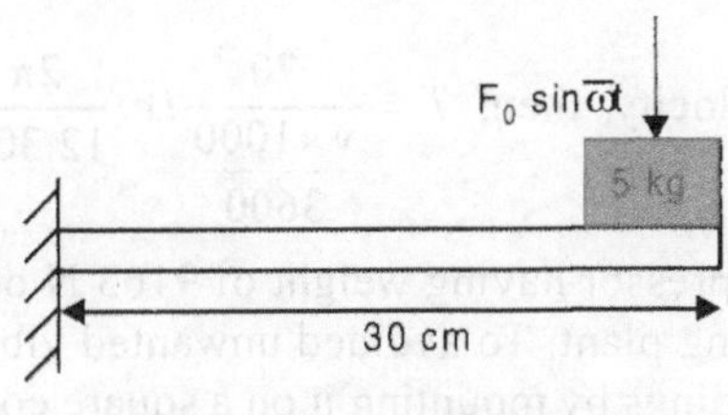

$$k = \frac{3EI}{L^3} = 2.22 \times 10^5 \text{ N/m}$$

$$\omega = \sqrt{\frac{k}{m}} = \sqrt{\frac{2.22\times10^5}{5}} = 210.82 \text{ rad/s}$$

Without damping, the free vibration motion is, $u(t) = \dfrac{F_0}{k(1-\beta^2)}\sin\bar{\omega}t$

$$\therefore\ 0.5\times10^{-3} = \frac{150}{2.22\times10^5(1-\beta^2)} \qquad \therefore\ (1-\beta^2) = \frac{150}{(0.5\times10^{-3}\times2.22\times10^5)} = 1.35$$

$\therefore\ \left(\dfrac{\bar{\omega}}{\omega}\right)^2 = -0.35$. Thus, excitation frequency cannot be less than natural frequency.

$$\therefore\ (\beta^2-1) = 1.35 \ \ i.e.,\ \beta = 0.591$$

$$\therefore\ \omega = \frac{\bar{\omega}}{\beta} = \frac{210.8}{0.591} = 355 \text{ rad/s}$$

Example 17.19 A racing car modeled as *SDOF* system vibrating in vertical direction due to road undulation. The road elevation is assumed to vary sinusoidally; with peak to trough difference of 0.2m and distance between peaks are 70m. Determine the (*i*) amplitude of vibration of the racing car at a speed of 12 kmph and (*ii*) most favorable speed. Assume system frequency as 2Hz and damping ratio as 0.15.

Time to travel one cycle, $T = \dfrac{70}{\dfrac{120\times1000}{3600}} = 2.1$ sec

$\therefore$ Excitation frequency, $\bar{\omega} = \dfrac{2\pi}{T} = 2.992$ rad/s

$$\omega = 2\pi\times2 = 12.5664 \text{ rad/s} \ \therefore\ \beta = \frac{\bar{\omega}}{\omega} = 0.2381,\ \ \zeta = 0.15.$$

Then, amplitude of vibration,

$$u = u_0\left[\frac{(1+2\zeta\beta)^2}{(1-\beta^2)^2+(2\zeta\beta)^2}\right]^{\frac{1}{2}} = 0.1\times\left[\frac{1+(2\times0.15\times0.2381)^2}{\left\{1-(0.2381)^2\right\}^2+(2\times0.15\times0.2381)^2}\right]^{\frac{1}{2}} = 0.105\text{m}$$

For most favorable speed, $\dfrac{d}{d\beta}\left(\dfrac{u}{u_0}\right) = 0$ which gives, $\beta = 0.9586$. Thus, $\bar{\omega} = 12.303 = \dfrac{2\pi}{T}$

If v is the required velocity, then, $T = \dfrac{70}{\dfrac{v\times1000}{3600}}$ *i.e*, $\dfrac{2\pi}{12.303} = \dfrac{70}{\dfrac{v\times1000}{3600}}$ $\therefore\ v = 493$ kmph

Example 17.20 A compressor having weight of 9163 N operates at 1800 rpm need to install on the floor of a manufacturing plant. To avoided unwanted vibration it was decided to isolate the compressor from the surroundings by mounting it on a square concrete block separated from rest of the floor. If the block has a size of 2 m × 2 m, then determine the depth required so that 75% reduction in transmission of forces to floor due to vibration can be achieved. Assume density of concrete as 23563 N/m^3 and vertical compression coefficient of underlying soil, k_z = 20.36 × 10^6 N/m^3.

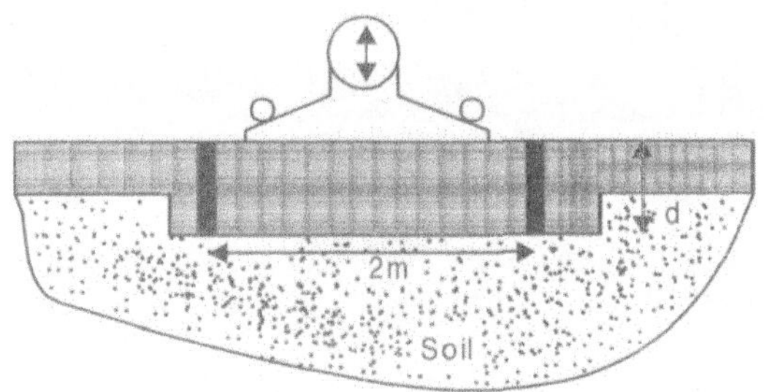

The system can be modeled as *SDOF* system with effective mass, $m + M$

$$= \frac{9163 + (2\times 2\times d)\times 23563}{9.81} = (934.1 + 9607.7d)\text{kg}$$

The spring constant of soil, $k = k_c \times$ bearing area

$$= 20.36 \times 10^6 \times (2 \times 2) = 81.44 \times 10^6 \text{ N/m}$$

Since, the transmissibility is relatively unaffected by damping in the isolated region, neglecting damping, $TR = \sqrt{\frac{1}{(1-\beta^2)^2}}$. Now considering negative root of above to yield positive *TR*,

$$1 - 0.75 = \frac{1}{\beta^2 - 1} \Rightarrow \beta = 2.24 \;\therefore\; \frac{\bar{\omega}}{\omega} = 2.24.$$ Then, natural frequency,

$$(\omega) = \left(\frac{1800\times 2\pi}{60}\right)\times\frac{1}{2.24} = 84.5 \text{ rad/sec} = \sqrt{\frac{k}{m_e}} = \sqrt{\frac{81.44\times 10^6}{934.1 + 9607.7d}} \quad \therefore\; d = 1.1\text{m}$$

Example 17.21 The static deflection of an electric motor of mass 400 kg supported by a system of four parallel springs and no damping is found to be 50 mm. If the electric motor has a rotating unbalance of 0.2 kg m, determine the amplitude of vibration at 200 rpm and the force transmitted to ground.

$$\delta_{st} = \frac{50.0}{1000} = \frac{m.g}{k} = \frac{400\times 9.81}{k}$$

$$\therefore \quad k = 78480 \text{ N/m Then, } \omega = \sqrt{\frac{k}{m}} = \sqrt{\frac{78480}{400}} = 14 \text{ rad/sec}$$

Frequency of forcing function, $\bar{\omega} = 200\times\frac{2\pi}{60} = 209.44$ rad/sec $\therefore\; \beta = \frac{\bar{\omega}}{\omega} = 14.96$ and $\zeta = 0$

Then, amplitude of vibration, $\rho = \frac{m_0 \cdot e\bar{\omega}^2}{k\sqrt{(1-\beta^2)^2 + (2\zeta\beta)^2}}$

$$= \frac{0.2}{78480}\,\frac{1}{\sqrt{\left(1-(14.96)^2\right)^2 + 0}} = 5.022\times 10^{-4}\text{ m}$$

And force transmitted due to this $= k.x = 78480 \times 50122 \times 10^{-4} = 39.416$ N

18

PLASTIC ANALYSIS OF STRUCTURES

18.1 INTRODUCTION

The success of the assumption of linear behavior in engineering analysis is obvious. But, it is equally obvious that many situations demand consideration of nonlinear behavior. The ultimate (collapse) load is becoming more significant to the desiginer due to devastating financial and social consequences of structural collapse particularly for extreme action like earthquake, wind, blast etc. The new codes are making it obligatory for a designer to limit the probability of collapse to a particular value (very low). For this, it is important to know the ultimate *i.e.*, collapse load. As linear behavior ceases long before collapse, so it is important to study the inelastic or plastic behaviour of structures. Idea of design on the basis of ultimate load rather than working stress method is more realistic. The idea has advantages *i.e.*, in terms of economy and simplicity. Since there is considerable reserve strength beyond the elastic limit, plastic design can make use of that judiciously. The attainments of the yield strength in most highly stressed fiber do not mean that serviceability ceased. Stress distribution is more realistic and rational. Moreover, the elastic condition of continuity is no longer needed, thus the analysis becomes simple than working stress method of evaluation. Thus, a plastic design approach is a realistic model of the actual behavior ensuring uniform margin of safety of structure and economy.

There are two sources of nonlinearity that exist in the anlaysis of solid continua, namely, the material nonlinearity and geometric nonlinearity. The geometric nonlinearity concerned with the changes in geometry having significant effect on the load deformation behavior. Whereas the material nonlinearity occurs when, for whatever reason, the stress strain behavior given by the constitutive relation is nonlinear. The plastic behaviour arises out of material nonlinearity is the subject of present chapter. In polycrystalline materials (like steel), the grains are mutually oriented in different directions. The crystal lattices have various points, linear and superficial defects. Impurities are also found on the grain boundaries. These factors affect the properties and behavior of the material under stress. The internal forces acting between individual atoms change under the effect of the external forces and the lattice is deformed. **Elastic behavior** of the material is the result of the static resistance of inter-atomic bonds. It appears as the change of shape and volume of the crystals *without* any impairment to these bonds. The deformation of the crystals depends on the direction of stress with respect to the lattice (as the individual crystals are anisotropic). **Plastic behavior** of material is caused by the formation of new bonds. Plastic deformation is the result of shift of dislocation and of their interaction. One part of the crystal slip along another part but the volume of the material remains unchanged. Such deformation is irreversible. Initially, there appear only local plastic deformations in the grains most favourably oriented for slips with regard to the direction of the stress. With increasing stress, plastic deformation spreads into the adjacent grains and into a large portion of the material. Strain-hardening takes place when the grain orientation is changed.

18.2 STRESS-STRAIN BEHAVIOR OF METAL

Figure 18.1 shows true elasto-plastic behaviour of a typical metal (annealed material). So long the stress is small, the material behaves **elastically** *i.e.*, the original size regains on removal of load (initial part of the curve straight line having slope E). **The proportional limit** is the point (Point A) on the curve where the linear stress-strain relation ceases to hold. The elastic range normally extends slightly beyond the proportional limit.

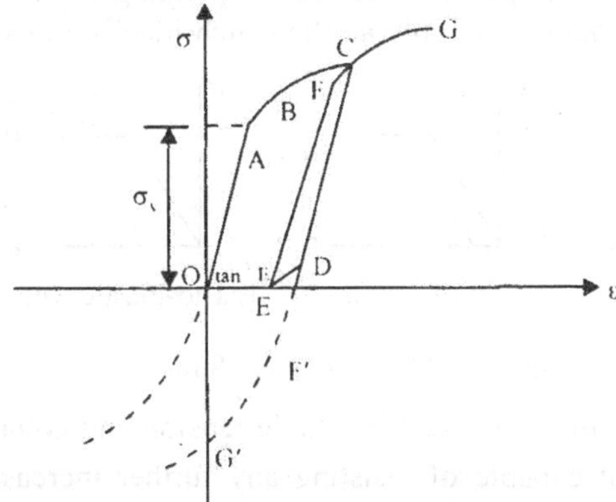

Fig. 18.1. The stress-strain diagram of a typical metal.

The yield point (Point B) and corresponding stress at that point *i.e.*, yield stress generally defined as that for which a specific small amuont of permanent deformation occurs. (Note that transition from elastic to plastic is gradual, owing to successive yielding of the individual crystal grains, thus location of point B largely a matter of convention. Normally, intersection of initial straight line OA and tangent of stress-strain curve slightly above B is taken. Beyond yield point, the stress continually increases with further plastic strain whereas slope of the curve (strain hardening rate) steadily decreases with increasing stress. If stressed upto some point C in the plastic range and the load is subsequently released, there is an elastic revovery (CD, nearly linear slope). It may be noted that OE is **permanent strain.** On reloading a new yield point, F is obtained and the loop thus formed is called hysteretic loop. The curve EFG represents the stress-strain relation when there is already a pre-stain of amount OE, the greater the pre-strain, the higher the new yield point and flatter the stress strain curve. If on complete unloading from tensile plastic state at C is reloaded in compression, the stress-strain curve follow the path DF′G′ where the yield point F′ correspond to a stress that is appreaciably smaller than earlier yield point in tension. This phenomenon is known as **Bauschinger effect** which is neglected in plasticity theory.

18.3 SOME USEFUL TERMINOLOGY

Ductile Metal: The material which undergoes large plastic deformation before failure.

Limit Design: A design based on any chosen limit of structural usefulness.

Plastic Design: A design based on ultimate load carrying capacity of the structure. The term 'plastic' is derive from the fact that the ultimate load is computed from knowledge of the strength in the plastic range.

Ultimate Load (P_u): The maximum load a structure will carry.

Plastification: The development of full plastic yield of cross-section.

Plastic Moment (M_p): Maximum moment of resistance of the fully yielded cross-section.

Plastic hinge: A yield section of a beam which acts as if it were hinged, except with a constant restraining moment.

Shape Factor: Ratio of Plastic moment (M_p) and Yield moment (M_y) *i.e.*, $\frac{M_P}{M_y}$.

18.4 ASSUMPTION IN PLASTIC ANALYSIS

The assumptions involve in plastic analysis (with a special reference to steel) are:

(*i*) Idealized stress-strain diagram is assumed depending on the experimental observations. Figure 18.2 show some commonly applied such idealizations.

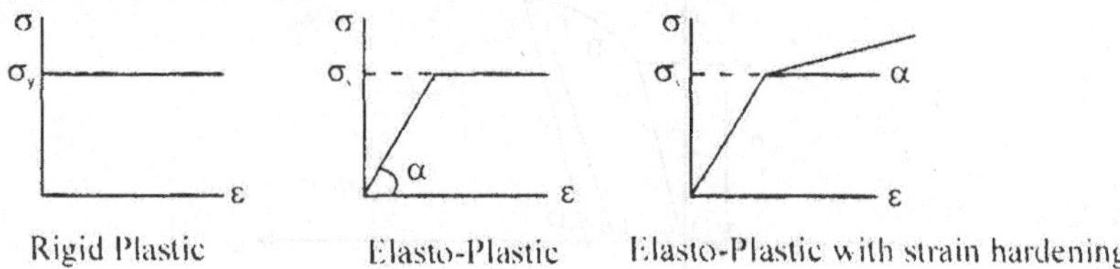

Fig. 18.2. Idealized Stress Strain Curve.

(*ii*) The material behavior is identical in simple tension and compression.

(*iii*) Plastic hinge is not capable of resisting any further increase in load (the plastic load bearing capacity is decided on this).

(*iv*) The deformations of structure upto the limit of plastic load bearing capacity are so small that the equilibrium of forces can be investigated for an undeformed structure *i.e.*, virtual work principle is applicable.

(*v*) The longitudinal strains of the cross-section are distributed linearly.

(*vi*) Condition of plasticity considers that the limit state develops when the eqivalent stress attains the yield strength value throughout the entire cross-section.

(*vii*) After the hinge mechanism is formed, the curvature of the elements between the hinges is supposed to remain unchanged.

(*viii*) Random imperfections are ignored.

(*ix*) The structural connections are rigid to the extent that they are capable of transferring the redistribution effect.

It may be noted that the Hook's Law, Superposition, Reciprocal theorem, Castigliano's least work theorem are not applicable for theory of plasticity.

18.5 FUNDAMENTALS OF PLASTIC ANALYSIS

To explain the various fundamental considerations, let us consider a simple three bar system as shown in figure 18.3. To investigate the behaviors of the members under axial forces, the idealized stress-strain behavior assumed is also shown in the same figure. All members have uniform cross-section. From figure 18.3, at equilibrium,

$$2T_1 + T_2 = P \tag{18.1}$$

and further consideration of compatibility (displacement of all the bars will be same) gives,

$$\frac{T_1 L_1}{AE} = \frac{T_2 L_2}{AE} \quad \text{or} \quad T_1 = T_2/2 \text{ so, } T_2 = P/2 \tag{18.2}$$

It may be noted that the load at which the structure will first yield when the forces in the bar which carry largest load will reach $\sigma_y A$. Thus, from Eqn. (18.2), one may obtain, $P_y = 2T_2 = 2\sigma_y A$ and the displacement at yield load will be, $\delta_y = \sigma_y L_2 = \frac{\sigma_y L}{2E}$.

At this stage, the structure is partially plastic. It will deform like a two bar truss system except that a constant force of $\sigma_y A$ will act at position of truss member 2. This situation is continued till the load reaches the yield value in other bars. Note that when the other two bars reach to their yield capacity, the equilibrium condition gives, $P = 3\sigma_y A$. Thus ultimate load carrying capacity of the system is, $P_u = 3\sigma_y A$. The associated load deflection behavior is shown at bottom right of figure 18.3. Now further increase of load causes instantaneous unrestricted deformation. From this simple analysis the essential features of plastic analysis can be noted *i.e.*, the equilibrium condition is satisfied at ultimate load and there is unrestricted plastic flow at ultimate load. It may be noted that how easily the ultimate load is obtained. This is due to the fact that the continuity need not be considered when the ultimate load is in the plastic range.

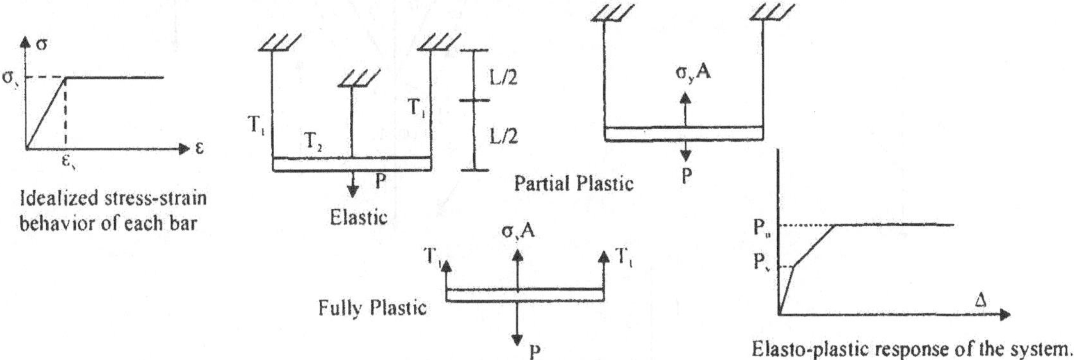

Fig. 18.3. Three Bar Systems

A second example of a three bar truss as shown in figure 18.4 is further considered. The idealized stress-strain behavior is also shown in the same figure. On application of load *P* on joint *A*, it moves to position *B*. Then, elastic strain on the vertical member is, $\varepsilon_v = \frac{\delta}{L}$ and it will become plastic when strain reaches to yield strain *i.e.*, $\varepsilon_v = \varepsilon_e$. And at this stage yield deformation of vertical member is $\delta_{ev} = \varepsilon_{ev} L$. Thus, force on vertical bar,

$$S_v = \sigma A = E\varepsilon_e A = \frac{\delta EA}{L} \qquad \text{for } \delta \le \delta_{ev}$$

$$= E\varepsilon_{ev} A = \frac{\delta_{ev} EA}{L} \qquad \text{for } \delta \ge \delta_{ev} \tag{18.3}$$

The corresponding load deformation behavior of the vertical bar at this stage is shown at bottom left of figure 18.5 (S_v *vs.* δ).

Now, let us consider the inclined bar *AG*. The elongation of inclined member *AG*=*BC*. IF *D* is the projection of *A* on straight line *BG* and if we assume for small value of δ, $BD = BC$ and $\phi = \alpha$, then elongation of $AG = \delta \cos\alpha$. Thus, strain in the inclined member is $\varepsilon_i = \frac{\delta \cos\alpha}{l/\cos\alpha} = \frac{\delta \cos^2\alpha}{l}$. When it reaches to plastic state *i.e.*, $\varepsilon_i = \varepsilon_e$, corresponding deformation of inclined member is obtained as,

$$\text{for} \quad \varepsilon_i = \varepsilon_e, \delta_e = \delta_{ei} \text{ thus, } \frac{\delta_{ei} \cos^2\alpha}{l} = \varepsilon_e \text{ i.e., } \delta_{ei} = \frac{\varepsilon_e l}{\cos^2\alpha} = \frac{\delta_e}{\cos^2\alpha} \tag{18.4}$$

as $\cos^2 \alpha < 1$ so, $\delta_{el} > \delta_{ev}$ which indicates that the deformation capacity of the inclined bar is more than the vertical bar. Thus, vertical bar becomes palstic first. Force in the inclined bars is,

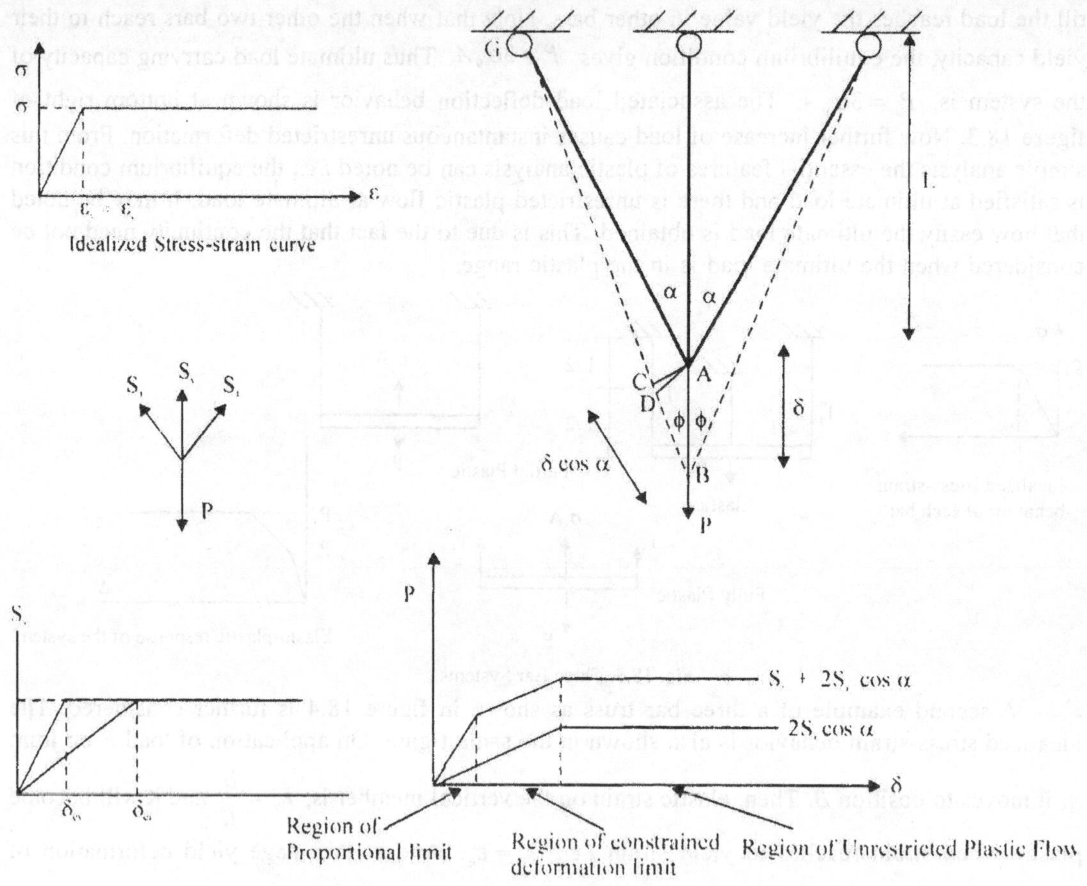

Fig. 18.4. Three Bar Truss and associated behaviour.

$$S_i = \sigma A = E\varepsilon, A = \frac{\delta AE\cos^2\alpha}{l} \quad \text{for } \delta \le \delta_{el}$$

$$= \frac{\delta_{el} AE\cos^2\alpha}{l} \quad \text{for } \delta \ge \delta_{el} \tag{18.5}$$

Equilibrium at joint A gives, $P = S_v + 2S_i \cos\alpha$. Thus,

$$P = \frac{AE}{L}\delta + 2\cos\alpha\frac{AE}{L}\delta\cos^2\alpha = \frac{AE}{L}(1+2\cos^3\alpha)\delta \quad \text{for } \delta \le \delta_{ev}$$

$$= \frac{AE}{L}\left(\delta_{ev} + 2\delta\cos^3\alpha\right) \quad \text{for } \delta_{ev} \le \delta \le \delta_{el} \tag{18.6}$$

$$= \frac{\delta_{ev}AE}{L} + 2\cos\alpha\frac{\delta_{el}AE\cos^2\alpha}{L} \quad \text{for } \delta \ge \delta_{el}$$

The associated force displacement diagram is shown at bottom right of figure 18.4.

18.6 FLEXURE OF BEAM

In previous section, we have tried to explain some of the considerations of plastic analysis of structures. However, those are only for tension or compression members. In present section, we will try to study the deformation of beams beyond elastic limit under bending. The assumptions of plain section remains plane before and after bending are valid and the idealized stress-strain curve is elasto-plastic as in figure 18.2 *i.e.*,

$$\sigma = E\varepsilon \qquad \text{for} \qquad 0 < \varepsilon < \varepsilon_y$$

$$= \sigma_y \qquad \text{for} \qquad \varepsilon_y < \varepsilon < \infty \tag{18.7}$$

Figure 18.5 shows the cross-section of a rectangular beam subjected to pure bending. From elastic theory,

$$\frac{M}{I} = \frac{E}{R} = \frac{\sigma_z}{y} \tag{18.8}$$

Recalling the Von-Misses Yield criteria *i.e.*,

$$(\sigma_1 - \sigma_2)^2 + (\sigma_2 - \sigma_3)^2 + (\sigma_3 - \sigma_1)^2 = 2\sigma_y^2 \quad \text{As, } \sigma_1 = \sigma_2 = 0; \sigma_3 = \sigma_z$$

$$\therefore \qquad 2\sigma_z^2 = 2\sigma_y^2 \quad i.e., \quad \sigma_z = \pm\,\sigma_y \tag{18.9}$$

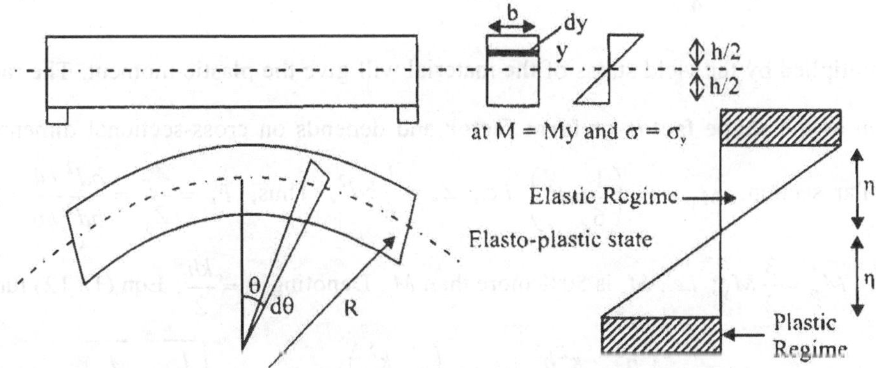

Fig. 18.5. A rectangular beam subjected to pure bending.

So, the plasticity begins when normal stress reaches the yield strength of the material under uniaxial load at extreme fibre *i.e.*, σ_y at $y = \pm\frac{h}{2}$ (*i.e.*, a localized yielding occur). Noting, M_y = Moment Capacity at first yielding, from equation (18.8) and (18.9) one gets,

$$\sigma_y = \frac{M}{I}\,y = \frac{M_y}{bh^3}\cdot 12 \cdot \frac{h}{2} = \frac{M_y}{6bh^2} \qquad \text{thus, } M_y = \frac{1}{6}\sigma_y bh^2 \tag{18.10}$$

At elasto-plastic state, some portion is under plastic stage, whereas inner portion of the section will be till elastic. Let the plasticity has spread upto a depth of ($h/2-\eta$) as shown in figure 18.5 *i.e.*,

$$\sigma_z = \frac{Ey}{R} \qquad \text{for} \qquad y \le \eta$$

$$= \sigma_y \qquad \text{for} \qquad y \ge \eta \tag{18.11}$$

The equilibrium condition gives,

$$M = \int_{-h/2}^{+h/2} (\sigma_z bdy)y = 2\left[\int_0^{\eta} \sigma_z (by)dy + \int_{\eta}^{h/2} \sigma_z (by)dy\right] = 2\left[\int_0^{\eta} \frac{Ey}{R}(by)dy + \int_{\eta}^{h/2} \sigma_y (by)dy\right]$$

$$= 2\left[\frac{Eb}{R}\cdot\frac{\eta^3}{3} + \frac{\sigma_y b}{2}\left(\frac{h^2}{4} - \eta^2\right)\right] = \frac{2}{3}\frac{Eb}{R}\eta^3 + \sigma_y b\left(\frac{h^2}{4} - \eta^2\right) = \frac{2}{3}b\eta^3\cdot\frac{E}{R} + \sigma_y b\left(\frac{h^2}{4} - \eta^2\right)$$

At elasto-plastic boundary, $\frac{E}{R} = \frac{\sigma_z}{y}$ *i.e.*, $\frac{E}{R} = \frac{\sigma_y}{\eta}$, then,

$$M = \frac{2}{3}b\eta^3\cdot\frac{\sigma_y}{\eta} + \sigma_y b\left(\frac{h^2}{4} - \eta^2\right) = \frac{\sigma_y bh^2}{4} - \frac{\sigma_y b\eta^2}{3} \quad \therefore\; M = \sigma_y b\left(\frac{h^2}{4} - \frac{\eta^2}{3}\right) \tag{18.12}$$

For fully plastic condition of the section,

$$M_p = 2\int_0^{h/2} \sigma_y bydy = 2\sigma_y b\cdot\frac{1}{2}\cdot\frac{h^2}{4} = \frac{1}{4}\sigma_y bh^2 = Z_p\sigma_y \tag{18.13}$$

In above, $Z_p = \frac{bh^2}{4}$ is termed as the plastic modulus of a section defines the quantity which, when multiplied by the yield stress of the material will give the plastic moment. The ratio of $\frac{M_p}{M_y}$ is known as the **shape factor** or **form factor** and depends on cross-sectional dimensions. For a rectangular section, $M_y = \sigma_y\left(\frac{1}{6}bd^2\right)$ *i.e.*, $Z_e = \frac{1}{6}bd^2$. Thus, $F_F = \frac{Z_p}{Z_y} = \frac{bd^2/4}{bd^2/6} = 1.5$. Which indicates $M_p = \frac{3}{2}M_y$; *i.e.*, M_p is 50% more than M_y. Denoting $\eta = \frac{kh}{2}$, Eqn (18.12) further yields,

$$M = \sigma_y b\left(\frac{h^2}{4} - \frac{k^2h^2}{4\times 3}\right) = M_p\left(1 - \frac{k^2}{3}\right) \quad \text{and} \quad k = \sqrt{3\left(1 - \frac{M}{M_0}\right)} \tag{18.14}$$

The shape factor for a general section can be readily derived. Let us consider a cross-section and associated stress distribution on it at fully plastic stage as shown in figure 18.6.

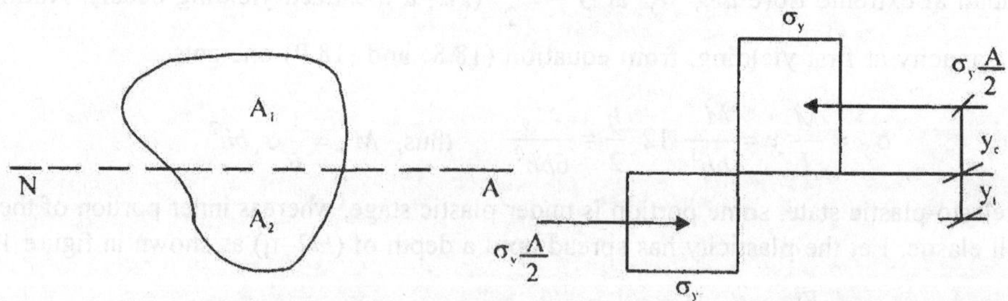

Fig. 18.6. A general cross-section and stress distribution on it at fully plastic stage.

Since, total compressive force is equal to total tensile force *i.e.*, $A_1\sigma_y = A_2\sigma_y$, then, $A_1 = A_2$. Hence neutral axis divides the cross-sectional area into two equal halves. Noting that y_c and y_t pass through centroids of A_1 and A_2 respectively, the plastic moment capacity of the section,

$$M_p = \frac{\sigma_y A}{2} y_c + \frac{\sigma_y A}{2} y_t = \frac{\sigma_y A}{2}(y_c + y_t) = \sigma_y \left[\frac{(y_c + y_t)A}{2}\right] = \sigma_y Z_p$$

In above, $Zp = (y_c + y_t)\frac{A}{2}$. Again, $M_y = \sigma_y Z_y$. Thus, $\frac{M_p}{M_y} = \frac{\sigma_y Z_p}{\sigma_y Z_y} = \frac{Z_p}{Z_y}$

The shape factor for some common symmetrical sections are shown in figure 18.7.

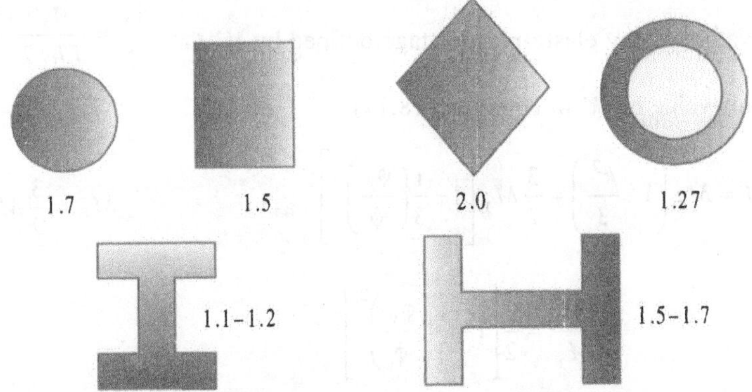

Fig. 18.7. The Shape Factor of common various cross-sections.

The Moment-Curvature Relationship

From the equations derived just earlier for curvature and moment, we are now in a position to derive the moment curvature relation of a section. To study the moment curvature relations in flexure, let us consider the rectangular beam section as shown in figure 18.8. The figure shows the stress and strain distribution along the cross-section when it is partly plastic and partly elastic (upto a depth of kh).

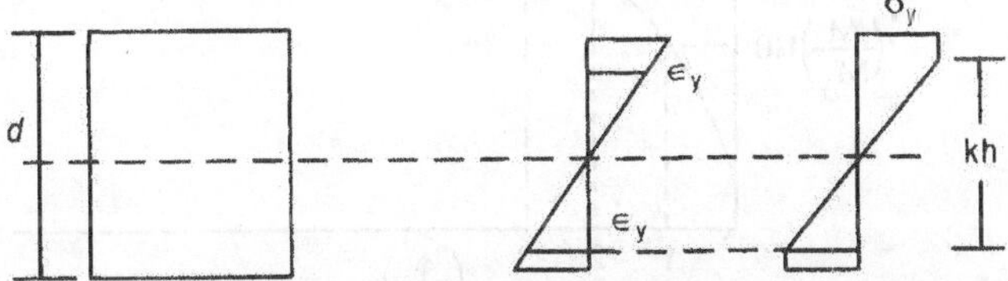

Fig. 18.8. Rectangular cross-section with elasto-plastic stress and strain distribution.

Usual moment curvature equation as given by equation (18.8) is valid over the elastic region.

At elasto-plastic boundary, $\frac{E}{R} = \frac{\sigma_y}{kh/2}$. Also using the relation $M = \frac{1}{4}\sigma_y bh^2\left(1 - \frac{k^2}{3}\right)$, $\sigma_y = \frac{4M}{bh^2(1-k^2/3)}$. Now, substituting this value of σ_y in the expression of $\frac{E}{R}$, one obtains

$$\frac{E}{R} = \frac{4M}{bh^2\left(1 - \frac{k^2}{3}\right)} \frac{2}{kh} = \frac{2M}{EIR(3-k^2)} \text{ as } I = \frac{bh^3}{12} \tag{18.15}$$

Let us now consider two strain diagrams for different conditions based on the load level *i.e.*, for incipient yielding stage and elasto-plastic stage (Figure 18.9)

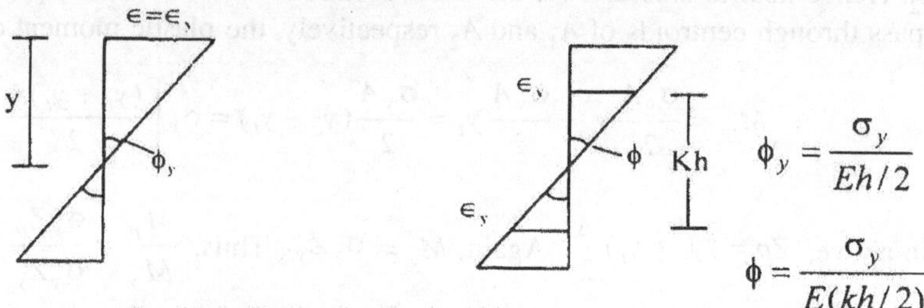

Fig. 18.9. Strain distribution diagrams.

ϕ_y is the curvature at any elasto-plastic stage defined by 'k' *i.e.*, $\dfrac{\phi_y}{\phi} = \dfrac{\sigma_y}{Eh/2} \dfrac{(kh/2)E}{\sigma_y} = k.$

Substituting this value of 'k' in equation (18.14),

$$M = M_p\left(1 - \frac{k^2}{3}\right) = \frac{3}{2}M_y\left[1 - \frac{1}{3}\left(\frac{\phi_y}{\phi}\right)^2\right] \quad \text{as} \quad M_p = \frac{3}{2}M_Y$$

so,
$$\frac{M}{M_y} = \frac{3}{2}\left[1 - \frac{1}{3}\left(\frac{\phi_y}{\phi}\right)^2\right] \tag{18.16}$$

The variation of ratio of elasto-plastic moment of resistance (M) to yielding moment (M_y) against the ratio of incipient yielding curvature (ϕ_y) to elasto-plastic curvature (ϕ) can be now easily obtained using Eqn. (18.16). Such a non-dimensional M-ϕ curve is shown in Fig. (18.10).

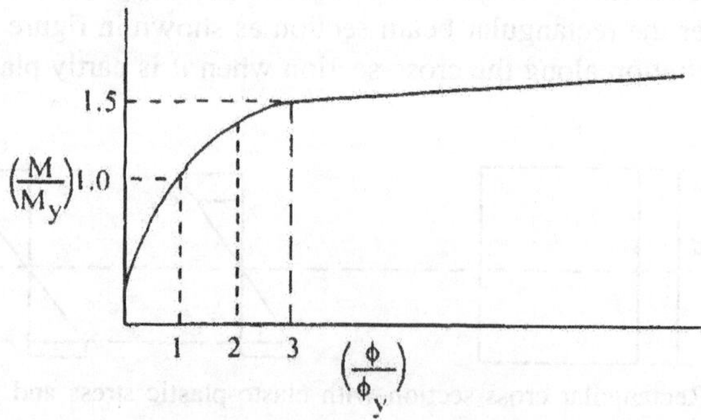

Fig. 18.10. M-ϕ curve at elasto-plastic stage.

The first point corresponds to incipient yielding stage. The stages 2 and 3 correspond to the stage, when max extreme fibre strain is $2\,\epsilon_y$ and $3\,\epsilon_y$, respectively. Stage 4 approached as a limit representing complete plastic yield of the cross-section. There is a 50% increase in strength above computed elastic limit due to plastification of the cross-section. This represents one of the sources of reserve strength beyond the elastic limit. It may be noted that the toughness (area under M - ϕ curve) is appreciably more at limiting plastic stage, in comparison with the first yielding.

Propagation of plasticity: Plastic Hinge

It may be observed from the moment curvature curve that as the moment reaches M_p, rotation occurs at constant moment (for ideal elasto-plastic materials) *i.e.*, the member acts as if it is hinged except the fact that a constant moment of value M_p exist at the section. Thus, a plastic hinge is a zone of yielding due to flexure in a structural member. The plastic hinges form at points of maximum moment. Thus, in a frame structure with prismatic members, it would be possible for plastic hinges to form at points of concentrated load, continuous support points, at the point of zero shear in span under distributed load.

Figure 18.11 shows the distribution of plastic hinge for an elasto-plastic material. At limiting stage at collapse the vertex of the parabolas are same and it passes through the origin. At this stage, a plastic hinge is said to be developed, the load carrying capacity of the beam is exhausted and turns into a mechanism. Beam theoretically breakdowns for rigidly plastic material without prior propagation of plastic zone along length of the beam.

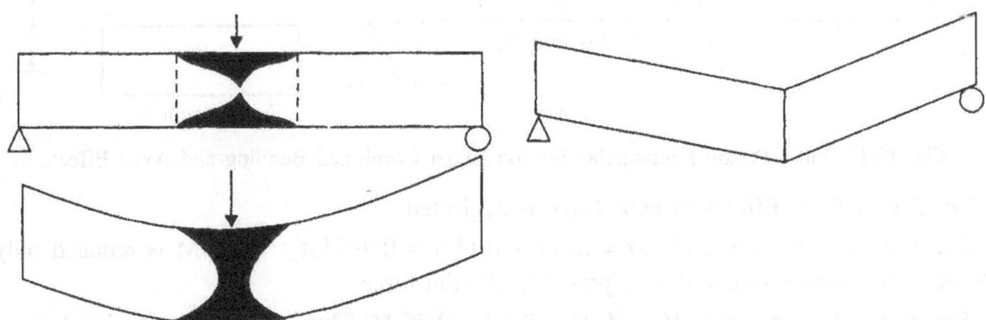

Figure 18.11 Distribution of plastic hinge

18.7 EFFECT OF AXIAL FORCE ON FULLY PLASTIC MOMENT

When a section is subjected to axial force *P*, then the plastic moment capacity is reduced. The reduction depends on shape of the section, length of the beam-column and intensity of compressive force. Due to axial force, yielding starts on the compression side prior to the tension side. In stress distribution, no longer compression area remains same with tension area. But, total force in bending tension must be equal to total force in bending compression. Hence, the stress distribution may be divided into two parts; one for bending another for axial compression. Figure 18.12 shows a ractangular section fully plastic under combined axial force *P* and bending *M*. Areas yielded in compression (direct and bending) and tension are shown separately in the figure. The entire section may be thought of as made up of areas under direct compression plus bending compression and tension. The resultant stress due to pure compression is equal to axial force *P i.e.*, $P = \sigma_y b(\alpha h) = \sigma_y b\alpha h$. If, the beam is subjected to direct compression only, then at plastic stage,

$$P_u = \sigma_y bd \quad \therefore \quad P = \alpha P_u, \alpha = \frac{P_u}{P} \tag{18.17}$$

P is called squash load of the section. The resultant of stress acting on the areas in bending is a couple equal to the applied moment *i.e.*,

$$M = \left[\sigma_y b \frac{(h - \alpha h)}{2}\right] \frac{(h + \alpha h)}{2} = \frac{\sigma_y b(h^2 - \alpha^2 h^2)}{4} = \frac{\sigma_y bh^2(1 - \alpha^2)}{4}$$

$$\therefore \quad M = M_p(1 - k^2) \tag{18.18}$$

So, M decreases with increases of α, *i.e.*, with increase of axial compressive force. Substituting α in eqn. (18.18)

$$M = M_p\left(1-\frac{P^2}{P_u^2}\right) \text{ i.e., } \frac{M}{M_p}+\left(\frac{P}{P_u}\right)^2 = 1 \tag{18.19}$$

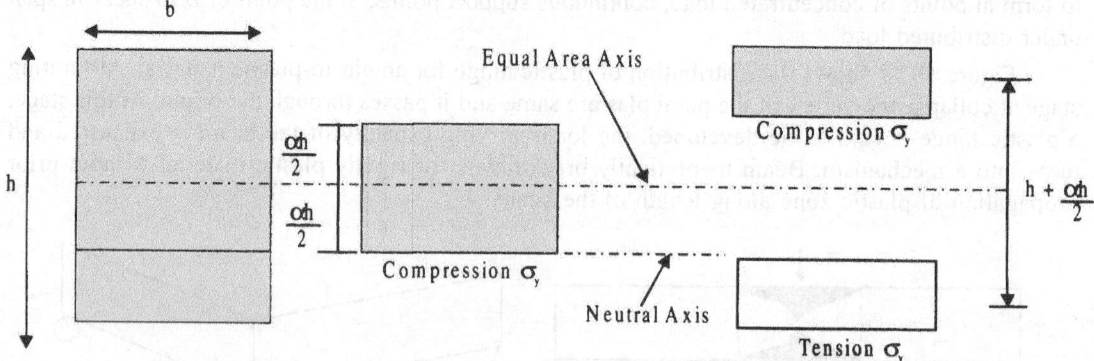

Fig. 18.12. Fully Plastic Rectangular Section Under Combined Bending and Axial Effect.

For $P/P_u < 0.15$: Effects of axial force is neglected.

For $P/P_u = 0.15$: $\alpha = 0.15$; $M = M_p(1 - 0.15^2) = 0.9775M_p$. Thus, M is reduced only by 2.25% of fully plastic moment due to presence of axial force.

For, $P/P_u = 0.5$: $\alpha = 0.5$; $M = M_p\ (1 - 0.5^2) = 0.75\ M_p$ Thus, 25% reduction *i.e.*, beam fails in flexure before reaching full momemt in presence of axial force.

Example 18.1: A simply supported beam 3 cm × 4 cm carries a concentrated load W at mid-point of span 4m. Determine 'W' and draw the shape of plastic zone, when strain at extreme fibre is $3\epsilon_y$. Draw moment curvature diagram along the beam. Also calculate maximum length of plastic hinge at collapse. Assume, $\sigma_y = 250$ Mpa.

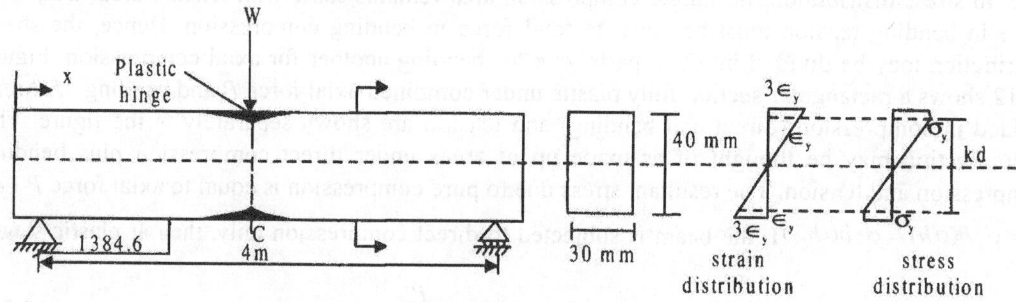

Fig. 18.13. The simply supported beam of example 18.1.

From strain distribution diagram, $\dfrac{3\epsilon_y}{(d/2)} = \dfrac{\epsilon_y}{(kd/2)}$ *i.e.*, $k = \dfrac{1}{3}$. Also, $M = \dfrac{Wx}{2}$ for $x \le \dfrac{l}{2}$

and $M_p = \dfrac{1}{4}\sigma_y bd^2$. Using the expressions of M and M_p in equation (18.14), $k = \sqrt{3\left(1-\dfrac{2Wx}{\sigma_y bd^2}\right)}$

Substituting, $x = \dfrac{4000}{2}$; $\sigma_y = 250$, $b = 30$, $h = 40$ and $k = 1/3$, we obtain, $W = 2889\ N$.

If fully elastic section is upto a distance 'α' from end, then $k = 1$ at $x \le \alpha$. Re-arranging the terms in the expression of k, $x = \left(1 - \frac{k^2}{3}\right)\frac{\sigma_y bd^2}{2W}$ *i.e.*, plot of x vs. k at $x > \alpha$ is an equation of parabola. So elasto plastic interface takes a shape of 'parabola'.

Substituting, $x = \alpha$, $k = 1$, $\sigma_y = 250$, $b = 30$, $d = 40$, $W = 2889$ N in above, $\alpha = 1384.6$mm, So plastic zone ends at (2000 – 1384.6) = 615.4 mm from centre of beam on both sides. From equation (18.15), $\phi = \frac{1}{R} = \frac{2M}{EI(3 - k^2)k}$ within elastic zone $0 \le x \le 1384.6$mm, $k = 1$ or, $\phi = \frac{M}{EI}$.

At mid span $k = 1/3$. Thus, $\phi = 2.0769 \frac{M}{EI}$. The curvature diagram is shown in figure 18.14.

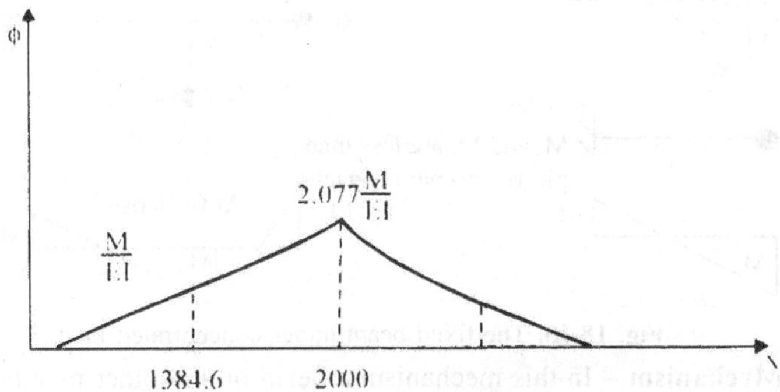

Fig. 18.14. Curvature diagram along length of beam.

Maximum length of plastic hinge:

At collapse $k = 0$ at mid-section. Thus, the collapse load is obtained by substituting $k = 0$ in expression of k *i.e.*, $W = 3000$N = 3kN. Now, Substituting $W = 3000$ N, $x = \alpha$, $k = 1$, $b = 30$, $d = 40$, $\sigma_y = 250$ as before, $\alpha = 1333.33$mm. So max length of plastic hinge is 2(2000 – 1333.33) = 1333.33 mm *i.e.*, approximately 33.33% of full length.

Example 18.2: Calculate the ultimate load, W for the beam in earlier problem if it is subjected to an axial thrust of 5W at ends.

$P_u = 250 \times 30 \times 40 = 300$ kN, $M_p = (250 \times 30 \times 40^2)/4 = 3$ kN-m

$P = 5W$; $M = (W \times 4)/4 = W$. Thus, substituting the values in the interaction diagram,

$$\left(\frac{5W}{300}\right)^2 + \left(\frac{W}{3}\right) = 1 \text{ or, } W = 3\text{kN}$$

18.8 COLLAPSE MECHANISMS

Before considering the methods of plastic analysis, let us first understand the basic considerations in collapse mechanism. The plastic structural analysis of beam and frame structures is concerned with the determination of ultimate load carrying capacity. The stress-strain relation may be idealized as elasto-plastic which admits only the elastic or plastic state. Furthermore, the elastic deformations are very small compare to theoretically infinite plastic deformations. The stress-strain relation thereby moment curvature relations can be further simplified to rigid-plastic (Fig. 18.2).

In our subsequent discussions, we will assume a rigid-plastic stress-strain behavior and also neglect the effect of axial and shear force.

The collapse mechanism can be defined as the conversion of whole or part of the structure into mechanism by the formation of plastic hinge. The various possible forms of mechanisms can be classified as : (*i*) Beam Mechanism (*ii*) Sway (Panel) Mechanism, (*iii*) Joint Mechanism and (*iv*) Combine or Composite Mechanism.

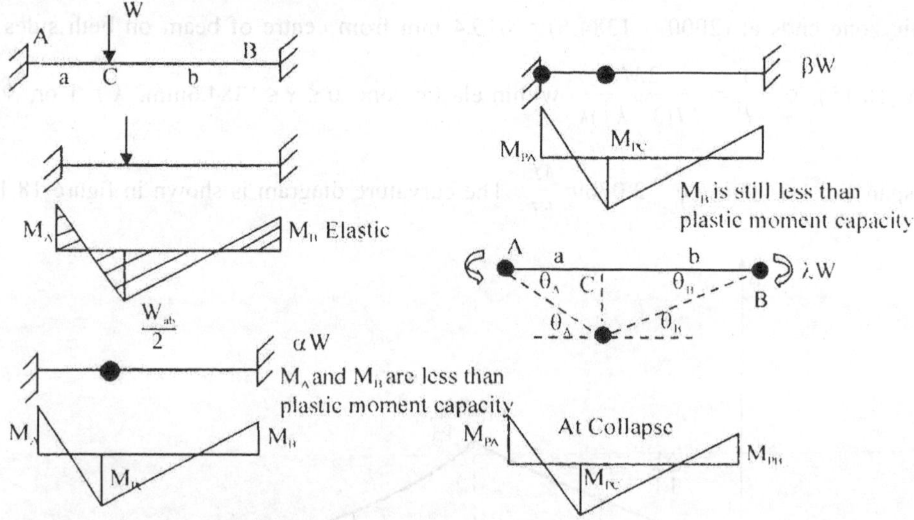

Fig. 18.15. The fixed beam under concentrated load.

(*i*) Beam Mechanism – In this mechanism a beam or any other member carrying transverse loads is converted into a mechanism due to formation of plastic hinges at two ends and at most critical section at some intermediate point *i.e.*, at point of zero shear.

Consider the beam *ABC* in Fig. 18.15 which supports a working load *W*, sufficiently low for stress everywhere to be within elastic range.

(*i*) As the load *W*, progressively increased to say αW, the bending moment at section *C* (say) reaches M_p and plastic hinge form at *C*. Now beam is statically indeterminate of degree one (originally two).

(*ii*) Further increase of load another plastic hinge at point *A* (say) at βW load. (Note that structure is now statically determinate).

(*iii*) As the load is further increased, the *BM* at *A* and *C* remain constant at full plastic moment M_p while that at *B* continues to grow until eventually at a load say λW, the moment at *B* also reaches M_p and hinge developed at *B* also.

As rigid-plastic behaviour is assumed, the beam will undergo unrestricted rotation *i.e.*, the structure becomes a mechanism. Now, the principle of virtual work *i.e.*, the external virtual work done (*EVW*) is equal to the internal virtual work (*IVW*) gives,

$$(\lambda W)\Delta = M_{pA}\theta_A + M_{pB}\theta_B + M_{pC}(\theta_A + \theta_B)$$

i.e.,
$$(\lambda W) = \frac{M_{pA}}{a} + \frac{M_{pB}}{b} + M_{pC}\left(\frac{1}{a} + \frac{1}{b}\right) \qquad (18.20)$$

Above gives an independent equation which can be used to determine the ultimate load factor λ.

(*ii*) Sway (Panel) Mechanism– The sway or panel mechanisms are concerned with the column members. It occurs due to formation of plastic hinges at the ends of the columns. With respect to

Figure 18.16 (*a*), if the collapse mechanism occurs due to side sway at jth story of the building frame, the principle of virtual work gives,

$$\Delta_j(P_1 + P_2 + ... + P_j) = \theta_j(M_{PAD} + M_{PBE} + M_{PCF} + M_{PDA} + M_{PEB} + M_{PFC}) \text{ where } \theta_j = \frac{\Delta_j}{h_j}$$

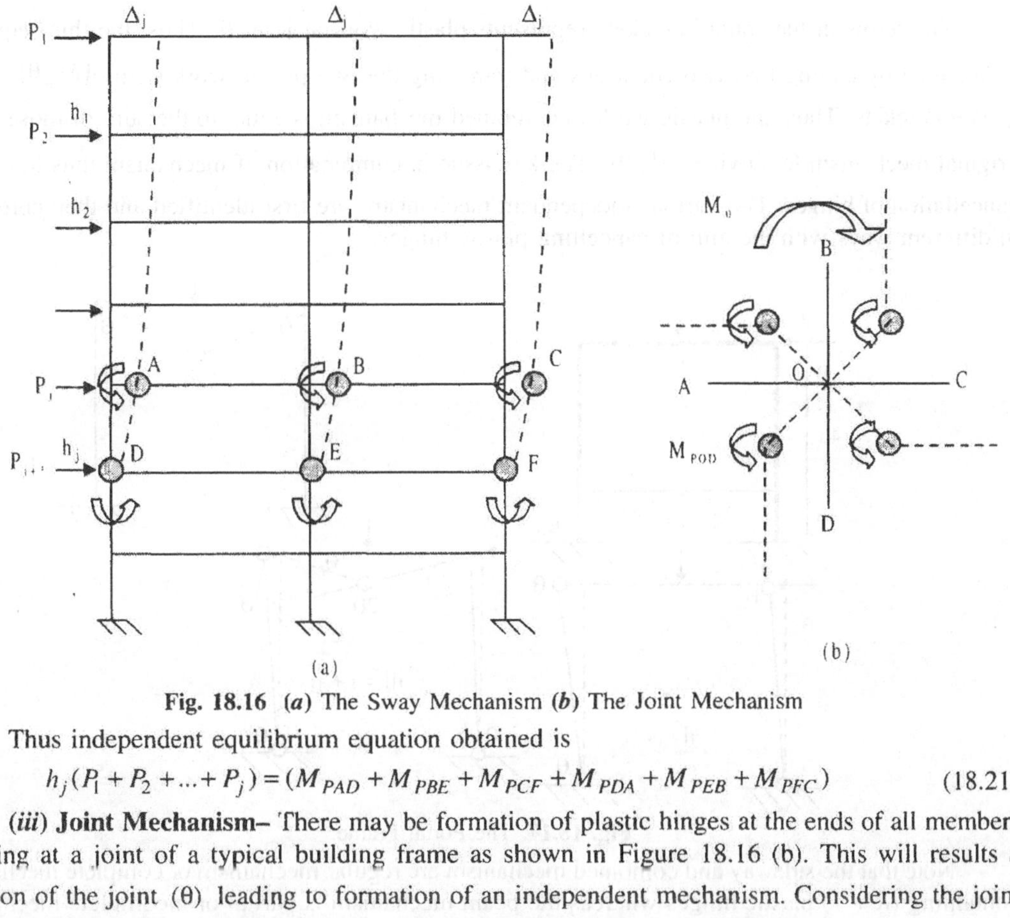

Fig. 18.16 (*a*) The Sway Mechanism (*b*) The Joint Mechanism

Thus independent equilibrium equation obtained is

$$h_j(P_1 + P_2 + ... + P_j) = (M_{PAD} + M_{PBE} + M_{PCF} + M_{PDA} + M_{PEB} + M_{PFC}) \quad (18.21)$$

***(iii)* Joint Mechanism–** There may be formation of plastic hinges at the ends of all members meeting at a joint of a typical building frame as shown in Figure 18.16 (b). This will results a rotation of the joint (θ) leading to formation of an independent mechanism. Considering the joint as shown in figure, if the external moment acts at the joint is M_0, the principle of virtual work gives,

$$M_0\theta = \theta(M_{POA} + M_{POB} + M_{POC} + M_{POD}) \text{ i.e., } M_0 = M_{POA} + M_{POB} + M_{POC} + M_{POD} \quad (18.22)$$

This provides an independent equation of equilibrium.

***(iv)* Combine or Composite Mechanism–** It may be noted that each independent collapse mechanism provides an independent equilibrium equation. Groups of mechanism are called independent if none of them can be obtained from a linear combination of others. It can be readily shown that for a given structure, the number of independent mechanisms, $M = N–R$, where N is the no. of possible plastic hinges and R is the no. of redundancies in the system. These M independent collapse mechanisms can be expressed as a combination of various other possible collapse mechanisms.

Example 18.3: The portal in figure 18.17 has uniform plastic moment capacity for all members. Investigate the various mechanisms to find the actual collapse load.

The two independent mechanisms as shown in (*i*) and (*ii*) of figure 18.17 are combined to form a third mechanism as shown in (*iii*). The virtual work equations for three mechanisms are :

(*i*) Beam Mechanism: $2P(3a\theta) = [M_P]\theta + 3M_p\theta$

(*ii*) Sidesway Mechanism: $P(4a\theta) = [M_p\theta] + 3M_p\theta$

(*iii*) Combined Mechanism: $P(4a\theta) + 2P(3a\theta) = 6M_p\theta$

The terms in the square brackets represents plastic work at joint B. Thus, the third equation is obtained by adding first two equations and removing the two plastic work terms $[M_p\theta]$ within square brackets. Thus, the plastic work in combined mechanism is equal to the sum of those in the original mechanism less twice $[M_p\theta]$. The key issue in combination of mechanism thus lies in the cancellation of hinges. The various independent mechanisms are first identified and then combined in different ways with the aim of cancelling plastic hinges.

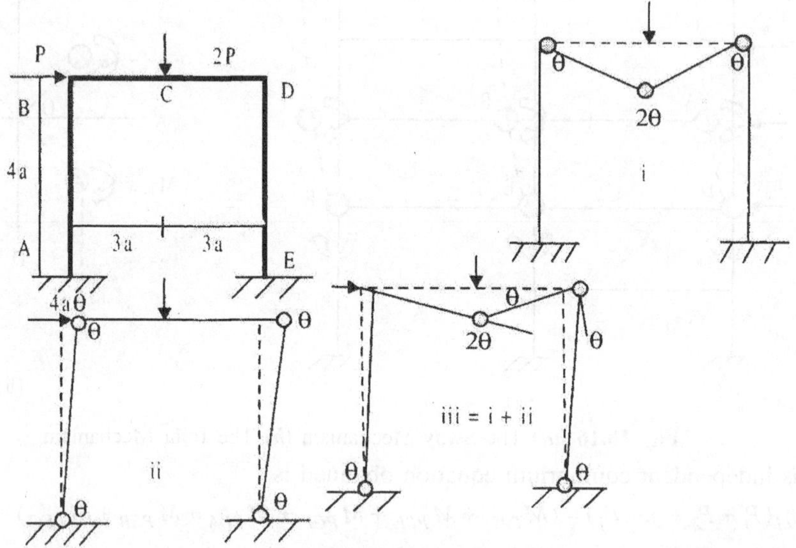

Fig. 18.17. The Portal Frame.

Note that the sideway and combined mechanism are regular mechanism or complete mechanism containing $(R + 1)$ plastic hinges whereas the beam mechanism is partial or incomplete mechanism containing less than $(R + 1)$ plastic hinges. A regular mechanism is statically determinate *i.e.*, the complete bending moment diagram can be obtained from applied load and known value of plastic moments at developed hinges.

18.9 PLASTIC ANALYSIS

With the background of fundamentals of plastic behavior of simple truss or moment curvature relation of beam bending beyond elastic limit, in this section, we introduce the principles of plastic analysis and then present those to analyze continuous beam and frame type structures. The basic difference between elastic and plastic analysis should be realized first. The elastic analysis needs a consideration of continuity, equilibrium and limiting stress condition whereas the plastic analysis method should satisfy the following :

(*i*) Mechanism condition *i.e.*, formation of sufficient plastic hinges to allow the structure to transfer into a mechanism.

(*ii*) Equilibrium conditions *i.e.*, the structure must be in equilibrium.

(*iii*) Yielding condition *i.e.*, the moment anywhere of the structure should be equal to or less than the fully plastic moment capacity of that section (M_p).

The conditions satisfied at elastic and plastic stage are further explained in figure 18.18.

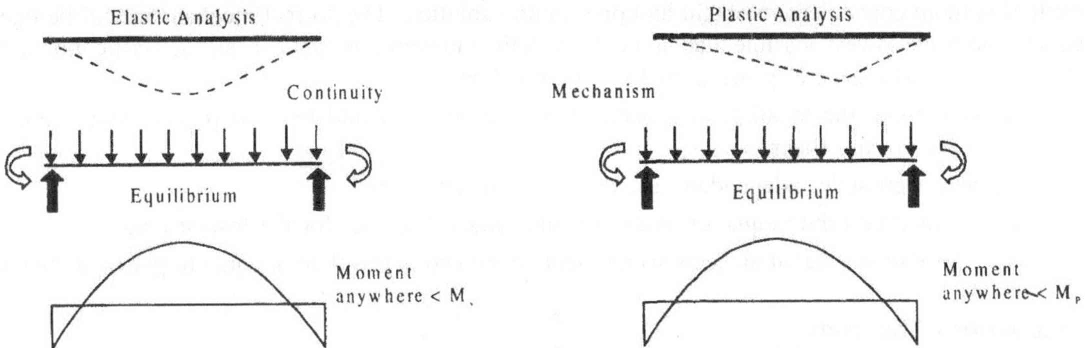

Fig. 18.18. Elastic and Plastic Analysis.

Method of Plastic Analysis:

Three necessary conditions of plastic analysis as discussed earlier are mechanism, equilibrium and yielding. It is not possible to solve three necessary conditions in one operation. However, the equilibrium conditions will be always satisfied. Now, a solution obtained on the basis of an assumed mechanism will give a load carrying capacity that is either correct or too high. On the other hand, one that is obtained by drawing a statical moment diagram that does not violate the plastic moment condition will either be correct or too low. Thus two approaches lead to two useful methods of analysis depending on particular conditions being satisfied.

(*i*) Statical or Equilibrium Method (Lower Bound Theory):

An equilibrium moment diagram is drawn such that moment at any section is equal to or less than the plastic moment capacity (M_p) of the section concerned. The resulting ultimate load is correct only if sufficient plastic hinges were formed at this stage to convert the system into a mechanism. The method leads to ***Lower Bound Theory*** which states that the magnitude of any loading which produces a statically admissible distribution of internal forces if a structure is smaller than or equal to the magnitude of the plastic limit load. When the load bearing capacity of the structure is exhausted, of all statically admissible distribution of internal forces, the actual distribution is that at which the work of the internal forces attains its lowest value.

The basis of the statical method is to find an equilibrium moment diagram in which the bending moment at any section of the structure is equal to or less than the fully plastic moment capacity of the section. The method can be briefly outlined as :

(*i*) Select redundant(s).

(*ii*) Draw bending moment diagram for determinate structure and also for the selected redundant.

(*iii*) Sketch composite bending moment diagram in such a way that a mechanism is formed.

(*iv*) Compute value of ultimate load by solving equilibrium equation.

(*v*) Check $M \leq M_p$.

(*ii*) Mechanical or Mechanism or Kinematic Method: Mechanisms are assumed and the resulting equilibrium equations are solved for the ultimate load value. The method leads to ***Upper Bound Theory*** *i.e.*, the magnitude of any loading which turns a structure into a kinematically admissible mechanism of collapse is larger than or equal to the magnitude of the plastic limit load. Any compatible plastic deformation pattern cannot stand up if the external work exceeds dissipation energy of the internal plastic moments. Thus, of all kinematically admissible mechanism of collapse the actual mechanism is that at which the work of the external forces attains its highest value.

As the number of redundant increases, the number of possible failure mechanisms also increases and it becomes difficult to construct the equilibrium moment diagram. For such case, mechanism method is more convenient to obtain an upper bound solution. The correct mechanism will be one which results the lowest possible load and for which the moment does not exceed the plastic moment at any section. The general process could be outlined as:

(*i*) Determine the location of possible plastic hinges – could be load points, connections, point of pure shear.

(*ii*) Select possible independent and composite mechanism.

(*iii*) Solve equilibrium equation (virtual displacement method) for the lowest load.

(*iv*) Check to see that at all sections moment developed is less than or equal to plastic moment M_p.

Uniqueness Theorem

The actual plastic limit load is that for which the static solution is in agreement with the kinematics one. The virtual work done by internal forces equals the work done by the external forces.

Example 18.4: Determine the ultimate load of the two span beam as shown in Fig. 18.19, assuming uniform plastic moment capacity.

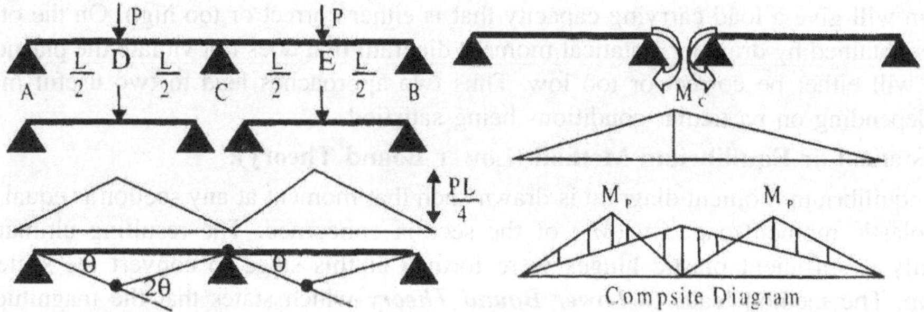

Fig. 18.19. The two span beam and associated mechanism of example 18.4.

Static Method

Taking moment at C as redundant, the composite diagram is shown at right side bottom of figure 18.19. For formation of collapse mechanism, at two section moments should reach M_p. Following figure thus, one can obtain,

$$\frac{PL}{4} - \frac{M_p}{2} = M_p, \qquad \text{Thus, } P_u = \frac{6M_p}{L}$$

Kinematic Method or Mechanism method:

Number of possible hinges $N = 2$ (C and D or E) and number of indeterminacy, $X = 1$. Thus, there is one possible independent mechanism as shown in figure. Work done by load $= \frac{P\theta L}{2}$ and work dissipated in the hinge, $Mp.\theta + Mp.2\theta$. The ultimate load can be obtained from work equation *i.e.*,

$$\frac{P\theta L}{2} = M_p\theta + M_p 2\theta \quad \therefore \quad P_U = \frac{6M_p}{L}$$

Example 18.5: A uniform beam is loaded as shown in the figure 18.20. Determine the plastic moment that will cause the beam to fail.

The beam is statically indeterminate to first degree. Thus, to form mechanism bending moment at any two points *i.e.*, *A*, *C*, *D* should be exactly equal to M_p.

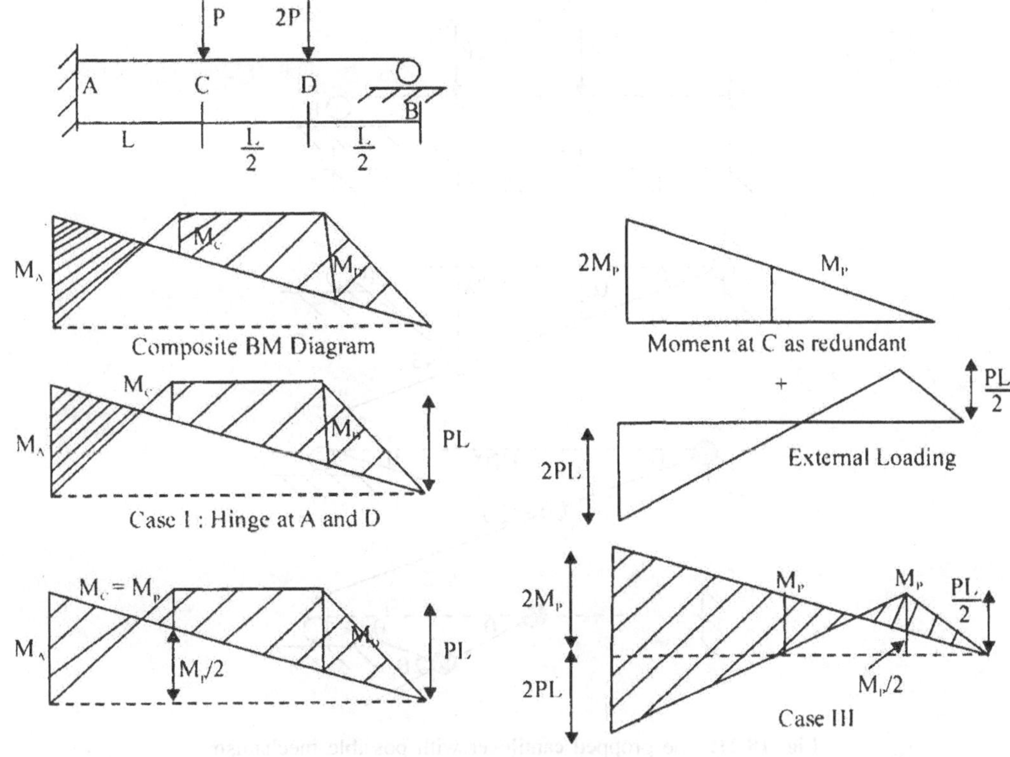

Case II: Hinge at A and C

Fig. 18.20. The propped cantilever with possible moment distribution of example 18.5.

Case I: Assuming plastic hinge at A and D gives, $M_A = M_D = M_p$. Thus, consideration of bending moment diagram for case I gives, $M_p + M_p/4 = PL$ *i.e.*, $P = 1.25M_p/L$.

Case II: Assuming plastic hinge at A and C: For this, $M_A = M_C = M_p$ and from diagram $M_p + M_p/2 = PL$ *i.e.* $P = 1.5\ M_p/L$

Case III. When partial failure due to hinge at C and D: For this, $M_C = M_D = M_p$, Considering M_C as redundant moment, the composite bending moment diagram is shown in figure. From the diagram we can write, $\dfrac{PL}{2} - \dfrac{M_P}{2} = M_P$ *i.e.*, $P = \dfrac{3M_P}{L}$.

The earlier problem is further solved by mechanism method.

The possible hinge locations are A, C and D, thus the number of hinges, $N = 3$ and number of indeterminacy, $X = 1$. Thus, possible number of independent mechanims are, $N - X = 2$. Three possible mechanisms *i.e.*, two indepedent and one partial failure are shown in figure 18.21. The work equation for

Mechanism 1 gives

$$P.\theta L + 2P.3/2\theta L = Mp.\theta + Mp.4\theta \ \text{ i.e., } P = 5\ Mp/4L$$

for mechanims 2, the work equation gives

$$P.\theta L + 2P.1/2\theta L = Mp.\theta + Mp.2\theta \ \text{ i.e., } P = 3\ Mp/2L$$

and for Mechanism 3, the work equation gives

$$2P.1/2\theta L = Mp.\theta + Mp.2\theta \ \text{ i.e., } P = 3\ Mp/L$$

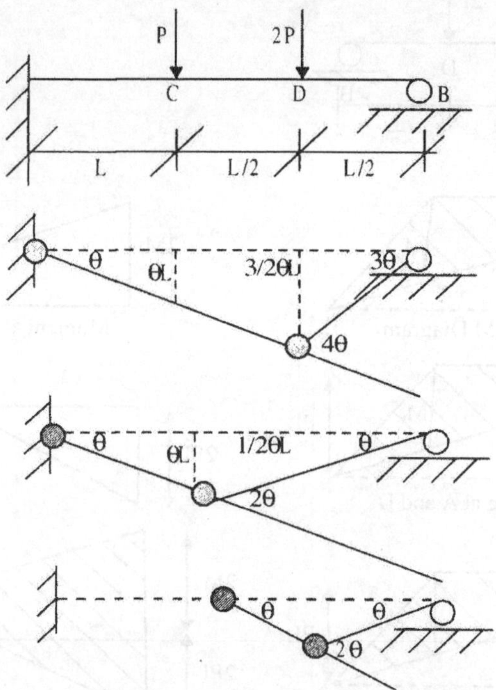

Fig. 18.21. The propped cantilever with possible mechanism.

The three values for P obtained by statical method agree with those obtained by mechanism method. In statical method, case II and III violates yield condition, thus mechanism cannot occur. In case II, $M_D > M_p$, which means collapse mechanism cannot occur unless plastic hinge formation is restricted at D. In case III, $M_A > M_p$, thus collapse mechanism 3 cannot occur. On the other hand, bending moment diagram in case I satisfies all three conditions (*a*) Plastic hinges at *A* and *D*, (*b*) Yielding – since at all point $M \le M_p$ and off course (*c*) Equilibrium. So, based on uniqueness theorem, the corresponding collapse load is $P = 1.25\ M_p/L$.

Example 18.6: A propped cantilever beam is subjected to distributed load of q per unit length as shown in figure 18.22. Find the ultimate load.

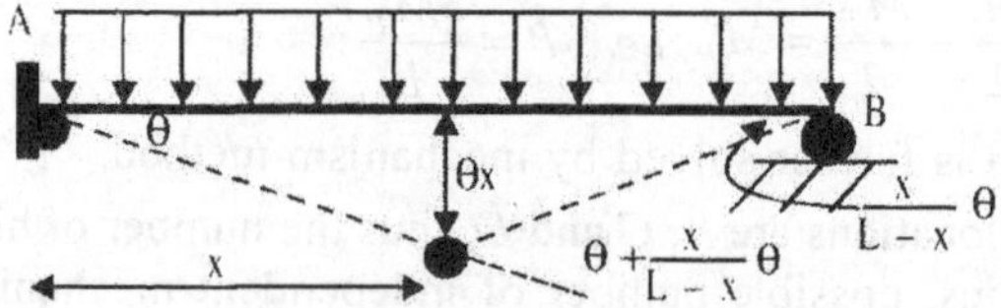

Fig. 18.22. The propped cantilever with distributed load.

The degree of indeterminacy is 1, thus minimum 2 hinges are required to form a mechanism. One hinge at A is obvious and another one will be between A and B. When members are subjected to distributed load, the precise locations of the plastic hinges are not known in advance. The correct value of the ultimate load is obtained by determining the distance to the point of maximum moment. The work equation can be solved to obtain the plastic moment value

$$\frac{1}{2} \times L \times \theta x \times q = M_p\left(\theta + \frac{L\theta}{L-x}\right) \therefore \quad M_p = qL\frac{x(L-x)}{2L-x}$$

By unsafe theory, the value of M_p corresponds to an arbitrarily assigned value of x is less than or at least equal to the plastic moment required to withstand load q. So, M_p is maximized with respect to x *i.e.*, $\frac{d}{dx}(M_p) = 0$, gives $x = 0.586L$ $\therefore$ $M_p = 0.0858\ qL^2$ or $q_u = \frac{M_p}{0.0858L^2}$

Example 18.7: Determine the plastic moment capacity of the beam shown in figure 18.23.

The degree of indeterminacy is 2, thus minimum 3 hinges are required to form a mechanism. The hinges at A and B are obvious, however third hinge location is not certain. But it could be anywhere between A and C. Assuming it at a distance of ξL from end A,

External work

$$\frac{\alpha L + \xi L\theta}{2}(L - \xi L)q + \frac{1}{2}\xi L\theta\xi Lq = WL\left(\xi\theta + \frac{\alpha}{2}\right) \text{where, } W = q\xi L$$

The work equation is $WL(\xi\theta + \frac{\alpha}{2})$

$$= M_p(\theta + \theta + \alpha + \alpha)\ i.e.,\ M_p = \frac{\xi(3 - 2\xi)}{8}WL$$

For M_p correspond to collapse mechanism,

$$\frac{dM_p}{d\xi} = 0 \text{ gives, } \xi = \frac{3}{4} \text{ thus, } M_p = \frac{9}{64}WL$$

Fig. 18.23. The Fixed Beam.

18.10 APPLICATION TO PORTAL FRAME

Example 18.8: Find the ultimate load for the portal frame shown in figure 18.24. Assume uniform plastic moment capacity.

If, P_u is the ultimate load, work equation of beam mechanism gives, $P_u.L\theta = M_p(\theta + 2\theta + \theta)$ i.e.

The work equation of sway mechanism gives, $rP_u.L\theta = M_p(\theta + \theta + \theta + \theta)$ *i.e.*, $P_u = \frac{M_p}{L}$

The work equation of combined mechanism gives $P_u.L\theta + rP_u.L\theta = M_p(\theta + 2\theta + 2\theta + \theta)$ *i.e.*, $P_u = 6M_p/(1+r)L$

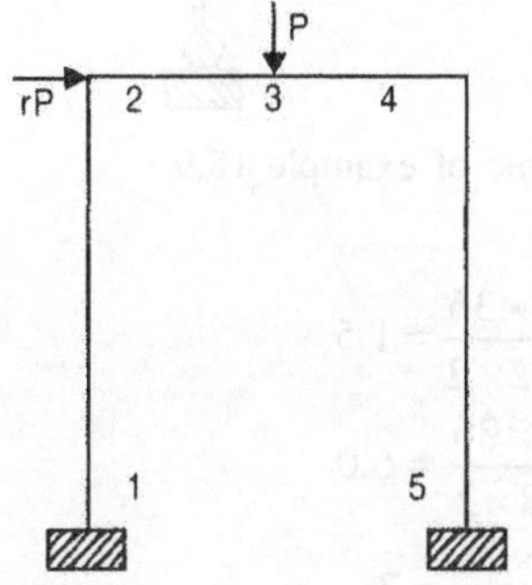

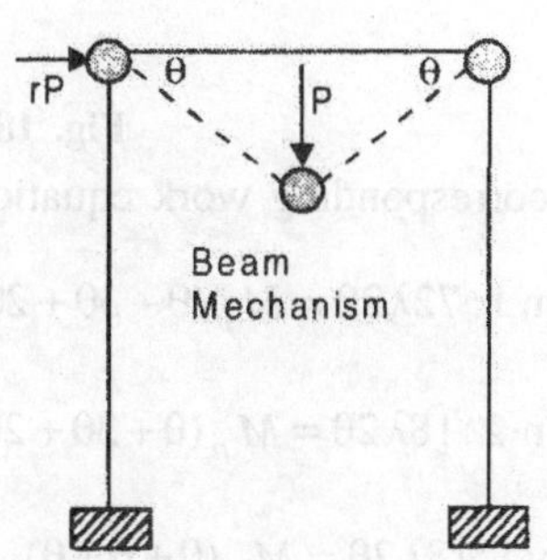

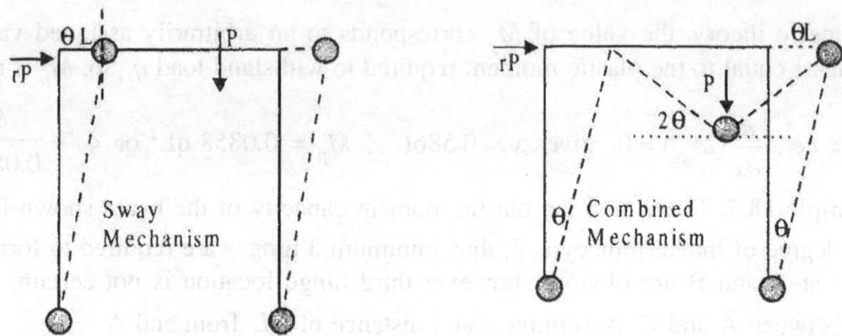

Fig. 18.24. The portal frame and associated mechanisms.

Thus, the ultimate load of the system depends on the value of r. For example, if $r = 2$, $P_u = 2M_p/L$ and for $r = 1$, $P_u = 3\ M_p/L$.

Example 18.9: Determine the load factor for the portal shown in figure 18.25.

The total numbers of possible hinge locations are 5 and degree of indeterminacy is 2, thus the number of independent mechanism 5 – 2 = 3. Three possible independent mechanisms and two combined mechanism are shown in the figure.

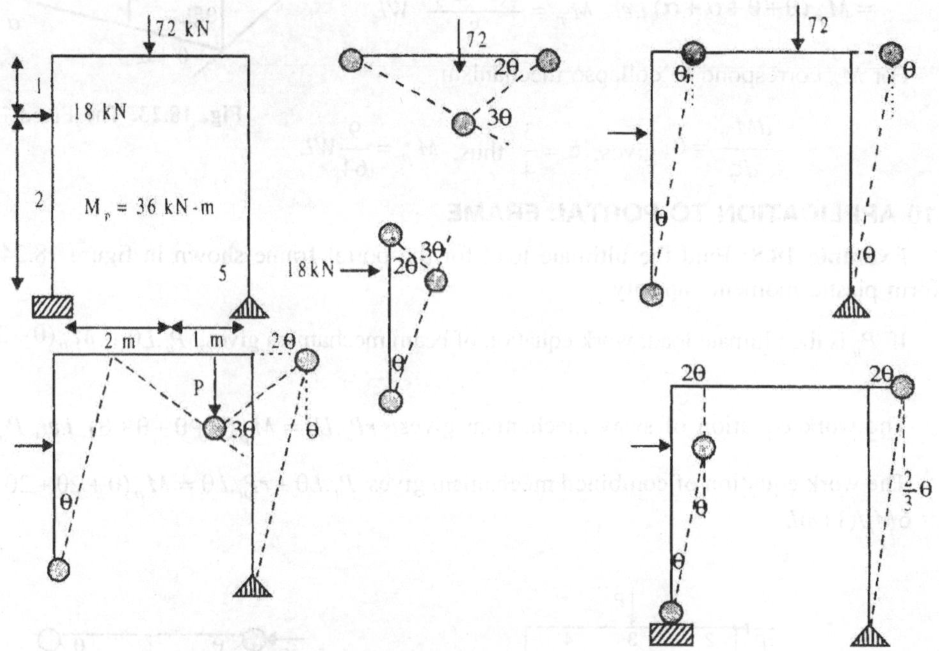

Fig. 18.25 (a) Portal frame of example 18.9.

The corresponding work equations are :

Beam 1: $72\lambda 2\theta = M_p(\theta + 3\theta + 2\theta)$ $\quad \therefore \quad \lambda = \dfrac{6\times 36}{72\times 2} = 1.5$

Beam 2: $18\lambda 2\theta = M_p(\theta + 3\theta + 2\theta)$ $\quad \therefore \quad \lambda = \dfrac{6\times 36}{18\times 2} = 6.0$

Sway 1: $18\lambda 2\theta = M_p(\theta + \theta + \theta)$ $\quad \therefore \quad \lambda = \dfrac{3\times 36}{18\times 2} = 3$

Sway 2: $18\lambda 2\theta = M_p(\theta+\theta+\frac{2}{3}\theta)$ $\therefore\ \lambda = \frac{8\times 36}{3\times 18\times 2} = 2.67$

Combined: $18\lambda 2\theta + 72\lambda 2\theta = M_p(\theta+3\theta+2\theta+\theta)$ $\therefore\ \lambda = \frac{7\times 36}{18\times 2+72\times 2} = 1.4$

Thus, the load factor $\lambda = 1.4$. The results are confirmed with statical method with associated bending moment diagram as shown in figure 18.25 (*b*).

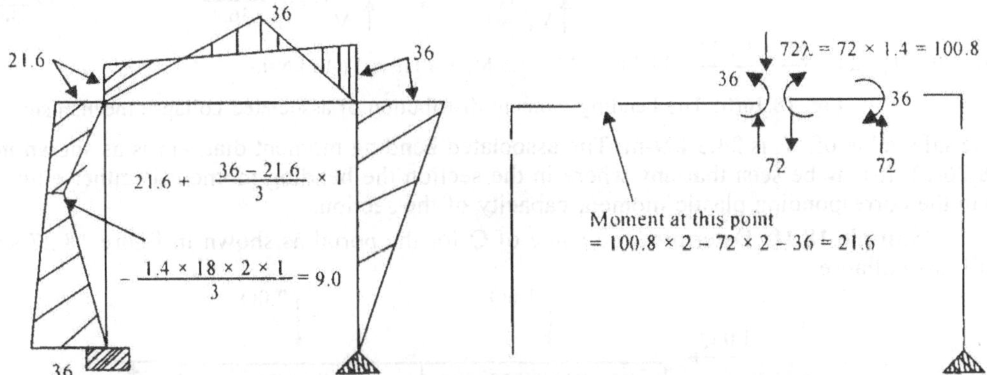

Fig. 18.25 (b) Moment correspond to collapse mechanism.

Example 18.10: Find allowable M_p to carry the load safely as shown in figure 18.26(a).

Various possible collapse mechanisms are shown in figure. Though, the failure mechanisms are same as example 18.8, formation of hinges occur earlier in the beam as $(M_P)_{beam} < (M_P)_{column}$.

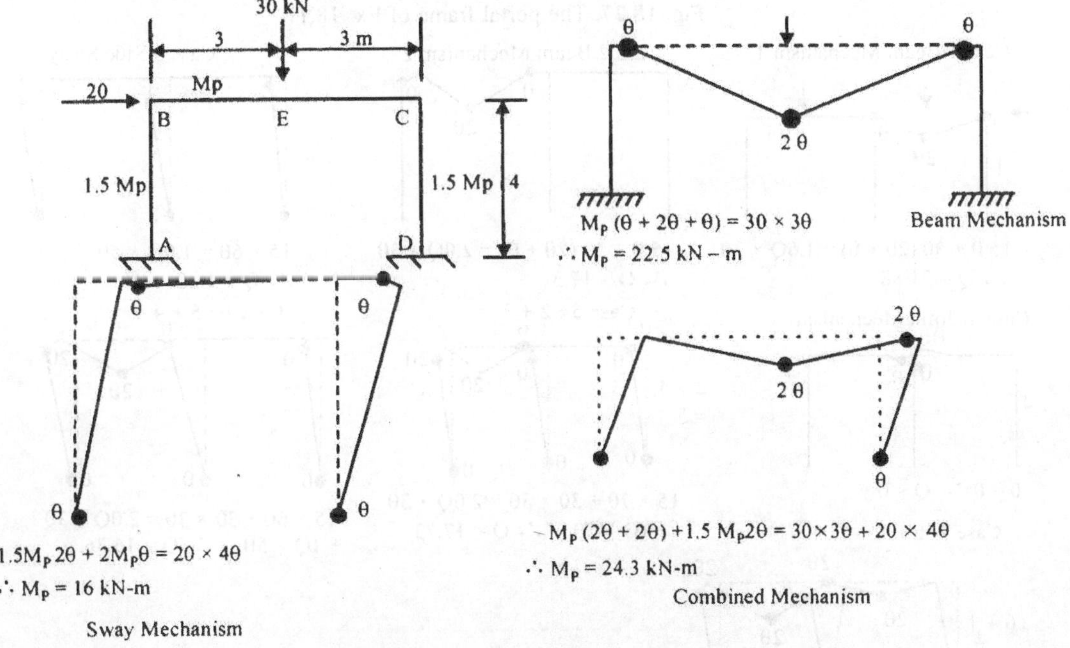

Fig. 18.26(a) Portal frame of Example 18.10.

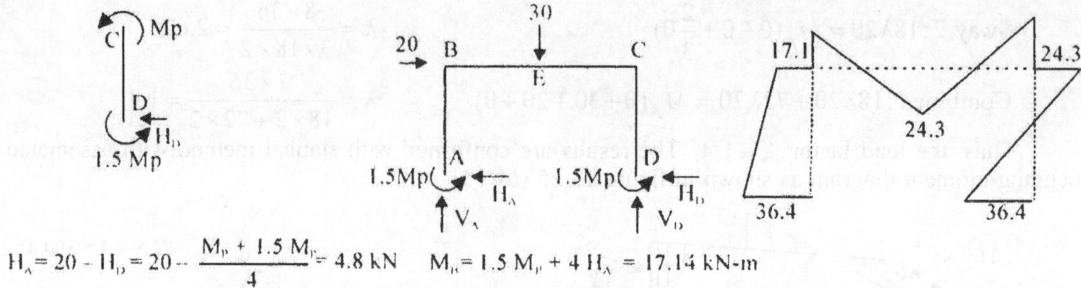

$H_A = 20 - H_D = 20 - \frac{M_p + 1.5\,M_p}{4} = 4.8$ kN $\quad M_E = 1.5\,M_p + 4\,H_A = 17.14$ kN-m

Fig. 18.26(b) The bending moment distribution of associated collapse mechanism.

The safe value of M_p is 24.3 kN-m. The associated bending moment diagram is as shown in figure 18.26(b). It may be seen that any where in the section the bending moment is either equal or less than the corresponding plastic moment capacity of the section.

Example 18.11: Determine the value of Q for the portal as shown in figure 18.27 so that it will just collapse.

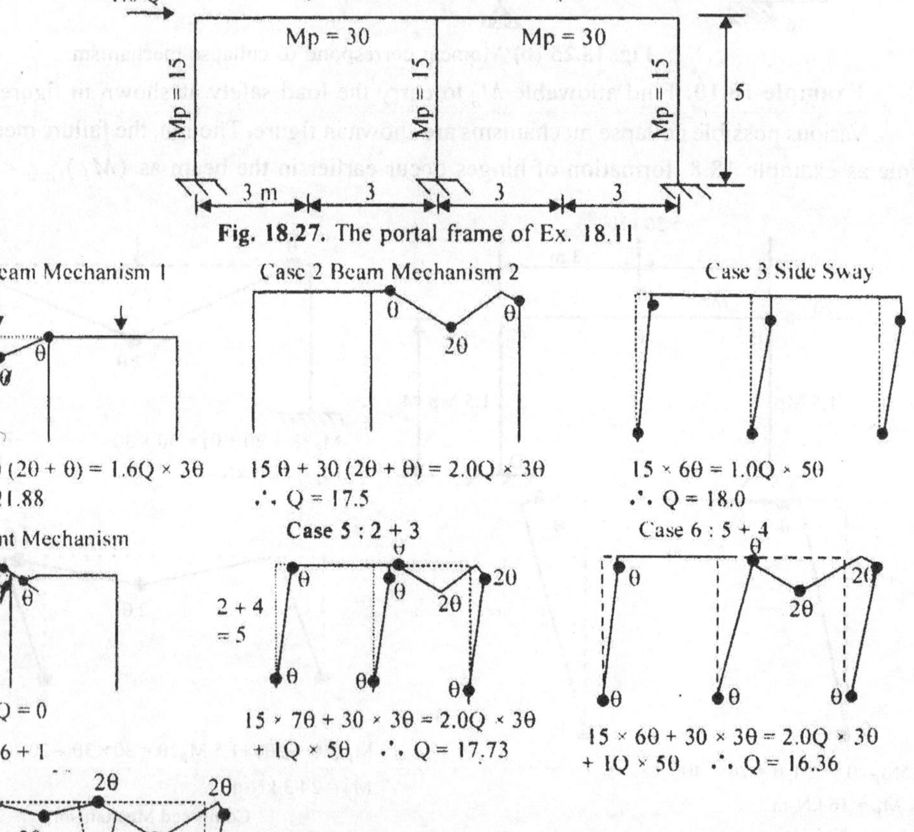

Fig. 18.27. The portal frame of Ex. 18.11

As we can see that mechanism 7 correspond to lowest value of the actual failure mechanism *i.e.*, $Q = 16.14$.

Example 18.12: Find the required plastic moment capacity of the members of the two storey frame as shown in figure 18.28.

Number of redundancies $R = 6$, $M = N - R = 14 - 6 = 8$. So there will be total eight independent mechanisms. These are mechanisms (*i*) to (*iv*) shown in figure and 4 joint rotations at *B*, *C*, *D* and *E*.

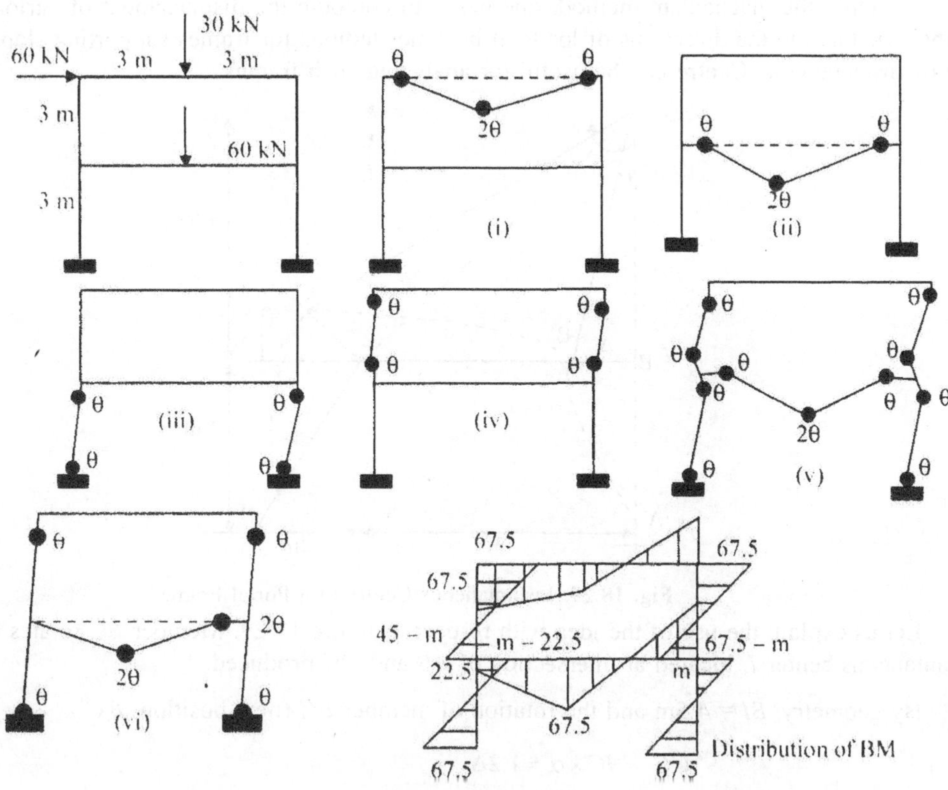

Fig. 18.28. The Two Storey Frame.

Thus, the work equation for various collapse mechanism are as follows :

Mechanism (*i*) $30 \times 3\theta = M_p(\theta + 2\theta + \theta)$ $\quad \therefore \; M_p = \frac{90}{4}$

Mechanism (*ii*) $60 \times 3\theta = M_p(\theta + 2\theta + \theta)$ $\quad \therefore \; M_p = \frac{180}{4}$

Mechanism (*iii*) $60 \times 3\theta = M_p(\theta + 2\theta + \theta)$ $\quad \therefore \; M_p = \frac{180}{4}$

Mechanism (*iv*) $60 \times 3\theta = M_p(\theta + 2\theta + \theta)$ $\quad \therefore \; M_p = \frac{180}{4}$

Mechanism (*v*) $60 \times 3\theta + 60 \times 6\theta = M_p(\theta + 6\theta + \theta + \theta + \theta + \theta)$ $\quad \therefore \; M_p = \frac{540}{12} = 45.0$

Mechanism (*vi*) $60 \times 3\theta + 60 \times 6\theta = M_p(\theta + \theta + 2\theta + 2\theta + \theta + \theta)$ $\quad \therefore \; M_p = \frac{540}{8} = 67.5$

case (*vi*) gives the highest value of the required plastic moment M_p = 67.5 kN-m. As the failure mechanism is partial, it is statically indeterminate. Thus, one cannot draw the actual bending moment diagram. However, assuming the bending moment at end *E* of member *ED* as *m*, the bending moment distribution can be drawn as shown in figure. It may be seen that for $0 \le m \le 67.5$, the bending moment exceeds nowhere the value of 67.5 kN-m. Thus, the uniqueness theorem says that required plastic moment, M_p = 67.5.

18.11 GEOMETRY OF MECHANISM (INSTANTANEOUS CENTRE)

To apply the mechanism method, one needs to compute the displacement of various points on the structures in the directions of loads. It becomes tedious for frames supporting sloping roof. Use of Instantaneous Centre can be useful for analyzing such frames.

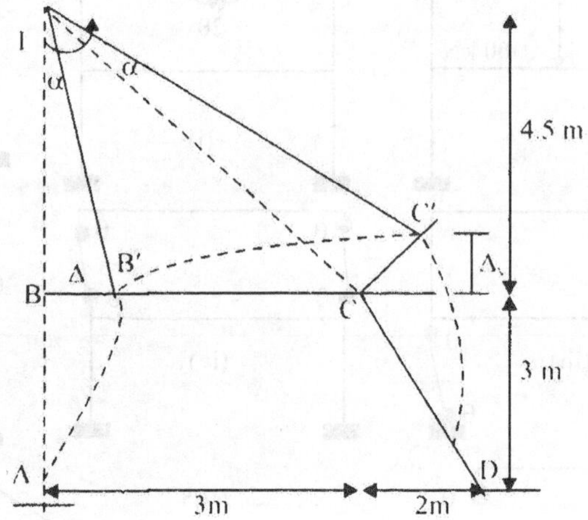

Fig. 18.29. Instantaneous Centre of a Portal Frame.

Let us explain the use of the idea with respect to figure 18.29. Member *BC* rotates about its instantaneous center *I*, located at intersection of *AB* and *DC* produced.

By geometry, *BI* = 4.5m and the rotation of member *BC* (new position *B'C'*), $\alpha = \Delta / 4.5$

$$CC' = IC \times \alpha = 1.2\Delta$$

and $$\Delta_{I'} = \sqrt{(1.2\Delta)^2 - \Delta^2} = 0.667\alpha.$$

Example 18.13. Determine the ultimate load *W* for the inclined portal as shown in figure 18.30

The associated beam and sway mechanism are shown in figure 18.30 (*a*) and the combined mechanism is shown in figure 18.30 (*b*). The corresponding work equations are:

Beam Mechanism: $W\theta\frac{L}{2} = M_p(\theta + 2\theta + \theta) \quad \therefore \quad W = \frac{8M_p}{L}$

Sway Mechanism: $W\theta\frac{L}{2} + WL\theta = M_p(\theta + 2\theta + 2\theta + \theta) \quad \therefore \quad W = \frac{4M_p}{L}$

Combined Mechanism: $W \times 3\theta\frac{L}{2} + WL\theta = M_p(\theta + 4\theta + 4\theta + \theta) \quad \therefore \quad W = \frac{4M_p}{L}$

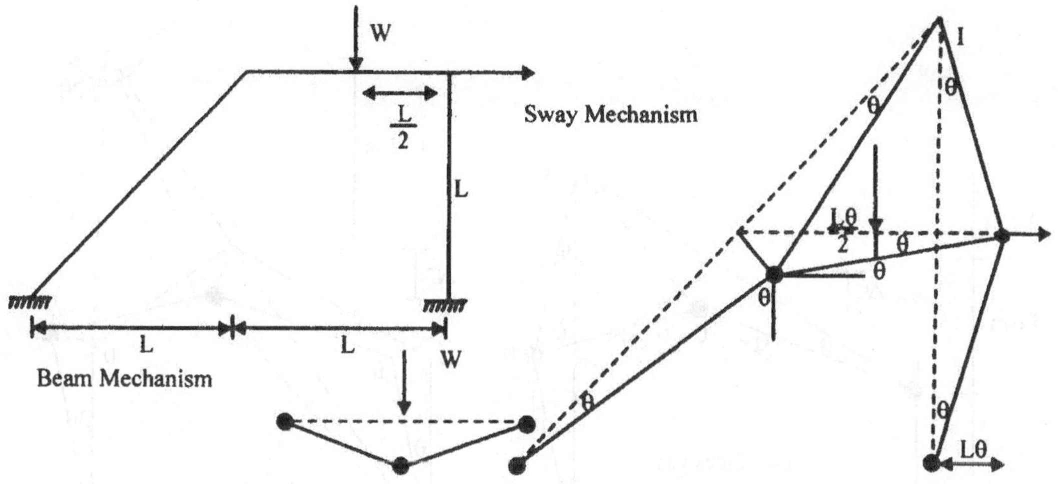

Fig. 18.30(*a*). Inclined Portal of Example 18.

Thus, the ultimate load is minimum of above *i.e.*, $\frac{4M_p}{L}$. Distribution of bending moment of the associated collapse mechanism is shown in figure 18.30(b).

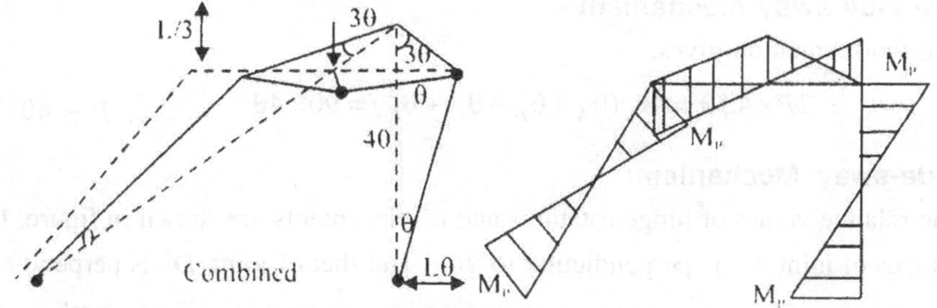

Fig. 18.30(*b*). The combined mechanism and bending moment distribution.

Example 18.14: Find the value of P which will just produce collapse of the gable frame as shown in figure 18.30. Assume plastic moment capacity for all members as 90 kN-m.

Number of plastic hinge is 5, thus there could be two possible independent mechanisms. The two independent mechanisms and one combined mechanism are shown in figure.

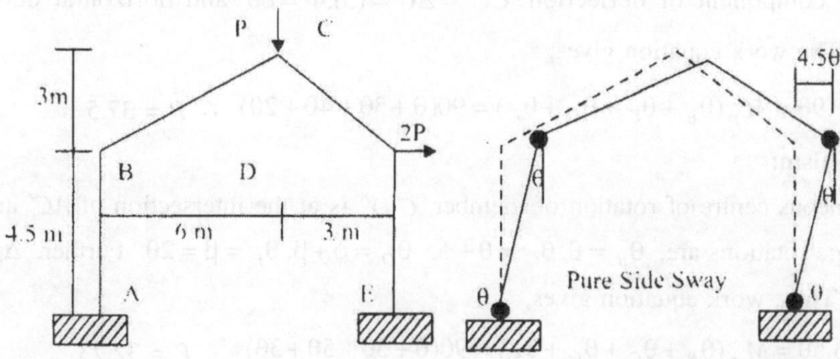

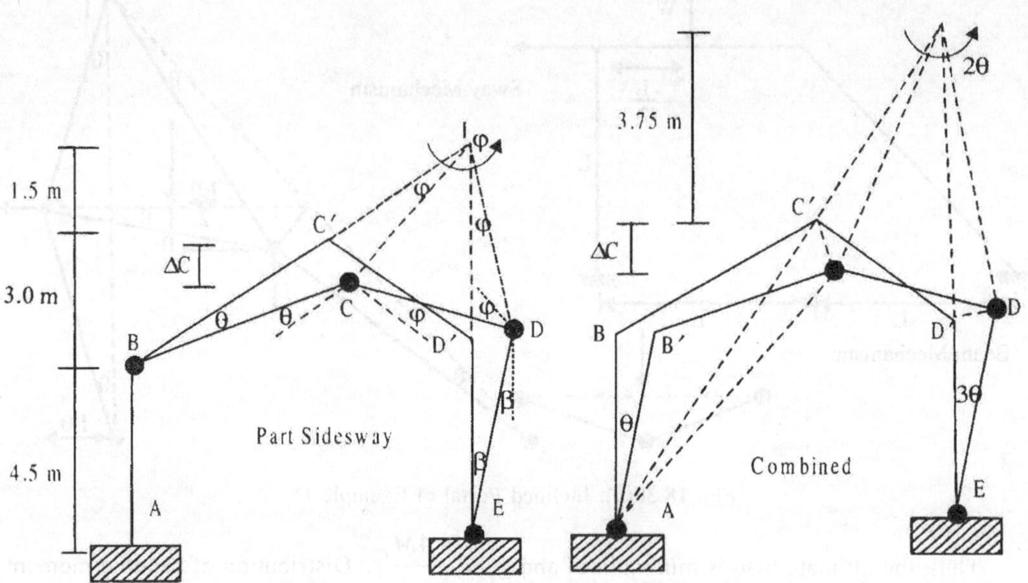

Fig. 18.31. The Gable Frame and associated collapse mechanisms.

For pure side sway mechanism

The work equation gives,

$$2P\times 4.5\theta = M_p(\theta_A+\theta_B+\theta_D+\theta_E) = 90\times 4\theta \qquad \therefore\ P = 40$$

Part Side-sway Mechanism:

The relative values of hinge rotations and displacements are shown in figure. It may be noted that the locus of joint C' is perpendicular to BC' and that of joint D' is perpendicular to ED'. If the rotation of BC' is taken as θ, then $CC' = \theta\times BC' = \phi\times IC'$ *i.e.*, $\phi = \dfrac{BC'}{IC'}\theta = 2\theta$.

Similarly, $DD' = \phi\times ID' = \beta ED'$ *i.e.*, $\beta = \dfrac{ID'}{ED'}\phi = \dfrac{4.5}{4.5}\phi = \phi$. Thus, the hinge rotations are, $\theta_B = \theta,\ \theta_C = \theta+\Phi = 3\theta,\ \theta_D = \Phi+\beta = 4\theta$ and $\theta_E = \beta = 2\theta$.

The vertical component of deflection $CC' = \Delta C = (3).\Phi = 6\theta$ and horizontal deflection $DD' = 4.5\beta = 9\theta$. The work equation gives,

$$P.6\theta + 2P.9\theta = M_p(\theta_B+\theta_C+\theta_D+\theta_E) = 90(\theta+3\theta+4\theta+2\theta) \quad \therefore\ P = 37.5$$

Combined Mechanism:

The instantaneous centre of rotation of member $C'D'$ is at the intersection of AC' and ED' produced. The hinge rotations are, $\theta_B = \theta, \theta_C = \theta+\phi,\ \theta_D = \phi+\beta, \theta_E = \beta = 2\theta$. Further, $\Delta_C = 6\theta$ and $DD' = 13.5\theta$. Thus, work equation gives,

$$P.6\theta + 2P.13.5\theta = M_p(\theta_B+\theta_C+\theta_D+\theta_E) = 90(\theta+3\theta+5\theta+3\theta) \quad \therefore\ P = 32.73$$

Thus, the maximum value of load P is 32.73 *kN*.

ANSWERS

Chapter 1.

1.2 (*a*)

Chapter 2.

2.2 (*i*) 18 – 1 = 17 (*ii*) (*c*) (*iii*) (*b*) (*iv*) (*b*).

2.3 (*i*) 1 (*ii*) 4 (*iii*) 0 (*iv*) 0 (*v*) 1 (*vi*) 0 (*vii*) 1 (*viii*) – 1 unstable (*ix*) 1 (*x*) 2 (*xi*) 0 (*xii*) 1 (*xiii*) 2 (*xiv*) 3 (xv) 4 (*xvi*) 7 (*xvii*) 1.

2.4 $V_A = 3\ kN \uparrow V_D = 5\ kN \uparrow\ \ H_D = 2\sqrt{3} \rightarrow$

2.5 $R_A = \dfrac{3P}{2} \uparrow R_D = \dfrac{3P}{2} \uparrow M_A = \overset{\curvearrowleft}{PL}$

2.6 $V_A = 11.16\ kN \uparrow H_A = 5$ kN $M_A = 34.15\ kNm \circlearrowright V_B = 2.5\ kN \uparrow H_D = 0\ \ H_B = 0$

2.7 $V_A = 0\ \ H_A = 5\ kN \leftarrow\ \ M_A = 10\ kNm \circlearrowleft\ \ V_E = 0\ \ H_E = 0$

2.8 $V_A = 0.86\ kN \downarrow\ \ V_B = 2.36\ kN \uparrow$

Chapter 3.

3.2 (*i*) (*c*) (*ii*) (*c*) (*iii*) $P_1 = 0$

3.3 $F_{CB} = W$ (*t*), $F_{BD} = -W$ (c) $F_{DA} = -W$ (c) $F_{AB} = W$ (*t*)

3.4 $F_{AB} = 10.1$ (*c*) *kN* $F_{AI} = 20.20$ (*t*) *kN*
$F_{IB} = 20.20$ (*c*) *kN* $F_{IH} = 20.20$ (*t*) *kN*
$F_{BH} = 8.66$ (*t*) *kN* $F_{BC} = 24.54$ (*c*) *kN*
$F_{HC} = 8.66$ (*c*) *kN* $F_{HG} = 28.87$ (*t*) *kN*
$F_{CG} = 14.43$ (*c*) *kN* $F_{DG} = 14.43$ (*t*) $F_{DF} = 14.43$ (*c*) *kN*
$F_{CD} = 21.65$ (*c*) *kN* $F_{GF} = 14.43$ (*t*) *kN*
$F_{DE} = 7.22$ (*c*) *kN*

3.5 $F_{DE} = 20$ (*t*) *kN* $F_{DC} = 10\sqrt{3}$ (*c*) *kN* $F_{CE} = 10$ (*t*) *kN* $F_{CB} = F_{CD} = 10\sqrt{3}$ (*c*) *kN*
$F_{EB} = 10$ (*c*) *kN* $F_{EA} = 30$ (*t*) *kN*

3.6 $F_{AH} = 12.95$ (*c*) *kN* $F_{AB} = 11.11$ (*t*) *kN*
$F_{BH} = 0$ $F_{BC} = 11.11$ (*t*) *kN*
$F_{HI} = 14.36$ (*c*) *kN* $F_{HC} = 2.59$ (*t*) *kN*
$F_{IC} = 8.66$ (*t*) *kN* $F_{IJ} = 8.49$ (*c*) *kN* $F_{CD} = 13.33$ (*t*) *kN*
$F_{ID} = 7.07$ (*c*) *kN* $F_{GF} = 5.56$ (*t*) *kN* $F_{FL} = 0$
$F_{EF} = 5.56$ (*t*) *kN* $F_{KL} = 7.18$ (*c*) *kN*
$F_{EL} = 1.29$ (*t*) *kN* $F_{KE} = 0.666$ (*c*) *kN*
$F_{JK} = 8.498$ (*c*) *kN* $F_{DK} = 2.357$ (*t*) *kN*
$F_{IJ} = 8.498$ (*c*) *kN* $F_{KJ} = 8.498$ (*t*) *kN*
$F_{JD} = 1.954$ (*t*) *kN* $F_{GF} = 5.56$ (*t*) *kN*
$F_{DE} = 6.660$ (*t*) *kN*

3.7 $F_a = 44.44\ kN$ (comp.) $F_d = 44.44\ kN$ (tension)
$F_b = 17.36\ kN$ (comp.) $F_C = 17.36\ kN$ (tension)

3.8 $F_{AB} = 14.144\ (t)\ kN$ $F_{AC} = 10\ (c)\ kN$
$F_{BC} = 14.144\ (c)\ kN$ $F_{BD} = 10\ (t)\ kN$
$F_{CD} = 20\sqrt{2}\ (t)\ kN$ $F_{CE} = 40\ (c)\ kN$
$F_{ED} = 7.07\ (t)\ kN$ $F_{DF} = 35.35\ (t)\ kN$
$F_{EF} = 5\ (c)\ kN$

3.9 $F_{OC} = 19.48\ (c)\ kN$ $F_{OB} = 1.777\ (c)\ kN$ $F_{OA} = 15.99\ (t)\ kN$

3.10 $F_{DB} = 190.96\ (c)\ kN$ $F_{DA} = 63.64\ (t)\ kN$ $F_{DC} = 95.96\ (t)\ kN$

3.11 $R_{cy} = 0$ $R_{by} = 20\ kN$ $R_{bz} = 17.33\ kN$
$R_{az} = -17.33\ kN$ $R_{ay} = -20\ kN$
$F_{da} = F_{df} = 0$ $F_{de} = -20 = F_{ed} = F_{eh}$
$F_{bc} = 20$ $F_{bf} = 0$

3.12 $t_{FB} = -\dfrac{3}{2}\dfrac{P}{L}$ $t_{FC} = \dfrac{5}{6}\dfrac{P}{L}$ $F_{EA} = -1.5\ P\ (c)$ $F_{ED} = 1.5\ P\ (t)$

$t_{FD} = \dfrac{2}{3}\dfrac{P}{L}$ $t_{EA} = -\dfrac{P}{L}$ $F_{EF} = -P\ (c)$ $F_{ED} = \dfrac{2P}{\sqrt{3}}(t)$

$t_{EF} = -\dfrac{2P}{L}$ $t_{ED} = \dfrac{P}{L}$ $F_{FB} = -2.25\ P\ (c)$ $F_{FC} = 1.25\ P\ (t)$

3.13 $F_{DF} = F_{ED} = F_{EB} = F_{EF} = 0$ $R_{CY} = 27.77\ kN.$
$F_{DA} = F_{DB} = 0$ $F_{AF} = -15.351\ (c)$ $R_{AX} = 0$ $R_{CZ} = 0$
$F_{AC} = 8.72\ (t)$ $F_{AB} = 2.473\ (t)$ $R_{BX} = 0$
$F_{BF} = -12.94\ (c)$ $R_{AY} = 11.11$ kN
$F_{BC} = 6.17\ (t)$ $F_{CF} = -30.948\ (c)$ $R_{BY} = 11.11$ kN

Chapter 4.

4.1 (*i*) (*d*) (*ii*) (*d*) (*iii*) (*b*) (*iv*) (*b*) (*v*) (*c*) (*vi*) (*a*)

4.2 $R_A = 3\ kN \uparrow$ $V_D = 5\ kN \uparrow$ $H_D = 2\sqrt{3}\ kN$
S.F. at $A = 3\ kN$, $B = 1\ kN$ $C_{left} = 1\ kN$ $C_{right} = -3\ kN$
$D_{left} = -3\ kN$ $D_{right} = +2\ kN$
B.M. at $A = 0$, $B = 4\ kNm$ $C = 6\ kNm$ $D = -2\ kNm$ $E = 0$

4.3 S.F. at $A = 3.33\ kN$ $B_{left} = 0.67\ kN$ $B_{right} = 5.67\ kN$
$C_{left} = 7.67\ kN$ $C_{right} = 4\ kN,$ $D = 0$
B.M. at $A = 0$ $B = 2.66\ kNm$ $C = -4\ kNm$ $D = 0$

4.4 S.F. at $A = 2.5\ kN,$ $B_{left} = 2.5$ $B_{right} = -2.5\ kN$
$D_{left} = -2.5\ kN$ $D_{right} = 0.625\ kN$ $E = 0.625\ kN$
B.M. at $A = 0$ $B = +5\ kNm$ $C = 0$ $D = -2.5\ kNm$ $E = 0$

4.5 $R_A = 10^t$ $R_B = 20^t$
S.F. at $A = 10\ t$ at $B = 20^t$
0 at 3.464 *m* from *A*
B.M. at $A = 0$ $B = 0$ *Max*^{m} B.M. = 23.094 *tm*

4.6 $R_B = 7.166\ kN$ $\theta = 42.48°$

Max^m B.M. $= 6.33\ kNm$

4.7 S.F. at $A = 43\ kN$ $B_{left} = 43\ kN$ $B_{right} = 57\ kN$

$C_{left} = 57\ kN$ $C_{right} = 54\ kN$ $D = 30\ kN$

$E_{left} = 30\ kN$ $E_{right} = 20\ kN$ $G = 56\ kN$

B.M. at $A = 0$ $B = +258\ kNm$ $C = -84\ kNm$

$D = 0$ $E = 120\ kNm$ $F = 0$ $G = -396\ kNm$

4.8 S.F. at $A = +16\ kN$ $B_{left} = +16\ kN$

$B_{right} = -24\ kN$ $C_{left} = -24\ kN$ $C_{right} = 40\ kN$ $D = 0$

Chapter 5.

5.1 (*a*) $M_A = -84\ kNm$ $M_B = -48$ $M_E = 0.00$

S.F. at A to $E = 12\ kN$

B to $F = 24\ kN$

A.F. at A to $B = 24\ kN$

5.1 (*b*) B.M. at $A = 0$ $B = 36\ kNm$ $C = 0$ $D = 0$

S.F. at A to $E = -12\ kN$

B to $F = 3\ kN$

F to $C = 21\ kN$

A.F. A to $B = 3\ kN$ D to $C = 21\ kN$

5.2 (*a*) B.M. at $A = 0$ $B = -10\ kNm$ $C = 0$

S.F. at A to $B = -10\ kN$

B to $C = +5\ kN$

A.F. D to $B = -15\ kN$

E to $C = +5\ kN$

5.2 (*b*) B.M. at $A = 0$ $B = -10\ kNm$ $C = -30\ kNm$ $D = -30\ kNm$

S.F. at $A = -10\ kN$ $B = -10\ kN$

A.F. at E to $C = -10\ kN$

5.3 (*a*) B.M. at $A = 0$ $B = +15\ kNm$

at $C = 0$ $D = -15\ kNm$ $E = 0.00$

S.F. at $A = 2.77\ kN$ $B_{left} = 2.77\ kN$ $B_{right} = 5\ kN$

A.F. at $A = 1.847\ kN$, $B_{left} = 1.847\ kN$ $B_{right} = 6.67\ kN$ $C = 6.67\ kN$

$D_{left} = 6.67\ kN$ $D_{right} = 7.8598\ kN$

5.3 (*b*) A.F. at $A = -5\ kN$

$B = -5\ kN$

$C, D = +\dfrac{30}{\sqrt{5}}$

S.F. at $A = 0$

$B = +5\ kN$

$C_{left} = +15\ kN$

$C_{right} = 3\sqrt{5}$ $D = -7\sqrt{5}\ kN$

B.M. at $A = 0$ $B = 0$ $C = +20\ kNm$ $D = 0$

Chapter 6.

6.2 (*i*) (*a*) (*ii*) (*c*)

6.3 Maximum Tension = 628.245 *t*
Tension at Back stay at B = 699.8973 *t*
Pressure on Pier B = 484.8233 *t*
Tension at Back stay at A = 699.899 *t*
Pressure on Pier A = 515.115 *t*

6.4 H = 312.5 *tf* Maximum tension in each cable = 336.572 *tf*
Required sectional area = 280.48 cm^2
Overturning force = 74.517 *tf*

6.5 V_D = 53.75 *tf* V_I = 51.25 *tf*

6.6 Maximum tension in cable = 13463 *tf*
S.F. at 25 *m* = 0
B.M. at 25 *m* = 156.25 *tm*

6.7 Max^{m} tension in cable = 80.777 *tf*
S.F. at 40 *m* from leftend = – 4.9
B.M. at 40 *m* = 0.00

6.8 Maximum tension = 131.94 *tf*
Maximum B.M. = 428.57 *tm*
Section Modulus = 35714 *cm*^{3}

6.9 B.M. at left quarter span = 13.90 *tm*
Maximum B.M. = 17.45 *tm*
at $\theta = \cos^{-1} 0.9284$

6.10 Maximum B.M
left half span = – 0.04201 *wR*
at θ = 19°
Right half span = + 0.10355 *wR*
at θ = 45°
Radial shear

at left quarter point $= \frac{W}{4}(\sin\theta - \cos\theta) = 0.0915W$ at $\theta = 60°$

A.F. $= \frac{W}{4}(\cos\theta + \sin\theta) = 0.3415\ W$ at $\theta = 60°$

6.11 $H = \frac{P}{2\sqrt{3}}$ Max^{m} B.M. $= \frac{PR}{\sqrt{3}}\left(\frac{2-\sqrt{3}}{2}\right)$

6.12 Radial shear at a distance R from left hinge = 0.1830 W
A.F. = 0.6830 W

6.13 H = 144,000 V_A = 288,000 V_B = 72,000
Normal Thrust at 4 *m* from support = 160992
S.F. = 0

6.14 H = 24$'$ B.M. at *tm*

from left support = 48 *tm*

S.F. = 5.3664

A.F. = 24.1488

Chapter 7.

7.2 (*i*) 2 *mm* (*ii*) (*b*) (*iii*) (*b*) (*iv*) (*b*) (*v*) (*c*) (*vi*) (*b*) (*vii*) (*c*) (*viii*) (*a*) (*ix*) (*d*) (*x*) (*d*)

7.3 (*ii*) $\frac{wL^3}{EI} \cdot \frac{1}{1152}$, $R_C = \frac{4W}{3}$.

7.4 $R_B = 7.5\ t$ $M_{AB} = 25\ tm$, $R_A = 12.5t$

Slope at $B = -\frac{41.67}{EI}$

7.5 (*i*) Deflection at $C = \frac{PL^3}{32\ EI} - \frac{ML^2}{16EI}$

Slope at $C = \frac{PL^2}{16\ EI} - \frac{ML}{6\ EI}$

(*ii*) Deflection under load $P = \frac{PL^3}{48\ EI} - \frac{ML^2}{16\ EI}$

7.6 Slope at $A = \frac{8.6108}{EI}$

deflection at $A = \frac{7.777}{EI}$

7.7 (*i*) B.M. $M_A = WL\frac{4}{27}$ $M_B = \frac{2WL}{27}$ $R_A = W \cdot \frac{20}{27}$ $R_B = \frac{7}{27} \cdot W$

(*ii*) deflection under load $= -0.0036579 \frac{WL^3}{EI}$

Maximum deflection $= \frac{WL^3}{EI} \times 0.004014$

(*iii*) $x = \frac{L}{5}, \frac{2L}{7}$

7.8 (*i*) -5.416×10^{-3} radian

(*ii*) 10 *mm* (*iii*) Maximum deflection = 15.03 *mm*

7.9 (*i*) Deflection under load = 11.85 *mm*

(*ii*) Maximum deflection occurs at $x = 3.266\ m$

Value = 12.872 *mm*

7.10 (*a*) slope at $B = -\frac{64}{3\ EI} = -5.33 \times 10^{-3}$ radian (*b*) deflection at $B = \frac{68.26}{EI} = 17.06\ mm$

(*c*) $R_B = 0.355\ t$ (*d*) Point of contraflexure = 1.03 *m*

7.11 (*a*) slope at $B = 0.2567\dfrac{Ph^2}{EI}$ (*b*) Deflection at $C = 0.1908\dfrac{Ph^3}{EI}$

7.12 (*i*) deflection under load $= \dfrac{Wa^2b^2}{3\,EI.L}$ (*ii*) slope under load $= \dfrac{Wab\,(b-a)}{3\,EI.L}$

(*iii*) Deflection under mid span $= \dfrac{Wa}{12EI}\left(\dfrac{6ab+3b^2-a^2}{4}\right)$

(*iv*) Support Reaction $= \dfrac{Wa\left(6ab+3b^2-a^2\right)}{L^3}$

7.13 (*i*) Deflection at middle $= \dfrac{5wL^4}{384\,EI}$ (*ii*) Deflection at middle $= \dfrac{wL^2l^2}{16\,EI}$

(*iii*) Ratio of $\dfrac{l}{L} = 2.19$

7.14 (*a*) .0349 *m* (*b*) $R = \dfrac{6}{17}wL$ **7.15** $M = \dfrac{PL}{4}$, Deflection at $D = \dfrac{PL^3}{162EI}$

Chapter 8.

8.2 (*i*) (*a*) (*ii*) (*a*) (*iii*) (*d*) (*iv*) (*b*)

8.3 26.553 *mm*

8.4 $R_H = 1.0626\,wL$

8.5 $\Delta_c = 3.5\dfrac{wL}{aE}$

8.6 $\Delta_V = 1.3\ cm$ $\Delta_H = 2.4\ cm$ $\Delta_\theta = 0.0135\ radian$

8.7 $\Delta = 0.008916\dfrac{PL^3}{EI}$

8.8 $V_D = 65.946\ kN$ $H_D = 23.784\ kN$
$V_A = 14.054\ kN$ $H_A = 16.216\ kN$

Chapter 9.

9.2 (*i*) (*d*) (*ii*) (*b*) (*iii*) (*a*) (*iv*) (*b*) (*v*) (*b*) (*vi*) (*c*) (*vii*) (*c*) (*viii*) (*b*).

9.3 Maximum Force in IJ = 8.08 *kN*

9.4 Maximum Value at 4 m from left support
S.F. = 32.4 *kN* B.M. = 207.36 *kNm*
Absolute maximum shear = 86.4 *kN*
Absolute Max^m B.M. = 216 *kNm*

9.5 Central ordinate of I.L. = 0.6652

9.6 I.L. for B.M. Max^m ordinate $= \dfrac{5}{3}$ at E and F
I.L. for Axial Thrust – Max^m ordinate = 0.9718

9.7 Max^m Value of tenstion = 764.23 *t*
Max^m Positive moment = 360 *tm*

Max^m Negative moment = 245 tm

Max^m Positive shear = 9.6 t

Negative shear = 9.6 t

Chapter 10.

10.2 (*i*) (*a*) (*ii*) (*c*) (*iii*) (*b*)

10.3 Slope at $B = \frac{wL^3}{48\ EI}$(clockwise)

deflection at $D = \frac{wL^4}{64\ EI}$(upward)

10.4 F_{CF} = 2.899 W (comp.)

10.5 Vertical deflection of joint C = 0.3646 mm downward

10.6 (*i*) Redundant force in CA = + 2071 kg

$F_{DA} = -1464.4$

$F_{DC} = -1464.4$

$F_{AB} = -1464.4$

$F_{AC} = +2071 = F_{DB}$

$F_{CB} = -1464.4$

(*ii*) $F_{DA} = -732.2$

$F_{DC} = -732.2$

$F_{DB} = +1035 = F_{AC}$

$F_{AB} = -732.2 = F_{CB}$

10.7 Slope at $B = \frac{PL^2}{12\ EI}$ (anticlockwise)

Deflection at $D = \frac{5}{8}.\frac{PL^3}{EI}$ (downward)

10.8 $M_B = 0$ $R_A = \frac{W}{2}$ $R_B = W$ $R_C = \frac{W}{2}$

Chapter 11.

11.2 (*i*) (*d*) (*ii*) (*a*) (*iii*) (*a*) (*iv*) (*d*) (*v*) (*d*) (*vi*) (*a*) (*vii*) (*c*) (*viii*) (*a*) (*ix*) (*a*) (*x*) (*d*) (*xi*) (*b*)

(*xii*) 5 kNm (*xiii*) $\frac{6\ EI}{L}$

11.3 $M_{AB} = 0$ $M_{BA} = -4.839\ kNm$ $M_{CB} = -5.8\ kNm$ $M_{DC} = -6.1\ kNm$

11.4 $M_{BA} = -1.2\ tm$ $M_{CB} = -6.851$ $M_{DC} = -9.174\ tm$

11.5 $M_{AB} = 1.8\ tm$ $M_{BA} = 3.6\ tm$ $M_{BC} = -3.6\ tm$

$M_{CB} = 3.6\ tm$ $M_{CD} = -3.6\ tm$ $M_{DC} = -1.8\ tm$

11.6 $M_{DB} = 40.208$ $M_{BD} = 80.416$ $M_{BA} = 39.58$

$M_{BC} = -120$ Slope $\theta_C = 0.02129$ radian

deflection $\Delta_C = 34.258\ mm \downarrow$

11.7 $\Delta = \frac{265.25}{EI}$

11.8 $M_{AB} = -16.829$ $M_{BA} = -13.1707$

$M_{BC} = +13.1706$ $M_{CB} = 13.1706$
$M_{CD} = -13.1706$ $M_{DC} = -16.829$

11.9 $M_{AO} = 100$ tm $M_{BO} = 50$ tm $M_{CO} = 50$ tm $M_{DO} = 0$

11.10 $M_{AB} = -7.72$ $M_{BA} = +3.76$ $M_{BC} = -2.89$
$M_{BD} = -0.87$ $M_{CB} = 0$ $M_{DB} = -0.43$

11.11 $M_{AC} = -108.11$ $M_{CA} = -378.38$ $M_{CB} = -621.62$ $M_{BC} = -351.35$

11.12 $M_{AB} = -16.75$ tm $M_{BA} = +5.00$ tm
$M_{BC} = -5.00$ $M_{CB} = +11.50$ tm

11.13 $M_{AB} = -13.26$ $M_{BA} = -9.25$ $M_{BC} = +9.25$
$M_{CB} = +9.25$ $M_{CD} = -9.25$ $M_{DC} = -13.26$

11.14 $M_{BC} = -4.72$ tm $M_{CB} = +2.849$

11.15 $H_A = 11.505\ t$ $H_D = 1.51\ t$
$V_D = 4.81$ $V_A = 3.19\ t$

11.16 $M_{BA} = +5.31$ tm $M_{AB} = +1.63$ tm
$M_{BC} = -5.31$ tm $M_{CB} = +3.61$ $M_{CD} = -3.61$ $M_{DC} = -2.08$

11.17 $M_{AD} = +15.40$ tm $M_{AB} = -15.40$ tm
$M_{BA} = +20.69$ tm $M_{BC} = -20.69$ tm
$M_{CB} = +20.69$ tm $M_{CD} = -20.69$ tm
$M_{BC} = +15.40$ tm $M_{DA} = -15.40$

11.18 $M_{AB} = +1.71$ tm $M_{BA} = +3.46$ tm $M_{BC} = -3.465$ tm
$M_{CB} = +2.83$ tm $M_{CE} = +1.06$ tm $M_{CD} = -3.89$ tm
$M_{DC} = 0.00$ tm $M_{EC} = -3.965$ $V_A = 20.64\ t$ $H_A = 6.47\ t$ $H_E = 12.42t$

11.19 $M_{AB} = -2.00$ tm $M_{BA} = -1.00$ $M_{BC} = +1.00$
$M_{CB} = -1.00$ $M_{CD} = +1.00$ $M_{DC} = +2.00$

11.20 $M_{AB} = -1.96$ tm $M_{BA} = -2.51$ tm $M_{BC} = +2.51$ tm $M_{CB} = +8.54$ tm
$M_{CE} = -12.00$ tm $M_{CD} = 3.46$ tm $M_{DC} = +1.01$ tm $V_A = 1.38\ t \downarrow$
$V_D = 3.38\ t \uparrow$ $H_A = 0.558\ t \leftarrow$ $H_D = 0.558 \rightarrow$

11.21 $M_{AB} = -0.58$ tm $M_{BA} = +1.84$ tm

11.22 $M_{AB} = -7.37$ tm $M_{BA} = +5.25$ tm $M_{CD} = -5.00$ tm

11.23 $M_{AB} = 0.96$ tm $M_{BA} = 2.44$ tm $M_{BC} = -2.44$ tm
$M_{CB} = 2.05$ tm $M_{CD} = -2.05$ tm $M_{DC} = 1.30$ tm

11.24 $M_{12} = -476.415$ kgm $M_{21} = -524.50$ kgm $M_{32} = +1190.48$
$M_{34} = +809.52$ kgm $M_{43} = 190.80$ kgm

11.25 $M_{AB} = 0.30$ tm MB_{BA} $+3.56$ $M_{CB} = 12.47$
$M_{CD} = -1.60$ $M_{CE} = -10.875$ $M_{DC} = -2.28$

11.26 $M_{AB} = -6.16$ tm $M_{BA} = -0.47$ tm
$M_{CB} = +7.49$ tm $M_{DC} = -6.38$ tm

11.27 $M_{AB} = -1.3053$ kNm $M_{BA} = 4.3764$ kNm
$M_{CB} = +6.4942$ kNm $M_{DC} = -5.2844$

Chapter 12

12.2 (*i*) (*a*) (*ii*) (*c*) (*iii*) (*c*)

12.3 $H = 21.25^t$ $\quad$ $B.M_{crown} = 0.00$ tm

B.M at quarter span = – 15.608 tm

12.4 $V_A = 75^t$ $\quad$ $V_B = 25^t$ $\quad$ $H = 62.5^t$

At quarter span S.F. = 0.00, A.F. = 67.375^t B.M. = 62.5 tm

12.5 $H = \dfrac{3wL^2}{16h}$

12.7 $V_A = V_B = 160$ kN $\quad$ $H = 162.95$ kN

Point of contraflexure = 11.07 m

Point of Max^m B.M. = 5.535 m, 14 m

Max^m B.M. = – 158.49 kNm, + 212.177

A.F. at load point = 225.80 kN, SF = 80.05 kN

Chapter 13

13.2 (*a*)

13.3 – 7.55 tm at A, – 5.31 tm at *B*

13.4 $M_A = +2.0838$ $\quad$ $M_B = -4.16592$

13.5 $M_A = -3.3555$ $\quad$ $M_B = -15.8505$

$M_C = -24.25$ $\quad$ $M_D = +16.36$ tm

13.6 $M_A = -2.7297$ tm

$M_C = -9.490$ tm

C.O.F. from *A* to *B* = – 1.19867

C.O.F. from *B* to *A* = – 0.31685 (B.M. sign convention)

Absolute stiffness at *A* = 90 64 tm

Absolute stiffness at *B* = 34287 tm

13.7 $M_B = 569.17$ kgm $\quad$ $M_C = 281.95$ kgm $\quad$ $M_D = -430.83$

13.8 $\dfrac{I_2}{I_1} = 2$

Chapter 14

14.2 (*i*) (*b*) (*ii*) (*c*) (*iii*) (*d*)

14.3 I.L ordinates at 1 m interval.

Span *AB* = 0, – .16, –.31, –.43, –.50, –.51, –.43, –.27,0

Span *BC* = 0, 0.26, 0.59, 0.98, 1.45

14.4 I.L. ordinates at 2.5 m interval 0, 0.367, 0.687, 0.914, 1

14.5 I.L. ordinates at lm interval

1, 0.672, 0.3754, 0.1412, 0 , 0.027, 0.0312, 0.0195

14.6 I.L. ordinates at span *AB* and *BC*, From *A*

1 m =.0439, 2m = 0.076 3m = .0615

4 m = 0, 5m = 0.125, 6m = 0.297, 6.5 = 0.3942

In Span *BC* and *CD*

From *D* 1m = 0.0482, 2 m = 0.0815

3m = 0.0846, 4m = 0.0426, 4.5 m = 0

From *C* 1m = 0.1316, 2m = 0.3079, 3m = 0.3079, 2.94 m = 0.4945, 3.5m = 0.6057

LANDMARKS IN THE HISTORY OF STRUCTURAL ENGINEERING

We have just now covered the fundamentals of structural engineering. Since structural engineering derives its basic principles from solid mechanics and strength of materials the great pioneering contributions of Archimedes (287-212 B.C), Leonard da Vinci (1452-1519) Galileo Galilei (1564-1642), Robert Hooke (1635-1703) Isaac Newton (1642-1727) and others should be gratefully acknowledged. The list below tabulates further contributions in structural analysis chronologically.

Year	Development or theory	Scientist/Engineer
1635-1703	Hooke's law	Robert Hooke
1654-1705	Deflection of an elastically bent bar	James Bernoulli
1667-1748	Principle of Virtual displacement	Johann Bernoulli
1700-1782	Theory of Elastic curves, Strain Energy for a beam.	Daniel Bernoulli
1707-1783	Application of minimal principle, Buckling of columns.	Leonard Euler
1736-1806	Theory of Friction, Stability of retaining walls	C.A. Coulomb
1736-1813	Generalised Forces and displacements	J.L. Lagrange
1773-1829	Elastic Modulus	Thomas Young
1785-1836	Fundamental Equations of Theory of Elasticity	L.M. Navier
1790-1888	Analysis of determinate trusses	A.F. Möbius and S. Whipple
1899-1864	Three Moment Equation, Conservation of Energy	B.P. Clapeyron
1821-1881	Graphical methods of analysis of structures	Karl Culmann
1830-1879	Theory of Reciprocal deflections	James Clark Maxwell
1867	Concept of influence Lines	E. Winkler
1835-1918	Conjugate beam method, circles of stress and strain	Otto Mohr
1847-1884	Theory of Least work	Alberto Castigliano
1873	Moment Area Method	Charles E Greene
1848-1931	Theorem of Compatibility	F. Engesser
1851-1921	Influence lines, tension coefficient method of analysis of trusses.	H. Müller Breslau
1854-1924	Analysis of Space Truss	August Föppl

1914-1915	Slope Deflection method	George A Maney
1919	Use of Singularity function for deflection of beams.	W.H. Macaulay
1924	Moment Distribution Method	Hardy Cross
1930	Column Analogy Method	Hardy Cross
1940	Relaxation Method	R.V. Southwell
1945-1960	Plastic Behaviour of Structures	J.F. Baker L.S. Beedle and others
1950	Matrix force method	H. Falkenheiner
1954	Concept of Stiffness in the analysis of Stiff	R.K. Livesley jointed frames
1955-56	Concept of Frame work analysis and continuum analysis	J.H. Argyris, M.T. Turner, R.W. Clough H.C. Martin and others
1960	Introduction to Finite Element terminology	R.W. Clough
1960-1988	Advances in Finite Element Method	Wilson, Bathe Zien Kiewicz and others

The list is obviously not exhaustive. It includes only significant contributions in the field of structural engineering as covered in this book.

BIBLIOGRAPHY

1. Anand, A.S. (1990) Theory of Structure. Satya Prakasan, New Delhi.
2. Balfour, James A.D. (1986) Computer Analysis of Structural Frameworks. Collins. London.
3. Bhatt, P. (1989) Problems in Structural Analysis by Matrix Method. Wheeler. New Delhi.
4. Chandrupatla, T.R. and A.D. Belegundu (1997) Introduction to Finite Elements in Engineering. Prentice Hall of India, New Delhi.
5. Dayaratnam, Pasala (1976) Analysis of Statically Determinate Structures. Affiliated East-West Press Pvt. Limited, New Delhi.
6. Dayaratnam, Pasala (1977) Analysis of Statically Indeterminate Structure. Affiliated East-West Press Pvt. Limited, New Delhi.
7. Dravid, P.S. (1962) Analysis of Continuous Beam and Rigid Frames. Asia Publishing House, New Delhi.
8. Ghali, A, A.M. Nevelle (1972) Structural Analysis (A Unified Classical and Matrix approach). 2nd Edition, Chapman and Hall, London and New York.
9. Gupta, S.P, G.S., Pandit, R Gupta. (1999) Theory of Structure, Vol II, Tata Magraw Hill, New Delhi.
10. Hibbeler, R.C. (1999). Structural Analysis. Prentice Hall International Inc.
11. Jain, Ashok. K. (1990) Elementary Structural Analysis. Nem Chand and Brothers, Roorkee, India.
12. Jenkins, W.M. (1969) Matrix and Digital Computer Methods in Structural Analysis. Mcgraw Hill London.
13. Jindal, R.L (1980). Indeterminate Structures. S. Chand and Company New Delhi.
14. Kennedy, John. B. and K.S. Madugula Murty (1990) Elastic Analysis of Structures—Classical and Matrix Methods. Harper and Row Publishers, New York.
15. Laible, Jeffery P. (1985). Structural Analysis. Holt Renehart and Winston, Japan.
16. Martin, H.C. and G.F. Cary (1975) Introduction to F.E. Analysis. Tata Macgraw Hill New Delhi.
17. Neal, B.G (1964) Structural Theorems and Their applications. Peragamon Press, London.
18. Parikh, S.K. (1995) Automated Optimum Design of R.C.C. Skeletons. Tata Macgraw Hill New Delhi.
19. Popov, Egor. P. (1973) Introduction to mechanics of solids. Prentice Hall of India Private Ltd, New Delhi.
20. Punmia, B.C., Ashok Jain and Arun Jain (1992) Strength of Materials. Laxmi Publications, New Delhi.
21. Punmia, B.C. Ashok Jain and Arun Jain (1992) Theory of structures. Laxmi Publications, New Delhi.
22. Reddy, C.S. (1996) Basic Structural Analysis. Tata McGraw Hill Publishing Company Limited.
23. Sack, R.L. (1984). Structural Analysis. Mcgraw Hill Book Company.
24. Shedd, Thomas Clark and Janison Vawter (1941) Theory of Simple structures. John Wiley and Sons. London
25. Sinha, N.C. and P.K. Gayen (1990) Advanced Theory of Structures. Dhampat Rai and Sons New Delhi - 6.
26. Smith, J.C. (1988) Structural Analysis. Harper and Row Publishers, New York.
27. Timoshenko, S. (1940). Strength of Materials, Part 1. Mcgraw Hill.
28. UtKu, S, C.H. Norris, J.B. Wilber, (1991). Elementary Structural Analysis. Mcgraw Hill, Inc.
29. Wang C.K. (1983). Intermediate Structural Analysis. Mcgraw Hill Book Company Singapore.
30. Wang, Chu Kia (1953). Statically Indeterminate Structures. Mcgraw Hill KogaKusha Ltd.
31. West, Hary. H (1989) Analysis of Structures. John Wiley & sons. New York.
32. Young, T.Y. (1986) Finite Element Structural Analysis. Prentice Hall International Inc. New Jersy.

ADDITIONAL BIBLIOGRAPHY

1. Hodge, P.C., Plastic Analysis of Structures, McGraw Hill, New York, 1959.
2. Logan D.L., A First Course in the Finite Element Method, Third Ed. Thomson, 2002.
3. Neal, B.G, The Plastic Methods of Structural Analysis, John Wiley, New York, 1956.
4. Rao S.S., The Finite Element Method in Engineering fourth Ed., Elsevier, 1st Indian Edition 2005.
5. Rao V. Dukkipati, Vibration Analysis, Narosa, New Delhi, 2004.
6. SP: 6(6) 1972 is Handbook for Structural Engineers: Application of Plastic theory in Design of Steel Structures.

INDEX